# OCEANIC ANGLERFISHES

# OCEANIC ANGLERFISHES

## EXTRAORDINARY DIVERSITY IN THE DEEP SEA

THEODORE W. PIETSCH

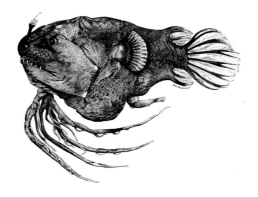

UNIVERSITY OF CALIFORNIA PRESS
*Berkeley   Los Angeles   London*

University of California Press, one of the most distinguished university presses in the United States, enriches lives around the world by advancing scholarship in the humanities, social sciences, and natural sciences. Its activities are supported by the UC Press Foundation and by philanthropic contributions from individuals and institutions. For more information, visit www.ucpress.edu.

University of California Press
Berkeley and Los Angeles, California

University of California Press, Ltd.
London, England

© 2009 by the Regents of the University of California
Published with the assistance of the University of Washington

The material in this volume is based on work supported by the U.S. National Science Foundation under Grants GB-40700, DEB 76-82279, DEB 78-26540, DEB 03-14637, T.W. Pietsch, principal investigator. Any opinions, findings, conclusions, or recommendations expressed in this work are those of the author and do not necessarily reflect the views of the Foundation.

Library of Congress Cataloging-in-Publication Data

Pietsch, Theodore W.
  Oceanic anglerfishes : extraordinary diversity in the deep sea / Theodore W. Pietsch.
      p.   cm.
  Includes bibliographical references and index.
  ISBN 978-0-520-25542-5 (case : alk. paper)  1. Anglerfishes.  2. Anglerfishes—Classification.   I. Title.

  QL637.9.L6P54   2009
  597'.62—dc22                                            2008048604

Manufactured in China
16  15  14  13  12  11  10  09
10  9  8  7  6  5  4  3  2  1

The paper used in this publication meets the minimum requirements of ANSI/NISO Z39.48-1992 (R 1997) (*Permanence of Paper*).♾

Cover illustration: Artwork by Ray Troll, 2007

Title-page illustration: *Linophryne brevibarbata*, 100-mm SL female, with an 18.5-mm SL parasitic male, BMNH 1995.1.18.4, a North Atlantic species represented in collections around the world by only six specimens. Drawing by Elisabeth Beyerholm; after Bertelsen (1980a).

*Dedicated to the memory of the late Erik Bertelsen*

# CONTENTS

PREFACE ix
ACKNOWLEDGMENTS xi

## PART I
## The Biology of Deep-Sea Anglerfishes

1. Introduction and Historical Perspective / 3
2. What Makes an Anglerfish? / 23
3. Biodiversity / 51
4. Evolutionary Relationships / 173
5. Geographic Distribution / 197
6. Bioluminescence and Luring / 229
7. Locomotion, Food, and Feeding / 253
8. Reproduction and Early Life History / 277

## PART II
## A Classification of Deep-Sea Anglerfishes

Introduction / 313
Families, Genera, and Species of the Ceratioidei / 319

REALLOCATION OF NOMINAL SPECIES OF THE CERATIOIDEI BASED ON FEMALES / 511
REALLOCATION OF NOMINAL SPECIES OF THE CERATIOIDEI BASED ON FREE-LIVING MALES / 519
SYMBOLIC CODES FOR INSTITUTIONAL COLLECTIONS / 521
GLOSSARY / 523
REFERENCES / 527
ILLUSTRATION CREDITS / 547
INDEX / 549

# PREFACE

My purpose in writing this book is to bring together a diverse and previously scattered array of facts and data surrounding an astonishing assemblage of deep-water oceanic fishes commonly referred to as the devilfishes or seadevils, and scientifically known as the deep-sea ceratioid anglerfishes. Although rarely seen by anyone but a small handful of specialists, these remarkable animals are surprisingly abundant and widespread geographically, contributing significantly to the biodiversity of the marine ecosystem. The 160 currently recognized species of ceratioids dwell primarily in the inky black oceanic midwaters below 1000 m, where their existence depends on their ability to lure prey with a uniquely modified dorsal-fin spine equipped with a bioluminescent bait. While females may attain a length of a meter or more, the males of all included taxa are dwarfed, reaching only a small fraction of the size of the females. Equipped with large, well-developed eyes, a highly developed sense of smell, and a specialized nipping device at the tips of its jaws, the male's sole purpose in life is to find a female. Once found, by a combination of visual and chemical cues, the male bites onto her body and in some cases remains there, only to become permanently joined to her body, the blood circulatory systems of the pair fusing so that the male becomes a sexual parasite on the female, and she, with ovaries, and now with testes as well, becomes a functional hermaphrodite. These adaptations alone, unique among fishes, and for that matter, among all animals, make these creatures one of the most intriguing groups of organisms for the lay person as well as the professional ichthyologist. They never fail to amaze any audience.

This monograph is the result of some 40 years of research and study, beginning in 1967 as a first-year graduate student at the University of Southern California, and greatly benefiting over the next three decades from a close collaboration with the famous Danish ichthyologist and world expert on ceratioids, the late Erik Bertelsen of the Zoological Museum of the University of Copenhagen. Working either jointly or apart, Bertel (as he always wanted to be called) and I embarked on a program of biological study of the various families, genera, and species that has finally reached culmination here in this book. It is the result of a critical review of the world literature on the Ceratioidei, as well as a detailed examination of approximately 7,095 specimens (including 6,310 females, 785 males, and all extant type material) made available by the curators and staffs of some 90 institutions located around the world. Following an introduction and a history of our knowledge of the Ceratioidei, from the earliest discoveries to the present, an analysis of the characters that best differentiate the taxa is provided. This is followed by an overview of the biological diversity found within the group, focusing primarily on the families and genera, within which most of the morphological variation is found. An analysis of evolutionary relationships is then presented, followed by a discussion of geographic and depth distributions in relation to environmental factors, including temperature, salinity, oxygen, light, currents, and food supply. The known facts concerning bioluminescence and the functional morphology of luring, a unique mode of energy capture so highly developed in these fishes, are then summarized, followed by what is known about ceratioid locomotion, food and feeding, and reproductive biology. Finally, in Part Two, a classification of the suborder is provided, along with definitions and a synopsis of the methods employed in this part of the study, followed by evidence for the recognition of the 160 species of ceratioids, including synonymies, diagnoses, descriptions, and keys to facilitate their identification. In some ways this is a biosystematic monograph, designed to satisfy the needs and interests of students and professionals in ichthyology; but to a greater extent I hope that these fascinating creatures have been presented in such a way that this book will be enjoyed by all those who find excitement in the wonders of the natural world.

# ACKNOWLEDGMENTS

Many people have generously given their help with this study by providing loans of specimens or by making collection data available. Many others have assisted by making the author feel welcome during visits to their respective institutions. To all of these people I extend my sincere gratitude: Rafael Bañón Díaz, Asociación Científica de Biología Marina, Vigo, Pontevedra, Galicia, Spain; Marcelo Melo, Auburn University, Auburn, Alabama; A.B. Stephenson, Auckland Museum, Auckland, New Zealand; Barbara A. Brown, American Museum of Natural History, New York; John R. Paxton, Jeffrey M. Leis, Mark McGrouther, and Amanda Hay, Australian Museum, Sydney; Lou Van Guelpen, Atlantic Reference Centre, Huntsman Marine laboratory, St. Andrews, Canada; Kwang-Tsao Shao and Hsuan-Ching Ho, Institute of Zoology, Academia Sinica, Taipei, Taiwan, Republic of China; Ralf Britz, Oliver A. Crimmen, James S. Maclaine, and Patrick D. Campbell, The Natural History Museum, London; Hiromitsu Endo, Kochi University, Kochi City, Japan; Jeffrey Landesman, Cabrillo Marine Aquarium, San Pedro, California; William N. Eschmeyer, Tomio Iwamoto, and David Catania, California Academy of Sciences, San Francisco; Juan Carlos Arronte, Centro de Experimentación Pesquera, Castropol (Asturias), Spain; Alastair Graham, Commonwealth Scientific and Industrial Research Organization, Hobart, Tasmania; Mary Anne Rogers, Brian L. Sidlauskas, Eric J. Hilton, Christian F. Kammerer, Field Museum of Natural History, Chicago; Ronald A. Fritzsche, Humboldt State University, Arcata, California; Kunio Amaoka, Mamoru Yabe, Kazuhiro Nakaya, and Mitsuomi Shimazaki, Hokkaido University, Hakodate, Japan; Aevar Petersen, Icelandic Institute of Natural History, Reykjavik, Iceland; Günther Behrmann, formerly of the Institut für Meeresforschung, Bremerhaven, Germany; Dmitriy L. Pitruk, Andrei Balanov, and Vladimir E. Kharin, Institute of Marine Biology, Russian Academy of Sciences, Vladivostok, Russia; Nakajima Masaaki, Ibaraki Nature Museum, Ibaraki, Japan; Nikolas V. Parin, Institute of Oceanography, Russian Academy of Sciences, Moscow; Matthias Stehmann, formerly of the Institut für Seefischerei, Hamburg, Germany; Robert J. Lavenberg, Jeffrey A. Seigel, and Richard F. Feeney, Natural History Museum of Los Angeles County, Los Angeles; Bruce H. Robison and Steven H.D. Haddock, Monterey Bay Aquarium Research Institute, Moss Landing, California; Karel F. Liem, Karsten E. Hartel and Andrew Williston, Museum of Comparative Zoology, Harvard University, Cambridge, Massachusetts; Jean-Claude Quéro, Musée d'Histoire Naturelle de La Rochelle, France; Manuel Biscoito, Museu Municipal do Funchal, Madeira, Portugal; Marie-Louise Bauchot, Guy Duhamel, Martine Desoutter, Patrice Pruvost, and Romain Causse, Muséum National d'Histoire Naturelle, Paris, France; Samuel Iglésias, Station de Biologie Marine du Museum National d'Histoire Naturelle et du Collège de France, Concarneau; Paulo A. Buckup, Marcelo F. G. de Brito, Paulo A. S. Costa, and Marcos A. L. Franco, Museu Nacional, Rio de Janeiro, Brazil; Gunnar Jónsson and Jónbjörn Pálsson, Marine Research Institute, Reykjavik, Iceland; José Lima de Figueiredo and Mário C. C. de Pinna, Museu de Zoologia da Universidade de São Paulo, São Paulo, Brazil; Masaki Miya, Natural History Museum and Institute, Chiba, Japan; Y.-M. Ju, National Museum of Marine Biology and Aquarium, Checheng, Pingtung, Taiwan, Republic of China; Clive D. Roberts, Chris D. Paulin, and Andrew L. Stewart, National Museum of New Zealand, Wellington; Geoffroy N. Swinney, National Museums of Scotland, Edinburgh; Martin F. Gomon and Dianne Bray, National Museum of Victoria, Melbourne, Australia; Helmut Wellendorf and Ernest Mikschi, Naturhistorisches Museum, Vienna, Austria; Keiichi Matsuura and Gento Shinohara, National Science Museum, Tokyo, Japan; Toshiro Saruwatari, Ocean Research Institute, University of Tokyo, Tokyo, Japan; Edith A. Widder, Ocean Research and Conservation Association, Fort Pierce, Florida; Douglas F. Markle, Oregon State University, Corvallis; Jeffrey W. Johnson, Queensland Museum, South Brisbane, Australia; Alexei M. Orlov, Russian Federal Research Institute of Fisheries and Oceanography, Moscow; Richard Winterbottom and Marty Rouse, Royal Ontario Museum, Toronto; M. Eric Anderson, South African Institute for Aquatic Biodiversity, Grahamstown; P. Alexander Hulley, Iziko Museums of Cape Town, South Africa; Ralph Foster and Steve Donnellan, South Australian Museum, Adelaide; Richard H. Rosenblatt, Cynthia Klepadlo, and H.J. Walker, Scripps Institution of Oceanography, University of California, La Jolla; Peter J. Herring, National Oceanography Centre, Southampton, England; John D. McEachran and R. Kathryn Vaughan, Texas Cooperative Wildlife Collection, Texas A & M University, College Station; George H. Burgess, Robert H. Robins, and Griffin Sheehy, Florida State Museum, University of Florida, Gainesville; Eric Taylor, University of British Columbia,

British Columbia, Canada; N. Justin Marshall, School of Biomedical Sciences, University of Queensland, Brisbane, Queensland, Australia; Bruce B. Collette, Susan L. Jewett, Jeffrey T. Williams, and Lisa F. Palmer, National Museum of Natural History, Washington, D.C.; J. Richard Dunn, Katherine P. Maslenikov, Christopher P. Kenaley, Michael J. Cooksey, Zachary H. Baldwin, and Kimberly A. Sawyer, University of Washington, Seattle; Tracey T. Sutton, Virginia Institution of Marine Science, College of William and Mary, Glouchester Point, Virginia; Barry Hutchins and Sue Morrison, Western Australian Museum of Natural History, Perth; Arcady V. Balushkin, V. V. Fedorov, Boris Sheiko, and Olga Voskoboinikova, Zoological Institute, Russian Academy of Sciences, St. Petersburg; Isaac Isbrücker, Instituut voor Taxonomische Zoölogie, Zoölogische Museum, Universiteit van Amsterdam, Netherlands; Horst Wilkens, Gudrun Schulze, Ralf Thiel, and Irina Eidus, Zoologisches Institut und Zoologisches Museum der Universität Hamburg, Hamburg; Hans-Joachim Paepke and Peter Bartsch, Museum für Naturkunde der Humboldt-Universität zu Berlin; Hans Kauri and Ingvar Byrkjedal, Zoological Museum, University of Bergen, Norway; Jørgen Nielsen, Peter Rask Møller, and Tomas Menne, Zoological Museum, University of Copenhagen, Denmark; Kazuo Sakamoto, Department of Zoology, University Museum, University of Tokyo, Japan.

For providing images and permission to reproduce illustrative materials, I thank Peter Bartsch, Elisabeth Beyerholm, Geert Brovad, Richard C. Brusca, John H. Caruso, Patricia Chaudhuri, A. H. Coleman, Oliver A. Crimmen, Peter David, Steinunn Einarsdottir, Richard Ellis, K. Elsman, Bryan Gim, Andrew Goodson, David B. Grobecker, Hsuan-Ching Ho, Elizabeth Anne Hoxie, Gunnar Jónsson, Christopher P. Kenaley, Rudie Kuiter, Michel Lamboeuf, Kyle Luckenbill, John G. Lundberg, James S. Maclaine, Caryl Maloof, N. Justin Marshall, Eileen C. Mathias, Peter Rask Møller, Joe Nakanishi, Jørgen Nielsen, Robert Nielsen, William T. O'Day, Jónbjörn Pálsson, Joanne M. Pavlak, John E. Randall, Richard H. Rosenblatt, Frank Schneidewind, David Shale, Matthias Stehmann, Gary T. Takeuchi, H. J. Walker, and Edith A. Widder. Sincere gratitude is extended to Steinunn Einarsdottir, granddaughter of Bjarni Saemundsson, for allowing me to reproduce the portrait of her grandfather. Special thanks to Ray Troll, Ketchikan, Alaska, for his extraordinary cover image; and to Zachary H. Baldwin for providing the scanning electron micrographs and the small figures of ceratioids that illustrate the synoptic keys to families and genera in Chapter Three. Thanks also to Secretary Maureen A. Donnelly and the American Society of Ichthyologists and Herpetologists for permission to reproduce copyrighted materials previously published in *Copeia*, and to Secretary Yoshiaki Kai and the Ichthyological Society of Japan for materials previously published in *Ichthyological Research*.

I am greatly indebted to Katherine P. Maslenikov for many years of expert curatorial assistance. I also thank J. Frank Morado and Pamela C. Jensen (NOAA Fisheries, Alaska Fisheries Science Center, Seattle, Washington) for their help in interpreting histological data; and James W. Orr and Duane E. Stevenson (both with NOAA Fisheries, Alaska Fisheries Science Center, Seattle), and William D. Anderson (Grice Marine Biological Station, College of Charleston), for their helpful critiques of portions of the manuscript. The entire draft manuscript was critically read by Christopher P. Kenaley and Zachary H. Baldwin. Finally, at the University of California Press, grateful thanks are extended to Charles R. Crumly, editor for science, for skillfully directing the publication of this volume, and to his publishing team for supervising its design and production: Scott Norton, project manager; Francisco Reinking, preproduction assistant; Meg Hannah, copyeditor; and Joanne Bowser, project manager, Aptara, Inc.

Support for the research presented here has come from numerous sources spread out over a 40-year period beginning in 1967 as a first-year graduate student at the University of Southern California. The bulk of the work has been generously supported by the U.S. National Science Foundation, most recently through grant DEB-0314637, T. W. Pietsch, principal investigator. I am extremely grateful to the foundation and to all its program officers and staff.

Finally, I thank David A. Armstrong, director of the School of Aquatic and Fishery Sciences, University of Washington, for providing the subsidy required to publish this volume.

*TWP*
*University of Washington*
*October, 2008*

# PART ONE

# THE BIOLOGY OF DEEP-SEA ANGLERFISHES

INTRODUCTION AND HISTORICAL PERSPECTIVE
    Historical Perspective

WHAT MAKES AN ANGLERFISH?
    Characters Shared by Both Sexes
    Characters Restricted to Females
    Characters Restricted to Males
    Characters Restricted to Larvae

BIODIVERSITY
    Suborder Ceratioidei Gill
    Family Centrophrynidae Bertelsen
    Family Ceratiidae Gill
    Family Himantolophidae Gill
    Family Diceratiidae Regan and Trewavas
    Family Melanocetidae Gill
    Family Thaumatichthyidae Smith and Radcliffe
    Family Oneirodidae Gill
    Family Caulophrynidae Goode and Bean
    Family Neoceratiidae Regan
    Family Gigantactinidae Boulenger
    Family Linophrynidae Regan

EVOLUTIONARY RELATIONSHIPS
    Ordinal Relationships
    Subordinal and Familial Relationships
    Key to the Major Subgroups of the Lophiiformes
    Relationships of Ceratioid Families and Genera
    Characters and Character States

    Comparisons with Previous Hypotheses
    Characters Restricted to Metamorphosed Males
    Characters Restricted to Larvae
    Conflicting Molecular Evidence
    Sexual Parasitism

GEOGRAPHIC DISTRIBUTION
    Seasonal Distribution
    Vertical Distribution
    Distribution Relative to Physical, Chemical, and Biological Parameters
    Geographical Distribution

BIOLUMINESCENCE AND LURING
    Internal Structure of the Esca
    Other Bioluminescent Structures
    Biological Significance of Luminous Structures

LOCOMOTION, FOOD, AND FEEDING
    Locomotion
    Food and Feeding
    Biomechanics of Feeding
    Specialized Feeding Mechanisms
    What Eats a Ceratioid?

REPRODUCTION AND EARLY LIFE HISTORY
    Family Accounts
    Mate Location and Species-Specific Selection
    Modes of Reproduction
    Reproductive Modes and Phylogeny

ONE

## Introduction and Historical Perspective

> The medieval imagination, rioting in strange imps and hobgoblins, could hardly have invented anything more malevolent in appearance than the ceratioids or deep-sea anglerfishes, sometime called black devils. Naturally enough, these black sea devils live in the kingdom of darkness, where they prowl about seeking whom they may devour. Many of them even carry a sort of torch, illumined with a phosphorescent glow, with which they lure their victims within reach of their devouring, traplike jaws.
>
> WILLIAM KING GREGORY AND GEORGE MILES CONRAD,<br>"The Evolution of the Pediculate Fishes," 1936:193

No place on earth can compete with the enormity of physical and biological constraints imposed on life in the deep oceanic midwaters. With temperatures near freezing, the absence of solar radiation, inconceivable pressures from the weight of water above, and biomass so low that meals are far and few between, it is almost inconceivable that animals could occupy this vast and forbidding habitat (Fig. 1). Yet fishes are there in surprising profusion, having adapted to these extreme limitations in a host of bizarre and unpredictable ways. Few groups, however, are as prolific and spectacular as the deep-sea ceratioid anglerfishes.

Ceratioids are part of a much larger assemblage—an order of teleost fishes called the Lophiiformes—nearly all of which share a peculiar and unique mode of feeding characterized most strikingly by the structure of the first dorsal-fin spine (called the "illicium"), placed out on the tip of the snout and modified to serve as a luring apparatus to attract prey. The 18 families, 66 genera, and approximately 323 living species of the Lophiiformes are distributed among five suborders (Pietsch, 1984): the Lophioidei (Caruso, 1981, 1983, 1985, 1986; Caruso and Bullis, 1976; Caruso and Suttkus, 1979), Antennarioidei (Last et al., 1983, 2007; Pietsch and Grobecker, 1987), Chaunacoidei (Caruso, 1989a, 1989b), Ogcocephaloidei (Ochiai and Mitani, 1956; Bradbury, 1967, 1980, 1988, 1999; Endo and Shinohara, 1999), and deep-sea Ceratioidei (Fig. 2). The most phylogenetically derived of these suborders is the Ceratioidei, distributed throughout the world's oceans below a depth of 300 m (Pietsch, 1984; Pietsch and Orr, 2007). With 160 species, it constitutes by far the most species-rich vertebrate taxon within the bathypelagic zone and below (Fig. 1), containing more than twice as many families and genera and more than three times the number of species as the whalefishes—suborder Cetomimoidei—the next most species-rich deep-sea vertebrate taxon (see Paxton, 1998; Herring, 2002). At the same time, new species are being added to the suborder at a steady if not increasing rate.

Members of the group differ remarkably from their less-derived, bottom-living relatives by having an extreme sexual dimorphism (shared by all contained taxa) and a unique mode of reproduction in which the males are dwarfed—those of some linophrynids, adults at 6 to 10 mm standard length, competing for the title of world's smallest mature vertebrates (see Winterbottom and Emery, 1981; Roberts, 1986; Weitzman and Vari, 1988; Kottelat and Vidthayanon, 1993; Watson and Walker, 2004; Pietsch, 2005b; Kottelat et al., 2006; Guinness World Records, 2007:41)—and attach themselves (either temporarily or permanently) to the bodies of relatively gigantic females (Figs. 3, 4). In *Ceratias holboelli*, the Northern Giant Seadevil, where the most extreme examples are found, females may be more than 60 times the length and about a half-a-million times as heavy as the males (Bertelsen, 1951; Pietsch, 1976, 1986b, 2005b). The males lack a luring apparatus, and those of most species are equipped with large well-developed eyes (Munk, 1964, 1966) and huge nostrils (Marshall, 1967a, 1967b), the latter apparently used for homing in on a female-emitted, species-specific chemical attractant (Bertelsen, 1951; Pietsch, 1976, 2005b; Munk, 1992). Normal jaw teeth of males are lost during metamorphosis but are replaced by a set of pincerlike denticles at the anterior tips of the jaws for grasping and holding fast to a prospective mate (Figs. 5, 6).

In some taxa, attachment is followed by fusion of epidermal and dermal tissues and, eventually, by an apparent connection of the circulatory systems so that the male becomes permanently dependent on the female for blood-transported nutrients, while the host female becomes a kind of self-fertilizing hermaphrodite (Regan, 1925a, 1925b, 1926; Parr, 1930b; Regan and Trewavas, 1932; Bertelsen, 1951; Pietsch, 1975a, 1976, 2005b; Munk and Bertelsen, 1983; Munk, 2000). Permanent attachment is usually accomplished by means of separate outgrowths from the snout and tip of the lower jaw of the male, both of which eventually fuse with the skin of the female. In some species, a papilla of female tissue protrudes into the mouth of the male, sometimes appearing to completely occlude the pharynx. The heads of some males become broadly fused to the skin of the female, extending from the tip of the

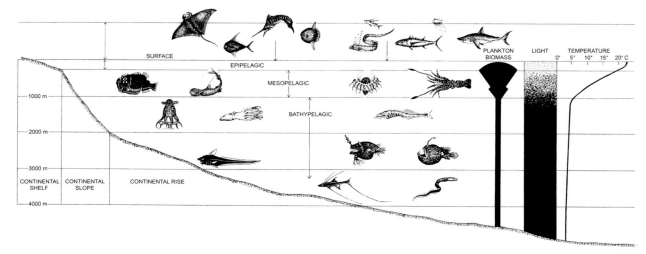

FIGURE 1. The ocean divided. For ease of discussion, oceanographers have divided the oceanic realm into a number of "zones" that are defined by both biological and physical-chemical parameters: the epipelagic zone extends from the surface to 200 m, a depth that corresponds on average to the margin of the continental slope, which in turn is approximately equivalent to the lower limit of photosynthesis, often called the euphotic zone; in terms of fishes, it provides habitat for many large, fast-swimming, predaceous forms like the tunas and mackerels, dolphinfishes, billfishes, and ocean-going sharks; it also serves as a nursery ground for tons of eggs and larvae of deeper-living forms. The mesopelagic zone lies between 200 and 1000 m, this greater depth corresponding to the limit of penetration of solar radiation, often called the twilight zone or disphotic zone; it supports mostly silvery fishes with large eyes and light-producing structures called photophores; communication between individuals and species is primarily mediated by vision and biological light. The bathypelagic zone extends from 1000 to 4000 m, below the extent of solar radiation, often called the aphotic zone; it contains mostly black fishes with small or degenerate eyes, poorly developed musculature, and weakly ossified skeletons; communication is mostly by way of waterborne chemicals and a highly developed sense of smell. The abyssopelagic zone includes all the deepest parts of the ocean below 4000 m, providing habitat for things like hagfishes, certain squaloid sharks, deep-sea skates, deep-sea cods, eelpouts, snailfishes, rattails, and more; communication is largely through sound production and well-developed hearing abilities. At the right of the diagram are represented the change of biomass with depth, the penetration of sunlight, and a typical temperature profile of a warm ocean. While the term "pelagic" refers to everything that lies within the water column, benthopelagic describes all those phenomena and things that are closely associated with the bottom, regardless of depth, and benthic is reserved strictly for those things that spend all or nearly all their time resting on the bottom. Epipelagic animals (top row) from left to right: giant devil-ray *(Manta)*, ocean bream *(Tarachtichthys)*, spearfish *(Tetrapturus)*, ocean sunfish *(Mola)*, ribbon-fish *(Regalecus)*, flying-fish, albacore tuna, Portuguese-man-of-war *(Physalia)*, and great white shark *(Carcharodon)*. Mesopelagic animals, left to right: hatchet-fish *(Agyropelecus)*, viperfish *(Chauliodus)*, coronate jellyfish *(Atolla)*, and squid *(Histioteuthis)*. Bathypelagic animals, left to right: *Vampyroteuthis*, midwater octopus *(Amphitretus)* both of which live also at mesopelagic levels, *Cyclothone*, and immediately below, two female ceratioid anglerfishes *(Linophryne* and *Melanocetus)*. Bottom-dwelling fishes, left to right: Rat-tail *(Nezumia)*, a benthopelagic form, tripodfish *(Bathypterois)*, a benthic form, and a deep-sea eel *(Synaphobranchus)*, a benthopelagic form. Modified after Marshall (1974).

lower jaw to the rear of the skull, appearing as if embedded or absorbed by their mate, while in others, the male is carried at the tip of an elongate, cylindrical stalk of female tissue. Increasing considerably in size once fused, their volume becoming much greater than free-living males of the same species, and being otherwise completely unable to acquire nutrients on their own, the males are considered to be parasites. They apparently remain alive and reproductively functional as long as the female lives, participating in repeated spawning events. A single male per female appears to be the rule in some taxa, but in others multiple attachments are relatively common, with as many as eight coupled to a single host (Saruwatari et al., 2001). Since its discovery more than 80 years ago (Saemundsson, 1922; Regan, 1925a, 1925b), the story of sexual parasitism in ceratioid anglerfishes has become a part of common ichthyological knowledge. However, the known facts concerning this remarkable reproductive mode have never been thoroughly or satisfactorily analyzed, despite the work of Bertelsen (1951) and more recently of Munk and Bertelsen (1983), Munk (2000), and Pietsch (2005b). The physiological mechanisms (endocrinological and immunological) that allow for sexual parasitism, which could be of significant biomedical importance, have never been explored.

Ceratioid anglerfishes differ further from their less-derived bottom-dwelling relatives in having a bacterial light organ that serves as bait to attract prey (for the possibility of bioluminescence in

FIGURE 2. The five suborders of anglerfishes that together form the teleost order Lophiiformes: (A) Lophioidei, containing the goosefishes or monkfishes, a single family, four genera, and 25 living species of shallow to deepwater, dorsoventrally flattened forms (represented here by *Lophiodes reticulatus*, 157 mm SL, UF 158902, dorsal and lateral views, photos by J. H. Caruso); (B–E) Antennarioidei, the frogfishes and handfishes, four families, 15 genera, and about 56 species of laterally compressed, shallow to moderately deepwater, benthic forms (represented by *Antennarius commerson*, 111 mm SL, UW 20983, photo by D. B. Grobecker; *Antennarius commerson*, about 150 mm SL, specimen not retained, photo by F. Schneidewind; *Antennarius striatus*, 150 mm SL, specimen not retained, photo by F. Schneidewind; *Sympterichthys politus*, specimen not retained, photo by R. Kuiter); (F) Chaunacoidei, the gapers, coffinfishes, or sea toads, a single family, at least two genera, and as many as 14 species of globose, deepwater benthic forms (*Chaunax umbrinus*, 305 mm SL, BPBM 17344, photo by J. E. Randall); (G) Ogcocephaloidei, the batfishes, a single family of 10 genera and about 68 species of dorsoventrally flattened, deepwater benthic forms (*Halieutaea retifera*, 102 mm SL, BPBM uncataloged, dorsal view, photo by J. E. Randall); and (H) Ceratioidei, the seadevils, containing 11 families, 35 genera, and 160 currently recognized species of globose to elongate, mesopelagic, bathypelagic, and abyssal-benthic forms (*Diceratias trilobus*, 86 mm SL, AMS I.31144-004, photo by T. W. Pietsch).

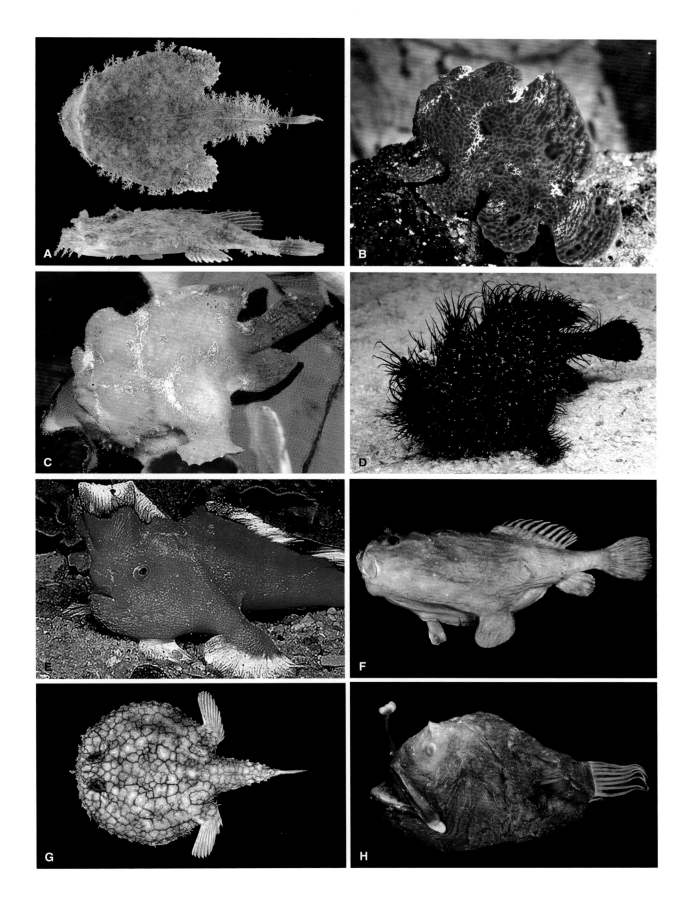

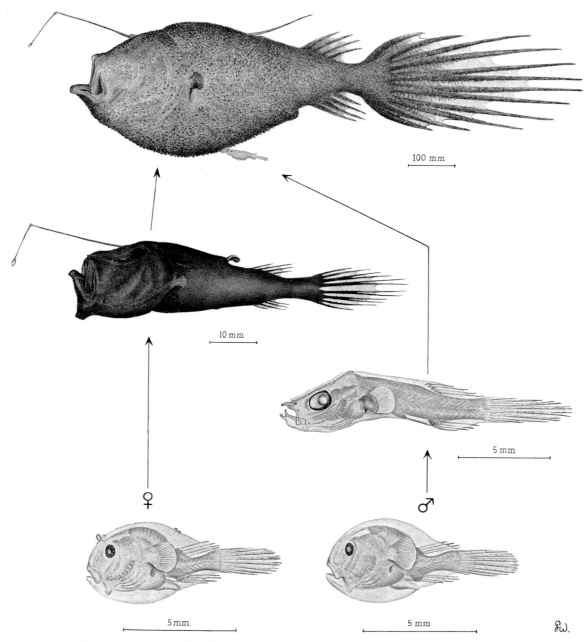

FIGURE 3. The life history of *Ceratias holboelli* Krøyer, the Northern Giant Seadevil: female and male larvae, shown at the bottom of the diagram, exist in the plankton of the epipelagic zone in roughly equal numbers; rudiments of the developing illicium are present even in the tiniest females, thus gender can be easily determined immediately after the eggs hatch; during metamorphosis the female rapidly increases in size, the illicium elongates, and the skin becomes darkly pigmented, while the eyes become reduced and eventually covered over with black skin; the male develops pincherlike denticular teeth on the tip of the snout and chin, and the eyes increase in relative size as he begins his search for a conspecific female; once found, the male attaches to the female, and the tissues of the two eventually fuse, resulting in permanent parasitic conjugation. Drawings by Poul H. Winther; after Bertelsen (1951).

batfishes, see Crane, 1968). Placed out on the tip of the illicium, the bait, or "esca" as it is technically called, is species specific; that is, in external appearance it differs without exception in all members of the suborder. Parr (1927) was the first to recognize the diagnostic value of the external morphology of escae in ceratioids, pointing out the need for a closer examination of individual variation in the structure of this organ. Regan and Trewavas (1932:3–4) agreed, employing small differences in escal morphology to introduce 45 oneirodid species as new to science (in the following quote, for "illicium," read "esca"):

It now appears that the Oneirodidae may attain the form of the adult when quite small, and that in this family, specimens only 15 to 20 mm long may show a definite and characteristic structure of the illicium. Although we cannot be sure that specimens that differ only in slight details of the illicium are specifically distinct, and we do not like basing new species on single specimens 15 mm in total length, we consider that to give specific names freely is probably the best way to describe the material now available.

Since that time, differences in the number, shape, and size of escal appendages and filaments, as well as variation in exter-

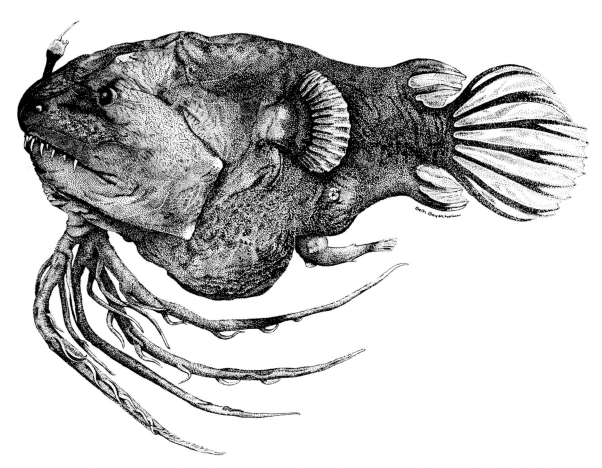

FIGURE 4. *Linophryne brevibarbata*, 100-mm SL female, with an 18.5-mm SL parasitic male, BMNH 1995.1.18.4, a North Atlantic species known from only six specimens. Drawing by Elisabeth Beyerholm; after Bertelsen (1980a).

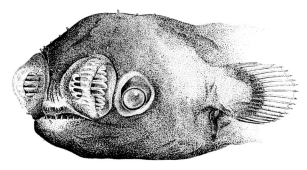

FIGURE 5. Head of a free-living male of the *Linophryne arborifera* group, 18.5 mm SL, BMNH 2004.7.5.1, showing the denticular teeth and extremely well-developed eyes and nostrils. Drawing by Robert Nielsen; after Bertelsen (1980a).

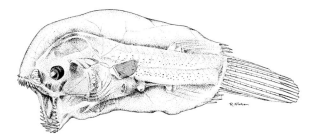

FIGURE 6. A free-living male of *Haplophryne mollis*, 13 mm SL, ZMUC (specimen sacrificed for histological study), with its mouth agape, indicating the effectiveness of the jaw apparatus in attaching to a female. Drawing by Robert Nielsen; after Munk and Bertelsen (1983).

nal escal pigment patterns, have been, for the most part, the sole basis on which new species have been described (e.g., see Pietsch, 1974a; Bertelsen et al., 1981; Bertelsen and Krefft, 1988).

The internal structure of ceratioid escae is infinitely more complex, involving a confusing array of bacteria-filled vesicles, light-absorbing pigment layers, reflecting tissues, tubular light-guiding structures, nerves, blood vessels, and smooth muscle fibers (Munk and Bertelsen, 1980; Munk, 1988, 1998, 1999; Herring and Munk, 1994; Munk and Herring, 1996; Munk et al., 1998). There is some evidence also that ceratioid escae contain pheromone-producing secretory glands that function to attract a conspecific male (Munk, 1992), but the true nature and adaptive significance of these structures and most of the other internal parts of escae are unknown.

In addition to the esca, all 22 currently recognized species of the ceratioid genus *Linophryne* (family Linophrynidae) bear an elaborate bioluminescent chin barbel, the light of which does not originate from symbiotic luminescent bacteria, but rather from a complex array of intrinsic, intracellular, paracrystalline photogenic granules; the bacteria-filled esca is ectodermal in origin, whereas the barbel light organ appears to be derived from the mesoderm (Hansen and Herring, 1977). This remarkable dual system, involving two entirely separate mechanisms of light production, is unique among animals.

FIGURE 7. Henrik Nikolai Krøyer (1799–1870), a painting by his stepson, the well-known Norwegian-Danish artist Peder Severin Krøyer (1851–1909), dated 1872. Courtesy of Peter Rask Møller, Jørgen Nielsen, and the Zoological Museum, University of Copenhagen.

In summary, ceratioid anglerfishes are among the most intriguing of all animals, possessing a host of spectacular morphological, behavioral, and physiological innovations found nowhere else. The suborder is taxonomically diverse: with 160 currently recognized species (and many more certain to be discovered in the future), it forms a major contribution to the biodiversity of the deep sea. It is exceedingly widespread geographically, occurring in deep waters of all major oceans and seas of the world, from high Arctic latitudes to the Southern Ocean; while some species appear to be almost cosmopolitan in distribution, many others have surprisingly small, restricted vertical and horizontal ranges. Their relative abundance, high species diversity, and trophic position as the top primary carnivores in meso- and bathypelagic communities make them important ecologically. Their unique mode of reproduction has significant biomedical implications to the fields of endocrinology and immunology. Yet, despite these many aspects of biological interest and importance, as well as a large amount of revisionary work published in the 1970s and early 1980s—including repeated attempts to resolve phylogenetic relationships—ceratioid anglerfishes have remained poorly known. Well short of providing all the answers, it is hoped that this monograph will have established a firm basis for future research and discovery.

## Historical Perspective

In 1833, following a severe storm, a strange and unknown fish was washed ashore near Godthaab, southwest Greenland. It was remarkable for its nearly spherical head and body, large mouth, the absence of pelvic fins, thick black skin sparsely set with large spine-bearing bony plates, restricted gill openings,

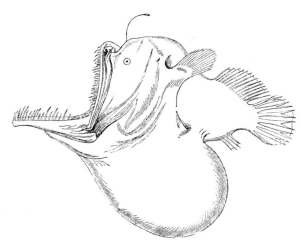

FIGURE 8. Albert Günther's holotype of *Melanocetus johnsonii*, female, 64 mm SL, BMNH 1864.7.18.6, brought to London from Madeira by James Yates Johnson (c. 1820–1900) and described by Günther in 1864:301: "It must be extremely rare, as the specimen entrusted to me by Mr. Johnson for description, and presented by him to the British Museum, is the only one which has ever come to the knowledge of naturalists." After Günther (1880).

and a conspicuous flexible appendage arising from a deep longitudinal groove on the forehead, terminating in a series of elongate filaments (Waterman, 1939a). The specimen was badly damaged by birds and decay, but fortunately it was preserved and given to Captain-Lieutenant Carl Peter Holbøll (1795–1856), who in turn sent it to Johannes Christopher Hagemann Reinhardt (1776–1845), then professor of zoology at the University of Copenhagen. In 1837, Reinhardt described the specimen and named it *Himantolophus groenlandicus*, the first representative of the suborder Ceratioidei known to science. Reinhardt noted its similarities to *Lophius* and *Antennarius* in body shape, in position and structure of the gill openings, and in the development of the large anterior cephalic fin spine, but the differences from these well-known shallow-water genera, especially the absence of pelvic fins, left him in doubt about its relationships (Bertelsen, 1951). In the end, however, he concluded that his new species was most probably related to those fishes that comprise the teleost order Pediculati (Regan, 1912).

The same Captain Holbøll who sent the first known specimen of *H. groenlandicus* to Reinhardt later obtained from deep waters off Greenland two additional ceratioids (originally called *barbugede Tudsefiske* or "bare-bellied toadfishes" by the Danes, because pelvic fins are absent in metamorphosed specimens of all known representatives). One of these specimens was described in 1845 by Danish ichthyologist and carcinologist Henrik Nikolai Krøyer (Fig. 7) as *Ceratias holboelli*. The other, sent by Holbøll to physiology and anatomy professor Daniel Frederick Eschricht (1798–1863), was eventually accessioned into the collections of the University of Copenhagen, where it remained unnoticed for the next 25 years.

In the meantime, Albert Carl Ludwig Gotthilf Günther (1830–1914), keeper of zoology at the British Museum in London, received a very different looking ceratioid from James Yates Johnson (c. 1820–1900) who brought it back from Madeira in 1864: "A fish which proves to be the type of a new genus, not only on account of its extraordinary form, but also on account of the absence of pelvic fins. In the latter respect it agrees with *Ceratias* from the coast of

FIGURE 9. Christian Frederik Lütken's holotype of *Himantolophus reinhardti*, female, 328 mm SL, ZMUC 66, a name now in the synonymy of *Himantolophus groenlandicus*; by any standard, one of the most accurate and detailed illustrations of a ceratioid ever published. After Lütken (1878a).

Greenland, from which, however, it differs in its dentition" (Günther, 1864:301). This new form, remarkable for its large pendulous belly—which when opened was found to contain, "rolled up spirally into a ball, a Scopeline fish" nearly twice the length of the angler—was described by Günther in that same year and named *Melanocetus johnsonii* in honor of its collector (Fig. 8).

Some time later, back in Denmark, yet another Danish professor, Christian Frederik Lütken (1827–1901), came across Eschricht's long forgotten Greenland specimen in the collections of the University of Copenhagen. Recognizing it as distinct from the three previously known genera, but most similar to Günther's *Melanocetus* in lacking skin spines, among other similarities, he described it in 1871 (1872 in English translation) under the name *Oneirodes eschrichtii*: "That the Greenland form is specifically distinct from the deep-sea Lophioid from Madeira *[Melanocetus]*, which has been so often mentioned, is seen at first glance. Their differences, notwithstanding their resemblance in many essential features, are very sharply marked; nay, I consider that it will even be admitted that they are great enough for the establishment of a generic distinction" (Lütken, 1872:330–331).

In many ways, Lütken's 1871 paper and another published in 1878—in which he described *Himantolophus reinhardti* based on a second known specimen of the genus, also from West Greenland—represent "remarkable pioneer work," not only on the Himantolophidae, but on ceratioids in general (Bertelsen and Krefft, 1988:10). In addition to providing exact and extremely detailed descriptions and illustrations of the specimens (Figs. 9, 10), Lütken (1878a) was the first to recognize that the four ceratioid genera known at that time, *Himantolophus*, *Ceratias*, *Melanocetus*, and *Oneirodes*, together represent a discrete taxon (which he called *Lophioidea apoda* or *les Cératiades*) within the then-recognized anglerfish order Pediculati: "Between all of these [genera] there is the nearest affinity; and they seem to form a very natural little group of deep-sea Lophioids, of weak vision and destitute of pelvic fins, within the great family of the Halibatrachi" (Lütken, 1871:70, 1872:340; see also Lütken, 1878b:342, 343). In this he preceded Theodore Nicholas Gill (1837–1914), who in 1873 included *Himantolophus*, *Ceratias*, and *Oneirodes* in the family Ceratiidae, the latter family-group name dating to Gill (1861). Although Reinhardt (1837), in describing the illicium of *Himantolophus*, certainly implied its use as a lure, Lütken (1878a) was the first to specifically state the probability that the first dorsal-fin spine (as in other lophiiform fishes) is used to attract prey (p. 325) and that the terminal bait or esca is bioluminescent (p. 313; but see also Willemoes-Suhm, 1876, and the somewhat later account of Collett, 1886:138, 142). He also provided the first osteological description of a ceratioid (Fig. 11), as well as a discussion of the importance of escal morphology, including the position, number, and shape of the various escal appendages, as characters that can be used to differentiate taxa (1878a:318; Fig. 10). Finally, he was the first to describe and illustrate a larval ceratioid, correctly referring it to *Himantolophus* (1878a:321–324; Fig. 10).

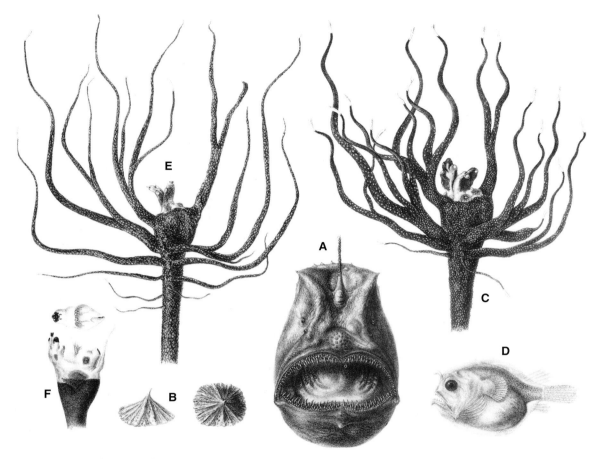

FIGURE 10. Illustrations taken from Lütken's publication on "Dybhavs-Tudsefiske" (deep-sea toadfishes): (A) Anterior view of the holotype of *Himantolophus reinhardti*, female, 328 mm SL, ZMUC 66; (B) lateral and dorsal views of large conical spines from the skin of *Himantolophus reinhardti*; (C) esca of the holotype of *Himantolophus reinhardti*; (D) a larval female of *Himantolophus* sp., about 18 mm SL; (E) esca of the holotype of *Himantolophus groenlandicus*, about 465 mm SL, ZMUC 65; (F) lateral and dorsal views of the esca of the holotype of *Oneirodes eschrichtii*, 153 mm SL, ZMUC 64, described by Lütken in 1871. After Lütken (1878a).

Following these initial discoveries, a few additional species were described based on single specimens caught incidentally in trawls or, more often, found dead on shore or floating on the surface. Notable among the latter are *Himantolophus appelii*, described in 1878 by Frank Edward Clarke (1849–1899) from a single specimen found stranded on a New Zealand beach (Fig. 12); and *Linophryne lucifer*, described in 1886 by Robert Collett (1842–1913) from a specimen found floating alive on the surface off Madeira, apparently incapacitated by ingestion of a large fish (Fig. 13). But it was not until the great oceanographic expeditions of the late nineteenth and early twentieth centuries that ceratioids become known in any reasonable numbers. The circumnavigation of the HMS *Challenger* (1872–1876) brought back half a dozen specimens, by far the largest single collection to date and the first specimens to be collected in their natural environment. From this material, Sir John Murray (1841–1914), an eminent deep-sea biologist and one of the founders of modern oceanography (Mills, 1989), described *Ceratias uranoscopus* in Charles Wyville Thomson's (1830–1882) *Voyage of the Challenger*, first published in 1877. Likewise, from this same collection, Günther, in his "Challenger" report on the deep-sea fishes of 1887, described *Melanocetus murrayi* and *Diceratias bispinosus*, recognizing overall eight genera and 11 ceratioid species based on the 13 specimens then known in collections around the world.

Material collected during the numerous scientific cruises of the U.S. Fish Commission Steamer *Albatross* (from 1883 to 1920) provided considerably larger collections (Hobart, 1999), from which Gill described *Cryptopsaras couesii* in 1883; George Brown Goode (1851–1896) and Tarleton Hoffman Bean (1846–1916), *Caulophryne jordani* in 1896; Samuel Walton Garman (1843–1927), *Dolopichthys allector* in 1899; Hugh McCormick Smith (1865–1941) and Lewis Radcliffe (1880–1950), *Dermatias platynogaster* and *Thaumatichthys pagidostomus*, both in 1912; and Charles Henry Gilbert (1859–1928), *Oneirodes acanthias* in 1915.

Samuel Garman's (1899) description of *Dolopichthys allector* that appeared in his "Report on an exploration off the west coasts of Mexico, Central and South America, and off the Galapagos Islands, in charge of Alexander Agassiz, by the U. S. Fish Commission steamer Albatross" is remarkable for its accuracy, detail, and beautiful illustrations (Figs. 14 through 16). With only a single small specimen available to him, Garman, through a series of careful dissections, was able to describe not only the complete skeleton, but the musculature as well, including that of the head, body, fins, and even the illicial apparatus. More surprising are his descriptions and figures of the gills and the viscera, the latter providing the earliest depiction of the internal organs of an anglerfish. In all of this he was well ahead of his contemporaries, and for completeness and accuracy he has rarely been eclipsed since.

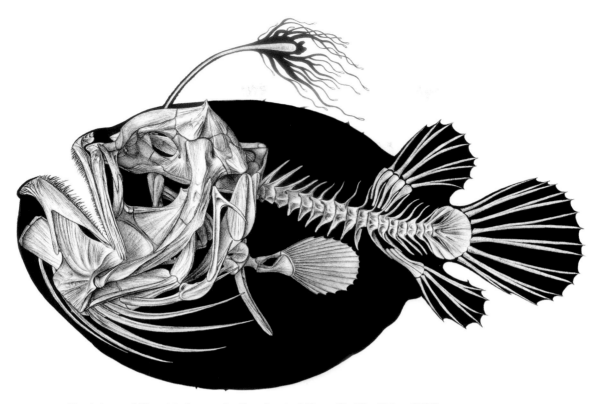

FIGURE 11. The skeleton of *Himantolophus groenlandicus*, female, 148 mm SL. After Lütken (1887).

August Bernhard Brauer (1863–1917), one-time director of the Zoological Museum of the Humboldt University of Berlin, in his 1902 report on the fishes collected during the German *Valdivia* Expedition of 1898–1899, described four new ceratioid species, recognizing overall a total of 23 species in three families, the Ceratiidae, Gigantactinidae, and Aceratiidae. One of these families, the Aceratiidae, was introduced by Brauer (1902) to contain some small anglerlike fishes that were similar to ceratiids and gigantactinids in the absence of pelvic fins and having restricted gill openings, but differed in lacking a cephalic luring apparatus. A decade later, Charles Tate Regan (1878–1943) of the British Museum of Natural History (Fig. 17), in revising the classification of anglerfishes based in part on an osteological examination, placed all those taxa lacking pelvic fins in a separate suborder, the Ceratioidei, which he divided into six families, including Brauer's Aceratiidae (Regan, 1912). Neither Brauer (1902) nor Regan (1912) was aware at the time that these little ceratioids without lures were actually the males of species based at that time only on females. The discovery of the remarkable sexual dimorphism that characterizes all ceratioids would have to wait until May 1924, when Regan realized that a small fish attached to the side of a large female *Ceratias holboelli* was in fact a "dwarfed parasitic male" (Regan, 1925a:41, 1925b:386; for more on dwarfed males and the history of the discovery of sexual parasitism, see Reproduction and Early Life History, Chapter Eight).

By far the greatest contribution to our knowledge of the Ceratioidei up until that time came from the Danish oceanographic expeditions aboard the Royal Danish Research Ship *Dana* under the direction of Johannes Schmidt (1877–1933). Prior to these efforts, there were only about 100 known specimens, distributed among 16 genera and 24 species—including the 40 or so specimens collected during the *Michael Sars* North Atlantic Deep-Sea Expedition of 1910, which were not described until 1944 by Einar Koefoed (but see John Murray and Johan Hjort, 1912; and Regan, 1926:3). However, the 1920–1922 *Dana* Expedition to the North Atlantic, Caribbean Sea, and Gulf of Panama alone took 217 specimens (excluding larvae). In reporting on this material, Regan (1925c, 1926) recognized 39 species, 27 of which were described as new, more than doubling the number of known species. The 1928–1930 Danish "Oceanographic Expedition Round the World," again under the leadership of Schmidt and funded this time by the Carlsberg Foundation, brought back 172 specimens, from which Regan, in coauthorship with Ethelwynn Trewavas (1900–1993; see Greenwood, 1994), described 79 new species, once again more than doubling the number of recognized forms (Regan and Trewavas, 1932). These descriptions based on the *Dana* collections—in addition to those published by Albert Eide Parr (1927, 1930, 1934), William Beebe (1926, 1932), Nikolai Andreevich Borodin (1930a, 1930b), Viktor Pietschmann (1926), John Roxbrough Norman (1930), Leonard Peter Schultz (1934), Alan Fraser-Brunner (1935), Gérard Belloc (1938), Wilbert McLeod Chapman (1939), Talbot Howe Waterman (1939b), Sadahiko Imai (1941), and Beebe and Jocelyn Crane (1947)—greatly increased the known ceratioid biodiversity, so that by the time of Erik Bertelsen's (1951) worldwide revision of the suborder, the number of described species had risen to 194. A good number of these, however, have since fallen into synonymy; for example, of the 106 species described by Regan, either alone or in coauthorship with Trewavas, only 41 are recognized today; of the 17 described by Parr, only four are still recognized; and of the 20 described by Beebe, alone or in coauthorship with Crane, only six are still recognized today. Despite this high attrition, however, new ceratioids continue to be discovered.

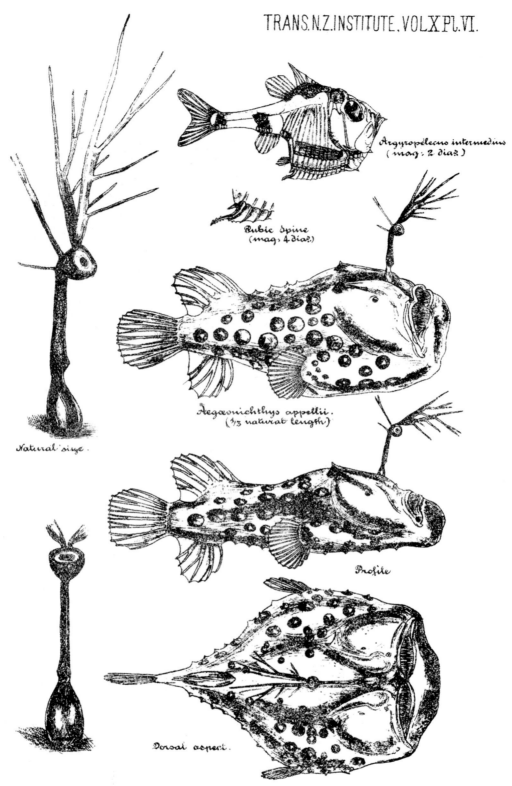

FIGURE 12. Frank Edward Clarke's illustrations of the holotype of *Aegoeonichthys appelii*, female, 287 mm SL (specimen not retained), now known as *Himantolophus appelii*: "The fish forms a new genus in the family of Pediculati and is truly 'a king among kings' in a class of fishes containing some of the most grotesque forms in nature." After Clarke (1878:243).

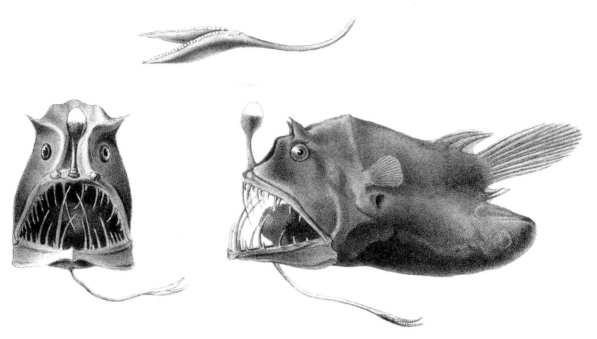

FIGURE 13. The holotype of *Linophryne lucifer*, female, 31.5 mm SL, ZMUO J.5560, discovered off the coast of Madeira by Captain P. Andresen, while on a voyage to the West Indies: "He was capturing turtle in his boat; there was a heavy swell, but the water was smooth. After a time he caught sight of this little black fish, which lay on the surface quite alive, but almost motionless, which was not surprising when it was discovered that it had just swallowed a fish longer than itself. It did not lie on its side, but was apparently unable to swim away. By getting the bailer under it, he lifted it out with ease, and in order to keep it fresh he gave up his search for turtle and rowed back to the ship, where it was placed in spirit for preservation" (Collett, 1886:143). After Collett (1886).

From an early date, it was believed that ceratioids were "degenerate" relatives of shallow-water, inshore anglerfishes, living much like *Lophius*, but at great depths in the open ocean, luring prey with their highly modified first dorsal-fin spine, while lying on soft muddy bottoms: "The Bathybial Sea-devils . . . are degraded forms of *Lophius*; they descend to the greatest depths of the oceans. Their bones are of an extremely light and thin texture, and frequently other parts of their organization, their integuments, muscles, and intestines, are equally loose in texture when the specimens are brought to the surface. In their habits they probably do not differ in degree from their surface representatives" (Günther, 1887:50).

When Reinhardt (1837) described the "frontal appendage" (illicium and esca) of *Himantolophus groenlandicus* as being the same as, but more highly developed than, that of *Lophius* and *Antennarius* (see Pietsch and Grobecker, 1987), he surely meant to imply its use as a lure to attract prey (Waterman, 1939a). Lütken (1871:62, 1872:334), however, as mentioned above, was the first to state this directly. In describing the bait at the tip of the first dorsal-fin spine of *Oneirodes eschrichtii*, he commented further on its resemblance to the head of a nereid polychaete worm (see Brusca and Brusca, 1990:383; Fig. 18): "I will not conceal that the whole arrangement has above all produced a 'mimetic' impression upon me, as if it were intended to resemble, e.g., the head of a Nereid; and I have been compelled to think of the old notions of the employment by the fishing-frog [*Lophius*] of its homologous frontal appendage as a means of attracting other fishes, which indeed, have given origin to its scientific specific name [*piscatorius*]."

Clarke (1878:243), in describing what is now called *Himantolophus appelii* (Fig. 12), agreed with Lütken's assessment: "The probable use of the tentacular appendage as an attractive lure, is beyond conjecture, as the habits of an allied fish (the angler), which is supplied with a far less complicated attachment, have been closely studied and provided to be 'a fish which angles for fish,' with a natural rod, line, and bait, and with certainly as deadly a 'creel' as any human disciple of Isaac Walton might have to which to relegate its captives."

While offering an alternative hypothesis for the function of the first dorsal-fin spine, Sir Wyville Thomson (1877:68), writing in reference to *Ceratias uranoscopus*, was explicit in describing its habitat as benthic (see also Goode, 1881:469, and Goode and Bean, 1896:490, where this same passage is repeated verbatim, without attribution):

> The presence of a fish of this group at so great a depth is of special interest. From its structure, and from the analogy of its nearest allies, there seems to be no reasonable doubt that it lives on the bottom. It is the habit of many of the family to lie hidden in the mud, with the long dorsal filament and its terminal soft expansion exposed. It has been imagined that the expansion is used as a bait to allure its prey, but it seems more likely that it is a sense-organ, intended to give notice of their approach.

A number of other early authors agreed with this bottom-living life-style (e.g., Goode, 1881:469; Gill, 1883b:284; Goode and Bean, 1896:490), but Henri Filhol (1885:81) went a step further when he published an unlikely picture of *Melanocetus johnsonii* in which several individuals (modeled after Günther, 1864; see Fig. 8) are shown buried up to their eyes in the mud, with their lures waving above their heads (Fig. 19). The fact that the nearly vertical mouth of this species opens upward at a sharp angle makes these fish look rather implausible buried tail-downward in the ooze of the ocean bottom (Waterman, 1939a).

Garman (1899:81), in his original description of *Dolopichthys allector* (Fig. 14), also remarked on the improbability of such fishes being active swimmers, calling them "degenerate pediculates

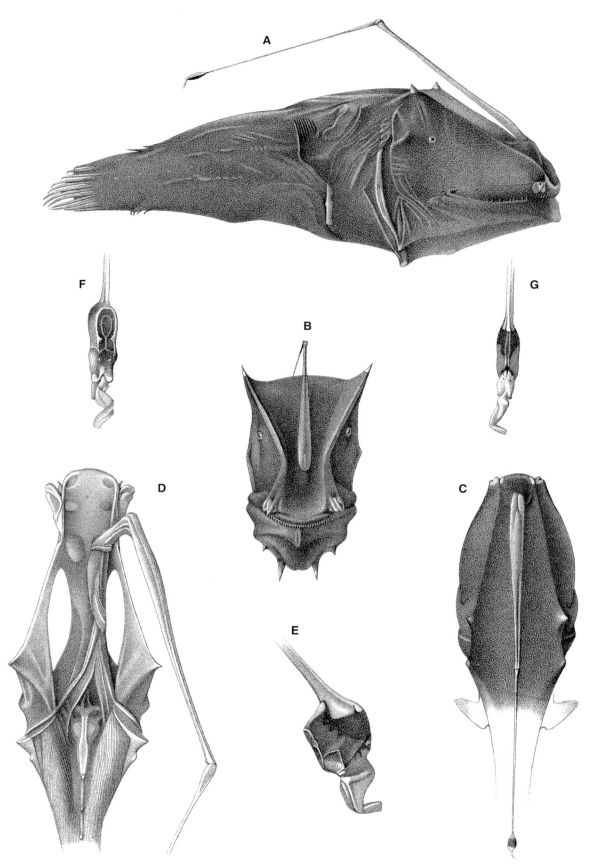

FIGURE 14. Illustrations from Samuel Garman's original description of the holotype of *Dolopichthys allector*, female, 61 mm SL, MCZ 28735: (A–C) lateral, anterior, and dorsal views; (D) head and body in dorsal view after removal of the skin, showing, among other things, the musculature of the illicial apparatus; (E–G) lateral, anterior, and posterior views of the esca. Lithograph by E. Meisel from drawings by A. M. Westergren; after Garman (1899).

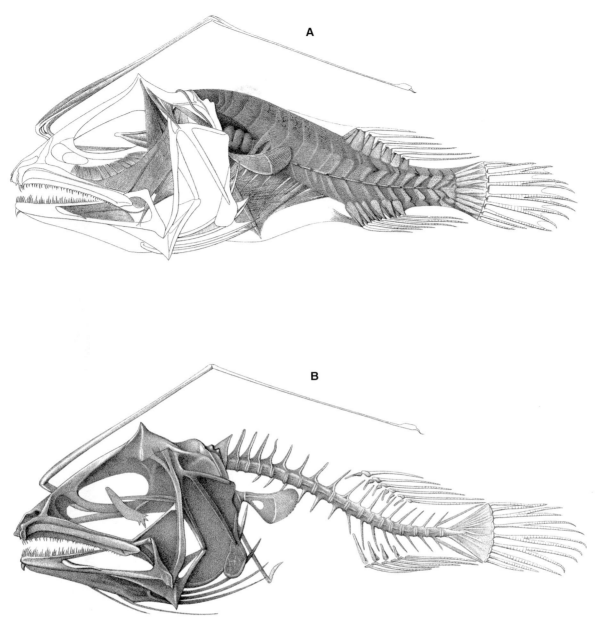

FIGURE 15. More illustrations from Samuel Garman's description of the holotype of *Dolopichthys allector*, female, 61 mm SL, MCZ 28735: (A) muscles and bones of the left side; (B) skeleton after removal of the soft tissues. Lithograph by E. Meisel from drawings by A. M. Westergren; after Garman (1899).

adapted to a life in the mud and ooze at great depths in the ocean, having fins more or less concealed in the skin and imperfectly suited to free progression off the bottom, and capturing prey by means of an illicium, a modification of the anterior dorsal spine." In his use of the word "illicium" for the first dorsal-fin spine, Garman (1899:15, 18, 75–77, 81–83) introduced a new term taken from the Latin, meaning "attraction" or "inducement." At the same time (1899:77, 82), he introduced to modern readers the related term "esca," the Latin word for "bait," which had been used rather frequently by classical Greek and Roman authors (e.g., Cicero, c. 45 BC; see Waterman, 1939a) as well as by the sixteenth-century naturalists (e.g., Rondelet, 1554; see Pietsch and Grobecker, 1987) in their descriptions of the feeding habits of the fishing-frogs *(Lophius piscatorius* and *Lophius budegassa)* of the Mediterranean. The term "illicium" was adopted rather quickly, first reinforced by Gill (1909) and somewhat later by Regan (1912), who immediately recognized its value, using it later throughout his two monographs on the Ceratioidei (Regan, 1926; Regan and Trewavas, 1932). In contrast, and for some unknown reason, the equally useful expression "esca" was slow to be accepted. In fact, no one after Garman (1899) seems to have used it, except Gill (1909) who gave brief mention, and subsequently by Waterman (1939a, 1939b) who reintroduced it and applied it to his original descriptions of *Gigantactis longicirra*, *Linophryne algibarbata*, and *Danaphryne nigrifilis*. But even then the term was not picked up for another decade, until Bertelsen (1951) used it throughout his worldwide revision of the suborder. Since that time, "esca," along with the previously accepted term "illicium," have become standards in describing anglerfish morphology.

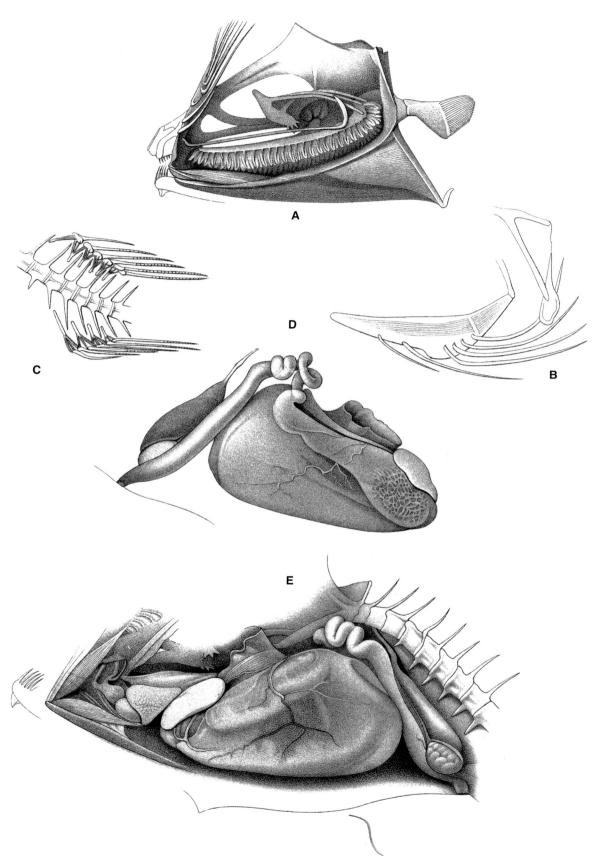

FIGURE 16. More illustrations from Samuel Garman's description of the holotype of *Dolopichthys allector*, female, 61 mm SL, MCZ 28735: (A) longitudinal section of the head, showing the gills and gill arches; (B) branchiostegal rays, hyoid apparatus, and opercular bones; (C) bones of the dorsal and anal fins; (D) stomach and intestine as viewed from the right side; (E) viscera as viewed from the left side. Lithograph by E. Meisel from drawings by A. M. Westergren; after Garman (1899).

FIGURE 17. Charles Tate Regan (1878–1943), renowned ichthyologist and director of the British Museum (Natural History), London. He singled out his discovery of dwarfed males and sexual parasitism in ceratioids as one of his major contributions to ichthyology, a finding that gave him special pleasure and satisfaction (see Burne and Norman, 1943:415). Courtesy of James Maclaine, Jamie Owen, and the Natural History Museum, London.

It was Brauer (1906) who first showed that the ceratioids were meso- and bathypelagic rather than bottom-living forms. He based his conclusion primarily on an analysis of discrete depth samples taken by a vertical closing trawl employed by the German research vessel *Valdivia* in 1898–1899, one of the earliest efforts to retrieve accurate depth distributions of organisms in deep oceanic waters. Brauer further pointed out that several ceratioids in the *Valdivia* collections were found with myctophiform and other pelagic fishes in their stomachs. We know now that ceratioid larvae generally occur at depths of less than 30 m and rarely below 200 m. The onset of metamorphosis leads to a rapid sinking to great depths, so that metamorphosis of females takes place at 2000 to 2500 m. Young juvenile females are most abundant at slightly shallower depths, 1500 to 2000 m, while older females seek slightly greater depths and are most common at about 2000 m (Bertelsen, 1951). In addition to the truly benthic ceratioid genus *Thaumatichthys* (see Bertelsen and Struhsaker, 1977), first discovered during the Danish *Galathea* Deep-Sea Expedition of 1950–1952, it is now understood that certain other ceratioid species associate with the bottom. Members of the diceratiid genera *Diceratias* and *Bufoceratias* are more often than not collected in bottom trawls, and benthic organisms have been found in their stomachs (Grey, 1959; Trunov, 1974; Uwate, 1979; Bertelsen, 1990; Anderson and Leslie, 2001; Pietsch et al., 2004:105). An unknown species of *Gigantactis* was recently videotaped from a submersible, swimming upside-down and apparently luring prey from soft muddy bottoms at a depth of 5000 m (Marzuola, 2002; Moore, 2002; see Locomotion, Food, and Feeding, Chapter Seven).

Of special importance to the understanding of ceratioid biology (as pointed out by Bertelsen, 1951) were the 1927 and 1930 publications of Albert Eide Parr (1890–1991). Based primarily on material collected during the oceanographic expeditions of the *Pawnee* and funded by the Bingham Oceanographic Foundation of Yale University, Parr (1927) was the first to demonstrate explicitly that escal morphology provides valuable characters for distinguishing species. He provided accurate descriptions and figures of this structure in introducing eight new species based on females, four of which are still recognized today (Fig. 20). In a subsequent publication, Parr (1930a) presented an osteological account of a specimen he called *Rhynchoceratias longipinnis*, comparing it to other forms then included in the Aceratiidae. Although his descriptions were detailed and accurate, "his attempt at a classification of the Aceratiids can only be regarded as evidence of the uncertainty of the correct placing of these fishes at that time" (Bertelsen, 1951:8). However, later in that same year, he (Parr, 1930b) found the key to the problem, showing that the specimen he had described earlier was a male with well-developed testes. He indicated the similarity between the known parasitic males and the aceratiids and suggested that all the latter are free-living stages of males, while all other ceratioids, which bear an illicium, are females:

> Shortly after having finished a study of the osteology of various genera of the aceratiid family of ceratioid deep-sea fishes, the writer [Parr] was, by general considerations, led to suspect that the family in question might in reality represent merely the free-living stages of the males which have heretofore only been known to science through Mr. C. Tate Regan's amazing discoveries of various more or less degenerate specimens attached in a presumably parasitic fashion to the skin of the much larger females. No such large females have ever been found in the family Aceratiidae; the latter, on the contrary, are exclusively known from a considerable number and variety of very small forms comparing very well with the parasitic males in regard to size. (Parr, 1930b:129)

Regan and Trewavas (1932) showed that it was now possible, based on Parr's (1930b) supposition, to place almost all the known free-living males within five of the 10 families then recognized on the basis of females (see Part Two, under Reallocation of Nominal Species of Ceratioids Based on Free-living Males). Their systematic revision of the suborder was based on a series of fundamental osteological characters common to all males and females of the same family, but which could not be used to establish taxonomic subunits within the families (Bertelsen, 1951). The free-living males were therefore referred to their own genera, separate from those based on females. Regan and Trewavas (1932:20) pointed out that two "male species," *Haplophryne mollis* and *Aceratias macrorhinus*, closely resemble the parasitic males of the "female genera" *Edriolychnus* and *Borophryne*, respectively. Because of the extreme sexual dimorphism, generic and specific divisions within the families were based on secondary sexual characters, with the result that the taxonomy of the females and males was radically different. In no case was there any agreement between the number of female and male genera and species within a given family. For example, two families, the Melanocetidae and Himantolophidae, each contained a "female genus" and two "male genera," and the number of species based on males was likewise greater than that of the females. On the other hand, Regan and Trewavas (1932:20) divided the Oneirodidae into 11 "female genera" containing 42 species, but only a single "male genus" with three species; two "male species" were placed within

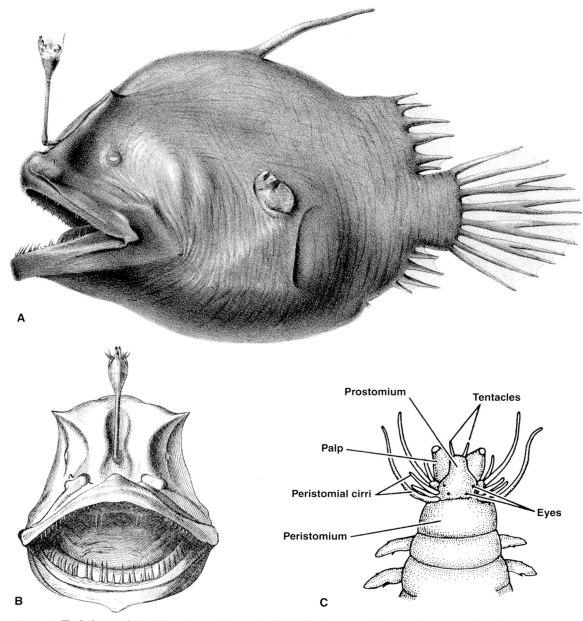

FIGURE 18. The holotype of *Oneirodes eschrichtii*, 153 mm SL, ZMUC 64, taken from Lütken's original description of 1871: (A, B) lateral and anterior views; (C) head of a nereid polychaete worm, after Brusca and Brusca (1990). Imagining that the esca might mimic the head of the worm, Lütken (1871) was the first to suggest mimicry in ceratioids as a means to attract prey.

a separate family, the Laevoceratiidae, as types of separate genera.

Such was the taxonomic confusion that faced Erik Bertelsen as he began his studies of the Ceratioidei in the early 1940s. Well aware that not all ceratioid species were known, he described the misfit between the number of genera and species as

> so great that we cannot ascribe it to faulty representation. So far as the free-living males are concerned, the nomenclature must be regarded as purely provisional. Both genera and species are exclusively defined by characters whose variation and consequent taxonomic value are entirely or almost unknown. On the whole, the female genera seem particularly well defined, but the separation of the species is often based on such small differences between so few specimens that it simply rests on a subjective estimate, whether they may be considered justifiable. (Bertelsen, 1951:8)

Erik Bertelsen's (Fig. 21) enormous contribution to our knowledge of ceratioid anglerfishes is founded primarily on a detailed examination of some 2400 larvae in the *Dana* collections. The objectives of his investigation, published in 1951, were to (1) sort the material into the lowest possible taxonomic units; (2) place these groups within the existing taxonomic hierarchy of juveniles and adults; (3) describe the ontogeny of the various constituent taxa; (4) resolve the relationship between metamorphosed free-living males and females (which had heretofore been placed in separate genera); and (5) add to the body of information regarding the extraordinary biology of these fishes, especially the related phenomena of male dwarfism and sexual parasitism. While impractical to list all of his findings here, a brief summary follows.

FIGURE 19. An imaginative view of a population of *Melanocetus johnsonii* living in the abyssal ooze published by Henri Filhol (1843–1902), professor of paleontology at the Jardin des Plantes, Paris, in his *La Vie au Fond des Mers (Life at the Bottom of the Seas)*. After Filhol (1885).

In Part One of his monograph, under the heading of "Ontogeny and Taxonomy," Bertelsen (1951:7–196) addressed the characters common to both sexes, showing that the osteological features that define families and genera can "as a rule" be determined in the larvae by examination of cleared and alizarin-stained specimens. He also showed that the pattern of subdermal pigmentation, which is genus specific, and in some taxa even species specific, is laid down in the smallest larvae and retained subdermally for some time following metamorphosis and may be recognized in most males and juvenile females. Thus, with this new body of information, coupled with an analysis of fin-ray counts, Bertelsen was able to link not only the larvae to families and genera based on females, but most of the free-living males as well.

As for secondary sexual characters, Bertelsen demonstrated among other things that the presence or absence of an illicium can be determined in the youngest of larvae. He found also that a size difference between the sexes becomes apparent at or immediately before metamorphosis, and that in early metamorphosis the shape of the body diverges, the head of the female increasing greatly in relative size, especially the jaw apparatus, while the male becomes more slender and its head and jaws decrease in relative size. He determined also that ceratioid males, even as larvae, show no indication of having light organs of any kind, while the light organ of the esca of females is laid down in the larval stage and is present in all metamorphosed females, with the exception of two genera, *Caulophryne* and *Neoceratias*. The eyes of the larvae are well developed and relatively largest in the smallest individuals. After metamorphosis, the eyes of the females of all families grow very little and are relatively tiny and vestigial in the largest adult specimens. In contrast, development of the eyes of the males is highly variable among families: they become vestigial in gigantactinids; small in centrophrynids and diceratiids; relatively large and provided with an aphakic space (a space anterior to the edge of the lens that has no focusing element, thought to enhance the forward binocular field of vision; see Schwab, 2004) in caulophrynids, melanocetids, himantolophids, and oneirodids; very large and bowl shaped, with a pupil diameter twice as large as the lens in ceratiids; and "telescopic" in linophrynids. The olfactory organs of the larvae and metamorphosed females are small, but in older larvae they are generally larger in males than in females. Except in the ceratiids (and perhaps neoceratiids), the olfactory organs of the males grow enormously during and after metamorphosis.

A detailed study of the ontogeny of ceratioids, and identification of the free-living males that had previously been placed in their own genera separate from the females, provided the materials for a full taxonomic revision of the suborder (Bertelsen, 1951). The number of families was reduced from 11 to 10: the family Laevoceratiidae was rejected, the Photocorynidae included in the Linophrynidae, and a new family, the Centrophrynidae, added. Hypotheses of evolutionary relationship were proposed for the first time: the Caulophrynidae was given the "most primitive position," based on, among other things, the discovery of pelvic fins in the larval stages. Seventeen genera were synonymized and five new ones erected (based on earlier described species), reducing the total number from 46 to 34. Three of the 34 genera were retained provisionally until such time that the males could be studied and

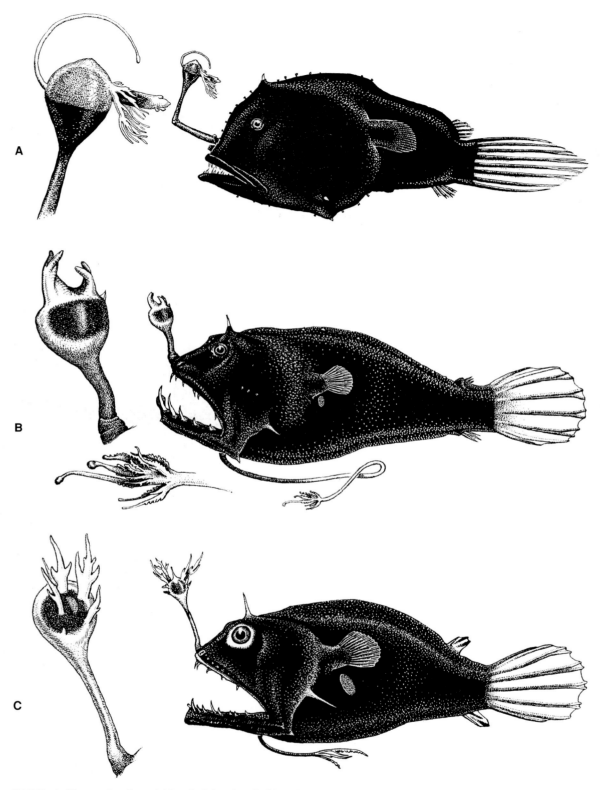

FIGURE 20. New species of ceratioid anglerfishes described by Albert Eide Parr (1890–1991) as a result of the Third Oceanographic Expedition of the yacht *Pawnee* under the direction of Harry Payne Bingham (1887–1955), philanthropist, sportsman, art patron, and founder of the Bingham Oceanographic Foundation of Yale University: (A) *Dolopichthys obtusus*, female, 13 mm SL, YPM 2028, now recognized as a junior synonym of *Oneirodes eschrichtii*; (B) *Linophryne coronata*, female, 33 mm SL, YPM 2005; (C) *Linophryne brevibarbis*, female, 25 mm SL, YPM 2001, a junior synonym of *Linophryne macrodon*. Drawings by Wilfrid S. Bronson; after Parr (1927).

FIGURE 21. Erik Bertelsen (1912–1993), celebrated ichthyologist, fisheries biologist, and the all-time world authority on the Ceratioidei, shown here having fun aboard the Russian Research Vessel *Vityaz*, 29 November 1988 (for a tribute to his life, see Nielsen, 1994). Courtesy of Jørgen Nielsen, Peter Rask Møller, and the Zoological Museum, University of Copenhagen.

identified. Of the more than 200 nominal species of ceratioids, which by 1951 had been reduced to 178 by various authors, Bertelsen rejected 61 and added two new ones, for a total of 119. Unable to revise certain species-rich genera (e.g., *Oneirodes*) for a lack of adequate comparative material, and acknowledging that a number of species based only on males must be regarded as provisional, he concluded that the known material of the suborder represents no more than 80 species.

Prior to Bertelsen's (1951) monograph, the scientific literature contained descriptions or mention of only about 50 ceratioid larvae. Among this material, representatives of four families were correctly referred to family, but only two genera and not a single species were correctly identified. In contrast, based on the approximately 2400 larvae in the *Dana* collections, Bertelsen recognized and described the early development of 30 taxa. Nineteen of these were identified with their respective species, three were referred to a group of species within a genus, seven could be identified only to genus, and one only to family. His descriptions of the larvae included representatives of all 10 families and 22 of the 34 genera then recognized.

Also prior to Bertelsen's (1951) work, the known free-living males were, with some few exceptions, correctly referred to four of the 10 families of the suborder, one was wrongly placed in a fifth family, and two were provisionally placed in the now discarded Laevoceratiidae. None was referred to a species based on females, and, apart from a few uncertain suppositions, only one was (but incorrectly) referred to a genus that also contained females. Of the 166 metamorphosing or metamorphosed free-living males in the *Dana* collections, 63 had been previously described by Regan (1926) and Regan and Trewavas (1932). In addition to representatives of the families Himantolophidae, Melanocetidae, Oneirodidae, and Linophrynidae to which Regan and Trewavas had already referred specimens, Bertelsen was able to assign free-living males to the Centrophrynidae and Gigantactinidae, the males of which had been unknown previously; as well as to the Caulophrynidae and Ceratiidae, whose males had previously been known only as parasites on females. Of the two remaining families, Neoceratiidae and Diceratiidae, the males of which had been previously unknown, he described a parasitic male of the first, and to the other he referred the free-living male *Laevoceratias liparis* that had been described by Parr (1927). The males examined by Bertelsen were divided into 27 taxa, all of which were referred to genera based on females; 14 of these were further identified to species based on females.

The second part of Bertelsen's (1951:197–251) monograph is devoted to an analysis of "Distribution, Ecology, and Biology." Thanks to the methodical and well-defined collecting protocols employed by the scientific crew of the *Dana* expeditions—but limited primarily to collections made in the North Atlantic—he was able to provide the first quantitative study of the seasonal, vertical, and horizontal distribution of ceratioids. At the same time, based on their anatomy, stomach contents, analyses of catch data, and the small and infrequent information on the behavior of specimens that lived a short time after capture, he provided an account of their ecology and general biology, including a discussion of their swimming powers, bioluminescence, food habits, and reproduction, the latter in special reference to sexual dimorphism and parasitism. In summary, Bertelsen's great work set the standard and laid the foundation for all future studies of this enigmatic group of fishes for the next half century.

In 1968, through the intermediacy of my major professor, Basil G. Nafpaktitis, who guided me with great care and expertise through the travails of graduate school, I was extremely fortunate to become closely associated with Erik Bertelsen, whose name was and still is synonymous with "Ceratioidei," at least among ichthyologists. Beginning at this early time, and during the 25 years that followed, Bertelsen and I undertook a program of systematic studies of ceratioid genera and families, sometimes working alone, independently, but more often together. In addition to a few small papers beginning in 1969, some of which described new ceratioid genera (Pietsch, 1969, 1972b, 1973), I reviewed the Centrophrynidae (1972a), based on new material collected from the Gulf of California. Following shortly thereafter, I published revisions of the oneirodid genera *Dolopichthys* (1972c), *Oneirodes* (1974a), *Lophodolos* (1974b), *Chaenophryne* (1975b), the so-called long-pectoraled oneirodid genera (i.e., *Leptacanthichthys*, *Chirophryne*, *Ctenochirichthys*, and *Puck*; 1978), and the families Caulophrynidae (1979) and Ceratiidae (1986b).

Bertelsen, during this same time period, produced independently seven revisionary papers on the Linophrynidae (1973, 1976, 1978, 1980a, 1980b, 1981, 1982), continuing a numbered series of publications on this family that he had started in 1965, the first in collaboration with the late Gerhard Krefft (Fig. 22). Also produced by Bertelsen during these years were two remarkably detailed and beautifully illustrated monographs: the first, a review of the genus *Thaumatichthys*, in coauthorship with Paul J. Struhsaker (1977), and second, a revision of the family Himantolophidae, with Krefft (1988).

As for Bertelsen and Pietsch together, our first collaboration was a review of the oneirodid genus *Spiniphryne* (1975), based on new material collected by the research cruises of the German FRV *Walther Herwig* off South America. This was followed

FIGURE 22. Erik Bertelsen with his long-time friend and collaborator Gerhard Krefft (1912–1993), ichthyologist, fisheries biologist, and founder of the great collection of marine fishes at the Institut für Seefischerei (ISH) in Hamburg, which was transferred to the Zoological Museum of the University of Hamburg in 1993 shortly after Krefft's death (for more on his life, see Stehmann and Hulley, 1994, and Stehmann, 1997). Photograph taken in Krefft's office at ISH, 12 December 1985; courtesy of Jørgen Nielsen, Peter Rask Møller, and the Zoological Museum, University of Copenhagen.

shortly thereafter by a review of the oneirodids resulting from those same expeditions (1977), a paper describing the ceratioid anglerfishes of Australia (1983), a revision of the family Gigantactinidae (with Robert J. Lavenberg, 1981), a "resurrection of the ceratioid anglerfish *Ceratias tentaculatus*" (1984), and three studies completed and published posthumously: a revision of the thaumatichthyid genus *Lasiognathus* (1996), a revision of the gigantactinid genus *Rhynchactis* (1998b), and a new species of *Gigantactis* (2002).

Despite this flurry of activity during the 1970s, 1980s, and 1990s, during which time all major taxa of the Ceratioidei were either reviewed or revised, and four genera and 66 species were described as new to science, we are not yet close to knowing the full diversity of ceratioids. Of the 160 species recognized in this monograph, 23, or about 14%, are still known only from the holotype, 53 (33%) are known from three female specimens or fewer, and only 84 species (52%) are represented by more than six females. Consequently, knowledge of individual, ontogenetic, and geographic variation is extremely limited. While the taxonomy of females is poorly known, that of males is much worse: only 29 of the 160 species are represented by males, and by far most of these are identified with species based on females only because they belong to monotypic genera or because they are parasitically attached to females; the hundreds of males in collections around the world, if identified at all, are still named only to genus. The rate of description of new species, while characterized by occasional large spikes that coincide with major revisions (e.g., Regan, 1925b; Regan and Trewavas, 1932; Pietsch, 1974a; Bertelsen, 1980a, 1980b, 1982; Bertelsen et al., 1981; Bertelsen and Krefft, 1988), has not leveled off despite steeply declining exploratory fishing activity around the world in the past 20 years (Bertelsen, 1982; Mihai-Bardan, 1982; Bertelsen and Pietsch, 1983, 1996, 1998b, 2002; Kharin, 1984, 1989; Balushkin and Fedorov, 1985, 1986; Leipertz and Pietsch, 1987; Balushkin and Trunov, 1988; Bertelsen and Krefft, 1988; Ni, 1988; Swinney and Pietsch, 1988; Orr, 1991; Gon, 1992; Stewart and Pietsch, 1998; Ho and Shao, 2004; Pietsch, 2004, 2005a, 2005b, 2007; Pietsch et al., 2004; Pietsch and Baldwin, 2006).

This monograph is by no means the last word on this extraordinary group of animals. The descriptive phase of our knowledge of ceratioids, and deep-sea biology in general, is not over and may well be still in its early stages. New species will continue to be discovered, and the filling of taxonomic gaps will provide new insights into phylogenetic relationships. The addition of new material will expand our knowledge of geographic distributions, both horizontal and vertical. Observations of freshly caught specimens, and in situ sightings from remote and manned submersibles will provide new information about behavior. Resolving the seemingly insurmountable problems of keeping ceratioids alive after capture will lead eventually to exhibits in public aquaria, which may in turn provide new insights into the unique reproductive modes of these fishes, sexual parasitism, and the intriguing biomedical implications of tissue compatibility.

Acknowledging that much more remains to be learned, I hope that what is presented in this volume is as correct and as complete as possible, and that where I have made decisions, they have been as objective as possible. Time will tell, and such errors as may come to light will be my responsibility alone.

# TWO

## What Makes an Anglerfish?

> At present we do not know certainly for any one species its vertical or horizontal distribution, the conditions most favorable for its existence, or the nature of its food; but we may be certain that these are different for the different species and that the differences in form, in the structure of the fins, in the size and shape of the mouth, in the dentition, and in the structure of the illicium, are related to different habits.
>
> CHARLES TATE REGAN,
> "The Pediculate Fishes," 1926:11

Anglerfishes differ radically from all other fishes. With the first dorsal-fin spine mounted on the snout and modified to serve as a luring apparatus, and gill openings narrowly constricted to form tubelike structures that open posteriorly behind the base of the pectoral fin, they can hardly be confused with anything else. The deep-sea suborder Ceratioidei is by far the largest, most highly derived, and certainly the least known of the five primary lineages of the order, which has come to be called the Lophiiformes. When compared with its less-derived lophiiform relatives, it too is easily distinguished. In fact, a single character complex is sufficient to diagnose the suborder: anglerfishes that display an extreme sexual dimorphism in which the males are dwarfed, reaching only a fraction of the size of the females. Although enough to encompass what is meant by "Ceratioidei," this single concise statement only begins to convey the enormous morphological uniqueness and diversity found within the group. The present chapter is intended to provide an overview of this diversity of form, to describe in some detail the various characters of biosystematic significance, and to demonstrate how they vary, not only among the various taxonomic subunits, but between genders as well.

While the families and genera of the Ceratioidei are well defined by combinations of the characters described below, the problems of differentiation are nearly all confined to the species level. The soft bodies and weakly ossified skeletons of ceratioids make nearly all morphometric characters traditionally utilized in teleost taxonomy useless in distinguishing the many morphologically similar species. This is especially true for those belonging to the species-rich genera, such as *Himantolophus*, *Oneirodes*, *Gigantactis*, and *Linophryne*, which together account for 70% of all recognized ceratioids. Values for body width and depth; head length, width, and depth; length of the upper and lower jaws, pectoral lobe, and bases of the unpaired fins; and length and depth of the caudal peduncle are often difficult to establish with accuracy, and so plastic and highly variable intraspecifically depending on the mode of preservation, that they are of very limited taxonomic importance.

With regard to female ceratioids, most frustrating has been the inability to discover characters of taxonomic value at the specific level other than those of the escal morphology. While the importance of this character complex cannot be overstated, the illicial apparatus of ceratioids is often damaged or lost upon capture, particularly in larger specimens, making positive identification difficult if not impossible. For example, nearly half the known specimens of *Ceratias* greater than 75 mm have lost the esca (Pietsch, 1986b:479), making assignment to one of the three recognized species of this genus next to impossible. The percentage is significantly higher in certain other genera like *Oneirodes*, in which a large number of damaged females in collections around the world remain unidentified (Pietsch, 1974a:107).

Ceratioid families are distinguished from one another primarily by osteological features, many of which are shared by members of both sexes: primarily the (1) presence or absence of certain cranial elements; (2) branchiostegal-ray counts; (3) fin-ray counts; and (4) number of pectoral radials. In contrast, differentiation of ceratioid genera is based primarily on characters present only in the females, the more important of which include the following: (1) shape of the skull and other bones of the head, including the development of head spines; (2) structure of the jaws, including dentition; (3) structure of the luring apparatus, including the basic pattern of escal appendages and filaments; (4) pigmentation of the skin; and (5) development of skin spines and spinules. Some of the distinguishing osteological characters of the females are shared with the males, for example, the shape of the opercular bones, which within some families (e.g., the Oneirodidae) show distinct and consistent intergeneric differences. On the other hand, certain structures restricted to the males, such as the denticular teeth, show distinct intergeneric differences that agree in full with distinctions based on characters of the females (Bertelsen, 1984:325). In addition to fin- and branchiostegal-ray counts, larvae are differentiated primarily on the basis of body shape, degree of skin inflation, size of the pectoral fins, and the pattern of subdermal pigmentation.

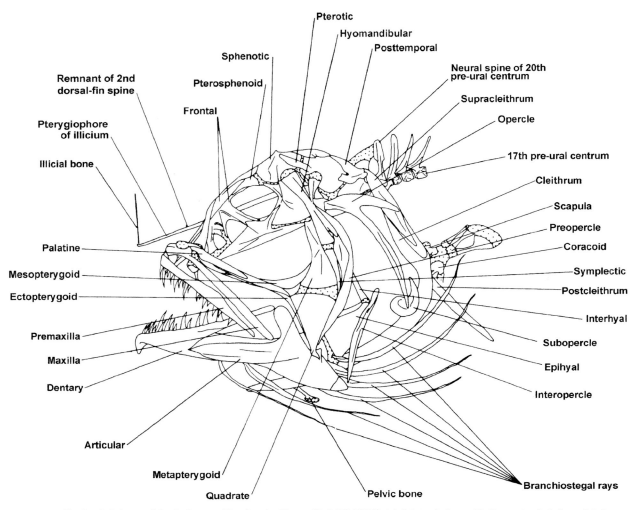

FIGURE 23. The head skeleton of *Oneirodes acanthias*, female, 71 mm SL, LACM 9960-4, left lateral view, with the pectoral girdle, pelvic bone, hyoid apparatus, and first four vertebrae in place. After Pietsch (1974a).

The species of *Linophryne* have been grouped into subgenera (Bertelsen, 1982) and those of *Himantolophus* (Bertelsen and Krefft, 1988), *Oneirodes* (Pietsch, 1974a), *Chaenophryne* (Pietsch, 1975b), and *Gigantactis* (Bertelsen et al., 1981) into species groups based on shared but relatively minor differences in one or more of the characters mentioned above. With very few exceptions, the separation of females into species is based on differences in the number, shape, size, and placement of the escal appendages and filaments (as well as, in some cases, variation in external escal pigment patterns), often combined with differences in illicial length and jaw-tooth counts. In the vast majority of the recognized species, no additional distinguishing features have been found, other than occasional small intrageneric differences in meristics (e.g., tooth and fin-ray counts) and minor osteological characters (e.g., the shape of the opercular bones and the development and dentition of the gill arches). A unique opportunity to check the validity of intrageneric distinctions based on escal characters is found in the genus *Linophryne*, in which the pattern of branching of the hyoid barbel of females shows distinct and very consistent differences between subgenera and species (see Bertelsen, 1980a, 1980b, 1982, 1984:325).

With some few exceptions, and despite major efforts, it has not been possible to establish characters that allow intrageneric identification of males. A few males found to differ from their supposed congeners in features restricted to males (e.g., the number and pattern of placement of denticular teeth; see Bertelsen, 1951:21) have been tentatively described as representatives of separate species. Detailed studies of males attached parasitically to identified females have not revealed characters that will allow intrageneric identification of free-living males (Bertelsen, 1984:325).

## Characters Shared by Both Sexes

### Head Skeleton

The shape, size, and relative position of the cranium and closely associated elements of the suspensorium, jaws, opercular apparatus, hyoid apparatus, and pectoral girdle of ceratioids vary enormously among families and genera, and even more so between males and females of the same species. Those of the genus *Oneirodes* are rather typical of females of the majority of the taxa (Fig. 23), but various extremes are found throughout the suborder. The crania of females of genera like *Ceratias* and *Cryptopsaras*, for example, are long and narrow, with the bones that suspend the jaws (the suspensorium) equally elongate and strongly directed anteriorly, resulting in relatively short jaws and a small, dorsally directed mouth (Fig. 24). In contrast, and

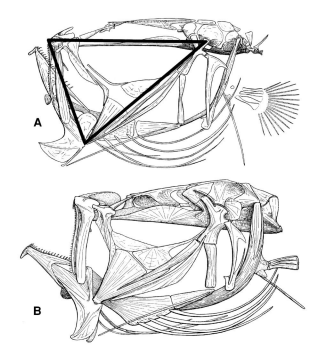

FIGURE 24. Head skeleton of females of *Cryptopsaras couesii*: (A) juvenile, 18 mm SL, emphasizing the acute angle formed between the cranium and the elements of the suspensorium, resulting in relatively short jaws and a small dorsally directed mouth; (B) adult, 160 mm SL. After Regan and Trewavas, 1932.

with almost every kind of arrangement in between, the suspensorium of genera like *Melanocetus* and *Caulophryne* is nearly twice the length of the cranium (Fig. 25A, B); but while that of *Melanocetus* is directed anteriorly, resulting in relatively long jaws and a dorsally directed mouth, that of *Caulophryne* is strongly directed posteriorly, resulting in extraordinarily long jaws and a huge oblique mouth. The relative proportions of the cranium, suspensorium, and jaws of most oneirodids are more or less equal, resulting in jaws of intermediate size and anteriorly directed mouths (Figs. 23, 25C, D). While great variation exists among females, the overall appearance of males is considerably more uniform (see Fig. 26).

The head skeleton of ceratioids differs further in the presence or absence of a number of elements: the supraethmoid is usually present, but very much reduced or absent in the Thaumatichthyidae, and absent in the Gigantactinidae (a tiny rudiment observed in a single specimen; see Bertelsen et al., 1981:10); parietals are usually present, but absent in the Himantolophidae and in metamorphosed females of the gigantactinid genus *Rhynchactis*; pterosphenoids are usually present, but reduced in the Diceratiidae and absent in the Ceratiidae, Caulophrynidae, Gigantactinidae, Linophrynidae, and the oneirodid genus *Lophodolos*; metapterygoids are usually present, but absent in the Neoceratiidae; mesopterygoids are usually present, but absent in the Neoceratiidae, Gigantactinidae, and Linophrynidae.

The dorsal end of the hyomandibular bone is distinctly bifurcated in most ceratioid taxa, forming two heads, both of which articulate with the cranium: an anterior head that fits within a cavity formed by the respective sphenotic and prootic bone and a posterior head that articulates on the lateral surface of the respective pterotic (Fig. 25A–D). In contrast, the hyomandibular bears only a single expanded head in the Neoceratiidae, Gigantactinidae, Linophrynidae, and the oneirodid genus *Bertella* (see Pietsch, 1973:199; Fig. 25E–F).

## Opercular Bones

The opercular bones of ceratioids are considerably reduced relative to those of other lophiiforms (see Pietsch, 1981). The opercle is almost always deeply incised posteriorly, the resulting dorsal arm considerably shorter than the lower. The relative lengths of the two arms of the opercle, the angle formed between them, and the extent to which they are united by bone vary widely among oneirodids, in many cases providing important diagnostic characters to distinguish the various genera of this family (Bertelsen, 1951; Pietsch, 1974a). Most divergent among the latter is *Chaenophryne*, in which the opercle is nearly triangular in shape, the posterior margin only slightly more concave than the anterior (Fig. 25C). A similarly shaped opercle is found in some linophrynids (see Bertelsen, 1951:162). In contrast to all other ceratioids, the upper arm of the opercle of thaumatichthyids is divided into several radiating branches, as few as 5 in adult specimens to as many as 13 in larvae (Bertelsen, 1951:119; Bertelsen and Struhsaker, 1977:12; Fig. 27).

In most ceratioids the subopercle is elongate, with a rounded ventral end and a narrow tapering dorsal projection, but it varies considerably in shape, especially among oneirodids, and adds significantly to the diagnostic value of this character complex among oneirodid genera (Bertelsen, 1951; Pietsch, 1974a). The anterodorsal margin of the subopercle is smooth and rounded in most ceratioids but bears a distinct, anteriorly directed spine in both sexes of the Centrophrynidae, Diceratiidae, Melanocetidae, and the ceratiid genus *Cryptopsaras*; a small subopercular spine is also present in the genus *Thaumatichthys* (Bertelsen and Struhsaker, 1977:14; Fig. 27) and in most specimens of the oneirodid genus *Chaenophryne* (Fig. 25C).

## Branchiostegal Rays

Nearly all ceratioids have 6 branchiostegal rays (Figs. 23–27), two attached anteriorly to the ventral or ventromedial margin of the ceratohyal and four attached posteriorly to the ventrolateral surface of the ceratohyal and respective epihyal (noted in the descriptive accounts as 2 + 4). This number is reduced in some families, the anterior-most element dropping out in some specimens of the Caulophrynidae and Neoceratiidae (producing a 1 + 4 pattern; see Bertelsen, 1951:158; Pietsch, 1979:8) and in all members of the Linophrynidae (very rarely both anterior branchiostegal rays are absent in linophrynids, producing a 0 + 4 pattern; Bertelsen, 1951:161). The number is very rarely increased to 7 (2 + 5) in some specimens of the genus *Gigantactis* (see Bertelsen et al., 1981:11).

## Caudal Skeleton

The caudal skeleton of ceratioids is similar to that of other lophiiform fishes in having a single hypural plate emanating from a single complex half-centrum (see Gosline, 1960; Rosen and Patterson, 1969; Pietsch, 1972a). The hypural plate, consisting of an unknown number of fused hypurals, is usually entire, but distinctly notched posteriorly in the Centrophrynidae, Neoceratiidae, the genus *Thaumatichthys*, and in some members of the Ceratiidae and Caulophrynidae

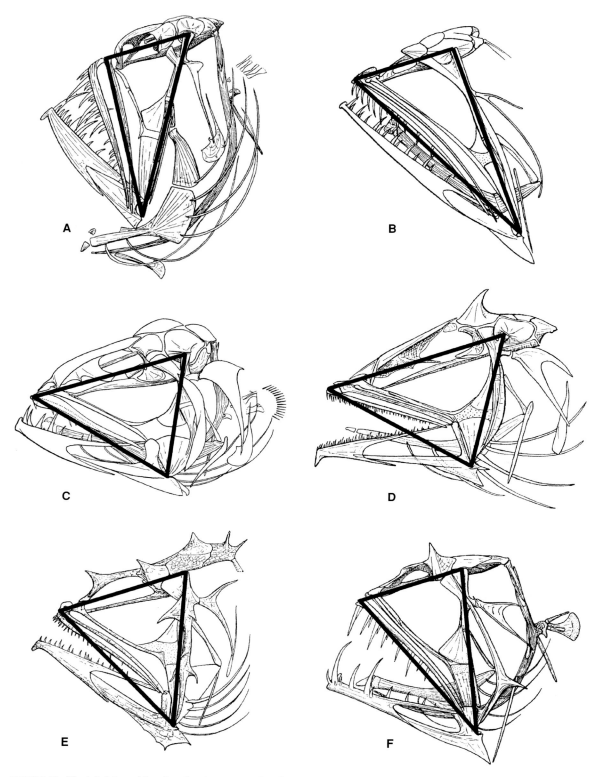

FIGURE 25. Head skeleton of females of various ceratioids, selected to emphasize the great morphological diversity among genera, and to show how the relative length of the suspensorium and the angle it forms with the cranium determines the size of the jaws and the angle of the opening of the mouth: (A) *Melanocetus johnsonii*, 22 mm SL, characterized by having a long suspensorium and long jaws, resulting in a large dorsally directed mouth; (B) *Caulophryne pelagica*, 57 mm SL, with an exceptionally long suspensorium and long jaws, resulting in an extremely large dorsally directed mouth; (C) *Chaenophryne draco*, 13 mm SL, a short suspensorium, short jaws, and an anteriorly directed mouth; (D) *Dolopichthys jubatus*, 24 mm SL, a short suspensorium, short jaws, and an anteriorly directed mouth; (E) *Photocorynus spiniceps*, 37 mm SL, elements combining to produce an oblique mouth opening; (F) *Linophryne racemifera*, 52 mm SL, elements combining to produce an oblique mouth opening. After Regan and Trewavas (1932).

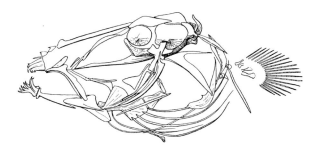

FIGURE 26. Head skeleton of *Melanocetus johnsonii*, male, 27 mm SL, showing a drastically different configuration of bony parts compared with that of a conspecific female (see Fig. 25A), making it rather easy to understand why early taxonomists assigned males and females to different fish families. After Regan and Trewavas (1932).

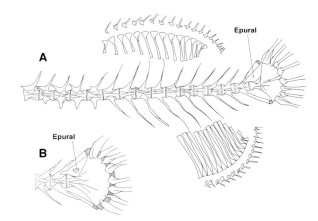

FIGURE 28. Vertebrae, caudal skeleton, and median fins of females of *Caulophryne jordani*, showing the presence of a tiny epural bone: (A) 68 mm SL, UW 22154; (B) 41 mm SL, LACM 6844-003. After Pietsch (1979).

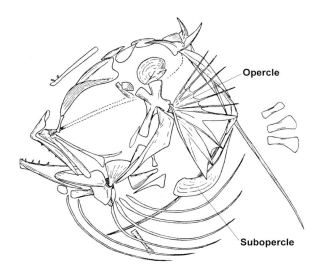

FIGURE 27. Head skeleton of *Thaumatichthys* sp., larval male, 18.5 mm SL, showing among other things the unique highly branched opercular bone. After Bertelsen (1951).

(Fig. 28). Epurals are absent in all ceratioids, except for a single tiny remnant in the Caulophrynidae. Bertelsen's (1984:326) report of an epural in the linophrynid genus *Photocorynus* could not be confirmed by reexamination of all available material.

## Pectoral Radials and Pelvic Bones

The number of pectoral radials (bony elements that support the rays of the pectoral fin) varies somewhat among ceratioid families. Most have 3, but there are 2 in the Caulophrynidae; 4 in the Centrophrynidae, Ceratiidae, and Melanocetidae; and 5 in the Gigantactinidae. In females of many of the families, the two ventral-most radials become fused to one another with growth, giving the appearance of one less element. For example, the 3 radials of diceratiids and at least some thaumatichthyids are often reduced to 2 (Regan and Trewavas, 1932:30; Bertelsen and Struhsaker, 1977:14), and the 4 of centrophrynids and melanocetids, reduced to 3 (Regan and Trewavas, 1932:27; Pietsch, 1972a:41; Pietsch and Van Duzer, 1980:67). In female ceratiids and gigantactinids, the third and fourth pectoral radials, respectively, become reduced with growth, the proximal end, in each case, absent in most specimens and the adjacent radials approaching or meeting each other proximally (Pietsch, 1972a:42; Bertelsen et al., 1981:14).

The pelvic bones of ceratioids are rudimentary, usually rod shaped or somewhat expanded distally, but triradiate in the Himantolophidae (Regan and Trewavas, 1932:31) and in some females of the oneirodid genus *Chaenophryne* (Pietsch, 1975b:79). They are further reduced or absent in the Neoceratiidae and Gigantactinidae, and absent in the Linophrynidae.

## Fins and Fin Rays

A summary of fin-ray counts shows great variability among ceratioid families and genera, especially in the dorsal and anal fins (Table 1). The dorsal fin consists of 3 to 8 rays in most families, but there are 12 to 17 in the Melanocetidae, 6 to 22 in the Caulophrynidae, and 11 to 13 in the Neoceratiidae. The dorsal fin of the two genera of the Ceratiidae each have 4 (very rarely 3 or 5) rays, but if the 2 or 3 embedded rays within the caruncles of these taxa are counted, there are 6 dorsal-fin rays in *Ceratias* and 7 in *Cryptopsaras*. The anal fin consists of 3 to 7 rays in most families, but there are 5 to 19 in the Caulophrynidae and 10 to 13 in the Neoceratiidae.

The caudal fin nearly always contains 9 rays, but the ninth or lowermost ray is reduced to less than one-half the length of the eighth ray in the Gigantactinidae, Linophrynidae, and the genus *Ceratias*. There are only 8 caudal-fin rays in the Caulophrynidae and the ceratiid genus *Cryptopsaras*, and 10 in some larval specimens of the Neoceratiidae. The pectoral fin usually contains from 12 to 23 rays, but there are 21 to 27 in the oneirodid genus *Pentherichthys* and 28 to 30 in the oneirodid genus *Ctenochirichthys*. Pelvic fins are absent in all ceratioids except in larval and newly metamorphosed specimens of the Caulophrynidae. On the basis of fin-ray counts alone, only two taxa, the Melanocetidae and oneirodid genus *Ctenochirichthys*, can be clearly distinguished from all other ceratioids (Table 1), but 3 rays in both the dorsal and anal fins almost always differentiate members of the Linophrynidae.

All fin rays of ceratioids are biserial and segmented. In nearly all taxa, the rays of the dorsal, anal, and pectoral fins are simple, but the posterior 2 to 4 rays of the dorsal and anal fins of the Himantolophidae are bifurcated (Bertelsen and Krefft, 1988:17). Branching of the rays of the caudal fin varies considerably among ceratioid families: almost always the outermost

TABLE 1
Fin-ray Counts of Families and Genera of the Ceratioidei

| Taxon | Dorsal Fin | Anal Fin | Caudal Fin | Pectoral Fin |
|---|---|---|---|---|
| Centrophrynidae | 6–7 | 5–6 | 9[a] | 15–16 |
|   *Centrophryne* | 6–7 | 5–6 | 9[a] | 15–16 |
| Ceratiidae | 4 (3, 5)[b] | 4 | 8–9 | 15–19 (14) |
|   *Ceratias* | 4 (3)[b] | 4 | 9[c] | 15–19 |
|   *Cryptopsaras* | 4 (5)[b] | 4 | 8[d] | 15–17 (14, 18) |
| Himantolophidae | 5–6 | 4 | 9[e] | 14–18 |
|   *Himantolophus* | 5–6 | 4 | 9[e] | 14–18 |
| Diceratiidae | 5–7 | 4 (5) | 9[e] | 13–16 |
|   *Diceratias* | 5–7 | 4 (5) | 9[e] | 13–16 |
|   *Bufoceratias* | 5–6 | 4 | 9[e] | 13–14 |
| Melanocetidae | 12–17 | 4 (3, 5) | 9[e] | 15–23 |
|   *Melanocetus* | 12–17 | 4 (3, 5) | 9[e] | 15–23 |
| Thaumatichthyidae | 6–7 | 4 | 9[a] | 14–16 |
|   *Lasiognathus* | 5 | 5 | 9[a] | 17–20 |
|   *Thaumatichthys* | 6–7 | 4 | 9[a] | 14–16 |
| Oneirodidae | 5–8 (4) | 4–7 (3) | 9[a] | 13–30 |
|   *Lophodolos* | 5–8 | 5–6 (4, 7) | 9[a] | 17–20 (21) |
|   *Pentherichthys* | 6–7 | 6 (5, 7) | 9[a] | 21–24 (25–27) |
|   *Chaenophryne* | 6–8 | 5–6 | 9[a] | 16–22 |
|   *Spiniphryne* | 6 | 4–5 | 9[a] | 15–17 |
|   *Oneirodes* | 5–7 | 4 (3, 5) | 9[a] | 14–18 (13, 19) |
|   *Dermatias* | 6 | 4 | 9[a] | 15–16 |
|   *Danaphryne* | 5–7 (8) | 5 (4) | 9[a] | 16–19 |
|   *Microlophichthys* | 5–7 | 5 (4, 6) | 9[a] | 18–20 (21–23) |
|   *Tyrannophryne* | 5 | 5 | 9[a] | 20 |
|   *Phyllorhinichthys* | 5 | 5 | 9[a] | 19–24 |
|   *Dolopichthys* | 5–7 (8) | 4–6 | 9[a] | 18–21 (17, 22) |
|   *Bertella* | 5–6 | 5 (4) | 9[a] | 18–22 |
|   *Puck* | 5 | 4 | 9[a] | 19 |
|   *Chirophryne* | 5 | 4 | 9[a] | 18–19 |
|   *Leptacanthichthys* | 5–6 (4) | 5 | 9[a] | 18–22 |
|   *Ctenochirichthys* | 6–7 | 4–5 | 9[a] | 28–30 |
| Caulophrynidae | 6–22 | 5–19 | 8[d] | 14–19 |
|   *Caulophryne* | 14–22 | 12–19 | 8[d] | 14–19 |
|   *Robia* | 6 | 5 | 8[d] | 17 |
| Neoceratiidae | 11–13 | 10–13 | 9–10[f] | 12–15 |
|   *Neoceratias* | 11–13 | 10–13 | 9–10[f] | 12–15 |
| Gigantactinidae | 3–9 (10) | 3–7 (8) | 9 | 14–22 |
|   *Gigantactis* | 5–9 (4, 10) | 4–7 (8) | 9[g] | 15–21 (14, 22) |
|   *Rhynchactis* | 3–4 (5) | 3–4 | 9[c] | 17–20 |
| Linophrynidae | 3 (4) | 3 (2, 4) | 9[c] | 12–19 |
|   *Photocorynus* | 3 (4) | 3 (4) | 9[c] | 15–17 |
|   *Haplophryne* | 3 | 3 | 9[c] | 15–16 |
|   *Acentrophryne* | 3 | 3 | 9[c] | 16–19 |
|   *Borophryne* | 3 | 3 | 9[c] | 16–18 |
|   *Linophryne* | 3 | 3 (2) | 9[c] | 12–19 |

NOTE: Rare counts, occurring in less than 2% of the known material, are in parentheses.
[a] 2 simple + 4 bifurcated + 3 simple.
[b] Plus two or three dorsal-fin rays embedded within caruncles of *Ceratias* and *Cryptopsaras*, respectively.
[c] 2 simple + 4 bifurcated + 3 simple; ninth ray reduced.
[d] 2 simple + 4 bifurcated + 2 simple.
[e] 1 simple + 6 bifurcated + 2 simple.
[f] 2 simple + 4 bifurcated + 3 simple; 10 rays in some larvae.
[g] All rays simple in females; 2 simple + 4 bifurcated + 3 simple in males; ninth ray reduced.

FIGURE 29. *Bufoceratias wedli*, about 27 mm SL, caught with a slurp gun from the submersible research vessel *Johnson Sea-Link II*, at 1000 m off St. Croix, Virgin Islands, showing the pattern of placement of acoustico-lateralis papillae on the head and body. Photographed aboard ship while still alive by N. J. Marshall.

rays are simple and the innermost rays are bifurcated. The pattern is 2 simple + 4 bifurcated + 3 simple in most families, but 1 + 6 + 2 in the Himantolophidae, Diceratiidae, and Melanocetidae (all caudal rays are simple in the single known diceratiid male), and 2 + 4 + 2 in the Caulophrynidae and the ceratiid genus *Cryptopsaras*. All the caudal-fin rays are simple in females of *Gigantactis*, but the 2 + 4 + 3 pattern characterizes males of this genus, as well as both sexes of the gigantactinid genus *Rhynchactis* (Bertelsen et al., 1981:14).

In contrast to all other ceratioids, the pectoral fins of both sexes of the Caulophrynidae are unusually large at all life stages (their length about 30% SL in females, 40% SL in males, and reaching well beyond the origin of the dorsal and anal fins in larvae), having broad, rounded, supporting lobes due primarily to a broadly expanded ventral-most pectoral radial (see Pietsch, 1979:11). Similarly, the pectoral fins of larval gigantactinids, but not those of juveniles or adults, are enormous, reaching to or beyond the base of the caudal fin (Bertelsen, 1951:148). The absence of pelvic fins in gigantactinids, so well developed in larval caulophrynids, however, readily separates the larvae of the two families.

## Acoustico-lateralis System

The elements of the lateral-line system are rather simple in ceratioids, not reaching "anything like the high development one finds in other abyssal fishes, e.g., macrourids and berycids" (Bertelsen, 1951:25). Rudiments of these structures appear as small papillae on the distended skin of the head and body of larvae, especially in members of the Linophrynidae. In juveniles and adults they are not sunk into canals but are instead exposed, raised above the surface of the head and body and sit-

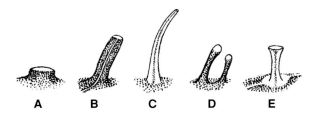

FIGURE 30. Morphological variation of the sensory papillae among various ceratioid genera: (A) *Melanocetus*; (B) *Oneirodes*; (C) *Caulophryne*; (D) *Neoceratias*; (E) *Photocorynus*. After Regan and Trewavas (1932).

uated at the tips of cutaneous papillae (Marshall, 1996:248, 250), the form and length of which vary among families. The papillae "may either be low or may project as tags, stalks, or filaments, pigmented or transparent, that bear the opaque, white sense-organs at their distal ends" (Regan and Trewavas, 1932:23; Fig. 29). The papillae are more or less distinctly connected in a series by narrow, usually unpigmented grooves (Marshall, 1996:248; Fig. 30). The lateral line is usually double, in many species extending out to the tips of the caudal-fin rays. Scattered, isolated organs outside the series may also occur on any part of the head or body. As a rule they are slightly better developed in females than in males. They are especially well developed in *Caulophryne* and *Neoceratias*, two genera that lack an escal light organ.

Although the pattern of placement of the lateral-line structures is complex (e.g., see Bertelsen et al., 1981:8; Fig. 29), with considerable variation among families, no characters of taxonomic significance have yet been found, despite detailed studies (Pietsch, 1969, 1972b, 1974a, 1974b).

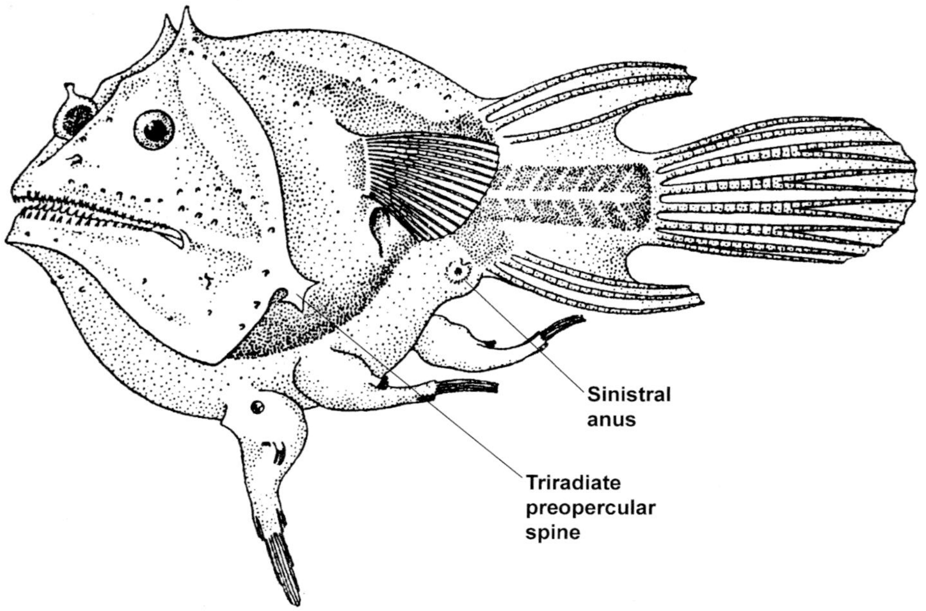

FIGURE 31. *Haplophryne mollis*, 50-mm SL female, with three parasitic males, 11 to 12 mm SL, ZMUC P921777, showing the unique shape of the preopercular spine of this species, and the sinistral anus characteristic of all members of the ceratioid family Linophrynidae. Drawing by W. P. C. Tenison; after Regan and Trewavas (1932).

### Anal Opening

In contrast to all other ceratioids, the anal opening, normally centered on the posteroventral midline, is displaced to the left (sinistral) in all members of the Linophrynidae, a feature evident in all metamorphosed males and females as well as larvae (Fig. 31). The cause of this displacement and its function, if any, are unknown.

## Characters Restricted to Females

### Body Shape

Female ceratioids vary considerably in the shape of the body and in the shape and size of the head and mouth. The body is short and deep, nearly globular in most families, but elongate and somewhat laterally compressed in the Centrophrynidae, Ceratiidae, Gigantactinidae, Neoceratiidae, and several oneirodid genera (e.g., *Pentherichthys*, *Spiniphryne*, and *Dolopichthys*). The body is also elongate in the Thaumatichthyidae, laterally compressed in *Lasiognathus*, but strongly depressed dorsoventrally in the bottom-dwelling genus *Thaumatichthys* (in sharp contrast to all other ceratioids). The length of the head is usually greater than 40% SL, but only about 25% SL in elongate taxa. The mouth is moderate to extremely large (compared to other lophiiforms), the length of the premaxilla as little as 10% SL in some gigantactinids but as great as 40% SL in some linophrynids. The cleft of the mouth is nearly horizontal in some families (e.g., the Centrophrynidae, Thaumatichthyidae, and Gigantactinidae) or oblique to almost vertical in others (e.g., the Ceratiidae and Melanocetidae). The caudal peduncle is exceptionally long and narrow in the Neoceratiidae and Gigantactinidae (length 20 to 30% SL, depth 5 to 10% SL), the dorsal and anal fins relatively well separated from the posterior margin of the hypural plate.

In contrast to the females, most male ceratioids are rather similar in shape, with elongate cylindrical bodies, the head usually less than 30% SL and the mouth small, the opening nearly horizontal. Males of the genus *Thaumatichthys*, however, are considerably more elongate than those of other ceratioids, the head less than 20% SL (see Bertelsen and Struhsaker, 1977:24).

### Frontal Bones

The shape of the frontals and their position relative to one another vary widely among female ceratioids. In the Himantolophidae, Diceratiidae, and Melanocetidae these bones are more or less triradiate and widely separated along their dorsal margins; they meet or approach one another on the midline only at their ventromedial extensions, where they are narrowly separated by cartilage from the supraethmoid anteriorly and the supraoccipital posteriorly, but make no direct bony contact with the parasphenoid. The frontals of all oneirodid genera examined, except those of *Lophodolos*, are also triradiate and widely separated dorsally, but the ventromedial extension of each frontal is bifurcated, the anterior branch narrowly separated by cartilage from the supraethmoid, the posterior branch similarly separated from the supraoccipital, but making direct bony contact as well with the parasphenoid (see Pietsch, 1974a:18). The frontals are widely separated but without ventromedial extensions in the Gigantactinidae, the genus *Thaumatichthys*, and the oneirodid genus *Lophodolos*. They are narrowly separated by cartilage in the Centrophrynidae but meet posteriorly along the midline in front of the supraoccipital in the Ceratiidae, Caulophrynidae, Neoceratiidae, and the linophrynid genus *Photocorynus*.

### Head Spines

The development of spines on the various bones of the head of female ceratioids varies considerably among families and genera, and only those elements that have significant value in identification are discussed here (Fig. 32). Sphenotic spines are well developed in the Himantolophidae, Diceratiidae, Oneirodidae (except for the genus *Chaenophryne*), Caulophrynidae, and Linophrynidae. In most cases, these spines seem to function in combination with well-developed quadrate and articular spines to protect the dorsal, lateral, and ventral aspects of

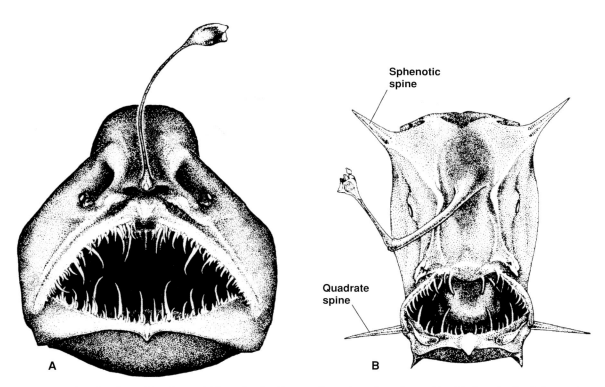

FIGURE 32. Anterior views of female ceratioids contrasting the smooth contour of a melanocetid, with the spiny head characteristic of most oneirodids: (A) *Melanocetus johnsonii*, 85 mm SL, LACM 31484-001 (drawing by Elizabeth Anne Hoxie); (B) *Chirophryne xenolophus*, 22 mm SL, SIO 70-306, with well-developed sphenotic and quadrate spines (drawing by Patricia Chaudhuri). After Pietsch and Orr (2007).

the head and body; exceptions include the Caulophrynidae and Linophrynidae, in which sphenotic spines are well developed but quadrate and articular spines rudimentary or absent. A symphysial spine of the lower jaw is well developed in most ceratioids, but rudimentary to absent in the Thaumatichthyidae, Neoceratiidae, Gigantactinidae, the oneirodid genera *Pentherichthys* and *Chaenophryne*, and some members of the genus *Linophryne*. A large, posteroventrally directed, preopercular spine is present in the linophrynid genera *Borophryne* and *Linophryne* (Fig. 25F); a similar spine in the linophrynid genus *Haplophryne* terminates in 3 to 5 radiating tips (Bertelsen, 1951:170; see Fig. 31). The linophrynid genus *Photocorynus* is unique in having well-developed spines on the sphenotics, epiotics, posttemporals, preopercles (the latter usually with a row of five or six alternating paired and unpaired spines; see Bertelsen, 1951:164; Fig. 25E), anterodorsal margins of the frontals, heads of the palatines, and symphysis of the lower jaw; in some specimens of *Photocorynus* the posteroventral margin of the articulars are notched to form a pair of short spines.

## Jaws and Dentition

The jaw mechanism of females of most ceratioids is similar to that of other lophiiform fishes in having a large gape and restricted gill-openings, combined with slender, sharply pointed, almost always depressible teeth, large muscular upper pharyngeals, and a highly extensible pharynx and stomach enabling engulfment of extremely large prey. The length of the jaws varies among genera from 50% SL to nearly 70% SL in some species of *Linophryne* and the oneirodid genus *Tyrannophryne* to less than 15% SL in the Gigantactinidae. The upper jaw is protrusible in most genera, but this ability is best developed in some oneirodids, and in the Ceratiidae that alone among ceratioids has a well-developed postmaxillary process of the premaxillae (see Bertelsen, 1951:127; Pietsch, 1986b:480). Highly specialized "trapping" or "snagging" mechanisms are developed in some genera, for example, greatly enlarged mobile premaxillae, with long hooked teeth in the Thaumatichthyidae (Bertelsen and Struhsaker, 1977:15), and a somewhat similar development of the lower jaw in the genus *Gigantactis* (Bertelsen et al., 1981:18; see Locomotion, Food, and Feeding, Chapter Seven). Extremely long, flexible, hooked teeth are present on the outer surface of the premaxillae and dentaries of the Neoceratiidae. Greatly reduced, toothless jaws covered with large glands of unknown function are present in the gigantactinid genus *Rhynchactis* (see Bertelsen et al., 1981:10).

Premaxillary and dentary teeth are present in all adult females except for those of the gigantactinid genus *Rhynchactis*. Vomerine teeth are present in females of most taxa, but lost with growth in some species of the Ceratiidae, the oneirodid genus *Bertella*, and in some species of the oneirodid genus *Dolopichthys*; and absent in the Himantolophidae, Thaumatichthyidae, Gigantactinidae, some members of the Ceratiidae, the oneirodid genera *Pentherichthys* and *Lophodolos*, and the linophrynid genera *Photocorynus* and *Haplophryne*. The palatine is toothless in all ceratioids. Teeth are present on the third pharyngobranchial of all females, and on the second pharyngobranchial of all those in which this bone is developed (reduced and toothless or absent in the Neoceratiidae, some species of *Oneirodes*, and the oneirodid genera *Lophodolos*, *Pentherichthys*, *Microlophichthys*, and *Bertella*; see Bertelsen, 1951:157; Pietsch, 1974a:25; Fig. 33). Other elements of the gill arches are toothless in females (and males) of most species, but teeth are present on epibranchial I in females of some

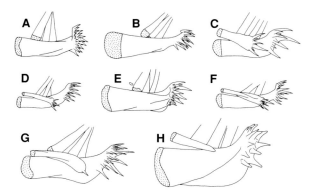

FIGURE 33. Upper pharyngeal bones of females of various species of the family Oneirodidae, showing reduction and loss of the second pharyngobranchial: (A) *Lophodolos acanthognathus*, 57 mm SL, ROM 27285; (B) *Pentherichthys atratus*, 119 mm SL, ISH 130/67; (C) *Chaenophryne draco*, 59 mm SL, LACM 9319-011; (D) *Danaphryne nigrifilis*, 82 mm SL, ISH 2658/71; (E) *Microlophicthys microlophus*, 99 mm SL, ROM 27286; (F) *Leptacanthichthys gracilispinis*, 54 mm SL, ROM 27284; (G) *Dolopichthys pullatus*, 76 mm SL, LACM 6723-033; (H) *Bertella idiomorpha*, 78 mm SL, LACM 30561-001. After Pietsch (1974a).

species of *Oneirodes* (*Oneirodes carlsbergi* and *O. luetkeni*), on epibranchial I and ceratobranchials I through IV in females of Centrophrynidae, and on the proximal ends of epibranchials I through IV and the full length of ceratobranchials I through IV in females of Himantolophidae. A few rudiments of ceratobranchial teeth are present in females of the linophrynid genus *Photocorynus* and males of the Himantolophidae (and possibly also males of the Centrophrynidae; see Bertelsen, 1983:313).

Within species the number of teeth in the jaws of females increases with standard length and varies greatly between species; for example, in each dentary there are 150 to nearly 250 teeth in young specimens (70 to 100 mm) of the Himantolophidae (Fig. 34) and nearly 300 in some specimens (70 to 115 mm) of the oneirodid genus *Dolopichthys* (Pietsch, 1972c:5); but the number rarely exceeds 25 in comparable-sized specimens of *Linophryne* (Fig. 35). Jaw teeth are slender, more or less recurved, and depressible except for the smallest and oldest teeth, which become fixed to the jaws with age (the anterior-most two or three premaxillary teeth of well-preserved females of the oneirodid genus *Spiniphryne* are also immobile). The length of the largest teeth is highly variable between species, and only in some cases comparable in size within genera; for example, the longest teeth are less than 10% SL in the Himantolophidae but approach 25% SL in some species of *Linophryne* (Figs. 34, 35).

The teeth of both upper and lower jaws of females are arranged in a more or less distinct pattern of two crossing series (Figs. 36, 37): (1) a greatly overlapping longitudinal series with teeth in each row becoming progressively larger posteriorly, and (2) a less overlapping transverse series with teeth increasing in length toward the inner side of the jaw (e.g., see Pietsch, 1972c:5; Bertelsen and Struhsaker, 1977:15; Bertelsen, 1980a:34). New and larger teeth are developed ontogenetically on the inner side of the jaw at the posterior end of the longitudinal series, thus keeping the number of series constant (and providing useful taxonomic characters), while increasing the number of transverse series. The smallest (outermost and oldest) teeth are gradually resorbed or lost, thus the total number of teeth does not continue to increase but remains constant in older specimens. In most genera, the pattern of tooth placement in the longitudinal series is more distinct and regular than in the transverse series, while in a few genera the transverse series is more distinct. The tooth pattern in the lower jaw of *Gigantactis* (Fig. 38), and in both upper and lower jaws of the Neoceratiidae, differs considerably from the description

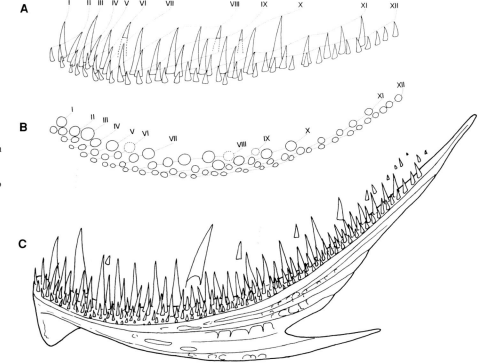

FIGURE 34. Lower jaw teeth of females of *Himantolophus*, with diagrammatic representations of tooth pattern: (A, B) *H. albinares*, 39 mm SL, ISH 3245/79, showing overlapping longitudinal series labeled I to XII (stippled teeth are those in development); (C) *H. groenlandicus*, 53.5 mm SL, ISH 2056/71. After Bertelsen and Krefft (1988).

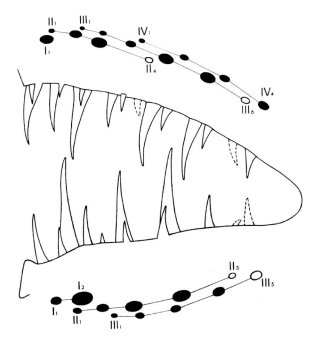

FIGURE 35. Jaw teeth of *Linophryne arborifera*, female, 44 mm SL, ISH 2009/71, with diagrammatic representations of tooth pattern forming overlapping longitudinal series labeled I to IV (upper jaw) and I to III (lower jaw) (stippled outlines and open dots indicate teeth in development). After Bertelsen (1980a).

above in having a longitudinal series of long teeth on the outer surface of the jaw, the outermost teeth being the largest (see Bertelsen, 1951:157; Bertelsen et al., 1981:5).

## Gills and Gill Filaments

The bones of the branchial arches of female ceratioids are somewhat reduced compared to those of other lophiiforms; most of the elements are weak and poorly ossified, and in some cases lost (Fig. 39). Pharyngobranchial I is usually absent, but present in the Centrophrynidae, Ceratiidae, Himantolophidae, Diceratiidae, and the oneirodid genera *Spiniphryne* and *Oneirodes*; pharyngobranchial II is usually present, but reduced or absent in some oneirodids (Fig. 33) and absent in the Neoceratiidae; and pharyngobranchial III is well developed in all taxa (especially in the Gigantactinidae and Neoceratiidae; see Bertelsen, 1951:156), while pharyngobranchial IV is invariably absent. Ceratobranchial V is toothless, reduced to a slender rod-shaped element in most families, but absent or represented only by one or two tiny ossified remnants in the Gigantactinidae (Bertelsen et al., 1981:12).

The extent to which portions of the gill arches are attached to one another and to the adjacent lateral and medial walls of the pharynx, and the distribution of gill filaments on the various elements of the arches, varies considerably among ceratioids, and only a small part of the complexity is described here. Epibranchials I through IV are closely bound to each other by connective tissue, the first nearly always free from the lateral wall of the pharynx (bound to the lateral wall of the pharynx in the Melanocetidae and Thaumatichthyidae), the fourth always bound to the medial wall of the pharynx; in the Himantolophidae, however, the proximal one-half to two-thirds of epibranchial III is free from the adjacent epibranchials. In nearly all families, the proximal one-fourth to two-thirds of ceratobranchial I is bound to the lateral wall of the pharynx; in the Ceratiidae, however, the full length of ceratobranchial I is bound to the lateral wall of the pharynx. In contrast to all other ceratioids, the distal one-third of ceratobranchial I of the Caulophrynidae is bound to the adjacent ceratobranchial II. Also in contrast to other ceratioids, the proximal ends of the ceratobranchials of the Neoceratiidae, Gigantactinidae, and Linophrynidae are bound to one another: the proximal one-quarter of ceratobranchial II bound to ceratobranchial III and the proximal one-third to one-half of ceratobranchial III bound to ceratobranchial IV. In all ceratioids, the full length of ceratobranchial IV, along with that of epibranchial IV, is bound to the medial wall of the pharynx, leaving no opening behind the fourth arch.

Gill filaments are present as holobranchs on gill arches II and III of females of all families and extend onto the proximal end of gill arch I as hemibranchs in the Centrophrynidae, Himantolophidae, Diceratiidae, Melanocetidae, Caulophrynidae, Neoceratiidae, and some oneirodids. Gill filaments are present as hemibranchs on gill arch IV of all families. A pseudobranch is greatly reduced (a few small gill filaments are present on the wall of the pharynx in some gigantactinids) or absent.

## Illicial Apparatus

The illicial apparatus of ceratioids, all that remains of a reduced spinous dorsal fin, is by far the single most important character complex for distinguishing females at all taxonomic levels (Fig. 40). It consists of two cephalic spines, both of which are supported by a single elongate illicial pterygiophore (basal bone of Bertelsen, 1943:189, 1951:17), with the all important bait, or esca, situated at the distal tip of the first cephalic spine.

ILLICIAL PTERYGIOPHORE: In females the pterygiophore of the illicium is elongate and slender, usually cylindrical, but sometimes bearing a compressed ventral keel; it ranges in length from approximately 5% SL in the Neoceratiidae to more than 100% SL in the Ceratiidae. Anteriorly the pterygiophore articulates with the illicial bone (except for the Neoceratiidae in which the latter element is absent), while its cartilaginous posterior end extends back beyond the anterior margin of the supraoccipital. In males the illicial pterygiophore is completely hidden beneath the skin of the head (the only known exception is a parasitic male of *Ceratias holboelli*, ZMUC P922481, in which the posterior end of the pterygiophore protrudes behind the head; Bertelsen, 1951:138); it usually varies in length between approximately 10 and 15% SL but reaches about 25% SL in the Ceratiidae. In males of the Ceratiidae, Himantolophidae, and Melanocetidae the anterior end of the illicial pterygiophore articulates with the upper denticular bone (e.g., see Bertelsen and Krefft, 1988:27; Fig. 41).

Unlike lophioid and antennarioid anglerfishes (but similar to some ogcocephaloids and perhaps some chaunacoids; see Bradbury, 1967:403), in which the illicial pterygiophore is relatively immobile, the pterygiophore of most female ceratioids can be protruded and retracted within a longitudinal groove on the dorsal surface of the cranium (Bertelsen, 1943:190, 1951:17; Fig. 42). In the anteriorly protruded position, the anterior end of the pterygiophore usually emerges from the skin of the head (completely covered by the skin of the head in the Neoceratiidae) from between the frontal bones and somewhat behind the tip of the snout, but its position varies among some taxa: it

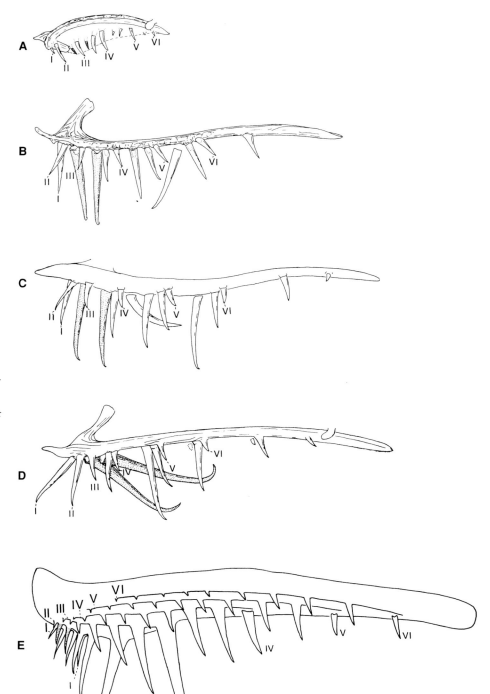

FIGURE 36. Left premaxillary tooth pattern of females of *Thaumatichthys*: (A) *T. binghami*, holotype, 36.5 mm, YPM 2015; (B) *T. binghami*, 70 mm, ZMUC P921948; (C) *T. axeli*, 85 mm, SIO 70-2265; (D) *T. pagidostomus*, holotype, 60 mm, USNM 72952; (E) *T. axeli*, holotype, 365 mm, ZMUC P92166 (diagrammatic). For comparison, the second and third teeth of series II are stippled. After Bertelsen and Struhsaker (1977).

emerges from between or somewhat behind the sphenotic spines in the oneirodid genus *Lophodolos*, behind the cranium in the diceratiid genus *Bufoceratias*, from the roof of the mouth between the anterior ends of the premaxillae in *Thaumatichthys*, and on the distal tip of the snout in *Gigantactis* (Fig. 43). In the retracted position, the posterior end of the pterygiophore usually extends only to the occipital region of the cranium, abutting against a vertically raised posterior portion of the supraoccipital in the Gigantactinidae, but protruding as a posteriorly directed, dorsal tentacle on the back between the head and soft-dorsal fin in the Ceratiidae, the thaumatichthyid genus *Lasiognathus*, and some species of the genus *Oneirodes* (see Bertelsen, 1951:18; Pietsch, 1974a:34, 1996:402, 2004:78).

SECOND DORSAL-FIN SPINE: The second dorsal-fin spine (second cephalic ray of Bertelsen, 1951:17) is minute in most ceratioids (apparently absent in the Neoceratiidae and in large adult females of the Ceratiidae), situated on the dorsal surface

34 WHAT MAKES AN ANGLERFISH?

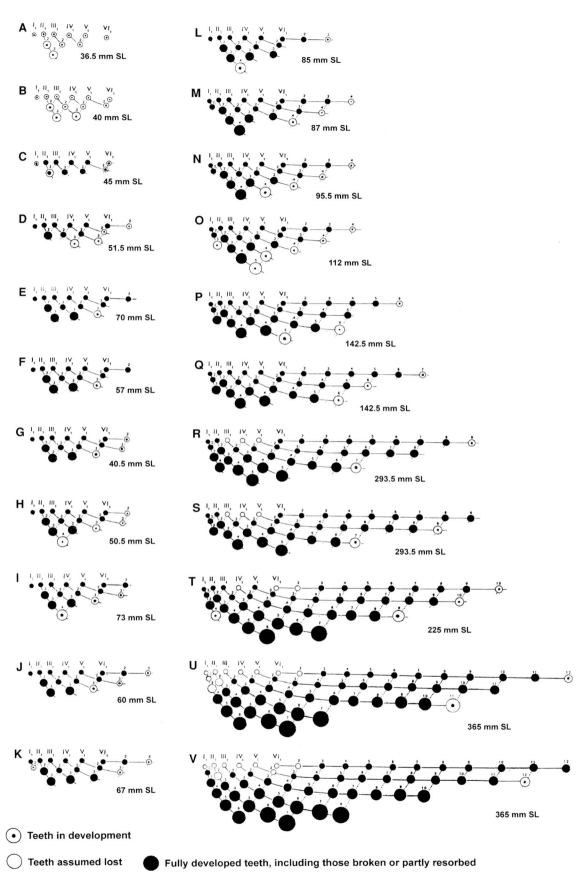

FIGURE 37. Diagrammatic representation resulting from Bertelsen's study of the premaxillary tooth pattern of a developmental series of females of *Thaumatichthys*, demonstrating among other things the remarkable attention to detail found in all his work: (A–I, K, M–T) *T. binghami*, 36.5 to 293.5 mm SL; (J) *T. pagidostomus*, holotype, 60 mm SL, USNM 72952; (L) *T. axeli*, 85 mm SL, SIO 70-2265; (U, V) Right and left premaxillae of *T. axeli*, holotype, 365 mm SL, ZMUC P92166. After Bertelsen and Struhsaker (1977).

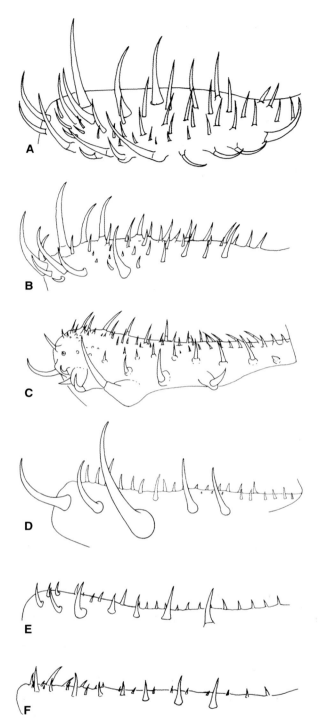

FIGURE 38. Lower jaw of females of species of *Gigantactis*, left side, showing variation in tooth size, number, and number of longitudinal series: (A) *G. longicirra*, 118 mm SL, ISH 2561/71; (B) *G. kreffti*, 252 mm SL, ISH 1099/71; (C) *G. meadi*, 207 mm SL, ISH 571/76; (D) *G. paxtoni*, 210 mm SL, SIOM uncataloged; (E) *G. macronema*, 354 mm SL, LACM 30599-020; (F) *G. microdontis*, 118 mm SL, LACM 9693-34. After Bertelsen et al. (1981).

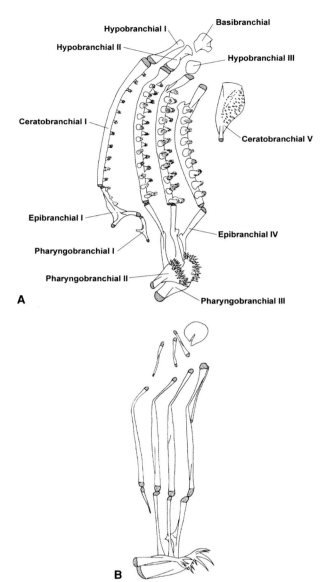

FIGURE 39. Gill arches of (A) a chaunacoid anglerfish, *Chaunax pictus*, 90 mm SL, UW 20770, compared to the much reduced gill arches of (B) a ceratioid, *Melanocetus murrayi*, 84 mm SL, LACM 31501-003. The ventral portion of the branchial basket is shown in dorsal view; the dorsal portion (epibranchials and pharyngobranchials) is folded back and shown in ventral view. Cartilage is stippled. After Pietsch (1981) and Pietsch and Van Duzer (1980).

of the illicial pterygiophore a short distance behind the articulation of the illicium, and nearly always embedded beneath the skin of the head. In larvae and juveniles of the Ceratiidae and Diceratiidae, however, the second dorsal-fin spine bears an esca-like gland (see below) that sometimes emerges slightly from the skin of the head but for the most part remains more or less sunk within a small pore just behind the base of the illicium (Fig. 44). In ceratioid males (except those of the Neoceratiidae), a tiny remnant of the second dorsal-fin spine can be found lying beneath the skin in the same position as that of the females (see Bertelsen, 1951:20).

ILLICIUM AND ESCA: The first dorsal-fin spine, or "illicium" (a Latin term meaning "attraction" or "inducement," first used by Garman, 1899:15, 18, 75–77, 81–83), is attached to the anterior tip of the illicial pterygiophore proximally and bears the esca distally. It is well developed in most ceratioids (its length usually between 25 and 100% SL), but minute (less than 5% SL) and almost totally enveloped by tissue of the esca in some genera (e.g., the ceratiid genus *Cryptopsaras*, the genus *Thaumatichthys*, and

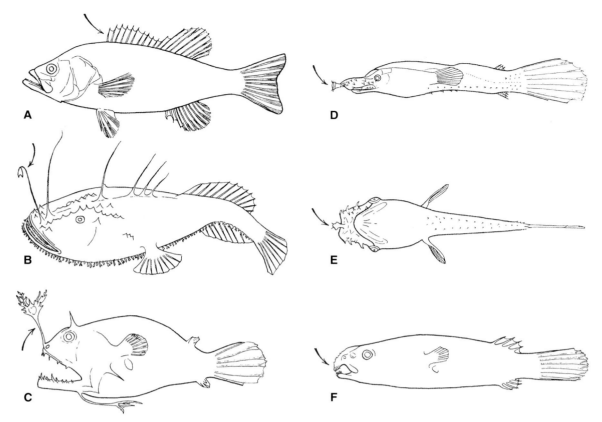

FIGURE 40. Evolutionary modification of the first dorsal-fin spine: (A) the spiny and soft-rayed dorsal fin of a largemouth bass, *Micropterus salmoides*; (B) anterior spines separated and displaced from the dorsal fin to the head and snout in the shallow-water goosefish *Lophius americanus*; (C) first dorsal-fin spine modified as a bioluminescent lure and placed out on the tip of the snout (remaining spines and rays of the dorsal fin greatly reduced) in *Linophryne macrodon*; (D, E) first dorsal-fin spine with luminous bait projecting from the roof of the mouth in *Thaumatichthys binghami*, lateral and ventral views; (F) luminous lure lost, the bony support for the first dorsal-fin spine (the latter drastically reduced to a tiny remnant) closely associated with the toothed upper denticular bone (used in part to attach to a female) in a free-living dwarfed male of the *Himantolophus brevirostris* group. After Parr (1932).

the linophrynid genera *Photocorynus* and *Haplophryne*) and absent in the Neoceratiidae. In other ceratioids it varies considerably in thickness and length, from short and stout (e.g., length less than 30% SL and width as great as 12% SL in some species of *Himantolophus*) to extremely long and threadlike (e.g., 225% SL in some members of the diceratiid genus *Bufoceratias*, almost 270% SL in the caulophrynid genus *Robia*, and nearly 500% SL in some species of *Gigantactis*). Within ceratioid species, however, illicial length is surprisingly constant relative to standard length. In ceratioid males, except those of the Neoceratiidae in which the illicium is absent in both sexes, a tiny remnant of this bone is present lying beneath the skin of the head at the anterior tip of the illicial pterygiophore (e.g., see Bertelsen, 1951:20).

The illicium of females of most ceratioids bears a cutaneous distal swelling, referred to as the "esca" or "escal bulb" (the Latin term for "bait," frequently used by classical Greek and Roman authors in reference to the feeding structures of Mediterranean anglerfishes, and reintroduced in modern times by Garman, 1899:77, 82), containing a globular, bacteria-filled light organ or photophore (from the Greek *photos*, meaning "light," and *phoros*, "a bearing") that opens to the outside by way of a small pore on its posterodorsal margin (the illicium is absent in the Neoceratiidae, the escal photophore is absent in the Caulophrynidae and the gigantactinid genus *Rhynchactis*). The escal bulb is more or less pear shaped in most ceratioids but otherwise varies in shape from nearly spherical (e.g., most members of the Linophrynidae) to elongate (e.g., most species of the genus *Gigantactis*). Its diameter varies from approximately 1% SL (e.g., in *Gigantactis macronema*) to 10 to 15% SL (e.g., in *Himantolophus macroceras*, *Diceratias trilobus*, and *Melanocetus eustalus*).

Pigmentation and the presence of dermal spinules in the skin of the illicium and esca vary considerably between species and with the age of specimens. The illicium is unpigmented in the linophrynid genus *Haplophryne* and juveniles of most other genera, with pigment spreading with growth from the illicial stem to the distal part of the escal bulb, so that in large specimens of some species only a narrow unpigmented field remains encircling the escal pore (e.g., see Bertelsen et al., 1981:29, 40). Filaments and other appendages may be present along the entire length of the illicium or may be restricted to the escal bulb (absent in juvenile *Ceratias*, some diceratiids and melanocetids, and the linophrynid genus *Photocorynus*). The number, position, and shape of the escal appendages show distinct interspecific differences, more often than not representing the only known distinguishing characters within genera (Fig. 45). In all cases, terminology describing the location and position of the various illicial and escal appendages and filaments (e.g., dorsal, anterior, posterior, and lateral) are derived from assuming a vertical orientation of the stem of the illicium; confusion caused by a possible twisting of the soft tissues on the illicial bone can be circumvented by determining the position of the escal pore, which in all cases is directed posteriorly. Appendages are usually present on the distal surface of the escal

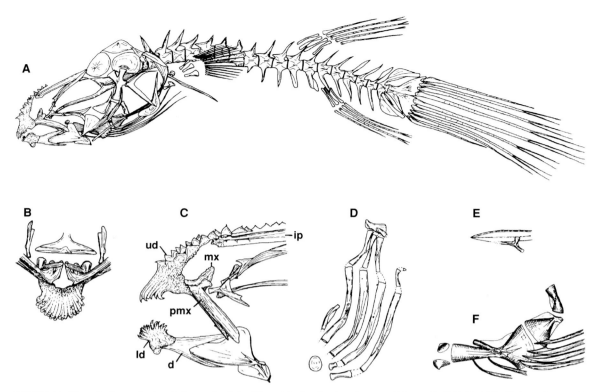

FIGURE 41. The osteology of a free-living male of the *Himantolophus brevirostris* group, 36 mm SL, ISH 3246/79: (A) lateral view of whole specimen, with gill arches removed; (B) ventral view of upper denticular bone and adjacent elements; (C) lateral view of jaws and adjacent bones; (D) gill arches; (E) triradiate pelvic bone and anterior tip of the left cleithrum; (F) lateral view of hyoid apparatus. d, dentary; ip, illicial pterygiophore; ld, lower denticular; mx, maxilla; pmx, premaxilla; ud, upper denticular. After Bertelsen and Krefft (1988).

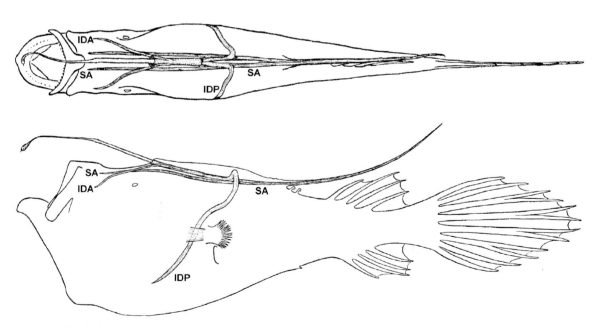

FIGURE 42. The highly mobile illicial apparatus of *Ceratias holboelli*, female, 185 mm SL, ZMUC 75, anterior and lateral views, showing the extrinsic musculature: the inclinator dorsalis anterior (IDA), responsible for vibrating and twisting movements of the illicial pterygiophore; inclinator dorsalis posterior (IDP), responsible for pulling the pterygiophore backward; and the supracarinales anterior (SA), responsible for anterior extension of the pterygiophore. After Bertelsen (1943, 1951).

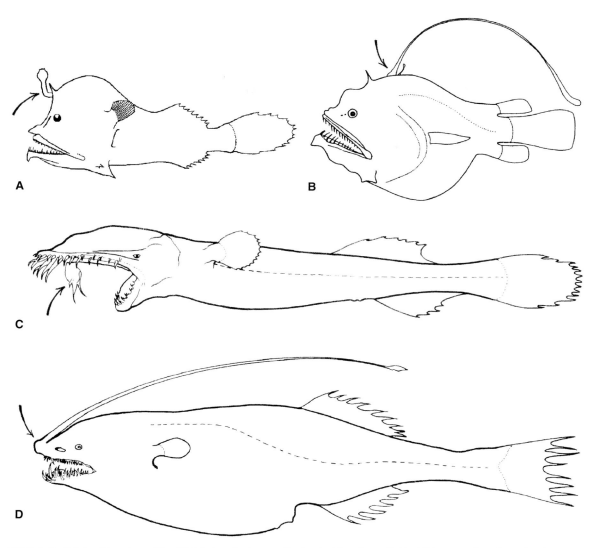

FIGURE 43. Points of insertion of the illicial pterygiophore on the head of female ceratioids: (A) between the sphenotic spines in *Lophodolos*; (B) behind the cranium in *Bufoceratias*; (C) from the roof of the mouth in *Thaumatichthys*; (D) on the distal tip of the snout in *Gigantactis*. Images courtesy of Michel Lamboeuf; copyright FAO, all rights reserved.

bulb anterior to the escal pore (exceptions include some diceratiids, melanocetids, linophrynids, and juvenile *Ceratias*), varying from simple conical or truncated prolongations of the escal bulb to complexly branched or otherwise subdivided structures. Appendages with internal light-guiding structures (that serve to restrict the transparency of the wall of the escal bulb) are present in species of most families (see Bioluminescence and Luring, Chapter Six). A posterior appendage emerging just below the escal pore is present in species of most taxa (but absent in the Ceratiidae, Gigantactinidae, and some species of the Himantolophidae, Diceratiidae, Melanocetidae, and Oneirodidae). Appendages on the lateral surface of the escal bulb and on the illicial stem are present in species of several families. The size and shape of the appendages vary greatly between species, but only slightly within species, ranging from low papilliform outgrowths, digitiform or slender filamentous structures that may be simple or branched, to compressed lobes or flanges with simple, fringed, or filamentous edges.

ABERRANT GENERA: Several ceratioid genera do not conform to the general description of the illicium and esca given above. In addition to the Neoceratiidae, which lacks the illicium, an escal photophore is absent in three ceratioid taxa: the caulophrynid genera *Caulophryne* and *Robia*, and the gigantactinid genus *Rhynchactis*. In the Caulophrynidae, the illicium is well developed, but its distal tip is unpigmented and only slightly inflated (see Pietsch, 1979:12). Although the esca of *Caulophryne* is more or less divided into stout filamentous branches, and that of *Robia* bears a few short filaments, there are no apparent specialized internal structures (e.g., a bacteria-filled central lumen; see Bioluminescence and Luring, Chapter Six). There is no evidence that these structures are bioluminescent, but observations of living specimens are lacking and a histological study has not been made.

In some species of the gigantactinid genus *Rhynchactis*, the posterior margin of the distal half of the illicium bears a series of appendages, each terminating in one or two unpigmented, tapering filaments, some with a tiny distal swelling (Bertelsen and Pietsch, 1998b:586). The proximal one-half to two-thirds of each appendage has two darkly pigmented bands bounding a single transparent band in between. Histologically, the walls of each appendage are without an internal reflecting layer; the central core is opaque, consisting of a dense concentration of

ILLICIAL MUSCULATURE: The illicial apparatus is controlled by five pairs of muscles: two intrinsic pairs and three extrinsic (Bertelsen, 1951:17; Figs. 42, 46). The intrinsic muscles are absent or rudimentary in males but consist of well-developed depressor and erector dorsalis muscles in females. The origins, insertions, and relative development of these muscles are highly variable among families and genera, depending for the most part on the development of the illicium. The intrinsic pairs provide for mobility of the illicium in the vertical plane relative to the illicial pterygiophore. The extrinsic illicial muscles consist of the supracarinales anterior, and anterior and posterior subdivisions of the inclinator dorsalis II, present in both sexes (see Winterbottom, 1974:283, 286). Here again, the origins, insertions, and development of these muscles are highly variable among females of families and genera, depending on the development of the illicium. The extrinsic muscles provide for sliding, vibrating, and twisting movements of the illicial apparatus in females, and movement of the denticular apparatus in males (e.g., see Bertelsen, 1951:17–20; Winterbottom, 1974:284; Bertelsen et al., 1981:18; Shimazaki and Nakaya, 2004:35).

## Other Bioluminescent Structures

In addition to the esca, structures that appear to have a light-producing function are present on the tip of the second dorsal-fin spine (second cephalic ray) of the Ceratiidae and Diceratiidae: an external bulb, appearing sessile on the head of ceratiid larvae (disappearing during metamorphosis; see Fig. 44), but mounted on a short stalk in diceratiid larvae and juveniles, which sinks beneath the skin of the head with age, but remains connected to the surface through a small pore. In larvae of the ceratiid genus *Cryptopsaras*, this structure is histologically similar to the escal photophore, with a glandular body opening to the outside by way of a distal pore, but without the vestibule, and completely covered by a pigmented wall (Munk and Herring, 1996:520; see also Bioluminescence and Luring, Chapter Six). The photophore-like structure of the second dorsal-fin spine has not been examined histologically in the Diceratiidae.

In the Ceratiidae, the anterior-most soft-dorsal rays are modified in a way similar to that of the illicium and second dorsal-fin spine: two club-shaped "caruncles" (a term first used by Günther, 1887:52) born on short, stalk-like fin rays in the genus *Ceratias*; and three, oval, more or less sessile "caruncles" in *Cryptopsaras* (Fig. 44). The caruncles decrease in size relative to standard length in *Ceratias* (tiny and degenerate in large specimens) but increase slightly in size in *Cryptopsaras*. The caruncles of *Cryptopsaras* are histologically similar to the photophore-like structure of the second dorsal-fin spine (Bertelsen, 1951:16). They are known to contain "dense populations of luminous bacteria" that can be discharged to the exterior by way of a distal pore (Hansen and Herring, 1977:104; Herring and Morin, 1978:324), the expulsion observed in a living specimen by Young and Roper (1977:247).

A hyoid barbel is present in *Centrophryne* and *Linophryne*. In *Centrophryne* the barbel is present in larvae and juveniles of both sexes, simple papilliform to digitform, disappearing in larger specimens, without photophores or other specialized internal structures (histologically examined by Pietsch, 1972a:24). In contrast, the barbel of *Linophryne*, which has been observed to luminous in six species (see Bioluminescence and Luring, Chapter Six), is present only in metamorphosed females and provides

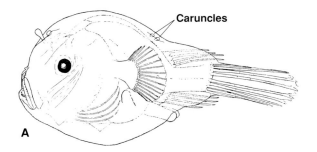
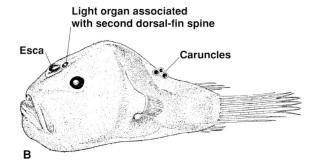
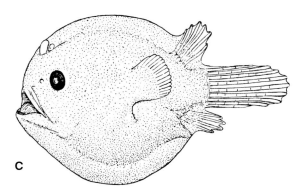
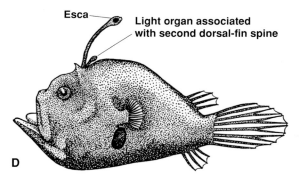

FIGURE 44. Female larvae and juveniles of ceratiids and diceratiids showing early development of accessory light organs of the second dorsal-fin spine and caruncles: (A) *Ceratias* sp., 7.6 mm SL, ZMUC P921133; (B) *Cryptopsaras couesii*, 9.6 mm SL, ZMUC P921291; (C) Diceratiidae sp., 10.5 mm SL, ZMUC P92676; (D) *Bufoceratias wedli*, 19 mm SL, BMNH 1930.1.12.1101. After Bertelsen (1951) and Norman (1930).

cells with large nuclei surrounded by blood vessels (Bertelsen et al., 1981:3). The terminal distal swellings contain numerous specialized cells that are possibly sensory in function. As in the Caulophrynidae, there is no evidence that these structures are bioluminescent.

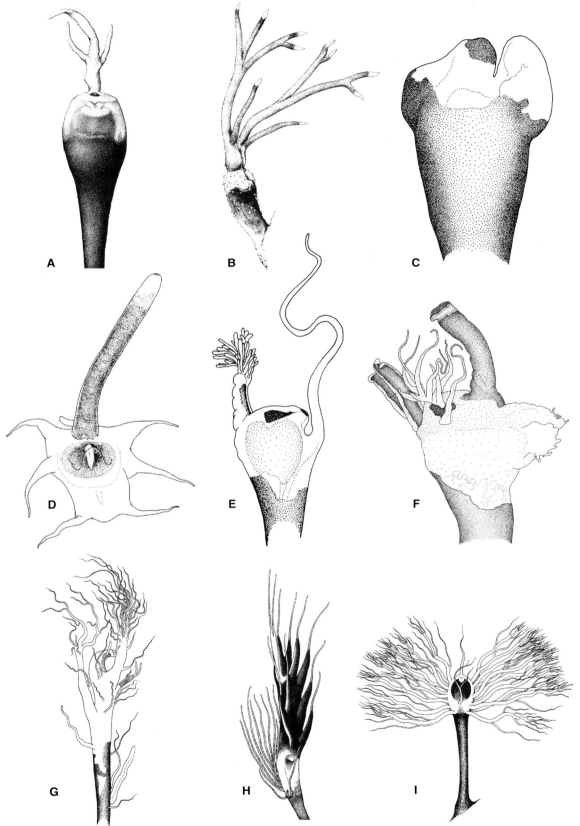

FIGURE 45. Escae of various female ceratioids selected to demonstrate the wide variety of form found within the suborder: (A) *Ceratias holboelli*, 590 mm SL, ZMUC P922184 (drawing by K. Elsman; after Bertelsen and Pietsch, 1984); (B) *Himantolophus cornifer*, holotype, 90 mm SL, MCZ 58858 (drawing by R. Nielsen; after Bertelsen and Krefft, 1988); (C) *Melanocetus johnsonii*, 75 mm SL, MCZ 49849 (after Pietsch and Van Duzer, 1980); (D) *Thaumatichthys binghami*, 142.5 mm SL, ZMUC P921950 (drawing by E. Bertelsen; after Bertelsen and Struhsaker, 1977); (E) *Oneirodes flagellifer*, holotype, 22 mm SL, ZMUC P9280 (after Pietsch, 1974a); (F) *Phyllorhinichthys micractis*, 96 mm SL, SIOM uncataloged (after Pietsch, 1972b); (G) *Caulophryne jordani*, 54 mm SL, LACM 33924-001 (after Pietsch, 1979); (H) *Gigantactis watermani*, holotype, 99 mm SL, ISH 2330/71 (drawing by K. Elsman; after Bertelsen et al., 1981); (I) *Linophryne racemifera*, 58 mm SL, LACM 25185-010 (drawing by R. Nielsen; after Bertelsen, 1982).

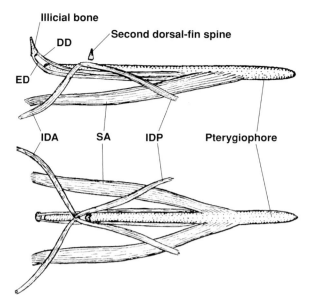

FIGURE 46. Illicial musculature of *Cryptopsaras couesii*, lateral and dorsal views, reconstructed from serial sections of a larval female, 6.6 mm TL: the depressor dorsalis (DD), responsible for drawing the illicium backward; the erector dorsalis (ED), responsible for extending the illicium forward; remaining muscles as described in Figure 42. After Bertelsen (1951).

characters of significant taxonomic importance at the species level (Fig. 47). In most species, it is single stemmed, but multistemmed in the subgenus *Rhizophryne*, and in *Linophryne algibarbata* and *L. polypogon*. In all species of the genus, the terminal branches of the barbel bear small spherical tubercles shown histologically to be photophores (see Hansen and Herring, 1977). For a detailed description of barbel characters in *Linophryne*, see Bertelsen (1982:86–98, table 4).

## Fins and Fin Rays

The rays of the unpaired fins are often greatly elongated (e.g., in females of the Caulophrynidae and Gigantactinidae) and often not interconnected, or interconnected with extremely thin, transparent, membranelike tissue. The caudal fin is rounded in nearly all families, but emarginate in the Gigantactinidae (except for the largest females of some species of *Gigantactis*; see Bertelsen et al., 1981:14). The caudal-fin rays of some large females of the Ceratiidae terminate in spherical skin-covered ossifications (mistaken for caruncle-like light organs by Bertelsen, 1951:142).

## Dermal Spines and Spinules

The skin is everywhere smooth and naked in females of the Caulophrynidae, Neoceratiidae, Linophrynidae, all oneirodids except *Spiniphryne*, and the thaumatichthyid genus *Lasiognathus* (excluding the two or three spinelike hooks attached to the esca). Minute, widely spaced spinules (only visible microscopically in cleared and stained specimens) are present in the skin of females of at least some species of *Melanocetus* and *Oneirodes*. The skin of the entire head, body, and fins, and often including the illicium and proximal part of the escal bulb as well, is covered with numerous, close-set dermal spinules in the Centrophrynidae, Ceratiidae (much larger and considerably less numerous in *Ceratias* than *Cryptopsaras*, especially in females greater than 200 mm), Diceratiidae, the oneirodid genus *Spiniphryne*, and the Gigantactinidae (except for all but the largest known females of *Rhynchactis*; Bertelsen et al., 1981:14, 67; Fig. 48). Similar close-set dermal spinules are present in metamorphosed females of *Thaumatichthys*, more or less restricted to the ventral surface of the body in smaller specimens, but covering the entire head, body, and fins of larger specimens. The family Himantolophidae is unique among ceratioids in having large, widely spaced, conical spines, with broad circular bases, scattered over the head and body (Fig. 49).

## Characters Restricted to Males

### Eyes

The eyes of ceratioid larvae of both sexes are large (their diameter approximately 10 to 18% SL) and morphologically similar in all taxa. Their growth is slow after metamorphosis, never attaining a diameter of more than approximately 3.0 mm and decreasing in relative size to less than 2% SL in the largest known females. Histologically there is no evidence of ocular degeneration in females (Brauer, 1908:184; Waterman, 1948:109; Munk, 1964:12), but as growth continues, the laterally directed eyes sink beneath a thin transparent layer of skin, greatly restricting the visual field. The lens is situated behind the iris and close to the retina, thus ocular function is probably reduced to mere light detection (Munk, 1964:12, 1966:58).

In contrast, the eyes of metamorphosed free-living males of most genera are moderately large, diameters ranging between about 7 and 10% SL. They are usually directed laterally, and spherical or slightly oval in shape, with a distinct aphakic space surrounding the lens. In the Ceratiidae, the eyes of males are exceptionally large and bowl shaped, with an extremely wide aphakic space and anterior sighting grooves (Figs. 50A, B, 51). On the other hand, the eyes of centrophrynid and especially gigantactinid males are surprisingly small (only about 3 to 5% SL in diameter in most specimens) and apparently have reduced ocular function. The eyes of neoceratiid males also appear to be especially small, but these males are so far known only as sexual parasites (Pietsch, 2005b:225), which invariably show signs of ocular degeneration, especially those of *Ceratias* and *Neoceratias* (see Reproduction and Early Life History, Chapter Eight). In the Linophrynidae, while similar in size to those of other ceratioid males, the eyes are directed anteriorly (most probably providing binocular vision), with an elongate axis and an enlarged lens (first described as "telescopic" by Brauer, 1908:184; a term adopted by Bertelsen, 1951:25), and without an aphakic space (see Munk, 1964, 1966; Figs. 50C, 51).

### Olfactory Organs

Olfactory organs are feebly developed in most female ceratioids, the two tiny nostrils on each side of the snout close set and usually mounted on the tip of a low papilla. A relatively large, elongate nasal papilla is present in females of the genus *Thaumatichthys*, the Gigantactinidae, and Neoceratiidae (Fig. 52), but associated nostrils and olfactory lamellae are poorly developed in the former two taxa and apparently absent in the latter (see Bertelsen, 1951:160). In sharp contrast, the nostrils of free-living ceratioid males are extremely large (larger in size relative to the head than in any other vertebrate; Bertelsen, 1951:25; Fig. 53), with the exception of ceratiid males in which

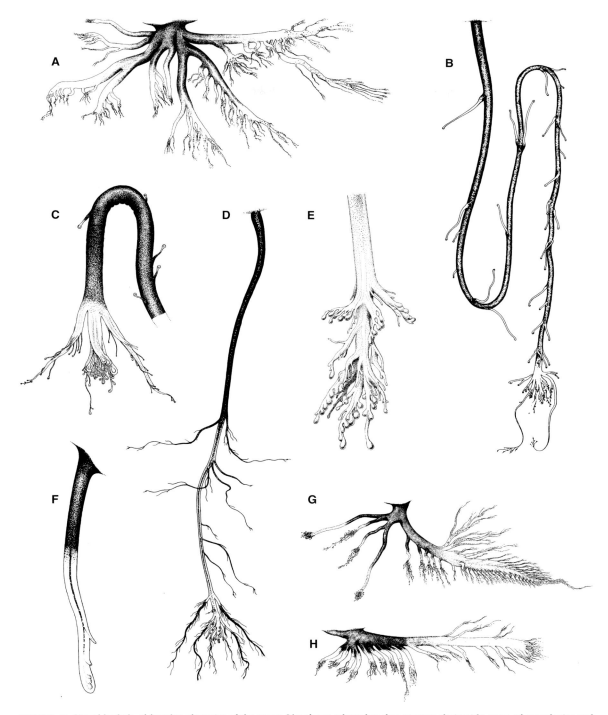

FIGURE 47. Hyoid barbels of females of species of the genus *Linophryne* selected to demonstrate their wide range of complexity and morphological variation: (A) *L. densiramus*, 43.5 mm SL, LACM 38440-001; (B) *L. coronata*, 51.5 mm SL, SIO 60-282; (C) *L. arcturi*, 51 mm SL, ISH 2132/71; (D) *L. coronata*, 36 mm SL, BMNH 2004.11.6.86; (E) *L. arcturi*, 34 mm SL, BMNH uncataloged; (F) *L. racemifera*, 58 mm SL, LACM 25185-010; (G) *L. pennibarbata*, holotype, 47 mm SL, USNM 219854; (H) *L. pennibarbata*, 36.5 mm SL, SIO 68-483. Drawings by R. Nielsen; after Bertelsen (1980a, 1982).

the olfactory organs are tiny, apparently degenerate, and functionless. Each nostril contains a series of large, bladelike olfactory lamellae (the number often providing important taxonomic distinctions between species and genera), covered with thin, more or less inflated and transparent skin in which the nostrils form large round or oval windowlike openings separated by a narrow bridge of skin (Bertelsen, 1951:25). The posterior nostril is usually somewhat larger than the anterior nostril; both are directed laterally in some taxa, but the anterior nostril is more or less directed anteriorly in the Centrophrynidae, Thaumatichthyidae, Oneirodidae, Caulophrynidae, Gigantactinidae, and Linophrynidae.

The olfactory organs of attached parasitic males show a tendency to degenerate. Marshall (1967a:58), in a now classic

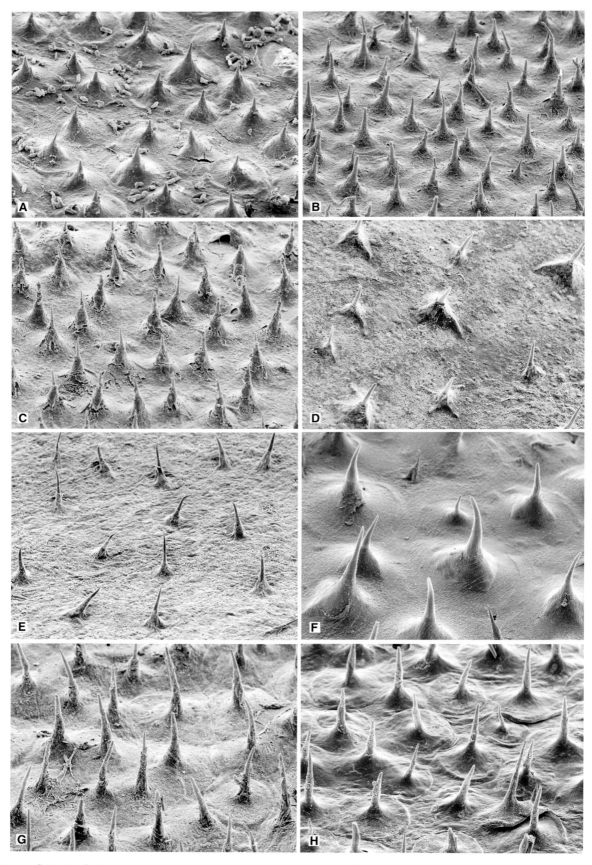

FIGURE 48. Scanning electron micrographs of dermal spinules of female ceratioids: (A) *Centrophryne spinulosa*, 141 mm SL, UW 117074; (B) *Ceratias* sp., 53 mm SL, UW 25504; (C) *Cryptopsaras couesii*, 142 mm SL, UW 47218; (D) *Diceratias pileatus*, 118 mm SL, UW 46525; (E) *Bufoceratias wedli*, 70 mm SL, UW 46524; (F) *Thaumatichthys pagidostomus*, 246 mm SL, ASIZP 63971; (G) *Spiniphryne gladisfenae*, 70 mm SL, UW 20824; (H) *Gigantactis vanhoeffeni*, 295 mm SL, UW 45951. Courtesy of Zachary H. Baldwin.

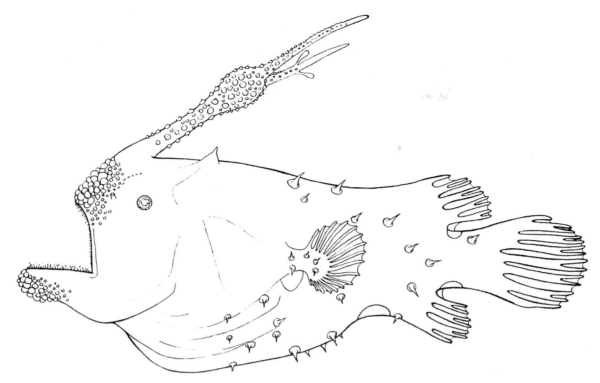

FIGURE 49. *Himantolophus azurlucens*, holotype, 98 mm SL, CAS-SU 46507, showing large, broad-based, widely spaced dermal skin spines. After Beebe and Crane (1947).

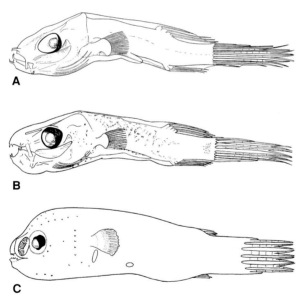

FIGURE 50. Free-living adult males of ceratioids selected for their well-developed eyes: (A) *Ceratias* sp., 10.8 mm SL, specimen sacrificed for histology (see Munk, 1964); (B) *Cryptopsaras couesii*, 10.2 mm SL, specimen sacrificed for histology (see Munk, 1964); (C) *Linophryne arborifera* group, 17 mm SL, ZMUC P921799. After Bertelsen (1951).

paper on the olfactory organs of bathypelagic fishes, demonstrated extreme sexual dimorphism not only in the external structure of the olfactory organs of ceratioids (the males being macrosmatic, the females microsmatic), but also in that part of the brain responsible for chemosensory function, the olfactory nerves, olfactory bulbs, and forebrain all comparatively larger in the males (see Figs. 52 and 53; for sexual dimorphism in the olfactory structures of lophiids and ogcocephalids, see Caruso, 1975:380; and Bradbury, 1988:4, 1999:263).

## Jaw Teeth and Denticular Bones

Premaxillary and dentary teeth, present in all females except adults of the gigantactinid genus *Rhynchactis*, are more or less lost during metamorphosis in all male ceratioids, except in the linophrynid genera *Photocorynus* and *Haplophryne*. Vomerine, palatine, and pharyngobranchial teeth are absent in all known males. Other elements of the gill arches are toothless in the males of all species, except for a few rudiments of ceratobranchial teeth in males of the Himantolophidae and possibly those of the Centrophrynidae (males of the latter family known only on the basis of three small specimens; Bertelsen, 1983:313).

The upper and lower jaws of males are relatively short (the lower jaw 15 to 20% SL in most specimens), becoming smaller relative to standard length during metamorphosis. The outer margins of the premaxillae and dentaries become conspicuously notched as jaw teeth are lost during metamorphosis. The premaxillae are gradually resorbed, eventually leaving only small anterior remnants in fully grown free-living males of most genera; the premaxillae are completely lost in the Ceratiidae. The anterior ends of the maxillae and dentaries of many genera develop into truncated bases to provide articular surfaces for the denticular bones.

Strongly hooked denticular teeth begin to develop in the skin near the anterior tip of the upper and lower jaws at metamorphosis in the males of all known species (apparently restricted to the lower jaw in the Neoceratiidae). In some taxa (e.g., ceratiids) these can be seen originating during metamorphosis by fusion of modified dermal spinules anterior to the toothed

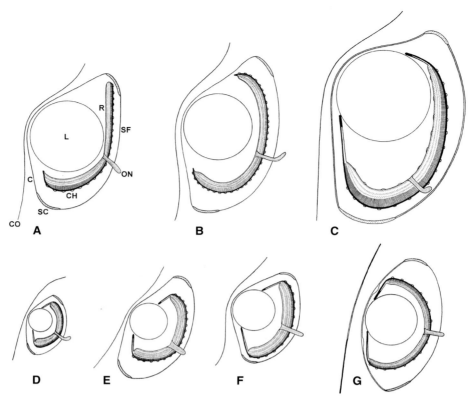

FIGURE 51. Diagrams of horizontal sections through the left eye of ceratioids: (A) *Ceratias* sp., free-living adult male, 10.8 mm SL (same specimen as shown in Fig. 50A); (B) *Cryptopsaras couesii*, free-living adult male, 10.2 mm SL (specimen shown in Fig. 50B); (C) *Linophryne arborifera* group, 17 mm SL, ZMUC P921799 (specimen shown in Fig. 50C); (D) *Cryptopsaras couesii*, larval male, 4 mm TL; (E) *Cryptopsaras couesii*, larval male, 9 mm TL; (F) *Cryptopsaras couesii*, larval female, 8 mm TL; (G) *Cryptopsaras couesii*, adult female, 70 mm SL. The eyes of free-living adult males of *Ceratias* and *Cryptopsaras* have a very wide field of binocular vision because of the large aphakic space, the peculiar form of the retina, and the horizontal sighting groove located ahead of each eye. Metamorphosed males of *Linophryne* have tubular eyes, also providing a wide binocular field of vision but, in addition, probably a greater sensitivity to light, a better judgment of distance, and perhaps stereopsis. In metamorphosed ceratiid females the visual field is restricted because of the position of the lens behind the iris and because the eyes are situated beneath the skin. For these reasons it is thought that the eyes of adult females are incapable of forming images and probably function only as light detectors. C, scleral cornea; CH, choroid; CO, corium of the skin of the head; L, lens; ON, optic nerve; R, retina; SC, scleral cartilage; SF, fibrous part of sclera. Text and figures after Munk (1964).

premaxillae and dentaries (see Bertelsen, 1951:21; Munk, 2000:315; Fig. 54). The denticular teeth eventually fuse at the base in most genera, to form an upper and lower denticular bone or tooth plate situated just anterior to the symphyses of the premaxillae and dentaries, respectively. The denticular teeth apparently remain mutually free in males of the Diceratiidae, some genera of the Oneirodidae, and the genus *Gigantactis*. The number and shape of the denticular teeth and the development of the denticular bones vary between species but show no distinct intrageneric variation. In males of the Ceratiidae, Himantolophidae, and Melanocetidae a posteromedial extension of the upper denticular bone articulates with the anterior end of the pterygiophore of the illicium, the latter element, along with the extrinsic muscles of the pterygiophore, thus participating in opening and closing the denticular jaws (Parr, 1930b:134; Bertelsen, 1951:22; Bertelsen and Krefft, 1988:26).

## Dermal Spinules

The skin is smooth and naked in males of the Centrophrynidae, Caulophrynidae, Neoceratiidae, Linophrynidae, the gigantactinid genus *Rhynchactis*, and all known genera of the Oneirodidae (however, males of the spiny-skinned oneirodid genus *Spiniphryne* are unknown), but everywhere covered with tiny, close-set dermal spinules in those of the Himantolophidae, Diceratiidae, and some members of the Melanocetidae and Gigantactinidae. Free-living males of the Ceratiidae are naked, but all known parasitic males of this family are spinulose. Males of the genus *Thaumatichthys* have numerous small spinules scattered over the body from the occipital region to the base of the caudal peduncle.

## Sexual Parasitism

The parasitic mode of reproduction is apparently obligatory in the Ceratiidae, Linophrynidae, and probably the Neoceratiidae. Males of the Himantolophidae, Diceratiidae, Melanocetidae, Gigantactinidae, and several of the better known oneirodid genera (for example, *Lophodolos*, *Chaenophryne*, *Oneirodes*, *Microlophichthys*, and *Dolopichthys*, each now known from well over 50 females) probably never become parasitic.

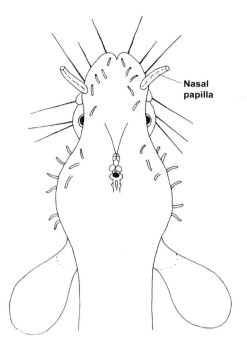

FIGURE 52. Diagrammatic dorsal view of *Neoceratias spinifer*, female, showing prominent nasal papilla. Note the relatively tiny brain compared to that of a free-living male shown in Figure 53B. Modified after Marshall (1971a).

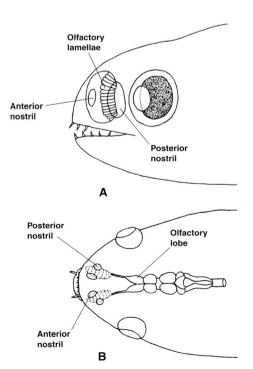

FIGURE 53. Brain and olfactory structures of free-living males of (A) *Linophryne* sp. and (B) *Oneirodes* sp. Note the relatively large brain, especially the olfactory lobes. Modified after Marshall (1967a).

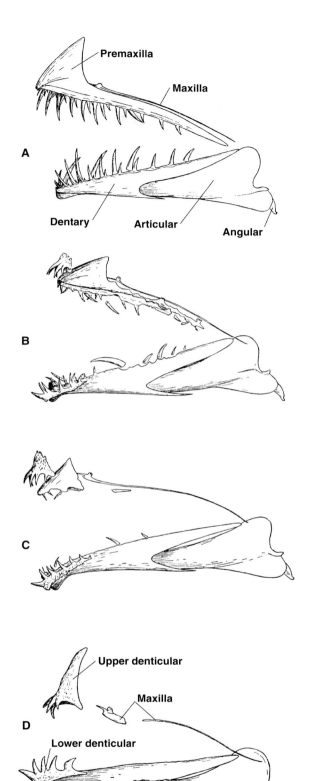

FIGURE 54. Jaws of free-living males of the *Linophryne arborifera* group, showing the progressive degeneration and loss of maxillae, premaxillae, and associated jaw teeth; and the simultaneous development of the upper and lower denticular bones, which originate during metamorphosis by fusion of modified dermal spinules located anterior to the toothed premaxillae and dentaries: (A) larval stage, 13 mm SL, ZMUC P921801; (B) early metamorphosis, 15 mm SL, ZMUC P921789; (C) late metamorphosis, 14 mm SL, ZMUC P921788; (D) postmetamorphosis, 17 mm SL, ZMUC P921787. All drawn to the same scale; modified after Bertelsen (1951).

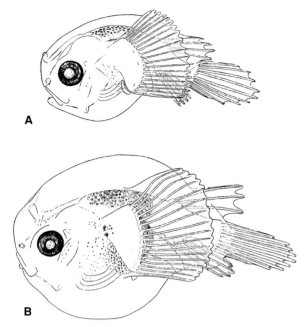

FIGURE 55. Larval females of *Rhynchactis* sp., showing the highly inflated transparent skin characteristic of most ceratioid larvae: (A) 6.0 mm TL, ZMUC P921734; (B) 7.2 mm SL, 9.5 mm TL, ZMUC P921753. After Bertelsen (1951).

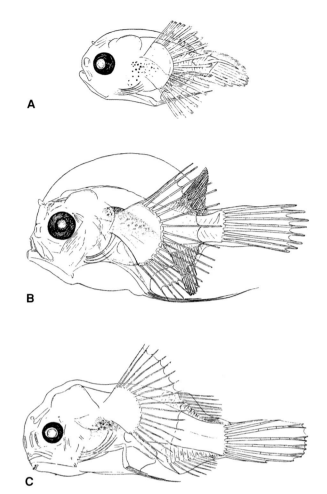

FIGURE 56. Developmental series of *Caulophryne* sp., showing the rudiment of the illicium at the tip of the snout and the unique presence of pelvic fins: (A) larva, 2.9 mm SL, 3.7 mm TL, ZMUC P92203; (B) larva, 6.6 mm SL, 9.5 mm TL, ZMUC P92192; (C) male in metamorphosis, 7.5 mm SL, 10 mm TL, ZMUC P92193. After Bertelsen (1951).

Spawning and fertilization is assumed to take place during a temporary sexual attachment that does not involve fusion of male and female tissues. Sexual parasitism is probably facultative in the Caulophrynidae and in the oneirodid genera *Bertella* and *Leptacanthichthys*. The remaining ceratioid families, Centrophrynidae and Thaumatichthyidae, are still so poorly known that little can be concluded concerning their mode of reproduction (see Reproduction and Early Life History, Chapter Eight).

## Characters Restricted to Larvae

### Body Size and Shape

Ceratioid larvae are generally small, the smallest known individuals measuring 2.0 to 3.5 mm total length (TL). In most genera, metamorphosis begins at about 8 to 10 mm SL, whereas in some taxa (Himantolophidae, Thaumatichthyidae, Gigantactinidae, and Linophrynidae) the larvae may reach lengths of 15 to 25 mm SL (Bertelsen, 1984:326). The head and body of both sexes are surrounded by inflated, balloonlike, transparent skin, which causes their shape to vary from nearly spherical, with the greatest width and depth of the body reaching 80 to 90% SL (Fig. 55), to elongate or pear shaped, with body depths of 40 to 60% SL. The inflation of the skin varies with preservation, but generally its greatest development is found in the Himantolophidae, Caulophrynidae, and Gigantactinidae, and least pronounced in the Ceratiidae, Oneirodidae, and Neoceratiidae. The larvae of most taxa may be described as short, more or less spherical, the length of the head (measured from the tip of the snout to the base of the pectoral-fin lobe) about 50% SL; those of the Oneirodidae and Linophrynidae may be categorized as elongate, the length of head generally about 45% SL; and those of the Neoceratiidae, long and slender, the length of the head 35 to 40% SL (Bertelsen, 1984:327). The vertebral column of larvae of the Ceratiidae is more strongly sigmoid in shape compared to those of other families, resulting in a characteristic hump-backed appearance.

### Illicium, Second Dorsal-Fin Spine, and Caruncles

With few exceptions, the primary sexually dimorphic features of ceratioids can be seen even in the youngest and newly hatched larvae that measure only 2.0 to 3.5 mm TL. The external second dorsal-fin spine of the Ceratiidae and Diceratiidae, and the caruncles of the Ceratiidae, are present as distinct rudiments in larval females of these taxa, but absent in the males. Similarly, a distinct, external, papilliform illicial rudiment is present in all larval females, but absent in males—the only possible exceptions to this rule are the larvae of the Caulophrynidae (16 known specimens) and Neoceratiidae (11 known specimens), among which no larvae without illicial rudiments have yet been found. In *Caulophryne*, in which metamorphosed females lack an escal bulb but have a well-developed illicium, the rudiment protrudes on the anterodorsal margin of head in the same position as in other ceratioid larvae (Fig. 56). In *Neoceratias*, in which the illicium is absent in metamorphosed females, all known larvae have an elongate, cylindrical illicial rudiment protruding on the tip of the snout just above the symphysis of the upper jaw, in a position unique among ceratioid larvae (Bertelsen, 1951:159, 1984:328).

## Pigmentation

The pattern of subdermal pigmentation of ceratioid larvae provides the most important character complex to differentiate species and genera. It is especially important in the identification of the smallest larvae (2.0 to 3.5 mm TL), at which stages no distinguishing characters other than pigmentation are usually developed. In most cases, the pattern of pigmentation is retained under the pigmented skin of postmetamorphic juveniles that have acquired adult characters (Bertelsen, 1951:12, 1984:327). In some taxa, for example the Neoceratiidae and some linophrynids (i.e., *Haplophryne* and *Linophryne* subgenera *Stephanophryne* and *Rhizophryne*; Bertelsen, 1984:327), the subdermal melanophores form distinct dorso- and ventrolateral rows along the body. But, generally speaking, the pigment pattern is usually separated into four primary, more or less well defined groups of melanophores: (1) peritoneal, (2) opercular, (3) dorsal, and (4) caudal-peduncular. In all species in which one of these primary groups occurs, the pigment is almost always laid down in the youngest larvae as a few small and scattered melanophores that during larval development gradually increase in size, number, and area covered. Additional groups of melanophores occur in some taxa, for example, on the base of the pectoral lobe in the gigantactinid genus *Rhynchactis*; internally, within the fin rays of the oneirodid genus *Pentherichthys*; on the posterior angle of the lower jaw in *Linophryne* subgenus *Stephanophryne* (containing only a single species, *Linophryne indica*); and on a dorsal swelling of the outer transparent skin in front of the dorsal fin in some larvae of the Himantolophidae (Bertelsen, 1984:327)

A complete lack of pigment is found in larval *Ceratias*, the linophrynid genus *Borophryne*, and some larvae of *Gigantactis* (Bertelsen, 1984:327). In addition to these three taxa, peritoneal pigment is absent only in the Neoceratiidae. In all other ceratioid larvae, peritoneal pigment is laid down on the dorsal margin of the peritoneum of the youngest larvae and gradually spreads with growth to the lateral and posterior margins.

Pigmentation of the opercular region varies greatly between taxa. It is absent or weakly developed in most genera, but dense and taking on characteristically different patterns in the Thaumatichthyidae, various genera of the Oneirodidae (e.g., *Oneirodes*, *Dolopichthys*, and *Microlophichthys*), and the ceratiid genus *Cryptopsaras* (Bertelsen, 1984:327).

In addition to the completely unpigmented larvae mentioned above, dorsal pigment is absent in the Caulophrynidae, Neoceratiidae, and Linophrynidae. In all other taxa it is laid down on the anterodorsal surface of the body. Differing between genera in density and coverage, it spreads laterally and posteriorly, in some cases reaching and overlapping the dorsal part of the peritoneal pigmentation and the bases of the dorsal and anal fins; in some larvae it becomes confluent with the pigment group of the caudal peduncle.

In occurrence, position, and development relative to total length, larval pigmentation on the caudal peduncle shows very distinct differences between genera (e.g., among oneirodids; Bertelsen, 1984:327), subgenera (e.g., within *Linophryne*; Bertelsen, 1984:327), and species groups (e.g., within the Himantolophidae and the oneirodid genus *Chaenophryne*; Bertelsen, 1984:327).

## Pectoral and Pelvic Fins

The larvae of the Caulophrynidae and Gigantactinidae differ distinctly from those of other ceratioids in the size of the pectoral fins, which have lengths of 40% to nearly 60% SL, when measured from the base of the pectoral-fin lobe. In all other families, pectoral-fin length is about 20 to 25% SL and certainly never exceeds 30% SL.

In contrast to all other ceratioids, the larvae and early metamorphic stages of *Caulophryne* have pelvic fins that consist of 3 or 4 well-developed fin rays (Fig. 56). The longest of these rays increases in relative length from about 45% SL in the smallest known larva (2.9 mm) to about 60% SL in the largest (7.8 mm). Pelvic rays are present also in the smaller of two known free-living metamorphosed males of this genus (7.5 to 11.0 mm), but in this specimen the length of the longest ray is reduced to about 28% SL. Pelvic-fin rays are absent in the three known parasitic males of *Caulophryne* (12 to 16 mm), as well as in all metamorphosed females (10 to 183 mm).

# THREE

# Biodiversity

> Much diversity is manifest in this group, extending to the compression of the head and body, the extent and direction of the mouth, the development of the dorsal and anal fins, and the modification of the rostral spine.
>
> THEODORE NICHOLAS GILL,
> "Angler Fishes: Their Kinds and Ways," 1909:577

The Ceratioidei, containing all the deep-sea anglerfishes, is by far the most species-rich vertebrate taxon within the bathypelagic realm, a vast, largely empty body of cold, nutrient-poor water that constitutes the world's oceans below 1000 m. When compared to other vertebrate groups that inhabit these waters, nothing comes close to anglerfishes for variety and number of forms. The present chapter is devoted to a demonstration of this surprisingly rich diversity, providing evidence for the recognition of 11 ceratioid families, 35 genera, and 160 species. It begins with a concise summary of the most significant distinguishing characters of the suborder (a "diagnosis," to use the scientific term) to serve as evidence that all subtaxa of the group are united in monophyly, that is, that all the diverse lineages of deep-sea anglerfishes evolved from a single common ancestor. This is followed by a brief synopsis of the families—an illustrated key of sorts that serves as a means of identification of each major lineage, while at the same time providing a better understanding of the family and generic accounts that follow.

## Suborder Ceratioidei Gill

### DISTINGUISHING CHARACTERS

Ceratioid fishes are recognized as one of five suborders of anglerfishes that together constitute the teleost order Lophiiformes. They are uniquely different and evolutionarily derived in a host of ways that clearly separate them from all other anglerfishes and, for that matter, from all other fishes. First and foremost, monophyly for the Ceratioidei is supported by extreme sexual dimorphism, an extraordinary difference between males and females of all included taxa that has no comparison in other organisms, at least not among vertebrate animals. This difference is much more than size. Along the evolutionary pathway toward male dwarfism, natural selective pressures have affected every organ system of the male body, resulting in an almost total structural reorganization. What can be most easily observed in the males, in addition to their small size, is the loss of the luring apparatus, the presence of denticular bones—found in no other fishes and used to grasp and hold fast to a prospective mate—and the greatly enlarged nostrils and eyes. But there are also the numerous unseen and poorly understood physiological and behavioral adaptations that allow for a unique mode of reproduction: the mechanisms that led to mate location and selection, male attachment to females, sexual parasitism, and the hormonal communication that in turn mediates simultaneous gamete maturation and subsequent spawning (see Reproduction and Early Life History, Chapter Eight).

Monophyly for the Ceratioidei is also supported by the mutual loss of a number of bony parts that are well developed in other anglerfishes. In their transition from a benthic life-style to a pelagic midwater existence, ceratioids as a group have lost the pelvic fins and have undergone a drastic repositioning and reduction in size of the pectoral fins, all resulting in the loss of the benthic ambulatory function that is so highly developed in the shallow-water, bottom-living ancestors of ceratioids, the goosefishes, frogfishes, batfishes, and their allies (see Fig. 1).

Finally, reflective of the nutrient-poor environment in which they live, ceratioid monophyly is supported by a general trend toward the reduction of body density through an overall decrease in skeletal ossification and the extent of muscle development, as well as the infusion of lipids throughout. In short, there is no reason to doubt that ceratioids have diverged and evolved together from a single common lophiiform ancestor. For more exhaustive diagnoses as well as descriptions of females, males, and larvae of the suborder, see Part Two: A Classification of Deep-Sea Anglerfishes.

The suborder contains 11 families, differentiated as follows:

### SYNOPSIS OF CERATIOID FAMILIES

The following key is provided to allow for the placement of ceratioids into the proper family. Each entry consists of a combination of features that together serve to differentiate each family. It works by progressively eliminating the most morphologically unique family; for that reason it should always be entered from the beginning. All character states listed for each family must correspond to the specimen being keyed; if not, the user should proceed to the next set of character states. The emphasis here, and throughout this chapter, is on the more common and much better known females, which are, with few exceptions, the defining entities for all ceratioid families, genera, and species. For a full overview of both genders, as well as larvae, including traditional dichotomous keys to the identification of all taxa, see Part Two: A Classification of Deep-Sea Anglerfishes.

Females with the body elongate and laterally compressed; the illicium absent; numerous elongate mobile teeth, attached to conical bony outgrowths, situated on the outer margins of the jaws; dorsal fin with 11 to 13 rays, anal fin with 10 to 13 rays; caudal fin large, distinctly rounded posteriorly; extremely rare, known from only 18 metamorphosed females, seven parasitic males; free-living males unknown:

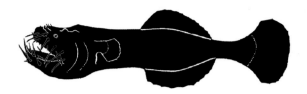

NEEDLEBEARD SEADEVILS, FAMILY NEOCERATIIDAE, A SINGLE GENUS AND SPECIES, P. 136

Females with the body elongate, compressed laterally (in *Lasiognathus*) or strongly depressed dorsoventrally (in *Thaumatichthys*); the illicium long, emerging from the forehead (in *Lasiognathus*), or extremely short, embedded within the esca and hanging from the roof of the mouth (in *Thaumatichthys*); the esca bearing 1 to 3 small bony hooks; the upper jaw extending anteriorly far beyond the lower jaw; relatively rare, only 44 known females:

WOLFTRAP SEADEVILS, FAMILY THAUMATICHTHYIDAE, TWO GENERA AND EIGHT SPECIES, P. 81

Females with the body elongate, laterally compressed; the illicium long and slender (but length highly variable, ranging from somewhat less than standard length to nearly five times standard length), emerging from the anteriormost tip of the snout, with (in *Gigantactis*) or without (in *Rhynchactis*) a bulbous distal light organ; length of the head less than 35% SL; mouth nearly horizontal; jaw teeth well developed (in *Gigantactis*) or absent (in *Rhynchactis*); epibranchial and ceratobranchial teeth absent; caudal peduncle unusually long and slender, length more than 20% SL; the caudal fin nearly always incised posteriorly, caudal rays usually highly elongate:

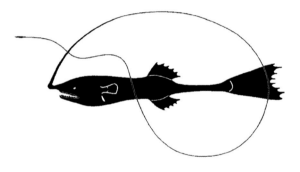

WHIPNOSE SEADEVILS, FAMILY GIGANTACTINIDAE, TWO GENERA AND 21 SPECIES, P. 138

Females with the body elongate, laterally compressed; the mouth almost vertical; illicium well developed (in *Ceratias*) or fully embedded within tissue of the esca (in *Cryptopsaras*); the pterygiophore of the illicium emerging anteriorly well behind the tip of the snout, and posteriorly on the back near the origin of the soft dorsal fin; 2 or 3 club-shaped caruncles on the dorsal midline just anterior to the soft dorsal fin; caudal fin rounded posteriorly, containing 8 well-developed fin rays (small remnant of a ninth ray present or absent); the best known ceratioid family, with well over 1300 females cataloged:

WARTY SEADEVILS, FAMILY CERATIIDAE, TWO GENERA AND FOUR SPECIES, P. 57

Females with the body elongate, laterally compressed; the mouth oblique to horizontal; the anterior end of the pterygiophore of the illicium emerging on the snout, the posterior end embedded beneath skin of the head; sphenotic spines absent, quadrate and articular spines present but greatly reduced; dorsal caruncles absent; numerous teeth present on gill arches; specimens less than 40 mm SL with a small digitiform barbel on the throat; skin everywhere covered with tiny close-set dermal denticles; relatively rare, only 41 known females:

PRICKLY SEADEVILS, FAMILY CENTROPHRYNIDAE, A SINGLE GENUS AND SPECIES, P. 55

Females with the body short and deep, more or less globular; the mouth large, the lower jaw usually extending posteriorly beyond the base of the pectoral-fin lobe; the illicium present, relatively short (less than 130 mm SL in *Caulophryne*) or unusually long (about 270 mm SL in *Robia*) with or without distal filaments, but lacking a bulbous bacteria-filled light organ; the pterygiophore of the illicium fully embedded beneath skin of head; jaw teeth unusually large, well-developed; epibranchial and ceratobranchial teeth absent; rays of the dorsal and anal fins apparently free (not interconnected by membrane) and usually long, length greater than 60% SL; dorsal-and anal-fin rays numerous (14 to 22 and 12 to 19, respectively, in *Caulophryne*) or relatively few (only 6 and 5, respectively, in *Robia*); caudal-fin rays eight; skin smooth and naked, without spines or dermal denticles; lateral-line structures unusually well developed, sense organs situated on the tips of elongate cutaneous papillae:

FANFIN SEADEVILS, FAMILY CAULOPHRYNIDAE, TWO GENERA AND FIVE SPECIES, P. 128

Females with the body short and deep, more or less globular; the illicium emerging from the snout (in *Diceratias*) or on the back well behind the head (in *Bufoceratias*); the skin covered with numerous, close-set dermal spinules; small specimens with a short, second, dorsal-fin spine, situated immediately posterior to the base of the illicium, bearing a distal luminous gland, withdrawn beneath the skin in larger specimens, its presence indicated by a small pore; jaw teeth large and well developed; epibranchial and ceratobranchial teeth absent; sphenotic, quadrate, and articular spines well developed:

DOUBLESPINE SEADEVILS, FAMILY DICERATIIDAE, TWO GENERA AND SIX SPECIES, P. 71

Females with the body short and deep, globular; the mouth large, opening oblique to nearly vertical; jaw teeth numerous and well developed; epibranchial and ceratobranchial teeth absent; vomer usually well-toothed, with a single row of as many as 12 teeth; the head smooth and rounded, without sphenotic, quadrate, or articular spines; illicium emerging on snout, supporting pterygiophore fully embedded beneath skin of head; the skin smooth, appearing naked, without dermal spines or spinules; dorsal fin unusually long, with 12 to 17 rays; anal fin relatively short, with only 3 to 5 rays; one of the best known ceratioid families, with more than 1200 females cataloged:

BLACK SEADEVILS, FAMILY MELANOCETIDAE, A SINGLE GENUS AND SIX SPECIES, P. 75

Females with the body short and deep, globular; sphenotic spines well developed, but quadrate, articular, angular, and preopercular spines absent; the lower jaw unusually blunt, extending anteriorly somewhat beyond the upper jaw; the illicium thick and stout, the esca unusually large and morphologically complex; the pterygiophore of the illicium fully embedded beneath skin of the head; low rounded wart-like papillae covering the snout and chin; the skin of specimens greater than 30 to 40 mm SL with large, widely spaced bony plates, each bearing a single median spine; jaw teeth numerous but short, arranged in several close-set longitudinal series, vomer unusually broad and toothless; epibranchial and ceratobranchial teeth well developed:

FOOTBALLFISHES, FAMILY HIMANTOLOPHIDAE, A SINGLE GENUS AND 18 SPECIES, P. 63

Females with the body short and deep, more or less globular; most species (genus *Linophryne*) with a conspicuous chin barbel; mouth especially large, opening oblique; the illicium well developed, emerging on snout (in *Acentrophryne*, *Borophryne*, and *Linophryne*) or reduced and nearly fully enveloped by tissue of the esca (in *Photocorynus* and *Haplophryne*); the pterygiophore of the illicium fully embedded beneath skin of head; sphenotic spines well developed, but quadrate and articular spines absent; the preopercle usually bearing one or more spines (absent in *Acentrophryne*); jaw teeth highly variable among genera: long and few, widely spaced, and arranged in several oblique longitudinal series (in *Acentrophryne*, *Borophryne*, and *Linophryne*), or unusually short and numerous, arranged in several close-set series (in *Haplophryne* and *Photocorynus*); the vomer usually bearing long, fanglike teeth (toothless in *Photocorynus* and *Haplophryne*); epibranchial and ceratobranchial teeth absent; the dorsal and anal fins each with only 3 rays (very rarely 2 or 4); branchiostegal rays 5 (rarely 4); the skin smooth and naked, without dermal spines or spinules; the anal opening displaced to the left side of the ventral midline:

LEFTVENT SEADEVILS, FAMILY LINOPHRYNIDAE, FIVE GENERA AND 27 SPECIES, P. 148

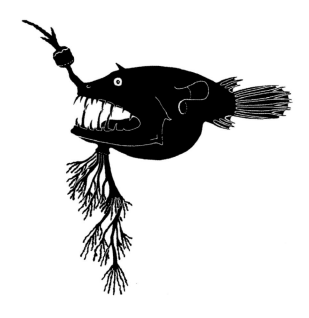

Females with the body short and deep to moderately elongate and laterally compressed; the mouth oblique to nearly horizontal; the illicium with a bulbous distal light organ; the pterygiophore of the illicium usually emerging anteriorly on the snout, but only rarely extending posteriorly on the back behind the head (genus *Oneirodes*); the top of the head armed with sharp sphenotic spines (short in *Ctenochirichthys*, absent in *Chaenophryne*); quadrate and articular spines usually well developed; the skin smooth and naked, without dermal spines or spinules (except in *Spiniphryne*):

DREAMERS, FAMILY ONEIRODIDAE, 16 GENERA AND 62 SPECIES, P. 87

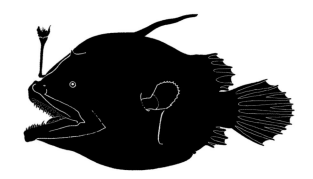

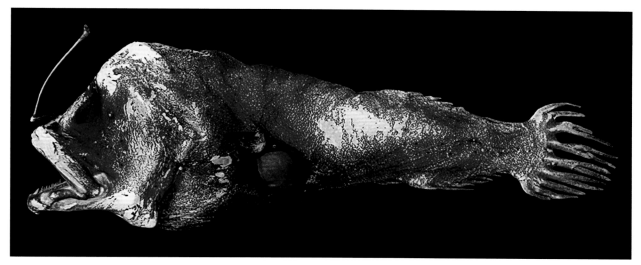

FIGURE 57. *Centrophryne spinulosa*, adult female, 136 mm SL, LACM 30379-001. Photo by T.W. Pietsch.

## FAMILY CENTROPHRYNIDAE Bertelsen, 1951
(Prickly Seadevils)

*Figures 57–60, Table 1*

Although now recognized as unique among ceratioid families, the Centrophrynidae has been confused with oneirodids, while, at the same time, having an overall close similarity to ceratiids, especially members of the genus *Ceratias*. *Centrophryne* was briefly described by Regan and Trewavas (1932) as a genus of the family Oneirodidae to contain two species: *Centrophryne spinulosa* Regan and Trewavas, 1932, represented by two female specimens; and *C. gladisfenae* (Beebe, 1932b), represented by a single female. The presence in *C. spinulosa* of 4 pectoral radials, a spine on the anterior margin of the subopercular bone, and a small digitiform barbel on the throat led Bertelsen (1951) to remove this species from the family Oneirodidae and place it in a new family, the Centrophrynidae. At the same time, he retained *C. gladisfenae* in the family Oneirodidae under a new generic name, *Spiniphryne*.

Regan (1926) referred a male in metamorphosis to *Rhynchoceratias leuchorhinus* Regan, 1925c (*Trematorhynchus leuchorhinus* of Regan and Trewavas, 1932), which was later shown by Bertelsen (1951) to differ from all other ceratioid males by the presence of a short hyoid barbel. A comparison of the subdermal pigment, number of fin rays, head skeleton, and pelvic bones with the known material of *Centrophryne*, confirmed the identity of the male as *C. spinulosa*. At the same time, Bertelsen (1951) described two larval specimens, a male and female, also bearing the characteristic digitiform barbel on the throat. Thus, prior to Pietsch's (1972a) revision of the family, the material of *C. spinulosa* recorded in the literature included only two small females, two larvae, and one male in metamorphosis.

Based on a nearly eightfold increase in the number of available specimens, including several large adult females and one additional metamorphosing male, Pietsch (1972a) reviewed the family, providing a more complete description of *C. spinulosa* and extending its known geographic range into the Atlantic and Indian oceans. A detailed comparative osteological analysis provided ample evidence to justify Bertelsen's (1951) removal of *C. spinulosa* from the Oneirodidae, and reallocation to a family of its own, the Centrophrynidae. Since Pietsch's (1972a) review, very little new information has come to light: Bertelsen and Quéro (1981) described an additional female from off the Canary Islands, and Bertelsen (1983) described the first known adult male from the South China Sea.

As for *Spiniphryne*, Pietsch's (1972a:18, 1974a:30) prediction that "study of adequate material, including a growth sequence of *Spiniphryne gladisfenae*, may prove this species and *C. spinulosa* to be congeneric" has proven to be incorrect. Based on an osteological examination of new material collected in the eastern tropical Atlantic, Bertelsen and Pietsch (1975) showed that the closest relatives of *Spiniphryne* lie within the family Oneirodidae, despite its superficial similarity to *Centrophryne*.

### DISTINGUISHING CHARACTERS

Centrophrynid females are elongate, laterally compressed forms, with relatively small anteriorly directed mouths that open almost horizontally. Their teeth are numerous but rather small compared to most ceratioids. Larvae and juveniles of both sexes are unique in having a small fingerlike barbel on the throat that gradually becomes smaller with growth, disappearing in males during metamorphosis, and in females once they reach about 40 mm standard length. Despite a detailed histological examination of this structure (see Pietsch, 1972a:24), no function has been assigned to this little appendage.

Perhaps even more curious are the gonads of female centrophrynids. Instead of paired structures, as are found in all other anglerfishes, and for that matter in nearly all vertebrates (hagfishes, lampreys, some elasmobranchs, and most birds are notable exceptions), metamorphosed females of this family are unique in having a single oval-shaped ovary (Pietsch, 1972a:24). Why this should be so is another mystery that defies explanation.

In addition to a host of other, mostly internal osteological characters (detailed in Part Two: A Classification of Deep-Sea Anglerfishes), female centrophrynids are characterized by the absence of sphenotic spines on the top of the head, elements that are a hallmark of several other ceratioid families, especially

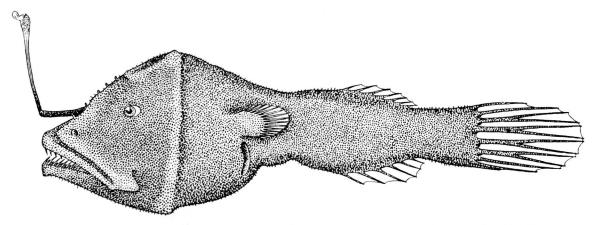

FIGURE 58. *Centrophryne spinulosa*, paralectotype, juvenile female, 33 mm SL, BMNH 1932.5.3.19. Drawing by W.P.C. Tenison; after Regan and Trewavas (1932).

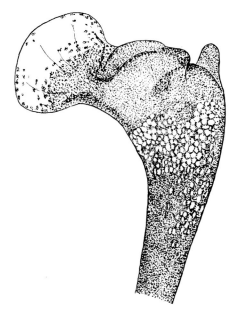

FIGURE 59. Esca of *Centrophryne spinulosa*, lectotype, juvenile female, 39 mm SL, ZMUC P92122. After Bertelsen (1951).

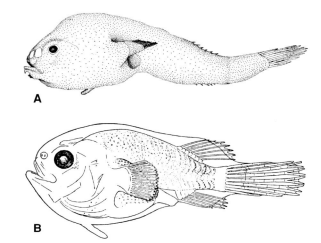

FIGURE 60. *Centrophryne spinulosa*: (A) male in late metamorphosis, 16 mm SL, LACM 30844-001 (after Pietsch, 1972a); (B) larval male, 7.2 mm SL, ZMUC P92153 (after Bertelsen, 1951).

oneirodids. They also have 4 pectoral radials, fusing to 3 in specimens greater than 150 mm. The skin is everywhere covered with numerous close-set dermal spinules, giving specimens a fine sandpapery feel (Fig. 48A). The esca is rather simple, consisting of a compressed, fan-shaped anterior appendage and a single, short, more or less compressed posterior appendage (Fig. 59). Finally, one last feature that helps to distinguish female centrophrynids from those of other ceratioids is found in the gills. A quick look inside the mouth reveals the presence of numerous, small, tooth-bearing plates along the anterior margin of all the gill arches (i.e., the first epibranchial and all four ceratobranchials), a feature found elsewhere among ceratioids only in members of the Himantolophidae. Why centrophrynids and himantolophids should have these gill teeth to the exclusion of all other ceratioids is unknown.

Centrophrynid males, known from only three juvenile specimens, differ from those of all other ceratioids in having the short barbel on the throat described above for females (Fig. 60A). They differ further in having a combination of unusually small eyes and relatively large olfactory organs, indicating that olfaction is the primary sensory mode used to locate conspecific females. Their denticular teeth set them apart as well: the upper denticular plate is roughly triangular in shape and bears a transverse series of 3 well-developed hooked denticles; while the lower denticular plate is crescent shaped, bearing a transverse series of 4 strong, symmetrically placed denticles, fused at the base. Unlike the females, the skin is naked, without dermal spinules. The males are free-living. There is no evidence of sexual parasitism (see Reproduction and Early Life History, Chapter Eight).

The larvae, known from only two specimens, are relatively short and deep, with the skin only moderately inflated (Fig. 60B). The presence of a short digitiform hyoid barbel easily distinguishes them from the larvae of all other ceratioids.

Fin ray counts, consistent in both genders as well as larvae, are as follows: dorsal-fin rays 6 or 7, anal-fin rays 5 or 6, pectoral-fin rays 15 or 16, caudal rays 9 (2 simple + 4 bifurcated + 3 simple) (Table 1).

The largest known female of the family is a 247-mm individual (ASIZP 59902) collected from off the east coast of

FIGURE 61. *Ceratias holboelli*, female, with parasitic male. Drawing by Richard Ellis, modeled after Bertelsen (1951).

Taiwan. The only known adult male (SIO 70-347), from the South China Sea, measures 12.8 mm.

DIVERSITY

As presently understood, the Centrophrynidae contains only a single genus:

GENUS *CENTROPHRYNE* REGAN AND TREWAVAS, 1932
(PRICKLY SEADEVILS)

Figures 57–60, Table 1

DISTINGUISHING CHARACTERS

This genus bears all the features of the family, as described above and detailed in Part Two: A Classification of Deep-Sea Anglerfishes.

ETYMOLOGY

The name chosen by Regan and Trewavas (1932) for this genus refers to the characteristic spiny skin of females: the Greek *kentron* and Latin *centrum* meaning "prickle" or "spine" (in addition to the midpoint of a circle), and the Greek *phryne*, meaning "toad"; hence, a "prickly toad."

DIVERSITY

The genus *Centrophryne* contains only a single species:

Prickly Seadevils, *Centrophryne spinulosa* Regan and Trewavas, 1932: 41 metamorphosed females (18 to 247 mm), one metamorphosed male (12.8 mm), two males in metamorphosis (11.5 to 16 mm), and two larvae (4.2 to 7.5 mm); Atlantic, Pacific, and Indian oceans; lectotype, ZMUC P92122, female, 39 mm, *Dana* station 3768(1), north of New Guinea, 1°20′S, 138°42′E, 4000 m of wire, 24 July 1929 (Regan and Trewavas, 1932:84; Bertelsen, 1951:125, 1983:313; Pietsch, 1972a:19; Bertelsen and Pietsch, 1975:10; Bertelsen and Quéro, 1981:89).

## FAMILY CERATIIDAE Gill, 1861
(Warty Seadevils)

Figures 61–67, Tables 1, 2

The ceratioid family Ceratiidae contains relatively elongate, large-sized anglerfishes, easily separated from members of allied families by having the cleft of the mouth vertical to strongly oblique, the posterior end of the pterygiophore of the illicium emerging from the dorsal midline of the trunk, and 2 or 3 caruncles on the back just anterior to the origin of the soft-dorsal fin. Two genera are recognized, *Ceratias* Krøyer and *Cryptopsaras* Gill. The genus *Ceratias* is one of the most common and best-known representatives of the lophiiform suborder Ceratioidei, now represented by at least 440 metamorphosed females (including numerous damaged specimens that cannot be identified to species). Since Krøyer's (1845, often erroneously dated 1844) original description, material of this genus has been described under eight generic and 13 specific names (see Table 2). All of these nominal forms, however, were placed by Bertelsen (1943, 1951) within the synonymy of the type species, *Ceratias holboelli* Krøyer (1845). Similarly, the genus *Cryptopsaras*, even better known than *Ceratias*, with some 980 metamorphosed females in collections around the world, contains seven nominal species (Table 2), all now placed (primarily by Bertelsen, 1951) in the synonymy of the type species, *Cryptopsaras couesii* Gill (1883b).

Pietsch (1986b) reviewed the Ceratiidae based on all known material, including examination of nearly all extant type specimens. The validity of *Ceratias tentaculatus* (Norman), resurrected from the synonymy of *C. holboelli* Krøyer by Bertelsen and Pietsch (1984), was further documented. In addition, Pietsch (1986b) provided evidence to support the resurrection of *C. uranoscopus* Murray (in Thomson, 1877), also from the synonymy of *C. holboelli*, thus bringing the total number of recognized species of the genus to three. A detailed examination of nearly all known material of *Cryptopsaras* indicated the presence of only a single nearly cosmopolitan species.

TABLE 2
Reallocation of Nominal Species of the Ceratiid Genera *Ceratias* and *Cryptopsaras* Based on Females

| Nominal Species | Currently Recognized Taxon |
| --- | --- |
| *Ceratias holboelli* Krøyer, 1845 | *Ceratias holboelli* Krøyer, 1845 |
| *Ceratias uranoscopus* Murray, in Thomson, 1877 | *Ceratias uranoscopus* Murray, in Thomson, 1877 |
| *Typlopsaras shufeldti* Gill, 1883b | *Ceratias uranoscopus* Murray, in Thomson, 1877 |
| *Cryptopsaras couesii* Gill, 1883b | *Cryptopsaras couesii* Gill, 1883b |
| *Ceratias carunculatus* Günther, 1887 | *Cryptopsaras couesii* Gill, 1883b |
| *Miopsaras myops* Gilbert, 1905 | *nomen dubium* |
| *Ceratias mitsukurii* Tanaka, 1908 | *Cryptopsaras couesii* Gill, 1883b |
| *Mancalias tentaculatus* Norman, 1930 | *Ceratias tentaculatus* (Norman, 1930) |
| *Cryptosparas normani* Regan and Trewavas, 1932 | *Cryptopsaras couesii* Gill, 1883b |
| *Cryptosparas pennifer* Regan and Trewavas, 1932 | *Cryptopsaras couesii* Gill, 1883b |
| *Cryptosparas valdiviae* Regan and Trewavas, 1932 | *Cryptopsaras couesii* Gill, 1883b |
| *Mancalias xenistius* Regan and Trewavas, 1932 | *Ceratias uranoscopus* Murray, in Thomson, 1877 |
| *Mancalias bifilis* Regan and Trewavas, 1932 | *Ceratias tentaculatus* (Norman, 1930) |
| *Mancalias uranoscopus triflos* Roule and Angel, 1933 | *Ceratias uranoscopus* Murray, in Thomson, 1877 |
| *Cryptosparas atlantidis* Barbour, 1941b | *Cryptopsaras couesii* Gill, 1883b |
| *Mancalias sessilis* Imai, 1941 | *Ceratias uranoscopus* Murray, in Thomson, 1877 |
| *Typhloceratias firthi* Barbour, 1942b | *nomen dubium* |
| *Parrichthys merrimani* Barbour, 1942b | *nomen dubium* |
| *Reganichthys giganteus* Bigelow and Barbour, 1944a | *Ceratias holboelli* Krøyer, 1845 |
| *Reganula gigantea*: Bigelow and Barbour, 1944b | *Ceratias holboelli* Krøyer, 1845 |
| *Mancalias kroyeri* Koefoed, 1944 | *Ceratias uranoscopus* Murray, in Thomson, 1877 |

### DISTINGUISHING CHARACTERS

Metamorphosed females of the family Ceratiidae are in many ways similar to those of centrophrynids. They have an elongate, laterally compressed head and body; sphenotic, quadrate, articular, angular, and preopercular spines are all absent; the illicial apparatus emerges on the snout from between the frontal bones; and their skin is everywhere covered with close-set dermal spinules (hypertrophied in *Ceratias*, especially in large females; Fig. 48B). But they differ most strikingly from centrophrynids, and from all other anglerfishes, in having 2 or 3 fleshy wartlike caruncles—bioluminescent glands associated with the anteriormost rays of the dorsal fin—located on the dorsal midline of the trunk. In further contrast to centrophrynids, their relatively small mouths are directed dorsally, the opening almost vertical. Another peculiar feature is the elongate pterygiophore of the illicium that slides back and forth within a deep narrow cranial trough that runs the full length of the dorsal surface of the skull, the posterior end of which is hidden within a hollow, cylindrical evagination of the skin, emerging from the dorsal midline just anterior to the caruncles (a feature present elsewhere only in some species of *Oneirodes* and the thaumatichthyid genus *Lasiognathus*; see Bertelsen, 1943:190, 1951:18–20).

In addition to numerous, mostly internal osteological characters given in Part Two: A Classification of Deep-Sea

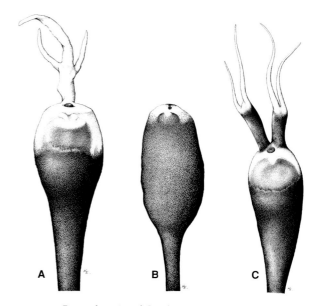

FIGURE 62. Escae of species of *Ceratias*, posterior views: (A) *C. holboelli*, 590 mm SL, ZMUC P922184; (B) *C. uranoscopus*, 173 mm SL, LACM 33313-002; (C) *C. tentaculatus*, 365 mm SL, ISH 230/71. Drawings by K. Elsman; after Pietsch (1986b).

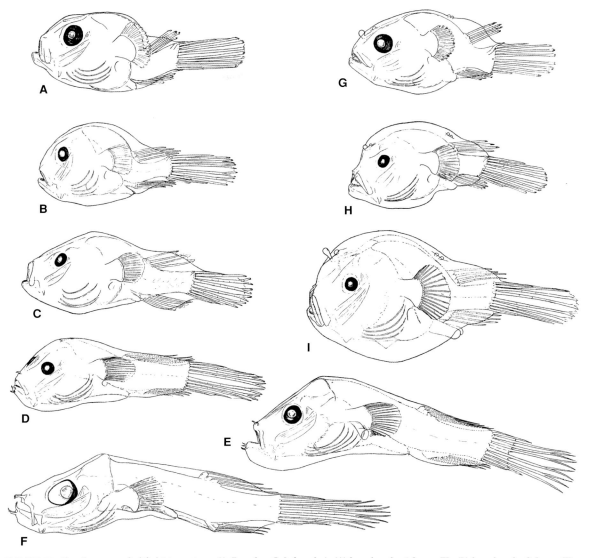

FIGURE 63. *Ceratias* sp., early life history stages (A–F, males; G–I, females): (A) larval male, 4.3 mm TL; (B) larval male, 8.5 mm TL; (C) larval male, 9.4 mm TL; (D) male in metamorphosis, 10.4 mm TL; (E) male in metamorphosis, 9.3 mm SL, 13.4 mm TL; (F) free-living adult male, 10.8 mm SL, 15.6 mm TL; (G) larval female, 4.5 mm TL; (H) larval female, 8.1 mm TL; (I) larval female, 7.6 mm SL, 11.9 mm TL. After Bertelsen (1951).

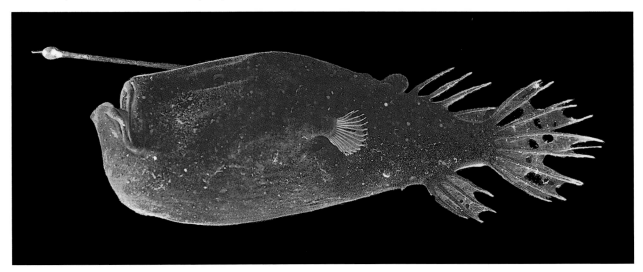

FIGURE 64. *Cryptopsaras couesii*, female, 34.5 mm SL, BMNH 2006.10.19.1. Photo by Edith A. Widder.

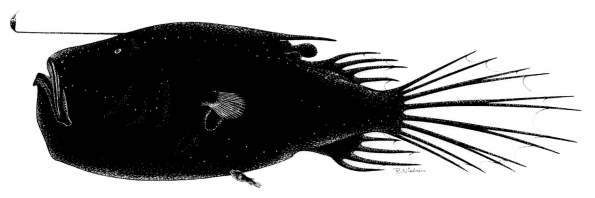

FIGURE 65. *Cryptopsaras couesii*, female, 34.5 mm SL, BMNH 2006.10.19.1. Drawing by R. Nielsen; after Pietsch (1986b).

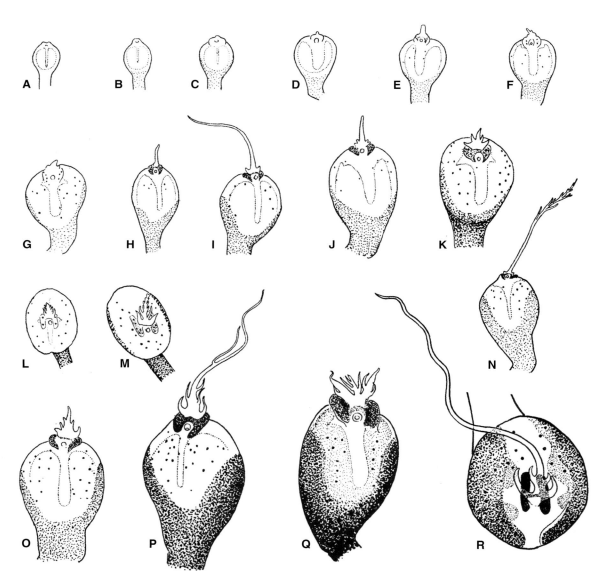

FIGURE 66. Escae of *Cryptopsaras couesii*, showing morphological variation and change with growth, posterior views: (A–E) early postmetamorphic stages, 9.6 to 9.8 mm SL; (F) 13.5 mm SL, 18 mm TL; (G) 19 mm SL; (H) 13.2 mm SL; (I) 20 mm SL; (J) 19.5 mm SL; (K) 24 mm SL; (L) 17.5 mm SL; (M) 27.5 mm SL; (N) 22 mm SL; (O) 28 mm SL; (P) 38 mm SL; (Q) 90 mm SL; (R) 122 mm SL. After Bertelsen (1951).

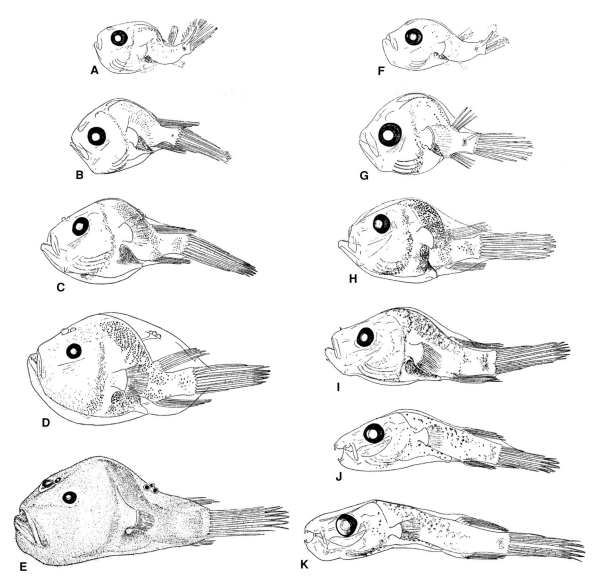

FIGURE 67. *Cryptopsaras couesii*, early life history stages (A–E, females; F–K, males): (A) larval female, 3.2 mm TL; (B) larval female, 4.1 mm TL; (C) larval female, 7.5 mm TL; (D) larval female, 11.8 mm TL; (E) female in metamorphosis, 9.6 mm SL, 15 mm TL; (F) larval male, 3.1 mm TL; (G) larval male, 4.5 mm TL; (H) larval male, 7.7 mm TL; (I) male in metamorphosis, 9.8 mm TL; (J) early postmetamorphic male, 11.3 mm TL; (K) free-living adult male, 10.2 mm SL, 14.3 mm TL. After Bertelsen (1951).

Anglerfishes, the second dorsal-fin spine of larval and juvenile female ceratiids bears a distal bioluminescent gland that is quickly reduced with further growth, and hidden beneath the skin just behind the base of the illicium in adults (Fig. 44A, B). There are 4 pectoral radials. The esca is oval in shape, with or without one or two distal appendages (Figs. 62, 66).

Free-living metamorphosed males differ strikingly from those of other ceratioids in having unusually large, bowl-shaped eyes, while the olfactory organs are minute (Figs. 50A, B, 51). A pair of large denticular teeth is present on the tip of the snout, fused at the base and articulating with the pterygiophore of the illicium. Two pairs of denticular teeth are present on the tip of the lower jaw. The skin is naked and unpigmented in juvenile free-living stages, but spinulose and darkly pigmented in parasitic stages. Males are obligatory sexual parasites as adults (see Reproduction and Early Life History, Chapter Eight).

The body of larval ceratiids is distinctly "hump-backed," the skin moderately inflated, and the mouth almost vertical (Figs. 63, 67). Sexual dimorphism is well developed, even in the smallest larvae, with the females having a distinct illicial rudiment and caruncles on the dorsal midline of the trunk. The pectoral fins are small, not reaching beyond the dorsal and anal fins.

Fin-ray counts are as follows: dorsal-fin rays 4 (rarely 3 or 5), excluding the 2 or 3 reduced rays enveloped by the caruncles; anal-fin rays 4; pectoral-fin rays 15 to 19 (rarely 14); caudal-fin rays 8 or 9 (8 in *Cryptopsaras*; the ninth or lowermost ray is reduced to a small remnant in *Ceratias*), 2 simple + 4 bifurcated + 2 (or 3) simple (Table 1).

Ceratiids represent the largest known ceratioids, the females of *Ceratias holboelli* attaining a standard length of at least 855 mm (BMNH 1953.2.25.1). All free-living males (about 175 known specimens) measure less than 20 mm, whereas parasitic males (107 known specimens, attached to 81 females) range from 8.0 to about 140 mm (see Pietsch, 1976:786, 2005b:223; and Reproduction and Early Life History, Chapter Eight).

SYNOPSIS OF GENERA OF THE FAMILY CERATIIDAE

The Ceratiidae contains two genera, differentiated as follows.

---

Females with the illicium long, considerably longer than the length of the escal bulb; 2 caruncles on the dorsal midline of the trunk just anterior to the origin of the soft-dorsal fin (reduced in larger specimens); the subopercle without an anterior spine; the ninth or lowermost caudal-fin ray reduced to a small remnant; one of the most common ceratioid genera, with well over 330 cataloged females:

DOUBLEWART SEADEVILS, GENUS *CERATIAS* KRØYER, 1845, THREE SPECIES, P. 62

Females with the illicium short, nearly completely enveloped by tissue of the escal bulb; 3 caruncles on the dorsal midline of the trunk just anterior to the origin of the soft-dorsal fin (well developed even in the largest known specimens); the subopercle with an anterior spine; the ninth or lowermost caudal-fin ray absent; the most common ceratioid genus, with over 1,000 cataloged females:

TRIPLEWART SEADEVILS, GENUS *CRYPTOPSARAS* GILL, 1883B, A SINGLE SPECIES, P. 63

---

GENUS *CERATIAS* KRØYER, 1845
(DOUBLEWART SEADEVILS)

*Figures 61–63, Tables 1, 2*

DISTINGUISHING CHARACTERS

The most striking difference between females of *Ceratias* and those of its sister genus *Cryptopsaras* lies in the structure of the luring apparatus. The pterygiophore of the illicium is unusually long in both genera, but the illicium itself, born on the distal tip of the pterygiophore, is long in *Ceratias*, many times longer than the length of the esca; whereas, in *Cryptopsaras* the illicium is extremely short, almost entirely embedded in the tissue of the esca. Also diagnostic and easily observed is the number of dorsal caruncles—2 in *Ceratias* and 3 in *Cryptopsaras* (see Figs. 44A, B; 63G–H; 67C–E). But for some unknown reason, and in sharp contrast to those of *Cryptopsaras*, the caruncles of *Ceratias* are gradually reduced in size with growth, becoming minute in females greater than 400 mm. For specimens with damaged or missing illicia and/or caruncles, a quick look at the caudal skeleton will always differentiate the two genera: like nearly all ceratioids, *Ceratias* has 9 caudal-fin rays, but the ninth or lowermost ray is reduced to a small basal remnant; whereas, in *Cryptopsaras* this element is absent leaving this genus with only 8 caudal-fin rays (Table 1). Finally, *Ceratias* differs from *Cryptopsaras* in lacking a spine on the anterodorsal margin of the subopercle.

In addition to a tiny remnant of the ninth caudal-fin ray and a spineless subopercle, metamorphosed males of *Ceratias* are distinguished from those of *Cryptopsaras* in having the two pairs of lower denticular teeth nearly equal in size. Larvae, males, and juvenile females lack subdermal pigment (Fig. 63).

ETYMOLOGY

Krøyer's (1845) *Ceratias*, from which we also get the family and subordinal names Ceratiidae and Ceratioidei, is derived from the Greek *keras* or *keratos*, meaning "horn" or "horned," in allusion to the lure projecting from the snout.

DIVERSITY

Three species are recognized, along with three nominal forms of doubtful validity (Table 2):

Northern Giant Seadevil, *Ceratias holboelli* Krøyer, 1845: 161 metamorphosed females (16.5 to 855 mm) and 19 parasitic males (35 to about 140 mm); Atlantic, Pacific, and Indian oceans; holotype, ZMUC 61, 680 mm, Southern Greenland, 0 to 340 m (Krøyer, 1845:639; Bertelsen, 1943:185, 1951:133; Pietsch, 1986b:484).

Stargazing Giant Seadevil, *Ceratias uranoscopus* Murray, in Thomson, 1877: 112 metamorphosed females (12 to 240 mm), and one parasitic male (22.5 mm); Atlantic, Pacific, and Indian oceans; holotype, BMNH 1887.12.7.15, 57 mm, *Challenger* station 89, between Canary and Cape Verde Islands, ca. 20°13′N, 20°13′W, 0 to 4392 m (Murray, in Thomson, 1877:70; Bertelsen, 1943:195, 196, 198–203; Pietsch, 1986b:485).

Southern Giant Seadevil, *Ceratias tentaculatus* (Norman, 1930): 57 metamorphosed females (6.2 to 640 mm), and one parasitic male (37 mm); except for three specimens from off South Africa, all of the known material has been collected from the Southern Ocean between approximately 35° and 68°S; holotype *(Mancalias tentaculatus)*, BMNH 1930.1.12.1100, 80 mm, *Discovery* station 114, 52°25′S, 9°50′E, 0 to 700 m, 12 November 1926 (Norman, 1930:355; Bertelsen, 1943:200; Bertelsen and Pietsch, 1984:45; Pietsch, 1986b:482).

*Miopsaras myops* Gilbert, 1905, *nomen dubium*: holotype, USNM 51637, 75 mm, *Albatross* station 4019, near Kauai Island,

Hawaiian Islands, 748 to 1325 m, 21 June 1902 (Gilbert, 1905:694, pl. 99; Pietsch, 1986b:487).

*Typhloceratias firthi* Barbour, 1942b, *nomen dubium*: holotype, MCZ 35771, 480 mm, Georges Bank, 275 to 366 m, 9 February 1927 (Barbour, 1942b:78; Pietsch, 1986b:487).

*Parrichthys merrimani* Barbour, 1942b, *nomen dubium*: holotype, YPM 2004, 35 mm, Crooked Island Passage, Bahamas, 2440 m of wire out, 20 March 1927 (Barbour, 1942b:84; Pietsch, 1986b:487).

## GENUS *CRYPTOPSARAS* Gill 1883 (TRIPLEWART SEADEVILS)

Figures 64–67, Table 1

### DISTINGUISHING CHARACTERS

The genus *Cryptopsaras* differs from *Ceratias* in having only 8 caudal-fin rays and a conspicuous spine on the anterodorsal margin of the subopercle. Metamorphosed females are further distinguished from those of *Ceratias* in having a tiny illicium, reduced to a small remnant (nearly fully enveloped by tissue of the esca), and 3 wartlike caruncles on the dorsal midline of the trunk just anterior to the origin of the soft-dorsal fin (Figs. 44B, 67C–E).

Metamorphosed males are further distinguished from those of *Ceratias* in having the anterior pair of lower denticular teeth considerably longer than the posterior pair. Larvae, males, and juvenile females are unique in having subdermal pigment on the gill cover, dorsal surface of the trunk, and caudal peduncle (Fig. 67).

### ETYMOLOGY

The name *Cryptopsaras*, coined by the well-known American ichthyologist Theodore N. Gill in 1883, is derived from the Greek *kryptos*, meaning "secret" or "hidden," and *psaras*, meaning "fisherman," in allusion to the tiny illicial bone of this genus, almost totally hidden within tissue of the esca.

### DIVERSITY

A single species is recognized, along with one nominal form of doubtful validity (Table 2):

Triplewart Seadevil, *Cryptopsaras couesii* Gill, 1883: 983 known metamorphosed females (5.5 to 358 mm), 74 parasitic males (8.0 to 99 mm), about 100 metamorphosed free-living males (7.5 to 10.5 mm), and about 350 larvae (2.5 to 15 mm TL); Atlantic, Pacific, and Indian oceans, holotype, USNM 33558, 30 mm, *Albatross* station 2101, western North Atlantic, 38°18′N, 68°24′W, 0 to 3085 m (Gill, 1883:284; Bertelsen, 1951:14, 17, 22, 23, 128, 129, 139, 145; Pietsch, 1986b:488).

Triplestar Seadevil, *Bathyceratias trilychnus* Beebe, 1934a, *nomen dubium*: new genus and species observed by Beebe from bathysphere, half-mile down off Bermuda, no specimen collected (Beebe, 1934a:211, 327, 1934b:191, 1934c:691; Bertelsen, 1951:130).

## FAMILY HIMANTOLOPHIDAE Gill, 1861 (Footballfishes)

Figures 68–74, Tables 1, 3

Himantolophids are among the most easily recognized taxa of the suborder. With their globose spiny bodies, short stout illicia, elaborately decorated escae, and protruding chins covered with numerous wartlike swellings, they can hardly be confused with anything else. The taxonomic history of the family begins with Reinhardt's (1837) description of *Himantolophus groenlandicus*, based on the first representative of the Ceratioidei known to science (Figs. 9–11). Reinhardt's discovery was followed some 40 years later by Lütken's (1878a, 1878b) description of a second specimen, which he called *H. reinhardti*; and, in that same year, by Clarke's (1878) description of *H. appelii* (Fig. 12). Over the next half century, the number of known specimens of the genus increased slowly. Regan (1926) listed only 10 metamorphosed females, two identified as *H. appelii* and eight as *H. groenlandicus*, the later including *H. reinhardti* Lütken, 1878a, and two additional species described by Tanaka, *Corynolophus sagamius* (1918a) and *C. globosus* (1918b), as junior synonyms. Regan and Trewavas (1932) increased the number of known metamorphosed females to 12, reporting a ninth specimen of *H. groenlandicus*, recognizing the two specimens identified as *H. appelii* by Regan (1926), and introducing a new species, *H. danae*, based on a single specimen from the South China Sea. By 1951, when Bertelsen published his monograph on the Ceratioidei, the number of known metamorphosed females had reached 32, and two additional species had been described: *H. ranoides* Barbour, 1942b; and *H. azurlucens* Beebe and Crane, 1947. By the time that Bertelsen and Krefft (1988) produced their revision of the family, the material had increased almost fivefold; Maul (1961) had drawn attention to the description of *H. compressus* (Osório, 1912) overlooked by preceding authors; and another two species had been introduced: *H. albinares* Maul, 1961; and *H. borealis* Kharin, 1984 (Table 3).

In the meantime, the first known metamorphosed males of *Himantolophus* were described by Regan (1925c) as species of two new genera, *Lipactis* and *Rhynchoceratias*, which he referred to the family Aceratiidae. After Parr (1930b) showed that the specimens that formed the basis of the Aceratiidae were males, Regan and Trewavas (1932) reallocated the two genera (by that time comprising eight specimens divided among five species; see Table 3) to the Himantolophidae. Synonymizing *Lipactis* and part of *Rhynchoceratias* with *Himantolophus*, Bertelsen (1951) referred most of the contained specimens to *H. groenlandicus*, tentatively retaining only *H. rostratus* (Regan, 1925c) for two aberrant specimens.

Although larvae of *Himantolophus* were first described by Lütken (1878a, 1878b), on the basis of two specimens collected from off Greenland, Bertelsen (1951:61, 265) presented complete ontogenetic series of both male and female larvae (311 specimens from the Danish *Dana* expeditions), separating them on the basis of pigmentation into two groups (later designated "Type A" and "Type B" by Bertelsen and Krefft, 1988:74) of uncertain taxonomic significance. Since then, a small number of additional larvae have been recorded as *Himantolophus* sp. by Tsukahara et al. (1974), Pietsch and Seigel (1980), and Bertelsen and Pietsch (1983).

In 1988, Bertelsen and Krefft published a remarkably detailed revision of the family Himantolophidae based on an examination of all known material. General descriptions of the females (about 150 specimens), males (48 specimens), and larvae (50 reexamined at random from 311 in the *Dana* collections) were presented, including descriptions of their morphology, osteology, dentition, and illicial apparatus, as well as vertical distribution, stomach contents, maturity stages, and functional anatomy. On the basis of illicial characters and other features of the females, Bertelsen and Krefft (1988) recognized 18 species (distributed among five "species groups"; see Part Two: A Classification of Deep-Sea Anglerfishes), including two

TABLE 3
Reallocation of Nominal Species of the Genus *Himantolophus*

| Nominal Species | Currently Recognized Taxon |
|---|---|
| **Based on Females** | |
| *Himantolophus groenlandicus* group | |
|   *Himantolophus groenlandicus* Reinhardt, 1837 | *Himantolophus groenlandicus* Reinhardt, 1837 |
|   *Himantolophus reinhardti* Lütken, 1878a | *Himantolophus groenlandicus* Reinhardt, 1837 |
|   *Corynolophus sagamius* Tanaka, 1918a | *Himantolophus sagamius* (Tanaka, 1918a) |
|   *Corynolophus globosus* Tanaka, 1918b | *Himantolophus sagamius* (Tanaka, 1918a) |
|   *Himantolophus danae* Regan and Trewavas, 1932 | *Himantolophus danae* Regan and Trewavas, 1932 |
|   *Himantolophus kainarae* Barbour, 1942b | *Himantolophus sagamius* (Tanaka, 1918a) |
|   *Himantolophus ranoides* Barbour, 1942b | *Himantolophus groenlandicus* Reinhardt, 1837 |
|   *Himantolophus crinitus* Bertelsen and Krefft, 1988 | *Himantolophus crinitus* Bertelsen and Krefft, 1988 |
|   *Himantolophus paucifilosus* Bertelsen and Krefft, 1988 | *Himantolophus paucifilosus* Bertelsen and Krefft, 1988 |
| *Himantolophus appelii* group | |
|   *Aegoeonichthys appelii* Clarke, 1878 | *Himantolophus appelii* (Clarke, 1878) |
| *Himantolophus nigricornis* group | |
|   *Himantolophus nigricornis* Bertelsen and Krefft, 1988 | *Himantolophus nigricornis* Bertelsen and Krefft, 1988 |
|   *Himantolophus melanolophus* Bertelsen and Krefft, 1988 | *Himantolophus melanolophus* Bertelsen and Krefft, 1988 |
|   *Himantolophus melanophus* Jónsson and Pálsson, 1999 | *Himantolophus melanolophus* Bertelsen and Krefft, 1988 |
| *Himantolophus albinares* group | |
|   *Himantolophus albinares* Maul, 1961 | *Himantolophus albinares* Maul, 1961 |
|   *Himantolophus borealis* Kharin, 1984 | *Himantolophus borealis* Kharin, 1984 |
|   *Himantolophus mauli* Bertelsen and Krefft, 1988 | *Himantolophus mauli* Bertelsen and Krefft, 1988 |
|   *Himantolophus pseudalbinares* Bertelsen and Krefft, 1988 | *Himantolophus pseudalbinares* Bertelsen and Krefft, 1988 |
|   *Himantolophus multifurcatus* Bertelsen and Krefft, 1988 | *Himantolophus multifurcatus* Bertelsen and Krefft, 1988 |
| *Himantolophus cornifer* group | |
|   *Corynophorus compressus* Osório, 1912 | *Himantolophus compressus* (Osório, 1912) |
|   *Himantolophus azurlucens* Beebe and Crane, 1947 | *Himantolophus azurlucens* Beebe and Crane, 1947 |
|   *Himantolophus cornifer* Bertelsen and Krefft, 1988 | *Himantolophus cornifer* Bertelsen and Krefft, 1988 |
|   *Himantolophus macroceras* Bertelsen and Krefft, 1988 | *Himantolophus macroceras* Bertelsen and Krefft, 1988 |
|   *Himantolophus macroceratoides* Bertelsen and Krefft, 1988 | *Himantolophus macroceratoides* Bertelsen and Krefft, 1988 |
| **Based on Males** | |
| *Himantolophus brevirostris* group | |
|   *Rhynchoceratias brevirostris* Regan, 1925c | *Himantolophus brevirostris* group Bertelsen and Krefft, 1988 |
|   *Lipactis tumidus* Regan, 1925c | *Himantolophus brevirostris* group Bertelsen and Krefft, 1988 |
|   *Rhynchoceratias oncorhynchus* Regan, 1925c | *Himantolophus brevirostris* group Bertelsen and Krefft, 1988 |
|   *Rhynchoceratias altirostris* Regan and Trewavas, 1932 | *Himantolophus brevirostris* group Bertelsen and Krefft, 1988 |
| *Himantolophus rostratus* group | |
|   *Rhynchoceratias rostratus* Regan, 1925c | *Himantolophus rostratus* group Bertelsen and Krefft, 1988 |

NOTE: All species groups after Bertelsen and Krefft, 1988.

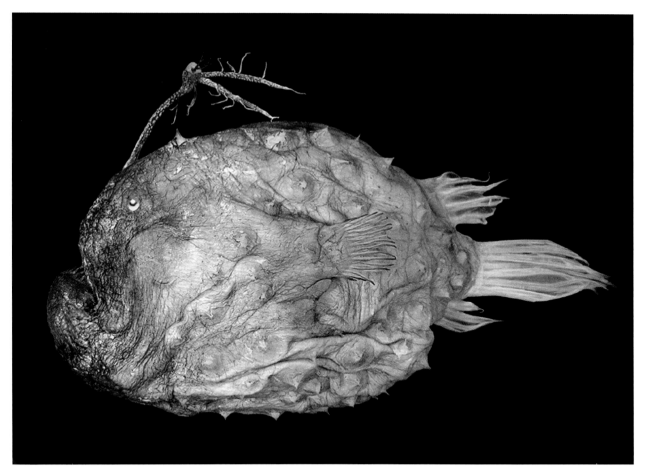

FIGURE 68. *Himantolophus appelii*, 124 mm SL, CSIRO H.5652-01. Photo by T. W. Pietsch.

species resurrected from synonymy and 10 species described as new. With the exception of *H. appelii*, characters that allow specific identification of males remain unknown. Of the 48 metamorphosed males available to Bertelsen and Krefft (1988), four were referred to *H. appelii* while the remaining 44 were separated into two species groups based on the development of the olfactory organs and denticular teeth. The account presented here follows closely the work of Bertelsen and Krefft (1988).

### DISTINGUISHING CHARACTERS

Metamorphosed females of the family Himantolophidae differ from those of other ceratioids in having an unusually prominent lower jaw that extends anteriorly beyond the snout (Figs. 49, 68–70); a broad toothless vomer; well-developed sphenotic spines; low, rounded, wartlike papillae covering the snout and chin; and, at least in larger specimens, broad-based, conical dermal spines, widely spaced and scattered over the head and body. While some smaller specimens (less than approximately 40 mm SL) may be confused with diceratiids or perhaps oneirodids, the presence of prominent teeth on all four gill arches will always separate them. The escae of himantolophids vary enormously in size and construction, perhaps more so than any other ceratioid taxon (Fig. 72).

Metamorphosed males are further differentiated by having a series of enlarged, more or less fused dermal spines above and posterior to the upper denticular bone (Fig. 41C). They are also characterized by having moderately sized, laterally directed eyes; the olfactory organs are large, the nostrils directed laterally (Fig. 73). There are 16 to 31 denticular teeth on the snout and 20 to 50 on the chin, each cluster of teeth fused at the base to form upper and lower denticular bones, respectively. The skin is densely covered with close-set dermal spinules. The males are free-living, apparently never becoming parasitic on females (see Reproduction and Early Life History, Chapter Eight).

Himantolophid larvae are short, almost spherical, the skin highly inflated (Fig. 74). The pectoral fins are of normal size, not reaching posteriorly beyond the dorsal and anal fins. Sexual dimorphism is evident even in the smallest larvae, the females bearing a small, club-shaped illicial rudiment protruding from the head. Unlike most other ceratioids, metamorphosis is delayed, the larvae attaining lengths of 17 to 22.5 mm SL, and metamorphosis taking place at lengths between about 20 and 33 mm SL (Bertelsen, 1984:326, 328; Bertelsen and Krefft, 1988:28).

Fin-ray counts, shared by all females, males, and larvae, are as follows: dorsal-fin rays 5 or 6, anal-fin rays 4, pectoral-fin rays 14 to 18, caudal-fin rays 9 (1 simple + 6 bifurcated + 2 simple) (Table 1).

The largest known female of the family is the holotype of *H. groenlandicus* from southwest Greenland (not preserved, except for the illicium; ZMUC 65), which measured about 465 mm standard length. There are several additional individuals in the range of 250 to 400 mm (Bertelsen and Krefft, 1988:14, 37).

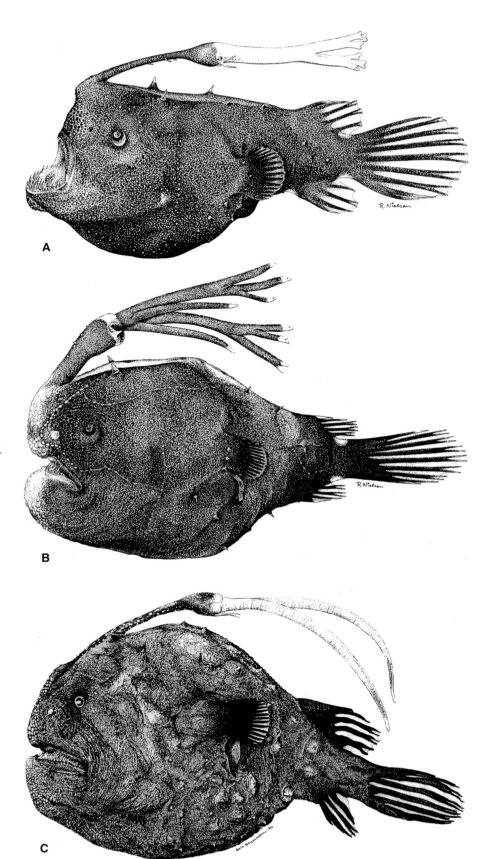

FIGURE 69. Species of *Himantolophus*: (A) *H. multifurcatus*, holotype, 48 mm SL, ISH 764/68 (drawing by R. Nielsen); (B) *H. cornifer*, paratype, 52 mm SL, MCZ 58176 (drawing by R. Nielsen); (C) *H. mauli*, paratype, 155 mm SL, ROM 267560 (drawing by E. Beyerholm). All after Bertelsen and Krefft (1988).

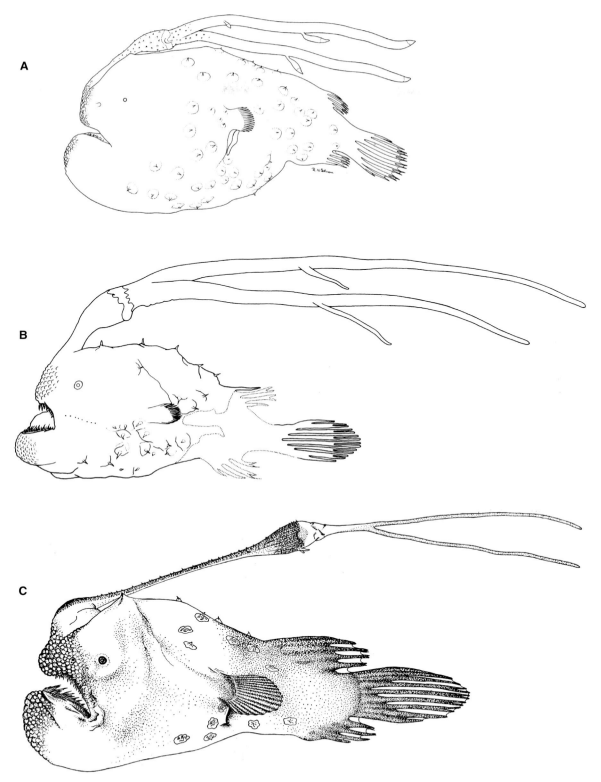

FIGURE 70. Species of *Himantolophus*: (A) *H. macroceratoides*, paratype, 180 mm SL, ZIN 47914 (drawing by R. Nielsen); (B) *H. macroceras*, paratype, 85 mm SL, ISH 722/68 (drawing by R. Nielsen); (C) *H. mauli*, holotype 100 mm SL, MMF 18291 (drawing by G.E. Maul; after Maul, 1961).

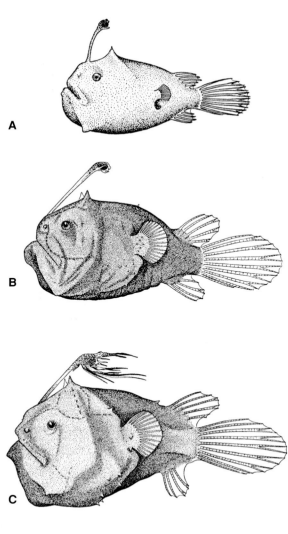

FIGURE 71. *Himantolophus*, females, showing ontogenetic change in body shape and development of the esca and skin spines: (A) *H.* sp., 33 mm SL (after Regan, 1926); (B) *H.* sp., 36 mm SL (after Regan and Trewavas, 1932); (C) *H. danae*, holotype, 39 mm SL, ZMUC P9262 (after Regan and Trewavas, 1932); (D) *H. groenlandicus*, the holotype of *H. reinhardti*, 328 mm SL, ZMUC 66 (after Regan, 1926). Drawings by W.P.C. Tenison.

Metamorphosed males range from 23 to 39 mm, the largest known larvae from 17 to 22.5 mm.

### DIVERSITY

The Himantolophidae contains a single genus:

### GENUS *HIMANTOLOPHUS* REINHARDT, 1837 (FOOTBALLFISHES)

*Figures 68–74, Tables 1, 3*

### DISTINGUISHING CHARACTERS

This genus bears all the features of the family, as described above and detailed in Part Two: A Classification of Deep-Sea Anglerfishes.

### ETYMOLOGY

The name *Himantolophus* is derived from the Greek *himas* or *himantos*, meaning a "leather strap," "thong," or "leash," in reference to the thick leathery illicium of himantolophids; and *lopho* or *lophio*, a "crest" or "tuft," in allusion to the baited illicium mounted on top of the head.

### DIVERSITY

Eighteen species based on females are currently recognized, distributed among five species groups (Table 3):

*HIMANTOLOPHUS GROENLANDICUS* GROUP BERTELSEN AND KREFFT, 1988:

*Himantolophus groenlandicus* Reinhardt, 1837: 143 metamorphosed females, 32 to 465 mm, including three tentatively identified specimens, 51.5 to 70 mm; widely distributed in the Atlantic Ocean, and apparently present also in the eastern Pacific and perhaps in the western Indian Ocean; holotype, ZMUC 65, c. 465 mm, only illicium preserved, found washed ashore, southwest Greenland, near Godthaab, 1833 (Reinhardt, 1837:116, pl. 4; Bertelsen, 1951:60, figs. 22–25; Maul, 1961:104, figs. 5–9; Bertelsen and Krefft, 1988:37, figs. 1–5, 6C, 11, 41; McEachran and Fechhelm, 1998:854, 856, fig.; Anderson and Leslie, 2001:8, fig. 6).

*Himantolophus sagamius* (Tanaka, 1918a): 25 metamorphosed females (32 to 380 mm), including two lost specimens (128 to 295 mm); widely distributed in the Pacific Ocean; holotype *(Corynolophus sagamius)*, originally SCMT 8201 (lost or mislaid at ZUMT), 200 mm, Sagami Sea, about 35°N, 139°E (Tanaka, 1918a:491, pl. 134, fig. 377; Bertelsen and Krefft, 1988:39, figs. 12, 41; Klepadlo et al., 2003:99, figs. 1–6; Kharin, 2006a:281, fig.).

*Himantolophus danae* Regan and Trewavas, 1932: a single metamorphosed female, 39 mm, the holotype, ZMUC P9262, 39 mm, *Dana* station 3714(1), South China Sea, 15°22′N, 115°20′E, 1000 m wire, 0245 h, 20 May 1929 (Regan and Trewavas, 1932:60, figs. 87, 88B, pl. 1, fig. 2; Bertelsen and Krefft, 1988:42, figs. 13, 41).

*Himantolophus crinitus* Bertelsen and Krefft, 1988: 11 metamorphosed females, 30 to 83 mm; eastern tropical Atlantic Ocean; holotype, USNM 229970, 83 mm, *La Rafaele* station 8, eastern tropical Atlantic, 09°10′N, 15°39′W, bottom trawl 600 to 610 m (Bertelsen and Krefft, 1988:43, figs. 14, 39).

*Himantolophus paucifilosus* Bertelsen and Krefft, 1988: 19 metamorphosed females, 32 to 163 mm; tropical Atlantic Ocean; holotype, ISH 393a/66, 111 mm, *Walther Herwig* 182/66, 10°46′N, 23°54′W, MT, 300 m, 2120 h, 16 May 1966 (Bertelsen and Krefft, 1988:45, figs. 15, 40).

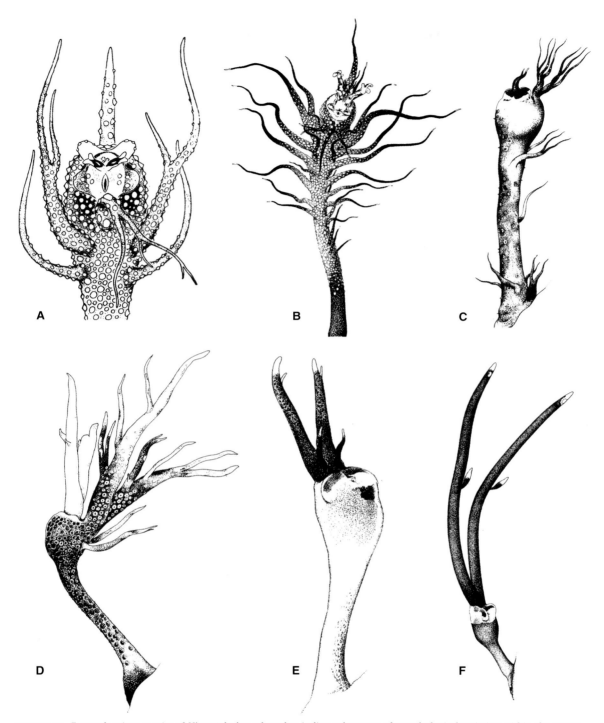

FIGURE 72. Escae of various species of *Himantolophus* selected to indicate the range of morphological variation within the genus: (A) *H. groenlandicus*, 61 mm SL, MMF 11141, posterior view; (B) *H. crinitus*, holotype, 83 mm SL, USNM 229970, posterior view; (C) *H. melanolophus*, paratype, 84 mm SL, USNM 235918, left lateral view; (D) *H. appelii*, 225 mm SL, ZIN 40510, left lateral view; (E) *H. cornifer*, paratype, 38 mm SL, LACM 33325-001, left lateral view; (F) *H. macroceratoides*, holotype, 50 mm SL, ISH 2400/71, left lateral view. Drawings by R. Nielsen; after Bertelsen and Krefft (1988).

*HIMANTOLOPHUS APPELII* GROUP BERTELSEN AND KREFFT, 1988:

*Himantolophus appelii* (Clarke, 1878): 115 metamorphosed females, 14 to 320 mm; southern parts of the Atlantic, Indian, and Pacific oceans; holotype, 287 mm, off Westland, New Zealand, about 43°S, 170°E, stranded, not preserved (Clarke, 1878:245, pl. 6; Günther, 1887:51; Waite, 1912:194–197, pl. 10; Bertelsen and Pietsch, 1983:82, fig. 4; Bertelsen and Krefft, 1988:47, figs. 16, 41; Meléndez and Kong, 1997:12, fig. 1; Stewart and Pietsch, 1998:6, fig. 3; Anderson and Leslie, 2001:6, figs. 3, 4).

*HIMANTOLOPHUS NIGRICORNIS* GROUP BERTELSEN AND KREFFT, 1988:

*Himantolophus nigricornis* Bertelsen and Krefft, 1988: three metamorphosed females, 145 to 195 mm; North Pacific; holotype, LACM 42697-1, 145 mm, *Velero IV*, San Clemente Basin,

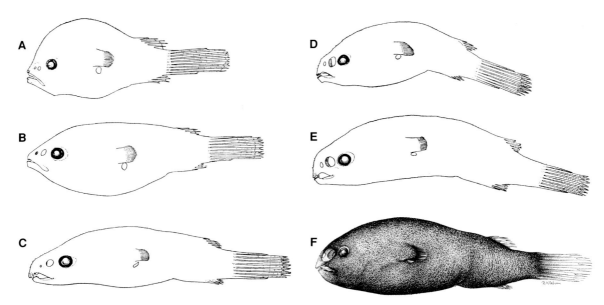

FIGURE 73. Males of *Himantolophus* sp. (A–C, *H. brevirostris* group; D–F, *H. rostratus* group): (A) lectotype of *Lipactis tumidus*, 21.5 mm SL, ZMUC P9261; (B) holotype of *Rhynchoceratias brevirostris*, 30.5 mm SL, ZMUC P9263; (C) lectotype of *R. oncorhynchus*, 34 mm SL, ZMUC P9264; (D) holotype of *R. rostratus*, 26.5 mm SL, ZMUC P9265; (E) *H. rostratus* group, 34.5 mm SL, ZMUC P92675; (F) *H. rostratus* group, 39 mm SL, UF 23267 (drawing by R. Nielsen). After Bertelsen (1951) and Bertelsen and Krefft (1988).

about 33°N, 118°W, in "deep water," 5 September 1980 (Bertelsen and Krefft, 1988:50, figs. 17, 41).

*Himantolophus melanolophus* Bertelsen and Krefft, 1988: six metamorphosed females, 35 to 150 mm; North Atlantic and West Pacific oceans; holotype, ISH 1212/74, 82 mm, *Anton Dohrn* station 32-II/74, North Atlantic, 9°31′N, 29°35′W, midwater trawl, 0 to 350 m, 2115 h, 20 July 1974 (Bertelsen and Krefft, 1988:52, figs. 18, 41; McEachran and Fechhelm, 1998:854, 857, fig.; Jónsson and Pálsson, 1999:200, figs. 3, 9).

### HIMANTOLOPHUS ALBINARES GROUP BERTELSEN AND KREFFT, 1988:

*Himantolophus albinares* Maul, 1961: 25 metamorphosed females, 28 to 190 mm; Atlantic and western South Pacific; holotype, MMF 2598, 190 mm, off Câmara de Lobos, Madeira, on long line (Maul, 1961:111, figs. 11–15, 1962a:8, figs. 1, 2; Trunov, 1981:51, fig. 5; Bertelsen and Krefft, 1988:54, figs. 19, 20; Jónsson and Pálsson, 1999:199, fig. 3; Anderson and Leslie, 2001:5, fig. 2; Moore et al., 2003:214).

*Himantolophus mauli* Bertelsen and Krefft, 1988: 20 metamorphosed females, 30.5 to 215 mm; North Atlantic Ocean; holotype, MMF 18291, 100 mm, off Câmara de Lobos, Madeira, August 1925 (Maul, 1961:96, figs. 2–4; Bertelsen and Krefft, 1988:57, figs. 7, 21, 22, 42; Rodríguez-Marín et al., 1996:69, fig. 2; Jónsson and Pálsson, 1999:200, fig. 3; Kukuev and Trunov, 2002:326; Moore et al., 2003:215).

*Himantolophus pseudalbinares* Bertelsen and Krefft, 1988: four metamorphosed females, 80 to 153 mm; southern parts of the western Pacific and western Indian oceans; holotype, ZIN 49711, 82 mm, *Fiolent* cruise 3-63, off Cape St. Francis, southwest of Port Elizabeth, South Africa, 35°01′S, 24°36.8′E, 1280 to 1300 m (Bertelsen and Krefft, 1988:59, figs. 23, 42; Stewart and Pietsch, 1998:8, fig. 4; Anderson and Leslie, 2001:9, fig. 7).

*Himantolophus multifurcatus* Bertelsen and Krefft, 1988: four metamorphosed females, 42 to 122 mm; North Atlantic Ocean; holotype, ISH 764/68, 48 mm, *Walther Herwig* station 14 II/68, 4°11′N, 24°37′W, MT, 180 to 200 m, 1 February 1968 (Bertelsen and Krefft, 1988:60, figs. 24, 25, 42; Donnelly and Gartner, 1990:77).

*Himantolophus borealis* Kharin, 1984: three metamorphosed females, 62 to 160 mm; off Honshu, Japan; holotype, ZIN 46021, 62 mm, *Mys Junony*, western North Pacific, off Honshu, Japan, 40°52.8′N, 142°20.8′E, BT, 1210 m (Kharin, 1984:663, fig. 1A, B; Bertelsen and Krefft, 1988:62, figs. 26, 42; Nakabo, 2002:476, figs.).

### HIMANTOLOPHUS CORNIFER GROUP BERTELSEN AND KREFFT, 1988:

*Himantolophus cornifer* Bertelsen and Krefft, 1988: eight metamorphosed females, 27 to 208 mm; Atlantic and Pacific oceans, with a single record from off South Africa; holotype, MCZ 58858, 90 mm, *Harbison* station 1047, Coral Sea, 12°31′S, 148°41′E, oblique MT, 0 to 1650 m, 3 December 1981 (Bertelsen and Krefft, 1988:64, figs. 27, 28, 43; Donnelly and Gartner, 1990:77; McEachran and Fechhelm, 1998:854, 855, fig.; Anderson and Leslie, 2001:8, fig. 5).

*Himantolophus macroceras* Bertelsen and Krefft, 1988: five metamorphosed females, 32 to 92 mm; tropical Atlantic; holotype, ISH 2329/71, 79 mm, *Walther Herwig* station 478/71, 1°04′N, 18°22′W, 2100 m, 12 April 1971 (Bertelsen and Krefft, 1988:68, figs. 29, 30, 43).

*Himantolophus macroceratoides* Bertelsen and Krefft, 1988: two metamorphosed females, 50 and 180 mm; eastern Atlantic and western Indian oceans; holotype, ISH 2400/71, 50 mm, *Walther Herwig* station 482-II/71, 4°36′N, 19°40′W, MT, 0 to 256 m, 13 April 1971 (Bertelsen and Krefft, 1988:70, figs. 31, 32, 43).

*Himantolophus azurlucens* Beebe and Crane, 1947: two metamorphosed females, 98 to 141 mm; eastern South Atlantic and Gulf of Panama; holotype, CAS-SU 46507, 98 mm, eastern Pacific *Zaca* Expedition station 228 T-1, 7°00′N, 79°16′W, about 900 m (Bertelsen and Krefft, 1988:72, figs. 33, 43; Trunov, 2001:543, fig.).

*Himantolophus compressus* (Osório, 1912): a single metamorphosed female, 130 mm, the holotype (*Corynophorus compressus*), off Portugal, about 38°20′N, 9°15′W, but lost in 1978 fire at the Museo Bocage, Lisbon (Bertelsen and Krefft, 1988:73, fig. 43).

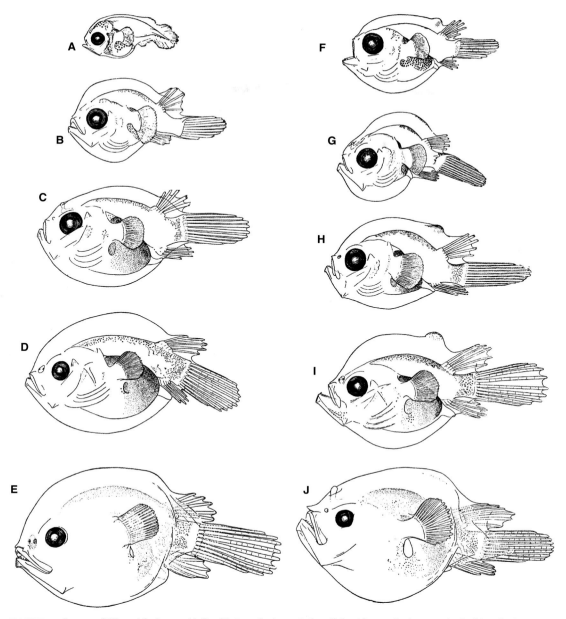

FIGURE 74. Larvae of *Himantolophus* sp. (A–E, with type A pigmentation; F–J, with type B pigmentation): (A) male, 2.4 mm TL; (B) male, 5.7 mm TL; (C) female, 6.0 mm SL, 9.0 mm TL; (D) male, 8.6 mm SL, 13.0 mm TL; (E) male, 17.2 mm SL, 24 mm TL; (F) male, 5.6 mm TL; (G) female, 5.0 mm TL; (H) male, 8.2 mm TL; (I) male, 7.1 mm SL, 10.7 mm TL; (J) female, 15.8 mm SL, 22 mm TL. After Bertelsen (1951).

## FAMILY DICERATIIDAE Regan and Trewavas, 1932 (Doublespine Seadevils)

*Figures 75–77, Table 1*

While appearing superficially like himantolophids or generalized oneirodids, especially members of the genus *Oneirodes*, metamorphosed females of the family Diceratiidae are unique among members of the suborder Ceratioidei in having a second light-bearing, dorsal-fin spine emerging from the head directly behind the insertion of the illicium. On that basis alone, they cannot be confused with any other family. The taxonomic history of the Diceratiidae begins with Günther's (1887) description of *Ceratias (Diceratias) bispinosus* based on a single female specimen collected by the HMS *Challenger* in the East Indies off Banda Island. Most likely unaware of Günther's (1887) publication, Alcock (1890:207) described a second female, from the Bay of Bengal, under the name *Paroneirodes glomerosus*, recognizing its similarity to *Oneirodes* Lütken (1871) but, at the same time, realizing the unique presence of "two clavate cephalic tentacles." Alcock (1899:57) later placed *Paroneirodes* in the synonymy of *Oneirodes*, while maintaining the subgeneric status of *Diceratias*. Emphasizing once again the significance of a "second ray of [the] spinous dorsal [situated] immediately behind the first," Regan (1912:287) raised *Diceratias* to generic status, while recognizing *Paroneirodes* as a junior synonym.

Somewhat later, Pietschmann (1926, 1930) discovered a very different looking diceratiid in the collections of the Natural History Museum in Vienna. Collected some 60 years earlier off the coast of Madeira, Pietschmann (1926, 1930) described the specimen as a new genus and species, naming it

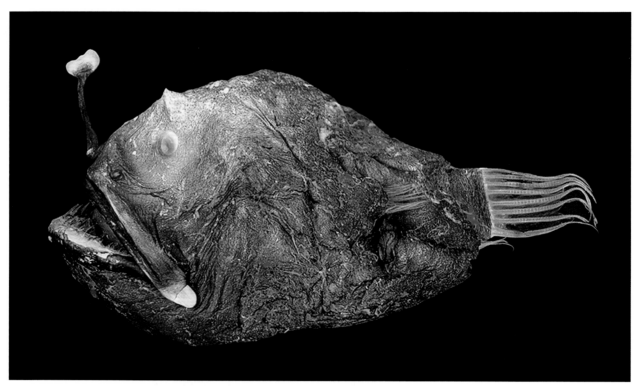

FIGURE 75. *Diceratias trilobus*, 86 mm SL, AMS I.31144-004. Photo by T. W. Pietsch.

*Phrynichthys wedli*. While sharing with *Diceratias* a second dorsal-fin spine lying just behind the first, *Phrynichthys* was unique in the much greater length of the first spine (illicium) and in the position of the two spines, emerging on the back at the rear of the skull rather than on the snout.

Unfortunately, Pietschmann's (1926) important contribution to diceratiid systematics was passed over by several subsequent authors including Norman (1930) who, had he known of Pietschmann's publication, would not have misidentified a small specimen of *P. wedli* from the eastern South Atlantic as *Paroneirodes glomerosus*, at the same time resurrecting *Paroneirodes* from the synonymy of *Diceratias*. In any case, with two rather distinct genera, differing in similar ways from all other known ceratioids, Regan and Trewavas (1932) established the family Diceratiidae to include *Diceratias* and *Paroneirodes*, with the latter containing *Phrynichthys* as a junior synonym. Contained also in the family Diceratiidae was *Caranactis pumilus*, a new genus and species erected by Regan and Trewavas (1932) to contain a single male specimen from the Indian Ocean subsequently referred to the genus *Oneirodes* by Bertelsen (1951).

Subsequent authors, including Bertelsen (1951) and Maul (1962a), followed Regan and Trewavas (1932) in recognizing *Phrynichthys* as a junior synonym of *Paroneirodes*. Karrer (1973b) went farther, regarding both *Paroneirodes* and *Phrynichthys* as junior synonyms of *Diceratias*. Uwate (1979), however, based on a detailed examination of all known material, was able to show conclusively that *Paroneirodes* is indeed synonymous with *Diceratias* as originally proposed by Regan (1912), the later genus now containing *Diceratias bispinosus* and a newly described species, *D. pileatus* Uwate, from the central Atlantic. At the same time, Uwate (1979) provided evidence to resurrect *Phrynichthys* from the synonymy of *Paroneirodes*, to include *Phrynichthys wedli* and another new species, *P. thele* Uwate, based on two specimens from the western central Pacific. However, Uwate (1979) and most other authors before and after (exceptions include Golvan, 1962:173; Eschmeyer, 1990:68; and Anderson and Leslie, 2001:10) failed to recognize Whitley's (1931:334) *Bufoceratias* as a replacement name for *Phrynichthys* Pietschmann, the later preoccupied by *Phrynichthys* Agassiz (1846), which was itself a replacement name for *Bufichthys* Swainson (1839), a synonym of *Synanceia* Bloch and Schneider (1801), family Scorpaenidae.

In the 30 years since Uwate's (1979) revision, little has been added to our knowledge of diceratiids. Two additional species, *D. trilobus* and *Bufoceratias shaoi*, have been described by Balushkin and Fedorov (1986) and Pietsch et al. (2004), respectively, bringing the total number to six (three species of *Diceratias* and three of *Bufoceratias*), and a large number of additional specimens of several species have been collected, including the first known metamorphosed male (Bertelsen, 1983). But the family is still based almost solely on metamorphosed juvenile females. Only one sexually mature female, one metamorphosed male, and two larvae are known. Differences in the morphology of the illicium and its supporting pterygiophore, various internal osteological features of the cranium, and length of the dermal spinules that cover the skin of the head and body are diagnostic at the generic level, but illicium length and escal morphology are the only diagnostic characters at the specific level. Consequently specimens of either genus with missing or damaged illicia are impossible to identify to species.

### DISTINGUISHING CHARACTERS

Metamorphosed females of the Diceratiidae are readily distinguished by having an externally exposed second cephalic spine, bearing a distal light organ and emerging from the head behind the base of the illicium. The structure is highly conspicuous in smaller specimens (less than about 30 mm; see Figs. 44C, D,

FIGURE 76. Species of *Bufoceratias*: (A) *B. wedli*, 96 mm SL, CSIRO H.2285-02 (photo by T.W. Pietsch); (B) *B. shaoi*, holotype, 101 mm SL, ASIZP 61796 (photo by H.-C. Ho).

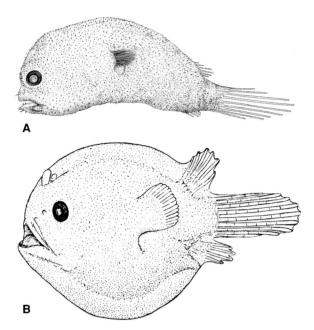

FIGURE 77. Diceratiids for which characters are unavailable for further identification to genus and species: (A) Diceratiidae sp., 14 mm SL, LACM 36091-4, the only known metamorphosed male of the family; (B) Diceratiidae sp., larval female, 10.5 mm SL, ZMUC P92676, one of only two known larvae of the family. After Bertelsen (1951, 1983).

77B), but withdrawn inside a narrow cavity in larger specimens. Like most oneirodids, female diceratiids have well-developed sphenotic spines; quadrate and articular spines are present, but angular and preopercular spines are absent. The jaws are subequal, the lower jaw extending anteriorly slightly beyond the upper jaw. There are 3 pectoral radials. Gill teeth are absent. The skin, including that of the illicium and the proximal half of the escal bulb, is everywhere covered with small close-set dermal spinules (Fig. 48D, E). The latter feature clearly separates diceratiid females from those of himantolophids and all oneirodids except for the spiny-skinned genus *Spiniphryne* (Fig. 48G).

The single known metamorphosed male (Fig. 77A), a juvenile specimen only 14 mm long, differs from those of all other ceratioids in a combination of characters that includes large, laterally directed eyes and nostrils, a single pair of denticular teeth on the snout, and two transverse series of denticular teeth on the chin, each containing 4 or 5 separate teeth. The skin is covered with small dermal spinules. The males are free-living, apparently never becoming parasitic (see Reproduction and Early Life History, Chapter Eight). Lacking all the features present in females that differentiate the two diceratiid genera, this little male can only be identified to family.

Diceratiid larvae are known from only two specimens, both females, 7 to 10.5 mm (Fig. 77B). Like the single known male described above, they cannot be identified to genus with any certainty. While the body is short and deep, nearly spherical, and the skin highly inflated, like those of many other ceratioid taxa, they are easily recognized by the presence of a second dorsal-fin spine, bearing a bulbous distal light organ. Subdermal pigmentation is evenly distributed over the body. The pectoral fins are of normal size, not extending posteriorly to the base of the dorsal and anal fins.

There are 5 to 7 dorsal-fin rays, 4 anal-fin rays (very rarely 5), 13 to 16 pectoral-fin rays, and 9 caudal rays (very rarely 8), 1 simple + 6 bifurcated + 2 simple (Table 1).

The largest known specimen of the family is a 275-mm female of *Diceratias pileatus* (BPBM 30655) found floating on the surface off Kona, Hawaii (Pietsch and Randall, 1987). The only known metamorphosed male (LACM 36091-4) measures 14 mm.

SYNOPSIS OF GENERA OF THE FAMILY DICERATIIDAE

The family contains two genera differentiated as follows.

---

Females with the anterior tip of the pterygiophore of the illicium emerging on the snout from between the frontal bones, the distance between the point of emergence and the symphysis of the upper jaw less than 16% SL; the posterior end of the pterygiophore, lying within an unusually deep cranial trough, embedded beneath skin of the head; the length of the illicium 26 to 47% SL; the dermal spinules of the skin relatively large, their length 230 to 380 μm:

DOUBLESPINE SEADEVILS, GENUS *DICERATIAS* GÜNTHER, 1887, THREE SPECIES, P. 75

Females with the pterygiophore of the illicium, lying within a relatively shallow cranial trough, fully embedded beneath the skin of the head, the illicium emerging from the dorsal surface of the head at the rear of the skull, the distance between the point of emergence and the symphysis of the upper jaw greater than 28% SL; the length of the illicium 25 to 225% SL; the dermal spinules of the skin relatively minute, their length 90 to 110 μm:

TOADY SEADEVILS, GENUS *BUFOCERATIAS* WHITLEY, 1931, THREE SPECIES, P. 75

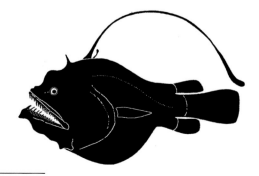

## GENUS *DICERATIAS* GÜNTHER, 1887 (DOUBLESPINE SEADEVILS)

*Figure 75, Table 1*

### DISTINGUISHING CHARACTERS

Females of *Diceratias* are distinguished from those of most species of *Bufoceratias* in having a short illicium, never reaching a length greater than 50% SL. A more striking difference, however, is the point of emergence of the luring apparatus on the head: the anterior tip of the pterygiophore of the illicium is exposed, emerging on the snout in *Diceratias*. In sharp contrast, the pterygiophore of *Bufoceratias* is concealed beneath the skin, the illicium emerging from the top of the head between the sphenotic spines in smaller specimens to well behind the rear of the skull in larger specimens. The illicial trough is unusually deep (Uwate, 1979:132). Although both genera are characterized by having the skin covered with numerous, close-set dermal spinules, those of *Diceratias* are significantly larger (Fig. 48D). The morphology of the esca of diceratiids varies from a simple rounded bulb without ornamentation of any kind to highly filamentous (Fig. 75). Males and larvae are unknown.

### ETYMOLOGY

The name *Diceratias* is formed from the Greek *di*, meaning "two" or "double," and *keras* or *keratos*, meaning "horn" or "horned," in allusion to the two dorsal-fin spines that are conspicuous in the larvae and young females of this genus.

### DIVERSITY

Three species are recognized as follows:

*Diceratias bispinosus* Günther, 1887: 10 metamorphosed females, 20 to 112 mm; Indo-west Pacific Ocean; holotype of *Ceratias (Diceratias) bispinosus*, BMNH 1887.12.7.14, 52 mm, *Challenger* station 194A, off Banda Island, 659 m (360 fathoms), 1200 to 1330 h, 29 September 1874 (Günther, 1887:50, 52, 53, pl. 11, fig. B; Uwate, 1979:139, figs. 16, 20).

*Diceratias pileatus* Uwate, 1979: 153 metamorphosed females, 20 to 275 mm; Atlantic, central Pacific, and eastern South Indian oceans; holotype, UF 23774, 82 mm, *Oregon II* station 10616, western North Atlantic, off Surinam, 7°37'N, 53°32'W, 0 to 722 m, 13 May 1969 (Uwate, 1979:140, figs. 1, 2, 4, 5, 7A, 8, 10A, 11A, 12, 17, 20; Fujii, 1983:259, fig.; Pietsch and Randall, 1987:419, figs. 1–3).

*Diceratias trilobus* Balushkin and Fedorov, 1986: 24 metamorphosed females, 47 to 140 mm; western Pacific and eastern Indian oceans; holotype, ZIN 47426, 122 mm, *Shantar*, trawl 28, E of Honshu, Japan, 38°20.7'N, 142°31.9'E, bottom trawl, 1211 to 1216 m, 28 March 1975 (Balushkin and Fedorov, 1986:855, fig. A; Pietsch et al., 2006:S98, figs. 1–3).

## GENUS *BUFOCERATIAS* WHITLEY, 1931 (TOADY SEADEVILS)

*Figure 76, Table 1*

### DISTINGUISHING CHARACTERS

Females of the genus *Bufoceratias* are highly unusual among anglerfishes in the placement of the luring apparatus. While it seems that natural selection in most anglers has favored an anterior position of the illicium and its supporting pterygiophore—to get the lure out as far as possible in front of the mouth, where it can seemingly do the most good—the opposite has occurred in *Bufoceratias*. In sharp contrast to all other lophiiform genera, the point of emergence of the illicium in adult females is on the back, near the posterior margin of the skull. Along with this surprising shift in position, the cranial trough within which the illicial pterygiophore lies is much shallower in this genus, and the pterygiophore itself much shorter, about half the length of that of its sister genus *Diceratias* (see Uwate, 1979:132, 134). No doubt related to these differences, the illicium of most species of *Bufoceratias* attains extraordinary lengths, as great as 225% SL. The only species of *Bufoceratias* with a short illicium (*Bufoceratias shaoi*; Pietsch et al., 2004), comparable in length to that of species of *Diceratias*, seems to compensate for this apparent disadvantage by having a huge, morphologically complex esca (see Fig. 76B).

### ETYMOLOGY

The name *Bufoceratias* is derived from the Latin *bufo* or *bufonis*, a "toad"; and the Greek *keras* or *keratos*, meaning "horn" or "horned," again in allusion to the hornlike lure projecting from the head.

### DIVERSITY

Three species are recognized as follows:

*Bufoceratias wedli* (Pietschmann, 1926): 82 metamorphosed females, 19 to 196 mm; Atlantic Ocean; holotype *(Phrynichthys wedli)*, NMW 3524, 35 mm, off Madeira, Steindachner collection, 1865 (Pietschmann, 1926:88; Uwate, 1979:142, figs. 3, 6, 7B, 9, 10B, 11B, 13, 18, 20; Machida and Yamakawa, 1990:60, figs. 1–3; Anderson and Leslie, 2001:10, figs. 8, 9; Pietsch et al., 2004:100, figs. 1, 3A, 5).

*Bufoceratias thele* (Uwate, 1979): 14 metamorphosed females, 22 to 204 mm; western Pacific Ocean; holotype *(Phrynichthys thele)*, LACM 36077-1, 32 mm, *Alpha Helix* station 155, Halmahara Sea, 0°38.6'S, 129°05.6'E, 680 to 850 m, 1210 to 1400 h, 22 May 1975 (Uwate, 1979:142, figs. 19, 20; Ni et al., 1989:89, 94, figs. 2, 3I, J; Pietsch et al., 2004:100, figs. 1, 3A, 5).

*Bufoceratias shaoi* Pietsch, Ho, and Chen, 2004: four metamorphosed females, 55 to 101 mm; western Pacific and western Indian oceans; holotype, ASIZP 61796, 101 mm, off northeast coast of Taiwan, 24°25 to 50'N, 122°00 to 10'E, bottom trawl, 0 to 800 m, 1999 (Pietsch et al., 2004:100, figs. 1, 3A, 5).

## FAMILY MELANOCETIDAE Gill, 1879b (Black Seadevils)

*Figures 78–84, Tables 1, 4*

The family Melanocetidae includes globose bathypelagic anglerfishes, easily separated from members of allied families by having a combination of features that includes 12 or more dorsal-fin rays, 3 to 5 anal-fin rays, and a huge dorsally directed mouth, equipped with numerous long fanglike teeth (Bertelsen, 1951; Pietsch and Van Duzer, 1980). The only currently recognized genus of the family was established by Günther (1864), with the description of *Melanocetus johnsonii*, based on a single female specimen collected in the Atlantic Ocean off Madeira. Since that time, 13 additional species based on females have been described (Table 4). From a comparison of the

TABLE 4
Reallocation of Nominal Species of the Genus *Melanocetus*

| Nominal Species | Currently Recognized Taxon |
| --- | --- |
| Based on Females | |
| *Melanocetus johnsonii* Günther, 1864 | *Melanocetus johnsonii* Günther, 1864 |
| *Melanocetus bispinossus* Günther, 1880 | *Nomen nudum* |
| *Melanocetus murrayi* Günther, 1887 | *Melanocetus murrayi* Günther, 1887 |
| *Melanocetus krechi* Brauer, 1902 | *Melanocetus johnsonii* Günther, 1864 |
| *Melanocetus vorax* Brauer, 1902 | *Melanocetus murrayi* Günther, 1887 |
| *Melanocetus rotundatus* Gilchrist, 1903 | *Melanocetus johnsonii* Günther, 1864 |
| *Melanocetus polyactis* Regan, 1925c | *Melanocetus polyactis* Regan, 1925c |
| *Melanocetus niger* Regan, 1925c | *Melanocetus niger* Regan, 1925c |
| *Melanocetus ferox* Regan, 1926 | *Melanocetus johnsonii* Günther, 1864 |
| *Melanocetus tumidus* Parr, 1927 | *Melanocetus murrayi* Günther, 1887 |
| *Melanocetus cirrifer* Regan and Trewavas, 1932 | *Melanocetus johnsonii* Günther, 1864 |
| *Melanocetus megalodontis* Beebe and Crane, 1947 | *Melanocetus johnsonii* Günther, 1864 |
| *Melanocetus eustalus* Pietsch and Van Duzer, 1980 | *Melanocetus eustalus* Pietsch and Van Duzer, 1980 |
| *Melanocetus rossi* Balushkin and Fedorov, 1981 | *Melanocetus rossi* Balushkin and Fedorov, 1981 |
| Based on Males | |
| *Rhynchoceratias acanthirostris* Parr, 1927 | *Melanocetus murrayi* Günther, 1887 |
| *Rhynchoceratias latirhinus* Parr, 1927 | *Melanocetus murrayi* Günther, 1887 |
| *Rhynchoceratias longipinnis* Parr, 1930a | *Melanocetus murrayi* Günther, 1887 |
| *Centrocetus spinulosus* Regan and Trewavas, 1932 | *Melanocetus johnsonii* Günther, 1864 |
| *Xenoceratias macracanthus*: Regan and Trewavas, 1932 | *Melanocetus johnsonii* Günther, 1864 |
| *Xenoceratias micracanthus* Regan and Trewavas, 1932 | *Melanocetus johnsonii* Günther, 1864 |
| *Xenoceratias heterorhynchus* Regan and Trewavas, 1932 | *Melanocetus johnsonii* Günther, 1864 |
| *Xenoceratias laevis* Regan and Trewavas, 1932 | *Melanocetus johnsonii* Günther, 1864 |
| *Xenoceratias brevirostris* Regan and Trewavas, 1932 | *Melanocetus johnsonii* Günther, 1864 |
| *Xenoceratias longirostris* Regan and Trewavas, 1932 | *Nomen dubium* |
| *Xenoceratias braueri* Koefoed, 1944 | *Melanocetus johnsonii* Günther, 1864 |
| *Xenoceratias regani* Koefoed, 1944 | *Melanocetus murrayi* Günther, 1887 |
| *Xenoceratias nudus* Beebe and Crane, 1947 | *Nomen dubium* |

characters used to distinguish these nominal forms, Bertelsen (1951:40–43) doubted that *M. krechi* and *M. cirrifer* could be maintained and that *M. ferox* and *M. niger* might be synonyms. *Melanocetus murrayi* and *M. johnsonii* were recognized as the only species known from the Atlantic; *M. niger*, *M. ferox*, and *M. polyactis* were considered forms restricted to the eastern tropical Pacific. Six larval specimens from the Gulf of Panama were assigned to *M. polyactis*. The remaining larvae, approximately 600 individuals, were separated into two groups, representing *M. murrayi* and *M. johnsonii*, on the basis of geographic distribution, fin-ray counts, and a comparison of larval and juvenile pigmentation. Despite these reallocations, Bertelsen (1951:43) made it clear that "the separation of the species is still very uncertain and future investigations and material will probably make it necessary to revise this synopsis."

At the time of Bertelsen's (1951) monograph on the Ceratioidei, 19 metamorphosed melanocetid males were known. Of these, 14 had been described as types of 13 separate species, and five were uncertainly placed (Table 4). On the basis of subdermal pigmentation, fin-ray counts, and geographic distribution,

FIGURE 78. *Melanocetus johnsonii*, 75-mm female, with a 23.5-mm attached male, BMNH 2004.6.3.2-3. Photo by Edith A. Widder.

Bertelsen (1951) synonymized seven of these 13 nominal forms with *M. johnsonii* and four with *M. murrayi*. The remaining two species based on males, *M. longirostris* and *M. nudus*, each differing slightly from the rest of the material, were tentatively retained (Table 4).

With the vast increase in the amount of material of *Melanocetus* made available in the 30 years following Bertelsen's (1951) work, Pietsch and Van Duzer (1980), in a revision of the family based on a study of some 600 specimens collected from all oceans, recognized five species based on females. Four of these were previously described forms: *M. johnsonii*, collected from all three major oceans of the world; *M. polyactis* and *M. niger*, both restricted to the eastern tropical Pacific; and *M. murrayi* from the Atlantic and Pacific. The fifth species, *M. eustalus* Pietsch and Van Duzer (1980), was described as new to science based on a single female collected in the eastern Pacific off Mazatlán, Mexico. One year later, Balushkin and Fedorov (1981) added another new species, *M. rossi*, based on a single female specimen from the Ross Sea, Antarctica.

Although the number of melanocetid males available to Pietsch and Van Duzer (1980) was nearly four times the amount examined by Bertelsen (1951), no new diagnostic characters were found. Of the 73 individuals (11.5 to 24 mm) examined, none of them could be satisfactorily identified to species based on females. As predicted by Bertelsen (1951), variation in the number of denticular teeth was found to be greater than previously thought and values given in his key overlap to a much greater extent than indicated. An attempt to utilize differences in larval pigmentation, thought to be more or less retained at least in the younger metamorphosed males, failed to separate the material into groups that could be associated with species based on females. Although Bertelsen's (1951) synonymies for nominal species based on males were retained by Pietsch and Van Duzer (1980), additional male specimens were listed as *Melanocetus* sp.

During the past two decades, Bertelsen and Pietsch (1983) reported on the melanocetids of Australia, Stewart and Pietsch (1998) on those found in New Zealand waters, and Anderson and Leslie (2001) on those of southern Africa, but no evidence has been presented in these or other publications during this time to question the taxonomic conclusions made by Pietsch and Van Duzer (1980).

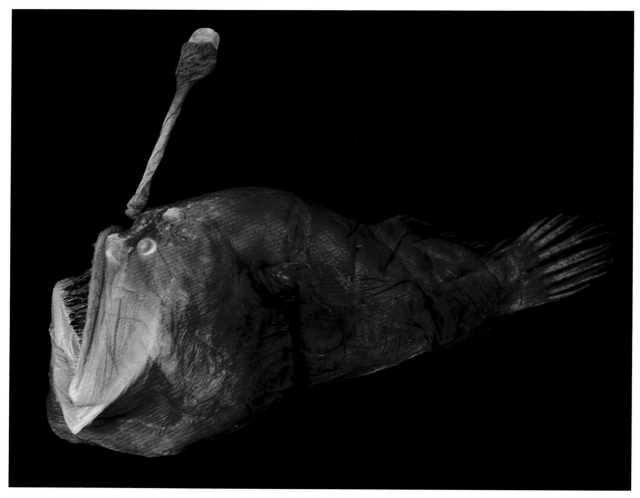

FIGURE 79. *Melanocetus eustalus*, 93 mm SL, SIO 55-229. Photo by T. W. Pietsch.

### DISTINGUISHING CHARACTERS

Members of the Melanocetidae are among the most easily recognized ceratioids. Both genders as well as larvae have a unique arrangement of the unpaired fins in which the dorsal contains more than three times the number of rays as the anal fin. The females of the family are short-bodied globular forms, characterized to a greater extent than many other ceratioids by having a large pendulous belly (Figs. 8, 78), an indication of their ability to engulf prey that is considerably larger than themselves. The mouth is large, well toothed, and dorsally directed, the opening almost vertical (Fig. 79). The head is smooth and rounded, unprotected by sphenotic spines; quadrate, articular, angular, and preopercular spines are also absent (Fig. 80). The esca is rather simple compared to many other ceratioids, often bearing laterally compressed anterior and posterior crests, but without elongate filaments or appendages (Fig. 81). When freshly caught the skin, except for the distal part of the esca, is velvety black. But like most all ceratioids and, for that matter, most bathypelagic fishes, the black fades rather quickly in preservative—the melanin and other pigments leached out by alcohol and otherwise broken down by photochemical bleaching—the specimens taking on a deep brownish red or even orange coloration. With the naked eye, the skin appears to be naked and smooth, but minute, widely spaced dermal spinules are present in at least some specimens (visible only microscopically in cleared and stained material; see Pietsch and Van Duzer, 1980:67).

Metamorphosed males have well-developed, laterally directed eyes, elliptical in shape, the pupil larger than the lens (Fig. 82). The olfactory organs are large, the nostrils inflated and directed laterally. Jaw teeth are absent. The upper denticular bears two or three semicircular series of strong recurved denticular teeth, fused with a median series of 3 to 9 enlarged dermal spines that articulate with the pterygiophore of the illicium. The lower denticular bears 10 to 23 recurved denticles, fused into a median and two lateral groups. The males are free-living, never becoming parasitic on females. Two examples of males attached to females in temporary attachment are known (Fig. 78; see Reproduction and Early Life History, Chapter Eight).

On first inspection, there appears to be nothing unusual about melanocetid larvae: the body is short, almost spherical; the skin is moderately inflated; and the pectoral fins are of normal size, not reaching beyond the dorsal and anal fins (Figs. 83, 84). But the unpaired fin-ray counts, consistent in the larvae and both genders, are all that is needed to distinguish them from all other anglerfishes. Sexual dimorphism is evident even in the tiniest larvae, the females bearing a small, club-shaped illicial rudiment protruding from the head. Metamorphosis begins at lengths of 8 to 10 mm (Bertelsen, 1951:45, 49; 1984:326, 328).

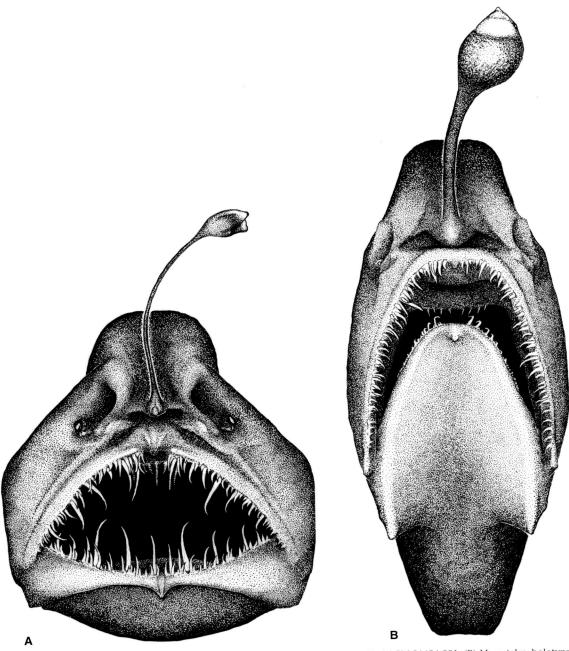

FIGURE 80. Species of *Melanocetus*, females, anterior views: (A) *M. johnsonii*, 85 mm SL, LACM 31484-001; (B) *M. eustalus*, holotype, 111 mm SL, LACM 30037-012. Drawings by E.A. Hoxie; after Pietsch and Van Duzer (1980).

There are 13 to 16 (rarely 12 or 17) dorsal-fin rays, 4 (rarely 3 or 5) anal-fin rays, 15 to 23 pectoral-fin rays, and 9 (1 simple + 6 bifurcated + 2 simple) caudal-fin rays (Table 1).

The largest known female is a 135-mm specimen of *M. johnsonii* (NMNZ P21373); the largest known metamorphosed male, also identified as *M. johnsonii*, measures 28 mm (the holotype of *Xenoceratias micracanthus*, ZMUC P9250).

### DIVERSITY

The family contains a single genus:

### GENUS *MELANOCETUS* GÜNTHER, 1864 (BLACK SEADEVILS)

*Figures 78–84, Tables 1, 4*

### DISTINGUISHING CHARACTERS

This genus has all the features of the family, as described above and detailed in Part Two: A Classification of Deep-Sea Anglerfishes.

### ETYMOLOGY

The name *Melanocetus*, chosen by Günther (1864) for his new genus and species from Madeira, is derived from the Greek *melas* or *melanos* meaning "black," and *ketos*, any large sea creature, but most often applied in reference to a whale.

### DIVERSITY

Six species are recognized based on females, and two of uncertain validity based on males (see Table 4):

BIODIVERSITY 79

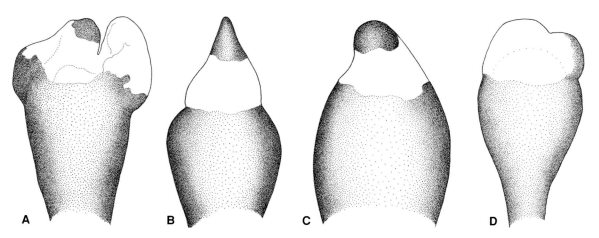

FIGURE 81. Escae of species of *Melanocetus*, left lateral views: (A) *M. johnsonii*, 75 mm SL, MCZ 49849; (B) *M. polyactis*, lectotype, 61 mm SL, ZMUC P9260; (C) *M. niger*, lectotype, 80 mm SL, ZMUC P9252; (D) *M. murrayi*, 120 mm SL, ISH 296/71. After Pietsch and Van Duzer (1980).

*Melanocetus johnsonii* Günther, 1864: 852 metamorphosed females, 10 to 154 mm; eight metamorphosed males, 15.5 to 28 mm; and 329 larvae, 2.5 to 17.5 mm TL; Atlantic, Pacific, and Indian oceans; holotype, BMNH 1864.7.18.6, female, 64 mm, off Madeira, 24 December 1863 (Günther, 1864:301, pl. 25; Bertelsen, 1951:48–53, figs. 13, 15, 16–19; Pietsch and Van Duzer, 1980:83).

*Melanocetus rossi* Balushkin and Fedorov, 1981: a single metamorphosed female, 118 mm, the holotype, ZIN 45349, 118 mm, fishery trawler *Babushkin Cape* trawl number 127, northern slope of Pennell Bank, Ross Sea, Antarctica, 74°46′S, 177°35′W, variable-depth otter trawl, 390 m, bottom depth 420 m, 22 March 1979 (Balushkin and Fedorov, 1981:79, figs. 1, 2; Pietsch, 1990:214; Balushkin and Fedorov, 2002:23, fig. 3).

*Melanocetus polyactis* Regan, 1925c: 32 metamorphosed females, 16.5 to 82 mm; three metamorphosed males, 9.0 to 19.5 mm; and seven larvae, 3.0 to 9.0 mm TL; eastern tropical Pacific Ocean; lectotype, ZMUC P9260, 61 mm, *Dana* station 1206(3), Gulf of Panama, 6°40′N, 80°47′W, 3500 m wire, 1845 h, 14 January 1922 (Regan, 1925c:565, 1926:34, pl. 8, fig. 2; Bertelsen, 1951:44, 54, 55; Pietsch and Van Duzer, 1980:77, figs. 21–24, 26, 30).

*Melanocetus niger* Regan, 1925c: 10 metamorphosed females, 12.5 to 80 mm; eastern tropical Pacific Ocean; lectotype, ZMUC P9252, 80 mm, *Dana* station 1208(4), Gulf of Panama, 6°48′N, 80°33′W, 3500 m wire, 0810 h, 16 January 1922 (Regan, 1925c:565; Bertelsen, 1951:44, 53; Pietsch and Van Duzer, 1980:78, figs. 21–24, 27, 30).

*Melanocetus eustalus* Pietsch and Van Duzer, 1980: four known females, 36 to 111 mm; eastern Pacific; holotype, LACM 30037-12, 111 mm, *Velero IV* station 11748, eastern Pacific off Mazatlán, Sinaloa, Mexico, 21°39′N, 106°58′W, 3-m Isaacs-Kidd midwater trawl, 0 to 1675 m, bottom depth 2820 m, 1320 to 2136 h, 11 November 1967 (Pietsch, 1972c:10, 1976:782, 783; Pietsch and Van Duzer, 1980:79, figs. 18, 28, 30).

*Melanocetus murrayi* Günther, 1887: 310 metamorphosed females, 13.5 to 124 mm; four metamorphosed males, 15 to 20 mm; and 77 larvae, 2.5 to 13 mm TL; Atlantic, Pacific, and Indian oceans; lectotype, BMNH 1887.12.7.17, female, 71 mm, *Challenger* station 106, central Atlantic, between St. Vincent and St. Paul's Rocks, 1°47′N, 24°26′W, 0 to 3386 m, 25 August 1873 (Günther, 1887:57, pl. 11, fig. A; Bertelsen, 1951:40, fig. 16; Pietsch and Van Duzer, 1980:81, figs. 2, 4, 5, 7–15, 16A, 19, 20, 29–31).

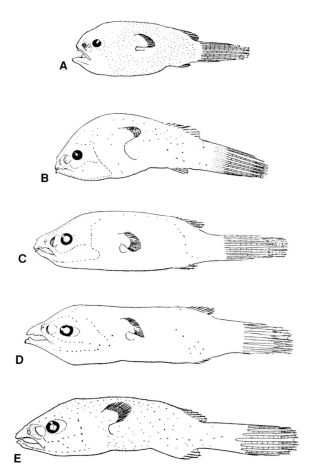

FIGURE 82. Free-living males of *Melanocetus* sp.: (A) lectotype of *Centrocetus spinulosus*, 15.5 mm SL, ZMUC P9246; (B) holotype of *Xenoceratias brevirostris*, 19 mm SL, ZMUC P9247; (C) holotype of *X. laevis*, 23 mm SL, ZMUC P9249; (D) syntype of *X. heterorhynchus*, 27 mm SL, ZMUC P9248; (E) syntype of *X. micracanthus*, 28 mm SL, ZMUC P9250. After Bertelsen (1951).

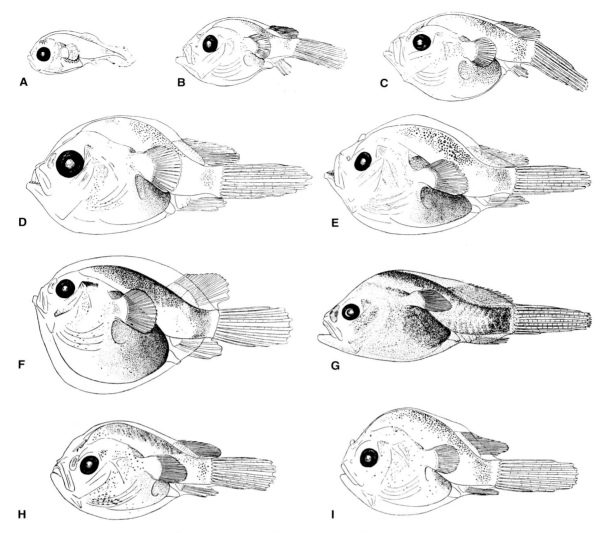

FIGURE 83. Larvae of *Melanocetus johnsonii*, showing the characteristic band of pigment on the caudal peduncle: (A) female, 3.3 mm TL; (B) female, 6.5 mm TL; (C) male, 8.5 mm TL; (D) male, 8.9 mm SL, 13 mm TL; (E) female, 11.9 mm SL, 17.0 mm TL; (F) male, 11.0 mm SL, 15.5 mm TL; (G) male in metamorphosis, 15.0 mm SL, 21.0 mm TL; (H) male, 8.5 mm SL, 12.3 mm TL; (I) female, 7.6 mm SL, 11 mm TL. After Bertelsen (1951).

*Melanocetus longirostris* (Regan and Trewavas, 1932), *nomen dubium*: a single metamorphosed male, 21 mm; holotype *(Xenoceratias longirostris)*, ZMUC P9259, 21 mm, *Dana* station 3751(7), north of New Guinea, 3°40′N, 137°53′E, 3000 m wire, 1240 h, 12 July 1929 (Regan and Trewavas, 1932:54, 55, fig. 80; Bertelsen, 1951:42–44, 54).

*Melanocetus nudus* (Beebe and Crane, 1947), *nomen dubium*: two metamorphosed males, 20 to 21.5 mm; holotype *(Xenoceratias nudus)*, CAS-SU 46495 (originally NYZS 28402), 21.5 mm, eastern Pacific *Zaca* Expedition station 210T-8, south of Cape Blanco, Costa Rica, 9°12′N, 85°10′W, 915 m, 27 February 1938 (Beebe and Crane, 1947:155, text fig. 2; Bertelsen, 1951:43, 44, 54, fig. 20).

## FAMILY THAUMATICHTHYIDAE Smith and Radcliffe, 1912 (Wolftrap Seadevils)

*Figures 85–95, Table 1*

Thaumatichthyids are among the most bizarre of anglerfishes, characterized by having large toothlike denticles associated with the esca, but more striking, an enormous upper jaw that extends far forward, each premaxilla bearing numerous long hooked teeth. The premaxillae, with their extraordinarily long teeth, are capable of flipping up and down to enclose, when in the ventral position, the much shorter lower jaw, forming a cagelike compartment within which prey are held prior to swallowing, reminiscent of the Venus flytrap *(Dionaea muscipula)* among carnivorous plants. The family was erected by Smith and Radcliffe (1912) to contain a new genus and species that they called *Thaumatichthys pagidostomus*, the trap-mouthed wonder fish. The holotype, a 60-mm female collected off Sulawesi in 1909, remained the only known representative of the genus until Parr (1927) discovered a second, somewhat smaller female (36.5 mm) in material collected off the Bahamas. This specimen, which Parr named *T. binghami*, differs from the type of *T. pagidostomus* in the development of the teeth and eyes, and, according to Parr (1927), in the number of dorsal-fin rays. Although no additional material was available, Regan and Trewavas (1932) placed Parr's *T. binghami* in a new genus, *Amacrodon*, based solely on the absence of a pair of long hooked premaxillary teeth found in *T. pagidostomus*. This reallocation was accepted by Parr (1934:7)

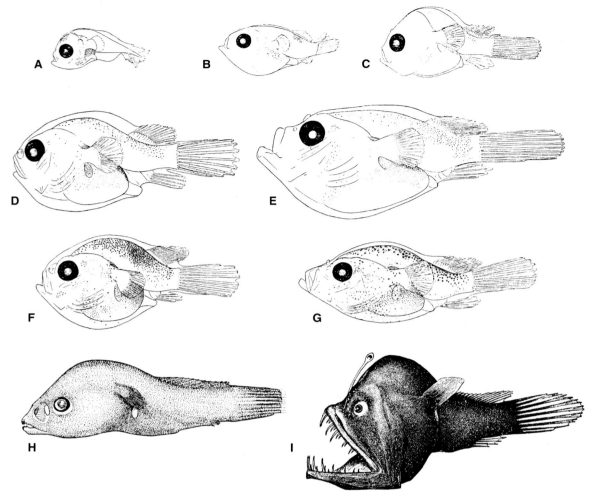

FIGURE 84. Larvae and juveniles of *Melanocetus murrayi*, distinguished from those of *M. johnsonii* by the absence of a pigment band on the caudal peduncle: (A) larval male, 3.0 mm TL; (B) larval female, 4.0 mm TL; (C) larval female, 6.3 mm TL; (D) larval male, 8.5 mm TL; (E) larval female, 9.0 mm SL, 13 mm TL; (F) larval male, 8.0 mm TL; (G) larval female, 9.0 mm TL; (H) juvenile male, 14 mm SL; (I) juvenile female, 14 mm SL (drawing by P. H. Winther). After Bertelsen (1951).

when he reported a second specimen of *"Amacrodon binghami"* (40 mm) captured in the same general area as the holotype.

Declared as a "living mouse-trap with bait . . . and altogether one of the oddest creatures in the teeming variety of the fish world," Anton Bruun (1953:177, 1956:177) announced the discovery of yet another thaumatichthyid, this one, a 365-mm female collected in the eastern tropical Pacific during the *Galathea* Deep-Sea Expedition of 1950–1952. In a short general description, along with a photograph, Bruun (1953:177, 1956:177) wrote that his specimen was "so different from any other [anglerfish] that we were justified in giving it the new generic and specific name *Galatheathauma axeli*." By this time, the material assigned to these three nominal genera, *Thaumatichthys*, *Amacrodon*, and *Galatheathauma*, included only eight specimens: the three holotypes, an additional specimen of *Amacrodon binghami* listed by Parr (1934), and four larvae from the *Dana* collections tentatively assigned to *T. pagidostomus* by Bertelsen (1951).

Over the next 25 years the number of specimens in collections around the world increased fourfold, representing tiny larvae of only 3 mm to large females reaching 365 mm, including one adolescent and three larval males. In a remarkably detailed examination of this material, Bertelsen and Struhsaker (1977) provided a general description, including the osteology, dentition, jaw mechanism, and illicial and escal morphology of the females, and a description of a previously unreported adult male. They found no reason to retain *Amacrodon* and *Galatheathauma* and concluded that *Thaumatichthys* contains three species: *T. pagidostomus*, *T. binghami*, and *T. axeli*.

In the meantime, Regan (1925c) described *Lasiognathus saccostoma*, a new genus and species based on a single 59-mm female collected by the *Dana* in the Caribbean Sea. While showing some obvious similarities to *Thaumatichthys*, Regan (1926) concluded that *Lasiognathus* was more closely allied with the oneirodid genus *Dolopichthys*, thus including both *Lasiognathus* and *Thaumatichthys* in the Oneirodidae and synonymizing Smith and Radcliffe's (1912) Thaumatichthyidae.

Based on a single 38-mm female collected by Beebe (1930, 1932a) off Bermuda, Regan and Trewavas (1932:90) described *L. beebei*, said to differ from the former species only by having the escal "hooks inserted directly on the bulb." Maul (1961, 1962b) later described three females of *Lasiognathus* from off Madeira: one 111-mm individual referred to "*Lasiognathus* sp." and two specimens described as the holotype (55.5 mm) and paratype (34 mm) of a new species, *L. ancistrophorus*, characterized by having 3 clawlike escal hooks born on an appendage, well separated from the escal bulb.

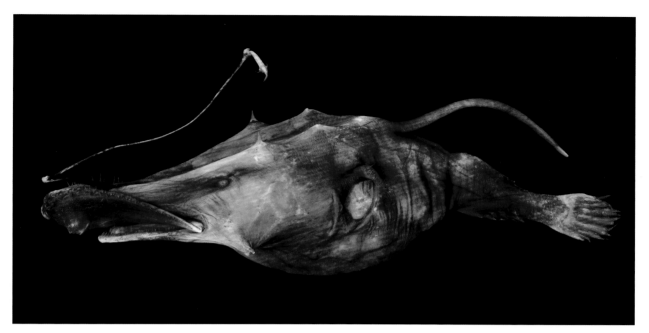

FIGURE 85. *Lasiognathus amphirhamphus*, holotype, 157 mm SL, BMNH 2003.11.16.12. Photo by T. W. Pietsch.

Nolan and Rosenblatt (1975) reviewed *Lasiognathus* on the basis of these five specimens from the Atlantic, and seven previously unreported specimens from the Pacific (five of which had lost the illicium). They synonymized *L. ancistrophorus* with *L. saccostoma*; referred an eastern Pacific specimen (97.5 mm) with an intact illicium to *L. saccostoma*; described the other Pacific specimen (94 mm) with an intact illicium as a new species, *L. waltoni*; and referred Maul's (1961) "*L.* sp." to *L. beebei*.

In a more recent revision, Bertelsen and Pietsch (1996) reexamined all previously recorded specimens and compared their escal characters with those of newly obtained individuals. They provided drawings of the escae of previously described specimens, and a new illustration reconstructing the esca of the holotype of *L. beebei* on the basis of its present somewhat shrunken condition, caused by the clearing and staining of the specimen by Gregory (1933). The final result was a new interpretation of the known material, in which four species were recognized, one of which was described as new. Finally, Pietsch (2005a) described a fifth species of *Lasiognathus* based on a single specimen collected from off Madeira, which differs from its congeners in having only 2 escal hooks.

#### DISTINGUISHING CHARACTERS

On casual inspection, the two genera of the family Thaumatichthyidae appear to have nothing to do with one another: *Lasiognathus* is laterally compressed, the illicium is well developed—emerging in typical anglerfish fashion from the snout—and its habitat is clearly pelagic; *Thaumatichthys*, on the other hand, is strongly depressed dorsoventrally, the illicium is drastically reduced—the esca hanging from the roof of the mouth—and it is without doubt a bottom-living creature. There are, however, several shared derived characters that serve to unite these two taxa to the exclusion of all other anglerfishes. First and foremost are the jaws. The elements of the upper jaw, the maxillae and premaxillae, are extraordinarily long, together extending forward far beyond the lower jaw. The bones do not come together anteriorly to unite on the midline as they do in other anglerfishes, but are instead widely separated, connected only by a broad elastic membrane. The premaxillary teeth are extremely long and somewhat hooked distally.

Thaumatichthyids are further unique among anglerfishes in having toothlike denticles associated with the esca, two or three sharp bony elements attached to the outer surface of the esca in *Lasiognathus*, and a single element of similar structure embedded within the tissue of the esca in *Thaumatichthys*. Finally a third shared feature of these two genera is a peculiar opercular bone (which is easily examined by removing the loose overlying skin): unlike all other ceratioids the dorsal part of this element is divided into two or more branches.

Nothing like this combination of structural peculiarities—the unique jaw structure, toothlike dermal denticles in the esca, and a divided opercle—exists elsewhere among lophiiform fishes, thus providing strong evidence that *Lasiognathus* and *Thaumatichthys* evolved together from a common ancestor, despite striking differences between the two in most all other character complexes. For details, as well as a diagnosis and description of thaumatichthyid males and larvae, which are known only for *Thaumatichthys*, see the generic accounts below as well as in Part Two: A Classification of Deep-Sea Anglerfishes.

Fin-ray counts vary little within the family: there are 5 to 7 dorsal-fin rays, 4 to 5 anal-fin rays, 14 to 20 pectoral-fin rays, and 9 (2 simple + 4 bifurcated + 3 simple) caudal-fin rays (Table 1).

Females of the two thaumatichthyid genera differ remarkably in size: the largest known specimen of *Lasiognathus* is only 157 mm (the holotype of *L. amphirhamphus*, BMNH 2003.11.16.12); whereas that of *Thaumatichthys* measures 365 mm (the holotype of *Galatheathauma axeli*, ZMUC P92166). Larvae and free-living males, known only for *Thaumatichthys*, attain lengths of 22.5 mm and 36 mm, respectively.

#### SYNOPSIS OF GENERA OF THE FAMILY THAUMATICHTHYIDAE

The Thaumatichthyidae contains two genera, differentiated as follows.

Females with the head narrow, compressed; the pterygiophore of the illicium exceptionally long, its anterior end emerging on the snout from between the frontal bones, its posterior end emerging on the back behind the head; the illicium long, with a well-developed terminal esca; esca with 2 or 3 hooked dermal denticles; the skin smooth and naked, dermal spinules absent:

SNAGGLETOOTH SEADEVILS, GENUS *LASIOGNATHUS* REGAN, 1925C, FIVE SPECIES, P. 84

Females with the head broad, depressed; the pterygiophore of the illicium relatively short, completely hidden beneath the skin of the head; the illicium short, nearly fully enveloped by tissue of the esca; the esca hanging from the roof of the mouth; esca with a single embedded dermal denticle; skin of the ventral and lateral surfaces of the head, body, and tail covered with close-set dermal spinules:

WONDERFISHES, GENUS *THAUMATICHTHYS* SMITH AND RADCLIFFE, 1912, THREE SPECIES, P. 85

## GENUS *LASIOGNATHUS* REGAN, 1925c (SNAGGLETOOTH SEADEVILS)

*Figures 85–88, Table 1*

### DISTINGUISHING CHARACTERS

Metamorphosed females of *Lasiognathus* can hardly be confused with anything else. In addition to having a narrow laterally compressed head and body; an enormous upper jaw equipped with numerous, long, mobile teeth; and a long illicial apparatus that emerges on the snout from between the frontal bones, the esca bears 2 or 3 large, toothlike dermal denticles (Figs. 86, 88). In most spiny skinned ceratioid genera (e.g., *Centrophryne*, *Ceratias*, *Cryptopsaras*, *Diceratias*, *Bufoceratias*, *Spiniphryne*, and *Gigantactis*), dermal spinules are present on the stem of the illicium and at the base of the esca, but in all cases these elements are tiny—nothing like the relatively huge, strongly curved, and specially arranged escal denticles of *Lasiognathus*. In addition to number (2 or 3 versus only 1), the escal denticles of *Lasiognathus* differ from that of *Thaumatichthys* in shape, pigmentation, and position. While the single denticle of the latter genus is transparent, pointed at both ends, and almost entirely embedded within the tissue of the esca (Fig. 93), those of *Lasiognathus* are darkly pigmented, and have a broad, hollow conical base attached superficially to the skin of the esca (Fig. 88), all features that make them more similar—and perhaps more obviously homologous—to skin spines than the denticle of *Thaumatichthys*. In *Lasiognathus* the escal denticles are either attached directly on the esca or on a filament that arises from the escal bulb. The function of this elaborately constructed assemblage of hooks can only be surmised. But rather than used for snagging prey, it seems more likely they serve to warn and thus fend off predators or, more likely still, they protect the esca from being damaged by the nibbling attacks of other fishes and invertebrates that mistake the esca for a meal.

In addition to the features described above, females of *Lasiognathus* have extremely well-developed sphenotic spines; quadrate and articular spines are present as well; the skin is smooth, without dermal spinules, a character that is found also in nearly all oneirodids. Finally, in sharp contrast to females of their sister genus *Thaumatichthys*, they display a full combination of features that indicate a pelagic existence at all life history stages (Bertelsen and Pietsch, 1996).

FIGURE 86. Esca of *Lasiognathus amphirhamphus*, holotype, 157 mm SL, BMNH 2003.11.16.12. Photo by T.W. Pietsch.

Males and larvae of *Lasiognathus* are unknown.

Fin-ray counts vary little within the genus: all known specimens have 5 dorsal- and 5 anal-fin rays; the pectoral-fin rays range from 17 to 20 (Table 1).

### ETYMOLOGY

The name *Lasiognathus*, proposed by Regan (1925c) for this strange ceratioid, is derived from the Greek *lasios*, meaning "hairy" or "wooly," and *gnathos*, meaning "jaw," thus drawing attention to the numerous long teeth of the upper jaw.

### DIVERSITY

Five species are recognized as follows:

*Lasiognathus saccostoma* Regan, 1925c: eight metamorphosed females, 30 to 77 mm; North Atlantic, and central Pacific off the Hawaiian Islands; holotype, ZMUC P92121, 59 mm, *Dana* station

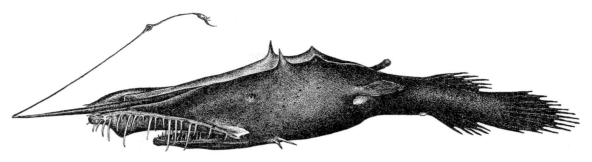

FIGURE 87. *Lasiognathus saccostoma*, 59 mm SL, ZMUC P92121, illicium rotated so that posteriorly directed escal hooks appear to be directly anteriorly. Drawing by P. H. Winther; after Bertelsen (1951).

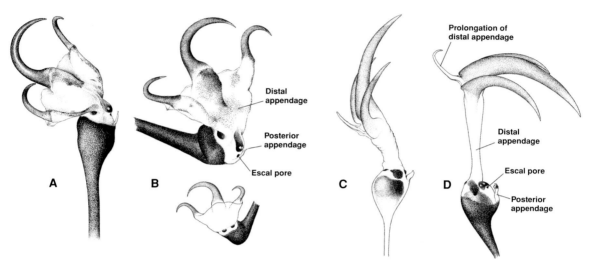

FIGURE 88. Escae of species of *Lasiognathus*: (A) *L. beebei*, 80 mm SL, MCZ 57779, left lateral view; (B) *L. beebei*, 112 mm SL, ISH 5542/79, posterolateral and anterior views; (C) *L. intermedius*, paratype, 26.5 mm SL, BMNH 1993.9.1.12-13, left lateral view; (D) *L. intermedius*, paratype, 29 mm SL, MCZ 49283, left lateral view. After Bertelsen and Pietsch (1996).

1217, Caribbean Sea, 18°50′N, 79°07′W, 4000 m of wire, 0630 h, 29 January 1922 (Regan, 1925c:563, 1926:31, pl. 7; Bertelsen, 1951:119, figs. 76, 77A; Maul, 1962b:39, figs. 4–6; Bertelsen and Pietsch, 1996:407, figs. 1, 5, 6; Pietsch, 2005a:79, 80, fig. 3).

*Lasiognathus beebei* Regan and Trewavas, 1932: five metamorphosed females, 27.5 to 112 mm; North Atlantic and central Pacific off the Hawaiian Islands; holotype, USNM 170956, 27.5 mm, off Bermuda, 0 to 1100 m (Regan and Trewavas, 1932:90; Bertelsen, 1951:119, fig. 77B; Maul, 1961:136, figs. 24–26; Nolan and Rosenblatt, 1975:65; Bertelsen and Pietsch, 1996:404, figs. 2, 6).

*Lasiognathus waltoni* Nolan and Rosenblatt, 1975: a single metamorphosed female, 94 mm; holotype, SIO 72-373, 94 mm, central North Pacific, 30°39.1′N, 155°23.4′W, 0 to 1350 m, 24 June 1972 (Nolan and Rosenblatt, 1975:64, figs. 4, 5; Bertelsen and Pietsch, 1996:406, figs. 3, 6).

*Lasiognathus intermedius* Bertelsen and Pietsch, 1996: seven metamorphosed females, 26.5 to 129 mm; Atlantic and eastern South Pacific; holotype, MCZ 57778, 31 mm, *Oceanus*, 34°18′N, 75°08′W, 0 to 1220 m (Nolan and Rosenblatt, 1975:62, figs. 1–3; Bertelsen and Pietsch, 1996:406, figs. 4, 6).

*Lasiognathus amphirhamphus* Pietsch, 2005a: a single known female, 157 mm; holotype, BMNH 2003.11.16.12, female, *Discovery*, station 10378-25, eastern North Atlantic, Madeira Abyssal Plain, off the SW coast of Madeira Island, 32°22′12″N, 29°50′42″W, 1200 to 1305 m, 9 June 1981 (Pietsch, 2005a:77, figs. 1, 2).

## GENUS *THAUMATICHTHYS* SMITH AND RADCLIFFE, 1912 (WONDERFISHES)

*Figures 89–95, Table 1*

### DISTINGUISHING CHARACTERS

No other anglerfish differs so strikingly from its close relatives than *Thaumatichthys*. The defining features of this enigmatic genus are nearly all the result of a shift from a pelagic way of life to a benthic or bottom-living existence, a life-style and habitat shared with no other ceratioid. Metamorphosed females differ most conspicuously in having a broad, dorsoventrally flattened head and anterior part of the body. The tiny eyes and olfactory organs are far apart, displaced posteroventrally to the distant corners of the mouth. The elements of the upper jaw extend far beyond the rather short lower jaw, their anterior ends widely separated from each other and hinged to the cranium in such a way that they can be flipped up and down. The premaxillae bear numerous long hooked teeth. The pterygiophore of the illicium is extremely short, completely hidden beneath the skin of the snout. The illicium is short as well, fully enveloped by the tissue of the esca, the latter bearing a single, partially embedded dermal denticle, and situated on the roof of the mouth, hanging from membranous tissue that interconnects the premaxillae (Figs. 89, 91B, 92, 93). The opercle is like that of no other

FIGURE 89. *Thaumatichthys axeli*, holotype, 365 mm SL, ZMUC P92166, originally described as *Galatheathauma axeli*, "the strange deep-sea angler-fish with the forked light organ in the mouth" (Bruun, 1953:175). Drawing by P.H. Winther; after Bruun (1953).

anglerfish, divided into 6 to 13 radiating branches (Fig. 27). The skin of the dorsal part of the head and body is smooth and naked, but the ventral and lateral surfaces are covered with small dermal spinules that extend out onto the caudal fin (Fig. 48F). In the shape of the head and position of the eyes and illicium,

*Thaumatichthys* differs from all other ceratioids, while only *Lasiognathus* has a similarly constructed upper jaw, denticles associated with the esca, and a multifurcated opercular bone.

Metamorphosed males of *Thaumatichthys* (four known specimens, 12 to 36 mm) differ from those of all other ceratioids in

FIGURE 90. *Thaumatichthys binghami*, 83 mm SL, UW 47537. Photo by Christopher P. Kenaley.

being considerably more slender and elongate (Fig. 94). The premaxillae are toothless, but unresorbed, bearing at their symphysis 4 or 5 separate denticles arranged in two transverse series. There are 7 denticles on the tip of the lower jaw, arranged in a transverse ventral series of 4 denticles and a dorsal series of 3 denticles. Like metamorphosed females, the dorsal fork of the opercle is divided into 6 or 7 radiating branches. The skin is covered with dermal spinules from the occipital region to the base of the caudal fin. There is no evidence that males become parasitically attached to females (see Reproduction and Early Life History, Chapter Eight).

The larvae of *Thaumatichthys* (five known specimens, 3.0 to 22.5 mm) can be rather easily identified by having a multiradiate opercle and a layer of subdermal melanophores that covers the entire surface of the head and body (Fig. 95). Like the adults of *Thaumatichthys*, but in contrast to those of *Lasiognathus*, the larvae of *Thaumatichthys* have 6 or 7 dorsal-fin rays, 4 anal-fin rays, and 14 to 16 pectoral-fin rays (Table 1).

### ETYMOLOGY

The name *Thaumatichthys* is derived from the Greek *thauma*, meaning "wonder" or "marvel," and *ichthys*, a "fish," thus providing the common name "wonderfishes."

### DIVERSITY

Three species are recognized as follows:

*Thaumatichthys pagidostomus* Smith and Radcliffe, 1912: three metamorphosed females, the holotype (60 mm) and two tentatively identified specimens (142 to 246 mm); western Pacific Ocean holotype, USNM 72952, 60 mm, *Albatross* station 5607, near Binang Unang Island, Gulf of Tomini, Sulawesi (Celebes), 00°04′S, 121°36′E, beam trawl, 1440 m, bottom of fine sand, 18 November 1909 (Smith and Radcliffe, 1912:580, pl. 72, figs. 1, 2; Bertelsen, 1951:121, figs. 78, 79; Bertelsen and Struhsaker, 1977:36, figs. 3, 7, 8, 13, 16).

*Thaumatichthys binghami* Parr, 1927: 39 metamorphosed females, 36.5 to 294 mm; Gulf of Mexico, Caribbean Sea, adjacent Atlantic waters off the Bahamas, to Rio de Janeiro, Brazil; holotype, YPM 2015, 36.5 mm, *Pawnee* station 25, 25°51′N, 76°37′W, pelagic net, 8000 feet of wire, 17 March 1927 (Parr, 1927:25, fig. 9; Regan and Trewavas, 1932:91; Bertelsen, 1951:122; Bertelsen and Struhsaker, 1977:36, figs. 1, 3–11, 13, 14, 16, 17, pls. 2, 3; McEachran and Fechhelm, 1998:871, fig.).

*Thaumatichthys axeli* (Bruun, 1953): two metamorphosed females, 85 to 365 mm; eastern tropical Pacific; holotype (*Galatheathauma axeli*), ZMUC P92166, 365 mm, *Galathea* station 716, eastern tropical Pacific, 09°23′N, 87°32′W, herring otter trawl, 3570 m, 6 May 1952 (Bruun, 1953:174, plate; Wolff, 1960:168, 175, 176–177, unnumbered color plate; Bertelsen and Struhsaker, 1977:37, figs. 2, 3, 6–8, 12, 13, 15, 16, 19, pl. 1; Gartner et al., 1997:133, fig. 4).

## FAMILY ONEIRODIDAE Gill, 1879a (Dreamers)

*Figures 96–147, Table 1*

The Oneirodidae is by far the largest, most complex, and certainly the least understood family of the suborder. With 16 genera and 62 species, it contains 40% of all recognized ceratioids. Of the 16 genera, five are currently represented by only one, two, or three females; only eight are represented by more than a dozen females. Males have been described for only seven genera, while larvae are known for only eight. But despite the rareness of most recognized taxa, new oneirodids continue to be discovered (see below).

The taxonomic history of the Oneirodidae begins with Lütken's (1871, 1872) description of a deep-sea anglerfish

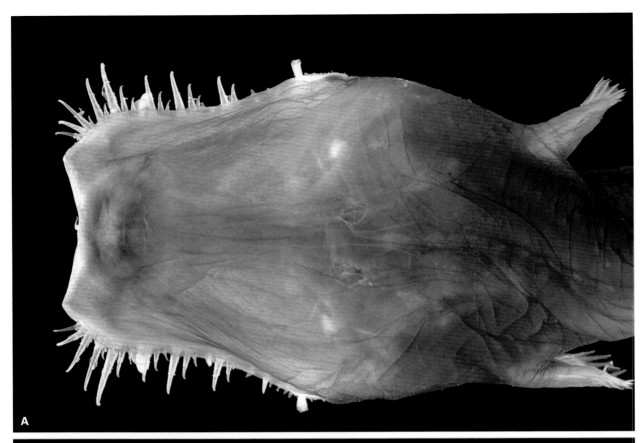

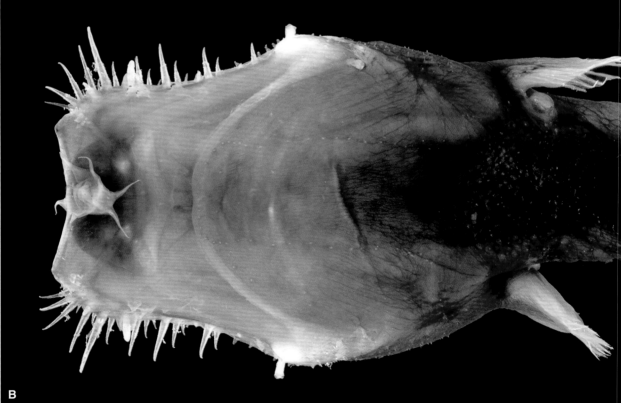

FIGURE 91. Head of *Thaumatichthys binghami*, 83 mm SL, UW 47537, dorsal (A) and ventral (B) views. Photos by C.P. Kenaley.

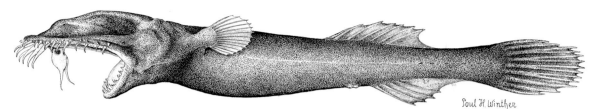

FIGURE 92. *Thaumatichthys axeli*, holotype, 365 mm SL, ZMUC P92166, originally described as *Galatheathauma axeli*. Drawing by P. H. Winther; after Bertelsen and Struhsaker (1977).

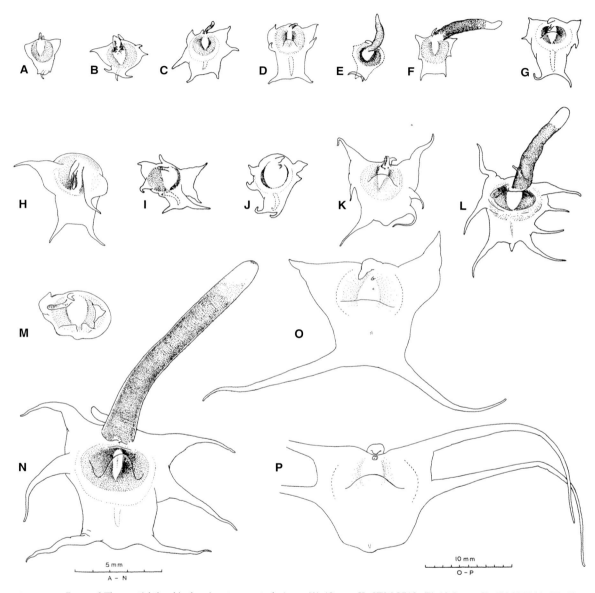

FIGURE 93. Escae of *Thaumatichthys binghami*, anteroventral views: (A) 40 mm SL, YPM 3713; (B) 40.5 mm SL, UF 230866; (C) 45 mm SL, TCWC; (D) 50.5 mm SL, UF 231363; (E) 51.5 mm SL, USNM 214570; (F) 58 mm SL, UF 231362; (G) 57 mm SL, UF 226927; (H) 67 mm SL, UF 230926; (I) 70 mm SL, ZMUC P921948; (J) 73 mm SL, ZMUC P921947; (K) 95.5 mm SL, UF 230930; (L) 87 mm SL, USNM 214571; (M) 112 mm SL, UF 227594; (N) 143 mm SL, ZMUC P921950; (O) 225 mm SL, UF 230958; (P) 294 mm SL, USNM 214571. After Bertelsen and Struhsaker (1977).

collected from the west coast of Greenland, which he named *Oneirodes eschrichtii* in honor of the celebrated Danish zoology professor Daniel Frederik Eschricht (1798–1863). The family remained monotypic until 1899, when Samuel Garman described *Dolopichthys* in his report on the fishes collected in the eastern tropical Pacific Ocean by the U.S. Fish Commission Steamer *Albatross*. Lloyd's description of *Lophodolos* followed in 1909, and shortly thereafter, Smith and Radcliffe's *Dermatias* in 1912.

The surprisingly large diversity of ceratioids in the deep-sea material brought back by the Danish research vessel *Dana* in the 1920s, resulted in a massive proliferation of new oneirodid

FIGURE 94. *Thaumatichthys* sp., juvenile male, 31 mm SL, ZMUC P921946. After Bertelsen and Struhsaker (1977).

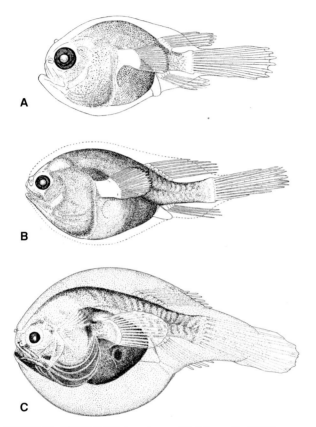

FIGURE 95. *Thaumatichthys* sp., larvae: (A) 6.5 mm SL, ZMUC P921956; (B) 17.2 mm SL, ZMUC P921731; (C) 22.6 mm SL, LACM 31107-001. After Bertelsen and Struhsaker (1977).

taxa, nearly all of which were described by Regan (1925c, 1926) and Regan and Trewavas (1932). Although they did their best to divide this material into well-defined subunits, and despite Lütken's detailed and accurate original description of the type genus *Oneirodes*, the systematics of the family became almost inextricably confused due primarily to a failure to distinguish correctly between *Oneirodes* and *Dolopichthys*. While maintaining *Oneirodes* as a monotypic genus, Regan and Trewavas (1932) provisionally recognized 43 species of *Dolopichthys*, distributed among five subgenera: *Dolopichthys*, *Dermatias*, *Microlophichthys*, *Leptacanthichthys*, and *Pentherichthys*, the three latter taxa described as new (Regan and Trewavas, 1932). Within their subgenus *Dermatias*, they included 29 species, 18 of which were represented by only one or two juvenile females less than 30 mm in standard length. Their subgenus *Dolopichthys* was recognized to contain seven species; *Microlophichthys*, four species; *Leptacanthichthys*, one species; and *Pentherichthys*, two species.

Such was the tangle of names and taxa that faced Bertelsen when he began his revision of the Oneirodidae in the 1940s. But with the advantage of additional material and a much better idea of individual morphological variation, Bertelsen (1951) was able to resolve most of the confusion by placing the subgenus *Dermatias* in synonymy with *Oneirodes* (subsequently resurrected by Pietsch and Kharin, 2004) and elevating the remaining four subgenera to generic status. Recognized also in his revision were *Chaenophryne*, described by Regan (1925c); *Chirophryne*, *Ctenochirichthys*, and *Tyrannophryne*, all introduced by Regan and Trewavas (1932); plus two new genera added by Bertelsen (1951), *Danaphryne* and *Spiniphryne*. The remaining three recognized genera of the family, *Phyllorhinichthys*, *Puck*, and *Bertella* were introduced by Pietsch (1969, 1973, and 1978, respectively).

In the meantime, Smith and Radcliffe (in Radcliffe, 1912) erected the Thaumatichthyidae to accommodate *Thaumatichthys*, a new genus that differed strikingly from all other ceratioids in having an elongate upper jaw, extending far out beyond the lower jaw, as well as a suite of remarkable features that indicated a benthic existence (see previous family account). Regan (1926), however, followed by Regan and Trewavas (1932), Bertelsen (1951), and Maul (1961, 1962b), chose not to recognize the Thaumatichthyidae, placing *Thaumatichthys* in the Oneirodidae, along with *Lasiognathus*, another genus with a highly elongate upper jaw that had been described earlier by Regan (1925c). Pietsch (1972a:18) later resurrected the Thaumatichthyidae to include both genera, stating that these two taxa "possess several important and unique characters that justify familial status." Bertelsen and Struhsaker (1977), however, compared the osteology of *Thaumatichthys* and *Lasiognathus*, pointing out that the latter appears more closely related to the Oneirodidae in several of the characters in which it differs most from *Thaumatichthys*. They (1977:34) concluded that "it becomes a subjective choice whether the genera *Lasiognathus* and *Thaumatichthys* both should be included in the Oneirodidae as Regan (1926) did, or placed together in Thaumatichthyidae as proposed by Pietsch (1972a), or whether each of them should be referred to a family of its own." At the same time, however, they cited the two unique features used by Pietsch (1972a) to diagnose the Thaumatichthyidae (premaxillae extending anteriorly far beyond the lower jaw, and enlarged dermal denticles associated with the esca) and added a third (dorsal portion of the opercle divided into two or more branches). In the end, they chose to retain the Thaumatichthyidae in the enlarged sense, as proposed by Pietsch (1972a) and supported here in the present analysis.

## DISTINGUISHING CHARACTERS

Oneirodids are so numerous and diverse, each of the 16 genera defined by its own set of unique features—with very few of these characters shared by any two or more genera—that it is almost impossible to provide a reasonable definition of the family. At present, there is only one known feature that serves to establish monophyly for this assemblage: metamorphosed females all have a narrow, spatulate, anterodorsally directed process that overlaps the posterolateral surface of the respective sphenotic (Pietsch and Orr, 2007:12). While not inclusive of all genera, a number of additional characters are helpful in understanding the family: each frontal has a prominent, bifurcated ventromedial extension (absent in *Lophodolos*); the parasphenoid has a pair of anterodorsal extensions that approach or overlap the posterior ventromedial extensions of the respective frontal (absent in *Lophodolos*); sphenotic, quadrate, and articular spines are usually well developed (absent in *Chaenophryne*); the skin is usually smooth and naked (well-developed dermal spinules are everywhere present in the skin of *Spiniphryne*).

Metamorphosed males are also nearly impossible to define as a family distinct from those of all other ceratioid families. While the following combination of features is helpful, the reader requiring more precise information is directed to the genera accounts below as well as the full diagnoses and descriptions provided in Part Two: A Classification of Deep-Sea Anglerfishes: the eyes are well developed and directed laterally, the diameter of the pupil greater than that of the lens. The olfactory organs are large, the anterior nostrils situated close together, their openings directed anteriorly; the posterior nostrils, usually larger than the eye, are directed laterally. The nasal area is usually pigmented and sometimes slightly inflated. The posterior end of the upper denticular is well separated from the anterior end of the pterygiophore of the illicium. The skin is smooth and naked, but males of the spiny-skinned oneirodid genus *Spiniphryne* are unknown. The males are free-living and nonparasitic, with two exceptions, both apparently representing facultative sexual parasitism: a single known attached pair of *Bertella idiomorpha* and another of *Leptacanthichthys gracilispinis* (see Reproduction and Early Life History, Chapter Eight).

There is no satisfactory combination of features that serve to adequately diagnose oneirodid larvae. As stated by Bertelsen (1951:71), it is "easier in practice to make the identification [of larvae] from the features characteristic of each of the many genera within the family." To a somewhat lesser extent, the same could also be said for the metamorphosed males and females.

Fin-ray counts vary widely in the family as a whole but range narrowly within most genera and all species, remaining consistent within both genders and larvae: dorsal-fin rays 5 to 8 (very rarely 4); anal-fin rays 4 to 7 (very rarely 3); pectoral-fin rays 13 to 30; caudal-fin rays 9 (2 simple + 4 bifurcated + 3 simple) (Table 1).

The largest known female is an unidentified, 370-mm specimen of *Oneirodes* (ISH 995/82), collected by the *Walther Herwig* during the 1982 expedition to the Mid-Atlantic Ridge. The largest known oneirodid male measures 16.5 mm (ZMUC).

The family contains 16 genera, differentiated as follows:

### SYNOPSIS OF GENERA OF THE FAMILY ONEIRODIDAE

The following key is provided to allow for the placement of oneirodids into the proper genus. Each entry consists of a combination of features that together serve to differentiate each genus. It works by progressively eliminating the most morphologically unique family; for that reason it should always be entered from the beginning. All character states listed for each genus must correspond to the specimen being keyed; if not, the user should proceed to the next set of character states. The emphasis here, and throughout this chapter, is on the more common and much better known females, which are, with few exceptions, the defining entities for all ceratioid families, genera, and species. For a full overview of both genders, as well as larvae, including traditional dichotomous keys to the identification of all taxa, see Part Two: A Classification of Deep-Sea Anglerfishes.

---

Females with the snout extremely short, the face more or less flat in lateral view, the profile almost vertical; the illicial apparatus emerging high on the forehead from between or somewhat behind extraordinarily well-developed sphenotic spines; the quadrate spine extremely well developed, much longer than the articular spine; the symphysial spine of the lower jaw extremely well developed; jaw teeth numerous, adults with 200 to 280 in the lower jaw; vomerine teeth absent; anal-fin rays 5 to 6, rarely 4 or 7:

PUGNOSE DREAMERS, GENUS *LOPHODOLOS* LLOYD, 1909A, TWO SPECIES, P. 96

Females with the lower jaw thickened anteriorly, without a ventrally directed spine at the symphysis; the dentaries forming a thick, broad, posteriorly directed flange immediately lateral to their union on the midline, the ventral margin of the lower jaw at the symphysis concave when viewed anteriorly; sphenotic spines small, but sharply pointed; quadrate and articular spines rudimentary; caudal-fin rays internally pigmented; lower jaw with 80 to nearly 200 teeth; vomerine teeth absent; anal-fin rays 6, rarely 5 or 7:

THICKJAW DREAMERS, GENUS *PENTHERICHTHYS* REGAN AND TREWAVAS, 1932, A SINGLE SPECIES, P. 97

Females with blunt protuberances on the dorsal surface of the head, but lacking sphenotic spines; quadrate and articular spines, and symphysial spine of the lower jaw, all rudimentary; the pterygiophore of the illicium unusually long, 70 to 82% SL (less than 50% SL in other oneirodids); the opercle more or less triangular in shape, only slightly concave along its posterior margin; the pelvic bones triradiate to broadly expanded distally; the skin smooth and naked, without dermal spinules, but unusually thick, almost leathery; the superficial bones of the head cancellous, having an almost bubbly appearance; lower jaw with 26 to 57 teeth; the vomer with 4 to 8 teeth; anal-fin rays 5 or 6:

SMOOTHHEAD DREAMERS, GENUS *CHAENOPHRYNE* REGAN, 1925C, FIVE SPECIES, P. 98

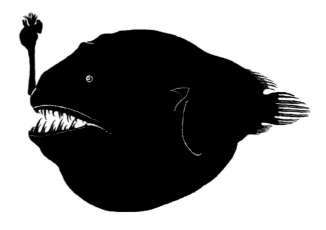

Females with the body unusually long and slender, laterally compressed; the dorsal surface of the head with sharp sphenotic spines; symphysial spine of the lower jaw well developed; quadrate and articular spines rudimentary; the skin everywhere covered with tiny close-set dermal spinules (in sharp contrast to all other oneirodids); lower jaw with 21 to 52 teeth; the vomer with 2 to 8 teeth; epibranchials and ceratobranchials toothless; anal-fin rays 5, rarely 4 or 6:

SPINY DREAMERS, GENUS SPINIPHRYNE BERTELSEN, 1951, TWO SPECIES, P. 102

Females with a blunt snout and short, highly convex frontals; sphenotic spines well developed, directed dorsolaterally; quadrate spine well developed; articular spine reduced, less than half the length of the quadrate spine; the lower jaw with a stout symphysial spine; the caudal peduncle unusually deep, greater than 20% SL; darkly pigmented skin of the caudal peduncle extending well past the base of the caudal fin; remarkably few teeth in the jaws, 20 to 32 in the upper jaw, 20 to 31 in the lower jaw; vomer with 4 to 6 teeth; anal-fin rays 4; extremely rare, only three known specimens:

FATTAIL DREAMERS, GENUS *DERMATIAS* SMITH AND RADCLIFFE, IN RADCLIFFE, 1912, A SINGLE SPECIES, P. 111

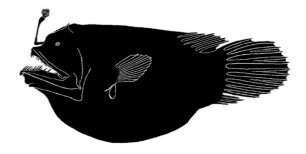

Females with the dorsal margin of the frontal bones convex; at least some species with the posterior end of the pterygiophore of the illicium emerging on the back, embedded within a cylindrical socklike evagination of the skin; sphenotic spines well developed; the lower jaw bearing a well-developed symphysial spine; the quadrate spine well developed, distinctly longer than the articular spine; the subopercle short and broad, the ventral end nearly circular; the caudal fin not covered by darkly pigmented skin except at the base; lower jaw with 18 to 160 teeth; vomer with 4 to 14 teeth; anal-fin rays 4, rarely 3 or 5; by far the most species-rich ceratioid genus:

COMMON DREAMERS, GENUS *ONEIRODES* LÜTKEN, 1871, 35 SPECIES, P. 105

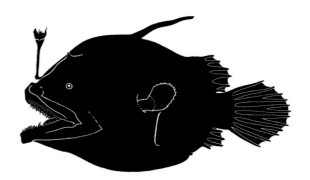

Females with an extremely large oblique mouth, the elements of the upper and lower jaws extending posteriorly far beyond the base of the pectoral fin and opercular opening; sphenotic spines well developed; quadrate spine reduced, but somewhat longer than the articular spine; the lower jaw with a well-developed symphysial spine; the caudal fin covered with darkly pigmented skin for some distance beyond the fin base; lower jaw with 38 to 110 teeth; vomerine teeth present or absent, apparently lost with growth; anal-fin rays 5, rarely 4; extremely rare, only three known specimens:

PUGNACIOUS DREAMERS, GENUS *TYRANNOPHRYNE* REGAN AND TREWAVAS, 1932, A SINGLE SPECIES, P. 116

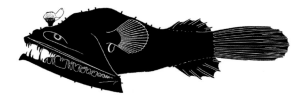

Females with 2 or 3 stout, nontapering appendages arising from the dorsal surface of the escal bulb (the posteriormost appendage sometimes exceedingly long); smaller specimens (less than about 100 mm) with a pair of unpigmented leaflike appendages on the snout; sphenotic spines well developed; quadrate spine well developed, much longer than the articular spine; the lower jaw with a small symphysial spine; darkly pigmented skin of the caudal peduncle extending well past the base of the caudal fin; remarkably few teeth in the jaws, 10 to 39 in the upper jaw, 8 to 21 in the lower jaw; vomer with 4 to 6 teeth; anal-fin rays 5:

LEAFYSNOUT DREAMERS, GENUS *PHYLLORHINICHTHYS* PIETSCH, 1969, TWO SPECIES, P. 115

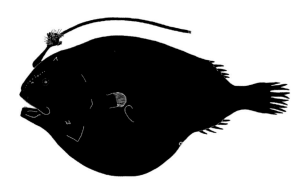

Females elongate and slender, not globular, with the mouth large, the cleft extending posteriorly past the eye; the illicium extremely short, less than twice the length of the escal bulb; sphenotic spines well developed; the quadrate spine well developed, longer than the articular spine; the lower jaw bearing a well-developed symphysial spine; jaw teeth numerous and extremely close-set, the upper jaw with about 160 to 320 teeth, the lower jaw with about 100 to 180 teeth; vomer with 4 to 12 teeth; anal-fin rays 5, rarely 4 or 6:

SHORTBAIT DREAMERS, GENUS *MICROLOPHICHTHYS* REGAN AND TREWAVAS, 1932, TWO SPECIES, P. 114

Females with mouth relatively small, the cleft terminating anterior to the eye; illicium relatively long, length 38 to 48% SL; esca unusually large, with a distal pair of stout, posteriorly directed, tentacle-like filaments, and a large, laterally compressed posterior appendage; sphenotic spines well developed; quadrate spine well developed, longer than the articular spine; the lower jaw with a well-developed symphysial spine; the dorsal end of the subopercle slender, tapering to a point; lower jaw with 30 to 105 teeth; vomer with 4 to 14 teeth; anal-fin rays 5, rarely 4:

DANA DREAMERS, GENUS *DANAPHRYNE* BERTELSEN, 1951, A SINGLE SPECIES, P. 112

Females with the frontal bones long and straight; sphenotic spines well developed; quadrate spine well developed, considerably longer than the articular spine; lower jaw with a well-developed symphysial spine; hyomandibula with a double head; subopercle long and narrow, the ventral end strongly oval; the pectoral-fin lobe short, shorter than the longest pectoral-fin rays; the esca with a large compressed posterior appendage, usually darkly pigmented; lower jaw with fewer than 35 to nearly 600 teeth; vomer with 0 to 14 teeth; second pharyngobranchial less than half the size of the third pharyngobranchial, the former bearing 1 to 4 teeth, the latter with 12 to 19 teeth; anal-fin rays 4 to 6:

LONGSNOUT DREAMERS, GENUS *DOLOPICHTHYS* GARMAN, 1899, SEVEN SPECIES, P. 118

Females with the frontal bones long and straight; sphenotic and quadrate spines extremely well developed; the distal end of the angular forming a well-developed spine; hyomandibula with a single head; the posteriormost rays of the dorsal and anal fins long, extending posteriorly well past the margin of the hypural plate; the pectoral-fin lobe short, shorter than the longest pectoral-fin rays; the esca with a single, short, unpigmented posterior appendage; the lower jaw with 60 to 250 teeth; vomerine teeth present in juveniles, but absent in adults; the second pharyngobranchial reduced and toothless, the third pharyngobranchial well-toothed; anal-fin rays 5, rarely 4:

SPIKEHEAD DREAMERS, GENUS *BERTELLA* PIETSCH, 1973, A SINGLE SPECIES, P. 121

Females with the pectoral-fin lobe extremely long and narrow (length greater than 15% SL), longer than the longest pectoral-fin rays; pectoral-fin rays 28 to 30, inserted along the dorsal margin of the pectoral-fin lobe; sphenotic, quadrate, and articular spines short, in some specimens not piercing skin; lower jaw with 16 to 31 teeth; vomer with 2 to 4 teeth; second and third pharyngobranchials present and well toothed; anal-fin rays 4 or 5; extremely rare, only two known females, a single juvenile, and two larval males:

COMBFIN DREAMERS, GENUS *CTENOCHIRICHTHYS* REGAN AND TREWAVAS, 1932, A SINGLE SPECIES, P. 128

Females with the pectoral-fin lobe long and narrow, longer than the longest pectoral-fin rays; pectoral-fin rays 18 to 22, inserted along the dorsal margin of the pectoral-fin lobe; sphenotic and quadrate spines well developed, the length of the latter less than the length of the articular spine; the dorsal profile of the frontal bones nearly straight; the illicium unusually slender; the esca with a single distal appendage; the lower jaw with 44 to 140 teeth; vomer with 6 to 16 teeth; second and third pharyngobranchials present and well toothed; anal-fin rays 5 to 6:

LIGHTLINE DREAMERS, GENUS *LEPTACANTHICHTHYS* REGAN AND TREWAVAS, 1932, A SINGLE SPECIES, P. 126

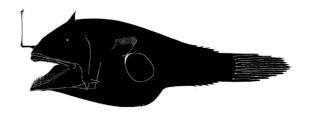

Females with the pectoral-fin lobe long and narrow, longer than the longest pectoral-fin rays; pectoral-fin rays 19 to 20, inserted along the dorsal margin of the pectoral-fin lobe; the mouth small, the gape not extending beyond the eye; the upper jaw extending slightly beyond the lower jaw; sphenotic and quadrate spines extremely well developed, the latter nearly six times the length of the articular spine; the subopercle long and slender, the upper end tapering to a point; the esca with a lateral filament; the lower jaw with 24 to 25 teeth; the vomer with 4 to 6 teeth; second and third pharyngobranchials present and well toothed; anal-fin rays 4; extremely rare, only four known metamorphosed females:

MISCHIEVOUS DREAMERS, GENUS *PUCK* PIETSCH, 1978, A SINGLE SPECIES, P. 124

Females with the pectoral-fin lobe long and narrow, longer than the longest pectoral-fin rays; pectoral-fin rays 18 to 19, inserted along the dorsal margin of the pectoral-fin lobe; the mouth large, the gape extending beyond the eye; the lower jaw protruding anteriorly beyond the upper jaw; sphenotic and quadrate spines extremely well developed, the latter four to nearly six times longer than the articular spine; the subopercle short and broad, the upper end rounded; the esca without a lateral filament; lower jaw with 34 to 40 teeth; vomer with 8 to 10 teeth; the second and third pharyngobranchials well developed and toothed; anal-fin rays 4; extremely rare, only four known metamorphosed females:

LONGHAND DREAMERS, GENUS *CHIROPHRYNE* REGAN AND TREWAVAS, 1932, A SINGLE SPECIES, P. 125

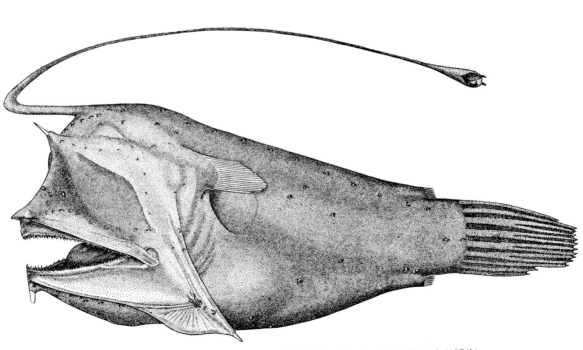

FIGURE 96. *Lophodolos indicus*, 58 mm SL, MCZ 47559. Drawing by Patricia Chaudhuri; after Pietsch (1974b).

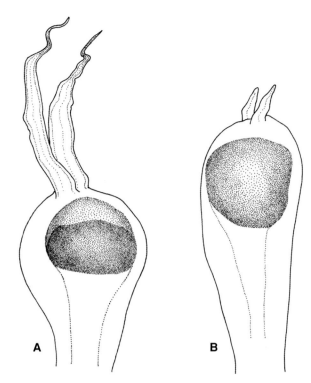

FIGURE 97. Escae of species of *Lophodolos*, left lateral views: (A) *L. acanthognathus*, 38 mm SL, LACM 10011-009; (B) *L. indicus*, 58 mm SL, MCZ 47559. Drawings by Patricia Chaudhuri; after Pietsch (1974b).

## GENUS *LOPHODOLOS* LLOYD, 1909a (PUGNOSE DREAMERS)

*Figures 96–98, Table 1*

The genus *Lophodolos* was erected by Lloyd (1909a) to include a single species, *Lophodolos indicus*, on the basis of a specimen collected from the Indian Ocean by the Royal Indian Museum Survey Ship *Investigator*. Since that time, three additional species have been described: *L. acanthognathus* Regan (1925c), to which some 149 specimens have been identified from the Atlantic and western Pacific oceans; *L. lyra* Beebe (1932b), synonymized with *L. acanthognathus* by Regan and Trewavas (1932); and *L. dinema* Regan and Trewavas (1932), synonymized with *L. indicus* by Pietsch (1974b).

Since the appearance of Bertelsen's (1951) monograph on the Ceratioidei, the number of metamorphosed female specimens of *Lophodolos* has more than doubled. But despite extensive new information gained from this increase in material, systematic study of the genus is by no means complete. Metamorphosed males are unknown; species are thus based solely on females. The separation of species is based on few characters, the most important being the morphology of the esca and the length of the illicium. But differences in these two characters merge in specimens less than 25 mm, making differentiation particularly difficult. Nevertheless, the material presently known appears to represent only two species (Pietsch, 1974b): *L. indicus* Lloyd (1909a) and *L. acanthognathus* Regan (1925c).

### DISTINGUISHING CHARACTERS

Unlike most female oneirodids, *Lophodolos* is relatively long and slender, but the forehead is thrust forward in a sense to produce

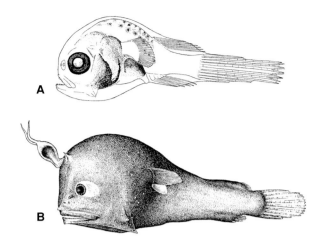

FIGURE 98. *Lophodolos* sp., early life-history stages: (A) *L.* sp., larval male, 5.0 mm SL; (B) *L. acanthognathus*, juvenile female, 8.5 mm SL, ZMUC P92923 (drawing by P. H. Winther). After Bertelsen (1951).

an extremely short snout and a concave face, hence the common name "Pugnose Dreamer." In addition to this rather strange combination of features, the mouth is large, the opening extending posteriorly past the eye; the jaws are equal anteriorly; the bony pterygiophore that supports the illicium emerges high on the head, from between or somewhat behind extraordinarily well developed sphenotic spines. The illicium itself is highly variable in length, ranging from as little as 10% SL in small females to nearly 140% SL in the largest known specimens. The quadrate bone, symphysis of the lower jaw, and posterior-most extent of the lower jaw all form extremely long sharp spines. Jaw teeth are close set and numerous, 200 to 280 in the lower jaw, but unlike most other oneirodids, the vomer is toothless. The skin is naked, without dermal spinules. The darkly pigmented skin of the caudal peduncle extends well past the base of the caudal fin. The esca bears a pair of unpigmented, bilaterally placed appendages arising from the distal surface (Fig. 97).

Metamorphosed males of *Lophodolos* are unknown, but larvae are represented by five specimens, two females and three males. They are readily separated from other oneirodid larvae by having a dark V-shaped patch of pigment on the gill cover (Fig. 98A). The dorsal pigmentation of the body extends only slightly beyond the anterior insertion of the dorsal fin, leaving the posterior part of the body, including the caudal peduncle, without pigment.

There are 5 to 8 dorsal-fin rays, the anterior-most ray reduced to a small stub; 5 to 6 (rarely 4 or 7) anal-fin rays; and 17 to 20 (rarely 21) pectoral-fin rays (Table 1).

### ETYMOLOGY

The name *Lophodolos* is derived from the Greek *lopho-* or *lophio-*, meaning a "mane," "crest," or "tuft"; plus the Greek *dolos*, meaning "artifice," "deceit," or "guile," the two roots combining to describe a deceitful little fish that beguiles its prey with a baited illicium mounted on top of the head.

### DIVERSITY

Two species and one nominal form of doubtful validity are recognized as follows:

*Lophodolos indicus* Lloyd, 1909a: 31 metamorphosed females, 9.5 to 77 mm; eastern Atlantic, Pacific, and Indian oceans;

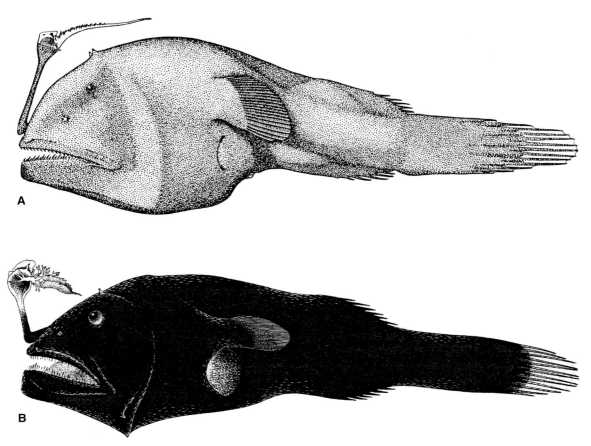

FIGURE 99. *Pentherichthys atratus*: (A) lectotype, 19 mm SL, ZMUC P92103 (drawing by W.P.C. Tenison; after Regan and Trewavas, 1932); (B) 19.5 mm SL, ISH 2058/71 (drawing by R. Nielsen).

holotype, ZSI 1024/1, 53 mm, *Investigator* station 307, off Kerala (formerly Travancore), southwest coast of India, 0 to 1624 m (Lloyd, 1909a:167, pl. 45, fig. 7; Regan and Trewavas, 1932:83, pl. 4, fig. 3; Bertelsen, 1951:108; Pietsch, 1974b:8, 13, figs. 1, 2, 4–9).

*Lophodolos acanthognathus* Regan, 1925c: 149 metamorphosed females (6 to 73 mm); five larvae, three males (7.0 to 8.0 mm TL), and three females (7.0 to 8.0 mm TL); Atlantic, Pacific, and Indian oceans; lectotype, ZMUC P92104, 12 mm, *Dana* station 1358(5), western North Atlantic, 28°15′N, 56°00′W, 3000 m wire, 1530 h, 2 June 1922 (Murray and Hjort, 1912:104, 614, fig. 90; Regan, 1925c:563; Beebe, 1932b:96, fig. 28; Regan and Trewavas, 1932:9, 11, 12, 83; Bertelsen, 1951:107, fig. 64; Pietsch, 1974a:17, 19, 21, 24–28, 82, 86–89, figs. 27, 33, 39E, 45, 51H, 53B, 104).

*Lophodolos biflagellatus* Koefoed, 1944, *nomen nudum*: this name was used by Koefoed in a manuscript dated 1918 (not seen by me), and later mentioned in published form (Koefoed, 1944:7) without application to a description or type (Pietsch, 1974b:17).

## GENUS *PENTHERICHTHYS* REGAN AND TREWAVAS, 1932 (THICKJAW DREAMERS)

*Figures 99–101, Table 1*

The earliest recorded specimen of the genus *Pentherichthys* is a 26-mm female collected by HMS *Discovery* in the North Atlantic, referred by Norman (1930:354) to *Dolopichthys allector* Garman, 1899. Norman provided a detailed description and illustration of the luring apparatus but remained unsure of his identification: "The structure of the bulb of the illicium of this specimen seems to be a little different to that of *D. allector*, and as I have been unable to detect any other differences, I have hesitated to describe a new species on the basis of a singe small specimen." Recognizing the uniqueness of the esca of Norman's specimen, Regan and Trewavas (1932:81) described it as the holotype of a new species, *Dolopichthys venustus*, and placed it together with another new species, *D. atratus*—represented by five tiny females, 16 to 19 mm, all collected by the *Dana* in the Gulf of Panama—in a new subgenus *Pentherichthys*.

As part of his almost total realignment of oneirodid taxa, Bertelsen (1951:102) elevated *Pentherichthys* to full generic status and, while no new metamorphosed females were added to the known material, he was able to distinguish males (four specimens) and larvae (19 specimens), based primarily on his discovery of the unique presence of melanophores embedded within the rays of the caudal fin. As for the two species, *Pentherichthys atratus* and *P. venustus*, the only discernable differences lie in the relative size and pattern of branching of the escal appendages, which led Bertelsen (1951:103) to doubt their validity: "As the first specimen *[P. venustus]* is somewhat larger than those used in founding *P. atratus* . . . it is possible they will be found to be synonyms." Since that time, additional females, including three described by Bertelsen and Pietsch (1977; see also Pietsch, 1974a), have further blurred the distinction between the two species, prompting the decision here to recognize only the type species of the genus, *P. atratus*.

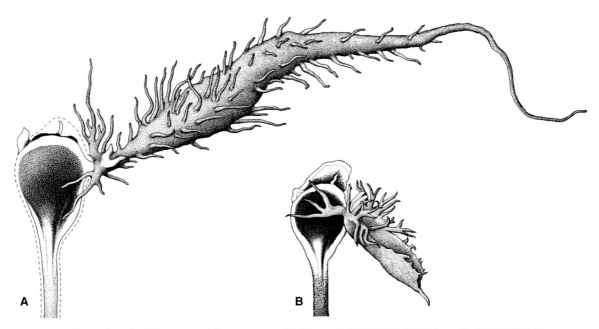

FIGURE 100. Escae of *Pentherichthys atratus*, left lateral views: (A) 92 mm SL, ISH 131/67; (B) 19.5 mm SL, ISH 2058/71. Drawings by K. Elsman; after Bertelsen and Pietsch (1977).

### DISTINGUISHING CHARACTERS

*Pentherichthys* has a rather strange and unique feature that sets it off clearly from all other anglerfishes: embedded within the rays of the caudal fin are large melanophores, spots of black pigment that can be easily detected with the naked eye in larvae and adults of both sexes (Bertelsen, 1951:103). In addition to these spots of pigment, the function of which, if any, is unknown, metamorphosed females of *Pentherichthys* differ in the structure of the lower jaw: the dentaries form a thick, broad, posteriorly directed flange immediately lateral to their union on the midline, so that instead of the ventrally directed symphysial spine found in most other oneirodids, the ventral margin of the lower jaw at the symphysis is concave when viewed anteriorly (Pietsch, 1974a:21).

Additional features of females of *Pentherichthys* that set them apart from those of other oneirodids include the absence of vomerine teeth, a character shared only with *Lophodolos*; sphenotic spines are well developed, but quadrate and articular spines are rudimentary; the skin is naked, without dermal spinules; and darkly pigmented skin of the caudal peduncle extends well past the base of the caudal fin. The esca bears a small conical or laterally compressed anterior papilla, a median distal crest, and a large tapering posterior appendage (Fig. 100). The posterior escal appendage bears in turn several to numerous short filaments along most of its length, and, on each side of its base, a large, simple to complexly branched, anteriorly directed appendage.

Metamorphosed males are rare—described on the basis of only six specimens, 7.5 to 13.5 mm—but readily identified by the melanophores embedded within the caudal-fin rays. They are also characterized by having the posterior nostril contiguous with the eye, the nasal area unpigmented, and 18 olfactory lamellae. There are 7 to 9 upper denticular teeth, fused at their base, and 4 lower denticular teeth (Bertelsen, 1951:102).

In addition to having internally pigmented caudal-fin rays, larvae of *Pentherichthys* are differentiated by having a few scattered melanophores on the gill cover, a restricted cluster of melanophores on the back terminating anterior to the origin of the dorsal fin, and only faint pigmentation on the posterior part of the caudal peduncle (Bertelsen, 1951:102; Fig. 101).

There are 6 to 7 dorsal-fin rays, 6 (rarely 5 or 7) anal-fin rays, and 21 to 27 pectoral-fin rays (Table 1).

### ETYMOLOGY

In naming *Pentherichthys*, Regan and Trewavas (1932) must have thought the six specimens included in their original description had a sad look about them: the Greek root *penthos* refers to "sorrow" or "mournfulness," thus, in combination with *ichthys*, a "mournful fish."

### DIVERSITY

A single species is recognized:

*Pentherichthys atratus* (Regan and Trewavas, 1932): 26 metamorphosed females (16.5 to 122 mm); six males, three in metamorphosis (7.5 to 10.5 mm), three metamorphosed (12 to 13.5 mm); and 19 larvae, seven females and 12 males (5.3 to 15 mm TL); Atlantic, Pacific, and eastern Indian oceans; lectotype *(Dolopichthys atratus)*, ZMUC P92103, 19 mm, *Dana* station 1208(6), Gulf of Panama, 6°48′N, 80°33′W, 2500 m wire, 0810 h, 16 January 1922 (Regan and Trewavas, 1932:81, figs. 2, 129, 130, pl. 3; Beebe and Crane, 1947:162; Bertelsen, 1951:105; Pietsch, 1974a:16–21, 24–29, 109, figs. 25, 38, 39H, 47, 48, 51G, 53C; Bertelsen and Pietsch, 1977:186, fig. 9; Bertelsen, 1984:331, fig. 170E).

### GENUS *CHAENOPHRYNE* REGAN, 1925c (SMOOTHHEAD DREAMERS)

*Figures 102–108, Tables 1, 5*

*Chaenophryne* includes globose bathypelagic anglerfishes, easily separated from members of allied genera by the absence

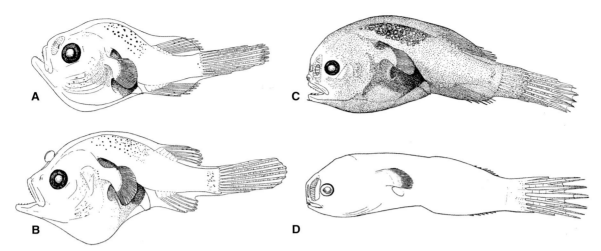

FIGURE 101. *Pentherichthys atratus*, early life-history stages: (A) larval male, 7.7 mm SL; (B) larval female, 10.6 mm SL; (C) metamorphosed male, 10.8 mm SL; (D) juvenile male, 13 mm SL. After Bertelsen (1951).

of sphenotic spines, a triangular opercular bone, with a slightly concave posterior margin, and peculiar highly cancellous bones (Bertelsen, 1951:109; Pietsch, 1975b:77, 79). Regan (1925c) established the genus with the description of *Chaenophryne longiceps* based on 14 specimens, only one of which actually represents the type species. The remaining 13 specimens were included in descriptions of 13 new species all introduced by Regan and Trewavas (1932): *C. bicornis*, *C. crenata*, *C. quadrifilis*, *C. haplactis*, *C. parviconus*, *C. atriconus*, *C. columnifera*, *C. melanodactylus*, *C. macractis*, *C. melanorhabdus*, *C. pterolophus*, *C. fimbriata*, and *C. ramifera*. Prior to Bertelsen's (1951) monograph on the Ceratioidei, four additional forms were described: *C. crossota* Beebe, 1932b; *C. draco* Beebe, 1932b; *C. intermedia* Belloc, 1938; and *C. pacis* Koefoed, 1944. The total number of nominal species is thus 18 (Table 5), 14 of which were originally based on one or two juvenile females less than 20 mm, none on more than nine specimens, and only one represented by a specimen larger than 24 mm.

From an examination of the extensive larval material in the *Dana* collections, Bertelsen (1951) was able to divide the then known material of *Chaenophryne* into two species groups based on the inner pigment layer and number of pectoral-fin rays. Within the *C. longiceps* group he tentatively recognized five species: *C. longiceps*, *C. quadrifilis*, *C. bicornis*, *C. crenata*, and *C. crossota*; and within the *C. draco* group, three species: *C. draco*, *C. parviconus*, and *C. ramifera*. *Chaenophryne fimbriata* and *C. intermedia* were considered synonyms of *C. ramifera* and all remaining available names were placed in the synonymy of *C. parviconus*.

With the advantage of considerably more comparative material, Pietsch (1975b) was able to show that the available female material of *Chaenophryne* represents four well-defined species (Table 5): *C. longiceps*, *C. draco*, and *C. ramifera*, each known from all three major oceans of the world; and *C. melanorhabdus*, apparently restricted to the continental slope of western North America. More recently, Pietsch (2007) added a fifth species based on two specimens collected in the eastern tropical Pacific off Peru. The separation of these five species is based almost entirely on the morphology of the esca. Significant differences exist, however, in the width of the escal bulb, the length of the illicium, jaw tooth counts, and fin-ray counts (for details, see the species accounts given in Part Two: A Classification of Deep-Sea Anglerfishes).

Despite a greater than fourfold increase in males since Bertelsen's (1951) revision, Pietsch (1975b) was only able to separate the known material two taxonomic units corresponding to the *C. longiceps* and the *C. draco* groups. The accounts provided here follow closely those provided by Pietsch (1975b, 2007).

### DISTINGUISHING CHARACTERS

*Chaenophryne* is unique in several ways that make it seem misplaced among oneirodids, appearing on superficial inspection to deserve a family of its own. While the females of other oneirodids have more or less well-developed sphenotic spines, the dorsal surface of the head of females of *Chaenophryne* is smooth and rounded—quadrate and articular spines as well as the symphysial spine of the lower jaw are rudimentary. The shape of the opercular bone is unique as well: instead of deeply forked, the posterior margin of this bone in *Chaenophryne* is only slightly concave, much like the opercle of some linophrynids (Fig. 25C). The pterygiophore of the illicium is exceptionally long in *Chaenophryne*, 70 to 82% SL compared to less than 50% SL in all other oneirodids (see Pietsch, 1974a:18). The pelvic bones are triradiate to broadly expanded distally, unlike those of other oneirodids but reminiscent of those of himantolophids (see Bertelsen and Krefft, 1988:19, 27). The teeth of females are rather few compared to those of many other oneirodids, only 21 to 51 in the upper jaw and 26 to 57 in the lower jaw; the vomer bears 4 to 8 teeth. The skin of *Chaenophryne* differs from other oneirodids and recalls that of himantolophids in being rather thick, almost leathery, accounting for the fact that specimens make it through the ordeal of capture in relatively good physical condition (the thin delicate skin of other oneirodids and most other ceratioids is almost always shredded and torn, specimens often completely skinned in the net). Finally, the bones of *Chaenophryne*, especially those closely associated with the external surface of the head, are highly cancellous, appearing bubbly (Pietsch, 1975b:77; Fig. 104), a condition not found in any other anglerfish, but similar to that of the chiasmodontid genera *Dysalotus* and *Kali* described by Johnson and Cohen (1974:14). The morphology of the esca is highly variable (Figs. 105, 106).

TABLE 5
Reallocation of Nominal Species of the Genus *Chaenophryne*

| *Nominal Species* | *Currently Recognized Taxon* |
|---|---|
| Based on Females | |
| *Chaenophryne longiceps* group Bertelsen, 1951 | |
|     *Chaenophryne longiceps* Regan, 1925c | *Chaenophryne longiceps* Regan, 1925c |
|     *Chaenophryne crossota* Beebe, 1932b | *Chaenophryne longiceps* Regan, 1925c |
|     *Chaenophryne bicornis* Regan and Trewavas, 1932 | *Chaenophryne longiceps* Regan, 1925c |
|     *Chaenophryne crenata* Regan and Trewavas, 1932 | *Chaenophryne longiceps* Regan, 1925c |
|     *Chaenophryne quadrifilis* Regan and Trewavas, 1932 | *Chaenophryne longiceps* Regan, 1925c |
| *Chaenophryne draco* group Bertelsen, 1951 | |
|     *Chaenophryne draco* Beebe, 1932b | *Chaenophryne draco* Beebe, 1932b |
|     *Chaenophryne atriconus* Regan and Trewavas, 1932 | *Chaenophryne draco* Beebe, 1932b |
|     *Chaenophryne columnifera* Regan and Trewavas, 1932 | *Chaenophryne draco* Beebe, 1932b |
|     *Chaenophryne fimbriata* Regan and Trewavas, 1932 | *Chaenophryne ramifera* Regan and Trewavas, 1932 |
|     *Chaenophryne haplactis* Regan and Trewavas, 1932 | *Chaenophryne longiceps* Regan, 1925c |
|     *Chaenophryne macractis* Regan and Trewavas, 1932 | *Chaenophryne draco* Beebe, 1932b |
|     *Chaenophryne melanodactylus* Regan and Trewavas, 1932 | *Chaenophryne draco* Beebe, 1932b |
|     *Chaenophryne melanorhabdus* Regan and Trewavas, 1932 | *Chaenophryne melanorhabdus* Regan and Trewavas, 1932 |
|     *Chaenophryne parviconus* Regan and Trewavas, 1932 | *Chaenophryne draco* Beebe, 1932b |
|     *Chaenophryne pterolophus* Regan and Trewavas, 1932 | *Chaenophryne melanorhabdus* Regan and Trewavas, 1932 |
|     *Chaenophryne ramifera* Regan and Trewavas, 1932 | *Chaenophryne ramifera* Regan and Trewavas, 1932 |
|     *Chaenophryne intermedia* Belloc, 1938 | *Chaenophryne ramifera* Regan and Trewavas, 1932 |
|     *Chaenophryne galeatus* Koefoed, 1944 | *Nomen nudum* |
|     *Chaenophryne pacis* Koefoed, 1944 | *Chaenophryne ramifera* Regan and Trewavas, 1932 |
|     *Chaenophryne quasiramifera* Pietsch, 2007 | *Chaenophryne quasiramifera* Pietsch, 2007 |
| Based on Males | |
| *Chaenophryne longiceps* group Bertelsen, 1951 | |
|     *Rhynchoceratias leuchorhinus* Regan, 1926, in part | *Chaenophryne longiceps* Regan, 1925c |
| *Chaenophryne draco* group Bertelsen, 1951 | |
|     *Rhynchoceratias leuchorhinus* Regan, 1926, in part | *Chaenophryne draco* group Bertelsen, 1951 |
|     *Trematorhynchus obliquidens* Regan and Trewavas, 1932 | *Chaenophryne draco* group Bertelsen, 1951 |
| *Chaenophryne* sp. Group? | |
|     *Trematorhynchus adipatus* Beebe and Crane, 1947 | *Chaenophryne* sp. |
|     *Trematorhynchus moderatus* Beebe and Crane, 1947 | *Chaenophryne* sp. |

In contrast to females, metamorphosed males of *Chaenophryne* are rather typical for oneirodids (Fig. 107E, 108E): the skin between the nostrils and between the posterior nostril and the eye is pigmented; there are 8 to 12 olfactory lamellae; the upper denticular bears 10 to 22 teeth, fused at their bases and forming a semicircular cluster, while the lower denticular has 13 to 31 teeth in two or three irregular series; the skin is black and naked, without a trace of dermal spinules; the shape of the opercular bones is the same as that of the females (Bertelsen, 1951:109). The males are free-living, apparently never becoming parasitic.

The larvae are rather easily identified by the pattern of pigmentation (Figs. 107, 108): that of the dorsal part of the body terminates behind the occipital region of the head, but extends posteriorly with increasing growth to the posterior margin of the base of the dorsal fin; pigmentation of the gill cover is weak to moderately strong; peduncular pigment is present or absent (see Bertelsen, 1951:109).

There are 6 to 8 dorsal-fin rays, 5 or 6 anal-fin rays, and 16 to 22 pectoral-fin rays (Table 1).

### ETYMOLOGY

The Greek root *chaeno* or *chaino* means "to gape," thus, Regan (1925c), in naming *Chaenophryne*, was apparently impressed with the large mouth of this genus; in combination with the Greek *phryne*, or "toad," the name implies a "gaping toad."

FIGURE 102. *Chaenophryne quasiramifera*, holotype, 157 mm, SIO 72-180. After Pietsch (2007).

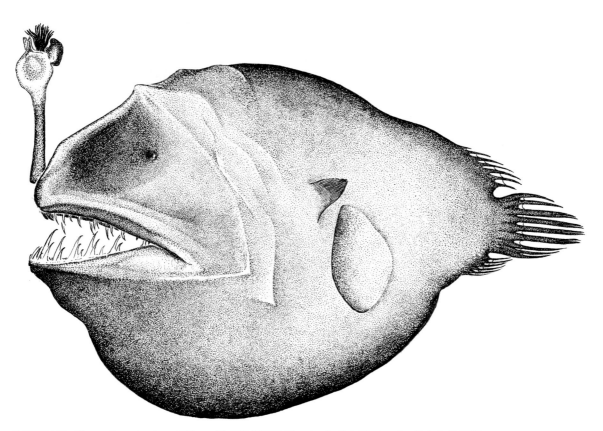

FIGURE 103. *Chaenophryne longiceps*, 102 mm SL, ISH 237/73. Drawing by E.A. Hoxie; after Pietsch (1975b).

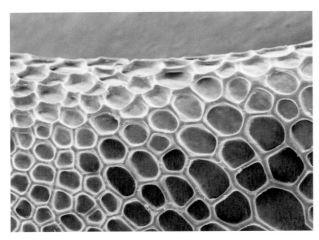

FIGURE 104. Scanning electronic micrograph of the dorsal-most margin of the sphenotic bone of *Chaenophryne melanorhabdus*, 90 mm SL, UW 48672, showing cancellous structure of the dermal skeletal elements unique to this genus. Courtesy of Zachary H. Baldwin.

### DIVERSITY

Five species and one nominal form of doubtful validity are divided among two species groups (Table 5):

CHAENOPHRYNE LONGICEPS GROUP BERTELSEN, 1951:

*Chaenophryne longiceps* Regan, 1925c: 84 metamorphosed females (10 to 245 mm); two females in metamorphosis (12 to 12.5 mm); seven metamorphosing, juvenile, and adult males (11 to 19.5 mm); and 106 larvae (62 females, 42 males, and two undetermined, 3.2 to 19 mm TL); Atlantic, Pacific, and Indian oceans; lectotype, ZMUC P92106, 20 mm, *Dana* station 1203(11), Gulf of Panama, 7°30′N, 79°19′W, 3000 m of wire, 1500 h, 11 January 1922 (Regan, 1925c:564; Parr, 1927:22, fig. 8; Regan and Trewavas, 1932:85, 86, figs. 14, 135; Bertelsen, 1951:71, 110–114, 269, figs. 30, 66, 68–71; Pietsch, 1975b:81, 82, figs. 1A, 3–7, 11, 12).

CHAENOPHRYNE DRACO GROUP BERTELSEN, 1951:

*Chaenophryne draco* Beebe, 1932b: 119 metamorphosed females, 11 to 123 mm; Atlantic, Pacific, and Indian oceans; holotype, USNM 170943 (originally NYZS 22396), 16.5 mm, Bermuda Oceanographic Expedition net 1181, 10 miles SE of Nonsuch, 32°12′N, 64°36′W, 1100 m, 15 August 1931 (Beebe, 1932b:84, fig. 22; Bertelsen, 1951:115, 116; Pietsch, 1975b:81, 87, figs. 1B, 2A, 3–5, 8, 11, 12).

*Chaenophryne melanorhabdus* Regan and Trewavas, 1932: 49 metamorphosed females, 11 to 102 mm; eastern Pacific Ocean; holotype, ZMUC P92117, 40 mm, *Dana* station 1203(14), Gulf of Panama, 7°30′N, 79°19′W, 2500 m wire, 2030 h, 11 January 1922 (Regan and Trewavas, 1932:85, 89, fig. 143; Bertelsen, 1951:117, fig. 74E; Pietsch, 1975b:81, 90, figs. 1C, 2B, 3–5, 9, 11, 12).

*Chaenophryne ramifera* Regan and Trewavas, 1932: 23 metamorphosed females, 13.5 to 87 mm; Atlantic, Pacific, and Indian oceans; holotype, ZMUC P92119, 17 mm, *Dana* station 3550(6), Gulf of Panama, 7°10′N, 78°15′W, 3000 m wire, 0145 h, 5 September 1928 (Regan and Trewavas, 1932:85, 90, fig. 146; Bertelsen, 1951:115, figs. 67D, 75; Pietsch, 1975b:81, 92, figs. 1D, 2C, 3–5, 9, 10–12; Bertelsen and Pietsch, 1977:186).

*Chaenophryne quasiramifera* Pietsch, 2007: two metamorphosed females, 45 to 98 mm; Peru-Chile Trench and Nasca

FIGURE 105. Esca of *Chaenophryne quasiramifera*, holotype, 157 mm SL, SIO 72-180, left lateral view. After Pietsch (2007).

Ridge; holotype, SIO 72-180, 98 mm, RV *Thomas Washington*, Cruise Southtow IV/MV 72-II, Station MV-72-II-23, Peru-Chile Trench, 20°19.2′ S, 71°14.9′ W, 3-m Isaacs-Kidd midwater trawl, 0 to 900 m, bottom depth 5856 m, 0023 to 0815 h, 3 May 1972 (Pietsch, 2007).

*Chaenophryne galeatus* Koefoed, 1944, *nomen nudum*: this name was used by Einar Koefoed in a manuscript dated 1918 (not seen by me), and later mentioned in published form (Koefoed, 1944:8) without application to a description or type (Pietsch, 1975b:94).

### GENUS *SPINIPHRYNE* BERTELSEN, 1951 (SPINY DREAMERS)

*Figures 109–111, Table 1*

The history of the genus *Spiniphryne* begins with Beebe's (1932b) description of *Dolopichthys gladisfenae*, based on a single, 40-mm female collected off Bermuda. Six months after Beebe's (1932b) publication, Regan and Trewavas (1932)

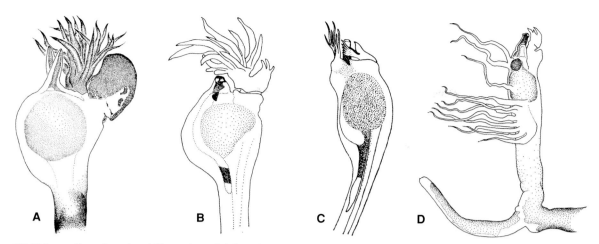

FIGURE 106. Esca of species of *Chaenophryne*, left lateral views: (A) *C. longiceps*, 102 mm SL, ISH 237/73; (B) *C. draco*, 12.5 mm SL, MCZ 49863; (C) *C. melanorhabdus*, holotype, 40 mm SL, ZMUC P92117; (D) *C. ramifera*, 36.5 mm SL, UF 229702. After Pietsch (1975b).

introduced *Centrophryne*, a new genus of the family Oneirodidae, to include *D. gladisfenae* and a new species that they named *Centrophryne spinulosa*. The presence of 4 pectoral radials, an anteriorly directed spine on the subopercle, and a small, digitiform hyoid barbel led Bertelsen (1951) to remove *C. spinulosa* from the Oneirodidae and place it in a new family, the Centrophrynidae. *Centrophryne gladisfenae* was retained by Bertelsen (1951) within the family Oneirodidae under a new generic designation, *Spiniphryne* (a senior synonym of *Bertelsenna* Whitley, 1954, an unacceptable replacement name for *Spiniphryne*). In a review of the family Centrophrynidae, Pietsch (1972a) retained *Spiniphryne* within the Oneirodidae following Bertelsen (1951), but predicted that sufficient material of *Spiniphryne gladisfenae* would show this species and *C. spinulosa* to be congeneric. However, an osteological examination by Bertelsen and Pietsch (1975), made possible by two additional females collected in 1971 in the tropical Atlantic, showed that the similarity between *Spiniphryne* and *Centrophryne* is superficial. The former is in some ways similar to *Oneirodes*, while the latter is quite distinct and appears to be most closely related to ceratiids (see Evolutionary Relationships, Chapter Four).

Since the review of Bertelsen and Pietsch (1975), several publications have added to our knowledge of *Spiniphryne*. Karrer (1976) reported an additional female (105 mm) from off Greenland; Pietsch and Seigel (1980) reported another female (18 mm) from the Banda Sea (the sixth known specimen of the genus and first record outside of the Atlantic); Moore et al. (2003) noted three additional females (14 to 36.5 mm) from off New England; and Pietsch and Baldwin (2006) added a second species based on two females collected from the central and eastern North Pacific Ocean. Although previously unreported records now bring the total number of identifiable specimens to 23, the genus is still based solely on metamorphosed females; males and larvae remain unknown. The account provided below is based almost solely on Bertelsen and Pietsch (1975) and Pietsch and Baldwin (2006).

### DISTINGUISHING CHARACTERS

Unlike most oneirodids, the body of metamorphosed females of *Spiniphryne* is elongate and slender, not globular (Figs. 109, 110). It is also the only spiny-skinned member of the family—the entire body and fins are covered with numerous close-set dermal spinules (tiny, but obvious without microscopic aid; Fig. 48G)—and for this reason may be confused with *Centrophryne* (family Centrophrynidae). Similar also to *Centrophryne* is the small head and anteriorly directed mouth, the latter rather small for an oneirodid, the cleft not extending past the eye. But *Spiniphryne* differs clearly in having well-developed sphenotic spines, toothless gill arches, 3 pectoral radials (versus 4 in *Centrophryne*), and numerous additional internal osteological characters detailed in Part Two: A Classification of Deep-Sea Anglerfishes. As in all ceratioids, the escal morphology of *Spiniphryne* is unique in consisting in part of 2 distal filaments or bulbous appendages, the filaments simple, the appendages usually covered with small digitiform papillae and clusters of tiny filaments around the base (see Fig. 111).

Males and larvae are unknown.

Fin-ray counts are as follows: 6 or 7 dorsal-fin rays, 5 (rarely 4 or 6) anal-fin rays, and 15 to 16 (rarely 17) pectoral-fin rays (Table 1).

### ETYMOLOGY

The Latin root *spina* or *spinula*, meaning "thorn" or "spine," combined with the Greek *phryne*, describes a "spiny toad."

### DIVERSITY

Two species are recognized as follows:

*Spiniphryne gladisfenae* (Beebe, 1932b): 21 metamorphosed females, 12.8 to 131 mm; Atlantic, western Pacific, and western Indian oceans; holotype *(Dolopichthys gladisfenae)*, USNM 170944 (originally NYZS 15490), 40 mm, female, Bermuda Oceanographic Expedition, net 639, 9.7 km south of Nonsuch Island, Bermuda, ca. 32°12′N, 64°36′W, 1280 m, 28 May 1930 (Beebe, 1932b:86–88; Regan and Trewavas, 1932:84; Bertelsen, 1951:122, fig. 81; Bertelsen and Pietsch, 1975:1–11, figs. 1–6, 1977:172; Pietsch and Baldwin, 2006:407, figs. 2, 4, 5, 7).

*Spiniphryne duhameli* Pietsch and Baldwin, 2006: two known specimens, 25.5 to 117 mm; eastern Pacific Ocean; holotype, SIO 60-239, female, 117 mm, RV *Spencer F. Baird*, Tethys Expedition, Station 17, eastern Pacific Ocean, 4°56.5′ to 5°28.0′N, 142°54.5′ to 143°10.0′W, 3-m Isaacs-Kidd midwater trawl, 0 to

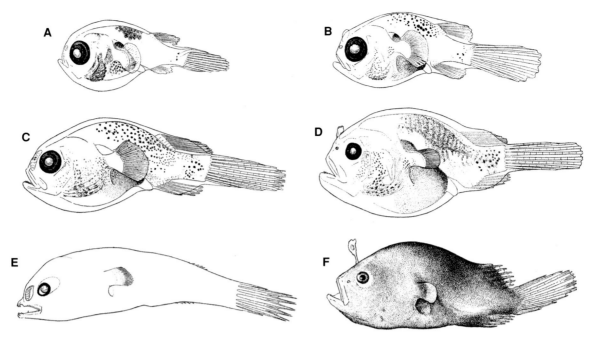

FIGURE 107. *Chaenophryne longiceps* group, early life-history stages: (A) larval male, 4.3 mm TL; (B) larval female, 8.1 mm TL; (C) larval male, 9.8 mm SL, 13.8 mm TL; (D) larval female, 13.3 mm SL, 19.0 mm TL; (E) juvenile male, 15.3 mm SL, 21 mm TL; (F) early postmetamorphic female, 13.2 mm SL, 18.0 mm TL (drawing by P.H. Winther). After Bertelsen (1951).

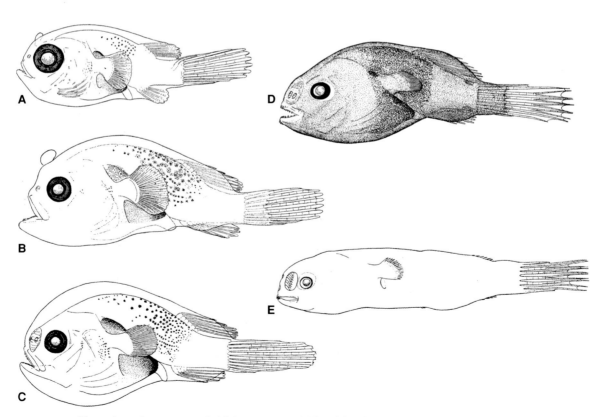

FIGURE 108. *Chaenophryne draco* group, early life-history stages: (A) larval female, 4.0 mm SL, 5.7 mm TL, ZMUC P92735; (B) larval female, 9.5 mm SL, 13.0 mm TL; (C) larval male, 9.0 mm SL, 12.4 mm TL; (D) early postmetamorphic male, 8.2 mm SL, 12.3 mm TL; (E) juvenile male, 15.3 mm SL, 20.5 mm TL. After Bertelsen (1951).

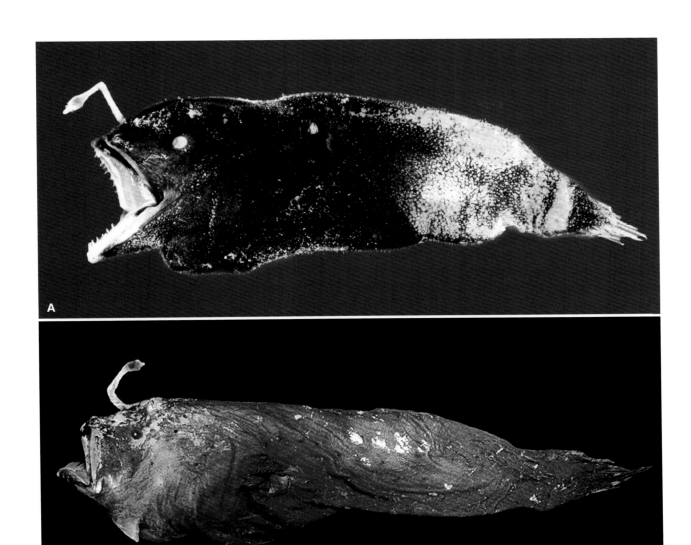

FIGURE 109. Species of *Spiniphryne*: (A) *S. gladisfenae*, 36.5 mm SL, MCZ 161504 (photo by C.P. Kenaley; courtesy of K. Hartel and the Museum of Comparative Zoology, Harvard University); (B) *S. duhameli*, holotype, 117 mm SL, SIO 60-239 (after Pietsch and Baldwin, 2006).

2500 m, 0452 to 1304 h, 6 July 1960 (Pietsch and Baldwin, 2006:405, figs. 1, 3, 5, 7).

### GENUS *ONEIRODES* LÜTKEN, 1871 (COMMON DREAMERS)

*Figures 112–115, Table 1*

*Oneirodes* is by far the largest genus of the suborder, with 35 currently recognized species—the next largest ceratioid genus is *Linophryne*, with 22 species, and after that *Gigantactis* and *Himantolophus*, each with 18—the result of a seemingly unending stream of original descriptions appearing on average almost every other year since the publication of Pietsch's (1974a) monograph, in which only 23 species were recognized. The taxonomic history of the genus begins with Lütken's (1871, 1872) description of *Oneirodes eschrichtii* based on a single specimen collected from the west coast of Greenland (Fig. 112). Since that time, and despite Lütken's detailed and accurate original description, the systematics of the genus has been confusing. Part of this confusion can be traced to an early failure to distinguish between *Oneirodes* and *Dolopichthys* Garman, 1899; problems that were not rectified until Bertelsen's (1951) revision of the suborder half a century later (see discussion of the family above).

Within the newly defined genus *Oneirodes*, Bertelsen (1951) listed 33 species, 28 of which were divided among three species groups. Within each of these groups

the separation of species is based on small details in the number, form, and relative size of the esca appendages. These differences are almost of the same dimensions as the individual and ontogenetic variation we find within well-defined species. Only four of the 28 species I place in these three groups are based on more than one specimen and none on more than four. The esca of the few specimens referred to the same species show differences, which do not seem essentially smaller than those used in the separation of the remaining species within the same group. As it is possible that each of the three groups embraces some few species, they may be designated, until closer examination of a larger material has been

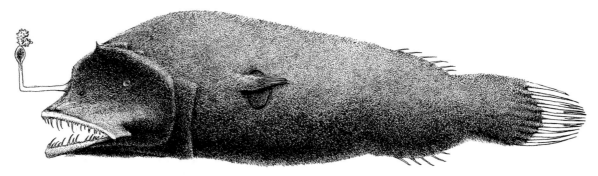

FIGURE 110. *Spiniphryne gladisfenae*, 63 mm SL, ISH 2734/71. Drawing by E.A. Hoxie; after Bertelsen and Pietsch (1975).

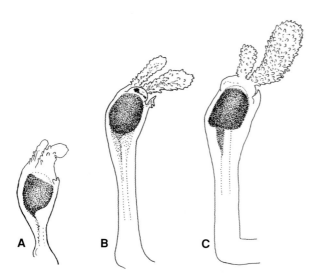

FIGURE 111. Escae of *Spiniphryne gladisfenae*, left lateral views: (A) 12.8 mm SL, BMNH 2004.8.24.7; (B) 49 mm SL, ISH 2131/71; (C) 63 mm SL, ISH 2734/71. After Bertelsen and Pietsch (1975).

made, as respectively: the *Oneirodes eschrichtii*, *flagellifer*, and *schmidti* groups. (Bertelsen, 1951:77)

With the advantage of a sixfold increase in the amount of material, Pietsch (1974a) revised the genus *Oneirodes* based on examination of approximately 450 metamorphosed male and female specimens, including all type material. Twenty-three species were recognized, seven of which were described as new. Eighteen nominal forms previously placed in the *O. eschrichtii* group were considered synonyms of *O. eschrichtii*. The remaining species of the *O. eschrichtii* group, *O. bulbosus*, *O. anisacanthus*, *O. heteronema*, and *O. theodoritissieri*, were considered valid. *Oneirodes inimicus* was a synonymized with *O. carlsbergi*, and *O. thysanophorus* with *O. flagellifer*. The *O. eschrichtii* and *O. flagellifer* groups were not recognized. The *O. schmidti* group, however, was retained to include a number of morphologically similar forms that are clearly differentiated from other species of *Oneirodes*: *O. mirus*, *O. schmidti*, *O. basili*, *O. theodoritissieri*, and additional unidentifiable material designated as *Oneirodes* sp. of *O. schmidti* group. Larvae, metamorphosing females, and males of all stages of development were listed as *Oneirodes* sp.

Despite the revision of Pietsch (1974a), the systematics of *Oneirodes* is by no means well understood. The separation of species is still based on very few characters, the most important being the morphology of the esca. Specimens with damaged or lost escae are usually impossible to identify. Characters that allow specific identification of males have not been found, thus species definitions are based solely on females. Even as restricted by Pietsch (1974a), *Oneirodes* is by far the largest of the 35 ceratioid genera, including almost a quarter of all known ceratioid species; yet, additional new forms remain to be described when adequate material becomes available, and the discovery of numerous additional species is predicted. While including descriptions of 14 additional species published in the last 35 years, the account provided here follows closely that of Pietsch (1974a).

### DISTINGUISHING CHARACTERS

Metamorphosed females of *Oneirodes* may be considered generalized oneirodids that exhibit few obvious external morphological features that clearly set them apart from other members of the family. The bodies of most species are short and deep, almost spherical, but some are narrow and elongate, approaching the shape of females of some other genera of the family, especially those of *Dolopichthys*. The illicium is usually short, not greater than about 33% SL, but exceeds 100% SL in some species. Tooth counts vary from 18 to 65 in the upper jaw, 18 to 160 in the lower jaw, and 4 to 14 on the vomer. The subopercle is usually short and broad, but narrow and elongate in some species. The skin appears smooth and naked, but at least some species have minute, widely spaced dermal spinules. But despite the great variation in most characters, the following combination of features more or less excludes all other oneirodids: sphenotic spines are well developed; the quadrate spine is also well developed and distinctly longer than the articular spine; the lower jaw bears a well-developed symphysial spine; the hyomandibula has a double head; darkly pigmented skin of the caudal peduncle terminates at the base of the caudal fin; the anal fin nearly always contains only 4 rays. Finally, in addition to these features, a quick glance at the esca will almost always allow accurate identification. Although the detailed morphology of the esca is extremely variable, all species of *Oneirodes* share a basic pattern of escal pigmentation and ornamentation that is unique for the genus: a distal arrow-shaped patch of pigment, flanked anteriorly and laterally by various filaments and appendages, and posteriorly by a low rounded terminal papillae, with an additional appendage arising posteriorly (see Figs. 113, 114).

Metamorphosed males of *Oneirodes* are unique in having the following combination of characters (Bertelsen, 1951:76, 77; Fig. 115E): The spaces between the anterior nostrils and

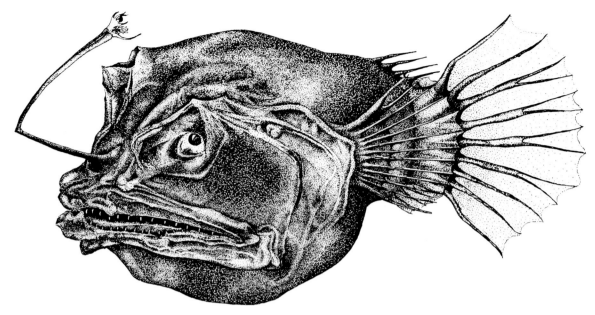

FIGURE 112. *Oneirodes eschrichtii*, 71 mm SL, LACM 31100-001. Drawing by J. Nakanishi; after Pietsch (1974a).

between the posterior nostril and the eye are pigmented. The septa between the anterior and posterior nostrils are unpigmented. There are 6 to 12 olfactory lamellae. There are 6 to 17 upper denticular teeth and 7 to 28 lower denticular teeth. The males are free-living, apparently never becoming parasitic on females.

Like the larvae of most ceratioids, those of *Oneirodes* can be distinguished by differences in subdermal pigmentation (Fig. 115): the anterior part of the body is distinctly pigmented, the melanophores distributed more or less evenly throughout, but terminating at the posterior insertion of the dorsal and anal fins, leaving the caudal peduncle unpigmented. The gill cover and the dorsal margin of the peritoneum are somewhat more darkly pigmented (Bertelsen, 1951:76, 77).

There are 5 to 7 dorsal-fin rays, 4 (very rarely 3 or 5) anal-fin rays, and 14 to 18 (very rarely 13 or 19) pectoral-fin rays (Table 1).

ETYMOLOGY

Lütken's (1871) *Oneirodes* is perhaps the most appropriate of the many early names assigned to deep-sea anglerfishes. It is derived from the Greek *oneiros*, meaning a "dream," "dreamlike," or "out of a dream," implying that this fish is so strange and marvelous that it could only be imagined in the dark of night during a state of unconsciousness.

DIVERSITY

Thirty-five recognized species compose by far the largest genus of the family as well as the suborder. Not surprisingly, it also has by far the broadest geographic range of any oneirodid genus and parallels that described above for the family as a whole:

*Oneirodes luetkeni* (Regan, 1925c): 58 metamorphosed females, 11.5 to 123 mm; Gulf of Panama and adjacent waters of the eastern Pacific Ocean; holotype *(Dolopichthys luetkeni)*, ZMUC P9287, 123 mm, *Dana* station 1203(10), 7°30′N, 79°19′W, 3500 m wire, bottom depth 2550 m, 1500 h, 1 November 1922 (Regan, 1925c:562; Beebe and Crane, 1947:159; Bertelsen, 1951:86, 87, figs. 31P–S, 40; Pietsch, 1974a:38, figs. 19, 28, 50, 61, 106).

*Oneirodes carlsbergi* (Regan and Trewavas, 1932): 30 metamorphosed females, 18 to 222 mm; tropical Atlantic and Pacific oceans between approximately 18°N and 9°S; lectotype *(Dolopichthys* [subgenus *Dermatias*] *carlsbergi)*, ZMUC P9285, 40 mm, *Dana* station 1206(7), Gulf of Panama, 6°40′N, 80°47′W, 1200 m wire, 1845 h, 15 January 1922 (Regan and Trewavas, 1932:76, fig. 115; Bertelsen, 1951:86, figs. 31M–O, 39; Pietsch, 1974a:39, figs. 62, 107; Bertelsen and Pietsch, 1977:172, fig. 1; Pietsch and Seigel, 1980:383; Jónsson and Pálsson, 1999:201, fig. 5).

*Oneirodes rosenblatti* Pietsch, 1974a: 21 metamorphosed females, 12.5 to 134 mm; eastern tropical Pacific Ocean; holotype, SIO 69-351, 94 mm, *Piquero* cruise 8, Gulf of Panama, 3°10′N, 84°10′W, 3-m Isaacs-Kidd midwater trawl, 0950 to 1453 h, 3 July 1969 (Pietsch, 1974a:41, figs. 63, 64, 108).

*Oneirodes eschrichtii* Lütken, 1871: 115 metamorphosed females, 10 to 213 mm; nearly cosmopolitan in the Atlantic, Pacific, and Indian oceans; holotype, ZMUC 64, 153 mm, west coast of Greenland (Lütken, 1871:56, figs. 1, 2, pl. 2, 1872:329, figs. 1, 2, pl. 9, 1949:34, figs. 13–17; Bertelsen and Pietsch, 1977:174, fig. 1; Pietsch and Seigel, 1980:383, fig. 3; Stearn and Pietsch, 1995:136, fig. 86; Swinney, 1995a:52, 55; Stewart and Pietsch, 1998:18, fig. 13; Jónsson and Pálsson, 1999:201, fig. 5; Anderson and Leslie, 2001:16, fig. 12B).

*Oneirodes sabex* Pietsch and Seigel, 1980: 24 metamorphosed females, 12 to 189 mm; southeast Asian, eastern Australian, and New Zealand waters; holotype, LACM 36116-3, 46 mm, *Alpha Helix* station 84, Banda Sea, 5°04.5′S, 130°12.0′E, 0 to 1500 m, 1400 to 2100 h, 28 April 1975 (Pietsch and Seigel, 1980:387, figs. 9, 10; Bertelsen and Pietsch, 1983:85, fig. 6; Ni, 1988:332, fig. 259; Meng et al., 1995:444, fig. 596; Stewart and Pietsch, 1998:17, 35, fig. 11).

*Oneirodes bulbosus* Chapman, 1939: 115 metamorphosed females, 30 to 160 mm; North Pacific Ocean and Bering Sea; holotype, USNM 108149, 57 mm, International Fisheries Commission station 1109C, 53°50′N, 133°54′W, 11 March

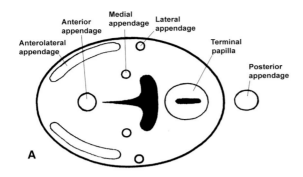

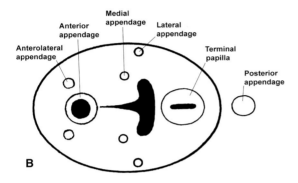

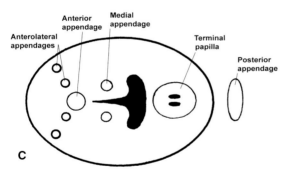

FIGURE 113. Escal appendage patterns of species of *Oneirodes*; diagrammatic representation of the dorsal surface of the escal bulb, indicating placement of appendages, papillae, and pigment. With one possible exception, each of the 37 recognized species of the genus falls easily within one of these patterns: (A) *O. luetkeni* and *O. rosenblatti*; (B) all species of the genus other than those with escae assigned to Pattern A and C; (C) the eight species of the *O. schmidti* group. Modified after Pietsch (1974a).

1934 (Chapman, 1939:538, fig. 70; Rass, 1955:334, 336; Pietsch, 1974a:52, figs. 71, 72, 109; Amaoka, 1983:250, 251, 325, fig. 141; Amaoka et al., 1995:113, fig. 173; Fedorov et al., 2003:55, fig. 83).

*Oneirodes anisacanthus* (Regan, 1925c): 11 metamorphosed females, 10.5 to 173 mm; North Atlantic Ocean; lectotype (*Dolopichthys anisacanthus*), ZMUC P9267, 27 mm, *Dana* station 1152(3), North Atlantic, 30°17′N, 20°44′W, 3000 m wire, 1930 h, 22 October 1922 (Regan, 1925c:562; Regan and Trewavas, 1932:72, text fig. 105, pl. 2, fig. 2; Maul, 1961:129, figs. 19–21; Pietsch, 1974a:54, figs. 73, 74, 106; Bertelsen and Pietsch, 1977:174, figs. 1, 2; Jónsson and Pálsson, 1999:201, fig. 5; Anderson and Leslie, 2001:15, fig. 12A).

*Oneirodes kreffti* Pietsch, 1974a: 38 metamorphosed females, 11 to 126 mm; southern latitudes of the Atlantic, western Pacific, and Indian oceans; holotype, ISH 1536/71, 50 mm, *Walther Herwig* station 431 III/71, 30°04′S, 5°22′E, CMBT-1600, 0 to 500 m, 2130 to 2252 h, 31 March 1971 (Pietsch, 1974a:57, figs. 75, 76, 107; Bertelsen and Pietsch, 1977:175, figs. 1, 3; Bertelsen and Pietsch, 1983:84, fig. 5; Stewart and Pietsch, 1998:17, figs. 10, 12; Anderson and Leslie, 2001:17, fig. 12D).

*Oneirodes posti* Bertelsen and Grobecker, 1980: three metamorphosed females, 36.5 to 135 mm; North Atlantic Ocean; holotype, ISH 3184/79, 112 mm, *Anton Dohrn* station 345/79, eastern North Atlantic, 35°24′N, 32°01′W, MT 1600, 0 to 1800 m (Bertelsen and Grobecker, 1980:63, fig. 1).

*Oneirodes myrionemus* Pietsch, 1974a: two metamorphosed females (43 to 121 mm), plus two tentatively identified specimens (76 to 137 mm); eastern North Atlantic; holotype, ISH 3100a/71, 43 mm, *Walther Herwig* station 512/71, 32°47′N, 16°24′W, CMBT-1600, 0 to 1800 m, 1945 to 2348 h, 22 April 1971 (Pietsch, 1974a:58, figs. 77, 78, 110; Bertelsen and Pietsch, 1977:176, fig. 1; Jónsson and Pálsson, 1999:202, fig. 5).

*Oneirodes clarkei* Swinney and Pietsch, 1988: a single known metamorphosed female, 119 mm; holotype, NMSZ 1986.005.1, 119 mm, *Challenger* cruise 14/83, station C83/2, eastern North Atlantic, 32°24′N, 16°51′W, RMT 10, two forward-facing spotlights on upper bar, 0 to 1500+ m, 2108 to 0032 h, 12 to 13 October 1983 (Swinney and Pietsch, 1988:1054, figs. 1–3, 4a).

*Oneirodes heteronema* (Regan and Trewavas, 1932): nine metamorphosed females, 13.5 to 119 mm; Gulf of Panama and Peru-Chile Trench; holotype (*Dolopichthys* [subgenus *Dermatias*] *heteronema*), ZMUC P92150, 13.5 mm, *Dana* station 1209(4), 7°15′N, 78°54′W, 2000 m wire, 1845 h, 17 January 1922 (Regan and Trewavas, 1932:72, fig. 106; Pietsch, 1974a:60, figs. 79, 80, 110).

*Oneirodes macrosteus* Pietsch, 1974a: 28 metamorphosed females, 11.5 to 185 mm; Atlantic Ocean; holotype, ROM 27262, 124 mm, *Brandal* tow 10, 49°00′N, 45°00′W, Engel trawl, 0 to 990 m, 1511 to 1830 h, 16 July 1968 (Pietsch, 1974a:61, figs. 81, 82, 110; Bertelsen and Pietsch, 1977:177, figs. 1, 4; Stearn and Pietsch, 1995:137, fig. 87; Jónsson and Pálsson, 1999:202, fig. 5).

*Oneirodes plagionema* Pietsch and Seigel, 1980: four metamorphosed females, 25 to 104 mm; Pacific Ocean; holotype, LACM 36114-2, 25 mm, *Alpha Helix* station 66, Banda Sea, 4°54′0S, 129°47.5′E, 1500 to 2000 m, 1515 to 1815 h, 17 April 1975 (Pietsch and Seigel, 1980:385, figs. 5, 6).

*Oneirodes cristatus* (Regan and Trewavas, 1932): three metamorphosed females, 20 to 165 mm; Banda and Celebes seas; lectotype (*Dolopichthys* [subgenus *Dermatias*] *cristatus*), ZMUC P9286, 165 mm, *Dana* station 3676(8), Banda Sea, 5°52′S, 131°14′E, 4000 m wire, bottom depth 7120 m, 0145 h, 23 March 1929 (Regan and Trewavas, 1932:67, fig. 93; Bertelsen, 1951:79, fig. 31C, D; Pietsch, 1974a:62, figs. 83, 108).

*Oneirodes pterurus* Pietsch and Seigel, 1980: a single metamorphosed female, 30 mm; holotype, LACM 36075-3, 30 mm, *Alpha Helix* station 121, Halmahera Sea, 0°41.7′S, 128°55.7′E, 1000 to 1400 m, 1730 to 1930 h, 16 May 1975 (Pietsch and Seigel, 1980:386, figs. 7, 8).

*Oneirodes acanthias* (Gilbert, 1915): 153 metamorphosed females, 11.5 to 167 mm; eastern North Pacific Ocean; holotype

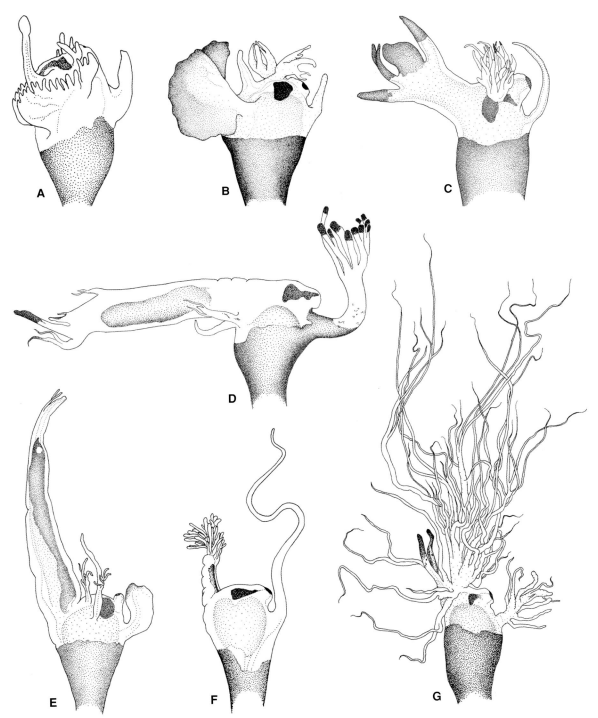

FIGURE 114. Escae of various species of *Oneirodes* selected to indicate the range of morphological variation within the genus, left lateral views: (A) *O. luetkeni*, 60 mm SL, LACM 31110-001; (B) *O. rosenblatti*, holotype, 94 mm SL, SIO 69-351; (C) *O. eschrichtii*, 118 mm SL, ISH 3048/71; (D) *O. macrosteus*, 182 mm SL, ISH 233/73; (E) *O. carlsbergi*, 62 mm SL, SIO 55-246; (F) *O. flagellifer*, holotype, 22 mm SL, ZMUC P9280; (G) *O. heteronema*, 91 mm SL, SIO 70-385. After Pietsch (1974a) and Bertelsen and Pietsch (1977).

*(Monoceratias acanthias)*, USNM 75825, 33 mm, *Albatross* station 4428, off Santa Cruz Island, southern California, 0 to 1629 m (Gilbert, 1915:379, pl. 22, fig. 24; Bolin and Myers, 1950:206–207; Bertelsen, 1951:85, fig. 38A, B; Lavenberg and Ebeling, 1967:195, fig. 5; Pietsch, 1974a:63, figs. 1–18, 30A, 84–86, 107).

*Oneirodes thompsoni* (Schultz, 1934): 108 metamorphosed females, 33 to 153 mm; North Pacific Ocean and Bering Sea; holotype, USNM 104495 (originally UW 2890, not UW 2530 as cited by Schultz, 1934), 33 mm, International Fisheries Commission haul 530, 54°13′N, 159°06′W, 2-m ring net, 0 to 900 m, 0858 h, 3 July 1931 (Schultz, 1934:66, figs. 1–4; Bertelsen,

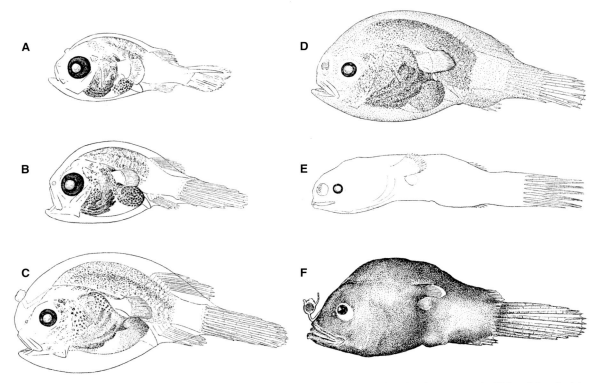

FIGURE 115. *Oneirodes* sp., early life history stages: (A) larval female, 3.0 mm TL; (B) larval male, 4.5 mm TL; (C) larval female, 8.0 mm SL, 12.0 mm TL, ZMUC P921073; (D) male in metamorphosis, 8.0 mm SL, 11.5 mm TL, ZMUC P9266; (E) adult male, 12.5 mm SL, ZMUC P921016; (F) early postmetamorphic female, 8.4 mm SL, 12.5 mm TL, ZMUC P921008 (drawing by P. H. Winther). After Bertelsen (1951).

1951:85, fig. 38C; Pietsch, 1974a:68, figs. 87, 88, 106; Amaoka, 1984:106, fig. 20, pl. 92-E; Mecklenburg et al., 2002:305, 308 [Alaska, in key]; Nakabo, 2002:472, figs. [Japan, in key]; Fedorov et al., 2003:55, fig. 84).

*Oneirodes notius* Pietsch, 1974a: 17 metamorphosed females, 30 to 150 mm; Southern Ocean; holotype, LACM 11165-9, 132 mm, *Eltanin* cruise 23, station 1615, 62°13'S, 95°39'W, 3-m Isaacs-Kidd midwater trawl, 0 to 1025 m, bottom depth 4914 m, 0610 to 0919 h, 9 April 1966 (Pietsch, 1974a:70, figs. 30B, 89, 90, 106; Bertelsen and Pietsch, 1977:177; Abe and Iwami, 1979:1, figs. 1–6).

*Oneirodes schistonema* Pietsch and Seigel, 1980: a single metamorphosed female, 74 mm; holotype, LACM 36036-3, 74 mm, *Alpha Helix* station 24, Banda Sea, 4°39.1'S, 129°53.7'E, 0 to 2000 m, 0115 to 0745 h, 28 March 1975 (Pietsch and Seigel, 1980:389, figs. 11, 12).

*Oneirodes flagellifer* (Regan and Trewavas, 1932): nine metamorphosed females, 10.5 to 22 mm; Indo-west Pacific Ocean; holotype *(Dolopichthys* [subgenus *Dermatias*] *flagellifer)*, ZMUC P9280, 22 mm, *Dana* station 3909(3), Indian Ocean off Sri Lanka, 5°21'N, 80°38'E, 3500 m wire, 1900 h, 22 November 1929 (Regan and Trewavas, 1932:74, fig. 111; Bertelsen, 1951:84, fig. 31J, K; Pietsch, 1974a:72, figs. 91, 110; Pietsch and Seigel, 1980:384).

*Oneirodes dicromischus* Pietsch, 1974a: two metamorphosed females, 21 to 35 mm; western and central Pacific Ocean; holotype, LACM 31463-1 35 mm, *Caride* cruise 3, station 59, central Pacific, 00°01'N, 139°06'W, 3-m Isaacs-Kidd midwater trawl, 0 to 840 m, 0816 h, 18 February 1969 (Pietsch, 1974a:73, figs. 92, 93, 108).

*Oneirodes thysanema* Pietsch and Seigel, 1980: two metamorphosed females, 13 to 26.5 mm; western North Atlantic and Banda Sea; holotype, USNM 207931, 26.5 mm, Ocean Acre cruise 7, station 13N, Bermuda, 32°18'N, 63°30'W, 3-m Isaacs-Kidd midwater trawl, 0 to 1500 m, 1430 to 1730 h, 8 September 1969 (Pietsch and Seigel, 1980:389, figs. 13, 14).

*Oneirodes haplonema* Stewart and Pietsch, 1998: two metamorphosed females, 35 to 116 mm; Tasman Sea, western South Pacific; holotype, NMNZ P-13409, 116 mm, off New Zealand, northern Challenger Plateau, 37°31.3'S, 169°31.9'E, 1132 to 1128 m, 23 February 1983 (Stewart and Pietsch, 1998:15, fig. 9).

*Oneirodes epithales* Orr, 1991: two metamorphosed females, 45 to 128 mm; western North Atlantic Ocean; holotype, ARC 8602571, 128 mm, vessel and cruise N067, off Newfoundland, 41°05'39"N, 56°25'33"W, 0 to 1829 m, 1 September 1986 (Orr, 1991:1024, figs. 1, 2).

*Oneirodes macronema* (Regan and Trewavas, 1932): two metamorphosed females, 16.5 to 27 mm; Caribbean Sea and off Oahu, Hawaii; holotype *(Dolopichthys* [subgenus *Dermatias*] *macronema)*, ZMUC P9282, 27 mm, *Dana* 1256(1), Caribbean Sea, 17°43'N, 64°56'W, 1000 m wire, 1920 h, 4 March 1922 (Regan and Trewavas, 1932:66, fig. 91; Pietsch, 1974a:75, figs. 96, 97, 110; Grobecker and Pietsch, 1978:547, figs. 1, 2).

*Oneirodes melanocauda* Bertelsen, 1951: five larval specimens: four females (6.5 to 15.5 mm) and one male (5.0 mm); Caribbean Sea, South China Sea, and Indian Ocean; holotype, ZMUC P9288, female, 15.5 mm, *Dana* station 3688(1), South China Sea, 6°55'N, 114°02'E, 4000 m wire, bottom depth 2900 m, 1700 h, 8 April 1929 (Bertelsen, 1951:76, 87, figs. 31L, 41; Pietsch, 1974a:76, fig. 108; perhaps the larvae of an undescribed ceratioid genus).

*Oneirodes pietschi* Ho and Shao, 2004: four metamorphosed females, 41.5 to 117 mm; North Pacific Ocean; holotype, ASIZP 61822, 100 mm, *Ocean Researcher I* station CD 191, off southwest coast of Taiwan, South China Sea, 21°22.18′N, 118°11.02′E, beam trawl, 1631 to 1635 m, 28 August 2002 (Ho and Shao, 2004:74, figs. 1–3).

*Oneirodes alius* Seigel and Pietsch, 1978: seven metamorphosed females, 10 to 38 mm; Halmahera Sea; holotype, LACM 36026-1, 38 mm, *Alpha Helix* station 122, Halmahera Sea, 0°36.3′S, 129°03.2′E, RMT-8, 575 to 600 m, 2240 to 2340 h, 16 May 1975 (Seigel and Pietsch, 1978:11, figs. 1, 2; Pietsch and Seigel, 1980:390, fig. 15; Orr, 1991:1025).

*Oneirodes micronema* Grobecker, 1978: two metamorphosed females, 17 to 89 mm; Banda Sea; holotype, LACM 36039-3, 89 mm, *Alpha Helix,* Southeast Asian Bioluminescence Expedition, station 113, Banda Sea, 5°07.5′S, 130°08.4′E, 650 to 1000 m, 2120 to 2255 h, 13 May 1975 (Grobecker, 1978:567, figs. 1, 2; Pietsch and Seigel, 1980:393, fig. 18).

*Oneirodes schmidti* (Regan and Trewavas, 1932): five metamorphosed females, 15.5 to 92 mm; Banda Sea; holotype, *Dolopichthys* (subgenus *Dermatias*) *schmidti*, ZMUC P9284, 32 mm, *Dana* station 3678(1), Banda Sea, 4°05′S, 128°16′E, 5000 m wire, bottom depth 4700 m, 1840 h, 24 March 1929 (Regan and Trewavas, 1932:75, fig. 113; Bertelsen, 1951:84, fig. 311; Pietsch, 1974a:78, figs. 98A, 99, 111; Pietsch and Seigel, 1980:391, figs. 16, 17).

*Oneirodes mirus* (Regan and Trewavas, 1932): a single metamorphosed female, 42 mm; holotype, *Dolopichthys* (subgenus *Dermatias*) *mirus*, ZMUC P9283, 42 mm, *Dana* station 3828 (10), off west coast of Sumatra, 1°22′N, 96°06.5′E, 3000 m wire, bottom depth 4980 m, 1600 h, 18 September 1929 (Regan and Trewavas, 1932:74, fig. 112; Bertelsen, 1951:84, fig. 31H; Grey, 1956a:246; Pietsch, 1974a:79, figs. 98B, 100, 111).

*Oneirodes basili* Pietsch, 1974a: three metamorphosed females, 95 to 159 mm; off southern California and Guadalupe Island, Mexico; holotype, LACM 30020-34, 95 mm, *Velero IV* station 11635, 28°08′N, 117°31′W, 3-m Isaacs-Kidd midwater trawl, 0 to 700 m; bottom depth 3520 to 3493 m, 2340 to 0430 h, 20 August 1967 (Pietsch, 1972a:42, 43, 45, fig. 24(5), 1974a:79, figs. 20, 29, 98C, 101, 111).

*Oneirodes theodoritissieri* Belloc, 1938: three metamorphosed females, 58 to 183 mm; eastern North Atlantic off Portuguese Guinea and the Cape Verde Islands; holotype, MHNLR P316-448, 64 mm, *President Theodore Tissier* cruise 5, station 733, Bissagos, Guinee-Bissau, 11°13′N, 17°26′W, Schmidt net, 1000 m wire, bottom depth 1460 m, 27 May 1936 (Belloc, 1938:303, figs. 23 to 25; Bertelsen, 1951:79; Grey, 1956a:245; Pietsch, 1974a:80, figs. 98D, 102, 111; Bertelsen and Pietsch, 1977:178).

*Oneirodes bradburyae* Grey, 1956: a single known metamorphosed female, 23.5 mm; holotype, USNM 164359, 23.5 mm, *Oregon* station 1028, Gulf of Mexico, 28°28′N, 87°18′W, 0 to 1426 m, 21 April 1954 (Grey, 1956b:245, fig. 2; Pietsch, 1974a:74, figs. 94, 95, 107; McEachran and Fechhelm, 1998:867, 869, fig.).

## GENUS *DERMATIAS* SMITH AND RADCLIFFE, IN RADCLIFFE, 1912 (FATTAIL DREAMERS)

*Figures 116, 117, Table 1*

The genus *Dermatias* and its type species *Dermatias platynogaster* were described and illustrated in accurate detail by Smith and Radcliffe (in Radcliffe, 1912) at a time when the Oneirodidae was represented by only three monotypic genera, *Oneirodes* Lütken, 1871; *Dolopichthys* Garman, 1899; and *Lophodolos* Lloyd, 1909a. Said to appear "more closely related to *Dolopichthys*," Smith and Radcliffe argued that the "form of the head and body and of the illicium and the better developed fins are distinctive." Since that time, only passing references to *Dermatias platynogaster* have appeared: Regan (1926), Parr (1927), and Regan and Trewavas (1932) included it in their monographic works on the Ceratioidei, but added nothing new. Bertelsen (1951), unable at the time to examine the holotype, placed it, along with 21 other nominal species, in his *Oneirodes eschrichtii* group, concluding that it was probably a synonym of *O. eschrichtii* Lütken, 1871. Pietsch (1974a), based on first-hand observation, went a step further and, without comment, formally placed it in the synonymy of *O. eschrichtii*. Thus, for the last 30 years the name *Dermatias platynogaster* has been buried and forgotten within the large and complex synonymy of *O. eschrichtii* (Pietsch, 1974a).

More recently, Kharin (1989) described the genus *Pietschichthys* and its type species *Pietschichthys horridus* on the basis of a single specimen collected from the Magellan Seamounts in the western North Pacific Ocean. In distinguishing *Pietschichthys* from other oneirodid genera, Kharin (1989:158) listed a number of characters, the most significant of which were features of the pectoral fin: "lobe of pectoral fin long and narrow, longer than its longest rays," thus apparently establishing an affinity of his new genus with the "long-pectoraled genera" of Pietsch (1978), which includes *Chirophryne*, *Leptacanthichthys*, and *Ctenochirichthys*, all originally described by Regan and Trewavas (1932), and *Puck* described by Pietsch (1978). Close examination of the holotype, however, showed that the pectoral lobe of *Pietschichthys* is in fact rather short, less than 9% SL, whereas the longest rays of the pectoral fin are about 20% SL (Pietsch and Kharin, 2004). Nevertheless, *Pietschichthys*, in other ways, was found to be unique among recognized oneirodid genera and, in addition, indistinguishable from Smith and Radcliffe's *Dermatias*. Superficially, *Dermatias* appears most similar to *Oneirodes*—possessing among other features, a surprisingly similar escal morphology—but it differs in a number of ways, the most significant of which are described below. Except for the addition of data describing a third specimen (Kharin and Pietsch, 2007), the following account is based solely on the review of Pietsch and Kharin (2004).

### DISTINGUISHING CHARACTERS

Metamorphosed females of *Dermatias* may be most easily confused with those of *Oneirodes*, but they differ from the latter and from those of all other genera of the family in having an unusually deep caudal peduncle; a blunt snout and short, highly convex frontal bones resulting in an extremely short head; and few teeth in the jaws, only 20 to 32 in the upper jaw and 20 to 31 in the lower jaw (Fig. 116). The mouth is small, the opening almost horizontal, and the cleft extending posteriorly slightly past the eye. The opercular opening is unusually large, while the illicium is rather short. Like most other genera of the family, *Dermatias* has well-developed sphenotic and quadrate spines, the lower jaw bears a stout symphysial spine, the posterior margin of the opercle is deeply notched, the subopercle is long and narrow, its dorsal end tapering to a point, its ventral end oval in shape; the skin is smooth and

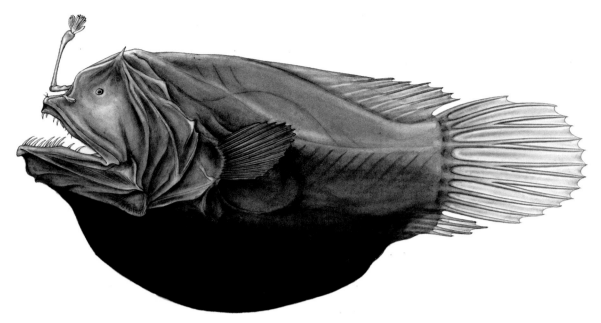

FIGURE 116. *Dermatias platynogaster*, holotype, 134 mm SL, USNM 70269. After Smith and Radcliffe (in Radcliffe, 1912).

apparently naked, without dermal spinules. The morphology of the escal is remarkably similar to that of some species of *Oneirodes* (Fig. 117).

Males and larvae are unknown.

There are 6 dorsal-fin rays, 4 anal-fin rays, and 15 to 16 pectoral-fin rays (Table 1).

ETYMOLOGY

Smith and Radcliffe (in Radcliffe, 1912:206) chose the name *Dermatias*, taken from the Greek *derma* or *dermatos*, meaning "skin" or "leather," to draw attention to the integument of their new genus, which they described as "naked, very loose, and soft."

DIVERSITY

The genus contains a single species:

*Dermatias platynogaster* Smith and Radcliffe, in Radcliffe, 1912: three metamorphosed females, 134 to 175 mm; western North Pacific Ocean; holotype, USNM 70269, 134 mm, *Albatross* station 5463, near Sialat Point Light, off east coast of Luzon, Philippine Islands, western North Pacific, 13°40'57"N, 123°57'45"E, beam trawl, 300 fathoms (549 m) (Smith and Radcliffe, in Radcliffe, 1912:206, 207, pl. 17, fig. 3; Bertelsen, 1951:79, 81; Pietsch, 1974a:44, 53, 102, table 9; Kharin, 1989:158–160, figs. 1, 2; Pietsch and Kharin, 2004:123, figs. 1–4).

GENUS *DANAPHRYNE* BERTELSEN, 1951
(DANA DREAMERS)

*Figures 118–120, Table 1*

The species for which Bertelsen (1951:101) established his genus *Danaphryne* was originally described by Regan and Trewavas (1932:67) as *Dolopichthys nigrifilis*, based on a single female specimen collected by the *Dana* in the South

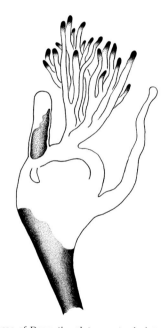

FIGURE 117. Escae of *Dermatias platynogaster*, holotype, 134 mm SL, USNM 70269, left lateral view. After Pietsch and Kharin (2004).

China Sea. Waterman (1939b:89), while recognizing a close similarity to *D. nigrifilis*, described *D. albifilosa*, again from a single specimen, this one collected by the *Atlantis* in the western North Atlantic. A decade or so latter, Bertelsen (1951:102) synonymized the two species, describing the series of differences cited by Waterman as "minor," and rejecting some as "only apparent and due to faulty description by Regan and Trewavas, whilst the reminder . . . all fall within the ontogenetic and individual variation one might expect to find." At the same time, Bertelsen (1951) discovered additional characters to warrant placement within a

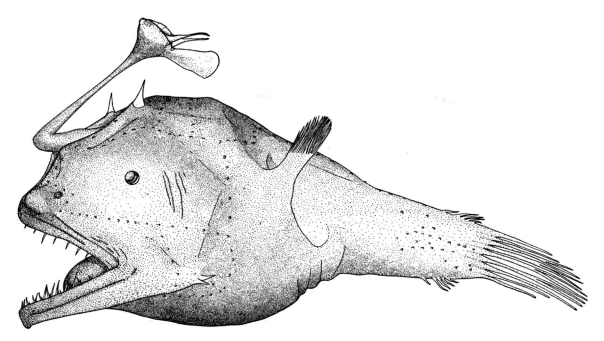

FIGURE 118. *Danaphryne nigrifilis*, the holotype of *Dolopichthys albifilosus*, 33 mm SL, MCZ 35067. Courtesy of the Museum of Comparative Zoology, Harvard University; after Waterman (1939b).

new genus, which he named *Danaphryne* in honor of the Danish research vessel that had been responsible for amassing by far the largest and most significant collection of deep-sea organisms up until that time.

Bertelsen and Pietsch (1977), in describing oneirodids collected during South Atlantic cruises of the German Fisheries Research Vessel *Walther Herwig*, added six additional females to the known material of *Danaphryne*, confirming the validity of a single widely distributed genus. Since that time, the number of metamorphosed females has increased to 20 (20 to 105 mm), in addition to four larvae (three females and one male). Metamorphosed males remain unknown.

## DISTINGUISHING CHARACTERS

At first glance, there is not much about *Danaphryne* to separate it from its congeners. The body of metamorphosed females is relatively short and globular to moderately fusiform; the mouth is on the small side, the cleft not extending past the eye; sphenotic and quadrate spines are well developed; and the lower jaw bears a well-developed symphysial spine—all features readily found within many other oneirodids (Fig. 118). However, when viewed anteriorly, females have an unusually narrow snout; the width of the ethmoid cartilage and vomer is considerably less than the distance between the anterolateral tips of the lateral ethmoids and frontals (see Pietsch, 1974a:16). The shape of the opercular bones is also unique: the upper fork of the opercle is unusually short compared to the lower fork, especially in smaller individuals; the subopercle is long and narrow, its dorsal end unusually long and slender, tapering to a point, its ventral end nearly circular, without an anterior spine or projection (see Bertelsen, 1951:102). The esca is especially large and expanded anterodorsally, bearing a distal pair of stout posteriorly directed tentacle-like filaments, and a large, laterally compressed posterior escal appendage, with a more or less spherical pigmented swelling on its ventral margin (Fig. 119).

Like the adults, larval females of *Danaphryne* are best distinguished from those of other genera of the family by the shape of the opercular bones: the opercle is notched posteriorly, its upper fork much broader than the lower fork; the dorsal end of the subopercle tapers to a point, its ventral end is broad and rounded, without a spine or projection on the anterior margin. Subdermal pigment covers the entire head and body except for the snout, pectoral-fin lobe, and caudal peduncle (Fig. 120). The pigment is heaviest on the anterior part of the lower jaw and along the ventral and posterior margin of the gill cover (Bertelsen, 1951:123).

Metamorphosed males are known.

There are 5 to 7 (rarely 8) dorsal-fin rays, 5 (rarely 4) anal-fin rays, and 16 to 19 pectoral-fin rays (Table 1).

## ETYMOLOGY

*Danaphryne* was named by Bertelsen (1951) to commemorate the extraordinary oceanographic contributions of the Royal Danish Research Vessel *Dana*.

## DIVERSITY

The genus contains a single species:

*Danaphryne nigrifilis* (Regan and Trewavas, 1932): 20 metamorphosed females (20 to 105 mm); and four larvae, three females (9 to 17 mm TL) and one male (9.5 mm); North Atlantic and western Pacific oceans; holotype *(Dolopichthys nigrifilis)*, ZMUC P92102, 24 mm, *Dana* station 3716, South China Sea, 19°18.5′N, 120°13′E, 3000 m wire, 1400 h, 1400 h, 22 May 1929 (Regan and Trewavas, 1932:67, fig. 92; Waterman, 1939b:89, figs. 5, 6; Bertelsen, 1951:102, 123, figs. 58, 59, 82, 83; Pietsch, 1974a:16–21, 24–28, 32, figs. 21, 31, 39A, 40, 51A, 52; Bertelsen and Pietsch, 1977:183, 185, fig. 8; Stearn and Pietsch, 1995:135, fig. 85).

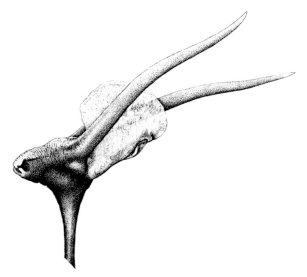

FIGURE 119. Esca of *Danaphryne nigrifilis*, 73 mm SL, ISH 2731/71. After Bertelsen and Pietsch (1977).

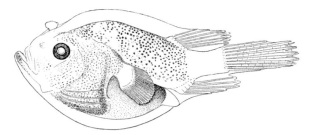

FIGURE 120. *Danaphryne nigrifilis*, larval female, 12.4 mm SL, ZMUC P921005. After Bertelsen (1951).

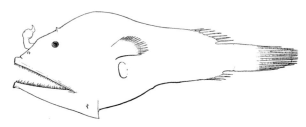

FIGURE 121. *Microlophichthys microlophus*, holotype, 26 mm SL, ZMUC P9292. After Regan and Trewavas (1932).

## GENUS *MICROLOPHICHTHYS* REGAN AND TREWAVAS, 1932 (SHORTBAIT DREAMERS)

*Figures 121–124, Table 1*

*Microlophichthys* was originally described by Regan and Trewavas (1932:77) as a subgenus of *Dolopichthys* to encompass four nominal species, each based on one to three small females. Later, given full generic status, the genus was revised by Bertelsen (1951:88) who placed all four species within the synonymy of the type species *Microlophichthys microlophus*. At the same time, he described larvae and males of *M. microlophus*, and introduced a new species, *M. andracanthus*, on the basis of a single male specimen. The osteology of the genus was described by Pietsch (1974a:15–29) and four additional females were reported by Bertelsen and Pietsch (1977:185), bringing the number of females known at the time to 22. Today, in collections around the world, *Microlophichthys* is represented by nearly 100 metamorphosed females and about the same number of larvae. Metamorphosed males, however, are still quite rare, known from only five specimens.

### DISTINGUISHING CHARACTERS

Metamorphosed females of *Microlophichthys* are rather easily recognized by the unusually short illicium, much of its length enveloped by tissue of the esca, the latter appearing to sit directly on the surface of the snout (Fig. 121). Also helpful in identifying females are unusually high toothcounts, which increase almost exponentially with the size of the specimen. In fact, the jaw teeth are so numerous, with many tiny teeth set so extremely close together, that counting is extremely difficult: there are about 160 to 320 teeth in the upper jaw, about 100 to 180 in the lower jaw, and 4 to 12 teeth on the vomer.

Like nearly all oneirodid genera, the shape of the opercular bones is significant in differentiating both females and males of *Microlophichthys*: the subopercle is short and broad, the dorsal end rounded or tapering to a blunt point, the ventral end nearly circular, without an anterior spine or projection. Finally, the esca is unique in having a truncated distal escal papilla and a more or less pointed compressed posterior escal appendage (Fig. 122).

Other than these few characters, females are similar to other oneirodids: the body is elongate and slender, not globular; the snout is relatively short, but the mouth is large, its cleft extending posteriorly well past the eye (Fig. 121). Sphenotic and quadrate spines are prominent, and the lower jaw bears a well-developed symphysial spine. The skin is smooth and naked, without dermal spinules, and darkly pigmented skin of the caudal peduncle extends well past the base of the caudal fin.

Metamorphosed males are best distinguished by the shape of the subopercle, which is short and broad, the dorsal end rounded (Fig. 123). There are 8 to 10 upper denticles and 8 lower denticles. The skin between the nostrils is only slightly pigmented, the posterior nostril well separated from the eye by a pigmented bridge of skin. The medial surface of the subopercle is darkly pigmented. The caudal peduncle bears a distinct subdermal group of melanophores (Bertelsen, 1951:75, 89).

The body of larvae is relatively elongate and slender (Fig. 124). The dorsal end of the subopercle is rounded, its medial surface with a dark spot of pigment. The branchiostegal pigment is faint. The caudal peduncle is pigmented, the melanophores forming a single isolated group on the lateral surface (Bertelsen, 1951:75, 89).

There are 5 to 7 dorsal-fin rays, 5 (rarely 4 or 6) anal-fin rays, and 18 to 23 pectoral-fin rays (Table 1).

### ETYMOLOGY

The name *Microlophichthys* was chosen by Regan and Trewavas (1932) to describe the unusually short illicium of this genus. It is derived from the Greek roots *mikros*, meaning "small"; *lophos*, a "crest" or "tuft"; and *ichthys*, a "fish"; thus, in this case, a "fish with a tiny lure."

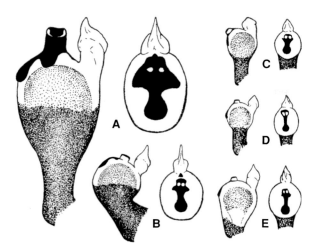

FIGURE 122. Escae of *Microlophichthys microlophus*: (A) 47.5 mm SL, ZMUC P92931; (B) holotype, 26 mm SL, ZMUC P9292; (C) syntype of *Dolopichthys implumis*, 15 mm SL, ZMUC P9291; (D) holotype of *Dolopichthys exiguus*, 13 mm SL, ZMUC P9289; (E) 16 mm SL, ZMUC P92935. After Bertelsen (1951).

FIGURE 123. *Microlophichthys andracanthus*, holotype, adult male, 16.5 mm SL, ZMUC P9293. After Bertelsen (1951).

## DIVERSITY

The genus contains two species, one based on females, males, and larvae; the other based solely on two adult males:

*Microlophichthys microlophus* (Regan, 1925c): 94 metamorphosed females (11.5 to 112 mm), five males (6.8 to 17 mm), and 99 larvae (63 females, 32 males, and four undetermined, 2.4 to 12.5 mm TL); widely distributed in the Atlantic, Pacific, and Indian oceans; holotype *(Dolopichthys microlophus)*, ZMUC P9292, female, 26 mm, *Dana* station 1159(3), western North Atlantic, 17°55′N, 24°35′W, 3000 m wire, 0450 h, 29 October 1921 (Regan, 1925c:563; Parr, 1927:20, fig. 7; Regan and Trewavas, 1932:77, 78, figs. 118, 119, 121, 122; Beebe and Crane, 1947:160, fig. 5; Bertelsen, 1951:90, figs. 42–46; Pietsch, 1974a:16, 24, 32, figs. 32, 39B, 41, 51B, 53A, 54, 56A; Bertelsen and Pietsch, 1977:185; Pietsch and Seigel, 1980:394).

*Microlophichthys andracanthus* Bertelsen, 1951: two adult males, 16 to 17 mm; Caribbean Sea and in eastern tropical Pacific; holotype, ZMUC P9293, 17 mm, *Dana* station 1269(1), Caribbean Sea, 17°13′N, 64°58′W, 4500 m wire, 1800 h, 15 March 1922 (based on the considerable differences between these two males and those of *M. microlophus*, they could represent an undescribed genus of the family; see Bertelsen, 1951:92, 93, fig. 47).

## GENUS *PHYLLORHINICHTHYS* PIETSCH, 1969 (LEAFYSNOUT DREAMERS)

*Figures 125–127, Table 1*

The deep-sea anglerfish genus *Phyllorhinichthys* was erected by Pietsch (1969) to contain a single species, *Phyllorhinichthys mi-*

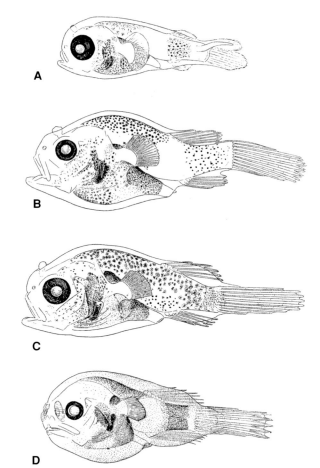

FIGURE 124. Larvae of *Microlophichthys microlophus*: (A) larval female, 4.0 mm TL; (B) larval female, 9.0 mm SL, 12.1 mm TL; (C) larval female, 9.0 mm SL, 12.5 mm TL, ZMUC P92976; (D) male in metamorphosis, 6.8 mm SL, 10.3 mm TL. After Bertelsen (1951).

*cractis*, described on the basis of a single 52-mm female, collected in the eastern North Pacific Ocean, off the northern end of Guadalupe Island, Mexico. Since that time, a second female (96 mm), from off Japan, was found in the collections of the Institute of Oceanology of the Russian Academy of Sciences, Moscow (since transferred to the Zoological Institute, St. Petersburg), and compared with the holotype (Pietsch, 1972b). More recently, three additional females (10.8 to 118 mm) were reported by Bertelsen and Pietsch (1977). Of these latter specimens, the largest differed significantly from the remaining four in details of escal morphology, especially in having a much longer distal escal appendage, variation attributed at that time to ontogenetic change. However, several newly discovered females, four of which represent the largest known specimens of the genus (120 to 132 mm), do not fit this pattern of growth and indicate instead the presence of a new species recently described by Pietsch (2004).

## DISTINGUISHING CHARACTERS

*Phyllorhinichthys* was originally distinguished from other oneirodid genera primarily by having an unusual pair of compressed, leaflike appendages on each side of the snout just anteroventral to the eye (Pietsch, 1969; Fig. 125A). Additional specimens, however, subsequently described by Pietsch

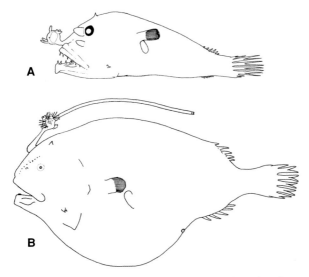

FIGURE 125. Species of *Phyllorhinichthys*, females: (A) *P. micractis*, 10.8 mm SL, BMNH 2001.2.1.75; (B) *P. balushkini*, holotype, 118 mm SL, ISH 536/73. After Bertelsen and Pietsch (1977) and Pietsch and Baldwin (2006).

(1972b) and Bertelsen and Pietsch (1977), showed these structures to be highly variable in size and development, becoming reduced with growth and eventually lost in the largest known specimens. Thus, while the "snout flaps" are still unique to *Phyllorhinichthys* and therefore useful in identification, additional characters are required to adequately define the genus. While having all those features that generally describe oneirodids—well-developed sphenotic and quadrate spines, a well-developed symphysial spine of the lower jaw, the hyomandibula with a double head, the posterior margin of the opercle deeply notched, the skin smooth and naked, without dermal spinules, and darkly pigmented skin of the caudal peduncle extending well past the base of the caudal fin—the snout of metamorphosed females of *Phyllorhinichthys* is unusually short; the dorsal margin of the frontal bones is strongly convex; and the illicium is short, nearly completely enveloped by the escal bulb to about three times the length of the escal bulb in the largest known specimens. Females are also characterized by having relatively few teeth in the jaws (upper jaw 10 to 39, lower jaw 8 to 21) and a unique escal morphology, consisting of 1 or 2 stout, internally pigmented anterior appendages; a single internally pigmented distal appendage; and a short, unpigmented, wedge-shaped posterior appendage (Figs. 126, 127).

Males and larvae are unknown.

There are 5 (rarely 6) dorsal-fin rays, 5 anal-fin rays, and 19 to 24 pectoral-fin rays (Table 1).

### ETYMOLOGY

The name *Phyllorhinichthys* is derived from the Greek *phyllon*, meaning "leaf"; *rhinos*, meaning "nose" or "snout"; and *ichthys*, a "fish," in allusion to the leaflike snout flaps characteristic of this genus.

### DIVERSITY

The genus contains two species:

*Phyllorhinichthys micractis* Pietsch, 1969: 11 metamorphosed females, 8.8 to 140 mm; Atlantic, Pacific, and Indian oceans;

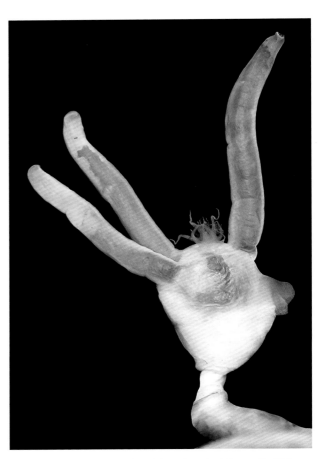

FIGURE 126. Esca of *Phyllorhinichthys micractis*, 120 mm SL, ARC 8602570, left lateral view. After Pietsch and Baldwin (2006).

holotype, LACM 9567-14, female, 52 mm, *Velero IV* cruise 832, station 11187, eastern North Pacific, 22 km off northern end of Guadalupe Island, Mexico, 29°16'30" to 29°35'30"N, 118°11'30" to 118°18'30"W, 3-m Isaacs-Kidd midwater trawl fished open between surface and 1050 m, bottom depth 2194 to 3429 m, 1315 to 2103 h, 3 August 1966 (Pietsch, 1969:365, figs. 1–4, 1972b:335, figs. 1–6, 2004:800, figs. 1, 4; Bertelsen and Pietsch, 1977:178, figs. 5–7).

*Phyllorhinichthys balushkini* Pietsch, 2004: seven metamorphosed females, 72 to 168 mm; Atlantic Ocean; holotype, ISH 536/73, female, 118 mm, *Walther Herwig*, "Overflow 73" Expedition, cruise 51, station 695/73, eastern North Atlantic, 55°43'N, 25°53'W, MT-1600, between the surface and 2600 m, bottom depth 3210 m, 1837 to 2130 hrs, 22 September 1973 (Bertelsen and Pietsch, 1977:178, figs. 5–7; Pietsch, 2004:801, figs. 2–4).

## GENUS *TYRANNOPHRYNE* REGAN AND TREWAVAS, 1932 (PUGNACIOUS DREAMERS)

*Figure 128, Table 1*

Certainly one of the more extraordinary members of the suborder, remarkable for its huge mouth and long jaws, *Tyrannophryne* and its type species, *Tyrannophryne pugnax*, were originally described by Regan and Trewavas (1932) on the basis of a single 12-mm female, collected in 1928 by the *Dana* off Tahiti. This tiny specimen remained the only known

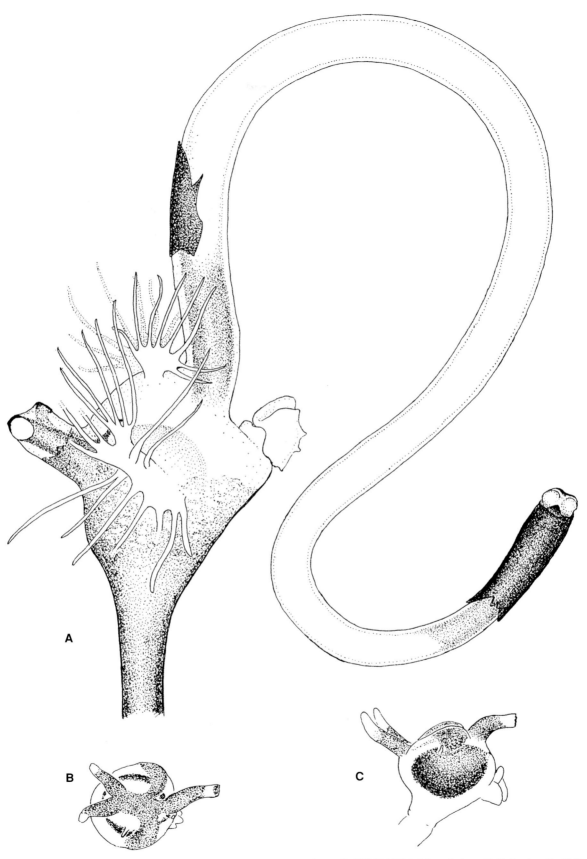

FIGURE 127. Escae of *Phyllorhinichthys*: (A) *P. balushkini*, holotype, 118 mm SL, ISH 536/73, left lateral view; (B, C) *P. micractis*, 10.8 mm SL, BMNH 2001.2.1.75, dorsal and left lateral views. After Bertelsen and Pietsch (1977).

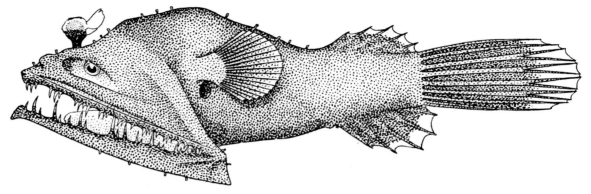

FIGURE 128. *Tyrannophryne pugnax*, holotype, 12 mm SL, ZMUC P9294. Drawing by W.P.C. Tenison; after Regan and Trewavas (1932).

individual until 1956 when a second small female (16.5 mm) was collected by the RV *Horizon* near the Marshall Islands; and, more recently (2006), a much larger female (50 mm), by commercial shrimpers off Taiwan. Unfortunately, with so few individuals available, and with males and larvae still unknown, very little is known about its internal anatomy and general biology.

### DISTINGUISHING CHARACTERS

In addition to the extremely large, oblique mouth—the elements of the upper and lower jaws extending posteriorly far beyond the base of the pectoral fin and opercular opening—metamorphosed females of *Tyrannophryne* are distinguished from those of all other genera of the family by having the following combination of features (Fig. 128): The body is elongate and slender, not globular, the snout relatively short. Vomerine teeth are present or absent, apparently lost with growth. The second pharyngobranchial is toothless and perhaps absent. Like most oneirodids, sphenotic and quadrate spines are prominent, and the lower jaw bears a well-developed symphysial spine, but, in addition, the angular bone is elongate and tapering, forming a long narrow spine, reminiscent of that of *Lophodolos*. The illicium is extremely short, almost totally enveloped by tissue of the esca in smaller specimens. The esca has a medial, distal streak of pigment, forked posteriorly (Fig. 128). A terminal escal papilla is absent. A large, laterally compressed posterior escal appendage is present, bearing a small distal cleft and a low anterior crest. The skin is naked, without dermal spinules. Darkly pigmented skin of the caudal peduncle extends well past the base of the caudal fin.

Larvae and metamorphosed males are unknown.

There are 5 dorsal-fin rays, 5 anal-fin rays, and 18 to 20 pectoral-fin rays (Table 1).

### ETYMOLOGY

Regan and Trewavas (1932), in naming their new genus, must have thought it looked rather menacing, combining the Greek roots *tyrannos* (or Latin *tyrannus*), meaning "tyrant," and *phryne*, to produce "tyrannical toad." The common name for the genus, Pugnacious Dreamer, is derived from the specific name of the type species of the genus, *T. pugnax*.

### DIVERSITY

A single species is recognized:

*Tyrannophryne pugnax* Regan and Trewavas, 1932: three metamorphosed females, 12 to 50 mm; western and central Pacific; holotype, ZMUC P9294, 12 mm, *Dana* station 3577(7), South Pacific near Tahiti, 18°49′S, 153°10′W, 4000 m wire, 0615 to 0815 h, 19 October 1928 (Regan and Trewavas, 1932:83, pl. 4, fig. 1; Pietsch, 1974a:109).

## GENUS *DOLOPICHTHYS* GARMAN, 1899 (LONGSNOUT DREAMERS)

*Figures 129–132, Table 1*

Prior to Bertelsen's (1951) monograph on the Ceratioidei, the genus *Dolopichthys* Garman, 1899, included a total of 43 described species assigned by Regan and Trewavas (1932) to five subgenera: *Dermatias*, *Microlophichthys*, *Dolopichthys*, *Leptacanthichthys*, and *Pentherichthys*. Within their subgenus *Dolopichthys*, Regan and Trewavas (1932) recognized 10 female specimens distributed among seven species: *Dolopichthys allector* Garman, 1899; *D. niger* Brauer, 1902; *D. danae* Regan, 1926; *D. longicornis* Parr, 1927; *D. pullatus* Regan and Trewavas, 1932; *D. jubatus* Regan and Trewavas, 1932; and *D. mucronatus* Regan and Trewavas, 1932. Bertelsen (1951) drastically amended this situation by placing the subgenus *Dermatias* in synonymy with the genus *Oneirodes* Lütken, 1871, and elevating the remaining four subgenera to generic status. Within the newly defined genus *Dolopichthys*, Bertelsen (1951) recognized four species, placing *D. mucronatus*, *D. pullatus*, and *D. jubatus* in the synonymy of *D. longicornis*.

In the 20 years between the publication of Bertelsen's (1951) monograph and Pietsch's (1972c) revision of *Dolopichthys*, the number of female specimens of the genus had increased more than sevenfold. Much of the new material was in excellent condition, making it possible to give a more thorough description, to say something about morphological variation and geographic distribution, and to discuss possible relationships within the genus. Despite this greatly increased knowledge, however, the taxonomy is by no means obvious. Characters that allow specific identification of males have not been found, thus the species are represented only by females. The separation of species is based on few characters, the most important being the morphology of the esca. Without the use of the esca, some specimens can be identified by the relative

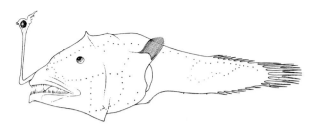

FIGURE 129. *Dolopichthys jubatus*, lectotype, 19.5 mm SL, ZMUC P9299. Drawing by R. Nielsen.

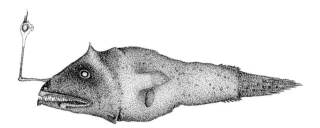

FIGURE 130. *Dolopichthys dinema*, holotype, 21.5 mm SL, BMNH 1970.11.25.1. After Pietsch (1972c).

length of the illicium or by the relative number of teeth in the jaws and on the vomer. But specimens with damaged or lost escae, especially those that are less than 20 mm, are particularly difficult to differentiate. Despite these problems, which are characteristic of deep-sea anglerfish systematics in general, *Dolopichthys* remains one of the best known of the 35 genera of the suborder Ceratioidei. The account provided here follows closely that of Pietsch (1972c).

### DISTINGUISHING CHARACTERS

Although oneirodids are generally thought of as being bulbous fishes, with short deep bodies, metamorphosed females of *Dolopichthys* are relatively long and slender (Figs. 129, 130). The frontal bones are long, linear, and placed more horizontally than those of their congeners, producing a long narrow snout. The mouth is large, the cleft extending posteriorly past the eye. Sphenotic and quadrate spines are well developed; the lower jaw bears a well-developed symphysial spine. The posterior margin of the opercle is deeply notched. The subopercle is long and narrow, its dorsal end slender, tapering to a point, its ventral end oval in shape. The second pharyngobranchial is reduced, but may bear as many as 6 teeth. Jaw teeth counts are highly variable: those in the upper jaw range from fewer than 25 teeth to over 400, while those in the lower jaw from fewer than 35 to nearly 600. The length of the illicium is highly variable (17.8 to 75.3% SL), becoming longer proportionately with growth. The esca bears a large compressed posterior appendage, consisting of a swollen basal part, darkly pigmented in some specimens; and a distal part consisting of a long tapering anterior filament usually, but not always, connected by a thin membrane to a shorter tapering posterior filament (Fig. 131). The skin is smooth and naked, without dermal spinules. Darkly pigmented skin of the caudal peduncle extends well past the base of the caudal fin.

Metamorphosed males are probably best distinguished by the shape of the opercular bone, which is rather broad, its posterior margin only slightly concave, the dorsal fork as long as and much broader than the ventral fork. The posterior nostril is contiguous to the eye. The nasal area is unpigmented. There are 10 to 11 olfactory lamellae. There are 5 to 8 upper denticular teeth and 8 to 10 lower denticular teeth (Bertelsen, 1951:96).

The larvae are unique in having the following combination of characters: The gill cover is heavily pigmented, especially along its ventral and posterior margins, the melanophores extending to some distance from the distal tips of the branchiostegal rays (Fig. 132). The dorsal pigment is restricted to the upper part of the body, reaching posteriorly slightly beyond the base of the dorsal fin in larger specimens. Caudal peduncle pigmentation is separated into dorsal, lateral, and ventral groups. The peritoneum is darkly pigmented (Bertelsen, 1951:96).

There are 5 to 7 (rarely 8) dorsal-fin rays, 4 to 6 anal-fin rays, and 18 to 21 (rarely 17 or 22) pectoral-fin rays (Table 1).

### ETYMOLOGY

The name *Dolopichthys* is derived from the Greek *dolops* or *dolopos*, meaning "ambusher," and *ichthys*, a "fish"; thus a fish that attacks by surprise.

### DIVERSITY

Seven species and two nominal forms of doubtful validity are recognized as follows:

*Dolopichthys pullatus* Regan and Trewavas, 1932: 45 metamorphosed females, 10 to 115 mm; Atlantic, Pacific, and Indian oceans; holotype (*Dolopichthys* [subgenus *Dolopichthys*] *pullatus*), ZMUC P92101, 34 mm, *Dana* station 3680(1), Molucca Sea, 2°22'S, 126°58.5'E, 5000 m wire, 1035 h, 27 March 1929 (Regan and Trewavas, 1932:79, fig. 123, pl. 3, fig. 1; Beebe and Crane, 1947:161, text fig. 6; Bertelsen, 1951:96–98, 100, 101, figs. 53d, 57; Pietsch, 1972c:7, figs. 4, 11; Bertelsen and Pietsch, 1977:185, 1983:87, fig. 8; Pietsch and Seigel, 1980:394; Stewart and Pietsch, 1998:13, fig. 8; Anderson and Leslie, 2001:14, fig. 11C).

*Dolopichthys longicornis* Parr, 1927: 26 metamorphosed females, 14 to 159 mm; Atlantic, Pacific, and Indian oceans; holotype, YPM 2008, 20 mm, *Pawnee* station 46, 21°46'N, 72°49'W, 3050 m wire, 4 April 1927 (Parr, 1927:18, fig. 6; Norman, 1930:354; Regan and Trewavas, 1932:79; Bertelsen, 1951:96, 100, 101, fig. 53e; Pietsch, 1972c:12, figs. 5, 11; Bertelsen and Pietsch, 1977:185; Pietsch and Seigel, 1980:394; Jónsson and Pálsson, 1999:200, fig. 4).

*Dolopichthys danae* Regan, 1926: six metamorphosed females, 20.5 to 115 mm; North Atlantic Ocean; holotype, ZMUC P9298, 75 mm, *Dana* station 1171(7), North Atlantic Ocean, 8°19'N, 44°35'W, 6000 m wire, 0800 h, 13 November 1921 (Regan, 1926:29, pl. 4, fig. 1; Regan and Trewavas, 1932:80, fig. 127; Bertelsen, 1951:96, 97, 100, 101, figs. 53f, 54d; Pietsch, 1972c:15, figs. 6, 11).

*Dolopichthys jubatus* Regan and Trewavas, 1932: eight metamorphosed females, 15 to 89 mm; Atlantic, Pacific, and Indian oceans; lectotype (*Dolopichthys* [subgenus *Dolopichthys*] *jubatus*), ZMUC P9299, 19.5 mm, *Dana* station 3920(4), Indian Ocean, 1°06'S, 62°25'E, 2500 m wire, 1830 h, 9 December 1929 (Regan and Trewavas, 1932:79, 80, fig. 126; Bertelsen, 1951:96,

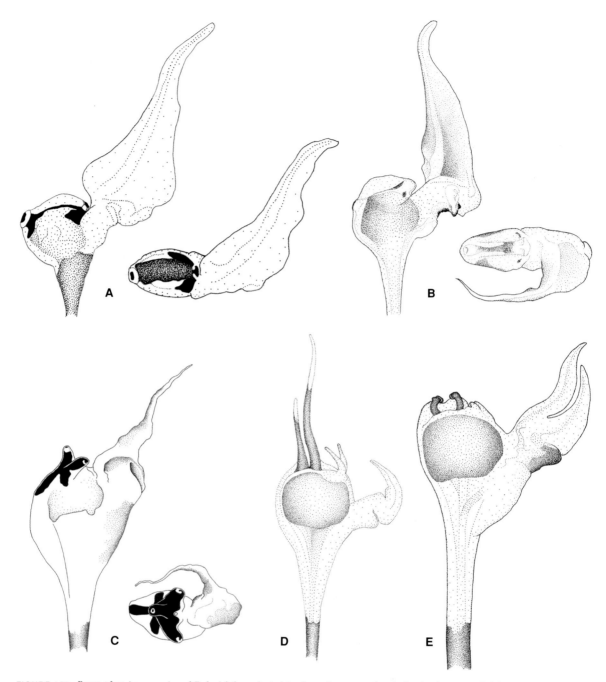

FIGURE 131. Escae of various species of *Dolopichthys* selected to show the range of variation in the genus, left lateral and dorsal views: (A) *D. pullatus*, 81 mm SL, ISH 2557/71; (B) *D. pullatus*, 93 mm SL, LACM 9261-018; (C) *D. karsteni*, holotype, 99 mm SL, MCZ 60991; (D) *D. dinema*, holotype, 21 mm SL, BMNH 1970.11.25.1; (E) *D. danae*, 20.5 mm SL, BMNH 2004.11.6.131. After Pietsch (1972c) and Leipertz and Pietsch (1987).

100, 101, fig. 53b, c; Pietsch, 1972c:17, figs. 7, 11; Bertelsen and Pietsch, 1977:186; Anderson and Leslie, 2001:13, fig. 11A).

*Dolopichthys allector* Garman, 1899: 15 metamorphosed females, 16 to 154 mm; Atlantic and eastern Pacific oceans; holotype, MCZ 28735, 61 mm, *Albatross* station 3371, Gulf of Panama, 5°26′20″N, 86°55′00″W, 0 to 1408 m (Garman, 1899:81, pls. 13–15, fig. 57; Regan, 1926:28; Beebe and Crane, 1947:161, text fig. 7; Bertelsen, 1951:100; Pietsch, 1972c:7, 19, figs. 8, 11; Bertelsen and Pietsch, 1977:186).

*Dolopichthys karsteni* Leipertz and Pietsch, 1987: eight metamorphosed females, 18 to 99 mm; western North Atlantic; holotype, MCZ 60991, 99 mm, *Knorr* cruise 98, station MOC 20-56(0), western North Atlantic, 39°28.0′N, 64°00.6′W, 0 to 1023 m, 30 September 1982 (Leipertz and Pietsch, 1987:406, fig. 1; Moore et al., 2003:215).

*Dolopichthys dinema* Pietsch, 1972c: two metamorphosed females, 21 to 22 mm; eastern tropical Atlantic Ocean; holotype, BMNH 1970.11.25.1, 21 mm, *Discovery II* station 6662-21, off coast of French Guinea, Africa, 10°50′N, 19°54′W, RMT

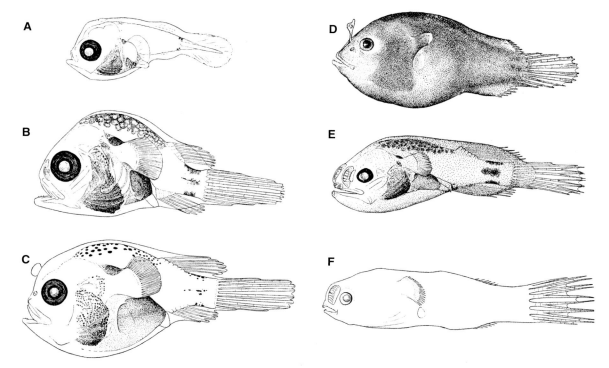

FIGURE 132. *Dolopichthys* sp., life-history stages: (A) larval female, 2.6 mm TL; (B) larval male, 5.4 mm SL, 8.0 mm TL, ZMUC P92818; (C) larval female, 7.0 mm SL, 10.4 mm TL; (D) female in metamorphosis, 10.7 mm SL, 14.3 mm TL (drawing by P. H. Winther); (E) male in metamorphosis, 7.7 mm SL, 11.0 mm TL; (F) adult male, 12.5 mm SL, 17.3 mm TL, ZMUC P92799. After Bertelsen (1951).

8/5, 0 to 700 m, 17 February 1968 (Pietsch, 1972c:7, 21, figs. 9–11).

*Dolopichthys niger* Brauer, 1902, *nomen dubium*: originally two metamorphosed females: 13-mm TL syntype from *Valdivia* station 237 apparently lost; extent syntype, ZMB 17707, 13 mm (15.5 mm TL), *Valdivia* station 173, Indian Ocean, 29°06′02″S, 89°39′00″E, 0 to 2500 m, bottom depth 3765 m (Brauer, 1902:292, 1906:316, pl. 15, fig. 6; Regan and Trewavas, 1932:79, fig. 125; Bertelsen, 1951:96, 100, 101, fig. 53a; Pietsch, 1972c:23).

*Dolopichthys cornutus* (Gilchrist and von Bonde, 1924), *nomen dubium*: a single metamorphosed female, holotype *(Oneirodes cornutus)*, 12 mm, South African waters, formerly in collections of the "Government Marine Survey," specimen discarded (Gilchrist and von Bonde, 1924:23, pl. 6, fig. 2; Regan, 1926:27; Smith, 1949:429, fig. 1231; Penrith, 1967:187, 188; Pietsch, 1972c:24).

## GENUS *BERTELLA* PIETSCH, 1973 (SPIKEHEAD DREAMERS)

*Figures 133–136, Table 1*

*Bertella* was originally described on the basis of eight females, collected from across the North Pacific Ocean, from Japan to the Gulf of California (Pietsch, 1973:197). Since then its osteology has been described and compared with that of other oneirodids (Pietsch, 1974a), a case of male-female attachment has been reported (Pietsch, 2005b:220), and new records have increased the number of known females to 34 and extended its range north to Kamchatka (Sheiko and Fedorov, 2000:24). But the genus still remains poorly known—larvae and free-living males have so far not been found.

### DISTINGUISHING CHARACTERS

Although showing marked similarity to the oneirodid genus *Dolopichthys*, *Bertella* displays a surprising difference in the way the elements of the suspensorium are attached to the cranium. Unlike all other oneirodids, but similar to gigantactinids, neoceratiids, and linophrynids, the dorsal end of the hyomandibular bone forms a single broad head that articulates with the ventrolateral surface of the respective sphenotic and pterotic. There has been some suggestion that this single-headed hyomandibula, in contrast to the double-headed hyomandibula of other oneirodids, might somehow be related to differences in trophic specializations (see Pietsch, 1973:198), but convincing arguments are lacking, and we are left only with speculation.

Like *Dolopichthys*, *Bertella* has a relatively long snout, the frontal bones long and linear in lateral profile, contrasting with the convex frontals of most other oneirodids (Fig. 134). Sphenotic, quadrate, and angular spines are all especially large and sharply pointed (Fig. 135). The lower jaw bears a well-developed symphysial spine. Vomerine teeth are present in juvenile specimens, but lost with growth in adults. The rear of the throat is armed with only a single pair of toothed pharyngobranchials (Pietsch, 1974a:25). Jaw teeth are numerous, but fewer and smaller than those of *Dolopichthys*, the upper jaw with 40 to 200 teeth, the lower jaw with 60 to 250. The skin is smooth and naked, without

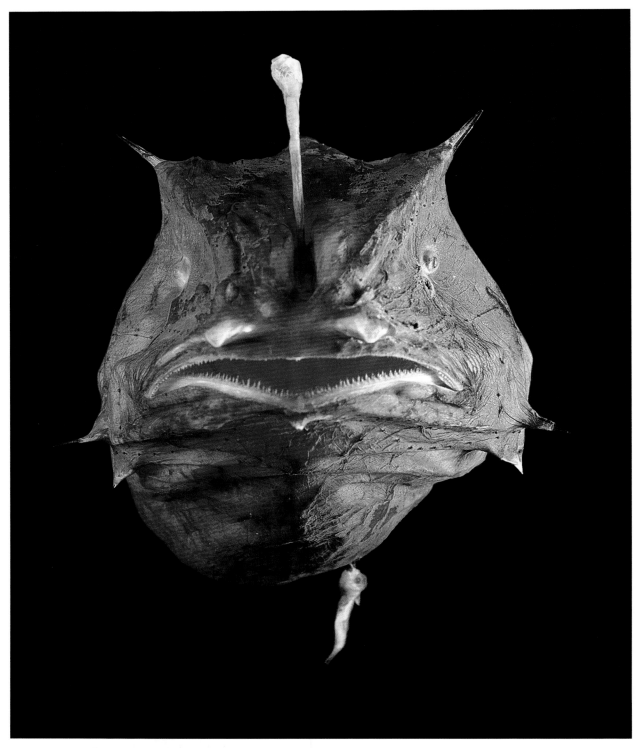

FIGURE 133. *Bertella idiomorpha*, 77-mm SL female, with an 11-mm SL parasitic male, UW 48712. Photo by Darlyne A. Murawski; courtesy of the National Geographic Image Collection.

dermal spinules. Darkly pigmented skin of the caudal peduncle extends well past the base of the caudal fin. The esca bears a single short unpigmented posterior appendage (Fig. 136).

There are 5 to 6 dorsal-fin rays, 5 (rarely 4) anal-fin rays, and 18 to 22 pectoral-fin rays (Table 1).

ETYMOLOGY

*Bertella* was named by Pietsch (1973) in honor of Erik Bertelsen (1912–1993), then curator of the *Dana* collections housed at Danmarks Fiskeri-og Havundersøgelser, Charlottenlund, Denmark, in recognition of his contributions to our knowledge of ceratioid fishes.

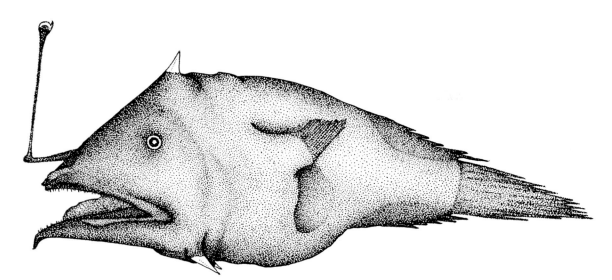

FIGURE 134. *Bertella idiomorpha*, holotype, 74 mm SL, LACM 30601-022. After Pietsch (1973).

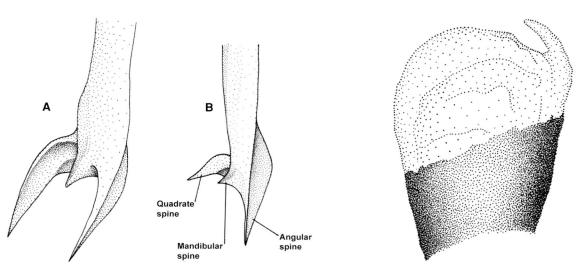

FIGURE 135. Ventral view of posterior end of lower jaw, left side, showing the arrangement and relative size of the quadrate, mandibular, and angular spines: (A) *Bertella idiomorpha*, holotype, 74 mm SL, LACM 30601-022; (B) *Dolopichthys pullatus*, 84 mm SL, LACM 9631-035. After Pietsch (1973).

FIGURE 136. Esca of *Bertella idiomorpha*, paratype, 84 mm SL, OS 1045, left lateral view. After Pietsch (1973).

### DIVERSITY

This genus contains a single species:

*Bertella idiomorpha* Pietsch, 1973: 34 metamorphosed females (11 to 101 mm) and one attached male (11 mm); North Pacific Ocean; holotype, LACM 30601-22, 74 mm, *Velero IV* station 12464, off Guadalupe Island, Mexico, 29°10'00"N, 118°28'15"W, 3-m Isaacs-Kidd midwater trawl, 0 to 940 m, bottom depth 2377 to 3475 m, 1719 to 2240 h, 15 November 1968 (Pietsch, 1973:194, figs. 1–6, 1974a:16–21, 24–26, 28, 30, 31, 33, 82–89, figs. 24, 36, 39G, 43, 51E, 56B, 103, 104; Amaoka, 1984:106, pl. 92F; Sheiko and Fedorov, 2000:24).

### "LONG-PECTORALED" ONEIRODID GENERA

Figures 137–147

Among the diverse members of the family Oneirodidae are four genera that are unique among ceratioids in having an unusually long pectoral-fin lobe that bears the fin rays along its dorsal margin: *Puck*, *Chirophryne*, *Leptacanthichthys*, and *Ctenochirichthys* (Fig. 137). *Chirophryne* and *Ctenochirichthys* were introduced by Regan and Trewavas (1932) as monotypic genera: *Chirophryne xenolophus*, described from a single metamorphosed female; and *Ctenochirichthys longimanus*, based on two metamorphosed females. Since the original publication, three additional specimens of *C. xenolophus* have been collected, one described osteologically by Pietsch (1974a:31) and the other two reported here for the first time. Three males of *C. longimanus*, two larvae and one adolescent specimen, were described by Bertelsen (1951:95) and Beebe and Crane (1947:166), respectively, but otherwise no additional material has become available.

*Leptacanthichthys* was introduced by Regan and Trewavas (1932) as one of five subgenera of the genus *Dolopichthys* Garman (1899) to include a single species, *Dolopichthys gracilispinis* Regan (1925c), based on two metamorphosed females. It was later given generic status by Bertelsen (1951:94), and Pietsch (1974a) described the osteology of

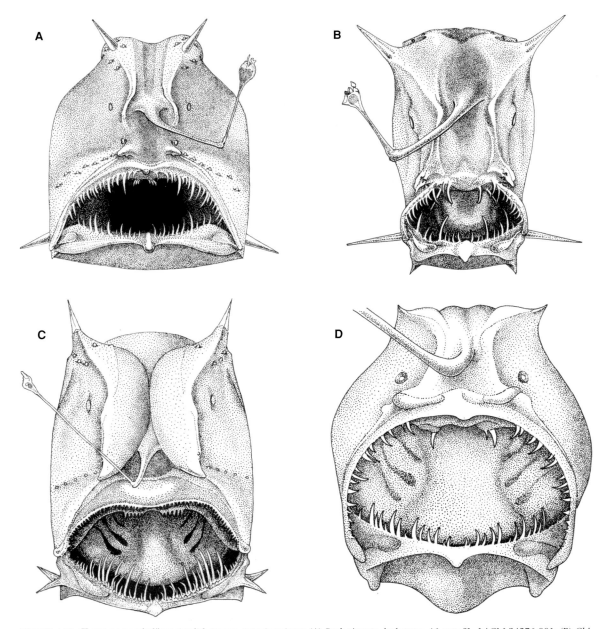

FIGURE 137. "Long-pectoraled" oneirodid genera, anterior views: (A) *Puck pinnata*, holotype, 46 mm SL, LACM 34276-001; (B) *Chirophryne xenolophus*, 22 mm SL, SIO 70-306; (C) *Leptacanthichthys gracilispinis*, 56 mm SL, LACM 33625-002; (D) *Ctenochirichthys longimanus*, paralectotype, 36.5 mm SL, BMNH 1932.5.3.20. Drawings by Patricia Chaudhuri; after Pietsch (1978).

the genus and discussed its phylogenetic relationship with other oneirodid genera. More recently, Pietsch (1976) reported a case of sexual parasitism in *Leptacanthichthys*, the first known occurrence of this mode of reproduction in the family Oneirodidae, and the first record of a male for the genus.

*Puck pinnata*, representing a fourth "long-pectoraled" oneirodid genus, was described by Pietsch (1978), on the basis of two metamorphosed females, and compared osteologically with the other members of this group. Since that time, nothing of significance has been added to our knowledge of this genus or the long-pectoraled assemblage as a whole. While the shared presence of an elongate pectoral-fin lobe points to monophyly for the group, the intrarelationships of the four as well as their placement among other oneirodids

remain unknown (see Evolutionary Relationships, Chapter Four).

GENUS *PUCK* PIETSCH, 1978 (MISCHIEVOUS DREAMERS)

Figures 137A, 138, 139, Table 1

DISTINGUISHING CHARACTERS

*Puck* has a bit of a bull-dog look to it, with an extremely short snout, and the frontal bones strongly convex in lateral profile (Fig. 138). At the same time, however, the body is rather elongate and stout, not globular. The mouth is small, its opening oblique, the cleft terminating well before the eye. The sphenotic spines are extremely large and sharply pointed. The quadrate spine is also

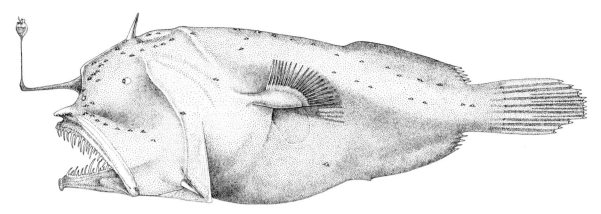

FIGURE 138. *Puck pinnata*, holotype, 46 mm SL, LACM 34276-001. Drawing by Patricia Chaudhuri; after Pietsch (1978).

unusually well developed, nearly six times the length of the articular spine. The lower jaw bears a prominent symphysial spine. There are 35 to 47 teeth in the upper jaw and 24 to 25 in the lower jaw. There are 3 teeth on each side of the vomer. The subopercle is small, elongate, and narrow throughout its length, its dorsal end tapering to a point, its ventral end rounded. The skin is naked, without dermal spinules. Darkly pigmented skin of the caudal peduncle extends well past the base of the caudal fin. The esca bears a stout, rounded, anterodorsally directed and internally pigmented anterior papilla and a similar, and posterodorsally directed, medial escal papilla, without internal pigment (Fig. 139). The distal ends of the anterior and medial papillae are darkly pigmented. An unpigmented compressed posterior escal appendage is present, bearing anterodorsally a lump of tissue of uncertain morphology. A tapering unpigmented lateral filament is present on each side, slightly less than the length of the escal bulb.

Males and larvae are unknown.

There are 5 dorsal-fin rays, 4 anal-fin rays, and 19 to 20 pectoral-fin rays (Table 1).

### ETYMOLOGY

The name *Puck* is taken from Germanic folklore, a minor order of mischievous devils, sprites, goblins, and demons; a devilish trickster.

### DIVERSITY

This genus contains a single species:

*Puck pinnata* Pietsch, 1978: four metamorphosed females, 38 to 81 mm; North Atlantic and Pacific oceans; holotype, LACM 34276-1, female, 46 mm, *Vityaz* cruise 19, station 3199, sample 123b, western North Pacific, 38°16′N, 152°34′E, 6-m diameter conical ring net fished open with 5350 m of wire, bottom depth 5420 to 5350 m, 0230 to 0545 h, 16 October 1954 (Pietsch, 1978:11, figs. 1–10, 13A, 18).

## GENUS *CHIROPHRYNE* REGAN AND TREWAVAS, 1932 (LONGHAND DREAMERS)

Figures 137B, 140, 141, Table 1

### DISTINGUISHING CHARACTERS

The body of metamorphosed females is rather short and stout, somewhat globular (Fig. 140). The snout is short, but

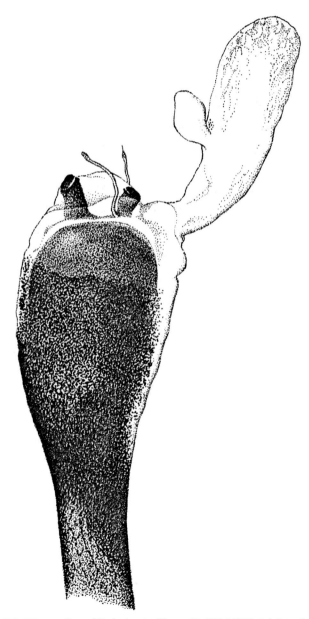

FIGURE 139. Esca of *Puck pinnata*, 38 mm SL, ISH 345/79, left lateral view. Drawing by R. Nielsen.

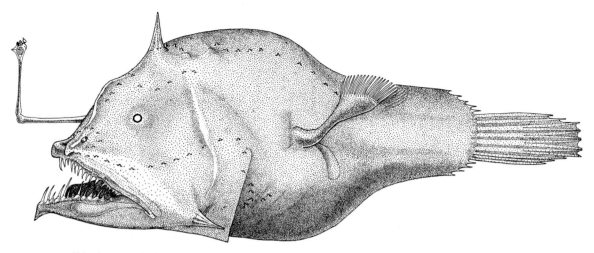

FIGURE 140. *Chirophryne xenolophus*, 22 mm SL, SIO 70-306. Drawing by Patricia Chaudhuri; after Pietsch (1978).

the mouth is large and oblique, the cleft extending slightly past the eye. The sphenotic spines are extremely large and sharply pointed. The quadrate spine is also very well developed, four to nearly six times longer than the articular spine. The lower jaw bears a small symphysial spine. There are 30 to 39 teeth in upper jaw, 34 to 40 in lower jaw. There are 8 to 10 teeth on the vomer. The subopercle is short and broad, its dorsal end rounded, its ventral end oval in shape. The skin is naked, without dermal spinules. Darkly pigmented skin of the caudal peduncle extends well past the base of the caudal fin. The escal bulb bears an unpigmented, tapering anterior appendage connected by a thin membrane to an internally pigmented anterodorsal appendage, darkly pigmented on its distal tip, except for a small circular unpigmented area on each side (Fig. 141). A pair of bilaterally placed, unpigmented medial escal appendages are present, along with an unpigmented compressed posterior escal appendage bearing distally a rounded lump of tissue that tapers to a point.

Males and larvae are unknown.

There are 5 to 6 dorsal-fin rays, 4 anal-fin rays, and 18 to 19 pectoral-fin rays (Table 1).

### ETYMOLOGY

The Greek root, *cheir*, meaning "hand," was chosen by Regan and Trewavas (1932), in combination with *phryne*, a "toad," in allusion to the elongate pectoral-fin lobe of this genus.

### DIVERSITY

A single species is recognized:

*Chirophryne xenolophus* Regan and Trewavas, 1932: four metamorphosed females, 11 to 42 mm; western North Atlantic and Pacific oceans; holotype, ZMUC P9296, 11 mm, *Dana* station 3731(12), South China Sea, 14°37′N, 119°52′E, 2500 m wire; 0200 h, 17 June 1929 (Regan and Trewavas, 1932:82, figs. 131, 132; Bertelsen, 1951:75, 94, fig. 50; Pietsch, 1974a:31, 32, 89, fig. 58, 1978:16, figs. 3C, 6D, 13C, 14, 15, 18; Pietsch and Seigel, 1980:394).

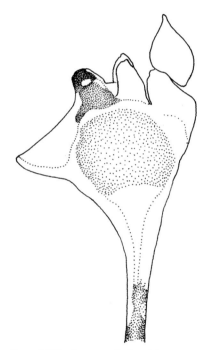

FIGURE 141. Esca of *Chirophryne xenolophus*, 22 mm SL, SIO 70-306. After Pietsch (1978).

## GENUS *LEPTACANTHICHTHYS* REGAN AND TREWAVAS, 1932 (LIGHTLINE DREAMERS)

*Figures 137C, 142–144, Table 1*

### DISTINGUISHING CHARACTERS

At first glance, *Leptacanthichthys* looks a lot like *Dolopichthys*, with its elongate, relatively slender body, long flat snout, and large horizontal mouth, but the rays of the pectoral fin articulating along the dorsal margin of the elongate pectoral-fin lobe clearly separate it (Fig. 142). Highly distinctive also is the unusually slender and flexible illicium as well as a unique arrangement of the quadrate and articular spines: in contrast

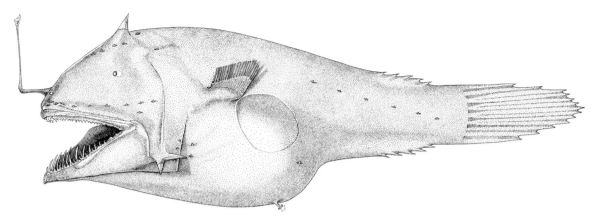

FIGURE 142. *Leptacanthichthys gracilispinis*, 56 mm SL, LACM 33625-002; stalk of tissue protruding from the belly bears the embedded bones of the upper jaws of a parasitic male (see Figure 144). Drawing by Patricia Chaudhuri; after Pietsch (1978).

to all other oneirodids, the articular spine of *Leptacanthichthys* is considerably longer and better developed than the quadrate spine (Fig. 137C). The following combination of features further helps to differentiate *Leptacanthichthys*: sphenotic spines are extremely well developed, but the symphysial spine of the lower jaw is rather small; there are 52 to 220 teeth in the upper jaw, 44 to 140 in the lower jaw, and 6 to 16 teeth on the vomer; the subopercle is short and broad, its dorsal end rounded to bluntly pointed, its ventral end rounded; the skin is naked, without dermal spinules; darkly pigmented skin of the caudal peduncle extends well past the base of the caudal fin; the esca bears a darkly pigmented streak on the dorsal surface and an unpigmented compressed posterior appendage (Fig. 143).

Metamorphosed males of *Leptacanthichthys*—known from only two specimens, a 13-mm free-living male and 7.5-mm individual in late metamorphosis, attached to a sexually mature female (see Reproduction and Early Life History, Chapter Eight)—are distinguished by the elongate pectoral-fin lobe, in combination with a short, broad opercular bone (Fig. 144). There are 6 lower denticular teeth and an unknown number of upper denticular teeth. The gill cover is pigmented, with slightly darker pigmentation along the posterior margin of the subopercle. The dorsal pigment is restricted to the upper part of the body, extending beneath the base of the dorsal fin and just past the anterior base of the anal fin, with a more heavily pigmented dorsal and ventral group of melanophores near the hypural plate.

Larvae are unknown.

There are 5 to 6 (rarely 4) dorsal-fin rays, 5 anal-fin rays, and 18 to 22 pectoral-fin rays (Table 1).

### ETYMOLOGY

The name *Leptacanthichthys* is derived from three Greek roots: *leptos*, meaning "thin" or "delicate"; *akantha*, "thorn" or "spine"; and *ichthys*, a "fish." Thus, a "thin-spined fish," alluding to the unusually thin, delicate illicium of this genus.

### DIVERSITY

The genus contains a single species:

*Leptacanthichthys gracilispinis* (Regan, 1925c): 24 metamorphosed females (10.5 to 103 mm), one adult free-living

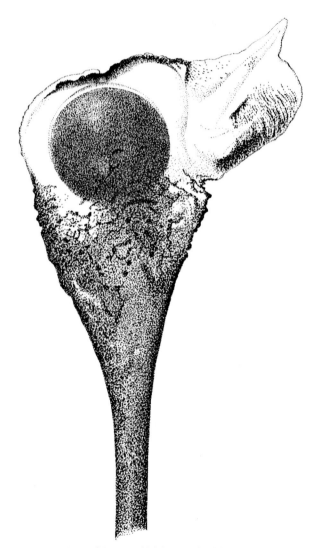

FIGURE 143. Esca of *Leptacanthichthys gracilispinis*, 72 mm SL, ISH 2980/79. Drawing by R. Nielsen.

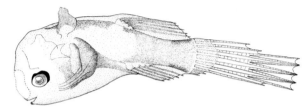

FIGURE 144. *Leptacanthichthys gracilispinis*, parasitic male, originally attached to 56-mm SL female, LACM 33625-002 (see Figure 142). After Pietsch (1976, 1978).

male (13 mm), and one parasitic male in late metamorphosis (7.5 mm); North Atlantic and Pacific oceans; lectotype (*Dolopichthys gracilispinis*), ZMUC P9295, 52 mm, *Dana* station 1206(3), Gulf of Panama, 6°40′N, 80°47′W, 3500 m wire, 1845 h, 14 January 1922 (Regan, 1925c:563; Regan and Trewavas, 1932:66, 80, fig. 128; Bertelsen, 1951:74, 94, fig. 49; Pietsch, 1974a:16–32, 82, 86–89, figs. 22, 34, 39c, 42, 51c, 55, 59, 103, 104, 1976:784, figs. 2–5, 1978:20, figs. 3D, 5B, 6E–H, 13D, 16–18; Bertelsen, 1986:1391, figs.; Swinney, 1995a: 52, 54).

## GENUS *CTENOCHIRICHTHYS* REGAN AND TREWAVAS, 1932 (COMBFIN DREAMERS)

*Figures 137D, 145–147, Table 1*

### DISTINGUISHING CHARACTERS

Of the four long-pectoraled oneirodid genera, *Ctenochirichthys* is certainly the most extreme in the development of the pectoral fin (Fig. 145). Although conspicuously long in metamorphosed males as well, the pectoral-fin lobe of females is greater than 15% SL, bearing 28 to 30 rays along its dorsal margin, more than any other ceratioid and approached only by some specimens of *Pentherichthys* (21 to 24, rarely 25 to 27; see Table 1). Distinctive also is the elongate, slender subopercular bone, tapering only slightly, the dorsal end rounded or somewhat squared off. Unlike *Puck* and especially *Chirophryne*, but similar to *Leptacanthichthys*, the body of females is elongate, not globular. The snout is short, but the mouth is moderate in size and oblique relative to the axis of the body, the cleft terminating anterior to or slightly beneath the eye. The sphenotic, quadrate, and articular spines are reduced (in some specimens not piercing the skin). The lower jaw bears a small symphysial spine. There are 17 to 37 teeth in the upper jaw, 16 to 31 teeth in the lower jaw, and 2 to 4 teeth on the vomer. The skin is naked, without dermal spinules. Darkly pigmented skin of the caudal peduncle extends well past the base of the caudal fin. The esca bears a short rounded anterior appendage; a darkly pigmented, raised band of tissue extending over the dorsomedial surface and down onto the sides of the bulb, with a circular unpigmented area on each side; and an unpigmented compressed tapering posterior appendage (Fig. 146).

In addition to the long pectoral-fin lobe and high number of pectoral-fin rays, metamorphosed males of *Ctenochirichthys* (only a single known juvenile male, 11.5 mm) differ from those of all other oneirodid genera in having 13 olfactory lamellae; 9 upper denticular teeth arranged in a single series; and lower denticular teeth arranged in two rows, the inner row with about 8 teeth, the outer row with 3 teeth.

The two known larvae (4.5 to 5 mm) are characterized most strikingly (in addition to the high number of pectoral-fin rays) by unique features of pigmentation: a dark spot medial to the subopercle in combination with the separation of pigment on the caudal peduncle into two distinct groupings (Fig. 147).

There are 6 or 7 dorsal-fin rays, 4 or 5 anal-fin rays, and 28 to 30 pectoral-fin rays (Table 1).

### ETYMOLOGY

Emphasizing the most distinctive feature of *Ctenochirichthys*, Regan and Trewavas (1932) combined the Greek roots, *kteis* or *ktenos*, meaning "comb"; *cheiros*, "hand" (and by extension, its evolutionary homolog, the pectoral fin); and *ichthys*, a "fish," to give us "Combfin Fish" or, in this case, "Combfin Dreamer."

### DIVERSITY

The genus contains a single species:

*Ctenochirichthys longimanus* Regan and Trewavas, 1932: two metamorphosed females (12.5 to 36.5 mm); a single free-living juvenile male (11.5 mm), and two larval males (4.5 to 5 mm); eastern North Atlantic and eastern tropical Pacific oceans; lectotype, ZMUC P9297, female, 12.5 mm, *Dana* station 3548(2), Gulf of Panama, 7°06′N, 79°55′W, 3000 m wire, 1030 h, 31 September 1928 (Regan and Trewavas, 1932:82, pl. 3, fig. 3; Bertelsen, 1951:75, 94, 95, figs. 51, 52; Pietsch, 1974a:31, 32, 89, 1978:14, figs. 3B, 6C, 8B, 11–13B, 18; Pequeño, 1989:43).

## FAMILY CAULOPHRYNIDAE Goode and Bean, 1896 (Fanfin Seadevils)

*Figures 148–152, Tables 1, 6*

The family Caulophrynidae includes globose bathypelagic anglerfishes, the females of which are easily distinguished from those of allied families by the absence of an expanded escal bulb, the presence of only 2 pectoral radials, extremely long dorsal- and anal-fin rays, only 8 caudal-fin rays, and neuromasts of the acoustico-lateralis system located at the tips of elongate filaments. *Caulophryne*, the first of two recognized genera of the family, was established by Goode and Bean (1896) with the description of *Caulophryne jordani* based on a single female specimen collected in 1887 off Long Island, New York, by the U.S. Fish Commission Steamer *Albatross*. From that time, and prior to Bertelsen's (1951) monographic treatment of the Ceratioidei, five additional members of the genus were described (Table 6), each based on a single female, and one with an attached male: *Caulophryne pelagica* (Brauer, 1902); *C. polynema* Regan, 1930b; *C. ramulosa* Regan and Trewavas, 1932; *C. acinosa* Regan and Trewavas, 1932; and *C. regani* (Roule and Angel, 1932). But despite this proliferation of new taxa, the amount of material available to Bertelsen (1951) remained remarkably small: only nine female specimens, one parasitic male, a male in metamorphosis, and 16 larvae. Not finding sufficient reason to subdivide this material into several species, Bertelsen (1951:37) provisionally recognized three subspecies: *C. jordani jordani*, representing a western North Atlantic subspecies; *C. jordani pelagica*, from the Indian and eastern Pacific oceans (including *C. ramulosa*

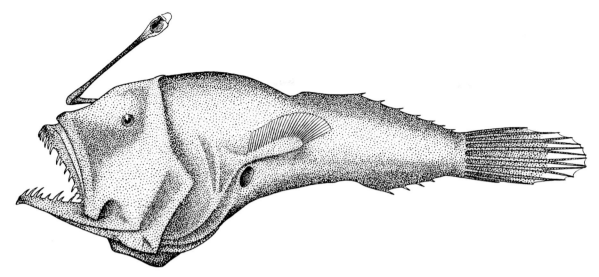

FIGURE 145. *Ctenochirichthys longimanus*, lectotype, 12.5 mm SL, ZMUC P9297. Drawing by W.P.C. Tenison; after Regan and Trewavas (1932).

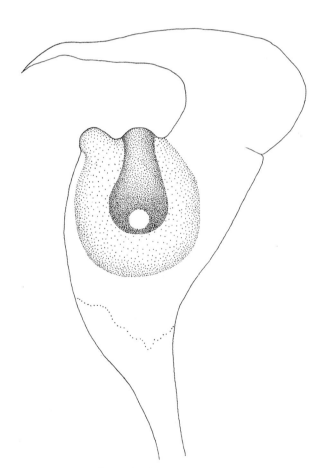

FIGURE 146. Esca of *Ctenochirichthys longimanus*, lectotype, 12.5 mm SL, ZMUC P9297. After Pietsch (1978).

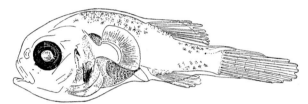

FIGURE 147. *Ctenochirichthys longimanus*, larval male, 6.5 mm TL, ZMUC P92794. After Bertelsen (1951).

and *C. acinosa* as junior synonyms); and *C. jordani polynema*, from the eastern North Atlantic (with *C. regani* as a junior synonym).

With the advantage of a nearly threefold increase in the number of known metamorphosed female specimens, Pietsch (1979) revised *Caulophryne* recognizing three species, distinguished by differences in dorsal- and anal-fin ray counts, tooth counts, illicial length, and illicial and escal morphology: *C. pelagica*, represented by six female specimens collected from the eastern and Indo-west Pacific oceans; *C. jordani*, 11 females from all three major oceans of the world; and *C. polynema*, known from nine females and one parasitically attached male from the Atlantic and eastern Pacific oceans. One additional specimen, a female with a parasitic male from the Banda Sea, could not be placed within the known material of any of these three recognized species and was thus identified by Pietsch (1979:18) as *Caulophryne* sp. This specimen was subsequently recorded by Pietsch and Seigel (1980:380) as *Caulophryne* sp. A.

While reinforcing Bertelsen's (1951) conclusions, that nearly all known females fall into one of three taxa, Pietsch (1979) was unable to add any new information concerning males, the available material remaining extremely rare: only four individuals, two free-living males in metamorphosis and two adults parasitically attached to females.

Since the revision by Pietsch (1979), the taxonomic situation has become more complicated by the discovery of three small metamorphosed females (10 to 11.5 mm) from the Sulu Sea identified by Pietsch and Seigel (1980:380) as *Caulophryne* sp. B. While differing significantly from the known material of the genus and thus probably representing a new species, formal description was postponed because of the small size of the specimens and their poor condition. Finally, and more recently, two additional species have been described: *C. bacescui* Mihai-Bardan, 1982, based on a single metamorphosed female specimen from Peruvian waters; and

## TABLE 6
Reallocation of Nominal Species of the Genus *Caulophryne* Based on Females

| Nominal Species | Currently Recognized Taxon |
| --- | --- |
| *Caulophryne jordani* Goode and Bean, 1896 | *Caulophryne jordani* Goode and Bean, 1896 |
| *Caulophryne setosus* Goode and Bean, 1896 | *Caulophryne jordani* Goode and Bean, 1896 |
| *Melanocetus pelagicus* Brauer, 1902 | *Caulophryne pelagica* (Brauer, 1902) |
| *Caulophryne polynema* Regan, 1930b | *Caulophryne polynema* Regan, 1930b |
| *Caulophryne ramulosa* Regan and Trewavas, 1932 | *Caulophryne pelagica* (Brauer, 1902) |
| *Caulophryne acinosa* Regan and Trewavas, 1932 | *Caulophryne pelagica* (Brauer, 1902) |
| *Ceratocaulophryne regani* Roule and Angel, 1932 | *Caulophryne jordani* Goode and Bean, 1896 |
| *Caulophryne radians* Fowler, 1936 | *Nomen nudum* |
| *Caulophryne racemosa* Monod, 1960 | *Nomen nudum* |
| *Caulophryne bacescui* Mihai-Bardan, 1982 | *Caulophryne bacescui* Mihai-Bardan, 1982 |
| *Caulophryne pietschi* Balushkin and Fedorov, 1985 | *Caulophryne pelagica* (Brauer, 1902) |

*C. pietschi* Balushkin and Fedorov, 1985, from the western South Pacific, also represented by a single metamorphosed female. A detailed comparison of the holotype of *C. bacescui* with all known material has reinforced its validity, while evidence is presented here to conclude that *C. pietschi* is a junior synonym of *C. pelagica* (see Part Two: A Classification of Deep-Sea Anglerfishes).

A second caulophrynid genus and its type species, *Robia legula*, was described by Pietsch (1979) and compared with *Caulophryne*. The single known female of this highly distinctive species was part of the extensive midwater collections made during the 1975 *Alpha Helix* Southeast Asian Bioluminescence Expedition. Since that time no additional females have become available, and males and larvae of *Robia* remain unknown.

The account presented here follows closely the revision of Pietsch (1979).

### DISTINGUISHING CHARACTERS

Metamorphosed females of the family Caulophrynidae are instantly recognized by having extremely long dorsal- and anal-fin rays, unlike any other anglerfish. Although the shortest of these rays are somewhat greater than 60% standard length in all specimens examined, their true length is difficult to estimate—all are thin, brittle, and nearly always broken off short upon capture (e.g., see Fig. 148). Unsupported by interconnecting membranes, but well innervated and equipped with erector and depressor dorsalis muscles, it is tempting to think that these rays might act like a network of sensory antennae—something akin to cat whiskers—continually sweeping independently forward and backward and side to side through the water, effectively creating a sphere of tactility around the body of the fish to monitor the nearby presence of predator or prey.

While these exceptionally long median fin rays are enough to distinguish caulophrynids from other anglerfishes, the family exhibits a number of additional diagnostic features: the maxilla, opercle, subopercle, posttemporal, and ventral portion of the cleithrum are all greatly reduced (Pietsch, 1979); there are only 2 pectoral-fin radials; the caudal fin contains only 8 rays (a character found elsewhere among ceratioids only in the ceratiid genus *Cryptopsaras*); the neuromasts of the acoustico-lateralis system are located at the tips of extremely long filaments (erroneously described by Goode and Bean, 1896:489, 496, as "numerous luminous filaments on head and body"); and the illicium lacks an expanded escal bulb, making it improbable that this structure is capable of producing bioluminescence.

Caulophrynids are further differentiated by having the following combination of character states: the body of metamorphosed females is short and globular; the mouth is large, its opening horizontal to slightly oblique, the lower jaw usually extending posteriorly beyond the base of the pectoral-fin lobe; sphenotic spines are present, but quadrate and articular spines are rudimentary, while angular and preopercular spines are absent; the lower jaw bears a well-developed symphysial spine; there are 20 to 45 teeth in the upper jaw, 12 to 34 in the lower jaw, and 1 to 5 teeth on the vomer; the skin is everywhere smooth and naked, without a trace of dermal spinules.

Metamorphosed males, known from only five specimens, all assigned to *Caulophryne* (two free-living and three parasitically attached to females), have large eyes (Fig. 56C). The olfactory organs are also large, the nostrils directed laterally in smaller specimens, the anterior nostrils close set and directed anteriorly in larger specimens. The upper denticular bone bears an irregular series of teeth along its ventral margin. The lower denticular is divided into three lobes, bearing 5, 8, and 5 teeth, respectively. The pectoral fins are large, their length about 40% SL. Pelvic fins are well developed in young free-living stages, but lost with growth in later stages, including parasitic adults. Males become parasitic, but they are probably facultative (see Reproduction and Early Life History, Chapter Eight).

Larvae are represented by 16 specimens, all assigned to *Caulophryne* (Fig. 56). They are characterized by having a short, rounded body, the skin highly inflated. The pectoral fins are

FIGURE 148. *Caulophryne pelagica*, 183 mm SL, BMNH 2000.1.14.106. Photo by David Shale.

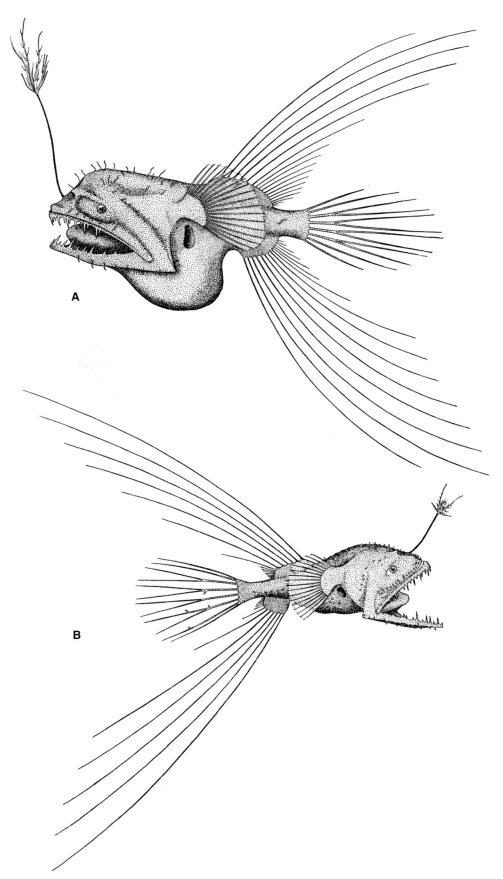

FIGURE 149. *Caulophryne pelagica*: (A) holotype of *Caulophryne ramulosa*, 57 mm, ZMUC P9245; (B) holotype of *Caulophryne acinosa*, 19 mm SL, ZMUC P9244. Drawings by W.P.C. Tenison; after Regan and Trewavas (1932).

unusually large, reaching posteriorly beyond the dorsal and anal fins. The pelvic fins are well developed, with 3 or 4 rays (lost during metamorphosis). Sexual dimorphism is apparently absent; all known specimens bear a rudiment of the illicium protruding from the anterodorsal margin of the head (Bertelsen, 1951:35; 1984:326, 328).

Fin-ray counts are as follows: dorsal-fin rays 6 to 22, anal-fin rays 5 to 19, pectoral-fin rays 14 to 19, and caudal-fin rays 8 (2 simple + 4 bifurcated + 2 simple) (Table 1).

The largest known female of the family is a 183-mm specimen of *C. pelagica* (BMNH 2000.1.14.106) collected in the Atlantic Ocean off the Cape Verde Islands. The two known free-living males, both in metamorphosis, measure 7.7 and 11 mm (ZMUC P92193, MCZ 69324). The three known parasitic males—one attached to the 142-mm holotype of *C. polynema* (BMNH 1930.2.7.1; see Regan, 1930b), another to a 137-mm specimen of *C. polynema* (MNHN 2001-0140), and the third attached to a 98-mm female identified here as *Caulophryne* sp. A (LACM 36025-001) following Pietsch and Seigel (1980:380)—measure 16, 15, and 12 mm, respectively.

### SYNOPSIS OF GENERA OF THE FAMILY CAULOPHRYNIDAE

The Caulophrynidae contains two genera, differentiated as follows.

---

Females with rays of unpaired fins relatively numerous and long: dorsal-fin rays 14 to 22, length of the longest rays greater than 70% SL; anal-fin rays 12 to 19, length of the longest rays greater than 60% SL; the illicium relatively short, less than 130% SL, that of some species bearing few to numerous slender filaments along its length, and usually darkly pigmented except near the esca; morphology of the esca highly variable, consisting of several branched appendages and/or numerous filaments, but without an expanded bacteria-filled light organ; upper jaw with 17 to 46 teeth, lower jaw with 12 to 34 teeth; vomer with 1 to 6 teeth; pectoral-fin rays 14 to 19:

FANFIN SEADEVILS, GENUS *CAULOPHRYNE* GOODE AND BEAN, 1896, FOUR SPECIES, P. 133

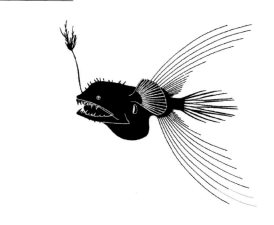

Females with rays of unpaired fins relatively few and short: dorsal-fin rays 6, length of the longest rays less than 65% SL; anal-fin rays 5, length of the longest rays less than 40% SL; the illicium relatively long, 268% SL in the only known specimen, the stem darkly pigmented except near the esca, without filaments along its length; the esca unpigmented and somewhat translucent, with two short lateral appendages, and a slightly more opaque distal tip bearing three short appendages, but without an expanded bacteria-filled light organ; upper jaw with 33 teeth, lower jaw with 31 teeth; vomer with 4 teeth; pectoral-fin rays 17; extremely rare, known only from a single metamorphosed female specimen:

LONG-LURE FANFINS, GENUS *ROBIA* PIETSCH, 1979, A SINGLE SPECIES, P. 135

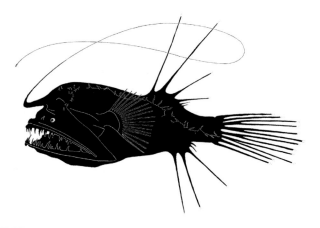

---

### GENUS *CAULOPHRYNE* GOODE AND BEAN, 1896 (FANFIN SEADEVILS)

Figures 148–151, Tables 1, 6

#### DISTINGUISHING CHARACTERS

*Caulophryne* differs from its sister genus *Robia* in a number of easily recognized characters: while the illicium is highly variable in length, 16 to 125.8% SL, it is considerably less than half the length of that of *Robia*; dorsal- and anal-fin ray counts are more than two or three times those of *Robia*; and the esca, highly variable among the species, consists of several branched appendages and/or numerous filaments—in some species the entire length of the illicium is covered with numerous translucent filaments—in contrast to the simple illicium and esca of *Robia*, lacking ornamentation of any kind (Fig. 150).

Characters that distinguish the males and larvae are provided in the family account above and in greater detail in Part Two: A Classification of Deep-Sea Anglerfishes (see Figs. 56, 151).

There are 14 to 22 dorsal-fin rays, 12 to 19 anal-fin rays, and 14 to 19 pectoral-fin rays (Table 1).

#### ETYMOLOGY

Unlike most authors who published in the older literature, Goode and Bean (1896:xxxi) provided a precise explanation for their choice of the name *Caulophryne*: "A *Phryne*-like fish with

FIGURE 150. Escae of *Caulophryne jordani*, left lateral views: (A) 130 mm SL, ISH 242/82; (B) 145 mm SL, ZMUC P922110. Drawings by K. Elsman.

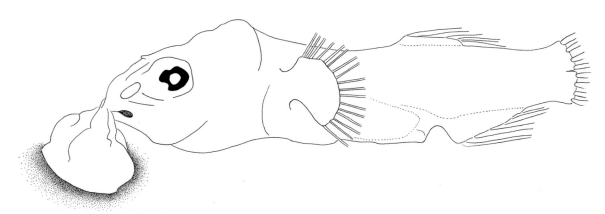

FIGURE 151. *Caulophryne* sp., 12-mm SL parasitic male, attached to a 98-mm SL female, LACM 36025-001. After Pietsch (1979).

the cephalic tentacle supported on a stemlike base," derived from the Greek *caulis* or *kaulos*, a "stem," and *phryne*, a "toad."

### DIVERSITY

The genus consists of five species; two forms labeled "A" and "B," respectively, that probably represent undescribed species; and three *nomina nuda* recognized as follows (Table 6):

*Caulophryne pelagica* (Brauer, 1902): 17 known specimens, 11 to 183 mm, Atlantic, Pacific, and Indian oceans; holotype *(Melanocetus pelagicus)*, ZMB 17711, 11 mm, *Valdivia* station 228, Indian Ocean, 2°38′S, 65°59′E, 0 to 2500 m, bottom depth 3460 m (Pietsch, 1979:14, figs. 16–18, 24).

*Caulophryne bacescui* Mihai-Bardan, 1982: a single known specimen, the holotype, NHMB 49922, 169 mm, *Anton Bruun* Cruise 11, Peru-Chile Trench, October 1965 (Băcescu, 1966:34,

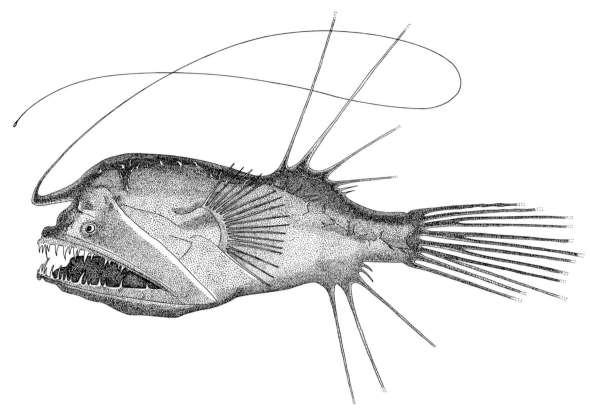

FIGURE 152. *Robia legula*, holotype, 41 mm SL, LACM 36024-1. Drawing by Caryl Maloof; after Pietsch (1979).

fig. 4; Mayer and Nalbant, 1972:164, fig. 5; Mihai-Bardan, 1982:17, figs. 1–3).

*Caulophryne jordani* Goode and Bean, 1896: 38 known specimens, 16 to 155 mm, Atlantic, Pacific, and Indian oceans; holotype, USNM 39265, 26 mm, *Albatross*, off Long Island, New York, 39°27′N, 71°15′W, 0 to 2335 m, 19 September 1887 (Pietsch, 1979:15, figs. 1–13, 16, 18, 24).

*Caulophryne polynema* Regan, 1930b: 14 known specimens, 11.5 to 142 mm, Atlantic and eastern Pacific oceans; holotype, BMNH 1930.2.7.1, female, 142 mm, with a 16-mm parasitic male, off Funchal Bay, Madeira, "long line, in deep water," 1 February 1929 (Pietsch, 1979:16, figs. 16, 18, 20, 21, 24).

*Caulophryne* sp. A, Pietsch and Seigel, 1980: a single metamorphosed female, 98 mm, with a 12-mm parasitic male; LACM 36025-1, *Alpha Helix*, Southeast Asian Bioluminescence Expedition station 37, Banda Sea, 4°56.3′S, 129°25.5′E, RMT-8, 0 to 2000 m, 1900 to 0155 h, 11 April 1975 (Pietsch, 1979:18, figs. 16, 18, 22, 23; Pietsch and Seigel, 1980:380).

*Caulophryne* sp. B, Pietsch and Seigel, 1980: three juvenile females, 10 to 11.5 mm: LACM 36112-1, 10 mm, *Alpha Helix*, Southeast Asian Bioluminescence Expedition station 183, Sulu Sea, 9°24.5′N, 122°12.3′E, RMT-8, 690 to 890 m, 0115 to 0215 h, 6 June 1975; LACM 36111-1, 10.5 mm, *Alpha Helix*, Southeast Asian Bioluminescence Expedition station 184, Sulu Sea, 9°18.0′N, 122°10.0′E, RMT-8, 480 to 550 m, 0355 to 0455 h, 6 June 1975; LACM 36109-2, 11.5 mm, *Alpha Helix*, Southeast Asian Bioluminescence Expedition station 193, Sulu Sea, 9°21.5′N, 122°14.7′E, RMT-8, 800 to 1100 m, 1020 to 1240 h, 7 June 1975 (Pietsch and Seigel, 1980:380, fig. 1).

*Caulophryne setosus* Goode and Bean, 1896, *nomen nudum*: this name appears in a figure caption and is entered in the index and list of plates of Goode and Bean's (1896:26, 541, pl. 121, fig. 409) original description of *C. jordani*, without application to a description or type. This error was caught by Jordan and Evermann (1898:2735): "plate named *C. setosus*, by slip in proof reading." USNM 39265, the holotype of *Caulophryne jordani*, is also listed under the name *setosus* in the "type files" of the Division of Fishes, National Museum of Natural History (Susan L. Jewett, personal communication, 14 September 1976; Pietsch, 1979:20).

*Caulophryne radians* Fowler, 1936, *nomen nudum*: this name appears in Fowler's (1936:1368) "Marine Fishes of West Africa," based on the collection of the American Museum Congo Expedition of 1909–1915. Original authorship is erroneously attributed to Regan, with *Caulophryne polynema* Regan and Trewavas listed as a synonym. There is no application to a description or type. The name *radians* is otherwise unknown in the Caulophrynidae and just what Fowler might have had in mind cannot now be established.

*Caulophryne racemosa* Monod, 1960, *nomen nudum*: This name appears in a figure that illustrates various stages of specialization of lophiiform pectoral radials. There is no application to a description or type. No doubt Monod (1960:687, fig. 80) meant to refer to *C. ramulosa*. His figure 80 was taken from Regan and Trewavas's (1932, fig. 58) illustration of the pectoral radials of the holotype of this nominal species (Pietsch, 1979:20).

GENUS *ROBIA* PIETSCH, 1979 (LONG-LURE FANFINS)

Figure 152, Table 1

DISTINGUISHING CHARACTERS

The genus *Robia* is known only on the basis of a single 41-mm female. It is therefore important to realize that additional

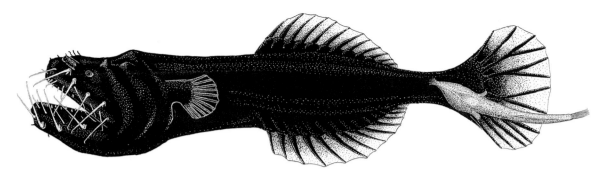

FIGURE 153. *Neoceratias spinifer*, 52-mm SL female, with 15.5-mm SL parasitic male, ZMUC P921726. After Bertelsen (1951).

specimens may well reduce the extent to which it differs from *Caulophryne*. Yet, with what we understand at present, its exceptionally long illicium (268.3% SL); only 6 and 5 dorsal- and anal-fin rays, respectively; and a morphologically simple esca, without filaments or appendages, clearly distinguish the two genera (Fig. 152).

Males and larvae are unknown.

There are 6 dorsal-fin rays, 5 anal-fin rays, and 17 pectoral-fin rays (Table 1).

ETYMOLOGY

This genus was named by Pietsch (1979) for Bruce H. Robison (better known to his friends as "Robie")—cruise leader of the 1975 *Alpha Helix* Southeast Asian Bioluminescence Expedition, during which the only known specimen of *Robia* was collected, and now senior scientist at the Monterey Bay Aquarium Research Institute—in recognition of his many contributions to our knowledge of midwater fishes.

DIVERSITY

The genus contains a single species:

*Robia legula* Pietsch, 1979: a single known specimen, 41 mm, the holotype, LACM 36024-1, *Alpha Helix*, Southeast Asian Bioluminescence Expedition station 81, Banda Sea, 4°56.5'S, 129°59.5'E, RMT-8 with closing device, 1000 to 1500 m, 0416 to 0616 h, 28 April 1975 (Pietsch, 1979:13, figs. 14, 15, 24).

## FAMILY NEOCERATIIDAE Regan, 1926
(Needlebeard Seadevils)

*Figures 153, 154, Table 1*

The family Neoceratiidae was erected by Regan (1926) to include a strange little ceratioid that Pappenheim (1914) called *Neoceratias spinifer*. The single known specimen, only 25 mm SL, was remarkable in several ways, not the least of which was the absence of an illicium and the presence of several series of long, hinged, needlelike teeth mounted outside the jaws. Regan (1926:39) thought it might be "related to the Gigantactinidae, because *Gigantactis* differs from other ceratioids in having the outer teeth larger than the inner, and moreover placed on the outer side of the jaws and inserted in muscular pads, a type of dentition from which that of *Neoceratias* might readily have been derived." Regan and Trewavas (1932:95), emphasizing the importance of the lack of an illicium, went a step further in considering that it might be "a male of some unknown gigantactinid."

Koefoed (1944), without comment on its sex or affinity among ceratioids, described a second, slightly larger specimen (43 mm) and, somewhat later, Bertelsen (1951) described a third (52 mm), this one with an attached parasitic male (15.5 mm), thus establishing the previously described specimens as females and adding the Neoceratiidae to ceratioid families known to practice sexual parasitism. While also describing neoceratiid larvae for the first time, Bertelsen (1951) used the 52-mm female and attached male to present a remarkably detailed osteology, the results of which strongly confirmed the rather isolated position of the Neoceratiidae among known ceratioid families.

Other than summaries of what is known about sexual parasitism (Pietsch, 1976, 2005b), including mention of new material, increasing the number of metamorphosed females to 18 (seven with a parasitic male), nothing of significance has been added to our knowledge of this family since Bertelsen's (1951) monograph.

DISTINGUISHING CHARACTERS

The monotypic family Neoceratiidae is perhaps the most aberrant of all ceratioids (Fig. 153). The lack of a luring apparatus, the single most important defining feature of an anglerfish, is baffling. Present without exception in all other members of the order Lophiiformes (some 323 living species) and, in the absence of any other known means by which this species might attract prey, the natural selective process responsible for this evolutionary loss is hard to imagine. How this animal acquires its food in the near sterile waters of the deep sea is beyond speculation.

In numerous other ways, the combination of characters that defines neoceratiids is unique: The body of metamorphosed females is elongate, slender, and laterally compressed (Fig. 153). The head is short, but the mouth is large, its cleft extending posteriorly well past the eye, its opening horizontal to slightly oblique. The eyes are minute, appearing degenerate, their diameter less than 2.5% SL. A large nasal papilla is present on each side of the snout; with nostrils and olfactory lamellae absent, the function of these papillae is unknown. The sphenotics are conical in shape, but without a distal spine. Quadrate, articular, angular, and preopercular spines are absent. The jaws are subequal, the lower extending anteriorly somewhat beyond the upper, and lacking a symphysial spine. The jaws bear an inner row of short, straight, widely spaced immobile teeth, 0 to 6 on each premaxilla and 10 to 21 on each dentary. The outer margins of the jaws bear prominent conical outgrowths, providing articular surfaces for two or three irregular series of long straight hinged teeth, strongly attached by connective tissue and well-developed

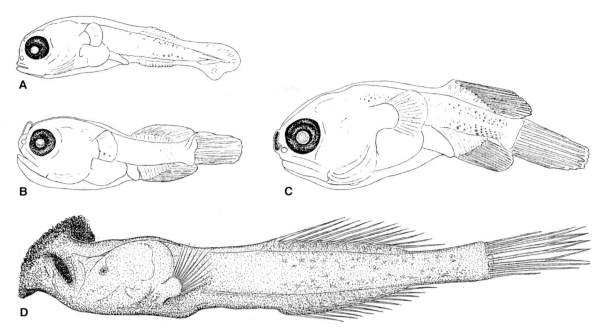

FIGURE 154. *Neoceratias spinifer* (A–C, larvae, sex indeterminate): (A) 4.5 mm TL; (B) 6.0 mm TL; (C) 6.3 mm SL, 8.5 mm TL, ZMUC P921725; (D) parasitic male, 15.5 mm SL, ZMUC P921726. After Bertelsen (1951).

musculature, and each with a tiny distal hook. There are 11 to 20 outer teeth on each premaxilla, and 18 to 20 on each dentary (all known material is in very poor condition, making accurate tooth counts difficult). The vomerine teeth are long and slender, one on each side. The pterygiophores that support the rays of the dorsal fin and, to a much greater extent, those that support the anal fin, are unusually long; the rays themselves are long as well, such that the fins appear to be mounted on prominent bases. The caudal peduncle is unusually long and slender, the caudal fin itself very broad, with the four innermost rays deeply bifurcated. The skin of the head and body lacks dermal spinules, but bears numerous elongate filaments (similar to those of caulophrynids), undoubtedly associated with the acoustico-lateralis system (not light organs as suggested by Koefoed, 1944:10).

Free-living males are unknown, but parasitic males are represented by seven individuals (11.5 to 18 mm). They differ from those of all other ceratioid families in having an unusually slender body (154D). The eyes and olfactory organs are degenerate. The upper denticular bone is apparently absent, the lower denticular bearing 3 elongate curved projections, each bifurcating distally. The skin is naked, without a trace of dermal spinules. Sexual parasitism is apparently obligatory (see Reproduction and Early Life History, Chapter Eight).

The larvae (11 known specimens, 3.7 to 9.8 mm) differ from those of all other ceratioid families in having an elongate, slender body, its depth 30 to 40% SL (Fig. 154). The skin is only slightly inflated. Sexual dimorphism is apparently absent; all known specimens have an elongate, cylindrical rudiment of the illicium protruding (in a position unique among ceratioid larvae) on tip of the snout, just above the symphysis of the upper jaw. The pectoral fins are relatively small, their length 15 to 20% SL. Metamorphosis begins at lengths of 8 to 10 mm (Bertelsen, 1951:159, 1984:326, 328).

There are 10 to 13 dorsal-fin rays, 10 to 13 anal-fin rays, 12 to 15 pectoral-fin rays, and 9 (2 unbranched + 4 branched + 3 unbranched) caudal-fin rays (10 caudal rays in some larvae; see Bertelsen, 1951:159) (Table 1).

The largest known female and male, 108 and 18 mm, respectively (SIO 70-336), are a parasitically attached couple, collected in 1970 by the RV *Melville* in the Philippine Sea off Luzon.

## DIVERSITY

The family contains a single genus:

### GENUS *NEOCERATIAS* PAPPENHEIM, 1914 (NEEDLEBEARD SEADEVILS)

*Figures 153, 154, Table 1*

#### DISTINGUISHING CHARACTERS

This genus has all the features of the family, as described above and detailed in Part Two: A Classification of Deep-Sea Anglerfishes.

#### ETYMOLOGY

In comparing his new genus to other ceratioids known at the time, Pappenheim (1914:198) was unsure of its position within the suborder: "I would like to mention under the above name [*Neoceratias spinifer*], a small fish of 25-mm length (without the caudal fin), whose systematic position I am unable to specify exactly, other than its affiliation with the Pediculati, which to me hardly seems doubtful" (translated from the German). But finding most similarity with the Ceratiidae, he chose the name *Neoceratias* to signify a new kind of ceratiid-like anglerfish.

#### DIVERSITY

The genus contains a single known species:

*Neoceratias spinifer* Pappenheim, 1914: 18 metamorphosed females (17 to 108 mm), seven parasitically attached males (8.5 to

18 mm), and 11 larvae (3.7 to 9.8 mm); Atlantic, Pacific, and Indian oceans; holotype, ZMB 19383, 25 mm, Deutschen Südpolar-Expedition, *Gauss*, tropical Atlantic near St. Helena, 12°11'S, 6°16'W, 0 to 2000 m, 4 September 1903 (Pappenheim, 1914:198, fig. 10; Regan and Trewavas, 1932:39, 95, fig. 153; Koefoed, 1944:9, pl. 2, fig. 5; Bertelsen, 1951:158, figs. 105–107; Pietsch, 1976:789, fig. 8; Bertelsen and Pietsch, 1983:93, figs. 14, 15).

## FAMILY GIGANTACTINIDAE Boulenger, 1904a
(Whipnose Seadevils)

*Figures 155–166, Table 1*

The Gigantactinidae is one of the most well defined and highly specialized families of deep-sea anglerfishes. The females are readily distinguished from those of the other 10 families of the suborder by having an elongate streamlined shape, a relatively small head and slender caudal peduncle, and a greatly prolonged illicium that reaches a length between one and five times standard length (Figs. 155–157, 161). The first of two gigantactinid genera to be discovered was established by Brauer (1902) with the description of *Gigantactis vanhoeffeni*, based on two female specimens collected in the Indian Ocean during the German *Valdivia* Expedition of 1898–1899. The family remained monotypic until 1925 when Regan described *Rhynchactis leptonema*, based on a single 42-mm female collected in the western tropical Atlantic by the *Dana* in 1921. In the same paper, Regan (1925c) recorded two additional specimens of *Gigantactis*, both with a well-preserved esca, describing them as holotypes of new species. Armed with this comparative material, and assuming that "the illicium is broken" in the only known specimen of *Rhynchactis*, Regan (1926:38) distinguished females of the two genera as follows: *Gigantactis* Brauer, 1902: "Snout produced in front of mouth; teeth moderately strong, depressible, curved inwards; about 1 to 3 series in upper jaw, 2 to 4 in lower, the outer teeth longest; similar teeth on upper pharyngeals; palate toothless; skin spinulose." *Rhynchactis* Regan, 1925c: "Snout truncated, not produced in front of mouth; teeth in jaws minute, in several series; on each side an external pair of anterior canines in front of upper jaw; strong teeth on upper pharyngeals; skin naked."

From the time of Regan's (1925c) publication to Bertelsen's (1951) monograph on the Ceratioidei, *Rhynchactis* remained monotypic, while eight new species were added to Brauer's (1902) *Gigantactis vanhoeffeni*, each based on a single female specimen: *G. macronema* Regan, 1925c; *G. gracilicauda* Regan, 1925c; *G. sexfilis* Regan and Trewavas, 1932; *G. exodon* Regan and Trewavas, 1932; *G. ovifer* Regan and Trewavas, 1932; *G. filibulbosus* Fraser-Brunner, 1935; *G. longicirra* Waterman, 1939b; and *G. perlatus* Beebe and Crane, 1947. A tenth species based on a single metamorphosed male not originally recognized as a gigantactinid, was described by Regan and Trewavas (1932) as *Teleotrema microphthalmus* and later referred to *Gigantactis* by Bertelsen (1951).

After examining the then known material of *Gigantactis*, Bertelsen (1951) concluded that the separation of nine nominal species based on a total of 11 metamorphosed females (30 to 100 mm) must be regarded as uncertain, that the five known metamorphosed males probably represent at least two species distinguished by the number and shape of the denticular teeth, and that the 233 larvae in the *Dana* collections could be divided into three groups based on differences in subdermal pigmentation and fin-ray counts. As for *Rhynchactis*, Bertelsen (1951) provided the first descriptions of larvae (23 specimens) and a metamorphosed male, but because additional records of metamorphosed females were unavailable at the time, the incomplete and misleading diagnosis of Regan (1926) was retained.

Bertelsen et al. (1981), in a detailed revision of the Gigantactinidae, examined all available material, consisting of about 500 specimens of *Gigantactis* (165 metamorphosed females, 43 free-living males, and about 300 larvae) and 36 specimens of *Rhynchactis* (five females, five males, and 26 larvae). The study confirmed that gigantactinids represent a highly specialized monophyletic family of the Ceratioidei, but also revealed that the two genera are remarkably and uniquely specialized. In *Gigantactis*, the illicium carries a terminal escal bulb, which, as in nearly all female ceratioids, has an internal globular gland containing luminous bacteria and bears on its external surface species-specific escal appendages and filaments. The jaws of *Gigantactis* females bear numerous teeth, which are greatly enlarged (especially those of the lower jaw), curved and partially placed on the outside of the jaw. In contrast, the illicium of the available females of *Rhynchactis* (in which the structure appears complete) has no terminal escal bulb; instead, the luring apparatus is a simple filament, without external ornamentation, or bears small secondary filaments along its length. The greatly reduced premaxillae of *Rhynchactis* each bear as many as 2 curved teeth, but the dentaries are toothless. A dense pavement of white papilliform glands covers the inner surface of the upper and lower jaws. Thus, it appears that the typical ceratioid feeding mode—luring prey by means of a luminous bacterial light organ and capturing it with the aid of well-developed jaw teeth—has been retained in *Gigantactis*, but replaced in *Rhynchactis* by a completely different, and at present unexplained, mechanism (for more on the feeding mechanism of gigantactinids, see Locomotion, Food, and Feeding, Chapter Seven).

Bertelsen et al. (1981) showed further that females of *Gigantactis* could be separated, like most other ceratioids, by differences in the morphology of the illicium and esca, but also by illicial length; the number, size, and pattern of placement of the jaw teeth; caudal fin morphology; and median fin-ray counts. Using these characters, they recognized 17 species. Five of these were previously described forms: *G. longicirra*, *G. vanhoeffeni*, *G. gracilicauda*, *G. macronema*, and *G. perlatus*; the remaining 12 species were described as new on the basis of recently collected material from all three major oceans of the world. The nominal species *G. ovifer* and *G. filibulbosus*, each represented only by a poorly preserved holotype, were regarded as *nomina dubia*. Since this work was published, Kharin (1984) described *G. balushkini* on the basis of a single 287-mm female collected in the western North Pacific off northern Honshu, Japan; and Bertelsen and Pietsch (2002) described *G. longicauda* from a single 114-mm female collected in the western North Atlantic in the Sargasso Sea.

Despite an eightfold increase in the number of known metamorphosed males, Bertelsen et al. (1981) were unable to find characters that would allow specific identification. The material, however, was divided into six groups based on differences in eye diameter, pigmentation, presence or absence of skin spines, and fin-ray counts: *G. longicirra* (recognized by the high number of dorsal-fin rays unique to this species) and five additional groups referred to as groups I through V. *Gigantactis* Male Group I included "the naked type" of Bertelsen (1951:153) as well as Parr's (1927) *Laevoceratias liparis* (tentatively included in the Diceratiidae by Bertelsen, 1951). Group II included *Teleotrema microphthalmus* Regan and Trewavas

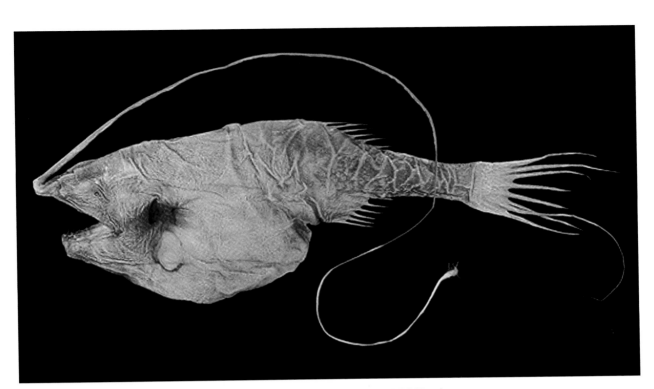

FIGURE 155. *Gigantactis gargantua*, paratype, 166 mm SL, LACM 9748-028. Photo by T.W. Pietsch.

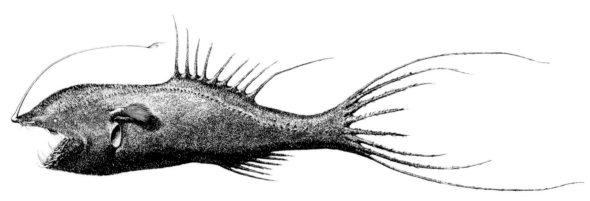

FIGURE 156. *Gigantactis longicirra*, 72.5 mm SL, SIO 60-215. Drawing by William Goodell; courtesy of Richard H. Rosenblatt, H.J. Walker, and the Scripps Institution of Oceanography.

(1932), and the remaining three groups were each based on one or two aberrant and previously undescribed specimens. Despite records of a number of large females (300 to 400 mm), no gravid females and no parasitic males have ever been found.

Although Bertelsen et al. (1981) brought the total number of known metamorphosed specimens of *Rhynchactis* to nine (five females and four males), and added a male in metamorphosis and two larvae, the material was still insufficient to determine whether the genus contained more than one species. Since that time, however, the number of available metamorphosed females of *Rhynchactis* has more than doubled, and four of the six newly recorded specimens have complete illicia. On the basis of this new material, Bertelsen and Pietsch (1998b) revised the genus, recognizing three species distinguished solely on differences in illicial morphology: the type species of the genus, *Rhynchactis leptonema* Regan, represented by the holotype collected from the western tropical Atlantic and a second specimen taken from off Hawaii; and two previously unrecorded species: *R. macrothrix*, described as new on the basis of three specimens from the North Atlantic and Indian oceans; and *R. microthrix*, a single specimen taken from the Indian Ocean.

With only a few minor changes and additions, the following account is based on the revisions of Bertelsen et al. (1981) and Bertelsen and Pietsch (1998b).

### DISTINGUISHING CHARACTERS

More so than other ceratioids, the body of female gigantactinids is slender, laterally compressed, and streamlined, with a relatively short head and a long, narrow caudal peduncle, a combination of features that indicates a much greater level of locomotory activity than is possible for short globose members of the suborder (Figs. 155–157, 161). The mouth is large, the

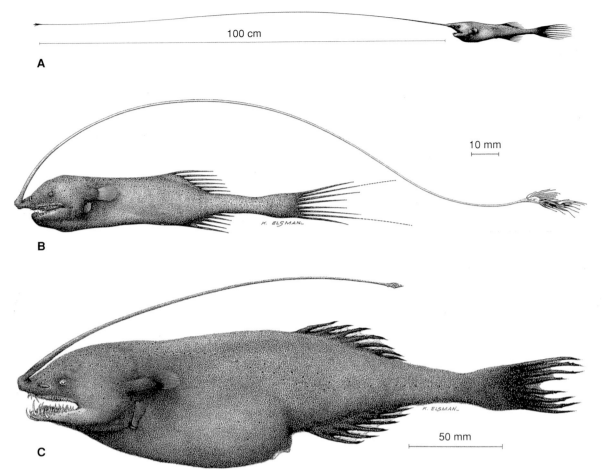

FIGURE 157. Species of *Gigantactis*: (A) *G. macronema*, 232 mm SL, ISH 1596/71; (B) *G. watermani*, holotype, 99 mm, ISH 2330/71; (C) *G. kreffti*, holotype, 252 mm SL, ISH 1099/71. Drawings by K. Elsman; after Bertelsen et al. (1981).

opening horizontal, the cleft extending past the eye. Although the pectoral fin of females is relatively small, the rays are supported by 5 pectoral radials, in contrast to 5 or fewer in other ceratioids. In further contrast to other ceratioids, the posterior margin of the caudal fin is deeply cleft in most species, often with one or two dorsal- and ventral-most rays greatly prolonged.

The pterygiophore of the illicium is unusually large and laterally compressed. The illicium itself is greatly prolonged, its length highly variable among species—ranging from 40 to 100% SL in *Gigantactis longicirra* to 340 to 490% SL in *G. macronema*—reaching its full relative length at a standard length of about 30 mm. The dentition of gigantactinids is highly variable (for details, see the generic accounts below). The second and third pharyngobranchials are extremely large, well ossified, and heavily toothed, lying in a forward position in the roof of the mouth. The organs of the acoustico-lateralis system are born on short stalks of pigmented skin, more or less distinctly connected in series by narrow unpigmented grooves (see Bertelsen et al., 1981:8, 10).

While several prominent structures of gigantactinids are hypertrophied (the illicial pterygiophore, the illicium itself, and the upper pharyngeals), females are generally characterized overall by an extreme reduction or loss of parts: in contrast to all other ceratioid families they lack a vomer, pterosphenoids, mesopterygoids, and ossified scapulae. The supraethmoid and parietals are greatly reduced or absent. Sphenotic, quadrate, articular, angular, and preopercular spines are absent. The preopercle, interopercle, opercle, and subopercle are all reduced to small, narrow struts of bone; the normal ligamentous connection between the interopercle and the angular is absent. The dentaries, attached to each other at the symphysis by thick elastic connective tissue, are simple, without posterior bifurcation and without a symphysial spine. The premaxillae are long and narrow, without ascending or postmaxillary processes; they are well ossified in *Gigantactis*, but greatly reduced in *Rhynchactis*. The maxillae are reduced to threadlike ossifications or absent.

Metamorphosed gigantactinid males are unique in having unusually small eyes, with diameters of only 3 to 5% SL, but the olfactory organs are large, their depth usually 8 to 10% SL (Fig. 160). The anterior nostrils are close together and directed anteriorly. The premaxillae are degenerate, but the maxillae are well developed. The denticular teeth are all or nearly all mutually free. The upper denticular bone bears 3 to 6 (rarely 2) teeth, without connection to the pterygiophore of the illicium. The lower denticular supports 4 to 7 (rarely 3) teeth. The skin is naked or densely covered with dermal spinules. The males are free-living, with no evidence that they ever become parasitic, but temporarily attached males have not yet been found (see Reproduction and Early Life History, Chapter Eight).

Gigantactinid larvae differ from those of other ceratioid families in having exceptionally large pectoral fins (length 45 to 55% SL), comparable only to those of the Caulophrynidae (Figs. 56, 161). They differ further in having a rather short, nearly spherical body. The skin is highly inflated. In contrast to their sister group, the neoceratiids, sexual dimorphism is evident even in the youngest larvae: the females bear a small, club-shaped illicial rudiment protruding from the head. Metamorphosis begins at 8 to 10 mm and subsequent developmental stages range in size from 9 to 20 mm (Bertelsen et al., 1981:62; Bertelsen, 1984:326).

There are 3 to 9 dorsal-fin rays, 3 to 7 anal-fin rays, 14 to 22 pectoral-fin rays, and 9 caudal-fin rays (the ventral-most ray extremely short and embedded within the skin surrounding the adjacent ray). The caudal-fin rays are all simple in females of *Gigantactis*, but there are 2 simple + 4 bifurcated + 3 simple in males of *Gigantactis* and in both sexes of *Rhynchactis* (Table 1).

Gigantactinids are among the largest known ceratioids, but surprisingly none seem to have attained sexual maturity. While some of the 30 or so females of *Gigantactis* greater than 200 mm (the largest is a 435-mm specimen of *Gigantactis elsmani*, ISH 1135/79) have relatively large ovaries, none contain ripe or ripening eggs. Eggs larger than about 0.5 mm in diameter have not been found (see Reproduction and Early Life History, Chapter Eight). At the same time, none of the more than 175 metamorphosed females in collections around the world are parasitized by males, thus it is assumed that males remain free-living. The largest known male, a member of the *Gigantactis* Male Group I, with ripe testes (BMNH 2004.9.12.168), measures 22 mm.

### SYNOPSIS OF GENERA OF THE FAMILY GIGANTACTINIDAE

The Gigantactinidae contains two genera differentiated as follows:

---

Females with the snout elongate and pointed, the illicium emerging anteriorly in front of the jaws from the tip of a greatly enlarged pterygiophore; the esca morphologically complex and highly variable, consisting of an expanded distal bulb containing a bacteria-filled light organ; the teeth of the lower jaw well developed and arranged in several rows, with one or two rows consisting of enlarged fangs placed on the outside of the jaw; the anterior bases of the dorsal and anal fins expanded, supported by unusually long radials; dorsal-fin rays 5 to 9 (rarely 4 or 10); anal-fin rays 4 to 7 (rarely 8); all rays of the caudal fin simple:

WHIPNOSE SEADEVILS, GENUS *GIGANTACTIS* BRAUER, 1902, 18 SPECIES, P. 141

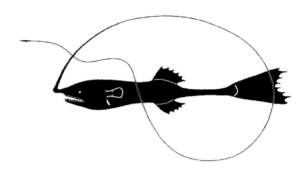

Females with the snout blunt and rounded, the illicium emerging behind the anterior margin of the upper jaw, the illicial pterygiophore fully embedded beneath skin of the head; a bulbous, terminal, escal light organ absent; the premaxillae greatly reduced, each bearing no more than two tiny teeth; the dentaries toothless or bearing no more than a few minute teeth; dorsal-fin rays 3 or 4 (rarely 5); anal-fin rays 3 or 4; innermost four rays of the caudal fin bifurcated; a rare genus, only 26 known female specimens:

TOOTHLESS SEADEVILS, GENUS *RHYNCHACTIS* REGAN, 1925C, THREE SPECIES, P. 146

---

### GENUS *GIGANTACTIS* BRAUER, 1902 (WHIPNOSE SEADEVILS)

*Figures 155–161, Table 1*

#### DISTINGUISHING CHARACTERS

While the females of both genera of the Gigantactinidae are elongate and slender, with an exceptionally large, laterally compressed illicial pterygiophore and an unusually long illicium, the two differ in a host of ways. The most striking differences are found in the structure of the luring apparatus and in the jaw dentition. The anterior end of the illicial pterygiophore of *Gigantactis* protrudes beyond the margin of the upper jaw, forming the tip of an elongate pointed snout. In contrast, the pterygiophore of *Rhynchactis* is completely concealed beneath the skin of the head and terminates behind the upper jaw, contributing to the formation of a much blunter, more rounded snout. *Gigantactis* has an elongate, club- or spindle-shaped escal bulb containing a relatively small, internal spherical photophore, bearing appendages in the form of papillae, filaments, and lobes; filaments along the stem of the illicium may also be present (Figs. 158, 159). On the other hand, *Rhynchactis* lacks an escal photophore as well as bioluminescent bacteria, in sharp contrast to all other ceratioid females except those of the Caulophrynidae, and the Neoceratiidae in which the illicium and esca are absent altogether.

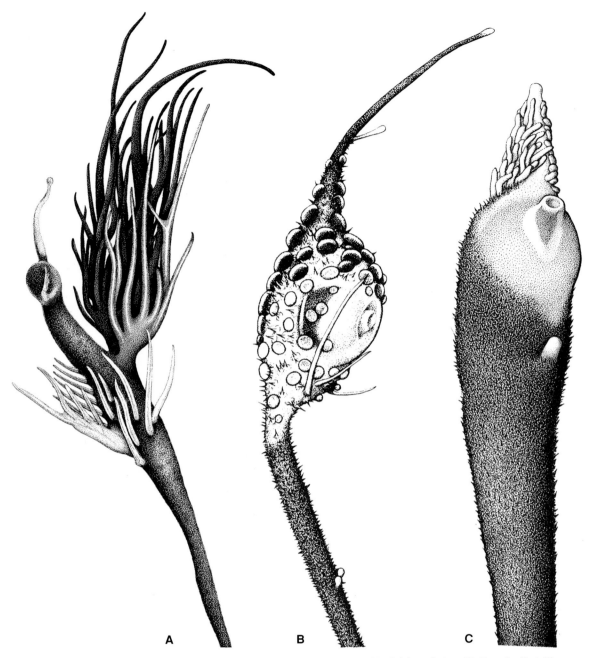

FIGURE 158. Escae of species of *Gigantactis*: (A) *G. longicirra*, 118 mm SL, ISH 2561/71, left lateral view; (B) *G. vanhoeffeni*, 38 mm, ISH 2188/71, left lateral view; (C) *G. krefffti*, holotype, 252 mm SL, ISH 1099/71, posterolateral view. Drawings by K. Elsman; after Bertelsen et al. (1981).

Unlike metamorphosed females of *Rhynchactis*, which are toothless except for a few premaxillary teeth retained in juveniles, *Gigantactis* has well-developed teeth on both the upper and lower jaws. Those of the premaxillae are small, depressible, and relatively few, arranged in one or two series. The dentary teeth are considerably larger, depressible, and more numerous, arranged in two to six, more or less distinct overlapping series, including an external series situated on the outer margin of the jaw (in contrast to all other ceratioids except for females of *Neoceratias*). Several anterior- and lateral-most dentary teeth are greatly enlarged into curved fangs. All the teeth are depressible and strongly curved, especially the large external fangs. The dentaries are connected anteriorly only by elastic ligaments; the right and left elements are mobile relative to one another, their long curved teeth capable of rotating inward from a widely outstretched open position to a situation in which they approach each other on the midline, within the cavity of the mouth (for details, see Locomotion, Food, and Feeding, Chapter Seven).

The elongate, streamlined body shape of metamorphosed females of *Gigantactis* is very similar in all species of the genus except for some variation in the relative length and depth of the caudal peduncle. The pterygiophores that support the rays of the dorsal and anal fins, especially the anterior-most elements in each series, are unusually long, giving the fins a

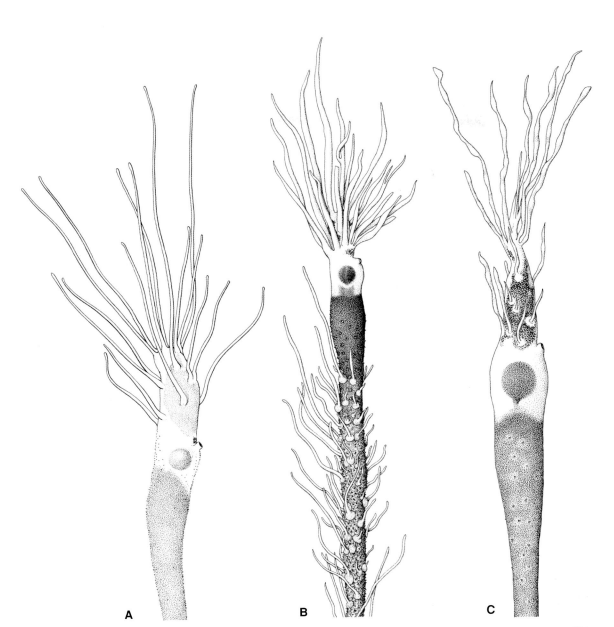

FIGURE 159. Escae of species of *Gigantactis*, left lateral views: (A) *G. macronema*, 232 mm SL, ISH 1596/71; (B) *G. macronema*, 354 mm SL, LACM 30559-020; (C) *G. savagei*, holotype, 150 mm SL, LACM 9706-041. Drawings by K. Elsman; after Bertelsen et al. (1981).

relatively large surface area. The caudal peduncle is long and slender, but broadened posteriorly by a very large hypural plate. The caudal fin itself is long, differing from all other ceratioids (including males of *Gigantactis* and both sexes of *Rhynchactis*) in having all the rays unbranched. The shape of the caudal fin and the extent of skin coverage on the caudal-fin rays show distinct interspecific variation (for details, see Part Two: A Classification of Deep-Sea Anglerfishes). Darkly pigmented skin covers each caudal-fin ray, the skin usually compressed, with broad extensions partially connecting the rays.

Numerous close-set dermal spinules cover the entire head, body, and fins of *Gigantactis*, extending out onto the illicium, and onto the esca in some species (visually obvious in females of all species, without microscopic aid). Minute dermal spinules are present in the skin of the largest known females of *Rhynchactis*, but the skin is smooth and naked in smaller specimens.

Metamorphosed males of *Gigantactis* are similar to those of other ceratioids in the shape of the body and fins, without significant intrageneric differences (Fig. 160). The eyes are smaller than those of all other ceratioids, their diameters 3.0 to 5.2% SL (approaching only those of *Centrophryne*, 5.5% SL). The anterior nostrils are close set, directed anteriorly, and slightly smaller than the posterior nostrils. The posterior nostrils are directed laterally. There 8 to 13 olfactory lamellae. The upper and lower denticular teeth are mutually free, hooked, compressed proximally, and very loosely attached to the symphyses of the maxillae and dentaries, respectively. There are 3 (rarely 2 or 4) upper denticular teeth and 4 (rarely 3 or 5) lower denticular teeth. Tooth counts vary without apparent relationship to the separation of specimens into groups.

Metamorphosed males differ significantly in pigmentation and in the occurrence of dermal spines, each of the following

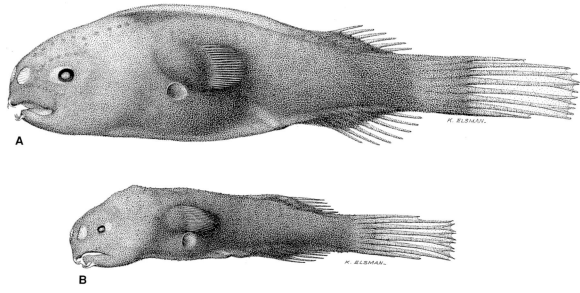

FIGURE 160. Free-living adult males of *Gigantactis* sp.: (A) Male Group I, 22 mm SL, BMNH 2004.9.12.168; (B) Male Group II, 14.5 mm SL, SIO 52-409. Drawings by K. Elsman; after Bertelsen et al. (1981).

combinations occurring: unpigmented and naked, unpigmented and spiny, pigmented and naked, and pigmented and spiny. Each combination shows a more or less distinct relationship to fin-ray counts, eye size, and subdermal pigmentation; and different combinations of these characters form the basis for the separation of *Gigantactis* males into six groups (originally proposed by Bertelsen et al., 1981:59; see accounts in Part Two: A Classification of Deep-Sea Anglerfishes).

The subdermal pigmentation of the larvae of *Gigantactis* is generally weak to completely absent but, when present, subdermal melanophores form an isolated dorsal group and are otherwise scattered on the dorsal margin of the peritoneum, on the dorsal and ventral margins of the caudal peduncle, and more rarely on the chin and underside of the head (Fig. 161). Based on combinations of subdermal pigmentation and the number of dorsal- and anal-fin rays, *Gigantactis* larvae are divided into four groups (following Bertelsen et al., 1981:62; see accounts in Part Two: A Classification of Deep-Sea Anglerfishes).

Fin-ray counts are as follows: dorsal-fin rays 5 to 9 (rarely 4 or 10), anal-fin rays 4 to 7 (rarely 8), and pectoral-fin rays 15 to 20 (rarely 14 or 22) (Table 1).

ETYMOLOGY

The name *Gigantactis* is derived from the Greek roots *gigas* or *gigantos*, meaning "gigantic"; and *aktis*, a "ray," in allusion to the unusually long first dorsal-fin spine that serves as a lure in this genus.

DIVERSITY

Eighteen species and two nominal forms of doubtful validity are based on females, five species groups are based on males, and four species groups are based on larvae:

*Gigantactis longicirra* Waterman, 1939b: 10 females, eight metamorphosed (34.5 to 221 mm), and one in metamorphosis (19.5 mm); four metamorphosed males (12 to 14.5 mm); and eight larvae (4.7 to 7.5 mm TL); Atlantic and eastern tropical Pacific oceans; holotype, MCZ 35065, 39 mm, *Atlantis* station 2894, western North Atlantic, 39°06′N, 70°16′W, closing net at 1000 m, bottom depth 2860 m, 20 July 1937 (Waterman, 1939b:82, figs. 1, 2, 1948:81, figs. 1–10; Bertelsen, 1951: 148, 150, 274, fig. 99E, F; Bertelsen et al., 1981:26, 62, 63, figs. 1A, 4F, 11A, 11D, 13A, 14A, 15A, 16, 17, 24–26, 66, 68).

*Gigantactis kreffti* Bertelsen, Pietsch, and Lavenberg, 1981: five metamorphosed females, 44 to 345 mm; eastern South Atlantic and western Pacific oceans; holotype, ISH 1099/71, 252 mm, *Walther Herwig* station 406/71, eastern South Atlantic, 39°19′S, 3°15′W, 0 to 2000 m, 19 March 1971 (Bertelsen et al., 1981:29, figs. 4D, 27–29, 65; Yamakawa, 1982:197, 362, fig. 123; Anderson and Leslie, 2001:23, fig. 15B).

*Gigantactis vanhoeffeni* Brauer, 1902: 90 known females, 86 metamorphosed (19 to 420 mm) and three in metamorphosis (16.5 to 21.5 mm); nearly cosmopolitan in the Atlantic, Pacific, and Indian oceans; lectotype, ZMB 17712, 35 mm, *Valdivia* station 239, off Zanzibar, 5°42′S, 43°36′E, open pelagic net, 2500 m (Brauer, 1902:296, 1908:103, 184, text fig. 1, pl. 31 [figs. 18–24], pl. 32 [figs. 1–6], pl. 34 [fig. 14], pl. 44 [fig. 1]; Regan, 1926:38; Parin and Golovan, 1976:271; Pietsch and Seigel, 1980:396; Bertelsen et al., 1981:31, figs. 1B, 4A, 5, 6, 8, 12, 19, 21, 30, 31, 64; Amaoka, 1983:116, 117, 198, figs. 69; Ni, 1988:329, fig. 257; Stearn and Pietsch, 1995:133, fig. 83; Anderson and Leslie, 2001:26, fig. 15E).

*Gigantactis meadi* Bertelsen, Pietsch, and Lavenberg, 1981: 20 females, 16 metamorphosed (35.5 to 353 mm) and four in late metamorphosis (19 to 21 mm); Southern Ocean; holotype, MCZ 52572, 306 mm, 34°14′S, 64°56′E (depth unknown) (Bertelsen et al., 1981:33, figs. 4B, 18, 20, 23, 32, 33, 64; Pequeño, 1989:44; Stewart and Pietsch, 1998:27, fig. 18; Anderson and Leslie, 2001:24, fig. 16).

*Gigantactis gibbsi* Bertelsen, Pietsch, and Lavenberg, 1981: four metamorphosed females (38 to 114 mm); North Atlantic Ocean; holotype, ZIN 44262, 50 mm, Gulf of Guinea, 2°01′N, 3°56′W, 0 to 465 m (Bertelsen et al., 1981:36, figs. 34, 64; Pequeño, 1989:44; Bertelsen, 1990:513).

*Gigantactis gracilicauda* Regan, 1925c: three metamorphosed females (21 to 82 mm); North Atlantic Ocean; holotype, ZMUC

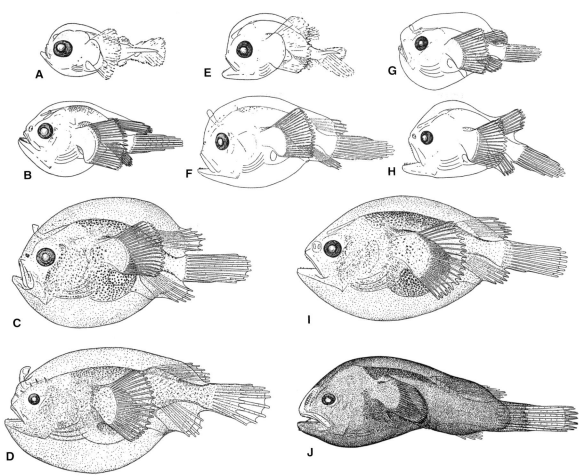

FIGURE 161. *Gigantactis* sp., early life-history stages: (A) larval male, 2.8 mm TL; (B) larval male, 8.0 mm TL; (C) larval female, 9.0 mm SL, 12.5 mm TL; (D) female in metamorphosis, 19.6 mm SL, 25 mm TL; (E) larval female, 6.3 mm TL; (F) larval male, 7.3 mm TL; (G) larval male, 4.0 mm TL; (H) larval female, 8.5 mm SL, 12.0 mm TL, ZMUC P921682; (I) male in metamorphosis, 13.0 mm SL, 17.0 mm TL; (J) male in metamorphosis, 12.2 mm SL, 17.0 mm TL. After Bertelsen (1951).

P92129, 82 mm, *Dana* station 1183(1), Caribbean Sea, 13°47′N, 61°26′W, open pelagic net, 4500 m wire, 1630 h, 24 November 1921 (Regan, 1925c:565, 1926:18, 19, 38, figs. 2, 12, pl. 10; Waterman, 1948:90, 93; Bertelsen, 1951:150; Pietsch, 1972a: 42, 45; Bertelsen et al., 1981:38, figs. 35, 64).

*Gigantactis paxtoni* Bertelsen, Pietsch, and Lavenberg, 1981: 18 metamorphosed females (50 to 305 mm); Indo-west Pacific Ocean; holotype, AMS I.20314-018, 237 mm, 100 km east of Broken Bay, New South Wales, 33°28′S, 152°33′E, 0 to 900 m over 4200 m, 14 December 1977 (Bertelsen et al., 1981:39, figs. 36–38, 64; Bertelsen and Pietsch, 1983:92, figs. 12, 13; Paulin, 1984:66, fig. 5; Stewart and Pietsch, 1998:26, fig. 17).

*Gigantactis perlatus* Beebe and Crane, 1947: 11 females, nine metamorphosed (23 to 223 mm) and two in metamorphosis (19 to 20 mm); Atlantic, Pacific, and eastern Indian oceans; holotype, CAS-SU 46487 (originally NYZS 28621), 32.5 mm, eastern Pacific *Zaca* Expedition station 225 T-1, off Jicaron Island, Panama, 7°08′N, 81°57′W, open pelagic net, 0 to 915 m 20 March 1938 (Beebe and Crane, 1947:167, text figs. 3, 13, pl. 2; Mead, 1958:133; Robins and Courtenay, 1958:151; Bertelsen et al., 1981:41, figs. 1C, 4C, 39, 40, 65; Amaoka, 1983:118, 119, 198, fig. 70; Anderson and Leslie, 2001:25, fig. 15D).

*Gigantactis elsmani* Bertelsen, Pietsch, and Lavenberg, 1981: 10 metamorphosed females (283 to 435 mm) and one tentatively identified female in metamorphosis (11.5 mm); Atlantic and Pacific oceans; holotype, ISH 1360/71, 384 mm, *Walther Herwig* station 459/71, 10°57′S, 11°20′W, 0 to 1900 m, 1818 to 2218 h, 7 April 1971 (Bertelsen et al., 1981:43, figs. 4E, 41, 42, 65; Amaoka, 1983:118, 119, 199, fig. 71; Fedorov, 1994:414, fig.; Anderson and Leslie, 2001:23, fig. 15A).

*Gigantactis golovani* Bertelsen, Pietsch, and Lavenberg, 1981: four metamorphosed females (25 to 179 mm); Atlantic Ocean; holotype, ISH 2250/71, 179 mm, *Walther Herwig* station 471-III/71, eastern tropical Atlantic, 2°27′S, 19°00′W, MT, 0 to 660 m, 2054 to 2225 h, 10 April 1971 (Bertelsen et al., 1981:44, figs. 4J, 43, 44, 65).

*Gigantactis gargantua* Bertelsen, Pietsch, and Lavenberg, 1981: 11 females, 10 metamorphosed (49 to 408 mm) and one in metamorphosis (25 mm); North Pacific and eastern South Indian oceans; holotype, LACM 6903-32, 408 mm, *Velero IV*, eastern Pacific, San Clemente Basin, 32°16′N, 117°43′W, 0 to 1250 m, bottom depth 1775 m, 21 February 1966 (Bertelsen et al., 1981:46, figs. 1D, 4H, 45, 46, 66; Amaoka, 1983:120, 121, 199, fig. 72; Ni, 1988:330, fig. 258; Meng et al., 1995:441, fig. 591).

*Gigantactis watermani* Bertelsen, Pietsch, and Lavenberg, 1981: two metamorphosed females (99 to 305 mm); eastern tropical Atlantic and western South Pacific; holotype, ISH 2330/71, 99 mm, *Walther Herwig* station 478/71, eastern tropical Atlantic, 1°04′N, 18°22′W, 0 to 2100 m, 1842 to 2245 h, 12 April 1971 (Bertelsen et al., 1981:49, figs. 47–49, 66).

*Gigantactis herwigi* Bertelsen, Pietsch, and Lavenberg, 1981: two metamorphosed females (105 to 262 mm); Atlantic Ocean; holotype, ISH 972/68, 262 mm, *Walther Herwig* station 17/68, tropical Atlantic, 4°43′S, 26°39′W, MT, 0 to 2000 m, 1155 to 1215 h, 4 February 1968 (Bertelsen et al., 1981:49, figs. 4G, 50, 51, 66).

*Gigantactis macronema* Regan, 1925c: 11 metamorphosed females (34 to 354 mm); Atlantic and eastern North Pacific oceans; holotype, ZMUC P92130, 98 mm, *Dana* station 1365(9), central North Atlantic, 31°47′N, 41°41′W, 5000 m wire, 1030 h, 8 June 1922 (Regan, 1925c:565; Regan and Trewavas, 1932:93, 94; Waterman, 1939b:84, 1948:130; Bertelsen, 1951:150, fig. 101; Pietsch, 1972a:29, 34, 35, 41, 42, 45; Bertelsen et al., 1981:50, figs. 1E, 4I, 22, 52, 53, 64; Fujii, 1983:264, fig.).

*Gigantactis savagei* Bertelsen, Pietsch, and Lavenberg, 1981: six females, two metamorphosed (56 to 150 mm), one in metamorphosis (19 mm), and three (33 to 44 mm) tentatively assigned to this species; eastern North Pacific; holotype, LACM 9706-41, 150 mm, *Velero IV*, Baja California, Cortez Bank, 31°40′N, 120°23′W, 0 to 650 m 31 July 1966 (Bertelsen et al., 1981:53, figs. 54, 55, 67).

*Gigantactis microdontis* Bertelsen, Pietsch, and Lavenberg, 1981: 12 females, 11 metamorphosed (25.5 to 310 mm) and one in late metamorphosis (19.5 mm); eastern Pacific; holotype, MCZ 52574, 66 mm, *Anton Bruun* cruise 18B, station 739B, off coast of Peru, 15°12′S, 75°44′W, 0 to 700 m, bottom depth 1060 m, 25 August 1966 (Bertelsen et al., 1981:54, figs. 56–58, 67).

*Gigantactis ios* Bertelsen, Pietsch, and Lavenberg, 1981: four metamorphosed female (38 to 81 mm); eastern North Atlantic; holotype, BMNH 1977.9.13.1, 57 mm, *Discovery* station 7856-48, eastern North Atlantic, southwest of Madeira, 29°49′N, 23°00′W, RMT, 1005 to 1250 m (Bertelsen et al., 1981:56, figs. 59, 57; Swinney, 1995a:52, 56; Jónsson and Pálsson, 1999:203, figs. 7, 11).

*Gigantactis longicauda* Bertelsen and Pietsch, 2002: a single metamorphosed female, the holotype, ISH 5539/79, 114 mm, *Anton Dohrn*, Sargasso Sea Expedition, cruise 210/92, station 98/79, western North Atlantic, 23°46′N, 58°59′W, MT-1600, 500 to 1200 (–0) m, bottom depth 5600 m, 28 March 1979 (Bertelsen and Pietsch, 2002:958, figs. 1, 2).

*Gigantactis ovifer* Regan and Trewavas, 1932, *nomen dubium*: a single metamorphosed female, the holotype, ZMUC P92131, 30 mm, *Dana* station 3731(12), South China Sea, 14°37′N, 119°52′E, 2500 m wire, 0200 h, 17 June 1929 (Regan and Trewavas, 1932:93, 95, fig. 152; Fraser-Brunner, 1935:326; Waterman, 1939b:85; Bertelsen, 1951:150; Pietsch, 1972a:42, 45; Bertelsen et al., 1981:57).

*Gigantactis filibulbosus* Fraser-Brunner, 1935, *nomen dubium*: a single metamorphosed female, the holotype, BMNH 1934.8.8.92, 25 mm, *Dynevor Castle*, Irish Atlantic slope, 53°15′N, 12°28′W, 0 to 320 m, July 1934 (Fraser-Brunner, 1935:326, fig. 5; Waterman, 1939b:85; Bertelsen, 1951:150; Robins and Courtenay, 1958:151; Wheeler, 1969:585; Maul, 1973:675; Bertelsen et al., 1981:57).

*Gigantactis* Male Group I: 10 metamorphosed males (15 to 22 mm); based in part on *Laevoceratias liparis* Parr, 1927; Atlantic and Pacific oceans; holotype, YPM 2013, 17 mm, 24°11′N, 75°37′W, 2440 m wire (Parr, 1927:33, fig. 13; Bertelsen, 1951:70, fig. 29; Bertelsen et al., 1981:59, figs. 61A, 68).

*Gigantactis* Male Group II: 22 metamorphosed males (10.5 to 15.5 mm); based in part on *Teleotrema microphthalmus* Regan and Trewavas, 1932; Atlantic, Pacific, and Indian oceans; holotype, ZMUC P92127, 16 mm, *Dana* station 4003(2), eastern tropical Atlantic, west of Sierra Leone, 8°26′N, 15°11′W, 5000 m wire, 1130 h, 9 March 1930 (Regan and Trewavas, 1932:93, fig. 149; Bertelsen, 1951:146, 152, 153, figs. 102C, 103E; Pietsch and Seigel, 1980:396; Bertelsen et al., 1981:60, figs. 61B, 68).

*Gigantactis* Male Group III: three metamorphosed males, 14 to 16 mm; central Pacific Ocean, leeward Oahu (Bertelsen et al., 1981:60, fig. 68).

*Gigantactis* Male Group IV: a single male in late metamorphosis, 16.5 mm; western tropical Pacific, Banda Sea (Pietsch and Seigel, 1980:396; Bertelsen et al., 1981:61, fig. 68).

*Gigantactis* Male Group V: a single male in early metamorphosis, 14.5 mm; central Pacific Ocean, leeward Oahu (Bertelsen et al., 1981:61, figs. 62, 68).

*Gigantactis* Larval Group A: probably containing larvae of species of the *G. vanhoeffeni* group (see Bertelsen, 1951:148, fig. 99A–D, I–J; Bertelsen et al., 1981:62, 63, fig. 68).

*Gigantactis* Larval Group B: probably containing larvae of species with dorsal subdermal pigmentation in metamorphosal stages very weak or absent (see Bertelsen, 1951:148, 150, fig. 99G–H; Bertelsen et al., 1981:62, 63, fig. 68).

*Gigantactis* Larval Group C: containing larvae of *G. longicirra*; eight specimens, 4.7 to 7.5 mm TL (see Bertelsen, 1951:148, 150, fig. 99E, F; Bertelsen et al., 1981:62, 63, fig. 68).

*Gigantactis* Larval Group D: containing larvae of one or more species of *Gigantactis* for which metamorphosal stages have not yet been identified (see Bertelsen et al., 1981:62, 63, fig. 68).

## GENUS *RHYNCHACTIS* REGAN, 1925c
## (TOOTHLESS SEADEVILS)

Figures 162–166, Table 1

### DISTINGUISHING CHARACTERS

In many ways the genus *Rhynchactis* looks like a pared down version of its sister genus *Gigantactis*. While its general appearance is similar—the body long and slender, with an exceptionally long caudal peduncle and caudal fin—it lacks the well-muscled robust look of *Gigantactis* (Fig. 162). Many structures are reduced or absent and this is especially evident in the jaws and dentition: each premaxilla is represented by only a small anterior remnant, bearing in smaller specimens no more than 2 tiny teeth; the maxillae are absent (but present in larvae); the lower jaw is greatly reduced in all metamorphosed females and the dentaries are toothless except for some tiny remnants in small specimens. Frontal and parietal bones are absent. The pterygiophore that supports the illicium is short, fully embedded within the skin of the head, and the illicium itself emerges from a blunt, rounded snout, slightly behind the anterior-most margin of the upper jaw. The illicium of metamorphosed females—in some specimens bearing elongate slender filaments of a complex structure and unknown function (Figs. 163, 164)—is longer than the body, remaining in approximately constant proportion to standard length during growth. The caudal fin is distinctly divided into upper and lower portions containing 4 and 5 caudal-fin rays, respectively. The skin is covered with minute spinules in larger specimens, but juveniles are naked. A bulbous, terminal, escal light organ is absent. For more details, see the comparison of gigantactinid genera above.

A dense pavement of strange, white, papillae-like glands, found in no other anglerfish, covers the inner surface of the upper and lower jaw of *Rhynchactis*, each gland with a more or

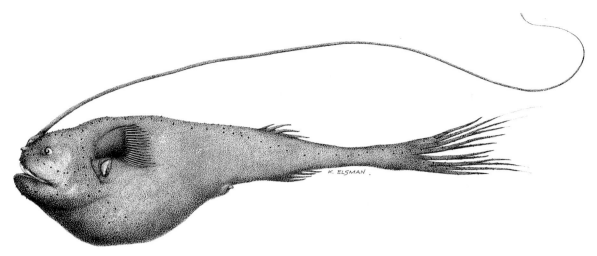

FIGURE 162. *Rhynchactis* sp., unidentifiable female, 126 mm SL, ISH 2560/71. Drawing by K. Elsman; after Bertelsen et al. (1981).

FIGURE 163. Distal portion of the illicium of *Rhynchactis macrothrix*, paratype, 27 mm SL, BMNH 2004.7.9.17. Drawing by K. Elsman; after Bertelsen et al. (1981).

less distinct central groove and outlined by pigmented skin (Fig. 165). Histologically each gland consists of a simple short tube, with a more or less pigmented wall, covered internally with large glandular cells that nearly fill the lumen of the tube (see Bertelsen et al., 1981:10). It is tempting to think that these glands are bioluminescent, but there is no direct evidence to support this idea and their true function is presently unknown.

Metamorphosed males are easily distinguished from those of *Gigantactis* (Fig. 166). In addition to the reduced number of dorsal- and anal-fin rays, there are 4 to 6 upper denticular teeth and 6 or 7 lower denticular teeth. At least some of the denticular teeth are paired, with broad conical bases. The diameter of the eye is 2.4% SL in the largest known specimens (17 to 18.5 mm). There are 13 to 15 olfactory lamellae. The depth of the nostrils is 10 to 12% SL. The skin is naked and weakly pigmented (subdermal pigment is distributed as described for larvae; see below).

The larvae of *Rhynchactis* differ from those of *Gigantactis* in having an extremely dense dorsal group of subdermal melanophores that is contiguous with the peritoneal pigment, but never extending back to the base of the dorsal fin (Fig. 55. The length of the pectoral fin is slightly longer than that of *Gigantactis*, measuring approximately 50 to 55% SL.

There are 3 or 4 (rarely 5) dorsal-fin rays, 3 or 4 anal-fin rays, 17 to 19 pectoral-fin rays (although 20 have been recorded in some larvae; see Bertelsen et al., 1981:67), and 9 (2 simple + 4 branched + 3 simple) caudal-fin rays (Table 1).

### ETYMOLOGY

Regan's (1925c) choice of a name for this taxon is better suited to its sister genus *Gigantactis*, being derived from the Greek *rhynch*, meaning a "beak" or "snout"; and *aktis* a "ray," in reference to the illicium that arises near the tip of the snout.

### DIVERSITY

Three species are recognized:

*Rhynchactis leptonema* Regan, 1925c: five metamorphosed females (42 to 118 mm; western tropical Atlantic, western and central Pacific; holotype, ZMUC P92133, 42 mm, *Dana* station 1171(11), western tropical Atlantic, 8°19′N, 44°35′W, 3000 m of wire, 0800 h, 13 November 1921 (Regan, 1925c:565; Regan and Trewavas, 1932:95; Bertelsen, 1951:153, fig. 104; Pietsch, 1972a:42, 45; Bertelsen et al., 1981:66, figs. 2, 3, 7, 9, 10, 11B, C, E, 13B, 14B, 15B, 63, 69; Bertelsen and Pietsch, 1998b:587, figs. 1, 7).

*Rhynchactis macrothrix* Bertelsen and Pietsch, 1998b: 10 metamorphosed females (27 to 152 mm); Atlantic, western Pacific,

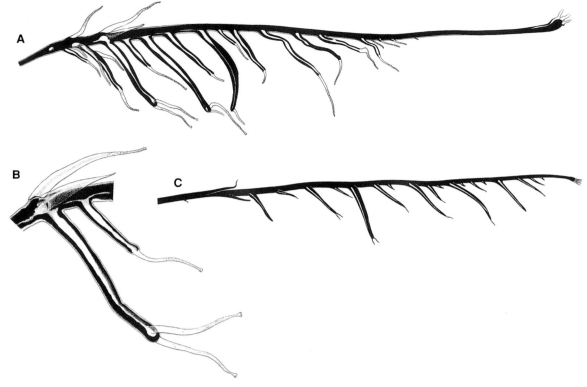

FIGURE 164. Distal portion of the illicium of *Rhynchactis macrothrix*: (A) holotype, 110 mm SL, ISH 605/74; (B) holotype, 110 mm SL, ISH 605/74, detail showing secondary illicial filaments at midlength of illicium; (C) paratype, 130 mm SL, SIOM uncataloged. Drawings by K. Elsman; after Bertelsen et al. (1981).

and western Indian oceans; holotype, ISH 605/74, 110 mm, *Anton Dohrn*, Gate Expedition, station 39/74, central Atlantic, 7°55′N, 32°41′W, 0 to 2000 m, bottom depth 4840 m, 21 July 1974 (Bertelsen and Pietsch, 1998b:587, figs. 2–4, 7).

*Rhynchactis microthrix* Bertelsen and Pietsch, 1998b: a single metamorphosed female, the holotype, MCZ 57516, 113 mm, *Anton Bruun* cruise VI, station 337B, western Indian Ocean, 00°14′S, 65°03′E, 0 to 2250 m, 28 May 1964 (Bertelsen and Pietsch, 1998b:588, figs. 5–7).

## FAMILY LINOPHRYNIDAE Regan, 1925c
(Leftvent Seadevils)

*Figures 167–185, Tables 1, 7*

Linophrynids are spectacular even among the most bizarre of their anglerfish relatives in having the largest mouths, with the largest and longest daggerlike teeth of any fish and perhaps of any vertebrate. In addition to an elaborately adorned esca that harbors symbiotic bioluminescent bacteria, most members of the family are uniquely equipped with a complex bioluminescent chin barbel. These features, coupled with low dorsal and anal fin-ray counts and a sinistral anus, easily differentiate female linophrynids from those of all other lophiiforms. The taxonomic history of the family begins with Collett's (1886:138) description of *Linophryne lucifer*, based on a "little black fish" found in May 1877 floating alive on the surface off Madeira, apparently incapacitated by the attempted ingestion of a large fish identified only as a "*Scopeloid* . . . not far from being half a length longer than the [anglerfish] itself." Following this brief introduction, the family remained monotypic until Regan (1925b, 1925c, 1926) and Regan and Trewavas (1932) reported on the ceratioids brought back by the *Dana*. Together these publications resulted in the addition of 16 new linophrynid species based on females, including *Photocorynus spiniceps* (1925b), *Borophryne apogon* (1925c), and *Acentrophryne longidens* (1926). Beebe (1926b, 1932b), Parr (1927), and Borodin (1930a), all working at about the same time on new material collected in the western North Atlantic, described an additional seven species; and somewhat later, Waterman (1939b), Imai (1941), and Beebe and Crane (1947) each added one more. So by the time of Bertelsen's (1951) great monograph on the Ceratioidei, in which he examined and compared all available material around the world, the total number of nominal species based on females had reached 26, of which Bertelsen recognized only 16.

In 1965, Bertelsen, in coauthorship with Gerhard Krefft, published the first of a series of eight papers on *Linophryne*, directing his efforts primarily toward females of the genus, but occasionally describing new records of parasitic males as they were discovered. By the time this work was completed in 1982, he had synonymized one species, resurrected three, and added six new species, bringing the total number of recognized linophrynids to 25, including Maul's (1961) *Linophryne maderensis*. Ofer Gon's (1992) description of *Linophryne andersoni* and Pietsch and Shimazaki's (2005) *Acentrophryne dolichonema* have now raised the number to 27.

In the meantime, while female linophrynids were being described in ever-increasing numbers, males too were being discovered. In 1902, August Brauer described four peculiar little fishes, correctly identified by him as deep-sea anglerfishes, but which differed from any known up to that time in lacking the

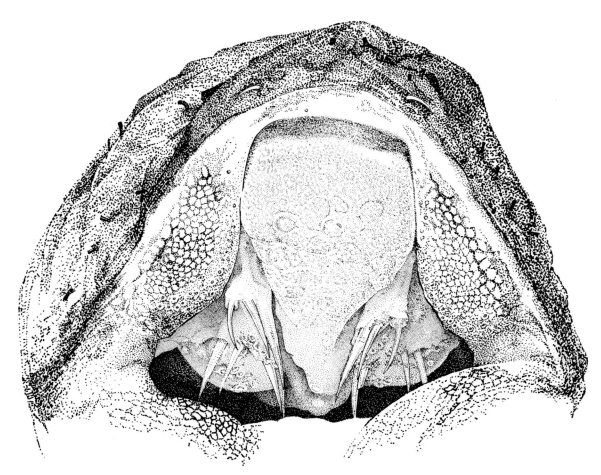

FIGURE 165. Mouth cavity of *Rhynchactis* sp., unidentifiable female, 60 mm SL, ISH 2332/71, showing oral glands of unknown function. Drawing by E. Beyerholm; after Bertelsen et al. (1981).

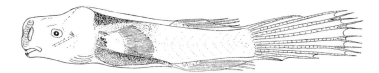

FIGURE 166. *Rhynchactis* sp., juvenile male, 18.5 mm SL, ZMUC P921732. After Bertelsen (1951).

angling apparatus and having greatly enlarged nostrils, large anteriorly directed eyes, and rostral teeth or denticles at the tip of the upper and lower jaws. Two species were recognized, *Aceratias mollis* and *A. macrorhinus*. Two of the three specimens assigned to the latter species, differing slightly from the third, were given subspecific rank, *A. macrorhinus indicus*; and all were subsequently placed in a family of their own, the Aceratiidae (Brauer, 1906). Brauer's family was later recognized by Regan (1912), but he divided the material into two genera: *Aceratias* Brauer, 1902, containing *A. macrorhinus* and *A. indicus*; and *Haplophryne*, a new genus erected to contain a single species, *H. mollis*. Additional genera and species were added to the Aceratiidae by Regan (1925b), but his discovery of parasitic males (Regan, 1925c) and the subsequent studies of "aceratiids" by Parr (1930a, 1930b) led to the surprising conclusion that the latter were nothing more than free-living ceratioid males.

Following the discovery of sexual parasitism, the numerous genera that had been erected to accommodate the diversity found among free-living males were redistributed among preexisting taxa based on females (see Table 7). Among these genera, *Aceratias* and *Haplophryne* were referred to the Linophrynidae by Regan and Trewavas (1932:111–113). Bertelsen (1951:166–168) went a step further by placing *Aceratias* in the synonymy of *Linophryne* and expressing agreement with Regan and Trewavas (1932:111) that *Haplophryne mollis* might represent males of the genus *Edriolychnus* Regan, 1925b. Conclusive demonstration of the latter was subsequently published by Bertelsen (1981) and Munk and Bertelsen (1983), along with convincing evidence that Brauer's (1902) *Aceratias indicus*, now recognized as *Linophryne indica* (Brauer), is a senior synonym of *Linophryne corymbifera* Regan and Trewavas, 1932.

Thanks to the detailed and careful work of Bertelsen (1951, 1982), adequate characters are now available to identify most free-living linophrynid males to genus, and to subgenus within *Linophryne* (see the keys provided in Part Two: A Classification of Deep-Sea Anglerfishes), but, as with all ceratioid genera that contain more than one species, specific identification is still not possible. *Photocorynus*, *Haplophryne*, *Borophryne*, and *Linophryne* are all relatively well represented in collections by metamorphosed females and free-living males, and each genus is known to have parasitic males. *Acentrophryne*, however, is represented by only five metamorphosed females; males and larvae are unknown. Thirty parasitized females of *Linophryne* are now present in collections around the world, representing only 12 of the 22 recognized species of the genus (see Reproduction and Early Life History, Chapter Eight). Clearly much about this family remains to be discovered. The account below closely follows the revision of Bertelsen (1982), with updates provided by Munk and Bertelsen (1983), Bertelsen (1986), Gon (1992), Pietsch (2005b), and Pietsch and Shimazaki (2005).

DISTINGUISHING CHARACTERS

The Linophrynidae is one of the most distinctive families of the suborder. Any one of three characters is sufficient to separate metamorphosed males and females as well as larvae from those of all other ceratioid taxa: all linophrynids have a reduced number of median-fin rays: only 3 (rarely 4) dorsal-fin rays and 3 (rarely 2 or 4) anal-fin rays; the number of branchiostegal rays is also reduced, only 5 (rarely 4); and all are characterized by a strange displacement of the vent or anal opening: normally positioned on the posteroventral midline, the intestine and ducts of the urogenital system terminate and open to the left of center (sinistral) in all members of the family (e.g., see Figs. 31, 171C, 172). With the exception of some flatfishes (order Pleuronectiformes), in which displacement to the right (e.g., all bothids) or left (most pleuronectids) is associated with full-body bilateral asymmetry (Kunio Amaoka, personal communication, 26 July 2007), this kind of asymmetry is apparently not found in any other group of fishes and, except for rare genetic anomalies recorded in the medical literature, apparently unknown in other vertebrates. The cause of this displacement and its function, if any, are unknown.

Aside from these three shared derived characters (and a few additional autapomorphic features of internal anatomy detailed in Part Two: A Classification of Deep-Sea Anglerfishes; see also Evolutionary Relationships, Chapter Four), morphological diversity with the family is so great that a good understanding of the group is best conveyed by the generic accounts provided below. But, in general, the body of metamorphosed females is short, more or less oval in shape to globular. The snout is relatively short as well, but the head is large, its length, when measured from the tip of the snout to the base of the pectoral fin, usually exceeding 50% SL. The mouth is also large, that of some species competing for the title of the largest mouth of any vertebrate. The jaws are more or less equal anteriorly, the lower jaw of some species, extending slightly beyond the upper. The lower jaw usually bears a symphysial spine (absent in the *Linophryne* subgenus *Stephanophryne*). Sphenotic spines are well developed, but quadrate and articular spines are absent. Like its close relatives, the neoceratiids and gigantactinids, the hyomandibular has a single head. The opercle is bifurcated, its posterior margin moderately concave. The subopercle is long and extremely slender, without a spine or projection on the anterior margin. The angular is produced to form a sharp spine in some taxa and, in contrast to all other ceratioids, the preopercle usually bears one or more prominent spines (absent in *Acentrophryne*). The skin is everywhere smooth and naked, without dermal spinules.

The dentition of linophrynid females is highly variable among genera: the teeth of *Acentrophryne*, *Borophryne*, and *Linophryne* are long and few, arranged in several oblique longitudinal series; while those of *Haplophryne* and *Photocorynus* are considerably shorter and more numerous, arranged in several series (for details, see the generic accounts below). The vomer is well toothed in *Acentrophryne*, *Borophryne*, and *Linophryne*, but toothless in *Photocorynus* and *Haplophryne*.

The pterygiophore of the illicium is unusually short—its anterior end hidden beneath the skin of the head or protruding slightly on the snout—and lacks a remnant of the second cephalic spine. The length of the illicium varies enormously, from especially short and almost totally embedded within the tissue of the esca in *Haplophryne*, *Photocorynus*, *Borophryne*, and several species of *Linophryne*; but reaching 35 to 40% SL in other species of *Linophryne* and about 70% SL in *Acentrophryne*. The morphology of the esca varies enormously as well, from simple, without adornment of any kind, to highly ornate (for details see the generic accounts below as well as Part Two: A Classification of Deep-Sea Anglerfishes). A bioluminescent hyoid barbel is present in *Linophryne*, but absent in all other genera.

Free-living and parasitic males are known for all five genera, except *Acentrophryne*. The eyes of metamorphosed free-living males are relatively large (their diameter 6 to 9% SL), more or less tubular, and directed anteriorly. The olfactory organs are inflated and moderately to strongly enlarged. There are 3 to 13 olfactory lamellae, but the number is variable among genera. Jaw teeth are present in the males of *Photocorynus* and *Haplophryne* but absent in those of *Borophryne* and *Linophryne*. The upper and lower denticular bones bear 3 to 7 and 2 to 13 teeth, respectively. Sphenotic spines are prominent in the males of *Borophryne* and in the *Linophryne* subgenus *Linophryne*, but absent in all other taxa. A hyoid barbel is absent in the males of all genera. The skin is smooth and naked, without dermal spinules. The denticular teeth, eyes, and olfactory organs of parasitic males appear somewhat degenerated; the belly is greatly inflated in mature specimens.

The larvae are more elongate than those of most other ceratioids, the length of the head generally about 45% SL. The skin of the head and body is moderately to highly inflated. The pectoral fins are relatively short. A dorsal group of subdermal melanophores is absent. A small rudiment of a hyoid barbel is present in the largest known female larvae of *Linophryne*. In contrast to the larvae of all other ceratioids, relatively well-developed, pointed sphenotic spines are present in *Borophryne* and in the *Linophryne* subgenus *Linophryne*.

The color of metamorphosed females is usually dark brown to black over the entire surface of the body, except for the escal appendages, the distal parts of the escal bulb, the barbel of *Linophryne*, and fin rays. The skin is everywhere unpigmented in *Haplophryne* (in contrast to all other metamorphosed ceratioid females). The free-living males of *Linophryne* are dark brown to black but unpigmented in all other linophrynid genera. The parasitic males of *Borophryne* and *Linophryne* are dark brown to black, but those of *Photocorynus* and *Haplophryne* are unpigmented.

There are 13 to 19 pectoral-fin rays and 9 (2 simple + 4 bifurcated + 3 simple) caudal-fin rays, the ninth very short in *Photocorynus*, about one-half the length of the eighth caudal-fin ray in all other genera (Table 1).

The largest known member of the family is a 275-mm specimen of *Linophryne lucifer* (ZMUC P922443, with a 29-mm parasitic male); several additional specimens of *Linophryne* are larger than 180 mm. The largest recorded females of the other four genera of the family range from 50 to 159 mm: 69 mm in *Photocorynus* (SIO 53-356), 159 mm in *Haplophryne* (NMNZ P.21248), 105 mm in *Acentrophryne* (HUMZ 175257), and 101 mm in *Borophryne*

(LACM 30053-10). The largest known free-living male of *Photocorynus spiniceps* is 8.6 mm (ZMUC P921727); maximum known lengths in the other genera are 16 to 21 mm. The largest known parasitic males are 7.3 mm in *Photocorynus* (ZMUC P92134), 15 mm in *Haplophryne* (AMS I.21365-8), 22 mm in *Borophryne* (LACM 30053-10), and 30 mm in *Linophryne* (IMB).

The family consists of five genera, three of which contain only a single species, while the fourth, *Acentrophryne*, contains two species, and the fifth, *Linophryne*, contains 22 species:

### SYNOPSIS OF THE GENERA OF THE FAMILY LINOPHRYNIDAE

The following key is provided to allow for the placement of linophrynids into the proper genus. Like the other keys in this chapter, each entry consists of a combination of features that together serve to differentiate each genus. It works by progressively eliminating the most morphologically unique genus; for that reason it should always be entered from the beginning. All character states listed for each genus must correspond to the specimen being keyed; if not, the user should proceed to the next set of character states. The emphasis here, and throughout this chapter, is on the more common and much better known females, which are, with few exceptions, the defining entities for all ceratioid families, genera, and species. For a full overview of both genders, as well as larvae, including traditional dichotomous keys to the identification of all taxa, see Part Two: A Classification of Deep-Sea Anglerfishes.

---

Females without a hyoid barbel; the illicium short, less than 10% SL, nearly fully embedded within the tissue of the esca; frontal, sphenotic, epiotic, and posttemporal spines present; the preopercle with 5 or 6 spines; jaw teeth numerous, but small and close-set, arranged in several overlapping series, large fangs absent; vomerine teeth absent; the esca nearly sessile on the snout, without appendages; the ventralmost caudal-fin ray extremely short; the skin only lightly pigmented; a rare ceratioid genus, with only 31 cataloged female specimens:

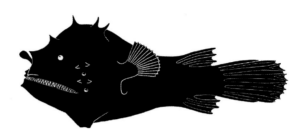

SPINYHEAD SEADEVILS, GENUS *PHOTOCORYNUS* REGAN, 1925B, A SINGLE SPECIES, P. 152

Females without a hyoid barbel; the illicium extremely short, nearly fully embedded within the tissue of the esca; epiotic and posttemporal spines absent; a single preopercular spine, divided distally to form 2 to 5 radiating cusps; jaw teeth numerous, but small and close-set, placed in several overlapping series, large fangs absent; vomerine teeth absent; the esca sessile on the snout, bearing a single, short, posterior escal appendage, divided distally into 2 to 6 short branches (simple in some juveniles); the lowermost caudal-fin ray about half the length of the adjacent ray; the skin without pigment, appearing white in preservation:

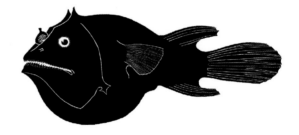

GHOSTLY SEADEVILS, GENUS *HAPLOPHRYNE* REGAN, 1912, A SINGLE SPECIES, P. 154

Females without a hyoid barbel; the illicium long, about 35 to 70% SL; epiotic and posttemporal spines absent; the preopercle sharply angled posteriorly, but without a spine on the posterior margin; jaw teeth few, but large and widely spaced, several forming prominent fangs; the vomer with 2 to 6 large teeth; the esca large, usually bearing a single, unpigmented distal appendage; the lowermost caudal-fin ray about half the length of the adjacent ray; the caudal peduncle unusually short, the dorsal and anal fins terminating nearly at the base of the caudal-fin rays; the skin darkly pigmented, except for distal parts of the esca; extremely rare, only five known female specimens:

FANGTOOTH SEADEVILS, GENUS *ACENTROPHRYNE* REGAN, 1926, TWO SPECIES, P. 157

Females without a hyoid barbel; the illicium short, about 20% SL; epiotic and posttemporal spines absent; the preopercle sharply angled posteriorly, bearing a prominent spine on the posterior margin; jaw teeth few, but large and widely spaced, several forming prominent fangs; the vomer with 0 to 4 teeth; the esca large (its diameter about 10% SL), bearing a prominent bifurcated terminal appendage and numerous fine filaments on each side; the lowermost caudal-fin ray about half the length of the adjacent ray; the skin darkly pigmented, except for distal parts of the esca and, except in some large specimens, the distal parts of the fins:

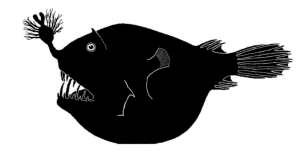

GREEDY SEADEVILS, GENUS *BOROPHRYNE* REGAN, 1925C, A SINGLE SPECIES, P. 158

Females with a conspicuous hyoid barbel; illicial length variable, less than 20% to nearly 40% SL; the anterodorsal margin of the frontals bearing a small spine in some species; epiotic and posttemporal spines absent; the preopercle sharply angled posteriorly, bearing a prominent spine on the posterior margin; a symphysial spine of the lower jaw present or absent; the angular and posteroventral part of the articular forming a sharp spine; jaw teeth relatively few (fewer than 30 on each premaxilla, fewer than 20 on each dentary), but large and widely spaced, several forming prominent fangs (some as long as 24% SL); the vomer with 2 to 7 teeth; escal morphology with extreme interspecific variation in the number, position, and shape of the escal appendages; similar variation in the length, pattern of branching, and shape of the bioluminescent structures of the hyoid barbel; the lowermost caudal-fin ray about half the length of the adjacent ray; the skin darkly pigmented, except for distal parts of the esca and barbel, and on the distal parts of the fins of all but some of the largest known females:

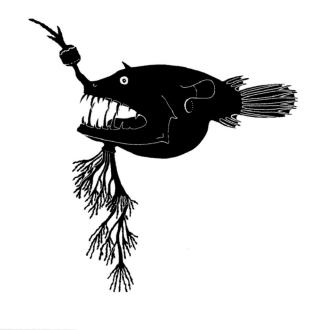

BEARDED SEADEVILS, GENUS *LINOPHRYNE* COLLETT, 1886, 22 SPECIES, P. 162

## GENUS *PHOTOCORYNUS* REGAN, 1925b (SPINYHEAD SEADEVILS)

*Figures 167–169, Table 1*

The genus *Photocorynus* figures largely in the history of ceratioid biology. Soon after Regan (1925a, 1925b) realized that the tiny fishes attached to much larger individuals of *Ceratias holboelli* were dwarfed parasitic males, a search of all available material revealed a second example of parasitism in a single 46-mm specimen from the Gulf of Panama that he named *Photocorynus spiniceps*. This discovery lead Regan (1925b:395) to assume that the sexual parasitic mode of reproduction was common to all ceratioids: "Seeing that dwarfed males, parasitic on the females and attached to them in the same way, are known to occur in two very distinct genera, it seems not unreasonable to suppose that the common ancestor of these genera had males of the same kind. Therefore it is probable that all the Ceratioid Fishes may have males of this type."

Although Regan (1925b) was not quite correct—we know now that reproduction in these fishes is considerably more complex (see Reproduction and Early Life History, Chapter Eight)—his conclusion opened the door to the realization that all the large ceratioids then known were females and that free-living males might be found within the large number of unidentified postlarval lophiiforms then present in the *Dana* collections.

In describing *Photocorynus*, Regan (1925b) identified a number of differences from all previously recognized genera that prompted him (Regan, 1926:16) to place the genus within a "well-marked" family of its own, the Photocorynidae, the skull of which he concluded "is less specialized than that of any other ceratioid." While admitting a possible close relationship with the Linophrynidae, Regan and Trewavas (1932:44) continued to recognize the family, but Bertelsen (1951:161) found no reason to do so, citing many osteological similarities between females of *Photocorynus* and linophrynids, as well as a surprisingly close resemblance between the free-living males of these taxa, especially the shared development of telescopic eyes.

In addition to reallocating *Photocorynus* to the Linophrynidae, Bertelsen (1951:161, 165) described and illustrated the first known free-living males, adding crucial osteological evidence of a close linophrynid affinity for the genus. Grey (1956a:270) described its geographic distribution, then thought to be restricted to the eastern tropical Pacific; Parin et al. (1973:146), Paulin et al. (1989:138), and Swinney (1995a:52, 57) each reported additional specimens, extending the known range of the genus into the Atlantic and Indian oceans; and Pietsch and

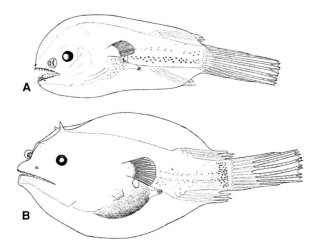

FIGURE 172. *Haplophryne mollis*: (A) metamorphosed free-living male, 13 mm SL, ZMUC P921901; (B) female in metamorphosis, 28 mm SL, ZMUC P921778. After Bertelsen (1951).

cannot be distinguished from those of the *Linophryne* subgenus *Rhizophryne*, and are thus included in an artificial assemblage of larvae here termed "Hyaloceratias" as first described and defined by Bertelsen (1951:180, 189).

Fin-ray counts are shared by both genders: the dorsal and anal fins each contain 3 rays; the pectoral fin, 15 or 16 rays; and the caudal-fin, 9 rays (2 simple + 4 bifurcated + 3 simple), the ninth ray about one-half the length of the eighth (Table 1).

### ETYMOLOGY

Regan's (11912) original description of *Haplophryne*, a single short phrase—"with depressible teeth and without nasal papilla"—meant to accommodate only one tiny male specimen, provided no explanation for the name. The Greek *haplos*, however, means "single" or "simple," so no doubt he thought this male was rather simple, with nothing very unusual about it. Thus, combined with the Greek route *phryne*, meaning "toad," we are left with a "simple toad."

### DIVERSITY

A single species is recognized:

*Haplophryne mollis* (Brauer, 1902): 88 metamorphosed and metamorphic females (13 to 159 mm), 43 parasitic males (8.9 to 15 mm), 20 free-living metamorphosed and metamorphic males (10 to 16 mm), and about 250 "Hyaloceratias" larvae (2.6 to 17.5 mm); Atlantic, Pacific, and Indian oceans; holotype, ZMB 17713, male, 13 mm, *Valdivia* station 175, Indian Ocean, 26°3'S, 93°43'E, open pelagic net, 0 to 2200 m, January 1899 (Brauer, 1902:297, 1906:324, pl. 16, fig. 10; Regan, 1925b:398, figs. 8, 9, 1926:14, 25, fig. 5, pl. 3, fig. 2; Regan and Trewavas, 1932:15, 104, figs. 3–6, 163, pl. 9, fig. 2; Bertelsen, 1951:168, figs. 108B, 111A, B, 112–115, 124A; Maul, 1961:155, figs. 27A, C, 32; Pietsch, 1976:782, 788).

## GENUS *ACENTROPHRYNE* REGAN, 1926 (FANGTOOTH SEADEVILS)

Figures 173–175, Table 1

*Acentrophryne*, one of the rarest of the 35 recognized ceratioid genera, represented by only five female specimens in collections around the world, was introduced by Regan (1926) to contain a single 50-mm female collected in 1922 by the *Dana* in the Gulf of Panama (Fig. 174). Although the original description was cited briefly by Regan and Trewavas (1932), the type species, which Regan (1926) had named *Acentrophryne longidens*, was not collected again until 1938, this time off Costa Rica, during the Eastern Pacific *Zaca* Expedition sponsored by the New York Zoological Society (Beebe and Crane, 1947:170). This second specimen of *Acentrophryne longidens*, only slightly smaller than the first (42 mm), together with the holotype, remained the only known representatives of the genus until December 2002 when three additional females were identified among collections made by Japanese colleagues off the coast of Peru. This new material was subsequently described as new to science and named *A. dolichonema* by Pietsch and Shimazaki (2005).

Ironically, this extremely rare ceratioid is the first member of the suborder ever to be recognized in the fossil record (Pietsch and Lavenberg, 1980). Although lophioid and antennarioid fishes are well known from Pliocene and Eocene deposits—*Lophius budegassa* Arambourg, Lower Pliocene of Algeria; *L. brachysomus* Agassiz, Eocene of Monte Bolca, Italy; *Histionotophorus bassani* (De Zigno), Monte Bolca; *Eosladenia caucasica* Bannikov, Middle Eocene of the northern Caucasus (for a summary, see Bannikov, 2004:420; Carnevale and Pietsch, 2006:454)—a fossil of *Acentrophryne* from the Late Miocene Puente Formation of southern California, described by Pietsch and Lavenberg (1980), is the first known fossilized ceratioid (Fig. 175). Easily recognized as a female ceratioid by the presence of an illicium and the absence of pelvic fins, the fossil clearly belongs to the family Linophrynidae in having 3 dorsal- and 3 anal-fin rays, 3 pectoral radials, and a single-headed hyomandibular bone; and to the genus *Acentrophryne* in the absence of a preopercular spine and in having an elongate illicium (Pietsch and Lavenberg, 1980). Since this discovery, about a dozen additional ceratioid fossils have been collected from this same formation during earth-moving activities associated with construction of the Metro Rail Red Line, Wilshire Boulevard/Vermont Avenue Subway Station, in Los Angeles. This material, donated by the Los Angeles Metropolitan Transportation Authority to the Department of Vertebrate Paleontology at the Los Angeles County Museum of Natural History, represents at least two families and five genera (see Carnevale et al., 2008). The Puente Formation and other southern California diatomite deposits are well known for their meso- and bathypelagic fishes (see Jordan, 1925; and Crane, 1966).

### DISTINGUISHING CHARACTERS

If it were not for the absence of a hyoid barbel and the lack of a spine on the posterior margin of the preopercular bone, *Acentrophryne* would be extremely difficult if not impossible to differentiate from *Linophryne* (Figs. 173, 174). The two, in all other ways, are nearly identical. The body of metamorphosed females is short and globular. The head and mouth are huge, the length of the lower jaw of some specimens approaching two-thirds the length of the body. The caudal peduncle is unusually short, the dorsal and anal fins terminating nearly at the base of the caudal-fin rays. Frontal, epiotic, and posttemporal spines are absent. A spine at the symphysis of the lower jaw is absent as well, but the sphenotic and posterior tip of the angular both form a prominent spine. Jaw teeth are relatively few—the largest known female (105 mm, HUMZ

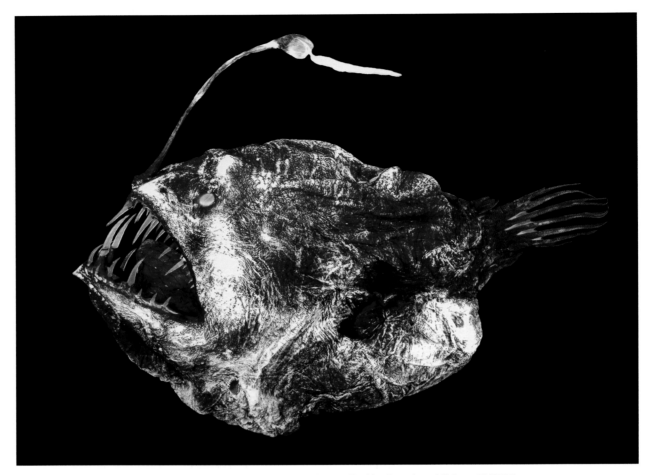

FIGURE 173. *Acentrophryne dolichonema*, holotype, HUMZ 175257, 105 mm SL. After Pietsch and Shimazaki (2005).

175257) has 25 or 26 teeth on each premaxilla and 16 on each dentary—but the size of some is extraordinary (Figs. 173, 174): the longest premaxillary tooth is almost 15% SL, the longest dentary tooth is about 20% SL. There are 2 to 6 vomerine teeth. The anterior end of the pterygiophore of the illicium protrudes slightly from the snout. The illicium itself is rather long, as great as 70% SL in some specimens. The escal bulb is relatively large and bears only a single, unpigmented distal appendage (Fig. 173). In contrast to *Photocorynus* and *Haplophryne*, the skin of the head and body is uniformly black to dark red brown, except for the distal part of the escal bulb and escal appendage.

Males and larvae are unknown.

The dorsal and anal fins each contain 3 rays, the pectoral fin, 16 to 19 rays. There are 9 caudal-fin rays, the ninth ray (the ventral-most) about one-half the length of the eighth caudal-fin ray (Table 1).

### ETYMOLOGY

To emphasize the absence of a preopercular spine, the single most important feature that distinguishes this genus from its close linophrynid relatives, Regan (1926) chose a combination of three Greek roots: the prefix *a*, meaning "without"; *kentron*, a "prickle" or "spine"; and *phryne*, a "toad"; hence a "spineless toad."

### DIVERSITY

Two species are recognized:

*Acentrophryne longidens* Regan, 1926: two metamorphosed females (42 to 50 mm), Gulf of Panama and Pacific coast of Costa Rica; holotype, ZMUC P921981, 50 mm, *Dana* station 1203(14), eastern tropical Pacific, Gulf of Panama, 7°30′N, 79°19′W, open conical ring-trawl, 2500 m wire out, bottom depth 2550 m, 2030 h, 11 January 1922 (Regan, 1926:23, pl. 1, fig. 2; Beebe and Crane, 1947:170, fig. 15; Pietsch and Lavenberg, 1980:906, figs. 1, 2; Pietsch and Shimazaki, 2005:247, fig. 1).

*Acentrophryne dolichonema* Pietsch and Shimazaki, 2005: three metamorphosed females (55 to 105 mm), off Peru; holotype, HUMZ 175257, 105 mm, *Shinkai-Maru*, eastern tropical Pacific, off Peru, 8°10.6 to 11.9′S, 80°32.0 to 32.4′W, Bacalao Trawlnet-586 MSK (nonclosing otter trawl), bottom depth 1061 to 1105 m, 27 April 2000 (Pietsch and Shimazaki, 2005:248, fig. 2).

## GENUS *BOROPHRYNE* REGAN, 1925C (GREEDY SEADEVILS)

*Figures 176–178, Table 1*

*Borophryne* was established by Regan (1925c:564) with a single short descriptive phrase: "As *Linophryne*, but without a barbel." Then based on only four females, all collected by the *Dana* in the Gulf of Panama, the single recognized species of the genus,

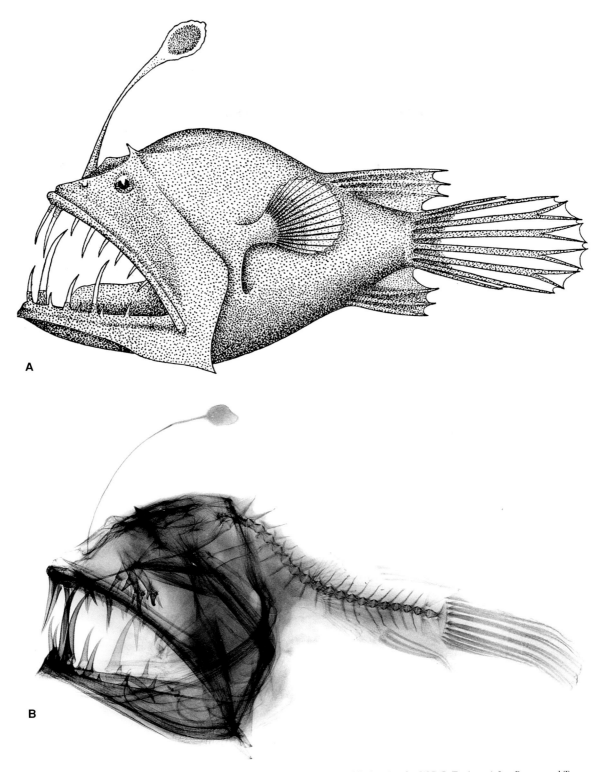

FIGURE 174. *Acentrophryne longidens*, holotype, 50 mm SL, ZMUC P921981: (A) drawing by W.P.C. Tenison (after Regan and Trewavas, 1932; number of dorsal- and anal-fin rays corrected by the author); (B) radiograph (after Pietsch and Lavenberg, 1980).

*Borophryne apogon*, is now represented in collections by some 39 females, 62 free-living males, 13 parasitic males, and 14 larvae, all of which were taken in a relatively small region of the eastern tropical Pacific Ocean. Parasitic males were first described by Regan and Trewavas (1932). Additional females, some with attached males, were subsequently reported by Beebe and Crane (1947), Brewer (1973), and Parin et al. (1973). Bertelsen (1951) added yet more females, but also assigned the first known larvae and free-living males to the known material. Bruun (1958) provided an explanation for the narrow geographic distribution of the species, then confined to the Gulf of Panama. Its reproductive biology was summarized by Pietsch

FIGURE 175. *Acentrophryne* sp., about 28 mm SL, from the Miocene of Chalk Hill, Los Angeles County, California, LACM-VP 117685, left lateral view. Photo by Gary T. Takeuchi; courtesy of the Department of Vertebrate Paleontology, Natural History Museum of Los Angeles County.

(2005b) and its sister-group relationship with *Linophryne* was reconfirmed by Pietsch and Orr (2007).

### DISTINGUISHING CHARACTERS

As so concisely indicated by Regan (1925c), metamorphosed females of *Borophryne* are very similar to those of *Linophryne*, differing in obvious ways only by the lack of a hyoid barbel. Metamorphosed females of the two genera are essentially the same in having a short globular body, with a huge head and mouth. The caudal peduncle is unusually short. The frontals bear a conspicuous, rounded, laterally compressed dorsal protuberance. The sphenotic spines are well developed in young specimens, but become proportionately smaller with growth and covered with skin in the largest known specimens. Epiotic and posttemporal spines are absent. The preopercle is angled posteriorly, its posterior margin bearing a strong pointed spine, a feature shared only with *Linophryne*. A symphysial spine is present on the lower jaw. The angular and posteroventral part of the articular each form a sharp spine. The maxillae are reduced and extremely slender. The pterygiophore of the illicium protrudes slightly from the snout, its exposed length rarely more than the diameter of the escal bulb. The length of the illicium is about 20% SL. The esca bears a large bifurcated terminal appendage and numerous fine filaments on each side (Fig. 177). The skin of the head and body is uniformly black or very dark red brown. The fins, however, are unpigmented in most small specimens, but pigmentation at the base of each gradually spreads with growth to cover the distal tips of the rays.

The dentition of females, including the placement and number and length of teeth, is very similar to that found in species of *Linophryne* (Bertelsen, 1980a): the jaw teeth are unusually well developed but few in number, arranged in three longitudinal, overlapping series on each premaxilla and dentary; there are 7 to 28 teeth on each premaxilla and 5 to 17 on each dentary, the longest of which measure about 25% SL. There are 0 to 4 vomerine teeth.

Metamorphosed males of *Borophryne* are similar to those of other linophrynid genera, but can be separated primarily by differences in dentition and pigmentation (Fig. 178E). The premaxillae are degenerate, more or less completely resorbed, bearing at most only a few tiny teeth; the dentaries are nearly toothless as well. The denticular bones, however, are strong and well toothed, connected basally and meeting in front of the mouth when the jaws are closed. There are 3 to 7 teeth on the upper denticular. The lower denticular bears an anteriorly directed median tooth and a row of 2 to 6 more or less posteriorly directed teeth on each side. Subdermal pigmentation is absent. The skin of free-living males is light brown, but weakly pigmented, and more or less transparent, closely surrounding the body; but that of parasitic males is black or dark red brown. The olfactory organs are unpigmented and strongly inflated, the length of the

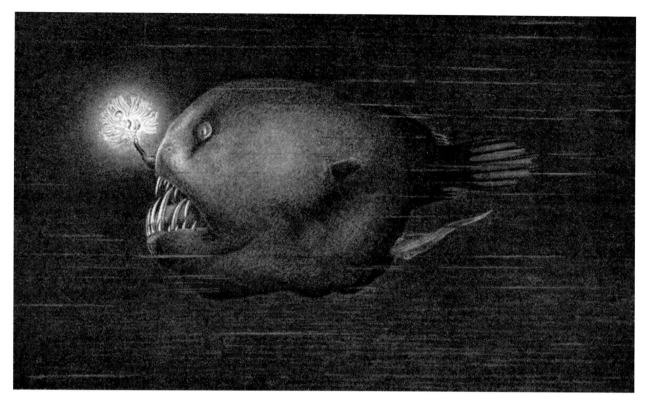

FIGURE 176. *Borophryne apogon*, 52-mm SL female, with 12-mm SL parasitic male, ZMUC P922322; "Deep-sea angler-fish, lamp on nose, with a dwarf male grown on to the ventral side" (Kramp, 1956:86). Drawing by P.H. Winther; after Kramp (1953, 1956).

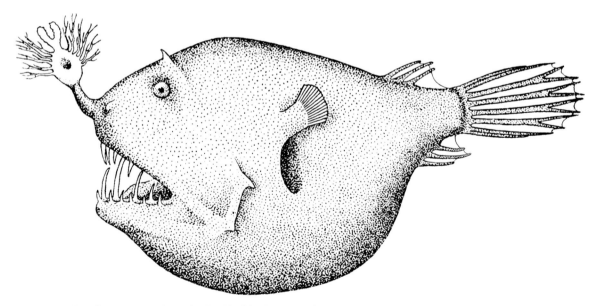

FIGURE 177. *Borophryne apogon*, 55 mm SL, ZMUC P92147. Drawing by W.P.C. Tenison; after Regan (1926).

posterior nostril almost twice the diameter of the eye. There are 7 to 9 olfactory lamellae. Sphenotic spines are well developed. The eyes and nostrils of parasitic males appear somewhat degenerate.

The larvae of *Borophryne* differ in having relatively long sphenotic spines, in combination with a short body and highly inflated, unpigmented, and more or less transparent skin (Fig. 178). The eye diameter is approximately 14% SL in smaller specimens, decreasing to about 8% in larger specimens.

The dorsal and anal fins each have 3 rays. There are 15 to 18 pectoral-fin rays and 9 caudal-fin rays, the ninth ray about one-half the length of the eighth (Table 1).

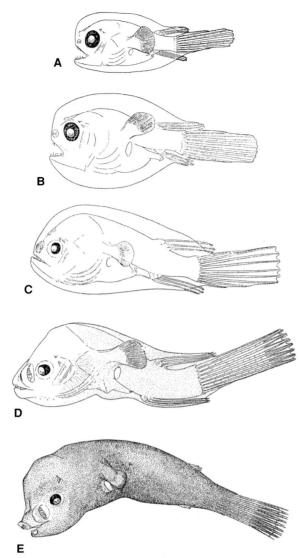

FIGURE 178. Males of *Borophryne apogon*: (A) larval male, 7.0 mm TL; (B) larval male, 4.3 mm SL, 7.0 mm TL, ZMUC P921759; (C) male in metamorphosis, 12 mm SL; (D) metamorphosed free-living male, 15 mm SL, ZMUC P921771; (E) parasitic male, 13 mm SL, ZMUC P92147. After Bertelsen (1951).

ETYMOLOGY

*Boros* is the Greek word for "greedy" or "glutinous," and this, combined with *phryne*, a "toad," provides the common name for this genus, the "Greedy Seadevil."

DIVERSITY

The genus contains a single species:

*Borophryne apogon* Regan, 1925c: 39 metamorphosed females (15 to 101 mm), 62 free-living males (11.0 to 17.5 mm), 13 parasitic males (9.5 to 22 mm), and 14 larvae (4.0 to 11 mm); eastern tropical Pacific; lectotype, ZMUC P92147, 55-mm female, with a 13-mm parasitic male, *Dana* station 1209(1), Gulf of Panama, 7°15′N, 78°54′W, 3500 m wire (Regan, 1925c:564; Regan and Trewavas, 1932:18, 20, 106, figs. 7, 8, 10A; Beebe and Crane, 1947:151, 171, pl. 2, fig. 4; Bertelsen, 1951:193, figs. 125, 126; Blache et al., 1970:453, fig. 1149; Parin et al., 1973:1467, fig. 37; Beltrán-León and Herrera, 2000:284, fig. 92).

## GENUS *LINOPHRYNE* COLLETT, 1886 (BEARDED SEADEVILS)

*Figures 179–185, Tables 1, 7*

*Linophryne* is probably the most easily recognized deep-sea anglerfish. Its big mouth, huge teeth, well-developed esca, and unique hyoid barbel combine to form what is most emblematic of a ceratioid. In fact, ceratioids in general have been popularized, more so than anything else, by Regan's (1926) famous illustration of *Linophryne arborifera*, which has been reproduced in the scientific as well as the popular literature more often than any other image of a ceratioid (Fig. 179). Following the introduction of the genus by Collett (1886), with his accurate and detailed description of *L. lucifer*, based on a single small female collected off Madeira (Fig. 13), additional species were not added to the genus until Regan's (1925c) discovery of unique material in the collections brought back by the *Dana*. From this time, and following in quick succession up until Bertelsen's (1951) monograph, 17 new species were added (Beebe, 1926b, 1932b; Parr, 1927; Regan and Trewavas, 1932; Waterman, 1939b; Imai, 1941; Beebe and Crane, 1947), each based on a single metamorphosed female (see Table 7). Bertelsen (1951), in the first monographic treatment of the genus, reduced the number to 12, synonymizing several forms as ontogenetic and intraspecific variants, and also dealing for the first time with larvae and free-living males, separating them into three groups, each with a characteristic pattern of subdermal pigmentation.

Following Maul's (1961) description of *Linophryne maderensis*, based on three small females, all recovered from the stomachs of the Black Scabbard Fish, *Aphanopus carbo*, Bertelsen initiated a series of eight papers on *Linophryne*. Intended to encompass a full revision of the genus, these articles were directed primarily toward females of the genus—including a redescription and report of additional material of the type species *L. lucifer* (Bertelsen and Krefft, 1965), descriptions of new species (Bertelsen, 1973, 1978, 1980b), and revisions of species groups within the genus (Bertelsen, 1980a, 1980b)—but occasionally describing new records of parasitic males as they were discovered (Bertelsen, 1976, 1978). By the time this work was completed, with a final summary revision of *Linophryne* published in 1982, he had synonymized one species, resurrected three, and added six new species, bringing the total number of recognized species to 21. In addition to full descriptive accounts of all species, along with comments on variation and geographic distribution, the number of sexually parasitized females was increased from two (*L. argyresca*, Regan and Trewavas, 1932; *L. arborifera*, Maul, 1961) to 16, representing 11 species. On the basis of features of the esca and hyoid barbel, coupled with subdermal pigmentation, the number of vomerine teeth, and development of sphenotic spines in the males, the genus was divided into three subgenera (*Stephanophryne*, *Rhizophryne*, and *Linophryne*), and an analysis of evolutionary relationships was proposed for the first time. Since 1982, the only significant additions to our knowledge of *Linophryne* have been Ofer Gon's (1992) description of *L. andersoni*, and analyses of reproductive modes and phylogenetic relationships by Pietsch (2005b) and Pietsch and Orr (2007), respectively.

DISTINGUISHING CHARACTERS

Females of the genus *Linophryne* differ from those of all other ceratioids by having an elongate, slender, bioluminescent hyoid barbel, a feature found elsewhere among anglerfishes only

TABLE 7
Reallocation of Nominal Species of the Genus *Linophryne*

| Nominal Species | Currently Recognized Taxon |
|---|---|
| **Based on Females** | |
| Subgenus *Stephanophryne* Bertelsen, 1982 | |
| *Linophryne corymbifera* Regan and Trewavas, 1932 | *Linophryne indica* (Brauer, 1902) |
| Subgenus *Rhizophryne* Bertelsen, 1982 | |
| *Linophryne arborifer* Regan, 1925c | *Linophryne arborifera* Regan, 1925c |
| *Linophryne brevibarbata* Beebe, 1932b | *Linophryne brevibarbata* Beebe, 1932b |
| *Linophryne eupogon* Regan and Trewavas, 1932 | *Linophryne arborifera* Regan, 1925c |
| *Linophryne densiramus* Imai, 1941 | *Linophryne densiramus* Imai, 1941 |
| *Linophryne quinqueramosus* Beebe and Crane, 1947 | *Linophryne quinqueramosus* Beebe and Crane, 1947 |
| *Linophryne pennibarbata* Bertelsen, 1980a | *Linophryne pennibarbata* Bertelsen, 1980a |
| *Linophryne parini* Bertelsen, 1980b | *Linophryne parini* Bertelsen, 1980b |
| *Linophryne andersoni* Gon, 1992 | *Linophryne andersoni* Gon, 1992 |
| Subgenus *Linophryne* Bertelsen, 1982 | |
| *Linophryne lucifer* Collett, 1886 | *Linophryne lucifer* Collett, 1886 |
| *Linophryne macrodon* Regan, 1925c | *Linophryne macrodon* Regan, 1925c |
| *Linophryne polypogon* Regan, 1925c | *Linophryne polypogon* Regan, 1925c |
| *Diabolidium arcturi* Beebe, 1926b | *Linophryne arcturi* (Beebe, 1926b) |
| *Linophryne coronata* Parr, 1927 | *Linophryne coronata* Parr, 1927 |
| *Linophryne bicornis* Parr, 1927 | *Linophryne bicornis* Parr, 1927 |
| *Linophryne brevibarbis* Parr, 1927 | *Linophryne macrodon* Regan, 1925c |
| *Linophryne longibarbata* Borodin, 1930a | *Linophryne coronata* Parr, 1927 |
| *Linophryne argyresca* Regan and Trewavas, 1932 | *Linophryne argyresca* Regan and Trewavas, 1932 |
| *Linophryne lucifera* Regan and Trewavas, 1932 | *Linophryne lucifer* Collett, 1886 |
| *Linophryne racemifera* Regan and Trewavas, 1932 | *Linophryne racemifera* Regan and Trewavas, 1932 |
| *Linophryne coronata diphlegma* Parr, 1934 | *Linophryne coronata* Parr, 1927 |
| *Linophryne algibarbata* Waterman, 1939b | *Linophryne algibarbata* Waterman, 1939b |
| *Linophryne maderensis* Maul, 1961 | *Linophryne maderensis* Maul, 1961 |
| *Linophryne sexfilis* Bertelsen, 1973 | *Linophryne sexfilis* Bertelsen, 1973 |
| *Linophryne trewavasae* Bertelsen, 1978 | *Linophryne trewavasae* Bertelsen, 1978 |
| *Linophryne escaramosa* Bertelsen, 1982 | *Linophryne escaramosa* Bertelsen, 1982 |
| *Linophryne bipennata* Bertelsen, 1982 | *Linophryne bipennata* Bertelsen, 1982 |
| *Linophryne digitopogon* Balushkin and Trunov, 1988 | *Linophryne lucifer* Collett, 1886 |
| **Based on Males** | |
| Subgenus *Stephanophryne* Bertelsen, 1982 | |
| *Aceratias macrorhinus indicus* Brauer, 1902 | *Linophryne indica* (Brauer, 1902) |
| Subgenus *Rhizophryne*? | |
| *Haplophryne simus* Borodin, 1930b | *Linophryne* subgenus *Rhizophryne* sp.? |
| Subgenus *Rhizophryne* Bertelsen, 1982 | |
| *Nannoceratias denticulatus* Regan and Trewavas, 1932 | *Linophryne* subgenus *Rhizophryne* sp. |
| Subgenus *Linophryne* Bertelsen, 1982 | |
| *Aceratias macrorhinus* Brauer, 1902 | *Linophryne* subgenus *Linophryne* sp. |
| *Haplophryne hudsonius* Beebe, 1929b | *Linophryne* subgenus *Linophryne* sp. |
| *Aceratias edentula* Beebe, 1932b | *Linophryne* subgenus *Linophryne* sp. |
| *Borophryne masculina* Parr, 1934 | *Linophryne* subgenus *Linophryne* sp. |

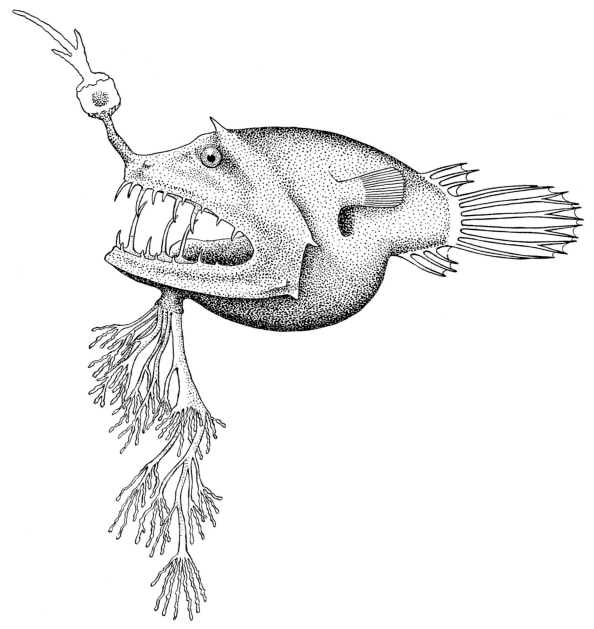

FIGURE 179. *Linophryne arborifera*, 49 mm SL, ZMUC P92139. Drawing by W.P.C. Tenison; after Regan (1926).

in *Centrophryne* (Figs. 179–182, 184). But in the latter genus the barbel is tiny and lacks photophores; it is present also in males and larvae, and reduced or absent in large metamorphosed females (Fig. 60). Like its closely related congeners, *Acentrophryne* and *Borophryne*, the body is short and globular in shape. The head and mouth are huge, the latter equipped with enormous fanglike teeth. The caudal peduncle is unusually short. Spines of the head are generally well developed: the sphenotic spines are large and sharply pointed; the anterodorsal margin of the frontal bears a small spine in some species, but in others it forms a conspicuous, rounded, laterally compressed protuberance; a symphysial spine of the lower jaw is present in the subgenera *Linophryne* and *Rhizophryne*, but absent in the subgenus *Stephanophryne*; and the angular and posteroventral part of the articular form a sharp spine. There is a single large preopercular spine, which among ceratioids is shared only with *Borophryne*. Epiotic and posttemporal spines are absent. The maxillae are reduced and extremely slender. The dentition of females, including the placement and the number and length of teeth, is extremely similar in all species (Bertelsen, 1980a; see Fig. 35): like those of *Acentrophryne* and *Borophryne*, the teeth are large, but few in number, fewer than 30 on each premaxilla and 20 on each dentary. Several teeth in both jaws are extremely well developed, the longest premaxillary tooth approaching 15% SL, the longest dentary tooth 24% SL.

The pterygiophore of the illicium protrudes slightly from the snout. The length of the illicium varies among species from less than 20% to nearly 40% SL. There is extreme interspecific variation in the number, position, and shape of the escal appendages, the longest usually less than 20% SL, but reaching 60% SL in some species (Fig. 183). Well-developed, internal, light-guiding escal structures are present in some species (see

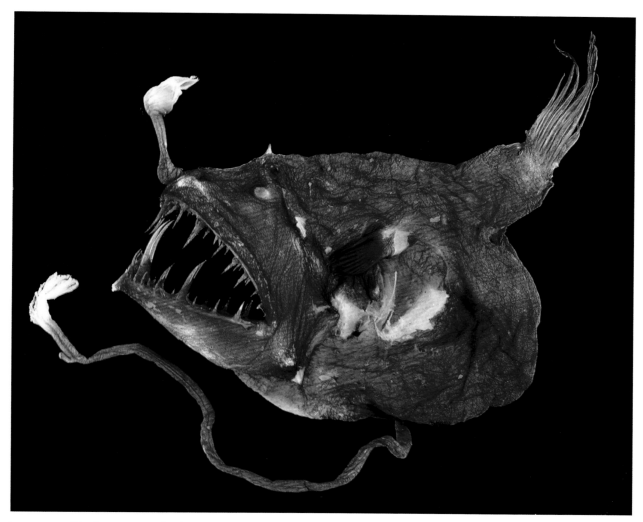

FIGURE 180. *Linophryne coronata*, female, 75 mm SL, MCZ 32307. Photo by C.P. Kenaley; courtesy of K. Hartel and the Museum of Comparative Zoology, Harvard University.

Bioluminescence and Luring, Chapter Six). There is similar extreme interspecific variation in the length, pattern of branching, and shape of the bioluminescent structures of the hyoid barbel, the length of the barbel varying from less than 50% SL in some species to 360% SL in others (Fig. 184). The skin of the head and body is uniformly black or dark red brown, but pigmentation of the fins and barbel is highly variable, absent in most small specimens, but gradually spreading toward the distal tips of the fin rays with growth.

Free-living metamorphosed males of *Linophryne*, while easy to identify to family by the "telescopic" eyes, large olfactory organs, and tiny dorsal and anal fins (each containing only 3 rays, very rarely 2), are only slightly more difficult to take to genus (Fig. 185D). The best characters are those of dentition and pigmentation. Upper and lower jaw teeth are few to absent, the premaxillae more or less completely resorbed. In contrast, the denticular teeth are well developed, meeting in front of the mouth, the upper denticular bearing 3 to 6 teeth, the lower denticular with a medial tooth lying just in front of a transverse series of 6 to 15 teeth. Subdermal pigmentation is usually well developed, rarely weak or absent, while the skin is darkly pigmented, brown or black. Sphenotic spines are present or absent. The olfactory organs are strongly inflated, larger than the eyes, containing 8 to 13 olfactory lamellae.

Although linophrynid larvae are easily separated from those of other ceratioid families by their more elongate body and low dorsal- and anal-fin ray counts, the larvae of *Linophryne* taken as a whole have no known shared characters that will differentiate them from those of other genera of the family (Fig. 185). The three subgenera of *Linophryne*, however, are defined primarily on the basis of intrageneric differences in larval characters (Bertelsen, 1982). Larvae of *Linophryne* subgenus *Linophryne* are unique among those of other ceratioids, except the linophrynid genus *Borophryne*, in having well-developed sphenotic spines; and unlike the larvae of *Borophryne*, which are completely unpigmented, they have a varying number of subdermal melanophores more or less restricted to the caudal peduncle. Sphenotic spines are absent or represented only by blunt rudiments in the larvae of *L.* subgenus *Rhizophryne* and *L.* subgenus *Stephanophryne*. Furthermore, these two subgenera share a pattern of subdermal pigmentation that consists of two lateral bands of melanophores along the sides of the body (Fig. 185). Larvae of *L.* subgenus *Stephanophryne* differ from those of other members of the genus in having additional groups of subdermal melanophores on the posterior angle of the lower

FIGURE 181. *Linophryne polypogon*, 33 mm SL, BMNH 2004.9.12.167. Photo by Peter David; after Pietsch and Orr (2007).

jaw and on the posterior tip of the peritoneum. However, this pigment pattern may not be developed in larvae of less than about 7.0 mm and like those of the subgenus *Rhizophryne*, along with larvae of the linophrynid genus *Haplophryne*, fall within an assemblage of inseparable linophrynid larvae termed "*Hyaloceratias*" (see Bertelsen, 1951:189).

There are 3 dorsal-fin rays, 3 (very rarely 2) anal-fin rays, 12 to 19 pectoral-fin rays, and 9 caudal-fin rays, the ninth ray about one-half the length of the eighth (Table 1).

The largest known specimen is a 275-mm female of *L. lucifer* (ZMUC P922443). The largest known free-living male is a 21-mm specimen of *L.* subgenus *Linophryne* sp. (the holotype of *Aceratias macrorhinus*, ZMB 17714). Of the 30 known parasitically attached males, 11 are greater than 21 mm (the largest is 30 mm, IMB; see Behrmann, 1977), indicating substantial growth after attachment. All have bellies swollen by enlarged testes, and some of the largest attached males also show distinct degeneration of olfactory organs and eyes.

ETYMOLOGY

*Linon* is the Greek word for "flax" or anything made of flax, such as a "cord," "rope," or a "net"; thus, this prefix, combined with *phryne*, a "toad," alludes to a toadlike fish that fishes with a net.

DIVERSITY

The genus is divided into three subgenera and 22 species (all based on females) as follows (Table 7):

LINOPHRYNE SUBGENUS *STEPHANOPHRYNE* BERTELSEN, 1982:

*Linophryne indica* (Brauer, 1902): 12 metamorphosed females (24 to 51 mm), two parasitic males (9.5 to 14.5 mm), 31 free-living males (11 to 16 mm), and 17 larvae (7.5 to 22 mm); Pacific and Indian oceans; lectotype, ZMB 17715, male, 16 mm, *Valdivia* station 230, 2°43′S, 61°12′E, 1500 m (Brauer, 1902:296, 1906:325, pl. 16, figs. 6–9; Regan, 1912:289, 1926:45; Regan and Trewavas, 1932:107, 110, 113, figs. 1, 169, 170, pl. 10; Bertelsen, 1951:174, 175, 177, fig. 116, 1978:30, fig. 3, 1981:2, figs. 1–4).

LINOPHRYNE SUBGENUS *RHIZOPHRYNE* BERTELSEN, 1982:

*Linophryne parini* Bertelsen, 1980b: a single metamorphosed female, the holotype, SIOM uncataloged, 64 mm, *Fiolent* cruise 3, station 126, off South Africa, 33°41′S, 27°26′E, bottom trawl, 1200 to 1220 m (Bertelsen, 1980b:234, fig. 1; Anderson and Leslie, 2001:28, fig. 19).

*Linophryne andersoni* Gon, 1992: a single metamorphosed female, the holotype, BPBM 24512, 32 mm, Townsend Cromwell cruise 46, station 9, western tropical Pacific, 11°49′N, 144°51′W, Cobb pelagic trawl, 0 to 50 m, 14 October 1969 (Gon, 1992:139, figs. 1–5).

*Linophryne quinqueramosa* Beebe and Crane, 1947: a single metamorphosed female, the holotype, CAS-SU 46506, 58 mm, eastern Pacific *Zaca* Expedition, station 234 T-1, Gulf of Panama, 7°24′N, 78°35′W, 910 m, 4 April 1938 (Beebe and Crane, 1947:174, fig. 17, pl. 3, fig. 5; Bertelsen, 1951:174, 176, 1980b:241, figs. 4, 6).

*Linophryne brevibarbata* Beebe, 1932b: six metamorphosed females (33 to 100 mm) and two parasitic males (13.6 to 18.5 mm); North Atlantic Ocean; holotype, USNM 170947, 33 mm, Bermuda Oceanographic Expedition, net 308, 32°12′N, 64°36′W, pelagic net (0 to) 1700 m (Beebe, 1932b:94, figs. 26, 27, 1937:207; Bertelsen, 1951:78, 1980a:59, figs. 3C, 4C, 5E, 5H, 13–15, 17, 1982:3, 96; Maul, 1961:146, fig. 29; Swinney, 1995a:52, 57; McEachran and Fechhelm, 1998:859, 861, fig.).

*Linophryne densiramus* Imai, 1941: 27 females (34 to 71 mm) and four parasitic males (9.0 to 17.5 mm); Atlantic, Pacific, and Indian oceans; holotype, ZUMT 55032, *Dainchi-Maru*, Suruga Bay, 21 October 1940 (Imai, 1941:247, figs. 14–17; Bertelsen, 1951:178, 1980a:55, figs. 3C, 4D, 5G, 11, 12, 17, 1982:3, 96; Parin et al., 1977; Bertelsen and Pietsch, 1983:94, fig. 16; Anderson and Leslie, 2001:27, fig. 17).

*Linophryne pennibarbata* Bertelsen, 1980a: 11 females, eight metamorphosed (36.5 to 47 mm) and three in metamorphosis (35 to 36.5 mm); North Atlantic and central Pacific; holotype, USNM 219854, 47 mm, Silver Bay station 3735, 29°58′N, 80°10′W, shrimp-trawl, 325 m (Bertelsen, 1980a:45, figs. 4A, 5A, 6, 7, 17, 1982:3 and 96, 1990:518).

*Linophryne arborifera* Regan, 1925c: 31 metamorphic and metamorphosed females (21 to 77 mm) and two parasitic males (14.5 to 15 mm); Atlantic Ocean; holotype, ZMUC P92139, 49 mm, *Dana* station 1161(3), central Atlantic Ocean, 14°52′N, 28°04′W, open pelagic net, 600 m of wire, 0345 h, 5 November 1921 (Regan, 1925c:564; Parr, 1927:5, 10, fig. 3; Regan and Trewavas, 1932:107, 111; Beebe, 1937:207; Bertelsen, 1951:174, 175, 178, figs. 117, 118, 124C, 1980a:50, figs. 1, 2, 3B, 4B, 5B, 8–10, 17, 18, 1982:92).

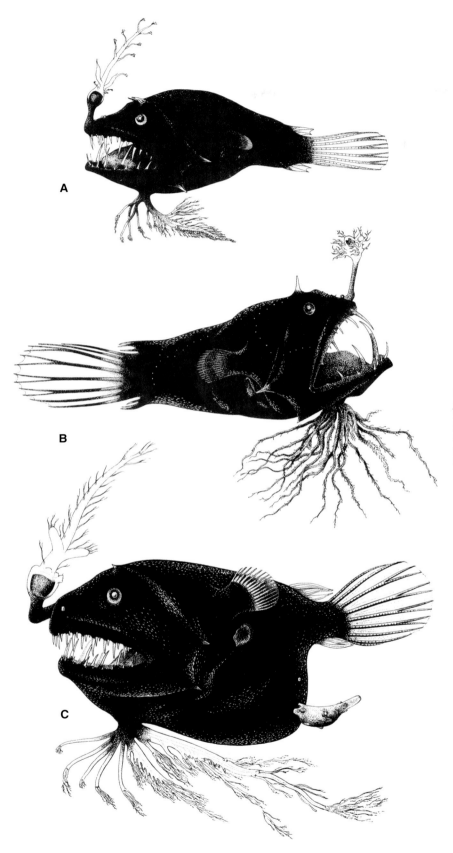

FIGURE 182. Species of *Linophryne*: (A) *L. pennibarbata*, holotype, 47 mm SL, USNM 219854; (B) *L. polypogon*, holotype, 32 mm SL, ZMUC P92145; (C) *L. arborifera*, 77-mm SL female, with 15-mm SL parasitic male, ISH 2736/71. Drawings by R. Nielsen; after Bertelsen (1980a, 1980b).

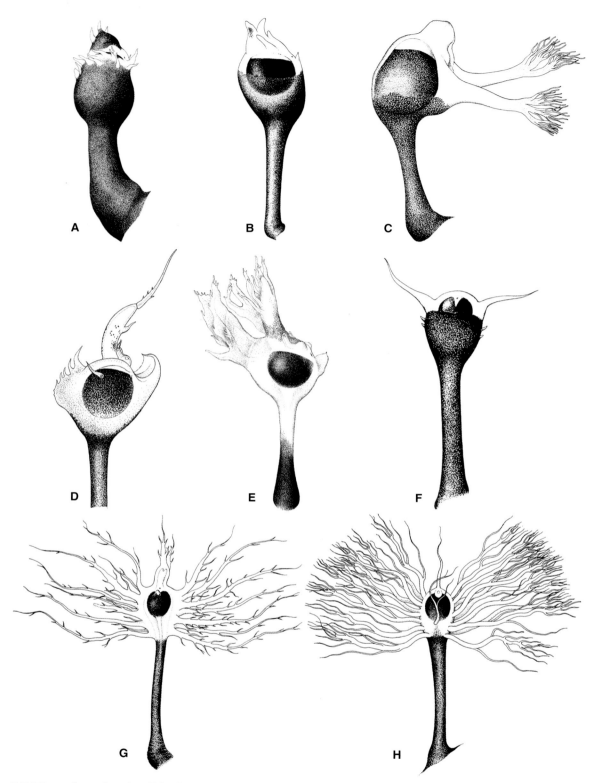

FIGURE 183. Escae of species of *Linophryne*, selected to show the range of morphological variation within the genus: (A) *L. lucifer*, 190 mm SL, ZIN uncataloged; (B) *L. coronata*, 105 mm SL, ROM 27258; (C) *L. parini*, holotype, 64 mm SL, ZIN uncataloged; (D) *L. arcturi*, 51 mm SL, ISH 2132/71; (E) *L. arcturi*, 34 mm SL, BMNH uncataloged; (F) *L. bicornis*, 180 mm SL, ZIN uncataloged; (G) *L. escaramosa*, holotype, 36.5 mm SL, LACM 42296-001; (H) *L. racemifer*, 58 mm SL, LACM 25185-010. Drawings by R. Nielsen; after Bertelsen (1980b, 1982).

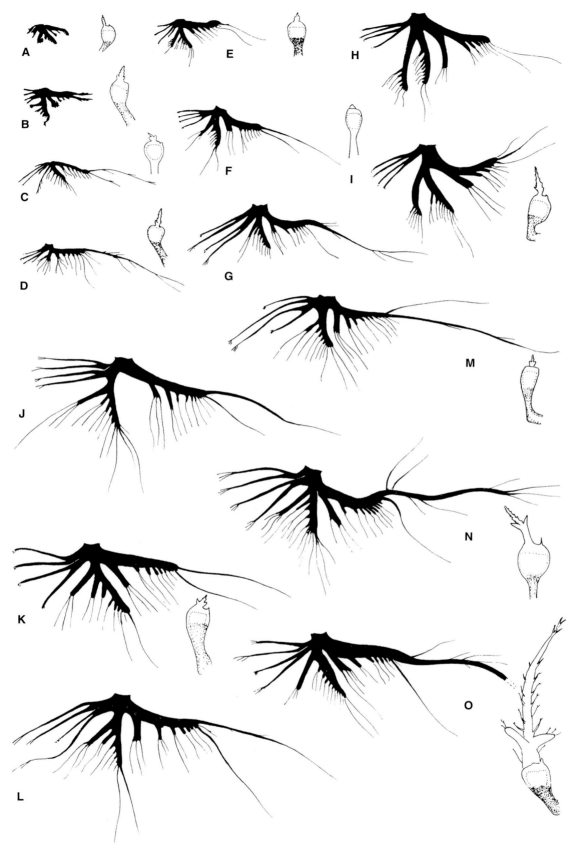

FIGURE 184. Diagrammatic developmental series of barbels and escae of *Linophryne arborifera*, showing pattern of placement of various parts, and the extent of morphological variation within the species: (A) 34.5 mm SL, SIO 63-552; (B) 35 mm SL, UF 221618; (C) holotype of *L. eupogon*, 30 mm SL, YPM 2029; (D) 41 mm SL, MCZ 14171; (E) 31 mm SL, BMNH 1998.9.8.11; (F) 30.5 mm SL, MCZ 51993; (G) 44 mm SL, ISH 2009/71; (H) 34 mm SL, ISH 75/67; (I) 35 mm SL, ISH 1753/71; (J) 30 mm SL, USNM 219675; (K) 32.5 mm SL, USNM 219674; (L) 31 mm SL, USNM 219676; (M) 44 mm SL, MCZ 49822; (N) holotype, 49 mm SL, ZMUC P92139; (O) 77 mm SL, ISH 2736/71. After Bertelsen (1980a).

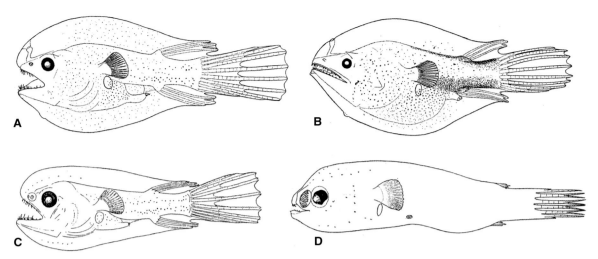

FIGURE 185. *Linophryne* sp., early life-history stages: (A) *L.* subgenus *Rhizophryne* sp., female in metamorphosis, 17.5 mm SL, ZMUC P921812; (B) holotype of *Cryptolychnus micractis*, female in metamorphosis, 32 mm SL, ZMUC P92140; (C) larval male, 13.5 mm SL; (D) metamorphosed free-living male, 17 mm SL, ZMUC P921799. After Bertelsen (1951).

*LINOPHRYNE* SUBGENUS *LINOPHRYNE* BERTELSEN, 1982:

*Linophryne arcturi* (Beebe, 1926b): three metamorphosed females (28.6 to 51 mm); eastern Atlantic and Gulf of Panama; holotype, CAS-SU 46505 (formerly NYZS 6333), 28.6 mm, *Arcturus* station 74 T-70, south of Cocos Island, Costa Rica, 4°50′N, 87°00′W, open pelagic net, 915 m (500 fathoms), 2 June 1925 (Beebe, 1926b:80, fig.; Regan and Trewavas, 1932:107; Beebe and Crane, 1947:173, fig. 16; Bertelsen, 1951:174, 183, 1982:65, 97, figs. 6, 7).

*Linophryne coronata* Parr, 1927: 22 metamorphosed females (29.5 to 225 mm), including three tentatively assigned specimens (29.5 to 51.5 mm); and four parasitic males (18 to 26 mm, one represented only by partial remains); North Atlantic and eastern Pacific oceans; holotype, YPM 2005, 33 mm, *Pawnee* station 39, Bahamas, 22°43′N, 74°23′W, open pelagic net, 8000 ft of wire, 29 March 1927 (Parr, 1927:13, fig. 4; Bertelsen, 1951:174, 176, 1973:68, 1976:10, figs. 2, 4, 1982:56, 97, figs. 3–6; Ponomarenko, 1959:83, fig.; Stearn and Pietsch, 1995:142, fig. 92; Jónsson and Pálsson, 1999:204, fig. 8).

*Linophryne lucifer* Collett, 1886: 29 females, 28 metamorphosed (30 to 275 mm) and one in late metamorphosis (22 mm); and six parasitic males (24 to 29 mm); North Atlantic and eastern South Indian Ocean; holotype, ZMUO J.5560, 31.5 mm, off Madeira, ca. 36°N, 20°W, found floating on the surface, May 1877 (Collett, 1886:138, pl. 15; Goode and Bean, 1896:496, pl. 121, fig. 408; Regan and Trewavas, 1932:107, 109; Bertelsen, 1951:174, 176, 1982:51, 97, figs. 1, 6; Bertelsen and Krefft, 1965:294, figs. 1–4; Balushkin and Trunov, 1988:62, fig. 1; Jónsson and Pálsson, 1999:204, fig. 8).

*Linophryne sexfilis* Bertelsen, 1973: a single metamorphosed female, the holotype, BMNH 1976.11.4.1, 38.5 mm, *Discovery* station 7856-50, eastern North Atlantic, 30°04′N, 23°00′W, RMT 8, 1250 to 1000 m, 5 April 1972 (Bertelsen, 1973:65, figs. 1, 2, 1982:92, 1986:1410, 1413, figs.).

*Linophryne argyresca* Regan and Trewavas, 1932: a single metamorphosed female, the holotype, ZMUC P92142, 61 mm, with a 12-mm parasitic male, *Dana* station 3904(1), eastern Indian Ocean, 5°18′N, 90°55′W, 3500 m wire, 1745 h, 18 November 1929 (Regan and Trewavas, 1932:19, 108, figs. 9, 167, 168, pl. 10, fig. 3; Bertelsen, 1951:175, 185, fig. 121E, 1982:54, 98, figs. 2, 6).

*Linophryne escaramosa* Bertelsen, 1982: a single known metamorphosed female, the holotype, LACM 42296-1, 36.5 mm, off the Hawaiian Islands, 21°10′ to 20′N, 158°10′ to 20′W, Isaacs-Kidd midwater trawl, oblique tow, 0 to 1350 m, 1134 to 1418 h, 27 May 1974 (Bertelsen, 1982:69, 95, figs. 8, 13).

*Linophryne bipennata* Bertelsen, 1982: a single metamorphosed female, the holotype, SIO 61-31, 37.5 mm, Monsoon Expedition, *Argo* station II-7, eastern South Indian Ocean, Java Trench, south of Bali, 12°05.9S, 115°26.2E, 3-m Isaacs-Kidd midwater trawl, 0 to 2000 m, 28 to 29 Oct 1960 (Bertelsen, 1982:71, 98, figs. 9, 13; Pietsch, 1999:2037).

*Linophryne maderensis* Maul, 1961: four metamorphosed females (29 to 105 mm) and one parasitic male (15 mm); eastern North Atlantic Ocean off Madeira; holotype, MMF 9094, 34 mm, off Madeira, from stomach of *Aphanopus carbo* (Maul, 1961:151, figs. 30, 31A, B; Bertelsen, 1982:73, 98, figs. 10, 13, 18, 1986:1409, 1412, figs.).

*Linophryne macrodon* Regan, 1925c: 10 metamorphosed females (19 to 91 mm) and one parasitic male (21.5 mm); western North Atlantic and Gulf of Panama; holotype, ZMUC P92144, 37 mm, *Dana* station 1208(4), Gulf of Panama, 6°48′N, 80°33′W, open pelagic net, 3500 m of wire, 0810 m, 16 January 1922 (Regan, 1925c:564; Parr, 1927:7, fig. 1; Bertelsen, 1951:175, 185, figs. 120B, 121C, 1982:76, 98, figs. 11–13; Pietsch, 1976:788).

*Linophryne polypogon* Regan, 1925c: three metamorphosed females (32 to 97 mm); eastern North Atlantic off Madeira and East China Sea; holotype, ZMUC P92145, 32 mm, *Dana* station 1142(6), 33°25′N, 16°58′W, open pelagic net, 5000 m of wire, 15 October 1921 (Regan, 1925c:565; Bertelsen, 1951:174, 183, figs. 120A, 121A, 1980b:243, figs. 5, 6; Ni, 1988:334, fig. 260; Bertelsen and Pietsch, 1998a:140).

*Linophryne racemifera* Regan and Trewavas, 1932: five metamorphosed females (25.5 to 81 mm); North Atlantic and eastern North Pacific; holotype, ZMUC P92146, 52 mm, *Dana* station 4009(6), western North Atlantic, south of Canaries, 24°36.5′N, 17°27′W, open pelagic net, 4000 m of wire, 0700 h, 18 March 1930 (Regan and Trewavas, 1932:108, figs. 165, 166, pl. 10, fig. 2; Bertelsen, 1951:175, 184, fig. 121B, 1982:80, figs. 13, 14; Swinney, 1995a:52, 57, 1995b:46, figs. 4, 5).

*Linophryne trewavasae* Bertelsen, 1978: a single metamorphosed female, the holotype, LACM 36116-5, 73.5 mm, with parasitic male (10.7 mm); Southeast Asian Bioluminescence Expedition, *Alpha Helix* station 84, Banda Sea, 5°04.5′S, 130°12.0′E, RMT-8 oblique haul, 0 to 1500 m, 28 April 1975 (Robison, 1976:40, color photograph; Bertelsen, 1978:26, figs. 1, 2; Pietsch and Seigel, 1980:397).

*Linophryne bicornis* Parr, 1927: five metamorphosed females (27 to 185 mm) and three parasitic males (18 to 30 mm); western North Atlantic and eastern South Indian oceans; holotype, YPM 2030, 27 mm, *Pawnee* station 59, Bermuda, 32°19′N, 64°32′W, open pelagic net, about 2500 m of wire, 21 April 1927 (Parr, 1927:10, fig. 2; Regan and Trewavas, 1932:107, 111; Bertelsen, 1951:184, 1982:82, fig. 15; Behrmann, 1977:93, figs. 1–4).

*Linophryne algibarbata* Waterman, 1939b: nine metamorphosed females (28 to 182 mm) and two parasitic males (23 to 29 mm); North Atlantic Ocean; holotype, MCZ 35066, 28 mm, *Atlantis* station 2894, 39°06′N, 70°16′W, closing net in 400 m, 20 July 1937 (Waterman, 1939b:85, figs. 3, 4; Koefoed, 1944:13, pl. 2, fig. 1; Bertelsen, 1951:174, 175, 1976:13, figs. 3, 4, 1980b:237, figs. 2, 3, 6; Behrmann, 1974:364, figs. 1, 2; Stearn and Pietsch, 1995:141, fig. 91; Jónsson and Pálsson, 1999:203, fig. 8).

# FOUR

# Evolutionary Relationships

> We may assume an ogcocephalid or chaunacid-like ancestral ceratioid which, from the benthic or littoral environment of its ancestors, has invaded the bathypelagic zone of the ocean. Probably this evolution passed through forms in which the adults were benthic, while the juveniles after metamorphosis continued the pelagic life of the larvae during adolescence as, for instance, found in the family Chaunacidae and as retained or reestablished in the [benthic] ceratioid genus *Thaumatichthys*. This move to a new adaptive zone has led to a dimorphism which separates the tasks of the two sexes, the females attaining adaptations to the bathypelagic conditions of the lophiiform feeding strategy by passive luring, the males being adapted solely to actively search for a sexual partner.
>
> ERIK BERTELSEN,
> "Ceratioidei: Development and Relationships," 1984:330

Under the extreme selective pressures imposed by meso- and bathypelagic environments, a vast array of bizarre and wonderful creatures has evolved, displaying behaviors and associated morphological adaptations that are almost beyond our ability to imagine. Among the highly diverse products of this evolution in the great oceanic depths, few are as surprising and spectacular as the deep-sea anglerfishes. Just knowing the full breadth of morphological and taxonomic diversity within this extraordinary taxon—which the preceding chapter is intended to provide—is enough to stimulate a host of intriguing questions: Why are there so many living species of ceratioids—more than twice as many families and genera and more than three times the number of species as the whalefishes of the suborder Cetomimoidei, the next most species-rich deep-sea vertebrate taxon? Can the "explosive" radiation of ceratioids, within all the world's oceans, across the tropics and from pole to pole, be attributed to one or more key innovations unique to their anatomy or life history? A quick glance at a sexually parasitized ceratioid immediately evokes wonder and amazement, not to mention an unending list of additional questions. How and why did this arrangement ever come to be? What natural selective constraints have led deep-sea ceratioids to develop male dwarfism and sexual parasitism rather than simple hermaphroditism, which is exhibited by so many other deep-sea fishes? Definitive answers to most of these questions are probably beyond our grasp, but, nevertheless, to move beyond guesswork (what my major professor used to call "armchair" biology) and begin to address the mysteries of ceratioid biology in a true scientific context, an understanding of the sequence of evolution of its major clades is essential. A rigorously tested systematic hypothesis of relationships allows for a precise reconstruction of aspects of character evolution, which in turn will cast light on broader evolutionary processes that have taken place within the Ceratioidei, and perhaps within the deep-ocean ecosystem as a whole. The present chapter is devoted to that effort.

## Ordinal Relationships

Anglerfishes have traditionally been allied with the toadfishes of the teleost order Batrachoidiformes (Fig. 186). This alignment dates back to Georges Cuvier's second edition of *Le Règne Animal* (1829:xi, 249), in which the then-known lophioids, antennarioids, and ogcocephaloids were placed together with the batrachoidids, in a group called the *acanthoptérygiens à pectorales pédiculées*, a name that Cuvier proposed in reference to the foot-like structure of the pectoral fins of these fishes. This phylogenetic arrangement, or a variation of it that placed anglerfishes and toadfishes side by side, was later followed by Valenciennes (1837:335) and numerous others (e.g., Cope, 1872:340; Jordan and Evermann, 1898:2712; Jordan and Sindo, 1902:361; Boulenger, 1904b:717; Jordan, 1905:542; Eaton et al., 1954:216), including Albert Günther (1861:178), who was the first to use the name Pediculati (see Gill, 1883a:552), which he Latinized from the original French vernacular of Cuvier (1829). Gill (1863:88), however, vehemently denied a close relationship between anglerfishes and toadfishes: "The genus *Batrachus*, referred to the Pediculati by Cuvier, has really little affinity to the true representatives of the group, and has been, by general consent, separated from them by all the more modern systematists." But just as sure of himself, Gill (1872:xli) later changed his mind: "The natural character of the association of forms combined therein [the Pediculati] is obvious, and has never been questioned, and the comparatively slight affinity with them of the Batrachoids, which were formerly combined with them, is now universally conceded.... Their relations are most intimate

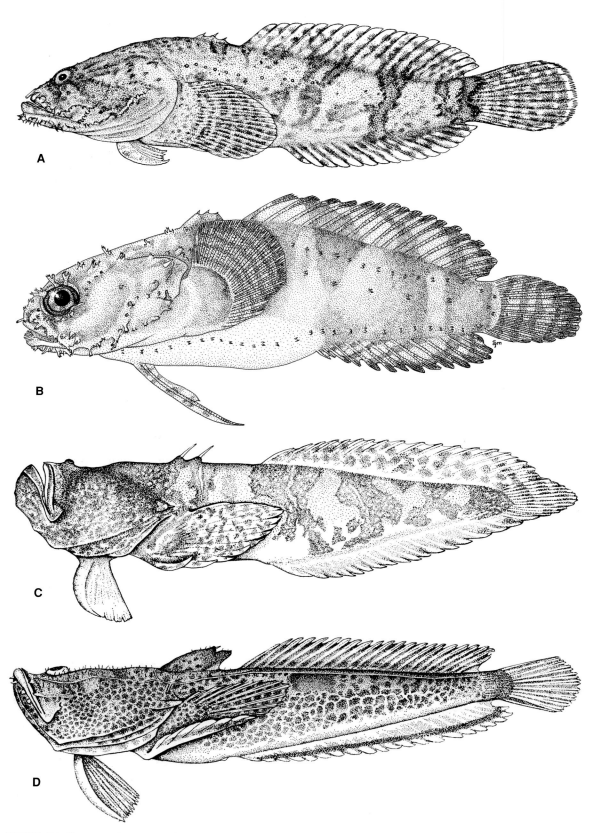

FIGURE 186. Representative species of the teleost family Batrachoididae. Although contradicted by recent molecular evidence, this group has been recognized by nearly all ichthyologists as the most likely sister taxon of lophiiform fishes: (A) *Potamobatrachus trispinosus*, holotype, 49.3 mm SL, MZUSP 4335 (drawing by K.H. Moore; after Collette, 1995); (B) *Halophryne hutchinsi*, 91.8 mm SL, USNM 150899 (drawing by S.G. Mondon; after Greenfield, 1998); (C) *Thalassophryne amazonica*, 68.2 mm SL, USNM 210867 (drawing by M.H. Carrington; after Collette, 1966); (D) *Daector schmitti*, paratype, 80.4 mm SL, CAS-SU 14949 (drawing by M.H. Carrington; after Collette, 1966, 1968).

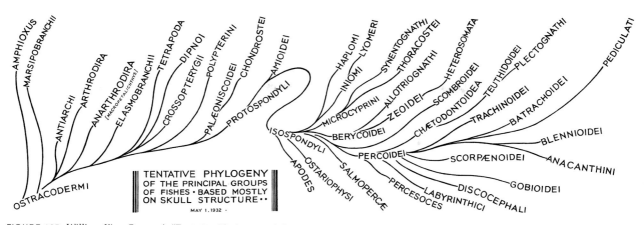

FIGURE 187. William King Gregory's "Tentative Phylogeny of the Principal Groups of Fishes," one of many evolutionary hypotheses published throughout the twentieth century that supports a close relationship between the toadfishes (Batrachoidei) and the anglerfishes (Pediculati), indicated by a shared branching point at the far right of the diagram. After Gregory (1933).

with the Batrachoid and Blennioid forms, and doubtless they have descended from the same common progenitors."

Like Cuvier (1829) and his disciple Valenciennes (1837), Regan (1912:278) initially believed anglerfishes and toadfishes to be so closely related that he included them as suborders of the Pediculati, a taxon recognized by him as an acanthopterygian order: "The Batrachoidea are here included in the Pediculati rather than in the Percomorphi, for it can hardly be the case that the resemblance in osteological characters, especially in the structure of the pectoral arch, is not due to real affinity." Later, however, in his review of *The Pediculate Fishes of the Suborder Ceratioidea*, Regan (1926:3) separated the Lophiiformes from the Batrachoidiformes (but kept them side by side, thereby implying a sister-group relationship), stating that "although the resemblances in the pectoral arch may be due to relationship, the differences in other characters are sufficient to keep them apart." Since that time, Regan's (1926) revised opinion has been almost universally accepted (e.g., Regan and Trewavas, 1932:25; Gregory, 1933:386, see Fig. 187, 1951:225; Gregory and Conrad, 1936:198, 200; Berg, 1940:498; Eaton et al., 1954:216; Monod, 1960:622; Greenwood et al., 1966:388; Rosen and Patterson, 1969:438; for a return to Regan's 1912 proposal, see Gosline, 1971:173).

In a paper titled "The Paracanthopterygii Revisited: Order and Disorder," Patterson and Rosen (1989:23) reaffirmed their earlier (see Rosen and Patterson, 1969:438) contention that the Batrachoidiformes and Lophiiformes are sister groups, citing new evidence derived from the dorsal gill-arch skeleton: in both these orders "the first pharyngobranchial and the suspensory tip of the first epibranchial are reduced or lost, and, if present, both are withdrawn laterally away from the second and third pharyngobranchials." In addition, Patterson and Rosen (1989:23) listed two other "probable" synapomorphies: (1) the ventral gill arches converge on a very short copula, which is ossified very feebly or not at all; and (2) the prezygapophyses of the first vertebra insert into hollow exoccipital bony tubes that are secondarily elongated, extending to or beyond the basioccipital condyle (Rosen, 1985:29).

Thus, in summary, batrachoidiforms are the only group ever shown to bear morphological evidence of a sister-group relationship with the lophiiforms (Pietsch, 1981:413): "the latter are surely monophyletic, the former less surely, but if so, the two are sister groups" (Patterson and Rosen, 1989:24). But very recent evidence of a different kind now indicates that perhaps anglerfishes have nothing at all to do with toadfishes but are most closely related instead to triggerfishes, boxfishes, puffers, and their allies, that is, to all those fishes that make up the teleost order Tetraodontiformes. The evidence for this is molecular. In a recent comprehensive study of teleost relationships based on 100 complete mitochondrial DNA sequences, Miya et al. (2003:131) showed that the commonly accepted position of the Lophiiformes among "relatively primitive groups within higher teleosts (Paracanthopterygii)" could not be supported. They found instead that lophiiforms were "confidently placed" within a crown group of teleosts that also included two tetraodontiform species *(Sufflamen fraenatus* and *Stephanolepis cirrhifer)*, plus *Antigonia capros* (Zeiformes: Caproidae). Holcroft (2004:8; see also Holcroft, 2005; Yamanoue et al., 2007), in testing hypotheses of tetraodontiform relationships using the *RAG1* gene, also suggested a more derived placement of lophiiforms, the one member of this order examined, *Lophius americanus*, clustering with *Siganus doliatus*; and these two taxa, along with *Antigonia capros*, forming the sister group of an assemblage containing, among other groups, a monophyletic Tetraodontiformes.

Miya et al. (2005), in a follow-up study of the relationships of the Batrachoidiformes, this time inferring relationships from a partitioned Bayesian analysis of 102 whole mitochondrial genomic sequences, verified the earlier findings (Miya et al., 2003), showing once again that the Lophiiformes form a sister-group relationship with a clade comprising tetraodontiforms, plus *Antigonia*, supported by 99 to 100% posterior probabilities (Miya et al., 2005:297). They concluded also that the least comprehensive monophyletic group that includes both Batrachoidiformes and Lophiiformes encompasses nearly the entire Percomorpha (Percomorpha minus Ophiiformes), strongly suggesting that the two groups diverged relatively basally within the Percomorpha. Bayesian analysis of the three data sets found no tree topology that is congruent with the above two hypotheses, indicating that the probabilities of the Batrachoidiformes representing a member of the Paracanthoptergii and the sister group of the Lophiiformes are less than 1/60,000 or 0.00002 in the Bayesian context (Miya et al., 2005:297). For more on the conflicting results of these two data sets, morphological and molecular, see below.

## Subordinal and Familial Relationships

That the order Lophiiformes itself constitutes a natural assemblage seems certain, based on at least six unique and morphologically complex, shared derived features (numbered 1 to 6 in

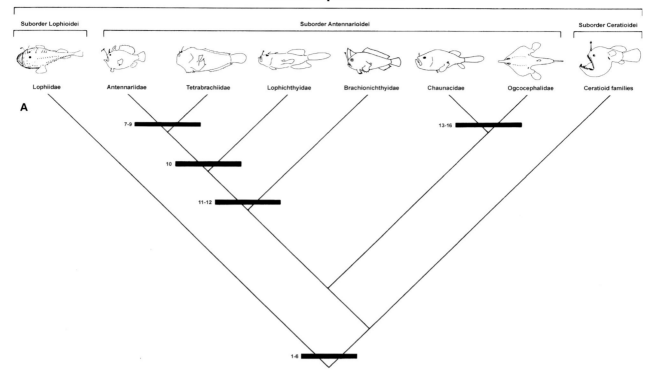
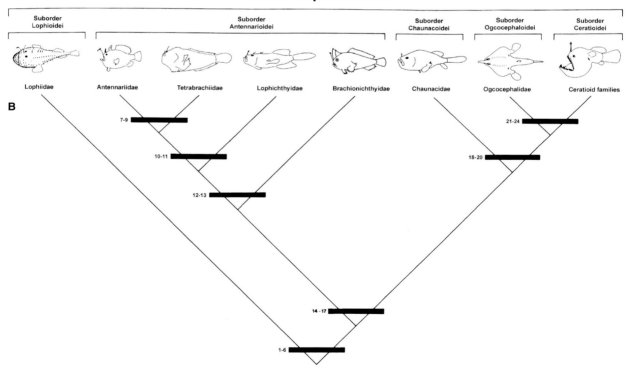

FIGURE 188. Cladistic hypotheses of relationship of the families of lophiiform fishes: (A) three suborders recognized, with the families of the Ceratioidei indicated as the sister group of the Antennarioidei (modified after Pietsch, 1981); (B) five suborders recognized, with the families of the Ceratioidei indicated as the sister group of the batfishes, family Ogcocephalidae (modified after Pietsch, 1984).

Fig. 188; a modification of Pietsch, 1981, 1984; Pietsch and Grobecker, 1987:268; see also Everly, 2002):

1. Spinous dorsal fin primitively of 6 spines, the anterior-most three of which are cephalic in position, the first modified to serve as a luring apparatus (involving numerous associated specializations, e.g., a medial depression of the anterior portion of the cranium, loss of the nasal bones [nasal of Rosen and Patterson, 1969, is the lateral ethmoid] and supraoccipital lateral-line commissure, and modifications of associated musculature and innervation)
2. Epiotics separated from the parietals and meeting on the midline posterior to the supraoccipital
3. Gill opening restricted to a small, elongate, tubelike opening situated immediately dorsal to, posterior to, or ventral to (rarely partly anterior to) the pectoral-fin base
4. A single hypural plate (sometimes deeply notched posteriorly) emanating from a single complex half-centrum (Rosen and Patterson, 1969:441)
5. Ventral-most pectoral radial considerably expanded distally (Pietsch, 1981:397, 411)
6. Eggs spawned in an oval or scroll-shaped mucous sheath (Rasquin, 1958; Pietsch and Grobecker, 1987:351)

Since Regan (1912), three major lophiiform taxa of equal rank have been recognized by nearly all authors. Listed here are these taxa, together with their currently recognized families (the 11 families of the bathypelagic Ceratioidei excluded; see Bertelsen, 1951:29; Pietsch, 1972a:18):

Suborder Lophioidei
Family Lophiidae

Suborder Antennarioidei
Family Antennariidae
Family Tetrabrachiidae
Family Lophichthyidae
Family Brachionichthyidae
Family Chaunacidae
Family Ogcocephalidae

Suborder Ceratioidei

Pietsch (1981:416) tested the validity of Regan's (1912) concept of three major lophiiform taxa by using cladistic analysis (see Hennig, 1950, 1966). In that study, serious difficulty was encountered in efforts to establish monophyly for the six families of Regan's (1912) Antennarioidei. Although a number of synapomorphic features were found to support a sister-group relationship between the four families Antennariidae through Brachionichthyidae, and between the families Chaunacidae and Ogcocephalidae, no convincing synapomorphy was found to link these two larger subgroups (Fig. 188A).

In a more recent attempt, Pietsch (1984:323) proposed a new hypothesis, which, in resolving the former difficulties, differed significantly from that previously published (Pietsch, 1981). In this revised cladogram (Fig. 188B), the suborder Antennarioidei is restricted to just four families: the Antennariidae, recognized as the sister group of the Tetrabrachiidae, these two families together forming the sister group of the Lophichthyidae, and this assemblage of three families forming the sister group of the Brachionichthyidae. These relationships are supported by a total of seven synapomorphies, most of which had been previously described by Pietsch (1981; numbered 7 through 13 in Fig. 188B):

7. Posteromedial process of the vomer emerging from the ventral surface as a laterally compressed, keel-like structure, its ventral margin (as seen in lateral view) strongly convex (Pietsch, 1981:397)
8. Postmaxillary process of the premaxilla spatulate (Pietsch, 1981:398)
9. Opercle similarly reduced in size (Pietsch, 1981:401)
10. Ectopterygoid triradiate, a dorsal process overlapping the medial surface of the metapterygoid (Pietsch, 1981:400)
11. Proximal end of hypobranchials II and III deeply bifurcate (Pietsch, 1981:407)
12. Interhyal with a medial, posterolaterally directed process that makes contact with the respective preopercle (Pietsch, 1981:400)
13. Illicial pterygiophore and pterygiophore of the third dorsal spine with highly compressed, bladelike dorsal expansions (Pietsch, 1981:410)

The present interpretation of lophiiform relationships differs further from any previously proposed hypothesis in considering the Antennarioidei (*sensu stricto*) to form the sister group of a much larger group that includes the Chaunacoidei (Pietsch, 1984), the Ogcocephaloidei (Pietsch, 1984), and the Ceratioidei. The Ogcocephaloidei are in turn recognized as the sister group of the Ceratioidei (Fig. 188B).

Monophyly for a group containing the suborders Antennarioidei, Chaunacoidei, Ogcocephaloidei, and Ceratioidei is supported by four synapomorphies, all previously identified by Pietsch (1984; numbered as they appear in Fig. 188B):

14. Eggs and larvae small (at all stages the eggs are considerably less than 50% the diameter of those of lophioids; the smallest larvae are certainly less than 50%, and probably less than 30%, the size of those of lophioids; size at transformation to the prejuvenile stage is less than 60% that of lophioids; Pietsch, 1984)
15. Head of larvae proportionately large relative to body (always greater than 45% SL, compared to less than 30% in lophioids; Pietsch, 1984)
16. Number of dorsal-fin spines reduced from a primitive 6 in lophioids to 3 or fewer (Pietsch, 1981:409)
17. Pharyngobranchial IV absent (present and well toothed in lophioids; Pietsch, 1981:401)

Monophyly for a group containing the suborders Chaunacoidei, Ogcocephaloidei, and Ceratioidei is supported by three synapomorphies (numbered as they appear in Fig. 188B):

18. Second dorsal-fin spine reduced and embedded beneath the skin of the head (Pietsch, 1981:410)
19. Interhyal simple, cylindrical, without a medial, posterolaterally directed process (present in lophioids and all antennarioids; Pietsch, 1981:400)

20. Gill filaments of gill arch I absent (but present on the proximal end of ceratobranchial I of some ceratioids; Bradbury, 1967:408; Pietsch, 1981:415)

Monophyly for a group containing the Ogcocephaloidei and Ceratioidei is supported by four synapomorphies (numbered as they appear in Fig. 188B):

21. Second dorsal-fin spine reduced to a small remnant (well developed in the ceratioid family Diceratiidae, and in all other lophiiforms; Bertelsen, 1951:17; Pietsch, 1981:410)
22. Third dorsal-fin spine and pterygiophore absent (present in all other lophiiforms; Bertelsen, 1951:17; Bradbury, 1967:401; Pietsch, 1981:410)
23. Epibranchial I simple, without ligamentous connection to epibranchial II (in batrachoidiforms and all other lophiiforms, epibranchial I bears a medial process ligamentously attached to the proximal tip of epibranchial II; Pietsch, 1981:401)
24. Posttemporal fused to the cranium (attached to the cranium in batrachoidiforms and all other lophiiforms in such a way that considerable movement in an anterodorsal-posteroventral plane is possible; Pietsch, 1981:411)

Of the possible cladograms that could be constructed on the basis of the morphological data provided by Pietsch (1981, 1984) and Pietsch and Grobecker (1987), the one shown here (Fig. 188B) is by far the most parsimonious (but see Pietsch, 1984:324, for a discussion of convergence or reversal of character states).

## Key to the Major Subgroups of the Lophiiformes

Plesiomorphic and autapomorphic features of the major subgroups of the Lophiiformes are incorporated into the following analytical key:

1A. Postcephalic, spinous dorsal-fin of 1 to 3 spines; pharyngobranchial IV present; cleithrum with prominent posterior spine; subopercle with large ascending process attached to anterior margin of ventral rami of opercle; pseudobranch well developed; eggs and larvae large; head of larvae small relative to body . . . . . . . . . . . . . . . . . . . . . . . . . . . . . . . . . Suborder Lophioidei
1B. Postcephalic, spinous dorsal-fin absent; pharyngobranchial IV absent; cleithral spine absent; subopercle with ascending process absent or reduced to a small projection detached from opercle; pseudobranch greatly reduced or absent; eggs and larvae small; head of larvae large relative to body . . . . . . . . . . . . . . . . . . . . . 2
2A. Spinous dorsal-fin of 3 spines emerging from dorsal surface of cranium; illicial pterygiophore and pterygiophore of third dorsal-fin spine with highly compressed, bladelike dorsal expansions; interhyal with a medial, posterolaterally directed process that comes into contact with the respective preopercle; interoapercle flat and broad (Suborder Antennarioidei) . . . . . . . . . . . . . . . 3
2B. Spinous dorsal-fin of 2 or 3 spines, but only anteriormost spine emerging from dorsal surface of cranium (spines II and III reduced and embedded beneath skin of head or lost); illicial pterygiophore and pterygiophore of third dorsal-fin spine without bladelike dorsal expansions; interhyal without a medial, posterolaterally directed process; interopercle elongate and narrow . . . . . . . . . . . . . . . . . . . . . . . . . . . . . . . . . 6
3A. Parietals meeting on midline dorsal to supraoccipital; ectopterygoid roughly oval in shape or absent; ceratobranchials I through III with 1 or more tooth plates; hypobranchial II simple, hypobranchial III absent; pectoral radials 2; pelvic fin of 1 spine and 4 rays . . . . . . . . . . . . . . . . . . . . . . . . . . . . . Family Brachionichthyidae
3B. Parietals well separated by supraoccipital; ectopterygoid triradiate, a dorsal process overlapping medial surface of metapterygoid; ceratobranchials I through IV toothless; hypobranchials II and III bifurcated proximally; pectoral radials 3; pelvic fin of 1 spine and 5 rays . . . . . . . . . 4
4A. Vomer wide, the width between lateral ethmoids nearly as great as that between lateral margins of sphenotics; vomer without posteromedial process; dorsal head of quadrate broad, the width equal to or greater than that of metapterygoid; postmaxillary process of premaxilla tapering to a point; opercle expanded posteriorly; pharyngobranchial and epibranchial of first arch toothed; bony connection between tips of haemal spines of fourteenth through sixteenth preural centra; pterygiophore of illicium elongate, greatly depressed and laterally expanded posteriorly . . . . . . . . . . . . . . . . . . . . . . . . . . . . . . . . . . . . . . Family Lophichthyidae
4B. Vomer narrow, the width between lateral ethmoids considerably less than that between lateral margins of sphenotics; posteromedial process of vomer emerging from ventral surface as a laterally compressed, keel-like structure, its ventral margin (as seen in lateral view) strongly convex; dorsal head of quadrate narrow, the width less than that of metapterygoid; postmaxillary process of premaxilla spatulate; opercle reduced in size; pharyngobranchial and epibranchial of first arch toothless; bony connection between tips of haemal spines absent; pterygiophore of illicium short, the posterior end cylindrical . . . . . . . . . . . . . . . . . . . . . . . . . . . . . . 5
5A. Eyes dorsal; dorsal-fin spines reduced; mouth small; pectoral fin double, the dorsal-most ray of ventral portion membranously attached to side of body; pectoral-fin lobe membranously attached to rays of pelvic fin; dorsal-fin rays 16 or 17, anal-fin rays 11 or 12 . . . . . . . . . . . . . . . . . . . . . . . . . . . . . Family Tetrabrachiidae
5B. Eyes lateral; dorsal-fin spines well developed; mouth large; pectoral fin single, the rays not membranously attached to side of body; pectoral-fin lobe not membranously attached to rays of pelvic fin; dorsal-fin rays 11 to 15, anal fin rays 6 to 9 . . . . . Family Antennariidae
6A. Second dorsal-fin spine elongate, embedded beneath skin of head; third dorsal-fin spine and pterygiophore present; epibranchial I with a medial process ligamentously attached to proximal tip of epibranchial II . . . . . . . . . . . . . . . . . . . . . . . . . . . . . Suborder Chaunacoidei

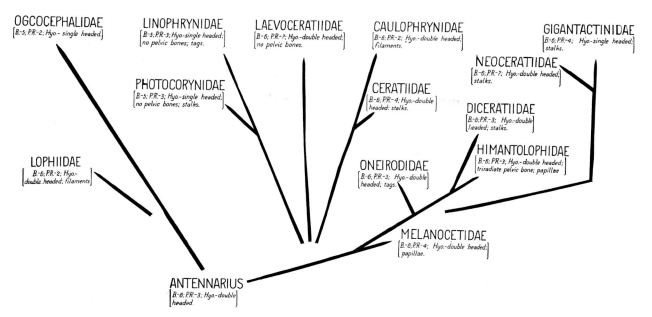

FIGURE 189. The earliest published branching diagram of relationships of ceratioid families, thought then to have been derived from an *Antennarius*-like ancestor, based almost exclusively on the conclusions of Regan (1926) and Regan and Trewavas (1932). Laevoceratiidae was then recognized to contain an assemblage of what we now know are ceratioid males. After Gregory and Conrad (1936).

6B. Second dorsal-fin spine reduced to a tiny remnant embedded beneath skin of head and lying on, or fused to, dorsal surface of pterygiophore just behind base of illicial bone; third dorsal-fin spine and pterygiophore absent; epibranchial I simple, without ligamentous attachment to epibranchial II .................. 7

7A. Palatine teeth usually present; ceratobranchial V toothed, expanded proximally; pelvic fins present; obvious sexual dimorphism absent, the males not dwarfed .........
..................... Suborder Ogcocephaloidei

7B. Palatine teeth absent; ceratobranchial V toothless, reduced to a slender rod-shaped element; pelvic fins absent (except in larval caulophrynids); sexual dimorphism strongly developed, the males dwarfed to a small fraction of size of females ............ Suborder Ceratioidei

## Relationships of Ceratioid Families and Genera

Based on the limited amount of ceratioid material available at the time, Regan and Trewavas (1932:26) thought it "probable that the families in which the males are parasitic form a natural group, but apart from this, the evidence from the skeleton would point to the Melanocetidae, Diceratiidae, and Himantolophidae as closely related families to be placed first, and the Photocorynidae and Linophrynidae [the former since synonymized with the latter] as forming a pair to be placed at the end of the series. The Oneirodidae are perhaps not very remote from the Diceratiidae, but the remaining families appear to be rather isolated."

While agreeing with almost everything proposed by Regan and Trewavas (1932), Gregory (1933, 1951) and Gregory and Conrad (1936) provided an elaborate scenario of how ceratioids might have evolved and diversified from shallow-water anglerfishes (lophioids and antennarioids). They illustrated their ideas with the first branching diagrams of anglerfishes ever published—"pictorial phylogenies of the pediculate fishes" that show the "principal lines of cleavage leading respectively to the true anglers *(Lophius)*, the sea-bats (ogcocephalids), the sea-mice (antennariids) and the deep-sea anglers (ceratioids)" (Gregory and Conrad, 1936:194, 198; Fig. 189); and the "divergent evolution" or "adaptive branching" of the deep-sea anglers (Gregory, 1933:405, 1951:312; Fig. 190).

Bertelsen (1951:28) too was inclined to agree with the general idea of family relationships proposed by Regan and Trewavas (1932), but he argued against the notion that families that exhibit sexual parasitism form a natural assemblage. Because the Caulophrynidae, the males of which become sexually parasitic on females, displays a number of primitive characters (e.g., caulophrynid larvae apparently lack sexual dimorphism in the luring apparatus (see below); adult females lack the bulbous bacteria-filled esca; and caulophrynid larvae retain pelvic fins, which are absent in larvae and adults of all other ceratioids), Bertelsen (1951:28) placed this family "first in the suborder." At the same time, he argued that linophrynids, which also have parasitic males, show "such a highly specialized condition that they must be placed last." For unstated reasons, Greenwood et al. (1966:397) implied a much closer relationship between the Caulophrynidae and Linophrynidae. Likewise, Pietsch (unpublished data) reported (at the 1975 Annual Meeting of the American Society of Ichthyologists and Herpetologists at Williamsburg, Virginia) numerous, apparently derived, character states shared by these two families as well as with the Gigantactinidae and Neoceratiidae. He presented a phylogeny that argued for a monophyletic origin of sexual parasitism within a lineage derived from some oneirodid-like ancestor (Pietsch, 1976; see also Pietsch, 1979).

In a more recent attempt to determine the phylogenetic relationships of ceratioid taxa, Bertelsen (1984:331) expressed frustration that "most of the derived osteological characters shared by two or more families are reduction states or loss of parts . . . and similarities among such characters may in many cases represent convergent developments. At the same time, most of the diagnostic family characters which represent new

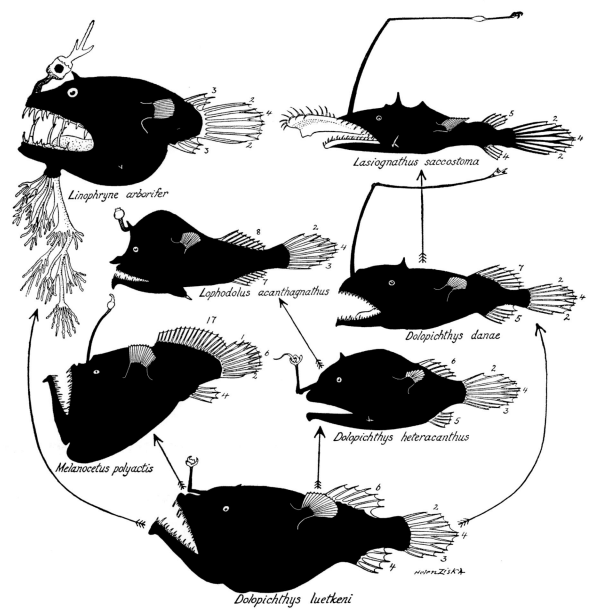

FIGURE 190. One of the earliest published branching diagrams of ceratioid relationships, showing oneirodids *(Dolopichthys)* as ancestral to melanocetids, linophrynids, and thaumatichthyids, based largely on the work of Regan (1926) and Regan and Trewavas (1932). Drawings by Helen Ziska; after Gregory (1933, 1951).

structures or specialization of organs are autapomorphic." Nevertheless, he was able to shed new light on the subject, analyzing a list of 30 characters, including 10 shared by both sexes, another 10 confined to metamorphosed females, six that describe only metamorphosed males, and another four that are restricted to larvae. Following a discussion of character-state evolution for each of the 30 characters, Bertelsen (1984:333) presented a tree that by his own admission "should be regarded only as a very schematic compilation of expressed views" (Fig. 191). He (1984:334) concluded by saying that "ceratioids are still very incompletely known and future studies on additional characters and as yet unknown forms may bring answers to at least some of the many questions about their phylogenetic relationships."

Thus, despite numerous efforts, the several hypotheses of phylogenetic relationship that have been proposed for the Ceratioidei in the past are unsatisfactory, mostly contradicting one another and containing one or more unresolved polytomies. It should be emphasized, however, that up until very recently (Pietsch and Orr, 2007) a rigorous cladistic analysis had never been attempted. Taxa had been grouped by inspection only; no computer-generated trees had ever been constructed. Now, however, Pietsch and Orr (2007), with the advantage of more than 20 years of additional accumulated data since Bertelsen's (1984) attempt, coupled with a reexamination of all previously identified characters—combined also with the results of analyses of new characters and the incorporation of morphological variation taken from newly discovered taxa—have provided the first computer-assisted cladistic analysis of relationships of ceratioid families and genera. Their work is followed here with only minor revisions.

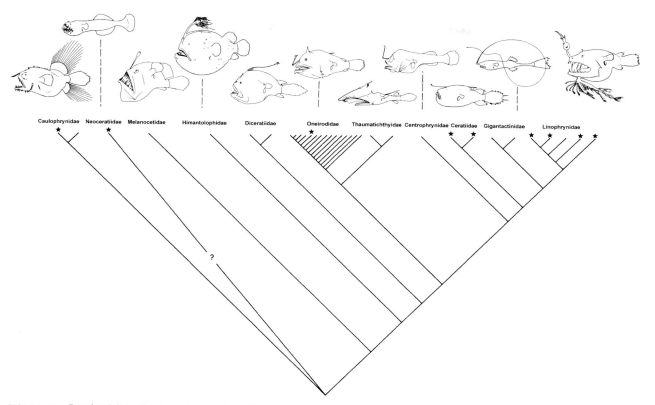

FIGURE 191. Bertelsen's hypothesis of relationships of the 11 families of the Ceratioidei, the stars indicating taxa known to have parasitic males. Note the unresolved position of the Neoceratiidae. After Bertelsen (1984).

This latest attempt to reconstruct the evolutionary history of deep-sea anglerfishes is based on an examination of all 66 known lophiiform genera, including cleared-and-stained osteological material of all except a few unavailable ogcocephaloid genera and five extremely rare ceratioid genera (*Dermatias, Tyrannophryne, Ctenochirichthys, Robia,* and *Acentrophryne*) known only from very few specimens (for lists of material examined osteologically, see the various papers of Bertelsen and Pietsch cited herein, especially Bertelsen, 1951; and Pietsch, 1972a, 1974a, 1981). A data matrix of 34 taxa (four outgroups and 30 ingroups, the five unavailable ceratioid genera excluded) and 71 characters applicable to metamorphosed females, plus another 17 characters applicable only to metamorphosed males and larvae (for a total of 88), was constructed (but note that ceratioid males and larvae are rare in collections and generally poorly described anatomically, the males unknown for 13 of the 35 ceratioid genera, and larvae unknown for 11 of the 35 ceratioid genera; see Bertelsen, 1984; Pietsch, 1984). Two separate analyses were conducted: one with only those characters applicable to metamorphosed females, and a second with all 88 characters, including those applicable to metamorphosed males and larvae. All characters were treated as unordered and unweighted and were polarized using the lophiiform suborders Lophioidei, Antennarioidei, Chaunacoidei, and Ogcocephaloidei as outgroups (Fig. 2). Characters for outgroup taxa were coded as a single state when all members of the suborder shared the same character state; when the state differed among subtaxa, the state was coded as polymorphic. All except 17 characters were binary. Character states that do not extend to or are unknown for a particular taxon are indicated in the data matrix by a question mark. Character states were coded 0 to 3 and indicated within paren-

theses after each respective character-state description. A matrix of character states for all taxa was published by Pietsch and Orr (2007) and is not repeated here. The matrix was analyzed with PAUP (v. 4.0b10, D.L. Swofford, PAUP*: Phylogenetic Analysis Using Parsimony and Other Methods, Sinauer Associates, Sunderland, MA, 2002), using the branch and bound algorithm, with accelerated transformation (ACCTRAN) to optimize characters. To evaluate branch support, a heuristic bootstrap analysis of 100 replicates was conducted, with simple addition sequence and TBR (tree bisection-reconnection) branch swapping options. Bremer decay values (Bremer, 1988) were calculated using TREEROT (v. 2, M.D. Sorenson, TREEROT.v2, Boston University, Boston, MA, 1999).

In addition to the extreme sexual dimorphism—including the presence of denticular bones in the males, used to grasp and hold fast to a prospective mate, as well as numerous associated features that allow for a unique mode of reproduction—monophyly for the Ceratioidei is supported by the loss of palatine teeth, the basihyal, and pelvic fins; a loss of the benthic ambulatory function and a consequent repositioning of the pectoral fins; and a general trend toward a reduction in body density by a loss of bony parts, an overall decrease in ossification and the extent of muscle development, and the infusion of lipids throughout.

The taxonomy and present classification of the Ceratioidei are based primarily on studies of metamorphosed females (only 22 of the 35 recognized ceratioid genera are represented by males). Except for larval stages and the few basic meristic and osteological characters shared by both sexes, synonymies, diagnoses, and descriptions require separate treatment of females and males. The families of the suborder form well-defined, highly distinct taxa, separated primarily by osteological characters, which more often than not are autapomorphic for each; the females of each family possess strikingly unique features that separate them from those of all other families (Bertelsen, 1984).

Likewise, the separation and definition of genera are based primarily on characters present only in females. However, some of the distinguishing meristic and osteological characters are shared with the males, such as fin-ray counts, which in some families show distinct intergeneric differences. The structures unique to the males, such as denticular teeth and nostril morphology, show distinct intergeneric differences, in full agreement with separations based on characters of the females. In most cases, however, it has not been possible to separate free-living males into taxa below the generic level, and studies of males attached to females have not revealed characters that will allow specific identification (Bertelsen, 1984).

## Characters and Character States

The following characters are divided into three categories: those that describe metamorphosed females (1 to 71), those restricted to metamorphosed males (72 to 81), and those that extend only to larvae (82 to 88). Within each category, characters are arranged by anatomical complex: cranium (1 to 14), suspensorium (15 to 18), opercular apparatus (19 to 24), jaws (25 to 35), hyoid and gill arches (36 to 47), axial skeleton and caudal fin (48 to 53), dorsal and anal fins (54 to 62), pectoral and pelvic fins (63 to 67), skin spines (68), additional soft structures unique to lophiiforms (69 to 71); sexual dimorphism (72), eyes (73 to 74), olfactory organs (75 to 76), jaws, teeth, and denticular bones (77 to 81); size and shape of head and body (82 to 84), body inflation (85), illicial apparatus (86), and pectoral and pelvic fins (87 to 88). For the tree derived from characters applicable to metamorphosed females, consistency and retention indices (CI and RI) were produced as a whole and for each character individually. The indices listed for characters of metamorphosed males and larvae were taken from the tree derived from all 88 characters combined. Both CI and RI are presented below after each character description.

### Characters of Metamorphosed Females

1. The supraethmoid is usually well developed in lophiiforms **(0),** but very much reduced or absent in the Thaumatichthyidae (Bertelsen and Struhsaker, 1977:9, figs. 1, 2), and absent in lophioids and the Gigantactinidae (a tiny rudiment observed in a single specimen of *Gigantactis*; Bertelsen et al., 1981:10, fig. 12) **(1)** (CI 0.33, RI 0.50).

2. An ossified vomer is present in nearly all lophiiforms **(0)**, but absent in juvenile and adult females of both genera of the Gigantactinidae **(1)** (1.00).

3. Vomerine teeth are well developed in nearly all lophiiforms **(0)**; lost with growth in the oneirodid genera *Dolopichthys* and *Bertella*, and in some species of *Ceratias* (Pietsch, 1986b:481, table 1) **(1);** and absent in some ogcocephaloids and all members of the Himantolophidae, Thaumatichthyidae, the oneirodid genera *Lophodolos* and *Pentherichthys*, and the linophrynid genera *Photocorynus* and *Haplophryne* **(2)** (0.33, 0.43).

4. The dorsal margin of the frontal bone is smooth in nearly all lophiiforms or interrupted in some lophioids by a series of short conical knobs or spines ("rugose ridge" of Caruso, 1985:873, fig. 3) **(0)**; a conspicuous, rounded, laterally compressed frontal protuberance is present in the linophrynid genera *Borophryne* and *Linophryne* (Regan and Trewavas, 1932:46, figs. 65, 66) **(1)**; the frontal protuberance forms a sharp spine in the linophrynid genera *Photocorynus* and *Haplophryne* (Regan and Trewavas, 1932:44, figs. 59, 62) **(2)** (0.67, 0.50).

5. The frontals of all the outgroups, as well as those of the Centrophrynidae, Ceratiidae, Caulophrynidae, Neoceratiidae, and the linophrynid genus *Photocorynus*, meet posteriorly on the midline (or are narrowly separated by cartilage) in front of the supraoccipital **(0)**; those of all remaining lophiiform taxa are widely separated along their dorsal margins **(1)** (0.25, 0.67).

6. The anterior end of the frontals of most lophiiforms are simple, more or less tapering, truncate, or slightly concave (Pietsch, 1981:397, figs. 4, 6, 15–19) **(0)**; but strongly bifurcate in the Ceratiidae, the thaumatichthyid genus *Lasiognathus*, the Oneirodidae, and Linophrynidae (Regan and Trewavas, 1932:40, 44, figs. 52, 61, 62, 65; Pietsch, 1974a:6, 18, figs. 1, 2, 4, 19–38; Bertelsen and Struhsaker, 1977:32, fig. 18) **(1)** (0.25, 0.77).

7. In most lophiiforms, the anterior extension of the frontal bone overlaps a small distal portion of the lateral ethmoid (Pietsch, 1981:397, figs. 4, 6, 15–19) **(0)**; in the thaumatichthyid genus *Lasiognathus*, the Oneirodidae, and Linophrynidae the frontal overlaps the full length of the lateral ethmoid (Regan and Trewavas, 1932:44, figs. 61, 62, 65; Pietsch, 1974a:6, 18, figs. 1, 2, 4, 19–38; Bertelsen and Struhsaker, 1977:32, fig. 18) **(1)** (0.33, 0.87).

8. The frontals of most lophiiforms, including all outgroup taxa, are relatively simple elongate bones, without ventromedial extensions **(0)**; those of the Himantolophidae, Diceratiidae, and Melanocetidae bear prominent ventromedial extensions that approach one another on the midline and are narrowly separated by cartilage from the supraethmoid anteriorly and the supraoccipital posteriorly, but make no contact with the parasphenoid (Pietsch and Van Duzer, 1980:62, figs. 1–4; Fig. 192) **(1)**; the ventromedial extension of each frontal is bifurcated in the thaumatichthyid genus *Lasiognathus* and all oneirodid genera (except *Lophodolos*), the anterior branch making contact with the supraethmoid, the posterior branch closely approaching or making contact with the supraoccipital as well as the parasphenoid (Pietsch, 1974a:6, figs. 1, 2, 30; Fig. 193) **(2)** (0.50, 0.87).

9. The parietals are present in nearly all lophiiforms **(0)**, but absent in the Himantolophidae and the gigantactinid genus *Rhynchactis* **(1)**; they are very much enlarged in the Centrophrynidae and Ceratiidae (Bertelsen, 1951:128, fig. 88B, C; Pietsch, 1972a:28, figs. 7, 8) **(2)** (0.67, 0.67).

10. Pterosphenoids are nearly always present **(0)**, but reduced in the Diceratiidae (Uwate, 1979:130, figs. 2, 3) **(1)** and absent in the Ceratiidae, the oneirodid genus

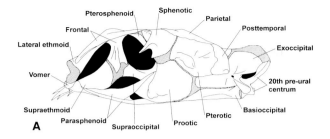
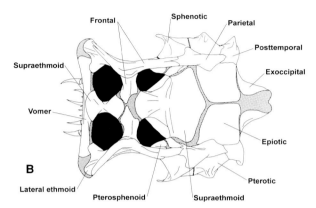

FIGURE 192. Cranium of *Melanocetus johnsonii*, 75 mm SL, LACM 32786-001, left lateral (A) and dorsal (B) views, showing, among other things, the simple ventromedial extension of the frontal bone, far removed from the parasphenoid. Cartilage stippled, open space rendered in solid black. After Pietsch and Van Duzer (1980).

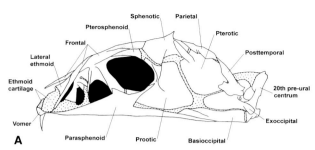
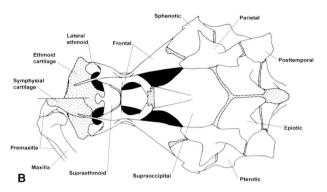

FIGURE 193. Cranium of *Oneirodes acanthias*, 71 mm SL, LACM 9960-004, left lateral (A) and dorsal (B) views, showing, among other things, the bifurcated ventromedial extension of the frontal bone, the posterior branch making direct contact with parasphenoid. Cartilage stippled, open space rendered in solid black. After Pietsch (1974a).

*Lophodolos*, the Caulophrynidae, Gigantactinidae, and Linophrynidae (Pietsch, 1974a:19, fig. 33) **(2)** (0.40, 0.70).

11. The parasphenoid is simple in nearly all lophiiforms **(0)**; in all oneirodid genera (except *Lophodolos*) this bone bears a pair of anterodorsal extensions each overlapping or approaching the distal end of the respective posterior ventromedial extension of the frontal (Pietsch, 1974a:6, 18, figs. 1, 2, 30; Fig. 193A) **(1)** (1.00).

12. Sphenotic spines are absent in most of the outgroups **(0)** but present in lophioids and in the Himantolophidae, Diceratiidae, the thaumatichthyid genus *Lasiognathus*, all oneirodid genera except *Chaenophryne* the Caulophrynidae, and Linophrynidae (Pietsch, 1974a:19, fig. 37, 1975b, fig. 1; Fig. 32) **(1)** (0.17, 0.55).

13. In nearly all lophiiforms the pterotic makes contact with, but bears no discrete process that overlaps, the respective sphenotic **(0)**; in the Oneirodidae a narrow, elongate, distally rounded, anterodorsally directed process overlaps the posterolateral surface of the sphenotic (Pietsch, 1974a:19, figs. 2, 28–37; Fig. 193A) **(1)**; in the Linophrynidae, the pterotic bears a similar, but tapering and distally pointed process **(2)** (1.00).

14. In nearly all lophiiforms, the supraoccipital is situated posterior relative to other elements of the cranium **(0)**; the supraoccipital is displaced anteriorly in metamorphosed females of the Gigantactinidae, most of its dorsal surface lying in the vertical plane, providing an abutment for the pterygiophore of the illicium (Bertelsen et al., 1981:10, fig. 11) **(1)**; it is displaced anteriorly and narrowly separated by cartilage from the distal ends of the posterior ventromedial extensions of the frontals in the Himantolophidae, Diceratiidae, Melanocetidae, and all oneirodid genera except *Lophodolos* (Bertelsen and Krefft, 1988:16, fig. 3; Pietsch, 1974a:4, 15, figs. 1–4; Pietsch and Van Duzer, 1980:63, figs. 1–7; Figs. 192, 193) **(2)** (0.50, 0.88).

15. The dorsal end of the hyomandibular bone is distinctly bifurcated, forming two heads in nearly all lophiiform taxa **(0)**; the hyomandibula bears only a single expanded dorsal head in the oneirodid genus *Bertella* and in the Neoceratiidae, Gigantactinidae, and Linophrynidae (Pietsch, 1973:199, fig. 1) **(1)** (0.50, 0.86).

16. Mesopterygoids are usually present **(0)**, but absent in the antennarioid families Tetrabrachiidae, Lophichthyidae, and Brachionichthyidae (Pietsch, 1981:393, 400, figs. 9, 22, 23), and in the ceratioid families Neoceratiidae, Gigantactinidae, and Linophrynidae **(1)** (1.00).

17. The palatines are well-developed toothed elements in all the outgroups, except for the Tetrabrachiidae and Brachionichthyidae (Pietsch, 1981:393, 400, figs. 9, 23), and some genera of the Ogcocephaloidei (Endo and Shinohara, 1999) **(0)**; they are reduced and toothless in all ceratioids **(1)** (1.00).

18. The quadrate of most lophiiforms usually bears no more than a small projection at its articulation with lower jaw **(0)**, but this projection forms a highly

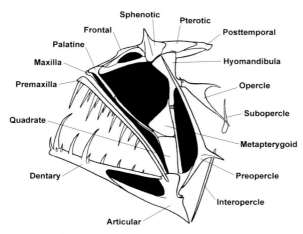

FIGURE 194. Head skeleton of *Linophryne racemifera*, 52 mm SL, ZMUC P92146, left lateral view, showing large open space (rendered in solid black) between the preopercle and remaining elements of the suspensorium. Pectoral girdle and hyoid apparatus not shown; open space rendered in solid black. After Regan and Trewavas (1932).

conspicuous, laterally directed spine in the thaumatichthyid genus *Lasiognathus* and most oneirodid genera (excluding *Pentherichthys, Chaenophryne, Spiniphryne,* and *Ctenochirichthys*) (Pietsch, 1974a:10, figs. 6, 40–47; Bertelsen and Struhsaker, 1977:32, fig. 18; Figs. 32, 135, 137) **(1)** (0.33, 0.80).

19. The preopercle is large and distinctly crescent shaped in nearly all lophiiforms **(0)**, straight in both genera of the Thaumatichthyidae (Bertelsen and Struhsaker, 1977:12, figs. 5A, 18) **(1)**; strongly bowed at midlength and extending posteriorly in the linophrynid genera *Haplophryne, Borophryne,* and *Linophryne,* leaving a large space between it and the remaining elements of the suspensorium (Regan and Trewavas, 1932:44, figs. 61, 66; Fig. 194) **(2)**; and very much reduced to a small strut of bone in the Gigantactinidae (Bertelsen et al., 1981:11, fig. 13) **(3)** (1.00).

20. The posterior margin of the preopercle is smooth and spineless in nearly all lophiiforms **(0)**, but bears a series of 4 to 6 short spines along its anterior, lateral, and posterior margins in the linophrynid genus *Photocorynus* (Regan and Trewavas, 1932:44, fig. 59); and a single, large, posteriorly directed spine in the linophrynid genera *Haplophryne, Borophryne,* and *Linophryne* (the preopercular spine of *Haplophryne* terminating in 3 to 5 radiating cusps; Regan and Trewavas, 1932:44, figs. 61, 66; Bertelsen, 1951:172, figs. 115, 118, 126; Figs. 31, 171, 177, 194) **(1)** (1.00).

21. The upper arm of the opercle is simple in nearly all lophiiforms **(0)**, but divided into 3 or more radiating ribs or branches in the Thaumatichthyidae (Bertelsen, 1951:118, figs. 77, 78; Bertelsen and Struhsaker, 1977:32–33, figs. 1, 4, 5, 18) **(1)** (1.00).

22. The anterodorsal margin of the subopercle bears a distinct anteriorly directed spine in most of the outgroups and in the Centrophrynidae, the ceratiid genus *Cryptopsaras,* the Diceratiidae, Melanocetidae, and the genus *Thaumatichthys* (Bertelsen, 1951:139, figs. 88B, 89A, C; Bertelsen and Struhsaker, 1977:14, figs. 4, 5A) **(0)**; a small subopercular spine or projection is present in some specimens of several oneirodid genera (*Chaenophryne*; Bertelsen, 1951:109, figs. 66, 67), but absent in ogcocephaloids, the antennarioid families Tetrabrachiidae and Brachionichthyidae, and all other lophiiforms **(1)** (0.20, 0.43).

23. The interopercle of nearly all lophiiforms is long and slender **(0)**, but extremely reduced to a small triangular bone in the Neoceratiidae and Gigantactinidae (Bertelsen, 1951:156, fig. 105; Bertelsen et al., 1981:11, fig. 13; Figs. 194–196) **(1)** (1.00).

24. An interopercular-mandibular ligament is present in nearly all lophiiforms **(0)**, but absent in both genera of the Gigantactinidae (Bertelsen et al., 1981:11, fig. 13) **(1)** (1.00).

25. The upper and lower jaws are more or less equal in length in nearly all lophiiforms **(0)**; in sharp contrast, the upper jaw extends anteriorly far beyond the lower in the Thaumatichthyidae, the distal ends of the premaxillae widely separated from each other and connected only by membranous connective tissue (Bertelsen and Struhsaker, 1977:11, 29, 31, figs. 1, 2, 17, 18; Fig. 195) **(1)** (1.00).

26. A rostral cartilage (symphysial cartilage of Bertelsen and Pietsch) of the upper jaw is present in nearly all lophiiforms **(0)**, but absent in the genus *Thaumatichthys,* the Gigantactinidae, and the linophrynid genera *Haplophryne, Borophryne,* and *Linophryne* (Bertelsen and Struhsaker, 1977:11, figs. 1, 2; Bertelsen et al., 1981:11, figs. 11, 13) **(1)** (0.33, 0.60).

27. The ascending process of the premaxilla of lophioids (as well as all batrachoidiform genera; Monod, 1960:665, figs. 47–49; Field, 1966:51, fig. 3; David W. Greenfield, personal communication, 23 March 2006) articulates with the toothed portion of this bone (i.e., autogenous; capable of considerable independent lateral movement) **(0)**, but fused to the latter in all other lophiiforms (Pietsch, 1981:399, fig. 20) **(1)** (1.00).

28. A postmaxillary process of the premaxilla is present in all the outgroups (Pietsch, 1981:399, fig. 20) and in the Ceratiidae (Lütken, 1878a:334, fig. 7; Regan and Trewavas, 1932:39, fig. 53; Bertelsen, 1951:128, fig. 88) **(0)**; it is absent in all other lophiiforms **(1)** (0.50, 0.80).

29. The maxillae are well developed in nearly all lophiiforms **(0)**; considerably reduced in the Caulophrynidae, Neoceratiidae, the genus *Gigantactis,* and the linophrynid genus *Photocorynus* **(1)**; and further reduced to a fine thread of bone or absent in the gigantactinid genus *Rhynchactis* and the linophrynid genera *Haplophryne, Borophryne,* and *Linophryne* (Bertelsen et al., 1981:10, fig. 13; Figs. 194, 196) **(2)** (0.67, 0.83).

30. A thick anterior-maxillomandibular ligament (labial cartilage of Le Danois, 1964; Pietsch, 1972a) is present in most lophiiforms (Pietsch, 1978:4, fig. 3) **(0)**, but very much reduced or absent in the Melanocetidae, Thaumatichthyidae, Caulophrynidae, Neoceratiidae, Gigantactinidae, and Linophrynidae **(1)** (0.50, 0.90).

31. The dentaries of nearly all lophiiforms are strongly bifurcated posteriorly **(0)**, but simple in the Neoceratiidae

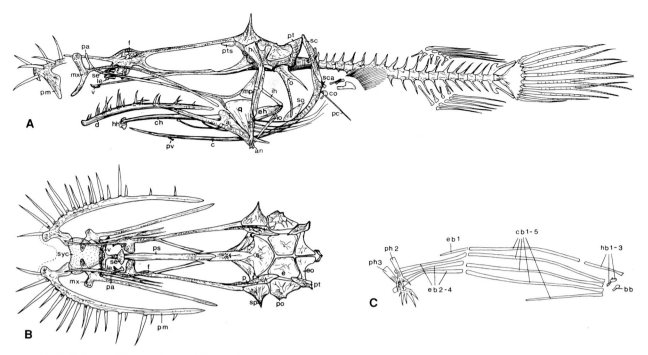

FIGURE 195. Skeleton of *Lasiognathus* sp., left lateral view, 40 mm SL, SIO 73-305, showing elements of upper jaw extending anteriorly far beyond lower jaw: (A) left lateral view; (B) dorsal view of head; (C) gill arches. an, angular; bb, basibranchial; c, cleithrum; cb, ceratobranchial; ch, ceratohyal; co, coracoid; d, dentary; e, epiotic; eb, epibranchial; eh, epihyal; eo, exoccipital; f, frontal; h, hyomandibula; hb, hypobranchial; hh, hypohyal; ih, interhyal; io, interopercle; le, lateral ethmoid; mp, metapterygoid; mx, maxilla; o, opercle; p, parietal; pa, palatine; pc, postcleithrum; ph, pharyngobranchial; pm, premaxilla; po, pterotic; ps, parasphenoid; pt, posttemporal; pts, pterosphenoid; pv, pelvic bone; q, quadrate; s, supraoccipital; sc, supracleithrum; sca, scapula; se, supraethmoid; so, subopercle; sp, sphenotic; syc, symphysial cartilage; v, vomer. After Bertelsen and Struhsaker (1977).

and Gigantactinidae (Bertelsen, 1951:156, fig. 105; Bertelsen et al., 1981:11, fig. 13; Figs. 194–196) **(1)** (1.00).

32. Jaw teeth vary considerably in size among the outgroups and most all ceratioids **(0)**, but never become as large and fanglike in proportion to the head and mouth as those of the linophrynid genera *Borophryne* and *Linophryne* (as well as those of the linophrynid genus *Acentrophryne*, unavailable for analysis; Regan, 1926:23, pl. 2, figs. 1–3, pl. 3, fig. 1; Bertelsen, 1951:172, 193, figs. 118, 126; Figs. 173, 174, 177, 179–182, 194) **(1)** (1.00).

33. The outermost lower-jaw teeth are smaller than the inner and mounted on the dorsal edge of the dentaries in nearly all lophiiforms **(0)**, but considerably larger than the inner and conspicuously attached to the lateral surface of the dentaries in the Neoceratiidae and the genus *Gigantactis* (jaw teeth are minute or absent in metamorphosed specimens of the gigantactinid genus *Rhynchactis*; Bertelsen, 1951:156, fig. 105; Bertelsen et al., 1981:11, figs. 5, 19, 23; Figs. 38, 196) **(1)** (1.00).

34. The posterior end of the lower jaw (articular and angular) extends posteriorly considerably beyond its articulation with the quadrate in nearly all lophiiforms **(0)**, but terminates at the articular-quadrate joint in the Thaumatichthyidae, Neoceratiidae, and Gigantactinidae (Bertelsen and Struhsaker, 1977:11, figs. 1, 9; Bertelsen et al., 1981:11, fig. 13; Figs. 194–196) **(1)** (0.50, 0.75).

35. The posteroventral margin of the articular is rounded in nearly all lophiiforms **(0)**, but greatly expanded and squared off in chaunacoids (Pietsch, 1981:403, fig. 24) and in the Ceratiidae (Regan and Trewavas, 1932:40, fig. 52; Bertelsen, 1951:128, fig. 88) **(1)** (0.50, 0.50).

36. The interhyal bears a medial, posterolaterally directed process in lophioids and antennarioids **(0)**, but this element is simple and cylindrical in all other lophiiforms (Pietsch, 1981:400, fig. 26) **(1)** (1.00).

37. Nearly all lophiiforms have 6 branchiostegal rays **(0)**; this number is reduced in some taxa, the anteriormost element dropping out in some specimens of the Caulophrynidae and Neoceratiidae, and in all members of the Linophrynidae **(1)** (1.00).

38. A basihyal is present in most of the outgroups **(0)**, but absent in some antennarioids (i.e., Brachionichthyidae), at least some ogcocephaloids, and all ceratioids (Pietsch, 1981:394, 400, fig. 10) **(1)** (1.00).

39. The first pharyngobranchial is present in most of the outgroup taxa and many ceratioids **(0)**, but present or absent in ogcocephaloids **(0/1)**, and absent in lophioids, the Melanocetidae, Thaumatichthyidae, all oneirodid genera except *Spiniphryne* and *Oneirodes*, the Caulophrynidae, Neoceratiidae, Gigantactinidae, and Linophrynidae (Pietsch, 1979:8, fig. 10; Pietsch and Van Duzer, 1980:66, fig. 13; Bertelsen et al., 1981:11, fig. 16) **(1)** (0.25, 0.67).

40. The third pharyngobranchial is large and well toothed in the outgroups and most ceratioid taxa **(0)**,

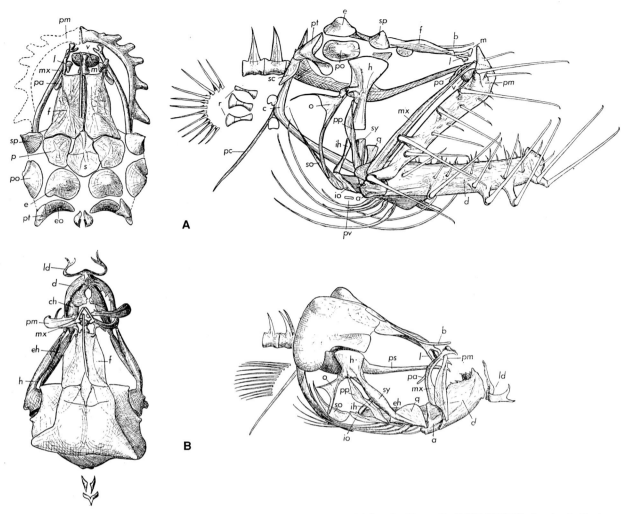

FIGURE 196. Head skeleton of *Neoceratias spinifer*, dorsal and right lateral views: (A) female, 52 mm SL, ZMUC P921726, showing teeth attached to the lateral margin of the dentary and premaxilla; (B) parasitic male, 15.5 mm SL, ZMUC P921726, showing uniquely shaped lower denticular bone. b, basal bone or pterygiophore of the illicium; ld, lower denticular bone; pp, preopercle; sy, symplectic; all other abbreviations as given in Figure 195. Modified after Bertelsen (1951).

but greatly hypertrophied and expanded distally in the Neoceratiidae and Gigantactinidae (Bertelsen, 1951:156, fig. 105A; Bertelsen et al., 1981:11, 17, figs. 9, 16; Fig. 197) **(1)** (1.00).

41. A fourth pharyngobranchial tooth plate is present and well developed in lophioids (Field, 1966:58, figs. 8, 9) **(0)**, but absent in all other lophiiforms (Pietsch, 1981:401, figs. 11, 28–32) **(1)** (1.00).

42. The first epibranchial is present in nearly all lophiiforms **(0)**, but absent in the genus *Thaumatichthys* and in the Gigantactinidae (Bertelsen and Struhsaker, 1977:14, fig. 5; Bertelsen et al., 1981:17, fig. 16; Figs. 197, 198) **(1)** (0.50, 0.50).

43. The first epibranchial of lophioids, antennarioids, and chaunacoids bears a medial process ligamentously attached to the proximal tip of the second epibranchial **(0)**; this element is simple and without ligamentous connection to the second epibranchial in ogcocephaloids and ceratioids (Pietsch, 1981:401, figs. 28–32) **(1)**. The first epibranchial is absent in the genus *Thaumatichthys* and in the Gigantactinidae (Fig. 197) (1.00).

44. The dorsal and ventral ends of the third and fourth ceratobranchials are more or less free of each other in nearly all lophiiforms **(0)**, but bound tightly to one another by connective tissue in the Neoceratiidae, Gigantactinidae, and Linophrynidae, greatly restricting the space between these two elements (Fig. 199) **(1)** (1.00).

45. The third hypobranchial is present in most lophiiforms **(0)**, but absent in the Himantolophidae, Caulophrynidae, Neoceratiidae, Gigantactinidae, and Linophrynidae (Bertelsen et al., 1981:12, fig. 16; Bertelsen and Krefft, 1988:16, fig. 4C) **(1)** (0.50, 0.88).

46. Among the outgroups, branchial teeth are present on the first three ceratobranchials (and usually the fourth) of chaunacoids, ogcocephaloids, some antennarioids, and the ceratioid families Centrophrynidae and Himantolophidae (Pietsch, 1972a:35, fig. 15;

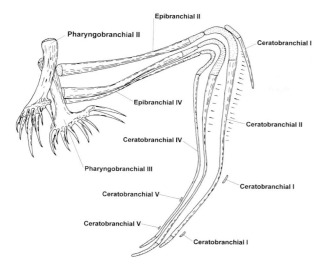

FIGURE 197. Gill arches of *Gigantactis longicirra*, 209 mm SL, ISH 973/71, showing extremely reduced ceratobranchials I and V, and hypertrophied pharyngobranchials II and III, the distal end of pharyngobranchial III greatly expanded. Modified after Bertelsen et al. (1981).

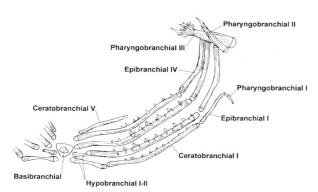

FIGURE 198. Gill arches of *Himantolophus groenlandicus*, 53.5 mm SL, ISH 2056/71, showing teeth on the proximal ends of the epibranchials and along nearly the full length of all four ceratobranchials. Modified after Bertelsen and Krefft (1988).

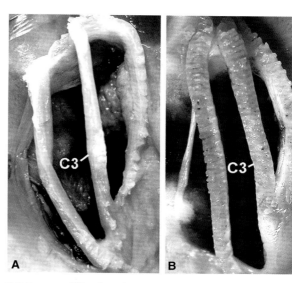

FIGURE 199. Gill arches of ceratioids, left posterior view, showing the extent to which the dorsal and ventral ends of the third (C3) and fourth ceratobranchials are bound together by connective tissue: (A) *Oneirodes thompsoni*, 143 mm SL, UW 43825, showing relatively large open space behind the third ceratobranchial; (B) *Gigantactis vanhoeffeni*, 290 mm SL, UW 46040, showing greatly restricted space behind the third ceratobranchial. Gill filaments removed for better clarity. After Pietsch and Orr (2007).

Bertelsen and Krefft, 1988:16, fig. 4C; Fig. 198) **(0)**; they are absent in all other lophiiforms **(1)** (0.33, 0.33).

47. Gill filaments are present on the first gill arch of lophioids and antennarioids **(0)**, but absent on this element in all other lophiiforms (present, however, on the ventral-most tip of the first ceratobranchial of some ceratioids; Bradbury, 1967:408; Pietsch, 1981:415) **(1)** (1.00).

48. Vertebral centra are short in nearly all lophiiforms, their greatest diameter approximately equal to their length (Pietsch, 1972a:37, figs. 16, 18; 1974a:12, fig. 12) **(0)**, but elongate in the Thaumatichthyidae, Neoceratiidae, and Gigantactinidae, their greatest diameter only about half their length (Bertelsen and Struhsaker, 1977:13, figs. 5, 18; Bertelsen et al., 1981:12, fig. 17; Figs. 195, 200) **(1)** (0.50, 0.75).

49. The caudal peduncle is relatively short in nearly all lophiiforms **(0)**, but exceptionally long and narrow in the Thaumatichthyidae, Neoceratiidae, and Gigantactinidae, the posterior insertion of the dorsal and anal fins well separated from the posterior margin of the hypural plate (Figs. 85, 87, 90, 92, 153, 155–157, 162) **(1)** (0.50, 0.67).

50. A single epural is present in all outgroup taxa and in the Caulophrynidae (Pietsch, 1979:9, fig. 11) **(0)**; epurals are absent in all other lophiiforms **(1)** (0.50, 0.75). Bertelsen's (1984:326) report of an epural in the linophrynid genus *Photocorynus* could not be confirmed by reexamination of all available material.

51. The caudal fin is rounded in nearly all lophiiforms **(0)**, but emarginate in females of both genera of the Gigantactinidae (Bertelsen et al., 1981:5, figs. 4, 63; Figs. 155–157, 162) **(1)** (1.00).

52. The caudal fin of lophiiforms nearly always contains 9 rays **(0)**, but the ninth or lowermost ray is reduced to less than one-half the length of the eighth ray in the genus *Ceratias*, the Gigantactinidae, and Linophrynidae **(1)**; there are only 8 caudal-fin rays in lophioids, the ceratiid genus *Cryptopsaras*, and in the Caulophrynidae **(2)** (0.33, 0.50).

53. The innermost six caudal-fin rays are bifurcated in nearly all the outgroups (seven or all nine caudal rays bifurcated in antennarioids) as well as in the ceratioid families Himantolophidae, Diceratiidae, and Melanocetidae **(0)**; the innermost four are bifurcated in all remaining ceratioid taxa **(1)**, except for females of the genus *Gigantactis* in which all nine caudal-fin rays are simple (Bertelsen et al., 1981:5, fig. 4) **(2)** (0.67, 0.86).

54. There are primitively 6 dorsal-fin spines in lophioids **(0)**, but 3 or fewer in all other lophiiforms (Pietsch, 1981:409, figs. 36–38) **(1)** (1.00).

55. The pterygiophore of the illicium is relatively small in nearly all lophiiforms **(0)**, but exceptionally well

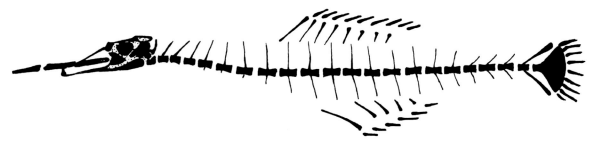

FIGURE 200. Diagrammatic representation of the axial skeleton of *Gigantactis longicirra*, female, 209 mm SL, ISH 973/71, showing, among other things, the elongate vertebral centra characteristic of members of the Gigantactinidae. After Bertelsen et al. (1981).

developed in the Gigantactinidae (Bertelsen et al., 1981:14, fig. 13) **(1)** (1.00).

56. The esca is nonluminescent in all the outgroups, and in the Caulophrynidae (Pietsch, 1979:12, figs. 15, 17, 19–22) and gigantactinid genus *Rhynchactis* (Bertelsen et al., 1981:3, figs. 2, 3; Bertelsen and Pietsch, 1998b:586, figs. 2–6) **(0)**; in all other ceratioids (except the Neoceratiidae, in which the illicium and esca have been lost; Bertelsen, 1951:156, fig. 105) the esca encloses an expanded central cavity containing bioluminescent bacteria (O'Day, 1974:4, figs. 3–6; Munk and Bertelsen, 1980:111, figs. 3–9; Munk, 1999:266, fig. 2; Fig. 201) **(1)** (0.33, 0.60).

57. The esca is a solid structure, with no central cavity or opening to the outside, in lophioids, antennarioids, chaunacoids, Caulophrynidae, and the gigantactinid genus *Rhynchactis* **(0)**; a tiny opening (the escal pore of Bertelsen, 1951:16) leading from a central cavity to the outside is present in ogcocephaloids (Bradbury, 1988:20) and in all other ceratioids (except the Neoceratiidae in which the illicium and esca have been lost; Brauer, 1904:18, fig. 1; Munk, 1999:266, fig. 2; Fig. 201) **(1)** (0.33, 0.50).

58. Large, toothlike dermal denticles embedded in the esca are absent in nearly all lophiiforms **(0)**, but present in both genera of the Thaumatichthyidae (Bertelsen and Struhsaker, 1977:21, figs. 10, 11; Bertelsen and Pietsch, 1996:402, figs. 2–5; Pietsch, 2005a:78, figs. 1–3; Fig. 86, 88, 93) **(1)** (1.00).

59. The cephalic second dorsal-fin spine is simple in nearly all lophiiforms **(0)**, but bears a more or less concealed (within a pore on the head just behind the base of the illicium) distal light organ in larvae of the Ceratiidae, and in larvae and juvenile females of the Diceratiidae (Bertelsen, 1951:16, 67, 127, figs. 28, 90G–I, 93C–E; Figs. 44, 63, 67, 77) **(1)** (0.50, 0.67).

60. A third cephalic dorsal-fin spine and pterygiophore are present in lophioids, antennarioids, and chaunacoids **(0)**, but absent in ogcocephaloids and ceratioids (Bertelsen, 1951:17; Bradbury, 1967:401; Pietsch, 1981:410, fig. 38) **(1)** (1.00).

61. The rays of the dorsal and anal fins are relatively short in nearly all metamorphosed lophiiforms **(0)**, but extremely long in both genera of the Caulophrynidae, in extreme cases exceeding 160% standard length (Pietsch, 1979:9, figs. 14, 20; Figs. 148, 149, 152) **(1)** (1.00).

62. Caruncles (light organs associated with dorsal-fin rays), absent in nearly all lophiiforms **(0)**, are present in both genera of the Ceratiidae (Brauer, 1908:103, pl. 32, fig. 17; Bertelsen, 1951:16, figs. 90, 93; Munk and Herring, 1996:517, figs. 1–4; Figs. 44, 63, 67) **(1)** (1.00).

63. The posttemporal of lophioids, antennarioids, and chaunacoids is attached to the cranium in such a way that considerable movement in an anterodorsal-posteroventral plane is possible **(0)**; this element is fused to the cranium in ogcocephaloids and all ceratioids (Pietsch, 1981:411, figs. 3–5, 15–19) **(1)** (1.00).

64. The pectoral lobe of nearly all lophiiforms is relatively short, shorter than the longest pectoral-fin rays **(0)**, but considerably longer than the longest pectoral-fin rays in the oneirodid genera *Puck*, *Leptacanthichthys*, *Chirophryne*, and *Ctenochirichthys* (Pietsch, 1978:7, figs. 8, 9, 11, 14, 16) **(1)** (1.00).

65. Most lophiiforms, including most outgroup taxa, have 3 pectoral radials **(0)**, but the Centrophrynidae, Ceratiidae, and Melanocetidae have 4 **(1)**; lophioids (small juveniles) and gigantactinids have 5 **(2)** (0.50, 0.60). Fusion of pectoral radials with increasing standard length is common in lophiiforms (the 5 radials present in juvenile lophioids fuse to 2 in adults; Pietsch, 1972a:41, fig. 23; Bertelsen and Struhsaker, 1977:14, fig. 6; Pietsch, 1979:11, fig. 13).

66. The pelvic bones of all the outgroups are well developed and expanded distally to form two heads, one bearing the spine and rays of the pelvic fin, the other making contact with its counterpart on the opposite side **(0)**; these bones are considerably reduced in ceratioids, lacking the pelvic spine and rays and ranging from triradiate in the Himantolophidae (Bertelsen and Krefft, 1988:19, fig. 5C) and some females of the oneirodid genus *Chaenophryne* (Pietsch, 1975b:79, fig. 2) to somewhat expanded distally, or slender and cylindrical throughout their length **(1)**; they are rudimentary or absent in the Neoceratiidae and Gigantactinidae, and absent in the Linophrynidae (Bertelsen, 1951:156, fig. 105A; Bertelsen et al., 1981:14, fig. 13; Fig. 196) **(2)** (1.00).

67. Pelvic fins are present in all the outgroups **(0)**, but absent in all juvenile and adult ceratioids of both sexes **(1)** (1.00).

68. The skin is covered with numerous, close-set, dermal spines or spinules in most of the outgroups (except lophioids and some antennarioids), and in the

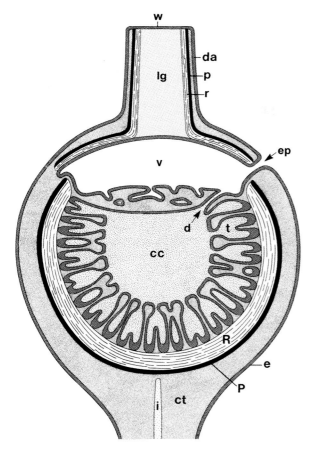

FIGURE 201. Diagrammatic medial section of a ceratioid esca, left lateral view. The light gland is enclosed within a lightproof capsule consisting of an inner reflecting layer (R) and an outer pigmented layer (P). A duct (d) connects the bacteria-filled central cavity (cc) of the light gland with the vestibule (v), which opens to the outside by way of an escal pore (ep) positioned on the posterior margin of the distal surface of the esca. The distal escal appendage (da) contains a single light guide (lg), with a terminal window (w). ct, connective tissue; e, epidermis; i, illicial bone; p, pigmented layer of light-guide wall; r, reflecting layer of light-guide wall. Modified after Munk (1999).

Centrophrynidae, Ceratiidae, Himantolophidae, Diceratiidae, the genus *Thaumatichthys*, the oneirodid genus *Spiniphryne*, and the Gigantactinidae (Regan, 1926:12, figs. 1, 3; Bertelsen and Krefft, 1988:21, figs. 1, 15) **(0)**; although tiny dermal spinules may be present in some specimens (detectable only microscopically in cleared and stained material; Pietsch, 1974a:29; Pietsch and Van Duzer, 1980:67), the skin is everywhere smooth and naked in all other lophiiforms **(1)** (0.20, 0.64).

69. In contrast to those of all other lophiiforms **(0)**, females of the genus *Thaumatichthys*, the Neoceratiidae, and Gigantactinidae have a large elongate nasal papilla (Bertelsen, 1951:160, fig. 106D) **(1)** (0.50, 0.67).

70. The opercular opening is primitively large in lophioids, extending not only behind the base of the pectoral fin, but in front of it as well (Caruso, 1985:873) **(0)**; the opercular opening is considerably more restricted and fully situated behind the base of the pectoral fin in all other lophiiforms (Pietsch and Grobecker, 1987:34, 349, pl. 16A) **(1)** (1.00).

71. The anal opening is situated on the ventral midline in nearly all lophiiforms **(0)**, but skewed to the left of the midline (i.e., sinistral) in all members of the Linophrynidae (males and larvae as well as females; Bertelsen, 1951:161, figs. 112, 115–117; Figs. 31, 169–172, 178, 185) **(1)** (1.00).

## Characters Restricted to Metamorphosed Males

72. Adult males and females of all outgroup taxa are similar in size **(0)**, whereas metamorphosed males of all ceratioids are dwarfed, reaching only a small fraction of the size of the females (Regan, 1925a:395, figs. 1–9; Parr, 1930b:129, figs. 1–7; Bertelsen, 1951:15) **(1)** (1.00).

73. The eyes of metamorphosed males of all outgroup taxa, as well as those of nearly all metamorphosed ceratioid males, are large, their diameters ranging between approximately 9.1 and 20.0% head length (Bertelsen, 1951:24) **(0)**; those of the Ceratiidae are relatively huge, with diameters greater than 28% head length (Fig. 50A, B) **(1)**; while those of the Centrophrynidae, Neoceratiidae, and Gigantactinidae are very much reduced, with diameters of 5.9 to 7.1% head length (Bertelsen, 1984:326, table 89; Figs. 60, 154, 160) **(2)** (0.67, 0.75).

74. The eyes of males of all the outgroups and most ceratioids are spherical **(0)**, but oval and bowl shaped in the Ceratiidae (Munk, 1964:5, 9, figs. 1A, B, 2A, B, D–G, 1966:22, fig. 10; Figs. 50A, B, 51) **(1)** and tubular (described as "telescopic" by Bertelsen, 1951:25) in the Linophrynidae (Munk, 1964:10, figs. 1C, 2C, 1966:31, figs. 17, 18; Figs. 50C, 51) **(2)** (1.00).

75. The olfactory organs of males of all the outgroups are small relative to head length **(0)**; those of nearly all metamorphosed ceratioid males are large, their greatest diameter ranging between approximately 12.5 and 21.7% head length (Bertelsen, 1951:25; Marshall, 1967a, 1967b) **(1)**; those of the Gigantactinidae are relatively huge, diameters greater than 30% head length (Fig. 160) **(2)**; but those of the Ceratiidae and Neoceratiidae are very much reduced and probably nonfunctional (Figs. 50, 154) (Bertelsen, 1984:326, table 89) **(3)** (0.75, 0.83).

76. The paired nostrils of males of all the outgroups and many ceratioids are similarly directed, either both laterally or both anteriorly **(0)**; in the Centrophrynidae, Thaumatichthyidae, Oneirodidae, Caulophrynidae, Gigantactinidae, and Linophrynidae the posterior nostrils are directed laterally, but the anterior nostrils are directed anteriorly (Bertelsen, 1984:326, table 89) **(1)** (0.33, 0.78).

77. The premaxillae of metamorphosed males of all outgroup taxa as well as those of nearly all ceratioids are well developed **(0)**; those of the linophrynid genera *Borophryne* and *Linophryne* are extremely reduced or absent (Bertelsen, 1951:161, 164, fig. 108; Fig. 54) **(1)** (1.00).

78. Jaw teeth are retained in adult males of all the outgroups and in those of the linophrynid genera

*Photocorynus* and *Haplophryne* (Bertelsen, 1951:21, fig. 5) **(0)**; they are lost during metamorphosis in those of all other ceratioids (Fig. 54) **(1)** (0.33, 0.60).

79. Denticular bones, a unique set of pincerlike denticles at the anterior tips of the jaws used for grasping and holding fast to a prospective mate (originating during metamorphosis by fusion of modified dermal spinules anterior to the toothed symphysis of the premaxillae and dentaries; Bertelsen, 1951:21, figs. 5, 6; Munk, 2000:315), are absent in all the outgroups **(0)**, but present in all metamorphosed ceratioid males (Bertelsen, 1984:326, table 89; Fig. 54) **(1)** (1.00).

80. An anterior medial ridge of consolidated dermal spinules is absent in nearly all Lophiiforms **(0)**; a series of fused dermal spinules form a conspicuous medial ridge on the snout of metamorphosed males of the Himantolophidae, Melanocetidae, and the oneirodid genus *Microlophichthys* (Bertelsen, 1951:22, 39, 93, figs. 14, 47; Bertelsen and Krefft, 1988:26, fig. 10A, C; Fig. 40) **(1)** (0.50, 0.50).

81. An upper denticular bone is absent or limited to the distal tip of the snout in nearly all lophiiforms **(0)**; the upper denticular bone makes contact with the anterior tip of the pterygiophore of the illicium of metamorphosed males of the Ceratiidae, Himantolophidae, and Melanocetidae (Parr, 1930a:7, figs. 2, 3, 1930b:132, fig. 6; Bertelsen, 1951:22, fig. 6; Bertelsen and Krefft, 1988:26, fig. 10A, C; Fig. 40) **(1)** (0.50, 0.67).

## Characters That Extend Only to Larvae

82. The eggs and larvae of lophioids are relatively large **(0)**, but small in all other lophiiforms (at all stages the eggs are considerably less than 50% the diameter of those of lophioids; the smallest larvae are certainly less than 50%, and probably less than 30%, the size of those of lophioids; size at transformation to the prejuvenile stage is less than 60% that of lophioids; Bertelsen, 1984:327, figs. 167–170; Pietsch, 1984:323, fig. 164) **(1)** (1.00).

83. The head of the larvae of lophioids is small relative to the body (less than 30% SL) **(0)**, but proportionately large in all other lophiiforms, always greater than 45% SL (Bertelsen, 1984:327, figs. 167–170; Pietsch, 1984:324, fig. 164) **(1)** (1.00).

84. Larvae are short and deep, nearly spherical in most outgroups and most ceratioids **(0)**, but elongate and slender in lophioids, most oneirodid genera (except *Lophodolos*), the Neoceratiidae, and the linophrynid genera *Haplophryne*, *Borophryne*, and *Linophryne* (larvae of *Photocorynus* and *Acentrophryne* are unknown; Bertelsen, 1984:327, figs. 167B–G, 170, table 89; Figs. 172, 178, 185) **(1)**; the larvae are distinctly "humpbacked" in the Ceratiidae (Bertelsen, 1984:327, fig. 168C–E, table 89; Figs. 63, 67) **(2)** (0.40, 0.73).

85. The skin of the larvae of most of the outgroups as well as those of the Himantolophidae, Thaumatichthyidae, Caulophrynidae, and Gigantactinidae is highly inflated (Bertelsen and Struhsaker, 1977:27, fig. 15; Bertelsen, 1984:327, figs. 167A, 168A, B, 169A, B, table 89; Fig. 55) **(0)**, but only moderately inflated in antennarioids and all other ceratioids **(1)** (0.17, 0.38).

86. Sexual dimorphism in the illicial apparatus is absent in all the outgroups and in the larvae of the ceratioid families Caulophrynidae and Neoceratiidae (Bertelsen, 1984:328, fig. 167A, B, table 89) **(0)**, but present in all other ceratioids **(1)** (0.33, 0.60).

87. The pectoral fins of larvae are large (the rays extending well beyond the origin of the dorsal and anal fins) in most outgroup taxa and in the ceratioid families Caulophrynidae and Gigantactinidae (Bertelsen, 1984:327, figs. 167A, 168A, B, table 89; Figs. 56, 161) **(0)**; they are relatively small in antennarioids and in all other ceratioids **(1)** (0.20, 0.33).

88. Pelvic-fin rays are present in the larvae of all the outgroups and in the ceratioid family Caulophrynidae (Bertelsen, 1984:327, fig. 167A, table 89; Fig. 56) **(0)**, but absent in those of all other ceratioids **(1)** (0.50, 0.75).

## Tree Based on Characters Applicable to Metamorphosed Females

The phylogenetic analysis produced five equally parsimonious trees, with a total length of 153, a consistency index of 0.5560, and a retention index of 0.7952. Differences between the trees were restricted to a single family, the relatively poorly understood Oneirodidae, which contains 16 genera and 62 species, nearly 40% of all recognized ceratioids. The strict consensus tree is presented in Figure 202. Monophyly of the Ceratioidei was confirmed, and all ceratioid genera were placed in currently recognized monophyletic families (i.e., as presented by Bertelsen, 1984). Characters without homoplasy (unique and unreversed within ceratioids) that support monophyly of the Ceratioidei, however, are surprisingly few: palatines reduced and toothless (character 17, state 1), basihyal absent (38, 1), and pelvic fins absent in metamorphosed specimens (67, 1). A clade comprising the Centrophrynidae (containing only *Centrophryne*) and Ceratiidae (with two genera, *Ceratias* and *Cryptopsaras*) is represented as the sister group of all other ceratioids. The sister-group relationship of the Centrophrynidae and Ceratiidae is supported by one nonhomoplastic character: parietals enlarged (9, 2). Monophyly of the Ceratiidae is supported by a single nonhomoplastic character: caruncles present (62, 1).

The Himantolophidae (containing only *Himantolophus*), Diceratiidae *(Diceratias* and *Bufoceratias),* and Melanocetidae *(Melanocetus)* diverge next in sequential stepwise fashion, the latter family forming the sister group of all remaining ceratioids. No nonhomoplastic character supports the relationships of each of these clades as basal to other ceratioids, and bootstrap and Bremer support is very low for each node. Members of the Diceratiidae share one nonhomoplastic character: pterosphenoid reduced (10, 1).

Two relatively large clades remain: one containing the Thaumatichthyidae *(Lasiognathus* and *Thaumatichthys)* and the Oneirodidae (only 13 of 16 genera available for analysis), and the other containing the Caulophrynidae (*Caulophryne*; a second genus, *Robia*, unavailable for analysis), Neoceratiidae *(Neoceratias),* Gigantactinidae (*Gigantactis* and *Rhynchactis),* and Linophrynidae (*Photocorynus*, *Haplophryne*, *Borophryne*, and *Linophryne*; a fifth genus, *Acentrophryne*, unavailable for analysis). Monophyly of the Thaumatichthyidae-Oneirodidae clade was poorly supported, having very low bootstrap sup-

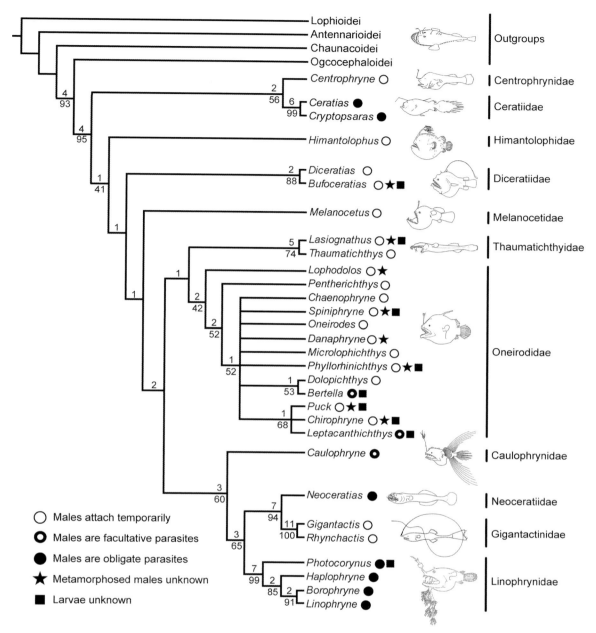

FIGURE 202. Strict consensus of five trees from a parsimony analysis of 71 morphological characters applicable to metamorphosed females for the genera of the Ceratioidei and four lophiiform outgroups. The number above the base of a node is the Bremer value, and the number below the node indicates bootstrap support greater than 40% for the respective node. Reproductive modes are plotted, and those genera for which metamorphosed males and/or larvae are unknown are indicated.

port and a Bremer value of 1. Monophyly of the Thaumatichthyidae *(Lasiognathus* and *Thaumatichthys),* however, is supported by four synapomorphies: preopercle straight (19, 1), upper arm of opercle with 3 or more radiating branches (21, 1), upper jaw extending anteriorly far beyond the lower jaw (25, 1), and esca with dermal denticles (58, 1). Monophyly of the Oneirodidae has low bootstrap and Bremer support, primarily because of homoplasy among character states of *Lophodolos,* the basal member of the family. All oneirodids share a narrow pterotic process that overlaps the respective sphenotic (13, 1), while all oneirodids except *Lophodolos* share one nonhomoplastic character: parasphenoid with a pair of anterodorsal extensions that approach or overlap the posterior ventromedial extensions of the respective frontal (11, 1). *Pentherichthys* diverges next, leaving the most derived clade among oneirodids as an unresolved polytomy comprising *Chaenophryne, Oneirodes, Spiniphryne, Danaphryne, Microlophichthys, Phyllorhinichthys,* a monophyletic *Dolopichthys* plus *Bertella* (supported by the ontogenetic loss of vomerine teeth; 3, 1), and an unresolved monophyletic triplet containing *Puck, Leptacanthichthys,* and *Chirophryne* (which share an elongate pectoral-fin lobe; 63, 1). In all five of the most parsimonious trees, the clades *Dolopichthys* plus *Bertella* and *Puck, Leptacanthichthys,* and *Chirophryne,* as well as the genera *Danaphryne, Microlophichthys,* and *Phyllorhinichthys,* formed a single large polytomy because of the absence of, as opposed to conflict between, characters to support any other relationship. However, *Chaenophryne*, *Spiniphryne*, and *Oneirodes* were always basal relative to other derived oneirodids, and *Chaenophryne* was basal-most in three of the five resolutions. In two of five resolutions, *Chaenophryne* formed a polytomy with *Spiniphryne* or a clade containing *Spiniphryne* plus *Oneirodes*.

Support for monophyly of the clade containing the Caulophrynidae, Neoceratiidae, Gigantactinidae, and Linophrynidae is moderate, having bootstrap support of 64% and a Bremer value of three. No nonhomoplastic character supports monophyly of the Caulophrynidae, Neoceratiidae, Gigantactinidae, and Linophrynidae clade. The Caulophrynidae (*Caulophryne* and *Robia*) is the sister group of the remaining three families. The Neoceratiidae (*Neoceratias*) and Gigantactinidae (*Gigantactis* and *Rhynchactis*) are sister groups that together form the sister group of the Linophrynidae (only four of five genera available). The latter clade of three families is supported by three characters: mesopterygoids absent (16, 1), dorsal and ventral ends of the third and fourth ceratobranchials tightly bound together by connective tissue (44, 1), and pelvic bones rudimentary or absent (66, 2). The clade comprising the Neoceratiidae and Gigantactinidae is strongly supported by four characters: interopercle extremely reduced (23, 1), dentary simple posteriorly (31, 1), outermost lower-jaw teeth large (33, 1), and the third pharyngobranchial enlarged and expanded distally (40, 1). Monophyly of the Gigantactinidae is the most strongly supported clade in this analysis, with bootstrap support at 100%, and seven synapomorphies: vomer absent (2, 1), supraoccipital anterior with dorsal surface in vertical plane (14, 1), preopercle reduced to a small strut (19, 3), interopercular-mandibular ligament absent (24, 1), caudal fin emarginate (51, 1), and pterygiophore of illicium exceptionally well developed (55, 1). Monophyly of the Linophrynidae is also well supported, with a bootstrap value of 99%. Among linophrynids, *Photocorynus* and *Haplophryne* diverge sequentially in stepwise fashion, the latter genus forming the sister group of *Borophryne* plus *Linophryne*, each clade with high bootstrap values and supported by nonhomoplastic characters. Monophyly of the Linophrynidae is supported by five characters: pterotic with tapered pointed process (13, 2), preopercle with a large posteriorly directed spine (20, 1), and anterior-most branchiostegal ray lost (37, 1). Monophyly of *Haplophryne*, *Borophryne*, and *Linophryne* is supported by one synapomorphy: preopercle bowed and extending posteriorly (19, 2). *Borophryne* and *Linophryne* share two characters: a conspicuous, rounded, laterally compressed frontal protuberance (4, 1), and greatly enlarged jaw teeth (32, 1).

## Tree Based on Characters of Metamorphosed Females, Males, and Larvae

This analysis produced 352 equally parsimonious trees, with a total length of 202, a consistency index of 0.5680, and a retention index of 0.7723. The strict consensus tree is presented in Figure 203. Differences between the trees were found among the deeper nodes and among derived oneirodid genera. Lack of resolution was present at the basal position of the tree where the Himantolophidae, Diceratiidae, and Melanocetidae, and a monophyletic Centrophrynidae and Ceratiidae formed a polytomy, as well as in the position of the Thaumatichthyidae in a polytomy with the Oneirodidae and the Caulophrynidae through Linophrynidae. The polytomies are the result of the lack of data rather than conflict among characters. Metamorphosed male and larval characters offered some support for monophyly of the Ceratioidei as a whole, as well as monophyly of some terminal taxa. Two characters of males supported monophyly of the Ceratioidei: an extreme sexual dimorphism in which males are dwarfed relative to females (character 72, 1), and denticular bones present (character 79, 1). Characters of males and larvae supported monophyly of the Ceratiidae: eyes of males huge (73, 1), eyes of males bowl shaped (74, 1), and larvae "humpbacked" (84, 1). Monophyly of the Gigantactinidae was supported by one additional male character: olfactory organs huge (75, 2). One character provided additional evidence for monophyly of the Linophrynidae, eyes of males tubular (74, 2), while within the Linophrynidae, *Borophryne* and *Linophryne* shared one male character: premaxillae greatly reduced or absent (77, 1).

## Comparisons with Previous Hypotheses

In some ways the relationships proposed here corroborate the findings of earlier studies of ceratioid evolution, but in more ways they are vastly different. Some of the similarities and more significant differences are summarized below, along with additional pertinent comments.

CENTROPHRYNIDAE AND CERATIIDAE: Despite Bertelsen's (1951:28) conclusion that the Centrophrynidae "shows no obviously close relationship to any other family," Pietsch (1972a:43, fig. 25) argued in support of a lineage containing the Centrophrynidae and Ceratiidae, listing 11 shared character states, most of which are incorporated here in this study. A sister-group relationship between these two families, however, was later challenged by both Pietsch (1979:23, figs. 25, 26) and Bertelsen (1984:333, fig. 171). A basal position among ceratioid families for either of these two families, as proposed here, has never been suggested before. Both taxa are deeply nested within the suborder in all earlier phylogenetic hypotheses.

HIMANTOLOPHIDAE, DICERATIIDAE, MELANOCETIDAE, THAUMATICHTHYIDAE, AND ONEIRODIDAE: The sequential stepwise divergence of these five families is not too surprising given that a similar arrangement has been proposed in nearly all previously published discussions of ceratioid relationships. In contrast to the present findings, however, all earlier proposals suggest that the Melanocetidae diverged first, followed in order by the Diceratiidae, Himantolophidae, and Oneirodidae (including the Thaumatichthyidae), according to Regan and Trewavas (1932); and the Himantolophidae, Diceratiidae, and Oneirodidae (including the Thaumatichthyidae), according to Bertelsen (1951, 1984) and Pietsch (1979). A monophyletic Himantolophidae, Melanocetidae, and Diceratiidae is supported by one character, ventromedial extensions of the frontal that make no contact with the parasphenoid (8, 1). No other nonhomoplastic character supports any other alternative resolution.

THAUMATICHTHYIDAE: Regan (1925b, 1926), followed by Regan and Trewavas (1932), Bertelsen (1951), and Maul (1961, 1962b), chose not to recognize Smith and Radcliffe's (1912) Thaumatichthyidae, placing the two relevant genera *Thaumatichthys* and *Lasiognathus* in the family Oneirodidae. Pietsch (1972a:18) resurrected the Thaumatichthyidae to include both genera, stating that these two taxa "possess several important and unique characters that justify familial status." Bertelsen and Struhsaker (1977), however, compared the osteology of *Thaumatichthys* and *Lasiognathus*, pointing out that the latter appears more closely related to the Oneirodidae in several of the characters

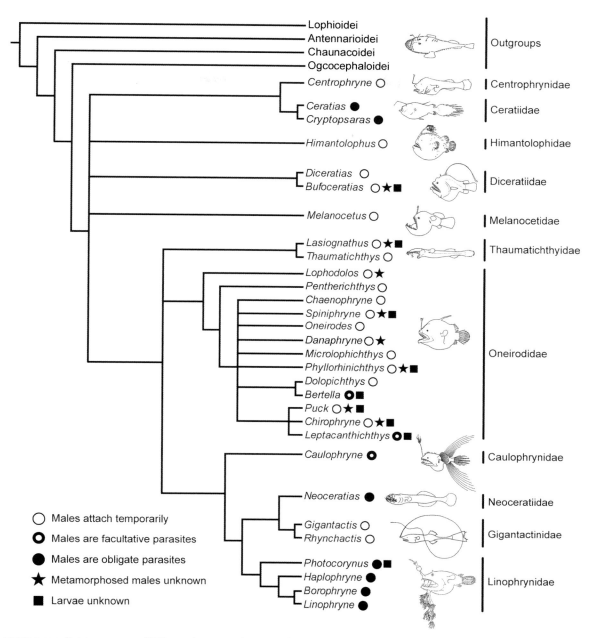

FIGURE 203. Strict consensus of 292 trees from a parsimony analysis of 88 morphological characters, applicable to metamorphosed females, metamorphosed males, and larvae, for the genera of the Ceratioidei and four lophiiform outgroups. Reproductive modes are plotted, and those genera for which metamorphosed males and/or larvae are unknown are indicated.

in which it differs most from *Thaumatichthys*. They (1977:34) concluded that "it becomes a subjective choice whether the genera *Lasiognathus* and *Thaumatichthys* both should be included in the Oneirodidae as Regan (1926) did, or placed together in Thaumatichthyidae as proposed by Pietsch (1972a), or whether each of them should be referred to a family of its own." At the same time, however, they cited the two unique features used by Pietsch (1972a) to diagnose the Thaumatichthyidae (premaxillae extending anteriorly far beyond lower jaw, and enlarged dermal denticles associated with the esca) and added a third (dorsal portion of opercle divided into two or more branches). In the end, they chose to retain the Thaumatichthyidae in the enlarged sense as proposed by Pietsch (1972a) and supported in the present analysis. It should be noted, however, that preliminary results of a molecular analysis of ceratioid evolution place *Lasiognathus* deep within the Oneirodidae, leaving *Thaumatichthys* as the only known genus of the Thaumatichthyidae (Masaki Miya, personal communication, 20 December 2005; see below).

No character in this analysis unequivocally supports a sister-group relationship of a Thaumatichthyidae composed of *Lasiognathus* and *Thaumatichthys* and the Oneirodidae. However, two characters show convergence between *Lasiognathus* (to the exclusion of *Thaumatichthys*) and most members of the Oneirodidae, and two other characters are convergent also with the Linophrynidae and Ceratiidae or both. Three of these characters are associated with the frontal bones: frontals with anterior bifurcation (6, 1; convergent also with the Ceratiidae and Linophrynidae), frontal overlapping the full length of the lateral ethmoid (7, 1; convergent also with the Linophrynidae), and ventromedial extension of the frontals bifurcate (8, 1); the fourth is the conspicuous, laterally directed quadrate spine (18, 1) found in *Lasiognathus* and most oneirodids. In contrast, four characters that are lost in the Caulophrynidae and Linophrynidae also support a close relationship of the Thaumatichthyidae with the Neoceratiidae and Gigantactinidae: articular and angular extending posteriorly (34, 1), vertebral centra elongate (48, 1), and caudal peduncle elongate (49, 1). In addition, one character, nasal papillae elongate (69, 1), is reversed in *Lasiognathus* as well as in the Caulophrynidae and Linophrynidae. Given the strength of the evidence for monophyly of the Thaumatichthyidae and absence of nonhomoplastic characters that unite *Lasiognathus* and the Oneirodidae to the exclusion of *Thaumatichthys*, the relationships described here are those best supported by the data.

ONEIRODIDAE: With 16 genera and 62 species, nearly 40% of all recognized ceratioids, the Oneirodidae is by far the largest, most complex, and certainly the least understood family of the suborder. Of the 16 genera, five are currently represented by only one, two, or three juvenile or adult females; only eight are represented by more than a dozen females. Males have been described for only seven genera, while larvae are known for only eight. Despite the rareness of most recognized taxa, however, new oneirodids are being discovered on a regular basis. The results of the present study provide the first evidence of monophyly for the family, although its more derived members remain in a large unresolved polytomy. In many ways, the relationships proposed here are remarkably similar to those suggested in a phenetic analysis of oneirodid genera published by Pietsch (1974a:86, fig. 103) but bear almost no resemblance to a cladistic attempt described in that same study (1974a:87, fig. 104).

CAULOPHRYNIDAE: Although the relationship is not supported by any nonhomoplastic characters, another surprising result of this study is the derived position of the Caulophrynidae. Bertelsen (1951, 1984) was convinced that this family is isolated from all other ceratioids, based primarily on three larval characters: (1) the presence of pelvic fins (lost in caulophrynids during metamorphosis; well developed in larvae and adults of all the outgroups, but absent in all other ceratioids), (2) the apparent absence of sexual dimorphism in rudiments of the illicium (all 16 known larval caulophrynids bear the beginnings of an illicial apparatus, a peculiarity also apparently found in neoceratiids; Bertelsen, 1984:328), and (3) the absence of a distal swelling of the illicial rudiments that would indicate the early development of a bioluminescent esca (the esca of adult caulophrynids is not a bulbous, bacteria-filled light organ as in other ceratioids, but a tuft of filaments like those found in the outgroups; Pietsch, 1979:12, figs. 15, 17, 19, 21). With respect to the ontogenetic implications of the latter two character states, Bertelsen (1984) held that the absence of an escal light organ in all life-history stages of caulophrynids is not due to secondary loss or reduction. All remaining ceratioids, being derived in these characters, thus cluster to the exclusion of the Caulophrynidae, leaving it in the most basal position in the suborder. The only prior suggestion that caulophrynids may be derived (aside from an unexplained juxtaposition of caulophrynids and linophrynids by Greenwood et al., 1966:397) was made by Pietsch (1979:22, fig. 25), who proposed an alignment with the Neoceratiidae and Gigantactinidae (see below). The results of this work indicate that the absence of these characters in the Caulophrynidae reflects a secondary loss.

NEOCERATIIDAE, GIGANTACTINIDAE, AND LINOPHRYNIDAE: In proposing the family Neoceratiidae to contain Pappenheim's (1914) *Neoceratias spinifer*, Regan (1926:39) wrote that it "may be related to the Gigantactinidae, because *Gigantactis* differs from other ceratioids in having the outer teeth larger than the inner, and moreover placed on the outer side of the jaws and inserted in muscular pads, a type of dentition from which that of *Neoceratias* might readily have been derived." Agreeing with this notion, Regan and Trewavas (1932:95), emphasizing the importance of the lack of an illicium, went a step further in considering that it might be "a male of some unknown gigantactinid." While quickly dismissing the latter proposal, Bertelsen (1951:28) concurred that the Neoceratiidae and Gigantactinidae "seem related and show a few approaches . . . to the linophrynids." That these three families form a clade was subsequently supported by Pietsch (1972a:44, fig. 25; 1979:23, fig. 26), but Bertelsen (1984:334, fig. 171), impressed with the apparent lack of sexual dimorphism in the illicium of neoceratiid larvae (a feature shared with caulophrynids; see above), tentatively proposed an unresolved trichotomy in which the Caulophrynidae and Neoceratiidae are basal to all remaining ceratioids.

The results described here include the first explicit demonstration of monophyly for the Linophrynidae as currently recognized (Bertelsen, 1951, 1984). Since Regan (1926) laid the foundation for the present classification of ceratioids, *Haplophryne*, *Borophryne*, and *Linophryne* have all been readily accepted as linophrynids, but the inclusion of *Photocorynus* has been more recent. When Regan (1925a, 1926:16) first described *Photocorynus*, he saw no affinity with linophrynids, placing the genus in a family of its own, the Photocorynidae, and describing its cranium as "less specialized than that of any other ceratioid, and more nearly approaching that of *Lophius* in essentials." While citing a number of features that seem to unite *Photocorynus* with the linophrynids, and which distinguish them from all other ceratioids, Regan and Trewavas (1932) continued to recognize the Photocorynidae. It was not until Bertelsen's (1951) in-depth comparative study of the Ceratioidei that a close relationship between *Photocorynus* and linophrynids was established. While no longer recognizing the Photocorynidae, Bertelsen placed *Photocorynus* basal to the other linophrynid genera, thus predicting the results described here. In further corroboration of the present findings, Bertelsen entered *Haplophryne* next in the sequence, thus implying a closer relationship between the terminal genera *Borophryne* and *Linophryne*.

## Characters Restricted to Metamorphosed Males

As mentioned above, sexual dimorphism is so strongly developed in ceratioids that separate taxonomic treatment of females and males is required. While some distinguishing meristic and osteological characters are shared with the males, such as fin-ray counts, families and genera are defined primarily on the basis of characters present only in females. Those few structures unique to the males, such as denticular teeth and nostril morphology, show distinct intergeneric differences (in full agreement with separations based on characters of the females), but most features, like those of the females, are autapomorphic and thus provide no evidence for reconstructing evolutionary history. In addition, ceratioid males are generally rare in collections, poorly described anatomically, and unknown for 13 of the 35 recognized genera. A thorough review of what is known about males produced only 10 characters that seemed useful to this study (most identified by Bertelsen, 1984:326, table 89). The addition of these features to the analysis provided support primarily for the monophyly of ceratioid families, including the Gigantactinidae, Ceratiidae, and Linophrynidae. While support for monophyly of the Ceratioidei as a whole is also provided, none of these characters is useful in resolving other deep nodes.

## Characters Restricted to Larvae

Larvae as well are relatively poorly known anatomically and available for only 24 of the 35 recognized ceratioid genera. Like metamorphosed males, they present few characters that can be used in phylogenetic studies (only seven; Pietsch, 1984). While providing additional support for monophyly of the Ceratioidei as a whole, larval characters offer no support for relationships within the suborder, and the absence of data for some taxa leads to a lack of resolution in basal clades.

## Conflicting Molecular Evidence

The results of a molecular study, still in its preliminary stages (Miya et al., in preparation), based on a partitioned Bayesian analysis of whole mitochondrial genome sequences of 47 lophiiform species, including representatives of all 11 ceratioid families, show very little resemblance to the hypothesis presented here. While characters for ceratioids are here polarized by outgroup comparison with nonceratioid lophiiforms, and those in turn by batrachoidiform fishes (Pietsch, 1981), the molecular results indicate that lophiiforms are deeply nested within the Perciformes, showing a close alignment with tetraodontiforms (Miya et al., 2003, 2005; Holcroft, 2004, 2005; Simmons and Miya, 2004; Yamanoue et al., 2007). While linophrynids hold a terminal position in the present proposal, the unpublished molecular findings show them basal to all other ceratioids. In further contrast to the hypothesis presented here, caulophrynids plus ceratiids diverge next, followed in stepwise fashion by gigantactinids, neoceratiids plus thaumatichthyids, centrophrynids, and oneirodids, the latter forming the sister group of a monophyletic assemblage that includes himantolophids, diceratiids, and melanocetids. Among these proposed relationships, only the clade containing the himantolophids, diceratiids, and melanocetids can be supported by aspects of our morphological data (i.e., the condition of the ventromedial extensions of the frontals, character 8). Clearly, considerably more work will be required to reconcile these two competing hypotheses.

## Sexual Parasitism

Bertelsen (1951:28) was the first to reject the idea that those ceratioids in which males become permanently and parasitically attached to females form a natural assemblage, and every study since then has corroborated this assumption (Pietsch, 1976, 2005b; Bertelsen, 1984; Shedlock et al., 2004). As currently understood, reproductive modes in lophiiform fishes exist in four states: (1) males never attach themselves to females, (2) males attach temporarily to females but never become parasitic, (3) parasitism is facultative in some taxa, and (4) parasitism is obligate in other taxa (Pietsch, 1976, 2005b). Obviously, the latter three states are derived relative to the first, but the results of this study provide no basis for further character-state transformation. While character state 1 describes the outgroup taxa identified in this study, all in-group taxa (i.e., ceratioids) are characterized by males that become attached to females (Pietsch, 2005b): those of the Centrophrynidae, Himantolophidae, Diceratiidae, Melanocetidae, Thaumatichthyidae, Gigantactinidae, and all the better known oneirodid genera except *Bertella* and *Leptacanthichthys* apparently attach themselves temporarily (character state 2); those of the Caulophrynidae and oneirodid genera *Bertella* and *Leptacanthichthys* are facultative parasites (state 3); and those of the Ceratiidae, Neoceratiidae, and Linophrynidae are obligate sexual parasites (state 4). When mapped on the strict consensus trees proposed here (Figs. 202, 203), these character states appear more or less scattered throughout the branches. In the most basal ceratioid clade proposed here, attached males have never been found in the Centrophrynidae (despite more than 40 known females, 18 to 247 mm), yet numerous examples of parasitized females are known for both genera of its sister family, the Ceratiidae (Pietsch, 2005b:223, table 1). On the other hand, the himantolophid-diceratiid-melanocetid lineages, all thought to reproduce by way of temporary nonparasitic attachment, are deeply nested within the suborder, while the thaumatichthyid-oneirodid clade contains primarily nonparasitic forms, but also at least two genera that employ facultative parasitism. Finally, the terminal assemblage, containing the Caulophrynidae through Linophrynidae, displays a mosaic of all three derived reproductive modes. Whether temporary attachment and facultative parasitism are precursors to obligate parasitism, or the former are more derived states of the latter, is thus still unknown.

The disjunct pattern of occurrence of sexual parasitism within ceratioids appears to be the result of independent acquisition among the various lineages rather than a repeated loss of this attribute within the suborder. Evidence to support this notion comes from the many differences in the precise nature of male-female attachment among the various taxa (for details, see Reproduction and Early Life History, Chapter Eight).

# FIVE

## Geographic Distribution

> To conceive of the habitat of a bottom-dwelling creature is not too difficult, but to keep in mind the three-dimensional spread of each mid-water species in a moving living space is practically impossible.
>
> NORMAN BERTRAM MARSHALL,
> *Aspects of Deep Sea Biology,* 1954:334

Ceratioid anglerfishes are widely distributed in all the major marine faunal realms of the world between extremes of 70°N and 75°S (Fig. 204). While most of the adults are restricted to lower mesopelagic and bathypelagic depths, those of some, for example *Oneirodes carlsbergi*, are concentrated at depths of only 300 or 400 m, while members of at least one genus *(Thaumatichthys)* have been captured off the bottom at depths in excess of 3600 m. At the same time, the eggs and larvae of all taxa are found in near-surface waters, thus the overall vertical and horizontal breadth of occurrence of these animals is truly vast. Taken by themselves, most of the 11 families and even some of the genera and species are almost as wide spread as the full distribution of all species of the suborder combined. Only species of the Diceratiidae, Thaumatichthyidae, and Neoceratiidae seem somewhat restricted. All three are unknown from the eastern Pacific, but they are also some of the least represented families in collections around the world. While some species are found nearly worldwide, for example *Cryptopsaras couesii*, others have surprisingly limited geographic ranges, for example *O. acanthias*, known only off the coast of southern California and Baja California, Mexico. In short, ceratioids are found wherever oceanic depths are adequate to provide suitable habitat. The only exception to this rule is the Mediterranean Sea, which is devoid of all anglerfishes except for two species of the shallow-water, benthic genus *Lophius* (Bertelsen, 1951; see also Caruso, 1986).

No thorough analysis of the distribution of ceratioids has been attempted since Bertelsen's (1951) great monograph on the suborder, in which he provided a detailed discussion of seasonal, vertical, and horizontal distributions, as well as distribution in relation to environmental factors, including temperature, salinity, oxygen, light, currents, and food supply. His analysis, based almost solely on the ceratioid larvae brought back by the *Dana* expeditions, is extraordinary for its detail and insight. The collecting efforts aboard the *Dana* were so effective that for numerous species the number of specimens known at the time was increased significantly. The collections were so geographically widespread and taken in such a large and regular series of hauls, from the surface to great depths, that the boundary limits, both vertical and horizontal, of a number of species could be determined for the first time. So complete were the hydrographic data taken for every haul that the correlation between the distribution of a species and the principle environmental factors could be studied. Finally, the protocols for collecting were so uniformly followed over the entire geographic study area that the catches from different positions, depths, and times could be compared quantitatively. Despite almost 60 years of additional oceanographic exploration, no comparable collections have been made since. Thus, Bertelsen's (1951) general conclusions, especially those that pertain to seasonal and vertical distributions, and distribution relative to physical, chemical, and biological parameters, still stand. Although the interested reader is encouraged to refer to Bertelsen's (1951) original publication, the following account is provided, much of it taken directly from his work.

### Seasonal Distribution

A detailed analysis of the number of larvae, adjusted for gear type and unit effort in the North Atlantic during each month of the year, showed that larvae of ceratioid species common in this area have a distinct annual frequency maximum. Despite the fact that adult ceratioids live at depths where temperature and light have little or no detectable annual rhythm, they have a spawning periodicity that is determined by season. The majority of the species are summer spawners, producing a maximum number of larvae in July and August. The only observed exceptions to this general rule are members of the family Linophrynidae, which spawn in the spring, with the maximum number of larvae found in March through April. No appreciable differences were found in the timing of larval maxima of the same species in different parts of the North Atlantic, but the frequency maximum for the larvae of the summer spawning species, taken as a whole, was somewhat later in the eastern

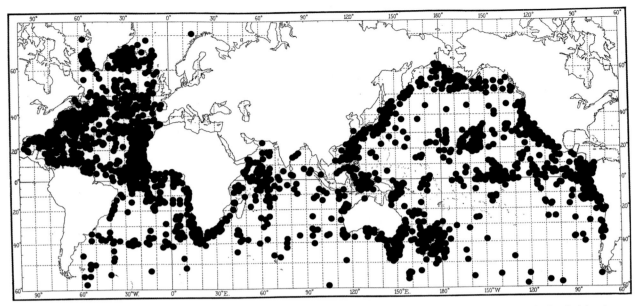

FIGURE 204. World distribution of all 160 species of the Ceratioidei. What appear to be sparsely populated regions in the South Atlantic, Central Indian Ocean, and central gyres of the North and South Pacific are rather most likely due to greatly reduced collecting efforts in those areas.

North Atlantic than in the western part of this ocean, and earliest in the Caribbean Sea and adjacent parts of the Atlantic. Specimens of the earliest spawning species contributed an increasing percentage of the total stock in the different areas as ordered above. The larval season for individual species and for the whole stock taken together appears to be more sharply defined in the eastern than in the western North Atlantic, and most extended in the Caribbean and adjacent parts of the Atlantic. The larval stage is of short duration, relatively large larvae about ready to begin metamorphosis appearing early in the larval season. The metamorphosis stage is also brief. Metamorphosing specimens occur in samples collected only within or shortly after the larval season.

## Vertical Distribution

### Ontogenetic Vertical Migration

Like most, if not all, midwater oceanic fishes, ceratioids undertake an enormous, two-way, ontogenetic vertical migration from great depths to near surface waters, and back down again, thus passing twice through a water column that is often as much as several thousand meters deep (Fig. 205). Spawning takes place at great depths, where the adults reside. Fertilization by the male, while attached to the female, takes place during the release of a large mucoid mass—or more typically, a continuous, scroll-shaped or ribbonlike sheath of gelatinous material, often referred to as an "egg raft" or "veil"—within which the eggs are embedded (a unique reproductive device characteristic of all lophiiform fishes; see Bertelsen, 1980a:66; Pietsch and Grobecker, 1980:551; 1987:351; Fig. 206). Acting like a sponge, the gelatinous material imbibes water, while also taking up waterborn sperm; thousands of tiny canals within the matrix leading to each egg facilitate fertilization. While serving to protect against predators—fishes that might otherwise feed on individual free-floating eggs, but are uninterested or unable to ingest the large jelly-like mass—the egg raft serves also to retain all the eggs in close proximity to the attached male, thus helping to ensure adequate fertilization. At the same time, the egg raft serves as a floatation device, carrying the fertilized eggs up into the relatively rich, sunlit waters of the epipelagic zone, where the eggs are released, eventually hatch, and the larvae undergo development. Bertelsen's (1951) analysis of the number of specimens collected at different depths in the North Atlantic, adjusted for gear type and unit effort, showed that the larvae of the summer-spawning species have their greatest frequency in less than 65 m below the surface, although they may occur also in less than 30 m, but only rarely below a depth of 200 m. On the other hand, linophrynid larvae have their greatest frequency between 100 and 200 m below the surface. In general, the smaller, younger larvae occur closer to the surface than the older, but this does not seem to apply to linophrynid larvae.

During the summer months, the onset of metamorphosis leads to a rapid descent (Fig. 205). The small metamorphosis stages and youngest juveniles of both sexes occur most frequently between 2000 and 2500 m (for definitions of terms used for the different stages of development, see Methods in Part Two: A Classification of Deep-Sea Anglerfishes). The metamorphosing or metamorphosed males are most abundant at about 2000 m, where they probably have the greatest chance of meeting mature females. Based on the relatively early development of the testes, those males that encounter a conspecific female will probably spawn during the next spawning season. After metamorphosis, females appear to seek somewhat shallower depths and are now most numerous between about 1500 and 2000 m below the surface. Judging by their growth rate, which appears to be very slow, at least in the first six months after metamorphosis, the females require more than one and perhaps several years to reach maturity. As maturity approaches, they again seem to descend somewhat and when fully mature have their greatest frequency probably at about 2000 m, where the free-living juvenile or adult males are also found (Bertelsen, 1951:220).

At all stages in their life cycle, ceratioid females, especially those of the more globose representatives such as *Himantolophus*,

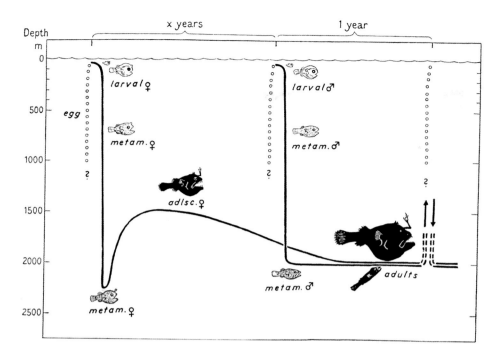

FIGURE 205. Diagrammatic and partially theoretical representation of the ontogenetic vertical migration of ceratioids. After Bertelsen (1951).

*Diceratias, Bufoceratias, Melanocetus,* and *Oneirodes,* are incapable of any prolonged horizontal locomotion and must be passively transported along to a considerable extent by water movements (see Locomotion, Food, and Feeding, Chapter Seven). The movements of water masses, and current gyres that help form and maintain these water masses, are important in the integration and concentration of populations of midwater organisms (Ebeling, 1962; Barnett, 1983, 1984; Sassa et al., 2002). While the broad geographic distributions of most ceratioids reflect this "planktonic" nature (see Parin, 1970:24), other members of the suborder somehow maintain restricted horizontal ranges. Such narrow geographic distributions may be maintained by ontogenetic vertical migrations that cause individuals of a population to pass through shallow and deep currents flowing in opposite directions. For that matter, such horizontal ranges may be altered by changes in the timing of vertical migration (Hardy and Günther, 1935; Ebeling, 1962).

To account for the restricted distribution of *Borophryne apogon* in the Gulf of Panama, Anton Bruun (1958) described a cycle in which larvae are carried westward, away from the Gulf, by the shallow equatorial current, and metamorphosis stages and juveniles seeking deeper layers are returned by the deeper, eastward flowing countercurrent. This hypothesis may explain the unusually high concentration and high degree of endemism of ceratioids in the eastern tropical Pacific as well as in the eastern tropical Atlantic (Pietsch, 1974a). In a similar way, Mackintosh (1937) and Moore (2002) suggested that the limits of distribution of Southern Ocean endemics may be controlled by vertical migrations between surface water, which has a northerly component of flow, and a southerly moving deep-water component. In the case of many Antarctic midwater fishes with epipelagic larval stages, shallow currents may carry eggs and larvae into lower latitudes, and deeper water movements return juveniles and adults to higher latitudes. For example, the ontogenetic vertical migration of *Oneirodes notius* in the Pacific sector of the Southern Ocean (see below), associated with shallow and deep waters flowing in opposite directions in this area, may explain the direct relationship between standard length of this species and latitude (see Pietsch, 1974a:95). According to McGinnis (1982) at least 14 Antarctic myctophids show this same size-latitude distribution, in which the juveniles occupy areas considerably north of conspecific adult populations. In an analogous way, this cyclic system may explain why ceratioid larvae are restricted to the warmer parts of the oceans between approximately 40°N and 35°S (as described by Bertelsen, 1951; Fig. 207), yet large females, some of which are gravid (e.g., see Bertelsen, 1951; Pietsch, 1974a; Bertelsen and Krefft, 1988), are found well outside these limits in the North Atlantic.

Thus, it seems that each distribution is "dependent upon a system of circulation, either an oceanic gyral or a current with associated countercurrents. These maintain both the water masses and the distribution" of the physicochemical parameters that relate to oceanic planktonic distributions (Brinton, 1962:198). The life cycles of geographically restricted species (such as *O. acanthias,* endemic to the transitional water of the California Current) must be therefore finely tuned to the dynamics of the waters in which they live.

## Vertical Distribution of Juveniles and Adults

Bertelsen (1951:217) found that the vertical distribution of metamorphosed ceratioids in different parts of the world oceans showed no distinct differences. A shift in the frequency maximum to somewhat shallower depths was found only in the Gulf of Panama. With a few known exceptions, most species seem to have nearly the same vertical distribution. Juveniles of *Cryptopsaras couesii,* for example, appear to be much more common at depths of less than 1000 m compared to those of other species; although the frequency maximum for this species is found between 1000 and 1500 m (Bertelsen, 1951; Pietsch, 1986b). *Melanocetus johnsonii* is most often collected between 500 and 1500 m, while its congener *M. murrayi* is a considerably deeper-dwelling species, the bulk of the known material collected between 1000 and 2500 m (see

FIGURE 206. *Linophryne arborifera*, female, 77-mm SL female, with 15-mm parasitic male hidden behind partially extruded gelatinous "egg raft," within which eggs are embedded, ISH 2736/71. Photo made aboard ship prior to preservation by Mattias Stehmann; after Bertelsen (1980a).

Pietsch and Van Duzer, 1980). The available data suggest that *Oneirodes carlsbergi* has an extremely wide vertical range compared to that of other ceratioids, and that it may be taken at relatively shallow depths: about 36% of the total known material (30 metamorphosed females), including the largest specimens, was collected by gear fished at maximum depths not exceeding 360 m; 72% was taken by nets fished above 1000 m; and two specimens (22.5 and 38.0 mm) were captured by closing trawls between 690 and 900 m (Pietsch, 1974a). The vertical distribution of the many rare species, that is, those known only from one or two specimens, shows that they generally inhabit deeper waters than the more abundant species, but the data do not indicate that their primary distribution lies within the relatively unexplored depths greater than 3000 m below the surface.

## Distribution Relative to Physical, Chemical, and Biological Parameters

Because of their limited powers of movement, without any appreciable active resistance to dispersal by water movement, the majority of ceratioids, both as larvae and adults, are not only widespread geographically (both horizontally and vertically), but appear as well to display a broad tolerance to the physicochemical and biological properties of the deep sea. Of the various parameters that typically affect marine distributions, Bertelsen (1951) was able to correlate ceratioid distributions with light, water temperature, salinity, oxygen content, currents, and food availability. He found that the vertical distribution of the larvae is determined primarily by light; the depths at which metamorphosis stages, juveniles, and adults are found, and the upper limit of their vertical distribution, are generally determined by temperature; while the lower limit of ceratioid distribution is probably determined by food availability. He found further that the fertile distribution of ceratioids—that is, the environmental conditions required for reproduction—seems determined by the temperature at the depths where the larvae have their frequency maximum and is indirectly dependent on the currents in the upper-water layers. The separation of species and populations with predominantly eastern or western distributions may be affected at least in part by currents.

The absence of ceratioids in the Mediterranean Sea is more difficult to explain. With an average depth of 1500 m, and a maximum depth of 5150 m off the southern coast of Greece, it is certainly deep enough to accommodate meso- and bathypelagic fishes (Cartes et al., 2004). The sill depth at the Strait of Gibraltar (only 14 km wide) is less than 100 m in a few places, but at least 300 m toward the center. This is quite shallow compared to the average depth of the Atlantic Ocean and the Mediterranean Sea on either side, but more than sufficient to allow epipelagic eggs and larvae to pass through. At the same time, the prevailing currents, which move at rates of 2 to 5 knots throughout the year, depending on the wind and tides, are from west to east: relatively fresh Atlantic water enters the Mediterranean through the Strait as a surface flow that is carried in to replace surface water that is evaporated at a high rate in the very arid eastern end of the Mediterranean Sea (Thurman, 1975; Longhurst, 2007). A countercurrent of dense, high-salinity water flows in the opposite direction, but at a depth well below the vertical distribution of ceratioid larvae (Morozov et al., 2002). Numerous larvae of a host of different ceratioid taxa have been collected just outside the Strait (Bertelsen, 1951; Fig. 207), yet not one larva has ever been reported from the relatively well-collected western Mediterranean. It has been hypothesized that high salinity and/or high temperature in the greater depths prevent the spreading of ceratioids into the Mediterranean (Bertelsen, 1951), but this does not explain the exclusion of epipelagic early life-history stages. The complete absence of antennarioid, chaunacoid, and ogcocephaloid anglerfishes in the Mediterranean, in stark contrast to lophioids (Caruso, 1986), which are very well established there, is equally puzzling.

## Geographical Distribution

The majority of the species of the suborder are known from such a small number of specimens that it is impossible to conclude anything about their geographic distribution. When Bertelsen (1951) published his monograph on the Ceratioidei, 39% of the 80 species recognized at the time were known from only a single specimen. Substantial collecting since then has doubled the number of species and significantly increased the number of specimens per species—14% of the 160 recognized species of the suborder are now known from a single specimen, 33% from three specimens or fewer, 48% from six specimens or fewer, and 63% from 12 or fewer (Table 8)—yet most are still very poorly known when compared to shallow-water, inshore species. On the other hand, most of those that are known in sufficiently greater numbers, for example, species represented in collections by more than 20 specimens (which account for 30% of the 160 recognized species), are recorded from all the major oceans of the world (Table 8). For the most part, therefore, once represented by sufficient numbers, most ceratioid species are proving to have very broad geographic distributions. There are, however, some exceptions to this general rule, perhaps the most striking of which have been found within the oneirodid genus *Oneirodes* (Pietsch, 1974a). These and other examples are described below by family.

Because the majority of collections of meso- and bathypelagic fishes are made with nonclosing nets, the actual depth of capture is unknown. Furthermore, because sample sizes are

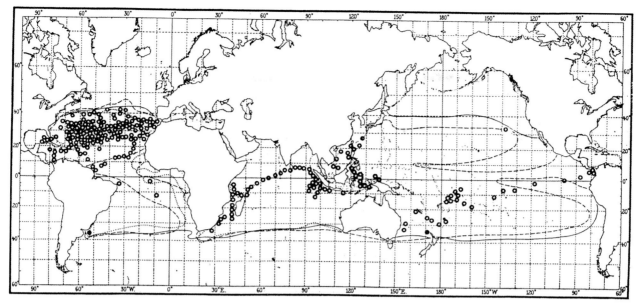

FIGURE 207. Geographic distribution of ceratioid larvae shown in relation to temperature isotherms, the solid line representing the 20°C surface average for August in the Northern Hemisphere and February in the Southern Hemisphere; the dashed and dotted lines representing 15 and 13°C averages at 100 and 200 m depth, respectively. Although based almost solely on material collected during the two circumnavigations of the Danish Research Vessel *Dana* in the 1920s (hence the obvious cruise tracks in the relatively poorly collected South Atlantic, Indian Ocean, and across the Pacific), very few ceratioid larvae have been reported since that time. Note the absence of ceratioid larvae in the Mediterranean Sea. After Bertelsen (1951).

small for most species, a statistical treatment of the nonclosing net data is usually impossible (but see Pietsch, 1974a). Ceratioids, however, are relatively rare; therefore, the chance is great that most specimens were caught at depths where gear is fished for the longest period of time. Thus, vertical distributions estimated with reference to the maximum depth reached by gear for each capture seem justifiable, but it is important to keep in mind that these data are only estimates of the actual habitats occupied in each case.

## Centrophrynidae

The family Centrophrynidae, containing only *Centrophryne spinulosa*, now represented in collections by 41 metamorphosed females, one metamorphosed male, two males in metamorphosis, and two larvae, has been collected in all three major oceans of the world (Pietsch, 1972a; Fig. 208). About a third of the material is from the Atlantic, where in the west it has been collected in the Gulf of Mexico, Caribbean Sea, and off the coast of Venezuela; and in the east from the Canary Islands and Cape Verde, south to about 18°S, 10°W. In the western Pacific it is known from Sagami Bay, Japan, south to Taiwan, the South China Sea, and off New Ireland in the Bismarck Archipelago. In the eastern Pacific it extends from off the Marquesas Islands eastward to Baja California, southern Mexico, and the Gulf of Panama. The only known metamorphosed male is from the South China Sea. A single larval male has been taken in the Mozambique Channel, Indian Ocean. The type locality is north of New Guinea.

Assuming that most of the specimens of *C. spinulosa* were taken at depth due to the relatively short period of time required for lowering and retrieving the trawl, the vertical distribution of metamorphosed females ranges from 590 to 2325 m. The two male specimens in metamorphosis were taken at depths of 750 and 1550 m. The single known metamorphosed male was taken somewhere between the surface and 1755 m; the two known larvae were collected somewhere between the surface and 35 m.

## Ceratiidae

Ceratiids are found at midwater depths worldwide, the distribution of the family taken as a whole nearly overlapping the geographic range of the entire suborder (Pietsch, 1986b; Figs. 209, 210). *Ceratias holboelli* has a broad distribution in all three major oceans of the world but is replaced in the Southern Ocean by its congener *C. tentaculatus* (Fig. 209). It is known from throughout the North Atlantic, ranging as far north as the Strait of Denmark at about 68°N and extending south to approximately 17°S off the coast of Brazil. In the Pacific, *C. holboelli* has been collected from off New South Wales, in the Coral and Celebes seas, from off the Bismarck Archipelago and Solomon Islands, from off Hokkaido, Japan, in the Bering Sea, off Oregon and California, and from off Oahu, Hawaiian Islands. Three records are known from the Indian Ocean, one from off Durban, South Africa, and the other two from the Arabian Sea.

Juveniles and adults of *C. holboelli* may be captured anywhere between approximately 150 and 3400 m, but the majority of specimens were captured between 400 and 2000 m. Of the material for which depth of capture is known (123 of 161 known specimens), 84% was taken by trawls that reached a maximum depth of 2000 m or less; 76% was taken by trawls that reached a maximum depth of 1100 m or less. At the upper end of its vertical range, 80% of the known material was taken at depths greater than 400 m. A number of large adult specimens have been found in relatively shallow water in high latitudes of the North Atlantic and in the Bering Sea, some of them from unknown depths, others in as little as 120 to 680 m. The 770-mm holotype of *Reganichthys giganteus* (MCZ 36042) was captured by otter trawl somewhere between the

TABLE 8
Geographic Distribution of Species of the Ceratioidei Arranged in Descending Order of the Number of Known Metamorphosed Females

| Species | WA | EA | WI | EI | WP | EP | SO | N |
|---|---|---|---|---|---|---|---|---|
| *Cryptopsaras couesii* | + | + | + | + | + | + | | 983 |
| *Melanocetus johnsonii* | + | + | + | + | + | + | | 852 |
| *Melanocetus murrayi* | + | + | + | + | + | + | | 310 |
| *Ceratias holboelli* | + | + | + | + | + | + | | 161 |
| *Diceratias pileatus* | + | + | | + | | + | | 153 |
| *Oneirodes acanthias* | | | | | | + | | 153 |
| *Lophodolos acanthognathus* | + | + | + | | + | + | | 149 |
| *Himantolophus groenlandicus* | + | + | ? | | | ? | | 143 |
| *Chaenophryne draco* | + | + | + | + | + | + | | 119 |
| *Himantolophus appelii* | + | + | + | + | + | + | | 115 |
| *Oneirodes bulbosus* | | | | | + | + | | 115 |
| *Oneirodes eschrichtii* | + | + | + | + | + | + | | 115 |
| *Ceratias uranoscopus* | + | + | + | + | + | + | | 112 |
| *Oneirodes thompsoni* | | | | | + | + | | 108 |
| *Microlophichthys microlophus* | + | + | + | + | + | + | | 94 |
| *Gigantactis vanhoeffeni* | + | + | + | | + | + | | 90 |
| *Haplophryne mollis* | + | + | | + | + | + | | 88 |
| *Chaenophryne longiceps* | + | + | + | + | + | + | | 84 |
| *Bufoceratias wedli* | + | + | | | | | | 82 |
| *Oneirodes luetkeni* | | | | | | + | | 58 |
| *Ceratias tentaculatus* | | | | | | | + | 57 |
| *Chaenophryne melanorhabdus* | | | | | | + | | 49 |
| *Dolopichthys pullatus* | + | + | + | | + | + | | 45 |
| *Centrophryne spinulosa* | + | + | + | | + | + | | 41 |
| *Borophryne apogon* | | | | | | + | | 39 |
| *Thaumatichthys binghami* | + | | | | | | | 39 |
| *Caulophryne jordani* | + | + | + | | + | + | | 38 |
| *Oneirodes krefft* | | + | + | + | + | | | 38 |
| *Bertella idiomorpha* | | | | | + | + | | 34 |
| *Melanocetus polyactis* | | | | | | + | | 32 |
| *Linophryne arborifera* | + | + | | | | | | 31 |
| *Lophodolos indicus* | | + | + | + | + | + | | 31 |
| *Photocorynus spiniceps* | + | + | | + | + | + | | 31 |
| *Oneirodes carlsbergi* | + | + | | | + | + | | 30 |
| *Linophryne lucifer* | + | + | | + | | | | 29 |
| *Oneirodes macrosteus* | + | + | | | | | | 28 |
| *Linophryne densiramus* | + | + | + | + | + | + | | 27 |
| *Dolopichthys longicornis* | + | + | + | | + | + | | 26 |
| *Pentherichthys atratus* | + | + | | + | + | + | | 26 |
| *Himantolophus albinares* | + | + | | | + | | | 25 |

TABLE 8 (continued)

| Species | WA | EA | WI | EI | WP | EP | SO | N |
|---|---|---|---|---|---|---|---|---|
| *Himantolophus sagamius* | | | | | + | + | | 25 |
| *Diceratias trilobus* | | | | + | + | | | 24 |
| *Leptacanthichthys gracilispinis* | + | + | | | + | + | | 24 |
| *Oneirodes sabex* | | | | | + | | | 24 |
| *Chaenophryne ramifera* | + | + | + | + | + | + | | 23 |
| *Linophryne coronata* | + | + | | | | + | | 22 |
| *Oneirodes rosenblatti* | | | | | | + | | 21 |
| *Spiniphryne gladisfenae* | + | + | + | | + | | | 21 |
| *Danaphryne nigrifilis* | + | + | | | + | + | | 20 |
| *Gigantactis meadi* | | | | | | | + | 20 |
| *Himantolophus mauli* | + | + | | | | | | 20 |
| *Himantolophus paucifilosus* | + | + | | | | | | 19 |
| *Gigantactis paxtoni* | | | + | | + | | | 18 |
| *Neoceratias spinifer* | + | + | + | + | + | + | | 18 |
| *Caulophryne pelagica* | + | + | + | | + | + | | 17 |
| *Oneirodes notius* | | | | | | | + | 17 |
| *Dolopichthys allector* | + | + | | | | + | | 15 |
| *Bufoceratias thele* | | | | | + | | | 14 |
| *Caulophryne polynema* | + | + | | | | + | | 14 |
| *Gigantactis microdontis* | | | | | | + | | 12 |
| *Linophryne indica* | | | + | + | + | + | | 12 |
| *Gigantactis gargantua* | | | | + | + | + | | 11 |
| *Gigantactis macronema* | + | + | | | | + | | 11 |
| *Gigantactis perlatus* | + | + | | + | + | + | | 11 |
| *Himantolophus crinitus* | | + | | | | | | 11 |
| *Linophryne pennibarbata* | + | + | | | + | + | | 11 |
| *Oneirodes anisacanthus* | + | + | | | | | | 11 |
| *Phyllorhinichthys micractis* | + | + | + | | + | + | | 11 |
| *Diceratias bispinosus* | | | | + | + | | | 10 |
| *Gigantactis elsmani* | + | + | | | + | + | | 10 |
| *Gigantactis longicirra* | + | + | | | | + | | 10 |
| *Linophryne macrodon* | + | | | | | + | | 10 |
| *Melanocetus niger* | | | | | | + | | 10 |
| *Rhynchactis macrothrix* | + | + | + | | + | | | 10 |
| *Linophryne algibarbata* | + | + | | | | | | 9 |
| *Oneirodes flagellifer* | | | | + | + | | | 9 |
| *Oneirodes heteronema* | | | | | | + | | 9 |
| *Dolopichthys jubatus* | | + | + | | | + | | 8 |
| *Dolopichthys karsteni* | + | + | | | | | | 8 |
| *Himantolophus cornifer* | + | + | | | + | + | | 8 |
| *Lasiognathus saccostoma* | + | + | | | | + | | 8 |

(Continued)

TABLE 8 (continued)

| Species | WA | EA | WI | EI | WP | EP | SO | N |
|---|---|---|---|---|---|---|---|---|
| *Lasiognathus intermedius* | + | | | | | + | | 7 |
| *Oneirodes alius* | | | | | + | | | 7 |
| *Phyllorhinichthys balushkini* | | + | | | | | | 7 |
| *Dolopichthys danae* | + | + | | | | | | 6 |
| *Gigantactis savagei* | | | | | ? | + | | 6 |
| *Himantolophus melanolophus* | + | + | + | | + | | | 6 |
| *Linophryne brevibarbata* | + | + | | | | | | 6 |
| *Gigantactis kreffti* | | + | | | + | | | 5 |
| *Himantolophus macroceras* | | + | | | | | | 5 |
| *Lasiognathus beebei* | + | + | | | | + | | 5 |
| *Linophryne bicornis* | + | | | + | | | | 5 |
| *Linophryne racemifera* | + | + | | | | + | | 5 |
| *Oneirodes melanocauda* | + | | + | | + | | | 5 |
| *Oneirodes schmidti* | | | | | + | | | 5 |
| *Rhynchactis leptonema* | + | | | | + | + | | 5 |
| *Bufoceratias shaoi* | | | + | | + | | | 4 |
| *Chirophryne xenolophus* | + | | | | + | + | | 4 |
| *Gigantactis gibbsi* | + | + | | | | | | 4 |
| *Gigantactis golovani* | + | + | | | | | | 4 |
| *Gigantactis ios* | | + | | | | | | 4 |
| *Himantolophus multifurcatus* | + | + | | | | | | 4 |
| *Himantolophus pseudalbinares* | | | | | | | + | 4 |
| *Linophryne maderensis* | | + | | | | | | 4 |
| *Melanocetus eustalus* | | | | | | + | | 4 |
| *Oneirodes pietschi* | | | | | + | | | 4 |
| *Oneirodes plagionema* | | | | | + | + | | 4 |
| *Puck pinnata* | + | | | | + | + | | 4 |
| *Acentrophryne dolichonema* | | | | | | + | | 3 |
| *Dermatias platynogaster* | | | | | + | | | 3 |
| *Gigantactis gracilicauda* | + | + | | | | | | 3 |
| *Himantolophus borealis* | | | | | + | | | 3 |
| *Himantolophus nigricornis* | | | | | + | + | | 3 |
| *Linophryne arcturi* | | + | | | | + | | 3 |
| *Linophryne polypogon* | | + | | | + | | | 3 |
| *Oneirodes basili* | | | | | | + | | 3 |
| *Oneirodes cristatus* | | | | | + | | | 3 |
| *Oneirodes posti* | + | + | | | | | | 3 |
| *Oneirodes theodoritissieri* | | + | | | | | | 3 |
| *Thaumatichthys pagidostomus* | | | | | + | | | 3 |
| *Tyrannophryne pugnax* | | | | | + | + | | 3 |
| *Acentrophryne longidens* | | | | | | + | | 2 |

TABLE 8 (continued)

| Species | WA | EA | WI | EI | WP | EP | SO | N |
|---|---|---|---|---|---|---|---|---|
| *Chaenophryne quasiramifera* | | | | | | + | | 2 |
| *Ctenochirichthys longimanus* | | + | | | | + | | 2 |
| *Dolopichthys dinema* | | + | | | | | | 2 |
| *Gigantactis herwigi* | + | + | | | | | | 2 |
| *Gigantactis watermani* | | + | | | + | | | 2 |
| *Himantolophus azurlucens* | | + | | | | + | | 2 |
| *Himantolophus macroceratoides* | | + | + | | | | | 2 |
| *Oneirodes dicromischus* | | | | | + | + | | 2 |
| *Oneirodes epithales* | + | | | | | | | 2 |
| *Oneirodes haplonema* | | | | | + | | | 2 |
| *Oneirodes macronema* | + | | | | | + | | 2 |
| *Oneirodes micronema* | | | | | + | | | 2 |
| *Oneirodes myrionemus* | | + | | | | | | 2 |
| *Oneirodes thysanema* | + | | | | + | | | 2 |
| *Spiniphryne duhameli* | | | | | | + | | 2 |
| *Thaumatichthys axeli* | | | | | | + | | 2 |
| *Caulophryne bacescui* | | | | | | + | | 1 |
| *Gigantactis longicauda* | + | | | | | | | 1 |
| *Himantolophus compressus* | | + | | | | | | 1 |
| *Himantolophus danae* | | | | | + | | | 1 |
| *Lasiognathus amphirhamphus* | | + | | | | | | 1 |
| *Lasiognathus waltoni* | | | | | | + | | 1 |
| *Linophryne andersoni* | | | | | + | | | 1 |
| *Linophryne argyresca* | | | | + | | | | 1 |
| *Linophryne bipennata* | | | | + | | | | 1 |
| *Linophryne escaramosa* | | | | | | + | | 1 |
| *Linophryne parini* | | | + | | | | | 1 |
| *Linophryne quinqueramosa* | | | | | | + | | 1 |
| *Linophryne sexfilis* | | + | | | | | | 1 |
| *Linophryne trewavasae* | | | | | + | | | 1 |
| *Melanocetus rossi* | | | | | | | + | 1 |
| *Oneirodes bradburyae* | + | | | | | | | 1 |
| *Oneirodes clarkei* | | + | | | | | | 1 |
| *Oneirodes mirus* | | | | | + | | | 1 |
| *Oneirodes pterurus* | | | | | + | | | 1 |
| *Oneirodes schistonema* | | | | | + | | | 1 |
| *Rhynchactis microthrix* | | | + | | | | | 1 |
| *Robia legula* | | | | | + | | | 1 |
| Totals | 75 | 84 | 35 | 29 | 79 | 82 | 4 | 5604 |

NOTE: WA, western Atlantic Ocean; EA, eastern Atlantic Ocean; WI, western Indian Ocean; EI, eastern Indian Ocean; WP, western Pacific Ocean; EP, eastern Pacific Ocean; SO, Southern Ocean; N, number of metamorphosed females.

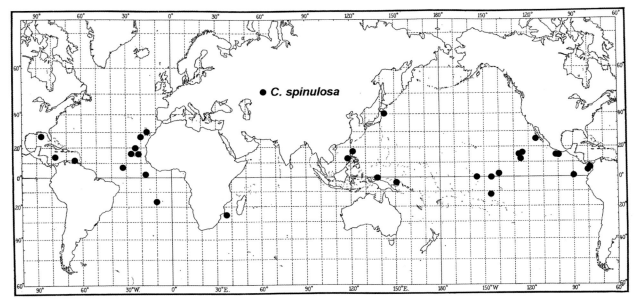

FIGURE 208. Distribution of *Centrophryne spinulosa*: Atlantic, Pacific, and Indian oceans. A single symbol may indicate more than one capture.

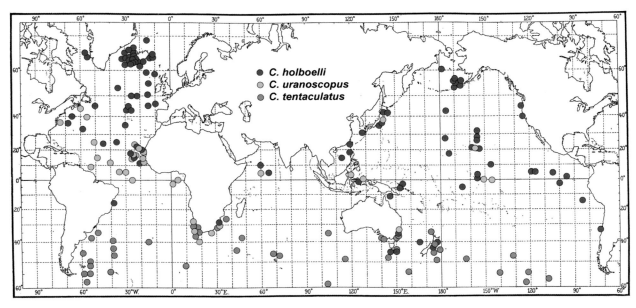

FIGURE 209. Distribution of species of *Ceratias*: *C. holboelli* (red), Atlantic, Pacific, and Indian oceans; *C. uranoscopus* (yellow), Atlantic, Pacific, and Indian oceans; *C. tentaculatus* (blue), Southern Ocean. A single symbol may indicate more than one capture.

surface and 230 m. The average maximum depth for all known captures was 1143 m.

*Ceratias uranoscopus* is broadly distributed, well represented in the Atlantic and Pacific, but known from the Indian Ocean on the basis of only two specimens, one from off Durban, South Africa, and another from the Arabian Sea (Fig. 209). Like its sympatric congener *C. holboelli*, it is replaced in the Southern Ocean by the third member of the genus, *C. tentaculatus*. In the Atlantic, *C. uranoscopus* ranges from off Nova Scotia in the west to approximately 40°S off Cape Town, South Africa, in the east. In the Pacific, it has been taken in the Halmahera and Celebes seas, off New South Wales, Australia, in the central tropical Pacific, and off leeward Oahu, Hawaiian Islands.

Metamorphosed females of this species may be taken anywhere between approximately 95 and 4000 m, but the majority of specimens were captured between 500 and 1000 m. Of the material for which data are available (79 of 112 known specimens), 94% was taken by fishing gear that reached a maximum depth of 2000 m or less; 85% was captured by gear fished in 1000 m or less. At the upper end of its vertical range, 89% of the material was taken in depths of 500 m or more. The average maximum depth for all known captures was 840 m. The largest known specimen, a 240-mm female with an attached parasitic male, was taken between the surface and 500 m.

Except for several specimens taken off South Africa and southern Mozambique (including a 534-mm specimen from off

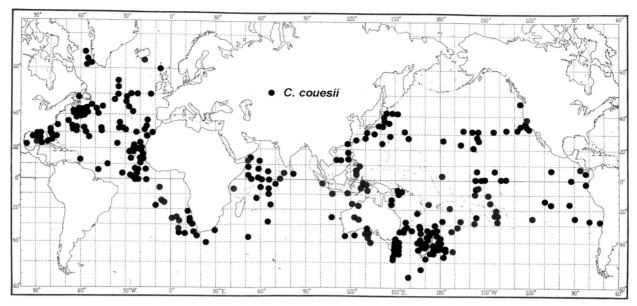

FIGURE 210. Distribution of *Cryptopsaras couesii*: Atlantic, Pacific, and Indian oceans. A single symbol may indicate more than one capture.

Saldanha Bay, South Africa; Penrith, 1967), all of the known material of *C. tentaculatus* has been collected within the Southern Ocean between approximately 35 and 68°S (Fig. 209). Although this species may be taken anywhere between approximately 100 and 2900 m, the majority of specimens were captured between 650 and 1500 m. Of the material for which depth of capture was recorded (41 of 57 known specimens), 90% was taken by fishing gear that reached a maximum depth of 2000 m or less; 81% was captured by gear fished at maximum depths of 1500 m or less. At the upper end of its vertical range, 86% of the known material was taken by trawls that reached maximum depths of greater than 650 m. A 470-mm specimen (ISH 657/71) was captured between 105 and 110 m. The average maximum depth for all known captures was 1194 m.

The geographic distribution of *Cryptopsaras couesii* is similar to the combined ranges of *Ceratias holboelli* and *C. uranoscopus*, occurring in all three major oceans of the world between approximately 64°N and 57°S (Pietsch, 1986b; Fig. 210). It occurs throughout the North Atlantic, ranging as far north as Iceland (a single record; Saemundsson, 1922) and from the Gulf of Mexico to the African coast. In the South Atlantic, however, it appears to be restricted to the eastern side, ranging as far south as approximately 40°S off the tip of Africa.

In the Pacific, *C. couesii* ranges from the Philippine and Molucca islands to the eastern tropical Pacific, and between Hokkaido, Japan, and the Oregon coast, to New Zealand and off northern Chile. In the Indian Ocean, this species is known from more than 80 specimens (primarily larvae and males taken from localities, nearly all of which appear to be associated with continental margins, but which in reality coincide with the route of the *Dana* Expedition of 1928–1930; see Knudsen et al., 1934).

Juvenile and adult females of *C. couesii* may be captured anywhere between approximately 75 and 4000 m. The majority of known specimens, however, were taken between 500 and 1250 m. Of the material for which depth of capture was known, 98% was taken by trawls that reached a maximum depth of 2000 m or less; 88% was collected by gear fished at 1250 m or less. At the upper end of its vertical range, 84% of the known material was taken at depths greater than 500 m. The average maximum depth for all known captures was 890 m.

## Himantolophidae

The distribution of the Himantolophidae was thoroughly analyzed by Bertelsen and Krefft (1988). Their conclusions are summarized below, augmented with range and depth extensions that reflect the records of numerous additional specimens collected during the last two decades (Figs. 211–214). Like most ceratioid families that contain multiple species, the small sample sizes for many of the included taxa, combined with the great variation in the intensity of fishing effort between geographic areas, significantly limit any meaningful comparison of their distributions. Also, like other families of the suborder, himantolophids are distributed primarily in the tropical and subtropical parts of the oceans. Larvae and metamorphosed males have not been found north of about 42°N and south of about 35°S, and the great majority of the metamorphosed females have been caught within these latitudinal limits. The only exceptions are a number of large females of *Himantolophus groenlandicus*, *H. albinares*, *H. mauli*, and *H. melanolophus* (greater than 100 mm) that have been recorded as far north as Iceland and Greenland close to the Polar Front, with a single record off the coast of Norway at 70°15′N representing the most northern locality of any ceratioid. Other notable exceptions include numerous large females of *H. appelii* collected between 35 and 56°S.

The best-represented species are *H. groenlandicus*, with about 143 known specimens, and *H. appelii*, with about 115. The next best represented species are *H. albinares* and *H. sagamius*, each with 25 specimens, and after that, *H. mauli*, with 20; *H. paucifilosus*, 19; *H. crinitus*, 11; *H. cornifer*, with 8; and the remaining species all represented by fewer than a half-dozen specimens (see Table 8). Each of the five species groups of the genus is widely distributed, probably circumglobal in occurrence, but few species within the groups have been recorded from more than one major ocean.

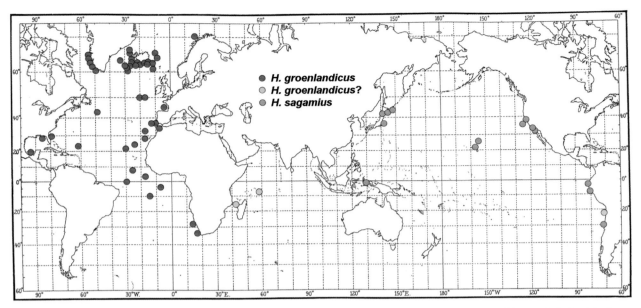

FIGURE 211. Distribution of two species of *Himantolophus*: *H. groenlandicus* (red), Atlantic Ocean, with questionable records (yellow) in the western Indian and eastern Pacific oceans; *H. sagamius* (blue), Pacific Ocean. A single symbol may indicate more than one capture.

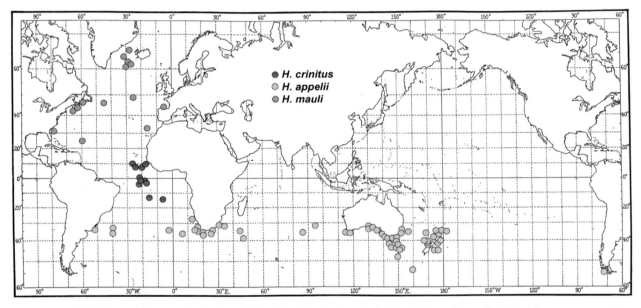

FIGURE 212. Distribution of three species of *Himantolophus*: *H. crinitus* (red), central Atlantic Ocean; *H. appelii* (yellow), Southern Ocean; *H. mauli* (blue), North Atlantic Ocean. A single symbol may indicate more than one capture.

Within the *H. groenlandicus* group, *H. groenlandicus* is widely distributed in the Atlantic, from off West Greenland, Iceland, and Norway in the north to Cape Town in the south, and from the Gulf of Mexico to the coasts of Europe and West Africa (Bertelsen and Krefft, 1988; Fig. 211). Most of the specimens have been collected far north of the known distribution of larvae and males and are thus assumed to be expatriates, that is, individuals living outside the area in which reproduction is possible. Long assumed to be endemic to the Atlantic, Meléndez and Kong (1997) reported a single female (340 mm, MNHNC P.6847) from the eastern South Pacific off Chile. Records of juvenile and larval specimens from the western Indian Ocean, which can be referred to the *H. groenlandicus* group, may also represent this species (for details, see Bertelsen and Krefft, 1988).

*Himantolophus sagamius*, found on both sides of the North Pacific, from Japan to California, as well as from off Ecuador and Chile, is the only species of the *H. groenlandicus* group known from the Pacific (Fig. 211). Three specimens collected in the Banda Sea and from off Hawaii, which are too small to be distinguished from *H. groenlandicus*, therefore most probably belong to this species (Bertelsen and Krefft, 1988). The only other identified member of this species group from Indonesian waters is the single known specimen of *H. danae*, collected in the South China Sea.

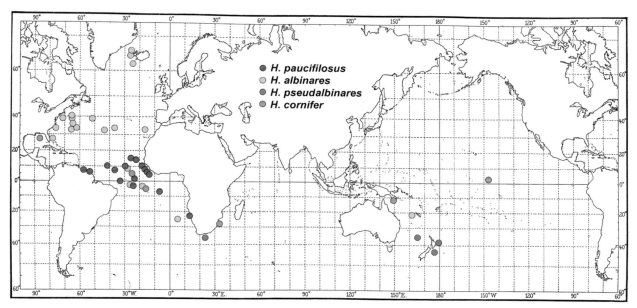

FIGURE 213. Distribution of four species of *Himantolophus*: *H. paucifilosus* (red), central and South Atlantic Ocean; *H. albinares* (yellow), Atlantic and western Pacific oceans; *H. pseudalbinares* (green), Southern Ocean; *H. cornifer* (blue), Atlantic, Pacific, and Indian oceans. A single symbol may indicate more than one capture.

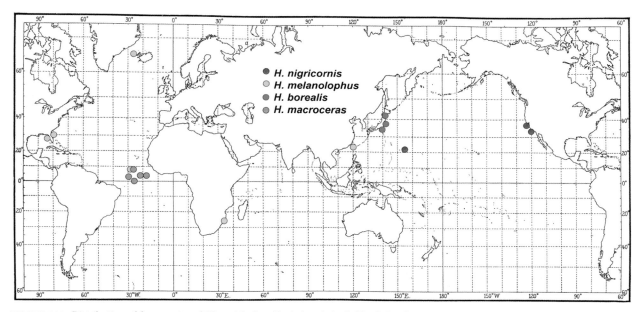

FIGURE 214. Distribution of four species of *Himantolophus*: *H. nigricornis* (red), North Pacific Ocean; *H. melanolophus* (yellow), Atlantic, Indian, and western Pacific oceans; *H. borealis* (green), western North Pacific Ocean; *H. macroceras* (blue), central Atlantic Ocean. A single symbol may indicate more than one capture.

The remaining two species of the *H. groenlandicus* group, *H. crinitus* and *H. paucifilosus*, have been found only in the Atlantic Ocean where their distributions overlap and are found as well within the known distribution of *H. groenlandicus* (Bertelsen and Krefft, 1988; Figs. 212, 213). Both are restricted, however, to the tropical part of the Atlantic, where records of *H. groenlandicus* are relatively rare. While *H. crinitus* has so far not been taken west of about 30°W, the range of *H. paucifilosus* extends from the African coast to waters off Venezuela, with a single record off Namibia at about 23°S.

*Himantolophus appelii* is unique among its congeners in being distributed along the southern subtropical convergence and restricted to latitudes below about 30°S (Bertelsen and Krefft, 1988; Fig. 212). It has been recorded from the western South Atlantic, off Uruguay, and across the Atlantic and Indian oceans to New Zealand, with a single record off the southernmost tip of Chile at about 56°S, 70°W (Meléndez and Kong, 1997). The absence of records from the central and eastern part of the South Pacific might be nothing more than the result of the relatively low level of research activity in this area.

GEOGRAPHIC DISTRIBUTION  209

The *H. nigricornis* group, with its two recognized species, is so poorly represented by specimens that hardly anything meaningful can be said about its distribution. Four of the six known specimens of *H. melanolophus* were collected in the Atlantic Ocean (Fig. 214): the holotype from the eastern equatorial Atlantic southwest of the Cape Verde Islands, the two paratypes from off the east and west coasts of Florida (Bertelsen and Krefft, 1988), and a fourth specimen reported from off Iceland (Jónsson and Pálsson, 1999). A fifth specimen was recently collected from off Taiwan (Hsuan-Ching Ho, personal communication, 26 July 2002) and a sixth from off southern Mozambique in the western Indian Ocean (M. Eric Anderson, personal communication, 20 November 2007). The three known specimens of *H. nigricornis* were all collected in the Pacific Ocean (Fig. 214): the holotype, and a specimen in the collections of the Scripps Institution of Oceanography, collected from off California; and the paratype from the central North Pacific near Wake Island (Bertelsen and Krefft, 1988).

Except for a single specimen of *H. pseudalbinares* caught slightly east of the Cape of Good Hope—with three additional records from off New Zealand and New Caledonia (Fig. 213)—no representative of the *H. albinares* group has been taken in the Indian Ocean (Bertelsen and Krefft, 1988). *Himantolophus borealis* has been collected only from the western North Pacific, all three known specimens from off the coast of Japan (Fig. 214). Of the remaining three species of the *H. albinares* group, *H. albinares*, with 25 females, is known from both sides of the Atlantic, extending as far north as Iceland (Jónsson and Pálsson, 1999) to approximately 24°S, 5°E, near Valdivia Bank (Bertelsen and Krefft, 1988; Fig. 213). A single western Pacific record of *H. albinares*, from off New Caledonia, was recently reported by Iglésias (2005). *Himantolophus mauli* (20 known females) appears to be restricted to the North Atlantic, its range extending from off Iceland at 66°N, south to about 25°N, and from off Florida in the west and northern Spain in the east (Bertelsen and Krefft, 1988; Fig. 212). The four known specimens of *H. multifurcatus* are all from the Atlantic, the holotype taken in the eastern equatorial Atlantic, the two paratypes from off the Atlantic coast of Florida, and a fourth specimen from the eastern Gulf of Mexico (Bertelsen and Krefft, 1988).

Two of the five species of the *H. cornifer* group are known from widely scattered localities, nearly all in the tropical regions of the world: *H. cornifer* is circumglobal, with the holotype from the Coral Sea, one record in the central Pacific, three from the central Atlantic, and two from the Gulf of Mexico (Bertelsen and Krefft, 1988; Fig. 213); an eighth specimen was recently reported from off South Africa (Anderson and Leslie, 2001). The two known specimens of *H. macroceratoides* were collected near the equator in the eastern Atlantic and western Indian oceans. In contrast, the five known specimens of *H. macroceras* are all from closely situated stations in the central equatorial Atlantic (Fig. 214). *Himantolophus azurlucens* is known from the holotype collected in the Gulf of Panama, and a second specimen from the eastern South Atlantic at about 32°S, 2°E, recently reported by Trunov (2001). Finally, the single known female of *H. compressus* was collected in the eastern North Atlantic, off the southwest coast of Spain.

Again like nearly all ceratioids, the vast majority of the specimens of *Himantolophus* have been caught in open nets; estimates of their vertical distribution are therefore necessarily based on the maximum depths reached by fishing gear. About 75% of the larvae were caught between the surface and 50 m, and most, if not all, of the approximately 5% caught in oblique or horizontal hauls that reached depths of more than 200 m were probably caught in lesser depths during the setting or retrieving of the nets (Bertelsen, 1951; Bertelsen and Krefft, 1988). At the same time, none of the specimens in metamorphosis or the 37 known metamorphosed males were caught in hauls with a maximum depth of less than 800 m, and only six metamorphosed males in hauls of less than 1000 m. On the other hand, metamorphosed females may occur as close to the surface as 170 to 200 m (only one has been taken from depths as shallow as 170; two at a maximum depth of 200 m). Of the approximately 80 specimens, within a size range of 20 to 100 mm and for which the maximum fishing is known, 70% were caught in depths of less than 800 m. While the vast majority of females of these sizes were caught in pelagic hauls far above the ocean bottom, most of the larger known females have been found washed up on shore or retrieved from whale stomachs, thus providing little information on the natural habitat of females approaching maximum size. Depth information is unfortunately lacking for the only known female with relatively large eggs (Bertelsen and Krefft, 1988), the 145-mm holotype of *H. nigricornis* caught in a midwater trawl in the San Clemente Basin off southern California, which has a maximum depth of about 2000 m. All the large females of *H. groenlandicus*, which make up the great majority of known specimens of the genus greater than 100 mm, were found either stranded or obtained from commercial fisheries conducted at or near the ocean bottom on coastal banks. Furthermore, they were all obtained far north or south of the area of distribution of the larvae and males, except for some few records at the border of this area (e.g., off Portugal, Madeira, and the Cape of Good Hope; see Bertelsen and Krefft, 1988). Several large females of *H. appelii*, a species nearly as common in collections as *H. groenlandicus*, have been found in the stomachs of sperm whales or, like *H. groenlandicus*, either stranded or caught in bottom trawls. This is also the case with the majority of the specimens greater than 150 mm, which represent 10 other species.

Thus, in summary, the vertical distribution exemplified by the known material of *Himantolophus* confirms the conclusions first proposed by Bertelsen (1951) to describe ceratioids in general: (1) prior to metamorphosis, the larvae of *Himantolophus* are epipelagic, occurring primarily at depths of less than 50 m below the surface; (2) at and during metamorphosis the young descend to depths of more than 800 m; (3) after metamorphosis the now mature males remain at meso- and bathypelagic depths, being most numerous in depths of more than 1000 m, while the females return to lesser depths. Immature females, 20 to 100 mm long, are caught most frequently at depths of 200 to 800 m, a vertical distribution that is similar to that of *Cryptopsaras couesii* (Bertelsen, 1951), and more shallow than that of any other ceratioid known in numbers sufficient for a comparison.

The limits of the horizontal distribution of the larvae, males, and young females of the Himantolophidae, to a region within the northern and southern subtropical convergences, is shared with these same life-history stages of all other members of the Ceratioidei, and the occurrence of large females far outside this region has been observed in representatives of several other families (i.e., the Ceratiidae, Oneirodidae, Gigantactinidae, and Linophrynidae). As argued by Bertelsen (1976, 1982), these large females are probably expatriates (individuals transported by currents beyond the environmental conditions required for reproduction), and their occurrence at relatively shallow depths, and thus within the reach of commercial fisheries, could be due to the lower temperatures of the upper water layers in these regions. The scarcity of females of

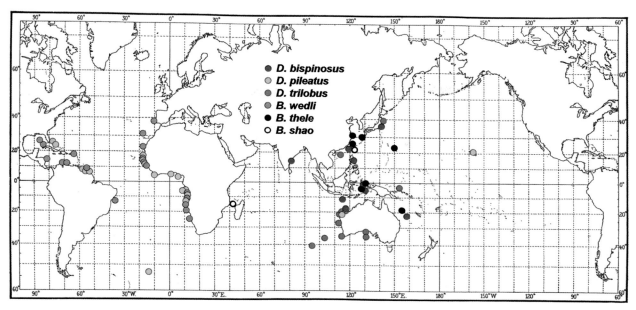

FIGURE 215. Distribution of six species of the family Diceratiidae: *Diceratias bispinosus* (red), Atlantic Ocean; *D. pileatus* (yellow), Atlantic, Indian, and central Pacific oceans; *D. trilobus* (green), eastern South Indian and western Pacific oceans; *Bufoceratias wedli* (blue), Atlantic Ocean; *B. thele* (black), western Pacific Ocean; *B. shaoi* (white), western Indian and western Pacific oceans. A single symbol may indicate more than one capture.

similar large sizes, within the area of distribution of the larvae and free-living males, is shared with most other ceratioid families and might indicate that they are less available for capture, being less abundant than the younger stages and possibly living in shallower, less accessible depths, for example, on or near the rough bottom of the continental slopes or mid-ocean ridges. Another possibility might be that these large sizes are reached only by expatriate females in connection with an inability to reach maturity, but this would not explain the almost total absence of ripe females in collections around the world (Bertelsen and Krefft, 1988).

## Diceratiidae

The family Diceratiidae is represented in all three major oceans of the world in a narrow belt between 40°N and 25°S (see Uwate, 1979; Pietsch et al., 2004; Fig. 215). To a much greater extent than other ceratioids (for which sufficient material is known), the geographic distributions of diceratiids follow the contours of high organic productivity. All known specimens have been collected near land, on the continental slope or shelf. None have been collected in the central gyres of any ocean (Fig. 215). Another area conspicuously lacking diceratiids is the eastern Pacific, with no member of the family ever having been taken east of the Hawaiian Islands, where a single specimen of *Diceratias pileatus* was reported by Pietsch and Randall (1987). All other Pacific records are from the extreme western margin of this ocean (Fig. 215): an individual of *D. bispinosus* from the Philippines and additional records of this species in equatorial waters of eastern Indonesia and in the Coral Sea off Brisbane, Australia; specimens of *D. trilobus* from off Japan, the South China Sea, and off New Ireland in the Bismarck Sea (Pietsch et al., 2006); and all known material of *Bufoceratias thele*, from the Ceram, Halmahera, and South and East China seas (Pietsch et al., 2004).

Based largely on collections made by the *Dana* in the 1920s, Regan and Trewavas (1932) and Bertelsen (1951) described the eastern tropical Pacific as especially rich in species and individuals of ceratioids. While the apparent absence of diceratiids in this region may be explained in part by inadequate sampling, it seems more likely that they are, for some unknown reason, restricted to other oceans.

On the basis of an analysis of all known material, Uwate (1979, fig. 20) was able to show that the two diceratiid genera are geographically sympatric in the tropical Atlantic and western Pacific oceans. He concluded further that the species within each genus are allopatric: *B. wedli* and *D. pileatus* are sympatric in the Atlantic, while *B. thele* and *D. bispinosus* are sympatric in the western Pacific. Realizing that in the Atlantic *B. wedli* has a filamentous esca while that of *D. pileatus* is relatively unadorned, and in the western Pacific, *D. bispinosus* has a filamentous esca while that of *B. thele* is simple, Uwate (1979) speculated that these character differences may function to reduce interspecific competition, perhaps important in partitioning food resources as well as attracting conspecific mates. Unfortunately, this nice story has been contradicted by Pietsch and Randall's (1987) discovery of a specimen of *D. pileatus* from Hawaii. It was thought also that Machida and Yamakawa's (1990) report of a specimen of *B. wedli* from the Okinawa Trough in the East China Sea was in opposition to Uwate's (1979) hypothesis, but recent examination of their specimen shows it to be a good example of *B. thele* (organic debris of some kind wound tightly around the base of the esca had been mistaken for escal appendages; see Machida and Yamakawa, 1990).

Most diceratiids have been collected by open midwater trawls, which indicate a broad vertical range for metamorphosed females, between 400 and 2300 m (Uwate, 1979). However, numerous specimens of both diceratiid genera have been taken recently in bottom trawls. Thus, in contrast to the midwater life-style generally assumed for most ceratioids, it seems evident that at least some diceratiids, especially larger individuals, are at least partly associated with the bottom (for more details, see Locomotion, Food, and Feeding, Chapter Seven).

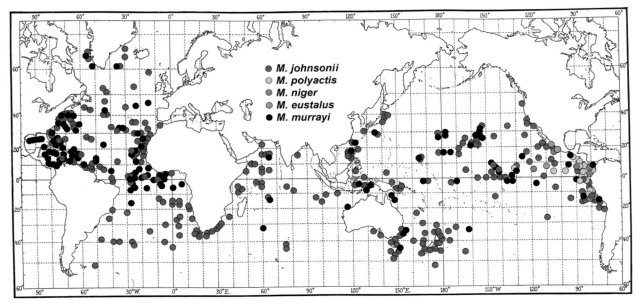

FIGURE 216. Distribution of five species of *Melanocetus*: *M. johnsonii* (red), Atlantic, Indian, and Pacific oceans; *M. polyactis* (yellow), eastern tropical Pacific Ocean; *M. niger* (green), eastern tropical Pacific Ocean; *M. eustalus* (blue), eastern tropical Pacific Ocean; *M. murrayi* (black), Atlantic, Indian, and Pacific oceans. A single symbol may indicate more than one capture.

### Melanocetidae

The family Melanocetidae is widely distributed throughout all three major oceans of the world in a broad belt limited by the Arctic and Antarctic Polar Fronts, with northern and southernmost records at approximately 66°N and 53°S (Fig. 216). It is present in the Caribbean and Gulf of Mexico but has not been collected in the Gulf of California or in the Mediterranean Sea. Two of the five species, each represented in collections by hundreds of specimens (see Table 8), are wide-ranging forms: *Melanocetus johnsonii*, found throughout the Atlantic, Pacific, and Indian oceans; and *M. murrayi* throughout the Atlantic and Pacific, but only four records from the Indian Ocean (see Pietsch and Van Duzer, 1980:83). The three remaining species, all represented by considerably fewer specimens, *M. polyactis* (32 known specimens), *M. niger* (10 specimens), and *M. eustalus* (one, plus three tentatively identified specimens) appear to be restricted to the eastern tropical Pacific Ocean (Fig. 216).

The vast majority of melanocetids were collected with non-closing nets, thus the actual depth of capture is unknown. Assuming, however, that most specimens were caught at depths where gear was fished for the longest period of time, members of the Melanocetidae may be taken anywhere between 250 m and some unknown lower depth limit exceeding 3000 m, but they are most commonly found between 500 and 2500 m. *Melanocetus johnsonii* is most often collected between 500 and 1500 m. *Melanocetus murrayi* is a considerably deeper-dwelling species; the bulk of the known material was collected between 1000 and 2500 m (see Pietsch and Van Duzer, 1980:84). The relatively thin integument and lighter, less well-ossified skeleton of *M. murrayi* most likely reflect the poorer trophic economies of these greater depths. The remaining species of the genus are so poorly represented in collections that their vertical distributions cannot be estimated.

### Thaumatichthyidae

Except for *Thaumatichthys binghami*, now represented by some 39 specimens, all collected in the western Atlantic, the species of the Thaumatichthyidae are so poorly represented in collections that little can be said about their horizontal ranges. As a whole, the genus *Lasiognathus* is now known from widely scattered localities in the Atlantic and Pacific oceans between approximately 45°N and 35°S (Bertelsen and Pietsch, 1996:407; Fig. 217), but its contained species are all extremely rare: *Lasiognathus beebei* is known from four specimens collected on both sides of the North Atlantic, and a fifth specimen from off Oahu, Hawaiian Islands. The single known specimen of *L. waltoni* is from the central Pacific, just north of Oahu. Five of the seven known specimens of *L. intermedius* were collected from the western North Atlantic, with single records from the eastern South Pacific and from off Cape Town, South Africa. *Lasiognathus saccostoma* is known from seven specimens captured on both sides of the North Atlantic, and one additional record from off the Hawaiian Islands in the central North Pacific.

*Thaumatichthys* is somewhat more broadly distributed than its sister genus, *Lasiognathus*, having a worldwide distribution in the tropical and subtropical parts of the oceans (Bertelsen and Struhsaker, 1977:28; Fig. 218). The holotype of *Thaumatichthys pagidostomus* and two additional specimens tentatively assigned to this species are all from the western Pacific, the former taken in the Gulf of Tomini, off Sulawesi (Celebes), the others from off Taiwan. The holotype of *T. axeli* and a second specimen referred to this species were collected in the eastern Pacific off the coast of Mexico at about 9 and 32°N, respectively. The known material of *Thaumatichthys binghami* was collected in the Gulf of Mexico, Caribbean Sea, and the adjacent waters surrounding the Bahamas. Finally, the genus is represented in the Indian Ocean by a single larval specimen collected at the southern end of the Mozambique Channel. Except for two larvae, captured some 725 to 800 km off the West African coast, all known specimens came from localities fewer than 400 km from land.

Species of the genus *Thaumatichthys* are the only truly benthic ceratioids. Except for five larvae and three of the smallest metamorphosed females, all taken with open pelagic nets

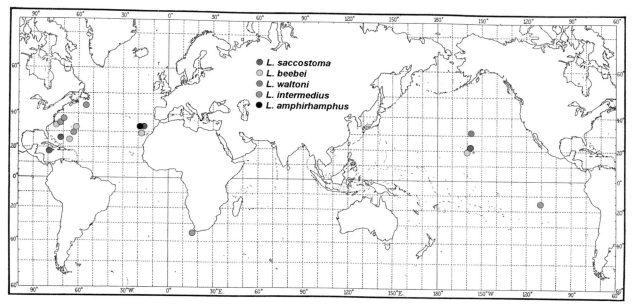

FIGURE 217. Distribution of five species of *Lasiognathus*: *L. saccostoma* (red), Atlantic and Pacific oceans; *L. beebei* (yellow), Atlantic and Pacific oceans; *L. waltoni* (green), central North Pacific Ocean; *L. intermedius* (blue), Atlantic and Pacific oceans; *L. amphirhamphus* (black), eastern North Atlantic Ocean. A single symbol may indicate more than one capture.

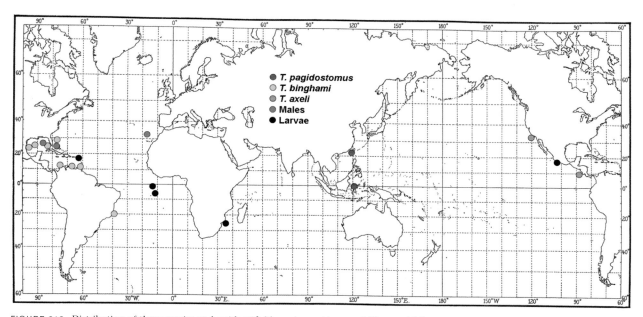

FIGURE 218. Distribution of three species and unidentifiable males and larvae of *Thaumatichthys*: *T. pagidostomus* (red), western Pacific Ocean; *T. binghami* (yellow), western Atlantic Ocean; *T. axeli* (blue), eastern North Pacific Ocean; unidentifiable males (green), North Atlantic Ocean; unidentifiable larvae (black) Atlantic, western Indian, and eastern North Pacific oceans. A single symbol may indicate more than one capture.

fished well off the bottom (between the surface and 1750 m), the remaining specimens of the genus were caught in bottom trawls. Of the 39 known females of *T. binghami* (40 to 294 mm) collected with bottom trawls, 37 were caught on the continental slope at depths between 1270 and 2193 m, one was captured at 1100 m, and another from the central Gulf of Mexico between 3151 and 3200 m. The two specimens of *T. axeli* were caught at abyssal depths of 3570 and 3595 to 3695 m, respectively. The holotype of *T. pagidostomus* was taken at 1440 m. The three smallest larvae (3 to 6.5 mm) were caught in nets fished open in maximum depths of about 20 to 100 m below the surface. The two larger larvae (18.5 to 22.6 mm) were captured in gear fished open at maximum depths of 1500 and 780 m, respectively. Finally, three of the metamorphosed males were collected with pelagic trawls fished open between the surface and 2000 m; the third male was taken with a bottom trawl at a depth of 1938 m.

Thus, it appears that the young larvae of *Thaumatichthys* are epipelagic, like those of other ceratioids. The older larvae and youngest metamorphosed females are meso- and bathypelagic, but at a length of 40 to 50 mm they become benthic, in contrast to all other known ceratioids.

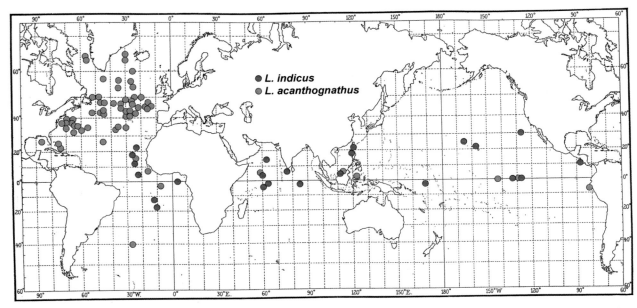

FIGURE 219. Distribution of two species of *Lophodolos*: *L. indicus* (red), eastern Atlantic, Indian, and Pacific oceans; *L. acanthognathus* (blue), Atlantic, Indian, and Pacific oceans. A single symbol may indicate more than one capture.

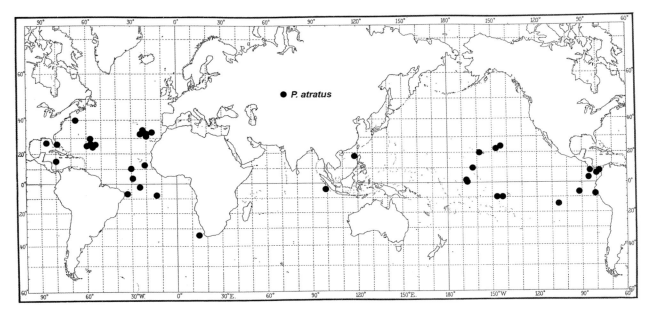

FIGURE 220. Distribution of *Pentherichthys atratus*: Atlantic, Indian, and Pacific oceans. A single symbol may indicate more than one capture.

## Oneirodidae

The vast majority of the 62 species recognized in the family Oneirodidae are so poorly represented in collections that nothing meaningful can be said about their geographic range or the depths they generally occupy. When viewed on the generic level, only six of the 16 genera of the family, each containing at least one species represented by 30 females or more, are considered worthy of detailed discussion: *Lophodolos*, *Chaenophryne*, *Oneirodes*, *Microlophichthys*, *Dolopichthys*, and *Bertella* (Table 8). These genera are dealt with below. For most of the remaining genera of the family, distributions are plotted (Figs. 219–229), but detailed information is referred to the species accounts provided in Part Two: A Classification of Deep-Sea Anglerfishes.

LOPHODOLOS: Two species of *Lophodolos* are currently recognized, the type species of the genus, *L. acanthognathus*, one of the best represented ceratioids in collections around the world, with 149 cataloged specimens; and *L. indicus*, with 31 specimens (Table 8). Both are known from all three major oceans of the world (Pietsch, 1974b; Fig. 219). *Lophodolos acanthognathus* is known from both sides of the Atlantic. Most of the material (including all type specimens) has been collected from the western side of this ocean, between about 22°N off the Cuban coast and about 66°N off the east coast of Greenland. In the eastern Atlantic the range extends from approximately 60°N to the continental slope of Africa at about 2°S, 9°W. A single record is known from the central South Atlantic at approximately 40°S, 28°W.

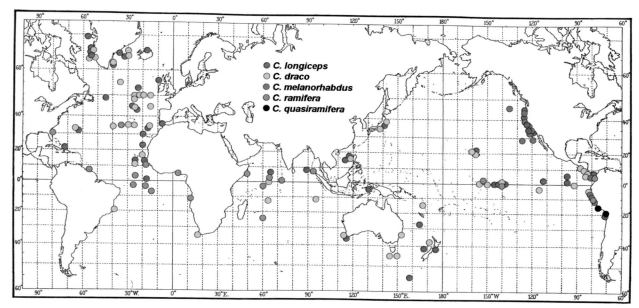

FIGURE 221. Distribution of five species of *Chaenophryne*: *C. longiceps* (red), Atlantic, Indian, and Pacific oceans; *C. draco* (yellow), Atlantic, Indian, and Pacific oceans; *C. melanorhabdus* (green), eastern Pacific Ocean; *C. ramifera* (blue), Atlantic, Indian, and Pacific oceans; *C. quasiramifera* (black), eastern South Pacific Ocean. A single symbol may indicate more than one capture.

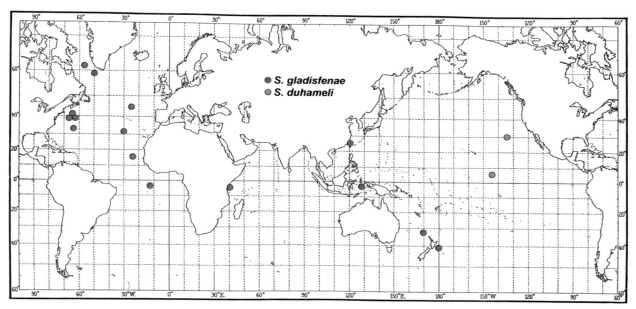

FIGURE 222. Distribution of two species of *Spiniphryne*: *S. gladisfenae* (red), Atlantic, Indian, and western Pacific oceans; *S. duhameli* (blue), eastern Pacific Ocean. A single symbol may indicate more than one capture.

In the Indo-Pacific, *L. acanthognathus* is represented by at least three specimens: one from the Bay of Bengal, Indian Ocean (at approximately 7°N, 59°E), and two from the South China and Celebes seas (Fig. 219). Three records are known from the eastern tropical Pacific: on the equator at 142°W and extending to the coast of Peru at about 8°S. The lectotype was collected from the western North Atlantic at 28°15'N, 56°00'W.

In the Atlantic Ocean, *L. indicus* appears to be restricted to the eastern side (Fig. 219). Seven specimens are known from off the continental slope of Africa from about 21°N, 21°W, east to the Gulf of Guinea and south to approximately 18°S, 10°W.

The remaining material is rather evenly distributed across the Indian and Pacific oceans between approximately 4°S and 30°N. The holotype was collected off the southwest coast of India.

The two species appear to have broadly overlapping vertical ranges. Based on maximum depths reached by fishing gear, metamorphosed specimens of both species were captured between approximately 650 m and an unknown lower limit. All specimens larger than 30 mm were captured by nets fished below 1000 m, and about half of these were captured by nets fished below 1500 m.

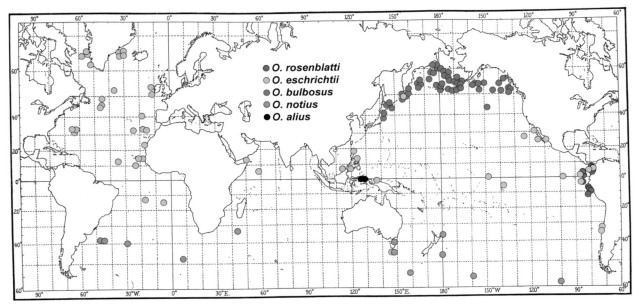

FIGURE 223. Distribution of five species of *Oneirodes*: *O. rosenblatti* (red), eastern tropical Pacific Ocean; *O. eschrichtii* (yellow), Atlantic, Indian, and Pacific oceans; *O. bulbosus* (green), North Pacific and Bering Sea; *O. notius* (blue), Southern Ocean; *O. alius* (black), western tropical Pacific Ocean. A single symbol may indicate more than one capture.

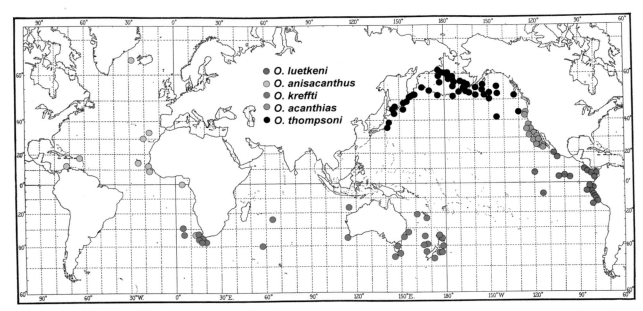

FIGURE 224. Distribution of five species of *Oneirodes*: *O. luetkeni* (red), eastern tropical Pacific Ocean; *O. anisacanthus* (yellow), Atlantic Ocean; *O. kreffti* (green), eastern South Atlantic, Indian, and western South Pacific oceans; *O. acanthias* (blue), eastern North Pacific Ocean; *O. thompsoni* (black), North Pacific Ocean and Bering Sea. A single symbol may indicate more than one capture.

CHAENOPHRYNE: The genus *Chaenophryne* contains five species: *Chaenophryne draco*, *C. longiceps*, and *C. ramifera*, known from 119, 84, and 23 female specimens, respectively, collected from all three major oceans of the world; *C. melanorhabdus*, 49 females, restricted to the eastern Pacific Ocean; and *C. quasiramifera*, two females, known only from the eastern tropical Pacific (Pietsch, 1975b, 2007; Fig. 221).

*Chaenophryne draco* is widely distributed on both sides of the Atlantic (Fig. 221): from Bermuda (the type locality), north to subarctic waters off West Greenland (at about 63°N), and extending south throughout temperate and tropical latitudes as far as Abrolhos Bank, Brazil, at 20°S, 39°W, to Cape Town, South Africa, at approximately 35°S, 18°E. It is also present throughout the central Indian Ocean and the temperate and tropical Pacific, from off Honshu, Japan; Taiwan; New South Wales, Australia; Tasmania; New Zealand; and east to Hawaii and through equatorial waters to the Gulf of Panama and south to the Peru-Chile Trench at about 20°S. Based on maximum depths reached by fishing gear, metamorphosed female specimens of *C. draco* are vertically distributed between approximately 350 m and some unknown lower limit. About 90% of the known material, however, was captured by gear

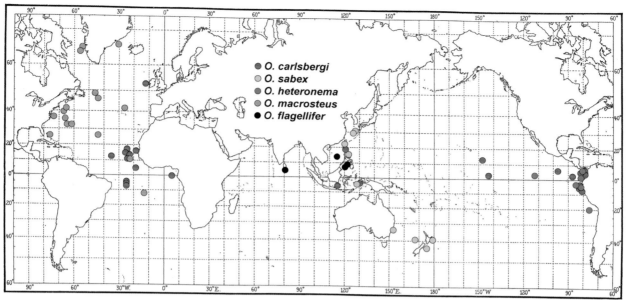

FIGURE 225. Distribution of five species of *Oneirodes*: *O. carlsbergi* (red), eastern Atlantic and Pacific oceans; *O. sabex* (yellow), western Pacific Ocean; *O. heteronema* (green), eastern tropical Pacific Ocean; *O. macrosteus* (blue), Atlantic Ocean; *O. flagellifer* (black), Indian and western Pacific oceans. A single symbol may indicate more than one capture.

reaching depths below 700 m. Overall, an analysis of available depth data indicates a concentration between 700 and 1500 m.

*Chaenophryne longiceps* has a similarly broad distribution, but in the Atlantic it has so far not been collected south of the equator (Fig. 221). Known records extend from the Caribbean and off Bermuda to West Greenland (as far north as 70°N), the northern coast of Iceland, and south throughout temperate and tropical waters to approximately 1°N. In the Indian Ocean it has been collected in the Arabian Sea (at approximately 7°N, 65°E) and the Bay of Bengal (9°N, 90°E), and east and south to off Cape Leeuwin, Western Australia. In the Pacific it extends from the South China Sea, south and east to New Caledonia, New Zealand, and off Macquarie Island (at approximately 55°S, 159°E); and east to Hawaii and throughout the eastern equatorial Pacific to the coast of the Americas from Oregon (46°N) to Chile (23°S). Based on maximum depths reached by fishing gear, metamorphosed female specimens of *C. longiceps* are vertically distributed between approximately 500 m and an unknown lower limit. About 90% of the known material was captured by gear fished below 850 m. The largest known specimens (100 to 245 mm) were all captured by nets fished below 950 m. Material is insufficient from any one geographic area for a more analytical treatment of distributional data.

In striking contrast to its two congeners described above, *C. melanorhabdus* appears to be restricted to the eastern Pacific Ocean (Fig. 221). Except for two tiny females taken from off Hawaii, all the known material falls within a relatively narrow distribution, extending along the western continental slope of North and Central America, from approximately 46°N, 135°W, in Pacific Subarctic Water, through the mixed transition zone of the California Current, into the eastern Pacific equatorial waters of the Gulf of Panama (type locality), and farther south to Peru at about 12°S. Based on maximum depths reached by fishing gear, metamorphosed female specimens of *C. melanorhabdus* are vertically distributed between approximately 200 m and some unknown lower limit. About 85% of the known material, including the largest known specimens (35 mm and larger) was collected by gear fished below 450 m.

An analysis of data for material collected in the transition zone of the California Current indicates a concentration between 300 and 1000 m (Pietsch, 1975).

*Chaenophryne ramifera*, although not as well represented in collections, has a wide geographic distribution in all three major oceans of the world, not unlike that of *C. draco* and *C. longiceps* (Pietsch, 1975b; Fig. 221). In the Atlantic, the known range lies between approximately 35°N and 12°S, including records off the coasts of Florida and Suriname in the west, and off Guinea-Bissau, the Gulf of Guinea, and Angola in the east. In the Indo-Pacific region, known records extend across the Indian Ocean between 9°N and 24°S, and farther east along the equator to the Gulf of Panama and off Peru at about 10°S. Based on maximum depth reached by fishing gear, metamorphosed female specimens of *C. ramifera* are distributed vertically between approximately 200 m and an unknown lower limit. Large specimens may be captured at relatively shallow depths: a 35-mm specimen was collected by gear fished above 200 m, a 36.5-mm specimen by gear fished above 550 m. Eighty-nine percent of the known material, however, was captured by gear fished below 550 m, and 46% by gear fished below 1000 m. Material is not sufficient from any one geographic area for a more analytical treatment of distributional data.

*Chaenophryne quasiramifera* is known from only two specimens, both collected in the eastern tropical Pacific Ocean (Fig. 221). The holotype was taken from the Peru-Chile Trench at approximately 20°S, 71°W, in gear fished open between the surface and 900 m, over a bottom depth of 5856 m. The paratype was captured over the Nasca Ridge, 16°S, 75°W, in gear fished open between the surface and 1130 m, over a bottom depth of 5307 m (Pietsch, 2007).

ONEIRODES: The genus *Oneirodes*, with 35 species, many of which are well represented in collections, provides a unique opportunity to analyze the geographic distribution of a ceratioid taxon in some detail. In general, it is found in the more productive waters of all three major oceans of the world, in a broad belt between approximately 66°N and 75°S (Figs. 223–225). It is present in the Gulfs of Aden, California, and

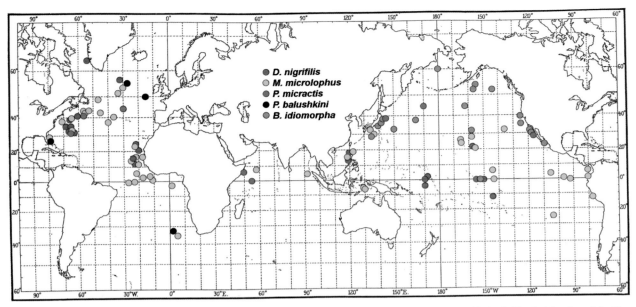

FIGURE 226. Distribution of species of *Danaphryne*, *Microlophichthys*, *Phyllorhinichthys*, and *Bertella*: *D. nigrifilis* (red), North Atlantic and Pacific oceans; *M. microlophus* (yellow), Atlantic, Indian, and Pacific oceans; *P. micractis* (green), western North Atlantic, western Indian, and Pacific oceans; *P. balushkini* (black), Atlantic Ocean; *B. idiomorpha* (blue), North Pacific Ocean. A single symbol may indicate more than one capture.

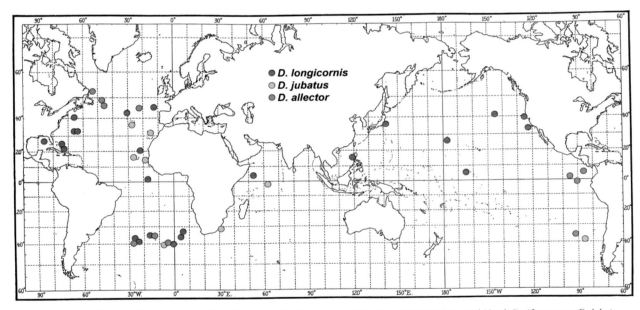

FIGURE 227. Distribution of three species of *Dolopichthys*: *D. longicornis* (red), Atlantic, western Indian, and North Pacific oceans; *D. jubatus* (yellow), Atlantic, western Indian, and eastern South Pacific oceans; *D. allector* (blue), Atlantic and eastern Pacific oceans. A single symbol may indicate more than one capture.

Mexico, and in the Caribbean Sea; but, like all other ceratioids, it is absent in the Mediterranean Sea (Bertelsen, 1951). Contrary to Bertelsen's (1951:223) prediction that "the majority of the ceratioid species will be found distributed in all the oceans," the species of *Oneirodes* are not cosmopolitan but are restricted for the most part to oceanic areas defined by distinct physical, chemical, and biological parameters (Pietsch, 1974a).

Twenty-two of the 35 recognized species of the genus are so poorly represented that the boundaries of their distributions are unknown; these are not dealt with further (Table 8). For the 15 remaining species, *Oneirodes acanthias*, *O. bulbosus*, *O. eschrichtii*, *O. thompsoni*, *O. luetkeni*, *O. krefffti*, *O. carlsbergi*, *O. sabex*, *O. rosenblatti*, *O. macrosteus*, *O. notius*, and *O. anisacanthus*, *O. heteronema*, *O. flagellifer*, and *O. alius*, horizontal ranges and, to a lesser extent, vertical ranges are known with some confidence. Ten of these 15 species, as well as the southern California population of *O. eschrichtii* (Pietsch, 1974a), appear to be limited to a single water mass; the remaining five cross one or more water mass boundaries. The vertical and horizontal distributions of these 15 better known species are discussed below.

Of those members of the genus that are limited to a single water mass, almost all occur in the North and eastern Pacific oceans where they inhabit four types of water: subarctic, transitional, quatorial, and subantarctic. The productively poor central water masses of the Pacific are avoided. Only the relatively rich eastern and western margins of the Atlantic Central Water Mass are inhabited by species of *Oneirodes* (Pietsch, 1974a).

*Oneirodes bulbosus* and *O. thompsoni*, two of the best-represented species of the genus, with 115 and 108 known females, respectively (Table 8), occur sympatrically across the Pacific Subarctic Water, including the Alaska Central Water (Alaskan Gyre) and Bering Sea (Western Gyre) (Figs. 223, 224). Neither species extends southward into the Transitional Water of the California Current. These species apparently occupy the same approximate depths, occurring between 500 and 1500 m, with peaks at 700 and 1000 m.

*Oneirodes acanthias*, with 153 known females, has the most restricted horizontal distribution of any ceratioid, found only in the mixed Transition Water of the eastern North Pacific off the coasts of California and Baja California (Fig. 224), where it appears to be concentrated between 500 and 1250 m. A morphologically distinct southern California population of *O. eschrichtii*, represented by six females, shares this same narrow distribution (see Pietsch, 1974a).

*Oneirodes luetkeni*, *O. rosenblatti*, and *O. heteronema* (except for a single record, see below) are confined to the more productive eastern part of the Pacific Equatorial Water Mass (Figs. 223–225). *Oneirodes luetkeni* (58 known females), formerly known only from the Gulf of Panama and adjacent waters of the eastern Pacific Ocean, is now known to extend along the west coast of the Americas to 20°21′N and to 15°20′S and as far west as about 110°W (Fig. 224). *Oneirodes rosenblatti* (21 known females) and *O. heteronema* (nine known females) are both somewhat more restricted, sharing ranges between approximately 7°22′N and 12°24′S (Figs. 223, 225). One individual of *O. heteronema* was collected just outside the boundary of the Pacific Equatorial Water Mass at approximately 20°S (Pietsch, 1974a). Although all three species are vertically sympatric between 750 and 2000 m, *O. luetkeni* ranges between 350 and 2000 m, with peaks at about 500 and 1250 m; *O. rosenblatti* appears to inhabit slightly deeper strata, between 700 and 3000 m, with a single peak at 1500 m; and *O. heteronema* occurs between 500 and 2000 m, with a single peak at 1250 m.

All known material of *Oneirodes krefftii* (38 females) was collected in southern latitudes of the eastern South Atlantic, Indian, and western Pacific oceans, between 0 and 180°E (Fig. 224). The northernmost record is off Port Hedland, Western Australia, at about 18°40′S; the southernmost record is on the Pacific side of the South Island, New Zealand, at about 44°52′S. Nearly all the specimens were taken in depths of less than 1000 m, many of them in bottom trawls.

*Oneirodes notius* (17 known females) is confined to subantarctic and Antarctic waters of both the Atlantic, Pacific, and Indian sectors of the Southern Ocean (Fig. 223). A 131-mm specimen of this species (ZIN 45350), collected in the Ross Sea, at 74°32′S, 178°19′E, represents the southernmost occurrence of a ceratioid. It appears to range vertically between 700 and 2000 m, with a single peak of concentration at roughly 1000 m.

Extending north into the region of mixing between North Atlantic Central and Arctic Intermediate Water masses, *O. macrosteus* (28 known females) occurs in the western part of the North Atlantic Central Water Mass, extending from the Bahamas, northward to the east coast of Greenland at approximately 66°N, 30°W (Fig. 225). Two specimens of this species, however, were collected more to the south, outside the boundary of the North Atlantic Central Water Mass, one south of the Cape Verde Islands at approximately 11°N, 21°W, and the other off Ascension Island at about 11°S, 11°W. Vertically, this species is found between approximately 1000 and 1500 m.

Crossing the boundary between the North and South Atlantic Central Water masses, *O. anisacanthus* (11 known females)—except for a single specimen collected in the far north between Greenland and Iceland at about 65°N, 29°W—inhabits the relatively productive eastern and western margins of the Atlantic, represented by two records in the Caribbean Sea, but more commonly in the east, extending from off Madeira, south into the Gulf of Guinea and off Cape Town, South Africa (Fig. 224).

Two other relatively well-represented species of *Oneirodes* commonly occur in and on both sides of the transition zone between the North and South Atlantic Central Water masses. One of these is *O. carlsbergi* (30 known females), which, in addition, extends across the tropical Pacific from the Java Sea and the Philippines to the Gulf of Panama (Fig. 225). Although, not yet known from the Indian Ocean, it most likely occurs there as well, thus representing the only known circumtropical species of the genus. Contradicting this hypothesis, however, is a single specimen of this species, the holotype of *Dolopichthys inimicus* (BMNH 1934.8.8.90), said to have been collected on the Irish Atlantic Slope. Fraser-Brunner (1935:324), who is responsible for this record, claims to have personally collected the specimen, along with some 62 additional species, during a two-week trip aboard a commercial trawler out of Swansea in July 1934. But being so far outside the known range of *O. carlsbergi*, it is difficult to believe the accuracy of his claim.

A second species of the genus commonly found in and on both sides of the transition zone between the North and South Atlantic Central Water masses is *O. eschrichtii* (115 known females). In fact, this species is by far the most broadly distributed member of the genus, ranging in the Atlantic from off Greenland to St. Helena (Fig. 223). Known also by several records in the western Indian Ocean, it is wide spread in the Pacific Ocean, ranging from Kamchatka, the Philippines, and Tasmania, to deep waters off the Americas, reaching south to central Chile. As mentioned above, a morphologically distinct population of *O. eschrichtii* occurs in the transition zone of the California Current (southern California population), and individual and ontogenetic variation in material worldwide, particularly in the morphology of the esca, is considerably greater in this species than in other members of the genus (see Pietsch, 1974a:51, figs. 66–70). It is thus quite possible that *O. eschrichtii* represents a complex of similar species that will be resolved only through detailed analysis of additional material from throughout its geographic distribution.

Vertically, *O. anisacanthus*, *O. carlsbergi*, and *O. eschrichtii* are all sympatric over a wide depth range between 600 and 2000 m. *Oneirodes carlsbergi* appears to have the broadest vertical distribution, between 100 and 2000 m, with its greatest concentration at roughly 300 m, which is surprisingly shallow for a ceratioid. The known material of *O. anisacanthus* is rather evenly distributed between 600 and 2000 m, whereas that of *O. eschrichtii* is spread out between 300 and 2600 m. Considering these differences in vertical distribution, the horizontal range of *O. carlsbergi* in the Atlantic might be better explained if reference is made to the 200-m isotherm for 14°C (Schroeder, 1963; Backus et al., 1965, 1970), which appears to be a better indicator of the meeting of the North and South Atlantic Central Water masses. This boundary runs from the coast of Africa at about 20°N west-southwestward to British Guiana. Although *O. anisacanthus* and *O. eschrichtii* commonly occur on both

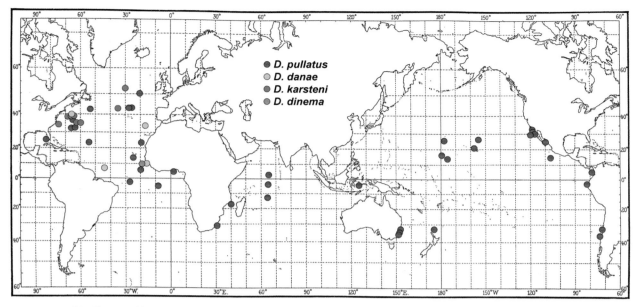

FIGURE 228. Distribution of four species of *Dolopichthys*: *D. pullatus* (red), Atlantic, Indian, and Pacific oceans; *D. danae* (yellow), North Atlantic Ocean; *D. karsteni* (green), North Atlantic Ocean; *D. dinema* (blue), eastern tropical Atlantic. A single symbol may indicate more than one capture.

sides of this boundary, *O. carlsbergi* does not and is thus limited (in the Atlantic) to the South Atlantic Central Water Mass. *Oneirodes anisacanthus* and *O. eschrichtii*, however, generally occupy deeper strata than *O. carlsbergi* and thus may cross the boundary of the North and South Atlantic Central Water masses at depths where physicochemical differences between these waters are appreciably smaller.

The horizontal distribution of three other species of *Oneirodes* should be mentioned. *Oneirodes sabex* (24 known species) is restricted to the western Pacific, where it ranges from the Okinawa Trough in the East China Sea, south to Taiwan, the Philippines, Banda Sea, off Sydney, Australia, and both sides of New Zealand (Fig. 225). Although represented by very few specimens, *O. flagellifer* (nine females) and *O. alius* (seven females), both appear to have restricted distributions, the former species known only from off Sri Lanka and the Celebes and South China seas, the latter species only from the Halmahara Sea (Figs. 223, 225).

To summarize what is known about the geographic distribution of *Oneirodes*, most of the better known species appear to follow the contours of organic productivity to a greater extent than most other ceratioids (Ebeling, 1962; Pietsch, 1974a), only rarely occurring in waters of low productivity. At the present time, the water mass concept—suggesting that the ranges of pelagic animals conform to water masses as defined by their temperature-salinity relationships—is sufficient to explain the distributions of many species of the genus. Most members appear to be limited to discrete water masses, in some cases, however, departing from this pattern at great depths where water-mass boundaries become weak. The distributional patterns exhibited by species of *Oneirodes* are therefore similar to those of many other groups of midwater organisms, thus conforming to zoogeographic regions proposed by Ebeling (1962), Baird (1971), and others. But, when compared with most other ceratioids, even closely related oneirodid genera such as *Chaenophryne*, *Microlophichthys*, and *Dolopichthys*, the horizontal ranges of many species of *Oneirodes* are considerably more restricted, yet their vertical ranges are approximately the same. Unlike most other ceratioids, some combination of physical, chemical, and resultant biological conditions restricts the ranges of many species of *Oneirodes*, but what those conditions might be is beyond speculation (Barnett, 1984; Sassa et al., 2002).

MICROLOPHICHTHYS: The genus *Microlophichthys* contains two species: *Microlophichthys microlophus*, based on females and now represented by some 94 females, five males, and about 110 larvae; and *M. andracanthus*, based on only two male specimens, one from the Caribbean Sea and the other from the tropical eastern Pacific. While nothing meaningful can be said about the latter relative to geographic distribution, *M. microlophus* is very broadly distributed in all three major oceans of the world (Fig. 226). It is well represented on both sides of the Atlantic between approximately 53°N and 36°S. It is also present on both sides of the North Indian Ocean and across the Pacific from the Philippines to the Gulf of Panama, between about 33°N and 26°S. But for some unknown reason, it appears to be absent in Australian and New Zealand waters. Unlike most ceratioids, many of the localities occur in the relatively nutrient-poor, central water masses. It appears to be a rather deep-dwelling species, most records for metamorphosed females indicating a range between 800 and 2200 m. Its broad geographic distribution may thus be explained by an ability to cross water-mass boundaries at greater depths where differences in salinity and temperature are less.

DOLOPICHTHYS: Like *M. microlophus*, but unlike the many species of *Oneirodes* that have surprisingly restricted distributions, the better known members of the genus *Dolopichthys* are found around the world. The two best-represented species, *Dolopichthys pullatus* (45 known females) and *D. longicornis* (26 known females), have been collected throughout all three oceans between about 50°N and 40°S in the Atlantic, 40°N and 38°S in the Pacific, and 3°N and 32°S in the Indian Ocean (Figs. 227, 228). *Dolopichthys allector* (15 females) has a similar distribution in the Atlantic and Pacific but has so far not been

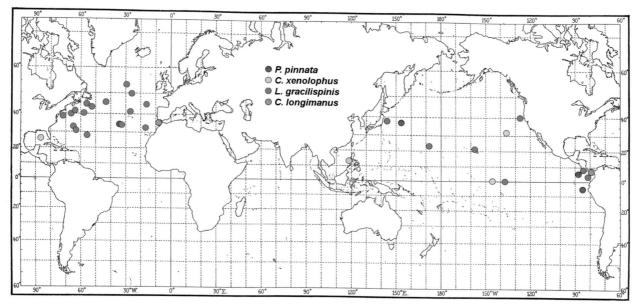

FIGURE 229. Distribution of species of "long-pectoraled" oneirodid genera: *Puck pinnata* (red), North Atlantic and Pacific oceans; *Chirophryne xenolophus* (yellow), Pacific Ocean; *Leptacanthichthys gracilispinis* (green), North Atlantic and Pacific oceans; *Ctenochirichthys longimanus* (blue), eastern North Atlantic and eastern tropical Pacific oceans. A single symbol may indicate more than one capture.

taken in the Indian Ocean (Fig. 227). Most of the material of *Dolopichthys jubatus* (eight females) is from the eastern Atlantic, but there are two records from the western Indian Ocean and a single record in the eastern South Pacific at about 39°S, 87°W (Fig. 227). The remaining species of the genus, represented by eight females or fewer, have all so far been taken only in the North Atlantic. Many individuals of the genus were taken from central water masses. As might be expected from these wide geographic distributions, the species of *Dolopichthys* are relatively deep-dwelling ceratioids. All the material of *D. pullatus* for which depth records are available was collected by nets that reached depths of at least 800 m, 68% was taken by nets fished below 1000 m, and 32% by nets fished below 1500 m. Concentrated at even greater depths, all the individuals of *D. longicornis* for which depth records are available were taken by nets that reached depths of at least 1375 m, and 77% was taken by nets fished at 2000 m or below.

BERTELLA: The monotypic genus *Bertella*, containing only *Bertella idiomorpha*, is unusual among ceratioids in being restricted geographically to the North Pacific Ocean (Fig. 226). Now represented in collections by some 34 females, it extends from off Japan to the California coast and from the Bering Sea to the equator at about 154°W. The available data indicate that it is a relatively shallow-living species. Of the material for which depth data are known, 66% was taken by trawls that reached a maximum of only 923 m, and 55% was taken by trawls that reached a maximum of only 679 m.

## Caulophrynidae

Caulophrynids are well represented in all three major oceans of the world between approximately 65°N and 50°S (Fig. 230). The best known species, *Caulophryne jordani*, with 38 females, is broadly distributed in the North Atlantic ranging from off the west coast of Greenland, off Iceland, and south to about 5°N; it has so far not been collected in the South Atlantic. There are two records in the western equatorial Indian Ocean, one off the coast of Somalia and another northwest of the Chagos Archipelago. In the Pacific Ocean, it ranges from off New South Wales, Australia, east and north to off southern California and Peru; it has so far not been collected in the western North Pacific. The species appears to have an unusually broad vertical distribution, with specimens taken as shallow as 275 m and others by gear fished at maximum depths of 3000 m; however, about 70% of the known material was taken by gear fished at maximum depths of 800 to 1625 m. Two specimens, both from the North Atlantic, were collected with a closing trawl between 1235 and 1260 m, and 1250 and 1510 m, respectively.

*Caulophryne pelagica*, with 17 known females, is well represented in the Pacific Ocean, ranging from Japan, Taiwan, the Halmahera Sea, and the Bounty Plateau southeast of New Zealand, to the Gulf of Alaska, off Baja California, and into the eastern tropical Pacific off Panama and Ecuador (Fig. 230). There are three records in the Indian Ocean, one off Somalia and two on the equator west of the Maldives, but only a single known specimen from the Atlantic, collected off the Cape Verde Islands. Several specimens were captured with gear fished at maximum depths of 500 m, and one with gear reaching a depth of 2500 m, but 70% of the known material was taken by gear fished at maximum depths of 900 to 1750 m. Two specimens, one tiny female (13 mm) from the Halmahera Sea, and the one specimen from the Atlantic (183 mm), were collected with closing trawls between 1250 and 1500 m, and 1500 m, respectively.

The 14 known specimens of *C. polynema* come from widely scattered localities in the Atlantic and eastern Pacific Ocean; it has so far not been taken in the Indian Ocean (Fig. 230). In the Atlantic it ranges from off Florida east to Madeira and south to Cape Verde, Zaire, and Namibia as far south as approximately 28°S. In the eastern Pacific, it has been collected off Oahu and off southern California and Baja. Four specimens were captured with gear fished at maximum depths of 585 to 834 m, and two with gear reaching a depth of 1920 to 2000 m, but all the remaining material was taken by gear fished at maximum depths of 1000 to 1250 m. Two specimens, both from the eastern North Atlantic (11.5 to 100 mm), were collected with a closing trawl between 900 and 1010 m, and 1000 and 1250 m, respectively.

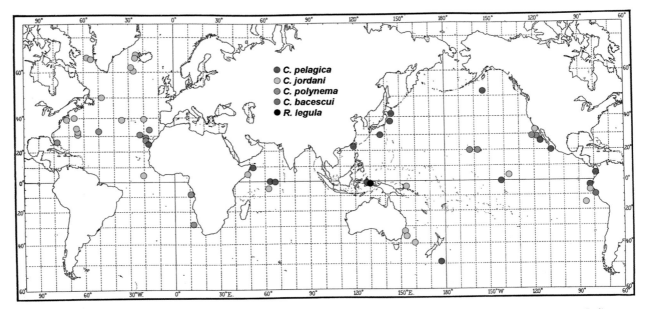

FIGURE 230. Distribution of five species of the family Caulophrynidae: *Caulophryne pelagica* (red), eastern North Atlantic, western Indian, and Pacific oceans; *C. jordani* (yellow), North Atlantic, western Indian, and Pacific oceans; *C. polynema* (blue), Atlantic and eastern North Pacific oceans; *C. bacescui* (green), eastern tropical Pacific Ocean; *Robia legula* (black), western tropical Pacific Ocean. A single symbol may indicate more than one capture.

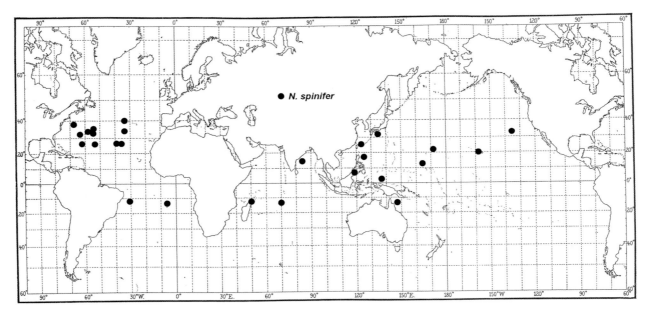

FIGURE 231. Distribution of *Neoceratias spinifer*: Atlantic, Indian, and Pacific oceans. A single symbol may indicate more than one capture.

## Neoceratiidae

*Neoceratias spinifer*, the only species of the Neoceratiidae, is known from all three major oceans of the world between about 40°N and 15°S (Fig. 231). In the Atlantic it ranges from off New England in the west and off St. Helena in the east. There are three known records in the Indian Ocean: from off the northern tip of Madagascar, south of Diego Garcia, and from the western Bay of Bengal. It is well represented in the western Pacific but known from only two specimens on the eastern side of this ocean: one from the Hawaiian Islands and another farther east at approximately 32°N, 136°W. It appears to be a relatively deep-living species: none of the material for which data are recorded was collected in gear fished at depths of less than 1000 m.

## Gigantactinidae

The family Gigantactinidae is widely distributed throughout all the major oceans of the world in a broad belt limited by the Arctic and Antarctic Polar Fronts, with northern- and southernmost records at approximately 63°N and 53°S, respectively (Figs. 232–235). The larvae and males, however, have been caught only in tropical and subtropical waters (as has been shown for all ceratioids; see Bertelsen, 1951), with northern- and southernmost records at about 48°N and 34°S, respectively (see Fig. 207). Gigantactinids are not found in the gulfs of California or Mexico, and, like all other ceratioids, they are unknown in the Mediterranean Sea (Bertelsen, 1951).

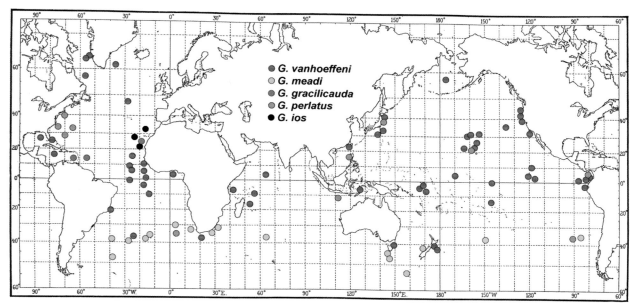

FIGURE 232. Distribution of five species of *Gigantactis*: *G. vanhoeffeni* (red), Atlantic, western Indian, and Pacific oceans; *G. meadi* (yellow), Southern Ocean; *G. gracilicauda* (green), North Atlantic Ocean; *G. perlatus* (blue), Atlantic, eastern Indian, and Pacific oceans; *G. ios* (black), eastern North Atlantic Ocean. A single symbol may indicate more than one capture.

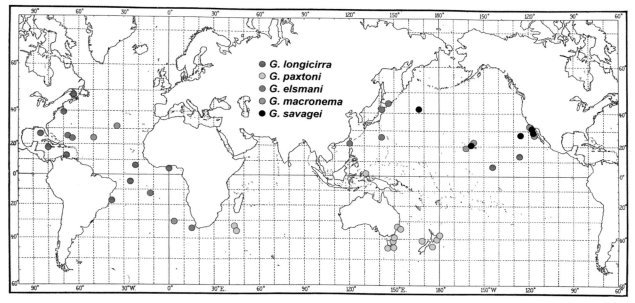

FIGURE 233. Distribution of five species of *Gigantactis*: *G. longicirra* (red), Atlantic and eastern North Pacific oceans; *G. paxtoni* (yellow), western South Indian and western South Pacific oceans; *G. elsmani* (green), Atlantic and western North Pacific oceans; *G. macronema* (blue), Atlantic and eastern North Pacific oceans; *G. savagei* (black), North Pacific Ocean. A single symbol may indicate more than one capture.

In common with most other ceratioid genera, gigantactinids are not particularly restricted to more organically productive waters and have typically broad geographic ranges, occurring throughout the central North Atlantic and central Pacific. This situation contrasts with that found for other ceratioid genera (e.g., *Oneirodes*, Pietsch, 1974a), whose species tend to avoid less productive regions and are, for the most part, restricted geographically into oceanic areas defined by distinct physiochemical and biological parameters (see above). With the exceptions of *Gigantactis meadi*, *G. gracilicauda*, and possibly *G. golovani*, no species known from more than three individuals is restricted to a single water mass.

Metamorphosing and metamorphosed members of the Gigantactinidae may be taken anywhere between 500 m and some unknown lower limit exceeding 3000 m. They are more commonly found, however, between roughly 1000 and 2500 m. Although some vertical separation of sympatric species may exist, such forms probably overlap in vertical range wherever they are found.

Four of the 21 recognized species of the family (*G. herwigi*, *G. watermani*, *G. longicauda*, and *Rhynchactis microthrix*) are so poorly represented in collections (all with fewer than three specimens) that the boundaries of their distributions are unknown; these are not dealt with further. Six species have

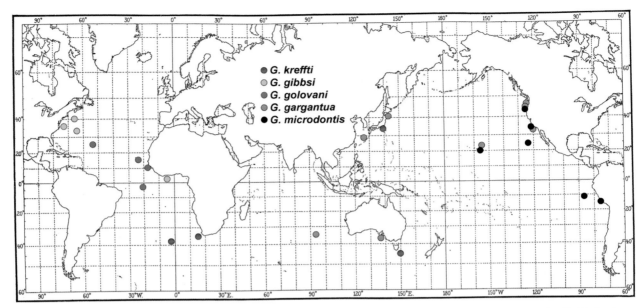

FIGURE 234. Distribution of five species of *Gigantactis*: *G. kreffti* (red), eastern South Atlantic and western Pacific oceans; *G. gibbsi* (yellow), North Atlantic Ocean; *G. golovani* (green), Atlantic Ocean; *G. gargantua* (blue), eastern South Indian and North Pacific oceans; *G. microdontis* (black), eastern Pacific Ocean. A single symbol may indicate more than one capture.

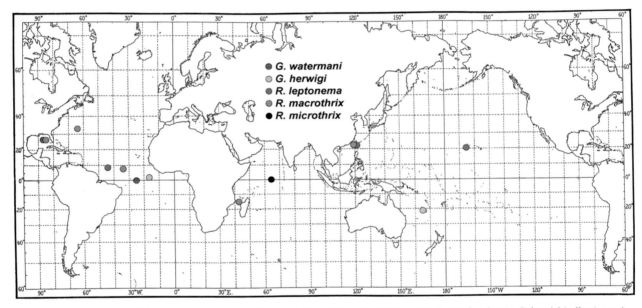

FIGURE 235. Distribution of five species of the family Gigantactinidae: *Gigantactis watermani* (red), North Atlantic; *G. herwigi* (yellow), tropical Atlantic and western South Pacific oceans; *Rhynchactis leptonema* (green), tropical Atlantic and North Pacific oceans; *R. macrothrix* (blue), North Atlantic, eastern South Indian, and western North Pacific oceans; *R. microthrix* (black), eastern tropical Indian Ocean. A single symbol may indicate more than one capture.

disjunct distributions: *G. macronema* (11 known females) and *G. longicirra* (10), from the Atlantic and eastern North Pacific (Fig. 233); *G. elsmani* (10), from the Atlantic and western North Pacific (Fig. 233); *R. macrothrix* (10), from the western North Atlantic, off Madagascar in the western Indian Ocean, and Taiwan (Fig. 235); *G. kreffti* (five), from the South Atlantic off Cape Town, South Africa, off Japan, and below Tasmania at about 44°S, 150°E (Fig. 234); and *R. leptonema* (five), the tropical western Atlantic, off Taiwan, and the Hawaiian Islands (Fig. 235). When these six species are better known, they will no doubt prove to be wide-ranging, if not cosmopolitan, forms.

Horizontal ranges and, to a lesser extent, vertical ranges for the remaining 11 species (*G. vanhoeffeni*, *G. meadi*, *G. paxtoni*, *G. gargantua*, *G. microdontis*, *G. perlatus*, *G. savagei*, *G. gibbsi*, *G. golovani*, *G. ios*, *G. gracilicauda*) are known with some confidence. Two of these have cosmopolitan distributions, four are known only from the Atlantic, two are Indo-Pacific, two are eastern Pacific, and one is endemic to the Southern Ocean.

COSMOPOLITAN SPECIES: *Gigantactis vanhoeffeni* (90 known females) and *G. perlatus* (11) are wide-ranging forms collected from all three major oceans of the world. *Gigantactis vanhoeffeni* is known from both sides of the Atlantic (including the Gulf of Mexico, Caribbean Sea, and Gulf of Guinea), extending from both sides of Greenland at approximately 64°N to the central South Atlantic at about 36°S, 23°W (Fig. 232). It is also well represented in the western Indian Ocean, and throughout the Pacific, where it ranges from the Banda Sea northward to Taiwan, Japan, and the Bering Sea, and southward to Tasmania and New Zealand, and east to the Hawaiian Islands, the coasts of Oregon and California, to the Gulf of Panama. *Gigantactis perlatus* has been collected off the east coast of North America, between about 29 and 40°N, and in the eastern South Atlantic at about 35°S, 4°E (Fig. 232). It has also been taken at several localities in the eastern Indian Ocean and across the Pacific, from the Philippines, Taiwan, Japan, the Hawaiian Islands, and Gulf of Panama, and in the eastern South Pacific at about 38°S, 91°W. Vertically, larger metamorphosed specimens of the two species coexist between approximately 800 m and some unknown lower limit (a 232-mm specimen of *G. vanhoeffeni* was taken in an otter trawl fished on the bottom at 3184 m). On the other hand, relatively small individuals have been captured with closing nets at considerably shallower depths: 32- and 24-mm females of *G. vanhoeffeni* between 550 and 815 m and 656 and 1000 m, respectively; and a 32-mm female of *G. perlatus* between 670 and 805 m.

ATLANTIC SPECIES: *Gigantactis gibbsi* (four females), *G. golovani* (four), *G. ios* (four), and *G. gracilicauda* (three) are each known only from the Atlantic Ocean. *Gigantactis gibbsi*, *G. golovani*, and *G. gracilicauda* are known from both sides of that ocean, while *G. ios* is perhaps restricted to the eastern side (Figs. 232, 234). Small sample sizes make it difficult to say with any certainty, but *G. gracilicauda*, having been taken in gear fished open at maximum depths of between 2000 and 2500 m, is perhaps a deeper-living species than most other gigantactinids. On the other hand, *G. golovani* may be a relatively shallow-living species: the 179-mm holotype was taken between the surface and 660 m, the 153-mm paratype between the surface and 1550 m.

INDO-PACIFIC SPECIES: *Gigantactis paxtoni* (18 females) is represented by more than a dozen individuals collected off New Zealand, the southeast coast of Australia, and Tasmania, the southernmost records nearing the northern boundary of the Subantarctic Water Mass (Fig. 233). It has also been collected in the Halmahara Sea, off the northwest coast of New Guinea; and in the western South Indian Ocean at approximately 35°S, 45°E. With the exception of a single specimen captured in a bottom trawl at 1210 to 1260 m, the material was collected by pelagic trawls fished open at maximum depths of 540 to 1500 m. These relatively shallow depths contrast with those of *G. meadi* (maximum depths between 1850 and 2000 m; see below), whose horizontal distribution appears to overlap slightly that of *G. paxtoni* (Fig. 232).

Most of the known material of *G. gargantua* (11) was collected from both sides of the North Pacific, extending from Japan to the Hawaiian Islands and the coasts of Oregon and California (Fig. 234). Two specimens have been taken in far southern waters, one off South Australia and another from the eastern South Indian Ocean at about 35°S, 93°E. All the known material was captured in open gear fished at maximum depths of 500 to 1300 m.

EASTERN PACIFIC SPECIES: *Gigantactis savagei* and *G. microdontis* are both more or less restricted to the eastern Pacific Ocean. *Gigantactis savagei* (six females) is known only from off southern California, the Hawaiian Islands, and a single record from the central North Pacific at about 42°N, 167°E (Fig. 233). *Gigantactis microdontis* (12) is known from the Hawaiian Islands and from off the coasts of Oregon and southern California, extending south into the eastern South Pacific off the coast of Peru and northern Chile (Fig. 234). Both species live sympatrically at relatively shallow depths (500 to 650 m) in the mixed Transition Water off southern California. In other regions, *G. savagei* appears to occupy slightly deeper levels, and *G. microdontis* maintains a rather shallow distribution; all material of the latter species was captured above 1200 m.

SOUTHERN OCEAN ENDEMIC: *Gigantactis meadi* (20 females) has a circumglobal distribution in Subantarctic Water, where it appears to have a relatively narrow, but relatively deep vertical range (Fig. 232). All specimens 87 mm and larger were taken in trawls fished open between 1850 and 2000 m.

At all stages in their life cycle, the more globose ceratioids, such as species of *Oneirodes*, are incapable of any prolonged horizontal locomotion and are to a considerable extent passively transported along by water movements. The movement of water masses and current gyres, which help to form and maintain these water masses, are no doubt important in the integration and concentration of distributions of these animals (Pietsch, 1974a). Gigantactinid females, on the other hand, at least among ceratioids, are probably the most hydrodynamically efficient and most active in terms of prolonged horizontal movement (see Locomotion, Food, and Feeding, Chapter Seven). Their considerably greater locomotory capabilities, as well as their tendency to inhabit deeper strata where physico-chemical differences between water masses are appreciably less, probably account for their broad horizontal as well as vertical ranges (Bertelsen et al., 1981).

## Linophrynidae

Four of the five genera of the family Linophrynidae, *Photocorynus*, *Haplophryne*, *Borophryne*, and *Linophryne*, are now represented by relatively large numbers of specimens, to the extent that a reasonable picture of their horizontal and vertical distributions can be drawn (Figs. 236–239). The fifth genus, *Acentrophryne*, is known from only five females, all collected in the eastern tropical Pacific, off the coast of Costa Rica, and in Gulf of Panama and Peru-Chile Trench (Fig. 236). The much better known genus *Borophryne*, with 39 females, 62 free-living males, and 14 larvae, also appears to be an eastern tropical Pacific endemic. It is most commonly found in the Gulf of Panama and nearby adjacent waters but ranges from approximately 28°N in the Gulf of California to about 6°S off the Pacific coast of Colombia (Fig. 236). All known metamorphosed material was collected with gear fished open between the surface and about 1750 m.

In contrast, the linophrynid genera *Photocorynus* (31 known females) and *Haplophryne* (88 females), each containing a single species, are found in all three major oceans of the world, spanning the breadth of both the Atlantic and Pacific oceans, yet each has only a single record in the Indian Ocean (Fig. 236). While *Photocorynus* is rather narrowly restricted between about 35°N and 13°S around the world, *Haplophryne* has a broader latitudinal range, found between 54°N and 30°S in the Atlantic, and 27°N and 45°S in the Pacific. *Photocorynus* appears to be relatively deep living: four metamorphosed females were captured in closing nets at depths of 990 and 1420 m. Of those caught in open nets, none were taken at maximum fishing depths of less than 1000 m. All known metamorphosed specimens of *Haplophryne* were caught in nonclosing pelagic trawls.

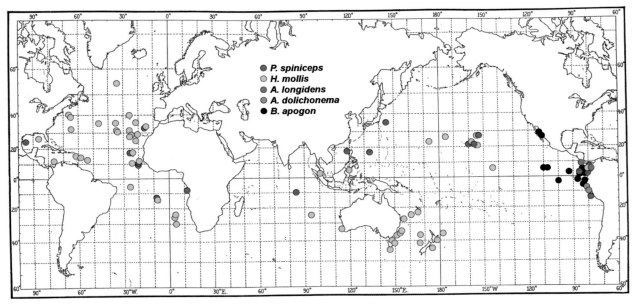

FIGURE 236. Distribution of species of *Photocorynus*, *Haplophryne*, *Acentrophryne*, and *Borophryne*: *P. spiniceps* (red), Atlantic, Indian, and Pacific oceans; *H. mollis* (yellow), Atlantic, Indian, and Pacific oceans; *A. longidens* (green), eastern tropical Pacific Ocean; *A. dolichonema* (blue), eastern tropical Pacific Ocean; *B. apogon* (black), eastern Pacific Ocean. A single symbol may indicate more than one capture.

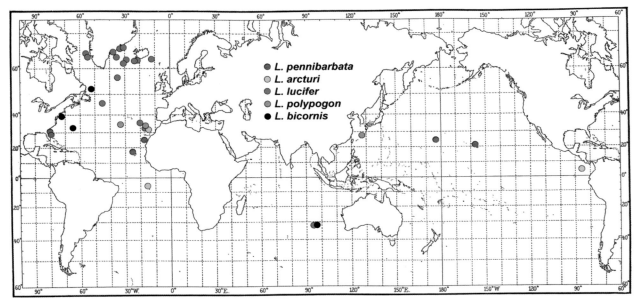

FIGURE 237. Distribution of five species of *Linophryne*: *L. pennibarbata* (red), North Atlantic and North Pacific oceans; *L. arcturi* (yellow), eastern Atlantic and eastern Pacific oceans; *L. lucifer* (green), North Atlantic Ocean; *L. polypogon* (blue), North Atlantic and western North Pacific oceans; *L. bicornis* (black), western North Atlantic and eastern South Indian oceans. A single symbol may indicate more than one capture.

Of the 88 known females, three specimens in metamorphosis were caught at maximum depths of 300 m, 10 others including six with parasitic males in maximum depths ranging from 550 to 900 m, and 12 others in nets with greater maximum fishing depth. With the exception of one specimen captured in less than 200 m, the 12 known free-living males came from hauls with maximum depths of more than 1500 m.

*Linophryne* is the last genus to be discussed. With 22 recognized species, it is by far the largest of the family and the second largest of the suborder (only *Oneirodes* is larger, with 35 species), but overall it is poorly represented in collections. The best-represented species is *Linophryne indica* (for which males and larvae have been identified, in contrast to all other species of the genus), with 60 specimens (Table 8). *Linophryne arborifera* is known from 31 metamorphosed females, and *L. lucifer*, *L. densiramus*, and *L. coronata*, 29, 27, and 22 females, respectively, but all the others are represented by fewer than a dozen specimens. For this reason, only a few preliminary estimates of their geographic distribution can be obtained.

Like those of all other ceratioids, the larvae, free-living males, and most metamorphosed females of *Linophryne* were caught beneath the warmer parts of the oceans between

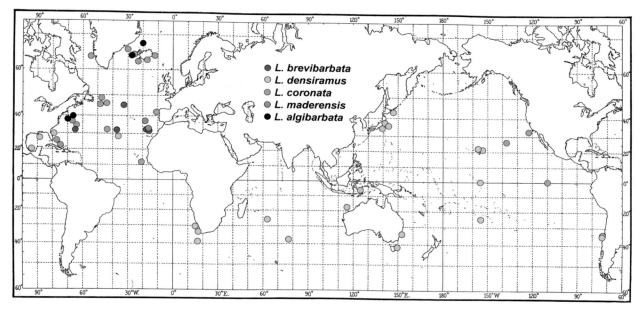

FIGURE 238. Distribution of five species of *Linophryne*: *L. brevibarbata* (red), North Atlantic Ocean; *L. densiramus* (yellow), Atlantic, Indian, and Pacific oceans; *L. coronata* (green), North Atlantic and eastern Pacific oceans; *L. maderensis* (blue), eastern North Atlantic Ocean; *L. algibarbata* (black), western North Atlantic Ocean. A single symbol may indicate more than one capture.

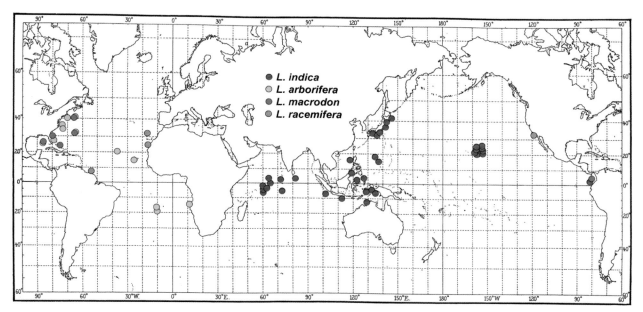

FIGURE 239. Distribution of four species of *Linophryne*: *L. indica* (red), Indian and Pacific oceans; *L. arborifera* (yellow), Atlantic Ocean; *L. macrodon* (green), western North Atlantic and eastern tropical Pacific oceans; *L. racemifera* (blue), North Atlantic and eastern North Pacific oceans. A single symbol may indicate more than one capture.

about 40°N and 40°S. A number of large, possibly expatriate females of *L. lucifer*, *L. coronata*, *L. algibarbata* and *L. bicornis* (some with parasitic males), however, have been found in the North Atlantic at latitudes higher than 66°N (see below). The records of metamorphosed females are scattered over the three major oceans of the world. That nearly 90% of the localities are less than 200 nautical miles from the nearest continental shelf or island is due primarily to more intensive collecting in these regions, especially in the Gulf of Mexico, off Florida, on the Newfoundland and Icelandic fishing banks, and off the island groups of Bermuda, Madeira, and Hawaii. However, this dominance is also of an order that might indicate some biological dependence on this environment (Bertelsen, 1982).

Eight of the 22 recognized species of the genus *Linophryne* (*L. andersoni*, *L. argyresca*, *L. bipennata*, *L. escaramosa*, *L. quinqueramosa*, *L. parini*, *L. sexfilis*, and *L. trewavasae*) are represented by a single specimen. Spread out evenly over the world's oceans—one from the Atlantic, one off South Africa, two from the eastern Indian Ocean, two from the western Pacific, and two from the eastern Pacific—nothing more can be said about their distribution. Eight species have disjunct distributions: *L. lucifer* (with 29 known females) and *L. bicornis* (five females) are primarily distributed in the North Atlantic, but each is represented by a single specimen collected in the eastern South Indian Ocean (Fig. 237); *L. polypogon* (three females) is known from the eastern North Atlantic, with a single record from the East China Sea (Fig. 237); *L. coronata* (22 females), *L. pennibarbata* (11), *L. macrodon* (10), *L. racemifera* (five), and *L. arcturi* (three) are all primarily distributed in the North Atlantic, but each is also represented in the eastern Pacific (Figs. 237–239). These eight species will no doubt eventually prove to be wide-ranging, if not cosmopolitan, species.

*Linophryne densiramus* (27 females) is the only wide-ranging species of the genus, with records in the North Atlantic, off South Africa, in the central South Indian Ocean, Indonesian waters, and across the Pacific from Tasmania, Australia, and Japan to the Hawaiian Islands and off southern Chile at about 36°S (Fig. 238). Four species appear to be confined to the Atlantic Ocean: *L. arborifera* (31), present on both sides of this ocean extending from about 40°N off New England to 18°S off Angola (Fig. 239); *L. algibarbata* (nine), so far collected only off New England and in high latitudes off Iceland and Greenland (Fig. 238); *L. brevibarbata* (six), across the North Atlantic between about 30° and 46°N (Fig. 238); and *L. maderensis* (four), known only from the eastern North Atlantic off Madeira (Fig. 238). Finally, the best-represented species of the genus, *L. indica* (12 females, 31 males, and 17 larvae), is an Indo-Pacific endemic, found in the western North Indian Ocean, in waters throughout Indonesia, Japan, the Hawaiian Islands, and the Gulf Panama (Fig. 239).

Bertelsen (1951) demonstrated that the larvae of *Linophryne* are most numerous in the upper 100 to 200 m, in contrast to the majority of other ceratioid larvae, which on the average are most abundant in the uppermost 100 m. But data describing the vertical distribution of metamorphosed specimens of this genus are limited because of small sample sizes for most species, and also by the common use of open pelagic nets, which collect continuously during both deployment and retrieval. Assuming that the bulk of the material was caught at the greatest possible depth reached by fishing gear, about 55% of the specimens were caught in nets fished between 1000 and 3000 m, about 45% in nets with a maximum fishing depth of 1000 m, and about 15% in less than 500 m. A few metamorphosed females have been caught at depths not exceeding 200 m. Approximately 25 metamorphosed females have been caught in bottom trawls. Fishing with large shrimp trawls in the Gulf of Mexico and Caribbean area has procured a small number of solitary females *(L. densiramus* and *L. racemifera)* of 40 to 82 mm in hauls made at 150 to 755 m. Considering the great amount of sampling carried out in these surveys and the semipelagic nature of the gear, these specimens might represent strays from the pelagic fauna of adjacent greater depths (Bertelsen, 1982). Several dozen females, including some with parasitic males, have been retrieved from the intensive fishery on the continental shelf off Newfoundland, Greenland, and Iceland. These specimens, representing four species *(L. lucifer, L. coronata, L. algibarbata,* and *L. bicornis)*, were caught at depths of about 310 to 660 m (the latter depth being near the maximum for this fishery) and are all among the largest specimens of *Linophryne* recorded (135 to 230 mm). Two specimens, a 190-mm *L. lucifer* and a 180-mm *L. bicornis*, were caught in bottom trawls by Russian research vessels at depths of 1220 to 1275 m on the western Australian Ridge in the eastern South Indian Ocean.

As mentioned above, the extremely large specimens of *Linophryne*, as well as several other ceratioid genera caught in the North Atlantic far north of their presumed spawning area, may be expatriates, and their occurrence on the bottom in relatively shallow water—and thus inside the reach of commercial fisheries—could be related to this possibility:

> Like most of the largest ceratioids, viz. specimens of *Ceratias holboelli* and *Himantolophus groenlandicus*, all these large specimens of *Linophryne* have drifted far north of their breeding area towards the continental shelves of the northern North Atlantic. Due to the lower temperatures in the upper water layers in this region, they are able to enter relatively shallow depths inside reach of the intensive commercial fishery off Newfoundland, Iceland, and East Greenland. The fact that no [sexually mature] adults of most of the numerous recognized species of ceratioids have been caught might indicate that many of them reach similar sizes in which they are less available for capture, probably due to a greater ability to avoid the relatively small fishing gears normally used in deep-sea research, less abundance than the younger stages, and possibly a tendency to live in greater and less-fished depths (Bertelsen, 1976:17; see also Bertelsen, 1982).

However, the records of similar-sized specimens of *Linophryne* (representing *L. lucifer* and *L. bicornis*) caught in bottom trawls in deep subtropical waters, and the fact that no specimens larger than about 100 mm have been caught in pelagic nets, make it seem probable that the larger individuals resort to benthic or benthopelagic environments, areas more abundant in suitable food than the poor meso- and bathypelagic zones (Bertelsen, 1982).

# SIX

## Bioluminescence and Luring

> From the tip of the snout there arose a stout stem supporting a branched, waxen-white affair, sprouting from a sheet of gold and with tiny luminous balls at the extremity of each branch. But the least believable thing was the mass of roots growing from the center of the lower jaw. A basal stem divided at once into several, and then, little by little, into a tangled mass of twigs and tentacles and tendrils—the mixed metaphor being quite intentional! This structure was longer than the entire body, with many scores of very small pink organs, several dozen diminutive lights, and the ability to contract and expand like a snarl of deadly medusa locks.
>
> WILLIAM BEEBE,
> "The Depths of the Sea," 1932a:85

The external morphology of the light organs of ceratioids—those aspects that provide the essential characters that delineate the numerous species of the suborder—have been described in Chapter Two, What Makes an Anglerfish? The internal anatomy, functional morphology, and luminescence of these structures are dealt with here, along with a discussion of the biological significance of bioluminescence in these fishes (Fig. 240). Specialized structures adapted to produce bioluminescence in ceratioids are restricted to females, the most conspicuous of these being the bait, or esca, located at the tip of the illicium. Other structures of females known or thought to luminesce include the esca-like bulb on the tip of the second dorsal-fin spine of diceratiids and ceratiids, the caruncles of ceratiids, and the jaw teeth, fin rays, dermal spines, and skin patches of various taxa. Beebe (1932b:101, fig. 29) thought that the somewhat swollen tips of the six innermost caudal-fin rays of a larval male of *Melanocetus murrayi* were "slightly but distinctly luminous in the fresh fish." Young and Roper (1977) described bioluminescent countershading in an attached pair of *Cryptopsaras couesii*, in which the skin of both female and male luminesced (see Luminescent Countershading, below). But apart from these two observations, neither light organs nor luminescence have ever been seen in ceratioid males, or larvae of either sex.

That the esca of deep-sea anglerfishes might be bioluminescent was first mentioned by Rudolph von Willemoes-Suhm (1847–1876), in a letter to his professor, Carl Theodor Ernst von Siebold (1804–1885), written while aboard the HMS *Challenger* in the Pacific in 1875 (see Büchner, 1973). While describing luminescence in various deep-sea fishes, Willemoes-Suhm (1876:lxxxi) wrote that "they are probably all phosphorescent, definitely so in *Sternoptyx*, which once, when the trawl came up at night, hung in the net like a shining star. The source of the light is probably the strange side organs [*Seitenorganen*, i.e., lateral photophores], which may be found in a similar fashion at the tip of the head barbel [illicium] of certain deep-sea lophioids." Lütken (1878a:313), in describing the silvery distal escal appendages of *Himantolophus*, agreed: "This specialization of the tips makes one think of a possible function, e.g., phosphorescence [*fosforescens*]. I mention this so that it can be studied further if a new opportunity arrives, but I lately recall that R. v. Willemoes-Suhm, in his 'letters from the Challenger-Expedition' . . . suggested the presence of a 'light-source' at the tip of the illicium in deep-sea lophioids."

The earliest recorded direct observation of bioluminescence in ceratioids is apparently that of William Beebe (1926b:80), in his original description of *Diabolidium arcturi*, now recognized as *Linophryne arcturi* (Fig. 241): "The white base of the candle-like organ [distal appendage of the esca] and all the longer teeth showed distinct luminescence in the dark room during the first three minutes after capture." Somewhat later, this time in *L. arborifera*, Beebe (1932a) again described glowing teeth—which he thought might be covered with a luminous mucus—as well as light emanating from the esca and hyoid barbel. A number of additional sightings were recorded by Beebe (1926b, 1926c, 1932a, 1934a; Beebe and Crane, 1947), but none are as well known as his eye-witness account and subsequent formal description of an anglerfish seen from the window of the bathysphere, as detailed in *Half Mile Down* (Beebe, 1934a:211; see also Beebe, 1934b, 1934c; Ellis, 2005:41; Fig. 242):

> One minute later, at 2470 feet, all my temporarily relaxed attention was aroused and focused on another splendid piece of luck. A tie rope had to be cut and in this brief interval of suspension, extended by my hurried order, a new anglerfish came out of all the ocean and hesitated long enough close to my window for me to make out its dominant characters. I am calling it the Three-starred Anglerfish, *Bathyceratias trilychnus*. It was close in many respects to the well-known genera *Ceratias* and *Cryptopsaras*, but the flattened angle of the mouth and the short, even teeth were quite different. It was six inches long, typically oval in outline, black, and with small eye. The fin rays were usual except that it had three tall tentacles or illicia, each tipped with a strong, pale yellow light organ. The light was clearly reflected on the upper side of the fish. In front of the dorsal fin were two pear-shaped organs

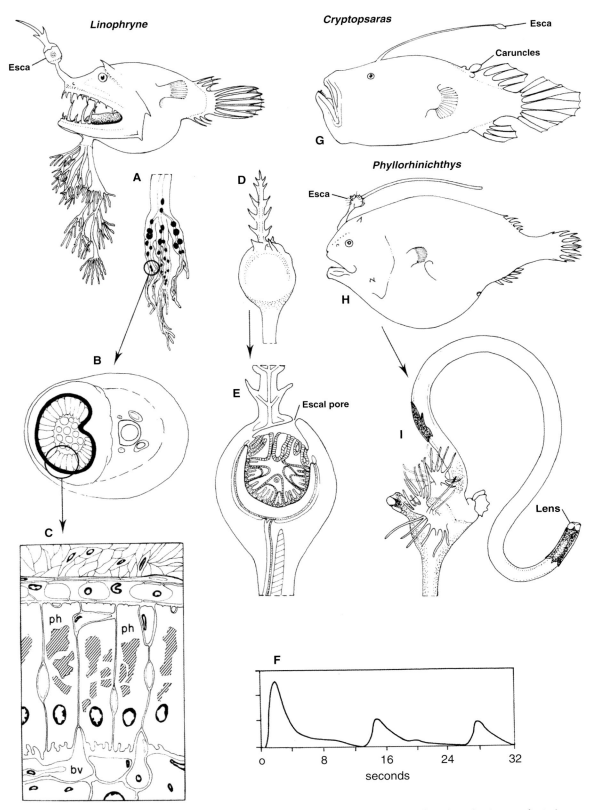

FIGURE 240. Summary of bioluminescent structures in ceratioids: Species of the genus *Linophryne* have luminescent bacteria within the esca and "self" luminescent tissue in the filaments of the hyoid barbel, shown here as small spherical bodies (A) in the barbel of *L. sexfilis*; a cross-section of a single barbel filament containing a single luminous body is shown in (B) and the paracrystalline structure of the material within the photogenic cells (ph) of the luminous body is shown in (C). The esca of *L. sexfilis* is shown in (D) and its appearance in sagittal section in (E). Luminescent bacteria are maintained within elaborate epithelial folds of the light organ, which opens to the outside by way of a small escal pore. Although rapid flashes of light are emitted, slow pulses of bioluminescence (F) are more often observed. In addition to the esca, members of the family Ceratiidae, represented here by *Cryptopsaras couesii* (G), have two or three luminous structures on the back just in front of the soft dorsal fin called caruncles. The esca of some ceratioids, for example *Phyllorhinichthys balushkini* (H), bears elaborate tubular light-guides (analogous to fiber optics), with light-emitting lenslike structures at the tip (I). Text and images modified after Herring and Morin (1978).

FIGURE 241. *Linophryne arcturi*, holotype, 28.5 mm SL, CAS-SU 46505, William Beebe's "Little Sea Devil of the Arcturus" (Beebe, 1926c:4), apparently the first ceratioid directly observed to emit escal bioluminescence. Drawing by Dwight Franklin; after Beebe (1926b, 1926c).

[caruncles] exactly like those of the common *Cryptopsaras*. No pioneer, peering at a Martian landscape, could ever have a greater thrill than did I at such an opportunity.

In all these early observations, escal luminescence was more or less discontinuous, indicating that it was under voluntary control. It almost always appeared in short flashes, each lasting only a few seconds, with longer intervals in between, and with varying strength. The longest duration of continuous luminescence observed was about half a minute, but in this case it was not spontaneous but produced by mechanical stimulation (Norman, 1930). Several observations showed that luminescence could be produced by various artificial means: stirring the water or adding more seawater, adding formalin to the water, touching the illicium, or squeezing the esca (e.g., Norman, 1930; Beebe and Crane, 1947; Bertelsen, 1951). Injections of adrenalin, however, shown to cause strong and prolonged luminescence in other fishes (e.g., Harvey, 1931, 1940; Bertelsen and Grøntved, 1949), produced no reaction in experiments with a living specimen of *Linophryne arborifera* (Harvey, 1931).

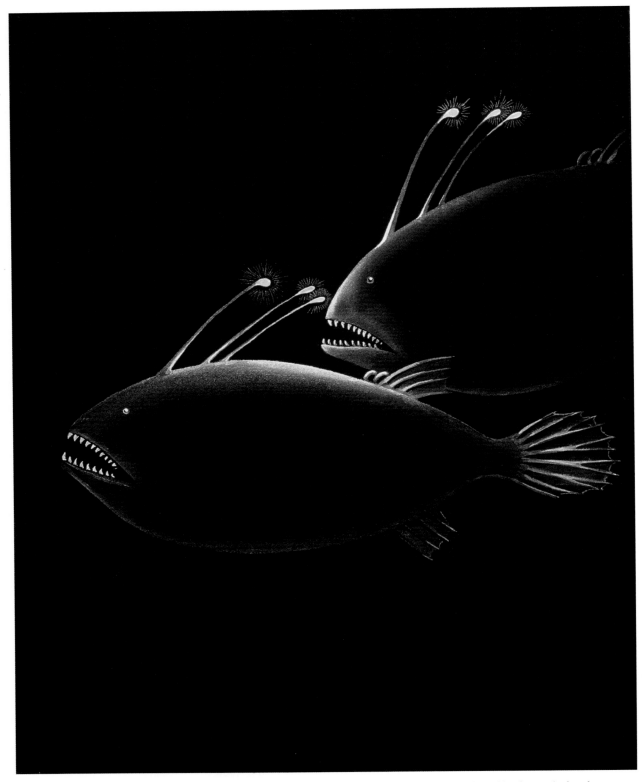

FIGURE 242. William Beebe's Three-starred Anglerfish, *Bathyceratias trilychnus*, as observed from the window of his famous bathysphere. Painting by Else Bostelmann; after Beebe (1934a, 1934b, 1934c); courtesy of the National Geographic Image Collection.

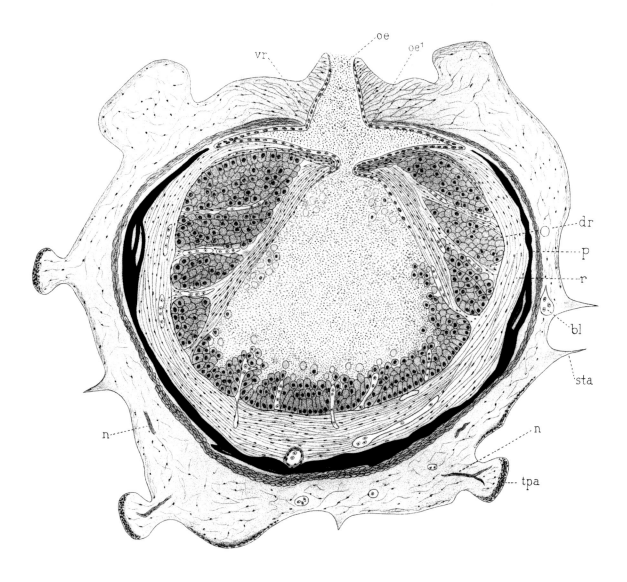

FIGURE 243. August Brauer's drawing of a cross-section though the esca of *Gigantactis vanhoeffeni*. Despite relatively primitive instruments available at the time, Brauer described what he saw in remarkable detail and with great accuracy. bl, *Blutgefäss* (blood vessel); dr, *Drüsenzellen* (glandular cells); n, *Nerv* (nerve); oe, *Oeffnung der Drüse* (glandular opening); p, *Pigment* (pigment layer); r, *Reflektor* (reflective tissue); sta, *Stachel* (dermal spinule); tpa, *Tastpapille* (mechanoreceptive papilla); vr, *Vorraum* (peripheral space). After Brauer (1908).

Adrenalin administered to several living specimens of *Cryptopsaras couesii*, aboard the *Dana* in 1947, stimulated swimming movements, but the esca only glowed in response to strong mechanical stimulus (Bertelsen, 1951).

In 1936, Waterman (1939a:71) observed escal luminescence in a living specimen of *Dolopichthys nigrifilis* (now *Danaphryne nigrifilis*) that had been taken aboard the Research Vessel *Atlantis* in the Sargasso Sea:

> Still feebly moving, this specimen was quickly removed from the net and placed in cold water. In the ship's dark-room, I was able to observe the luminescence of the escal light organ during the ten minutes or so that the fish remained alive. This remained dark until the fish was quite actively stimulated mechanically. When it did luminesce, the light was confined within the esca, which was transparent and more or less free from pigment distally, and no trace of luminous secretion was observed to be extruded from the external pore of the organ. The light was bluish green in color like that of the majority of luminous organisms and lasted for six seconds beginning dimly and rising to a peak in a second or so, maintaining the maximum intensity for several seconds, and then slowly fading out.

Bertelsen (1951) summarized what was then known about bioluminescence in ceratioids, citing various observations of light production in living specimens and describing the few histological examinations made to date of structures thought to be luminescent. But the surprising internal complexity of these organs, especially that of the esca, remained unappreciated.

August Brauer (1904, 1908) was the first to conduct histological studies of the illicial apparatus of ceratioids. In the anatomical section of his well-known work on the deep-sea fishes collected by the German research vessel *Valdivia*, he described the internal escal photophores of *Gigantactis* and *Oneirodes* (*Dolopichthys* of Brauer, 1902, 1904, 1906) as a hollow, spherical, glandular structure located within the escal bulb, surrounded by an inner layer of reflecting cells and a heavily pigmented outer capsule (Fig. 243). The lumen of the gland was said to be filled with small bodies that Brauer considered to be secretory granules produced by the cells of the gland itself. Some of these granules could be seen breaking down and releasing from their cytoplasm similar granules into the cavity of the organ (1904, 1908). The lumen opened into a more distal cavity, which in turn communicated with the exterior by a canal that opened on the posterodorsal midline of the escal bulb. Because of its general structure and especially because of the presence of the external opening, Brauer believed this organ to be a *Spitzdrüse*, that is, a gland whose secretions were forcibly ejected, implying that the organ emitted light only when the secretion of the gland was squeezed out into the surrounding water. However, no indication of muscles within the esca that might serve to force material out through the opening could be found. Nor was Brauer able to find evidence of innervation to the light organ, although branches of an illicial nerve were found associated with supposed sensory papillae of the esca. On the other hand, he was able to show convincingly that the surroundings of the gland are highly vascularized.

From similar studies of the fine structure of the esca of *Ceratias*, Dahlgren (1928) concluded that the secretory granules observed by Brauer (1904, 1908) were luminous bacteria living symbiotically within the cavity of the esca. The symbionts were thought to supply light for the purpose of attracting prey, in exchange for nutrients provided by a kind of culture medium, maintained within the luminous gland and produced by the activity of the secretory cells (a relationship that had been described previously in the cephalopod genus *Sepiola* and in a number of teleosts; e.g., Pierantoni, 1917; Harvey, 1922; Yasaki, 1928; see also Yasaki and Haneda, 1936). Despite some belief to the contrary—"According to my own test cultivation and my observation of the luminescence of the luminous organs of this fish *[Himantolophus groenlandicus]*, I have come to the conclusion that the luminescence . . . is not due to symbiotic bacteria" (Haneda, 1968:3)—Dahlgren's (1928) results have since been universally confirmed and greatly expanded, by both light and electron microscopy, in several additional ceratioid taxa by a number of authors (notably Bassot, 1966; Hulet and Musil, 1968; O'Day, 1974; and Hansen and Herring, 1977; see Table 9). The presence of bacterial luciferase in escal extracts has been shown as well (Leisman et al., 1980). But the definitive work on the structure and function of the ceratioid esca is that of Danish anatomist Ole Munk, who either alone (1988, 1992, 1998, 1999) or in collaboration (Munk and Bertelsen, 1980; Herring and Munk, 1994; Munk and Herring, 1996; Munk et al., 1998) provided almost all that is known on this subject. Although the interested reader is encouraged to refer to Munk's original publications, especially his 1999 review paper on "the escal photophore of ceratioids," the following summary is provided, taken primarily from his extraordinary work.

### Internal Structure of the Esca

#### The Escal Photophore

With exception of the Caulophrynidae and Neoceratiidae, all ceratioid females have an escal light organ that first begins to function toward the end of metamorphosis (Bertelsen, 1951). Of the 160 described species of the suborder, the escae of only about 20 have been studied by light or electron microscopy or both. The results of these studies show an amazing complexity and interspecific variation in morphology, involving a confusing array of bacteria-filled vesicles, heavily pigmented light-absorbing layers, silvery light-reflecting structures, nerves, blood vessels, smooth muscle fibers, various so-called accessory escal glands, and tubular modifications that serve to guide light, analogous to human-designed fiber-optics. Basically, however, the esca consists of a globular, bacteria-filled, light-producing organ or photophore (from the Greek *photos*, meaning "light," and *phoros*, "a bearing") that opens to the outside by way of a small pore on its posterodorsal surface (Fig. 244). The photophore appears through the outer translucent wall of the esca as a spherical inner capsule (the "cup" of Munk, 1999:266), darkly pigmented except for a distal transparent field of varying size and shape (Fig. 245A). The inner capsule is supported by a cartilaginous swelling of the distal end of the illicial bone (Fig. 245B). The walls of the pigmented part of the photophore consist of two layers: (1) a heavily pigmented external layer containing large blood vessels and, in most specimens examined, strands of smooth muscle fibers; and (2) an internal reflecting layer consisting of cells that contain large, plate-shaped crystals that appear silvery in fresh unfixed specimens (Figs. 245B, 246). Within the walls of the photophore, numerous tubular glands open into a central lumen, with muscle fibers between the tubules in some species (see Hansen and Herring, 1977); the glands and the lumen contain rod-shaped bioluminescent bacteria. The distal aperture of the photophore is transparent in life, circular to crescent shaped, and often forms a deep cleft that extends into a pigmented portion of the photophore. Ventral to the distal transparent field, a more or less empty sacklike structure (the vestibule), with transparent walls, is narrowly connected with the lumen of the glandular body and opens to the outside through the escal pore (Fig. 246).

#### Musculature

Although somehow missed by Brauer (1908), perhaps because of poor or improper fixation, a layer of smooth muscle lies between the light gland and the inside of the reflecting layer of the spherical inner capsule. In at least some species of *Melanocetus*, *Gigantactis*, and *Linophryne*, the musculature consists of a thin sheet with up to three layers of tangentially oriented smooth muscle cells. In *Himantolophus albinares*, however, the layer is considerably thicker, containing up to 10 layers of cells (see Munk, 1999). In all these taxa, radiating strands of smooth muscle cells may also be present in the connective tissue between the glandular tubules, especially in the peripheral parts of the light gland. Additional complexity and locations of intra-escal smooth muscle are found in these and other taxa examined (e.g., a ring-shaped sphincter that encloses the entire spherical inner bulb in ceratiids; Munk, 1999). Although Bertelsen (1951) thought it unlikely, it does seem possible that strong contractions of these various muscle-cell layers may serve to expel at least part of the contents of the light gland out of the escal pore and into the surrounding water, as first postulated by Brauer (1908).

#### Blood Vessels and Nerves

Although no one has yet investigated escal vascularization and innervation based on vascular injection techniques and special staining methods for nerve fibers, the primary blood vessels and their larger branches of the esca and illicium of two species, *Chaenophryne draco* and *Oneirodes eschrichtii*, have been identified and followed in serial sections (Munk and Bertelsen,

TABLE 9
Observations of Bioluminescence and Histological Studies of the Esca of Various Ceratioid Taxa Arranged by Family

| Genus | Species | Observation | Source |
|---|---|---|---|
| Ceratiidae | | | |
| "Bathyceratias | trilychnus" | Luminescence | Beebe, 1934a, 1934b, 1934c |
| Ceratias | holboelli | Histology | Munk, 1992 |
| Ceratias | tentaculatus | Histology | Munk, 1992 |
| Ceratias | uranoscopus | Histology | Munk, 1992 |
| Ceratias | sp. | Histology | Dahlgren, 1928 |
| Ceratias | sp. | Histology | Bassot, 1966 |
| Cryptopsaras | couesii | Luminescence | Norman, 1930 |
| Cryptopsaras | couesii | Luminescence and histology | Bertelsen, 1951 |
| Cryptopsaras | couesii | Luminescence | Herring, 1983 |
| Cryptopsaras | couesii | Histology | Munk and Herring, 1996 |
| Cryptopsaras | couesii | Luminescence | Herring et al., 1997 |
| Himantolophidae | | | |
| Himantolophus | albinares | Luminescence and histology | Bertelsen and Krefft, 1988 |
| Himantolophus | albinares | Luminescence and histology | Munk, 1999 |
| Himantolophus | azurlucens | Luminescence | Beebe and Crane, 1947 |
| Himantolophus | groenlandicus | Luminescence and histology | Bertelsen and Krefft, 1988 |
| Himantolophus | paucifilosus | Histology | Bertelsen and Krefft, 1988 |
| Himantolophus | sagamius | Luminescence | Haneda, 1968 |
| Melanocetidae | | | |
| "Melanocetus | sp." | Luminescence | Beebe, 1934a |
| Melanocetus | sp. | Histology | Munk et al., 1998 |
| Melanocetus | eustalus[a] | Luminescence | Pietsch, 1972c |
| Melanocetus | johnsonii[b] | Luminescence | Beebe and Crane, 1947 |
| Melanocetus | johnsonii | Histology | Bassot, 1966 |
| Melanocetus | murrayi | Histology | Hulet and Musil, 1968 |
| Melanocetus | murrayi | Histology | Munk et al., 1998 |
| Oneirodidae | | | |
| Lophodolos | sp. | Luminescence | Herring, 1983 |
| Chaenophryne | draco[c] | Luminescence | Beebe and Crane, 1947 |
| Chaenophryne | draco | Histology | Munk and Bertelsen, 1980 |
| Chaenophryne | draco | Histology | Munk, 1999 |
| Oneirodes | acanthias | Histology | O'Day, 1974 |
| Oneirodes | eschrichtii | Histology | Munk, 1988, 1998, 1999 |

(Continued)

TABLE 9 (continued)

| Genus | Species | Observation | Source |
|---|---|---|---|
| Oneirodes | luetkeni | Histology | Bertelsen, 1951 |
| Oneirodes | sp. | Luminescence | Beebe, 1926c |
| Oneirodes | sp. | Luminescence | Herring, 1983 |
| Danaphryne | nigrifilis[d] | Luminescence | Waterman, 1939a |
| Phyllorhinichthys | micractis[e] | Histology | Munk, 1992, 1998 |
| Dolopichthys | longicornis | Luminescence | Herring and Munk, 1994 |
| Dolopichthys | niger[f] | Histology | Brauer, 1908 |
| Dolopichthys | sp. | Luminescence | Herring et al., 1997 |
| Gigantactinidae | | | |
| Gigantactis | vanhoeffeni | Histology | Brauer, 1904, 1908 |
| Gigantactis | vanhoeffeni | Histology | Munk, 1999 |
| Gigantactis | sp. | Histology | Bassot, 1966 |
| Gigantactis | sp. | Luminescence | Herring, 1983 |
| Linophrynidae | | | |
| Haplophryne | mollis | Luminescence | Herring, 1983 |
| Haplophryne | mollis | Luminescence and histology | Herring and Munk, 1994 |
| Haplophryne | mollis | Luminescence | Herring et al., 1997 |
| Linophryne | arborifera | Luminescence | Beebe, 1932a |
| Linophryne | arborifera | Luminescence and histology | Hansen and Herring, 1977 |
| Linophryne | arborifera | Luminescence | Herring, 1983 |
| Linophryne | arcturi[g] | Luminescence | Beebe, 1926b, 1926c |
| Linophryne | arcturi | Luminescence | Beebe and Crane, 1947 |
| Linophryne | coronata | Histology | Munk, 1998, 1999 |
| Linophryne | indica[h] | Luminescence | Herring, 1983 |
| Linophryne | sexfilis | Luminescence | Hansen and Herring, 1977 |

NOTE: Modified after Munk (1999).
[a]Misidentified as *Melanocetus ferox*.
[b]As *Melanocetus ferox*.
[c]As *Chaenophryne parviconus*.
[d]As *Dolopichthys nigrifilis*.
[e]Based in part on *Phyllorhinichthys balushkini*.
[f]As *Oneirodes niger*.
[g]As *Diabolidium arcturi*.
[h]As *Linophryne corymbifera*.

1980; Munk, 1988). In both taxa, the proximal part of the illicial stem contains a single median artery and vein that run along the posterior margin of the illicium (Munk and Bertelsen, 1980). Branches of these vessels supply the light gland itself as well as other parts of the esca.

Thin nerve fibers, which originate from thicker, longitudinally oriented nerves situated within the illicium, are present in the connective tissue around the light gland and within the outer escal filaments and appendages. The proximal part of the illicial stem of *Chaenophryne draco* contains a single pair of nerves, one running along each side of the vein (Munk and Bertelsen, 1980). The myelin sheaths of the nerve fibers are easily observed by light microscopy in plastic sections stained with toluidine blue. It is likely that some of the myelinated fibers in the esca are sensory (see Waterman, 1939a, 1948), but dendrites of putative sensory neurons have never been identified, nor has any specialized type of receptor been found situated in the skin or elsewhere.

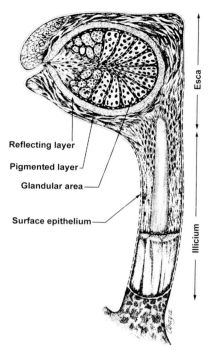

FIGURE 244. Diagrammatic representation of a sagittal section through the illicium and esca of *Melanocetus murrayi*. After Hulet and Musil (1968).

In the specimen of *O. eschrichtii* examined (Munk, 1988, 1998), thin unmyelinated nerve fibers with varicosities containing putative synaptic vesicles are associated with bundles of smooth muscle cells that lie just outside the spherical inner capsule in the dorsal part of the esca (Munk, 1999). These nerve fibers are probably visceral motor axons that innervate the smooth muscle cells.

### Light-guiding and Reflecting Structures

Light-guiding structures, most always located distal to the vestibule of the photophore, are present in most, if not all species (observed in all specimens in which the esca was examined prior to formalin fixation and/or by histological examination; Munk, 1998, 1999). In their simplest form (e.g., in *Ceratias*), the light-guiding structures consist of a circular, ventrally concave, mirrorlike reflecting shield. In most species they are divided into a number of lobes (e.g., two or four in *Linophryne* and some species of *Himantolophus*; three in some members of the Oneirodidae), all more or less covered externally with pigment. The lobes are often raised into conical or tubular, simple or unbranched structures lying within the variously shaped distal appendages anterior to the escal pore (Fig. 247).

More complex light-guiding appendages vary from conical prolongations of the escal bulb (short as in *Melanocetus* or very long as in some species of *Gigantactis*), or one or more structures that vary in length and shape from short, papilliform, or truncated (e.g., the "terminal papilla" of the Oneirodidae) to extremely long, cylindrical, tapering tubes (e.g., in some species of *Himantolophus* and the oneirodid genus *Phyllorhinichthys*; see Munk, 1998; see Figs. 72, 127). They may be simple or branched (e.g., as in the genus *Linophryne* and some species of *Himantolophus*), transparent or more or less pigmented, and naked or covered with dermal spines.

Light-guiding appendages are nearly always restricted to the distal surface of the escal bulb, anterior to the escal pore. The only exception is found in species of the oneirodid genus *Chaenophryne*, which have a subcutaneous "descending appendage" that extends down along the anterior margin of the illicium (Pietsch, 1975b; Munk and Bertelsen, 1980; Figs. 105, 106B–D, 248). The length of the descending appendage varies greatly among species: usually no greater than the length of the escal bulb in *Chaenophryne longiceps*, but often extending the full length of the illicium in *C. draco* and *C. melanorhabdus* (Fig. 248). In *C. ramifera* and *C. quasiramifera*, this structure is extraordinarily well developed, extending in large specimens not only along the entire length of the illicium, but protruding as well from the point at which the illicium articulates with the illicial pterygiophore, forming a free anterodorsally directed prolongation of the descending appendage that in some specimens is nearly as long as the illicium itself (see Pietsch, 1975b, 2007; Fig. 106D).

The internal anatomy of light-guiding tubes is essentially the same in all taxa in which they are found. Basically, they consist of an internal central core of transparent tissue, surrounded by silvery reflecting walls, with or without an external layer of dark pigment (some are translucent). The distal tip of the light-guiding tubes is transparent or equipped with one or more window- or lenslike openings in the pigmented wall (e.g., Figs. 127, 248). Located at these openings, and present in the wall between the lumen of the glandular body and the vestibule of the photophore, are a number of large, transparent, protein-rich cells that probably have refracting and color-filtering functions (Munk, 1988, 1998, 1999).

### Accessory Escal Glands

A few ceratioids (e.g., *Ceratias* and *Phyllorhinichthys*) possess what Munk (1992:33) called "accessory escal glands," located beneath the skin around the upper part of the illicium and the proximal part of the inner escal capsule. The presence of ducts opening on the surface of the esca indicates that these glands develop in the same way as the escal light gland, that is, from invaginations of epidermal cells (see Ontogenetic Development of the Escal Photophore, below). More than anything else, these accessory glands resemble typical exocrine glands, but their biological function remains unknown (Munk, 1992). There is no evidence to suggest that they are bioluminescent. Their location suggests instead that their function might be somehow supplementary to that of the escal light gland, perhaps by producing a chemical secretion that attracts prey or perhaps even a conspecific male (Munk, 1992; for evidence of chemical attractants produced by escae of shallow-water anglerfishes, see Combs, 1973; Pietsch and Grobecker, 1987; and Bradbury, 1988).

### Ontogenetic Development of the Escal Photophore

Studies of early stages of development of the esca have shown that bacteria are absent in the light glands of larval ceratioids, and very few or absent even in small metamorphosed females (Munk and Herring, 1996; Munk et al., 1998). Histological examinations of larval females of *Cryptopsaras* and *Melanocetus* show that the light gland proper originates from an invagination of epidermal cells from the distal surface of the roughly spherical escal primordium and that the intraglandular lumen, vestibule, and the duct that opens to the exterior probably

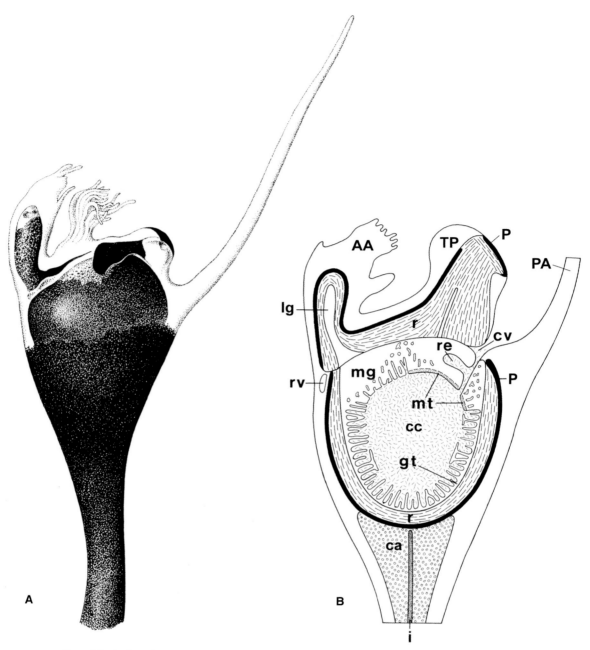

FIGURE 245. Esca of *Oneirodes eschrichtii*, 26 mm SL, left lateral views: (A) external view, showing escal appendages emanating from distal surface (from left to right): an anterior appendage, containing an internal, pigmented, light-guiding tubule, with an unpigmented distal "window"; paired medial appendages; a rounded terminal papilla; and an elongate tapering posterior appendage; (B) diagrammatic sagittal section, showing the anterior appendage (AA), with internal light-guiding tubule (lg); terminal papilla (TP), with darkly pigmented distal surface (P), and posterior appendage (PA). ca, cartilage; cc, central cavity; cv, caudal part of ring-shaped vestibule; gt, glandular tissue; i, illicium; mg, modified glandular tissue consisting almost exclusively of darkly stained cells; mt, modified glandular tissue with luminal cells; r, reflecting layer; re, recess in canal connecting central escal cavity with caudal part of vestibule; rv, rostral part of ring-shaped vestibule. Modified after Munk (1999).

originate from the breakdown of the central cells within the originally solid mass of epithelial cells. Rapid and marked changes in the structure of the epithelium of the light gland and the contents of the lumen occur during early growth. For example, even in the smallest metamorphosed specimens of *Melanocetus* and *Haplophryne*, the light gland has the same basic structure as that of adults: a central escal cavity from which branched ducts ramify into a system of radial tubules (see Munk, 1999).

## Escal Luminescence and Its Control

Numerous observations of escal luminescence in a host of different ceratioid taxa are now available (see Table 9; Fig. 249). According to these accounts, the color of the light varies from pink or purple, to white, yellow, orange, yellowish green, blue, and bluish green, depending on the species. The light emitted from the escae of several species has been found to peak within the 470- to 490-nm range, which corresponds well to the peak of the emission spectra of most luminous bacteria, falling within the range

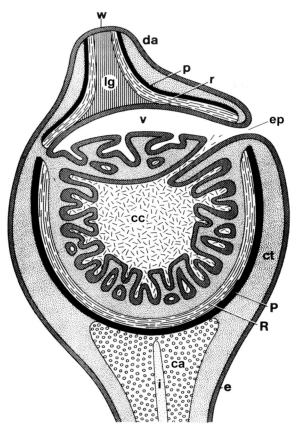

FIGURE 246. Diagrammatic sagittal section through a generalized esca, with a single light guide (lg) located within a distal appendage (da). The escal light gland is enclosed in a lightproof, cup-shaped structure consisting of an inner reflecting layer (R) and an outer pigmented layer (P). A tubular extension of the same kind of layers in the distal wall of the esca forms a dorsally directed light guide (lg) with a terminal window (w). A duct connects the central cavity (cc) of the light gland with the vestibule (v), which, in turn, opens to the exterior by way of a short duct leading to the escal pore (ep), situated posteriorly on the distal surface of the escal bulb. ca, cartilage; ct, connective tissue; e, epidermis; i, illicium; p, pigmented layer of light-guide core; r, reflecting layer of light-guide wall. Text and image after Munk (1998).

FIGURE 247. Esca of *Linophryne arborifera*, 35 mm SL, ISH 1753/71, showing the development of light-guiding tubules embedded within the escal appendages: "Inside the distal appendage an opaque tubular structure can be detected in well-preserved specimens. As observed by the author [Bertelsen] in freshly caught specimens . . . , this structure has very bright silvery walls and projects from a four-lobed, leaf-shaped basal plate covering the distal part of the escal bulb above the unpigmented field of the luminous gland, surrounding the escal pore; from this base it tapers into a cylindrical tube with a pair of side-branches stretching into the stout proximal pair of the appendage. The tip of each of the three tubes is transparent and may in the unpreserved specimens appear as a bright-yellow, circular window. . . . in some cases . . . escal structures of this type act as light guides, the light from the luminous gland being emitted partly through the windows at the tip of the tubes, and partly beneath the leaf-like basal plate after reflection in the silvery concave undersides of its lobes" (Bertelsen, 1980a:39). Drawing by R. Nielsen; after Bertelsen (1980a).

of 475 to 510 nm (Herring and Morin, 1978; Herring, 1983; Fig. 250). Blue green luminescence achieves maximum transmission in oceanic water and is the light to which the eyes of the majority of deep-sea animals are most sensitive (Herring, 1983).

One of the most detailed observations of escal light production is that of Hansen and Herring (1977:106), based on a living specimen of *Linophryne brevibarbata* (BMNH 1995.1.18.4, reported as *L. arborifera*), studied while aboard the RRS *Discovery* in the North Atlantic in March 1973: "A bright turquoise light was observed from the esca when the fish was handled, in the form of a series of pulses with rise time to maximum intensity of about 1 sec and a duration of about 10–12 sec" (Fig. 251). In this and most other observations of escal luminescence the light appears to emanate from the sides of the semitransparent distal portion of the escal bulb below the distal shield, and from the windowlike apertures of light-guiding appendages (Hansen and Herring, 1977; Munk and Bertelsen, 1980; Munk, 1988). Light production appears to be restricted to the bacteria present in the tubular glands and lumen of the glandular body, but the way in which light is generated and controlled by the escal light gland is not understood (Herring and Morin, 1978; Herring et al., 1997). Although small nerves are occasionally seen to enter the pigmented outer layer of the capsule that surrounds the light gland, nerve endings associated with the glandular epithelium have so far not been found (Munk, 1999); thus, it seems unlikely that there is any direct nervous control. Alternatively, the intensity of light may be controlled by modifying the supply of oxygen via the blood (as first suggested by Waterman, 1939a; see also Harvey, 1940; Waterman, 1948; Bertelsen, 1951; Munk, 1988), and also perhaps by the amount of glandular secretion made available to the bacteria (no ceratioid luminescence has ever been produced in bacterial cultures, i.e., in the absence of unknown factors most likely produced by the glandular tissue; see O'Day, 1974; Leisman et al., 1980; Herring and Munk, 1994). The amounts of oxygen and glandular secretion are in turn probably controlled by contractions of the intra-escal muscles, which would cause a mixing of bacteria and glandular secretion

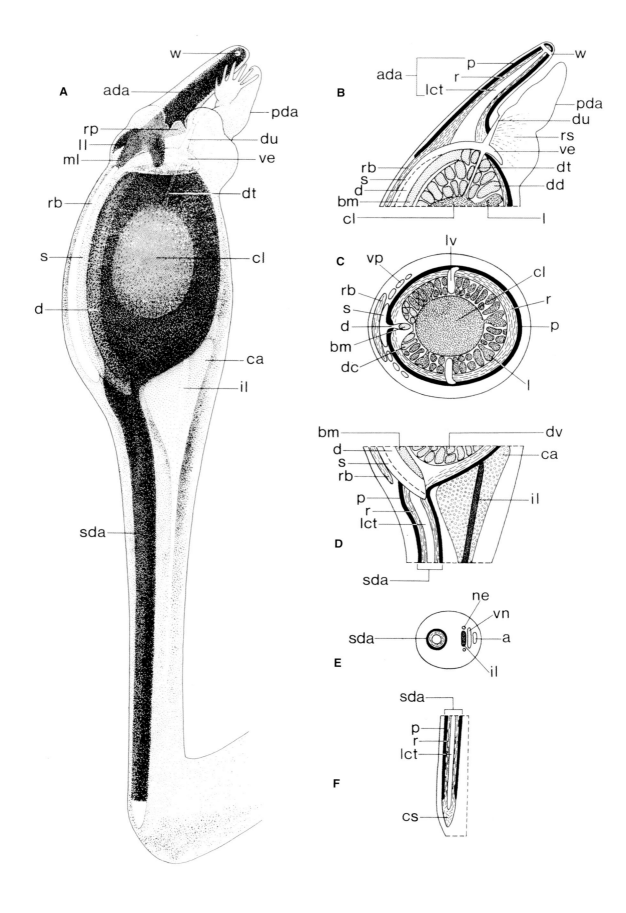

(Hansen and Herring, 1977). Visibility of the light is probably regulated by contractions of the ring-shaped smooth muscle that is situated along the edge of the distal aperture of the photophore (Hansen and Herring, 1977), or alternatively, it may be mediated by pressure on the glandular body, caused either by contraction of the muscles of the wall or by a widening of its many large blood vessels. By either mechanism, the resulting pressure would push the glandular body in a distal direction (the base of the inner bulb being fixed to the cartilage of the tip of the illicial bone), causing a widening of the elastic, unpigmented distal part of the escal bulb, and a flattening of the vestibule, thus bringing the luminous contents of the glandular body in close contact with the base of the light-guiding structures (Munk, 1998). Light would then be allowed to pass through the completely transparent core of these structures, being in part reflected inward and upward by the more or less concave, mirrorlike, inner surfaces and thus passing through the tubular light-guides to their transparent distal tips.

The strength and quality of the light produced by the esca may also be controlled, at least in some taxa, by movement of internal pigment layers. By whatever mechanism, the distal aperture of the pigmented outer wall of the escal capsule is apparently capable of rapidly opening and closing, thus resulting in pulses of emitted light (Herring and Morin, 1978). Pietsch and Van Duzer (1980:80) described an example of this control of escal light production by migrating layers of pigmented tissue in a specimen of *Melanocetus eustalus* (LACM 30037-12, misidentified as *M. ferox*; see Fig. 252) that was maintained alive for several minutes after capture:

> The bulb of the esca glowed continuously with a bright golden-orange light. The amount of light actually emitted, however, appeared to be controlled by an up and down movement of the darkly pigmented, inner wall of the photophore of the esca. The glowing bulb was almost entirely covered and uncovered four or five times within a period of at least 1 min. A mechanism of this kind would provide a rapid means of extinguishing light that may not only attract potential mates and prey, but also predators.

The color of the light produced by the esca is probably modified in passage through the modified cells in the distal part of the glandular body and at the outlets. Appendages that lack light-guiding structures may be diffusely illuminated by light entering at the base. Filaments with external reflectors (e.g., the silvery tips of the escal appendages of *Himantolophus*; see Bertelsen and Krefft, 1988) are probably illuminated by reflecting light that originates from the openings of the light-guiding structures.

### Bacterial Symbionts and Their Source

The bacteria found within the esca are rodlike in shape, Gram negative, and without spores, capsules, or flagella (O'Day, 1974).

Ultrastructural features of the bacteria include a double-layered cell wall and mesosomal invaginations of the plasma membrane (Figs. 253, 254). Although the microorganisms apparently grow on seawater-nutrient broth or agar (O'Day, 1974; Leisman et al., 1980), no luminescence has been observed under these conditions. Nevertheless, it is generally assumed that these symbiotic bacteria are responsible for light production in ceratioid escae. Evidence indicates that the escal light gland does not become functional until the final stages of metamorphosis, when the intraglandular lumen and the duct to the outside have developed. It also appears likely that bacteria from the surrounding seawater gain access to the light gland by way of the escal pore and duct that leads into the interior (but see the work of Haygood et al., 1992, summarized below). Unlike the bacterial light organ of the squid *Euprymna scolopes*, however, in which colonization of the gland by bacteria from the surrounding seawater is apparently assisted by microvilli and cilia on the epithelium covering the outer surface of the juvenile light gland (McFall-Ngai and Ruby, 1991; Ruby and McFall-Ngai, 1992), there is no evidence of such devices in ceratioids that might assist in the assumed transfer of bacteria from the outside.

Recent studies suggest that each ceratioid species may have its own particular species of bacterium. In a comparison of sequences of the 16S ribosomal RNA gene of luminous bacterial symbionts from the escae of *Cryptopsaras couesii* and *Melanocetus johnsonii*, with each other and with a sequence obtained from a strain of *Photobacterium phosphoreum* (the culturable light-organ symbiont of the deep-sea teleost *Opisthoproctus grimaldii*), Haygood et al. (1992; see also Haygood, 1993; Haygood and Distel, 1993) showed that the anglerfish symbionts group with *P. phosphoreum* but are not phylogenetically closely related to the latter (Fig. 255). Instead, the ceratioid symbionts appear to represent a unique lineage related to the free-living, luminous, marine bacteria genus *Vibrio*. More surprising yet, the two ceratioid symbionts differ from each other at least on the species level, implying that the bacteria are species specific throughout the suborder. Realizing that these new strains of *Vibrio*-like symbionts have not so far been recognized as free-living organisms in seawater, it is difficult to understand how the ceratioid esca acquires its symbionts (Herring, 1993). Electron microscopy has shown that no bacteria are present prior to the development of the duct that connects the escal glandular lumen with the exterior, and that the esca of even small, recently metamorphosed ceratioids contains no bacteria (Munk et al., 1998) or only a very small number of cells (Munk, 1988; Herring and Munk, 1994); these studies suggest that bacteria gain access from the exterior by way of the escal pore. But Haygood (1993) has proposed that bacteria discharged from the esca during spawning may somehow be transferred to the eggs. If so, the bacteria may perhaps thrive on the outside of the larvae in the mucus secreted by the skin, until the escal

---

FIGURE 248. Illicium and esca of *Chaenophryne draco*, 34 mm SL, intended in part to demonstrate the extreme complexity of this structure: (A) external, left lateral view; (B) sagittal section of distal part of esca, with anterior and posterior distal appendages; (C) horizontal cross section through center of esca; (D) sagittal section of proximal part of esca, showing distal tip of illicial bone and proximal end of subcutaneous descending escal appendage; (E) horizontal cross section through illicium; (F) sagittal section through distal end of subcutaneous descending escal appendage. a, artery; ada, anterior distal appendage; bm, band-shaped mass of darkly stained cells in deep portion of diverticulum; ca, cartilage; cl, central lumen of esca; cs, cone-shaped mass of cells; d, deep portion of diverticulum of vestibule; dc, darkly stained cells in anterior part of esca glandular tissue; dd, darkly stained cells in upper (dorsal) part of esca glandular tissue; dt, duct connecting central lumen of esca with vestibule; du, duct connecting vestibule with exterior; dv, darkly stained cells in lower (ventral) part of esca glandular tissue; il, illicial bone; l, lumen of esca glandular tubule; lct, light connective tissue; ll, lateral lobe of leaflike structure; lv, lateral vein draining esca glandular tissue; ml, median anterior lobe of leaflike structure; ne, nerve; p, pigmented layer; pda, posterior distal appendage; r, reflecting layer; rb, reflecting band; rp, low unpigmented reflecting protuberance; rs, reflecting streaks; s, superficial portion of diverticulum of vestibule; sda, subcutaneous descending appendage; ve, vestibule; vn, vein; vp, vascular plexus; w, window in anterior distal appendage. Drawings by E. Beyerholm; after Munk and Bertelsen (1980).

FIGURE 249. Esca of *Linophryne macrodon*, 28.5 mm SL, MCZ 164217. Photo of living specimen made aboard ship by C. P. Kenaley.

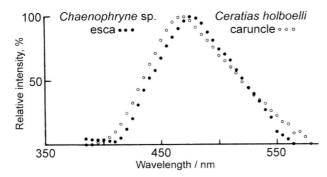

FIGURE 250. Bioluminescence emission spectra from caruncle exudate of *Ceratias holboelli* and from the esca of an unidentified species of *Chaenophryne*, showing emission maximum at 470 and 475 nm; nearly identical values have been obtained from the escae of other ceratioids, including *Cryptopsaras couesii*, *Haplophryne mollis*, and identified species of *Lophodolos*, *Oneirodes*, and *Gigantactis*. After Herring (1983).

pore and intraglandular lumen develop at the time of metamorphosis (Munk, 1999).

## Other Bioluminescent Structures

### Second Dorsal-Fin Spine

In addition to the esca, a bulbous structure that looks very much like an escal photophore, but which has never been seen to luminesce, is present on the tip of the second dorsal-fin spine (second cephalic ray of Bertelsen, 1951) of females of the Ceratiidae and Diceratiidae. In female ceratiids, it is externally visible only in the larvae, situated more or less sessile on the head immediately behind the slightly larger primordium of the developing esca (Fig. 44A, B). During metamorphosis, it is withdrawn beneath the skin, remaining open to the surface via a small pore in young juveniles, but eventually losing its connection with the rudimentary second

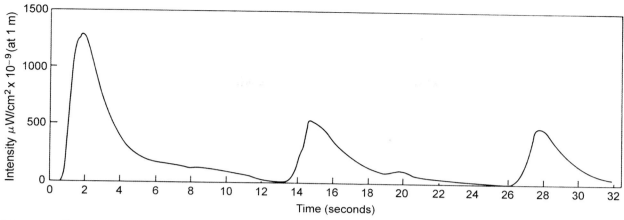

FIGURE 251. Continuous record, over a 32-s period, of spontaneous (i.e., unintentionally stimulated), bright turquoise-colored bioluminescence from the esca of a handheld, living specimen of *Linophryne brevibarbata* (100 mm SL, BMNH 1995.1.18.4; whole specimen illustrated in Fig. 4), showing three bursts of light emission, with a rise time to maximum intensity of about 1 s and a duration of about 10 to 12 s. "The light appeared from all around the edge of the dorsal silvered cap of the esca and up the silvered tubule in the distal appendage" (Hansen and Herring, 1977:106). After Hansen and Herring (1977).

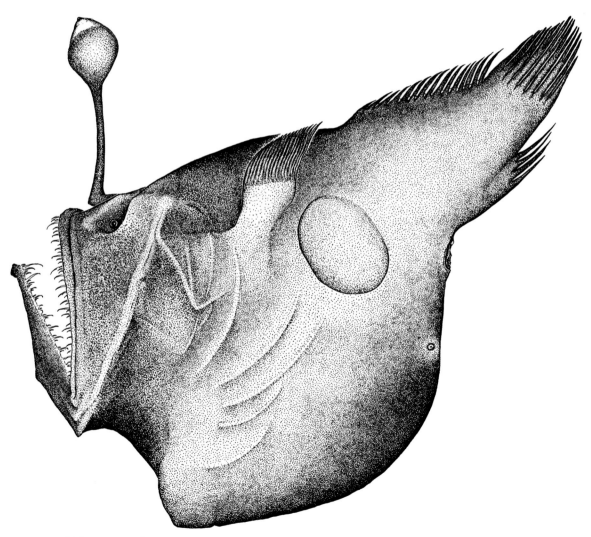

FIGURE 252. *Melanocetus eustalus*, holotype, 111 mm SL, LACM 30037-012, a species with perhaps the largest esca in proportion to the body of any ceratioid (see Fig. 80B). Drawings by E. A. Hoxie; after Pietsch and Van Duzer (1980).

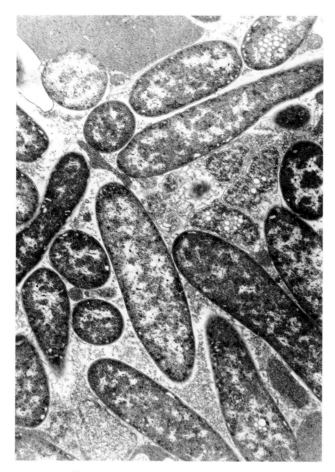

FIGURE 253. Transmission electron micrograph of a section through the central lumen of the esca of *Oneirodes acanthias*, showing numerous tightly clustered, aflagellate, rod-shaped bacteria. Courtesy of William T. O'Day; after Pietsch (1974a).

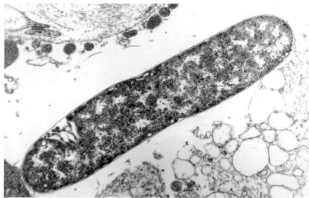

FIGURE 254. Transmission electron micrograph of a section through the central lumen of the esca of *Oneirodes acanthias*, showing a single bacterium. Courtesy of William T. O'Day.

dorsal-fin spine and disappearing with further growth (Bertelsen, 1951). In juvenile diceratiids it is found at the tip of a short stalk that emerges behind the insertion of the pterygiophore of the illicium and persists as an external bulb in young adults (Fig. 44C, D). With subsequent growth it tends to sink beneath the skin of the head but remains connected to the second dorsal-fin spine and to the surface through a small pore, even in large adults.

In the ceratiid genus *Cryptopsaras*, the photophore of the second dorsal-fin spine is histologically similar to that of the esca (but only a single larval specimen has been examined; see Munk and Herring, 1996). Like the esca, the glandular epithelium originates from an invagination of epidermal cells from the distal surface of a spherical primordium (Bertelsen, 1951; Munk and Herring, 1996). Once formed, the internal glandular body remains open to the outside by way of a distal pore, but a vestibule and light-guiding structures are absent, and bacteria have not been found. A transparent distal aperture is absent as well, the entire structure covered instead by a heavily pigmented outer wall. If, at any later stage in development, this esca-like structure acquires bacteria and becomes luminescent, it seems likely that only a secretion of luminous fluid would be visible to the outside (see Caruncles, below). Clearly more work remains to be done before function can be assigned; for example, the photophore-like structure of the second dorsal-fin spine of the Diceratiidae, which, unlike that of the Ceratiidae, persists into adulthood, has never been examined histologically.

## Caruncles

In the Ceratiidae, the anterior-most dorsal-fin rays are modified in a way similar to that of the illicium and second dorsal-fin spine: 2 club-shaped "caruncles" (a term apparently first used by Günther, 1887:52) born on short stalks in the genus *Ceratias*; and 3, oval, more or less sessile caruncles in *Cryptopsaras* (Fig. 44A, B). The caruncles decrease in size relative to standard length in *Ceratias* (tiny and degenerate in specimens greater than about 400 mm) but increase slightly in size in *Cryptopsaras*. The caruncles of *Cryptopsaras* are histologically similar to the photophore of the esca (Brauer, 1908; Fig. 256) and even more so to that of the second dorsal-fin spine (Bertelsen, 1951). They differ from the esca but resemble the latter in having the inner wall of the gland and the surrounding skin entirely covered with dense pigment, so that luminescence visible to the outside would be possible only by the emission of a secretion through the opening of the gland. Luminescence from the caruncles was observed for the first time, in a living female of *Cryptopsaras couesii*, by Bertelsen (1951:239) aboard the *Dana* in 1947:

> Spontaneous luminescence was not observed nor did adrenalin injection produce any reaction from the glands, but on pressing the large median, dorsal caruncle a whitish yellow secretion poured out into the water like a thread and quickly broke and spread into numerous points of light, which could be observed scattered over the basin used for about a minute. The experiment was repeated five times during about one hour, and all with the same result. The last just after life appeared to be extinct.

Since Bertelsen's (1951) early observation, histological studies have now shown that the caruncles of *C. couesii* contain "dense populations of luminous bacteria" that can be discharged to the exterior by way of a distal pore (Hansen and Herring, 1977:104; Herring and Morin, 1978:324). Spontaneous expulsion of luminous fluid from the caruncles has been observed by Young and Roper (1977) in a living specimen of this species.

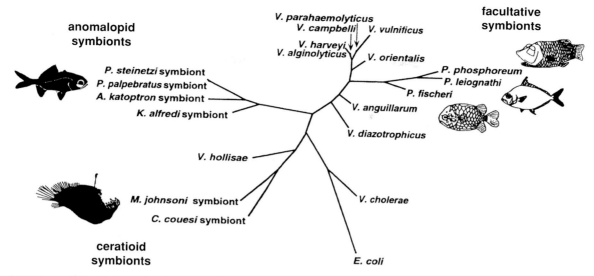

FIGURE 255. Phylogenetic relationships among luminous bacterial symbionts and other strains of *Vibrio*, based on parsimony analysis of small subunit rRNA sequences; representative hosts are illustrated next to their respective symbionts. Each of the two ceratioids examined (*Melanocetus johnsonii* and *Cryptopsaras couesii*) are shown to have unique symbionts that form a monophyletic group within a *Vibrio* clade, indicating that the symbionts have been evolving and diverging along with their hosts for a long time. After Haygood and Distel (1993).

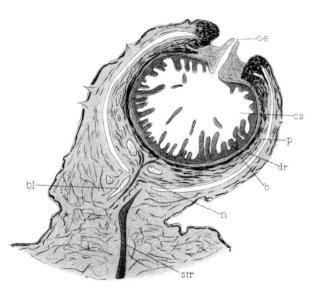

FIGURE 256. August Brauer's drawing of a sagittal section though a caruncle of *Cryptopsaras couesii*. b, Bindegewebige Huelle (envelope of connective tissue); bl, Blutgefäss (blood vessel); cs, Centraler Sinus (central cavity); dr, Drüsenzellen (glandular cells); n, Nerv (nerve); oe, Oeffnung der Drüse (glandular opening); p, Pigment (pigment layer); str, Flossenstrahl (dorsal-fin ray). After Brauer (1908).

## Hyoid Barbels

Among ceratioids a hyoid barbel is found only in *Centrophryne* and *Linophryne*. In *Centrophryne* the barbel is present in larvae and juveniles of both sexes: a simple, papilliform to digitform structure, present in females less than about 45 mm, but reduced to a minute protuberance or absent in larger specimens, distinguishable in some large individuals only because it is somewhat more darkly pigmented than the surrounding skin. It lacks photophores or other specialized internal structures. Sagittal sections of the barbel of a 42-mm female revealed a highly convoluted outer pigmented layer, surrounding an inner core of connective tissue (Pietsch, 1972a:24; fig. 5). Within the connective tissue at the distal tip of the barbel there appears to be a blood space or sinus, but there are no nerves or any other structure that might allude to its function.

In contrast, the barbel of *Linophryne* is a highly complex structure, both internally and externally, present only in metamorphosed females and providing characters of significant taxonomic importance at the species level (Bertelsen, 1980a, 1980b, 1982). In all species of the genus, the terminal branches of the barbel bear small spherical tubercles (Fig. 257), which were presumed to be "phosphorescent" as early as 1886, when Robert Collett (1886:142; see Fig. 13) introduced the genus as new to science. In his original description of *Linophryne lucifer*, Collett wrote that the "guttural tentacle" (i.e., barbel) is "thinner than the cephalic spine [illicium] and divides itself at the end into two short pointed blades. . . . Whilst the tentacle [barbel] otherwise is black, the inner edges of these blades are white, like the upper half of the snout tentacle [esca], and are furnished with a row of around papillae, about 30 on each, resembling a chain of pearls. These small bodies undoubtedly have a use, either as organs of sense or as the source of the phosphorescent light."

That the barbel of *Linophryne* is indeed bioluminescent was not confirmed, however, until 1932 when William Beebe (1932a:85–86) published his observations of the glowing barbel in a living specimen of *L. arborifera*: "This structure was longer than the entire body, with many scores of very small pink organs, several dozen diminutive lights, and the ability to contract and expand like a snarl of deadly medusa locks." Unaware of Beebe's (1932a) observation, Koefoed (1944:13) described the barbel of *L. algibarbata* as follows:

> Below the symphysis of the mandible, behind its process, sit four barbels [emanating from a common base] whose length at least answers to the length of the body. They are richly ramified and are like the stinging tentacles of Siphonophores; some of them are short and simple, others are long and richly ramified. The barbels are brown pigmented nearest to the mandible, but for the greater part of their length they are colourless and transparent so that an axis can be seen in them, which both in the stem and in the branches—particularly in their ends—finishes

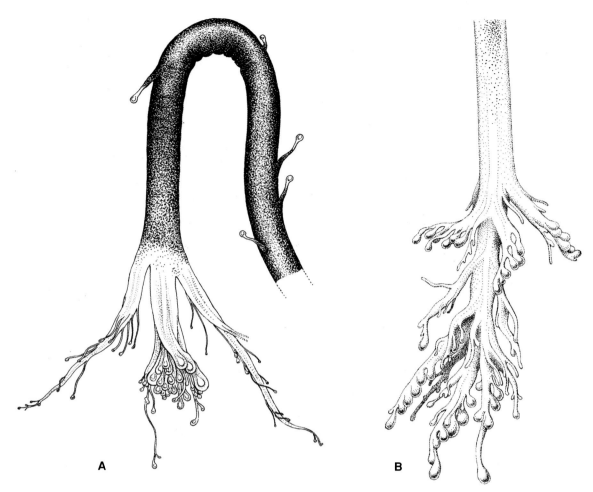

FIGURE 257. Hyoid barbels of *Linophryne arcturi*: (A) 51 mm SL, ISH 2132/71; (B) 34 mm SL, BMNH uncataloged. Drawings by R. Nielsen; after Bertelsen (1982).

in transparent knots. Each of the four chief stems ends in such a big knot. The knots contain, as far as can be seen through dyeing with Alum carmine and clarifying in glycerine, a layer of glandular cells which radiate from a lumen in the middle and are highest towards the vertex of the knot. They are therefore most likely organs of light.

More recently, the luminous function of the barbel of *Linophryne* has been confirmed by observations of light production in living specimens of six species. In a series of experiments performed by Hansen and Herring (1977) aboard the RRS *Discovery* in the North Atlantic, induced light production was recorded in *L. sexfilis*, *L. brevibarbata* (reported as *L. arborifera*), *L. indica* (as *L.* sp. A), and *L. trewavasae* (as *L.* sp. B). The barbels of all four species failed to emit light in response to immersion in adrenalin and to mechanical and electrical stimulation. But when washed with a dilute solution of hydrogen peroxide, all three primary branches of the barbel of *L. sexfilis* (BMNH 1976.11.4.1) emitted a glow that lasted about 30 seconds (Hansen and Herring, 1977). When similarly stimulated, the barbel of *L. brevibarbata* (BMNH 1995.1.18.4) gave off a steady blue glow, which persisted for one or two minutes; when examined with a binocular microscope, the light could be seen to originate from numerous tiny spherical tubercles at the tips of the terminal barbel branches. Again, when stimulated with hydrogen peroxide, the barbels of *L. indica* (LACM 36046-11) and *L. trewavasae* (LACM 36116-5) gave off a simi-

lar steady blue glow, and likewise the light was clearly observed to come from the tubercles themselves and not from the tissues of the barbel filaments. Subsequent histological studies of the barbel of *L. brevibarbata* clearly identified the spherical tubercles as photophores (Hansen and Herring, 1977).

Barbel luminescence has been witnessed firsthand in two additional species of *Linophryne*: a 33-mm specimen of *L. polypogon* (BMNH 2004.9.12.167) came aboard the RRS *Discovery* in the eastern central Atlantic in June 1981, with both esca and barbel glowing; an excellent photograph, taken by the late Peter David of the Southhampton Oceanography Centre, Southhampton, England, was published by Bertelsen (1986) and again by Bertelsen and Pietsch (1998a; see Fig. 181). In a similar observation, luminescence was recorded in a 39-mm specimen of *L. macrodon* (MCZ 164217), collected by the RV *Delaware* on Georges Bank in June 2004: the barbel (as well as the esca) gave off a continuous, steady white glow from the time of capture until the specimen was placed in formalin about 60 minutes later (Christopher P. Kenaley, personal communication, 3 June 2004; Fig. 258).

The barbel photophores are stalked, sessile, or embedded in the terminal branches of the barbel (Fig. 257). Their fine structure differs distinctly from that of any other known photophore (Hansen and Herring, 1977). Instead of originating from symbiotic luminous bacteria, which are absent in the barbels of all specimens examined, light appears to emanate from

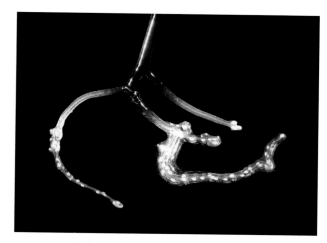

FIGURE 258. Hyoid barbel of *Linophryne macrodon*, 28.5 mm SL, MCZ 164217. Photo of living specimen made aboard ship by C.P. Kenaley.

a complex array of intracellular, paracrystalline photogenic granules bathed in a matrix of blood capillaries (Hansen and Herring, 1977; Fig. 259). The barbel photophores are probably mesodermally derived, in contrast to nearly all other known photophores, including ceratioid escal photophores, which are ectodermal in origin (Hansen and Herring, 1977).

Unlike the barbels of stomiiform fishes, which are clearly ectodermally derived and always contain distinct nerves (e.g., see Hansen, 1970), the barbel photophores of *Linophryne* seem to lack nerves altogether. Hansen and Herring (1977) were unable to identify nerves in the barbel photophores or in the branches of the barbel itself. Thus, innervation, if present at all, must be very diffuse and most probably related to the vascular system. In fact, the structure of the barbel light organs seems to be so intimately linked with the blood supply that an ontogenetic association with the cells and endothelia of the vascular system is probable. But, overall, a morphological interpretation of the barbel photophores is difficult; they are not readily comparable with any other light organs in fishes. A superficial similarity exists between the barbels of *Linophryne* and the barbels of some stomiiform fishes, especially the stomiid subfamily Melanostomiinae, but, in addition to the apparent lack of innervation, the histological structure of the organs as well as the ultrastructure of the photocytes differs significantly (Hansen and Herring, 1977; Fig. 259).

Extrapolating from the anatomy described above, control of barbel light production is most likely mediated by the blood supply. Realizing that control of luminescence in most all other luminous fishes is believed to be nervous, and the barbels of stomiiforms always contain prominent nerves, the virtual absence of major nerve fibers in the barbels of *Linophryne* is very unusual. On the other hand, the complex vascular network and abundant blood supply may indicate not only a means of supplying oxygen for the oxidative processes of luminescence, but also the possibility of control of such processes (Hansen and Herring, 1977).

### Teeth, Dermal Spines, Fin Rays, and Skin Patches

In addition to complex luminescent organs associated morphologically with spines and rays of the dorsal fin (the escae and caruncles), and the unique hyoid barbel of *Linophryne*, various published observations of light production have been

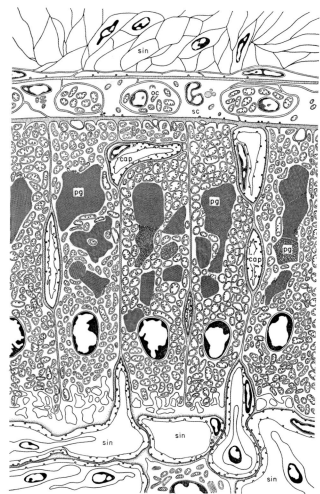

FIGURE 259. Diagrammatic representation of the organization of photocytes and associated tissues in a barbel light organ of *Linophryne arborifera*, 77 mm SL, ISH 2736/71. cap, capillary; oc, "ovoid" cell; pg, photocyte; sc, supportive cell; sin, sinusoid. After Hansen and Herring (1977).

recorded in other structures. Coincidently, William Beebe is responsible for nearly all these observations: in addition to glowing teeth in various taxa (*Linophryne arcturi*, 1926b, 1926c; *Oneirodes* sp., 1926c; *Chaenophryne* sp., 1926c, 1934c; *L. arborifera*, Beebe, 1932a; *Melanocetus* sp., Beebe, 1934a), which he thought might be due to a coating of luminescent mucous, he described an elaborate assortment of light-producing structures in his original description of *Himantolophus azurlucens* (Beebe and Crane, 1947:156; Fig. 260):

Body with conspicuous, sharp, sparsely scattered spines. Three of these, two on the left and one on the right upper posterior back, have luminous turquoise blue bases. A patch, not a spine, of the same color on upper, and another on lower base of peduncle; five luminous-based spines in a loose group on belly in front of anal fin; a triangular patch at anterior base of anal fin; others on the upper and lower caudal rays and on the anterior anal ray. In day-light all are brilliant turquoise blue, and in the darkroom we got pale blue luminescence from several of the spine bases and from two of the patches. In the dead preserved fish all trace of blue color is lost, the areas being distinguishable only by a slightly paler color of the tissue. Much of the illicium and its tentacles and many of the snout and chin papillae were distinguishable as pale grayish-white in

FIGURE 260. *Himantolophus azurlucens*, holotype, 98 mm SL, CAS-SU 46507, showing parts described by Beebe and Crane as luminescent. After Beebe and Crane (1947).

the dark, and several times before the fish expired we detected flashes of yellow light, possibly from the facial papillae and illicium stem, but strongly from the swollen distal end of the illicium club, at the base of the tentacles.

Although Beebe's observations of glowing teeth in ceratioids have not been reported since—nor has there been another case of "rhythmic flashing" from the mouth of a ceratioid, as witnessed in *Chaenophryne* by Herring and Morin, 1978:325—there is some additional evidence to verify some of the observations he described in *Himantolophus*: a 115-mm specimen of *Himantolophus albinares* (MCZ 161503), which remained alive for some 45 minutes after capture, appeared to emit white light from patches of skin on the anterior margin of the anterior-most ray of the dorsal fin, the dorsal margin of the caudal peduncle, and the lowermost ray of the caudal fin; light production was continuous and steady from the time the specimen came aboard ship until it was placed in formalin (Christopher P. Kenaley, personal communication, 30 July 2002). Similarly, a 39-mm specimen of *L. macrodon* (MCZ 164217) remained alive for some 60 minutes after capture, during which time the esca and barbel, and the distal tips of all the fin rays (dorsal, anal, caudal, and pectoral) gave off a continuous, steady white glow (Christopher P. Kenaley, personal communication, 3 June 2004).

## Luminescent Countershading

Young and Roper (1977:247) observed luminescence from the skin of a living female *Cryptopsaras couesii* (146 mm, with a 17-mm parasitic male; USNM 219906), caught at night off Oahu, Hawaii, between the surface and 200 m: "The glow was very directional, which suggested a counter-shading function, although it was directed posteriorly. We could not detect the source of the luminescence, but it appeared to originate from the skin; where the skin was abraded or purposely cut, there was no luminescence. Except for the anteriorly placed, blunt lower jaw, all of the black skin, including that on the fin rays and on the dwarf male, luminesced."

When placed in an aquarium, the fish immediately took on a head-up, vertical position in the water, directing the low-intensity glow downward (Fig. 261). Tested with specialized apparatus, it was found to be capable of adjusting this countershading luminescence to match the intensity of downward-directed light: "when the overhead light was turned on the fish was darkly silhouetted, but it rapidly increased its luminescence until it virtually disappeared from view." Through a series of trials in which the intensity of the overhead light was altered, the fish continued to match the overhead illumination perfectly, while on the last two trials its luminosity was slightly greater than the overhead illumination:

> After 10 min the fish discharged luminous material from the caruncles, turned head downward, and beat its tail vigorously for 3 min while its head pressed against the bottom; finally it lay motionless on the bottom. Although the intensity of the overhead illumination was decreased, the fish did not resume the head-up position. The fish was preserved while still alive nearly 2 h later after considerable additional handling and observation.

Although there is no reason to doubt Young and Roper's (1977) detailed description, no such luminescence, nor any organs or structures that might be assumed to provide this function, has ever been recorded in any other ceratioid.

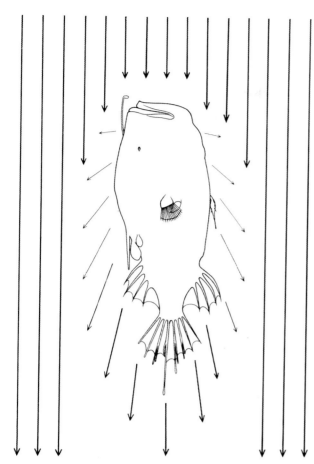

FIGURE 261. Diagrammatic representation of a living specimen of *Cryptopsaras couesii*, 146-mm SL female, with a 17-mm SL parasitic male, USNM 219906, positioned head-up in an experimental apparatus, emitting downward-directed low-intensity luminescence in response to light from above. After Young and Roper (1977).

## *In Situ* Observations from Submersibles

There are only two published sightings of ceratioids in situ, both made from remotely operated vehicles (ROVs): one or more individuals of an unidentified species of *Gigantactis* reported by Moore (2002), and a single unidentified individual of *Oneirodes* described by Luck and Pietsch (2008). In a video recording of the latter, the esca and most of the distal tips of the fin rays are distinctly illuminated. While it appears that these structures, especially the esca, are producing their own light, it turns out that the powerful lights of the ROV overwhelm and mask any biological light that might be coming from organisms at depth. What can be observed on the video therefore is only light from the ROV reflected back by the unpigmented distal surface of the esca and tips of the fin rays (for a lengthy description of these two observations, see Locomotion, Food, and Feeding, Chapter Seven).

## Biological Significance of Luminous Structures

Nearly all we know about the biology of ceratioids is based on indirect evidence—biological function and behavior extrapolated almost solely from morphology, a sometimes criticized approach often referred to as "armchair biology." The very few sightings of these animals *in situ* from the observation windows of submersibles have added little to our general understanding (see Locomotion, Food, and Feeding, Chapter Seven). There seems no question, however, that the illicium with its terminal bioluminescent bait serves as a lure to attract prey. Successful luring behavior has now been observed and recorded countless times in shallow-water lophiiform relatives, primarily of the families Lophiidae, Antennariidae, and Ogcocephalidae (e.g., see Wilson, 1937; Pietsch and Grobecker, 1987); there is no reason to doubt that ceratioids obtain their food in the same way. The illicial apparatus is superbly constructed for this function in nearly all anglerfishes, but that of ceratioids is extraordinary in its anatomical complexity and function. The ceratioid esca is capable of (1) attracting other organisms through its bioluminescence and external embellishment of appendages, filaments, and pigmentation, and perhaps also by secretion of chemical attractants; (2) detecting the close presence of other organisms through its nervous supply; and (3) a wide range of precisely controlled movement by an array of both intrinsic and extrinsic muscles that insert on the illicial pterygiophore and illicial bone. So well adapted for a luring mode of energy capture, the typical ceratioid female—with its short, almost spherical body; huge head and mouth, bearing numerous large teeth; and weak body musculature, incapable of providing for sustained swimming—has been described as a passively floating, baited trap, wriggling its bait to entice prey toward its cavernous mouth, and engulfing would-be predators by a powerful suction made possible by a sudden and violent expansion of the oral and opercular cavities (Waterman, 1948; Bertelsen, 1951; Bertelsen and Pietsch, 1998a).

The tremendous interspecific morphological variation in external escal morphology, combined with an even greater amount of internal variation, especially in the extent to which various reflecting and light-guiding structures are developed, obviously implies a great variety of luminous display (Munk, 1999). But the biological significance of this diversity is unknown. It has been suggested by a number of workers that ceratioid escae mimic small organisms that share the same water column. Certainly this is the case in at least some shallow-water anglerfishes. The esca of *Antennarius maculatus*, for example, shows an uncanny resemblance to a small fish (Pietsch and Grobecker, 1978:369, fig. 1, 1987:89, 311, fig. 28; see also Pietsch and Grobecker, 1990), while that of *Phyllophryne scortea* appears to mimic a pontogeneiid amphipod of the genus *Gondogeneia*: the esca is "a perfect copy of a swimming amphipod common along the south coast of Australia, but what makes this particularly interesting is that *P. scortea* feeds primarily on gobies of the genus *Nesogobius*, the several common South Australian species of which feed primarily on amphipods" (Rudie H. Kuiter, personal communication, 26 August 1984). Among deep-sea anglerfishes, Lütken (1872:334), in describing the esca of *Oneirodes eschrichtii*, wrote, "I am, of course, not in a position to indicate the purpose of this singular structure; but I will not conceal that the whole arrangement has above all produced a 'mimetic' impression upon me." Garman (1899:82) made a similar statement in his original description of *Dolopichthys allector*: "The esca evidently simulates the appearance of certain prey." Pietsch (1974a) went further by suggesting that by being different (i.e., by mimicking different kinds of organisms), the escae of closely related ceratioid taxa (such as species of *Oneirodes*) may attract different sets of prey organisms and, in doing so, may effectively partition food resources. In this way, closely related, especially sympatric, forms with similar needs may avoid or at least reduce interspecific competition for food. Realizing, however, the general scarcity of trophic resources in the lower mesopelagic and bathypelagic zones, this partitioning seems very unlikely. Rather it seems that any fish living at these great depths, where other organisms of a size worth the effort to consume are few and far between, could not, under any circumstances, afford to let a meal go by. The ability to eat anything and everything, even prey items of equal, or in some cases of greater size than the predator, seems most advantageous. The few analyses of stomach contents of ceratioids that have been carried out seem to support this contention. Bertelsen (1951:241), in looking at diets among juveniles and adults of all available ceratioid taxa, concluded that "one cannot find any distinct difference in their choice of food." Similarly Pietsch (1974a), in an examination of stomach contents of all available female material of *Oneirodes*, found a diverse array of organisms, including chaetognaths, amphipods, copepods, squid, and various kinds of fishes, but failed to find differences between species in the kinds of prey taken. Bertelsen and Krefft (1988) came to the same conclusion in their detailed examination of the stomach contents of *Himantolophus*. In the only known departure from these general conclusions, Pietsch (1972c) was able to correlate food and feeding habits with some morphological trends, including differences in escal morphology, among species of the oneirodid genus *Dolopichthys*. Of the six species of *Dolopichthys* recognized at the time, *Dolopichthys pullatus* had the longest jaws, greatest head depth, shortest illicium, greatest number of teeth, and simplest escal morphology and was thought to be the least derived member of the genus. On the other hand, *D. allector*, in which these character states were essentially reversed, was considered to be most derived. An analysis of the stomach contents of available material of these species (stomachs of five specimens of each contained food items) showed a dissimilarity in food items captured (stomach contents showing no signs of digestion were considered net contaminates and not included in the analysis). Eighty percent of the stomachs of *D. pullatus* contained only the remains of fish and crustaceans; the remaining 20% contained squid. Fish were not found in the stomach of any specimen of *D. allector*. Food items of the latter species were restricted to squid (some identifiable as *Bathyteuthis abyssicola*) and small, unidentifiable crustaceans. Speculating on small sample sizes, Pietsch (1972b) concluded that *D. pullatus*, lying near the base of the evolutionary trends, is better adapted to preying on fish; whereas, *D. allector*, at the apex of the trends, is better adapted to capturing squid. Extrapolating further, it might be assumed that the different escal morphologies of these two species might have something to do with attracting different organisms.

Another possible explanation for the species specificity and extreme morphological complexity of the ceratioid esca is the attraction of a conspecific male, an idea suggested previously by a number of authors (e.g., Waterman, 1948; Bertelsen, 1951; Munk, 1964, 1966; Marshall, 1971a, 1971b; Pietsch, 1974a, 1976, 2005b). Most fully metamorphosed male ceratioids, unlike other deep-dwelling midwater fishes, have well-developed eyes (Munk, 1964, 1966; Marshall, 1971a; Pietsch, 2005b). In addition, the olfactory organs of most ceratioid males are extremely large relative to their body size (Bertelsen, 1951; Marshall, 1967; Pietsch, 1976, 2005b). It seems probable that a species-specific chemical communicant excreted by the female is the initial stimulus for the male to begin his search for a mate (Bertelsen, 1951). At close range the species-specific escal morphology may be the necessary behavioral cue required before the male (in forms in which the males apparently attach only for a short time, as well as in those in which the sexes become permanently attached; see Reproduction and Early Life History, Chapter Eight) will attach to the female and undergo subsequent spawning: "Even if the eyes of the male are well developed, they can hardly observe and recognize a specific mate at any great distance, but the esca with its light organ and [species-] specific attachments may presumably function as a distinguishing mark which the males can recognize when they come sufficiently near" (Bertelsen, 1951:249). Munk and Bertelsen (1980:127), however, argued against the idea that specific features of the escae mediate pair formation: "The possibility that these specific characteristic organs aid sexual recognition seems contradicted by the fact that at least in some of the deep-sea anglerfish families the eyes of dwarf males are very poorly developed." While it is true that adult males of certain families have small eyes (e.g., the Centrophrynidae and Gigantactinidae), it seems rather extreme to reject the hypothesis as unlikely for all taxa. Otherwise, it is difficult to explain the well-developed eyes of males of other taxa (e.g., the Ceratiidae; for more discussion, see Reproduction and Early Life History, Chapter Eight).

The elements that provide support and mobility for the esca vary considerably among ceratioids as well, the differences being most obvious in the relative length of the illicial bone and illicial pterygiophore. In some species, the illicial bone is extremely short, almost totally enveloped by tissue of the esca. The most extreme example is found in *Haplophryne mollis* (and somewhat less so in *Phyllorhinichthys micractis*, *Microlophichthys microlophus*, and *Photocorynus spiniceps*), in which the esca is relatively large, lying sessile on the snout between the frontal bones, with very restricted mobility. In other taxa, for example *Melanocetus* and *Himantolophus*, the illicial pterygiophore is relatively short but bears a considerably longer illicial bone. An illicial apparatus of this construction, as found also in most shallow-water anglerfishes (e.g., see Pietsch and Grobecker, 1987), is capable of providing almost unlimited mobility. In one of the very few published observations of luring behavior in a living ceratioid, Beebe and Crane (1947:158) described great flexibility in the illicium of *Himantolophus azurlucens*: "the entire illicium was occasionally thrown forward, until the stem almost touched the snout, the slender remaining tentacle waving back and forth as the stem moved." Bertelsen and Krefft (1988:32) described similar movements of the illicium of living specimens of *H. groenlandicus* and *H. albinares*. In an unusually healthy specimen of *H. groenlandicus* (ISH 264/73) they were able to record heretofore unobserved details:

> The illicium was swept back and forth in periods [of] about six times a minute through the full arc of about 180° between its extreme posterior and anterior positions. The movements were in beat with the respiratory movements of the gill cover. Throughout these active periods the slender, silver-tipped [escal] appendages were waving with snakelike twists. We were unable to determine whether this twisting was caused solely by the produced turbulence of the water or included active movements of these appendages. Furthermore it was observed in this specimen that the short pair of light guiding distal appendages vibrated very distinctly when irritated by touch.

An even greater breadth and variety of movement of the luring apparatus is found in *Gigantactis*. The members of this genus are characterized by having an extremely long illicium, in some species (e.g., *Gigantactis macronema*) attaining a length of nearly five times the standard length (Bertelsen et al., 1981; Fig. 157). In contrast, the illicial pterygiophore is relatively short, and despite its restricted mobility relative to the cranium (only a few millimeters, even in the largest individuals), it is equipped with moderately developed extrinsic muscles (the supracarinales anterior and the anterior and posterior subdivisions of the inclinator dorsalis II; see Bertelsen et al., 1981; see Locomotion, Food, and Feeding, Chapter Seven). The extrinsic illicial musculature of *Gigantactis* does not provide for gross movement in the anteroposterior and lateral directions, as is the case in most other ceratioids (especially *Ceratias* and *Oneirodes*, see Bertelsen, 1943; Pietsch, 1974a, respectively). Instead, it is used to produce vibration that passes out along the stiff illicial bone to the esca and escal filaments. That most of this vibration is transferred to the surrounding water by the esca and its filaments rather than by the stem of the illicium has been confirmed through direct observation (R. J. Lavenberg, in Bertelsen et al., 1981:19). Immediately upon capture, the 408-mm holotype of *G. gargantua* was placed live in an aquarium. Several whiplike, backward and forward thrusts of the entire illicium were followed by moderately strong vibrations of the esca, with no apparent movement of the illicium, although quick, rapid contractions could be felt throughout its length. This vibratory action, combined with the bioluminescence of the bait, appears to be the most important mechanism of attracting prey; to what extent sweeping movements of the entire illicium are used in luring is unknown.

In yet another variation in the morphology of the illicial apparatus, the illicial pterygiophore of some ceratioids is exceedingly long, nearly equal to standard length in ceratiids, some thaumatichthyids (*Lasiognathus*), and some oneirodids (*Oneirodes*). As first described by Bertelsen (1943; see also Bertelsen, 1951:18, fig. 4) in *Ceratias holboelli*, the illicial pterygiophore of these taxa is capable of sliding back and forth on top of the head so that in the extended position the illicium and terminal esca extend far forward in front of the mouth, while in the fully retracted position this bone extends posteriorly, enveloped in a "sock" of tissue that appears as a "dorsal tentacle," reaching well beyond the insertion of the dorsal fin (Fig. 42). Although never observed in nature, direct extrapolation of function from morphology indicates an ability to greatly increase the distance between the twinkling bait and the waiting jaws, thereby greatly enhancing the luring capability of these fishes. While extension and retraction of the illicial pterygiophore is mediated by elongate protractor and retractor muscles in *C. holboelli* (Bertelsen, 1951), a unique arrangement of the illicial musculature of *Cryptopsaras couesii*, in which the protractors and retractors are wound around the pterygiophore in opposite directions, suggests that this longitudinal movement takes place by rotation (see Shimazaki and Nakaya, 2004).

Smooth muscle fibers found within the esca of several species indicate that the esca may be capable of ejecting luminous material out into the water, perhaps to attract prey or males, but more likely to confuse predators. In addition to describing continuous luminescence from the esca of a specimen of *Himantolophus sagamius* that was kept alive in an aquarium for eight days, Haneda (1968:3) wrote that "when the body of the fish was irritated, luminous mucus was ejected into the surrounding water from the pore situated at the front part of the main luminous organ. The range of the released luminous cloud was then almost equal to the body length of the fish, while the main and secondary luminous organs were still emitting light." Despite numerous attempts to duplicate this event in freshly caught specimens, and although stimulation of the esca of a live specimen of *H. groenlandicus* resulted in distinct contractions of the escal bulb (Bertelsen and Krefft, 1988), ejection of luminous fluid has not been reported in any other ceratioid. While Haneda's (1968) specimen was found floating on the surface, all other specimens examined for this function came from relatively deep trawls and "may have ejected all the secretion available for this during their struggle in the net" (Bertelsen and Krefft, 1988:33).

The dual bioluminescent system of *Linophryne* is unique—nothing like it has been found in any other organism. As emphasized by Hansen and Herring (1977:122), who are responsible for nearly all the work that has been done on this subject, "it is remarkable that it involves two entirely separate systems of light production, one being an open organ harbouring luminous bacteria, while the other is a closed organ with intrinsic, intracellular luminescence." It is also remarkable that the two systems "seem to be derived from different germinal layers, the esca being clearly ectodermal in origin and the photophores of the barbel probably mesodermally derived." While mesodermal light organs are not unknown, they are very rare, having been so far described only in the scopelarchid fish *Benthalbella infans* (Merrett et al., 1973), in which the luminous organs are derived from muscle tissue.

The natural selective pressure responsible for the evolutionary origin of separate luminous systems can only be guessed in the absence of any certain functional information about both escal and barbel luminescence from a biological point of view (Hansen and Herring, 1977). Hansen and Herring (1977) did find significant spectral differences in emission maxima from the two organs, but they concluded that while these differences may reflect their different developmental origins they are probably not of any functional significance. Bertelsen (1951:28) found the linophrynids to be "not only morphologically but also biologically distinct from all other ceratioids." Phylogenetic analysis indicates also that the Linophrynidae occupies a highly derived position within the suborder (see Evolutionary Relationships, Chapter Four)—the development of a complex barbel in only one of the five linophrynid genera, which, when compared with other ceratioid genera, is by far one of the most successful in terms of diversity (22 currently recognized species, second only to *Oneirodes*, by far the largest genus of the suborder, with 35 species), probably represents a specialization of considerable adaptive value that has evolved recently and rapidly in ceratioid evolution (Hansen and Herring, 1977).

It might be assumed that the barbel, equipped with what must appear in life as a cluster of tiny "dancing lights beneath the female," is an adaptation to attract a particular type of prey organism (Hansen and Herring, 1977:123), but an analysis of stomach contents of all available material of the genus (Bertelsen, 1951) failed to show any significant dietary differences when compared to other ceratioids in which a barbel is absent. Successfully reproducing males probably become parasitically attached to females in all species of *Linophryne* (Pietsch, 2005b), but because other linophrynids also have parasitic males, the notion that the luminous barbel may serve as a species-specific sexual attractant in this one genus seems unlikely (Hansen and Herring, 1977). The function of the esca is generally believed to be the attraction of prey and a means of species recognition, but there is also some evidence that a bright flash from the esca may serve to confuse or dazzle a potential predator. Such a function may also be assigned to the barbels, whose structure seems to indicate that they are capable of a faster reaction and/or a more rapidly repeated flashing than the esca, which is essentially a continuously glowing organ. However, convincing evidence is lacking, and it is unknown whether the function of barbel luminescence simply reinforces that of escal luminescence or whether it serves some additional or entirely separate purpose.

# SEVEN

## Locomotion, Food, and Feeding

> We have enough specimens . . . to have made some good guesses as to the kind of life they lead in the numbing and lightless waters of the very deep ocean. We are aided in this study of the deep-sea anglers by their relationship to the well-known frogfish, the angler of the shallow seas.
>
> CLARENCE PURVIS IDYLL,
> *Abyss*, 1964:197

Inhabiting such great depths in the vastness of the world oceans, living ceratioid anglerfishes are totally inaccessible by all reasonable means of observation. Their behavior is therefore almost completely unknown, and what we may assume about their life-style is a combination of "armchair biology" and a few scattered observations of individuals that have survived for minutes or hours aboard ship immediately after capture (in some highly unusual cases, ceratioids have been keep alive for several days in the laboratory; e.g., see Cowles and Childress, 1995:1633). But these opportunities to witness behavior firsthand have been few and far between, and in all cases the animals have been abnormally and severely affected by the immense stress of capture. The rapid release of pressure and increasing temperatures—endured by these animals in fishing gear during their quick ascent from the extreme cold and enormous weight of water at meso- and bathypelagic depths—wreak havoc on physiological processes. Moreover, the churning and grinding of dead and dying organisms in the cod end of the net more often than not cause considerable mechanical damage (Cowles and Childress, 1995:1631)—skin is torn away, delicate structures like illicia and escae are lost, and bones are broken. Ceratioids exposed to these conditions thus provide little or no information about their normal mode of existence.

Direct observations of these animals *in situ* made from manned submersibles or by cameras attached to remotely operated vehicles have not helped very much. Population sizes of these and other midwater organisms are so small, with solitary individuals so widely spaced, that the chances of encountering an anglerfish at depth are exceedingly small—a distinct disadvantage for those wanting to study ceratioids in their natural habitat, but a seemingly impossible situation for ceratioid males seeking conspecific mates. By knowing the number of ceratioids caught per unit effort aboard cruises of the *Dana*, Bertelsen (1951:249) provided a "rough estimate" of how scattered metamorphosed females might be at the depths where most are found:

From the speed of the boat whilst fishing and the diameter of the mouth of the net, we come to the result that on average there is scarcely 30 m between the single specimens in the fertile area of the distribution of the ceratioids in 1000 to 2000 m, where the stock is densest. This seems perhaps less than might be expected from the relatively small number of known specimens, but when we consider that this number applies to all specimens of all species, and how small a percentage of the stock is made up of the mature unoccupied [solitary] females, there must be, in spite of the considerable excess of the males, a great distance as a rule between such a female and the nearest, free-living male of the same species.

While their relative rarity makes observations of ceratioids from submersibles unlikely, the immense disruption caused by such vehicles is also a likely factor that impedes their direct observation. The noise and vibration given off by the motors, the illumination of search lights, and the approach of underwater gear are easily detected at great distances by midwater organisms—once described by my former major professor as "a freight train in the night that can be heard and felt by midwater fishes and invertebrates a city block away" (Basil G. Nafpaktitis, personal communication, February 1969)—affording plenty of time for them to move slowly and silently out of visible range. Consequently, in all the many years of underwater exploration, there have been only two published observations of living ceratioids at depth: the spectacular videos of upside-down gigantactinids taken in the eastern North Pacific by the ROV *Jason* (Marzuola, 2002; Moore, 2002), and footage of an oneirodid recorded off Monterey, California, by the ROV *Tiburon* (Luck and Pietsch, 2008), both examples described below.

In general then, the overall picture of how ceratioids make their living must be based on extrapolation from observations made on more readily accessible, shallow-water anglerfishes (e.g., see Pietsch and Grobecker, 1987), combined with what can be concluded from their external morphology, internal anatomy, early life history, and stomach contents, in the context of what we know about the environmental conditions under which they live (Bertelsen, 1951).

Because the larval stages of ceratioids occupy a very different habitat than the adults, and because metamorphosed females and males are so strikingly different in their morphology and behavior, any discussion of locomotion and of food and feeding must treat these life history stages separately.

## Locomotion

### Locomotion in Larvae

Ceratioid larvae live in water of relatively high temperature and low viscosity. Like the adults, they have no swimbladder, and apart from a liver that is rather well developed, especially in the older larvae, they show no deposits of fat or oil that might serve to reduce their specific gravity. On the other hand, the gelatinous tissue beneath the inflated skin, which is more or less present in all ceratioid larvae, most likely provides some buoyancy (see Reproduction and Early Life History, Chapter Eight). The unusually large pectoral fins of some larvae (those of the Caulophrynidae and the gigantactinid genus *Rhynchactis*; Figs. 55, 56, 161) must increase their buoyancy significantly, but most larvae are obliged to counterbalance the tendency to sink by continuously swimming. However, the nearly spherical shape of most ceratioid larvae, combined with their weakly developed fins, suggests that they are rather poor swimmers at best.

During metamorphosis, ceratioids of both sexes quickly descend to great depths (Fig. 205). How much of this activity is accomplished actively by swimming under the influence of changing light and temperature requirements is unknown. Generally speaking, older larvae occur at greater depths than the younger individuals. This stratification is probably due at least in part to shifting requirements but may also be connected with an increase in specific gravity as the musculature and especially the bones are developing. During metamorphosis a great reduction in the gelatinous subdermal tissue that surrounds the larvae also takes place, and it is possible that the resulting increase in density, accompanied by a decrease in the activity of the larva during metamorphosis, is partly the cause of the sinking.

### Locomotion in Metamorphosed Females

For most ceratioid taxa, the swimming powers of the females are significantly reduced during metamorphosis and later development. The head grows to a large size relative to the trunk and tail, the skeleton remains poorly ossified and the musculature weakly developed, thus helping to reduce the specific gravity. The dorsal and anal fins of most species are only slightly developed and often concealed under the skin; the pectoral fins, especially in comparison to those of other lophiiform fishes, are small and weak; and the pelvic fins are absent. As described by Waterman (1948:116, 143), even females of the genus *Gigantactis*, which, with their elongate slender bodies, must be among the best ceratioid swimmers, are probably relatively sedentary:

> Despite its elongate and slim body, there are several lines of evidence that imply that *Gigantactis* is a weak swimmer and that active locomotion does not make up a large part of its behavior pattern. First, the body musculature, which plays the major locomotory role in most good swimmers . . . , is weak and not well developed. Then the structure of the fins and their musculature is not that of highly effective propulsive organs. The caudal fin, for example, with its long trailing rays, resembles more the drooping comet tail of certain gold-fish varieties than it does the sturdy falciform tail of wide-ranging active fishes like the bonito. Furthermore, the position and anatomy of the pectoral fins indicate that these are not used as paddles for vigorous swimming. When they are important in propulsion, the pectorals in other fishes usually lie just ahead of and below the center of gravity of the body and have the adductor musculature more powerfully developed than that of the abductors [thereby leaving these fins with little leverage for propulsive force]. . . . But in *Gigantactis* the fin lies above the longitudinal axis of the body on which the center of gravity usually occurs; also the pectoral abductors considerably exceed the adductors in size. Finally, the pectoral girdle, which must anchor the fin to the rest of the body, is quite attenuate and weakly supported in this fish.

Published reports of females observed alive in aquaria immediately after capture are extremely rare. Perhaps the earliest is that of Norman (1930:355) who described swimming movements of a living female *Cryptopsaras couesii* that came aboard the *Discovery* in June 1926: "It was observed continually to beat the water with the pectoral fins, suggesting that this action was in some way connected with respiration." Bertelsen (1951:238) witnessed similar behavior in a living specimen of the same species brought aboard the *Dana* during a cruise in summer 1947: "I made the same observation [as Norman, 1930] and noted further that these movements produced a forward current in the aquarium, at the bottom of which the fish rested."

Beebe and Crane (1947:153), in their report of the ceratioids collected during the eastern Pacific expeditions of the New York Zoological Society, described similar behavior in a freshly captured ceratioid that they named *Melanocetus megalodontis* (now recognized as a junior synonym of *M. johnsonii*): "The fish lived for two hours after which we preserved it. In the dark room we caught a fairly strong orange gleam, given forth three times, but could not be sure the illicium was the source. It swam actively but slowly about, with alternative movements of the pectorals. When the caudal moved from side to side it wagged the whole fish."

Later, in that same publication, Beebe and Crane (1947:158, 173), in describing their newly discovered *Himantolophus azurlucens*, wrote the following:

> This Blue-lighted Anglerfish was brought up at 2:30 p.m. and was very much alive, swimming around its large dish, keeping upright and twice biting the finger of the senior author as he turned it over. The entire illicium was occasionally thrown forward, until the stem almost touched the snout, the slender remaining tentacle waving back and forth as the stem moved. It swam almost entirely by movements of the caudal fin but turned with the help of the pectorals. It lived until 5 o'clock in ice water, then expired slowly, and only lost its shape when put into preservative. The stretched mouth and the greatly distended gill arches gave it a wholly abnormal contour.

Still later, Beebe and Crane (1947:173) described a sexually parasitized female of *Borophryne apogon*:

> One of the two females with attached males was alive as it reached the surface, and in an aquarium swam strongly and easily, using the caudal fin for propulsion and the pectorals only for turning. Her eyes moved slightly as she turned. The illicium was usually extended well forward, but several times it was jerked back and forth, and once flattened into its groove. The jaws moved through a slight arc, but never closed on account of the length of the teeth. Under a hand lens, slight muscular twitches could be detected on the part of the parasitic male attached to her ventral surface.

Cowles and Childress (1995:1633), while measuring the aerobic metabolism of deep-sea fishes aboard ship in the central

FIGURE 262. Frame grab from a video taken by the ROV *Jason* at 5000 m in the North Pacific Ocean, showing *Gigantactis* sp. drifting upside down, with its illicium extended and esca held just off the bottom, as if attempting to lure prey from the soft muddy substrate. The two shadows on the bottom behind the fish are caused by lights on the ROV. Image courtesy of Jon Moore; after Moore (2002).

North Pacific off Hawaii, had several opportunities to retrieve specimens of *Melanocetus johnsonii* that appeared to be unharmed, and to observe them in a "healthy" state for several days at a time:

> During the experiment the anglerfish swam slowly, using primarily the caudal fin with some auxiliary motions of the pectoral fins; or else hung motionless in the water. The fish's physical condition could be assessed by a gentle touch to the body at the end of the experiment. In healthy specimens this elicited a slow swimming response if the touch was to the lateral or posterior body, or often a fast strike if the touch was on or near the esca.

Although, for the most part, female ceratioids drift passively in the water—a position most likely assumed for very long periods of time, while waiting patiently for a meal—they are able to move a short distance rather quickly by wriggling forward with the body and tail (Bertelsen 1951:238). This was only an assumption when first suggested by Bertelsen, but we now know that it is true from two *in situ* observations made from submersibles. The first of these was recorded at about 5000 m in the eastern North Pacific Ocean, about half way between Hawaii and the California coast (27°53′N, 141°59′W), during ROV *Jason* dive 271, on 28 September 1999 (Moore, 2002). Three females of an unidentified species of *Gigantactis*, each approximately 250 mm SL, were videotaped at different times during a 10-hour period (whether three individuals were involved or only one or two seen intermittently is unknown). Representing the first recorded behavior of a ceratioid in its natural environment, the following is taken almost verbatim from the excellent description provided by Moore (2002:1144; see also Marzuola, 2002). All three individuals were first observed drifting motionless in the water at the periphery of the visual field. In striking contrast to all we have assumed about these fishes in the past, they were closely associated with the bottom, positioned upside down, with the body held rigid, the unpaired fins completely splayed out, the mouth somewhat open, and the illicium held stiffly in a slight arc out in front (Fig. 262). The esca was held several centimeters above the bottom as the fish drifted along with the current. Instead of dangling a glowing lure in the open water column to attract prey, behavior that we would expect to see, the videos give the distinct impression that the anglerfish were trolling for small invertebrates on the seafloor. Numerous small burrows were visible in the soft muddy bottom just below the fish (Moore, 2002:1144).

None of the fish moved until the ROV was approximately 2 to 4 m away, at which point each individual exhibited the same escape response. While still oriented upside-down, each fish performed an initial fast-start escape response (the classic "C-start," which describes the shape of the body at its maximum curvature; Westneat et al., 1998:3041), swimming away in front of the ROV. As the fish moved off, the illicium swung under the body. For the first 5 to 6 seconds, the fish undulated, in sinusoidal anguilliform fashion, about three-quarters of the body length at a rate of two beats per second. After this period, the rate of undulations decreased to one beat per second. By this time, the fish was about 1 m above the ROV (which was typically positioned approximately 1 m above the seafloor). In one of the three individuals observed, the rate of undulations slowed to one beat per second, despite the fact that continued movement of the ROV brought it even closer to the fish. By about 15 seconds after the initial C-start, each fish had moved out of view.

While it is possible that the lethargic drifting behavior recorded in *Gigantactis* was a response to the noise, vibration, and/or lights of the ROV—the fish perhaps mesmerized or otherwise incapacitated by the intrusion, or, on the other hand, "playing possum" to avoid predation—Moore (2002:1145) argued that "the lack of any motion [of the fish] at the dim edge of the lit field [as the ROV first approached] suggests a continuation of behavior from the dark region beyond into the brightly lit field of view," despite the increased illumination and on-coming mechanical disruption. It seemed very unlikely also that either the noise or the lights would invoke an upside-down orientation, given that many other submersible observations of meso- and bathypelagic fishes have failed to note such unusual behavior. That these individuals not only drifted upside down while presumably foraging for benthic organisms, but also clearly maintained this posture while attempting to escape from the ROV, was a further surprise. There is no hint of this kind of behavior in gigantactinids, or, for that matter, in any ceratioid, based on their general morphology. It may very well be that this behavior is displayed only when the fish is near the bottom, but the use of the esca as a lure to attract benthic prey, or even possibly to probe the substrate to flush burrowing prey out of hiding, is yet another amazing feature of the biology of these extraordinary fishes.

In a second example, the ROV *Tiburon*, designed, built, and operated by the Monterey Bay Aquarium Research Institute (MBARI, 1997), captured 24.4 minutes of footage of a living ceratioid. The video was taken off the coast of Monterey, California (36.329°N, 122.899°W), on 6 June 2005, at a depth of 1474 m and a water temperature of 2.8°C. The fish was not captured, but identified from the video as either *Oneirodes acanthias* or *O. eschrichtii*, based on geographic distribution and observable features of the escal morphology (Pietsch, 1974a:44, 63).

During the approximately 44,000 frames in which the anglerfish was clearly visible, it exhibited a full range of behaviors from passive drifting to rapid "burst" swimming (Luck and Pietsch, 2008; Fig. 263). When first approached by the ROV, the fish remained motionless, drifting slowly with the current, maintaining this behavior intermittently during approximately 73.4% of the recorded time. While drifting passively, the orientation of the fish relative to the horizontal plane varied almost continually, the animal going through a kind of three-dimensional, slow-motion tumble, from right-side up (as we perceive a fish) to upside down, and from snout-up to down, and back again. But despite this variability, it appeared to be neutrally buoyant, requiring no effort to maintain its relative position in the water column.

Approximately 24.6% of the time, the anglerfish slowly beat its pectoral fins, creating very little if any body displacement (Luck and Pietsch, 2008). During these pectoral movements, the fish was almost always oriented with its snout directed upward. Occasionally, however, while weakly oscillating its pectoral fins, it drifted slowly into a head-down position; but twice it followed this movement with a quick point-turn that reoriented the fish 180°, so that it again faced upward.

On four occasions, the angler accelerated rapidly away from the camera, apparently in response to the approach of the ROV. Each of these burst swimming events was marked by an initial C-start escape response, followed immediately by a rapid increase in caudal-fin beat frequency (Fig. 263). After about 5 seconds, the angler in every case quickly decelerated but continued to swim in a more or less straight line at greatly reduced speed for an additional 15 seconds or so. In total, the fish spent approximately 1.9% of the time in burst swimming.

The illicium of the anglerfish was most often folded back onto the dorsal surface of the head and body. On two occasions, however, the angler rotated the structure away from its body nearly 180° so that the esca was brought forward and held out in front of the mouth. It was in this latter configuration that the animal was initially sighted by scientists conducting the dive. Each time the illicium was deployed, it was held in a fully extended position for about 30 seconds, without any discernible wriggling or vibration. It took approximately 7 seconds to rotate forward and slightly longer, 10 to 12 seconds, to bring it back to the nonluring position. In total, the illicium was fully deployed about 4.1% of the time.

When the ROV neared the fish, the esca and most of the distal tips of the fin rays were distinctly illuminated (Fig. 263A–C, 264). Excited initially to think this might be the first example of ceratioid bioluminescence observed *in situ* since William Beebe's numerous sightings from his famous bathysphere (see Bioluminescence and Luring, Chapter Six), it was disappointing to learn later that the powerful lights of the *Tiburon* overwhelm and mask any biological light that might be coming from organisms at depth. While parts of this individual of *Oneirodes* sp. may well have been luminescing at the time, what can be seen on the video is only light from the ROV reflected back by the unpigmented distal surface of the esca and tips of the fin rays (Bruce H. Robison and Peter J. Herring, personal communication, 27 April 2007 and 3 August 2007, respectively).

The overall level of activity and the major locomotory behaviors recorded in *Oneirodes* are consistent with those of *Gigantactis* described above: in both cases, the fishes were initially at rest and totally immobile, drifting passively with the current; they did not begin to move until approached by the ROV, at which time they displayed the classic C-start escape response and acceleration; and both decelerated significantly after 5 or 6 seconds of swimming, quickly resuming the same quiescent behavior as seen initially.

Moore (2002:1145) suggested that this pattern of behavior was in keeping with the standard energetic profile of a lie-in-wait predator (Kitchell, 1983; Cowles and Childress, 1995; Gartner et al., 1997). In the lower meso- and the bathypelagic zones, food availability is extremely low and stands, without a doubt, as the single most important factor in limiting the size of populations and individual organisms. Energy saving behavior is thus of obvious value. Lie-in-wait predators are particularly well suited to conserve energy because they effectively forage while remaining lethargic and, on average, devote just 2% of their energy intake to swimming activities (Kitchell, 1983). That an individual of *Oneirodes* sp. was initially sighted drifting quiescently with its illicium extended (Fig. 263A), that it demonstrated a significant lack of stamina for high-energy activity in its escape response, and that it remained completely motionless for nearly three-quarters of a 24.4-minute period

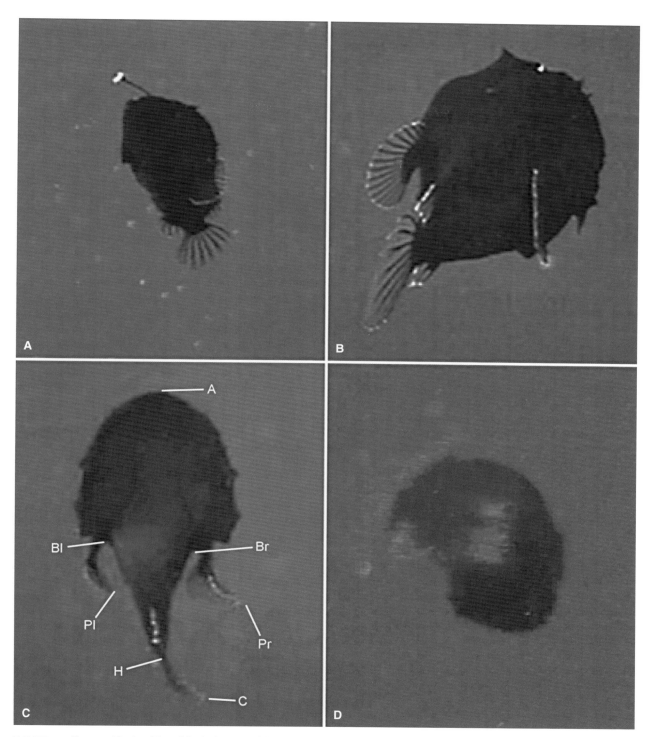

FIGURE 263. Frames of *in situ* video of *Oneirodes* sp. used for kinematical analyses (light appearing to emanate from the esca and unpigmented tips of fin rays is reflection from lights of the ROV): (A) initial sighting with illicium fully extended; (B) passive drifting with body and extended fins immobile; (C) anglerfish in dorsal view, showing points tracked in kinematical analyses: A, anterior-most margin of upper jaw; $B_l$ and $B_r$, base of left and right pectoral fin, respectively; C, distal tip of longest observable caudal-fin ray; H, posterior-most margin of hypural plate; $P_l$ and $P_r$, distal tips of left and right pectoral-fin rays, respectively; (D) initial fast-start escape response that marked the beginning of each burst swim. Courtesy of Bruce H. Robison and the Monterey Bay Aquarium Research Institute; after Luck and Pietsch (2008).

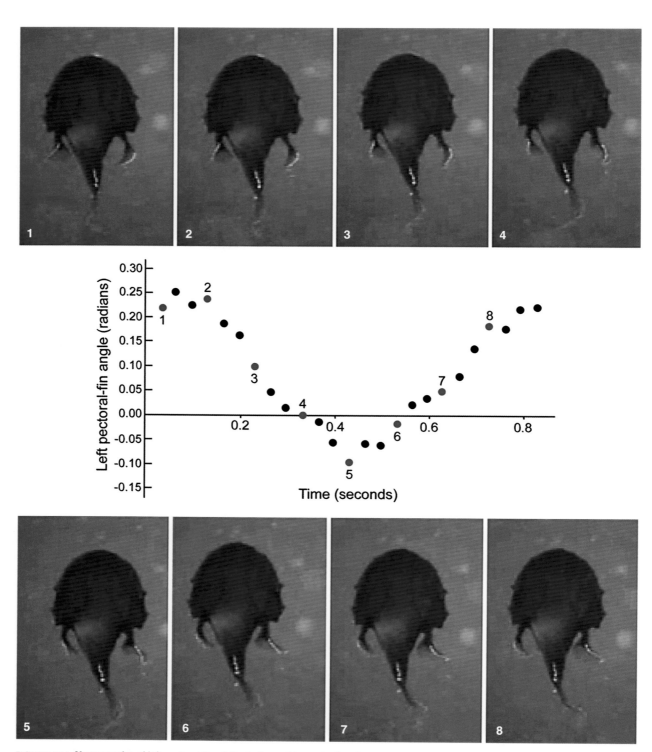

FIGURE 264. Vector angles of left pectoral fin of *Oneirodes* sp. (shown in dorsal view), recorded during a slow-swimming sequence (plotted as a function of time), showing in-phase pectoral-fin oscillation. Courtesy of Bruce H. Robison and the Monterey Bay Aquarium Research Institute; after Luck and Pietsch (2008).

strongly supports earlier speculation that this foraging strategy is utilized by oneirodids and is probably characteristic of all females of the suborder (Luck and Pietsch, 2008).

While the anglerfish drifted passively during most of the observation period, it also spent at least 24.6% of the time with its pectoral fins in motion. Cowles and Childress (1995) documented this "sculling" behavior in a series of rare laboratory observations of *Melanocetus johnsonii*, but the exact biological significance of sculling remains unknown. Several authors have suggested that the steady, continuous pectoral oscillations that characterize this behavior are related to respiration (Norman, 1930:355; Bertelsen, 1951:238). Additionally, sculling may help ceratioids maintain a relatively neutral position in the water column despite up- or downwelling currents (Luck and Pietsch, 2008). At the same time, however, the video of *Oneirodes* sp. included several sculling sequences that featured clear vertical translation of the body, thus the maintenance of body position probably does not explain sculling in all instances. It also appears unlikely that pectoral-fin motion is required to prevent the animal from sinking, because in some passive drifting sequences in which water turbulence was low (i.e., no apparent translation of particulate matter suspended in the water column), the angler held its vertical position.

One of the more interesting kinematic results obtained from the analysis of the video of *Oneirodes* sp. was the high degree of in-phase character between the pectoral fins in both the slow-swimming and directional change sequences (Luck and Pietsch, 2008; Fig. 264). This phenomenon was particularly pronounced in the slow-swimming sequence—in which the pectoral-fin angles reached local minima at the same time—but was slightly less pronounced in the directional change sequence. Nevertheless, the pectoral fins predominantly oscillated synchronously in both sequences; such in-phase character is relatively rare among undulatory swimmers (Thomas L. Daniel, personal communication, 15 April 2006). Synchronous pectoral-fin motion has been observed in some species of the fish families Ostraciidae and Scaridae, particularly at slow to medium swimming speeds (Hove et al., 2001; Korsmeyer et al., 2002). But unlike these taxa, ceratioids use their body and caudal-fin as the primary propulsive surfaces, thus the biological significance of in-phase pectoral-fin oscillation remains uncertain.

The passive, lie-in-wait, predatory life-style revealed by these *in situ* videos of *Gigantactis* sp. and *Oneirodes* sp. is undoubtedly characteristic of most if not all pelagic deep-sea ceratioids. Realizing that their closest relatives—shallow-water benthic anglerfishes—are among the most highly evolved sit-and-wait predators (see Pietsch and Grobecker, 1987), the lineage that led to the remarkable diversity of forms found today within the Ceratioidei was left with little choice of life-style as it diversified and invaded the great oceanic depths. Constraints of this kind, demanded by phylogenetic history, have been largely responsible for the breadth of form and function that characterize the group.

## Locomotion in Metamorphosed Males

In sharp contrast to the females, the swimming powers of the males apparently increase during metamorphosis. Relative to the head, which usually becomes somewhat elongate and pointed, the body increases in length and becomes more or less laterally compressed, while the dorsal and anal fins become almost completely enveloped by skin (Bertelsen, 1951). These changes, along with the development of a rather powerful caudal fin, appear to provide in the adult male a rather efficient swimming mechanism required for the arduous task of finding a proper female.

Free-living ceratioid males are very rarely taken alive. Small and delicate, they appear to be even more intolerant than females of the physiological and mechanical trauma of being taken by nets from the great depths they occupy as adults. In fact, Marshall (1974:5, 1979:258) seems to be the only one ever to record the behavior of a living male. Thinking that the males are rather good swimmers compared to the females—as suggested earlier by Bertelsen (1951) based on morphology—and that the quick escape response of the males might be better developed as well, he set out to test the hypothesis: "When I gently held the tail-fin of a male ceratioid in a pair of forceps he wriggled hard as though to escape. Using thumb and forefinger, I did the same to several females, but they promptly turned round and tried to bite me." From this simple experiment, Marshall categorized male ceratioids as having a positive escape response and females a negative escape response:

> Indeed, the females of most species have relatively short luminous lures, a globular or deep-bodied form, and fins nicely related to the centre of gravity, features that are admirably fitted to a quick rounding on their prey. After all, a female cannot always expect prey to swim both towards her flashing lure and straight for her large and gently smiling jaws. Prey may approach from behind, where, incidentally, the body is studded with free-ending lateral-line organs, able to detect nearby disturbances in the water (Marshall, 1974:5).

Bertelsen and Krefft (1988:33) reported similar behavior in two females of *Himantolophus groenlandicus*, each kept alive for about 24 hours aboard the *Walther Herwig* in 1982. A slight touch of the tail elicited an immediate and nearly explosive counter attack, but similar contact with the illicium and head gave no comparable reaction. But, aside from the single brief observation by Marshall (1974) described above, no additional direct information about locomotion in living males is available. With males having little or no interest in obtaining prey, and every effort directed instead toward the single ultimate goal of locating a conspecific mate, it is reasonable to conclude that they would be relatively well adapted for swimming. With the prospect of predators most likely coming from behind, it would be difficult to expect anything but a strong positive escape response.

## Locomotion by Jet Propulsion

Although never observed directly, it seems quite probable that ceratioids, both males and females as well larvae, also move about by jet propulsion, a kind of locomotion used generally for swimming and as an escape response in a wide variety of aquatic animals (see Pietsch and Grobecker, 1987:349). In lophiiform fishes, jet propulsion is made possible by the narrowly restricted, tubelike opercular openings located behind the pectoral-fin bases, which are characteristic of all members of the order (Pietsch, 1981; Fig. 265). Best described in shallow-water anglerfishes of the family Antennariidae (Gregory, 1928; Gregory and Conrad, 1936; Pietsch and Grobecker, 1987), the biomechanics of jet propulsion appear to be essentially the same as those involved in respiration, except that all of the movements are made with considerably more vigor. The propulsive force is generated by sucking a large quantity of water into the enlarged oral and opercular cavities and then forcing the water out through the small, restricted gill openings.

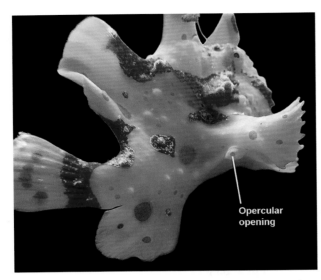

FIGURE 265. *Antennarius maculatus*, jet-propelling in open water by forcing water out through tiny restricted opercular openings situated beneath the pectoral-fin lobe; although never observed directly, there is every reason to believe that ceratioids, which have similar tubular opercular opening, move about in the same way. Photo by Andrew Goodson.

In antennariid anglerfishes, a rapid, posteriorly directed discharge from these jetlike openings results in forward progress at speeds that vary between 0.20 and 1.25 body lengths per second (Pietsch and Grobecker, 1987).

As with just about everything else concerning deep-sea anglerfishes, Bertelsen (1951:238) was the first to suggest that ceratioids might get about by jet propulsion:

> It has been observed that the Sargassum fish *Pterophryne histrio [Histrio histrio]* sometimes moves chiefly by means of the respiration current, which by means of the large mouth and gill cavity they press rapidly like a jet through the narrow gill-openings. . . . It seems reasonable to believe that the ceratioid females, with mouth and gill cavity even more [highly] developed, whilst their other swimming powers are greatly reduced, use this method of propulsion to an even greater extent. Possibly the forward current of the pectorals serves as a kind of brake on respiratory current and thus enables the fish, lurking for booty, to remain comparatively still in the water.

## Food and Feeding

What a fish eats as part of its normal diet in the wild is often difficult to determine, especially in fishes that are trawled up from great depths. While undergoing what can only be imagined as a horrible death in the confined space of the cod end of a net, fishes, in their attempt to escape, bite and sometimes swallow each other. At the same time, in their agony, but also under the pressure exerted by the mass of tightly packed organisms, they often regurgitate the contents of their stomachs, thus adding material to the mix that may well be subsequently ingested by other fishes. Whole organisms as well as partially digested pieces of this and that can thus become thoroughly jumbled among the captured lot. Small fishes and invertebrates newly captured may well end up in the stomachs of larger fishes, thus appearing as part of their normal diets. What might have been in the stomach of a viperfish may well end up in the stomach of a ceratioid and vice versa. The extent to which this "net-feeding" occurs in meso- and bathypelagic fishes is not well understood (Hopkins and Baird, 1975, 1977; Lancraft and Robison, 1980; Gartner et al., 1997). The only direct evidence for or against comes from the experimental work of Lancraft and Robison (1980) who introduced bogus food items into a net before launching it, and then examined the stomach contents of captured fishes after recovery. Their results indicated that contamination of stomach contents can and does occur, but the degree to which it occurs, and thus the magnitude of the bias imparted to dietary studies, is highly variable. Factors such as time at depth and speed of the trawl play important roles. Trawls sent to lower meso- and bathypelagic depths, where ceratioids are most commonly found, tend to stay down for a long time and are usually pulled more slowly to avoid damage to the fragile catch. Fishes under these circumstances succumb to the ordeal of capture more slowly—they remain alive and active for a longer time, thus increasing the chances and extent to which net-feeding occurs. Moreover, ceratioids, with their huge mouths and unusually short gullets—the stomach separated from the throat by little more than a large sphincter muscle—seem particularly susceptible to intermittent swallowing and vomiting while under extreme stress. In short, food studies based on stomach content analyses of meso- and bathypelagic fishes, especially those studies that propose large and all-encompassing conclusions, must be treated with caution. Lacking any other means to assess diet, however, much of the following discussion is based on examinations of the stomachs of preserved specimens.

### Food and Feeding in Larvae

Bertelsen (1951) examined the stomachs of about 100 larvae in the *Dana* collections and found that about half of these contained no identifiable traces of prey. The stomach contents of the remaining specimens were very much the same in all species, and thus a detailed study of a very large number of specimens was not undertaken. The examined material consisted almost exclusively of copepods and chaetognaths—not at all surprising, considering that these early life history stages live and eat predominately within the upper reaches of the epipelagic zone. Copepods, the majority belonging to the genus *Coryceus*, were found in the stomachs of larval *Cryptopsaras*, *Himantolophus*, *Melanocetus*, *Chaenophryne*, *Oneirodes*, *Microlophichthys*, *Dolopichthys*, *Caulophryne*, and *Gigantactis*. In a single case, the remains of some indeterminable larval stages of shrimps were found in a larva of *Chaenophryne* (Bertelsen, 1951).

Individuals of the chaetognath genus *Sagitta* were found in larvae of all the same genera noted above except *Caulophryne* (of which only a single specimen was examined), and also in the only available specimen of *Diceratias* as well as in all "*Hyaloceratias*" examined. *Sagitta* was found very rarely in larvae less than 10 mm in total length, except in the linophrynids examined. This is probably because the larvae of this family acquire strong teeth very early in their development, whereas the teeth in the earlier stages of other ceratioids are not suited to grasp prey of this kind (Bertelsen, 1951). The only other kinds of food identified were some small pteropods of the genus *Limacina* in a single larva of *Melanocetus* and some few larvae of *Oneirodes*. As in most other fish larvae, the eyes of ceratioid larvae are well developed, and prey are undoubtedly detected visually.

### Food and Feeding in Metamorphosed Females

Despite a large number of examinations, only scattered and incomplete information is available on the diets of female ceratioids. While it might be assumed that this lack of data could be remedied by a full-scale survey of preserved material in collections around the world, such an effort would quickly and invariably lead to frustration. First of all, the majority of stomachs examined are empty. This is not too surprising, given that ceratioids, and meso- and bathypelagic fishes in general, often regurgitate their meals when captured in nets (see the discussion of net-feeding, above). But it is also a reflection of the reduced energetic needs of deepwater organisms. Childress (1971) found that certain crustaceans living between the depths of 900 and 1300 m off California had a surprisingly low respiratory rate, about one-tenth of that of crustaceans taken in less than 400 m. If an animal's rate of respiration provides a measure of its food requirements, as suggested by Macdonald (1975), the food needed by the deeper-living crustaceans should be about one-tenth that of the shallower living species. This magnitude of reduction in metabolism led Macdonald (1975:313) to argue further that an animal such as *Ceratias holboelli* "may only require prey equal to 10 per cent of its own body weight at three month intervals or longer." Thus, for deep-sea anglerfishes, it may well be feast followed by long periods of famine. Eating only infrequently, combined with a sedentary life-style and an ability to consume relatively large, energy-rich food resources (whole fishes, through macrophagy), seems to be the only successful strategy in the deep-sea (see Arrington et al., 2002).

While most specimens have empty stomachs, many "food" items found in the buccal cavity and stomach of ceratioids are probably net contaminates, as mentioned above. Nevertheless, results from numerous examinations show that female ceratioids are predominantly piscivorous, their stomachs containing the remains of many different kinds of meso- and bathypelagic fishes, but traces of several different groups of invertebrates indicate a wide assortment of these generally smaller organisms as well (e.g., Saemundsson, 1922, 1927; Regan, 1926; Beebe and Crane, 1947; Maul, 1949; Bertelsen, 1951; Fitch and Lavenberg, 1968; Pietsch, 1972c, 1974a; Nolan and Rosenblatt, 1975; Uwate, 1979; Bertelsen and Krefft, 1988; Pietsch et al., 2004). Among invertebrates, cephalopods and crustaceans dominate, the latter including representatives of the Penaeidae, Gnathophausiidae, Amphipoda, and Ostracoda. Some coelenterates have also been found, as well as holothurians, the latter taken from stomachs of the benthic genus *Thaumatichthys* (Bertelsen and Struhsaker, 1977).

Ceratioids have long been categorized as midwater animals, occupying depths far off the bottom (Brauer, 1906; Bertelsen, 1951), but recent findings in stomachs indicate a benthic existence, at least for some taxa. For example, while most diceratiids have been collected by open midwater trawls, indicating a broad vertical range for metamorphosed females, between 400 and 2300 m (Uwate, 1979), Grey (1959) recorded an adult specimen of *Diceratias pileatus* (*D. bispinosus* of Grey, 1959) caught in gear designed for bottom-fishing. Subsequently, Trunov (1974) reported three specimens of *Bufoceratias wedli* from bottom trawls, and Anderson and Leslie (2001) reported a 117-mm specimen of *B. wedli* (SAM 29918) taken in a crab trap at 800 m. Seven specimens of *Bufoceratias* in the HUMZ collections (two *B. wedli* and five *B. thele*) were taken off the bottom, three of the four known specimens of *B. shaoi* were collected in bottom trawls (Pietsch et al., 2004), and finally, all 20 known specimens of *D. trilobus* were trawled off the bottom at depths of 503 to 1216 m (Pietsch et al., 2006). An analysis of 51 diceratiid stomachs (25 of which were empty) reported by Uwate (1979) revealed a variety of partially digested food items including fish and fish scales, coelenterates, various crustaceans, polychaetes, gastropods, and sea urchins (tests and spines). Remains of polychaetes, gastropods, and especially sea urchins indicate benthic feeding. Thus, in contrast to the midwater life-style generally assumed for most all ceratioids (known exceptions include the benthic genus *Thaumatichthys* and perhaps the recently discovered upside-down, bottom-swimming individuals of *Gigantactis*; Bertelsen and Struhsaker, 1977; Moore, 2002), it seems evident that at least some diceratiids, especially larger individuals, are at least partly associated with the bottom (Bertelsen, 1990; Anderson and Leslie, 2001).

Ceratioids at least occasionally take extremely large prey, sometimes with fatal results. Several examples of females found floating on the surface with large food organisms caught in the jaws and throat have been recorded; most likely these individuals have been brought to the surface by the expanding swimbladder of the prey. The earliest of these is Günther's (1864:301) original description of the type specimen of *Melanocetus johnsonii*. When first discovered, its "belly was very much distended, and contained, rolled up spirally into a ball, a Scopeline fish [*Lampanyctus crocodilus*, according to Regan, 1913], which measured about 7 1/2 inches in length and about 1 inch in depth" (Fig. 8). Next is Collett's (1886:142) account of *Linophryne lucifer*, based on a "little black fish" found floating alive on the surface off Madeira, apparently incapacitated by the ingestion of a large fish identified only as a "scopeloid, not far from being half a length longer than the [anglerfish] itself" (Fig. 13). Regan (1913) recorded another two cases, very similar to Günther's report mentioned above, and both involving *M. johnsonii*. Finally, in the most spectacular example, Paxton and Lavenberg (1973) described a 112-mm female of *Diceratias trilobus*, found floating on the surface off the island of New Ireland, Papua New Guinea, with a more than 369-mm rattail of the genus *Ventrifossa* protruding from is mouth (Fig. 266). Both specimens were dead, but they were otherwise in relatively good condition. The rattail had been engulfed to a point just behind the tips of the pectoral fins, its head distending the stomach and body wall of the angler posteriorly and ventrolaterally to such a degree that the caudal fin was no longer the most posterior part of the body. With such a huge mass filling the entire mouth and body cavity of the angler, Paxton and Lavenberg (1973:47) speculated that both fishes "probably died because they could not pass enough water over the gills to obtain sufficient oxygen." Because the jaw teeth of anglerfishes are constructed to fold inward—to allow prey to slide off backward into the rear of the throat to be swallowed—but not outward, there is no mechanism to allow for the ejection of large prey items once they are engulfed; thus, "the trial and error method of learning maximum food size must be severely limited" (Paxton and Lavenberg, 1973:50).

Predation by female ceratioids is assumed to take place by way of illicial luring, as is characteristic of most all lophiiform fishes (see Bioluminescence and Luring, Chapter Six), but there are some indications that feeding may occasionally occur on unattracted or unattractable prey items. The presence of fish scales or other fish fragments of individuals far too large to be swallowed whole may indicate occasional feeding as scavengers, but it could also be the result of net contamination (Bertelsen, 1951). No stomach contents have been found in the few known representatives of the Neoceratiidae, a family that lacks the illicium; nor are there any records of stomach contents in the Caulophrynidae and the gigantactinid genus *Rhynchactis*, both of which lack the escal photophore. How members of these taxa find their food and what they eat is unknown.

FIGURE 266. *Diceratias trilobus*, female, 112 mm SL, AMS I. 15602-001, found floating on the surface off the island of New Ireland, Papua New Guinea, with a 369+-mm SL specimen of *Ventrifossa johnboborum* protruding from its mouth. Photo by C. V. Turner; courtesy of John R. Paxton and the Australian Museum; after Paxton and Lavenberg (1973).

Thus, to summarize what we know about diet from stomach content analysis, female ceratioids utilize a broad diversity of mechanisms to obtain their prey—either through a wide variety of different angling devices (as well as feeding mechanisms, see below) or in ways not understood—yet there is no convincing evidence of interspecific differences in their choice of food (Bertelsen, 1951; Marshall, 1979). Based on the assumption that the esca simulates the appearance of certain prey, it has been suggested that different escal morphologies may attract different kinds of prey (Pietsch, 1974a:98): "By being morphologically different (i.e., by mimicking different kinds of organisms), the esca of closely related forms (such as species of *Oneirodes*) may attract different kinds of prey, and in so doing, split up the food resource. In this way, closely related sympatric forms with similar needs may avoid competition for food."

In support of this hypothesis, but based on very small sample sizes, Pietsch (1972c) was able to correlate food and feeding habits with some morphological trends, including differences in escal morphology, among species of the oneirodid genus *Dolopichthys* (for details, see Bioluminescence and Luring, Chapter Six). Later, however, in a much more detailed analysis of stomach contents of all available material of *Oneirodes*, no differences were found between species in the kinds of prey taken (stomachs contained a diverse array of organisms, including chaetognaths, amphipods, copepods, squid, and various kinds of fishes; see Pietsch, 1974a). But, again, sample sizes of each species examined were small, and net contamination may have biased the study. Nevertheless, prey selection in ceratioids may yet be shown by a more detailed statistical analysis of stomach contents of larger sample sizes, when and if they become available. Based on current knowledge, however, it seems clear that animals at great oceanic depths cannot afford to let a meal go by, no matter how large or small. There is apparently no choice for them other than the widest possible variety of food resources (Marshall, 1971b, 1979).

The attraction of prey through illicial luring is achieved by a combination of bioluminescence and escal movements that simulate the appearance and motion produced by organisms that normally form part of the diet of the prey (Pietsch, 1974a; Pietsch and Grobecker, 1978). By means of the illicial musculature, the esca and its various filamentous appendages can be made to wriggle and vibrate and, depending on the length of the illicium and protrusibility of the illicial pterygiophore, can be brought into a variety of swinging and sweeping movements (probably most similar to the swimming and respiratory movements of various small crustaceans).

The complex structure of the escal photophore and the associated light-guiding structures, as well as the observations of its bioluminescence, indicate that the intensity of the light can be regulated from a steady glow to a sudden burst of illumination (Bertelsen, 1951; Munk, 1999). The quality of the luminescence can be subdivided into a variety of "light effects": (1) bright spots of light emanating from the tips of light-guiding appendages and openings situated below the mirrorlike distal shield; (2) glow from semitransparent appendages and parts of the escal bulb; and (3) reflections from various silvery surfaces of these escal structures (see Bioluminescence and Luring, Chapter Six). This array of light-producing elements may simulate the bioluminescence produced by one or more small organisms as they might appear in nature, their bodies and limbs partly hidden in darkness, and their visibility changing with their movements. A luring device such as this may provide a more efficient attractant than a single spot of light.

How female ceratioids detect their prey is still open to speculation. It is unlikely that the small, laterally directed eyes, with their inability to produce a well-defined image, could play any significant role (see Munk, 1964, 1966). In species that have a short illicium, and those with an illicium borne on a retractable illicial pterygiophore, the acoustico-lateralis system of the head might serve as a detection apparatus, but it seems unlikely that this system could provide sufficient distance and directional signals required by those species with long illicia (e.g., *Gigantactis*, some species of *Oneirodes*, and the caulophrynid genus *Robia*). Although unsupported by direct observation, or by histological evidence of sensory structures, further study of the innervation of the esca might indicate that this organ and its various filaments and appendages are sensitive to touch or to pressure waves produced by moving prey. Similarly, the existence of gustatory structures located on the snout or escal appendages cannot be excluded.

## Food and Feeding in Metamorphosed Males

With few exceptions, feeding and growth of free-living ceratioid males stop at metamorphosis. The pincherlike denticular jaws, with their small, relatively few, and inflexible teeth, appear poorly suited for seizing prey or producing the amount of suction required to pull prey into their mouths (see Biomechanics of Feeding, below). Except for some prolongation of the body (which could be accounted for by a similar reduction in body depth and width), distinct growth during this stage has been observed only in males of the Himantolophidae and Melanocetidae; in correlation with this fact, stomach contents, consisting of chaetognaths and small crustaceans, have been found only in some metamorphosed males of these two families. Substantial growth of males after parasitic attachment to their respective females has been observed in ceratiids and some linophrynids (see Reproduction and Early Life History, Chapter Eight).

## Biomechanics of Feeding

The jaws and associated elements of the feeding apparatus of most female ceratioids are similar to those of other lophiiforms. Basically, the mechanism consists of a large gape and narrow, restricted gill openings, combined with pointed, hinged teeth; large, muscular upper pharyngeals; and a highly extensible pharynx and stomach, allowing engulfment of extremely large prey. The length of the jaws varies between genera, from 50 to nearly 75% SL in some species of *Linophryne* and the oneirodid genus *Tyrannophryne*, to less than 15% SL in the Gigantactinidae. The upper jaw is protrusible in most genera, but this is best developed in the ceratiids and some oneirodids. Highly specialized "snapping" or "gripping" mechanisms are developed in some genera (e.g., greatly enlarged mobile premaxillae, with long hooked teeth in *Thaumatichthys*, and a somewhat similar development of the lower jaw in *Gigantactis*; see below). Extremely long, flexible hooked teeth are mounted on the outer surface of the premaxillae and dentaries of *Neoceratias*. Greatly reduced, toothless jaws covered with large glands of unknown function are present in the gigantactinid genus *Rhynchactis*.

In contrast to females, male ceratioids have much shorter jaws—the lower jaw only about 15 to 20% SL in most specimens—becoming smaller relative to standard length during metamorphosis. Jaw teeth are generally lost during metamorphosis, and, as a consequence, the outer margins of the premaxillae and dentaries are characteristically notched. The premaxillae are gradually resorbed, leaving only small portions of the anterior heads of these bones in full-grown free-living specimens of some genera. The premaxillae are completely lost in males of the Ceratiidae. The anterior tips of the maxillae and dentaries of many genera develop into truncated bases to support the denticular bones. A posteromedial extension of the upper denticular bone of the Ceratiidae and Melanocetidae is modified to articulate with the anterior end of the pterygiophore of the illicium, the extrinsic muscles of the pterygiophore thus participating in the opening and closing of the jaws (Parr, 1930a, 1930b; see Reproduction and Early Life History, Chapter Eight).

## Gape-and-suck Feeding

Almost all female ceratioids and, for that matter, the vast majority of teleosts, are gape-and-suck feeders. They engulf prey by creating negative pressure (suction pressure) inside the mouth (Ballintijn and Hughes, 1965; Alexander, 1970; Liem, 1970), which results from a large increase in volume produced by rapid expansion of the oral and opercular cavities. The amount of expansion, as well as the rate at which it is produced, is crucial to the effectiveness of gape-and-suck feeding, and anglerfishes are surprisingly good at maximizing both volume increase and speed (see e.g., Whitmee, 1875; Gill, 1909; Beebe, 1933; Barbour, 1942a; Gudger, 1945a, 1945b; Gordon, 1955; Schultz, 1957). Judging by direct cinematographic analysis of feeding events in shallow-water anglerfishes of the genus *Antennarius*, the amount of oral expansion during a single feeding event is huge, considerably greater than that of other teleosts examined (see Grobecker and Pietsch, 1979; Pietsch and Grobecker, 1987). The much larger and more expansible oral and opercular cavities of most female deep-sea anglerfishes indicate an even greater, more effective sucking function, adapted to pull large prey inside the barrier of pointed, depressible jaw teeth and into the reach of the large teeth of the muscular upper pharyngeals (Figs. 33, 39, 165, 197). The alternating forward and backward movement of the upper pharyngeals transports the prey through the expansible pharynx and forces the material through the short esophagus into the extremely expansible stomach. While gape-and-suck feeding in ceratioids is undoubtedly more highly developed than in their shallow-water relatives, the basic biomechanics of lower-jaw depression, upper-jaw protrusion, and lateral expansion of the suspensoria and opercular bones are essentially the same as described in *Antennarius*. For a more detailed account, the interested reader is encouraged to consult the work of Pietsch and Grobecker (1987).

## Specialized Feeding Mechanisms

Aside from the vast majority of ceratioids that employ suction feeding, there are a few deep-sea anglers with highly specialized jaw mechanisms that are evidently adapted for seizing prey by grasping or snagging, without relying on negative pressure. Outstanding among these taxa are members of the families Gigantactinidae and Thaumatichthyidae. While these are all equally interesting, considerably more attention is given below to gigantactinids, which have been studied in much greater detail than the others. An overview of the muscles involved in feeding is given initially as a basis for understanding the functional aspects that follow.

## Myology of the Feeding Mechanism of the Gigantactinidae

The musculature of the feeding mechanism of gigantactinid females is similar to that described by Field (1966) for the much less-derived, shallow-water anglerfish, *Lophius piscatorius*. Differences are mainly found in the degree of development of various muscles, and in the reduction and loss of muscle segments due to a corresponding reduction and loss of bony parts. The muscles of the illicial apparatus of *Gigantactis* were first studied by Brauer (1908) and later redescribed by Waterman (1948) who added an incredibly detailed discussion and figures of the superficial muscles of the cheek, the pectoral girdle, and upper pharyngeal musculature (Figs. 267–269). Bertelsen et al. (1981)

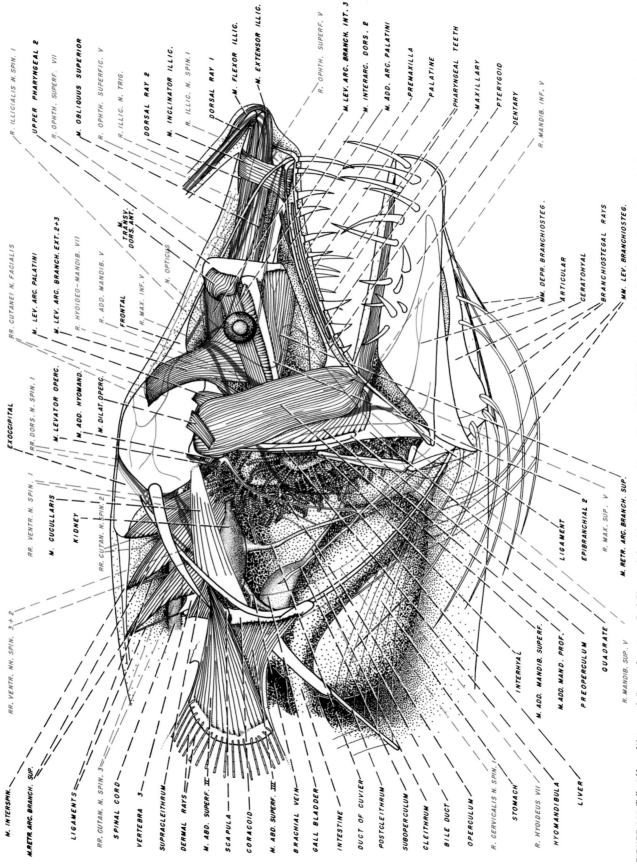

FIGURE 267. Talbot Howe Waterman's "general anatomy" of *Gigantactis longicirra*, holotype, 39 mm SL, MCZ 35065; lateral view of the head, after skin and superficial connective tissue had been dissected away; the nervous system is shown in red. After Waterman (1948).

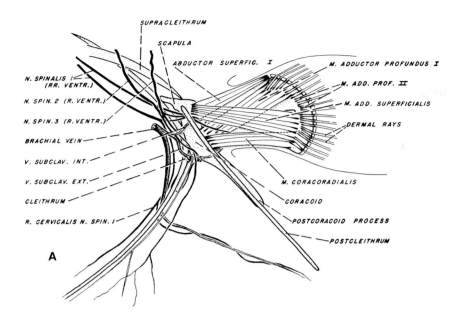

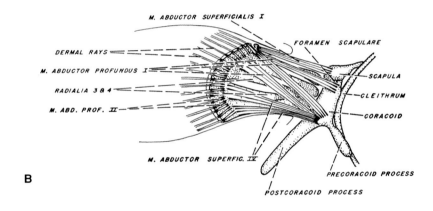

FIGURE 268. Anatomy of the pectoral fin and girdle of *Gigantactis longicirra*, holotype, 39 mm SL, MCZ 35065: (A) medial view of right pectoral fin and associated structures; (B) lateral view of right pectoral fin and girdle. After Waterman (1948).

took up the subject again, adding an account of the previously undescribed musculature of the lower jaw and floor of the mouth and throat.

CHEEK MUSCLES (Fig. 270): Section $A_1$ of the adductor mandibulae, defined by its dorsal position and insertion on the maxilla (Winterbottom, 1974), is absent in gigantactinids, corresponding to the severe reduction or loss of this upper jaw bone. In *Gigantactis*, section $A_2$ (the adductor mandibulae superficialis of Waterman, 1948:95) has a broad origin on the posterodorsal margin of the hyomandibula and a narrow insertion on the posterodorsal margin of the articular; it does not share a myocomma anteriorly with the posterior fibers of Section $A_w$ (see discussion of the lower jaw, below). Section $A_3$ (the adductor mandibulae profundus of Waterman, 1948:95) lies medial to $A_2$, originating broadly on the quadrate, the anteroventral margin of the hyomandibula, and metapterygoid. A small dorsal subdivision originating on the sphenotic is thought to be part of $A_3$. All of these muscles are essentially the same in *Rhynchactis*, the sister genus of *Gigantactis*.

The levator arcus palatini has its origin on the sphenotic and its insertion on the lateral face of the hyomandibula. A narrow dilatator operculi originates on the sphenotic and inserts on the proximal tip of the opercle. The levator operculi is also narrow, originating on the pterotic and inserting on the upper fork of the opercle. These are all essentially the same in *Rhynchactis*.

LOWER JAW (Fig. 271): In *Gigantactis*, section $A_w$ of the adductor mandibulae (articulodentary of Field, 1966:54; see also Winterbottom, 1974:242) is extremely well developed, but has lost all connection with section $A_2$, becoming purely an intrinsic lower jaw muscle. It covers nearly the entire medial surface of the lower jaw, stretching between the dentary and articular, its oblique fibers running in an anterodorsal-posteroventral direction. In lateral view, it can be seen extending well below the ventral margin of the lower jaw. In the material of *Rhynchactis* examined, no trace of this muscle could be found. This absence corresponds with the narrow, extremely reduced and toothless bones of the lower jaw of the latter genus.

FLOOR OF THE MOUTH AND THROAT (Fig. 271): The intermandibularis of *Gigantactis* is a relatively long, narrow muscle that has a broad insertion on the ventral margin of the

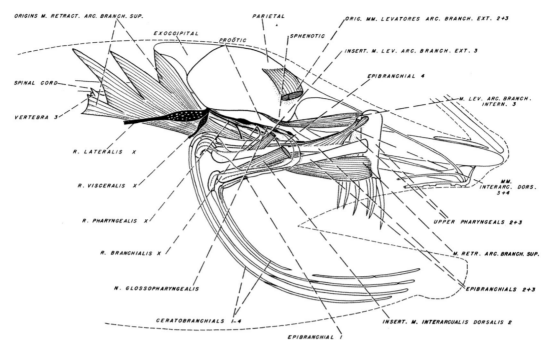

FIGURE 269. Gill arches and other elements associated with the hypertrophied upper pharyngeals of *Gigantactis longicirra*, holotype, 39 mm SL, MCZ 35065: "Despite the phylogenetic trend toward simplification of these structures in teleosts, a highly developed mechanism to aid in quieting and swallowing large prey has been evolved from these dorsal branchial elements" (Waterman, 1948:120). After Waterman (1948).

dentaries just behind the symphysis. Its transverse fibers pass ventral to a thick, crescent-shaped pad of elastic connective tissue that lies between the two halves of the lower jaw preventing them from meeting on the midline. No trace of this muscle could be found in *Rhynchactis*.

The protractor hyoidei of *Gigantactis* has a broad origin on the ventral-most margin of the dentary overlapping the insertion of the intermandibularis muscle. Each protractor hyoidei narrows posteriorly, approaching its counterpart from the other side and inserting on a fascia near the midline. From this narrow insertion, each protractor hyoidei widens farther posteriorly, attaching broadly to the respective ceratohyal. Only traces of this muscle could be found in *Rhynchactis*.

The sternohyoideus muscle of *Gigantactis*, originating on the cleithrum, splits into two sections: a considerably larger, lateral section inserts on the respective hypohyal; the medial section (not shown in Fig. 271A) passes to the distal tip of ceratobranchial IV. This muscle has essentially the same morphology in *Rhynchactis* but is slightly larger than that of *Gigantactis* of a similar standard length.

The hypaxial musculature forms a broad insertion along most of the posterior margin of the lower half of the cleithrum in both gigantactinid genera.

UPPER PHARYNGEALS (Fig. 272): As described by Waterman (1948:96), the largest and most complex musculature of *Gigantactis*, except for the body musculature, is the system that operates the hypertrophied pharyngobranchials and epibranchials of the second and third branchial arches. These muscles are essentially the same in both gigantactinid genera; *Rhynchactis* is mentioned below only when known differences or additions occur.

The levatores externi muscles (levatores arcuum branchialium externi of Waterman, 1948:97), of which only two can be differentiated, originate together on the parietal. The larger, more posterior of these muscles divides distally to insert broadly on epibranchials II and III. The smaller muscle, in contrast to the usual situation (see Winterbottom, 1974:250), and not mentioned by Waterman (1948), does not insert on an epibranchial but passes ventrally between epibranchials II and III to insert on the posterior margin of pharyngobranchial III just at the base of the pharyngobranchial teeth. The levatores externi muscles appear to be similar in *Rhynchactis* except that, in the absence of the parietal bones, they originate on the medial margin of the pterotic.

The levatores interni (levatores arcuum branchialium interni of Waterman, 1948:97), of which only one can be differentiated (most probably those that serve pharyngobranchials II and III have fused to form a single muscle mass; Waterman, 1948:97), originates on the prootic and sphenotic. It passes dorsal to the retractor dorsalis and medial to the levator externus muscles to insert on the proximal ends of pharyngobranchials II and III.

Three obliqui dorsales muscles are present in *Gigantactis*. That serving the second arch (interarcualis dorsalis 2 of Waterman, 1948:97) is especially well developed, originating on the dorsolateral surface of pharyngobranchial II and inserting on the proximal-lateral surface of epibranchial II.

The transversi dorsales originates partly on the parasphenoid and partly on a fascia near the midline. It passes medial to the levator externus and under the levator internus to form a large, bulbous insertion on the dorsolateral surface of pharyngobranchial II.

The retractor dorsalis (retractor arcuum branchialium of Waterman, 1948:97) is an enormous muscle that originates on the three anterior-most vertebrae and part of the fourth. Insertion is on the posteroventral margin of pharyngobranchial

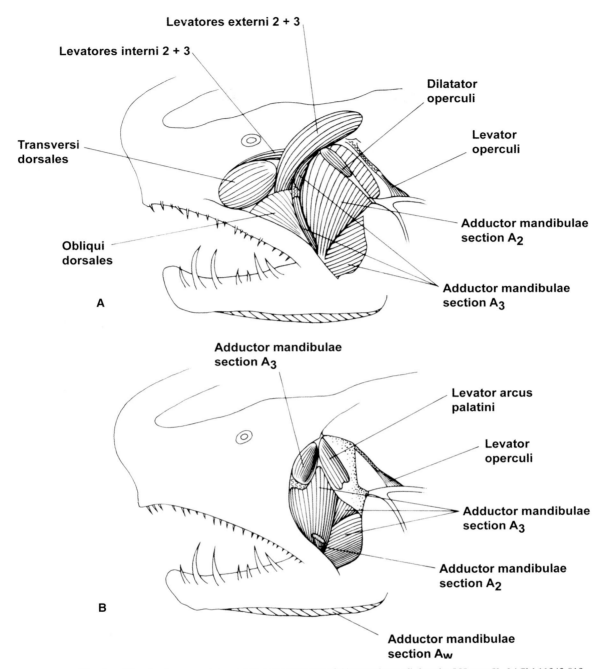

FIGURE 270. Muscles of the cheek, upper pharyngeals, and lower jaw of *Gigantactis meadi*, female, 353 mm SL, LACM 11242-012: (A) superficial musculature; (B) portions of superficial musculature removed. After Bertelsen et al. (1981).

III. In conflict with Waterman's (1948:97) description, no fibers of the retractor dorsalis insert on pharyngobranchial II.

ILLICIAL MUSCULATURE (Fig. 273): The illicial apparatus is controlled by five pairs of muscles: two intrinsic pairs, the depressor and erector dorsalis I (flexor and extensor, respectively, of Bertelsen, 1951); and three extrinsic pairs, the supracarinales anterior (exertor of Bertelsen, 1951), and anterior and posterior subdivisions of the inclinator dorsalis II (inclinator and retractor, respectively, of Bertelsen, 1951). The origins and insertions of these muscles were accurately described by Waterman (1948) and Bertelsen (1951). Compared to most other ceratioids, the extrinsic illicial muscles of gigantactinids are small. In contrast, the two intrinsic muscle pairs are unusually large.

## Functional Morphology of the Feeding Mechanism of Gigantactinid Females

*GIGANTACTIS*: Despite its restricted mobility relative to the cranium (only a few millimeters even in the largest individuals), the pterygiophore of the illicium of *Gigantactis* is equipped with moderately developed extrinsic muscles: the supracarinales anterior and the anterior and posterior subdivisions of the inclinator dorsalis II (Fig. 273). The extrinsic illicial musculature of *Gigantactis* does not provide for gross movement in the anteroposterior or lateral directions, as is the case in most other ceratioids (particularly *Ceratias* and *Oneirodes*, see Bertelsen, 1943, and Pietsch, 1974a, respectively). Instead, it is used to produce vibration that passes out along the stiff illicial bone

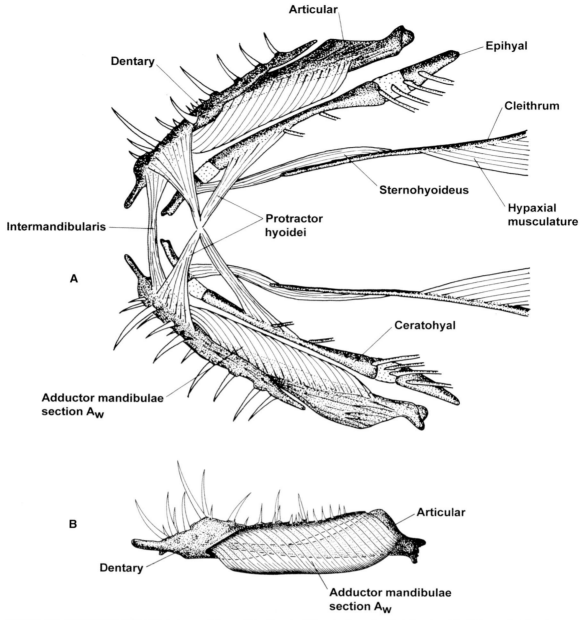

FIGURE 271. Musculature of the lower jaw and floor of the throat of *Gigantactis vanhoeffeni*, females: (A) 67.5 mm SL, ZMUC P921972, ventral view; (B) medial view of lower jaw showing intrinsic lower jaw muscles, 152 mm SL, ISH 802/68, left side reversed. After Bertelsen et al. (1981).

to the esca and escal filaments. That most of this vibration is transferred to the surrounding water by the esca and its filaments rather than by the stem of the illicium has been confirmed through direct observation (Bertelsen et al., 1981:19): "Immediately upon capture, the 408-mm holotype of *Gigantactis gargantua* was placed live in an aquarium. Several whiplike, backward and forward thrusts of the entire illicium were followed by moderately strong vibrations of the esca, with no apparent movement of the illicium, although quick, rapid contractions could be felt throughout its length."

This vibratory action, combined with the bioluminescence of the bait, appears to be the most important mechanism of attracting prey; to what extent sweeping movements of the entire illicium are used in luring is unknown. Feeding on unattracted (or unattractable) prey may occasionally occur, but how female ceratioids detect their attracted (or unattracted) prey is still open to speculation. In *Gigantactis*, this is especially difficult to understand with the large predator-to-prey distance resulting from the exceptionally long fishing apparatus (Fig. 157). It is unlikely that the small, laterally positioned eyes, lacking stereoscopic vision and an ability to produce a well-defined image (Munk, 1964, 1966), play any role in prey detection. The acoustico-lateralis system of the head, well developed in most ceratioids, may function in this way in those species in which prey is brought up close to the head, but it seems questionable that the system could provide sufficient distance and directional signals at the distance required by *Gigantactis*. Although unsupported by direct observation, the innervation of the esca

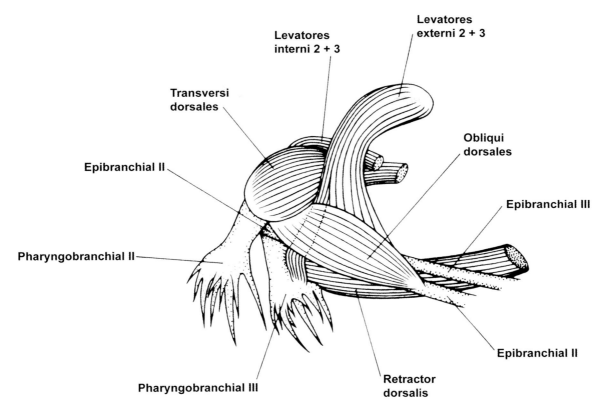

FIGURE 272. Musculature of the upper pharyngeals of *Gigantactis meadi*, female, 353 mm SL, LACM 11242-012, right side reversed. After Bertelsen et al. (1981).

described by Brauer (1908) and Waterman (1948) indicates that this organ and its filaments are sensitive to touch or to pressure waves produced by moving prey. It is hypothesized that once prey has been attracted to the bait, it is seized by a sudden, forward lunge, for which the streamlined body and powerful caudal fin of *Gigantactis* seem well adapted (Bertelsen, 1951).

The jaw mechanism of *Gigantactis* is similar to that of *Thaumatichthys* in that it allows the seizure of prey to be made by means of long hooked teeth placed outside the mouth, rather than through suction as is the case in nearly all other anglerfishes (Bertelsen, 1951; Bertelsen and Struhsaker, 1977; Grobecker and Pietsch, 1979; Pietsch and Grobecker, 1987). Although it is the lower jaw of *Gigantactis*, in contrast to the upper jaw of *Thaumatichthys*, that is specialized in this way, the principle is the same. In both mechanisms, the bones of the right and left sides of the jaw are free at the symphysis, connected only by elastic ligaments; each bone can be twisted relative to the other in such a way that their long, curved teeth can be rotated inward from a widely outstretched, open position to a situation in which the teeth of the opposite side approach each other and overlap within the cavity of the mouth (Bertelsen et al., 1981:19, fig. 23; see below). In both, the relatively small teeth in the opposing jaw (the upper in *Gigantactis*, the lower in *Thaumatichthys*) play a secondary role in seizing prey. It seems apparent that in *Gigantactis* the prey is snagged by the outstretched, recurved dentary teeth and brought into the buccal cavity within reach of the huge, upper pharyngeal teeth by a sudden, inward twist of the lower jaw (possibly aided by negative pressure created by a sudden expansion of the buccal and opercular cavities). The elastic connection between the rami of the lower jaw may allow for asymmetrical opening and closing of the mouth, much like the feeding mechanism of a snake. This alternating side to side adduction of the mandible, in association with the hooked and hinged jaw teeth, would facilitate the transport of prey items back toward the reach of the upper pharyngeal teeth. The extreme development and forward position of the upper pharyngeal teeth contribute to the efficiency with which the prey is seized and transferred to the stomach. Morphological evidence as well as direct observation on the living holotype of *G. gargantua* shows that the upper pharyngeals work in pairs on each side, each pair alternating with the other, and thrusting forward and pulling back with the teeth extended, like the pedals of a bicycle (Bertelsen et al., 1981).

Unlike other ceratioids, numerous individuals of *Gigantactis* have been captured with their stomachs everted; this eversion presumably occurs as a reaction to the stress of capture but could also indicate a mechanism by which *Gigantactis* is able to void the stomach of unwanted food. A stomach content analysis of all available material of *Gigantactis* was largely unsuccessful. In the majority of specimens examined, the stomach was either empty or everted. The stomachs of only nine individuals contained organisms, most of which showed little or no evidence of digestion and could well have been swallowed while in the net (see Bertelsen et al., 1981).

*RHYNCHACTIS*: As in *Gigantactis*, the restricted mobility of the pterygiophore of the illicium indicates that the primary function of the extrinsic illicial muscles is to create vibratory movement of the illicium and esca. In the absence of an escal bulb providing a site for the maintenance of bioluminescent bacteria, it seems apparent that the *Rhynchactis* esca

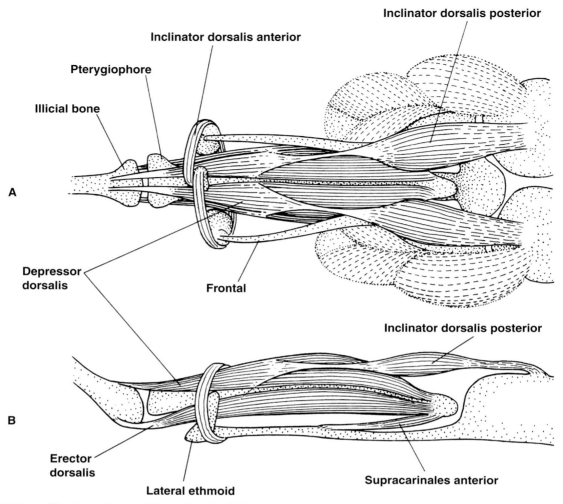

FIGURE 273. Illicial musculature of *Gigantactis vanhoeffeni*, female, 109 mm SL, ISH 2331/71, dorsal (A) and left lateral (B) views. After Bertelsen et al. (1981).

cannot produce light. If so, perhaps the attraction of prey is based solely on pressure waves produced by pterygial vibration, in addition to other movements of the illicium. The possibility of luminescence, however, cannot be completely disregarded.

In *Rhynchactis*, the reduction of nearly all the elements of the jaws, the loss of jaw teeth, and the reduced musculature reflect a very different kind of feeding mechanism than that found in *Gigantactis*. Prey attracted by the elongate illicial apparatus of both genera is far beyond the reach of the jaws. However, the lack of effective jaw teeth and less slender body of *Rhynchactis* make it unlikely that this genus is able to reach out and seize prey by a sudden forward dart as is supposed for *Gigantactis*. It seems more likely that the curious oral glands are in some way involved in luring the prey the remaining distance from the bait to within reach of the jaws and powerful pharyngeal teeth (Fig. 165). The inadequate preservation of the glands in all females examined does not allow conclusions about their function; bioluminescence cannot be excluded, nor can the more likely function of secretion of some chemical attractant. If the latter is true, and if the esca has lost the ability to produce light, then perhaps *Rhynchactis* is adapted to feeding on prey that are not attracted by light. Unfortunately, nothing is known about diet; the stomachs of all available females of *Rhynchactis* are empty.

## Functional Morphology of the Feeding Mechanism of Thaumatichthyid Females

As indicated above, the feeding mechanisms of the two thaumatichthyid genera, *Thaumatichthys* and *Lasiognathus*, are in many ways remarkably similar to that of *Gigantactis*. In fact, the similarities are so great that a detailed comparison of the two is given here as an extraordinary example of convergent evolution between two relatively unrelated groups of organisms. Most of the following is taken almost verbatim from the work of Bertelsen and Struhsaker (1977).

In his description of *Galatheathauma axeli* (*Thaumatichthys axeli*), Bruun (1956:177) assumed that juvenile and adult females of *Thaumatichthys* are benthic:

> There is a possibility that our fish was caught as the trawl was on its way up; but if that were so, it is strange that it had never been caught before. It seems to me to be far more probable that we here have a deep-sea angler-fish which lives close to or on the bottom. There it need only lie with its jaw open, leaving the large light organ with its two fine extensions to

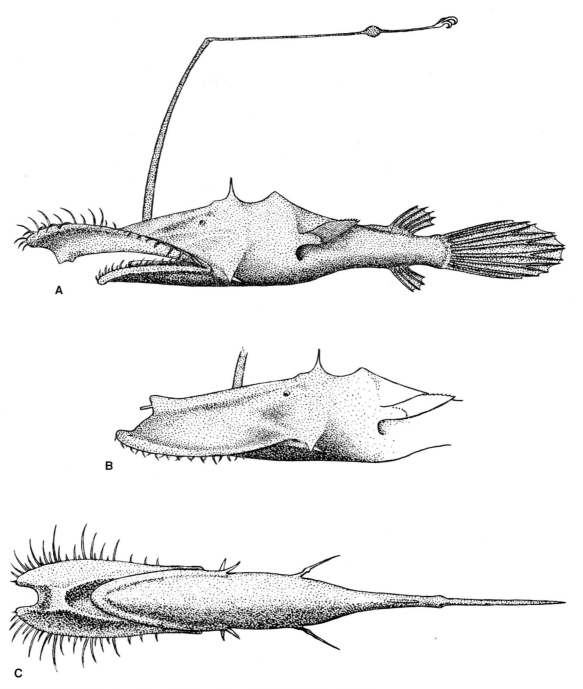

FIGURE 274. *Lasiognathus* saccostoma, holotype, 59 mm SL, ZMUC P92121: (A) left lateral view, showing upper jaw in the open position, with premaxillary teeth directed dorsally; (B) left lateral view of head, showing upper jaw in the closed position, with teeth of each premaxillae interdigitating and directed medially; (C) ventral view, showing extent to which the upper jaw extends anteriorly beyond the lower jaw. Drawings by W.P.C. Tenison; after Regan (1926).

lure the fish or prawn to it; then as soon as the prey is within reach of the long teeth the jaws will shut, the corners of the upper jaw falling down and trapping the prey against the lower jaw.

This assumption has now been supported by numerous records of specimens caught in bottom trawls and by the continued absence of catches in pelagic nets except for larvae and young juveniles (Bertelsen and Struhsaker, 1977). Furthermore, as discussed below, several features of the morphology of *Thaumatichthys* may be regarded as adaptations to benthic life. As a final proof of the correctness of this assumption, stomach contents of unquestionable benthic origin have been found in a specimen of *Thaumatichthys binghami*.

Most of the preserved specimens have been examined for stomach contents, but, except for the largest known specimen of *T. binghami* (294 mm SL, USNM 214571), they were all empty. The stomach of the latter individual contained three well-preserved specimens (50 to 80 mm long) of the holothurian

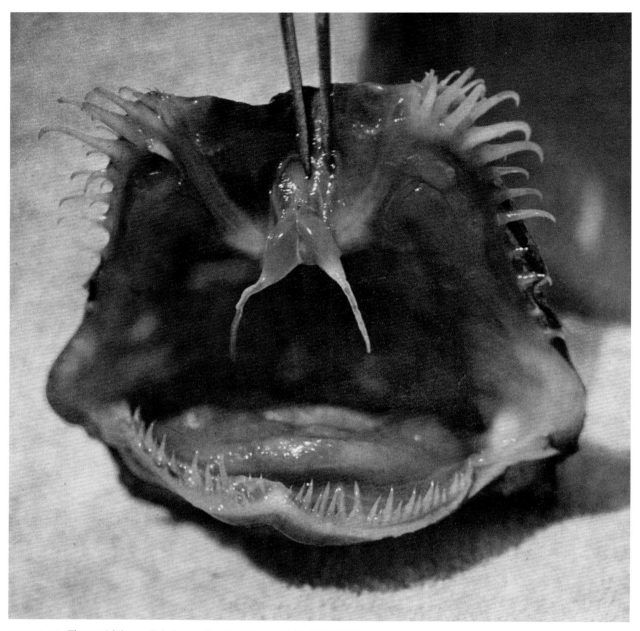

FIGURE 275. *Thaumatichthys axeli*, holotype, female, 365 mm SL, ZMUC P92166, anterior view, showing the characteristic forked esca protruding from the roof of the mouth; photographed aboard ship just after capture. After Bertelsen and Struhsaker (1977).

*Benthodytes typica*, a species that occurs in abyssal depths and the anatomy of which certainly suggests a benthic existence. A piece of pelagic seaweed of the genus *Sargassum* and some other indeterminable plant remnants were found embedded between the holothurians (Bertelsen and Struhsaker, 1977).

The extraordinary jaw mechanisms of *Thaumatichthys* and its sister genus *Lasiognathus* have been mentioned by several authors. Smith and Radcliffe (1912:581), who described the first known specimen of *Thaumatichthys*, chose the name *pagidostomus*, meaning "trap-mouth": "It would appear that the cavernous, elastic mouth is a trap into which the food is lured and dispatched." Regan (1926:31) illustrated the similar jaw mechanism of *Lasiognathus* with opened and closed premaxillae and hypothesized that prey are "evidently enclosed by the downward movement of the premaxillae, the bristlelike teeth being used not for piercing, but to close the aperture of the pouch" (Fig. 274). And, finally, Bruun (1956:177), in his original description, called the *Galathea* specimen "a living 'mouse-trap' with bait" (Fig. 275).

While in *Gigantactis* the major structural innovations of the feeding apparatus are found in the lower jaw, it is the upper jaw that is highly modified in thaumatichthyids. The anterior tip of each premaxilla of *Thaumatichthys* is connected to the heads of the maxilla and palatine bone of the opposite side by a highly elastic premaxillary ligament (Figs. 276, 277). These two ligaments, one on each side, cross each other just in front of the esca. The more slender anterior parts of the ligaments support the edge of a thin membrane that connects the premaxillae. Posteriorly the ligaments become much thicker and appear inflated where they pass dorsal to extraordinarily long

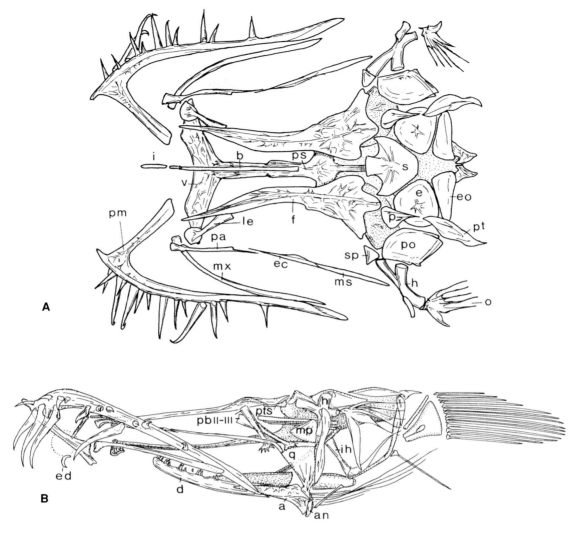

FIGURE 276. Skeleton of the head of *Thaumatichthys binghami*, female, 70 mm SL, ZMUC P921948, dorsal (A) and lateral (B) views. a, articular; an, angular; b, pterygiophore of illicium; d, dentary; e, epiotic; ec, ectopterygoid; ed, denticle of esca; eo, exoccipital; f, frontal; h, hyomandibular; i, illicium; ih, interhyal; le, lateralethmoid; mp, metapterygoid; ms, mesopterygoid; mx, maxillary; o, opercle; p, parietal; pa, palatine; pb II-III, pharyngobranchials; pm, premaxillary; po, pterotic; pro, prootic; ps, parasphenoid; pt, posttemporal; pts, pterosphenoid; q, quadrate; s, supraoccipital; sp, sphenotic; v, vomer. After Bertelsen and Struhsaker (1977).

articular processes of the premaxillae. The distal ends of the premaxillary processes are loosely attached to the maxillae by ligaments and also to semispherical, anteriorly directed projections of the ethmoid cartilage by connective tissue. In relation to the longitudinal axis of the head, the premaxillae are capable of rotating nearly 180° when they are moved from the open to the closed position (Fig. 277). The articular processes of the premaxillae point obliquely downward and inward when the upper jaws are lifted, and upward and outward when they are lowered. In both positions, the elastic premaxillary ligaments pull the tips of the premaxillae toward each other, thus contributing to the opening as well as the closing of the mechanism. These movements can be easily demonstrated and observed directly in preserved specimens that have retained the elasticity of the premaxillary ligaments. When the premaxillae are pushed from the open position in which they are usually preserved, the ligaments are stretched until a certain intermediate position is reached, at which point they cause the jaws to snap together into the closed position. The same movements can be demonstrated in well-preserved specimens of the closely related genus *Lasiognathus* (Figs. 85, 195).

The long and extremely well-developed upper jaw muscles (levator maxillae superioris), which insert posteriorly on the hyomandibulae and anteriorly on the proximal ends of the maxillae (Fig. 277A), are primarily responsible for the "spring" of the "trap" (Bertelsen and Struhsaker, 1977:30). The contraction of these muscles causes a backward and slightly laterally directed pull on the maxillae, which, directed by the cartilages and ligaments, is translated into an upward and outward pull on the articular processes of the premaxillae in such a way that these long processes, like a pair of handles, turn the jaws downward into the closed position. As no muscles are present to account for the lifting of the premaxillae, it must be assumed that the relapse into this open position is accomplished by the elastic ligaments, bones, and cartilages of the mechanism when the levator muscles are relaxed.

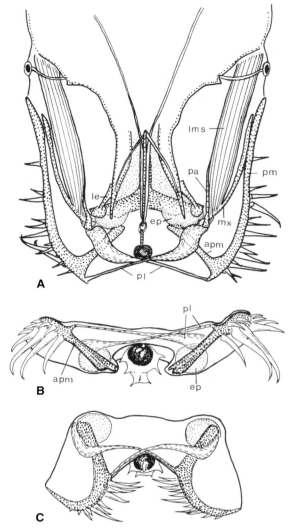

FIGURE 277. Jaw mechanism of *Thaumatichthys binghami*: (A) dorsal view, showing the premaxillae in the open position; (B) anterior view, with the premaxillae in the open position; (C) anterior view, with the premaxillae in the closed position. apm, articular process of premaxillary; ep, projection of ethmoid cartilage; lms, levator maxillae superioris muscle; pl, premaxillary ligaments; remaining abbreviations as given in Figure 276. After Bertelsen and Struhsaker (1977).

This highly specialized jaw mechanism permits the upper and lower jaws to function independently. With the premaxillae in an open position ready to snatch prey, the lower jaw is free to open and close in connection with respiration function, and when a prey item is trapped by the long teeth of the closed premaxillae it can be transported farther into the mouth within reach of the pharyngeal teeth by movements of the lower jaw.

By means of the intrinsic muscles of the illicium (which probably function primarily to produce vibration of the lure), the esca of *Thaumatichthys* can be raised to a more exposed position under (adults) or in front of (juveniles) the concave anterior edge of the membrane that connects the premaxillae. Alternatively, these same muscles allow the esca to be moved down and backward toward the roof of the mouth. When the premaxillae are closed, the esca is completely hidden and protected (Fig. 277C), with the possible exception of the greatly prolonged median appendage present in some specimens (Figs. 93, 275, 277).

The eyes, olfactory organs, and enlarged cephalic sensory papillae of thaumatichthyid females all seem well positioned on the head to detect prey or predators near or on the seafloor. The small eyes on the edge of the depressed head may appear well adapted to scan the bottom close to the jaws, but most probably they function merely as light detectors because the lens, as in other female ceratioids, is situated close to the retina and probably cannot produce clear images (Munk, 1964, 1966). This seems to exclude the possibility that light from the esca is used directly in search of food. The downward-directed nasal papillae and enlarged sensory papillae of the subopercular series of the acoustico-lateralis system, as well as the enlarged papillae on the roof of the mouth, are all located in optimal positions for the detection of prey on the seafloor.

That the only known stomach contents of *Thaumatichthys* consist of holothurians (which might not be expected to be attracted by light) as well as of quite large algal remnants confirms the impression obtained through examination of the food of other ceratioids (see above) that, in spite of their highly specialized feeding mechanisms, these fishes tend to be omnivorous. By means of these unique and bizarre mechanisms, they are able to secure prey that are otherwise difficult to obtain, but they will swallow anything that might serve as food.

## What Eats a Ceratioid?

Most of the information on predation of ceratioids comes from the rather extensive studies of stomach contents of the Black

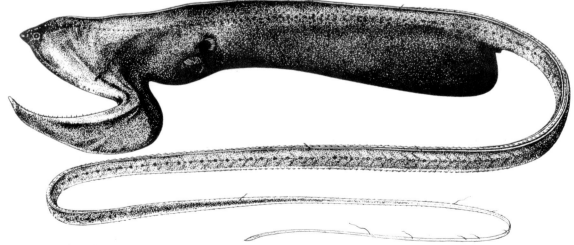

FIGURE 278. *Saccopharynx lavenbergi*, holotype, female, 950 mm TL, LACM 30532-001. After Nielsen and Bertelsen (1985).

Scabbard Fish (*Aphanopus carbo*) and the Longnose Lancetfish (*Alepisaurus ferox*) caught at depths of 600 to 800 m off Madeira (e.g., Maul, 1961, 1962a; Bertelsen and Krefft, 1988). Some additional data from the stomachs of western North Atlantic tuna (*Thunnus* spp.) and lancetfish (*Alepisaurus* spp.) are also available (see Matthews et al., 1977). The stomachs of these deep-dwelling predators contained a considerable number of ceratioids representing several species and most of the known families (mainly young females less than approximately 150 mm SL, including several in late larval stages). In addition, a number of large specimens of *Ceratias holboelli*, *C. tentaculatus*, *Cryptopsaras couesii*, *Himantolophus sagamius*, and *H. appelii* have been retrieved from the stomachs of sperm whales (*Physeter catodon*) caught in the Azores, and off South Africa, but also in subpolar waters of both hemispheres (see Clarke, 1950, 1956; Penrith, 1967; Bertelsen and Krefft, 1988).

Finally, the most interesting of the few examples of ceratioids found as prey for other fishes is one found in the stomach of a Deep-sea Swallower *Saccopharynx lavenbergi* (family Saccopharynidae), about 1070 mm total length (LACM 30385-25), captured off San Clemente Island, southern California, by the RV *Velero IV*, 26 to 27 March 1969, at a maximum depth of about 1000 m (Fig. 278). In addition to four partly digested specimens of the lanternfish, *Triphoturus mexicanus* (approximately 20 to 26 mm SL; family Myctophidae), the stomach of the swallower contained the disarticulated bones of a large ceratioid anglerfish. By comparing these bones, especially the opercles, hyomandibulae, lower jaw bones, and teeth, with those of readily available cleared-and-stained ceratioid material, the angler was found to be a member of the oneirodid genus *Oneirodes*, approximately 100 mm SL. Although specific identification was impossible, the species was most likely *Oneirodes acanthias* or *O. eschrichtii*, by far the two most commonly collected ceratioids off southern California. Clearly not a case of net-feeding, nearly all the bones of the skull were accounted for, including upper and lower jaws, the suspensoria, opercular bones, most of the branchial apparatus, and the pterygiophore of the illicium. The vertebral column, pectoral girdles, and caudal skeleton were absent, perhaps having been ejected from the stomach of the swallower at the time of capture.

Extremely little is known about the feeding habits of *Saccopharynx* (Nielsen and Bertelsen, 1985). In contrast to the closely related gulper eels (family Eurypharyngidae) that feed primarily on crustaceans and other smaller forms, saccopharyngids are thought to feed largely on fishes (Böhlke, 1966; Bertelsen et al., 1989). Like most ceratioids, members of both families have an enormously developed gape even as juveniles, able to swallow relatively large food items from the start of their deep-sea life (Mead et al., 1964). But, also like ceratioids, most preserved specimens have empty stomachs, which Nielsen and Bertelsen (1985) hypothesized is probably due to a tendency to vomit at capture, as indicated by Beebe's (1932b) observation of a live specimen that ejected a half-digested fish when brought on deck. In any case, this example of *Oneirodes* consumed by *Saccopharynx* is the only known record of a ceratioid taken from the stomach of a nonmigrating, fellow associate of the bathypelagic fauna.

# EIGHT

## Reproduction and Early Life History

> But to be driven by impelling odor headlong upon a mate so gigantic, in such immense and forbidding darkness, and willfully to eat a hole in her soft side, to feel the gradually increasing transfusion of her blood through one's veins, to lose everything that marked one as other than a worm, to become a brainless, senseless thing that was a fish—this is sheer fiction, beyond all belief unless we have seen the proof of it.
>
> WILLIAM BEEBE,
> "*Ceratias*—Siren of the Deep," 1938:52

In 1922, the Icelandic fisheries biologist Bjarni Saemundsson (Fig. 279) published a remarkable description of two small fish attached by their snouts to the belly of a large female deep-sea anglerfish identified as *Ceratias holboelli* (Fig. 280). Not recognizing them as males, Saemundsson (1922) described them as the young of the same species: "I can form no idea of how, or when, the larvae, or young, become attached to the mother; I cannot believe that the male fastens the egg to the female. This remains a puzzle for some future researcher to solve" (Saemundsson, 1922:164, translated from the Danish by H. Grönwold, in Regan, 1925b:387–388). A short time later, Regan (1925a, 1925b) dissected a small fish attached to a newly discovered female of *C. holboelli* and concluded that it must be a male parasitic on the female (Fig. 281). The male fish is "merely an appendage of the female, and entirely dependent on her for nutrition, . . . so perfect and complete is the union of husband and wife that one may almost be sure that their genital glands ripen simultaneously, and it is perhaps not too fanciful to think that the female may possibly be able to control the seminal discharge of the male and to ensure that it takes place at the right time for fertilization of her eggs" (Regan, 1925b:396–397).

Referring to them as "dwarfed males" that represent a heretofore unknown and "unique type of parasitism," Regan (1925b:397) searched for additional examples among the ceratioids collected by the Danish *Dana* Expedition of 1920–1922, discovering two additional attached males, both linophrynids, *Photocorynus spiniceps* and *Haplophryne mollis* (*Edriolychnus schmidti* of Regan, 1925b). Finding this highly specialized reproductive mode in what were considered to be widely divergent taxa, Regan (1925b:395) concluded that parasitic males are probably characteristic of all ceratioid fishes: "So far as is known . . . all the free-swimming ceratioids are females. Free-swimming males have not yet been found." He further summarized his extraordinary discovery as follows:

> The evolution of dwarfed males, parasitic on the females, in the Ceratioid Fishes, is intelligible if consideration be given to the conditions of life of these fishes. They are carnivorous, and are necessarily much inferior in numbers to the more active fishes that they prey on and attract by their luminous lure. They are wide-ranging, solitary and sluggish, and they float about in the darkness of the middle depths of the ocean. In such circumstances it would not be surprising if the difficulty experienced by the mature fish in finding mates had led the males to change their manner of life completely, in order to ensure the continuance of the race.
>
> I believe, then, that the first step was a change of habits; immature males formed the habit of attaching themselves to the females, preferably those approaching maturity, at the first opportunity that occurred. The ultimate result was that the males became dwarfed and parasitic.
>
> The structural peculiarities of the male—its small size, the outgrowths that unite it to the female, the absence of a lure and of teeth, the vestigial condition of the alimentary system—are all obviously adaptive. The evolution of these peculiarities must have been intimately related to, and even determined by, the changed activities of the male fish (Regan, 1925b:397–398).

Shortly thereafter, Parr (1930b) found that the ceratioid family Aceratiidae, the contents of which were long thought to be taxonomically distinct from all other ceratioids because of their extremely small size and the absence of the luring apparatus, consisted entirely of free-living stages of males, heretofore recognized only in the attached form. Thus it became possible for the first time to assign the free-living males to the same genera and often to the same species as the females (see Reallocation of Nominal Species of Ceratioids Based on Free-living Males in Part 2). In leading up to this discovery, Parr (1930a, 1930b) was the first to show also that rostral structures found in the males are homologous to the bony support for the illicium of the females, and that these structures are primarily responsible for opening and closing the pincherlike denticular jaws that allow the males to grab and hold fast to the females (Figs. 282, 283). Thus, assuming that males are at least partially attracted to the females by the bioluminescent display afforded by the illicium, the very structures that bring the sexes together are also those that allow for attachment. "An arrangement of this kind might certainly be regarded as an

FIGURE 279. Bjarni Saemundsson (1867–1940), Icelandic fisheries biologist and for 35 years (1905–1940) sole curator of what is now the Icelandic Institute of Natural History, Reykjavík. Courtesy of Steinunn Einarsdottir and Gunnar Jónsson.

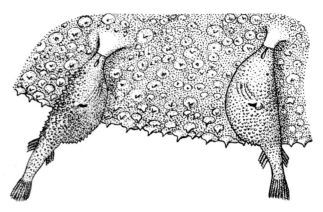

FIGURE 280. The earliest known examples of parasitic males (60 and 65 mm SL), thought initially by Bjarni Saemundsson to be the young of the species that somehow became attached to the mother, a 690-mm SL female of *Ceratias holboelli*, IINH uncataloged and ZMUC P922480. After Regan (1925b), based on a drawing published by Saemundsson (1922).

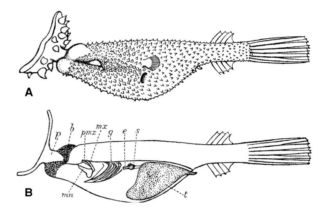

FIGURE 281. Parasitic male of *Ceratias holboelli*, 75 mm SL, attached to a 670-mm SL female, BMNH 1924.12.29.2, left lateral view: (A) external view; (B) internal view. b, outgrowth of tissue at point of contact of male and female; e, esophagus; g, gills; mn, mandible; mx, maxilla; p, papilla of female; pmx, premaxilla; s, stomach; t, testes. After Regan (1925b).

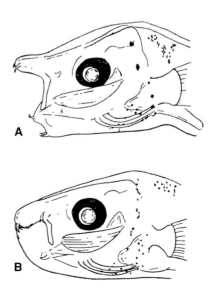

FIGURE 282. Head of *Cryptopsaras couesii*, free-living juvenile male, with mouth open (A) and closed (B). After Bertelsen (1951).

example of extreme phylogenetic economy of morphological parts" (Parr, 1930b:134; see also Parr, 1932:9).

Parr (1930b) supposed further that all the free-living metamorphosed males that had previously been placed in the anomalous family Aceratiidae sooner or later become attached parasitically to females. However, in support of earlier assumptions made by Regan and Trewavas (1932:21), Waterman (1939a:75) suggested that perhaps "in the more primitive families of the sub-order, in which no attached males have yet been found, that instead of actually growing fast to the females, . . . the males merely nip on to the females for a relatively short period." The supposition that a temporary attachment might occur in those species whose males do not become parasitic was further supported by Bertelsen (1951:257), who found evidence that the males of some ceratioids never become parasitic: whereas free-living, metamorphosed males of families in which parasitic males have been found become attached when immature and do not feed after metamorphosis,

> free-living himantolophid males grow essentially after metamorphosis and, according to the contents of their stomachs, melanocetid males must also be able to feed at this stage. In these and also the families Oneirodidae and Gigantactinidae, the testes of the free-living males attain a very considerable size and it is most probable that males of these four families never become parasitic, but only attach themselves for a time to the female by means of their denticular apparatus.

278  REPRODUCTION AND EARLY LIFE HISTORY

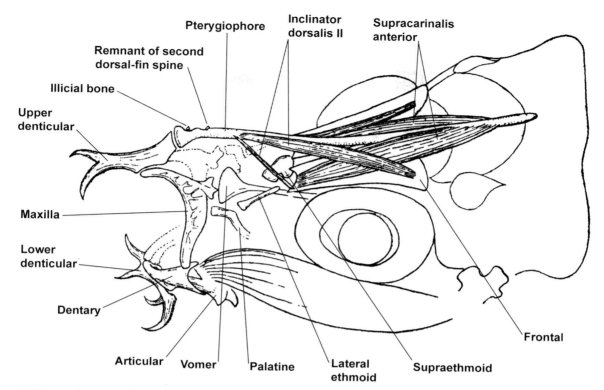

FIGURE 283. Anatomy of the head of a free-living adult male of *Cryptopsaras couesii*, dorsolateral view, showing among other things how the pterygiophore of the illicium makes contact with, and controls the movement of, the upper denticular bone, allowing the male to bite and hold fast to a female. After Bertelsen (1951).

Bertelsen (1951:250) found further that the males of those families that never become parasitic have an especially well-developed, toothed "denticular apparatus" (originating during metamorphosis by fusion of modified dermal spinules anterior to the toothed premaxillae and dentaries; see Munk, 2000; Fig. 54) on the tips of the snout and chin that he believed must be used for "a temporary attachment to the female without subsequent fusion."

The physiological aspects of the attachment of males to females in deep-sea anglerfishes have been only superficially explored. Regan (1925b, 1926) reported that longitudinal sections of tissue taken from the junction of an attached male and female *Ceratias holboelli* seemed to demonstrate a complete blending of tissues: "The highly vascularized fibrous tissue of the outgrowths of the male are continuous with that of . . . the female; in both, the general direction of the fibers and of the blood vessels is longitudinal, i.e., leading from one fish to the other, and it seems almost certain that the blood-systems of the two fishes are continuous and that the male is nourished by the female" (Regan, 1926:12). Subsequent work by Regan and Trewavas (1932, see Fig. 284), Bertelsen (1951), Shoemaker (1958), Pietsch (1976), Munk and Bertelsen (1983), and Munk (2000) fully substantiated the earlier findings. "In the neighborhood of the fusion area the skin of the female is rendered spongy by the development of a system of thin-walled irregular intercommunicating blood-sinuses, which are continuous with similar sinuses in the skin and underlying tissues of the snout and jaws of the male" (Regan and Trewavas, 1932:15). With this information, in addition to finding a poorly developed alimentary canal in the attached males, Regan (1925b, 1926) and Regan and Trewavas (1932) maintained that the attached males were nutritional parasites on the females. Furthermore, since all attached males known at the time retained lateral openings to the mouth just behind the area of attachment, and all males examined had a well-developed heart and gills, it was also concluded that "the male does not depend on the blood of the female for oxygen, but only for nutritive materials" (Regan, 1926:14). But, while Regan (1925b:392) and others maintained that the vascular systems of both sexes unite, Waterman (1939a:77) suggested that nourishment of the male may be accomplished by a placenta-like arrangement of male-female tissues in the area of fusion, a notion that was accepted without evidence by Norman and Greenwood (1975:274). That a placenta might exist was considered by Regan (1925b:13), but he thought it "unlikely, for a placenta is essentially an attachment between two individuals that separate later on; but when the attachment is permanent such a complication is unnecessary and need not be expected."

By whatever means energy is acquired, Parr (1930b, 1932) suggested that the nutriment received by the males from the blood of the females may not be as important as originally suggested by Regan (1925b). "The enormous size of the liver [found in attached males], particularly in comparison with the greatly reduced alimentary canal, is, under the circumstances, much more easily understood in relation to the food storage properties of this organ than in relation to its digestive functions, and it may not be unreasonable to assume that its own liver, not the blood of the female, may be the most important factor in sustaining the life and further growth and maturing of the male after it becomes attached to a female" (Parr, 1930b:135). Parr (1930b:135, 1932:10) believed further that the relationship between male and female should not be

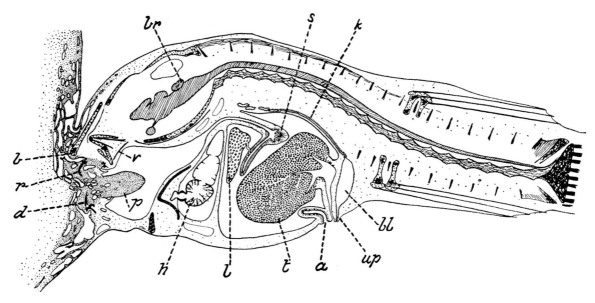

FIGURE 284. One of three parasitic males of *Haplophryne mollis*, 12 mm SL, attached to a 50-mm SL female, ZMUC P921777, composite sagittal section. a, anus; b, basal bone; bl, urinary bladder; br, brain; d, dentary; h, heart; k, kidney; l, liver; p, papilla of female tissue protruding into mouth of male; r, rostral denticle; s, stomach; t, testes; up, urinary papilla; v, vomer. After Regan and Trewavas (1932).

thought of as a case of true parasitism but simply a protracted mating that may last a few months, at least not longer than the male is able to maintain itself on its own resources.

Bertelsen (1951), however, in opposition to Parr (1930b, 1932), presented considerable evidence to show that at least in *Ceratias holboelli*, (1) attachment is probably of considerable duration; (2) males grow considerably after attachment, increasing their weight several hundred times; and (3) after metamorphosis, but before attachment, males do not increase in body weight, and the liver decreases somewhat in size during this period. These facts indicate that free-living metamorphosed males of this species do not eat on their own but are sustained by the food stored in the liver, both in the free-living stage after metamorphosis and in the period from attachment until effective connection with the blood system of the female has been made. Finding no reason to doubt that attached males of other ceratioids are also nourished in the same manner as *Ceratias*, Bertelsen (1951:245) concluded that "in all species and presumably in all families where attached males are found, the males are true parasites."

Shoemaker (1958:145) made histological sections through the region of attachment of one of three males attached to a female of *Cryptopsaras couesii* and concluded that the circulatory systems of the two were in fact fused: "It was quite clear that blood sinuses in this area communicated directly with blood vessels of both male and female." Wickler (1961) described the structure and function of the angling apparatus of ceratioids and discussed some problems of pair formation. Pietsch and Nafpaktitis (1971) described an anomalous case of a sexually mature male of *Melanocetus johnsonii* (Melanocetidae) attached, but without tissue fusion, to a juvenile female *Centrophryne spinulosa* (Centrophrynidae). Olsson (1974) reexamined the histological sections of the attached male of *Haplophryne mollis* (a senior synonym of *Edriolychnus schmidti*) that had been described and figured earlier by Regan and Trewavas (1932:15–17), with respect to the endocrine glands. Results showed that those organs involved in regulation of various internal physiological factors, as well as those involved in adaptation to external factors, appear to be poorly developed or reduced, whereas mechanisms regulating reproductive activities seem to be well developed. Pietsch (1976) summarized what was known about sexual parasitism to date, describing male-female attachment in a number of previously unreported specimens and taxa, including several examples of surprisingly small parasitized females (for example, a 15.5-mm female of *Cryptopsaras couesii*, with a 9.8-mm attached male; see Pietsch, 1975a:38), but more importantly providing evidence that parasitism may be facultative in some ceratioids. Finally, Munk and Bertelsen (1983) provided the first detailed histological examination of parasitic attachment in a ceratioid, supporting the assumption that, at least in *H. mollis*, a connection is established between the vascular systems of male and female. This conclusion was reaffirmed by Munk (2000:322), but he cautioned that, despite a high probability, there is as yet "no critical proof of a real continuity between the female blood vascular system and that of the parasitic male" in any ceratioid.

In summarizing his discovery of sexual parasitism in deep-sea anglerfishes, Regan (1926:14) proposed the following scenario:

> The reason why the Ceratioids, alone amongst Vertebrates, have males of this kind is evident. They are necessarily few in numbers in comparison with the more active fishes on which they prey, and they lead a solitary life, floating about in the darkness of the middle depths of the ocean. Under such circumstances it would be very difficult for a mature fish to find a mate, but this difficulty appears to some extent to have been got over by the males, soon after they are hatched, when they are relatively numerous, attaching themselves to the females, if they are fortunate enough to meet them, and remaining attached throughout life. In all probability the males are incapable of free development, and it is likely that the great majority of them fail to find a female and perish, although another possibility has been suggested to me, namely, that the post-larval fish that find and become attached to females develop into males, and those that do not into females.

The question of sex change in ceratioids was dispelled by Bertelsen (1951), who demonstrated for the first time that males and females of most species can be readily distinguished in the smallest larvae (2 to 3 mm in total length) by the early development of the illicial apparatus, appearing as a small undifferentiated papilla on the snout of females. Of 2366 larvae examined, the sex of only 142 (6%) was ranked as uncertain (Bertelsen, 1951:248). Although these collections contain slightly more larval females than males, correcting for the type of gear used, and calculating on the basis of how many specimens of each sex would be caught if the fishing had been equally great at all depths, males amount to an average of 56% in areas around the world where ceratioids are most abundant. However, assuming that the females represent several year-groups and the free-living males only one, and considering further that females, with their self-advertising bioluminescence, are under greater predation pressure, Bertelsen (1951:249) surmised that for every female there are at least 15 to 30 metamorphosed males, the great majority of which would be mature or ready to attach themselves.

Bertelsen (1951), in a review of the literature and a worldwide search of collections, listed only 20 examples of females with parasitic males. In a similar survey made 25 years later, Pietsch (1976) raised the number to 45. Now, after another 30 years, the number has increased to 165 (slightly more than the 153 reported by Pietsch, 2005b:215; see Table 10). Most of these additional records are the result of a more intense search of preexisting collections, but a few are newly discovered and noteworthy. These are described below, along with information for each ceratioid family on taxonomic content, available material, occurrence of sexual parasitism, gravid females and ripe males, the development of eyes and nostrils of the males, the ability of males to capture and ingest food independently of the female, occurrences of multiple attachments of males to a single female, and the nature of the fusion between coupled males and females.

In the family accounts given below, sexually mature or gravid individuals are assumed to be those whose gonads are obviously larger than those of other conspecific individuals of a similar standard length. Because ceratioids are still extremely rare and curators nearly always reluctant to sacrifice specimens, the developmental state of gonads has been subjected to little or no histological verification.

## Family Accounts

### Centrophrynidae

The family Centrophrynidae, containing only *Centrophryne spinulosa* Regan and Trewavas, 1932, is now known on the basis of at least 41 metamorphosed females, three juvenile males, and two larvae. In contrast to the paired gonads of all other lophiiform fishes, female centrophrynids possess only a single oval-shaped ovary (Pietsch, 1972a). Unlike the epithelial folds that line the inner walls of the ovaries of other ceratioids (Bertelsen, 1951), the lumen of the ovary of centrophrynids (three specimens examined, 136 to 209 mm; Pietsch, 1972a) is filled with villi-like projections of the epithelium. Whole cross-sections throughout the length of the ovary show small undeveloped oocytes embedded within the villi-like projections (Pietsch, 1972a:24). The two largest known females, a 230-mm specimen from the eastern North Atlantic (MHNLR P-811; Bertelsen and Quéro, 1981) and a 247-mm specimen from off Taiwan (ASIZP 59902), both appear to be immature, although the ovaries of the latter contain numerous developing eggs, measuring approximately 0.2 to 0.4 mm in diameter (Hsuan-Ching Ho, personal communication, 10 July 2004). The three known males (11.5 to 16 mm), two in metamorphosis and one recently metamorphosed (12.8 mm, SIO 70-347), are also immature, with testes less than 4% SL. Nevertheless, the upper and lower denticulars of these young males are well developed, each bearing 3 or 4 curved, slightly hooked teeth (Bertelsen, 1983:314). The olfactory organs are relatively large, the greatest diameter of the posterior nostril of the 12.8-mm specimen measuring 7.7% SL, but the eyes are small in all three specimens compared to other ceratioid males, their diameter less than 4.5% SL (Fig. 285).

There is no evidence to support sexual parasitism in this family, but the possibility cannot be ruled out. Of particular interest in this regard is the discovery of a female specimen of *C. spinulosa* (168 mm) with an attached male of *Melanocetus johnsonii* (Pietsch and Nafpaktitis, 1971). The male is sexually mature, with large ripe testes, but there is no evidence of fusion of male-female tissue. Whether this bizarre case of possible mistaken identity occurred naturally, or whether it occurred accidentally in the net, despite the small statistical probability of such an event, is unknown (Pietsch, 1976).

### Ceratiidae

The family Ceratiidae contains two genera and four species (Pietsch, 1986b). *Ceratias* Krøyer, 1845, with three species, is one of the most common and best-known ceratioids, now represented by more than 442 metamorphosed females, about 75 free-living males, 37 parasitic males (attached to 35 females), and about 160 larvae (Table 10). The genus *Cryptopsaras* Gill, 1883, with a single species, is even better known, with more than 980 metamorphosed females, at least 100 free-living males, 75 parasitic males (attached to 47 females), and about 350 larvae. As pointed out by Bertelsen (1951:245) and reconfirmed here, all free-living *Ceratias* males are small, less than 20 mm, whereas the vast majority of attached males are considerably larger, greater than 25 mm (for the few exceptions to this general rule, see Table 10). Most weigh several times that of the largest known free-living male, and a few have attained quite remarkable sizes, exceeding 140 mm. In contrast to those of *Ceratias*, the size ranges of free-living and attached *Cryptopsaras* males overlap slightly: the largest known free-living male measures 10.5 mm (14.3 mm TL; Bertelsen, 1951:144), whereas known attached males range from 9.8 (attached to a 15.5-mm female; Pietsch, 1975a, 1976) to 99 mm (Table 10).

No free-living ceratiid male with large testes has ever been found, yet large ripe testes have been described in several attached males (Regan, 1925a, 1925b, Fig. 281B; Saemundsson, 1939; Bertelsen, 1943, 1951; Olsson, 1974): those of three UW specimens (reported by Pietsch, 2005b; see Table 10) range from 7.3 mm to about 30 mm long (32.4 to 37.5% SL). Histological examination of the testes of two of these specimens (UW 21774, UW 21775; 20 and 34 mm, respectively) showed evidence of resorption, this indicating a recent spawning event. All known gravid females have an attached parasitic male (Bertelsen, 1943, 1951; Fast, 1957; Shoemaker, 1958; Mead et al., 1964; Fitch, 1973), and the ovaries of one of these, a 580-mm specimen (ZMUC P922481) first described by Saemundsson (1939), contain about five million ripening eggs measuring approximately 0.2 to 0.4 mm in diameter—upon capture, Saemundsson (1939) noted that "ripe sperm" were

TABLE 10
Records of Attached Males of the Ceratioidei in Collections Around the World, Arranged by Family

| Species | Female | Male(s) | Author and Date | Catalog Number |
|---|---|---|---|---|
| Ceratiidae | | | | |
| *Ceratias holboelli* | 690 mm | 2 (65–70 mm) | Saemundsson, 1922 | IINH/ZMUC P922480[a] |
| *Ceratias holboelli* | 720 mm | 75 mm (detached) | Regan, 1925a, 1925b | BMNH 1924.12.29.1-2 |
| *Ceratias holboelli* | 580 mm | 86 mm | Saemundsson, 1939 | ZMUC P922481 |
| *Ceratias holboelli* | 770 mm | 118 mm | Bigelow and Barbour, 1944a | MCZ 36042[b] |
| *Ceratias holboelli* | 735 mm | 75 mm | Bertelsen, 1951 | MRIR[c] |
| *Ceratias holboelli* | 1100 mm TL | 80 mm TL | Krefft, 1961 | NMH 1-1959 |
| *Ceratias holboelli* | 1270 mm TL | 160 mm TL | Krefft, 1961 | Apparently lost |
| *Ceratias holboelli* | 640 mm | 70 mm | Blacker, 1972 | BMNH 1970.10.28.20 |
| *Ceratias holboelli* | 1000 mm TL | ? | Du Buit et al., 1980 | MNHN, misplaced |
| *Ceratias holboelli* | 670 mm | 97.5 mm | Ni, 1988 | ECFR |
| *Ceratias holboelli* | 595 mm | 70 mm | Amaoka et al., 1995 | HUMZ 77841 |
| *Ceratias holboelli* | 855 mm | 85 mm | Pietsch, 2005b | BMNH 1953.2.25.1 |
| *Ceratias holboelli* | 560 mm | 35 mm | Pietsch, 2005b | CSIRO H.2746-01 |
| *Ceratias holboelli* | 690 mm | 105 mm | Pietsch, 2005b | HUMZ 95300 |
| *Ceratias holboelli* | 1210 mm TL | 190 mm TL | Pietsch, 2005b | MRIR[d] |
| *Ceratias holboelli* | 980 mm TL | 90 mm TL | Pietsch, 2005b | MRIR[c] |
| *Ceratias holboelli* | 1200 mm TL | ? | Pietsch, 2005b | MRIR[c] |
| *Ceratias holboelli* | 900 mm TL | 70 mm TL | Pietsch, 2005b | MRIR 94-34 |
| *Ceratias tentaculatus* | 550 mm | 37 mm | Pietsch, 2005b | CSIRO H.3101-01 |
| *Ceratias uranoscopus* | 240 mm | 22.5 mm | Pietsch, 1986b | LACM 33376-3 |
| *Ceratias* sp. | 850 mm TL | 38 mm TL | Beebe, 1938 | Apparently lost |
| *Ceratias* sp. | ? | ? | Matsubara, 1955 | Lost; see Abe, 1967 |
| *Ceratias* sp. | 650 mm | 58 mm | Abe, 1967 | Apparently lost[e] |
| *Ceratias* sp. | 460 mm | 32 mm TL | Abe, 1967 | Apparently lost[e] |

| | | | | |
|---|---|---|---|---|
| Ceratias sp. | 700 mm | 37 mm | Penrith, 1967 | SAM 29607 |
| Ceratias sp. | 525 mm | 37 mm | Fitch, 1973 | LACM 33718-1 |
| Ceratias sp. | 212 mm | 9.2 mm | Pietsch, 1986b | UF 25149 |
| Ceratias sp. | 370 mm | 5-mm remnant | Pietsch, 2005b | CAS 60358 |
| Ceratias sp. | discarded | 73 mm | Pietsch, 2005b | CAS 82498 |
| Ceratias sp. | 275 mm | 24.5 mm | Pietsch, 2005b | LACM 36955-1 |
| Ceratias sp. | 660 mm | 46 mm | Pietsch, 2005b | MNHN 2004-1518 |
| Ceratias sp. | 740 mm | 60 mm | Pietsch, 2005b | MTF |
| Ceratias sp. | 790 mm | 2 (65–89 mm) | Pietsch, 2005b | NMNZ P.40886 |
| Ceratias sp. | 553 mm | 46 mm | Pietsch, 2005b | UW 22322 |
| Ceratias sp. | 750 mm | 115 mm | Pietsch, 2005b | ZMH 21014 |
| Cryptopsaras couesii | 290 mm | 12 mm | Tanaka, 1908; Barbour, 1941b | MCZ 29855 |
| Cryptopsaras couesii | 276 mm | 12 mm | Abe and Nakamura, 1954 | Apparently lost[e] |
| Cryptopsaras couesii | 213 mm | 2 (27–27.5 mm) | Fast, 1957 | CAS-SU 49556 |
| Cryptopsaras couesii | 176 mm | 3 (16–37 mm) | Shoemaker, 1958 | USNM 177939[b] |
| Cryptopsaras couesii | 272 mm | 74 mm | Ueno, 1966; Ueno and Abe, 1967 | HUMZ 70815 |
| Cryptopsaras couesii | 322 mm | 41 mm | Penrith, 1967 | SAM 23587 |
| Cryptopsaras couesii | 356 mm | 73 mm | Penrith, 1967 | SAM 24283 |
| Cryptopsaras couesii | 77 mm | 15 mm | Pietsch, 1975a | BMNH 2004.6.29.4-5 |
| Cryptopsaras couesii | 173 mm | 35 mm | Pietsch, 1975a | LACM 33621-1 |
| Cryptopsaras couesii | 15.5 mm | 9.8 mm | Pietsch, 1975a | USNM 234867 |
| Cryptopsaras couesii | 146 mm | 17 mm | Young and Roper, 1977 | USNM 219906 |
| Cryptopsaras couesii | 300 mm | 4 (33–60 mm) | Abe and Funabashi, 1992 | INM 9201 |
| Cryptopsaras couesii | 348 mm | 3 (54–60 mm) | Amaoka et al., 1995 | HUMZ 124596 |
| Cryptopsaras couesii | 311 mm | 40 mm TL | Anderson and Leslie, 2001 | SAM 34480 |
| Cryptopsaras couesii | 343 mm | 2 (88–120 mm TL) | Anderson and Leslie, 2001 | SAM 34481 |
| Cryptopsaras couesii | 316 mm | 8 (35–56 mm) | Saruwatari et al., 2001 | ORI |
| Cryptopsaras couesii | 45 mm | 10 mm | Pietsch, 2005b | ARC 8707665 |
| Cryptopsaras couesii | 215 mm | 44 mm | Pietsch, 2005b | CAS 73320 |

*(Continued)*

TABLE 10 (continued)

| Species | Female | Male(s) | Author and Date | Catalog Number |
|---|---|---|---|---|
| *Cryptopsaras couesii* | 195 mm | 34 mm | Pietsch, 2005b | CAS 76509 |
| *Cryptopsaras couesii* | 322 mm | 51 mm | Pietsch, 2005b | CSIRO H.2532-02 |
| *Cryptopsaras couesii* | 268 mm | 36 mm | Pietsch, 2005b | CSIRO H.4391-01 |
| *Cryptopsaras couesii* | 324 mm | 48 mm | Pietsch, 2005b | HUMZ 69165 |
| *Cryptopsaras couesii* | 345 mm | 6 (18–44 mm) | Pietsch, 2005b | HUMZ 69166 |
| *Cryptopsaras couesii* | 308 mm | 4 (78–88 mm) | Pietsch, 2005b | HUMZ 73014 |
| *Cryptopsaras couesii* | 320 mm | 48 mm | Pietsch, 2005b | HUMZ 98265 |
| *Cryptopsaras couesii* | 162 mm | 24.5 mm | Pietsch, 2005b | MCZ 164112 |
| *Cryptopsaras couesii* | 390 mm | 2 (47–99 mm) | Pietsch, 2005b | NMNZ P.34960 |
| *Cryptopsaras couesii* | 252 mm | 62 mm | Pietsch, 2005b | OSUO 12056 |
| *Cryptopsaras couesii* | 302 mm | 50 mm | Pietsch, 2005b | SIO 95-30 |
| *Cryptopsaras couesii* | 202 mm | 20 mm | Pietsch, 2005b | SIO 97-52 |
| *Cryptopsaras couesii* | 94 mm | 2 (8.0–14.5 mm) | Pietsch, 2005b | UF 25157 |
| *Cryptopsaras couesii* | 164 mm | 49 mm | Pietsch, 2005b | UF 25160 |
| *Cryptopsaras couesii* | 108 mm | 20.5 mm | Pietsch, 2005b | UF 25164 |
| *Cryptopsaras couesii* | 132 mm | 17 mm | Pietsch, 2005b | USNM 219906 |
| *Cryptopsaras couesii* | 215 mm | 20 mm | Pietsch, 2005b | UW 21774 |
| *Cryptopsaras couesii* | 240 mm | 34 mm | Pietsch, 2005b | UW 21775 |
| *Cryptopsaras couesii* | 230 mm | 80 mm | Pietsch, 2005b | UW 46112 |
| *Cryptopsaras couesii* | 210 mm | 2 (21–28 mm) | Unpublished | CSIRO T.646-01 |
| *Cryptopsaras couesii* | 230 mm | 11 mm | Unpublished | NMNZ P.17798 |
| *Cryptopsaras couesii* | 255 mm | 11 mm | Unpublished | NMNZ P.23797 |
| *Cryptopsaras couesii* | 210 mm | Small remnant | Unpublished | NMNZ P.23888 |
| *Cryptopsaras couesii* | 300 mm | 31 mm | Unpublished | NMNZ P.24933 |
| *Cryptopsaras couesii* | 320 mm | 26 mm | Unpublished | NMNZ P.25942 |
| *Cryptopsaras couesii* | 81 mm | 9 mm | Unpublished | NSMT-P 40226 |
| *Cryptopsaras couesii* | 122 mm | 9.5 mm | Unpublished | UF 23798 |

| | | | |
|---|---|---|---|
| *Cryptopsaras couesii* | 68 mm | 10.5 mm | Unpublished | UF 167357 |
| *Cryptopsaras couesii* | 207 mm | 2 (55–62 mm) | Unpublished | UW 48058 |
| Melanocetidae | | | | |
| *Melanocetus johnsonii* | 75 mm | 23.5 mm | Pietsch, 2005b | BMNH 2004.6.3.2-3 |
| *Melanocetus murrayi* | 73 mm | 15 mm | Pietsch, 2005b | BSKU 57842 |
| Oneirodidae | | | | |
| *Bertella idiomorpha* | 77 mm | 11 mm | Pietsch, 2005b | UW 48712 |
| *Leptacanthichthys gracilispinis* | 56 mm | 7.5 mm | Pietsch, 1976 | LACM 33625-2 |
| Caulophrynidae | | | | |
| *Caulophryne polynema* | 142 mm | 16 mm | Regan, 1930a, 1930b | BMNH 1930.2.7.1 |
| *Caulophryne polynema* | 137 mm | 15 mm | Pietsch, 2005b | MNHN 2001-0140 |
| *Caulophryne* sp. | 98 mm | 12 mm | Pietsch, 1979 | LACM 36025-1 |
| Neoceratiidae | | | | |
| *Neoceratias spinifer* | 52 mm | 15.5 mm | Bertelsen, 1951 | ZMUC P921726 |
| *Neoceratias spinifer* | 86 mm | 17.5 mm | Pietsch, 1976 | SIOM |
| *Neoceratias spinifer* | 42 mm | 8.5 mm | Pietsch, 1976 | LACM 34271-1 |
| *Neoceratias spinifer* | 108 mm | 18 mm | Pietsch, 1976 | SIO 70-336 |
| *Neoceratias spinifer* | 77 mm | 12.5 mm | Bertelsen and Pietsch, 1983 | AMS I.20908-2 |
| *Neoceratias spinifer* | 74 mm | 12.5 mm | Munk, 2000 | ISH 5546/79 |
| *Neoceratias spinifer* | 67.5 mm | 17.5 mm | Pietsch, 2005b | SIO 68-478 |

*(Continued)*

TABLE 10 (continued)

| Species | Female | Male(s) | Author and Date | Catalog Number |
|---|---|---|---|---|
| Linophrynidae | | | | |
| *Photocorynus spiniceps* | 46 mm | 7.3 mm | Regan, 1925b | ZMUC P92134 |
| *Photocorynus spiniceps* | 50.5 mm | 7.0 mm | Pietsch, 2005b | ISH 1913/71 |
| *Photocorynus spiniceps* | 46 mm | 6.2 mm | Pietsch, 2005b | SIO 70-326 |
| *Photocorynus spiniceps* | 49 mm | 6.5 mm | Pietsch, 2005b | SIO 70-346 |
| *Haplophryne mollis* | 48 mm | 10 mm | Regan, 1925b | ZMUC P92135 |
| *Haplophryne mollis* | 50 mm | 3 (11–12 mm) | Regan and Trewavas, 1932 | ZMUC P921777[f] |
| *Haplophryne mollis* | 34 mm | 11 mm | Regan and Trewavas, 1932 | ZMUC P92138 |
| *Haplophryne mollis* | 38 mm | 2 (9.5–10 mm) | Pietsch, 1976 | SOC[g] |
| *Haplophryne mollis* | 40 mm | 2 (10.5–11 mm) | Pietsch, 1976 | BMNH 2004.8.17.48-50 |
| *Haplophryne mollis* | 70 mm | 2 (11–12.5 mm) | Pietsch, 1976 | LACM 11235-25 |
| *Haplophryne mollis* | 62 mm | 2 (12–15 mm) | Munk and Bertelsen, 1983 | AMS I.21365-8[h] |
| *Haplophryne mollis* | 52.5 mm | 2 (11.5–12 mm) | Bertelsen and Pietsch, 1983 | AMS I.20071-1 |
| *Haplophryne mollis* | 46 mm | 11 mm | Bertelsen and Pietsch, 1983 | AMS I.20314-14 |
| *Haplophryne mollis* | 39 mm | 12.5 mm | Bertelsen and Pietsch, 1983 | AMS I.20315-9 |
| *Haplophryne mollis* | 48 mm | 2 (10.5–12 mm) | Bertelsen and Pietsch, 1983 | AMS I.21364-3 |
| *Haplophryne mollis* | 54 mm | 6 (8.9–10.5 mm) | Stewart and Pietsch, 1998 | NMNZ P.26070 |
| *Haplophryne mollis* | 58 mm | 2 (11.6–14.2 mm) | Stewart and Pietsch, 1998 | NMNZ P.24927 |
| *Haplophryne mollis* | 60 mm | 11 mm | Stewart and Pietsch, 1998 | NMNZ P.24164 |
| *Haplophryne mollis* | 159 mm | 2 (9.8–10.8 mm) | Stewart and Pietsch, 1998 | NMNZ P.21248 |
| *Haplophryne mollis* | 60 mm | 14 mm | Pietsch, 2005b | CSIRO H.3286-01 |
| *Haplophryne mollis* | 47 mm | 13 mm | Pietsch, 2005b | MCZ 59223 |
| *Haplophryne mollis* | 45 mm | 12 mm | Pietsch, 2005b | MNHN 2003-2032 |
| *Haplophryne mollis* | 48 mm | 12 mm | Pietsch, 2005b | MNHN 2004-0810 |
| *Haplophryne mollis* | 45 mm | 13.6 mm | Pietsch, 2005b | NMNZ P.8045 |
| *Haplophryne mollis* | 51 mm | 11.6 mm | Pietsch, 2005b | NMNZ P.35370 |

| | | | | |
|---|---|---|---|---|
| Haplophryne mollis | 70 mm | 2 (11.4–11.5 mm) | Pietsch, 2005b | NMNZ P.36807 |
| Haplophryne mollis | 52 mm | 2 (10–10.3 mm) | Pietsch, 2005b | NMV A.5924 |
| Haplophryne mollis | 35 mm | 2 (10.5–11.5 mm) | Pietsch, 2005b | SOC[g] |
| Haplophryne mollis | 54 mm | 9 mm | Unpublished | CSIRO T.680-01 |
| Borophryne apogon | 55 mm | 13 mm (detached) | Regan and Trewavas, 1932 | ZMUC P92147 |
| Borophryne apogon | 47 mm | 14.5 mm | Regan and Trewavas, 1932 | BMNH 1932.5.3.38 |
| Borophryne apogon | 50 mm | 11 mm (detached) | Beebe and Crane, 1947 | AMNH 211332 |
| Borophryne apogon | 51.5 mm | 10.5 mm | Beebe and Crane, 1947 | Apparently lost |
| Borophryne apogon | 65 mm | 14 mm | Bertelsen, 1951 | ZMUC P921755 |
| Borophryne apogon | 53 mm | 12 mm | Bertelsen, 1951 | ZMUC P921756 |
| Borophryne apogon | 52 mm | 12 mm | Kramp, 1953 | ZMUC P922322 |
| Borophryne apogon | 56 mm | 9.5 mm | Kramp, 1953 | ZMUC P922324 |
| Borophryne apogon | 60 mm | Tiny remnant | Kramp, 1953 | ZMUC P922325 |
| Borophryne apogon | 101 mm | 2 (16–22 mm) | Pietsch, 1976 | LACM 30053-10 |
| Borophryne apogon | 89 mm | 20 mm | Pietsch, 1976 | SIO 68-112 |
| Borophryne apogon | 65.5 mm | 12 mm | Pietsch, 2005b | USNM 326570 |
| Linophryne algibarbata | 182 mm | 29 mm | Behrmann, 1974 | IMB |
| Linophryne algibarbata | 155 mm | 23 mm | Jónsson and Pálsson, 1999 | IINH |
| Linophryne arborifera | 77 mm | 15 mm | Bertelsen, 1980a | ISH 2736/71 |
| Linophryne arborifera | 63 mm | 14.5 mm | Unpublished | MCZ 164735 |
| Linophryne argyresca | 61 mm | 12 mm | Regan and Trewavas, 1932 | ZMUC P92142 |
| Linophryne bicornis | 185 mm | 30 mm | Behrmann, 1977 | IMB |
| Linophryne bicornis | 180 mm | 18 mm | Bertelsen, 1982 | SIOM |
| Linophryne bicornis | 101 mm | 19 mm | Moore et al., 2003 | MCZ 138063 |
| Linophryne brevibarbata | 86 mm | 13.6 mm | Maul, 1961 | MM SLF 18214 |
| Linophryne brevibarbata | 100 mm | 18.5 mm | Bertelsen, 1980a | BMNH 1995.1.18.4 |
| Linophryne coronata | 225 mm | Remnant | Ponomarenko, 1959 | PINRO |
| Linophryne coronata | 219 mm | 26 mm | Bertelsen, 1976 | MRIR[c] |
| Linophryne coronata | 290 mm TL | ? | Jónsson et al., 1986a | MRIR[c] |

*(Continued)*

TABLE 10 (continued)

| Species | Female | Male(s) | Author and Date | Catalog Number |
|---|---|---|---|---|
| Linophryne coronata | 152 mm | 18 mm | Pietsch, 2005b | ICMB |
| Linophryne densiramus | 68 mm | 9.0 mm | Parin et al., 1977 | SIOM |
| Linophryne densiramus | 71 mm | 17 mm | Anderson and Leslie, 2001 | SAIAB 54760 |
| Linophryne densiramus | 68 mm | 13.5 mm | Pietsch, 2005b | CSIRO H.3210-02 |
| Linophryne densiramus | 70 mm | 17.5 mm | Pietsch, 2005b | SAIAB 63715 |
| Linophryne indica | 42 mm | 9.5 mm | Bertelsen, 1978 | LACM 36046-11 |
| Linophryne indica | 51 mm | 14.5 mm | Bertelsen, 1981 | SIO 70-306 |
| Linophryne lucifer | 230 mm | 24 mm | Jónsson, 1967b | MRIR[c] |
| Linophryne lucifer | 190 mm | 24.5 mm | Bertelsen, 1982 | SIOM |
| Linophryne lucifer | 190 mm TL | 30 mm TL | Pietsch, 2005b | MRIR |
| Linophryne lucifer | 180 mm TL | 40 mm TL | Pietsch, 2005b | MRIR |
| Linophryne lucifer | 174 mm | 23.5 mm | Pietsch, 2005b | ZMUC P922290 |
| Linophryne lucifer | 275 mm | 29 mm | Pietsch, 2005b | ZMUC P922443 |
| Linophryne macrodon | 91 mm | 21.5 mm | Pietsch, 1976 | UF 233292 |
| Linophryne maderensis | 105 mm | 15 mm | Pietsch, 2005b | ZMB 33308[i] |
| Linophryne trewavasae | 73.5 mm | 10.7 mm | Bertelsen, 1978 | LACM 36116-5 |
| Linophryne sp. | ? | ? | Jónsson et al., 1986a | MRIR[c] |

NOTE: Records for the Melanocetidae represent temporary attachment, in which male and female tissues are not fused, while those of other families represent examples of permanent sexual parasitism, in which fusion between male and female is evident. Specimens listed as Ceratias sp. have lost the illicium and/or esca and therefore cannot be fully identified.

[a] One of two males removed, sent to Copenhagen, and cataloged ZMUC.
[b] With evidence of an additional male lost prior to capture.
[c] Discarded (Gunnar Jónsson, personal communication, 19 April 2004).
[d] Stuffed and mounted.
[e] Part of the late Tokiharu Abe's private collection, whereabouts unknown (Kazuo Sakamoto, personal communication, 7 June 2004).
[f] One of three males sacrificed by Regan and Trewavas (1932) for histological study.
[g] Peter Herring collection, Southampton Oceanography Centre, Southampton, England.
[h] With scar left behind by a third male apparently lost prior to capture; males sacrificed by Munk and Bertelsen (1983) for histological study.
[i] Specimen not examined by me (Peter Bartsch, personal communication, 16 February 2005).

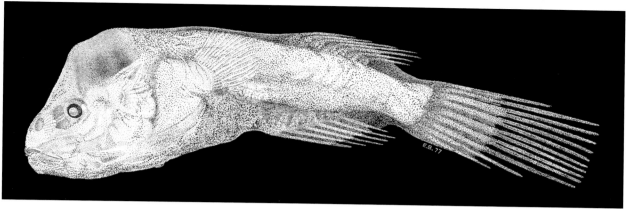

FIGURE 285. Free-living male of *Centrophryne spinulosa*, 12.8 mm SL, SIO 70-347. After Bertelsen (1983).

shed from the parasitic male (see Bertelsen, 1943:189). These data taken from both males and females reaffirm the idea that sexual maturity is never attained in members of this family unless stimulated by the attachment of a male.

The eyes of metamorphosed free-living males are unusually large in ceratiids, each having a prominent crescent-shaped aphakic space (Bertelsen, 1951:127; Munk, 1964:5–10, 1966:28–30; Fig. 286A, B), but they quickly degenerate upon attachment to a female (Fig. 287). The nostrils of ceratiid males, however, are minute, in marked contrast to those of all other ceratioids (Fig. 286A, B). The general assumption that pair formation in ceratioids is mediated by a species-specific pheromone emitted by the female and tracked by the male does not appear to apply to this family.

The denticular jaw apparatus of metamorphosed ceratiid males is well developed, consisting of a pair of upper and two pairs of lower teeth, each elongate and slightly hooked distally, appearing quite capable of nipping onto a female, but not especially well suited for prey capture (Bertelsen, 1951:137, 144; Figs. 282, 283, 286A, B). The alimentary canal is rather poorly developed (Bertelsen, 1951:131, 245; Fig. 288). None of the specimens examined by Bertelsen (1951:245) had food in its stomach. The few millimeters that the males increase in length during and after metamorphosis seem to result from a stretching of the body rather than any increase in body weight, and the liver decreases somewhat in size during this period. Bertelsen (1951:245) thus concluded that free-living metamorphosed males of this family do not eat.

It was long assumed that female ceratioids, before acquiring a parasitic male, must mature to an adult stage that is of considerable size in some taxa, especially in ceratiids: "A short time . . . after metamorphosis the males become mature or ready for attachment and in these stages are more numerous than the much larger and undoubtedly much older adult females" (Bertelsen, 1951:257; see also Regan, 1925b). But we now know that females, at least in *Cryptopsaras couesii*, may become sexually parasitized at almost any size once past metamorphosis. Examples of small parasitized individuals include a 15.5-mm female, with a 9.8-mm male (USNM 234867; Fig. 289); a 45-mm female, with a 10-mm male (ARC 8707665); and a 77-mm female, with a 15-mm male (BMNH 2004.6.29.4-5). In these three couples, the ovaries are as small as those found in nonparasitized females of a similar size, whereas the testes of the males are well developed, occupying more than half the volume of the coelomic cavity (1.7 mm long or 17% SL in ARC 8707665). Histological examination of the testes of the 10-mm

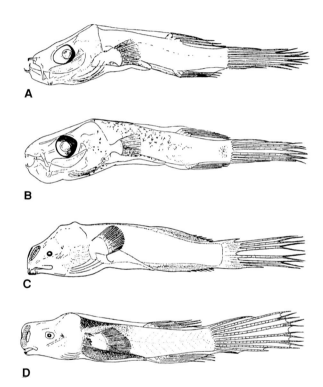

FIGURE 286. Free-living males of ceratiids, characterized by having small nostrils, but extremely well-developed eyes; and gigantactinids, with small eyes, but extremely large, well-developed nostrils: (A) *Ceratias* sp., 10.8 mm SL, specimen sacrificed for histology (see Munk, 1964); (B) *Cryptopsaras couesii*, 10.2 mm SL, specimen sacrificed for histology (see Munk, 1964); (C) *Gigantactis* male group II, 14.5 mm SL, ZMUC P921533; and (D) *Rhynchactis* sp., 18.5 mm SL, ZMUC P921732. After Bertelsen (1951).

male shows moderate resorption, thus indicating a recent spawning event. The members of the smallest attached pair appeared to be quite young, perhaps six months and certainly less than 12 months old.

Bertelsen (1951:250) speculated that "in one way or another, males are prevented from fixing themselves to females occupied already." In two out of the three examples of females with more than one attached male known at that time (Table 10), "the size, degree of development and stage of degeneration, are so much the same that we may believe the attachment to have taken place at the same time." Shoemaker (1958), how-

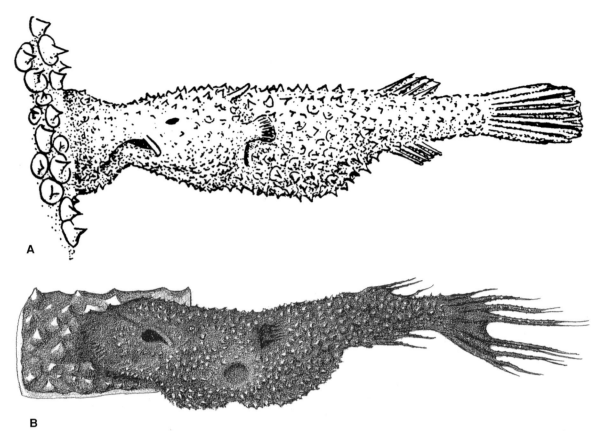

FIGURE 287. Parasitic males of *Ceratias holboelli*: (A) 86-mm SL, attached to a 580-mm SL female, ZMUC P922481 (after Bertelsen, 1943); (B) 70 mm SL, one of two males attached to the earliest discovered parasitized female, 690 mm SL, ZMUC P922480, originally described by Saemundsson, 1922 (artist unknown; courtesy of J. Nielsen and the Zoological Museum, University of Copenhagen).

ever, described a 176-mm *C. couesii* (USNM 177939) with three parasitic males that differed greatly in size; the two smaller males, 16 and 21 mm, no doubt had become attached more recently than the largest, 37-mm male (Fig. 290). Since that time, numerous cases of multiple attachments of males at widely varying stages of development have been found in *Cryptopsaras* (as well as in some linophrynids; see Table 10). The current record is now held by a 316-mm female with eight males ranging in size from 35 to 56 mm (Saruwatari et al., 2001).

The parasitic males of *Ceratias* are invariably attached to the belly of the female on or close to the ventral midline and somewhat anterior to the anus. Those of *Cryptopsaras* are usually found on the belly, most often off-set somewhat to the left or right, but may also be placed almost anywhere on the body. Although it is difficult to say in all cases, males more often than not attach themselves upside down with respect to the surface of the female, and they are almost invariably directed anteriorly as if they approached their mate from behind (see Figs. 65, 289, 290).

The details of attachment of all known parasitic males of ceratiids is similar to those first described in *Ceratias* by Regan (1925b:390; Figs. 280, 281, 291A):

> In front of the mouth the snout and chin are produced forwards into outgrowths, which unite in front of the end of the lower jaw, although a groove on each side indicates the limits between them.... Anteriorly these outgrowths end in a swollen ring of tissue covered with naked skin, which is thicker above and below than at the sides, indicating that it is a continuation of the outgrowths of the male fish; this ring surrounds and is united to a thick stalk that projects from the female.

As the male grows and the extent of fusion increases, a large expansion of tissue eventually displaces the male away from the point of attachment so that a considerable distance is established between the surface of the female and the tips of the jaws of the male. In some couplings of *Ceratias*, a nipple-like outgrowth of female tissue partially fills the mouth cavity of the male (Bertelsen, 1951:138), but this outgrowth is never as well developed as that of the linophrynid genus *Haplophryne* (see below). In *Cryptopsaras*, more often than not, tissue fusion extends along the side of the male that comes to lie against the surface of the female, completely closing the mouth opening on that side, but always leaving the opposite side open to the gills and opercular openings. Fast (1957:240), in his description of two males attached to a 213-mm female of *Cryptopsaras couesii* (CAS-SU 49556), found it impossible to force air through the openings of the mouth and out past the opercle, "indicating that the mass of female tissue which is taken into the mouth for attachment not only blocks off the alimentary canal but also the path of the respiratory currents." However, upon re-examination of Fast's (1957) specimens, a narrow probe was easily passed from the mouth and out through the gill opening (Pietsch, 2005b:224). In contrast to the point of attach-

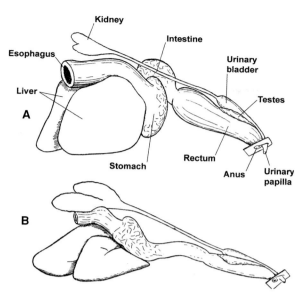

FIGURE 288. Viscera of males of *Ceratias* sp.: (A) in metamorphosis, 9.3 mm SL (whole specimen shown in Figure 63E); (B) free-living juvenile, 11.4 mm SL. After Bertelsen (1951).

ment of *Ceratias* males, conspicuous flattened, circular scars ("expanded discs," as described and figured by Shoemaker, 1958:143; see Fig. 290) surround the point of attachment of those of *Cryptopsaras* (Fig. 291A).

## Himantolophidae

The family Himantolophidae, containing a single genus (*Himantolophus* Reinhardt, 1837) and 18 species distributed among five species groups, is represented in collections around the world by more than 150 females, 48 free-living males, and 311 larvae (Bertelsen and Krefft, 1988). Except for the presence of white circular scars found on the bodies of five females, which Maul (1961:115) argued might be "the result of an injury caused by a male that had grown fast there and has for some reason become suddenly detached"—but which Bertelsen and Krefft (1988) suggested might be caused by parasitic copepods—there is no evidence of sexual parasitism in this family.

Contrary to reports of ovaries containing ripe or nearly ripe eggs in a number of large females (Regan and Trewavas, 1932; Bertelsen, 1951; Mead et al., 1964), Bertelsen and Krefft (1988:25), based on a detailed examination of all known material, stated that "no fully mature females have been found." Eggs with diameters of more than 0.10 mm were found in only four specimens (112 to 180 mm), and only one of these (145 mm; LACM 42697-1) appeared to be near maturity. In the remaining larger females examined (90 to 328 mm), egg diameters varied between 0.05 and 0.10 mm and ovaries measured 9 to 31% SL in length and 6 to 13% SL in width. All smaller females examined (40 to 80 mm) had immature ovaries less than 10% SL in length and 5% SL in width. That several large females of *Himantolophus groenlandicus* from far northern Atlantic waters have spent ovaries (Saemundsson, 1927:168; Bertelsen, 1951:26) also appears to be unfounded. Those of several large specimens (185 to 328 mm), all caught in northern latitudes and assumed to be expatriates (but see Jónsson and Pálsson, 1999:205), are somewhat enlarged (length 17 to 30% SL) but compressed (less than about 3% SL in width), containing only minute eggs in early developmental stages (Bertelsen and Krefft, 1988).

Contrary to the development of known females, numerous free-living himantolophid males have large well-developed testes (Regan and Trewavas, 1932; Bertelsen, 1951; Mead et al., 1964). As described by Bertelsen (1951), the testes show a distinct increase in relative size during late larval stages, become well developed in metamorphic stages, and nearly fill the body cavity in metamorphosed specimens. Those of three specimens (30.5 to 34 mm) examined microscopically by Bertelsen and Krefft (1988:78) were filled with mature spermatozoa with elongate condensed nuclei and distinct tails. None of the known males appeared to be spent.

Similar to that of melanocetids and ceratiids, as described by Parr (1930b) and Bertelsen (1951), the denticular jaw apparatus of himantolophid males appears well adapted for nipping the skin of the females. It also seems well adapted for predation (Bertelsen and Krefft, 1988:26; Fig. 292A), supporting the contention of Bertelsen (1951) that males of this family are able to feed on their own beyond metamorphosis. However, the latter assumption was not supported by the more detailed review of Bertelsen and Krefft (1988:30), who found no evidence of continued postmetamorphic feeding of the males; even the existence of any postmetamorphic growth appears questionable. Of 38 metamorphosed males examined by them, the stomachs of 30 were distinctly reduced in size, their walls thin and transparent, and completely empty. Of the remaining seven specimens, the stomachs of two (27 to 32 mm) contained an unidentifiable white substance (very similar to material found in the stomachs of a number of smaller metamorphosing males), and one of these two also contained the remains of chaetognath setae. While this may be evidence of postmetamorphic feeding, it seems more likely that the observed food remains in the stomachs of these specimens originated from a last feeding in epi- or upper mesopelagic layers prior to full metamorphosis and descent into deeper waters where they were captured. Thus, Bertelsen and Krefft (1988:30) were unable to provide convincing proof that metamorphosed himantolophid males are able to eat on their own: "The possibility cannot be excluded that the store of nutrients in the liver and other tissues of the metamorphic males might be sufficient for the change in body proportions and the final development of the testes during and after metamorphosis."

Himantolophid males reach a greater adult size than free-living males of any other known ceratioid, the largest among the approximately 38 known metamorphosed specimens measuring 39 mm (UF 232657; Bertelsen and Krefft, 1988:78; Fig. 73F). The olfactory organs are as well developed as those of most other ceratioid males, increasing significantly in relative size during metamorphosis, but the eyes are relatively small, actually decreasing slightly in relative diameter with increasing standard length (Figs. 73, 292A). Contrary to the general assumption that pair formation in ceratioids is mediated at least in part by vision (Munk, 1966), the relatively small eyes of himantolophid males may not be capable of assisting in mate location and recognition (Bertelsen and Krefft, 1988:26, 34). For more on reproduction in himantolophids, see Bertelsen and Krefft (1988).

## Diceratiidae

The Diceratiidae, containing two genera (*Diceratias* Günther, 1887; and *Bufoceratias* Whitley, 1931) and six species, is now

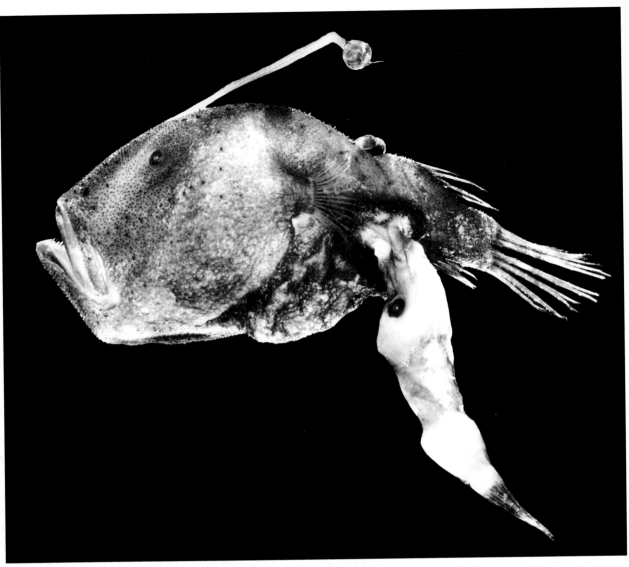

FIGURE 289. *Cryptopsaras couesii*, 15.5-mm SL female, with a 9.8-mm SL male, USNM 234867, the smallest known parasitized ceratioid. Photo by A. H. Coleman; after Pietsch (1975a).

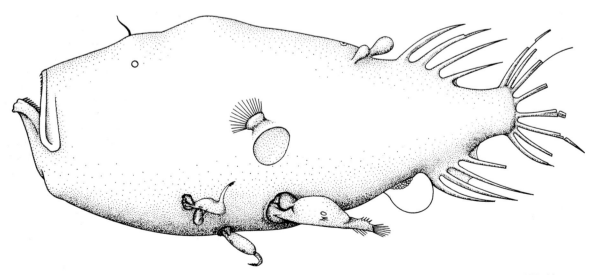

FIGURE 290. *Cryptopsaras couesii*, a 176-mm SL female, with three parasitic males, 16, 21, and 37 mm SL, USNM 177939. After Shoemaker (1958).

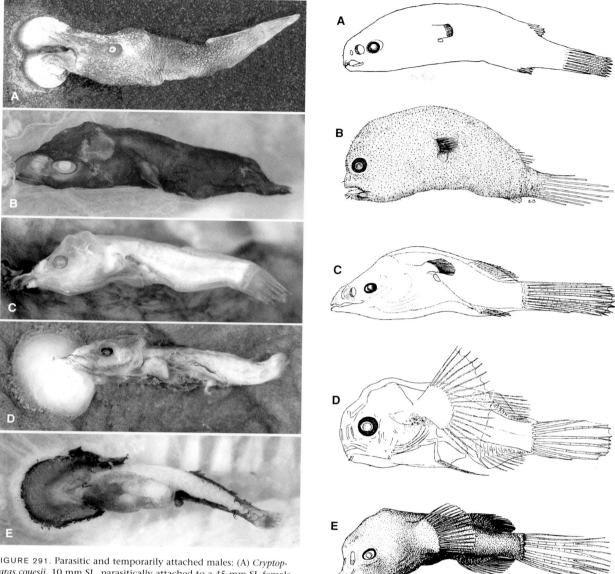

FIGURE 291. Parasitic and temporarily attached males: (A) *Cryptopsaras couesii*, 10 mm SL, parasitically attached to a 45-mm SL female, ARC 8707665; (B) *Melanocetus murrayi*, 15 mm SL, temporarily attached (without tissue fusion) to a 73-mm SL female, BSKU 57842; (C) *Bertella idiomorpha*, 11 mm SL, parasitically attached to a 77-mm SL female, UW 48712; (D) *Caulophryne polynema*, 15 mm SL, parasitically attached to a 137-mm SL female, MNHN 2001-140; (E) *Neoceratias spinifer*, 18 mm SL, parasitically attached to a 108-mm SL female, SIO 70-336. Modified after Pietsch (2005b).

FIGURE 292. Free-living males: (A) *Himantolophus* sp., 34.5 mm SL, ZMUC P92675; (B) Diceratiidae sp., 14 mm SL, LACM 36091-4; (C) *Melanocetus* sp., 20 mm SL, ZMUC P92458; (D) *Caulophryne* sp., 7.5 mm SL, ZMUC P92193, in late metamorphosis, showing presence of pelvic fins; (E) *Caulophryne* sp., 11 mm SL, MCZ 69324, in late metamorphosis, pelvic fins lost. After Bertelsen (1951, 1983).

known on the basis of at least 200 females, a single metamorphosed free-living male, and two larvae (Uwate, 1979; Balushkin and Fedorov, 1986; Pietsch et al., 2004). Although still very poorly known, there is no evidence of sexual parasitism in this family (Pietsch, 1976; Bertelsen, 1983). The only known sexually mature diceratiid is a 235-mm female of *Diceratias pileatus* (UF 233052), the ovaries of which each contain $10^4$ to $10^5$ pear-shaped eggs, measuring 0.3 to 0.7 mm in diameter (Uwate, 1979). The ovaries of the largest known specimen of the family, a 275-mm female of *D. pileatus* (BPBM 30655), are large, their length approximately 80 mm or 29% SL, but they contain no eggs and appear to be spent (Pietsch and Randall, 1987). The ovaries of the largest known specimen of *Bufoceratias thele* (200 mm, ASIZP 65075) contain numerous developing eggs, 0.16 to 0.28 mm in diameter (Hsuan-Ching Ho, personal communication, 5 May 2004). Judging by the small olfactory organs and testes, the position and small size of the denticular teeth, and the retention of a few tiny larval teeth, the single known male (14 mm, LACM 36091-4) is a juvenile (Bertelsen, 1983:312; Fig. 292B). The eyes are relatively well developed, measuring 1.2 mm in diameter (8.6% SL), each with a narrow aphakic space surrounding the lens.

### Melanocetidae

The Melanocetidae, containing a single genus (*Melanocetus* Günther, 1864) and six species (Pietsch and Van Duzer, 1980), is now known from more than 1200 females, at least 120 free-living males, and about 400 larvae. Until very recently, the only evidence of sexual parasitism in this family was an anomalous case of a 20-mm male *Melanocetus johnsonii* attached to the upper lip of a 168-mm female *Centrophryne spinulosa* (LACM 30843-1; Pietsch and Nafpaktitis, 1971; see Centrophrynidae, above). However, the attachment did not involve fusion of male-female tissue and is therefore not considered to be a parasitic association. Two similar couplings in *Melanocetus*, but between sexes of the same taxon, have recently been discovered. One of these, a 23.5-mm male attached to a 75-mm female *M. johnsonii* (BMNH 2004.6.3.2-3, Fig. 78), was collected in the eastern North Atlantic off Ireland in 1999 by the RRS *Discovery* in a cruise partially funded by the British Broadcasting Corporation for the celebrated *Blue Planet* video series. The second example is part of collections made by Hiromitsu Endo, aboard the RV *Tansei-Maru*, west of Okinawa, in April 2002: a 15-mm male attached to a 73-mm female *M. murrayi* (BSKU 57842, Fig. 291B). Both males are only loosely attached, with tissue of the female pinched by the tightly closed denticular jaws of the male, the BMNH example hanging from the middle of the belly of the female, the BSKU specimen attached to the right side of the head of the female, just beneath the sphenotic bone. In both cases, it does not appear that any fusion of male and female tissue has taken place. Either the connections of the two were so recent that the tissues did not have time to fuse, or, more likely, these are the first and only known examples of a nonparasitic coupling—male ceratioids caught in the act of temporary attachment.

The gonads of the smaller of the two melanocetid couples are not especially large: the testes are only about 1.9 mm long (12.7% SL), but histological examination shows them to be at an early stage of resorption (therefore indicating a recent spawning event), with a small number of sperm heads evident; the ovaries are about 12 mm long (16.4% SL) and contain thousands of tiny eggs, each approximately 0.20 to 0.25 mm in diameter. The testes of the larger pair are slightly smaller relative to standard length, about 2.7 mm long (11.5% SL), but histological examination shows numerous mature spermatozoa; the ovaries are huge, about 35 mm long (46.7% SL), containing thousands of slightly dehydrated eggs that are among the largest ever recorded for a ceratioid, each approximately 0.7 to 0.8 mm in diameter.

Metamorphosed free-living males of *Melanocetus* with large ripe testes have been described (Pietsch, 1976:783), but personal examination of the ovaries of numerous females, including the largest known individuals of the family (up to 135 mm), failed to identify any females with eggs greater than about 0.10 mm in diameter (examination of more than 60 females recently collected from off Taiwan produced the same result; Hsuan-Ching Ho, personal communication, 25 August 2003). Metamorphosed males continue to feed after metamorphosis: food items have been found in their stomachs and considerable growth continues after stores in the liver have become exhausted (Bertelsen, 1951; Pietsch, 1976). This ability to take prey after metamorphosis is indicated by the relatively large, heavily toothed upper and lower denticular bones of adult *Melanocetus* males (Bertelsen, 1951:39, 250; Pietsch, 1976:783; Fig. 293). The eyes and nostrils of the males are large and well developed (Bertelsen, 1951:48; see Fig. 292C). For more details on reproduction in melanocetids, see Pietsch (1976).

## Thaumatichthyidae

The family Thaumatichthyidae, containing two genera (*Lasiognathus* Regan, 1925c; and *Thaumatichthys* Smith and

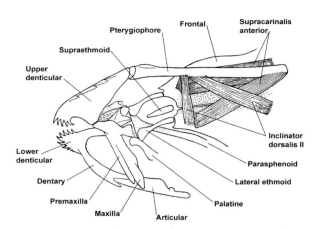

FIGURE 293. Anatomy of the snout of a free-living male of *Melanocetus murrayi*, the holotype of *Rhynchoceratias longipinnis*, 16 mm SL, YPM 2592, showing among other things the large, heavily toothed upper and lower denticular bones. Modified after Parr (1930a, 1930b).

Radcliffe, 1912) and eight species, is now represented by 65 known females, four free-living metamorphosed males, and five larvae (Bertelsen and Struhsaker, 1977; Bertelsen and Pietsch, 1996; Pietsch, 2005b; Figs. 94, 95, 294). There is no evidence of sexual parasitism in this family, and all known specimens are sexually immature. In the largest known females of *Thaumatichthys* (225 to 365 mm), the ovaries consist of a pair of relatively thin-walled sacks, each approximately 50 mm long and 15 mm wide in the 365-mm holotype of *Thaumatichthys axeli* (ZMUC P92166), and about 6 by 30 mm in a 294-mm specimen of *T. binghami* (USNM 21471). In both these females the slightly folded walls of the ovaries are covered with a single layer of oocytes 0.05 to 0.10 mm in diameter (Bertelsen and Struhsaker, 1977:23). The testes of the metamorphosed males (31 and 36 mm; ZMUC P921946, UW 47520) are immature: smaller and of unequal length (12.8 to 16.7% SL) in the 36-mm specimen, about 21% SL in the 31-mm specimen (Bertelsen and Struhsaker, 1977:26). The upper and lower denticulars of the males are extraordinarily well developed, each bearing long, curved, distally hooked teeth (Bertelsen and Struhsaker, 1977:24; Fig. 294C). The olfactory organs are relatively large as well, the length of each posterior nostril about 8 to 10% SL, each containing 13 olfactory lamellae (Fig. 294A, B). The diameter of the pigmented part of the eye is 1.6 to 2.0 mm (5.2 to 5.6% SL); the diameter of the transparent outer coat of the eyeball is 2.1 to 2.5 mm (6.8 to 6.9% SL).

## Oneirodidae

The Oneirodidae is by far the largest, most complex, and certainly the least understood family of the suborder. With 16 genera and 62 species, it contains 40% of all recognized ceratioids. Of the 16 genera, five are currently represented by only one, two, or three juvenile or adult females; only eight are represented by more than a dozen females. Males have been described for only seven genera, while larvae are known for only eight. But despite the rareness of most recognized taxa, new oneirodids continue to be discovered (e.g., Orr, 1991; Stewart and Pietsch, 1998; Ho and Shao, 2004; Pietsch, 2004, 2007; Pietsch and Baldwin, 2006).

The known material of the family has increased dramatically in the last quarter of a century. There are now at least 1200 females, about 164 free-living males, and 697 larvae in

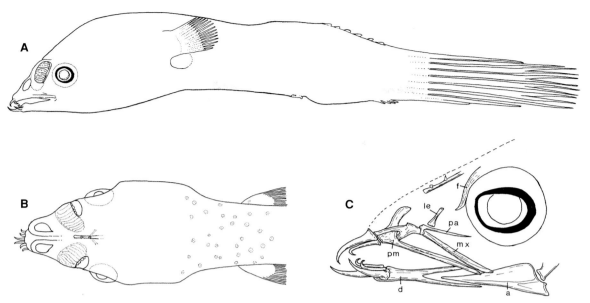

FIGURE 294. Free-living male of *Thaumatichthys* sp., 31 mm SL, ZMUC P921946, showing the well-developed eyes, olfactory organs, and upper and lower denticular bones: (A) left lateral view; (B) dorsal view of head; (C) skeleton of the snout. a, angular; d, dentary; f, frontal; le, lateral ethmoid; mx, maxilla; pa, palatine; pm, premaxilla. After Bertelsen and Struhsaker (1977).

collections around the world. Among this material are two attached males, both representing the only known males of their respective genus and species: a 7.5-mm male attached to a 56-mm female of *Leptacanthichthys gracilispinis* (LACM 33625-2) described by Pietsch (1976:784; 1978:19; Figs. 142, 144) and an 11-mm male attached to a 77-mm female of *Bertella idiomorpha* described by Pietsch (2005b:221; see below).

Prior to the discovery of these two attached males, there was no evidence to conclude that males of oneirodids ever become sexually parasitic. Food items have not been found in the stomach of any oneirodid male, but, nevertheless, the males of at least some genera grow considerably after metamorphosis. For example, metamorphosis stages of *Oneirodes* measure 6.0 to 9.5 mm; whereas, the largest known metamorphosed specimen is 16.5 mm long (Pietsch, 1974a, 1976). These data indicate that oneirodid males, all equipped with large, well-developed eyes and nostrils (Fig. 295), probably continue to feed after metamorphosis and are thus not nutritionally dependent on a parasitic association with the female. The especially well-developed denticular apparatus of males of this family has been thought to be used for temporary attachment without subsequent tissue fusion. The discovery of parasitized female oneirodids is therefore surprising and of considerable interest.

Except for a few details, the circumstances surrounding the attached male of *Bertella idiomorpha* (Figs. 291C, 296) are very similar to those described by Pietsch (1976, 1978) for that of *Leptacanthichthys gracilispinis*. Both are connected by a narrow cylindrical stalk of tissue protruding from the midbelly of the female. The length of the stalk of *Bertella* is considerably less than that of *Leptacanthichthys*: 1.6 mm (2.1% SL) and 2.2 mm (3.9% SL), respectively. Whereas only the tip of the upper jaw of the male of *Leptacanthichthys* was embedded in the stalk of female tissue (since torn away, but the bones of the upper jaw and ethmoid region remain attached to female; Pietsch, 1976:784; Fig. 144), the tips of both upper

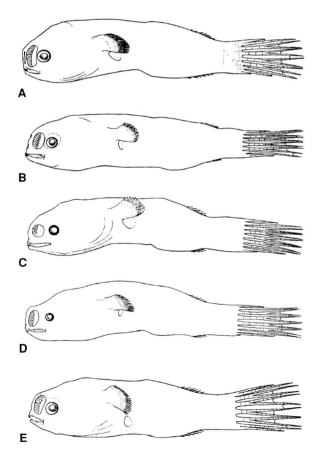

FIGURE 295. Free-living males of oneirodids: (A) *Pentherichthys atratus*, 13 mm SL, ZMUC P921113; (B) *Chaenophryne draco* group, 14 mm SL, ZMUC P92686; (C) *Oneirodes* sp., 12.5 mm SL, ZMUC P921016; (D) *Microlophichthys andracanthus*, 16.5 mm SL, ZMUC P9293; (E) *Dolopichthys* sp., 12.5 mm SL, ZMUC P92799. After Bertelsen (1951).

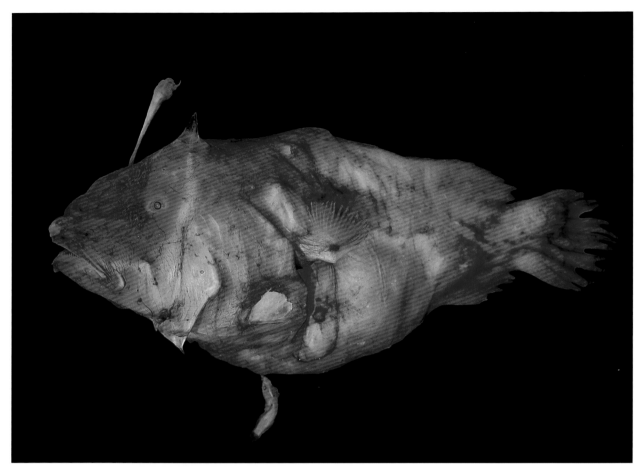

FIGURE 296. *Bertella idiomorpha*, 77-mm SL female, with an 11-mm SL parasitic male, UW 48712. After Pietsch (2005b).

and lower jaws are fused with female tissue in *Bertella*. In the latter, small openings to the mouth and opercular cavities of the male are retained on both sides; in both couplings, there is no papilla of female tissue extending into the mouth of the male.

The male of *Leptacanthichthys*, at 7.5 mm, is in late metamorphosis, with subdermal pigment well developed and skin only lightly pigmented (Pietsch, 1976:785; Fig. 144). The somewhat larger male of *Bertella*, at 11 mm, appears to be slightly more mature, lacking any visible subdermal pigment, but the dermal and subdermal pigmentation overall of both male and female has largely disappeared as a result of photochemical bleaching over time. Juvenile and adult males of most all known ceratioids are greater than approximately 10 mm (exceptions to this general rule are found primarily in the family Linophrynidae; see below), their subdermal pigmentation has disappeared, and the skin is darkly pigmented in all known genera except *Haplophryne* (see Bertelsen, 1951:10, 167; Figs. 170, 172).

The viscera of the two oneirodid males are similar (Pietsch, 1976:785), each having a moderately sized liver occupying a third or less of the volume of the coelomic cavity. The paired testes are small (their length 0.9 mm or only about 8.2% SL in *Bertella*) compared to those of a ripe male (greater than 20% SL and filling most of the abdominal cavity in some specimens; for example, see Jespersen, 1984). Serial sections through the entire length of both testes of the *Leptacanthichthys* male revealed only small undifferentiated spermatocytes. In contrast, the gonads of both females are well developed; the left ovary of *Leptacanthichthys* is filled with approximately 7,500 eggs, measuring approximately 0.4 to 0.6 mm in diameter; those of *Bertella* contain a similar number of slightly larger eggs (among the largest known for any ceratioid), approximately 0.6 to 0.8 mm in diameter (but now somewhat dehydrated, the eggs were undoubtedly larger in life).

Several facts indicate that these two couplings represent permanent parasitic associations of male and female rather than temporary attachments: (1) definite evidence of tissue fusion between male and female exists in both cases; (2) although not examined histologically in *Bertella*, the stalk of female tissue of *Leptacanthichthys*, to which the male was attached, is full of intercommunicating vessels (Pietsch, 1976:785) capable of transferring blood and nutrients to the male (much like those described in the area of fusion of parasitic males of other ceratioids; see Regan and Trewavas, 1932; Munk and Bertelsen, 1983); and (3) as shown by their small size, unpigmented skin, and undeveloped testes, the males are sexually immature; metamorphosed males of ceratioid families that exhibit sexual parasitism usually become attached while immature, whereas temporary attachment should not be expected before the male has reached sexual maturity.

An examination of the ovaries of all known material of *Leptacanthichthys gracilispinis* (24 metamorphosed females, 10.5 to 103 mm) turned up one additional female (ROM 27284, 54 mm) approaching sexual maturity, with ovaries 8.5 mm long (15.7% SL), containing numerous eggs approximately 0.2 mm in diameter (Pietsch, 1976). A similar examination of the known material of *Bertella idiomorpha* (34 metamorphosed females, 11 to 101 mm) revealed two specimens with large ovaries containing ripening eggs: an 84-mm female (OS 1045) with ovaries approximately 15 mm long (17.8% SL) that appear to be partially spent (about one-third of the volume of each ovary consists of a tight cluster of numerous eggs, each approximately 0.3 mm in diameter; Pietsch, 1973); and a 101-mm female (SIO 92-175) with ovaries about 15 mm long (14.8% SL), containing numerous eggs, each approximately 0.2 mm in diameter.

## Caulophrynidae

The family Caulophrynidae, now containing two genera (*Caulophryne* Goode and Bean, 1896; and *Robia* Pietsch, 1979) and five species, is currently represented in collections by some 60 females, five males, and 16 larvae (Bertelsen, 1951; Pietsch, 1979). Two of the males are free-living: a 7.5-mm specimen in an early stage of metamorphosis (ZMUC P92193, Fig. 292D) and an 11-mm specimen in late metamorphosis (MCZ 69324, Fig. 292E). The remaining three are parasitically attached to females (Table 10). Two of the latter are assigned to *Caulophryne polynema*: the holotype of the species, a 16-mm male attached to a 142-mm female (BMNH 1930.2.7.1) described in detail by Regan (1930b); and a previously unrecorded pair, a 15-mm male attached to a 137-mm female (MNHN 2001-0140, Fig. 291D). The third known example of sexual parasitism in caulophrynids is a 12-mm male attached to a 98-mm female (LACM 36025-1, Fig. 297A) identified by Pietsch (1979:18) as *Caulophryne* sp. (*Caulophryne* sp. A of Pietsch and Seigel, 1980:380).

The gonads of the BMNH pair have not been examined directly, but the "abdomen [of the male] is somewhat swollen, no doubt by the developing testes" (Regan, 1930b:194). Those of the LACM pair are large in each case: the right ovary of the female is 22 mm long (22.4% SL) and contains numerous eggs approximately 0.3 mm in diameter; the testes of the male are 2.3 mm long (19.2% SL). Unfortunately, the male attached to MNHN 2001-0140 has been eviscerated, perhaps by damage upon capture, but more likely by later dissection. The length of the ovaries of the female is about 23% SL, each containing thousands of tiny eggs, approximately 0.15 to 0.20 mm in diameter.

The ovaries of a number of large nonparasitized females of *C. jordani* are well developed, indicating that females may become sexually mature in the absence of a male: the length of those of a 109-mm specimen from off Newfoundland (ROM 27250) are nearly 30% SL and filled with numerous eggs measuring approximately 0.2 mm in diameter (Pietsch, 1976); lengths of those of four specimens from off Iceland (MRIR) are 16.8% SL in a 119-mm specimen, 26.9% SL in a 130-mm specimen, 27.4% SL in a 153-mm specimen (containing thousands of tiny eggs), and 22.3% SL in a 155-mm specimen. The testes of the two known free-living males are small and apparently immature.

In all three known cases of sexual parasitism, the male is joined to the female on the belly (just slightly to the left or right of the midline and facing forward in the same direction as the female as if it approached the female from behind) by means of separate outgrowths of tissue from the snout and lower jaw that unite with the distal surface of a prominent, unpigmented, conical or hemispherical outgrowth from the female (diameter nearly 5% SL in the 137-mm female; see Pietsch, 1979:20; Figs. 291D, 297A). Thus, the heads of the males are neither embedded in, nor broadly attached to, the

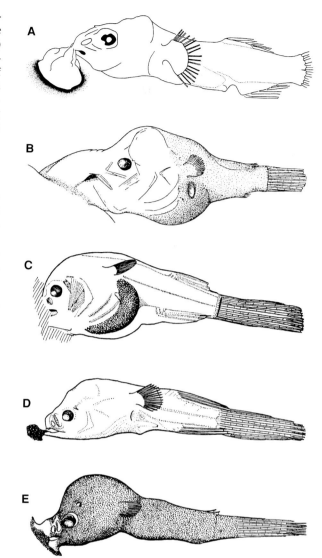

FIGURE 297. Parasitic males: (A) *Caulophryne* sp., 12 mm SL, attached to a 98-mm SL female, LACM 36025-1 (after Pietsch, 1979); (B) *Photocorynus spiniceps*, 7.4 mm SL, attached to a 46-mm SL female, ZMUC P92133 (after Bertelsen, 1951); (C) *Haplophryne mollis*, 12 mm SL, one of three attached to a 50-mm SL female, ZMUC P921777 (after Regan and Trewavas, 1932); (D) *Borophryne apogon*, 14.5 mm SL, attached to a 47-mm SL female, BMNH 1932.5.3.38 (after Regan and Trewavas, 1932); (E) *Linophryne argyresca*, 12 mm SL, attached to a 61-mm SL female, ZMUC P92142 (after Regan and Trewavas, 1932).

surface of the female, but rather a considerable distance is maintained between the two by newly formed tissue. The males are upside down relative to the females in the BMNH and LACM specimens, but right-side up in the MNHN specimen. In all three examples, prominent openings to the mouth and opercular cavities are retained on both sides, and there is no papilla of female tissue extending into the mouth of the male, characteristic of couplings in some other taxa (for example, *Haplophryne*; see below). A comparison of all three attached males shows no significant morphological differences. Relative to other ceratioid males, those of caulophrynids have large eyes, with a conspicuous aphakic space anterior to the lens; the olfactory organs are also large, at least in free-living stages, but appear to degenerate quickly upon attachment to a female. For more details, see Regan (1930b) and Pietsch (1976, 1979).

## Neoceratiidae

The family Neoceratiidae, containing only *Neoceratias spinifer* Pappenheim, 1914, is now represented by 18 metamorphosed females, seven males, and 11 larvae. All of the known males are parasitically attached to females (Fig. 298, Table 10). At least one of the parasitized females appears to be sexually mature or very close to it: an 86-mm female (IOAN) with ovaries 14% SL, containing numerous eggs approximately 0.6 mm in diameter (Pietsch, 1976:789). Others are in various stages of development: a 77-mm female (AMS I.20908-002) with ovaries 14% SL, containing numerous small eggs, the largest of which measure 0.15 mm in diameter; a 42-mm specimen (LACM 34271-1) with ovaries only 4% SL but containing numerous eggs, the largest of which are 0.2 mm in diameter; and, finally, the largest known specimen, a 108-mm female (SIO 70-336) with eggs 0.25 to 0.30 mm in diameter. The 15.5-mm male attached to a 52-mm female (ZMUC P921726) has testes that "are in the process of development but not remarkably large" (Bertelsen, 1951:161). Jespersen (1984), however, found the testes of a 12.5-mm attached male (ISH 5546/79) to be exceptionally large, filling most of the abdominal cavity and containing all spermatogenetic stages, spermatozoa being present in the lumina of the testes as well as in the sperm ducts. The testes of a 17.5-mm male attached to a 67.5-mm female (SIO 68-478) are 1.8 mm long (10.3% SL); those of a 18-mm male attached to the largest known specimen of the genus (108 mm, SIO 70-336) are 3.7 mm long (20.6% SL). Histological examination of the testes of the latter specimen revealed the presence of sperm, but also early resorption, indicating a recent spawning event.

The peculiar denticular jaw apparatus of *Neoceratias* males appears to be poorly adapted for use in prey capture and, for that matter, seemingly unsuited for grasping a female as well: the upper element is apparently absent, the lower trifurcated, each elongate arm flattened and bifurcated distally, and the lateral arms curved anteriorly (see Bertelsen, 1951:157; Pietsch, 1976:790; Fig. 299). The eyes and olfactory structures are in an advanced state of degeneration in all the known males. In every case, the males are upside down and facing forward (with respect to the female), attached on the side and toward the posterior end of their mate, either on the caudal peduncle or between the bases of the dorsal and anal fins. One of the attached males (SIO 70-336) is located slightly anterior to the base of the anal fin, just above the anus.

In nearly all previously described examples of parasitically attached males, lateral openings to the pharynx leading to the gills and opercular openings have been found behind the area of attachment. Because the heart and gills of attached males are well developed and show no signs of degeneration even in individuals that have obviously been attached to their hosts for a considerable amount of time (Munk, 2000:321), it has been suggested that the male does not depend on the blood of the female for oxygen, but only for nutrition (Regan, 1925b:396, 1926:14; Bertelsen, 1951:245). Although one of the seven known attached males of *N. spinifer* (ZMUC P921726, Fig. 298A) has retained a large opening to the pharynx on each side (through which the teeth of the jaws can be clearly observed), the flattened dorsal surface of the head of the remaining six is broadly affixed to the side of the female (appearing as if embedded in or absorbed by the female), leaving no lateral openings into the mouth cavity (Figs. 291E, 298B). The gills of these six males, however, are as well developed as those of the seventh (and those of free-living males of other ceratioid families), perhaps indicating that sufficient oxygen is not available via the blood of the female and that this gas is extracted from water that is pumped in and out through the opercular opening (Pietsch, 1976:790; Munk, 2000:317; the gills of parasitic males may also remain well developed to accommodate for the elimination of carbon dioxide and nitrogenous waste products, the maintenance of acid-base and mineral balance, and passive diffusion of water). It might be assumed also that these latter six males represent older examples of parasitism in which the process of fusion is more complete and that over time the seventh specimen would have eventually reached a similar morphology, including complete closure of the mouth. This complete blockage of the pharynx is in sharp contrast to the situation in all other ceratioid taxa, except for the linophrynids *Haplophryne mollis* and *Photocorynus spiniceps,* in which a similar occlusion appears to be present in some but not all known specimens (see below). Fast's (1957) claim that the mouth is fully occluded in an attached male of *Cryptopsaras couesii* (CAS-SU 49556) could not be verified by reexamination.

## Gigantactinidae

The family Gigantactinidae, containing two genera (*Gigantactis* Brauer, 1902; and *Rhynchactis* Regan, 1925c) and 21 species, is represented by at least 200 females, 50 free-living males, and 333 larvae (Bertelsen et al., 1981; Bertelsen and Pietsch, 1998b). Gigantactinids are among the largest known ceratioids. Some of the 30 or so females of *Gigantactis* greater than 200 mm have relatively large ovaries, but none appear to be fully mature; eggs greater than about 0.5 mm in diameter have not been found. The ovaries of a 136-mm female of *Rhynchactis macrothrix* (ASIZP 61797) are about 40 mm long (30% SL) and 29 mm wide (21% SL wide) and contain about 150,000 eggs, each 0.3 to 0.5 mm in diameter (Hsuan-Ching Ho, personal communication, 3 May 2004). A number of *Gigantactis* males with well-developed testes, "so large that they seem to be near maturity," have been described (Bertelsen, 1951:153).

None of the females in collections around the world are parasitized by males, thus it is assumed that males remain free-living. The largest of the known *Gigantactis* males in metamorphosis is 14.5 mm, whereas metamorphosed individuals are between 10.5 and 22 mm. This relatively large postmetamorphic increase in size indicates that these males continue to grow after metamorphosis, yet there is no evidence that postmetamorphic males are able to feed on their own; the stomachs of all metamorphosed males examined were empty (Bertelsen et al., 1981). Compared to most other ceratioids, gigantactinid males have extremely small eyes (Fig. 286C, D)—diameters in most specimens are not more than 5% SL, but their nostrils are well developed and it is thus assumed that they rely almost exclusively on olfaction to locate potential mates. The denticular jaw apparatus is well developed, consisting of long hooked teeth, 2 to 4 upper and 3 to 5 lower (Bertelsen et al., 1981:11; Figs. 160, 300).

## Linophrynidae

The Linophrynidae includes five genera and 27 species (Bertelsen, 1951, 1980a, 1980b, 1982; Balushkin and Trunov, 1988; Gon, 1992). Three of the genera, *Photocorynus* Regan, 1925b, *Haplophryne* Regan, 1912, and *Borophryne* Regan, 1925c are

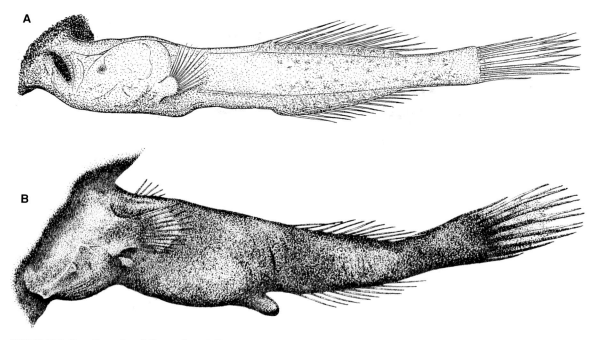

FIGURE 298. Parasitic males of *Neoceratias spinifer*: (A) 15.5 mm SL, attached to a 52-mm SL female, ZMUC P921726 (after Bertelsen, 1951); (B) 12.5 mm SL, attached to a 74-mm SL female, ISH 5546/79 (after Munk, 2000).

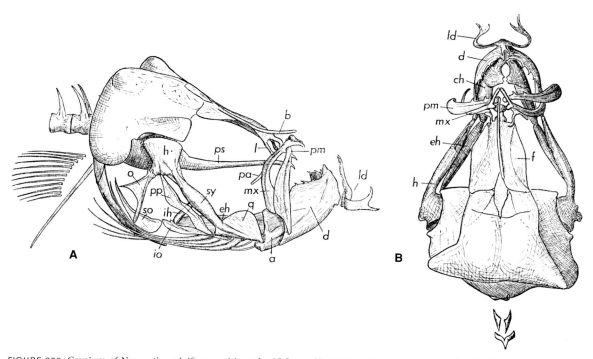

FIGURE 299. Cranium of *Neoceratias spinifer*, parasitic male, 15.5 mm SL, attached to a 52-mm SL female, ZMUC P921726, showing among other things the peculiar shaped lower denticle bone: (A) right lateral view; (B) dorsal view. a, angular; b, basal bone or pterygiophore of the illicium; ch, ceratohyal; d, dentary; eh, epihyal; f, frontal; h, hyomandibula; ih, interhyal; io, interopercle; l, lateral ethmoid; ld, lower denticular bone; mx, maxilla; o, opercle; pa, palatine; pm, premaxilla; pp, preopercle; ps, parasphenoid; q, quadrate; so, subopercle; sy, symplectic. Modified after Bertelsen (1951).

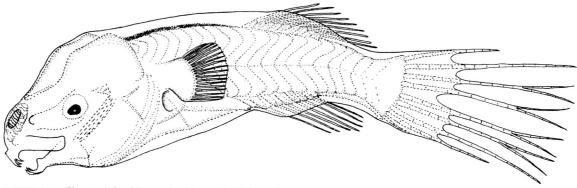

FIGURE 300. *Gigantactis longicirra*, male, 14 mm SL, UF 227412. After Bertelsen et al. (1981).

monotypic; *Acentrophryne* Regan, 1926, contains two species; and the fifth genus, *Linophryne* Collett, 1886, currently contains 22 species. *Photocorynus*, *Haplophryne*, *Borophryne*, and *Linophryne* are all relatively well represented in collections by metamorphosed females and free-living males, and each genus is known to have parasitic males. *Acentrophryne*, however, is known only from five metamorphosed female specimens, all of which have small undeveloped ovaries (see Pietsch and Shimazaki, 2005).

Bertelsen (1951:247) found that the testes of all free-living linophrynid males are "less developed" than in any parasitically attached male. Males with large testes containing "numerous spermatozoa" have been found only in attached males (Regan and Trewavas, 1932:17; Bertelsen, 1976:10, 13, 15). The only known females with well-developed ovaries, many of which contain eggs that "seem almost ripe," carry an attached male (Bertelsen, 1951:26, 1976:10, 13, 1980a:66; Mead et al., 1964). One of these females, a 77-mm *Linophryne arborifera*, with a 15-mm parasitic male, was found to have numerous eggs embedded in a gelatinous mass (the so-called egg raft, or veil, a reproductive device characteristic of all lophiiform fishes; see Pietsch and Grobecker, 1987:351) protruding from the genital opening; the eggs, 0.6 to 0.8 mm in diameter, are among the largest known of any ceratioid (Bertelsen, 1980a:66, Fig. 206).

An examination of stomachs of the majority of free-living males in the *Dana* collections led Bertelsen (1951:244) to consider it "extremely probable" that linophrynid males are unable to feed during their free-living stage after metamorphosis. The "short and stout" denticulars of the upper and lower jaws of these males do not "appear suitable for prey capture" (Bertelsen, 1951:161, 250; Figs. 54, 301). The eyes of linophrynid males are very well developed and unique among ceratioids in being tubular: Bertelsen (1951:161; Figs. 50C, 51, 297B–E, 301) described them as "telescopic, directed somewhat forward" (see also Munk, 1964:10, 1966:31). The nostrils of linophrynid males are also large and well developed (see Figs. 5, 6, 301).

*Photocorynus*, containing only *P. spiniceps*, is now known from about 31 metamorphosed females, two free-living metamorphosed males, and four parasitic males. One of the parasitic males, a 7.3-mm specimen attached on the face just above the right eye of the 46-mm holotype of *P. spiniceps* (ZMUC P92134, Fig. 297B), was described and figured by Regan (1925b; see also Bertelsen, 1951:166). The remaining three couples were reported by Pietsch (2005b:228): ISH 1913/71, a 50.5-mm female, with a 7-mm male attached to

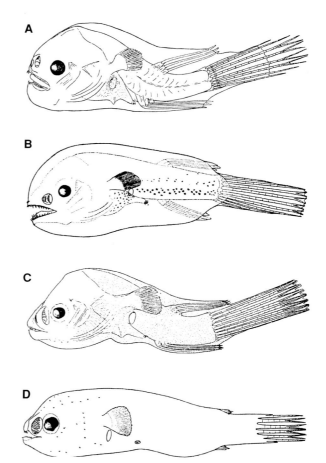

FIGURE 301. Free-living males of linophrynids: (A) *Photocorynus spiniceps*, 8.6 mm SL, ZMUC P921727; (B) *Haplophryne mollis*, 13 mm SL, ZMUC P921901; (C) *Borophryne apogon*, 15 mm SL, ZMUC P921771; (D) *Linophryne* sp., 17 mm SL, ZMUC P921799. After Bertelsen (1951).

the belly; SIO 70-326, a 46-mm female, with a 6.2-mm male attached just behind the head above the pectoral fin, slightly to the right of the dorsal midline (Figs. 302, 303); and SIO 70-346, a 49-mm female, with a 6.5-mm male in almost the same position, but slightly to the left of the dorsal midline. The ovaries of the ISH specimen are large and contain ripening eggs 0.4 to 0.5 mm in diameter, while the three remaining parasitized females appear to be immature with egg diameters of less than 0.1 mm. The belly of the 7.3-mm male attached

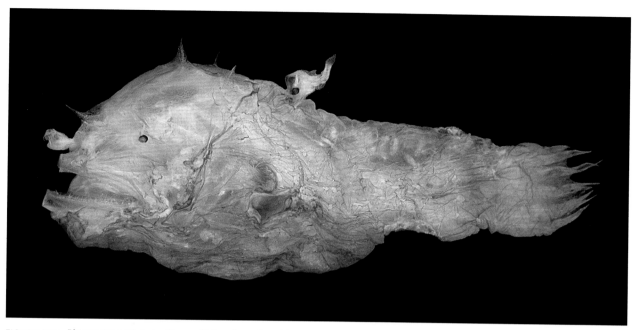

FIGURE 302. *Photocorynus spiniceps*, 46-mm SL female, with a 6.2-mm SL parasitic male, SIO 70-326. After Pietsch (2005b).

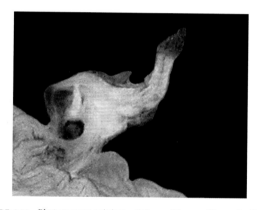

FIGURE 303. *Photocorynus spiniceps*, 6.2-mm SL parasitic male, SIO 70-326, the smallest known sexually mature vertebrate. After Pietsch (2005b).

to the holotype "is greatly inflated by the testes" (Bertelsen, 1951:166; Fig. 297B). The three remaining parasitic males are also mature or at least close to maturity, their testes distinctly enlarged. Those of the two SIO specimens (6.2 to 6.5 mm) are especially well developed, their greatest length about 1.4 and 1.3 mm (22.6 and 20.0% SL, respectively), each containing large amounts of developing spermatozoa, but without evidence of flagellated sperm. Together with the 7.5-mm parasitic male of *Leptacanthichthys gracilispinis* described above, the 6.2- and 7.4-mm parasitic males of *P. spiniceps* are the smallest known within the suborder, and, if regarded as adults, which histological evidence seems to indicate in the case of *Photocorynus*, they are among the world's smallest known sexually mature vertebrates as defined in terms of length, volume, and weight (see Winterbottom and Emery, 1981; see also Roberts, 1986; Weitzman and Vari, 1988; Kottelat and Vidthayanon, 1993; Watson and Walker, 2004; Guinness Book, 2007:41).

*Haplophryne*, containing only *H. mollis*, is known from about 88 metamorphosed females, 16 metamorphosed free-living males, 43 parasitic males, and 22 larvae (Fig. 172). Twenty-five of the females are parasitized: 11 of them carry only a single male, but 11 have two males, one has three (Figs. 31, 170, 171), and another has six (see Table 10). In contrast to *Borophryne* and *Linophryne*, in which males are nearly always found upside down, facing forward, and attached to the belly close to the anus, those of *Haplophryne* may be found facing in any and all directions, almost anywhere on the head and trunk, and even, in one case, on the esca of the female (Bertelsen and Pietsch, 1983:96; Munk and Bertelsen, 1983:50; see Fig. 304). In contrast also to the parasitic males of all other ceratioids (except for a few specimens of *Ceratias*; see above), a prominent nipplelike papilla of tissue projects from the female, more or less filling the mouth of the male (see Regan and Trewavas, 1932:15; Fig. 284). Munk and Bertelsen (1983:57, 71) speculated that this papilla, if developed prior to actual fusion of male and female tissues, "may facilitate the earliest stages of attachment by procuring a firmer hold for the male teeth." At the same time, however, the papilla, in filling the mouth cavity of the male, would seem to "represent a hindrance for the establishment of effective respiratory currents across the male gills." The majority of the attached males, however, have retained an opening to the pharynx on each side, although one has lost the opening on one side, and three are so deeply embedded in female tissue that the mouth has become completely closed (ZMUC P92138; NMNZ P.26070, Andrew L. Stewart, personal communication, 8 June 2004).

*Borophryne*, containing only *Borophryne apogon*, is represented in collections around the world by about 39 metamorphosed females, 62 metamorphosed males (13 of which are parasitically attached to females), and six larvae. Eleven of the females carry a single parasitic male, all attached in nearly the same place on the ventral midline just anterior to the anus, all upside down and directed forward with respect to the female (Table 10). A twelfth female, the largest known specimen of the genus (101 mm, LACM 30053-10), has two

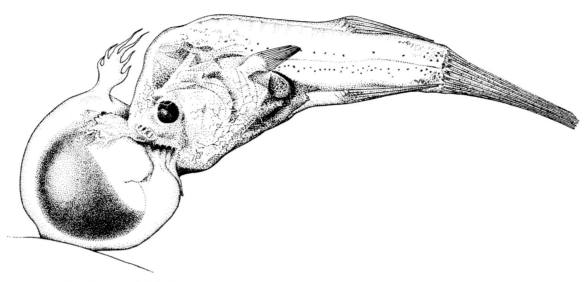

FIGURE 304. Parasitic male of *Haplophryne mollis*, 12 mm SL, fused to the distal surface of the esca of a 62-mm SL female, AMS I.21365-8. Drawing by Robert Nielsen; after Munk and Bertelsen (1983).

attached males (Fig. 305), both facing forward, a larger (22 mm), more heavily pigmented male situated upside down on the belly close to the anus, and a smaller (16 mm), much less pigmented male, placed right-side up, more posteriorly and slightly to the right side, at the base of the anal fin (see Pietsch, 1976:788). Both of these males have large testes: those of the smaller specimen, which appears to have attached itself to the female much more recently than the other, are about 2.8 mm long (17.5% SL); those of the larger male are 4.7 mm long (21.4% SL). Histological examination of the testes of the latter specimen showed evidence of late resorption, indicating a recent spawning event.

The size range of free-living *Borophryne* males is 11 to 17.5 mm (Figs. 178, 301C); that of the attached males is 10.5 to 22 mm. Two of the parasitic males of *Borophryne*, both described by Regan and Trewavas (1932:18), are attached only by the tip of their lower jaw, leaving the upper jaw and its denticles more or less free (Fig. 297D); but the remaining examples are attached by both upper and lower jaws, in all cases leaving prominent openings on each side that lead into their mouths and opercular cavities. A papilla of tissue projecting from the female into the mouth of the male, as described above in *Haplophryne*, is absent.

*Linophryne* is now represented in collections by a total of some 170 metamorphosed females, of which 29 carry a single parasitic male. In addition, there are about 110 known free-living males and about 80 larvae. Bertelsen (1982:99–100) summarized the known facts surrounding sexual parasitism in this genus; his discussion follows, with only a few minor additions and updates: The 30 parasitized *Linophryne* females now known represent 12 of the 22 recognized species (Table 10). Each of these has a single male, in contrast to the linophrynid genera *Haplophryne* and *Borophryne* (and the ceratiid genera *Ceratias* and *Cryptopsaras*) in which females with two or more males are known. In all known cases, the male is directed forward with respect to the female and attached in nearly the same position on the ventral midline of the female, somewhat in front of and below the sinistral anus; with only one or two exceptions, all are attached upside down with respect to the female (Figs. 4, 306–308). This is again in contrast to the linophrynid genera *Photocorynus* and *Haplophryne* (and the ceratiid genus *Cryptopsaras*) in which males may attach in any direction and almost anywhere on the head and body of the female (Munk and Bertelsen, 1983:50; Figs. 302, 304). In all known examples, the males are attached by both upper and lower jaws, leaving prominent openings on each side that lead into their mouths and opercular cavities; there is no papilla of tissue projecting from the female into the mouth of the male (Figs. 297E, 309).

Approximately half of all the known females of the genus greater than 150 mm (seven out of 13) are parasitized, about one-third of the specimens ranging from 50 to 140 mm (eight out of 25) are parasitized, and only a single female (42 mm) among the about 75 specimens less than 50 mm carries a parasitic male. Parasitic males considerably larger (21.5 to 30 mm) than any of the nearly 200 known metamorphosed free-living males of *Linophryne* (10.5 to 19.5 mm; see Fig. 301D) have been observed in five species (*Linophryne lucifer, L. coronata, L. algibarbata, L. bicornis,* and *L. macrodon*; see Table 10), confirming that these males are true parasites receiving nourishment from the females (Bertelsen, 1976:16, 1978:31).

## Mate Location and Species-Specific Selection

It has generally been assumed that male ceratioids locate conspecific females by olfactory or visual cues, or more likely by a combination of both these senses. First stated explicitly by Bertelsen (1951:249), these ideas are supported by the fact that the eyes and olfactory structures of males are generally very well developed in free-living stages but degenerate rapidly after parasitic attachment (Bertelsen, 1951:246; Pietsch, 1974a:98; Munk, 2000:323):

> The considerable and often enormous development of either eyes or olfactory organs or both must undoubtedly be regarded as an adaptation to the difficult task of the male to find a female. The large olfactory organs of the males indicate that the females give off specific odors. It may be assumed that the slow females leave behind them a scent, which will spread very slowly in the still water of the depths and may be perceived and followed by the searching males. Even if the eyes of the male are well developed, they can hardly observe and recognize a specific mate at any great distance, but the esca with its light organ and [species-] specific attachments

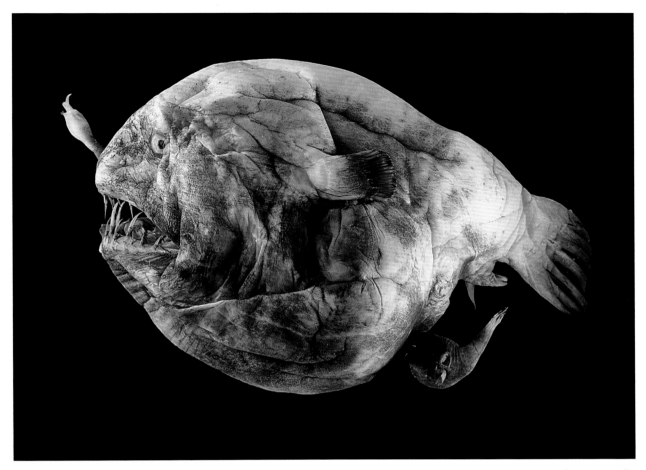

FIGURE 305. *Borophryne apogon*, 101-mm SL female, with two parasitic males, 16 and 22 mm SL. Photo by Darlyne A. Murawski; courtesy of the National Geographic Image Collection.

may presumably function as a distinguishing mark which the males can recognize when they come sufficiently near (Bertelsen, 1951:249).

This dual mechanism for mate location and species-specific selection—evolved to prevent wastage of gametes in an environment that is vast and dark, and where population densities of forms with restricted activities are low—probably functions in most ceratioids, in which both eyes and olfactory structures of the free-living males are well developed. But whether vision plays a significant role remains uncertain in *Himantolophus*, in which the eyes of metamorphosed males are relatively small and actually decrease slightly in diameter with increasing standard length. Furthermore, it is highly unlikely that vision and olfaction together mediate coupling in the ceratiid genera *Ceratias* and *Cryptopsaras*, in which the nostrils are surprisingly small and undeveloped (Fig. 286A, B); and in *Centrophryne*, and especially the gigantactinid genera *Gigantactis* and *Rhynchactis*, in which the eyes are very much reduced (Figs. 285, 286C, D). However, in apparent compensation for a reduced olfactory sense, the eyes of adult free-living ceratiid males are especially large and peculiarly specialized, having an unusually wide binocular field of vision made possible by large aphakic spaces, a curved rostral part of the retina, and prominent sighting grooves (Munk, 1964:5–10: Figs. 50, 51). The unique accessory bioluminescent structures (dorsal caruncles; Bertelsen, 1951:16) of ceratiid females probably also play a role in what must be primarily a visually mediated mechanism of pair-bonding in this family. Adult free-living gigantactinid males, on the other hand, have minute eyes, yet their olfactory structures are exceptionally large, even among ceratioids, indicating that the detection of a specific-specific pheromone emitted by the female is the dominant mode of solving the problem of coupling in this family. The eyes of centrophrynid males are also small, but there is nothing about these males that would seem to compensate for this apparent deficiency. Finally, the mechanism by which males and females of *Neoceratias* find each other remains a mystery. Although free-living males are unknown, the eyes and nostrils of all known parasitic males of this genus are especially small and degenerate, and besides these apparent deficiencies, the females lack an illicium and esca and any other known bioluminescent structure (see Bertelsen, 1951:158; Figs. 154, 298).

## Modes of Reproduction

### Obligatory Sexual Parasitism

All evidence indicates that the sexual parasitic mode of reproduction is obligatory in some ceratioids (Table 11). Examination of available specimens of five genera (*Ceratias*, *Cryptopsaras*, *Haplophryne*, *Borophryne*, and *Linophryne*) in two families (Ceratiidae and Linophrynidae) has shown that free-living males and nonparasitized females never have well-developed gonads. Males thus apparently never mature unless they are in

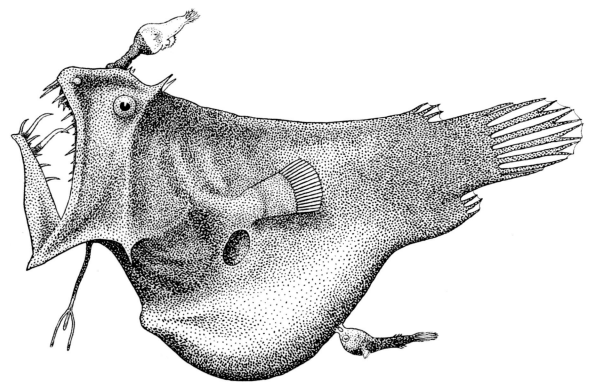

FIGURE 306. *Linophryne argyresca*, holotype, 61-mm SL female, with a 12-mm parasitic male, ZMUC P92142. Drawing by W. P. C. Tenison; after Regan and Trewavas (1932).

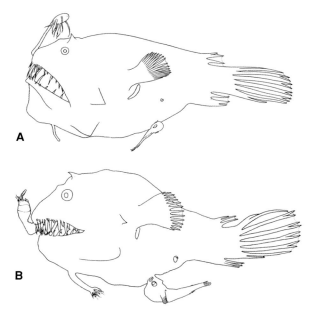

FIGURE 307. Species of *Linophryne*: (A) *L. trewavasae*, 73.5-mm SL female, with 10.7-mm SL parasitic male, LACM 36116-005; (B) *L. indica*, 42-mm SL female, with 9.5 mm SL parasitic male, LACM 36046-011. After Bertelsen (1978).

parasitic association with a female, and, failing to locate a conspecific female within the first few months of their lives, they die. Likewise, females never become gravid until stimulated by the permanent parasitic attachment of a male. That sexual maturity is determined not by size or age in these fishes, but by parasitic sexual association, may well be unique among animals. The jaw apparatus of free-living males of these taxa seems to be unsuited to serve in prey capture, and the alimentary canal is undeveloped, indicating that the males do not feed after metamorphosis and thus are fully dependent on a parasitic association with a female for long-term survival. These two lines of evidence suggest that spawning and fertilization in members of these families occur only during a permanent parasitic association of male and female. The question then arises: why do we have so few parasitized females of these obligatory forms? Members of the genus *Ceratias* (all three species combined) are among the most commonly collected ceratioids, now known from more than 440 metamorphosed females, but only 35, or 7.9%, of these specimens carry a parasitic male (see Table 10; Fig. 310). *Cryptopsaras couesii* is even better represented in collections, with more than 980 metamorphosed females, but only 47, or 4.8%, are parasitized. Assuming that these numbers reflect the true structure of populations at meso- and bathypelagic depths, they indicate a remarkably low percentage of individuals participating in reproduction at any one time. Furthermore, considering that females, at least in the case of *C. couesii*, are receptive to males at an extremely young age, beginning at a standard length of 15 mm and extending to the largest known individuals, just under 400 mm (see Pietsch, 1975a; Table 10), it is reasonable to expect a higher percentage of parasitized females. It seems highly unlikely that more than 90% of the females are purely vegetative, contributing nothing to successive generations. Under these circumstances, successful recruitment on an annual basis is difficult to understand.

The ratio of parasitized to nonparasitized females of *Photocorynus spiniceps*, as represented in preserved collections, is only slightly more equitable: of the 31 known metamorphosed fe-

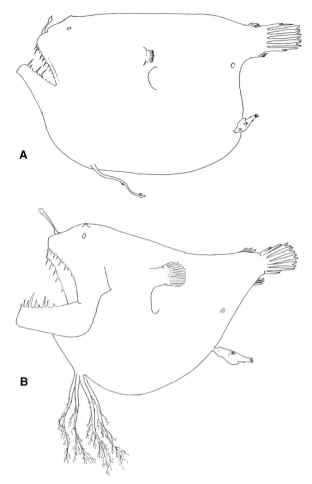

FIGURE 308. Species of *Linophryne*: (A) *L. lucifer*, 230-mm SL female, with 24-mm SL parasitic male, MRIR uncataloged; (B) *L. algibarbata*, 182-mm SL female, with 29-mm SL parasitic male, IMB uncataloged. After Bertelsen (1976).

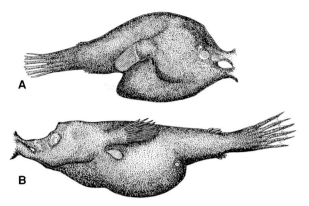

FIGURE 309. Parasitic males of *Linophryne*, with body cavities greatly expanded by enlarged testes (females shown in Figure 308. (A) *L. lucifer*, 24 mm SL, attached to a 230-mm SL female, MRIR uncataloged; (B) *L. algibarbata*, 29 mm SL, attached to a 182-mm SL female, IMB uncataloged. After Bertelsen (1976).

## Nonparasitic Attachment

It seems evident also that males and females of some ceratioids never become associated parasitically (Table 11). The Himantolophidae, Diceratiidae, Melanocetidae, Gigantactinidae, and most oneirodid genera, although now well represented in collections, have yielded no evidence of male parasitism. Except in the Melanocetidae, some females of these groups have been found to contain large ovaries, with eggs visible to the naked eye; and in most all these taxa (the Melanocetidae included, but not the Diceratiidae), free-living males with large testes have been reported. Either sex is thus able to attain sexual maturity without the presence of the other. Although food items have been found in the stomachs of melanocetid males only, these and males of the Himantolophidae, Gigantactinidae, and the oneirodid genus *Oneirodes* undergo a postmetamorphic increase in length of 7 to 12 mm, indicating that they are able to sustain themselves independent of the female after energy stores in their liver have been exhausted. That spawning takes place during a temporary sexual attachment, not involving fusion of male and female tissues, is supported by the presence in all these males of especially well-toothed upper and lower denticular bones that are presumably also effective in capturing prey. That collections around the world are nearly devoid of examples of temporary sexual couplings is no doubt because males are able to release themselves rather easily from their mates when startled by on-coming fishing gear. The only known exceptions are two pairs of *Melanocetus*, in which the males are firmly attached, but without tissue fusion.

## Facultative Sexual Parasitism

The possibility that sexual parasitism may be facultative in some ceratioids, as proposed by Pietsch (1976:791), was reaffirmed by Pietsch (2005b:230; Table 11). The known material of the Caulophrynidae and the oneirodid genera *Leptacanthichthys* and *Bertella* each includes at least two gravid females, one that is parasitized by a male and one that is not. Unless the nonparasitized females have somehow lost their males, leaving behind no trace, they are capable of reaching sexual maturity either with or without the presence of an attached male. These three taxa, therefore, cannot be reasonably placed within either of the two categories treated above—obligatory parasitism or

males, 12.9% carry an attached male. Counts for *Haplophryne mollis* and *Borophryne apogon* are even better, with about 28% and 33% of known females parasitized, respectively. However, even in the best situation, these numbers imply that two-thirds or more of all metamorphosed females are living out their lives as solitary nonreproducing individuals.

Although sufficient data are still unavailable to say with certainty, *Neoceratias* probably also reproduces by obligatory sexual parasitism. Of the 18 known metamorphosed females of this genus, seven carry a parasitic male. At least one of these parasitized females is sexually mature or very close to it, and several others are in various stages of development; the 11 remaining nonparasitized females all have small ovaries without visible eggs. Although free-living males are unknown, the testes of at least one of the attached males have been shown to be exceptionally large, containing all spermatogenetic stages, with spermatozoa present in the lumina of the testes as well as in the sperm ducts. The denticular apparatus of males of *Neoceratias* is highly modified and appears essentially useless in prey capture, indicating that, like ceratiid and linophrynid males, those of *Neoceratias* cannot feed and therefore depend solely on a parasitic association with a female for postmetamorphic survival.

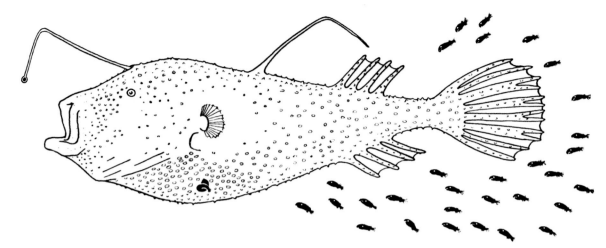

FIGURE 310. Fanciful image of *Ceratias holboelli* being attacked by males: "This sketch of Holboell's Angler shows the parasitic, dwarf male attached to her side. A swarm of free-living males is following, drawn by the scent of the female and waiting for a chance to attach themselves—on the side, below the body, or even between the eyes" (Beebe, 1938:51). After Beebe (1938).

nonparasitic temporary attachment. It is thus likely that fertilization of eggs can occur either during a temporary attachment or during a permanent parasitic association of male and female. Males of facultative forms probably attach to females whenever the two meet, regardless of sexual readiness. If both partners are in a state of readiness at the time of attachment, spawning and fertilization take place, after which the male releases his hold on the female and is then presumably capable of beginning a new search for another mate. If one or both partners are not ready to spawn, the male remains attached until spawning can take place. The longer the male remains attached to the female, the greater are his chances of becoming fused and establishing a permanent parasitic association.

The remaining ceratioid families, Thaumatichthyidae and Centrophrynidae, are still so poorly known that nothing can be concluded relative to their mode of reproduction. There is no evidence of sexual parasitism in either family. Although the known material contains large females in each case (225 to 365 and 230 to 247 mm, respectively), all appear to be immature. The known males of these taxa (only two and three, respectively) are also immature.

## Reproductive Modes and Phylogeny

Despite major efforts over the past 35 years, no satisfactory phylogeny of ceratioid anglerfishes has been available to help understand the evolution of sexual parasitism. But a rigorous cladistic analysis based on morphology has now been published by Pietsch and Orr (2007) and is presented here with only minor revisions (see Evolutionary Relationships, Chapter Four). At the same time, several recent and on-going attempts to determine ceratioid relationships through analysis of DNA are providing good resolution (Shedlock et al., 2004; Miya et al., in preparation). Results of the two approaches, morphological and molecular, differ markedly from each other, and from previously published hypotheses, yet some clear similarities exist. Of special interest is the lack of support for a monophyletic origin of sexual parasitism. The data in all cases suggest instead an abrupt appearance and subsequent loss of sexual parasitism during ceratioid evolution, followed by a secondary gain in the Oneirodidae. The apparent plasticity of this life history strategy is more consistent with the multiple gains and losses of parasitism required by several earlier proposals (e.g., that of Bertelsen, 1984). In summary, no matter how we

TABLE 11
Modes of Reproduction in Ceratioid Anglerfishes

| *Obligatory* | *Nonparasitic* | *Facultative* |
|---|---|---|
| Attached males common, always well fused to female | Attached males rarely found, never fused to female | Attached males rarely found, always fused to female |
| Gravid females never found without attached males | Gravid females often found without attached males | Gravid females are known with or without attached males |
| Free-living males always with small undeveloped testes | Free-living males often found with large, well-developed testes | Free-living males are found with large, well-developed testes |
| Gut of free-living males always empty | Gut of free-living males found to contain food | Gut of free-living males found to contain food |

look at ceratioid phylogeny, whether based on comparative morphology or molecular analysis, results indicate that sexual parasitism has evolved multiple times independently within the suborder.

These findings are consistent with the many differences evident in the precise nature of attachment among the various taxa: males attached to the apex of an unpigmented conical growth from the female in *Caulophryne*; male fused to the tip of a cylindrical stalk of female tissue in the oneirodid genera *Bertella* and *Leptacanthichthys*; nearly always single males attached invariably on the belly in *Ceratias*, but multiple males found almost anywhere on the body in *Cryptopsaras*; broadly attached males, with fully occluded mouths in *Neoceratias*; multiple males common in *Haplophryne*, attached anywhere on the head and body, and always involving a papilla of female tissue that fills the mouth of the male; and, finally, single males always the rule in *Linophryne*, almost always attached upside down at nearly the same spot on the ventral midline of the female. Having been established independently three and possibly as many as five times within the suborder, it seems evident, yet difficult to believe, that sexual parasitism in ceratioid anglerfishes, with all its extreme complexity of morphological, physiological, and behavioral adaptations, is a considerably less drastic evolutionary event than might be supposed. When viewed in this light, it is perhaps surprising that this remarkably successful reproductive strategy has not evolved in other vertebrate taxa that have come to occupy the deep sea. At the same time, the absence of hermaphroditism among ceratioids, so widespread and highly developed in other deep-sea taxa (e.g., see Mead et al., 1964), most likely reflects the overall effectiveness of sexual attachment and parasitism, a marvelous adaptation that has evolved in no other organism.

PART TWO

# A CLASSIFICATION OF DEEP-SEA ANGLERFISHES

## INTRODUCTION

## FAMILIES, GENERA, AND SPECIES OF THE CERATIOIDEI

### FAMILY CENTROPHRYNIDAE BERTELSEN, 1951 (PRICKLY SEADEVILS)

**Genus *Centrophryne* Regan and Trewavas, 1932 (Prickly Seadevils)**

*Centrophryne spinulosa* Regan and Trewavas, 1932

### FAMILY CERATIIDAE GILL, 1861 (WARTY SEADEVILS)

**Genus *Ceratias* Krøyer, 1845 (Doublewart Seadevils)**

*Ceratias holboelli* Krøyer, 1845
*Ceratias uranoscopus* Murray, in Thomson, 1877
*Ceratias tentaculatus* (Norman, 1930)

**Genus *Cryptopsaras* Gill, 1883b (Triplewart Seadevils)**

*Cryptopsaras couesii* Gill, 1883b

### FAMILY HIMANTOLOPHIDAE GILL, 1861 (FOOTBALLFISHES)

**Genus *Himantolophus* Reinhardt, 1837 (Footballfishes)**

***Himantolophus groenlandicus* Group Bertelsen and Krefft, 1988**

*Himantolophus groenlandicus* Reinhardt, 1837
*Himantolophus sagamius* (Tanaka, 1918a)
*Himantolophus danae* Regan and Trewavas, 1932
*Himantolophus crinitus* Bertelsen and Krefft, 1988
*Himantolophus paucifilosus* Bertelsen and Krefft, 1988

***Himantolophus appelii* Group Bertelsen and Krefft, 1988**

*Himantolophus appelii* (Clarke, 1878)

***Himantolophus nigricornis* Group Bertelsen and Krefft, 1988**

*Himantolophus nigricornis* Bertelsen and Krefft, 1988
*Himantolophus melanolophus* Bertelsen and Krefft, 1988

***Himantolophus albinares* Group Bertelsen and Krefft, 1988**

*Himantolophus albinares* Maul, 1961
*Himantolophus mauli* Bertelsen and Krefft, 1988
*Himantolophus pseudalbinares* Bertelsen and Krefft, 1988
*Himantolophus multifurcatus* Bertelsen and Krefft, 1988
*Himantolophus borealis* Kharin, 1984

***Himantolophus cornifer* Group Bertelsen and Krefft, 1988**

*Himantolophus cornifer* Bertelsen and Krefft, 1988
*Himantolophus macroceras* Bertelsen and Krefft, 1988
*Himantolophus macroceratoides* Bertelsen and Krefft, 1988
*Himantolophus azurlucens* Beebe and Crane, 1947
*Himantolophus compressus* (Osório, 1912)

***Himantolophus rostratus* Group Bertelsen and Krefft, 1988**
***Himantolophus appelii* Group Bertelsen and Krefft, 1988**
***Himantolophus brevirostris* Group Bertelsen and Krefft, 1988**

### FAMILY DICERATIIDAE REGAN AND TREWAVAS, 1932 (DOUBLESPINE SEADEVILS)

**Genus *Diceratias* Günther, 1887 (Doublespine Seadevils)**

*Diceratias bispinosus* Günther, 1887
*Diceratias pileatus* Uwate, 1979
*Diceratias trilobus* Balushkin and Fedorov, 1986

**Genus *Bufoceratias* Whitley, 1931 (Toady Seadevils)**

*Bufoceratias wedli* (Pietschmann, 1926)
*Bufoceratias thele* (Uwate, 1979)
*Bufoceratias shaoi* Pietsch, Ho, and Chen, 2004

### FAMILY MELANOCETIDAE GILL, 1879B (BLACK SEADEVILS)

**Genus *Melanocetus* Günther, 1864 (Black Seadevils)**

*Melanocetus johnsonii* Günther, 1864
*Melanocetus rossi* Balushkin and Fedorov, 1981
*Melanocetus polyactis* Regan, 1925c
*Melanocetus niger* Regan, 1925c
*Melanocetus eustalus* Pietsch and van Duzer, 1980
*Melanocetus murrayi* Günther, 1887

## FAMILY THAUMATICHTHYIDAE SMITH AND RADCLIFFE, 1912 (WOLFTRAP SEADEVILS)

### Genus *Lasiognathus* Regan, 1925c (Snaggletooth Seadevils)

*Lasiognathus beebei* Regan and Trewavas, 1932
*Lasiognathus waltoni* Nolan and Rosenblatt, 1975
*Lasiognathus intermedius* Bertelsen and Pietsch, 1996
*Lasiognathus saccostoma* Regan, 1925c
*Lasiognathus amphirhamphus* Pietsch, 2005a

### Genus *Thaumatichthys* Smith and Radcliffe, 1912 (Wonderfishes)

*Thaumatichthys pagidostomus* Smith and Radcliffe, 1912
*Thaumatichthys binghami* Parr, 1927
*Thaumatichthys axeli* (Bruun, 1953)

## FAMILY ONEIRODIDAE GILL, 1879A (DREAMERS)

### Genus *Lophodolos* Lloyd, 1909a (Pugnose Dreamers)

*Lophodolos indicus* Lloyd, 1909a
*Lophodolos acanthognathus* Regan, 1925c

### Genus *Pentherichthys* Regan and Trewavas, 1932 (Thickjaw Dreamers)

*Pentherichthys atratus* Regan and Trewavas, 1932

### Genus *Chaenophryne* Regan, 1925c (Smoothhead Dreamers)

*Chaenophryne longiceps* Group Bertelsen, 1951

*Chaenophryne longiceps* Regan, 1925c

*Chaenophryne draco* Group Bertelsen, 1951

*Chaenophryne draco* Beebe, 1932b
*Chaenophryne melanorhabdus* Regan and Trewavas, 1932
*Chaenophryne ramifera* Regan and Trewavas, 1932
*Chaenophryne quasiramifera* Pietsch, 2007

### Genus *Spiniphryne* Bertelsen, 1951 (Spiny Dreamers)

*Spiniphryne gladisfenae* (Beebe, 1932b)
*Spiniphryne duhameli* Pietsch and Baldwin, 2006

### Genus Oneirodes Lütken, 1871 (Common Dreamers)

*Oneirodes luetkeni* (Regan, 1925c)
*Oneirodes carlsbergi* (Regan and Trewavas, 1932)
*Oneirodes rosenblatti* Pietsch, 1974a
*Oneirodes eschrichtii* Lütken, 1871
*Oneirodes sabex* Pietsch and Seigel, 1980
*Oneirodes bulbosus* Chapman, 1939
*Oneirodes anisacanthus* (Regan, 1925c)
*Oneirodes kreffti* Pietsch, 1974a
*Oneirodes posti* Bertelsen and Grobecker, 1980
*Oneirodes myrionemus* Pietsch, 1974a
*Oneirodes clarkei* Swinney and Pietsch, 1988
*Oneirodes heteronema* (Regan and Trewavas, 1932)
*Oneirodes macrosteus* Pietsch, 1974a
*Oneirodes plagionema* Pietsch and Seigel, 1980
*Oneirodes cristatus* (Regan and Trewavas, 1932)
*Oneirodes pterurus* Pietsch and Seigel, 1980
*Oneirodes acanthias* (Gilbert, 1915)
*Oneirodes thompsoni* (Schultz, 1934)
*Oneirodes notius* Pietsch, 1974a
*Oneirodes schistonema* Pietsch and Seigel, 1980
*Oneirodes flagellifer* (Regan and Trewavas, 1932)
*Oneirodes dicromischus* Pietsch, 1974a
*Oneirodes thysanema* Pietsch and Seigel, 1980
*Oneirodes haplonema* Stewart and Pietsch, 1998
*Oneirodes epithales* Orr, 1991
*Oneirodes macronema* (Regan and Trewavas, 1932)
*Oneirodes melanocauda* Bertelsen, 1951

*Oneirodes schmidti* Group Bertelsen, 1951

*Oneirodes alius* Seigel and Pietsch, 1978
*Oneirodes micronema* Grobecker, 1978
*Oneirodes schmidti* (Regan and Trewavas, 1932)
*Oneirodes mirus* (Regan and Trewavas, 1932)
*Oneirodes basili* Pietsch, 1974a
*Oneirodes theodoritissieri* Belloc, 1938
*Oneirodes pietschi* Ho and Shao, 2004
*Oneirodes bradburyae* Grey, 1956b

### Genus *Dermatias* Smith and Radcliffe, in Radcliffe, 1912 (Fattail Dreamers)

*Dermatias platynogaster* Smith and Radcliffe, in Radcliffe, 1912

### Genus *Danaphryne* Bertelsen, 1951 (Dana Dreamers)

*Danaphryne nigrifilis* (Regan and Trewavas, 1932)

### Genus *Microlophichthys* Regan and Trewavas, 1932 (Shortbait Dreamers)

*Microlophichthys microlophus* (Regan, 1925c)
*Microlophichthys andracanthus* Bertelsen, 1951

### Genus *Phyllorhinichthys* Pietsch, 1969 (Leafysnout Dreamers)

*Phyllorhinichthys micractis* Pietsch, 1969
*Phyllorhinichthys balushkini* Pietsch, 2004

### Genus *Tyrannophryne* Regan and Trewavas, 1932
### Pugnacious Dreamers

*Tyrannophryne pugnax* Regan and Trewavas, 1932

### Genus *Dolopichthys* Garman, 1899 (Longsnout Dreamers)

*Dolopichthys pullatus* Regan and Trewavas, 1932
*Dolopichthys longicornis* Parr, 1927
*Dolopichthys danae* Regan, 1926
*Dolopichthys jubatus* Regan and Trewavas, 1932
*Dolopichthys allector* Garman, 1899
*Dolopichthys karsteni* Leipertz and Pietsch, 1987
*Dolopichthys dinema* Pietsch, 1972c

### Genus *Bertella* Pietsch, 1973 (Spikehead Dreamers)

*Bertella idiomorpha* Pietsch, 1973

### "Long-pectoraled" Oneirodid Genera

### Genus *Puck* Pietsch, 1978 (Mischievous Dreamers)

*Puck pinnata* Pietsch, 1978

### Genus *Chirophryne* Regan and Trewavas, 1932 (Longhand Dreamers)

*Chirophryne xenolophus* Regan and Trewavas, 1932

### Genus *Leptacanthichthys* Regan and Trewavas, 1932 (Lightline Dreamers)

*Leptacanthichthys gracilispinis* (Regan, 1925c)

### Genus *Ctenochirichthys* Regan and Trewavas, 1932 (Combfin Dreamers)

*Ctenochirichthys longimanus* Regan and Trewavas, 1932

## FAMILY CAULOPHRYNIDAE GOODE AND BEAN, 1896 (FANFIN SEADEVILS)

### Genus *Caulophryne* Goode and Bean, 1896 (Fanfin Seadevils)

*Caulophryne pelagica* (Brauer, 1902)
*Caulophryne bacescui* Mihai-Bardan, 1982
*Caulophryne jordani* Goode and Bean, 1896
*Caulophryne polynema* Regan, 1930b
*Caulophryne* sp. A
*Caulophryne* sp. B
**Genus *Robia* Pietsch, 1979 (Longlure Fanfins)**
*Robia legula* Pietsch, 1979

# FAMILY NEOCERATIIDAE REGAN, 1926 (NEEDLEBEARD SEADEVILS)
**Genus *Neoceratias* Pappenheim, 1914 (Needlebeard Seadevils)**
*Neoceratias spinifer* Pappenheim, 1914

# FAMILY GIGANTACTINIDAE BOULENGER, 1904A (WHIPNOSE SEADEVILS)
**Genus *Gigantactis* Brauer, 1902 (Whipnose Seadevils)**
*Gigantactis longicirra* Waterman, 1939b
*Gigantactis kreffti* Bertelsen, Pietsch, and Lavenberg, 1981
*Gigantactis vanhoeffeni* Brauer, 1902
*Gigantactis meadi* Bertelsen, Pietsch, and Lavenberg, 1981
*Gigantactis gibbsi* Bertelsen, Pietsch, and Lavenberg, 1981
*Gigantactis gracilicauda* Regan, 1925c
*Gigantactis paxtoni* Bertelsen, Pietsch, and Lavenberg, 1981
*Gigantactis perlatus* Beebe and Crane, 1947
*Gigantactis elsmani* Bertelsen, Pietsch, and Lavenberg, 1981
*Gigantactis golovani* Bertelsen, Pietsch, and Lavenberg, 1981
*Gigantactis gargantua* Bertelsen, Pietsch, and Lavenberg, 1981
*Gigantactis watermani* Bertelsen, Pietsch, and Lavenberg, 1981
*Gigantactis herwigi* Bertelsen, Pietsch, and Lavenberg, 1981
*Gigantactis macronema* Regan, 1925c
*Gigantactis savagei* Bertelsen, Pietsch, and Lavenberg, 1981
*Gigantactis microdontis* Bertelsen, Pietsch, and Lavenberg, 1981
*Gigantactis ios* Bertelsen, Pietsch, and Lavenberg, 1981
*Gigantactis longicauda* Bertelsen and Pietsch, 2003
**Gigantactis Males**
*Gigantactis* Male Group I
*Gigantactis* Male Group II
*Gigantactis* Male Group III
*Gigantactis* Male Group IV
*Gigantactis* Male Group V

*Gigantactis* **Larvae**
*Gigantactis* Larval Group A
*Gigantactis* Larval Group B
*Gigantactis* Larval Group C
*Gigantactis* Larval Group D
**Genus *Rhynchactis* Regan, 1925c (Toothless Seadevils)**
*Rhynchactis leptonema* Regan, 1925c
*Rhynchactis macrothrix* Bertelsen and Pietsch, 1998
*Rhynchactis microthrix* Bertelsen and Pietsch, 1998

# FAMILY LINOPHRYNIDAE REGAN, 1925C (LEFTVENT SEADEVILS)
**Genus Photocorynus Regan, 1925b (Spinyhead Seadevils)**
*Photocorynus spiniceps* Regan, 1925b
**Genus *Haplophryne* Regan, 1912 (Ghostly Seadevils)**
*Haplophryne mollis* (Brauer, 1902)
**Genus *Acentrophryne* Regan, 1926 (Fangtooth Seadevils)**
*Acentrophryne longidens* Regan, 1926
*Acentrophryne dolichonema* Pietsch and Shimazaki, 2005
**Genus *Borophryne* Regan, 1925c (Greedy Seadevils)**
*Borophryne apogon* Regan, 1925c
**Genus *Linophryne* Collett, 1886 (Bearded Seadevils)**
*Linophryne* **Subgenus *Stephanophryne* Bertelsen, 1982**
*Linophryne indica* (Brauer, 1902)
*Linophryne* **Subgenus *Rhizophryne* Bertelsen, 1982**
*Linophryne parini* Bertelsen, 1980b
*Linophryne andersoni* Gon, 1992
*Linophryne quinqueramosa* Beebe and Crane, 1947
*Linophryne brevibarbata* Beebe, 1932b
*Linophryne densiramus* Imai, 1941
*Linophryne pennibarbata* Bertelsen, 1980
*Linophryne arborifera* Regan, 1925c
*Linophryne* **Subgenus *Linophryne* Bertelsen, 1982**
*Linophryne arcturi* (Beebe, 1926b)
*Linophryne coronata* Parr, 1927
*Linophryne lucifer* Collett, 1886
*Linophryne sexfilis* Bertelsen, 1973
*Linophryne argyresca* Regan and Trewavas, 1932
*Linophryne escaramosa* Bertelsen, 1982
*Linophryne bipennata* Bertelsen, 1982
*Linophryne maderensis* Maul, 1961
*Linophryne macrodon* Regan, 1925c
*Linophryne polypogon* Regan, 1925c
*Linophryne racemifera* Regan and Trewavas, 1932
*Linophryne trewavasae* Bertelsen, 1978
*Linophryne bicornis* Parr, 1927
*Linophryne algibarbata* Waterman, 1939b

10B. Skin covered with numerous close-set dermal spinules .................................................. 12
11A. Dorsal- and anal-fin rays 3 (rarely 2 or 4); branchiostegal rays 4 or 5 .................. Linophrynidae, p. 477
11B. Dorsal-fin rays more than 4, anal-fin rays 4 to 7; branchiostegal rays 6 .......................... 13
12A. Ceratobranchials toothless; branchiostegal rays narrow and cylindrical; 3 pectoral radials ................ Oneirodidae (genus *Spiniphryne*), p. 392
12B. Ceratobranchials well toothed; branchiostegal rays broad and laterally compressed; 4 pectoral radials (fusing to 3 in specimens greater than about 100 mm) .................. Centrophrynidae, p. 319
13A. Snout and chin rounded and blunt; bones of lower jaw enlarged and expanded ventrally, chin large, protruding anteriorly beyond snout ............. Himantolophidae (females less than 30 to 40 mm), p. 331
13B. Snout and chin more or less pointed; bones of lower jaw narrow; chin small, more or less equal to anterior extent of snout .................... Oneirodidae, p. 376

## Free-living Males

1A. Anal-fin rays greater than 9 ........................ 2
1B. Anal-fin rays fewer than 9 ......................... 3
2A. Lower denticular with approximately 9 to 18 teeth; pelvic fins present in larval and early metamorphic stages; dorsal-fin rays 14 to 22; anal-fin rays 12 to 19 ........... ............................ Caulophrynidae, p. 443

NOTE Males of the caulophrynid genus *Robia*, the single known female of which has dorsal-fin rays 6 and anal-fin rays 5, are unknown.

2B. Lower denticular trifurcate, each branch with a double hook; pelvic fins absent; dorsal-fin rays 11 to 13; anal-fin rays 10 to 13 .................. Neoceratiidae, p. 450
3A. Olfactory organs small; eyes large, bowl-shaped ....... ............................... Ceratiidae, p. 322
3B. Olfactory organs large; eyes not bowl-shaped ........ 4
4A. Dorsal-fin rays 12 to 17 ......... Melanocetidae, p. 359
4B. Dorsal-fin rays fewer than 11 ....................... 5
5A. Dorsal-fin rays fewer than 5 ........................ 6
5B. Dorsal-fin rays 5 to 8 ............................. 7
6A. Eyes large (diameter 6 to 9% SL), slightly tubular, directed more or less anteriorly; dorsal- and anal-fin rays 3 (rarely 2 or 4) .................. Linophrynidae, p. 477
6B. Eyes small (diameter 3 to 5% SL) spherical, not tubular, directed laterally; dorsal-fin rays 4 (rarely 3), anal-fin rays 3 to 4 .............................. Gigantactinidae (genus *Rhynchactis*), p. 475
7A. Eyes small, diameter 5% SL or less .................. 8
7B. Eyes large, diameter greater than 5% SL ............. 9
8A. A small digitiform hyoid barbel; bases of denticular teeth fused to form dorsal and ventral denticular tooth plates ............................. Centrophrynidae, p. 319
8B. Hyoid barbel absent; all bases of denticular teeth mutually free, not forming discrete tooth plates .............. ............................. Gigantactinidae, p. 452
9A. Skin completely covered with well-developed dermal spinules; anterior nostrils opening laterally ........... 10
9B. Skin naked, dermal spinules absent or extremely small and scattered; anterior nostrils opening anteriorly near end of snout ............................................ 11
10A. Tip of snout with more than 10 denticular teeth, all fused at base ...................... Himantolophidae, p. 331
10B. Tip of snout with only 2 separate denticular teeth ..... ................................ Diceratiidae, p. 352
11A. Body unusually long and slender, snout pointed; skin with small, but distinct dermal spinules scattered over surface of body; upper part of opercle divided into several radiating branches ............. Thaumatichthyidae, p. 368
11B. Body stout, snout blunt; skin naked or appearing so, dermal spinules absent or microscopic in size; upper part of opercle undivided .............. Oneirodidae, p. 376

NOTE Males of nine of the 16 recognized oneirodid genera are unknown, including *Spiniphryne*, the females of which have spinulose skin.

## Larvae

1A. Pelvic fins present ........... Caulophrynidae, p. 443
1B. Pelvic fins absent .................................. 2
2A. Pectoral fin large, reaching to or beyond dorsal and anal fins .......................... Gigantactinidae, p. 452
2B. Pectoral fin small, not reaching to dorsal and anal fins . 3
3A. Dorsal fin with more than 10 rays .................. 4
3B. Dorsal fin with fewer than 9 rays ................... 5
4A. Anal-fin rays 10 to 13 .......... Neoceratiidae, p. 450
4B. Anal-fin rays 3 to 5 ............ Melanocetidae, p. 359
5A. Body short, strongly arched, giving a "humpbacked" appearance; mouth subvertical; females with caruncles on back ............................ Ceratiidae, p. 322
5B. Axis of body straight or only slightly curved, body not "humpbacked"; mouth nearly horizontal; caruncles absent ............................................ 6
6A. Branchiostegal rays 4 or 5; dorsal- and anal-fin rays 3 (rarely 2 or 4) ............... Linophrynidae, p. 477
6B. Branchiostegal rays 6; dorsal-fin rays 5 to 7 (rarely 4), anal-fin rays 4 to 7 (rarely 3) ....................... 7
7A. A small hyoid barbel present .... Centrophrynidae, p. 319
7B. Hyoid barbel absent ................................ 8
8A. Subdermal melanophores present everywhere on the head and body, more or less evenly distributed on body ... 9
8B. Subdermal melanophores not present everywhere, arranged in distinct groups on body ............. 10
9A. Subdermal melanophores evenly distributed on body; upper part of opercle simple; females with an illicium-like second dorsal-fin spine .......... Diceratiidae, p. 352
9B. Subdermal melanophores with slight concentrations along margin of gillcover, peritoneum, and back; upper part of opercle divided into a series of radiating branches; females with second dorsal-fin spine rudimentary and concealed under skin ...... Thaumatichthyidae, p. 368
10A. Skin on head and body anterior to dorsal and anal fins highly inflated; dorsal-fin rays 5 or 6; anal-fin rays 4 .... ........................... Himantolophidae, p. 331
10B. Skin on head and body anterior to dorsal and anal fins moderately inflated; dorsal-fin rays 5 to 8 (very rarely 4); anal-fin rays 4 to 7 (very rarely 3) ...... Oneirodidae, p. 376

# Families, Genera, and Species of the Ceratioidei

## Family Centrophrynidae Bertelsen, 1951 (Prickly Seadevils)

Figures 48, 57–60, 202, 203, 208, 285; Tables 1, 8

Type genus *Centrophryne* Regan and Trewavas, 1932

Aceratiidae Regan, 1926:42 (in part; family of suborder Ceratioidea to contain a free-living male of genus *Rhynchoceratias* reassigned to genus *Centrophryne* by Bertelsen, 1951:125, 126).

Eurostrinae Parr, 1930a:20 (in part; subfamily of family Aceratiidae to contain a free-living male of genus *Rhynchoceratias* reassigned to genus *Centrophryne* by Bertelsen, 1951:125, 126).

Centrophrynidae Bertelsen, 1951:124 (family of suborder Ceratioidei to contain genus *Centrophryne*).

### Distinguishing Characters

Larvae and juveniles of both sexes of the family Centrophrynidae are unique among ceratioids in having a small digitiform hyoid barbel (present elsewhere in the suborder only in females of the linophrynid genus *Linophryne*).

Metamorphosed females of the family are further distinguished from those of all other ceratioid families in having a single oval-shaped ovary (ovaries are paired in all other ceratioid families; see Pietsch, 1972a:24, fig. 5). They differ further in having the following combination of character states: A supraethmoid is present. The frontals are narrowly separated by cartilage along the dorsal midline, each without a ventromedial extension. Parietals, pterosphenoids, metapterygoids, and mesopterygoids are present. Sphenotic spines are absent. The hyomandibula has a double head. They have 2 hypohyals and 6 (2 + 4) branchiostegal rays. The opercle is bifurcated; the dorsal fork is short, less than 50% the length of the ventral fork. The subopercle is long and slender, at least as long as the ventral fork of the opercle, with a slender tapering upper end. The lower end of the subopercle bears a well-developed spine on its anterior margin (less conspicuous in large females). Quadrate and articular spines are present but minute. Angular and preopercular spines are absent. The jaws are equal anteriorly, the lower jaw bearing a well-developed symphysial spine. A postmaxillary process of the premaxilla is absent. The anterior-maxillomandibular ligament is long and well developed. The first pharyngobranchial is present and suspensory in function. The second and third pharyngobranchials are well developed and toothed. The fourth pharyngobranchial is absent. The first, second, and third hypobranchials are well ossified, but there is only a single ossified basibranchial. Teeth are present on the first epibranchial and all four ceratobranchials. Epurals are absent. The hypural plate is deeply notched posteriorly. The pterygiophore of the illicium bears a small ossified remnant of the second cephalic spine. The escal bulb contains a central lumen with an escal pore leading to the exterior. The esca is without toothlike denticles. The posteroventral process of the coracoid is absent. Four pectoral radials are present, fusing to 3 in specimens greater than 150 mm. Pelvic bones are present, each only slightly expanded distally. Fin ray counts are as follows: dorsal-fin rays 6 or 7; anal-fin rays 5 or 6; pectoral-fin rays 15 or 16; pelvic fins absent; caudal rays 9 (2 simple + 4 bifurcated + 3 simple) (Table 1). The skin is everywhere covered with numerous, close-set dermal spinules. Pyloric caeca are absent.

Males of the family Centrophrynidae, known only from three specimens (one adult and two in metamorphosis), differ from those of all other ceratioids in having a short hyoid barbel situated just behind the symphysis of the lower jaw. They are further unique in having the following combination of character states: The eyes are unusually small, each without an aphakic space. The olfactory organs are relatively large; the anterior nostril, about half the size of the posterior nostril, is directed anteriorly. The upper denticular plate is roughly triangular in shape and bears a transverse series of 3 well-developed hooked denticles. The lower denticular plate is crescent shaped, bearing a transverse series of 4 strong, symmetrically placed denticles, fused at the base. Unlike the females, the skin is naked, without dermal spinules (Bertelsen, 1983:313, fig. 2). The males are free-living. There is no evidence of sexual parasitism (see Pietsch, 2005b; and Reproduction and Early Life History, Chapter Eight).

Larvae of the family Centrophrynidae, known from only two specimens (a male and female), are relatively short and deep, with the skin only moderately inflated, and a short, digitiform hyoid barbel. The pectoral fins are moderate in size, not reaching to the base of the dorsal and anal fins. Pelvic fins are absent (Bertelsen, 1951:126, fig. 85A, 1984:329, fig. 168F).

The osteological features cited here and elsewhere in this account of the Centrophrynidae are based on examinations of two juvenile females, one larval male, and one adult male (Bertelsen, 1951:124, fig. 84, Bertelsen, 1983:313–314, fig. 3; Pietsch, 1972a:26–43, figs. 7–24).

### Description

The body of metamorphosed females is long and slender, not globular, its depth approximately 35 to 40% SL. The maxilla terminates below the eye in smaller specimens (less than approximately 150 mm), but extends posteriorly beyond the eye in larger specimens. The oral valve is well developed, lining the inside of both upper and lower jaws. A short hyoid barbel is present in larvae and juvenile specimens less than 50 mm, but is reduced to a minute protuberance or lost in larger specimens (Pietsch, 1972a:24, fig. 4). Two nostrils are present on each side of the snout, located at the end of a single short tube. The teeth are slender, recurved, and all depressible, large and small intermixed in both jaws. The teeth of the lower jaw are generally larger but less numerous than those of the upper jaw. The number of teeth in the lower jaw ranges from 23 to 96, in the upper jaw, 32 to 153. There is an increase in the number of teeth in both jaws with increasing standard length up to approximately 80 mm, but a decrease in number with further growth. The number of vomerine teeth is 0 to 9. The first epibranchial bears 3 or 4, broad-based tooth plates; 5 to 16 similar tooth plates are present on all four ceratobranchials (see Pietsch, 1972a:36, fig. 15). The first epibranchial is free, not bound to the wall of the pharynx by connective tissue. The four epibranchials are closely bound together and the epibranchial and ceratobranchial of the fourth arch are bound to the wall of the pharynx. There is no opening behind the fourth arch. The proximal one-third to one-half of the first ceratobranchial is bound to the wall of the pharynx; the distal two-thirds to one-half are free. The distal end of the first ceratobranchial is free, not bound by connective tissue to the adjacent ceratobranchial of the second arch. The proximal one-quarter to one-half of the second, third, and fourth ceratobranchials are free, not bound together by connective tissue. Gill filaments are present on the proximal tips of the second and third epibranchials, on the proximal tip of the first ceratobranchial, and along the full length of the second, third, and fourth ceratobranchials. A pseudobranch is absent. The length of the illicium of females measures 18.7 to 26.0% SL. The anterior end of the pterygiophore of the illicium is exposed, emerging on the snout, its proximal end concealed under the skin. The esca bears a compressed, fan-shaped anterior appendage and a single, short, more or less compressed posterior appendage. Neuromasts of the acoustico-lateralis system are located at the tips of low cutaneous papillae, the pattern of placement as described for other ceratioids (Pietsch, 1969, 1972b, 1974a, 1974b).

The two known males in metamorphosis (11.5 to 16 mm) differ somewhat: the skin of the smaller specimen is slightly inflated, faintly pigmented, and semitransparent, the subdermal larval pigmentation showing through. It bears a short, stout barbel on the throat, without pigment on the distal tip. The depth of the body is approximately 40% SL, the head approximately 45% SL. Larval teeth are present in the lower jaw, but absent in the upper jaw. The denticles, preformed as small papillae, consist of a pair and perhaps a median denticle on the tip of the snout, and 4 somewhat larger denticles on the chin. The anterior nostril is slightly larger than the posterior nostril, its depth approximately 14% SL (about 1.5 times the diameter of the eye). There are 7 olfactory lamellae. The liver and intestines are well developed, but the testes are small and undeveloped (Bertelsen, 1951:127, figs. 84B, 87).

The larger male in late metamorphosis has the characters of the smaller male, except the skin is darkly pigmented and not inflated (Pietsch, 1972a:22, fig. 2). The skin between the anterior nostrils and between the anterior and posterior nostrils is pigmented. The distance between the tip of the snout and the anterior edge of the eye is 15.6% SL. Only traces of larval teeth are present on the lower jaw near its symphysis. The denticles are well ossified but not fused basally; there are 3 on the snout and 4 embedded in the skin below the symphysis of the lower jaw. The olfactory lamellae are not countable. A well-developed symphysial spine is on the lower jaw. A small anterior spine is present on the subopercle. The dorsoventral length of the anterior and posterior nostrils is 5.0% SL and 6.9% SL, respectively. The hyoid barbel is darkly pigmented, not tipped with white; its length is approximately 1.0 mm or 6.3% SL. The distance from the tip of the symphysial spine of the lower jaw to the base of the hyoid barbel is 5.0 mm or 31.3% SL. The testes are only slightly larger than those of the 11.5-mm male (see Bertelsen, 1951:126, fig. 87).

In the single known adult male (12.8 mm), parietal bones are clearly present. The hyomandibula has a double head. The opercle is bifurcate, the length of the upper fork about 75% of the length of the lower fork. The upper part of the subopercle is slender and tapering to a fine point. The anterior margin of the lower part of the subopercle bears a sharp spine. The pectoral radials are unossified. Larval teeth are absent, the edge of the dentary irregularly resorbed. A triangular upper denticular plate is present on tip of the snout, bearing a transverse series of 3 well-developed hooked denticles. A crescent-shaped lower denticular plate is present on the chin, bearing a transverse series of 4 strong, symmetrically placed denticles, fused at the base. The eyes are small, 0.55 mm or 4.3% SL in diameter, without an aphakic space. The olfactory organs are relatively large, the posterior nostril measuring 0.9 mm or 7.7% SL in vertical diameter, the anterior nostril about half the size of the posterior nostril. There are 7 or 8 olfactory lamellae. A short hyoid barbel is present, situated about 34% SL behind the tip of the lower jaw. The skin is naked, without dermal spinules. The testes are small and immature, each measuring about 0.5 by 0.3 mm (Bertelsen, 1983:313, figs. 2, 3).

The skin of the two known larval specimens (a female and male, 4.2 and 7.5 mm) is inflated, but pigmentation is very faint, the melanophores not sharply separated. A short, digitiform hyoid barbel is present. The pectoral fins are of normal size, their length terminating well before the insertion of the dorsal and anal fins. Pelvic fins are absent. The male has a few melanophores behind the eye and on the upper part of the opercle; a faint group of melanophores extends from the subopercle down to the barbel. Pigmentation of the body increases in strength posteriorly, sharply delimited by the unpigmented posterior portion of the caudal peduncle. The diameter of the eye is approximately 10% SL. The diameter of the olfactory organ is approximately a third that of the eye (Bertelsen, 1951:126, fig. 85A).

Except for the illicial rudiment and slightly smaller olfactory organ, the smaller larval female is similar to the male, although the pigmentation is slightly weaker. There are no melanophores behind the eye and only a very few are present on the opercular region (Bertelsen, 1951:127).

The color of metamorphosed females is dark reddish brown to black over the entire head, body, and fins. The tip of the anterior escal appendage is white, with large scattered melanophores (Pietsch, 1972a:22, fig. 3). The adult male is light brown and semitransparent; subdermal pigment is very faint, without any distinct concentrations.

The largest known female of the family is a 247-mm individual (ASIZP 59902) collected from off the east coast of Taiwan. The only known adult male (SIO 70-347), from the South China Sea, measures 12.8 mm.

Diversity

A single genus:

Genus *Centrophryne* Regan and Trewavas, 1932 (Prickly Seadevils)

Figures 48, 57–60, 202, 203, 208, 285; Tables 1, 8

Females

*Centrophryne* Regan and Trewavas, 1932:84, pl. 4, fig. 2 (in part; type species *Centrophryne spinulosa* Regan and Trewavas, 1932, by subsequent designation of Burton, 1933).

Males

*Rhynchoceratias* Regan, 1926:44 (in part; a male in metamorphosis misidentified as

*Rhynchoceratias* subsequently referred to *Centrophryne* by Bertelsen, 1951; type species *Rhynchoceratias brevirostris* Regan, 1925c, by subsequent designation of Fowler, 1936).

Distinguishing Characters and Description

As given for the family.

Diversity

A single species.

CENTROPHRYNE SPINULOSA REGAN AND TREWAVAS, 1932 (PRICKLY SEADEVILS)

Figures 48, 57–60, 208, 285 Table 8

Females

*Centrophryne spinulosa* Regan and Trewavas, 1932:84, pl. 4, fig. 2 (original description, two specimens, lectotype ZMUC P92122, designated by Bertelsen, 1951). Fowler, 1949:159 (after Regan and Trewavas, 1932). Bertelsen, 1951:125, figs. 84, 85 (subordinal revision, comparison of all known material, designation of lectotype, elevated to family status). Beaufort and Briggs, 1962:253, fig. 61 (after Regan and Trewavas, 1932). Pietsch and Nafpaktitis, 1971:322, figs. 1–4 (female with attached male of *Melanocetus johnsonii*). Pietsch, 1972a:19, figs. 1–25 (revision of family; description, comparison of all known material, osteology, in key). Brewer, 1973:25 (Gulf of California). Parin et al., 1973:144, fig. 37 (additional specimen, eastern South Pacific). Nielsen, 1974:97 (listed in type catalog). Bertelsen and Pietsch, 1975:10 (not an oneirodid; Pietsch and Seigel, 1980:395 (reference to type locality). Bertelsen and Quéro, 1981:89, fig. 1 (additional specimen, eastern North Atlantic). Bertelsen, 1983:313, figs. 2, 3 (first known adult male). Bertelsen and Pietsch, 1983:89 (reference to type locality). Bertelsen, 1984: 329, fig. 168F (early life history, phylogeny). Bertelsen, 1986:1401, figs. (eastern North Atlantic). Bertelsen, 1990:509 (synonymy). Meng et al., 1995:442, fig. 592 (China). Stewart and Pietsch, 1998:21 (reference to type locality). Pietsch, 1999:2034, fig. (western central Pacific). Anderson and Leslie, 2001:19 (after Bertelsen, 1951, and Pietsch, 1972a; larval male redescribed). Pietsch, 2002b:1066, fig. (western central Atlantic). Love et al., 2005:59 (eastern North Pacific). Pietsch, 2005b:214, 215, 218, 222, 223, fig. 14 (reproduction). Pietsch and Orr, 2007:3, 6, 8, figs. 2A, 4A, 5A (phylogenetic relationships).

Males

*Rhynchoceratias leucorhinus*: Regan, 1926: 44 (in part; a male in metamorphosis misidentified as *Rhynchoceratias leucorhinus* subsequently referred to *Centrophryne spinulosa* by Bertelsen, 1951).

Material

Forty-one metamorphosed females (18 to 247 mm), one metamorphosed male (12.8 mm), two males in metamorphosis (11.5 to 16 mm), and two larvae (4.2 to 7.5 mm):

Lectotype of *Centrophryne spinulosa*: ZMUC P92122, female, 39 mm, *Dana* station 3768(1), north of New Guinea, 1°20′S, 138°42′E, 4000 m of wire, 24 July 1929.

Paralectotype of *Centrophryne spinulosa*: BMNH 1932.5.3.19, female, 33 mm, *Dana* station 3549(4), Gulf of Panama, 7°16′N, 78°30′W, 4000 m of wire, 4 September 1928.

Additional material: ASIZP, six females (70 to 247 mm); BMNH, three females (72 to 118 mm); ISH, seven females (59 to 137 mm); LACM, five females (25 to 209 mm), one male in late metamorphosis (16 mm); MCZ, four females (27.5 to 150 mm); MHNLR, one female (230 mm); NSMT, one female (18 mm); SIO, nine females (27 to 182 mm), one metamorphosed male (12.8 mm); SIOM, one female (230 mm); UF, one female (53 mm); UW, one female (22.5 mm); ZMUC, one male in metamorphosis (11.5 mm), one larval female (4.2 mm), one larval male (7.5 mm).

Distinguishing Characters and Description

As given for the genus and family.

Distribution

*Centrophryne spinulosa* has been collected in all three major oceans of the world (Fig. 208). About a third of the known material is from the Atlantic, where it ranges from the Gulf of Mexico, the Caribbean Sea, and off the coast of Venezuela, eastward to Cape Verde and the Canary Islands, and south to about 10°W, 18°S. In the Pacific it ranges from Japan, south to Taiwan, the South China Sea, off New Ireland in the Bismarck Archipelago, and eastward across the equatorial Pacific to Baja California, southern Mexico, and the Gulf of Panama. A single larval male has been taken in the Mozambique Channel, Indian Ocean. Based on maximum depths reached by fishing gear, metamorphosed specimens may be taken anywhere between 590 and 2325 m. The two known larvae were collected somewhere between the surface and 35 m.

Comments

The small hyoid barbel of *Centrophryne spinulosa* is a curious structure that cannot be explained in functional terms. It was described by Bertelsen (1951:15) as "a simple, papilla-shaped, undifferentiated part of the epithelium, found in larvae and metamorphosed individuals of both sexes. It is shaped like a finger and comparatively longest in the larvae and the adolescent . . . male, wart-shaped in the metamorphosed females." The barbel is present in the 39-mm lectotype of *C. spinulosa*, but, as pointed out by Bertelsen (1951), it is vestigial and difficult to find in most metamorphosed females. Of the remaining known material, the barbel is present in all the smallest females (to approximately 50 mm), in the 16-mm male in late metamorphosis (Pietsch, 1972a:23, fig. 2), and just barely recognizable in the only known adult male. In the 42-mm female (LACM 30882-1), the barbel is 1.2 mm long and located just behind the angle of the jaws, approximately 13.0 mm from the tip of the symphysial spine of the lower jaw. In the 25-mm metamorphosed female (LACM 30886-1), the barbel is 1.5 mm long and about 9.0 mm from the tip of the symphysial spine of the lower jaw. The tip of the barbel is white in the 11.5-mm metamorphic male described by Bertelsen (1951:127), but darkly pigmented in all other known specimens. In females greater than about 50 mm, the hyoid barbel is reduced to a minute protuberance or absent; when present, it is distinguishable only because it is slightly darker than the surrounding skin.

Sagittal sections of the hyoid barbel of the 42-mm female reveal a highly convoluted, outer pigmented layer surrounding an inner core of connective tissue (Pietsch, 1972a:24, fig. 4). Within the connective tissue at the distal tip of the barbel there appears to be a blood space or sinus. There are no nerves or any other structure that might allude to the function of the hyoid barbel. The only other ceratioid taxon with a hyoid barbel is the genus *Linophryne* (family Linophrynidae), where it occurs only in the females.

The characters of the only known adult male of *C. spinulosa*, as described and figured by Bertelsen (1983), are in good agreement with those of the previously recorded males: a young metamorphosal stage of 11.5 mm (Bertelsen, 1951:126–127, figs. 85A, 87) and a somewhat older metamorphosal stage of 16 mm (Pietsch, 1972a:22–23, fig. 2). However, it differs especially in the position and development of the denticular teeth and in reduction of the hyoid barbel. In the smaller metamorphosal stage, the denticular teeth are unossified and represented by small papillae situated some distance behind the tip of the jaws; no loss of larval teeth has occurred. In the 16-mm specimen the denticles are formed but still placed behind the tip of the jaws and not fused at the base; some larval teeth are present. In both metamorphosing specimens the barbel is digitiform, its length three to four times its width.

In its small eyes, the number of fin rays and denticular teeth, and in overall general appearance, the adult male is extremely similar to certain males of *Gigantactis* (see Biodiversity, Chapter Three), differing only in bearing a rudiment of the barbel and in a few characters that are difficult to observe, such as shape of the head of the hyomandibula, presence of a subopercular spine, and fusion of the bases of the denticular teeth. This striking similarity is apparently the result of evolutionary convergence between the two taxa, rather than an indication of phylogenetic relationship (see Evolutionary Relationships, Chapter Four).

## Family Ceratiidae Gill, 1861 (Warty Seadevils)

Figures 3, 24, 42, 44–46, 48, 50, 51, 61–67, 202, 203, 209, 210, 240, 250, 256, 261, 280–283, 286–291, 310; Tables 1, 2, 8–10

Type genus *Ceratias* Krøyer, 1845

Ceratianae Gill, 1861:47 (subfamily of family Lophioidae to contain genus *Ceratias*).

Ceratioidae Gill, 1863:89, 90 (family of suborder Pediculati to contain genus *Ceratias*).

Ceratiaeformes Bleeker, 1865:6 (subfamily of family Chironecteoidei to contain genus *Ceratias*).

Ceratiidae Gill, 1872:2 (family of order Pediculati to contain genus *Ceratias*).

Ceratiinae Gill, 1879a:217 (subfamily of family Ceratiidae to contain genus *Ceratias*). Gill, 1879b:227 (subfamily of family Ceratiidae to contain genera *Ceratias* and *Mancalias*).

### Distinguishing Characters

Metamorphosed females of the family Ceratiidae are distinguished from those of all other ceratioid families by having a well-developed postmaxillary process of the premaxilla. The posterior end of the articular is greatly expanded and squared off ventrally, sometimes terminating in an anterodorsally directed spine (Bertelsen, 1951:128, figs. 88, 89). Two or 3 fleshy caruncles (modified dorsal-fin rays, each bearing a bioluminescent gland; see Brauer, 1908:105, pl. 32, fig. 17; Bertelsen, 1951:16, fig. 88B) are present on the dorsal midline of the trunk just anterior to the origin of the soft dorsal fin. The posterior end of the pterygiophore of the illicium emerges from the dorsal midline just anterior to the caruncles (the later feature present elsewhere only in one species of *Oneirodes* and the thaumatichthyid genus *Lasiognathus*; see Bertelsen, 1943:190, 1951:18 to 20, fig. 4).

Metamorphosed females are further distinguished by having the following combination of character states: The supraethmoid is present. The frontals meet on the midline, each without a ventromedial extension. The parietals are large, considerably longer than wide, each overlapping the respective frontal, sphenotic, pterotic, and epiotic. The medial margin of each parietal contributes to the formation of a cranial trough, the latter deep and narrow, running the full length of the dorsal surface of the skull. Sphenotic spines are absent. The pterosphenoid is absent (erroneously said to be present by Pietsch, 1986b:480). The metapterygoid and mesopterygoid are present. The hyomandibula bears a double head. There are 2 hypohyals and 6 (2 + 4) branchiostegal rays. The opercle is bifurcate. The subopercle is elongate, an anterior spine or projection present or absent. Quadrate, articular, angular, and preopercular spines are all absent. The jaws are equal anteriorly, the lower jaw with a well-developed symphysial spine. An anterior-maxillomandibular ligament is present. The first pharyngobranchial is present or absent. The second and third pharyngobranchials are well developed and heavily toothed. The fourth pharyngobranchial is absent. Ossified hypobranchials are present, but ossified basibranchials are absent. The epibranchials and ceratobranchials are toothless. Epurals are absent. The posterior margin of the hypural plate is deeply notched posteriorly. The second dorsal-fin spine bears a distal bioluminescent gland, more or less concealed beneath the skin just behind the base of the illicium in larger specimens. The escal bulb contains a central lumen, with an escal pore that leads to the exterior. Toothlike escal denticles are absent. A posteroventral process of the coracoid is present *(Ceratias)* or absent *(Cryptopsaras)*. There are 4 pectoral radials. The pelvic bones are cylindrical, without a distal expansion. Fin-ray counts are as follows: dorsal-fin rays (excluding the 2 or 3 reduced rays enveloped by the caruncles) 4 (rarely 3 or 5); anal-fin rays 4; pectoral-fin rays 15 to 19 (rarely 14); caudal-fin rays 8 or 9 (8 in *Cryptopsaras*; the ninth or lowermost ray is reduced to a small remnant in *Ceratias*), 2 simple + 4 bifurcated + 2 (or 3) simple (Table 1). The skin is everywhere covered with close-set dermal spinules or spines (the spines are hypertrophied in *Ceratias*, especially in large females). The ovaries are paired. Two small pyloric caecae are present (see Clarke, 1950:10, fig. 3).

Metamorphosed males of the Ceratiidae are distinguished from those of all other ceratioid families by having the following combination of character states: The eyes of free-living juvenile stages are large, bowl shaped, and directed laterally; the axis is short and the pupil much larger than the lens. The olfactory organs are minute. A pair of large denticular teeth is present on the tip of the snout, fused at the base and articulating with the pterygiophore of the illicium (Bertelsen, 1951:129, fig. 89C, D). Two pairs of denticular teeth are present on the tip of the lower jaw. The premaxillae are degenerate. Jaw teeth are absent. Caruncles are absent. Fin-ray counts are the same as given above for metamorphosed females. The skin is naked and unpigmented in juvenile free-living stages, but spinulose and darkly pigmented in parasitic stages. Males are obligatory sexual parasites as adults (see Reproduction and Early Life History, Chapter Eight).

Larval ceratiids are distinguished from those of other ceratioid families by having the following combination of character states: The body is distinctly "humpbacked," the skin moderately inflated, and the mouth subvertical. Sexual dimorphism is well developed, with females having a distinct illicial rudiment and caruncles present on the dorsal midline of the trunk. The pectoral fins are small, not reaching beyond the dorsal and anal fins. Pelvic fins are absent. Fin-ray counts are the same as those given for metamorphosed females. Metamorphosis begins at 8 to 10 mm (Bertelsen, 1951:135, 140, figs. 90, 93, 1984: 329, fig. 168C–E).

The osteological features cited here and elsewhere in this account of the Ceratiidae are based on examinations of all life history stages of both *Ceratias* and *Cryptopsaras* (Lütken, 1878a:326–337, figs. 2–8, 1887:327–334; Regan, 1912:286–287, fig. 6B, 1926:19, fig. 11; Regan and Trewavas, 1932:39–42, figs. 51–55; Gregory, 1933: 401–404, 407, fig. 273; Bertelsen, 1951: 130–131, figs. 88, 89; Pietsch, 1972a:25, 29, 31, 34, 35, 38, 39, 41, 42, 45).

### Description

Metamorphosed females are relatively long and slender, and laterally compressed. Head length is approximately 40% SL. The greatest body depth is approximately 50% SL. The mouth is large, the cleft vertical to strongly oblique, not extending past the eye. Jaw teeth are slender, recurved, and depressible, those in the lower jaw considerably larger and slightly more numerous than those in the upper jaw. Vomerine teeth are present or absent (see Pietsch, 1986b:487, table 1). The first epibranchial is free from the wall of the pharynx. The 4 epibranchials are closely bound together with connective tissue. The epibranchial and ceratobranchial of the fourth arch are bound to the wall of the pharynx, without an opening behind. The full length of the first ceratobranchial is bound to the wall of the pharynx by connective tissue. The distal end of the first ceratobranchial is free, not bound by connective tissue to the adjacent second ceratobranchial. The proximal one-quarter to one-half of the second, third, and fourth ceratobranchials is free, not bound together by connective tissue. Gill filaments are present on the second, third, and fourth epibranchials and on the full length of the second, third, and fourth ceratobranchials. A pseudobranch is absent. The illicium is long *(Ceratias)* or extremely reduced and nearly totally enveloped by tissue of the esca *(Cryptopsaras)*. The anterior end of the pterygiophore of the illicium is exposed, emerging on the head between the frontal bones just posterior to the eye, its posterior end protruding on the dorsal midline of the trunk just anterior to the caruncles. The escal bulb is oval in shape, with or without 1 or 2 distal appendages. The neuromasts of the acousticolateralis system are located at the tips of short cutaneous papillae, the pattern of placement as described for other ceratioids (Pietsch, 1969, 1972b, 1974a, 1974b).

The color of females in preservative is dark red brown to black over the entire surface of the body (except for the distal portion of the escal bulb) and oral cavity. The skin is unpigmented in juvenile males, but darkly pigmented in parasitic stages. Larval pigmentation is present or absent.

For additional description of males and larvae, see the generic accounts below.

Ceratiids represent the largest known ceratioids, the females of *Ceratias holboelli* attaining a standard length of at least 855 mm (BMNH 1953.2.25.1). All free-living males (at least 175 known specimens) measure less than 20 mm, whereas parasitic males (112 known specimens, attached to 82 females) range from 8.0 to about 140 mm (see Pietsch, 1976:786, 2005b:223; Reproduction and Early Life History, Chapter Eight, Table 10).

Diversity

Two genera, differentiated as follows:

Keys to Genera of the Family Ceratiidae
FEMALES

1A. Illicium long, considerably longer than length of escal bulb; 2 caruncles on dorsal midline of trunk just anterior to origin of soft-dorsal fin; subopercle without anterior spine . . . .
. . . . . . . . . . . . . . . . . . . . . . *Ceratias* Krøyer, 1845, p. 323

NOTE Three species; Atlantic, Pacific, Indian, and Southern oceans

1B. Illicium short, nearly completely enveloped by tissue of escal bulb; 3 caruncles on dorsal midline of trunk just anterior to origin of soft-dorsal fin; subopercle with an anterior spine . . . . . . . . . . . . . . *Cryptopsaras* Gill, 1883b, p. 328

NOTE A single species; Atlantic, Pacific, and Indian oceans

FREE-LIVING MALES

1A. Head and body without subdermal pigment; two pairs of denticles on lower jaw nearly equal in length, posterior pair emerging immediately behind the other . . . . . . . . *Ceratias* Krøyer, 1845, p. 323

NOTE An unknown number of species represented and at least 75 known specimens; Atlantic, Pacific, Indian, and Southern oceans

1B. Subdermal pigment present on gill cover, dorsal surface of trunk, and caudal peduncle; posterior pair of denticles of lower jaw distinctly shorter than anterior pair, emerging some distance behind the latter . . . .
. . . . . . . . . . . . . . . . . . . *Cryptopsaras* Gill, 1883b, p. 328

NOTE A single species and at least 100 known specimens; Atlantic, Pacific, and Indian oceans

LARVAE

1A. Head and body without subdermal pigment; females with 2 caruncles; caudal-fin rays 9 . . . . . . . . . *Ceratias* Krøyer, 1845, p. 323

NOTE An unknown number of species represented and about 160 known specimens; Atlantic, Pacific, Indian, and Southern oceans

1B. A band of subdermal pigment extending like a collar around the posterior margin of the head; subdermal pigment present also on dorsal surface of trunk and caudal peduncle; females with 3 caruncles; caudal-fin rays 8 . . . . . . . . . . . . *Cryptopsaras* Gill, 1883b, p. 328

NOTE A single species and about 350 known specimens; Atlantic, Pacific, and Indian oceans

Genus *Ceratias* Krøyer, 1845 (Doublewart Seadevils)

Figures 3, 42, 44, 45, 48, 50, 51, 61–63, 202, 203, 209, 250, 280, 281, 286–288, 310; Tables 1, 2, 8–10

Females

*Ceratias* Krøyer, 1845:639 (type species *Ceratias holboelli* Krøyer, 1845, by monotypy).

*Caratias* Rink, 1877:434 (erroneous spelling of *Ceratias*, therefore taking the same type species *Ceratias holboelli* Krøyer, 1845, by monotypy).

*Mancalias* Gill, 1879b:227, 228 (type species *Ceratias uranoscopus* Murray, in Thomson, 1877, by original designation and monotypy).

*Typlopsaras* Gill, 1883b:284 (type species *Typlopsaras shufeldti* Gill, 1883b [a synonym of *C. uranoscopus* Murray, in Thomson], by monotypy).

*Miopsaras* Gilbert, 1905:694, pl. 99 (type species *Miopsaras myops* Gilbert, 1905 [*nomen dubium*; see Pietsch, 1986b:487], by monotypy); Jordan and Jordan, 1922:90 (after Gilbert, 1905).

*Typhlosparas* Regan, 1926:37 (emendation of *Typlopsaras* Gill, 1883b, therefore taking the same type species).

*Myosparas* Regan, 1926:37 (emendation of *Miopsaras* Gilbert, 1905, therefore taking the same type species).

*Myopsaras* Roule and Angel, 1933:58 (emendation of *Miopsaras* Gilbert, 1905, therefore taking the same type species).

*Typhlopsarus* Barbour, 1942b:78, 86 (emendation of *Typlopsaras* Gill, 1883b, therefore taking the same type species).

*Typhloceratias* Barbour, 1942b:78 (type species *Typhloceratias firthi* Barbour, 1942b [*nomen dubium*; see Pietsch, 1986b:487], by monotypy).

*Parrichthys* Barbour, 1942b:84 (type species *Parrichthys merrimani* Barbour, 1942b [*nomen dubium*; see Pietsch, 1986b:487], by monotypy).

*Reganichthys* Bigelow and Barbour, 1944a:10 (type species *Reganichthys giganteus* Bigelow and Barbour, 1944a [a synonym of *C. holboelli* Krøyer], by monotypy); preoccupied by *Reganichthys* Ogilby, a junior synonym of *Glaucosoma*, family Glaucosomatidae).

*Reganula* Bigelow and Barbour, 1944b:123 (replacement name for *Reganichthys* Bigelow and Barbour, 1944a, preoccupied, taking the same type species).

*Mancalius* Beebe and Crane, 1947:169 (emendation of *Mancalias* Gill, 1879b, therefore taking the same type species).

Distinguishing Characters

The genus *Ceratias* differs from *Cryptopsaras*, the only other recognized genus of the family, in having 9 caudal-fin rays, the ninth or ventral-most ray reduced to a small remnant; and in lacking a spine on the anterodorsal margin of the subopercle (Bertelsen, 1951:132, figs. 88A, C, 89B, D, E). Metamorphosed females of the genus are further distinguished from those of *Cryptopsaras* in having a long illicium (19.0 to 28.2% SL) and only 2 club-shaped caruncles (minute in specimens greater than approximately 400 mm) on the dorsal midline of the trunk just anterior to the origin of the soft-dorsal fin. Metamorphosed males of the genus are further distinguished from those of *Cryptopsaras* in having the two pairs of lower denticular teeth nearly equal in size. Larvae, males, and juvenile females lack subdermal pigment.

Description

The esca of metamorphosed females is somewhat variable, with or without 1 or 2 distal appendages. If present, the escal appendages are simple or bear 1 to 8 lateral filaments. The number of teeth in the lower jaw ranges from 44 to 102. Vomerine teeth are sometimes absent, but there may be as many as 8. Fin-ray counts are as follows: dorsal-fin rays 4 (excluding those embedded within the caruncles), very rarely 3 (only one of 196 specimens counted); anal-fin rays 4; pectoral-fin rays 15 to 19, usually 17 or 18 (Table 1). The skin of juvenile females is covered with minute, close-set spinules, that of large adults with large, conical dermal spines, each with a broad rounded base.

The skin of metamorphosed, free-living males is naked. The denticles are faintly barbed. There are two pairs of lower denticulars nearly equal in size, one pair situated immediately behind the other. Caruncles are absent (Bertelsen, 1951:132). Adult males are parasitic, the skin spinulose, the denticles, eyes, and gut degenerated (Bertelsen, 1951:132).

Larval females are distinguished in having 2 dorsal caruncles (see Bertelsen, 1951:132, fig. 90G–I).

Females are dark red brown to black over the entire surface of the body (except for the distal portion of the escal bulb) and oral cavity. The skin is unpigmented in juvenile males, but darkly pigmented in parasitic stages. Subdermal pigment is absent. The larvae are unpigmented (Bertelsen, 1951:132).

Females are the largest known ceratioids, attaining a standard length of at least 855 mm (BMNH 1953.2.25.1). All known free-living males are small, less than 12 mm. Parasitically attached males become considerably larger, ranging from 9.2 to about 140 mm, and having a weight several times greater than that of the largest known free-living males (see Pietsch, 1976:786, 2005b:223).

## Key to Females of Species of the Genus *Ceratias*

1A. Esca with a pair of distal appendages; vomerine teeth present . . . . . . . . . . . . . . . . . . . . . . . . . *Ceratias tentaculatus* (Norman, 1930), p. 327

   NOTE Fifty-seven specimens, 6.2 to 640 mm; Southern Ocean

1B. Esca with not more than 1 distal appendage; vomerine teeth present or absent . . . . . . . . . . . . . . . . . . . . 2

2A. Esca with a single distal appendage; illicium length 15.1 to 37.8% SL; vomerine teeth nearly always present in metamorphosed specimens less than approximately 80 mm, only occasionally present in larger individuals . . . . . . . . . . . . . *Ceratias holboelli* Krøyer, 1845, p. 324

   NOTE One hundred and sixty-one specimens, 16.5 to 855 mm; Atlantic, Pacific, and Indian oceans

2B. Esca without distal appendages; illicium length 14.0 to 28.8% SL; vomerine teeth absent . . . . . . . . . . . . . . . . . . . *Ceratias uranoscopus* Murray, in Thomson, 1877, p. 326

   NOTE One hundred and twelve specimens, 12 to 240 mm; Atlantic, Pacific, and Indian oceans

## *CERATIAS HOLBOELLI* KRØYER, 1845
(NORTHERN GIANT SEADEVILS)

Figures 3, 42, 45, 61, 62, 209, 250, 280, 281, 287, 310; Tables 2, 8–10

### Females and Parasitic Males

*Ceratias holboelli* Krøyer, 1845:639 (original description, single specimen). Gaimard, 1852, pl. 9 (figured). Reinhardt, 1857:23 (listed for Greenland). Günther, 1861:205 (after Krøyer, 1845). Lütken, 1878a:324, 326–337, figs. 2–8 (osteology). Gill, 1879a:217 (after Krøyer, 1845). Gill, 1879b:227 (after Krøyer, 1845). Jordan and Gilbert, 1883:847 (after Krøyer, 1845). Jordan, 1885:926 (listed). Günther, 1887:52 (additional material, off Greenland). Goode and Bean, 1896:489, pl. 117, fig. 399 (after Krøyer, 1845). Saemundsson, 1922:163, 165, fig. (discovery of attached specimens, thought to be juveniles). Regan, 1925a:41, pl. 2 (attached specimens shown to be parasitic males). Regan, 1925b:387, figs. 1–4 (attached specimens shown to be parasitic males). Beebe, 1926a:79, figs. (sexual parasitism). Regan, 1926:12, 34, text figs. 1, 3, 4, 20 (description after Krøyer, 1845, and Saemundsson, 1922). Schmidt, 1926:266–267, figs. 6, 7 (after Krøyer, 1845, and Regan, 1925a, 1925b, 1926). Regan, 1927b:3, postcards M 16, M 17 (popular account). Regan, 1930a:215, figs. 2, 3 (sexual parasitism, semipopular account). Regan and Trewavas, 1932:96 (after Krøyer, 1845, and Regan, 1926). Ehrenbaum, 1936:164, figs. 137, 137a (seas of northern Europe, sexual parasitism; figures after Regan, 1926, and Saemundsson, 1922). Schnakenbeck, 1936:7 (additional material, Iceland). Beebe, 1938:50, figs. (additional specimen, northern edge of Georges Bank). Waterman, 1939a:65, 74, 77 (historical review, parasitism). Rauther, 1941:131, figs. 1–3 (illicial anatomy). Bertelsen, 1943:185, figs. 1–4 (revision based on all known material). Saemundsson, 1949:47 (additional material, Iceland). Bertelsen, 1951:133, figs. 88A, C, 89B, D, E, 90, 91A–D, F–H, 92 (in part, revision based on all known material). Einarsson, 1952:90 (biology; after Bertelsen, 1951). Brandes et al., 1953:47 (two additional specimens, Iceland). Albuquerque, 1954–1956:1067, figs. 438, 438A (after Fowler, 1936; figures after Goode and Bean, 1896, and Saemundsson, 1922). Krefft, 1954:78, figs. (additional specimen, Iceland; figures after Bertelsen, 1951). Marshall, 1954:270, 305, 306, 339, fig. XII, 4 (life history; figure after Bertelsen, 1951). Matsubara, 1955:1349, 1350, 1352, 1353, 1357, figs. 525, 530A, 532 (review, ceratioids of Japan; figures after Regan and Trewavas, 1932, Imai, 1941, and Bertelsen, 1951). Brandes et al., 1956:30 (two additional specimens, distant northern seas). Clarke, 1956:260 (from stomachs of sperm whales, Azores). Grey, 1956a:261 (in part; synonymy, distribution). Brandes et al., 1957:54 (two additional specimens, distant northern seas). Krefft, 1960:71 (additional specimen, Faroe-Iceland Ridge). Krefft, 1961:90 (two females each with parasitic male, Iceland and West Greenland). Wickler, 1961:382, 385, fig. 3 (sexual parasitism). Bowen, 1963:91, figs. 1–3 (from stomach of sperm whale; esca damaged). Idyll, 1964:202, 279 (popular account). Mead et al., 1964:588 (reproduction). Munk, 1964:9 figs. 1A, 2A (histology of eyes of free-living males). Munk, 1966:22, 28, figs. 14A, 15B (ocular anatomy of free-living males). Abe, 1967:797, figs. 1, 2 (additional parasitized females; northern Japan). Blacker, 1967:186 (additional specimen, off Newfoundland). Krefft, 1967:186 (additional specimen, Faroe-Iceland Ridge). Rass, 1967:243 (distribution, Pacific Ocean). Blacker, 1968:186 (additional specimen, distant northern seas). Jónsson, 1968a:211 (additional specimen, east Greenland). Briggs, 1960:179 (worldwide distribution). Jónsson, 1970:278 (additional specimen, Iceland). Marshall, 1971a:41, 45, 95 (eyes of males, temperature and depth distribution). Blacker, 1972:195 (female with parasitic male, Iceland). Jones, 1972:24, fig. 1 (North Atlantic). Pietsch, 1972a:25, 29, 31, 34, 35, 38, 39, 41, 45 (eggs, after Mead et al., 1964; osteological comments). Rice, 1972:158 (from stomach of a sperm whale, off Richmond, California). Blacker, 1973:222 (additional specimen, Iceland). Karrer, 1973a:88 (additional specimen, Newfoundland). Maul, 1973:674 (eastern North Atlantic, synonymy). Nielsen, 1974:97 (listed in type catalog). Parin et al., 1974:124 (western South Atlantic). Martin, 1975:1, fig. 1 (leeward Oahu; parasitized by hydroid *Hydrichthys pietschi* sp. nov.). Pietsch, 1976:786 (reproduction). Jónsson et al., 1977:181 (additional specimen, Iceland). Jónsson et al., 1978:182 (two additional specimens, Iceland). Hubbs et al., 1979:13 (listed for California). Jónsson et al., 1979:229 (additional specimen, Iceland). Marshall, 1979:269, 338, 383, 390, 391, 449 (reduced food requirements, buoyancy, eyes and vision, fecundity). Du Buit et al., 1980 (female with parasitic male, off Hebrides). Leisman et al., 1980:1272 (bioluminescence). Jónsson et al., 1981:193 (additional specimen, Iceland). Ayling and Cox, 1982:140, fig. (New Zealand). Yamakawa, 1982:199, 363, fig. 125 (additional specimen, Kyushu-Palau Ridge). Bertelsen and Pietsch, 1983:89, fig. 10A, B (in part, additional material, New South Wales). Herring, 1983:202, table 9 (bioluminescence emission spectra). Amaoka, 1984:108, pl. 93F (Japan). Bertelsen, 1984:329, fig. 168E (early life history, phylogeny). Bertelsen and Pietsch, 1984:44, figs. 2, 3 (resurrection of *Ceratias tentaculatus*). Jónsson et al., 1984:205 (nine additional specimens, Iceland). Jónsson et al., 1985:185 (two additional specimens, Iceland). Paulin and Stewart, 1985:27 (New Zealand). Bertelsen, 1986:1403, figs. (eastern North Atlantic.

Jónsson et al., 1986a:186 (five additional specimens, Iceland). Jónsson et al., 1986b:160 (two additional specimens, Iceland). Jónsson et al., 1986c:628 (two additional specimens, Iceland). Pietsch, 1986a:374, fig. 105.1 (South Africa). Pietsch, 1986b:484, figs. 2, 3, 6 (family revision; comparison of all known material, resurrection of *C. uranoscopus*). Jónsson et al., 1987:360 (three additional specimens, Iceland). Tarakanov and Balushkin, 1987:38, figs. 1–5 (additional material, description; esca, caruncles, and skin spines figured; in key). Ni, 1988:325, fig. 255 (additional specimen, East China Sea). Quéro and Vayne, 1988:173, figs. 3–5, 7 (additional material, eastern North Atlantic). Du Buit et al., 1989:192, fig. 1 (North Atlantic distribution). Paulin et al., 1989:137, fig. 64.3b (New Zealand). Bertelsen, 1990:510 (synonymy, eastern tropical Atlantic). Haygood et al., 1992:149 (gene sequences from bioluminescent bacterial symbionts). Munk, 1992:34, figs. 2, 3 (accessory escal gland). Nielsen and Bertelsen, 1992:62, fig. 3 (North Atlantic). Andriyashev and Chernova, 1994:103 (listed for Arctic seas). Amaoka et al., 1995:114, fig. 175 (northern Japan). Meng et al., 1995:439, fig. 588 (China). Stearn and Pietsch, 1995:131, fig. 81 (five additional specimens, Greenland). Ellis, 1996:268, 270, figs. (popular account). Shedlock et al., 1997:396, 398, figs. 1, 2 (DNA extracted from formalin-fixed specimens). Bertelsen and Pietsch, 1998a:141 (popular account). Stewart and Pietsch, 1998:23, fig. 15 (four additional specimens, New Zealand). Jónsson and Pálsson, 1999:202, fig. 6 (Iceland). Munk, 1999:267, 280 (bioluminescence). Pietsch, 1999:2035 (western central Pacific). Sheiko and Fedorov, 2000:24 (Kamchatka). Anderson and Leslie, 2001:20, fig. 14A (after Pietsch, 1986b). Wagner, 2001:121, 127 (sensory brain areas). Kukuev and Trunov, 2002:326 (North Atlantic). Mecklenburg et al., 2002:302, 303 (Alaska, in key). Nakabo, 2002:468, figs. (Japan, in key). Pietsch, 2002a:274, fig. 148 (Gulf of Maine). Pietsch, 2002b:1067, fig. (western central Atlantic). Moore et al., 2003:216 (off New England). Love et al., 2005:61 (eastern North Pacific). Pietsch, 2005b:207, 209, 211, 213–215, figs. 3, 4 (reproduction). Kharin, 2006b:409, fig. 1 (Russian waters). Pietsch and Orr, 2007:5 (sexual dimorphism; phylogenetic relationships).

*Caratias holboellii*: Rink, 1877:434 (emendation, Greenland).

*Ceratias holbolli*: Jordan and Evermann, 1896:510 (after Krøyer, 1845; emendation of specific name). Jordan and Evermann, 1900, fig. 954 (figured). Gill, 1909:578, fig. 12 (after Krøyer, 1845). Regan, 1925a:41, pl. 2 (additional specimen, Iceland). Regan, 1925b:387, figs. 1–4 (sexual parasitism). Kujawa, 1964:301 (Flemish Cape). Utrecht, 1957:72, figs. 1, 2 (comparison with additional specimen identified as *Ceratias* sp.; off Durban, South Africa).

*Ceratias holboelli holboelli*: Bertelsen, 1943:203 (subspecies). Bertelsen, 1951:133, footnote (northern subspecies). Bertelsen and Pietsch, 1983:90 (northern subspecies).

*Reganichthys giganteus* Bigelow and Barbour, 1944a:10, pls. 4–6 (original description, single specimen).

*Reganula gigantea*: Bigelow and Barbour, 1944b:123 (new combination; after Bigelow and Barbour, 1944a).

Free-living Males and Larvae

Unknown.

Material

One hundred and sixty-one metamorphosed females (16.5 to 855 mm) and 19 parasitic males (35 to about 140 mm):
Holotype of *Ceratias holboelli*: ZMUC 61, 680 mm, South Greenland, 0 to 340 m.
Holotype of *Reganichthys giganteus*: MCZ 36042, 770-mm female with 108-mm parasitic male, 19.3 km south of Mt. Desert Rock, off Portland, Maine, otter trawl in 0 to 230 m, 1 October 1943.
Additional females and parasitic males: AIM, one (173 mm); AMS, one (168 mm); ASIZP, one (172 mm); BMNH, 10 (20 to 855 mm; 640-mm female with 70-mm parasitic male; 720-mm female with 75-mm parasitic male; 855-mm female with 85-mm parasitic male; BSKU, five (131 to 282 mm); ECFR, two (140 to 670 mm; 670-mm specimen with 97.5-mm parasitic male); CSIRO, two (500 to 560 mm; 560-mm female with 35-mm parasitic male); HSU, one (635 mm); HUMZ, 10 (76 to 690 mm); 595-mm female with 70-mm parasitic male; 690-mm female with 105-mm parasitic male); IINH, one (690-mm female with two parasitic males, 65 to 70 mm; one removed and now at ZMUC); ISH, 12 (30 to 240 mm); LACM, two (23 to 35 mm); MCZ, eight (16.5 to 770 mm; 770-mm female with 118-mm parasitic male); MNHN, one (1000 mm TL, with parasitic male of unknown length); MNRJ, one (167 mm); MRIR, six (690 to 735 mm; 735-mm female with 75-mm parasitic male; four specimens, 980 to 1210 mm TL, each with a parasitic male); MTF, three (111 to 550 mm); NMNZ, three (55 to 138 mm); NSMT, one (79 mm); OS, one (500 mm); SAIAB, one (170 mm); SIO, 38 (11 to 62 mm); SIOM, one (78 mm); SWFSC, one (26 mm); USNM, two (89 to 565 mm); UW, eight (455 to 655 mm); ZIN, five (117 to 489 mm); ZMH, one (1100-mm TL female with 80-mm TL parasitic male); ZMUB, one (39 mm); ZMUC, 29 (61 to 700 mm, 580-mm female with 86-mm parasitic male).

Distinguishing Characters

A species of *Ceratias*, the metamorphosed females of which are distinguished from those of all other described species of the genus in having a single distal escal appendage; illicium length 14.5 to 37.8% SL; vomerine teeth (2 to 6) nearly always present in metamorphosed specimens less than approximately 80 mm, but rarely present in larger specimens.

Description

Proximal one-half to two-thirds of escal bulb darkly pigmented, distal portion oval in shape, proximal portion tapering into stem of illicium; a single slender escal appendage arising just anterior to escal pore, usually simple but occasionally with as many as 3 short filaments on each side; escal pore situated at apex of bulb, nearly always raised on a pigmented papilla; at least 1, but usually 2 to 6 vomerine teeth present in all known material less than 80 mm (except for a single 57.5-mm specimen, ISH 2007/71), and in five of 19 examined specimens larger than 80 mm.

Additional description as given for the genus and family.

Distribution

*Ceratias holboelli* has a broad distribution in all three major oceans of the world, but is excluded from the Southern Ocean by its congener *C. tentaculatus* (Fig. 209). It is known from throughout the North Atlantic, ranging as far north as the Greenland Sea above Iceland at about 68°N and extending south to approximately 8°S in the central Atlantic and off Rio de Janeiro, Brazil, in the west. In has also been collected on both sides and across the Pacific, from off southern Tasmania, New Zealand, New South Wales, the Coral and Celebes seas, the Bismarck Archipelago, and Solomon Islands, from off Taiwan, Japan, the Bering Sea, and the Aleutian Islands, off Oregon and California, the Hawaiian Islands, and east to the Peru-Chile Trench at about 24°S, 72°W. In the Indian Ocean, it is known from off Durban, South Africa, and several localities in the Arabian Sea.

Juvenile and adults of *C. holboelli* may be captured anywhere between approximately 150 and 3400 m, but the majority of specimens were captured between 400 and 2000 m. Of the material for which depth of capture is known (133 of 161 recorded specimens), 84% was taken by trawls that reached a maximum depth of 2000 m or less; 76% was taken by trawls that reached a maximum depth of 1100 m or less. At the upper end of its vertical range, 80% of the known material was taken in 400 m or more. A number of large adult specimens have been found in relatively shallow water in high latitudes of the North Atlantic and in the Bering Sea, some of them from unknown depths, others in as little as 120 to 680 m. The 770-mm holotype of *Reganichthys giganteus* (MCZ 36042) was captured by otter trawl somewhere between the surface and 230 m. The average maximum depth for all known captures was 1143 m.

Comments

*Ceratias holboelli* has had a long and confusing taxonomic history (see Comments, below).

CERATIAS URANOSCOPUS MURRAY, IN THOMSON, 1877 (STARGAZING SEADEVILS)

Figures 62, 209; Tables 2, 8–10

Females and Parasitic Males

*Ceratias uranoscopus* Murray, in Thomson, 1877:70, fig. 20 (original description, single specimen). Bertelsen, 1943:195, 196, 198–203 (a synonym of *Ceratias holboelli*). Lütken, 1878a:324 (after Murray, in Thomson, 1877). Günther, 1887:52, 54, pl. 11, fig. C (description). Bertelsen and Pietsch, 1984:49, fig. 2 (in synonymy of *C. holboelli*, distribution). Pietsch, 1986b:485, figs. 3–6 (resurrection from synonymy of *C. holboelli*, comparison with all known material). Bertelsen, 1990:510 (synonymy, eastern tropical Atlantic). Munk, 1992:34, figs. 1, 4 (accessory escal gland). Ellis, 1996:273 (popular account). Munk, 1999:267, 280 (bioluminescence). Pietsch, 1999:2035 (western central Pacific). Anderson and Leslie, 2001:21, fig. 14C (after Pietsch, 1986b; additional material). Nakabo, 2002:468, 1496, figs. (Japan, in key). Pietsch, 2002b:1067 (western central Atlantic). Menezes et al., 2003:65 (after Pietsch, 1986b). Moore et al., 2003:216 (off New England). Pietsch, 2005b:209 (reproduction). Senou et al., 2006:424 (Sagami Sea).

*Mancalias uranoscopus*: Gill, 1879b:228 (new combination; after Murray, in Thomson, 1877). Jordan and Gilbert, 1883:848 (after Murray, in Thomson, 1877, and Gill, 1879b). Jordan, 1885:926 (listed). Goode and Bean, 1896:490 (after Murray, in Thomson, 1877, and Gill, 1879b). Jordan and Evermann, 1896:510 (after Murray, in Thomson, 1877, and Gill, 1879b). Jordan and Evermann, 1898:2729 (after Murray, in Thomson, 1877, and Gill, 1879b). Gill, 1909:579 (reference to original description). Regan, 1926:37, fig. 21 (in part; description after Murray, in Thomson, 1877, Günther, 1887; figure after Goode and Bean, 1896). Regan and Trewavas, 1932:99 (in part, North Atlantic). Barbour, 1942b:77, pl. 9, upper figure (figure after Murray, in Thomson, 1877). Imai, 1942:43, figs. 4, 5 (Suruga Bay). Bertelsen, 1943:195, 198, 200 (in synonymy of *Ceratias holboelli*). Bertelsen, 1951:133 (after Bertelsen, 1943). Bassot, 1966:584, pl. 8, fig. 1 (bioluminescence). Bertelsen and Pietsch, 1984:51 (possibly distinct from *C. holboelli*).

*Typlopsaras shufeldti* Gill, 1883b:284 (original description, single specimen). *Ceratias shufeldti*: Günther, 1887:52, 54 (new combination; after Gill, 1883b). *Ceratias (Mancalias) uranoscopus*: Lütken, 1894:78 (new combination). *Ceratias (Typlopsaras) schufeldti*: Lütken, 1894:78 (new combination, emendation of specific name). *Mancalias shufeldti*: Goode and Bean, 1896:490, pl. 119, fig. 401 (new combination, description). Jordan and Evermann, 1896:510 (after Gill, 1883b, and Günther, 1887). Jordan and Evermann, 1898:2730 (after Jordan and Evermann, 1896). Jordan and Evermann, 1900, fig. 955 (figured). Gill, 1909:579, fig. 14 (after Gill, 1883b, and Günther, 1887). Regan, 1926:37 (in synonymy of *Ceratias uranoscopus*).

*Mancalias xenistius* Regan and Trewavas, 1932:99, pl. 6, fig. 2 (original description, single specimen). Bertelsen, 1943:200, 202 (a synonym of *Ceratias holboelli*). Nielsen, 1974:97 (listed in type catalog).

*Mancalias uranoscopus triflos* Roule and Angel, 1933:57, pl. 3, fig. 27 (original description, single specimen). Bertelsen, 1943:200, 202 (a synonym of *Ceratias holboelli*).

*Mancalias (Ceratias) uranoscopus triflos* Roule, 1934:203, color plate 12 (semipopular account; after Roule and Angel, 1933).

*Typhlopsarus shufeldti*: Barbour, 1942b:78, 86, pl. 9, lower figure (new combination, photograph of holotype, emendation of generic name).

*Mancalias sessilis* Imai, 1941:245, figs. 12, 13 (original description, single specimen, 28 mm, existence of holotype unknown, originally in collections of Mitui Institute of Marine Biology, Sagami Sea, east of Hashima, 1220 m wire out, 14 November 1937). Matsubara, 1955:1357, fig. 532 (in synonymy of *Ceratias holboelli*; figure after Imai, 1941).

*Ceratias holboelli xenistius*: Bertelsen, 1943:203 (subspecies).

*Ceratias holboelli sessilis*: Bertelsen, 1943:203 (subspecies).

*Mancalias kroyeri* Koefoed, 1944:11, pl. 3, fig. 4 (original description, single specimen). Frøiland, 1979:150 (listed in type catalog).

*Mancalius uranoscopus*: Beebe and Crane, 1947:169 (four specimens, eastern tropical Pacific, but identifications unconfirmed, material not seen; emendation of generic name). Matsubara, 1955:1357 (in synonymy of *Ceratias holboelli*).

*Mancalius uranoscopus typhlops*: Beebe and Crane, 1947:170 (perhaps young of *M. tentaculatus* Norman, emendation of subspecific name).

*Mancalias uranoscopus typhlos*: Bertelsen, 1951:134 (in synonymy of *Ceratias holboelli*, emendation of subspecific name).

*Ceratias murrayi*: Bertelsen, 1990:511 (*nomen nudum*; an error for *Ceratias uranoscopus* Murray, in Thomson, 1877).

Free-living Males and Larvae

Unknown.

Material

One hundred and twelve metamorphosed females (12 to 240 mm) and one parasitic male (22.5 mm):

Holotype of *Ceratias uranoscopus*: BMNH 1887.12.7.15, 57 mm, *Challenger* Expedition, station 89, between Canary and Cape Verde Islands, 22°18′N, 22°02′W, 0 to 4392 m, 23 July 1873.

Holotype of *Typlopsaras shufeldti*: USNM 33552, ca. 50 mm, *Albatross* station 2099, off east coast U.S., 37°12′N, 69°39′W, 0 to 3841 m, 2 October 1883.

Holotype of *Mancalias xenistius*: ZMUC P92124, 20 mm, *Dana* station 3690(1), South China Sea, 8°02′N, 109°36.5′E, 1000 m wire out, 10 April 1929.

Holotype of *Mancalias uranoscopus triflos*: MOM Galerie, Level 3, Window 55, 34 mm, station 3414, Fosse Sigsbee, SE of Nova Scotia on the route to the Azores, 0 to 4000 m, 10 August 1913.

Holotype of *Mancalias kroyeri*: ZMUB 4299, 92 mm, *Michael Sars*, North Atlantic Deep-Sea Expedition 1910, station 42, Canary Islands, 28°02′N, 14°17′W, young-fish trawl, 900 m wire out, 23 to 24 May 1910.

Additional material: AMS, one (135 mm); BMNH, 20 (20.5 to 162 mm); HUMZ, one (126 mm); ISH, eight (30.5 to 74 mm); LACM, 38 (12.5 to 240 mm); 240-mm female with 22.5-mm parasitic male); MCZ, 25 (12 to 129 mm); NSMT, one (97 mm); SAIAB, two (27 to 47 mm); SAM, two (59 to 65 mm); SIO, one (91 mm); UF, five (37.5 to 149 mm); UW, one (56 mm); ZMUC, two (51 to 81 mm).

Distinguishing Characters

A species of *Ceratias*, the metamorphosed females of which are distinguished from those of all other described species of the genus in having a simple esca, without distal escal appendages; illicium length 14.0 to 28.8% SL; vomerine teeth invariably absent.

Description

Escal bulb nearly always darkly pigmented except for distal tip, distal portion oval in shape, proximal portion tapering into stem of illicium; bulb terminating without appendages and without elevation of small pigment spot surrounding escal pore; vomerine teeth absent in all known material.

Additional description as given for genus and family.

Distribution

*Ceratias uranoscopus* is broadly distributed, well represented in the Atlantic and

Pacific, but known from the Indian Ocean on the basis of only two specimens, one from off Durban, South Africa, and the other in the Arabian Sea (Fig. 209). Like its sympatric congener *C. holboelli*, it is excluded from the Southern Ocean by the third member of the genus *C. tentaculatus*. In the Atlantic, *C. uranoscopus* ranges from off Nova Scotia, Florida, the eastern Caribbean Sea and across to the eastern Atlantic, where it extends from Canary Island south to approximately 40°S off Cape Town, South Africa. In the Pacific, it has been taken from off Japan and south to the Halmahera and Celebes seas, off New South Wales, Australia, and east to the central tropical Pacific and off the Hawaiian Islands.

Metamorphosed females of this species may be taken anywhere between approximately 95 and 4000 m, but the majority of specimens were captured between 500 and 1000 m. Of the material for which data were available (89 of 112 known specimens), 94% was taken by fishing gear that reached a maximum depth of 2000 m or less; 85% was captured by gear fished in 1000 m or less. At the upper end of its vertical range, 89% of the material was taken in depths of 500 m or more. The average maximum depth for all known captures was 840 m. The largest known specimen, a 240-mm female with an attached parasitic male, was taken between the surface and 500 m.

Comments

*Ceratias uranoscopus* has long been confused with *C. holboelli*. Because no small individuals (less than approximately 90 mm) of *Ceratias* with a single escal appendage had ever been collected, and large specimens (greater than approximately 100 mm) with an unadorned esca were unknown, material of *C. uranoscopus* was universally recognized as the young of *C. holboelli*. In recent years, however, 16 juvenile specimens, ranging in size from 16.5 to 88 mm, have been discovered bearing the diagnostic, single escal appendage of *C. holboelli*. At the same time, a number of large individuals (97 to 240 mm) without escal appendages have been collected. Thus, with the presence of a series of specimens, completely overlapping in size with the known material of *C. holboelli* and clearly distinguishable from the later on the basis of escal morphology, it became evident that a previously unrecognized species of *Ceratias* must exist. As additional evidence accumulated (see below), it further became evident that the name *C. uranoscopus* Murray must be resurrected from the synonymy of *C. holboelli* to accommodate a third member of the genus (Pietsch, 1986b).

Comparison of these newly collected specimens with all known material of the genus revealed two additional characters that confirmed the validity of *C. uranoscopus* (Pietsch, 1986b). The first of these is a difference in the length of the illicium, that of *C. holboelli* being slightly longer than that of *C. uranoscopus*. The second is the presence or absence of vomerine teeth, these being present in nearly all juvenile individuals of *C. holboelli* less than approximately 80 mm, but absent in all known material of *C. uranoscopus*.

CERATIAS TENTACULATUS (NORMAN, 1930) (SOUTHERN GIANT SEADEVILS)

Figures 62, 209; Tables 2, 8–10

Females and Parasitic Males

*Mancalias tentaculatus* Norman, 1930: 355, fig. 45 (original description, single specimen). Regan and Trewavas, 1932: 100, fig. 158D (after Norman, 1930). Bertelsen, 1943:200, fig. 5 (tentative recognition as *Ceratias tentaculatus*, based on comparison of all known material). Bertelsen, 1951:134 (in synonymy of *C. holboelli*).
*Mancalias bifilis* Regan and Trewavas, 1932:100, fig. 158C, pl. 6, fig. 1 (original description, single specimen). Bertelsen, 1943:200 (a possible subspecies of *Ceratias holboelli*). Bertelsen, 1951:134 (in synonymy of *Ceratias holboelli*). Whitley, 1956:413 (a synonym of "*C. holbolli tentaculatus*"; New Zealand). Nielsen, 1974: 97 (listed in type catalog).
*Ceratias tentaculatus*: Bertelsen, 1943: 202, 203 (new combination). Bertelsen, 1951:134 (in synonymy of *Ceratias holboelli*). Bertelsen and Pietsch, 1984:45, figs. 1, 2 (resurrection from synonymy of *C. holboelli*). Pietsch, 1986a:374, fig. 105.2 (South Africa). Pietsch, 1986b:482, figs. 1, 6 (review based on all known material). Tarakanov and Balushkin, 1987:37, figs. 1–5 (additional material, description; esca, caruncles, and skin spines figured; in key). Paulin et al., 1989:137, fig. 64.3a (New Zealand). Pietsch, 1990:210, fig. 1 (Southern Ocean; after Pietsch, 1986b). Drioli and Vignes, 1992:19, figs. 1, 2 (additional specimen, Argentine sea). Munk, 1992:35 (accessory escal gland). Smale et al., 1995:70, pl. 36A (otolith described, figured). Ellis, 1996:273 (popular account). Stewart and Pietsch, 1998:22, fig. 14 (eight additional specimens, New Zealand). Munk, 1999:267, 280 (bioluminescence). Anderson and Leslie, 2001:20, fig. 14B (after Pietsch, 1986b; additional material). Pietsch, 2005b:209 (reproduction).
*Ceratias holboelli*: Clarke, 1950:3, figs. 1–3, 4a–c, pl. 1 (misidentifications, additional specimens, Antarctic Ocean). Penrith, 1967:185 (misidentifications, additional material, off Cape Town).
*Ceratias holboelli tentaculatus*: Bertelsen, 1951:133, footnote (southern subspecies). Fonseca, 1968:491, 493, fig. 38 (additional record, off Peru, specimen not seen). Whitley, 1968:88 (listed).

*Ceratias holbolli tentaculatus*: Whitley, 1956:413 (listed; New Zealand).

Free-living Males and Larvae

Unknown.

Material

Fifty-seven metamorphosed females (6.2 to 640 mm) and one parasitic male (37 mm):

Holotype of *Ceratias tentaculatus*: BMNH 1930.1.12.1100, 80 mm, *Discovery* station 114, 52°25′S, 9°50′E, 0 to 700 m, 12 November 1926.

Holotype of *Mancalias bifilis*: ZMUC P92123, 75 mm, *Dana* station 3642(1), east of New Zealand, 46°43′S, 176°08′E, 3000 m wire out, 9 January 1929.

Additional material: BMNH, two (15 to 430 mm, larger specimen from stomach of a sperm whale); CSIRO, six (230 to 550 mm; 550-mm female with 37-mm parasitic male); ISH, 12 (90 to 470 mm); LACM, 10 (83 to 276 mm); MACN, one (272 mm); MCZ, one (19.5 mm); MNHN, two (233 to 420 mm); NMNZ, eight (60 to 510 mm); NMV, three (29 to 640 mm); SAM, four (6.2 to 534 mm, 534-mm specimen taken from sperm whale stomach); SIO, two (31.5 to 45 mm); UW, one (460 mm); ZIN, two (94 to 478 mm); ZMUC, one (400 mm).

Distinguishing Characters

A species of *Ceratias*, the metamorphosed females of which are distinguished from those of all other described species of the genus in having a pair of distal escal appendages; illicium length 19.1 to 28.2% SL; vomerine teeth (2 to 6) present in all but smallest known specimens.

Description

Proximal one-half to two-thirds of escal bulb darkly pigmented, distal portion oval in shape, proximal portion tapering into stem of illicium; 2 slender escal appendages arising just anterior to escal pore, each appendage nearly always bifurcate, trifurcate, or multibranched; escal pore, situated approximately at apex of bulb, raised on a pigmented papilla in all known material 75 mm and larger. Vomerine teeth (2 to 6) present in all examined material, except for recently metamorphosed specimens (of all material examined only three, 6.2 to 19.5 mm, lacked vomerine teeth).

Additional description as given for the genus and family.

Distribution

Except for several small specimens from off Durban and Delagoa Bay, South Africa (6.2 to 15 mm), and off southern Mozambique (in addition to a 534-mm specimen from off Saldanha Bay, South

Africa; Penrith, 1967), all of the known material of *Ceratias tentaculatus* has been collected from numerous localities across all sectors of the Southern Ocean between approximately 35 and 68°S (Fig. 209). Although this species may be taken anywhere between approximately 100 and 2900 m, the majority of specimens were captured between 650 and 1500 m. Of the material for which depth of capture was recorded (41 of 57 known specimens), 90% was taken by fishing gear that reached a maximum depth of 2000 m or less; 81% was captured by gear fished at maximum depths of 1500 m or less. At the upper end of its vertical range, 86% of the known material was taken by trawls that reached maximum depths of greater than 650 m. A 470-mm specimen (ISH 657/71) was captured between 105 and 110 m. The average maximum depth for all known captures was 1194 m.

Comments

Although nearly all recent authors have treated *Ceratias* as a monotypic genus, the possibility of the existence of a northern and southern species has remained. Bertelsen (1943) was able to show that the isolated, so-called free dorsal spine of *Ceratias* is the posterior end of the movable pterygiophore of the illicium, which when pulled forward lies hidden under the skin of the back and that the presence or relative length of this structure could therefore not be used to distinguish species. Primarily on this evidence, the genus *Mancalias* Gill (1879b) was synonymized with *Ceratias,* and the nominal species of these genera were reduced to two: *Ceratias holboelli* Krøyer (1845) and *C. tentaculatus* (Norman, 1930). The separation of these two species was based on differences in the morphology of the esca, coupled with geographic distribution: the two adult *C. holboelli* with well-preserved escae known at that time had a single escal appendage and were collected from the North Atlantic, whereas the two, or possibly three, known *C. tentaculatus* had 2 escal appendages and were taken in southern waters beyond the tropics (Bertelsen, 1943). By 1951, additional material led Bertelsen (1951) to conclude that the number and composition of the escal appendages were more variable and of less taxonomic value than previously supposed. Furthermore, since the distribution of the larvae did not seem to support the existence of a separate southern species, Bertelsen (1951) synonymized the two. In the same publication, however, hearing of two additional specimens, one from the North Atlantic with a single escal appendage and one from the Southern Ocean (Clarke, 1950) having 2, Bertelsen (1951:133) made a final recommendation in footnote that until additional material becomes available, two subspecies should be recognized, a northern *C. holboelli holboelli* and a southern *C. h. tentaculatus*.

By the mid-1980s, the number of specimens of southern ocean *Ceratias* with 2 escal appendages had increased by nearly sevenfold (excluding eight with damaged or lost escae of the 31 specimens then known). This new material came primarily from two sources: a 534-mm specimen recovered from the stomach of a sperm whale captured in the South Atlantic off Cape Town, described by Penrith (1967); and 12 large females (90 to 470 mm) collected during the South Atlantic cruises of the FRV *Walther Herwig.* Citing these additional specimens, Bertelsen and Pietsch (1984) presented evidence to resurrect *C. tentaculatus* from the synonymy of *C. holboelli.* The validity of this Southern Ocean endemic was supported further by another 10 specimens described by Pietsch (1986b): a 19.5-mm female from off Cape Town, South Africa; and nine well-preserved females, 83 to 275 mm, captured in the Pacific sector of the Southern Ocean below 54°S latitude.

*Ceratias* Species

Females and Parasitic Males

The following nominal species are based on females that cannot be identified to species because of the loss of the illicium; none have vomerine teeth. They all are based on material collected from geographic areas where two species of *Ceratias* occur sympatrically and thus cannot reasonably be synonymized with any recognized species of the genus:

*Miopsaras myops* Gilbert, 1905:695, pl. 99 (*nomen dubium*; holotype USNM 51637, 75 mm, *Albatross* station 4019, deep-sea off Kauai, Hawaiian Islands, 748 to 1325 m, 21 June 1902).

*Typhloceratias firthi* Barbour, 1942b:78, pl. 10, 11, figs. 2–4 (*nomen dubium*; holotype MCZ 35771, 480 mm, southeast portion of Georges Bank, 275 to 366 m, 9 February 1927).

*Parrichthys merrimani* Barbour, 1942b: 84, pl. 11, fig. 1 (*nomen dubium*; holotype YPM 2004, 35 mm, Crooked Island Passage, Bahamas, 2440 m wire out, 20 March 1927).

The following list of material examined (109 metamorphosed females, 8.5 to 630 mm; and 12 parasitic males, 9.2 to 115 mm) includes unidentifiable females that have lost the illicium and/or esca: AMS, one (22 mm); BMNH, 20 (10 to 75 mm); CAS, two (370-mm female with 5-mm remnant of parasitic male; discarded female of unknown length with 73-mm parasitic male); HUMZ, one (361 mm); ISH, 11 (35.5 to 182 mm); LACM, 32 (8.5 to 630 mm; 275-mm female with 24.5-mm parasitic male, 525-mm female with 37-mm parasitic male); MCZ, four (10.5 to 71 mm); MNHN, one (660-mm female with 46-mm parasitic male); MTF, one (740-mm female with 60-mm parasitic male); NMNZ, 13 (64 to 500 mm; 790-mm female with two parasitic males, 65 to 89 mm); SAM, one (700-mm female with 37-mm parasitic male); SIOM, seven (10 to 31 mm); UF, six (20 to 212 mm, 212-mm female with 9.2-mm parasitic male); USNM, one (565 mm); UW, four (455 to 553 mm, 553-mm female with 46-mm parasitic male); WAM, two (480 to 570 mm); ZMH, one (750-mm female with 115-mm parasitic male); ZMUC, one (46 mm).

Free-living Males and Larvae

For a list of unidentifiable males and larvae, see Bertelsen (1951:134, 271, table 9).

GENUS *CRYPTOPSARAS* GILL, 1883B (TRIPLEWART SEADEVILS)

Figures 24, 44, 46, 48, 50, 51, 64–67, 202, 203, 210, 240, 250, 256, 261, 282, 283, 286, 289, 290, 291; Tables 1, 2, 8–10

Females, Males, and Larvae

*Cryptopsaras* Gill, 1883b:284 (type species *Cryptopsaras couesii* Gill, 1883b, by monotypy).

*Paraceratias* Tanaka, 1908:18, pl. 2, fig. 3 (type species *Ceratias* [*Paraceratias*] *mitsukurii* Tanaka, 1908 [a synonym of *Cryptopsaras couesii* Gill], by monotypy).

*Cryptosparas* Regan, 1925b:393 (emendation of *Cryptopsaras* Gill, 1883b, followed by numerous subsequent authors, therefore taking the same type species).

*Bathyceratias* Beebe, 1934a:211, 327, fig. 117 (new genus as seen from bathysphere, half-mile down, no specimens collected; type species *Bathyceratias trilychnus* Beebe, 1934a, by monotypy). Beebe, 1934b: 191, fig. (after Beebe, 1934a). Beebe, 1934c:691, pl. 7 (popular account). Bertelsen, 1951:130 ("to be received with caution," *nomen dubium*).

*Cryptosarus* Barbour, 1942b:79 (emendation of *Cryptopsaras* Gill, 1883b, therefore taking the same type species).

*Cryptosaras* Fowler, 1949:160 (emendation of *Cryptopsaras* Gill, 1883b, therefore taking the same type species).

*Cryptopsarus* Fitch and Lavenberg, 1968:129 (emendation of *Cryptopsaras* Gill, 1883b, therefore taking the same type species).

Distinguishing Characters

The genus *Cryptopsaras* differs from *Ceratias*, the only other recognized genus of the family, in having 8 caudal-fin rays and a conspicuous spine on the anterodorsal margin of the subopercle (Bertelsen, 1951:138, figs. 88B, 89A, C).

Metamorphosed females of the genus are further distinguished from those of *Ceratias* in having a tiny illicium, reduced to a small remnant (nearly fully enveloped by tissue of the esca), and 2 club-shaped caruncles on the dorsal midline of the trunk just anterior to the origin of the soft-dorsal fin.

Metamorphosed males are further distinguished from those of *Ceratias* in having the anterior pair of lower denticular teeth considerably longer than the posterior pair.

Larvae, males, and juvenile females are unique in having subdermal pigment on the gill cover, dorsal surface of trunk, and caudal peduncle.

### Description

The escal bulb of metamorphosed females bears a single distal filament (laid down as a tiny papilla in metamorphosing females as small as approximately 10 mm), with one to several pairs of smaller filaments arising from the base. The number of basal escal filaments, and length and branching of the distal filament, is highly variable (see Bertelsen, 1951:142, fig. 95). The number of teeth in the lower jaw ranges from 42 to 68. Vomerine teeth range from 2 to 10. Fin-ray counts are as follows: dorsal-fin rays 4 (excluding those embedded in the caruncles), rarely 5 (only three of 311 specimens counted); anal-fin rays 4; pectoral-fin rays 14 to 18, usually 15 to 17 (Table 1). The skin is covered with close-set dermal spinules. Larval females have a distinct longitudinal row of 3 caruncles (see Bertelsen, 1951:140, fig. 93A–E).

The skin of metamorphosed free-living males is naked. There are two pairs of denticles on the lower denticular; the basal portion of the anterior pair is prolonged and directed anteriorly when the mouth is closed. The posterior pair is vertical and considerably shorter than the anterior. Caruncles are absent (Bertelsen, 1951:138). Adult males are parasitic, the skin is spinulose, and the eyes are degenerate (Bertelsen, 1951:138).

Females are dark red brown to black over the entire surface of the body (except for the distal portion of the escal bulb) and oral cavity. The skin is unpigmented in juvenile males, but darkly pigmented in parasitic stages. Subdermal pigment is present, melanophores grouped as in the larvae (Bertelsen, 1951:138).

The larvae are characterized by having a band of pigment on the posterolateral margin of the head, extending from the occipital region, continuing along the margin of the gill cover and meeting anteriorly on the isthmus. Larger specimens have a lateral group of melanophores spreading from the base of the anal fin and becoming continuous with a dorsal group of melanophores. There is an isolated group of melanophores on the caudal peduncle. The peritoneum is pigmented dorsally (Bertelsen, 1951:138).

Females attain a standard length of at least 358 mm. The largest known free-living male measures 10.5 mm, while parasitic males range from 8.0 to 99 mm (see Pietsch, 1976:786, 2005b:223, table 1).

## *CRYPTOPSARAS COUESII* GILL, 1883B (TRIPLEWART SEADEVILS)

Figures 64–67, 210; Tables 2, 8–10

### Females, Males, and Larvae

*Cryptopsaras couesii* Gill, 1883b:284 (original description, single specimen). Günther, 1887:55 (after Gill, 1883b). Goode and Bean, 1896:491, pl. 119, fig. 402 (notes on holotype). Jordan and Evermann, 1896:510 (listed). Jordan and Evermann, 1898:2731, fig. 956 (after Gill, 1883b). Jordan and Evermann, 1900, fig. 956 (figured). Gill, 1909:579, fig. 13 (after Gill, 1883b). Roule, 1919:76, pl. 5, fig. 6 (additional specimen, Azores). Barnard, 1927:1004, 1005 (west coast of South Africa). Oshima, 1934:108, fig. (additional specimen, Japan). Bertelsen, 1943:195 (comparison with other ceratiids, distribution). Smith, 1949: 428, fig. 1229 (South Africa). Bertelsen, 1951:14, 17, 22, 23, 128, 129, 139, 145, figs. 2, 3, 6b, 7, 88b, 89a, c, 93–97 (revision based on all known material). Abe and Nakamura, 1954:95 (northern Japan). Grey, 1955:299 (Bermuda). Matsubara, 1955: 1349, 1350, 1353, 1357, 1358, figs. 524, 530F, 533 (in synonymy of *Ceratias holboelli*; figures after Barbour, 1941b, Imai, 1942, Bertelsen, 1951). Grey, 1956a:264 (synonymy, distribution). Brandes et al., 1957:54 (two additional specimens, North Atlantic). Fast, 1957:237 (Monterey Bay, California). Shoemaker, 1958:143, fig. 1 (Gulf of Mexico). Krefft, 1960:71 (additional specimen, northwest Iceland). Okada, 1961:155, figs. 1–3 (Japan). Maul, 1962a:26 (Madeira). Munk, 1964:5, figs. 1B, 2B, 2D–G (histology of eyes). Bussing, 1965:223 (off Chile). Ueno, 1966:536, 538, 539, fig. 2E (female with parasitic male; off Kushiro, Hokkaido). Munk, 1966:22, 28, figs. 10, 14B, 15A (ocular anatomy). Penrith, 1967:186 (South Africa). Ueno and Abe, 1967:274, figs. 1, 2 (after Ueno, 1966). Kobayashi et al., 1968:12, pl. 2, figs. A, B (after Ueno, 1966, and Ueno and Abe, 1967). Briggs, 1960:179 (worldwide distribution). Kubota and Uyeno, 1970:25 (from stomach of *Alepisaurus ferox*). Marshall, 1971a:41, 45 (eyes of males). Pietsch, 1972a:31, 38, 39, 45 (osteological comments). Jónsson, 1973:220 (additional specimen, Iceland). Kubota, 1973:235 fig. 3 (from stomach of *A. ferox*). Maul, 1973:674 (eastern North Atlantic; synonymy). Parin et al., 1973:144, fig. 37 (additional specimens, eastern South Pacific). Parin et al., 1974:124 (western South Atlantic). Jónsson, 1975:207 (additional specimen, Iceland). Karrer, 1975:78 (Gulf of Aden). Parin, 1975:323 (faunal study). Pietsch, 1975a:38 (sexual parasitism). Parin and Golovan, 1976:271 (off West Africa). Pietsch, 1976:786 (reproduction). Jónsson et al., 1977:181 (additional specimens, Iceland). Herring and Morin, 1978:322, fig. 9.12 (bioluminescence). Jónsson et al., 1978:182 (three additional specimens, Iceland). Hubbs et al., 1979:13 (listed for California). Jónsson et al., 1979:229 (three additional specimens, Iceland). Marshall, 1979:351, 383, 390, 417, 452, 488, 489, fig. 146 (caruncle luminescence, eyes and vision, lateral-line structures, larval growth rates, vertical distribution). Blacker, 1980:250 (additional specimen, eastern North Atlantic). Jónsson et al., 1980:248 (two additional specimens, Iceland and east Greenland). Leisman et al., 1980:1272 (bioluminescence). Pietsch and Seigel, 1980:396 (Philippine Archipelago, in key). Jónsson et al., 1981:193 (four additional specimens, Iceland and east Greenland). Ayling and Cox, 1982:140, fig. (New Zealand). Yamakawa, 1982:199, 362, fig. 124 (additional specimen, Kyushu-Palau Ridge). Bertelsen and Pietsch, 1983:89, 91, fig. 11 (Australia, in key). Fujii, 1983:265, fig. (three additional specimens, Surinam). Herring, 1983:202, table 9 (bioluminescence emission spectra). Jónsson et al., 1983:235 (additional specimen, Iceland). Amaoka, 1984:108, pl. 93G (Japan). Bertelsen, 1984:329, fig. 168C, D (early life history, phylogeny). Jónsson et al., 1985:185 (additional specimen, Iceland). Paulin and Stewart, 1985:27 (New Zealand). Bertelsen, 1986:1404, figs. (eastern North Atlantic). Jónsson et al., 1986a:186 (three additional specimens, Iceland). Jónsson et al., 1986c: 628 (four additional specimens, Iceland). Pietsch, 1986a:375, fig. 105.3 (South Africa). Pietsch, 1986b:488, figs. 7, 8 (family revision, comparison of all known material). Jónsson et al., 1987:360 (additional specimen, Iceland). Minchin, 1988:313 (additional specimen, eastern North Atlantic). Quéro and Vayne, 1988:173, figs. 1, 2, 5, 6 (additional material, eastern North Atlantic). Du Buit et al., 1989:192, fig. 1 (North Atlantic distribution). Paulin et al., 1989:136, fig. 64.2b (New Zealand). Amaoka, 1990:202, fig. 142 (New Zealand). Bertelsen, 1990:511 (synonymy, eastern tropical Atlantic). Abe and Funabashi, 1992:1, figs. (female with four parasitic males; off Ibaraki Prefecture, Japan). Haygood et al., 1992:149, 151, figs. 2–4 (gene sequences from bioluminescent bacterial symbionts). Nielsen and Bertelsen, 1992:62, fig. 4 (North Atlantic). Haygood, 1993:198, 203, fig. 6 (bioluminescent bacterial symbionts). Haygood and Distal, 1993:154, figs. 1, 2 (bioluminescent bacterial symbionts). Andriyashev and Chernova, 1994:103 (listed for Arctic seas). Amaoka et al., 1995:114, fig. 176 (northern Japan). Meng et al., 1995:440, fig. 589 (China). Smale et al., 1995:70, pl. 36B (otolith described, figured). Stearn and Pietsch, 1995:132, fig. 82 (additional specimen, Greenland). Swinney, 1995a:52, 56 (additional specimen, off Madeira). Ellis, 1996:273, fig. (popular account). Munk and Herring, 1996:517, figs. 1–4 (development and histology of escae and caruncles). Herring et al., 1997:52 (bioluminescent signaling). McEachran and Fechhelm, 1998:850, fig. (Gulf of Mexico).

Stewart and Pietsch, 1998:25, fig. 16 (73 specimens, New Zealand). Jónsson and Pálsson, 1999:203, fig. 6 (Iceland). Munk, 1999:267, 272, 274, 275, 277, 282, figs. 4C, 5E, 7, 12 (bioluminescence). Pietsch, 1999:2035 (western central Pacific). Trunov, 1999:493 (Atlantic sector of the Subantarctic Ocean). Anderson and Leslie, 2001:22, fig. 14C (after Pietsch, 1986b; additional material). Saruwatari et al., 2001:82, figs. 1–3 (female with eight parasitic males, sibling analysis). Mecklenburg et al., 2002:302 (Alaska, in key). Nakabo, 2002:468, figs. (Japan, in key). Pietsch, 2002b:1067 (western central Atlantic). Menezes et al., 2003:65 (after Fujii, 1983). Moore et al., 2003:216 (off New England). Shimazaki and Nakaya, 2004:33, figs. 1–9 (functional anatomy of the luring apparatus). Love et al., 2005:61 (eastern North Pacific). Pietsch, 2005b:209, 214, 223, 224, 226, 232, figs. 8D, 15B, 16 (reproduction). Shinohara et al., 2005:421 (off Ryukyu Islands). Woodland, 2005:251, fig. 5.41C (sexual parasitism). Kharin, 2006b:410, fig. 2 (Russian waters). Senou et al., 2006:424 (Sagami Sea). Kharin and Milovankin, 2007:112, fig. (western North Pacific). Pietsch and Orr, 2007:3, 6, 8, 18, figs. 2B, 4C, 5C, 15B (phylogenetic relationships).

*Ceratias carunculatus* Günther, 1887: 52, 55, pl. 11, fig. D (original description, single specimen). Lütken, 1894:79 (after Günther, 1887). Goode and Bean, 1896: 491 (in synonymy of *Cryptopsaras couesii*). Regan, 1926:36 (in synonymy of *Cryptosparas mitsukurii*).

*Ceratias (Cryptopsaras) couesii*: Lütken, 1894:79 (new combination).

*Ceratias couesi*: Brauer, 1906:317, pl. 15, fig. 7 (new combination, Gulf of Aden). Brauer, 1908:103, 105, 147, 184, pl. 32, fig. 17, pl. 34, fig. 17 (histology of caruncle and eye). Murray and Hjort, 1912:92, 608, 614, 627, fig. 466 (after Brauer, 1906). Saemundsson, 1922:160, pl. 3 (Iceland). Borodin, 1931:83 (central North Atlantic). Ehrenbaum, 1936:164 (seas of northern Europe). Fowler, 1936:1137, fig. 477 (West Africa). Saemundsson, 1949:47 (additional specimen, Iceland). Harvey, 1952:531, 532, fig. 178 (bioluminescence). Jónsson, 1967a:55, 58 (additional material, Iceland).

*Ceratias (Paraceratias) mitsukurii* Tanaka, 1908:18, pl. 2, fig. 3 (original description, single female with parasitic male). Barbour, 1941a:175 (type in MCZ).

*Paraceratias mitsukurii*: Tanaka, 1911:30, pl. 8, fig. 25 (after Tanaka, 1908). Jordan et al., 1913:427 (after Tanaka, 1908). Barbour, 1941a:175 (type in MCZ). Barbour, 1941b: 7, pl. 5, figs. 1, 2 (reference to Barbour, 1941a; figured). Matsubara, 1955:1357 (in synonymy of *C. couesii*).

*Cryptosparas couesi*: Regan, 1925b:393, fig. 5 (visceral anatomy, emendation of generic name). Regan, 1926:35, pl. 9, fig. 2 (additional material). Parr, 1927:29 (Bahamas). Norman, 1930:354, fig. 44 (eastern central and South Atlantic). Regan and Trewavas, 1932:96, fig. 154 (comparison of all known material). Beebe, 1937:207 (Bermuda). Roule, 1934:199, 202, fig. (semipopular account). Barbour, 1941b:11, 14, pl. 4, fig. 2, pl. 6, figs. 5, 6 (comparison with *Cryptosparas atlantidis* Barbour, 1941b). Kuronuma, 1941:64 (additional specimen, Japan). Rauther, 1941:137 (illicial anatomy, comparison with *Ceratias holboelli*). Imai, 1942:42, fig. 3 (Suruga Bay, Japan). Koefoed, 1944:11 (central North Atlantic). Albuquerque, 1954–1956:1069, fig. 439 (after Fowler, 1936).

*Cryptosparas mitsukurii*: Regan, 1926:36 (new combination, comparison with allied nominal species). Regan and Trewavas, 1932:98 (after Tanaka, 1908, and Regan, 1926). Barbour, 1941b:11, 14, pl. 5 (comparison with *Cryptosparas atlantidis* Barbour, refigured from the holotype).

*Cryptosparas carunculatus*: Regan and Trewavas, 1932:98, fig. 155 (after Günther, 1887; esca figured). Barbour, 1941b:11, 14, pl. 3, fig. 2, pl. 6, figs. 7, 8 (comparison with *C. atlantidis* Barbour, 1941b). Matsubara, 1955:1357 (in synonymy of *C. couesii*).

*Cryptosparas valdiviae* Regan and Trewavas, 1932:98, fig. 156 [original description, single specimen, holotype apparently lost, not in BMNH or in ZMB (Hans-Joachim Paepke, personal communication, 9 February 1978, 4 December 1984), 67 mm, *Valdivia* station 270, Gulf of Aden, 13°01′N, 47°11′E, 0 to 1840 m].

*Cryptosparas pennifer* Regan and Trewavas, 1932:98, fig. 157 (original description, three specimens, lectotype ZMUC P92125, designated by Pietsch, 1986b). Nielsen, 1974:97 (listed in type catalog).

*Cryptosparas normani* Regan and Trewavas, 1932:98 (original description, single specimen). Barbour, 1941b:11, 14, pl. 4, fig. 1, pl. 6, figs. 1, 2 (comparison with *Cryptosparas atlantidis* Barbour). Imai, 1942:39, figs. 1, 2 (Suruga Bay, Japan). Beebe and Crane, 1947:168, fig. 14 (Galapagos Islands). Matsubara, 1955:1357, 1358, fig. 533 (in synonymy of *C. couesii*; figure after Imai, 1942).

*Cryptopsaras pennifer*: Whitley, 1934, unpaged (emendation of generic name; listed). Whitley, 1956:413 (listed; New Zealand). Whitley, 1958:49, fig. 12 (after Regan and Trewavas, 1932; figure by Fraser-Brunner). Whitley, 1968:88 (listed).

*Ceratias mitsukurii*: Barbour, 1941a:175 (new combination, type in MCZ).

*Cryptosparas atlantidis* Barbour, 1941b: 11, 14, pl. 3, fig. 1, pl. 6, figs. 3, 4 (original description, single specimen).

*Cryptosarus mutsukurii*: Barbour, 1942b:79 (reference to type specimen, emendation of generic and specific names).

*Ceratias holboelli* (non Krøyer): Koefoed, 1944:10, pl. 2, fig. 2 (eastern central Atlantic).

*Cryptosaras couesii*: Fowler, 1949:160 (Atlantic, Pacific).

*Cryptopsaras couesi pennifer*: Bertelsen, 1951:143 (subspecies for three type specimens of *Cryptosparas pennifer* Regan and Trewavas, 1932; New Zealand and Tasman Sea).

*Cryptosarus couesii*: Fitch and Lavenberg, 1968:129, fig. 71 (distinguishing characters, natural history).

### Material

Nine hundred and eighty-three metamorphosed females (5.5 to 358 mm), 74 parasitic males (8.0 to 99 mm), about 100 metamorphosed free-living males (7.5 to 10.5 mm), and about 350 larvae (2.5 to 15 mm TL):

Holotype of *Cryptopsaras couesii*: USNM 33558, 30 mm, *Albatross* station 2101, western North Atlantic, northwest of Bermuda, 38°18′N, 68°24′W, 0 to 3085 m, 3 October 1883.

Holotype of *Ceratias carunculatus*: BMNH 1887.12.7.16, 28 mm, *Challenger* Expedition, station 232, south of Yeddo, Japan, 35°11′N, 139°28′E, 0 to 630 m, 12 May 1875.

Holotype of *Ceratias (Paraceratias) mitsukurii*: MCZ 29855, 290-mm female with 12-mm parasitic male, Yodomi, Sagami Bay, Japan, 0 to 1464 m, May 1907, Alan Owston.

Lectotype of *Cryptosparas pennifer*: ZMUC P92125, 27 mm, *Dana* station 3630(2), northeast of New Zealand, 34°24′S, 178°42.5′E, 2000 m wire, 17 December 1928.

Paralectotypes of *Cryptosparas pennifer*: BMNH 1932.5.3.29, 24 mm, *Dana* station 3653(9), Tasman Sea, 33°30.5′S, 165°53′E, 2000 m wire, 26 January 1929. ZMUC P92126, 17 mm, *Dana* station 3651(5), NW of New Zealand, 35°36′S, 171°52′E, 1000 m wire, 22 November 1929.

Holotype of *Cryptosparas normani*: BMNH 1930.1.12.1095, 50 mm, *Discovery* station 87, off South Africa, 33°53′S, 9°26′E, 0 to 1000 m, 25 June 1926.

Holotype of *Cryptosparas atlantidis*: MCZ 31650, 83 mm, *Atlantis* station 143, Azores, 50°00′N, 35°20′W, 0 to 915 m, 2 September 1928.

Additional material: AIM, 5 (45 to 167 mm); AMS, four (8 to 152 mm); ARC, 97 (standard lengths not recorded; 45-mm female with 10-mm parasitic male); ASIZP, three (38 to 169 mm); BMNH, 93 females (9.5 to 305 mm; 77-mm female with 15-mm parasitic male), 10 free-living males (5.8 to 9.5 mm); CAS, three (195 to 215 mm; 195-mm female with 34-mm parasitic male; 213-mm female with two parasitic males, 27 to 27.5 mm; 215-mm female with 44-mm parasitic male); CSIRO, 19 (34 to 328 mm; 210-mm female with two parasitic males, 21 to 28 mm; 268-mm female with 36-mm parasitic male; 322-mm female with 51-mm parasitic male); FMNH, four (16 to 44.5 mm); HUMZ, 24 (40 to 348 mm;

272-mm female with 74-mm parasitic male; 308-mm female with four parasitic males, 78 to 88 mm; 320-mm female with 48-mm parasitic male; 324-mm female with 48-mm parasitic male; 345-mm female with six parasitic males, 18 to 44 mm; 348-mm female with three parasitic males, 54 to 60 mm); IIPB, one (190 mm); INM, one (300-mm female with four parasitic males, 33 to 60 mm); ISH, 23 (21 to 215 mm); LACM, 93 (7 to 180 mm, 173-mm female with 35-mm parasitic male); MCZ, 280 (5.5 to 290 mm); 290-mm female with 12-mm parasitic male; 162-mm female with 24.5-mm parasitic male); MNHN, 17 (12 to 85 mm); NMNZ, 68 (10 to 390 mm, 210-mm female with small remnant of parasitic male; 230-mm female with 11-mm parasitic male; 255-mm female with 11-mm parasitic male; 300-mm female with 31-mm parasitic male; 320-mm female with 26-mm parasitic male; 390-mm female with two parasitic males, 47 to 99 mm); NMSZ, one (9 mm); NMV, 12 (10 to 225 mm); NSMT, 12 (12 to 223 mm, 81-mm female with 9-mm parasitic male); ORI, one (316-mm female with eight parasitic males, 35 to 56 mm); OS, one (252-mm female with 62-mm parasitic male); ROM, two (98 to 198 mm); SAIAB, two (53 to 123 mm); SAM, 19 (9.5 to 358 mm); 311-mm female with 40-mm TL parasitic male; 322-mm female with 41-mm parasitic male; 343-mm female with two parasitic males, 88 to 120 mm TL; 356-mm female with 73-mm parasitic male); SIO, 43 (11.5 to 300 mm); 202-mm female with 20-mm parasitic male, 302-mm female with 50-mm parasitic male); SIOM, 15 (26 to 113 mm); TCWC, one (59 mm); UF, 24 (20 to 164.5 mm); 68-mm female with 10.5-mm parasitic male; 94-mm female with two parasitic males, 8.0 to 14.5 mm; 108-mm female with 20.5-mm parasitic male; 122-mm female with 9.5-mm parasitic male; 164.5-mm female with 49-mm parasitic male); USNM, 69 (12 to 176 mm); 15.5-mm female with 9.8 mm parasitic male); 132-mm female with 17-mm parasitic male; 146-mm female with 17-mm parasitic male; 176-mm female with three parasitic males, 16 to 37 mm); UW, 10 (14 to 240 mm); 207-mm female with two parasitic males, 55 to 62 mm; 215-mm female with 20-mm parasitic male; 230-mm female with 80-mm parasitic male; 240-mm female with 34-mm parasitic male); ZIN, nine (standard lengths not recorded); ZMA, one (320 mm); ZMUC, 62 (10.5 to 210 mm).

### Distinguishing Characters and Description

As given for the genus and family.

### Distribution

The geographic distribution of *Cryptosparas couesii* is similar to the combined ranges of *Ceratias holboelli* and *C. uranoscopus*, occurring in all three major oceans of the world between approximately 64°N and 57°S (Fig. 210). It occurs throughout the North Atlantic, ranging as far north as Iceland (first recorded there by Saemundsson, 1922) and southern Greenland, and from the Gulf of Mexico and Caribbean Sea to the African coast, with numerous records from the relatively nutrient-poor Central Atlantic Water Mass. In the South Atlantic, however, it appears to be restricted to the eastern side (but this could well be the result of little collecting in this area), ranging as far south as approximately 40°S off the tip of South Africa.

In the Pacific, *C. couesii* is represented by numerous localities throughout Indonesian waters, extending north to Japan, and south to Australia, Tasmania, and New Zealand. It ranges further across the central Pacific between about 35°N and 30°S, to Hawaii, north to the Washington, Oregon, and California coasts, and south to Chile. In the Indian Ocean, it is known from more than 80 specimens (including numerous larvae and males; see Bertelsen, 1951:272), mostly taken on the western side of that ocean, but there are several records as well from off Sumatra, Java, and western Australia.

Juvenile and adult females of *C. couesii* may be captured anywhere between approximately 75 and 4000 m. The majority of the known specimens, however, were taken between 500 and 1250 m. Of the material for which depth of capture was known, 98% was taken by trawls that reached a maximum depth of 2000 m or less; 88% was collected by gear fished at 1250 m or less. At the upper end of its vertical range, 84% of the known material was taken at depths greater than 500 m. The average maximum depth for all known captures was 890 m.

### *Cryptopsaras* Species

The following nominal species, one of three described by William Beebe as new to science, based on a one-time observation from the port of his famous bathysphere off Nonsuch Island, Bermuda, in August 1934, at a depth of 753 m, is herein assigned to *Cryptopsaras* as a *nomen dubium*:

On the present dive, Number Thirty, we had the good fortune to observe three new fish so clearly and for such an appreciable length of time, that I shall again chance the accusation of proposing for them *nomina nuda*. I have only our recorded observations, very carefully prepared outline drawings and colored paintings for evidence, but in lieu of physical types I feel the unusual conditions, and possible future continued explorations, warrant the definitive establishment of these few fish (Beebe, 1934b:190).

Beebe's new ceratioid was differentiated in part by having 3 illicia, each tipped with a bioluminescent esca (see Bioluminescence and Luring, Chapter Six; Fig. 242). The anatomical improbability of a compound luring apparatus such as this, coupled with the absence of a specimen, makes it likely that Beebe's observation was a figment of his imagination, "to be received with caution" (Bertelsen, 1951: 130; see also Ellis, 2005:41).

#### Female

*Bathyceratias trilychnus* Beebe, 1934a: 211, 327, fig. 117 (new genus and species as observed by Beebe from bathysphere, half-mile down off Bermuda, no specimen collected). Beebe, 1934b:191, fig. (after Beebe, 1934a). Beebe, 1934c:691, pl. 7 (popular account). Bertelsen, 1951:130 (of doubtful validity). Ellis, 2005:41, fig. (Beebe's "cryptozoological fishes").

## Family Himantolophidae Gill, 1861 (Footballfishes)

Figures 9–12, 34, 40, 41, 45, 49, 68–74, 198, 202, 203, 211–214, 260, 292; Tables 1, 3, 8, 9

### Type Genus *Himantolophus* Reinhardt, 1837

Himantolophinae Gill, 1861:47 (subfamily of family Lophioidae to contain genus *Himantolophus*). Gill, 1879a:217, 218 (subfamily of family Ceratiidae to contain genera *Himantolophus* and *Corynolophus*).

Himantolophiformes Bleeker, 1865:6 (subfamily of family Chironecteoidei to contain genus *Himantolophus*).

Aegaeonichthyinae Gill, 1879b:227, 228 (subfamily of family Ceratiidae to contain genus *Aegaeonichthys*).

Himantolophidae Regan, 1912:285, 287 (family of division Ceratiiformes to contain genera *Aegeonichthys*, *Diceratias*, *Dolopichthys*, *Himantolophus*, *Linophryne*, *Oneirodes*, and *Paroneirodes*).

Aceratiidae Parr, 1927:30 (in part; family of suborder Ceratioidea to contain genera *Lipactis*, *Aceratias*, *Rhynchoceratias*, and *Laevoceratias*, all based on free-living males, those of *Lipactis* and *Rhynchoceratias* subsequently reassigned in part to genus *Himantolophus* by Regan and Trewavas, 1932).

Eurostrinae Parr, 1930a:20 (in part; subfamily of family Aceratiidae to contain genera *Aceratias* and *Rhynchoceratias*, both based on free-living males, those of *Rhynchoceratias* subsequently reassigned in part to genus *Himantolophus* by Regan and Trewavas, 1932).

### Distinguishing Characters

Metamorphosed females and males of the family Himantolophidae are distinguished from those of all other ceratioid families by lacking parietal bones throughout life (lost during metamorphosis in females of the gigantactinid genus *Rhynchactis*; see Bertelsen et al., 1981:10, fig. 13) and in having a broad toothless vomer and triradiate pelvic bones (triradiate pelvic bones are also found in some specimens of the oneirodid genus *Chaenophryne*; see Pietsch, 1975b:79, fig. 2).

Metamorphosed females and males are further unique in having the following combination of character states: The supraethmoid, pterosphenoid, metapterygoid, and mesopterygoid are all present. The hyomandibula has a double head. There are 2 hypohyals and 6 (2 + 4) branchiostegal rays. The opercle is bifurcate. The subopercle has a slender tapering dorsal prolongation, but the ventral end is rounded, without an anterior spine or projection. Quadrate, articular, angular, and preopercular spines are absent. A postmaxillary process of the premaxilla is absent. The first pharyngobranchial is reduced in females, but rudimentary in males. The second and third pharyngobranchials are present, well toothed in females, but toothless in males. The fourth pharyngobranchial is absent, the first epibranchial is present, the third hypobranchial is absent, and there is only a single ossified basibranchial. Epibranchial and ceratobranchial teeth are present in females, but there are only minute remnants of ceratobranchial teeth present in males; see below. Epurals are absent. The hypural plate is entire, without a posterior notch. The posteroventral process of the coracoid is absent. There are 3 pectoral radials. Fin-ray counts are as follows: dorsal-fin rays 5 or 6; anal-fin rays 4; pectoral-fin rays 14 to 18; pelvic fins absent; caudal-fin rays 9 (1 simple + 6 bifurcated + 2 simple), the ninth or ventral-most caudal ray is well developed, more than half the length of the eighth (Table 1). The skin is everywhere covered with dermal spines or spinules, the spines hypertrophied in females (see below).

Metamorphosed females are further differentiated by having low, rounded wartlike papillae covering the snout and chin; and large conical dermal spines, with extremely broad rounded bases, widely spaced and scattered over the head and body. Metamorphosed females are also unique in having the following combination of character states: The frontal bones are widely separated, but without ventromedial extensions. Sphenotic spines are present. The jaws are subequal, the lower jaw extending anteriorly beyond the upper. The lower jaw has a well-developed symphysial spine. The anterior-maxillomandibular ligament is present. The pterygiophore of the illicium bears a small ossified rudiment of the second cephalic spine. The escal bulb has a central lumen, with a pore leading to the outside. The esca is without toothlike denticles. The ovaries are paired. Pyloric caecae are absent.

Metamorphosed males are further differentiated by having a series of enlarged dermal spines above and posterior to the upper denticular bone. They are also unique in having the following combination of character states: the eyes are positioned laterally and are slightly oval in shape, with a narrow aphakic space in front of the lens. The olfactory organs are large, the nostrils directed laterally. There are 16 to 31 denticular teeth on the snout and 20 to 50 on the chin, each cluster of teeth fused at the base to form upper and lower denticular bones, respectively. Fin-ray counts are as given for metamorphosed females. The skin is densely covered with close-set dermal spinules. They are free-living, never becoming parasitic on females, but temporarily attached males are so far unknown (see Pietsch, 2005b:219; and Reproduction and Early Life History, Chapter Eight).

Himantolophid larvae differ from those of other ceratioids in having the following combination of character states: The body is short, almost spherical. The skin is highly inflated. The pectoral fins are of normal size, not reaching posteriorly beyond the dorsal and anal fins. Pelvic fins are absent. Sexual dimorphism is evident, the females bearing a small, club-shaped illicial rudiment protruding from the head. Fin-ray counts are as given for metamorphosed females. Unlike most other ceratioids, metamorphosis is delayed, the larvae attaining lengths of 17 to 22.5 mm, and metamorphosis taking place at lengths between about 20 and 33 mm (Bertelsen, 1984:326, 328, fig. 169A, B; Bertelsen and Krefft, 1988:28).

The osteological features cited here and elsewhere in this account of the Himantolophidae are based on detailed examinations of both males and females (see Lütken, 1887:325, figure and plate; Regan and Trewavas, 1932:31–33, figs. 31–36; Gregory, 1933:407, fig. 278; Bertelsen, 1951:56, 57, fig. 22; Maul, 1961:98, 99, 105, 108, 114, 119, 121, figs. 3, 5, 6, 8, 13, 16, 17; Pietsch, 1972a:29, 31, 35, 36, 43–45; Bertelsen and Krefft, 1988:16–21, figs. 2–6).

Description

Metamorphosed females are characterized by having a short stout body, best described as globose, the depth (in the best-preserved specimens) 60 to 70% SL. The head is about 60% SL. The mouth is large, the opening oblique to almost vertical, the cleft extending below or slightly behind the eye. The snout and chin are unusually blunt, padded with thick skin, and covered with wartlike papillae, forming thick fleshy lips. The lower jaw is unusually thick and heavy, protruding slightly in front of the upper jaw, with a well-developed symphysial spine. The oral valve is weakly developed. The nostrils are set on low rounded papillae. The jaw teeth are relatively short, arranged in several oblique, longitudinal series, with 10 to more than 20 such series in the lower jaw. There are about 35 to 150 teeth on each side of the upper jaw, about 40 to 210 on each side of the lower jaw. The longest teeth in the lower jaw are 3 to 8% SL, those of the upper jaw somewhat shorter. Vomerine teeth are absent. Epibranchial and ceratobranchial teeth are present: 1 or more tooth plates on the proximal tips of all four epibranchials and 8 to 20 tooth plates along the full length of all four ceratobranchials (see Bertelsen and Krefft, 1988:16, fig. 4C). The first epibranchial is free from the wall of the pharynx, but closely bound to the second epibranchial. The fourth epibranchial and ceratobranchial are bound to the wall of the pharynx, leaving no opening behind the fourth arch. The proximal one-half to two-thirds of the first ceratobranchial is bound to the wall of the pharynx, but the distal one-half to one-third is free. The distal end of the first ceratobranchial is also free, not bound by connective tissue to the adjacent second ceratobranchial. The proximal one-quarter to one-half of ceratobranchials II to IV are free, not bound together by connective tissue. Gill filaments are present on epibranchials II to IV, the proximal one-half to two-thirds of ceratobranchial I, and the full length of ceratobranchials II to IV. A pseudobranch is absent. The illicium is unusually stout and thick, its length 26 to 86% SL. The pterygiophore of the illicium is short, its anterior end nearly completely covered by skin of the head and not reaching beyond the anteriormost margin of the snout when protruded. The escal bulb is short, oval in shape, with well-developed distal light-guiding appendages. The neuromasts of the acoustico-lateralis system are set on low rounded papillae, the pattern of placement as described for other ceratioids (Pietsch, 1969, 1972b, 1974a, 1974b).

Recently metamorphosed females, measuring 30 to about 34 mm, differ from larger specimens in having a shorter illicium and shorter escal appendages, in lacking dermal spines and papillae on snout and chin, and in having fewer jaw teeth.

Metamorphosed males have relatively well-developed eyes, their diameter 5.5 to 8.7% SL. The olfactory organs are large, the nostrils directed laterally. The posterior nostril is slightly larger than the anterior, its greatest diameter 3.3 to 7.6% SL, and well separated from the eye and from the anterior nostril by pigmented skin. The nasal area is pigmented, but not inflated. There are 10 to 17 olfactory lamellae. The maxillae and premaxillae are not reduced, but jaw teeth are absent. The upper denticular bone bears 16 to 31 recurved teeth arranged in a semicircular, fan-shaped series, with some shorter denticles within. The upper denticular is fused dorsally with a number of large dermal spinules of the snout forming a conspicuous median ridge between the large olfactory organs, and attached to anterior end of the illicial pterygiophore. The lower denticular bone bears 20 to 50 recurved teeth, more or less distinctly separated into a median and a pair of lateral groups (Bertelsen and Krefft, 1988:26, 75, figs. 10, 35, 36).

The skin of the head and body of the larvae is greatly inflated, giving a nearly spherical appearance. A conspicuous, pigmented dorsal hump is sometimes present just in front of the dorsal fin (Bertelsen and Krefft, 1988:74, fig. 34). The subdermal pigmentation consists of a dorsal group of melanophores well separated from the peritoneal pigmentation and a group of melanophores on the caudal peduncle.

The color in preservation is black to dark brown over the entire surface of the head and body. The females of some species have whitish or lightly pigmented areas on the chin, snout, and upper surface of the head or body. One or more shiny white patches are sometimes present on the ventral and/or dorsal midline of the body, situated at or on the bases of one or more of the median fins. The pigmentation of the illicium and fin rays increases with the size of specimens and is highly variable among species.

The largest known female is the holotype of *Himantolophus groenlandicus* from southwest Greenland (not preserved, except for the illicium; ZMUC 65), which measured about 465 mm. There are several additional individuals in the range of 250 to 400 mm (Bertelsen and Krefft, 1988:14, 37). Metamorphosed males measure between 23 and 39 mm, and the largest larvae range from 17 to 22.5 mm.

Diversity

A single genus:

Genus *Himantolophus* Reinhardt, 1837 (Footballfishes)

Figures 9–12, 34, 40, 41, 45, 49, 68–74, 198, 202, 203, 211–214, 260, 292; Tables 1, 3, 8, 9

Females

*Himantolophus* Reinhardt, 1837:116, pl. 4 (type species *Himantolophus groenlandicus* Reinhardt, 1837, by monotypy).

*Aegoeonichthys* Clarke, 1878:245, pl. 6 (type species *Aegoeonichthys appelii* Clarke, 1878, by monotypy).

*Corynolophus* Gill, 1879b:228 (type species *Himantolophus reinhardtii* Lütken, 1878a, by monotypy).

*Aegaeonichthys* Gill, 1879b:228 (emendation; type species *Aegoeonichthys appelii* Clarke, 1878, by monotypy).

*Aegonichthys* Girard, 1893:604, 610 (emendation; type species *Aegoeonichthys appelii* Clarke, 1878, by monotypy). Golvan, 1962:172 (emendation; type species *Aegoeonichthys appelii* Clarke, 1878, by monotypy).

*Corynophorus* Osório, 1912:90, figs. 1, 2 (type species *Corynophorus compressus* Osório, 1912, by monotypy).

*Aegeonichthys* Regan, 1912:288 (emendation; type species *Aegoeonichthys appelii* Clarke, 1878, by monotypy).

Males

*Lipactis* Regan, 1925c:566 (type species *Lipactis tumidus* Regan, 1925c, by monotypy).

*Rhynchoceratias* Regan, 1925c:566 (in part; type species *Rhynchoceratias brevirostris* Regan, 1925c, by subsequent designation of Fowler, 1936).

Distinguishing Characters

With the diagnostic and descriptive features of the family.

Comments

Within the Ceratioidei, genera and subgenera have nearly always been designated for taxa that have, or are assumed to have, distinguishing osteological and/or other characters shared by its females, males, and larvae. In contrast, the subdivisions of the single recognized genus of the family Himantolophidae recognized here are based solely on females, thus the term "species group" is used below, following Bertelsen and Krefft (1988). The characters that separate *Himantolophus appelii* from the other species of the genus are of the same order as those on which the species groups are based, thus this species is recognized as a species group unto itself.

With the exception of four males that are assumed to represent *H. appelii* based on geographic distribution, it has not been possible to identify males to species based on females. For this reason, the males are treated separately in the following account, where they are separated into two groups each containing an unknown number of species. The names of the five nominal species based solely on males are retained for their primary types; if it should be found possible in the future to refer them to species based on females, any or all of these names may turn out to be senior synonyms.

A similar separate account is given for the larvae and those metamorphic specimens that have not yet developed specific characters. They are separated below into two groups, designated types A and B, following the descriptions provided by Bertelsen (1951) and Bertelsen and Krefft (1988).

Diversity

Five species groups are recognized based on females and three based on males, following the revision of Bertelsen and Krefft (1988) and differentiated as follows:

Keys to Species Groups of the Genus *Himantolophus*

FEMALES

The following key will differentiate juvenile and adult females greater than approximately 32 mm.

1A. Posterior escal appendage present .................... 2
1B. Posterior escal appendage absent; distal escal appendage long, 16 to 92% SL in specimens 30 to 75 mm, 32 to 214% in larger specimens .... ...... *Himantolophus cornifer* group Bertelsen and Krefft, 1988, p. 346

   NOTE  Five species; Atlantic, Pacific, and Indian oceans

2A. Distal escal appendage long (8 to 52% SL in specimens 30 to 75 mm, about 24 to 82% SL in larger specimens), nearly as long or longer than posterior escal appendage in specimens 30 to 75 mm, twice as long in larger specimens ............... ..... *Himantolophus albinares* group Bertelsen and Krefft, 1988, p. 341

   NOTE  Five species; Atlantic and western Pacific oceans

2B. Distal escal appendage short (0.4 to 5% SL in specimens 30 to 75 mm, about 1.0 to 20% SL in larger specimens), distinctly shorter than posterior escal appendage (less than one-half in specimens 30 to 75 mm, less than two-thirds in larger specimens) .................... 3

3A. Proximal half of distal escal appendage simple, undivided; base of escal bulb without appendages; a distal pair of appendages on illicial stem situated 1.5 to two times diameter of esca bulb below base of posterior escal appendage ................. .... *Himantolophus nigricornis* group Bertelsen and Krefft, 1988, p. 340

   NOTE  Two species; Atlantic and Pacific oceans

3B. Distal escal appendage bifurcated at base; base of escal bulb with a distal pair of illicial appendages, less than diameter of escal bulb below base of posterior escal appendage ...... 4

4A. Posterior escal appendage bifurcated at base, each primary branch with an anterior series of 2 to 7 side branches; distal swellings of escal bulb not distinctly divided into 4 lobes ....... ...... *Himantolophus appelii* group Bertelsen and Krefft, 1988, p. 339

   NOTE  A single species; southern parts of the Atlantic, Indian, and western Pacific oceans

4B. Posterior escal appendage with (1) proximal part undivided and distal part simple or bifurcated, or (2) divided near base to form 3 simple filaments, or (3) represented by 2 or 3

separate undivided filaments; base of distal escal appendage surrounded by 4 well-developed escal lobes . . . . . .
. . . *Himantolophus groenlandicus* group Bertelsen and Krefft, 1988, p. 334

NOTE   Five species; Atlantic, Pacific and western Indian oceans

MALES

1A. Olfactory lamellae 14 to 17 (greatest diameter of posterior nostril 3.5 to 6.4% SL; length of snout 11 to 19% SL) . . . . . . . . . . . . . . . . . . . . . . . .
. . . . . *Himantolophus rostratus* group Bertelsen and Krefft, 1988, p. 349

NOTE   An unknown number of species; tropical and subtropical Atlantic and Pacific oceans

1B. Olfactory lamellae 10 to 13 . . . . . 2
2A. Greatest diameter of posterior nostril 5.7 to 6.7% SL; length of snout 17 to 18% SL . . . . . . . . . . . . . . . . . . . . .
. . . . . . . *Himantolophus appelii* group Bertelsen and Krefft, 1988, p. 350

NOTE   Probably only a single species; off South Africa and New South Wales

2B. Greatest diameter of posterior nostril 3.3 to 5.5% SL; length of snout 13 to 17% SL . . . . . . . . . . . . . . . . . . . . .
. . . *Himantolophus brevirostris* group Bertelsen and Krefft, 1988, p. 350

NOTE   An unknown number of species; Atlantic, western Pacific, and Indian oceans

*Himantolophus groenlandicus* Group Bertelsen and Krefft, 1988

Females

*Himantolophus groenlandicus* Group Bertelsen and Krefft, 1988:36 (new species group). Meléndez and Kong, 1997:11, 13 (after Bertelsen and Krefft, 1988; Chile, in key).

Males and Larvae

Males are included within the *Himantolophus brevirostris* group; larvae belong to pigment pattern type A (see Bertelsen, 1951:61, fig. 23; Bertelsen and Krefft, 1988:74, 78, fig. 35B, 45).

Distinguishing Characters

Females of the *Himantolophus groenlandicus* group differ from those of other species of the genus in having the following combination of character states: The light-guiding distal escal appendage is divided at the base, its total length 0.4 to 2% SL, shorter than the diameter of the escal bulb in specimens less than 110 mm, 1 to 11% SL in larger specimens. Each primary branch of the distal escal appendage has a simple or bifurcated tip and 0 to 2 tiny papilliform side branches. The base of the distal appendage is surrounded by 4 escal lobes of about equal size. An anterior escal appendage is present in most species (but rudimentary or absent in *H. paucifilosus*), its length 1.5 to 42% SL. The posterior escal appendage is longer (about 1.5 to 34% SL) than the distal escal appendage; its proximal part is undivided, with a simple bifurcated tip, or it is trifurcated (but never bifurcated) at its base (in some specimens it is represented by a transverse series of 2 to 3 separate, simple filaments). Two to 23 posterolateral appendages are present on and below the base of the escal bulb, the distal-most pair measuring 5.9 to 49% SL, longer than the posterior escal appendage, and emerging just below the base of the bulb. Small dermal spinules are present on the stem of the illicium, the escal bulb, and the escal appendages of specimens greater than 33 mm. The dermal papillae of the snout and chin are low and rather indistinct. The skin is devoid of "white patches." The caudal-fin rays are white or only faintly pigmented, with dark tips in specimens less than 70 to 100 mm, but covered with dark pigment in larger specimens.

Key to Females of Species of the *Himantolophus groenlandicus* Group

The following key will differentiate females from a size of about 32 mm, except for *Himantolophus groenlandicus* and *H. sagamius* in which the distinguishing character of the length of the distal escal appendage is not developed in specimens less than about 100 mm.

1A. Anterior escal appendage present . . . . . . . . . . . . . . . . . . . . . . . . 2
1B. Anterior escal appendage absent or rudimentary; 2 to 4 appendages on and below escal bulb . . . . . . . . . . .
. . . . . . . *Himantolophus paucifilosus* Bertelsen and Krefft, 1988, p. 338

NOTE   Nineteen known specimens, 32 to 163 mm; tropical Atlantic Ocean

2A. Posterior escal appendage trifurcated at base (or represented by a transverse series of 2 or 3 simple filaments); 8 to 32 illicial appendages
. . . . . . . . . . *Himantolophus crinitus* Bertelsen and Krefft, 1988, p. 338

NOTE   Eleven known specimens, 30 to 83 mm; eastern tropical Atlantic Ocean

2B. Proximal part of posterior escal appendage undivided, distal part simple or bifurcated; 2 to 11 illicial appendages . . . . . . . . . . . . . . . . 3
3A. Distal escal lobes pointed; length of anterior and posterior escal appendages about 30% SL in 39-mm specimen . . . . *Himantolophus danae* Regan and Trewavas, 1932, p. 337

NOTE   A single known specimen, 39 mm; South China Sea

3B. Escal lobes blunt; length of anterior and posterior escal appendages less than 15% SL in specimens less than 40 mm . . . . . . . . . . . . . . . . . . . . . 4
4A. Length of distal escal appendage 1 to 2% SL in specimens less than 200 mm, 1.0 to 4.2% in larger specimens . . . .
. . . . . . . *Himantolophus groenlandicus* Reinhardt, 1837, p. 334

NOTE   One hundred and forty-three known specimens, 32 to 465 mm; Atlantic and perhaps eastern Pacific and western Indian oceans

4B. Length of distal escal appendage 6.2 to 11% (3.1%?) SL in specimens larger than 100 mm . . . . . . . . . . . .
. . . . . . . . . . *Himantolophus sagamius* (Tanaka, 1918a), p. 336

NOTE   Twenty-five known specimens, 32 to 380 mm; Pacific Ocean

HIMANTOLOPHUS GROENLANDICUS REINHARDT, 1837

Figures 9–11, 34, 71, 72, 198, 211; Tables 3, 8, 9

Females

*Himantolophus groenlandicus* Reinhardt, 1837:116, pl. 4 (original description, single specimen, only illicium preserved; earliest description of a ceratioid fish, comparison with *Lophius* and *Antennarius*). Reinhardt, 1857:24 (listed for Greenland). Rink, 1877:434 (Greenland). Lütken, 1878a:320, pl. 2 (comparison with *Himantolophus reinhardti* Lütken, 1878a). Jordan, 1885:927 (listed). Günther, 1887:51 (additional specimen, off Greenland). Girard, 1893:605, pl. 1, pl. 2, figs. 4, 4A (additional specimen, illicium lost, description, *H. reinhardti* a synonym of *H. groenlandicus*). Lütken, 1894:80 (after Girard, 1893). Regan, 1926:40, fig. 23 (in part, additional specimen, comparison with all known material; *Corynolophus sagamius* Tanaka, 1918a, and *Corynolophus globosus* Tanaka, 1918b, erroneously included as synonyms of *H. groenlandicus*). Regan and Trewavas, 1932:31, 59, figs. 31–33, 88, pl. 1 (in part, additional specimen, osteology). Nobre, 1935:233, pl. 32, fig. 105; pl. 32a, fig. 105a, b (two additional specimens). Ehrenbaum, 1936:163 (Iceland and Greenland). Bertelsen, 1951:60, figs. 22–25 [in part, description of males and larvae, comparison with all known material; *H. danae* Regan and Trewavas, 1932, *H. globosus* (Tanaka, 1918b), *H. kainarae* Barbour, 1942b, and *H. ranoides* Barbour, 1942b, included as synonyms of *H. groenlandicus*]. Einarsson, 1952:90 (biology; after Bertelsen, 1951). Albuquerque, 1954–1956: 1059, figs. 433 (Portugal; figure after Lütken, 1878a, 1878b). Matsubara, 1955: 1349, 1351, 1358, 1359, figs. 526, 534 (misidentifications; Japan; figures after Bertelsen, 1951). Brandes et al., 1956:30

(additional specimen). Briggs, 1960:179 (worldwide distribution). Lozano y Rey, 1960:465, fig. 147 (additional specimen). Maul, 1961:104, figs. 5–9 (three additional specimens). Krefft, 1961:90 (additional specimen, Iceland). Krefft, 1963:83 (additional specimen, Iceland). Leim and Scott, 1966:426 (additional specimen). Chen et al., 1967:22, fig. 11 (misidentification; specimen now lost, but probably *H. danae*; Taiwan). Krefft, 1967:186 (additional specimen, Iceland). Jónsson, 1970:278 (additional specimen, Iceland). Marshall, 1971a:46, 50, 95, fig. 16 (eyes and vision, olfaction, temperature and depth distribution). Pietsch, 1972a:29, 31, 35, 36, 45 (osteological comments). Jónsson, 1973:220 (additional specimen, Iceland). Maul, 1973:668 (synonymy). Nielsen, 1974:87 (listed type catalog). Blacker, 1977:185 (additional specimen). Jónsson et al., 1978:182 (additional specimen, Iceland). Jónsson et al., 1979:229 (additional specimen, Iceland). Marshall, 1979:132, 405, 451, figs. 49, 142 (figure after Bertelsen, 1951; brain and olfactory structures, eggs and larvae). Amaoka, 1984:105, pl. 92B (misidentification, Japan). Jónsson et al., 1984:205 (additional specimen, Iceland). Jónsson et al., 1985:185 (two additional specimens, Iceland). Bertelsen, 1986:1378, 1380, figs. (eastern North Atlantic; in key). Jónsson et al., 1986a:186 (three additional specimens, Iceland). Jónsson et al., 1986b:160 (two additional specimens, Iceland). Jónsson et al., 1986c:628 (two additional specimens, Iceland). Lloris, 1986: 246, fig. 128 (additional specimen, off Namibia). Jónsson et al., 1987:360, fig. (two additional specimens, Iceland). Bertelsen and Krefft, 1988:37, figs. 1–5, 11, 41 [24 additional specimens, family revision based on all known material; *H. sagamius* (Tanaka, 1918a) and *H. danae* Regan and Trewavas, 1932, resurrected from synonymy of *H. groenlandicus*]. Lee, 1988:22, fig. 7 (misidentification, after Chen et al., 1967). Pequeño, 1989:43 (Chile; specimen not seen, probably *H. appelii*). Bertelsen, 1990:494 (eastern tropical Atlantic). Nielsen and Bertelsen, 1992:62, fig. 2 (North Atlantic). Lee, 1993:184, fig. (misidentification; description and figure after Chen et al., 1967; Taiwan). Andriyashev and Chernova, 1994:102 (listed for Arctic seas and adjacent waters). Amaoka et al., 1995:112, fig. 172 (misidentification, northern Japan). Meng et al., 1995:442, fig. 594 (misidentification, China). Smale et al., 1995:70, pl. 36C (otolith described, figured). Swinney, 1995a:52, 53 (additional specimen, off Madeira). Meléndez and Kong, 1997:12, 13 (Chile, in key). Quigley and Flannery, 1997:442 (Irish Waters). Shedlock et al., 1997:396, fig. 1 (DNA extracted from formalin-fixed specimens). McEachran and Fechhelm, 1998:854, 856, fig. (Gulf of Mexico, in key). Jónsson and Pálsson, 1999:199, fig. 3 (Iceland). Munk, 1999:267, 278, 281, 282 (biolumines-cence). Anderson and Leslie, 2001:8, fig. 6 (SAM 23120, erroneously reported as SAM 23129 by Bertelsen and Krefft, 1988). Nakabo, 2002:476, figs. (misidentification, Japan, in key). Pietsch, 2002b:1061, fig. (western central Atlantic). Moore et al., 2003:215 (off New England). Pietsch, 2005b:219 (reproduction). Pietsch and Orr, 2007:15, fig. 12B (gill arches, phylogenetic relationships).

*Himantolophus reinhardti* Lütken, 1878a: 320, pls. 1, 2 (original description, single specimen, dead at surface). Jordan, 1885: 927 (listed). Günther, 1887:51 (additional specimen, off Greenland). Lütken, 1887:325, fig. 1, pl. 1 (additional specimen, osteology). Lütken, 1894:79 (after Lütken, 1878a, 1878b). Saemundsson, 1908:27 (two additional specimens, Iceland). Holt and Byrne, 1909:195, pls. 1, 2 (additional specimen). Osório, 1909:18, pl. 1, figs. 5, 6 (additional specimen). Williamson, 1911:51, pl. 3 (additional specimen). Barnard, 1927:1006, pl. 37, fig. 2 (additional specimen). Ehrenbaum, 1936:163 (seas of northern Europe). Fowler, 1936:1142, fig. 481 (after Lütken, 1878a, 1878b). Saemundsson, 1949:46 (seven additional specimens, Iceland). Amanieu and Cazaux, 1962:76, fig. 2 (additional specimen). Nielsen, 1974:87 (listed in type catalog).

*Corynolophus reinhardti*: Gill, 1879b: 228 (new genus to contain *Himantolophus reinhardti* Lütken, 1878a). Goode and Bean, 1896:494, pl. 120, fig. 405 (after Lütken, 1878a, and Gill, 1879b). Jordan et al., 1913:427 (in part; Japanese localities pertain to *H. sagamius*).

*Himantolophus ranoides* Barbour, 1942b: 83, pls. 13, 14 (original description, two specimens).

Males

Males of this species are included within the *Himantolophus brevirostris* group (Bertelsen and Krefft, 1988), see below.

Material

One hundred and forty-three metamorphosed females (32 to 465 mm), including two tentatively identified specimens (51.5 to 70 mm):

Holotype of *Himantolophus groenlandicus*: ZMUC 65, c. 465 mm, only illicium preserved, found washed ashore, southwest Greenland, near Godthaab, 1833.

Holotype of *Himantolophus reinhardti*: ZMUC 66, 328 mm, found dead at surface, Sukkertoppen, West Greenland.

Holotype of *Himantolophus ranoides*: MCZ 33928, 375 mm, south of Sable Island, Nova Scotia, beam trawl, "100 fathoms" (183 m), Frank E. Firth, March 1935.

Paratype of *Himantolophus ranoides*: MCZ 35773, 320 mm, Atlantic Slope, "probably off George's Bank," Frank E. Firth, 1942.

Additional females: BMNH, six (49 to 350 mm); IIPB, one (120 mm); ISH, 14 (39 to 265 mm); MCZ, one (37 mm); MMF, two (61 to 365 mm); MNHN, two (280 to 320 mm); MNHNC, one (340 mm); MRIR, 93 (75 to 430 mm); NMSZ, one (69 mm); SAM, one (207 mm); UF, three (32 to 36 mm); USNM, three (32 to 75 mm); ZMUC, nine (146 to 240 mm).

Tentatively identified females: SIOM, one (70 mm); MNHN, one (51.5 mm).

Distinguishing Characters

Metamorphosed females of *Himantolophus groenlandicus* differ from those of all other species of the *H. groenlandicus* group in having the following combination of character states: length of anterior escal appendage 2.7 to 16% SL in specimens less than 100 mm, 16 to 42% SL in larger specimens; distal escal appendages 0.9 to 2.1% SL in specimens less than 200 mm, 1.0 to 4.2% SL in larger specimens; distal escal lobes blunt; posterior escal appendage undivided proximally, simple or divided distally into 2 to 6 branches, its total length 4.6 to 20% SL in specimens less than 100 mm, 15 to 34% SL in larger specimens; 2 to 13 (rarely fewer than 4) posterolateral appendages situated on and closely below base of escal bulb, all or nearly all arranged in pairs, simple or divided into 2 to 9 branches, distal and longest pair 14 to 41% SL in specimens less than 100 mm, about 23 to 51% SL in larger specimens.

Description

Illicium and escal appendages relatively shortest in juveniles; length of illicium averaging about 49% SL in 34.5- to 106-mm specimens, followed by a slight decrease in relative length in larger specimens; average relative length of distal escal appendage remaining constant (about 1.5% SL) in specimens less than 200 mm, increasing to about 4% SL in the largest known specimens; average lengths of anterior and posterior escal appendages and longest illicial appendage showing gradual positive allometric growth; diameter of esca decreasing from a mean of about 7.5% SL in smaller specimens to about 5% in larger specimens.

Anterior escal appendage simple in most specimens, rarely divided into more than 2 branches; posterior esca appendage simple or bifid in about half known specimens, but divided into 8 or 9 branches in some largest known specimens; illicial appendages more or less regularly arranged in pairs, tending to increase in number with growth, nearly all simple, except in distal-most and longest pair, which may bear several branches; except for extreme tips, which are bright silvery in fresh specimens, all escal appendages darkly pigmented; stem of illicium, escal bulb, and all appendages covered with small dermal spinules; papillae of snout and chin low and indistinct; 1 to 5 large spines on each lobe of pectoral fins, 4 to about 60 similar spines on each side of body, numbers increasing with growth; smallest known specimens, up to 32 mm, naked; head and body dark gray to blackish brown, fin rays more or less grayish brown, black on extreme tips; dorsal-fin rays 5; anal-fin rays 4; pectoral-fin rays 14 to 18.

Additional description as given for the species group, genus, and family.

### Distribution

*Himantolophus groenlandicus* is widely distributed in the Atlantic Ocean, with more than 90 records in the far north from West Greenland and Iceland to off Norway at about 70°N, 17°E (Fig. 211). It extends as well to both sides of temperate and tropical Atlantic, including the Gulf of Mexico, and farther south to Cape Town, South Africa. Meléndez and Kong (1997; see also Pequeño, 1989) reported a single female (340 mm, MNHNC P.6847) from the eastern South Pacific off Chile. Juvenile and larval specimens from the western Indian Ocean, which key out to the *H. groenlandicus* group, may also represent this species (see below).

A majority of the metamorphosed specimens caught in open pelagic nets in tropical or subtropical waters came from maximum depths between 200 and 800 m, while most of the large specimens recorded from northern waters were found stranded or were caught in benthic trawls in less than 200 m.

### Comments

*Himantolophus groenlandicus* is the best-known species of the genus, represented by more than a third of all the recorded specimens of the family. It is most similar to *H. sagamius*, distinguished from this species only at sizes of more than 100 mm. Because *H. sagamius* is so far known only from the Pacific, smaller Atlantic specimens with the shared characters of these two species are assumed to be *H. groenlandicus* (Bertelsen and Krefft, 1988).

The identity of two specimens from the Indian Ocean (SIOM uncataloged and MNHN 1986-334) remains somewhat doubtful. Considering that they came from the western part of this ocean (from off the Seychelles and Mozambique Channel) and that *H. groenlandicus* is known from off the tip of South Africa, they are tentatively referred to this species following the conclusions of Bertelsen and Krefft (1988:39).

The 43-mm female collected off Taiwan recorded as *H. groenlandicus* by Chen et al. (1967) has been lost (Bertelsen and Krefft, 1988:39). From the description and illustration it can be referred to the *H. groenlandicus* group, but not to any of its species as recognized and described here.

### HIMANTOLOPHUS SAGAMIUS (TANAKA, 1918A)

Figure 211; Tables 3, 8, 9

#### Females

*Corynolophus reinhardti*: Tanaka, 1908:22, pl. 1, fig. 5 (new combination; three specimens from off Japan, one described and illustrated). Jordan et al., 1913:427 (in part; Greenland locality pertains to *Himantolophus groenlandicus*).

*Corynolophus sagamius* Tanaka, 1918a: 491, pl. 134, fig. 377 (original description, single specimen). Matsubara, 1955:1359 (in synonymy of *Himantolophus groenlandicus*).

*Corynolophus globosus* Tanaka, 1918b: 529, pl. 139, fig. 388 (original description, single specimen). Matsubara, 1955:1359 (in synonymy of *Himantolophus groenlandicus*).

*Himantolophus groenlandicus*: Regan, 1926:40 (in part; *Corynolophus sagamius* Tanaka, 1918a, and *C. globosus* Tanaka, 1918b, referred to *Himantolophus groenlandicus*). Bertelsen, 1951:60 (in part; after Regan, 1926). Matsubara, 1955:1349, 1351, 1358, 1359, figs. 526, 534 (misidentification; *Corynolophus sagamius* Tanaka, 1918a, and *C. globosus* Tanaka, 1918b, erroneously included in synonymy of *H. groenlandicus*; figures after Bertelsen, 1951). Haneda, 1968:1, figs. 1–5 (misidentification, bioluminescence). Amaoka, 1984:105, pl. 92B (misidentification, Japan). Amaoka et al., 1995:112, fig. 172 (misidentification, northern Japan). Nakabo, 2002:476, figs. (misidentification, Japan, in key). Senou et al., 2006:424 (misidentification, Sagami Sea).

*Himantolophus kainarae*: Barbour, 1942b: 82 (based on a misspelling of "kaibarae," in mistaken reference to original description of *Himantolophus sagamius* Tanaka, 1918a; see legend to Tanaka's plate 134).

*Himantolophus globosus*: Barbour, 1942b:82, pl. 12 (new combination; one of three specimens referred to *Corynolophus reinhardti* by Tanaka, 1908).

*Himantolophus sagamius*: Bertelsen and Krefft, 1988:37, fig. 12 (new combination; resurrected from synonymy of *Himantolophus groenlandicus*, comparison with all known material). Landesman, 1990:6, figures (south of San Clemente Island, southern California). Meléndez and Kong, 1997:12, 14 (Chile, in key). Munk, 1999:267, 278, 282 (bioluminescence). Pietsch, 1999:2029 (western central Pacific). Klepadlo et al., 2003:99, figs. 1–6 (additional specimen, southern California; notes on visceral anatomy). Love et al., 2005:61 (eastern North Pacific). Kharin, 2006a:281, fig. (Kuril-Kamchatka trough).

*Himantolophus* sp.: Lea, 1988:180 (two additional specimens; eastern North Pacific, off California).

#### Material

Twenty-five metamorphosed females (32 to 380 mm), including two lost specimens (128 to 295 mm) and three tentatively identified juveniles (32 to 40 mm):

Holotype of *Corynolophus sagamius*: originally SCMT 8201 (lost or mislaid at ZUMT; Kazuo Sakamoto, personal communication, 3 July 2006), 200 mm, Sagami Sea, about 35°N, 139°E.

Holotype of *Corynolophus globosus*: originally SCMT 8460 (lost or mislaid at ZUMT; Kazuo Sakamoto, personal communication, 3 July 2006), 274 mm, Misaki, Sagami Bay, Kumakichi Aoki.

Additional material: CAS, one (180 mm); CMA, one (350 mm); HUMZ, six (134 to 370 mm); IMBV, one (186 mm); ISH, two (160 to 240 mm); LACM, one (111 mm); MCZ, one (231 mm); MNHNC, one (111 mm); NSMT, two (69 to 73 mm); SIO, two (36 to 380 mm).

Specimens now lost: 356 mm TL, about 295 mm SL (see Haneda, 1968); 128 mm SL, T. Abe to E. Bertelsen, personal communication (see Bertelsen and Krefft, 1988:40).

Tentatively identified juveniles: SIOM, one (40 mm); LACM, two (32 to 37 mm).

#### Distinguishing Characters

Adult females (111 to 380 mm) of *Himantolophus sagamius* differ from those of all other species of the *H. groenlandicus* group in having the following combination of character states: length of anterior escal appendage 11 to 38% SL; distal escal appendage 6.2 to 11% (3.1%?) SL; distal lobes of esca blunt; posterior escal appendage 12 to 41% SL, simple or with bifid tip; two to four pairs of posterolateral appendages situated on and below base of escal bulb, length of distal and longest pair 30 to 45% SL.

#### Description

Within the size range of 111 to 274 mm, no distinct changes evident in relative lengths of illicium and escal appendages or in diameter of escal bulb with increasing size of specimens (the low value for length of the distal escal appendage of ISH 18/55a, obtained from the stomach of a sperm whale, may be due to shrinkage, and the short posterior escal appendage of MCZ 29854 appears to have been broken). Each primary branch of distal escal appendage with a simple or bifurcated tip and a pair of short side branches; anterior escal appendage simple in most specimens, bifurcated distally in ISH 18/55b, and an additional short appendage above and/or below anterior escal appendage

## HIMANTOLOPHUS NIGRICORNIS BERTELSEN AND KREFFT, 1988

Figure 214; Tables 3, 8

### Females

*Himantolophus nigricornis* Bertelsen and Krefft, 1988:50, fig. 17 (original description, two specimens). Pietsch, 1999:2029 (western central Pacific). Love et al., 2005:61 (eastern North Pacific).

### Material

Three metamorphosed females (145 to 195 mm):

Holotype of *Himantolophus nigricornis*: LACM 42697-1, 145 mm, *Velero IV*, San Clemente Basin, about 33°N, 118°W, in "deep water," 5 September 1980.

Paratype of *Himantolophus nigricornis*: ZIN 47913, 195 mm, *Vityaz* station 6176, 21°03′N, 167°09.7′E, conical net, 0 to 2500 m.

Additional material: SIO, one (195 mm).

### Distinguishing Characters

Metamorphosed females of *Himantolophus nigricornis* differ from those of the only other species of the *H. nigricornis* group in having the following combination of character states: length of illicium 70 to 87% SL; each primary branch of posterior escal appendage simple, total length 4.0 to 4.5% SL; a single pair of posterolateral appendages on stem of illicium, length 1.0 to 2.1% SL.

### Description

Known material extremely similar in nearly all characters; distal and posterior escal appendages darkly pigmented except at tips; undivided proximal half of short distal escal appendage conical in shape, distal part bifurcated, branches simple in holotype, extreme distal tips secondarily bifurcated in paratype; posterior escal appendage bifurcated near base, each branch simple; single pair of filamentous illicial appendages situated two or three times the diameter of escal bulb below base of posterior escal appendages; small dermal spinules present on stem of illicium and base of escal bulb; papillae of snout and chin well developed, those situated medially unpigmented or nearly so; skin of body and fin rays uniformly dark; 3 to 7 large dermal spines present on lobe of pectoral fin, about 50 spines on each side of body; dorsal-fin rays 5; anal-fin rays 4; pectoral-fin rays 16 or 17.

Additional description as given for the species group, genus, and family.

### Distribution

The three known specimens of *Himantolophus nigricornis* were all collected in the Pacific Ocean, the holotype and SIO specimen from off the coast of California, and the paratype from the central North Pacific near Wake Island (Fig. 214).

## HIMANTOLOPHUS MELANOLOPHUS BERTELSEN AND KREFFT, 1988

Figures 72, 214; Tables 3, 8

### Females

*Himantolophus melanolophus* Bertelsen and Krefft, 1988:52, fig. 18 (original description, three females). McEachran and Fechhelm, 1998:854, 857, fig. (Gulf of Mexico, in key). Pietsch, 2002b:1061 (western central Atlantic).

*Himantolophus melanophus*: Jónsson and Pálsson, 1999:200, figs. 3, 9 (misspelling of specific name, Iceland).

### Material

Six metamorphosed females (35 to 150 mm):

Holotype of *Himantolophus melanolophus*: ISH 1212/74, 82 mm, *Anton Dohrn* station 32-II/74, eastern tropical Atlantic, 9°31′N, 29°35′W, midwater trawl, 0 to 350 m, 2115 h, 20 July 1974.

Paratypes of *Himantolophus melanolophus*: UF 40805, 35 mm, *Delaware II* station 53, 29°30′N, 80°05′W, BT, 411 m, 30 May 1984; USNM 235918, 94 mm, *Libra*, Gulf of Mexico, about 28°N, 87°W, BT, 360 to 550 m, April 1982.

Additional material: MRIR, one (150 mm); NMNS, one (40 mm); SAIAB, one (48 mm).

### Distinguishing Characters

Metamorphosed females of *Himantolophus melanolophus* differ from those of the only other species of the *H. nigricornis* group in having the following combination of character states: length of illicium 40 to 41% SL in specimens 82 to 94 mm (23% SL in 35-mm specimen); each primary branch of posterior escal appendage bifurcating once or twice to form 2 to 5 slender branches, its total length 11 to 13% SL; 5 to 7 appendages on stem and base of illicium, the longest 5.7 to 12% SL.

### Description

Diameter of esca 6.7 to 8.6% SL; distal escal appendages very similar in holotype and paratypes, with proximal undivided part nearly cylindrical, each pair of distal branches simple, all darkly pigmented except for distal tips; posterior escal appendages repeatedly bifurcated, twice in holotype, two or three times in paratypes; posterior escal appendages darkly pigmented in holotype and largest paratype, only pigmented below second bifurcation in smaller paratype; holotype with a distal pair of illicial appendages located one to nearly twice diameter of escal bulb below base of posterior escal appendage, a similar but smaller proximal pair near base of illicium, and a single unpaired appendage at base of illicium; paratypes with illicial appendages similar to those of holotype but with 1 or 2 additional unpaired appendages on illicial stem, some bifurcated or trifurcated; small dermal spinules present on illicial stem and base of escal bulb; papillae of snout and chin well developed, those situated medially only lightly pigmented; indistinct lightly pigmented dorsal and ventral spots present on caudal peduncle; caudal-fin rays black in two largest specimens, pigmented distally and spotted proximally in 35-mm paratype; 5 or 6 large dermal spines present on base of each pectoral fin, about 30 spines on each side of body of holotype and largest paratype, 5 spines on each side of body of smaller paratype; dorsal-fin rays 5; anal-fin rays 4; pectoral-fin rays 15 to 17.

### Distribution

The holotype of *Himantolophus melanolophus* was captured in the eastern equatorial Atlantic just southwest of the Cape Verde Islands (Fig. 214). The two paratypes were collected off the east and west coasts of Florida (Bertelsen and Krefft, 1988:53). Jónsson and Pálsson (1999:200) reported a fourth specimen from off Iceland; a fifth, was recently collected from off Taiwan (Hsuan-Ching Ho, personal communication, 26 July 2002); and a still more recent capture was made off southern Mozambique in the western Indian Ocean (Eric Anderson, personal communication, 20 November 2007). The holotype was collected with an open midwater trawl between the surface and 350 m, while the remaining specimens were all taken by bottom trawls at depths between 360 and 880 m.

### Comments

*Himantolophus melanolophus* differs from *H. nigricornis* primarily in the length of the illicium, the pattern of branching of the posterior escal appendages and illicial appendages, and in the number of illicial filaments. Four of the five known specimens of *H. melanolophus* are considerably smaller than the three known individuals of *H. nigricornis*, but on the basis of ontogenetic changes observed in other members of the genus none of the differences mentioned above can be explained by differences in size.

## *Himantolophus albinares* Group Bertelsen and Krefft, 1988

### Females

*Himantolophus albinares* Group Bertelsen and Krefft, 1988:53 (new species group).

### Males and Larvae

Males belong to the *Himantolophus rostratus* group; larvae belong to pigment pattern Type B (see Bertelsen, 1951:61, fig. 23; Bertelsen and Krefft, 1988:74, 77, figs. 34, 35D, 36, 45).

### Distinguishing Characters

Metamorphosed females of the *Himantolophus albinares* group differ from those of the other species of the genus in having the following combination of character states: The distal light-guiding escal appendage is bifurcated at less than 1% SL to 32% SL above its base; each main branch is simple or with 1 to 3 bifurcations, its total length greater than the diameter of the escal bulb in specimens more than 30 mm, 8 to 52% SL in specimens 30 to 70 mm, 24 to 82% SL in larger specimens. A more or less distinct pair of posterolateral, distal swellings is present on the escal bulb. An anterior escal appendage is absent. The posterior escal appendage is divided near its base into 2 primary branches, each simple, or bifurcating distally once or twice. The total length of the posterior escal appendage is 3.5 to 23% SL, about as long as the distal escal appendage in smaller specimens, much shorter than the distal escal appendage in specimens greater than 70 mm. There are 0 to 9 posterolateral appendages on the stem of the illicium; the distal-most pair, when present, is located near the base of the escal bulb, just below the base of the posterior escal appendage, the longest 0 to 23% SL. Most species of the group have small dermal spinules on the stem of the illicium; in some species, the spinules extend onto the surface of the escal bulb and distal escal appendage. The papillae of the snout and chin are well developed (except in specimens less than about 40 mm). "White patches" of skin are present or absent. The caudal-fin rays are unpigmented distally, irregularly spotted proximally in juvenile specimens (less than about 50 mm), but darkly pigmented in larger specimens.

### Diversity

The *Himantolophus albinares* group contains five species:

### Key to Females of the *Himantolophus albinares* Group

1A. Each primary branch of distal escal appendage simple ............ 2
1B. Each primary branch of distal escal appendage with 1 to 3 bifurcations ........................ 3
2A. Length of illicium 56 to 87% SL; length of distal escal appendage 35 to 40% SL in specimens 30 to 38 mm, 78 to 82% SL in specimens 86 to 155 mm .... *Himantolophus mauli* Bertelsen and Krefft, 1988, p. 343

NOTE Twenty known specimens, 30.5 to 215 mm; North Atlantic Ocean

2B. Length of illicium 36 to 54% (61%?) SL; length of distal escal appendage 8 to 13% SL in specimens 34 to 66 mm, 25 to 74% SL in specimens 75 to 190 mm .................. ......... *Himantolophus albinares* Maul, 1961, p. 342

NOTE Twenty-five known specimens, 28 to 190 mm; Atlantic and western South Pacific oceans

3A. Each primary branch of distal escal appendage with a single bifurcation .......................... 4
3B. Each primary branch of distal escal appendage with 2 or 3 bifurcations ...... *Himantolophus multifurcatus* Bertelsen and Krefft, 1988, p. 344

NOTE Four known specimens, 42 to 122 mm; North Atlantic Ocean

4A. Length of undivided part of distal escal appendage 7% SL; length of illicium 52% SL ................. ...... *Himantolophus pseudalbinares* Bertelsen and Krefft, 1988, p. 344

NOTE Four known specimen, 80 to 153 mm; western Pacific and western Indian oceans

4B. Length of undivided part of distal escal appendage 17 to 18% SL; length of illicium 26 to 32% SL ......... .......... *Himantolophus borealis* Kharin, 1984, p. 345

NOTE Three known specimens, 62 to 160 mm; off Japan

## HIMANTOLOPHUS ALBINARES MAUL, 1961

Figures 34, 213; Tables 3, 8, 9

### Females

*Himantolophus albinares* Maul, 1961: 111, figs. 11–15 (original description, single specimen). Maul, 1962a:8, figs. 1, 2 (additional specimen). Maul, 1973:668 (synonymy; after Maul, 1961). Trunov, 1981:51, fig. 5 (additional specimen). Bertelsen, 1986:1378, figs. (eastern North Atlantic; in key). Bertelsen and Krefft, 1988:54, figs. 19, 20 (additional specimens, description, comparison with all known material). Bertelsen, 1990:495 (eastern tropical Atlantic; synonymy). Jónsson and Pálsson, 1999:199, fig. 3 (Iceland). Munk, 1999:267, 268, 271, 274, 275, 279, 281, 282, figs. 3C, 4A, B, 8D, 9 (bioluminescence). Anderson and Leslie, 2001:5, fig. 2 (after Trunov, 1981). Pietsch, 2002b:1061 (western central Atlantic). Moore et al., 2003:214 (off New England). Iglésias, 2005:191, figs. 1–3 (first record from the Pacific).

### Material

Twenty-five metamorphosed females (28 to 190 mm):

Holotype of *Himantolophus albinares*: MMF 2598, 190 mm, off Câmara de Lobos, Madeira, on long line, probably 1933 or 1934.

Additional material: MMF, one (34 mm); ISH, eight (28 to 104 mm); MCZ, six (45 to 115 mm); MNHN one, (153 mm); MRIR four, (155 to 190 mm); SIOM, one (141 mm); UF, one (29 mm); USNM, two (66 to 72 mm; 72-mm specimen cleared and stained).

### Distinguishing Characters

Females of *Himantolophus albinares* differ from those of other species of the *H. albinares* group in having the following combination of character states: length of illicium 36 to 54% (61%?) SL in specimens greater than 30 mm (34 to 190 mm); distal escal appendage deeply cleft, undivided part 1.0 to 9.4% SL; each primary branch of distal escal appendage simple, darkly pigmented except on distal tip, its total length 8 to 13% SL in specimens 34 to 66 mm, 25 to 74% SL in specimens 75 to 190 mm; length of posterior escal appendage 8.5 to 21% SL in specimens 30 to 190 mm, pigmentation faint or absent; 2 to 9 posterolateral appendages on illicial stem, simple or branched at tip, longest 7 to 23% SL in specimens greater than 30 mm; escal bulb and appendages without dermal spinules.

### Description

Smallest known specimen (28 mm) with least number of illicial appendages, and shortest illicial appendages and posterior escal appendage; two smallest specimens (28 to 29 mm) with shortest illicium and illicial appendages; a general but irregular trend toward an increase in relative length of escal appendages with increasing SL among larger specimens (34 to 190 mm); distal escal appendage nearly or completely covered with pigment, except on extreme distal tip, in specimens more than 30 mm; distal escal appendage of now bleached holotype with "some brownish but not dense pigmentation" (Maul, 1961:115); posterior escal appendage and illicial appendages white or with scattered melanophores; posterior escal appendage bifurcated near base, each primary branch simple or divided into 2 to 4 filamentous branches near tip; distal pair of illicial appendages situated about diameter of escal bulb below posterior escal appendage; escal and illicial appendages naked; stem of illicium with small dermal spinules in specimens larger than 50 mm, base of escal bulb with dermal spinules in 190-mm holotype; skin naked in specimens less than 35 mm, larger specimens with 2 to 7 large spines on each pectoral lobe, 4 to 9 spines on each side of body in 34- and 39-mm specimens, increasing to about 15 to 40 in larger specimens; papillae of snout and chin well developed except in juvenile specimens less than 30 mm.

According to the description and illustration provided by Trunov (1981:51, fig. 5), the 141-mm SIOM specimen is extremely similar to the holotype except in the length of the illicium, which Trunov recorded as 61% SL. Measured on Trunov's illustration, Bertelsen and Krefft (1988:55) found the length to be about 54% SL, which might indicate a slight difference in their method of measuring.

The known specimens differ somewhat in pigmentation: four specimens, 39 to 80 mm, with nearly uniform brownish black skin, only slightly lighter on medial surface of papillated area of snout and chin; in remaining specimens these medial areas white or only very faintly pigmented, and more or less distinct white patches present on dorsal and ventral side of caudal peduncle; five of these specimens (including holotype and specimen described by Trunov, 1981) with similar white spots on or in front of anterior-most dorsal- and anal-fin rays, as well as on dorsal- and ventral-most caudal-fin rays; white olfactory papillae (for which this species was named; see Maul, 1961:115) present in holotype and only three other specimens (60, 66, and 104 mm); in four smallest specimens (28 to 39 mm), only dorsal- and ventral-most caudal-fin ray pigmented, pigment restricted to first and distal part of remaining rays of dorsal and anal fins; in larger specimens, all fin rays darkly pigmented; except in four specimens (39 to 80 mm), dorsal surface of body faintly pigmented or nearly white from base of illicium to origin of dorsal fin; in two smallest specimens, posterior part of this field raised into a dorsal hump with a dark central spot, similar to pigmented hump characteristic of "Type B" larvae (see Fig. 74); faint remains of this spot can be traced in some larger specimens; dorsal-fin rays 5; anal-fin rays 4; pectoral-fin rays 15 to 17.

### Sexual Maturity

Each ovary of the holotype is approximately 45 mm long, 29 mm in greatest width, and 8 mm thick, containing numerous eggs of about 0.17 mm in diameter (Maul, 1961:115).

### Distribution

Except for one specimen taken off New Caledonia in the western South Pacific (Iglésias, 2005), all the known material of *Himantolophus albinares* is from the Atlantic, scattered across both sides of that ocean, and extending as far north as Iceland (Jónsson and Pálsson, 1999:199, fig. 3) to approximately 24°S near Valdivia Bank in the south (Fig. 213). Two specimens were caught on long lines at unknown depths and another in a bottom trawl at 1423 m, but all the remaining individuals were taken in open trawls fished at maximum depths of 330 to 1950 m, most of them in trawls with a maximum depth of more than 500 m. This might indicate a preference for a somewhat greater depth than the species of the *H. groenlandicus* group.

## HIMANTOLOPHUS MAULI BERTELSEN AND KREFFT, 1988

Figures 69, 70, 212; Tables 3, 8

### Females

*Himantolophus compressus*: Maul, 1961: 96, figs. 2–4 (single specimen, MMF 18291, misidentification). Maul, 1973:668 (synonymy; after Maul, 1961).

*Himantolophus mauli* Bertelsen and Krefft, 1988:57, figs. 21, 22 (original description, seven specimens). Rodríguez-Marín et al., 1996:69, fig. 2 (additional specimen, Flemish Cap bank, east of Newfoundland). Jónsson and Pálsson, 1999: 200, fig. 3 (11 additional specimens, Iceland). Kukuev and Trunov, 2002:326 (additional specimen, off Greenland). Pietsch, 2002b:1061 (western central Atlantic). Moore et al., 2003:215 (off New England). Arronte and Pietsch, 2007:85, figs. 1, 2 (additional specimen, Cantabrian Sea, eastern North Atlantic).

### Material

Twenty metamorphosed females (30.5 to 215 mm):

Holotype of *Himantolophus mauli*: MMF 18291, 100 mm, off Câmara de Lobos, Madeira, August 1925.

Paratypes of *Himantolophus mauli*: ARC 860-1638, 111 mm, 47°42′N, 59°28′W, 430 m; ISH 682/79, 38 mm, *Anton Dohrn* station 170-I/79, 25°09′N, 58°07′W, MT-1600, 300 to 1200 m, 4 April 1979; ROM 26760, 155 mm, 42°21′N, 65°08′W, BT, 411 m; ROM 37721, 86 mm, *Shirane-Maru*, 42°44′N, 63°32′W, MT, 765 m; USNM 231672, 30.5 mm, *Pelican* station 43, 30°03′N, 80°03′W, ET, 410 to 440 m, 2 May 1956; ZMUC P922195, 98 mm, *Walther Herwig* station 390/ 82, 49°48.4′N, 26°32.8′W, MT-1600, 500 m, bottom depth 1800 m, 16 June 1982.

Additional material: IEOS, one (105 mm); MMA, one (192 mm); MRIR, 11 (125 to 215 mm).

### Distinguishing Characters

Metamorphosed females of *Himantolophus mauli* differ from those of other species of the *H. albinares* group in having the following combination of character states: length of illicium 56 to 87% SL; distal escal appendage bifurcated some distance from base (distance to bifurcation 7.1 to 22 % SL in specimens 38 to 155 mm); each primary branch simple, darkly pigmented in smallest specimens (30.5 to 38 mm), very faintly pigmented in larger specimens, its total length 35 to 40% SL in specimens 30 to 38 mm, 78 to 83% in specimens 86 to 155 mm; posterior escal appendage unpigmented, its length 3.5 to 13% SL; 0 to 5 posterolateral appendages on stem of illicium, longest appendage 0 to 6.3% SL; escal bulb and appendages without dermal spinules.

### Description

Distal portion of distal escal appendage of 98-mm specimen torn off just above bifurcation and most skin of distal half of stem of illicium lost, leaving short remains of 2 filamentous illicial appendages; only 1 branch of distal escal appendage present in 86-mm specimen; a scar 21% SL above its base appears to represent position of lost branch; illicial appendages absent in 100-mm holotype and two smallest paratypes (30.5 and 38 mm); posterior escal appendage bifurcating at or near base, each branch bifurcating again in 30.5-mm paratype, each branch bifurcating once or twice in 38-mm specimen, primary branches simple in five largest specimens; except for distal tips, 2 branches of distal escal appendage of two smallest specimens appear nearly black due to strong pigmentation of wall of internal light-guiding structures, those of larger specimens unpigmented except for light dusky external pigmentation; posterior escal appendage and illicial appendages, when present, unpigmented; all escal appendages naked; small dermal spinules present on stem of illicium in five specimens (86 to 115 mm) and on base of escal bulb in four larger specimens (98 to 115 mm); except for a single spine on each pectoral lobe in 30.5-mm specimen, skin of two smallest specimens naked; holotype with 12 to 15 spines on each side of body, but none on pectoral lobes; four larger paratypes (86 to 155 mm) with about 30 to 35 spines including 2 to 5 on each pectoral lobe; papillae of snout and chin well developed in all specimens; color of holotype originally "rather light with irregular, large patches of brown" (Maul, 1961:98), but now completely bleached; skin of head and body of paratypes uniformly dark brown to black, without white or faintly pigmented patches; rays of all fins black except proximal part of median rays of caudal fin unpigmented in two smallest specimens; dorsal-fin rays 5; anal-fin rays 4; pectoral-fin rays 15 to 17.

### Sexual Maturity

The ovaries of the 155-mm paratype are about 25 mm long, 12 mm wide, and 5 mm thick, with eggs of about 0.1 mm in diameter.

### Distribution

All known specimens of *Himantolophus mauli* were collected in the North Atlantic extending from off the west coast of Iceland at 66°N and south to 25°N, including records from both sides of that ocean (Bertelsen and Krefft, 1988:59, fig. 42; Arronte and Pietsch, 2007:86; Fig. 212). They were all caught in bottom trawls and open pelagic nets, five of them in maximum fishing depths of 400 to 750 m.

### Comments

The holotype of *Himantolophus mauli* (100 mm) was originally identified by Maul (1961) as *H. compressus* (Osório, 1912), a species then known only from the holotype of a slightly larger size (130 mm). In addition to similarities in body shape and coloration, Maul's assumption was based on the relatively long illicium of the two specimens, considerably greater than that of any other specimen of *Himantolophus* known at that time. The taxonomic significance of escal morphology, however, was not understood prior to the revisionary work of Bertelsen and Krefft (1988): it is now evident that the absence of a posterior escal appendage in *H. compressus* excludes the possibility of conspecificity with Maul's specimen.

*Himantolophus mauli* is very similar to *H. albinares*, but differs from the latter in the relative length of the illicium, the morphology of the distal escal appendage and longest illicial appendages, and pigmentation of the head and body. When examined separately, however, each of the characters on which *H. mauli* is based might appear to be possible extreme values of the intraspecific variation found in *H. albinares*, but the fact that they are shared by all the specimens of *H. mauli* indicates that the two species are valid.

## *HIMANTOLOPHUS PSEUDALBINARES* BERTELSEN AND KREFFT, 1988

Figure 213; Tables 3, 8

### Females

*Himantolophus pseudalbinares* Bertelsen and Krefft, 1988:59, fig. 23 (original description, single specimen). Stewart and Pietsch, 1998:8, fig. 4 (additional specimen, description, New Zealand). Anderson and Leslie, 2001:9, fig. 7 (after Bertelsen and Krefft, 1988).

### Material

Four metamorphosed females (80 to 148 mm), including one tentatively identified specimen (153 mm):

Holotype of *Himantolophus pseudalbinares*: ZIN 49711, 82 mm, *Fiolent*, cruise 3-63, off Cape St. Francis, southwest of Port Elizabeth, South Africa, 35°01'S, 24°36.8'E, 1280 to 1300 m.

Additional material: AIM, one (99 mm); NMNZ, two (80 to 148 mm).

### Distinguishing Characters

Metamorphosed females of *Himantolophus pseudalbinares* differ from those of other species of the *H. albinares* group in having the following combination of character states: length of illicium 52.0 to 66.7% SL; undivided proximal part of distal escal appendage 6.7 to 10.0% SL, tip of each primary branch bifurcated, darkly pigmented except on tip, 27% SL in total length; posterior escal appendage 13% SL, proximal part pigmented, distal part of each primary branch bifurcated; 5 posterolateral appendages on stem of illicium, longest (distal-most pair) 10% SL; escal bulb and appendages without dermal spinules.

### Description

Diameter of escal bulb 8.7 to 9.1 % SL; escal bulb with a posterior pair of distal swellings; holotype with distal-most pair of illicial appendages situated about diameter of escal bulb below posterior escal appendage; a second pair of illicial appendages, length less than 2% SL; a tiny fifth filament present at about middle of illicial stem; illicial appendages absent in second known specimen; illicium and esca of holotype without small dermal spinules; spinules present on illicium of second known specimen; each pectoral lobe, about 25 on each side of body; papillae of snout and chin well developed; pigmentation of papillated area slightly lighter than other parts of head and body; skin of body without white patches; dorsal-fin rays 5; anal-fin rays 4; pectoral-fin rays 16.

### Distribution

The holotype of *Himantolophus pseudalbinares* was caught in a bottom trawl on the continental slope off Cape St. Francis, southwest of Port Elizabeth, South Africa, at a depth of about 1300 m (Fig. 213). The three remaining specimens are all from New Zealand waters: the 99-mm specimen taken off Mahia Peninsula, Hawke Bay, on the North Island of New Zealand, was caught in an open pelagic trawl fished at a maximum depth of 1138 m; the 80-mm specimens from off eastern Chatham Rise, in a bottom trawl at 1040 m; and the 148-mm specimen from Lord Howe Rise in an open trawl fished between the surface and 955 m.

### Comments

*Himantolophus pseudalbinares* differs from *H. albinares* only in having the tips of the primary branches of the distal escal appendage bifurcated. The assumption by Bertelsen and Krefft (1988:60, 84, fig. 42), that this represents a specific difference, was based partly on geographic distribution: the only known specimen of *H. pseudalbinares*, from off South Africa, was far outside the North Atlantic distribution of *H. albinares*. The possibility, however, that these two species are one and the same was proposed recently by Iglésias (2005), on the basis of a well-preserved female specimen from off New Caledonia, having 1 primary branch of the distal escal appendage bifurcated, the other simple. Assuming that *H. pseudalbinares* was known only from the holotype, with information on morphological variation therefore absent, and given that the available material of *H. albinares* shows "extremely large variation in measurements and ornamentation of the illicial apparatus," the intermediate nature of the New Caledonian specimen prompted Iglésias (2005:193, figs. 1, 2) to synonymize the two species. Since that time, however, three additional specimens, all with bifurcated distal escal appendages, have been identified in the collections of the Auckland Institute and Museum, Auckland, and the National Museum of New Zealand, Wellington. In light of this additional material, reinforcing the difference in escal morphology between the two populations, it seems best to recognize both as valid species until such time that additional specimens indicate otherwise.

## *HIMANTOLOPHUS MULTIFURCATUS* BERTELSEN AND KREFFT, 1988

Figure 69; Tables 3, 8

### Females

*Himantolophus multifurcatus* Bertelsen and Krefft, 1988:60, figs. 24, 25 (original description, three specimens). Donnelly and Gartner, 1990:77 (additional specimen, eastern Gulf of Mexico). Pietsch, 2002b:1061 (western central Atlantic).

### Material

Four metamorphosed females (42 to 122 mm):

Holotype of *Himantolophus multifurcatus*: ISH 764/68, 48 mm, *Walther Herwig* station 14 II/68, 4°11'N, 24°37'W, MT, 180 to 200 m, 1 February 1968.

Paratypes of *Himantolophus multifurcatus*: USNM 229975, 43 mm, *Combat* station C-42, 29°58'N, 80°10'W, 366 m, 17 August 1956; USNM 229976, 122 mm, Pea Patch Expedition, about 30°N, 80°W, March 1962.

Additional specimen: UF, one (42 mm).

### Distinguishing Characters

Metamorphosed females of *Himantolophus multifurcatus* differ from those of other species of the *H. albinares* group in having the following combination of character states: length of illicium 50 to 69% SL; undivided proximal part of distal escal appendage 18 to 32% SL, each primary branch with 2 or 3 bifurcations, unpigmented, 34 to 52% SL in total length; posterior escal appendage unpigmented, 5.6 to 7.0% SL; 0 to 4 posterolateral appendages on stem of illicium, longest (distal-most) pair 0 to 4.2% SL; escal bulb and distal escal appendage of largest known specimen with small scattered dermal spinules.

### Description

Holotype with each primary branch of distal escal appendage distinctly trifurcated, each secondary branch cleft at distal tip; paratypes with each primary branch of distal escal appendage bifurcated; 43-mm paratype with extreme tip of each secondary branch of distal escal appendage bifurcated or trifurcated; 122-mm paratype with 3 or 4 secondary branches of distal escal appendage divided into tiny tertiary branches; posterior escal appendage bifurcated some distance above base; holotype with 3 unpaired illicial appendages, distal-most emerging from stem of illicium about 1.5 times diameter of escal bulb below base of posterior escal appendage; 43-mm paratype with two pairs of illicial appendages, distal-most pair situated about diameter of bulb below base of posterior escal appendage; 122-mm paratype with illicial appendages absent; illicial appendages unpigmented except for some scattered melanophores in skin of distal escal appendage; escal structures and stem of illicium without small dermal spinules in two smaller specimens (43 to 48 mm); dermal spinules present in 122-mm paratype, distal-most scattered on undivided part of distal escal appendage and on proximal part of one of its primary branches; 2 to 4 large spines present on each pectoral lobe, 7 or 8 on each side of body of two smaller specimens, about 30 on each side of body of larger paratype; papillae of snout and chin well developed; skin on papillated area as well as other parts of body darkly pigmented except for small white spots behind dorsal fin and in front and behind anal fin of holotype; dorsal-fin rays 5; anal-fin rays 4; pectoral-fin rays 16 or 17.

### Sexual Maturity

The ovaries of the 122-mm paratype measure 22 by 10 by 4 mm and contain eggs with diameters less than 0.1 mm.

### Distribution

The holotype was caught in the eastern equatorial Atlantic, the two paratypes from off Florida, and a fourth specimen from the eastern Gulf of Mexico. The maximum fishing depths for the three smaller specimens were between 200 and 365 m; depth of capture of the 122-mm paratype is unknown.

### Comments

*Himantolophus multifurcatus* has the characters of the *H. albinares* group, but differs distinctly from the other members of the group in the characters of the distal light-guiding escal appendage.

## HIMANTOLOPHUS BOREALIS KHARIN, 1984

Figure 214; Tables 3, 8

### Females

*Himantolophus borealis* Kharin, 1984:663, fig. 1A, B, (original description, single specimen). Bertelsen and Krefft, 1988:62, figs. 26, 42 (additional specimen, Japan). Nakabo, 2002:476, figs. (Japan, in key).

### Material

Three metamorphosed females (62 to 160 mm):

Holotype of *Himantolophus borealis*: ZIN 46021, 62 mm, *Mys Junony*, western North Pacific, off Honshu, Japan, 40°52.8'N, 142°20.8'E, BT, 1210 m.

Additional specimen: SIO, one (112 mm); color photograph of a third specimen, about 160 mm (200 mm total length), northeast of Honshu, Japan, 600 m, Tokiharu Abe to E. Bertelsen, no other data available, existence of specimen unknown.

### Distinguishing Characters

Metamorphosed females of *Himantolophus borealis* differ from those of other species of the *H. albinares* group in having the following combination of character states: illicium short (26 to 32% SL); distal escal appendage 22 to 24% SL, distal one-fourth to one-third of length bifurcated twice, darkly pigmented except for extreme distal tips; posterior escal appendage 4.8 to 7.1% SL, bifurcated distally; 2 posterolateral appendages at base of escal bulb; small dermal spinules present on escal bulb and distal appendage.

### Description

The following description is based solely on the 112-mm specimen (SIO 77-157): distal swellings of unpigmented distal part of escal bulb absent (present at base of distal escal appendage of most species of the genus); undivided part of distal escal appendage stout, nearly cylindrical, diameter about half diameter of escal bulb; each branch of bifurcated part of distal escal appendage with bifurcated distal tip, resulting in 4 short, close-set conical terminations of nearly equal size, each with a small white tip; distal one-third of posterior escal appendage bifurcated; each primary branch of posterior escal appendage and paired illicial appendage with 2 to 5 tiny distal branches; paired illicial appendage emerging from base of escal bulb, about two-thirds diameter of bulb below base of posterior escal appendage; 4 to 6 large spines present on base of pectoral lobe, about 30 on each side of body; papillae of snout and chin well developed, darkly pigmented; head, body, and fin rays brownish black; dorsal-fin rays 5; anal-fin rays 4; pectoral-fin rays 16.

The illicial characters of the 160-mm specimen (reported by Bertelsen and Krefft, 1988:63, on the basis of a photograph provided by the late Tokiharu Abe of Tokyo) are very similar to those of the specimen described above. The bifurcation of the black posterior escal appendage is distinct, but the photo is not sharp enough to show its secondary distal branches or the presence or absence of the illicial appendages.

The 62-mm holotype is similar to the specimens described above in the relative dimensions of the illicium and distal and posterior escal appendages, but differs in the following details: (1) the primary bifurcation of the distal escal appendage is deeper than the secondary bifurcation, resulting in a distinct separation of the distal part of the appendage forming 2 branches, each with a pair of tiny terminal branches; (2) the illicial appendages are absent; and (3) only 3 dorsal-fin rays are present. According to the original description and illustrations (Kharin, 1984:663, fig. 1A, B), the holotype of *Himantolophus borealis* differs further from all other members of the *H. albinares* group in having a simple posterior escal appendage and a short appendage that emerges from the left side of the central part of the escal bulb. However, reexamination of the specimen (by E. Karmowskaya, Zoological Institute, St. Petersburg; see Bertelsen and Krefft, 1988:63) showed that the posterior escal appendage "is slightly bifurcated" and there is "no trace of additional appendages on the bulb."

### Sexual Maturity

The ovaries of the 112-mm specimen measure about 22 by 15 by 5 mm and contain eggs measuring 0.1 to 0.2 mm in diameter.

### Distribution

All three known specimens of *Himantolophus borealis* were caught off the east coast of Honshu, Japan (Bertelsen and Krefft, 1988:64; Fig. 214).

### Comments

Based on the shape and size of the distal escal appendage, the presence of a posterior escal appendage, and the development of papillae on the snout and chin, *Himantolophus borealis* is a typical member of the *H. albinares* group. Within this species group, it is most similar to *H. multifurcatus*, but differs distinctly in the relative length of the illicium and in the pigmentation of the distal escal appendage.

The differences between the holotype and the SIO specimen in the branching of the distal escal appendage and in the development of the illicial appendage are of an order similar to the intraspecific variation observed in other species of the genus. However, considering that the three specimens here referred to *H. borealis* were caught at nearly the same locality and are the only representatives of the *H. albinares* group known from the Pacific, it seems unlikely that they should represent more than one species. The "appendage" observed on the left side of the escal bulb of the holotype, but not found upon reexamination, was assumed by Bertelsen and Krefft (1988) to be a "skin-flap" caused by abrasion during capture. Considering that the two additional specimens referred to *H. borealis* as well as the several hundred specimens of the genus examined by Bertelsen and Krefft (1988) have 5 (or very rarely 6) dorsal-fin rays, it is furthermore assumed, that the existence of only 3 dorsal-fin rays in the holotype is due to damage or abnormal development.

## *Himantolophus cornifer* Group Bertelsen and Krefft, 1988

### Females

*Himantolophus cornifer* Group Bertelsen and Krefft, 1988:64 (new species group).

### Males and Larvae

Males probably belong to the *Himantolophus rostratus* group; the larvae of some and possibly all species belong to pigment pattern Type B (see Bertelsen, 1951:61, fig. 23; Bertelsen and Krefft, 1988:74, 77, figs. 34, 35D, 36, 45).

### Distinguishing Characters

Metamorphosed females of the *Himantolophus cornifer* group differ from those of the other species of the genus in lacking a posterior escal appendage and in having the following combination of character states: The distal escal appendage is bifurcated at or very close to its base, each primary branch bearing a single unbranched appendage emerging from the posterior margin (with two exceptions, the paratype of *H. macroceratoides* and the holotype of *H. azurlucens*). The total length of the distal escal appendage is 16 to 214% SL in specimens greater than 30 mm. A more or less distinct pair of swellings is present just posterior to the base of the distal escal appendage. No additional appendages are present on the esca or on the stem of the illicium. Small dermal spinules are present on the esca and on the distal escal appendage in specimens greater than about 55 mm. The papillae of the snout and chin are well developed. "White patches" on the body are present or absent. The caudal-fin rays are unpigmented distally, spotted proximally in juveniles less than about 50 mm, but black in larger specimens.

### Diversity

The *Himantolophus cornifer* group contains five species:

### Key to Females of Species of the *Himantolophus cornifer* Group

1A. Length of illicium 75% SL . . . . . . . .
  . . . . . . . . *Himantolophus compressus*
  (Osório, 1912), p. 349

  NOTE A single known specimen, 130 mm; eastern North Atlantic Ocean

1B. Length of illicium 29 to 44% SL . . .
  . . . . . . . . . . . . . . . . . . . . . . . . . . . . 2

2A. Each primary branch of distal escal appendage with 1 to 3 distal bifurcations . . . . . . . . . . . . . . . . . . . . . .
  . . . . . . . . . . *Himantolophus cornifer*
  Bertelsen and Krefft, 1988, p. 346

  NOTE Eight known specimens, 27 to 208 mm; Atlantic, Pacific, and western Indian oceans

2B. Each primary branch of distal escal appendage simple . . . . . . . . . . . . 3

3A. Side branches of distal escal appendage cylindrical and long (15 to 34% SL in specimens 59 to 92 mm) . . . . *Himantolophus macroceras*
  Bertelsen and Krefft, 1988, p. 347

  NOTE Five known specimens, 32 to 92 mm; tropical Atlantic Ocean

3B. Side branches of distal escal appendage fusiform or club shaped and short (3 to 10% SL in specimens 50 to 180 mm) . . . . . . . . . . . . . . . . 4

4A. Distal escal appendage 92 to 100% SL, each primary branch with 1 or 2 fusiform side branches emerging 22 to 49% SL above base . . . . . . . . . .
  . . . . . *Himantolophus macroceratoides*
  Bertelsen and Krefft, 1988, p. 348

  NOTE Two known specimens, 50 to 180 mm; eastern Atlantic and western Indian oceans

4B. Distal escal appendage about 30% SL, 1 primary branch simple, the other with 2 club-shaped side branches emerging about 10% SL above base
  . . . . . . . . *Himantolophus azurlucens*
  Beebe and Crane, 1947, p. 348

  NOTE Two known specimens, 98 to 141 mm; eastern South Atlantic and Gulf of Panama

## *HIMANTOLOPHUS CORNIFER* BERTELSEN AND KREFFT, 1988

Figures 45, 69, 72, 213; Tables 3, 8

### Females

*Himantolophus cornifer* Bertelsen and Krefft, 1988:64, figs. 27, 28 (original description, five specimens). Donnelly and Gartner, 1990:77 (additional specimen, eastern Gulf of Mexico). McEachran and Fechhelm, 1998:854, 855, fig. (Gulf of Mexico, in key). Pietsch, 1999:2029 (western central Pacific). Anderson and Leslie, 2001:8, fig. 5 (additional specimen, South Africa). Pietsch, 2002b:1061 (western central Atlantic).

### Material

Eight metamorphosed females (27 to 208 mm):

Holotype of *Himantolophus cornifer*: MCZ 58858, 90 mm, *Harbison* station 1047, Coral Sea, 12°31'S, 148°41'E, oblique MT, 0 to 1650 m, 3 December 1981.

Paratypes of *Himantolophus cornifer*: ISH 801/68, 39 mm, *Walther Herwig* station 14-III/68, 4°08'N, 24°41'W, MT, 580 to 600 m, 1 February 1968; ISH 2129/71, 27 mm, *Walther Herwig* station 467/71, 5°30'S, 16°28'W, MT, 1850 to 1900 m, 9 April 1971; LACM 33325-1, 38 mm, *Townsend Cromwell* 47-52, 3°03.5'N, 145°00'W, MT, 450 m, 6 February 1970; MCZ 58176, 52 mm, *Chain* 60, RHB 1310, 26°12'N, 87°54'W, oblique MT, 0 to 2150 m, 23 June 1966.

Additional females: IFM, one (63 mm); SAM, one (208 mm); UF, one (42 mm).

### Distinguishing Characters

Metamorphosed females of *Himantolophus cornifer* differ from those of other species of the *H. cornifer* group in having distally bifurcated primary branches of the distal escal appendage and in having the following combination of character states: length of illicium 29 to 35% SL; each primary branch of distal escal appendage with a posterior side branch situated (2.6 to 11% SL) near base; length of distal escal appendage increasing from 9% SL in 27-mm specimen to 54% SL in 52-mm specimen and 72% SL in 90-mm specimen; length of posterior side branch of distal escal appendage increasing from 4.3% SL in 27-mm specimen to about 30% SL in 52 to 90 mm specimens; distal escal appendage without small dermal spinules (present but restricted to base of esca in 90-mm specimen).

Description

Distal escal appendage darkly pigmented except for distal tips in specimens 38 to 90 mm, pigmented only at base in 27-mm juvenile; primary branches of distal escal appendage emerging less than 2.5% SL from base of appendage, distal part of each bifurcating to form 3 or 4 papilliform tips in 27-mm specimen, 2 digitiform branches in 38-mm specimen, each primary branch bifurcating twice in 90-mm holotype and 39-mm paratype, a third bifurcation of some of ultimate tips present in 52-mm paratype; side branches emerging from near base of distal escal appendage simple, slightly tapering; small dermal spinules present on stem of illicium and base of distal escal appendage of holotype, on illicial stem and base of escal bulb in 52-mm paratype, absent on illicium in smaller paratypes; no large dermal spines in the 27-mm specimen, 3 or 4 on each pectoral-fin lobe of 38 to 90 mm specimens, 8 on each side of body in 52-mm specimen of 52 mm, about 30 in holotype; papillae of snout and chin well developed; white or faintly pigmented fields of skin present on anterior side of stem of illicium, on papillated part of snout and chin, along dorsal margin of body to base of dorsal fin, and on caudal peduncle behind dorsal and anal fins; a distinct spot of melanophores is present on a slight swelling in front of dorsal fin in 27-mm juvenile; caudal-fin rays white, with scattered melanophores in 27 to 39 mm specimens, darkly pigmented in two larger specimens; dorsal-fin rays 5; anal-fin rays 4; pectoral-fin rays 16 to 18.

Sexual Maturity

The ovaries of the holotype are immature, about 10 mm in length.

Distribution

Seven of the eight known records of *Himantolophus cornifer* are widely scattered in the tropical and subtropical parts of the oceans, one from the central Pacific, the holotype from the Coral Sea, three from the central Atlantic, and two from the Gulf of Mexico (Fig. 213). An eighth specimen was recently reported from off South Africa (Anderson and Leslie, 2001). All were caught in pelagic nets fished open with maximum fishing depths varying from 450 to 2100 m.

*HIMANTOLOPHUS MACROCERAS*
BERTELSEN AND KREFFT, 1988

Figures 70, 214; Tables 3, 8

Females

*Himantolophus macroceras* Bertelsen and Krefft, 1988:68, figs. 29, 30 (original description, five specimens).

Material

Five metamorphosed females (32 to 92 mm):

Holotype of *Himantolophus macroceras*: ISH 2329/71, 79 mm, *Walther Herwig* station 478/71, 1°04′N, 18°22′W, 2100 m, 12 April 1971.

Paratype of *Himantolophus macroceras*: BMNH 1986.9.22.3, 32 mm, *Discovery* station 8568-5, 2°39′N, 23°10′W, 800 to 600 m, 3 August 1974; ISH 1863/66, 59 mm, *Walther Herwig* station 186/66, 1°24′S, 25°58′W, 170 to 330 m, 19 May 1966; ISH 722/68, 85 mm, *Walther Herwig* station 13-II/68, 8°21′N, 24°10′W, 500 to 520 m, 31 January 1968 (this specimen was in an early stage of decomposition when found in the net and may have been left behind from a haul taken about 24 hours previously at *Walther Herwig* station 12-II/68, 12°16′N, 23°05′W, in about 2000 m); MCZ 58177, 92 mm, *Chain*, cruise 35, RHB 969, 0°24′N, 27°32′W, 300 m, 19 February 1963.

Distinguishing Characters

Metamorphosed females of *Himantolophus macroceras* differ from those of the other species of the *H. cornifer* group in having the following combination of character states: length of illicium 30 to 44% SL; each primary branch of distal escal appendage unbranched, with a single side branch emerging from posterior margin at varying distances from base (7.2% SL in 32-mm specimen, 17 to 80% SL in 59- to 92-mm specimens); total length of distal escal appendage 18% SL in 32-mm specimen, 59 to 214% SL in 59- to 92-mm specimens; length of side branch of distal escal appendage 5% SL in 32-mm specimen, 15 to 34% SL in 59- to 92-mm specimens; each side branch cylindrical or slightly tapering toward blunt tip; distal escal appendage, except for tips, covered with dermal spinules in specimens 59 mm and larger.

Description

Primary branches of distal escal appendage separated less than 5% SL from base, each with an undivided distal part and a single, cylindrical or slightly tapering, posterior side branch; distal escal appendage darkly pigmented except for tips in all specimens; illicium and distal escal appendage naked in 32-mm specimen, covered with small dermal spinules except for tips in holotype and 59 to 92 mm paratypes; in the poorly preserved 85-mm specimens, remains of spinules present on stem of illicium, escal bulb, and proximal part of distal escal appendage (skin of distal part of distal escal appendage lost); except for a single large dermal spine on each pectoral-fin lobe, skin of 32-mm paratype naked, 4 to 7 spines on pectoral lobe of four larger specimens; each side of body with about 10 dermal spines in 59-mm specimen, 25 to 30 on each side of body in remaining paratypes; papillae of snout and chin well developed; white or faintly pigmented fields of skin present on anterior margin of stem of illicium, on snout and chin, and as two very distinct spots behind dorsal and anal fins; two smallest paratypes with faintly pigmented dorsal streak in front of dorsal fin, surrounding an indistinct darker patch (most probably representing the dorsal spot present in "Type B" larvae; see below); caudal fin of 32-mm specimen white except for dorsal-most and ventral-most rays, all fin rays black in larger specimens; dorsal-fin rays 5; anal-fin rays 4; pectoral-fin rays 17 or 18.

Sexual Maturity

The ovaries of the 92-mm paratype are small and immature.

Distribution

The five known specimens of *Himantolophus macroceras* were all caught within a remarkably small region of the eastern equatorial Atlantic between 1°S and 8°N and 18° and 26°W (Fig. 214). Except for the 32-mm juvenile, which was caught in a closing net between 600 and 800 m, all were caught in open pelagic nets with maximum fishing depths of 300 and 2100 m.

Comments

*Himantolophus macroceras* differs distinctly from *H. cornifer* in having simple primary branches of the distal escal appendage (those of *H. cornifer* are bifurcated or trifurcated) and in having the distal escal appendage almost completely covered with dermal spinules in specimens larger than 50 mm. Among the known specimens of *H. macroceras*, the 85-mm specimen differs from the remaining material in the relative length of the distal escal appendage and the position of its side branches, but this is assumed to be individual, possibly ontogenetic intraspecific variation, considering the other general similarities and the fact that the specimen was caught in nearly the same locality as the others. The intestine and ovaries of this specimen were lost.

The escal diameter of the 85-mm specimen, nearly 15% SL, is the largest recorded for any known ceratioid anglerfish. Furthermore, the distal light-guiding escal appendages of the four largest known specimens of *H. macroceras* are longer in relative length (59 to 214% SL) than those of any other known ceratioid, the previous record being nearly 58% SL in *Phyllorhinichthys balushkini*, recorded by Bertelsen and Pietsch (1977) and Pietsch (2004).

### HIMANTOLOPHUS MACROCERATOIDES BERTELSEN AND KREFFT, 1988

Figures 70, 72; Tables 3, 8

#### Females

*Himantolophus macroceratoides* Bertelsen and Krefft, 1988:70, figs. 31, 32 (original description, two specimens).

#### Material

Two metamorphosed females (50 and 180 mm):

Holotype of *Himantolophus macroceratoides*: ISH 2400/71, 50 mm, *Walther Herwig* station 482-II/71, 4°36′N, 19°40′W, MT, 0 to 256 m, 13 April 1971.

Paratype of *Himantolophus macroceratoides*: ZIN 47914, 180 mm, *Zwiezda Kryma*, cruise 6-1, 0°54.4′N, 56°32′E, 800 to 900 m.

#### Distinguishing Characters

Metamorphosed females of *Himantolophus macroceratoides* differ from those of other species of the *H. cornifer* group in having the following combination of character states: length of illicium 33 to 38% SL; each primary branch of distal escal appendage with a posterior side branch emerging 42 to 49% SL from base (left primary branch of 180-mm specimen with an additional small side branch at 22% SL from base); total length of distal appendage 92 to 100% SL, length of side branches 9 to 10% SL (length of small additional side branch 3.8% SL); all side branches fusiform; small dermal spinules on distal appendage absent or restricted to base.

#### Description

Length of illicium 33 to 38% SL; length of distal escal appendage 92 to 100% SL; length of posterior side branch of each distal escal appendage 9 to 10% SL, emerging 42 to 49% SL above base of distal escal appendage (paratype with an additional side branch, length 3.8% SL, emerging 22% SL above left primary branch of distal escal appendage); all side branches fusiform, lengths three to five times their greatest diameter; diameter of escal bulb 9.8 to 10% SL; distal escal appendage and its side branches black except for distal tips; holotype with a transparent, windowlike spot at base of white tip of primary branches; illicium and distal escal appendage of holotype naked, small scattered spinules present on escal bulb and base of distal escal appendage of paratype; 5 or 6 large spines on each pectoral lobe; holotype and paratype with about 16 and 50 large spines, respectively, on each side of body; papillae of snout and chin well developed; papillated area and median part of the back of holotype somewhat more lightly pigmented than brownish black sides of body and fin rays; paratype uniformly brownish black; dorsal-fin rays 5; anal-fin rays 4; pectoral-fin rays 16 or 17.

#### Sexual Maturity

The ovaries of the paratype are enlarged, each about 45 mm long, 25 mm wide, and 15 mm thick, containing numerous eggs 0.20 to 0.25 mm in diameter.

#### Distribution

The holotype of *Himantolophus macroceratoides* was taken in the eastern tropical Atlantic off the coast of West Africa in an open pelagic net with a maximum fishing depth of 256 m; the paratype is from the western equatorial Indian Ocean, collected in a bottom trawl at 800 to 900 m (Bertelsen and Krefft, 1988:71, fig. 43).

#### Comments

*Himantolophus macroceratoides* is very similar to *H. macroceras*, but the differences in size and shape of the side branches of the distal escal appendage and the complete or nearly complete absence of dermal spinules on the distal escal appendage are of a magnitude to warrant specific recognition (Bertelsen and Krefft, 1988:72).

### HIMANTOLOPHUS AZURLUCENS BEEBE and CRANE, 1947

Figures 49, 260; Tables 3, 8, 9

#### Females

*Himantolophus azurlucens* Beebe and Crane, 1947:155, text figs. 3, 4, plate 1, figs. 1, 2 (original description, single specimen). Harvey, 1952:530, 531, fig. 178 (bioluminescence). Mead, 1958:133 (type transferred to SU). Fonseca, 1968:487, 493, fig. 37 (additional record, off Peru, specimen not seen). Munk, 1999:267, 281 (bioluminescence). Trunov, 2001:543, fig. (additional specimen, South Atlantic).

#### Material

Two metamorphosed females (98 to 141 mm):

Holotype of *Himantolophus azurlucens*: CAS-SU 46507, 98 mm, eastern Pacific *Zaca* Expedition station 228, trawl 1, east of Cape Mala, Panama, 7°00′N, 79°16′W, about 900 m (500 fathoms), 25 March 1938.

Additional specimen: ZIN, one (141 mm).

#### Distinguishing Characters

*Himantolophus azurlucens* differs from other species of the *H. cornifer* group in having the following combination of character states: length of illicium 32% SL; a pair of side branches on left primary branch of distal escal appendage emerging about 8% SL from base; length of right distal escal appendage 23% SL, left distal escal appendage 32% SL; side branches club shaped, length 3 to 4% SL; except for distal tips, primary branches of distal escal appendage covered with small dermal spinules.

#### Description

Length of illicium 32% SL; length of primary branches of distal escal appendage about 23% (right side) and 32% (left side) SL; a pair of club-shaped side branches on right primary branch (length 3 to 4% SL), emerging about 8% SL above base of distal escal appendage; diameter of escal bulb 9% SL; distal escal appendage pigmented except at tip, silvery in freshly caught specimen; stem of illicium, bulb of esca, and distal escal appendage covered with small spinules; 5 to 7 large spines on each pectoral lobe, about 20 on each side of body; papillae of snout and chin well developed; central part of papillated area white or weakly pigmented; patches of unpigmented skin present behind dorsal fin, in front of and behind anal fin, and on first and ninth caudal-fin ray; dorsal-fin rays 5; anal-fin rays 4; pectoral-fin rays 17.

For more description, including observations of the live holotype, see Beebe and Crane (1947:156–158).

#### Sexual Maturity

The ovaries of the holotype are small (approximately 9 by 6 mm) and immature.

#### Distribution

The two known specimens of *Himantolophus azurlucens* are from widely spaced localities: the 98-mm holotype was collected in the Gulf of Panama, between the surface and 900 m; the 141-mm specimen is from the South Atlantic at approximately 31°S, 2°E, captured between the surface and 1100 m.

### Comments

*Himantolophus azurlucens* appears to be most similar to *H. macroceras* and *H. macroceratoides*, but differs from both in having a much shorter distal escal appendage. It differs further from *H. macroceras* in the size and shape of the side branches of the distal escal appendage and from *H. macroceratoides* in the development of spinules on the esca and distal escal appendage. The asymmetrical development of the side branches of the distal escal appendage is attributed to intraspecific variation (Bertelsen and Krefft, 1988:73).

## HIMANTOLOPHUS COMPRESSUS (OSÓRIO, 1912)

Tables 3, 8

### Females

*Corynophorus compressus* Osório, 1912: 90, figs. 1, 2 (original description, single specimen). Nobre, 1935:235, pl. 32a, fig. 105c (after Osório, 1912).
*Corinophorus compressus*: Nobre, 1935, pl. 33, fig. 106 (erroneous spelling of generic name).
*Himantolophus compressus*: Maul, 1961: 96, fig. 4 (new combination, description after Osório, 1912). Bertelsen, 1986:1378, 1379, figs. (eastern North Atlantic; in key).

### Material

A single metamorphosed female (130 mm):
Holotype of *Himantolophus compressus*: 130 mm, off Portugal, about 38°20′N, 9°15′W, but lost in 1978 fire at the Museo Bocage, Lisbon; description below based in part on examination by E. Bertelsen made in October 1973.

### Distinguishing Characters

The only known specimen of *Himantolophus compressus* differs from those of other species of the *H. cornifer* group in having a relatively long illicium, 75% SL; the length and shape of the distal escal appendage is unknown.

### Description

At the time of Bertelsen's examination of the holotype (October 1973), the illicium and esca were well preserved: diameter of escal bulb 6.4% SL, bearing an unpigmented, distal pair of escal swellings, without a trace of a posterior escal appendage; distal escal appendage torn off, only about 5 mm of its undivided base retained; stem of illicium and escal bulb covered with dermal spinules, but none present on retained remnant of distal escal appendage; body with numerous large spines (not counted by Bertelsen; indistinct in illustrations published by Osório, 1912, and Maul, 1961); papillae of snout and chin well developed; skin of head, body, and fin rays uniformly brownish black; dorsal-fin rays 5; anal-fin rays 4; pectoral-fin rays 18.

### Distribution

The only known specimen of *Himantolophus compressus* was caught in the eastern North Atlantic off the coast of Portugal.

### Comments

The lack of a posterior escal appendage in the only known specimen of *Himantolophus compressus* indicates its membership in the *H. cornifer* group. It differs significantly from all other species of this group in having a relatively long illicium (75% SL versus 30 to 44% SL). This feature alone justifies the retention of a separate species, despite the lack of knowledge of the shape and length of the distal escal appendage, the most significant distinguishing character of the group.

## *Himantolophus rostratus* Group Bertelsen and Krefft, 1988

Figure 73

### Males

*Rhynchoceratias rostratus* Regan, 1925c: 567 (original description, single specimen). Regan, 1926:44, fig. 25d (two additional specimens, description). Regan, 1927a:4, postcard M 15a (popular account). Regan and Trewavas, 1932:62 (additional specimen, description after Regan, 1925c, 1926). Nielsen, 1974:88 (listed in type catalog). Bertelsen and Krefft, 1988:75, figs. 35C, 45).
*Himantolophus rostratus*: Bertelsen, 1951:66, fig. 25D, E, 26F (new combination; species based only on males, two specimens).
*Himantolophus* sp.: Pietsch and Seigel, 1980:381 (in part; unidentifiable male, LACM 36124-1).
*Himantolophus rostratus* Group Bertelsen and Krefft, 1988:77, figs. 35D, 45 (new species group, based only on males; 17 specimens, comparison with all known material, in key). Pietsch, 1999:2029 (western central Pacific). Pietsch, 2002b: 1061 (western central Atlantic). Pietsch, 2005b:215 (reproduction).

### Material

Twenty metamorphosed males (23 to 39 mm):

Holotype of *H. rostratus*: ZMUC P9265, 26.5 mm, *Dana* 1209(1), Gulf of Panama, 7°15′N, 78°54′W, ring-trawl, 3500 m wire, 1845 h, 17 January 1922.
Additional material: ISH, four (30.5 to 35 mm); LACM, five (25 to 35 mm; 31-mm specimen recorded as *Himantolophus* sp. by Pietsch and Seigel, 1980:381); SIO, three (26 to 30 mm); SIOM, one (37 mm); UF, one (39 mm); USNM, four (23 to 30.5 mm); ZMUC, one (34 mm; recorded as *Rhynchoceratias rostratus* by Regan and Trewavas, 1932:62; as *Himantolophus rostratus* by Bertelsen, 1951:66, figs. 25E, 26F).

### Distinguishing Characters

Metamorphosed males of the *Himantolophus rostratus* group differ from those of all other species of the genus in having the following combination of character states: 14 to 17 olfactory lamellae; greatest diameter of posterior nostril 3.5 to 6.4% SL; length of snout 11 to 18% SL; upper denticular teeth 19 to 30; lower denticular teeth about 23 to 50.

### Description

The range of variation of counts and measurements, along with average values, are given by Bertelsen and Krefft, 1988:76, table 16. Fourteen olfactory lamellae were found only in a single specimen (LACM 31465-3) and only in the left organ, while 16 were counted on the right side; 15 lamellae were counted in a second specimen from the same haul. The relative size of the eye shows a slight decreasing trend with increasing SL; the greatest number of lower denticles (50) was found in the largest male (39 mm, UF 232657). Except for these two features, variation in the characters examined seems independent of standard length within the available observed range of size (23 to 39 mm). The two 35-mm specimens (LACM 31465-3) appear to be recently metamorphosed, showing some retention of larval teeth and lightly pigmented skin; one of the specimens has a distinct concentration of pigment in front of the dorsal fin and traces of this dorsal spot can be observed in the other.

### Sexual Maturity

All the specimens have well-developed testes, large in specimens greater than 25 mm, somewhat smaller in the four smaller specimens (23 to 25 mm). The testes of three specimens (30.5 to 34 mm) sectioned for microscopic examination by Bertelsen and Krefft (1988:78) were found to be filled with mature spermatozoa, with elongate, condensed nuclei and distinct tails.

### Distribution

Capture localities of the males of the *Himantolophus rostratus* group are widely scattered over the tropical and subtropical parts of the Atlantic and Pacific oceans including the Caribbean Sea and eastern Indonesian waters (Bertelsen and Krefft, 1988:78, fig. 45).

### Comments

The presence of a dorsal pigment spot in juvenile males of the *Himantolophus rostratus* group indicates that the group corresponds to larvae with pigmentation "pattern B," which, as discussed below, are assumed to represent the *H. albinares* and *H. cornifer* groups. The geographic distribution of these males does not contradict this assumption. Except for *H. compressus* (Osório, 1912), the original descriptions of all the species of these two groups based on females postdates *H. rostratus* (Regan, 1925c), which means, in all likelihood, *H. rostratus* will represent a senior synonym if and when specific identity with females can be shown.

## *Himantolophus appelii* Group Bertelsen and Krefft, 1988

### HIMANTOLOPHUS APPELII (CLARKE, 1878)

#### Males

*Himantolophus appelii*: Bertelsen and Krefft, 1988:75, figs. 35C, 45, table 16 (first records of males, four specimens, comparison with all known material, in key). Anderson and Leslie, 2001:6, fig. 4 (two males, SAM 30981, erroneously cited as SAM 27805 by Bertelsen and Krefft, 1988).

#### Material

Four metamorphosed males (23.5 to 27 mm): AMS, two (23.5 to 27 mm); SAM, two (24.5 to 26 mm).

#### Distinguishing Characters

Metamorphosed males of *Himantolophus appelii* differ from those of other species of the genus in having the following combination of character states: olfactory lamellae 11 to 13; greatest diameter of posterior nostril 4.1 to 6.7% SL; length of snout 14 to 18% SL; upper denticular teeth about 20 to 25; lower denticular teeth about 22 to 32 (see Bertelsen and Krefft, 1988:75, table 16).

#### Description

The two AMS specimens differ slightly in proportions and counts: except for the diameter of the posterior nostril (5.7 to 6.7% SL), which like the eye (6.0 to 7.6% SL) is relatively largest in the smaller specimen (23.5 mm), the larger values in the diagnosis refer to the larger specimen (27 mm); in both AMS specimens, 5 or 6 upper denticular teeth are placed within the outer semicircular series; the lower denticular teeth form a median and two lateral, but indistinctly separated, groups; olfactory lamellae 11 or 12; length of snout 17 to 18% SL; upper denticular teeth about 20 to 24; lower denticular teeth 22 to 30; body depth 27% SL in the smaller specimen, 30% SL in the larger specimen.

The two SAM specimens differ from the AMS specimens in having the following character states: olfactory lamellae 12 or 13, greatest diameter of posterior nostril 4.1 to 5.8% SL; length of snout 14 to 16% SL; upper denticular teeth about 25; lower denticular teeth about 28 to 32; both are juveniles having incompletely fused denticular teeth, some retention of larval teeth, deep bodies 45 to 46% SL, relatively large eyes 7.3 to 8.1 % SL, and medium-sized testes.

#### Sexual Maturity

Histological sections of the testes of the two AMS specimens indicate that both are fully mature (Bertelsen and Krefft, 1988:75).

#### Distribution

The two AMS specimens were caught in the same haul, taken in the Tasman Sea, northeast of Sydney, Australia; the two SAM specimens, from the same haul as well, were captured off Durban, South Africa (Bertelsen and Krefft, 1988:76, fig. 45).

#### Comments

The two AMS males were identified as *Himantolophus appelii* by Bertelsen and Krefft (1988:75) primarily on the basis of geographic distribution: both were caught in an area where *H. appelii* is the only known representative of the genus. The locality of the two SAM specimens is near the northern border of the known distribution of *H. appelii* but well south of records of all other specimens of the genus, except for the holotype of *H. pseudalbinares* and a single stranded specimen of *H. groenlandicus* (SAM 23129).

These identifications are supported further by the fact that, like the females of this species, the males differ from those of other members of the genus in characters of a level that corresponds to species groups. They are well separated from males of the *H. rostratus* group in the number of olfactory lamellae, and from those of the *H. brevirostris* group in the length of the snout and size of the posterior nostril. Viewed anteriorly, the lower denticular bones of the males recognized here as *H. appelii* are somewhat broader and less deep than those of the *H. rostratus* and *H. brevirostris* groups. That the two SAM specimens are juveniles most probably accounts for their having a somewhat shorter snout and smaller posterior nostril.

## *Himantolophus brevirostris* Group Bertelsen and Krefft, 1988

Figure 73

### Males

*Lipactis tumidus* Regan, 1925c:566 (original description, two specimens, lectotype ZMUC P9261, designated by Bertelsen and Krefft, 1988). Regan, 1926:43, pl. 12, fig. 2 (after Regan, 1925c; figured). Norman, 1930:357 (after Regan, 1926; additional specimen). Regan and Trewavas, 1932:61 (after Regan, 1925c, 1926; additional specimen). Bertelsen, 1951:59, 60, figs. 23E, 25A (description, comparison with all known material, referred to *Himantolophus groenlandicus*). Nielsen, 1974:88 (listed in type catalog). Bertelsen and Krefft, 1988:79 (tentatively assigned to the *H. brevirostris* group).

*Rhynchoceratias brevirostris* Regan, 1925c:566 (original description, single specimen). Regan, 1926:43, fig. 25A, pl. 13, fig. 1 (after Regan, 1925c; figured). Norman, 1930:357 (after Regan, 1926; additional specimen, described as *Rhynchoceratias altirostris* by Regan and Trewavas, 1932). Regan and Trewavas, 1932:62 (after Regan, 1925c, 1926; in key). Bertelsen, 1951:59, 60, fig. 25B (description, comparison with all known material, referred to *Himantolophus groenlandicus*). Nielsen, 1974:88 (listed in type catalog). Bertelsen and Krefft, 1988:78 (assigned to the *H. brevirostris* group).

*Rhynchoceratias oncorhynchus* Regan, 1925c:566 (original description, two specimens, lectotype ZMUC P9264, designated by Bertelsen and Krefft, 1988). Regan, 1926:44, fig. 25C, pl. 13, fig. 3 (after Regan, 1925c; additional specimens, figured). Regan and Trewavas, 1932:62 (after Regan, 1925c, 1926; in key). Nielsen, 1974:88 (listed in type catalog).

*Rhynchoceratias altirostris* Regan and Trewavas, 1932:62, fig. 90 (original description, single specimen, previously identified as *Rhynchoceratias brevirostris* by Norman, 1930). Bertelsen, 1951:59, 60, 61 (description, comparison with all known material, referred to *Himantolophus groenlandicus*). Bertelsen and Krefft, 1988:79 (assigned to the *H. brevirostris* group).

*Rhynchoceratias onchorhynchus*: Bertelsen, 1951:59, 60, 61, fig. 25C (misspelling of specific name, description, comparison with all known material, referred to *H. groenlandicus*). Bertelsen and Krefft, 1988:78 (assigned to the *H. brevirostris* group).

*Himantolophus groenlandicus*: Bertelsen, 1951:60 (nominal species of *Rhynchoceratias* based only on males placed in synonymy of *Himantolophus groenlandicus*; description, comparison of all known material).

*Rhynchoceratias ogcorrhynchus*: Günther and Deckert, 1956:179, 181, figs. 123, 124F (misspelling of generic and specific names, homology of first dorsal-fin spine and upper denticular of free-living male, popular account).

*Himantolophus* sp.: Pietsch and Seigel, 1980:381 (in part; three unidentifiable males, LACM 36040-3, LACM 36046-8).

*Himantolophus brevirostris* Group Bertelsen and Krefft, 1988:78, figs. 35A, B, 45 (new species group, based only on males; 17 specimens, comparison of all known material, in key). Stewart and Pietsch, 1998:9, fig. 5 (additional specimen, description, New Zealand; figure after Regan, 1926). Pietsch, 1999:2029 (western central Pacific). Pietsch, 2002b:1061 (western central Atlantic). Pietsch, 2005b:215 (reproduction).

Material

Eighteen metamorphosed males (27 to 38 mm) and 10 tentatively identified specimens in metamorphosis (20 to 26 mm):

Holotype of *Rhynchoceratias brevirostris* Regan, 1925c: ZMUC P9263, 30.5 mm, *Dana* station 1171(7), North Atlantic, 8°19′N, 44°35′W, ring-trawl, 6000 m wire, 0800 h, 13 November 1921.

Lectotype of *Rhynchoceratias oncorhynchus* Regan, 1925c: ZMUC P9264, 34 mm, *Dana* station 1171(7), North Atlantic, 8°19′N, 44°35′W, ring-trawl, 6000 m wire, 0800 h, 13 November 1921.

Paralectotype of *Rhynchoceratias oncorhynchus* Regan, 1925c: BMNH 1925.8.11.46, 30 mm, *Dana* station 1189(2), Caribbean Sea, south of the Virgin Islands, 17°58.5′N, 64°41′W, 4000 m wire, 1440 h, 8 December 1921.

Lectotype of *Lipactis tumidus* Regan, 1925c: ZMUC P9261, 21.5 mm, *Dana* station 1186(3), North Atlantic, 17°54′N, 64°54′W, 4000 m wire, 1815 h, 30 November 1921.

Paralectotype of *Lipactis tumidus* Regan, 1925c: BMNH 1925.8.11.42, 20.5 mm, *Dana* station 1171(10), North Atlantic, 8°19′N, 44°35′W, 4000 m wire, 0800 h, 13 November 1921.

Holotype of *Rhynchoceratias altirostris* Regan and Trewavas, 1932: BMNH 1930.1.12.1103, 30 mm, *Discovery* station 298, North Atlantic, 13°02′N, 21°35′W, 900 to 1200 m, 29 August 1927.

Additional material: BMNH, three (30 to 33 mm); ISH, six (28.5 to 36 mm); LACM, three (27 to 33 mm); MCZ, three (28.5 to 32 mm); NMNZ, one (29 mm); SIOM, one (32.5 mm); USNM, one (38 mm).

Tentatively identified males in metamorphosis: ISH, two (22 to 22.5 mm); LACM, two (25 to 26 mm); MCZ, two (24.5 to 25 mm); SIO, two (20 to 23.5 mm); VIMS, one (25 mm).

Distinguishing Characters

Metamorphosed males of the *Himantolophus brevirostris* group differ from those of all other species of the genus in having the following combination of character states: olfactory lamellae 10 to 13; greatest diameter of posterior nostril 3.3 to 5.5% SL; length of snout 13 to 17% SL; upper denticular teeth about 16 to 31; lower denticular teeth about 20 to 38.

Description

The range of variation of counts and measurements, along with average values, are given by Bertelsen and Krefft, 1988:76, table 16. Ten olfactory lamellae were found only in one specimen (ISH 3246/79b) and only on one side. The smallest eyes (5.5 to 5.6% SL) were found in the two largest known specimens (36 to 38 mm). The shortest snouts (13 to 14% SL) were found in MCZ 55373 and the holotype of *Himantolophus altirostris*, which both appear to be juveniles, judging from their great body depth (38 to 50% SL) and weak development of denticular teeth. Except for these two features, variation in the characters examined seems independent of standard length. All specimens have darkly pigmented skin, without a concentration of melanophores in front of the dorsal fin.

The tentatively identified males in metamorphosis have deep bodies (greatest depth 48 to 77% SL), larval teeth, rudimentary denticular teeth, relatively large eyes, short snouts, small olfactory organs containing 10 to 12 rudimentary lamellae, faintly pigmented skin without a trace of a pigmented swelling in front of the dorsal fin. Their stomachs are well developed and some appear to contain remains of food.

Sexual Maturity

All the metamorphosed males have large well-developed testes, which, when sectioned in two specimens (28.5 mm, ISH 3243/79; 34 mm, ISH 921/68), were found to be filled with mature spermatozoa, with elongate, condensed nuclei and distinct tails (Bertelsen and Krefft, 1988:79). The testes of the tentatively identified males in metamorphosis are relatively large as well, but smaller than those of the metamorphosed males; when examined histologically, the testes of a 25-mm specimen (MCZ 58354) were found to be filled with uniform stages of spermatocysts.

Distribution

With the exception of one specimen from the eastern Indian Ocean, one from off New Zealand, and two from Indonesian waters, all metamorphosed males of the *Himantolophus brevirostris* group have been caught in the central and North Atlantic. Six of the metamorphic males came from the same part of the Atlantic, while four were caught in the western and central Pacific (Bertelsen and Krefft, 1988:79, fig. 45).

Comments

Males of the *Himantolophus brevirostris* group differ distinctly from those of the *H. rostratus* group only in the number of olfactory lamellae, but it seems most unlikely that this difference is caused by random or ontogenetic variation within the examined material. The variation in other characters is relatively large in both groups and the differences in the average values too small and based on too few specimens to be statistically significant.

If, as assumed above, the *H. rostratus* group represents males of the *H. albinares* and *H. cornifer* groups, it is most likely that the *H. brevirostris* group represents males of the *H. groenlandicus* group. The dominant geographic distribution in the central and North Atlantic and the fact that no dorsal pigment spot was observed in any specimen of these groups are in agreement with this assumption. The absence of this pigment spot, which, if present, would be very distinct in the faintly pigmented skin of the metamorphic males, is the primary reason to include the specimens of these stages in the *H. brevirostris* group.

If in the future, specific identity of males and females can be determined, it seems likely that the holotypes of the four nominal species based on males (*H. brevirostris*, *H. oncorhynchus*, *H. altirostris*, and *H. tumidus*) will be found to represent the most common species, *H. groenlandicus* (as assumed by Bertelsen, 1951:59). The possibility cannot be excluded, however, that one or more of these names might be found to represent senior synonyms for other species of the *H. groenlandicus* group.

*Himantolophus* Larvae

Figure 74

Bertelsen (1951) divided the known larval material of *Himantolophus* into two groups, which he designated "Type A" and "Type B." Type B larvae were characterized by having a distinct spot of pigment in front of the dorsal fin, in most specimens raised on a prominent swelling of the skin; Type A larvae were characterized by the absence of both pigment spot and swelling (for full description, see Bertelsen, 1951:61, fig. 23). About 70% of the 311 specimens examined by Bertelsen (1951) were designated Type A. Among the 50 specimens in the random sample of larvae examined by Bertelsen and Krefft (1988:74–75) about two-thirds had the Type B pigment spot. The largest of these larvae (24 mm, ISH 970/68), in which the dorsal swelling of the skin is especially well developed, is illustrated by Bertelsen and Krefft (1988:74, fig. 34).

Comments

With very few exceptions, the inflated skin of ceratioid larvae is completely unpigmented; thus the predorsal pigment spot found only in the Himantolophidae is assumed to be a derived character state. Furthermore it might be assumed that it represents a synapomorphy shared by the larvae of all the members of the more derived species groups of the family.

In all metamorphosed specimens greater than about 35 mm, pigmentation of the skin is too dense to distinguish the presence or absence of any remains of the black spot in front of the dorsal fin that distinctly separates larval and metamorphic specimens of *Himantolophus* into two groups. The available material contains a total of 23 females less than 35 mm (20 to 34.5 mm) in which the escal appendages have developed to the extent that specific identification is possible. These specimens represent eight of the 18 species recognized by Bertelsen and Krefft (1988): nine specimens (30 to 34.5 mm) representing three of the five members of the *Himantolophus groenlandicus* group; eight specimens of *H. appelii* (20 to 32 mm); three of *H. albinares* and one of *H. mauli* (28 to 34 mm) of the *H. albinares* group; and one each of *H. cornifer* (27 mm) and *H. macroceras* (32 mm) of the *H. cornifer* group. The smallest known representative of the *H. nigricornis* group is 35 mm, but it has completely black skin hiding any possible remains of the predorsal pigment spot. Several of the specimens of the *H. groenlandicus* group and *H. appelii* are sufficiently weakly pigmented to show that they lack the pigment spot. Among the four specimens of the *H. albinares* group, the pigment spot is very distinct in the two smallest individuals (see Bertelsen and Krefft, 1988:56, fig. 20) and present, but less distinct in the 34-mm specimen. In the 30.5-mm specimen of *H. mauli*, the skin is too heavily pigmented to distinguish remains of the pigment spot. Finally, the pigment spot is present in both representatives of the *H. cornifer* group, distinct in *H. cornifer*, and present but less conspicuous in *H. macroceras*.

The absence of the predorsal pigment spot in larval *H. appelii* is confirmed by the fact that this pigmentation is lacking in all the larvae that have been caught in the Tasman Sea, south of 30°S, where this species may be assumed to be the only representative of the genus (eight larvae, 3 to 6 mm TL, listed by Bertelsen, 1951:265; and two larvae, 14 to 15 mm recorded by Bertelsen and Pietsch, 1983:83).

The majority and possibly all of the Type A larvae caught north of 30°S in all oceans may represent the *H. groenlandicus* group, but the fact that the presence or absence of the pigment spot is unknown in the less common *H. nigricornis* group makes it possible that larvae of this group may be included. Similarly it may be assumed that the specimens with the dark pigment spot (Type B) represent species of either the *H. albinares* or the *H. cornifer* group, but might include larvae of the *H. nigricornis* group as well.

## Family Diceratiidae Regan and Trewavas, 1932 (Doublespine Seadevils)

Figures 2, 29, 43, 44, 48, 75–77, 202, 203, 215, 266, 292; Tables 1, 8

### Type genus *Diceratias* Günther, 1887

Diceratiidae Regan and Trewavas, 1932:4, 30, 57 (family of suborder Ceratioidea to contain genera *Diceratias*, *Paroneirodes*, and *Caranactis*).

Aeschynichthyidae Golvan, 1962:173 (unjustified replacement name for Diceratiidae based on Ogilby's, 1907:25, erection of *Aeschynichthys* to replace *Diceratias* Günther, the latter assumed to be preoccupied by *Diceratia* Oken, a genus of mollusks).

### Distinguishing Characters

Metamorphosed females of the family Diceratiidae are distinguished from those of all other ceratioid families by having an externally exposed second cephalic spine, bearing a distal light organ and emerging from the head behind the base of the first dorsal-fin spine (illicium). The structure is highly conspicuous in smaller specimens (less than about 30 mm), but withdrawn inside a narrow cavity in larger specimens (see Uwate, 1979:134, fig. 11).

Metamorphosed females are further differentiated by having the following combination of character states: The supraethmoid is present. The frontals are widely separated, each with a prominent ventromedial extension. The parietals are present. The sphenotic spines are well developed, but the pterosphenoid is reduced. The metapterygoid and mesopterygoid are present. The hyomandibula bears a double head. There are 2 hypohyals and 6 (2 + 4) branchiostegal rays. The opercle is bifurcate. The subopercle is without a posterior notch; the anterior margin of the ventral part bears a well-developed spine. Quadrate and articular spines are present, but angular and preopercular spines are absent. The jaws are subequal, the lower jaw extending anteriorly slightly beyond the upper jaw. The lower jaw bears a well-developed symphysial spine. A postmaxillary process of the premaxilla is absent. The anterior-maxillomandibular ligament is present. The pharyngobranchial of the first arch is rudimentary, while those of the second and third arches are well developed and toothed. The fourth pharyngobranchial is absent. The first epibranchial is well developed. There are 3 hypobranchials and a single ossified basibranchial. Epibranchial and ceratobranchial teeth are absent. Epurals are absent. The hypural plate is entire, without a posterior notch. The pterygiophore of the illicium bears 2 spines, the illicium and a second cephalic spine. The escal bulb has a central lumen and a pore to the outside. The esca is without toothlike denticles. A posteroventral process of the coracoid is absent. There are 3 pectoral radials. The pelvic bones are slightly expanded distally. There are 5 to 7 dorsal-fin rays, 4 anal-fin rays (very rarely 5; of 69 specimens counted, only one, the holotype of *Diceratias trilobus*, has 5 anal rays; see Balushkin and Fedorov, 1986), and 13 to 16 pectoral-fin rays (Table 1). Pelvic fins are absent. There are 9 caudal rays (very rarely 8; of 69 specimens counted, only one had 8 caudal rays), 1 simple + 6 bifurcated + 2 simple. The skin, including that of the illicium and proximal half of the escal bulb, is everywhere covered with small dermal spinules. The ovaries are paired. Pyloric caeca are absent.

The single known metamorphosed male (a juvenile specimen, 14 mm, LACM 36091-4) differs from those of all other ceratioid families in having the following combination of character states: The eyes and nostrils are large, directed laterally. The parietals, hyomandibular, subopercle, pectoral radials, and pelvic bones are as described for females. There is a single pair of denticular teeth on the snout and two transverse series of denticular teeth on the chin, each containing 4 or 5 separate teeth. Fin-ray counts are as given for metamorphosed females. The skin is covered with small dermal spinules (Bertelsen, 1983:310, fig. 1). The males are free-living, apparently never becoming parasitic (see Reproduction and Early Life History, Chapter Eight).

The larvae (only two known specimens, both females, 7 to 10.5 mm; ZMUC P922538, P92676) differ from those of all other ceratioid families in having the following combination of character states: The body is short, globose, and nearly spherical. The length of the head is approximately 60% SL. The skin is highly inflated. The pectoral fins are of normal size, not extending posteriorly to the base of the dorsal and anal fins. Pelvic fins are absent. Fin-ray counts are as given for metamorphosed females (Bertelsen, 1951:67, 69, fig. 28; 1984:330, fig. 169E).

The osteological features cited here and elsewhere in this account of the Diceratiidae are based on examinations of females of both *Diceratias* and *Bufoceratias* (Regan, 1912:286, fig. 6A, 1926:21, fig. 13; Regan and Trewavas, 1932:30–31, figs. 29, 30; Bertelsen, 1951:67, fig. 27; Pietsch, 1972a:31, 43, 45; Uwate, 1979:130–137, figs. 1–13), as well as a single male identified only to family (Bertelsen, 1983:310–312, fig. 1).

### Description

The body of metamorphosed females is short, globular, its depth approximately 50% SL (Figs. 2, 75, 76). The mouth is large, the cleft extending past the eye, the opening oblique. The oral valve is well developed, lining the inside of both the upper and lower jaws. There are two nostrils on each side of the snout at the end of a single short tube. The jaw teeth are slender, recurved, and depressible, arranged in overlapping sets (as described for other ceratioids; see Pietsch, 1972b). There are 14 to 65 teeth in the lower jaw and 12 to 99 in the upper jaw. There are 4 to 15 teeth on the vomer. The first epibranchial is free from the wall of the pharynx. All four epibranchials are closely bound together by connective tissue. The fourth epibranchial and ceratobranchial are bound to the wall of the pharynx, leaving no opening behind the fourth arch. The proximal two-thirds of the first ceratobranchial are bound to the wall of the pharynx, but the distal one-third is free. The distal end of the first ceratobranchial is free, not bound by connective tissue to the adjacent second ceratobranchial. The proximal one-quarter to one-half of ceratobranchials II to IV are free, not bound together by connective tissue. Gill filaments are absent on the epibranchials, but present on the proximal tip of the first ceratobranchial, the full length of the second and third ceratobranchials, and the distal three-quarters of the fourth ceratobranchial. A pseudobranch is absent. The length of the illicium of females is highly variable, 26 to 47% SL in *Diceratias*, 83 to 225% SL in *Bufoceratias*. The anterior end of the pterygiophore of the illicium is exposed, emerging on the snout (*Diceratias*), or concealed beneath the skin of the head, the illicium emerging on the back at the rear of the skull (*Bufoceratias*). The posterior end of the pterygiophore of the illicium is concealed beneath the skin of the head. There is a second cephalic spine (second dorsal-fin spine), with a distal light organ, emerging from the dorsal surface of the head just behind the base of the illicium. The second cephalic spine tends to sink beneath the skin of the head with age, but remains connected to the surface through a small pore. The lumen of the escal bulb is connected to the outside by a pore located on the posterior margin of the base of the terminal escal papilla. The internal pigment of the escal lumen is visible in lateral view, while the basal half of the escal bulb is usually covered with dark pigment. There are numerous, small, rounded, darkly pigmented papillae on the head and body associated with the acoustico-lateralis system, each with an unpigmented distal tip (Regan and Trewavas, 1932:23); the pattern of placement is as described for other ceratioids (Pietsch, 1969, 1972b, 1974a, 1974b).

The single known metamorphosed male (a juvenile specimen from the Halmahera Sea, 14 mm, LACM 36091-4; see Bertelsen, 1983:312, fig. 1) has relatively large eyes, about 1.2 mm (8.6% SL) in diameter, with a narrow aphakic space surrounding the lens. The olfactory organs are well separated from the eye, the vertical diameter of the posterior nostrils are about 0.5 mm, slightly larger than the anterior nostrils. The number of olfactory lamellae is less than 10 (no exact count possible). The frontals are broad, meeting on the midline. Crescent shaped parietals are present, but they are relatively small, their anterior tips just touching the posterior margin of the frontals. The opercle is bifurcate, the dorsal fork nearly as long (95%) as the ventral fork. The dorsal part of the subopercle is slender and tapering to a fine point; the ventral part is elongate and rounded, with a well-developed spine on the anterior margin. There are 6 dorsal-, 4 anal-, and 15 pectoral-fin rays. The caudal fin contains 9 rays, the ninth well developed and nearly one-half the length of the longest medial rays. All the caudal-fin rays are simple. The testes are oval in shape, about 2.0 mm in length and 0.9 mm in greatest width.

The premaxillae and dentaries of the male have irregularly resorbed edges. There are few larval teeth, only 2 to 4 on each premaxilla and 1 or 2 on each dentary. There is a pair of recurved denticular teeth on the snout lying slightly posterior to the symphysis of the upper jaw, each about 0.25 mm in length. There are 9 similar denticular teeth lying slightly behind the tip of the lower jaw, 8 of which are arranged in a regular symmetrical pattern consisting of an anterior and posterior transverse series of 4 teeth in each series. The ninth lower denticular tooth is the smallest, placed asymmetrically to the right of the lower series. The slender distal part of each of the four largest denticular teeth is about 0.25 to 0.30 mm in length (lying medial to the anterior series and lateral to the posterior series), emerging in an obtuse angle from a stout, nearly cylindrical base. All the denticular teeth are mutually free, without expanded connecting bases (see Bertelsen, 1983:310, fig. 1).

The pterygiophore of the illicium of the male is subdermal, its length 2.5 mm or 18% SL. The anterior end of the pterygiophore lies near the tip of the snout, the posterior end is connected to the anterior edge of the frontals by relatively strong extrinsic muscles (supracarinales anterior; see Bertelsen, 1951:17, figs. 3, 4, 1983: 312, fig. 1). An irregularly shaped rudiment of the second cephalic spine lies slightly posterior to the middle of the pterygiophore, connected to the anterior edge of the parietals by retractor muscles (posterior inclinatores dorsalis).

The skin of the male is everywhere covered with tiny conical dermal spinules, those on the tip of the snout and chin slightly larger, more sharply pointed, and more closely spaced. The rounded basal plates of the largest spinules are about 0.15 to 0.2 mm in diameter.

The larvae (two known specimens, both females, 7 to 10.5 mm; ZMUC P922538, P92676) are extremely similar despite the large difference in size. The eye diameter is about 1.1 to 1.2 mm, relatively larger in the smaller specimen. The skin is inflated, forming an almost perfect sphere. The head is very large, its length more than 50% SL, but the mouth is relatively small. Both specimens are females, with relatively large rudiments of 2 cephalic spines: the illicium arising just in front of the eyes, its length almost equal to the diameter of the eye; and the second cephalic spine, arising just behind the first, about one-half as large. The overall color of the head and body is light gray brown. There are tiny melanophores of almost uniform density over the entire body. Only the illicium and distal part of the fins are unpigmented. The second cephalic spine is pigmented with the same density as the rest of the skin. The inner pigmentation of the body is visible through the skin, consisting of very small branched melanophores, arranged without distinct groups. The dorsal surface is slightly darker than the belly. The melanophores are grouped slightly more densely along the margins between the myomeres. There are 5 to 6 dorsal-fin rays, 4 anal-fin rays, 14 to 15 pectoral-fin rays, and 9 caudal-fin rays (Bertelsen, 1951:67, 69, fig. 28, 1984:327, fig. 169E; Table 1).

The color of metamorphosed specimens is dark brown to black over the entire surface of the head, body, and oral cavity, except for the distal portion of the escal bulb. The dorsal, anal, and caudal fins are unpigmented in females less than about 50 mm. The skin of the male is brownish black, except for that associated with the olfactory organs and tip of the snout. The subdermal pigmentation is light, without distinct concentrations of melanophores.

The largest known specimen of the family is a 275-mm female of *Diceratias pileatus* (BPBM 30655) found floating on the surface off Kona, Hawaii. The only known metamorphosed male (LACM 36091-4) measures 14 mm.

Diversity

Two genera, differentiated as follows:

Key to Females of Genera of the Family Diceratiidae

The following key will differentiate metamorphosed female specimens only. Metamorphosis occurs between 10 and 15 mm, after which all taxa of the family can be readily determined.

1A. Pterygiophore of illicium emerging from snout, distance between point of emergence and symphysis of upper jaw less than 16% SL (Fig. 75) .................... *Diceratias* Günther, 1887, p. 354

   NOTE Three species; Atlantic, western Pacific, and Indian oceans

1B. Pterygiophore of illicium embedded beneath skin of head, illicium emerging from dorsal surface of head at rear of skull, distance between point of emergence and symphysis of upper jaw greater than 28% SL (Fig. 76) . . . . . . . . . . . *Bufoceratias* Whitley, 1931, p. 356

NOTE  Three species; Atlantic, western Pacific, and Indian oceans

## Genus *Diceratias* Günther, 1887
(Doublespine Seadevils)

Figures 2, 44, 48, 75, 202, 203, 215, 266; Tables 1, 8

### Females and Larvae

*Ceratias* Günther, 1887:52, 53, pl. 11, fig. B (in part; type species *Ceratias holboelli* Krøyer, 1845, by monotypy). Alcock, 1899:56 (in part; includes *Diceratias*, *Mancalias*, *Typhloceratias*, and *Cryptopsaras*).

*Diceratias* Günther, 1887:52, 53, pl. 11, fig. B (type species *Ceratias* [subgenus *Diceratias*] *bispinosus* Günther, 1887, by monotypy). Regan, 1926:20, 21, 42, figs. 13, 24 (elevation of subgenus to generic rank; includes *Paroneirodes* and *Aeschynichthys*). Karrer, 1973b:246, figs. 28, 29 (includes *Paroneirodes* and *Phrynichthys*).

*Paroneirodes* Alcock, 1890:206, pl. 9, fig. 6 (type species *Paroneirodes glomerosus* Alcock, 1890, by monotypy).

*Onirodes* Alcock, 1899:57 (in part; emendation of *Oneirodes*, therefore taking the same type species *Oneirodes eschrichtii* Lütken, 1871, by monotypy).

*Aeschynichthys* Ogilby, 1907:25 (unjustified replacement name for *Diceratias* Günther, the latter assumed by Ogilby to be preoccupied by *Diceratia* Oken, a genus of mollusks; taking the same type species *Ceratias* [subgenus *Diceratias*] *bispinosus* Günther, 1887). Golvan, 1962:173 (after Ogilby, 1907).

*Paraneirodes* Maurin et al., 1970:21 (erroneous spelling of *Paroneirodes*, therefore taking the same type species *Paroneirodes glomerosus* Alcock, 1890).

### Males

Unknown.

### Distinguishing Characters

Females of the genus *Diceratias* are distinguished from those of *Bufoceratias* (the only other genus of the family) in having the following character states: The length of the illicium is 26 to 47% SL. The anterior tip of the pterygiophore of the illicium is exposed, emerging on the snout. The distance between the point of emergence of the pterygiophore and the symphysis of the upper jaw is 7 to 15% SL. The anterior margin of the supraethmoid forms an angle of approximately 52° with the horizontal plane of the cranium (see Uwate, 1979:132, fig. 2). The illicial trough is deep (Uwate, 1979:132, fig. 5). The dermal spinules are large, their length 230 to 380 µm.

### Description

As given for the family.

### Diversity

Three species as follows:

### Key to Females of Species of the Genus *Diceratias*

1A. Esca with anterior and posterior appendages well developed; terminal escal papilla absent or only moderately developed . . . . . . . . . . . . . . 2
1B. Esca with anterior and posterior escal appendages absent; terminal escal papilla large, bulbous, with a small pointed distal appendage in specimens greater than about 100 mm . . . . . . . . . . . . . . . *Diceratias pileatus* Uwate, 1979, p. 355

NOTE  One hundred and fifty-three known specimens, 20 to 275 mm; Atlantic, central Pacific, and eastern South Indian oceans

2A. Esca width and length approximately equal; anterior escal appendage bulbous, sometimes branched, posterior escal appendage with a row of filaments running dorsoventrally . . . . . . . . . . . . . . . . . . . . . . . . . . *Diceratias bispinosus* Günther, 1887, p. 354

NOTE  Ten known specimens, 20 to 77 mm; Indo-west Pacific Ocean

2B. Esca width more than 1.5 times length; anterior and posterior escal appendages each bearing a slender terminal filament . . . . . . . . . . . . . . . . . . . . . . . . . . *Diceratias trilobus* Balushkin and Fedorov, 1986, p. 356

NOTE  Twenty-four known specimens, 47 to 140 mm; western Pacific and eastern Indian oceans

## *DICERATIAS BISPINOSUS* GÜNTHER, 1887

Figure 215; Table 8

### Females

*Ceratias* (*Diceratias*) *bispinosus* Günther, 1887:50, 52, 53, pl. 11, fig. B (original description, single specimen). Lütken, 1894:78 (after Günther, 1887). Alcock, 1899:56 (additional specimen, off Malabar, ZSI F-14008). Fowler, 1936:1343 (after Günther, 1887).

*Paroneirodes glomerosus* Alcock, 1890:206, pl. 9, fig. 6 (original description, single specimen). Lütken, 1894:78 (after Alcock, 1890). Alcock, 1896:318 (after Alcock, 1890). Goode and Bean, 1896:493, pl. 119, fig. 404 (after Alcock, 1890). Gill, 1909:579, 580, fig. 15 (after Alcock, 1890). Grey, 1956a:242 (synonymy, distribution). Menon and Yazdani, 1963:165 (holotype listed in type catalog). Menon and Rama Rao, 1975:48 (after Menon and Yazdani, 1963).

*Ceratias bispinosus*: Alcock, 1896:318 (after Günther, 1887). Goode and Bean, 1896:488, 489 (after Günther, 1887). Alcock, 1900:353, pl. 35, fig. 2 (figured; after Alcock, 1899). Annandale and Jenkins, 1910:17 (Indian seas). Fowler, 1936:1343 (after Günther, 1887).

*Onirodes glomerosus*: Alcock, 1899:57 (new combination, emendation of *Oneirodes*; description after Alcock, 1890). Alcock, 1900:353, pl. 28, fig. 4 (figured; after Alcock, 1899). Alcock, 1902:236, 240, 247, fig. 32 (eyes and vision, luring function of first dorsal-fin spine, gill-arch morphology).

*Oneirodes glomerosus*: Annandale and Jenkins, 1910:17 (new combination; Indian seas). Grey, 1956a:243 (in synonymy of *Paroneirodes glomerosus*).

*Diceratias bispinosus*: Regan, 1912:286, 287, fig. 6A (new combination; *Diceratias* raised to generic status). Regan, 1926:20, 42, figs. 13, 24 (after Günther, 1887, and Alcock, 1899, 1900). Regan and Trewavas, 1932:30, 31, 58, figs. 29A, 30, 85A (after Regan, 1926). Bertelsen, 1951:67, 69 (description after Günther, 1887, and Alcock, 1899; first known larval diceratiids, description). Grey, 1956a:242 (synonymy, distribution). Beaufort and Briggs, 1962:246, fig. 58 (description after Regan, 1926, and Regan and Trewavas, 1932; figure after Alcock, 1899). Munro, 1967:578, pl. 77 (New Guinea). Pietsch, 1972a:31, 45 (osteological comments). Paxton and Lavenberg, 1973:47, figs. 1, 2 (misidentification, specimen is *Diceratias trilobus*; feeding mortality). Parin and Golovan, 1976:271 (off West Africa). Uwate, 1979:139, figs. 16, 20 (comparison with all known material). Pietsch and Seigel, 1980:382 (Halmahara Sea, in key; after Uwate, 1979). Bertelsen and Pietsch, 1983:83 (Indonesian waters). Balushkin and Fedorov, 1986:855, fig. B (comparison with congeners, in key). Ni et al., 1989:88, 93, figs. 1, 3A–C (misidentifications, material is *D. trilobus*, two additional specimens, South China Sea). Stewart and Pietsch, 1998:10 (Indonesian waters). Pietsch, 1999:2030 (western central Pacific).

*Dicerathias glomerulosus*: Regan, 1926:42 (new combination, emendation of specific name; description after Alcock, 1890, 1899).

*Paroneirodes glomerulosus*: Bertelsen, 1951:70 (reference to known material, in key; erroneous spelling of specific name after Regan, 1926:42).

### Males and Larvae

Unknown.

Material

Ten metamorphosed females (20 to 77 mm):

Holotype of *Ceratias (Diceratias) bispinosus*: BMNH 1887.12.7.14, 52 mm, *Challenger* station 194A, off Banda Island, 659 m (360 fathoms), 1200 to 1330 h, 29 September 1874.

Holotype of *Paroneirodes glomerosus*: ZSI F-12840, 21 mm, *Investigator* station 103, Bay of Bengal, off Madras, 15°14′N, 81°09′E, 2306 m (1260 fathoms), 2 April 1890.

Additional material: ASIZP, one (28 mm); CSIRO, one (77 mm); LACM, one (20 mm); MNHN, two (30 to 35 mm); NSMT, one (34 mm); SIO, one (21 mm); USNM, one (25.5 mm).

Distinguishing Characters

Metamorphosed females of *Diceratias bispinosus* differ from those of all other species of the genus in having a relative small esca, width about equal to length; a low rounded terminal escal papilla; anterior and posterior escal appendages well developed, both usually bearing small secondary filaments.

Description

Esca with terminal papilla low and rounded to absent; anterior escal appendage bulbous, sometimes branched; posterior escal appendage laterally compressed, with a row of as many as 5 small secondary filaments along outer margin; escal pore at posterobasal margin of terminal papilla; number of teeth in lower jaw 14 to 65, in upper jaw 18 to 99; vomerine teeth 4 to 9.

Additional description as given for the genus and family.

Distribution

The holotype of *Paroneirodes glomerosus* was collected in the Bay of Bengal, off Madras, in the eastern Indian Ocean, but all the remaining material is from the western Pacific where it ranges from Japan, Taiwan, the Philippines, and the Halmahera and Banda seas, to New Ireland in the Bismarck Archipelago in the east and off Rowley Shoals, western Australia, in the west (Fig. 215). A 20-mm specimen (LACM 36075-1) was captured in a closing trawl between 1000 and 1400 m. All the remaining specimens were taken in open nets fished at maximum depths of 533 to 2306 m.

## DICERATIAS PILEATUS UWATE, 1979

Figures 48, 215; Table 8

Females

*Diceratias bispinosus*: Grey, 1959:225, 226, fig. 1 (misidentification, single specimen, FMNH 64543).

*Diceratias pileatus* Uwate, 1979:140, figs. 1, 2, 4, 5, 7A, 8, 10A, 11A, 12, 17, 20 (original description, 32 metamorphosed females; osteology, comparison with all known material). Bertelsen, 1983:313 (evidence that diceratiid males do not become sexual parasitic on females). Fujii, 1983:259, fig. (13 additional specimens, off Surinam and French Guiana). Balushkin and Fedorov, 1986:855, fig. C (comparison with congeners, in key). Pietsch and Randall, 1987:419, figs. 1–3 (additional specimen, first Pacific record). Ni et al., 1989:87, 90, fig. 3D, E (esca figured; after Uwate, 1979). Bertelsen, 1990: 496 (synonymy, biology; after Uwate, 1979). Pietsch, 2002b:1062, fig. (western central Atlantic). Pietsch, 2005b:220, 215 (reproduction).

*Oneirodes notius*: Figueiredo et al., 2002:145, color photograph (misidentification, single specimen, MZUSP 78220); Menezes et al., 2003 (misidentification, after Figueiredo et al., 2002).

Males and Larvae

Unknown.

Material

One hundred and fifty-three metamorphosed females (20 to 275 mm):

Holotype of *Diceratias pileatus*: UF 23774, 82 mm, *Oregon II* station 10616, western North Atlantic, off Surinam, 7°37′N, 53°32′W, 0 to 722 m, 13 May 1969.

Paratypes of *Diceratias pileatus*: FMNH 64543, 59.5 mm, *Oregon* station 2010, off Surinam, 7°44′N, 54°40′W, 640 m. FMNH 83822, 100 mm, *Oregon* station 2011, off Surinam, 7°46′N, 54°36′W, 730 m. LACM 37441-1, 49 mm, *Poltawa* station 70, Golovan No. 18, off Guinea, 9°55′N, 17°33′W, 750 to 780 m. SIOM uncataloged, two (33 to 36.5 mm), *Poltawa* station 70, Golovan No. 18, off Guinea, 9°55′N, 17°33′W, 750 to 780 m. SIOM uncataloged, three (138 to 197 mm), *Zwiezda Kryma* station 276, Golovan No. 23, off Guinea Bissau, 12°00′N, 17°32′W, 1380 to 1400 m. UF 23771, 109 mm, *Oregon II* station 10604, off Surinam, 7°49′N, 54°22′W, 731 m, 10 May 1969. UF 23772, nine (45 to 78.5 mm), *Oregon II* station 10603, off Surinam, 7°48′N, 54°25′W, 644 m, 10 May 1969. UF 23773, two (45.5 to 63.5 mm), *Oregon II* station 10606, off Surinam, 7°41′N, 53°48′W, 677 m, 5 October 1969. UF 23775, 96.5 mm, *Oregon II* station 10614, off Surinam, 7°06′N, 52°44′W, 686 m, 13 May 1969. UF 23776, two (82 to 84 mm), *Oregon II* station 10620, off Surinam, 7°47′N, 54°05′W, 768 m, 15 May 1969. UF 23777, four (96 to 114 mm), *Oregon II* station 10620, off Surinam, 7°47′N, 54°05′W, 768 m, 13 May 1969. UF 215166, 45.5 mm, *Pillsbury* station 74, off Liberia, 4°20′ to 4°30′N, 9°26′ to 9°22′W, 732 m, 4 June 1964. UF 221660, 21 mm, *Pillsbury* station 309, off Nigeria, 4°15′ to 4°12′N, 4°27′ to 4°28′E, 1281 to 1318 m, 26 May 1965. UF 233052, 235 mm, *Columbus Iselin* station 145, Bahamas, 24°15.6′ to 24°13.5′N, 77°18.8′ to 77°18.5′W, 1430 m, 3 February 1974. USNM 217527, 41 mm, *La Rafale*, Guinean Trawling Survey I, Transect 5, station 8, off Guinea, 9°32′N, 16°20′W, 0 to 400 m, 1 December 1963.

Additional material: AMS, one (58 mm); BMNH, one (45 mm); BPBM, one (275 mm); CAS, 42 (35 to 50 mm); FMNH, 17 (29.5 to 116 mm); HUMZ, 20 (59 to 140 mm); MNHN, three (24 to 116 mm); NSMT, 20 (27 to 167 mm); SAIAB, 12 (20 to 67 mm); UW, three (38 to 118 mm); ZMUC, one (38 mm).

Distinguishing Characters

Metamorphosed females of *Diceratias pileatus* differ from those of all other species of the genus in having a relative small esca, width about equal to length; a conspicuous bulbous terminal escal papilla; anterior and posterior escal appendages absent.

Description

Esca with a bulbous posteriorly oriented terminal papilla, rounded in smaller specimens, becoming pointed by 100 mm, often capped with a small pointed terminal appendage; escal pore at posterobasal margin of terminal papilla; number of teeth in lower jaw 25 to 65, in upper jaw 27 to 82; vomerine teeth 4 to 14.

Additional description as given for the genus and family.

Sexual Maturity

The only known sexually mature diceratiid is a 235-mm female of *Diceratias pileatus* (UF 233052) reported by Uwate (1979:141). The ovaries of this specimen each contain $10^4$ to $10^5$ pear-shaped eggs, measuring 0.3 to 0.7 mm in diameter. The ovaries of the largest known specimen of the family, a 275-mm female of *D. pileatus* (BPBM 30655), are large, their length approximately 80 mm or 29% SL, but they contain no eggs and appear to be spent (Pietsch and Randall, 1987).

Distribution

Except for a single specimen trawled off Port Hedland, Western Australia, and another found floating on the surface off Kona, Hawaii (Pietsch and Randall, 1987), all of the known material of *Diceratias pileatus* is from the Atlantic Ocean (Uwate, 1979:141, fig. 20). In the western Atlantic, it has been captured in the Bahamas and adjacent waters, but all remaining records (39 females) are from off French Guiana and Surinam. On the eastern side of the

Atlantic, it is known from a number of localities along the West African coast, ranging from off Guinea Bissau to Angola (74 females), as far south as about 13°S. A large majority of the specimens (about 113) were collected on the bottom at depths of 660 and 910 m. All the remaining specimens were taken in pelagic nets fished open, one individual (USNM 217527, 41 mm) in gear that reached a maximum depth of only 400 m, the remaining in gear that reached maximum depths of 640–1430 m.

### DICERATIAS TRILOBUS BALUSHKIN AND FEDOROV, 1986

Figures 2, 75, 215, 266; Table 8

#### Females

*Diceratias bispinosus*: Paxton and Lavenberg, 1973:47, figs. 1, 2 (misidentification, AMS I.15602-001; Bismarck Sea off New Ireland, feeding mortality). Uwate, 1979:139 (misidentification, after Paxton and Lavenberg, 1973).

*Diceratias trilobus* Balushkin and Fedorov, 1986:855, fig. A (original description, single specimen, in key). Pietsch et al., 2006:98, figs. 1–3 (additional material from off Australia and eastern Indian Ocean). Pietsch and Orr, 2007:3, fig. 2D (phylogenetic relationships).

#### Males and Larvae

Unknown.

#### Material

Twenty-four metamorphosed females (47 to 140 mm):

Holotype of *Diceratias trilobus*: ZIN 47426, 122 mm, *Shantar* trawl 28, 38°20.7'N, 142°31.9'E, 1211 to 1216 m, 28 March 1975.

Additional material: AMS, four (80 to 114 mm); CSIRO, 15 (47 to 140 mm); SCFR, two (110 to 126 mm); SFU, two (79 to 134 mm).

#### Distinguishing Characters

Metamorphosed females of *Diceratias trilobus* differ from those of all other species of the genus in having an unusually large, laterally compressed esca, greatest width slightly more than 1.5 times its length (9.6 to 10.5% SL); a rounded terminal escal papilla; and anterior and posterior escal appendages well developed, each usually bearing 1 or more, tiny, slender terminal filaments.

#### Description

Esca laterally compressed, with large, rounded anterior and posterior appendages; distal margin of anterior and posterior escal appendages smooth, usually with 1 to 6 tiny, slender filaments; posterior escal appendage slightly longer than anterior appendage, length, including slender terminal filament, more than one-half width of escal bulb; tiny, close-set, dermal spinules of skin extending out onto esca, covering entire dorsal and lateral surfaces, including distal filaments, in largest known specimens; proximal half of escal bulb, and distal half of slender terminal filaments of some specimens, darkly pigmented, remaining parts of esca unpigmented; escal pore at posterobasal margin of a rounded terminal papilla.

Measurements in percent of standard length: head length 40.0 to 50.1; head width 27.0 to 37.2; head depth 50.1 to 55.7; premaxilla length 41.2 to 47.5; lower jaw length 54.4 to 60.7; illicium length 27.1 to 33.7; greatest width of esca 9.6 to 10.5; least width of esca 2.7 to 4.3; length of esca 5.8 to 7.9; longest tooth in upper jaw 6.1 to 7.3. Number of teeth in upper jaw 67 to 111, in lower jaw 51 to 68; vomerine teeth 6 to 10; dorsal-fin rays 5 or 6, anal-fin rays 4 or 5, pectoral-fin rays 14 or 15.

Additional description as for the genus and family.

#### Distribution

The known material of *Diceratias trilobus* ranges from the type locality off Honshu, Japan, south into temperate waters of Australia and the eastern Indian Ocean (Fig. 215). Except for the 114-mm specimen (AMS I.15602-001), found dead on the surface in the Bismarck Sea off New Ireland (*D. bispinosus* of Paxton and Lavenberg, 1973), all the material was trawled off the bottom at depths of 503 to 1216 m.

### *Diceratias* Species

Figure 77

#### Larvae

*Diceratias bispinosus*: Bertelsen, 1951:69, figs. 27, 28 (first known larval diceratiids, two specimens, description, osteology). Uwate, 1979:140 (after Bertelsen, 1951).

#### Material

Two larval females (6.5 to 10.5 mm, 10 to 15 mm TL): ZMUC P922538, 6.5 mm, *Dana* station 3770(4), Celebes Sea, 0°54'S, 137°29'E, 300 m of wire, 2345 h, 25 July 1929. ZMUC P92676, 10.5 mm, *Dana* station 3912(3), Indian Ocean, 6°52'S, 79°30'E, 300 m of wire, 0050 h, 24 November 1929.

#### Description

As given for the genus and family.

#### Comments

These two larval diceratiids, first identified and described by Bertelsen (1951:69, figs. 27, 28), have 2 distinct cephalic rays, clearing placing them within the Diceratiidae. The anterior insertion of the illicium on the snout and the distinct well-developed socket on the posterior margin of the hyomandibula for the articulation of the opercle identify them as *Diceratias* (see Uwate, 1979:133, figs. 8, 9). The escal morphology of the larval stages is not differentiated, thus specific determination is not possible, but the capture localities of these two specimens (Celebes Sea and Indian Ocean) coincide with the known geographic distribution of *Diceratias bispinosus* (western Pacific and Indian oceans) and thus probably belong to this species (identified as such by Bertelsen, 1951).

### Genus *Bufoceratias* Whitley, 1931
(Toady Seadevils)

Figures 29, 44, 48, 76, 202, 203, 215; Tables 1, 8

#### Females

*Phrynichthys* Pietschmann, 1926:88 (type species *Phrynichthys wedli* Pietschmann, 1926, by monotypy). Golvan, 1962:173 (name replaced by *Bufoceratias* following Whitley, 1931). Eschmeyer, 1990:313 (*Phrynichthys* "incorrectly treated as valid" following Whitley, 1931, and Golvan, 1962).

*Paroneirodes* Norman, 1930:356, fig. 46 (in part; based on misidentification; type species *Paroneirodes glomerosus* Alcock, 1890, by monotypy).

*Bufoceratias* Whitley, 1931:334 (replacement name for *Phrynichthys* Pietschmann, 1926, preoccupied by *Phrynichthys* Agassiz, 1846, a replacement name for *Bufichthys* Swainson, 1839, and a junior synonym of *Synanceia* Bloch and Schneider; therefore taking the same type species *Phrynichthys wedli* Pietschmann, 1926). Golvan, 1962:173 (replacement name for *Phrynichthys* following Whitley, 1931). Eschmeyer, 1990:68 (valid, after Whitley, 1931). Anderson and Leslie, 2001:10 (valid, after Whitley, 1931; diagnosis).

#### Males and Larvae

Unknown.

#### Distinguishing Characters

Females of the genus *Bufoceratias* are distinguished from those of *Diceratias* (the only other known genus of the family) in having the following character states: The length of the illicium is 25 to 225% SL. The anterior tip of the pterygiophore of the illicium is concealed beneath the skin. The illicium emerges from the dorsal surface of the head at the rear of the skull; the distance from the point of emergence to the symphysis of the upper jaw is 29 to 61% SL. The anterior margin of the supraethmoid forms an angle of approximately 65 to 74° with the horizontal plane of the cranium (see Uwate, 1979: 132, figs. 2, 3). The illicial trough is relatively shallow (Uwate, 1979: 132, figs. 5, 6). The dermal spinules are minute, their length 90 to 110 μm.

Description

As given for the family.

Diversity

Three species as follows:

Key to Metamorphosed Females of Species of the Genus *Bufoceratias*

1A. Illicium short, 25 to 40% SL; esca large, with numerous, long slender filaments ....... *Bufoceratias shaoi* Pietsch, Ho, and Chen, 2004, p. 358

NOTE   Four known specimens, 55 to 101 mm; western Pacific and western Indian oceans

1B. Illicium long, 83 to 225% SL; esca small, with or without short filaments ..................... 2

2A. Esca with a small rounded, sometimes pointed terminal papilla; anterior, posterior, and lateral escal appendages absent ............. ............... *Bufoceratias thele* (Uwate, 1979), p. 358

NOTE   Fourteen known specimens, 22 to 204 mm; western Pacific Ocean

2B. Esca with a low terminal papilla, becoming elongate by 60 mm; anterior, posterior, and lateral escal appendages present ............. ............... *Bufoceratias wedli* (Pietschmann, 1926), p. 357

NOTE   Eighty-two known specimens, 19 to 196 mm; Atlantic Ocean

*BUFOCERATIAS WEDLI* (PIETSCHMANN, 1926)

Figures 29, 44, 48, 76, 215; Table 8

Females

*Phrynichthys wedli* Pietschmann, 1926:88 (original description, single specimen). Pietschmann, 1930:419, fig. (description after Pietschmann, 1926; figured). Fowler, 1936:1344, fig. 564 (description and figure after Pietschmann, 1926). Uwate, 1979:142, figs. 3, 6, 7B, 9, 10B, 11B, 13, 18, 20 (comparison with all known material). Costa, 1980:89, fig. (additional specimen, off Portugal). Lloris, 1981:54, figs. 1–4 (additional specimen, Namibia). Fujii, 1983:260, fig. (additional specimen, off Surinam). Bertelsen, 1986:1381, fig. (eastern North Atlantic). Lloris, 1986:244, figs. 125–127 (four additional specimens, Namibia). Pietsch, 1986a:377, fig. 109.1 (eastern South Atlantic; figure after Pietschmann, 1930). Ni et al., 1989:87, 90, fig. 3F–H (esca figured; after Uwate, 1979). Bertelsen, 1990:496 (synonymy, biology; after Uwate, 1979). Machida and Yamakawa, 1990:60, figs. 1–3 (additional specimen, East China Sea; after Yamakawa, 1984). Marshall, 1996:239, 248, 250, figs. 1, 2, 7 (lateral line anatomy). Montgomery and Pankhurst, 1997:338, fig. 5 (photograph). Shedlock et al., 1997:396, fig. 1 (DNA extracted from formalin-fixed specimens). Bertelsen, 1994:141, color plate (popular account). Bertelsen and Pietsch, 1998a:141, color photograph (after Bertelsen, 1994). Bertelsen and Pietsch, 1998a:141, color photograph (popular account). McEachran and Fechhelm, 1998:852, fig. (Gulf of Mexico). Nakabo, 2002:475, figs. (Japan, in key). Pietsch, 2002b:1062 (western central Atlantic). Shinohara et al., 2005:420 (off Ryukyu Islands).

*Paroneirodes glomerosus*: Norman, 1930:356, fig. 46 (in part; misidentification of additional specimen, BMNH 1930.1.12.1101, South Atlantic, specimen referred to *Phrynichthys wedli* by Uwate, 1979). Regan and Trewavas, 1932:31, 38, figs. 29B, 85B (description after Norman, 1930). Grey, 1956a:242 (synonymy, distribution; description after Norman, 1930).

*Paroneirodes wedli*: Regan and Trewavas, 1932:58 (new combination; description after Pietschmann, 1926). Fowler, 1936:1364 (after Regan and Trewavas, 1932). Grey, 1956a:243 (synonymy, distribution; description after Pietschmann, 1926). Grey, 1959:225, 227 (description, single specimen, FMNH 64469, western Atlantic). Bertelsen, 1951:70 (listed; after Pietschmann, 1926). Albuquerque, 1954–1956:1057, fig. 432 (after Fowler, 1936; Portugal). Pietsch, 1972a:31 (osteological comments). Trunov, 1974:145, figs. 1, 2 (three additional specimens, eastern South Atlantic off Namibia).

*Diceratias glomerosus*: Fowler, 1936:1344, fig. 563 (misidentification; figure after Norman, 1930).

*Paroneirodes glomerulosus*: Maul, 1962a:12, figs. 4–6 (in part; misidentification of two specimens referred to *Paroneirodes wedli* by Uwate, 1979; erroneous spelling of specific name after Regan, 1926; eastern Atlantic). Maul, 1973:669 (after Maul, 1962a). Parin and Golovan, 1976:271 (off West Africa).

*Paraneirodes wedli*: Maurin et al., 1970:21 (additional specimen, eastern Atlantic; erroneous spelling of generic name).

*Diceratias wedli*: Karrer, 1973b:246, figs. 28, 29 (new combination; description of four additional specimens, ZMB 22194, South Atlantic).

*Bufoceratias wedli*: Anderson and Leslie, 2001:10, figs. 8, 9 (new combination; additional material, Namibia). Pietsch et al., 2004:103, figs. 2, 3, 5 (generic revision, additional material). Pietsch and Orr, 2007:3, 18, figs. 2E, 15D (phylogenetic relationships).

Males and Larvae

Unknown.

Material

Eighty-two metamorphosed females (19 to 196 mm):

Holotype of *Phrynichthys wedli*: NMW 3524, 35 mm, off Madeira, Steindachner collection, 1865.

Additional material: ARIK, three (32 to 49 mm); BMNH, one (19 mm); BSKU, one (72 mm); CAS, five (20 to 110 mm); FMNH, eight (27 to 62 mm); HUMZ, three (56 to 142 mm); IFAN, two (25 to 30.5 mm); IIPB, five (30 to 134 mm). LACM, three (44 to 77 mm); MBL, one (70 mm); MNHN, eight (24 to 106 mm); MNRJ, three (24 to 196 mm); NSMT, one (35 mm); SAIAB, five (29 to 117 mm); SAM, three (69 to 117 mm); SIOM, nine (29 to 178 mm); UF, nine (26 to 91 mm); USNM, one (140 mm); UW, one (70 mm); ZIN, one (132 mm); ZMB, four (78 to 117 mm); ZMUC, two (94 to 117 mm).

Distinguishing Characters

Metamorphosed females of *Bufoceratias wedli* differ from those of *B. shaoi* in having a much smaller esca and a longer illicium (83 to 225% SL versus 25 to 40% SL) and from those of *B. thele* in having distinct anterior, posterior, and lateral escal appendages.

Description

Esca with a small terminal papilla, rounded in smaller specimens, becoming elongate and cylindrical by 60 mm (Pietsch et al., 2004:104, fig. 3B); escal pore at posterobasal margin of terminal papilla; a pair of filamentous anterior escal appendages developing by 40 mm, remaining separate at base; a single bifurcated filamentous posterior escal appendage; a pair of lateral escal appendages (usually branched) developing by 40 mm; increased branching and elongation evident in all escal appendages by 50 mm; number of teeth in lower jaw 16 to 44, in upper jaw 21 to 65; vomerine teeth 7 to 15; dorsal-fin rays 5 or 6; anal-fin rays 4; pectoral-fin rays 13 or 14 (Table 1).

Additional description as given for the genus and family.

### Distribution

As concluded by Uwate (1979, fig. 20), *Bufoceratias wedli* is restricted to the Atlantic Ocean (Machida and Yamakawa's, 1990, report of a specimen from the East China Sea, is a misidentification; see below). In the western Atlantic it is known from the Gulf of Mexico and Caribbean Sea, and southward to coastal waters off Surinam and southern Brazil. In the eastern Atlantic it ranges from off Portugal to coastal waters of Namibia, as far south as about 24°S (Pietsch et al., 2004:105; Fig. 215). All of the specimens were collected with open trawls: four individuals (43 to 178 mm) in nets fished at maximum depths of 1200 to 1750 m, the remaining specimens by nets fished at maximum depths of 400 to 1000 m. Several specimens were taken in bottom trawls.

### BUFOCERATIAS THELE (UWATE, 1979)

Figure 215; Table 8

#### Females

*Phrynichthys thele* Uwate, 1979:142, figs. 19, 20 (original description, two specimens). Pietsch and Seigel, 1980:382 (Ceram and Halmahera seas, in key; after Uwate, 1979). Bertelsen and Pietsch, 1983:83 (Indonesian waters). Ni, 1988: 327, fig. 256 (two additional specimens, East China Sea). Ni et al., 1989:89, 94, figs. 2, 3I, J (after Ni, 1988). Machida and Yamakawa, 1990:63 (comparison with *Bufoceratias wedli*). Meng et al., 1995:440, fig. 590 (China). Stewart and Pietsch, 1998:10 (expected to occur in New Zealand waters). Pietsch, 1999:2030 (western central Pacific).

*Phrynichthys* sp.: Yamakawa, 1984:289, 385, fig. 204 (additional specimen, BSKU 29877, Okinawa Trough).

*Phrynichthys wedli*: Machida and Yamakawa, 1990:60, figs. 1–3 (misidentification, single specimen, BSKU 29877, Okinawa Trough, East China Sea).

*Bufoceratias thele*: Pietsch et al., 2004: 104, figs. 4, 5 (generic revision, additional material). Pietsch, 2005b:220, 215 (reproduction).

#### Males and Larvae

Unknown.

#### Material

Fourteen metamorphosed females (22 to 204 mm):

Holotype of *Phrynichthys thele*: LACM 36077-1, 32 mm, *Alpha Helix* station 155, Halmahara Sea, 0°38.6′S, 129°05.6′E, 680 to 850 m, 1210 to 1400 h, 22 May 1975.

Paratype of *Phrynichthys thele*: LACM 36076-1, 22 mm, *Alpha Helix* station 26, Ceram Sea, 2°46.0′S, 127°53.7′E, 0 to 1500 m, 0150 to 0800 h, 31 March 1975.

Additional material: ASIZP, two (200 to 204 mm); BSKU, one (72 mm; Yamakawa, 1984); CSIRO H.2285-02, one (97 mm, with a paired cluster of extra illicial filaments below base of esca); HUMZ, six (36 to 103 mm); SCFR, two (115 to 162 mm).

#### Distinguishing Characters

Metamorphosed females of *Bufoceratias thele* differ from those of *B. shaoi* in having a much smaller esca and a longer illicium (112 to 143% SL versus 25 to 40% SL), and from those of *B. wedli* in lacking escal appendages (Uwate, 1979, fig. 19; Pietsch et al., 2004:104, fig. 4).

#### Description

Esca with a small rounded, sometimes pointed terminal papilla arising from distal surface; escal pore at postero-basal margin of terminal papilla; anterior, posterior, and lateral escal appendages absent (Uwate, 1979, fig. 19); number of teeth in lower jaw 15 to 48, in upper jaw 12 to 49; vomerine teeth 4 to 10; dorsal-fin rays 5 or 6; anal-fin rays 4; pectoral-fin rays 13 or 14 (Table 1).

Additional description as given for the genus and family.

#### Distribution

*Bufoceratias thele* is known only from the western Pacific Ocean, ranging from the East China Sea, south to Taiwan, the Ceram and Halmahera seas, and off the Great Barrier Reef, Australia (Pietsch et al., 2004:105; Fig. 215). The holotype was collected between 680 and 850 m, the paratype somewhere between the surface and 1500 m, the BSKU specimen between 780 and 810 m (Machida and Yamakawa, 1990), and the two specimens reported by Ni (1988) between the surface and 950 m. The six HUMZ specimens were all taken in bottom trawls at depths of 996 to 999 m.

### BUFOCERATIAS SHAOI PIETSCH, HO, AND CHEN, 2004

Figures 76, 215; Table 8

#### Females

*Phrynichthys* sp.: Uwate, 1979: 143, fig. 21 (unidentified specimen, MNHN 1977-304, unique among known material of the genus). Machida and Yamakawa, 1990:63 (after Uwate, 1979; thought to be *Bufoceratias wedli* with regenerated illicial apparatus).

*Bufoceratias shaoi* Pietsch, Ho, and Chen, 2004:100, figs. 1, 3A, 5 (original description, four specimens). Pietsch and Orr, 2007:3, fig. 2F (phylogenetic relationships).

#### Males and Larvae

Unknown.

#### Material

Four metamorphosed females (55 to 101 mm):

Holotype of *Bufoceratias shaoi*: ASIZP 61796, 101 mm, off northeast coast of Taiwan, 24°25 to 50′N, 122°00 to 10′E, bottom trawl, 0 to 800 m, 1999.

Paratypes of *Bufoceratias shaoi*: ASIZP 59952, 2 (56 to 75 mm), off northeast coast of Taiwan, 24°55′N, 122°04′E, bottom trawl, 0 to 650 m, 20 March 1998; MNHN 1977-304, 55 mm, Mozambique Channel, 17°36′ to 22°25′S, 42°59′ to 43°56.5′E, 0 to 1200 m.

#### Distinguishing Characters

Metamorphosed females of *Bufoceratias shaoi* differ from those of *B. wedli* and *B. thele* in having a considerably shorter illicium (25 to 40% SL versus 83 to 225% SL) and a much larger and more complex esca.

#### Description

Esca with an elongate, unpigmented terminal papilla, cylindrical and truncated in 55-mm paratype (MNHN 1977-304), gradually tapering to a point in remaining specimens; escal pore situated at postero-abasal margin of terminal papilla; anterior escal appendage divided into several secondary branches, each branch bearing numerous slender filaments; a pair of slender filaments emerging from escal bulb adjacent to origin of anterior escal appendage (perhaps a basal bifurcation of the anterolateral appendages); a pair of short unbranched anterolateral appendages; a pair of lateral escal appendages, each divided into 4 or 5 secondary branches, each bearing numerous long, slender filaments; increased branching and elongation of appendages and filaments with increasing size of specimens; length from base of escal bulb to tip of longest filaments 23 to 48% SL; proximal parts of all appendages and filaments lightly pigmented, distal ends unpigmented; number of teeth in lower jaw 25 to 30, in upper jaw 34 to 42; vomerine teeth 8 to 11; dorsal-fin rays 5 or 6; anal-fin rays 4; pectoral-fin rays 13 or 14 (Table 1).

#### Distribution

The holotype and two paratypes of *Bufoceratias shaoi* were collected near Guei-san Island off the northeast coast of Taiwan in bottom trawls fished at depths of 800 m (ASIZP 61796) and 500 to 650 m

(ASIZP 59952), respectively (Fig. 215). The fourth known specimen (MNHN 1977 to 304) was captured in the Mozambique Channel, western Indian Ocean, in an open trawl fished at a maximum depth of 1200 m (Pietsch et al., 2004:105, fig. 5).

Diceratiidae Species

Figure 77

Metamorphosed Male

*Himantolophus* sp.: Pietsch and Seigel, 1980:381 (misidentification).

Material

A single juvenile male: LACM 36091-4, 14 mm, Southeast Asian Bioluminescence Expedition, *Alpha Helix* station 142, Halmahera Sea, 0°10.5′S, 128°33.3′E, RMT 8 with closing device, 750 to 1000 m, 20 May 1975.

Comments

There is no way to accurately determine the generic affinity of this only known male diceratiid (Bertelsen, 1983: 312, fig. 1). *Diceratias* and *Bufoceratias* are distinguished by characters unique to the females, primarily the length and position of the illicium; none of the other differences (angle of the supraethmoid, depth of the illicial trough, and size of the dermal spinules) noted by Uwate (1979: 138–141) can be assumed to be shared with the males. The subdermal pigmentation of the two known larvae of the family, described and identified by Bertelsen (1951:69, fig. 28) as *Diceratias*, is in agreement with that of the male, but with larvae and males of *Bufoceratias* unknown, the pattern of pigmentation could well be similar in the two genera. The geographic distribution of the known material does not provide help: both genera are represented in the Halmahera Sea where the single known male was collected (Pietsch and Seigel, 1980:381–382).

No sexually parasitized female of the family has ever been found. The largest known female of the family, a solitary and apparently mature female *Diceratias pileatus*, with large ovaries containing numerous eggs 0.3 to 0.7 mm in diameter (Uwate, 1979:141), is good evidence that sexual parasitism does not occur in this family (Pietsch, 2005b). The assumption that diceratiid males do not become parasitically attached to females seems supported further by the general morphology and spinulose skin of the single male described here, being remarkably similar to the males of the Himantolophidae and Melanocetidae, which undoubtedly are nonparasitic (Bertelsen, 1951:244; Pietsch, 1976:791; Pietsch, 2005b:218–220).

*Caranactis pumilus*, based on a single metamorphosed male and originally described as a diceratiid by Regan and Trewavas (1932), was referred to the genus *Oneirodes* by Bertelsen (1951:68, 73, 74). The poorly preserved holotype of *Laevoceratias liparis* Parr, 1927, was tentatively included in the Diceratiidae by Bertelsen (1951:70, fig. 29), based on Parr's observation of an anterior spine on the subopercle (A. E. Parr to E. Bertelsen, personal communication; see Bertelsen, 1951:68). Examination of the holotype by Uwate (1979), however, revealed a suite of characters (subopercular spine absent, pelvic bones absent, 5 pectoral radials, dorsal-fin rays 5, anal-fin rays 6, and caudal-fin rays 9) consistent with that of the Gigantactinidae, prompting a reallocation to that family. Additional confirmation of Uwate's reallocation of *L. liparis* to the Gigantactinidae was provided by Bertelsen et al. (1981:59).

## Family Melanocetidae Gill, 1879b (Black Seadevils)

Figures 1, 8, 19, 25, 26, 30, 32, 39, 45, 78–84, 190, 192, 202, 203, 216, 244, 252, 291–293; Tables 1, 4, 8–10

Type genus *Melanocetus* Günther, 1864

Melanocetinae Gill, 1879b:227, 228 (subfamily of family Ceratiidae to contain genus *Melanocetus*).

Melanocetidae Regan, 1912:285, 288 (family of division Ceratiiformes to contain genera *Melanocetus* and *Liocetus*).

Aceratiidae Parr, 1927:30 (in part; family of suborder Ceratioidea to contain genera *Lipactis*, *Aceratias*, *Rhynchoceratias*, and *Laevoceratias*, all based on free-living males, those of *Rhynchoceratias* reassigned in part to genus *Xenoceratias* by Regan and Trewavas, 1932, and subsequently to genus *Melanocetus* by Bertelsen, 1951).

Eurostrinae Parr, 1930a:20 (in part; subfamily of family Aceratiidae to contain genera *Aceratias* and *Rhynchoceratias*, both based on free-living males, those of *Rhynchoceratias* reassigned in part to genus *Xenoceratias* by Regan and Trewavas, 1932, and subsequently to *Melanocetus* by Bertelsen, 1951).

Melanocoetidae Smith, 1949:429 (revised spelling of Melanocetidae; family of order Pediculati to contain genus *Melanocoetus*).

Distinguishing Characters

Metamorphosed males and females of the family Melanocetidae are distinguished from those of all other ceratioid families by having a combination of 13 to 16 (rarely 12 or 17) dorsal-fin rays and 4 (rarely 3 or 5) anal-fin rays (Table 1).

Metamorphosed females are further differentiated by having the following combination of character states: The supraethmoid is present. The frontals are widely separated, each with a prominent ventromedial extension. The parietals are present. Sphenotic spines are absent. The pterosphenoid, metapterygoid, and mesopterygoid are present. The hyomandibula bears a double head. There are 2 hypohyals and 6 (2 + 4) branchiostegal rays. The opercle is bifurcate, the dorsal fork reduced. The subopercle is long and slender, as long as the lower fork of the opercle in some specimens; the ventral part bears a well-developed spine on the anterior margin. Quadrate, articular, angular, and preopercular spines are absent. The jaws are equal anteriorly. The lower jaw bears a well-developed symphysial spine. The postmaxillary process of the premaxilla, and anterior-maxillomandibular ligament, are absent. The first and fourth pharyngobranchials are absent, but the second and third pharyngobranchials are well developed and toothed. The first epibranchial is present but reduced. There are 3 well-developed hypobranchials and a single ossified basibranchial. Epibranchial and ceratobranchial teeth are absent. Epurals are absent. The hypural plate is entire, without a posterior notch. The pterygiophore of the illicium bears a small ossified remnant of the second cephalic spine. The escal bulb has a central lumen and a pore to the outside. The esca is without toothlike denticles. A posteroventral process of the coracoid is absent. There are 4 pectoral radials, fusing to 3 with growth. The pelvic bones are expanded distally. There are 15 to 23 pectoral-fin rays (Table 1). Pelvic fins are absent. There are 9 (1 simple + 6 bifurcated + 2 simple) caudal-fin rays. The skin appears smooth and naked, but minute, widely spaced dermal spinules are present in at least some specimens (only visible microscopically in cleared and stained material; see Pietsch and Van Duzer, 1980:67). The ovaries are paired. Pyloric caecae are absent.

Metamorphosed males are further differentiated in having the following combination of character states: The eyes are directed laterally, elliptical in shape, and the pupil is larger than the lens. The olfactory organs are large, the nostrils inflated and directed laterally. The nasal area is unpigmented. There are 12 to 24 olfactory lamellae. Jaw teeth are absent. The upper denticular bears two or three semicircular series of strong recurved denticular teeth, fused with a median series of 3 to 9 enlarged dermal spines that articulate with the pterygiophore of the illicium. The lower denticular bears 10 to 23 recurved denticles, fused into a median and two lateral groups. The skin is spinulose or naked. Fin-ray counts are as given for metamorphosed females. The males are free-living and never become parasitic. Two examples of males attached to females in temporary coupling are known (see Reproduction and Early Life History, Chapter Eight).

The larvae are further differentiated in having the following combination of character states: The body is short, almost spherical. The skin is moderately inflated. The pectoral fins are of normal size, not reaching beyond the dorsal and anal fins. Pelvic fins are absent. Sexual dimorphism is evident, the females bearing a small, club-shaped illicial rudiment protruding from the head. Fin-ray counts are as given for metamorphosed females. Metamorphosis begins at lengths of 8 to 10 mm (Bertelsen, 1951:45, 49, figs. 16A–G, 17A–F; 1984:326, 328, fig. 169C, D).

The osteological features cited here and elsewhere in this account of the Melanocetidae are based on examinations of both females and males (see Regan, 1912:286, fig. C, 1926:18, fig. 10; Regan and Trewavas, 1932:27–30, figs. 19–28; Gregory, 1933: 401–405, 407, 409, 410, figs. 272, 281; Bertelsen, 1951:39, figs. 13, 14; Pietsch, 1972a: 29, 34–36, 38, 42–45; Pietsch and Van Duzer, 1980:61–67, figs. 1–16).

Description

The body of metamorphosed females is short and deep, globular, the depth 60 to 75% SL (but often appearing highly compressed due apparently to deformation following capture). The head is short, the mouth large, its opening oblique to nearly vertical and the cleft not extending past the eye. The jaws are equal anteriorly. The oral valves are only weakly developed. There are two nostrils on each side of the snout, situated on the distal surface of a rounded papilla. The eye is small and subcutaneous, appearing through a circular translucent area of the integument, within a shallow orbital pit formed between the sphenotic and frontal bones. The teeth are slender, recurved, and depressible, some slightly hooked distally, those in the lower jaw less numerous (except in some small specimens, less than approximately 20 mm) but slightly longer than those in the upper jaw. There are 29 to 178 teeth in the upper jaw and 32 to 142 in the lower jaw. The longest tooth in the lower jaw measures 6.9 to 25.0% SL. There are 0 to 12 vomerine teeth. The first epibranchial and the proximal one-half of the first ceratobranchial are bound to the wall of the pharynx by connective tissue. All four epibranchials are closely bound together. The fourth epibranchial and ceratobranchial are bound to the wall of the pharynx, leaving no opening behind the fourth arch. The proximal one-half of the first ceratobranchial is bound to the wall of the pharynx, while the distal half is free, not bound by connective tissue to the adjacent second ceratobranchial. The proximal one-quarter to one-half of ceratobranchials II to IV are not bound together by connective tissue. Gill filaments are absent on the epibranchials, but present on the proximal tip of ceratobranchial I and the full length of ceratobranchials II to IV. A pseudobranch is absent. The length of illicium is 23.1 to 60.8% SL. The anterior-most tip of the pterygiophore of the illicium is exposed, emerging on the snout between the eyes, the posterior end concealed under the skin. The escal bulb is simple, usually with a rounded or conical distal prolongation, and often with posterior and anterior crests. Elongate cylindrical escal appendages and filaments are absent. The neuromasts of the acoustico-lateralis system are located at the tips of low cutaneous papillae, the pattern of placement as described for other ceratioids (Pietsch, 1969, 1972b, 1974a, 1974b).

For description of the males and larvae, see Bertelsen (1951:38–56).

The color of metamorphosed females is dark brown to black over the entire surface of the head and body (except for the distal portion of the escal bulb), and the oral cavity. All the fins are white in specimens less than approximately 40 mm (except for the caudal-fin rays of juvenile specimens of *Melanocetus murrayi*; Bertelsen, 1951:47, fig. 16I). The external pigmentation of metamorphosed males is as described for females (except the nasal area is unpigmented); the subdermal pigmentation of males variable (see Bertelsen, 1951:42; Pietsch and Van Duzer, 1980:83).

The largest known female is a 135-mm specimen of *M. johnsonii* (NMNZ P21373); the largest known metamorphosed male, also identified as *M. johnsonii*, measures 28 mm (ZMUC P9250, the holotype of *Xenoceratias micracanthus*).

Diversity

A single genus:

### Genus *Melanocetus* Günther, 1864 (Black Seadevils)

Figures 1, 8, 19, 25, 26, 30, 32, 39, 45, 78–84, 190, 192, 202, 203, 216, 244, 252, 291–293;

Tables 1, 4, 8–10

Females

*Melanocetus* Günther, 1864:301, pl. 25 (type species *Melanocetus johnsonii* Günther, 1864, by monotypy).

*Ceratias* Günther, 1887:53 (in part; type species *Ceratias holboelli* Krøyer, 1845, by monotypy).

*Melanocetus* (subgenus *Liocetus*) Günther, 1887:56, pl. 11, fig. A (type species *Melanocetus murrayi* Günther, 1887, by monotypy).

*Liocetus* Goode and Bean, 1896:495, pl. 120, fig. 407 (type species *Melanocetus murrayi* Günther, 1887, by monotypy).

*Melanocoetus* Smith, 1949:429 (erroneous spelling of *Melanocetus*, therefore taking the same type species, *Melanocetus johnsonii* Günther, 1864).

*Linocetus* Bertelsen, 1951:40, 44 (erroneous spelling of *Liocetus*, therefore taking the same type species, *Melanocetus murrayi* Günther, 1887).

*Melanocetes* Wagner, 2001:118 (erroneous spelling of *Melanocetus*, therefore taking the same type species, *Melanocetus johnsonii* Günther, 1864).

Males

*Rhynchoceratias* Parr, 1927:30, figs. 11, 12 (in part; type species *Rhynchoceratias brevirostris* Regan, 1925c, by subsequent designation of Fowler, 1936).

*Centrocetus* Regan and Trewavas, 1932:53, fig. 79 (type species *Centrocetus spinulosus* Regan and Trewavas, 1932, by monotypy).

*Xenoceratias* Regan and Trewavas, 1932:53, fig. 79 (type species *Xenoceratias longirostris* Regan and Trewavas, 1932, by subsequent designation of Fowler, 1936).

Distinguishing Characters and Description

As given for the family.

Diversity

Six species are recognized based on females, and two of uncertain validity based on males:

Key to Females of Species of the Genus *Melanocetus*

The following key will differentiate juvenile and adult female specimens greater than 20 mm. *Melanocetus rossi* Balushkin and Fedorov (1981), based on a single metamorphosed female in poor condition, is omitted from the key below. It appears to be very similar to *M. johnsonii* and may well prove to be a synonym of this species.

1A. Width of escal bulb 11.3% SL in 111-mm specimen; longest tooth in lower jaw 5.9% SL in 111-mm specimen . . . . . . . . . *Melanocetus eustalus* Pietsch and Van Duzer, 1980, p. 366

   NOTE   The holotype, plus three tentatively assigned specimens, 36 to 111 mm; eastern tropical Pacific Ocean

1B. Width of escal bulb less than 10% SL; longest tooth in lower jaw 6.9 to 25.0% SL . . . . . . . . . . . . . . . . . . . . 2

2A. Anterior margin of vomer deeply concave; least outside width between frontals 9.1 to 17.8% SL; number of teeth in lower jaw 46 to 142 (more than 60 in specimens 25 mm and larger); width of escal bulb 1.9 to 5.1% SL (less than 3% SL in specimens greater than 50 mm) . . . . . . . . . . . . . . . . . . *Melanocetus murrayi* Günther, 1887, p. 366

   NOTE   Three hundred and ten specimens, 13.5 to 124 mm; Atlantic, Pacific, and Indian oceans

2B. Anterior margin of vomer nearly straight; least outside width between frontals 13.5 to 28.6% SL; number of teeth in lower jaw 32 to 90; width of escal bulb 3.8 to 8.6% SL (more than 4% SL in specimens greater than 50 mm) .................. 3

3A. Longest tooth in lower jaw 8.4 to 25.0% SL; esca with compressed posterior and (usually) anterior crests; distribution nearly cosmopolitan ...... *Melanocetus johnsonii* Günther, 1864, p. 361

NOTE  Eight hundred and fifty-two specimens, 10 to 154 mm; Atlantic, Pacific, and Indian oceans

3B. Longest tooth in lower jaw 6.9 to 13.1% SL; esca without posterior or anterior crests; distribution restricted to eastern tropical Pacific ....... 4

4A. Number of teeth in upper jaw 62 to 120, in lower jaw 58 to 90; width of escal bulb 5.2 to 8.5% SL; length of illicium 34.6 to 56.0% SL; escal bulb with a conical distal prolongation occasionally pigmented on tip ...... ............ *Melanocetus polyactis* Regan, 1925c, p. 364

NOTE  Thirty-two known specimens, 16.5 to 82 mm; eastern tropical Pacific Ocean

4B. Number of teeth in upper jaw 51 to 87, in lower jaw 37 to 57; width of escal bulb 3.8 to 5.0% SL; length of illicium 29.8 to 38.8% SL; escal bulb with a low, rounded distal prolongation usually darkly pigmented on tip ............ *Melanocetus niger* Regan, 1925c, p. 365

NOTE  Ten known specimens, 12.5 to 80 mm; eastern tropical Pacific Ocean

Key to Males of Species of the Genus *Melanocetus*

Although characters that allow specific identification of males are not well defined, the following key is reproduced in slightly modified form after Bertelsen (1951:44).

1A. Posterior nostril contiguous with eye ............................ 2
1B. Posterior nostril well separated from eye .......................... 3
2A. Upper denticular with a median series of 8 to 11 teeth; lower denticular with 12 to 24 teeth; dorsal-fin rays 13 to 15; pectoral-fin rays 17 to 21 ...... ............ *Melanocetus johnsonii* Günther, 1864, p. 361

NOTE  Eight known specimens, 15.5 to 28 mm; Atlantic, Pacific, and Indian oceans

2B. Upper denticular with a median series of 6 teeth; lower denticular with 22 teeth; dorsal-fin rays 16 or 17; pectoral-fin rays 18 to 23 ........... ............ *Melanocetus polyactis* Regan, 1925c, p. 364

NOTE  Three known specimens, 9.0 to 19.5 mm; eastern tropical Pacific Ocean

2C. Upper denticular with a median series of 3 to 5 teeth; lower denticular with 10 to 13 teeth; dorsal-fin rays 12 to 14; pectoral-fin rays 15 to 18 ............ *Melanocetus murrayi* Günther, 1887, p. 366

NOTE  Four known specimens, 15 to 20 mm; Atlantic and Pacific oceans

3A. Skin spinulose ................ ......... *Melanocetus longirostris* (Regan and Trewavas, 1932), p. 368

NOTE  A single known specimen, 21 mm; western tropical Pacific Ocean

3B. Skin naked ..... *Melanocetus nudus* (Beebe and Crane, 1947), p. 368

NOTE  Two known specimens, 20.0 to 21.5 mm; eastern tropical Pacific Ocean

Key to Larvae of Species of the Genus *Melanocetus*

As with the males of this genus, characters that allow specific identification of larvae are not well defined. The following key, which will separate specimens greater than 3.5 mm TL, is slightly modified from that of Bertelsen (1951:44).

1A. A distinct lateral group of melanophores on caudal peduncle, distinctly separated from dorsal pigmentation in smaller specimens, more or less connected with dorsal pigmentation in larger specimens ........ ............ *Melanocetus johnsonii* Günther, 1864, p. 361

NOTE  Three hundred and twenty-nine known specimens, 2.5 to 17.5 mm TL; Atlantic, Pacific, and Indian oceans

1B. Caudal peduncle unpigmented or, in larger specimens, peduncle reached by spreading dorsal pigmentation ........................... 2
2A. Gill cover with many large melanophores; dorsal-fin rays 16 or 17; pectoral-fin rays 18 to 23 ........... ............ *Melanocetus polyactis* Regan, 1925c, p. 364

NOTE  Seven known specimens, 3 to 9 mm TL; eastern tropical Pacific Ocean

2B. Gill cover only faintly pigmented; dorsal-fin rays 12 to 14; pectoral-fin rays 15 to 18 ................. ............ *Melanocetus murrayi* Günther, 1887, p. 366

NOTE  Seventy-seven known specimens, 2.5 to 13 mm TL; Atlantic and Pacific oceans

## *MELANOCETUS JOHNSONII* GÜNTHER, 1864

Figures 8, 19, 25, 26, 32, 45, 78, 80, 81, 83, 192, 216; Tables 4, 8–10

### Females

*Melanocetus johnsonii* Günther, 1864: 301, pl. 25 (original description, single specimen). Lütken, 1871:64, 74 (comparison with *Oneirodes eschrichtii*). Lütken, 1872:329 (after Lütken, 1871). Günther, 1880:473, fig. 211 (after Günther, 1864). Günther, 1887:56 (after Günther, 1864; comparison with *M. murrayi*. Vaillant, 1888:346 (after Günther, 1864). Goode and Bean, 1896:488, 494, pl. 120, fig. 406 (description after Günther, 1864). Gill, 1909:582, 584, fig. 20 (after Günther, 1864, and Goode and Bean, 1896). Regan, 1912:286, fig. 6C (cranial osteology). Regan, 1913:1096 (description of additional specimen, natural history). Regan, 1926:18, 32, 33, fig. 10 (description of additional material, cranial osteology; *M. krechi* and *M. rotundatus* synonyms). Parr, 1927:29 (description of additional specimen). Regan, 1927a:3, postcard M 11 (popular account). Norman, 1930:354 (additional record). Regan, 1930a:212, fig. 1 (semipopular account). Regan and Trewavas, 1932:27, 49, figs. 19–21, 22A, 22B, 72, 73 (description of additional material, osteology and esca figured, in key). Fowler, 1936:1143, 1346, 1363 (description after Günther, 1864, Regan, 1926, and Norman, 1930). Norman, 1939:114 (additional material). Koefoed, 1944:3, pl. 1, fig. 1 (description of additional specimen, comparison with *M. murrayi*). Beebe and Crane, 1947:152 (description of additional specimen, color). Fowler, 1949:158 (listed). Bertelsen, 1951:7, 40, 43, 48, figs. 13, 15, 17–19 (description of females, males, and larvae; comparison with all known material; in key). Albuquerque, 1954–1956: 1055, fig. 431 (Portugal). Marshall, 1954:15, 114, 142, fig. V, 17c (deep sea biology; figure after Bertelsen, 1951). Grey, 1956a:235 (synonymy, distribution). Whitley, 1956:413 (listed; New Zealand). Briggs, 1960:179 (worldwide distribution). Monod, 1960:687, fig. 80 (pectoral radials). Maul, 1961:91, fig. 1 (description of additional material). Maul, 1962a:6 (description, additional material). Struhsaker, 1962:841 (description, additional specimen, skin spines). Bussing, 1965:222 (additional specimen). Fitch and Lavenberg, 1968:127, fig. 70 (distinguishing characters, natural history). Bassot, 1966:584, pl. 8, fig. 2 (bioluminescence). Whitley, 1968:88 (listed). Clarke and Herring, 1971:201, fig. 186.4 (figure after Brauer, 1906). Pietsch, 1972a: 29, 35, 36, 38, 45 (osteological comments). Brewer, 1973:25 (Gulf of California). Maul, 1973:667 (synonymy, after Bertelsen, 1951). Parin et al., 1973:141, figs. 37, 38 (eastern South Pacific). Parin et al., 1974: 123 (western South Atlantic). Parin and

Golovan, 1976:270 (off West Africa). Hubbs et al., 1979:13 (listed for California). Kotthaus, 1979:50 (additional records, off Goa, Indian Ocean). Marshall, 1979:256, 452, 489 (stomachs contents, larval growth rates, vertical distribution). Leisman et al., 1980:1272 (bioluminescence). Pietsch and Van Duzer, 1980:71, figs. 1, 3, 6, 16B, 17, 19–21, 25, 30, 31 (family revision, comparison of all known material). Pietsch and Seigel, 1980:381 (additional material, Philippine Archipelago). Ayling and Cox, 1982:137, fig. (New Zealand). Amaoka, 1983:116, 117, 197, fig. 68 (northeastern Sea of Japan). Bertelsen and Pietsch, 1983: 81, fig. 2 (new material, Australian waters). Fujii, 1983:258, fig. (two additional specimens, off Surinam). Amaoka, 1984:108, pl. 93I (Japan). Paulin and Stewart, 1985:28 (New Zealand). Bertelsen, 1986:1376, figs. (eastern North Atlantic; in key). Lloris, 1986:242, fig. 124 (two additional specimens, Namibia). Pietsch, 1986a:375–376, fig. 107.1 (southern Africa, description; figures after Gilchrist, 1903, and Pietsch and Van Duzer, 1980). Ni, 1988:324, fig. 254 (four additional specimens, East China Sea). Paulin et al., 1989:140, fig. 69.1 (New Zealand). Pequeño, 1989:43 (Chile). Bertelsen, 1990:492 (eastern tropical Atlantic; synonymy). Haygood et al., 1992:149, 151, figs. 1, 3, 4 (gene sequences from bioluminescent bacterial symbionts). Nielsen and Bertelsen, 1992:62, fig. 1 (North Atlantic). Haygood, 1993:198, 203, fig. 6 (bioluminescent bacterial symbionts). Haygood and Distal, 1993:154, figs. 1, 2 (bioluminescent bacterial symbionts). Herring, 1993:110, fig. (bioluminescence). Cowles and Childress, 1995:1631–1638 (aerobic metabolism). Stearn and Pietsch, 1995:139, fig. 89 (additional specimen, Greenland). Amaoka et al., 1995:112, fig. 171 (after Amaoka, 1983; northern Japan). Meng et al., 1995:439, fig. 587 (China). Swinney, 1995a:52, 53 (additional material, Madeira). Ellis, 1996: 279, fig. (deep Atlantic). McEachran and Fechhelm, 1998:864, 865, fig. (Gulf of Mexico, in key). Munk et al., 1998:1321 (development and histology of escae). Stewart and Pietsch, 1998:5, fig. 2 (new material, New Zealand; figure after Bertelsen, 1951). Jónsson and Pálsson, 1999: 199, fig. 2 (Iceland). Munk, 1999:267, 268, 271, 272, 274, figs. 1, 5B (bioluminescence). Pietsch, 1999:2028 (western central Pacific). Trunov, 1999:493 (Discovery Seamount, Atlantic sector of the Subantarctic Ocean). Anderson and Leslie, 2001:4, fig. 1A (additional material, southern Africa). Nakabo, 2002:474, figs. (Japan, in key). Pietsch, 2002b:1059, fig. (western central Atlantic). Menezes et al., 2003:65 (after Pietsch and Van Duzer, 1980, and Fujii, 1983). Moore et al., 2003: 214 (off New England). Love et al., 2005: 59 (eastern North Pacific). Pietsch, 2005b: 209, 214, 215, 218, 223, 232, fig. 10 (reproduction). Pietsch and Orr, 2007:12, figs. 7A, 8A (phylogenetic relationships).

*Melanocetus johnstoni*: Filhol, 1885:80, 81, 96, fig. 27 (Günther, 1864; depiction as benthic, burrowing in mud).

*Ceratias johnsonii*: Günther, 1887:53 (comparison with *Ceratias [Diceratias] bispinosus*).

*Melanocetus krechi* Brauer, 1902:293 (original description, single specimen). Brauer, 1906:319, pl. 15, figs. 1, 2 (description after Brauer, 1902). Gill, 1909:583, fig. 21 (after Brauer, 1902, 1906). Murray and Hjort, 1912:87, 610, 614, 618, 627 (in part; additional specimen, misidentification). Borodin, 1931:84 (additional specimen, misidentification). Regan and Trewavas, 1932:49, 52, fig. 74 (misidentification, in key). Fowler, 1936:1143, 1144 (description after Brauer, 1902, 1906; in key). Bertelsen, 1951:40 (comparison with all known material).

*Melanocetus rotundatus* Gilchrist, 1903:206, pl. 15 (original description, two specimens, both lost; see Pietsch and Van Duzer, 1980:76). Thompson, 1918:155 (after Gilchrist, 1903). Gilchrist and Thompson, 1917:417 (after Gilchrist, 1903). Barnard, 1927:1007, pl. 37, fig. 5 (after Gilchrist, 1903). Bertelsen, 1951:48 (in synonymy of *Melanocetus johnsonii*). Lloris, 1986:243 (a synonym of *M. johnsonii*; after Pietsch and Van Duzer, 1980).

*Melanocoetus rotundatus*: Smith, 1949: 429, fig. 1232 (after Gilchrist, 1903; emendation of generic name). Penrith, 1967: 187, 188 (type material lost, a synonym of *Melanocetus johnsonii*).

*Melanocetus ferox* Regan, 1926:33, pl. 9, fig. 1 (original description, single specimen). Schmidt, 1926:263, fig. 3 (after Regan, 1926). Regan and Trewavas, 1932:49, 52, fig. 75 (in part, only holotype; additional material referred to *Melanocetus polyactis* by Pietsch and Van Duzer, 1980; in key). Roule, 1934:203, 205, fig. (semipopular account). Beebe and Crane, 1947:152 (in part, only holotype). Bertelsen, 1951:44, 53 (in part, only holotype; comparison with all known material, in key). Harvey, 1952:530 (bioluminescence). Grey, 1956a:237 (synonymy, distribution). Pietsch, 1972b:10 (misidentification; specimen made holotype of *M. eustalus* by Pietsch and Van Duzer, 1980). Brewer, 1973:25 (misidentification after Pietsch, 1972b). Nielsen, 1974:86 (listed in type catalog). Munk, 1999:267 (bioluminescence).

*Melanocetus cirrifer* Regan and Trewavas, 1932:76A, 77, pl. 2, fig. 1 (original description, single specimen, with three questionably assigned specimens). Bertelsen, 1951:44, 53 (description, comparison with all known material, in key). Grey, 1956a:237 (synonymy, distribution). Günther and Deckert, 1956: 135, fig. 92 (popular account; figure after Regan and Trewavas, 1932). Nielsen, 1974:86 (listed in type catalog).

*Melanocetus niger*: Gregory, 1933:400, 402, fig. 272 (misidentification, osteology).

*Melanocetus megalodontis* Beebe and Crane, 1947:152, fig. 1 (original description, single specimen). Bertelsen, 1951:43, 48 (description, comparison with all known material, in key). Grey, 1956a:235 (synonymy, distribution). Mead, 1958:133 (type transferred to SU).

*Melanocetus* sp.: Roule and Angel, 1930:121, pl. 6 fig. 159 (additional material).

*Melanocetes johnsoni*: Wagner, 2001: 119, fig. 2 (sensory brain areas).

Males

*Centrocetus spinulosus*: Regan and Trewavas, 1932:53, fig. 79 (original description, two specimens; lectotype ZMUC P9246, designated by Pietsch and Van Duzer, 1980:73). Bertelsen, 1951:42, 43, 48, 52, 53, fig. 19A (description, comparison with other melanocetid males, in synonymy of *Melanocetus johnsonii*). Nielsen, 1974:86 (listed in type catalog). Pietsch and Van Duzer, 1980:70, 73 (after Bertelsen, 1951).

*Xenoceratias macracanthus*: Regan and Trewavas, 1932:11, 12 (listed; an error for *Xenoceratias micracanthus* Regan and Trewavas, 1932:54). Pietsch and Van Duzer, 1980:73 (in synonymy of *M. johnsonii*).

*Xenoceratias micracanthus* Regan and Trewavas, 1932:54, 55, fig. 81 (original description, single specimen). Fowler, 1936: 1364 (type species designation). Bertelsen, 1951:42, 43, 48, 52, 53, fig. 19E (description, comparison with other melanocetid males, in synonymy of *Melanocetus johnsonii*). Nielsen, 1974:87 (listed in type catalog). Pietsch and Van Duzer, 1980:73 (after Bertelsen, 1951).

*Xenoceratias heterorhynchus* Regan and Trewavas, 1932:54, 56, fig. 82 (original description, two specimens). Bertelsen, 1951:42, 43, 48, 52, 53, fig. 19D (description, comparison with other melanocetid males, in synonymy of *Melanocetus johnsonii*). Grey, 1956a:236 (synonymy, distribution). Nielsen, 1974:87 (listed in type catalog). Pietsch and Van Duzer, 1980:74 (after Bertelsen, 1951).

*Xenoceratias laevis* Regan and Trewavas, 1932:54, 57, fig. 83 (original description, single specimen). Bertelsen, 1951:42, 43, 48, 52, fig. 19C (description, comparison with other melanocetid males, in synonymy of *Melanocetus johnsonii*). Nielsen, 1974:87 (listed in type catalog). Pietsch and Van Duzer, 1980:74 (after Bertelsen, 1951).

*Xenoceratias brevirostris* Regan and Trewavas, 1932:54, 57, fig. 84 (original description, single specimen). Bertelsen, 1951:42, 43, 48, 52, 53, fig. 19B (description, comparison with other melanocetid males, a synonym of *Melanocetus johnsonii*). Nielsen, 1974:86 (listed in type catalog). Pietsch and Van Duzer, 1980:74 (after Bertelsen, 1951).

*Xenoceratias braueri* Koefoed, 1944:6, fig. 2 (original description, single specimen).

Bertelsen, 1951:42, 43, 48, 52, 53, fig. 19B (description, comparison with other melanocetid males, a synonym of *Melanocetus johnsonii*). Frøiland, 1979:151 (listed in type catalog). Pietsch and Van Duzer, 1980:74 (after Bertelsen, 1951).

*Centrocetus spinulosa*: Bertelsen, 1951:40 (emendation of specific name).

*Melanocetus johnsoni*: Bertelsen, 1951:44, 48, fig. 17C, D, F–H (description of females, males, and larvae; comparison with all known material, synonymy, distribution, in key). Grey, 1956a:236 (synonymy, distribution). Maul, 1962b:36, fig. 2 (description of additional specimen). Maul, 1973:667 (synonymy, after Bertelsen, 1951).

### Material

Eight hundred and fifty-two metamorphosed females (10 to 154 mm), eight metamorphosed males (15.5 to 28 mm), and 329 larvae (2.5 to 17.5 mm TL):

Holotype of *Melanocetus johnsonii*: BMNH 1864.7.18.6, female, 64 mm, off Madeira, 24 December 1863.

Holotype of *Melanocetus krechi*: ZMB 17688, female, 45 mm, *Valdivia* station 239, Indian Ocean, 5°42′S, 43°36′E, 0 to 2500 m, 13 March 1899.

Holotype of *Melanocetus ferox*: ZMUC P9257, female, 78 mm, *Dana* station 1208(14), Gulf of Panama, 6°48′N, 80°33′W, 3100 m wire, 16 January 1922.

Holotype of *Melanocetus cirrifer*: ZMUC P9258, female, 25.5 mm, *Dana* station 3678(2), Banda Sea, 4°05′S, 128°16′E, and 4000 m wire, 24 March 1929.

Holotype of *Melanocetus megalodontis*: CAS-SU 46488 (originally NYZS 25791), female, 25.5 mm, Templeton-Crocker Expedition station 165 T-3, eastern tropical Pacific, 20°36′N, 115°07′W, 0 to 915 m, 17 May 1936.

Lectotype of *Centrocetus spinulosus*: ZMUC P9246, male, 15.5 mm, *Dana* station 3847(2), Indian Ocean, 12°02′S, 96°43′E, 3000 m wire, 11 October 1929.

Syntypes of *Xenoceratias micracanthus*: ZMUC P9250, male, 28 mm, *Dana* station 4000(8), eastern tropical Atlantic, 0°31′S, 11°02′W, 4000 m wire, 4 March 1930; BMNH 1932.5.3.9, male, 18 mm, *Dana* station 4007(6), eastern tropical Atlantic, 18°22′N, 18°14′W, 4000 m wire, 15 March 1930.

Syntypes of *Xenoceratias heterorhynchus*: ZMUC P9248, male, 27 mm, *Dana* station 3716(2), South China Sea, 19°18.5′N, 120°13′E, 3000 m wire, 22 May 1929; BMNH 1932.5.3.10, male, 27 mm, *Dana* station 3716(2), South China Sea, 19°18.5′N, 120°13′E, 3000 m wire, 22 May 1929.

Holotype of *Xenoceratias laevis*: ZMUC P9249, male, 23 mm, *Dana* station 3731(13), South China Sea, 14°37′N, 119°52′E, 2000 m wire, 17 June 1929.

Holotype of *Xenoceratias brevirostris*: ZMUC P9247, male, 19 mm, *Dana* station 3739(8), Celebes Sea, 3°20′N, 123°50′E, 3000 m wire, 2 July 1929.

Holotype of *Xenoceratias braueri*: ZMUB 4309, male, 18.5 mm, *Michael Sars*, North Atlantic Deep-Sea Expedition 1910, station 53, central North Atlantic, 34°59′N, 33°01′W, 2600 m wire, 8 to 9 June 1910.

Additional females: AIM, four (16 to 85 mm); AMS, 104 (11 to 105 mm); ASIZP, two (62 to 103 mm); BMNH, 55 (13 to 97 mm); ECFR, two (45 to 90.5 mm); CSIRO, 24 (16 to 63 mm); FMNH, three (13.5 to 64); HUMZ, 35 (23 to 122 mm); IIPB, two (37 to 41); ISH, 89 (15 to 119 mm); LACM, 76 (13.5 to 83 mm); MCZ, 90 (12 to 75 mm); MMF, 18 (15 to 101 mm); MNHN, eight (21 to 39 mm); MNRJ, three (30 to 80 mm); MRIR, two (105 to 150 mm); NMNZ, 33 (15 to 135 mm); NMSZ, 13 (10 to 45.5 mm); NMV, 16 (12 to 110 mm); NHRM, one (22 mm); NSMT, two (15 to 72 mm); ROM, four (15 to 56 mm); SAIAB, four (15 to 83 mm); SAM, 21 (10 to 57 mm); SIO, 122 (11.5 to 154 mm); SIOM, 35 (12 to 75 mm); UF, 10 (13 to 82 mm); USNM, 19 (12 to 85 mm); UW, five (13 to 133 mm); ZIN, two (59 to 77 mm); ZMUB, one (109 mm); ZMUC, 41 (11.5 to 122 mm).

Larvae: MCZ, 20 (3.5 to 9.0 mm); ZMUC, 144 females, 133 males, 32 indeterminate (2.5 to 17.5 mm TL). See Bertelsen, 1951:263–265.

### Distinguishing Characters

Metamorphosed females of *Melanocetus johnsonii* are distinguished from those of all other described species of the genus in having the anterior margin of the vomer nearly straight (see Pietsch and Van Duzer, 1980:61, fig. 1); least outside width between frontals 13.5 to 28.6% SL; number of teeth in upper jaw 48 to 134, in lower jaw 32 to 78; length of longest tooth in lower jaw 8.4 to 25.0% SL; width of pectoral-fin lobe 10.7 to 17.8% SL; width of escal bulb 4.3 to 8.6% SL; length of illicium 32.4 to 60.8% SL; esca with posterior and usually anterior crests; skin with minute spinules over most of body; integument relatively thick (1.55 mm).

Metamorphosed males of *M. johnsonii* differ from those of all other described species of the genus in having the upper denticular with 9 to 13 ventrally directed anterior teeth and a posteromedial series of 8 to 11 teeth; lower denticular with 12 to 24 teeth; posterior nostril contiguous with eye; as many as 24 olfactory lamellae; skin naked or spinulose (Bertelsen, 1951:48, fig. 19).

Larvae of *M. johnsonii* are distinguished from those of all other described species of the genus in having the anterior body musculature unpigmented; dorsal pigment extending posteriorly to beneath middle of base of dorsal fin; a separate lateral group of peduncular pigment in smaller specimens, connecting in larger specimens with dorsal pigmentation; branchial pigmentation strong; branchiostegal, cephalic, sphenotic, preopercular, and pectoral pigment usually present (Bertelsen, 1951:48, fig. 17).

### Description

Escal bulb slightly compressed, with a low rounded or conical distal prolongation nearly always darkly pigmented on tip; a compressed posterior crest usually darkly pigmented, becoming larger and more conspicuous with growth; a considerably smaller compressed anterior crest present in some specimens; integument relatively thick (cross sections measuring 1.55 mm in thickness), not easily torn, usually retaining heavy pigmentation during fixation and preservation; vomerine teeth 2 to 12; dorsal-fin rays 13 to 15 (rarely 16); anal-fin rays 4 (very rarely 3 or 5); pectoral-fin rays 17 to 22 (rarely 23).

Additional description as given for the genus and family. For description of males and larvae, see Bertelsen (1951:49–53, figs. 17, 19).

### Distribution

*Melanocetus johnsonii* is nearly cosmopolitan around the world, with numerous records throughout the Atlantic, Pacific, and Indian oceans between approximately 66°N and 53°S (Fig. 216). Compared to its congener *M. murrayi*, it appears to occupy relatively shallow depths: about 65% of the material (for which data are available) was captured by open nets fished at maximum depths of 1000 m; 88% of the material can be accounted for by gear fished above 1500 m, and 98% by gear fished above 2100 m.

### Comments

*Melanocetus krechi* Brauer (1902) was synonymized with *M. johnsonii* by Regan (1926), resurrected by Regan and Trewavas (1932), and tentatively synonymized again with *M. johnsonii* by Bertelsen (1951). From the description and figure given by Brauer (1902, 1906), and based now on a much greater knowledge of variation within the genus, there can be little doubt that this nominal form is correctly placed within the synonymy of *M. johnsonii*.

*Melanocetus ferox* was described from a single specimen (78 mm) collected in the Gulf of Panama (Regan, 1926). Two additional specimens of this nominal species were listed by Regan and Trewavas (1932). A thorough comparison of all known material led Bertelsen (1951, table 4) to suspect that *M. ferox* might represent individual variation of *M. niger*. The holotype of *M. ferox*, however, has relatively long lower-jaw teeth (longest, 12.0% SL; see Pietsch and VanDuzer, 1980:76, fig. 21). In this, and in all other morphometric and meristic characters cited by Pietsch and Van Duzer (1980), it fits well within the material here recognized as *M. johnsonii*.

Although the esca of the holotype is in poor condition, traces of a posterior crest remain. For these reasons *M. ferox* is here retained within the synonymy of *M. johnsonii* following Pietsch and Van Duzer (1980:72). The two additional specimens identified as *M. ferox* by Regan and Trewavas (1932) (ZMUC P92210, 30.5 mm; BMNH 1932.5.3.6, 42 mm) have short jaw teeth; in this and in other ways they fit well within the material of *M. polyactis* (see below; Pietsch and Van Duzer, 1980:77).

*Melanocetus cirrifer* Regan and Trewavas (1932), described on the basis of two small females, was tentatively maintained by Bertelsen (1951) because of supposed differences in escal morphology and pigmentation that can now easily be shown to be part of the variation found within *M. johnsonii*. *Melanocetus megalodontis* Beebe and Crane (1947), based on a single specimen, was distinguished from all other species of the genus by "the character of the illicium; in the great length and robustness of the fangs . . . and in the shortness of the lower jaw." However, some specimens of *M. johnsonii* have longer teeth and some individuals of several other species of *Melanocetus* have as short a lower jaw (Bertelsen, 1951:41, table 4). Furthermore, as predicted by Bertelsen (1951:48), the "peculiar minute distal flaps" of the esca are artifacts. In all ways the holotype of *M. megalodontis* fits well within the variation now known to occur within *M. johnsonii*. Thus, these nominal forms, *M. cirrifer* and *M. megalodontis*, are here included within the synonymy of *M. johnsonii* following Pietsch and Van Duzer (1980:72).

Finally, the holotype and paratype of *M. rotundatus* Gilchrist (1903) have been lost. The circumstances of their demise are the same as for the holotype of *Dolopichthys cornutus* described elsewhere (Pietsch, 1972b:24; see also Barnard, 1927, and Penrith, 1967). Although Gilchrist's (1903) original description is poor, the figure provided by him shows rather long jaw teeth, a long illicium bearing a relatively large escal bulb, and a large pectoral-fin lobe. This combination of characters makes it nearly certain that *M. rotundatus* is a synonym of *M. johnsonii* (see Penrith, 1967).

### MELANOCETUS ROSSI BALUSHKIN AND FEDOROV, 1981

Tables 4, 8

#### Females

*Melanocetus rossi* Balushkin and Fedorov, 1981:79, figs. 1, 2 (original description, single specimen). Pietsch, 1990:214 (description of holotype after Balushkin and Fedorov, 1981). Balushkin and Fedorov, 2002:23, fig. 3 (additional description of holotype, figured).

#### Males and Larvae

Unknown.

#### Material

A single metamorphosed female (118 mm):

Holotype of *Melanocetus rossi*: ZIN 45349, 118 mm, fishery trawler *Babushkin Cape* trawl number 127, northern slope of Pennell Bank, Ross Sea, Antarctica, 74°46′S, 177°35′W, variable-depth otter trawl, 390 m, bottom depth 420 m, 22 March 1979.

#### Distinguishing Characters

*Melanocetus rossi*, known only from a single poorly preserved desiccated female, apparently differs from those of other species of the genus in having the following combination of character states: anterior margin of vomer nearly straight; least outside width between frontals 16.9% SL; number of teeth in upper jaw 33, in lower jaw 48; length of longest tooth in lower jaw 8.6% SL; width of pectoral fin lobe 10.4% SL; width of escal bulb 3.7% SL; length of illicium 29.2% SL; esca with a medial crest, anterior and posterior escal crests absent; integument relatively thick.

#### Description

Females with esca laterally compressed (perhaps the result of desiccation), bearing a distinct unpigmented medial crest, anterior and posterior crests absent; integument apparently as described for *Melanocetus johnsonii*; vomerine teeth 6; dorsal-fin rays 14; anal-fin rays 5; pectoral-fin rays 20 (Balushkin and Fedorov, 1981:79).

Additional description as given for the genus and family.

#### Distribution

*Melanocetus rossi* is known from a single female specimen collected in the Ross Sea, Antarctica, with a variable-depth otter trawl in 390 m, over a bottom depth of 420 m (Balushkin and Fedorov, 1981).

#### Comments

Based on Balushkin and Fedorov's (1981) original description, as well as personal examination of the holotype and only known specimen, *Melanocetus rossi* appears to be very similar to *M. johnsonii*, differing only in the absence of anterior and posterior escal crests, and in having a smaller escal bulb (width 3.7% SL versus 4.3 to 8.6% SL in *M. johnsonii*) and a slightly shorter illicium (length 29.2% SL versus 32.4 to 60.8% SL in *M. johnsonii*). These small differences, however, could well be the result of poor preservation and desiccation of the holotype. The differences may also be related to the greatly distended stomach of the holotype (Balushkin and Fedorov, 1981:79; 2002:23, fig. 3) found to contain 11 large specimens of *Pleuragramma antarcticum* (see below), which may have affected the accuracy of measurements. In any case, *M. rossi* is probably a synonym of *M. johnsonii*, but this determination must await the discovery of additional material.

The belly of the holotype and only known specimen of *M. rossi* is greatly distended due to ingestion of a huge meal of the Antarctic Silverfish *P. antarcticum* Boulenger. Dissection of the stomach revealed nine partially digested specimens of this rather common circum-Antarctic nototheniid (pelagic in 0 to 700 m; see DeWitt et al., 1990), plus parts of the vertebral column of two additional specimens. The length of each of the 11 specimens was considerably greater than that of the anglerfish itself (Balushkin and Fedorov, 1981, 2002).

### MELANOCETUS POLYACTIS REGAN, 1925C

Figures 81, 216; Tables 4, 8

#### Females

*Melanocetus polyactis* Regan, 1925c:565 (original description, three specimens; lectotype ZMUC P9260, designated by Bertelsen, 1951). Regan, 1926:34, pl. 8, fig. 2 (description after Regan, 1925c). Regan and Trewavas, 1932:53, fig. 78 (listed, after Regan, 1925c, 1926). Bertelsen, 1951:44, 54–55 (description, additional material, two males, six larvae; comparison with all known material, in key). Grey, 1956a:238 (synonymy, distribution). Nielsen, 1974:86 (listed in type catalog). Pietsch and Van Duzer, 1980:77, figs. 21–24, 26, 30 (family revision, comparison of all known material).

*Melanocetus niger*: Regan, 1925c:565 (in part, three of 11 syntypes, all from Gulf of Panama; see Comments). Regan, 1926:33, pl. 8, fig. 1 (in part, description, four additional specimens). Regan and Trewavas, 1932:53, fig. 76B (in part, listed; after Regan, 1925c, 1926). Bertelsen, 1951:44, 53 (in part, description, comparison with all known material, in key). Grey, 1956a:237 (in part, synonymy, distribution).

*Melanocetus ferox*: Regan and Trewavas, 1932:49, 52, fig. 75 (in part, nontype material only, in key). Bertelsen, 1951:44, 53 (in part, nontype material only; comparison with all known material, in key). Grey, 1956a:237 (in part, after Bertelsen, 1951; synonymy, distribution).

#### Males

*Rhynchoceratias rostratus*: Regan, 1926:44 (in part, misidentification).

*Rhynchoceratias leucorhinus*: Regan, 1926:44 (in part, misidentification).

#### Material

Thirty-two metamorphosed females (10 to 82 mm; including nine tentatively referred to this species), three metamorphosed males (9.0 to 19.5 mm), and seven larvae (3 to 9 mm TL):

Lectotype of *Melanocetus polyactis*: ZMUC P9260, 61 mm, *Dana* station 1206(3), Gulf of Panama, 6°40′N, 80°47′W, 3500 m wire, 1845 h, 14 January 1922.

Paralectotypes of *Melanocetus polyactis*: BMNH 1925.8.11.32, 42 mm, *Dana* station 1209(1), Gulf of Panama, 7°15′N, 78°54′W, 3500 m wire, 1845 h, 17 January 1922. ZMUC P92155, 25 mm, *Dana* station 1209(4), Gulf of Panama, 7°15′N, 78°54′W, 2000 m wire, 1845 h, 17 January 1922.

Paralectotypes of *Melanocetus niger*: BMNH 1925.8.11.30, 26 mm, *Dana* station 1205(2), Gulf of Panama, 6°49′N, 80°25′W, 1000 m wire, 0420 h, 14 January 1922. ZMUC P9251, 29 mm, *Dana* station 1203(14), Gulf of Panama, 7°30′N, 79°19′W, 2500 m wire, 2030 h, 11 January 1922. ZMUC P9253, 47 mm, *Dana* station 1208(4), Gulf of Panama, 6°48′N, 80°33′W, 3500 m wire, 0810 h, 16 January 1922.

Additional females: BMNH, one (42 mm); HUMZ, two (38 to 60 mm); LACM, five (16.5 to 35 mm); SIO, six (17 to 82 mm); SIOM, one (33 mm); ZMUC, two (26 to 30.5 mm).

The following metamorphosed females, all collected from the eastern tropical Pacific, are tentatively referred to *Melanocetus polyactis*: LACM, eight (17 to 20 mm); SIO, 1 (10 mm).

Metamorphosed males: SIO, one (9 mm); ZMUC, two (16 to 19.5 mm).

Larvae: SIO, one male (4.5 mm); ZMUC, two males, four females (3 to 9 mm TL). See Bertelsen (1951:55).

Distinguishing Characters

Metamorphosed females of *Melanocetus polyactis* differ from those of all other described species of the genus in having the following combination of characters states: anterior margin of vomer nearly straight; least outside width between frontals 18.0 to 26.0% SL; number of teeth in upper jaw 62 to 120, in lower jaw 58 to 90; length of longest tooth in lower jaw 9.3 to 13.1% SL; width of pectoral-fin lobe 10.9 to 16.0% SL; width of escal bulb 5.2 to 8.5% SL; length of illicium 34.6 to 56.0% SL; esca with a conical distal prolongation, crests absent; integument relatively thick.

Metamorphosed males of *M. polyactis* are distinguished from those of all other described species of the genus in having the upper denticular with 19 ventrally directed anterior teeth and a posteromedial series of 6 teeth; the lower denticular with 22 teeth; posterior nostril contiguous to eye; olfactory lamellae 16; skin with a few dermal spinules (Bertelsen, 1951:54, fig. 21C).

Larvae of *M. polyactis* differ from those of all other described species of the genus in having skin with scattered melanophores; caudal peduncle unpigmented in smaller specimens, dorsal pigment spreading to dorsal half of peduncle in larger specimens; branchiostegal pigment well developed; cephalic, sphenotic, and branchial pigmentation usually present (Bertelsen, 1951:54, fig. 21A, B).

Description

Escal bulb rounded, not compressed, bearing a conical distal prolongation nearly always slightly constricted at base, and usually as long as or longer than length of escal bulb, pigmented on tip in some specimens, posterior and anterior crests absent; integument as in *M. johnsonii*; vomerine teeth 0 to 10; dorsal-fin rays 14 to 17; anal-fin rays 4 (very rarely 5); pectoral-fin rays 17 to 21 (rarely 22 or 23).

Additional description as given for the genus and family. For description of males and larvae, see Bertelsen (1951:55–56, fig. 21).

Distribution

*Melanocetus polyactis* appears to be restricted to the eastern tropical Pacific Ocean where 23 (plus another nine specimens tentatively identified specimens) have been collected between 13°N and 13°S and as far west as 131°W (Fig. 216). Approximately 67% of the material was captured by open nets fished at maximum depths of 1000 to 2200 m.

Comments

*Melanocetus polyactis* is most easily confused with *M. niger*. Both forms are similar in having exceptionally short lower-jaw teeth, but differ significantly in the number of upper- and lower-jaw teeth, escal-bulb width, and illicial length. Part of the material originally listed as *M. niger* has been reallocated to *M. polyactis* (see Comments below). Also included with the material of *M. polyactis* are two specimens (ZMUC P92210, 30.5 mm; BMNH 1932.5.3.6, 42 mm) previously identified as *M. ferox* by Regan and Trewavas (1932).

## *MELANOCETUS NIGER* REGAN, 1925C

Figures 81, 216; Tables 4, 8

Females

*Melanocetus niger* Regan, 1925c:565 (original description, in part, four of 11 syntypes, all from Gulf of Panama; lectotype ZMUC P9252, designated by Pietsch and Van Duzer, 1980; see Comments below); Regan, 1926:33, pl. 8, fig. 1 (in part, description, four additional females). Regan and Trewavas, 1932:53, fig. 76B (in part; listed after Regan, 1925c, 1926). Beebe and Crane, 1947:153 (in part, description of four additional females). Bertelsen, 1951:44, 53 (in part, description, comparison with all known material, in key). Grey, 1956a:237 (in part, synonymy, distribution). Ueno, 1971:102 (listed for Hokkaido, Japan). Parin et al., 1973:142, figs. 37, 38 (eastern South Pacific). Nielsen, 1974:86 (listed in type catalog). Pietsch and Van Duzer, 1980:78, figs. 21–24, 27, 30 (family revision, comparison of all known material). Pequeño, 1989:43 (Chile). Nakabo, 2002:1497 (reference to erroneous Japanese record published by Ueno, 1971).

Males and Larvae

Unknown.

Material

Ten metamorphosed females (12.5 to 80 mm):

Lectotype of *Melanocetus niger*: ZMUC P9252, 80 mm, *Dana* station 1208(4), Gulf of Panama, 6°48′N, 80°33′W, 3500 m wire, 0810 h, 16 January 1922.

Paralectotypes of *Melanocetus niger*: BMNH 1925.8.11.29, 47 mm, *Dana* station 1208(4), Gulf of Panama, 6°48′N, 80°33′W, 3500 m wire, 0810 h, 16 January 1922. ZMUC P9254, 22 mm, *Dana* station 1208(5), Gulf of Panama, 6°48′N, 80°33′W, 3000 m wire, 16 January 1922. ZMUC P9256, 37 mm, *Dana* station 1209(1), Gulf of Panama, 7°15′N, 78°54′W, 3500 m wire, 17 January 1922.

Additional material: SIOM, one (77 mm); SIO, four (12.5 to 62 mm); ZMUC, one (42 mm).

Distinguishing Characters

Metamorphosed females of *Melanocetus niger* differ from those of all other described species of the genus in having the following combination of characters: anterior margin of vomer nearly straight; least outside width between frontals 14.3 to 24.3% SL; number of teeth in upper jaw 51 to 87, in lower jaw 37 to 57; longest tooth in lower jaw 6.9 to 10.5% SL; width of pectoral-fin lobe 9.1 to 13.5% SL; width of escal-bulb 3.8 to 5.0% SL; length of illicium 29.8 to 38.8% SL; esca without crests; integument relatively thick.

Description

Females with escal bulb rounded, not compressed, bearing a low rounded or conical distal prolongation nearly always pigmented on tip; anterior and posterior crests absent; integument as described for *Melanocetus johnsonii*; vomerine teeth 0 to 4; dorsal-fin rays 14 or 15; anal-fin rays 4; pectoral-fin rays 18 to 21.

Additional description as given for the genus and family.

Distribution

All known specimens of *Melanocetus niger* were collected in the Gulf of Panama and adjacent waters of the eastern tropical

Pacific Ocean, ranging off the west coast of the Americas between 18°N and 12°S, and extending as far west as approximately 119°W (Fig. 216). Ninety percent of the material was captured by open nets fished at maximum depths of 1200 to 2000 m.

### Comments

*Melanocetus niger* was described briefly by Regan (1925c) from seven specimens collected in the Gulf of Panama without type designation and without listing individual sizes, station numbers, or other means of identification. Regan (1926) added four more specimens without providing means of separating the original seven. All 11 specimens bear labels indicating "cotype" status and all are treated here as part of the original type material. Pietsch and Van Duzer (1980) designated one of these 11 specimens as the lectotype (ZMUC P9252, 80 mm), three were referred to *M. polyactis* (BMNH 1925.8.11.30, 26 mm; ZMUC P9251, 29 mm; ZMUC P9253, 47 mm), three unidentifiable specimens were listed as *Melanocetus* sp. (ZMUC P9255, 13.5 mm; BMNH 1925.8.11.31, 14 mm; BMNH 1925.8.11.28, 43 mm), and one is unaccounted for and presumed lost (*Dana* station 1209(3), 37-mm TL). The remaining three specimens are recognized as *M. niger*.

### MELANOCETUS EUSTALUS PIETSCH AND VAN DUZER, 1980

Figures 79, 80, 216, 252; Tables 4, 8, 9

#### Females

*Melanocetus ferox*: Pietsch, 1972b:10 (misidentification, bioluminescence). Brewer, 1973:25 (after Pietsch, 1972b, distribution).

*Melanocetus* sp.: Pietsch, 1976:782, 783 (reproduction).

*Melanocetus eustalus* Pietsch and Van Duzer, 1980:79, figs. 18, 28, 30 (original description, single specimen; family revision, comparison of all known material).

*Melanocetus eustales*: Pietsch and Orr, 2007:3, fig. 2G (erroneous spelling of specific name).

#### Males and Larvae

Unknown.

#### Material

Four metamorphosed females: the holotype (111 mm) and three tentatively identified specimens (36 to 93 mm):

Holotype of *Melanocetus eustalus*: LACM 30037-12, 111 mm, *Velero IV* station 11748, eastern Pacific off Mazatlán, Sinaloa, Mexico, 21°39'N, 106°58'W, 3-m Isaacs-Kidd midwater trawl, 0 to 1675 m, bottom depth 2820 m, 1320 to 2136 h, 11 November 1967.

Additional females: SIO, three (36 to 93 mm).

#### Distinguishing Characters

Metamorphosed females of *Melanocetus eustalus* differ from those of all other described members of the genus in having the following combination of character states (based solely on the holotype): anterior margin of vomer nearly straight; least outside width between frontals 18.0% SL; number of teeth in upper jaw 91, in lower jaw 60; longest tooth in lower jaw 5.9% SL; length of illicium 30.6% SL; width of pectoral-fin lobe 9.9% SL; width of escal bulb 11.3% SL; esca without crests; integument relatively thick.

#### Description

Female holotype with escal bulb large (length 14.4% SL, width 11.3% SL), slightly compressed, with a low conical distal prolongation, pigment absent; posterior and anterior crests absent; integument as described for *Melanocetus johnsonii*; gill opening exceptionally large, greatest diameter 23.4% SL; vomerine teeth 8; dorsal-fin rays 15; anal-fin rays 4; pectoral-fin rays 16.

Tentatively referred specimens similar to holotype, but with a slightly smaller esca, longer illicium and lower-jaw teeth, and a greater number of jaw teeth: least outside width between frontals 16.7 to 22.6% SL; number of teeth in upper jaw 77 to 141, in lower jaw 60 to 85; longest tooth in lower jaw 8.6 to 13.3% SL; length of illicium 46.2 to 49.1% SL; width of pectoral-fin lobe 8.6 to 13.9% SL; length of escal bulb 9.2 to 15.1% SL; width of escal bulb 8.9 to 10.9% SL; vomerine teeth 0 to 6; dorsal-fin rays 14 to 16; anal-fin rays 4; pectoral-fin rays 18.

#### Distribution

The four known specimens of *Melanocetus eustalus* were all collected in the eastern Pacific Ocean, two off the coast of Mexico at 18° and 22°N, and two in open water at approximately 9°N, 110°W (Fig. 216). The 55-mm female (SIO 73-247) was collected in an open net fished at a maximum depth of 750 m, the remaining specimens with gear fished open at maximum depths of 1610 to 1675 m.

### MELANOCETUS MURRAYI GÜNTHER, 1887

Figures 39, 81, 84, 216, 244, 291, 293; Tables 4, 8, 9

#### Females

*Melanocetus murrayi* Günther, 1887:57, pl. 11, fig. A (original description, two specimens, lectotype BMNH 1887.12.7.17, designated by Regan, 1926). Regan, 1926:32 (description, additional material, in key). Parr, 1927:27 (description, additional material). Regan and Trewavas, 1932:27, 49, figs. 22C, 23, 71 (description, additional material; pectoral radials, pelvic bone, and escae figured; in key). Beebe, 1932b:99, figs. 29, 30 (description of postlarvae). Parr, 1934:7 (listed). Whitley, 1934, unpaged (listed). Fowler, 1936:1144, 1346, 1363, fig. 483 (after Günther, 1887, and Regan, 1926; in key). Koefoed, 1944:3, 5 (description, comparison, additional material). Fowler, 1949:158 (listed). Bertelsen, 1951:40, fig. 16 (description of females, males, and larvae; comparison with all known material, in key). Albuquerque, 1954–1956:1056 (Portugal). Grey, 1955:299 (additional material, color). Matsubara, 1955:1352, fig. 528 (jaw mechanism of free-living male; figure after Parr, 1930a). Grey, 1956a:234 (synonymy, distribution). Monod, 1960:687, fig. 80 (pectoral radials). Wolff, 1967:187, fig. (after Bertelsen, 1951). Marshall, 1971a:50, fig. 20 (olfaction). Pietsch, 1972a:34, 38 (osteological comments). Maul, 1973:667 (synonymy, after Bertelsen, 1951). Parin et al., 1974:123 (western South Atlantic). Marshall, 1979:405, fig. 142 (olfactory organs and brain). Pietsch and Van Duzer, 1980:81, figs. 2, 4, 5, 7–15, 16A, 19, 20, 29–31 (family revision, comparison of all known material, in key). Pietsch and Seigel, 1980:381 (additional material, Philippine Archipelago). Bertelsen and Pietsch, 1983:81, fig. 3 (additional material, Australian waters). Bertelsen, 1984:330, fig. 169D (early life history, phylogeny). Yamakawa, 1984:289, 384, fig. 203 (additional specimen, Okinawa Trough). Bertelsen, 1986:1377, figs. (eastern North Atlantic, in key). Bertelsen, 1990:493 (eastern tropical Atlantic; synonymy). Stearn and Pietsch, 1995:140, fig. 90 (additional specimen, Greenland). McEachran and Fechhelm, 1998:864, 866, fig. (Gulf of Mexico, in key). Munk et al., 1998:1321, figs. 1–5 (development and histology of escae). Munk, 1999:267, 268, 271, 272, 274, 277, 278, 281, figs. 3E, 8A, B (bioluminescence). Pietsch, 1999:2028 (western central Pacific). Anderson and Leslie, 2001:4, fig. 1B (additional material, southern Africa). Nakabo, 2002:474, figs. (Japan, in key). Pietsch, 2002b:1059 (western central Atlantic). Menezes et al., 2003:65 (after Pietsch and Van Duzer, 1980). Moore et al., 2003:214 (off New England). Pietsch, 2005b:209, 213, 215, 218, 232, figs. 5, 8B (reproduction). Shinohara et al., 2005:420 (off Ryukyu Islands).

*Melanocetus bispinossus*: Günther, 1880:473 (*nomen nudum*, but no doubt an erroneous reference to *Diceratias bispinosus* Günther). Goode and Bean, 1896:495 (in synonymy).

*Melanocetus* (*Liocetus*) *murrayi*: Günther, 1887:56 (new subgenus). Lütken, 1894:79 (after Günther, 1887).

*Liocetus murrayi*: Goode and Bean, 1896:495, pl. 120, fig. 407 (new combination; after Günther, 1887). Gill, 1909:583, 584, fig. 22 (after Günther, 1887, and Goode and Bean, 1896).

*Melanocetus vorax* Brauer, 1902:294 (original description, single specimen).

Brauer, 1906:320, pl. 15, fig. 4 (description after Brauer, 1902). Fowler, 1936:1143, 1144 (description after Brauer, 1902, 1906; in key).

*Melanocetus johnsoni*: Brauer, 1906:318, pl. 15, fig. 3 (misidentification). Murray and Hjort, 1912:609, 614, 618, fig. 469 (in part, misidentification). Regan, 1926:33 (in part, misidentification). Fowler, 1936, fig. 482 (figure after Brauer, 1906). Harvey, 1952:531, fig. 178 (bioluminescence).

*Melanocetus krechi*: Murray and Hjort, 1912:87, 610, 614, 618, 627 (in part, misidentification).

*Melanocetus tumidus* Parr, 1927:28, fig. 10 (original description, single juvenile specimen). Regan and Trewavas, 1932:49 (reference to original description). Grey, 1956a:239 (synonymy, distribution; a young female *Melanocetus murrayi*).

*Melanocetus niger*: Parr, 1927:29 (misidentification). Beebe, 1929a:18 (misidentification).

Males

*Rhynchoceratias acanthirostris* Parr, 1927:31, fig. 11 (original description, single specimen). Parr, 1930b:130, 134 (anatomy, life history). Bertelsen, 1951:45 (in synonymy of *Melanocetus murrayi*). Pietsch and Van Duzer, 1980:81 (after Bertelsen, 1951).

*Rhynchoceratias latirhinus* Parr, 1927:32, 33, fig. 12 (original description, single specimen). Bertelsen, 1951:45 (in synonymy of *Melanocetus murrayi*). Pietsch and Van Duzer, 1980:81 (after Bertelsen, 1951).

*Rhynchoceratias longipinnis* Parr, 1930a:7, figs. 2–5 (original description, single specimen, osteology). Bertelsen, 1951:39, 45 (in synonymy of *Melanocetus murrayi*). Pietsch and Van Duzer, 1980:81 (after Bertelsen, 1951).

*Xenoceratias acanthirostris*: Regan and Trewavas, 1932:54, 55 (new combination, description after Parr, 1927; in key). Parr, 1937:63 (after Parr, 1927). Bertelsen, 1951:42, 43, 45, 47 (description; comparison with other melanocetid males, in synonymy of *Melanocetus murrayi*). Pietsch and Van Duzer, 1980:81 (after Bertelsen, 1951).

*Xenoceratias longipinnis*: Regan and Trewavas, 1932:54, 56 (new combination, description after Parr, 1930a; in key). Parr, 1937:63 (after Parr, 1930a). Bertelsen, 1951:42, 43, 45, 47 (description, comparison with other melanocetid males, in synonymy of *Melanocetus murrayi*). Pietsch and Van Duzer, 1980:81 (after Bertelsen, 1951).

*Xenoceratias latirhinus*: Regan and Trewavas, 1932:54, 57 (new combination, description after Parr, 1927; in key). Parr, 1937:63 (after Parr, 1927). Bertelsen, 1951:42, 43, 45, 47 (description, comparison with other melanocetid males, in synonymy of *Melanocetus murrayi*). Pietsch and Van Duzer, 1980:81 (after Bertelsen, 1951).

*Xenoceratias regani* Koefoed, 1944:4, 6, pl. 1, fig. 6 (original description, single specimen). Bertelsen, 1951:45, 48 (description, comparison with other melanocetid males, in synonymy of *Melanocetus murrayi*). Pietsch and Van Duzer, 1980:82 (after Bertelsen, 1951).

*Melanocetus murrayi*: Bertelsen, 1951:44, figs. 16A, D, F, H (synonymy, description, comparison with all known material, in key). Grey, 1956a:235 (synonymy, distribution). Maul, 1962b:37, fig. 3 (description of additional specimen). Maul, 1973:667 (synonymy; after Bertelsen, 1951).

Material

Three hundred and ten metamorphosed females (13.5 to 124 mm), 4 metamorphosed males (15 to 20 mm), and 77 larvae (2.5 to 13 mm TL):

Lectotype of *Melanocetus murrayi*: BMNH 1887.12.7.17, female, 71 mm, *Challenger* station 106, central Atlantic, between St. Vincent and St. Paul's Rocks, 1°47′N, 24°26′W, 0 to 3386 m, 25 August 1873.

Paralectotype of *Melanocetus murrayi*: BMNH 1887.12.7.18, female, 24 mm, *Challenger* station 348, central Atlantic, between Ascension Island and St. Vincent, 03°10′N, 14°51′W, 0 to 4484 m, 9 April 1876.

Holotype of *Melanocetus vorax*: ZMB 17710, female, 85 mm, *Valdivia* station 63, Gulf of Guinea, 2°00′N, 8°04′W, 0 to 2492 m, 26 September 1898.

Holotype of *Melanocetus tumidus*: YPM 2022, female, 15 mm, *Pawnee*, Third Oceanographic Expedition, station 11, western North Atlantic, 23°58′N, 77°26′W, 2135 m of wire, 2 March 1927.

Holotype of *Rhynchoceratias acanthirostris*: YPM 2011, male, 20 mm, *Pawnee*, Third Oceanographic Expedition, station 22, western North Atlantic, 23°37′N, 77°15′W, 2135 m of wire, 12 March 1927.

Holotype of *Rhynchoceratias latirhinus*: YPM 2012, male, 15 mm, *Pawnee*, Third Oceanographic Expedition, station 33, western North Atlantic, 24°11′N, 75°37′W, 2440 m of wire, 22 March 1927.

Holotype of *Rhynchoceratias longipinnis*: YPM 2592, male, 16 mm, *Pawnee*, Third Oceanographic Expedition, station 59, Bermuda, 32°19′N, 64°32′W, 2440 m of wire, 21 April 1927.

Holotype of *Xenoceratias regani*: ZMUB 4311, male, 20 mm, *Michael Sars*, North Atlantic Deep-Sea Expedition station 53, central North Atlantic, 34°59′N, 33°01′W, 2600 m of wire, 8 to 9 June 1910.

Additional females: AMS, two (63 to 74); ASIZP, 87 (13.5 to 79 mm); BMNH, nine (17 to 80 mm); BSKU, two (42 to 115 mm); CAS, three (14.5 to 51 mm); CSIRO, one (60 mm); FMNH four, (14 to 33.5 mm); ISH, 38 (15 to 124 mm); LACM, 14 (13 to 84 mm); MCZ, 34 (14 to 84 mm); NMNZ, one (53 mm); SAM, one (30 mm); SIO, 45 (12 to 90 mm); SIOM, five (14 to 56 mm); UF, 26 (13.5 to 99 mm); USNM, 14 (15 to 78 mm); UW, five (15 to 45 mm); VIMS, one (33 mm); ZMUC, 14 (14 to 80 mm).

Larvae: MCZ, 3 (8 to 11 mm); ZMUC, 24 females, 35 males, 15 indeterminate (2.5 to 13 mm TL). See Bertelsen (1951:263).

Distinguishing Characters

Metamorphosed females of *Melanocetus murrayi* differ from those of all other described species of the genus in having the anterior margin of the vomer deeply concave; least outside width between frontals 9.1 to 17.8% SL; number of teeth in upper jaw 34 to 178, in lower jaw 46 to 142; length of longest tooth in lower jaw 7.7 to 16.7% SL; width of pectoral-fin lobe 6.1 to 8.9% SL; width of escal bulb 1.9 to 5.1% SL; length of illicium 23.1 to 37.2% SL; esca with crests minute or absent; skin with minute spinules restricted to caudal peduncle; integument relatively thin (0.48 mm).

Metamorphosed males of *M. murrayi* differ from those of all other described species of the genus in having the upper denticular with 9 to 12 ventrally directed anterior teeth and a posteromedial series of 3 to 5 teeth; lower denticular with 10 to 13 teeth; posterior nostril contiguous to eye; olfactory lamellae 12 to 14; skin naked or with a few scattered dermal spinules (Bertelsen, 1951:45, fig. 16H).

Larvae of *M. murrayi* differ from those of all other described species of the genus in having a dorsal group of melanophores extending from the anterior part of body musculature to or beyond the posterior margin of base of dorsal fin (except in specimens less than 3 mm TL); caudal peduncle unpigmented in smaller specimens, peduncle reached by spreading dorsal pigment in larger specimens; branchial and branchiostegal pigmentation weak or absent; cephalic, sphenotic, preopercular, and pectoral pigment usually present (Bertelsen, 1951:45, fig. 16A–G).

Description

Escal bulb rounded, not compressed, bearing a low, rounded distal prolongation usually unpigmented on tip; posterior and anterior escal crests minute or absent; integument thin (cross sections measure 0.48 mm in thickness), easily torn, pigment readily lost during fixation and preservation, often transparent, especially in gill region and over branchiostegal rays; vomerine teeth 0 to 10; dorsal-fin rays 12 to 14; anal-fin rays 4 (very rarely 3 or 5); pectoral rays 15 to 19 (rarely 20).

Additional description as given for the genus and family. For description of males and larvae, see Bertelsen (1951:45–48, fig. 16).

Distribution

*Melanocetus murrayi* is nearly cosmopolitan around the world, with numerous records throughout the Atlantic, Pacific, and Indian oceans between approximately

64°N and 43°S (Fig. 216). Compared with its congener, *M. johnsonii*, it is a deep-living species: only 10% of the material (for which data was available) was captured in open nets fished at maximum depths of less than 1000 m. Approximately 60% of the material was taken by gear fished at maximum depths of 1500 m or below, and 55% by gear fished at 2000 m or below. The relatively thin integument of *M. murrayi* (less than one-third the thickness of that of its congeners) as well as a lighter, less well-ossified skeleton apparently reflects the poorer trophic economies of these greater depths (Pietsch and Van Duzer, 1980:84).

### Comments

*Melanocetus vorax* Brauer (1902) was tentatively synonymized with *M. murrayi* by Regan (1926). This decision was confirmed by Regan and Trewavas (1932) and later by Bertelsen (1951). *Melanocetus tumidus* Parr, 1927 (not mentioned by Bertelsen, 1951), was based on a single metamorphosing female (15 mm) that fits well within the larval and metamorphosing material of *M. murrayi* in lacking pigment on the caudal peduncle and in having a faintly pigmented gill cover. This nominal form is thus included within the synonymy of *M. murrayi* following Pietsch and Van Duzer (1980:81).

### *Melanocetus* Species

The following two nominal species based on males are distinguished from other known males of *Melanocetus* in having the posterior nostril well separated from the eye. They are probably not specifically distinct from each other and are most likely the males of one of the above recognized species based on females (Bertelsen, 1951:54, fig. 20).

### *MELANOCETUS LONGIROSTRIS* (REGAN AND TREWAVAS, 1932), NOMEN DUBIUM

#### Males

*Xenoceratias longirostris* Regan and Trewavas, 1932:54, 55, fig. 80 (original description, single specimen). Nielsen, 1974:86 (listed in type catalog).
*Melanocetus longirostris*: Bertelsen, 1951:42–44, 54 (new combination, comparison with all known material, in key). Grey, 1956a:238 (synonymy; distribution after Bertelsen, 1951).

#### Females and Larvae

Unknown.

#### Material

A single metamorphosed male (21 mm): Holotype of *Xenoceratias longirostris*: ZMUC P9259, 21 mm, *Dana* station 3751(7), north of New Guinea, 3°40'N, 137°53'E, 3000 m wire, 1240 h, 12 July 1929.

#### Distinguishing Characters

A single known male distinguished from those of other species of the genus by a combination of features including upper denticular with 8 ventrally directed anterior teeth and a dorsomedial series of 10 teeth; lower denticular with 16 or 17 teeth; posterior nostril well separated from eye; olfactory lamellae 20; skin covered with tiny dermal spinules (Bertelsen, 1951:54).

### *MELANOCETUS NUDUS* (BEEBE AND CRANE, 1947), NOMEN DUBIUM

#### Males

*Xenoceratias nudus* Beebe and Crane, 1947:155, text fig. 2 (original description, single specimen). Mead, 1958:134 (type transferred to SU).
*Melanocetus nudus*: Bertelsen, 1951:43–44, 54, fig. 20 (new combination, description of additional specimen, comparison with all known material, in key). Grey, 1956a:238 (synonymy, distribution after Bertelsen, 1951).

#### Females and Larvae

Unknown.

#### Material

Two metamorphosed males (20 to 21.5 mm):
Holotype of *Xenoceratias nudus*: CAS-SU 46495 (originally NYZS 28402), 21.5 mm, eastern Pacific *Zaca* Expedition station 210T-8, south of Cape Blanco, Costa Rica, 9°12'N, 85°10'W, 915 m, 27 February 1938.
Additional Specimen: ZMUC, one (20 mm). See Bertelsen, 1951:54.

#### Distinguishing Characters

Metamorphosed males distinguished from those of other species of the genus by a combination of features including upper denticular with 10 to 12 ventrally directed anterior teeth and a dorsomedial series of about 7 teeth; lower denticular with 17 or 18 teeth; posterior nostril well separated from eye; olfactory lamellae 17; skin naked, without dermal spinules (Bertelsen, 1951:54, fig. 20).

For more on males and larvae of *Melanocetus*, see Bertelsen (1951:38–56, figs. 13–17, 19–21).

## Family Thaumatichthyidae Smith and Radcliffe, 1912 (Wolftrap Seadevils)

Figures 27, 36, 37, 40, 43, 45, 48, 85–89, 93–95, 190, 195, 202, 203, 217, 218, 274–277; Tables 1, 8

Type Genus *Thaumatichthys* Smith and Radcliffe, 1912

Thaumatichthyidae Smith and Radcliffe, 1912:579 (family of Pediculati to contain genus *Thaumatichthys*). Pietsch, 1972a:18 (family of Ceratioidei to include *Thaumatichthys* and *Lasiognathus*, resurrected from synonymy of Oneirodidae).
Oneirodidae Regan, 1925c:563, 1926:25, 31 (in part; family of suborder Ceratioidea to contain genera *Thaumatichthys* and *Lasiognathus* reassigned to Thaumatichthyidae by Pietsch, 1972a:18).
Galatheathaumatidae Whitley, 1970:246 (family of order Lophiiformes to contain *Galatheathauma*).

### Distinguishing Characters

Metamorphosed females of the family Thaumatichthyidae are distinguished from those of all other ceratioid families by having elongate premaxillary that extend anteriorly far beyond the lower jaw, the bones widely separated anteriorly at the symphysis, but connected by a broad elastic membrane; the premaxillary teeth are extremely long and curved or hooked; the esca bears 1 to 3 large toothlike dermal denticles; and the opercle is bifurcate, the dorsal fork divided into 2 or more branches.

Metamorphosed females are further differentiated by having the following combination of character states: The supraethmoid is well developed (*Lasiognathus*) or very much reduced or absent (*Thaumatichthys*). The frontals are long, narrow, and widely separated, ventromedial extensions present (*Lasiognathus*) or absent (*Thaumatichthys*). The parietals are present. The sphenotics are large, with an extremely well developed spine (*Lasiognathus*), or small, conical, without a spine (*Thaumatichthys*). The pterosphenoid, metapterygoid, and mesopterygoid are present. The hyomandibula has a double head. There are 2 hypohyals and 6 (2 + 4) branchiostegal rays. The subopercle is long and narrow (*Thaumatichthys*) or short and oval (*Lasiognathus*), the posterior margin of the dorsal part without an indentation, the ventral part with (*Thaumatichthys*) or without (*Lasiognathus*) a spine or projection on the anterodorsal margin. Quadrate and articular spines are well developed (*Lasiognathus*) or rudimentary (*Thaumatichthys*). Angular and preopercular spines are absent. The lower jaw is without a symphysial spine. A postmaxillary process of the premaxilla is absent. The anterior-maxillomandibular ligament is reduced (*Thaumatichthys*) or absent (*Lasiognathus*). The first and fourth pharyngobranchials are absent, but those of the second and third arches are well developed and toothed. There are 3 hypobranchials but only a single ossified basibranchial. Epibranchial and ceratobranchial teeth are absent. Epurals are absent. The posterior margin of the hypural plate is entire *(Lasiognathus)* or deeply notched *(Thaumatichthys)*. The pterygiophore of the illicium bears a small ossified remnant of the second cephalic spine. The escal bulb contains a central lumen and a pore to the outside. A posteroventral process of the coracoid is absent. There are 3 pectoral radials. The pelvic bones are cylindrical, only slightly expanded distally. There are 5 to 7 dorsal-fin rays, 4 to 5 anal-fin rays, 14 to 20 pectoral-fin rays, and 9 (2 simple + 4 bifurcated + 3 simple) caudal-fin rays (Table 1). The skin is everywhere naked, dermal spinules absent *(Lasiognathus)*, or present on the ventral surface of the head and body of metamorphosed specimens *(Thaumatichthys)*. The ovaries are paired. Pyloric caecae are absent.

Males and larvae are known only for *Thaumatichthys*; for a diagnosis and description, see the generic accounts below.

The osteological features cited here and elsewhere in this account of the Thaumatichthyidae are based on examinations of females of *Lasiognathus*, and females and adolescent and larval males of *Thaumatichthys* (Gregory, 1933:402, 404, fig. 275, 276B; Bertelsen, 1951:118, 120, figs. 77, 79; Bertelsen and Struhsaker, 1977:9–18, 24–26, 29–33, figs. 1–9, 14, 17, 18).

### Description

The two genera of the Thaumatichthyidae are so morphologically divergent that additional description is relegated to separate generic accounts given below.

### Comments

Regan (1925c, 1926), followed by Regan and Trewavas (1932), Bertelsen (1951), and Maul (1961, 1962b), included *Lasiognathus* in the family Oneirodidae, together with *Thaumatichthys*, the latter originally placed in a family of its own, the Thaumatichthyidae Smith and Radcliffe (1912). Pietsch (1972a) resurrected the Thaumatichthyidae to include both *Lasiognathus* and *Thaumatichthys*. Bertelsen and Struhsaker (1977) compared the osteology of *Thaumatichthys* and *Lasiognathus*, pointing out that the latter appears more closely related to the Oneirodidae in several of the characters in which it differs from *Thaumatichthys*. Bertelsen and Struhsaker (1977:34) noted, therefore, that "it becomes a subjective choice whether the genera *Lasiognathus* and *Thaumatichthys* both should be included in the Oneirodidae as Regan (1926) did, or placed together in Thaumatichthyidae as proposed by Pietsch (1972a), or whether each of them should be referred to a family of its own." At the same time, however, they cited the two unique features used by Pietsch (1972a) to diagnose the Thaumatichthyidae (premaxillae extending anteriorly far beyond lower jaw, and enlarged dermal denticles associated with the esca) and added a third (dorsal portion of opercle divided into 2 or more branches). In the end, they chose to retain the Thaumatichthyidae in the enlarged sense as proposed by Pietsch (1972a).

### Diversity

Two genera, differentiated as follows:

### Key to Females of Genera of the Thaumatichthyidae

1A. Head narrow; pterygiophore of illicium long, anterior end emerging on snout from between frontal bones; illicium long, greater than 35% SL; esca at tip of illicium, bearing 2 or 3 large toothlike denticles; skin naked, dermal spinules absent; dorsal-fin rays 5; anal-fin rays 5 . . . . . . . *Lasiognathus* Regan, 1925c, p. 369

NOTE  Five known species; Atlantic and Pacific oceans

1B. Head broad, depressed; pterygiophore of illicium short, completely hidden beneath skin of head; illicium short, nearly fully enveloped by tissue of esca; esca sessile on roof of mouth, with 1 deeply embedded dermal denticle; dermal spinules present in skin of ventral surface of head and body; dorsal-fin rays 6 or 7; anal-fin rays 4 . . . . . *Thaumatichthys* Smith and Radcliffe, 1912, p. 373

NOTE  Three known species; Atlantic, Pacific, and Indian oceans

## Genus *Lasiognathus* Regan, 1925c (Snaggletooth Seadevils)

Figures 85–88, 195, 202, 203, 217, 274; Tables 1, 8

### Females

*Lasiognathus* Regan, 1925c:563 (type species *Lasiognathus saccostoma* Regan, 1925c, by monotypy).

### Males and Larvae

Unknown.

### Distinguishing Characters

Metamorphosed females of the genus *Lasiognathus* are distinguished from those of *Thaumatichthys*, the only other genus of the family, in having a narrow, laterally compressed head. The pterygiophore of the illicium is long, its anterior end emerging on the snout from between the frontal bones. The illicium is also long, greater than 35% SL. The esca bears 2 or 3 large hooklike denticles. The ventral fork of the opercle is slender and simple; the dorsal fork is expanded and more or less distinctly divided into 2 (sometimes 3) supporting ribs. The skin is naked. There are 5 dorsal- and 5 anal-fin rays (Table 1). They are pelagic at all life-history stages (Bertelsen and Pietsch, 1996).

Metamorphosed females differ further from those of *Thaumatichthys* in having the following character states: The supraethmoid is well developed. The sphenotics are large, with extremely well developed spines. The frontals bear posterior ventromedial extensions that connect with the parasphenoid. The symphysial cartilage of the upper jaw is large and well developed. The subopercle is without a spine or projection on the anterodorsal margin. The first epibranchial is present. The third (ventral-most) pectoral radial is much larger than second and shows no evidence of fusion to the second radial. The pelvic bones are closely spaced, their distal ends free, not connected by cartilage. The posterior margin of the hypural is entire. The skin beneath the eye and the epithelium of the roof of the mouth are without enlarged sensory papillae.

### Description

The body of metamorphosed females is slender and elongate. The head is extremely large, its length about 60% SL (measured to the posterior-most margin of the cleithrum). The mouth is also large, the cleft extending posteriorly slightly beyond the eye. Sphenotic, quadrate, and articular spines are extremely well developed, the articular spine longer than the quadrate spine. The pterygiophore of the illicium is about 85% SL, extending posteriorly when retracted to form a dorsal tentacle. The illicium measures 38 to 53% SL and lies within a deep groove on the dorsal surface of the head and between the epaxial musculature of the anterior part of the body. The pectoral-fin lobe is small and relatively short and broad, the third (ventral-most) pectoral radial considerably larger than the second. The first epibranchial is bound to the wall of the pharynx. The proximal one-half to two-thirds of the first ceratobranchial is bound to the wall of the pharynx. The distal end of the first ceratobranchial is free, not bound by connective tissue to the adjacent second ceratobranchial. The proximal one-quarter to one-half of ceratobranchials II to IV is free, not bound together by connective tissue. The skin is naked and darkly pigmented, covering the rays of the caudal fin.

The premaxillae are separated anteriorly, connected only by membrane (capable of flipping dorsally and ventrally, enclosing the lower jaw in the latter position), and extending anteriorly far beyond the lower jaw. The articular processes of the premaxillae are short, not reaching the ethmoid cartilage, but attached to the anterolateral projection of a well-developed, anteriorly notched symphysial (rostral) cartilage. The upper jaw is connected posteriorly to the lower jaw by thin ligaments (the anterior-maxillomandibular ligament is absent). The dorsal and ventromedial processes of the head of the maxillae are extremely large and prolonged, clasping the lateral surface of the symphysial cartilage.

The premaxillae bear numerous, long hooked teeth, roughly grouped in an oblique series of 1 + 1 + 2 + 3 + 3 + 2 + 2 + 1, increasing in length anteromedially (the arrangement of teeth in regular longitudinal series is not so distinctly developed as in *Thaumatichthys*). The dentary teeth are similar to those of the premaxillae but are somewhat shorter. The vomer and palatine are toothless. Pharyngobranchials II and III are well toothed (see Bertelsen and Struhsaker, 1977, fig. 18C). The escal denticles or hooks are usually darkly pigmented, especially in larger specimens; each has a broad, hollow, conical base attached superficially to the tissue of the esca and borne on a short, fan-shaped appendage or an elongate cylindrical appendage arising from the distal surface of the escal bulb.

### Diversity

Five species are recognized as follows:

### Key to Females of Species of *Lasiognathus*

1A. Escal hooks borne on a short, transverse, fan-shaped, distal appendage ............. *Lasiognathus beebei* Regan and Trewavas, 1932, p. 370

NOTE Five known specimens, 27.5 to 112 mm; North Atlantic and central Pacific oceans

1B. Escal hooks borne on an elongate, cylindrical, distal appendage .... 2

2A. Escal bulb with a membranous anterior crest; prolongation of distal appendage absent ............. *Lasiognathus waltoni* Nolan and Rosenblatt, 1975, p. 370

NOTE A single known specimen, 94 mm; central North Pacific Ocean

2B. Escal bulb without membranous anterior crest; prolongation of distal appendage present ........... 3

3A. Distal escal appendage with a short, cylindrical prolongation (length 5.2 to 8.6% SL) emerging anteriorly from bases of escal hooks, without lateral serrations or filaments; posterior escal appendage cylindrical ........ *Lasiognathus intermedius* Bertelsen and Pietsch, 1996, p. 371

NOTE Seven known specimens, 26.5 to 129 mm; Atlantic and eastern South Pacific oceans

3B. Distal escal appendage with an elongate, compressed prolongation (length 2.9 to 20.8% SL) emerging anteriorly from bases of escal hooks, bearing lateral serrations and/or tiny filaments on distal tip; posterior escal appendage broad, laterally compressed .................. 4

4A. Three darkly pigmented escal hooks; length of distal escal appendage 8.0 to 12.5% SL (in specimens ranging from 30 to 77 mm); length of prolongation of distal escal appendage increasing with standard length from 7.7% SL (in a 30-mm specimen), to 10.8% SL (55.5 mm), to 20.8% SL (77 mm); prolongation of distal escal appendage bearing numerous lateral serrations and distal filaments .................. *Lasiognathus saccostoma* Regan, 1925c, p. 371

NOTE Eight known specimens, 30 to 77 mm; North Atlantic and central Pacific oceans

4B. Two lightly pigmented escal hooks; length of distal escal appendage 3.8% SL (in a 157-mm specimen); length of prolongation of distal escal appendage 2.9% SL (in 157-mm holotype); prolongation of distal escal appendage without lateral serrations, but bearing a series of 6 tiny distal filaments ................ *Lasiognathus amphirhamphus* Pietsch, 2005a, p. 372

NOTE A single known specimen, 157 mm; eastern North Atlantic Ocean

### *LASIOGNATHUS BEEBEI* REGAN AND TREWAVAS, 1932

Figures 88, 217; Table 8

#### Females

*Lasiognathus beebei* Regan and Trewavas, 1932:90 (original description, taking a specimen collected and illustrated by Beebe, 1930, 1932a, as holotype). Gregory, 1933:404, figs. 275–277 (osteology of holotype). Bertelsen, 1951:119, fig. 77B (after Regan and Trewavas, 1932, and Gregory, 1933). Mead, 1958:133 (type transferred to USNM). Nolan and Rosenblatt, 1975:65 (inclusion of "*L.* sp." Maul, 1961). Bertelsen and Pietsch, 1996:404, figs. 2, 6 (additional material, comparison with all known material). Shedlock et al., 1997:396, fig. 1 (DNA extracted from formalin-fixed specimens). Pietsch, 2002b:1065 (western central Atlantic).

*Lasiognathus* sp.: Maul, 1961:136, figs. 24–26 (additional specimen, MMF 12839, 111 mm, off Madeira). Maul, 1973:672 (after Maul, 1961; eastern North Atlantic).

#### Males and Larvae

Unknown.

#### Material

Five metamorphosed females (27.5 to 112 mm):

Holotype of *Lasiognathus beebei*: USNM 170956, 27.5 mm, Bermuda Oceanographic Expedition, station 9804, net 71, Bermuda, near Nonsuch Island, 0 to 1100 m, 1929.

Additional material: ISH, one (112 mm); LACM, one (36 mm); MCZ, one (80 mm); MMF, one (111 mm; "*L.* sp." of Maul, 1961, from stomach of *Aphanopus carbo*).

#### Distinguishing Characters

Escal bulb without membranous anterior crest; distal escal appendage compressed, transversely fan shaped, bearing 3 bony escal hooks along posterior margin, without elongate prolongation emerging from bases of escal hooks; posterior escal appendage cylindrical, tapering to a point.

#### Description

Length of illicium 43 to 44% SL in two smallest specimens (27.5 to 31 mm), 47 to 50% in three larger specimens (80 to 112 mm); posterior escal appendage 0.5 to 2.5% SL, emerging from below escal pore; thin edge of distal appendage behind and between bases of escal hooks more or less folded; escal hooks unpigmented in two smallest specimens (27.5 to 31 mm), black in three larger specimens (80 to 112 mm); medial (largest) escal hook 5 to 9% SL in two smallest specimens, 2.5 to 3.5% SL in larger specimens; length of distal appendage, measured from surface of escal bulb to bases of escal hooks, about 1.3 to 2.8% SL.

Additional description as given for the genus and family.

#### Distribution

*Lasiognathus beebei* is known from four specimens collected on both sides of the North Atlantic, between approximately 25° and 34°N, including records from off Bermuda and Madeira; a fifth specimen was collected from off Oahu, Hawaiian Islands (Fig. 217). One specimen (MCZ 57779) was taken in a closing trawl between 800 and 1050 m; the remaining material was obtained with gear fished open between the surface and 1100 m.

### *LASIOGNATHUS WALTONI* NOLAN AND ROSENBLATT, 1975

Figure 217; Table 8

#### Females

*Lasiognathus waltoni* Nolan and Rosenblatt, 1975:64, figs. 4, 5 (original description, single specimen). Bertelsen and Pietsch, 1996:406, figs. 3, 6 (generic revision, comparison with all known material).

#### Males and Larvae

Unknown.

#### Material

A single metamorphosed female (94 mm):

Holotype of *Lasiognathus waltoni*: SIO 72-373, 94 mm, *Melville*, cruise Cato I, station A-6, central North Pacific, 30°39.3′N, 155°18.1′W, 0 to 1350 m, 3-m Isaacs-Kidd midwater trawl, bottom depth 5661 m, 24 June 1972.

Distinguishing Characters

Escal bulb with a membranous anterior crest; distal escal appendage an elongate cylindrical stalk, flattened and expanded distally, bearing slender filaments on periphery, without elongate prolongation emerging from bases of escal hooks; posterior escal appendage broad, laterally compressed.

Description

Length of illicium 38% SL; anterior crest of escal bulb with rounded outer margin and broad base; posterior escal appendage a compressed flange 2.1% SL in length, with irregularly rounded edge and broad base, emerging from below escal pore; distal escal appendage with swelling at tip carrying escal hooks, its length to base of hooks 12.2% SL, expanded and flattened distally, bearing 9 elongate slender filaments (proximal four filaments divided to base), more or less equally spaced around periphery, without prolongation arising anteriorly from bases of hooks; escal hooks black, length of medial (largest) hook about 2% SL.

Additional description as given for the genus and family.

Distribution

The single known specimen of *L. waltoni* was collected in the central North Pacific, just north of Oahu, by gear fished open between the surface and 1350 m (Fig. 217).

*LASIOGNATHUS INTERMEDIUS* BERTELSEN AND PIETSCH, 1996

Figures 88, 217; Table 8

Females

*Lasiognathus saccostoma*: Nolan and Rosenblatt, 1975:62, figs. 1–3 (misidentification, only record from the Pacific, SIO 69-342, 97.5 mm).

*Lasiognathus intermedius* Bertelsen and Pietsch, 1996:406, figs. 4, 6 (original description, seven specimens). Anderson and Leslie, 2001:18, fig. 13 (reference to paratype, SAM 32819; description and figure after Bertelsen and Pietsch, 1996). Pietsch, 2002b:1065 (western central Atlantic). Moore et al., 2003:216, fig. 36 (reference to type material, photograph of holotype).

Males and Larvae

Unknown.

Material

Seven metamorphosed females (26.5 to 129 mm):

Holotype of *Lasiognathus intermedius*: MCZ 57778, 31 mm, *Oceanus*, cruise 49, Florida Current, Cape Hatteras section, 34°32′N, 75°26′W, 0 to 1050 m, 12 August 1978.

Paratypes of *Lasiognathus intermedius*: BMNH 1993.9.1.12-13, 2 (both 26.5 mm), *Discovery* station 8281-37, 31°48′N, 63°37′W, 1240 to 1265 m, 19 March 1973. MCZ 49283, 29 mm, *Atlantis*, cruise 71, 35°21′N, 68°14′W, 0 to 1050 m, 24 September 1972. ROM 27288, 42 mm, 44°00′N, 57°52′W, 0 to 1000 m. SAM 32819, 129 mm, eastern Agulhas Bank, South Africa, 35°09.8′S, 23°37.0′E, 975 to 1000 m, 20 October 1992. SIO 69-342, 97.5 mm, *Thomas Washington*, cruise Piquero 5, station 46, 17°42′S, 110°20′W, 0 to 1100 m, 29 March 1969.

Distinguishing Characters

Escal bulb without membranous anterior crest; distal escal appendage an elongate cylindrical stalk, with a small slender prolongation emerging anteriorly from bases of escal hooks, without lateral serrations or filaments; posterior escal appendage compressed distally, constricted at base.

Description

Length of illicium 41 to 53% SL; posterior escal appendage 1.3 to 3.1% SL, emerging from below escal pore (not obvious in partially dried, 42-mm specimen, ROM 27288; with somewhat broader base in 97.5-mm specimen, SIO 69-342); length of distal escal appendage, measured from surface of escal bulb to bases of escal hooks (ca. 3.6% SL, SIO 69-342) 5.7 to 15.7% SL; prolongation of distal escal appendage (ca. 4% SL, SIO 69-342) 5.2 to 8.6% SL; escal hooks weakly pigmented; length of medial (largest) escal hook decreasing proportionately with standard length, 21 to 24% SL in three specimens measuring 26.5 to 29 mm, 10.6% SL in 31-mm holotype, 14.3% SL in 42-mm specimen, but only about 2% SL in 97.5-mm Pacific specimen (SIO 69-342).

Additional description as given for the genus and family.

Distribution

Five of the seven known specimens of *L. intermedius* were collected from the western North Atlantic, between approximately 31 and 34°N, and 58 and 68°W, with single records each from off Cape Town, South Africa, and the eastern South Pacific at about 17°S, 110°W (Fig. 217). Two specimens (BMNH 1993.9.1.12-13) were collected in a closing trawl between 1240 and 1265 m; the remaining material was collected in gear fished open between the surface and 1220 m.

*LASIOGNATHUS SACCOSTOMA* REGAN, 1925C

Figure 87, 217, 274; Table 8

Females

*Lasiognathus saccostoma* Regan, 1925c: 563 (original description, single specimen). Regan, 1926:31, pl. 7 (after Regan, 1925c; holotype figured). Schmidt, 1926:262, fig. 1 (after Regan, 1925c, 1926). Regan, 1927a:4, postcard M 14 (popular account). Parr, 1930a:4 (illicial anatomy; after Regan, 1926). Regan and Trewavas, 1932:90 (after Regan, 1925c). Roule, 1934:201, 204, fig. (semipopular account). Ehrenbaum, 1936: 165, fig. 138 (figure after Regan, 1926). Bertelsen, 1951:119, figs. 76, 77A (after Regan, 1925c). Marshall, 1954:114, fig. V, 17d (deep sea biology; figure after Bertelsen, 1951). Idyll, 1964:199, 202, fig. 10-5 (popular account). Marshall, 1966, pl. 7 (figured). Wolff, 1967:181, fig. (figure after Bertelsen, 1951). Nielsen, 1974:96 (in type catalog). Bertelsen, 1986:1400, figs. (eastern North Atlantic). Swinney, 1995a:52, 55 (additional material, Madeira). Bertelsen and Pietsch, 1996:407, figs. 1, 5, 6 (additional material, comparison with all known material). Pietsch, 2002b:1065 (western central Atlantic).

*Lasiognathus ancistrophorus* Maul, 1962b:39, figs. 4–6 (original description, two specimens). Maul, 1973:672 (eastern North Atlantic).

Males and Larvae

Unknown.

Material

Eight metamorphosed females (30 to 77 mm):

Holotype of *Lasiognathus saccostoma*: ZMUC P92121, 59 mm, *Dana* station 1217(1), Caribbean Sea, 18°50′N, 79°07′W, 4000 m wire, 0630 h, 29 January 1922.

Holotype of *Lasiognathus ancistrophorus*: MMF 19019, 55.5 mm, *Discovery* station 4742, off Madeira, 32°42′N, 16°32′W, 3-m Isaacs-Kidd midwater trawl, 0 to 1700 m, 20 September 1961.

Paratype of *Lasiognathus ancistrophorus*: MMF 19020, 34 mm, *Discovery* station 4742, data as given for holotype.

Additional material: ISH, one (30 mm); NMSZ, three (34.5 to 50 mm); SIO, one (77 mm).

Distinguishing Characters

Escal bulb without membranous anterior crest; distal escal appendage an elongate, cylindrical stalk, with a slender, elongate prolongation emerging anteriorly from bases of escal hooks, bearing numerous lateral serrations or filaments; posterior escal appendage laterally compressed and rounded.

### Description

Length of illicium 39 to 46% SL; posterior escal appendage a compressed flange, length 1.4 to 4.4% SL, with irregularly rounded edge and broad base, emerging from below escal pore; distal escal appendage with swelling at tip carrying escal hooks, its length to base of hooks 8.0 to 12.5% SL; a prolongation of distal appendage present in front of hooks, tapering and flattened at base where it is as broad as head of distal appendage; lateral edges serrated at base, gradually followed by slender filaments along lateral margins and on tip; its length increasing with standard length (in the three specimens in which it is complete) from 7.7% SL (in 30-mm specimen), 10.8% SL (55.5 mm), and 20.8% SL (77 mm); prolongation of distal appendage lost in paratype of *Lasiognathus ancistrophorus*; escal hooks black, relative length of medial (largest) hook decreasing with age from about 10.3% SL in 30-mm specimen to 3.6% SL in 77-mm specimen.

Additional description as given for the genus and family.

### Distribution

*Lasiognathus saccostoma* is known from both sides of the North Atlantic, between approximately 18 and 32°N, 16 and 79°W; one additional record is from the central North Pacific, off the Hawaiian Islands (Fig. 217). All the material was obtained in gear fished open between the surface and about 2000 m.

## *LASIOGNATHUS AMPHIRHAMPHUS* PIETSCH, 2005A

Figures 85, 86, 217; Table 8

### Females

*Lasiognathus amphirhamphus* Pietsch, 2005a: figs. (original description, single specimen). Pietsch, 2005b:215, 222 (reproduction). Pietsch and Orr, 2007:3, fig. 2H (phylogenetic relationships).

### Males and Larvae

Unknown.

### Material

A single known female (157 mm): Holotype of *Lasiognathus amphirhamphus*: BMNH 2003.11.16.12, female, 157 mm, *Discovery* station 10378-25, eastern central Atlantic, Madeira Abyssal Plain, 32°22′12″N, 29°50′42″W, 1200 to 1305 m, 9 June 1981.

### Distinguishing Characters

Escal bulb without membranous anterior crest; distal escal appendage an elongate, cylindrical stalk, with an expanded distal end and an elongate, compressed prolongation emerging anteriorly from bases of escal hooks, without lateral serration, but bearing tiny filaments at distal tip; prolongation of distal escal appendage extremely short (2.9% SL versus 5.2 to 20.8% SL in congeners); 2 lightly pigmented, bony, hooklike denticles embedded in esca; posterior escal appendage laterally compressed, with a rounded posterior margin and pointed distal tip.

### Description

Length of illicium 76 mm (48.4% SL); length of cylindrical cutaneous sheath surrounding posterior extension of illicial pterygiophore 62 mm (39.5% SL). Posterior escal appendage a broad-based, highly compressed flange, length 3.1 mm (2.0% SL, measured from proximal insertion on escal bulb), with a rounded posterior margin and pointed distal tip, emerging from below escal pore. Distal escal appendage with distal swelling bearing a strongly curved escal hook on each lateral margin, its length to base of hooks 6.0 mm (3.8% SL). A posterodorsally directed projection arising from posterior margin of distal appendage between hooks, conical in lateral view, laterally compressed in posterior view. A highly compressed prolongation of distal appendage arising anterior to base of hooks, length 4.6 mm (2.9% SL), expanded at base where it is somewhat broader than head of distal appendage, tapering distally; lateral edges smooth, without serrations, distal end terminating in 6 tiny filaments. Escal hooks lightly pigmented, each about 2.2 mm long (1.4% SL).

Measurements in percent of standard length: tip of upper jaw to posterior-most margin of preopercle 44.6; tip of upper jaw to anterior-most margin of opercular opening 56.7; distance between tips of sphenotic spines 14.3; tip of sphenotic spine to posttemporal 11.8; tip of sphenotic spine to tip of quadrate spine 23.2; length of sphenotic spine 2.6; length of quadrate spine 1.8; length of premaxilla 30.9; length of lower jaw 28.3; longest tooth in upper jaw 4.3; longest tooth in lower jaw 3.2. Total number of teeth on premaxillae 122, on dentaries 94; vomerine teeth absent; dorsal- and anal-fin rays 5; pectoral-fin rays 19; pelvic fins absent; caudal-fin rays 9.

Additional description as given for the genus and family.

### Distribution

*Lasiognathus amphirhamphus* is known from a single female specimen collected in the eastern central Atlantic Ocean, on the Madeira Abyssal Plain, off the southwest coast of Madeira, in a closing trawl fished between 1200 and 1305 m (Fig. 217).

### Comments

Like the vast majority of ceratioid anglerfishes, examination of the available specimens of *Lasiognathus* has revealed no characters that allow separation of the species other than those of the esca and its appendages. The differences in escal morphology, however, strongly indicate the existence of five species.

The validity of *Lasiognathus beebei* Regan and Trewavas (1932), and the reallocation of the specimen described as "*L.* sp." (Maul, 1961) to *L. beebei* by Nolan and Rosenblatt (1975), are supported by the three new specimens reported by Bertelsen and Pietsch (1996). Besides showing the known shared characters, the new material revealed two additional escal characters: the slender, pointed shape of the posterior appendage and the absence of a prolongation of the distal appendage.

The 77-mm specimen from the Pacific (SIO 79-345) appears to be a more advanced stage of the same species as the 55.5-mm holotype of *L. ancistrophorus*. Following Nolan and Rosenblatt's (1975) assumption that *L. ancistrophorus* is a junior synonym of *L. saccostoma*, the observed ontogenetic variation of the species appears to be almost continuous from 55.5 to 94 mm.

The possibility exists that the small specimens of *L. intermedius* (26.5 to 42 mm), each with a more or less stalked distal escal appendage and a short undeveloped prolongation of the distal appendage (lacking lateral filaments), might be juveniles of *L. saccostoma*. This is contradicted, however, by the 30-mm specimen of *L. saccostoma* (ISH 5541/79), which, in spite of its small size, has well-developed lateral serrations or filaments on the prolongation of the distal appendage.

The assumption that the specimens with a short prolongation of the distal escal appendage, lacking lateral serrations or filaments, represent a distinct species is supported by the presence of a number of additional shared escal characters: the laterally compressed posterior escal appendage with a blunt tip, constricted at its base; the club-shaped distal appendage; and the lightly pigmented escal hooks. The similarity between the escae of the five North Atlantic specimens (26.5 to 42 mm) is very distinct. The 97.5-mm Pacific specimen (SIO 69-342), which Nolan and Rosenblatt (1975) referred to *L. saccostoma*, has a somewhat more slender, distal escal appendage and a compressed posterior appendage that appears somewhat broader at the base than those mentioned above. However, it seems more likely that these differences are due to ontogenetic and/or geographic variation rather than to assume the identity of this species with the largest known specimens (55.5 to 77 mm) here referred to *L. saccostoma*.

*Lasiognathus amphirhamphus* differs from all previously recognized species of the genus in having only 2 (instead of 3) bony hooks in the esca. In having an elongate, compressed prolongation of the distal escal appendage, and a broad, laterally compressed posterior escal appendage, it is most similar to *L. saccostoma*. In addition to the lightly pigmented and reduced number of escal hooks, it differs from the latter species in having a considerably shorter distal escal appendage (3.8% SL in a 157-mm specimen versus 8.0 to 12.5% SL in specimens ranging from 30 to 77 mm); and a simple (without lateral serrations), much shorter prolongation of the distal escal appendage. The length of the prolongation of the distal escal appendage of *L. saccostoma* increases significantly with growth (from 7.7% SL in a 30-mm specimen, to 10.8% SL in a 55.5-mm specimen, and 20.8% SL in a 77-mm specimen), whereas that of the much larger holotype of *L. amphirhamphus* is considerably shorter than all of these specimens at only 2.9% SL.

It might be argued that the holotype of *L. amphirhamphus* is simply a specimen of *L. saccostoma* that has lost the medial escal hook at some time prior to capture, but it seems highly unlikely that this could have happened without leaving a trace, even given ample time for healing and possible tissue regeneration. The escae of ceratioid anglerfishes are surprisingly resistant to damage, and, other than immediate injury caused by trauma in the net at the time of capture, examples are very rarely observed. When they are damaged, it is always obvious on casual inspection (e.g., see Pietsch, 1974a:100, fig. 116). If the absence of the medial escal hook was the only distinguishing feature, the existence of a fifth species of *Lasiognathus* might be suspect, but in combination with several additional unique escal characters, the evidence for its validity is convincing.

### Genus *Thaumatichthys* Smith and Radcliffe, 1912 (Wonderfishes)

Figures 27, 36, 37, 40, 43, 45, 48, 89, 93–95, 190, 202, 203, 218, 275, 277; Tables 1, 8

#### Females, Males, and Larvae

*Thaumatichthys* Smith and Radcliffe, 1912:579, pl. 72, figs. 1, 2 (type species *Thaumatichthys pagidostomus* Smith and Radcliffe, 1912, by monotypy).

*Amacrodon* Regan and Trewavas, 1932:91 (type species *Thaumatichthys binghami* Parr, 1927, by monotypy).

*Galatheathauma* Bruun, 1953:174, unnumbered plate (type species *Galatheathauma axeli* Bruun, 1953, by monotypy). Whitley, 1970:246 (new family).

#### Distinguishing Characters

Metamorphosed females of the genus *Thaumatichthys* are distinguished from those of *Lasiognathus*, the only other genus of the family, in having a conspicuously broad, dorsoventrally depressed head, the eyes and olfactory organs displaced posteroventrally to the corner of the mouth. The pterygiophore of the illicium is short, completely hidden beneath the skin of the head. The illicium is also extremely short, fully enveloped by tissue of the esca. The esca is sessile on the roof of the mouth, bearing a single deeply embedded dermal denticle. The opercle is divided into 6 to 13 radiating branches. The skin of the ventral part of the head and body is covered with small dermal spinules. There are 6 to 7 dorsal-fin rays and 4 anal-fin rays. Metamorphosed females display numerous adaptations to a benthic existence (see Bertelsen and Struhsaker, 1977:29, 30).

Metamorphosed females differ further from those of *Lasiognathus* in having the following character states: An ossified supraethmoid is absent in juveniles, but represented by a small remnant in adults. The sphenotics are small, conical, without a spine. The frontals lack posterior ventromedial extensions. A symphysial cartilage of the upper jaw is absent. The subopercle bears a spine or projection on its anterodorsal margin. The first epibranchial is absent. The two ventral-most pectoral radials are nearly equal in size, fusing together with age. The pelvic bones are widely spaced, their distal ends connected by a band of cartilage. The hypural plate is deeply notched posteriorly. Enlarged sensory papillae are present beneath the eye and on the roof of the mouth.

Metamorphosed males of *Thaumatichthys* (four known specimens, 12 to 36 mm, the following diagnosis based on the two largest specimens, 31 to 36 mm) differ from those of all other ceratioids in having the following combination of character states: The premaxillae are toothless, but unresorbed, bearing at their symphysis 4 or 5 separate denticles arranged in two transverse series, a ventral series of 2 and a dorsal series of 2 or 3; the ventral series is shorter and more strongly hooked than the dorsal series. There are 7 denticles on the tip of lower jaw, arranged in a transverse series of 4 denticles and a dorsal series of 3 denticles; the dorsal series is shorter and more strongly hooked than the ventral series. The dorsal fork of the opercle is divided into 6 or 7 radiating branches. Like metamorphosed females of *Thaumatichthys*, but in contrast to those of *Lasiognathus*, the male of *Thaumatichthys* lacks a supraethmoid. There are 6 dorsal-fin rays and 4 anal-fin rays. The skin is covered with dermal spinules from the occipital region to the base of the caudal fin.

The larvae of *Thaumatichthys* (five known specimens, 3.0 to 22.5 mm) differ from those of all other ceratioids in having a multiradiate opercle and a layer of subdermal melanophores that covers the entire surface of the head and body. Like the adults of *Thaumatichthys*, but in contrast to those of *Lasiognathus*, the larvae of *Thaumatichthys* have 6 or 7 dorsal-fin rays and 4 anal-fin rays.

#### Description

The body of metamorphosed females is somewhat longer and more slender than that of most other female ceratioids. The anterior portion of the body just behind the head is depressed, while the posterior portion is cylindrical. The head is broad and highly depressed, its length about 35% SL (measured to the posteriormost margin of the cleithrum). The mouth is large, the cleft extending posteriorly slightly beyond the eye. Sphenotic, quadrate, and articular spines are rudimentary. The pterygiophore of the illicium is unusually short (only 14.3% SL in a 69.9-mm specimen of *Thaumatichthys binghami*) and completely hidden beneath the skin of the head. The illicial bone is also extremely short (only about 3.2% SL in a 69.9-mm specimen of *T. binghami*), directed ventrally, and emerging anteriorly from the roof of the mouth (i.e., from the ventral margin of membrane that connects the premaxillae, see below). The escal denticle is deeply embedded, curved, and sharply pointed at both ends. The pectoral-fin lobe is small, relatively short and broad, the first (dorsal-most) radial simple and rodlike, without a distal expansion. The second and third pectoral radials are nearly equal in size; their widely expanded distal ends are fused together in a 69.9-mm specimen of *T. binghami*; the fusion is nearly complete in a 143-mm specimen of *T. binghami*; the second and third pectoral radials are completely fused to each other in a 294 mm specimen of *T. binghami* and in the 365-mm holotype of *T. axeli*. The first epibranchial is bound to the wall of the pharynx. All four epibranchials are closely bound together. The proximal one-half to two-thirds of the first ceratobranchial is bound to the wall of the pharynx, the distal one-half to one-third free. The fourth epibranchial and ceratobranchial are bound to the wall of the pharynx, leaving no opening behind the fourth arch. Gill filaments are absent on the epibranchials, but present on the full length of ceratobranchials II to IV. A pseudobranch is absent. Pyloric caeca are absent. The ventral surface of the head and body is covered with numerous small dermal spinules, increasing in number and size with growth; they are restricted in smaller specimens to the opercular region of the head and ventral surface of the body (except for the caudal peduncle where they reach the dorsal midline), but in the largest known specimens they extend to the ventral surface of the lower jaw, the lateral surfaces of body, and the rays of the anal and caudal fins. The skin is everywhere dark brown to black; the roof of the mouth and extending over lower jaw is brown, while the inner walls of the mouth cavity are nearly or completely unpigmented. Darkly pigmented skin covers the rays of the dorsal, anal, and caudal fins.

The premaxillae of females are strong, well ossified, and widely separated anteriorly, where they are connected only by membrane; they are hinged to the ethmoid region of the cranium in such a way as to allow them to flip dorsally and ventrally, enclosing the lower jaw in the latter position. The ascending process of each premaxilla forms a slightly upturned spine, lying in front of an extremely long, posterolaterally directed articular process. Each ascending process is connected by an elastic band of tissue to the anterior end of the maxilla and palatine of the opposite side. The maxillae are thin and slender, each with a small nearly undivided head attached by ligaments to the head of the respective palatine. The dentaries are relatively weakly developed, without a symphysial spine; there is a small anterior-maxillomandibular present posteriorly that connects the dentary to the respective premaxilla and maxilla (for a full description of the structure and function of the premaxillae and associated anatomy, see Bertelsen and Struhsaker, 1977:11, 12, 29, 30).

The premaxillae of females bear long, slender, hooked teeth, arranged in a very distinctive pattern, consisting of six overlapping, oblique longitudinal series (see Figs. 36, 37): the first tooth of each of six series is spaced along the edge of the jaw in young specimens; the second tooth of each series develops slightly in front of and ventromedial to the first tooth of the following series; the third tooth of each series develops in front of and inside the second tooth of the following series, and so forth, forming in this way oblique transverse series that cross the longitudinal series. In each series, longitudinal as well as transverse, the teeth increase in length in both posterior and medial directions. The larger teeth of series II to IV are especially well hooked (for a detailed description of individual, ontogenetic, and specific variation in the premaxillary teeth of *Thaumatichthys*, see Bertelsen and Struhsaker, 1977:15, figs. 7, 8).

The dentary teeth of females are shorter and less curved than those of the premaxillae; they increase in number and size with age, varying from 9 to 10, the largest measuring 1.2 to 1.5 mm in the smallest known specimens of *T. binghami*; 22 to 30 and 5.6 to 10.3 mm in the three largest specimens of *T. binghami*; and 39 to 40 and 10.5 mm in the holotype of *T. axeli*. The dentary teeth are arranged in distinct anteromedial series, each of 2 to 4 teeth of increasing length, with the smallest positioned outermost. The vomer and palatine are toothless. The second and third pharyngobranchials are well toothed.

The body of metamorphosed males (four known specimens, 15 to 36 mm, the following description based on the two largest specimens, 31 to 36 mm) is long and slender. The snout is pointed, with hooked denticles at the tip (Fig. 94). There are 4 or 5 upper denticles, arranged in two transverse series, a dorsal (outer) series of 2 or 3 and a ventral (inner) series of 2: those of the dorsal pair are somewhat larger (3.3 to 4.2% SL) and curved ventrolaterally; those of the ventral pair are more strongly hooked, with their distal ends curved ventrally and their sharp tips thus pointing posteriorly, nearly parallel to their shafts. The premaxillae are toothless, each with a well-developed articular process, without any trace of resorption. The maxillae are similarly well developed, the head of each bearing two processes, a shorter medial process that extends beneath the articular process of the respective premaxilla, and a lateral process that lies close to the outer surface, reaching the base of the uppermost denticles. There are 7 lower denticles arranged on the chin in two bilateral, symmetrical series: a ventral (outer) row of 4 large denticles, with dorsolaterally curved pointed tips; and a dorsal (inner) series of 3 shorter denticles. The somewhat expanded bases of the 7 lower denticles are fused and attached to the anterior tips of the dentaries by connective tissue. The dentaries are toothless, without a trace of tooth sockets.

The opercle and subopercle of the males is similar in shape to those of the females: the dorsal part of the opercle is divided into 6 or 7 slender radiating branches; the dorsal part of the subopercle is slender, tapering to a point, the ventral part truncate, with an anterodorsal spine. The pelvic bones are as described for females. The diameter of the pigmented part of the eye is 1.5 to 1.6 mm (4.2 to 5.2% SL); the diameter of the transparent outer coat of the eyeball is 2.0 to 2.1 mm (5.6 to 6.8% SL). The olfactory organs are unusually large, the posterior nostril much larger than the anterior nostril, its length 2.6 to 3.0 mm (7.2 to 9.7% SL), each with 13 olfactory lamellae. There are 6 dorsal-fin rays, 4 anal-fin rays, 15 pectoral-fin rays, and 9 (2 simple + 4 bifurcated + 3 simple) caudal-fin rays. The skin is dark brown, with numerous small spinules scattered over the body from the occipital region to the base of the caudal fin. The head and body are covered with a layer of light dusky subdermal pigment of nearly uniform density. The testes are immature, each about 4.2 to 6.5 mm in length (11.7 to 21.0% SL) and 1.1 to 1.2 mm (3.1 to 3.9% SL) in width.

The larvae of *Thaumatichthys* have a layer of subdermal pigment covering the entire surface of the head and body, with concentrations along the margin of the gill cover, on the peritoneum, and on the dorsal surface of the trunk. The pigment is less concentrated to absent on the pectoral-fin lobe and at the base of the caudal fin. Small scattered melanophores are present in the outer transparent skin. The skin of the largest known larva (22.6 mm) is somewhat more densely pigmented (Fig. 95).

For a more detailed description of *Thaumatichthys*, including osteology, dentition, illicium and esca, sensory structures, functional morphology of the feeding mechanism, and early life-history stages, see Bertelsen (1951:121, fig. 80A, B, table 25), Bertelsen and Struhsaker (1977:27, 37, fig. 15), and Bertelsen (1984:326, fig. 169).

Diversity

Three species are recognized as follows:

Key to Females of Species of *Thaumatichthys*

1A. Length of premaxillae 33% SL (in 60 mm specimen); anterior premaxillary teeth long: I-1:3.8, II-1:3.0, II-2:6.5, II-3:7.0, III-1:1.5, III-2:3.3, III-3:4.3, IV-1:1.4, IV-2:2.4, V-1:1.0, V-2:3.4 mm . . . . . . . . . . . . . . . . . . . . . . . . . . . *Thaumatichthys pagidostomus* Smith and Radcliffe, 1912, p. 375

   NOTE  Three known specimens, 60 to 246 mm; western Pacific Ocean

1B. Length of premaxillae 23.5 to 27.0% (average 25.4%) SL (in specimens 40 mm and larger), anterior premaxillary teeth short: I-1: 2.2 to 3.5, II-1: 1.1 to 2.3, II-2: 3.0 to 5.6, II-3: 3.8 to 5.8, III-1: 1.1 to 1.8, III-2: 0.8 to 2.3, III-3: 1.8 to 3.4, IV-1: 0.6 to 1.5, IV-2: 0.9 to 2.3, V-1: 0.6 to 1.4, V-2: 0.8 to 2.6 mm . . . . . . . . . . . . . . . . . . . . 2

2A. Esca with two or three pairs of laterally directed lobes, which in larger specimens are prolonged into tapering tentacle-like filaments; dorsalmost medial escal appendage digitiform and tapering . . . . . . . . . . . . . . . . . *Thaumatichthys binghami* Parr, 1927, p. 375

   NOTE  Thirty-nine known specimens, 36.5 to 294 mm; western Atlantic Ocean

2B. Esca with a single pair of laterally directed lobes, which in larger specimens are prolonged into tapering tentacle-like filaments; dorsalmost medial escal appendage wartlike and untapering . . . . *Thaumatichthys axeli* Bruun, 1953, p. 376

   NOTE  Two known specimens, 85 to 365 mm; eastern Pacific Ocean

### *THAUMATICHTHYS PAGIDOSTOMUS* SMITH AND RADCLIFFE, 1912

Figures 36, 37, 48, 218; Table 8

#### Females

*Thaumatichthys pagidostomus* Smith and Radcliffe, 1912:580, pl. 72, figs. 1, 2 (original description, single specimen). Regan, 1926:31, fig. 19 (description after Smith and Radcliffe, 1912). Parr, 1927:24 (comparison with *Thaumatichthys binghami*, in key). Parr, 1930a:3 footnote (illicial anatomy). Regan and Trewavas, 1932:91 (after Smith and Radcliffe, 1912). Bertelsen, 1951:121, figs. 78, 79 (description and figures of holotype after Smith and Radcliffe, 1912; osteology, description, and figures of four larvae, referred to *T. binghami* and *T.* sp. by Bertelsen and Struhsaker, 1977). Günther and Deckert, 1956:177, fig. 120 (popular account; figure after Smith and Radcliffe, 1912). Mead et al., 1964:587 (reproduction). Bertelsen and Struhsaker, 1977:36, figs. 3, 7, 8, 13, 16 (generic revision, description, comparison with all known material). Pietsch and Seigel, 1980:395 (reference to holotype). Bertelsen and Pietsch, 1983:89 (reference to holotype). Meng et al., 1995:442, fig. 593 (China). Stewart and Pietsch, 1998:21 (reference to holotype). Pietsch, 1999:2033 (western central Pacific). Anderson and Leslie, 2001:18 (reference to larval male from off Mozambique).

#### Males and Larvae

Unknown.

#### Material

Three metamorphosed females: the holotype (60 mm), plus two tentatively identified specimens (142 to 246 mm):

Holotype of *Thaumatichthys pagidostomus*: USNM 72952, 60 mm, *Albatross* station 5607, near Binang Unang Island, Gulf of Tomini, Sulawesi (Celebes), 00°04′S, 121°36′E, beam trawl, 1440 m, bottom of fine sand, 18 November 1909.

Tentatively identified females: ASIZP, two (147 to 246 mm).

#### Distinguishing Characters

Based on the holotype: length of premaxillae 33% SL (in 60 mm specimen); anterior premaxillary teeth long: I-1: 3.8, II-1: 3.0, II-2: 6.5, II-3: 7.0, III-1: 1.5, III-2: 3.3, III-3: 4.3, IV-1: 1.4, IV-2: 2.4, V-1: 1.0, V-2: 3.4 mm; esca of holotype and only known specimen damaged, only inner bulb, with a loosely attached curved denticle (2.1 mm long) remains (see Comments below).

#### Description

As given for the genus and family.

#### Distribution

The holotype of *Thaumatichthys pagidostomus* was collected in the Gulf of Tomini, off Sulawesi (approximately on the equator at 121°E), with a beam trawl at 1440 m (Bertelsen and Struhsaker, 1977:28; Fig. 218). The two tentatively identified specimens are from off Taiwan, taken with shrimp trawls on the bottom between approximately 300 and 600 m.

#### Comments

Smith and Radcliffe (1912:580) described the illicium and esca of *Thaumatichthys pagidostomus* as "a slender pedicel bearing at its tip a sharp toothlike process that curves backward." In their illustration, the esca is pictured as a small, simple spherical structure, with the denticle protruding at the tip. However, the fact that no escal lobes or appendages are mentioned or illustrated does not necessarily indicate that the esca of this species differs from those of other species of *Thaumatichthys*. In Smith and Radcliffe's (1912, pl. 72, fig. 2) drawing, the esca has a diameter of about 1% SL, or somewhat less than the present diameter of the inner bulb, which is now 1.0 mm or 1.7% SL, a difference that might indicate that the damage to the esca occurred during capture (Bertelsen and Struhsaker, 1977:22).

The two tentatively identified females, both collected off Taiwan, are listed here as *T. pagidostomus* based on little more than geographic distribution. While they compare well with the holotype of *T. pagidostomus* in the length of the teeth (the longest anterior premaxillae teeth measuring 10.5 to 12.0 mm or about 4.9 to 7.4% SL), the premaxillae of both specimens are short (only about 23% SL), comparable to those of *T. binghami* (see below). Accurate identification must therefore await the discovery of additional material, especially individuals of a size range between that of the 60-mm holotype and these two larger specimens.

### *THAUMATICHTHYS BINGHAMI* PARR, 1927

Figures 36, 37, 40, 45, 93, 218, 276, 277; Table 8

#### Females

*Thaumatichthys binghami* Parr, 1927:25, fig. 9 (original description, single specimen). Parr, 1930a:3, fig. 1 (illicial anatomy; after Parr, 1927). McEachran and Fechhelm, 1998:871, fig. (Gulf of Mexico). Pietsch, 2002b:1065, fig. (western central Atlantic). Pietsch, 2005b:215, 222 (reproduction). Pietsch and Orr, 2007:3, fig. 2I (phylogenetic relationships).

*Amacrodon binghami*: Regan and Trewavas, 1932:91 (new combination, description after Parr, 1927). Parr, 1934:7 (second known specimen, listed). Bertelsen, 1951:122 (after Regan and Trewavas, 1932). Günther and Deckert, 1956:181, fig. 124D, E (popular account; figure after Parr, 1927). Bertelsen and Struhsaker, 1977:36, figs. 1, 3–11, 13, 14, 16, 17, pls. 2, 3 (generic revision, osteology, description, comparison with all known material).

#### Males and Larvae

Unknown.

#### Material

Thirty-nine metamorphosed females (36.5 to 294 mm):

Holotype of *Thaumatichthys binghami*: YPM 2015, 36.5 mm, *Pawnee*, Third Oceanographic Expedition, station 25, 25°51′N, 76°37′W, pelagic net, 8000 ft wire, 17 March 1927.

Additional metamorphosed females: LACM, one (56 mm); MNRJ, one (166 mm); TCWC, three (45 to 128 mm); UF, 22 (40 to 235 mm); USNM, three (51.5 to 294 mm); UW, one (83 mm); VIMS, three (49 to 65.5 mm); YPM, one (40 mm); ZMUC, three (70 to 142 mm).

#### Distinguishing Characters

Length of premaxillae 23.5 to 27.0% (average 25.4%) SL (in specimens 40 mm and larger), anterior premaxillary teeth short: I-1: 2.2 to 3.5, II-1: 1.1 to 2.3, II-2: 3.0 to 5.6, II-3: 3.8 to 5.8, III-1: 1.1 to 1.8, III-2: 0.8 to 2.3, III-3: 1.8 to 3.4, IV-1: 0.6 to 1.5, IV-2: 0.9 to 2.3, V-1: 0.6 to 1.4, V-2: 0.8 to 2.6 mm (Fig. 37); esca with two or three pairs of laterally directed lobes, which in larger specimens are prolonged into tapering tentacle-like filaments; dorsal-most medial escal appendage digitiform and tapering.

#### Description

As given for the genus and family.

#### Distribution

The known metamorphosed females of *Thaumatichthys binghami* were collected in the western Atlantic, all but one specimen west of 60°W and between approximately 10° and 30°N extending from the Gulf of Mexico, Caribbean, adjacent waters off the Bahamas, and south to Trinidad, with a single record from off Rio de Janeiro, Brazil (Bertelsen and Struhsaker, 1977; Fig. 218). Three of the smaller specimens (36.5 to 45 mm) were caught pelagically at intermediate depths, but 21 females, including most of the larger specimens (40.5 to 294 mm) were obtained by bottom trawls fished at 1100 to 3200 m: 18 of these were caught on the continental slope at depths between 1270 and 2193 m; one at 1100 m, and a single specimen from the central Gulf of Mexico at 3151 to 3200 m.

## THAUMATICHTHYS AXELI (BRUUN, 1953)

Figures 36, 37, 89, 218, 275; Table 8

### Females

*Galatheathauma axeli* Bruun, 1953: 174, plate (original description, single specimen, in Danish). Bruun, 1956:174, plate (description, single specimen, in English). Wolff, 1960:168, 175, 176–177, unnumbered color plate (animals from *Galathea* station 716, painting by P. H. Winther). Hjortaa, 1961:169, two unnumbered figs. (model for exhibition, cast from holotype). Wolff, 1961:137, 154, 159, color pl. 9 (after Wolff, 1960). Idyll, 1964:297, fig. 14-4 (popular account; figure after Bruun, 1953, 1956). Mead et al., 1964:587 (reproduction). Munk, 1964:10 (ocular anatomy). Whitley, 1970:246 (new family). Nielsen, 1974:96 (in type catalog).

*Thaumatichthys (Galatheathauma) axeli*: Wolff, 1967:282–283, color plate (new combination; plate after Wolff, 1960, 1961).

*Thaumatichthys axeli*: Bertelsen and Struhsaker, 1977:37, figs. 2, 3, 6–8, 12, 13, 15, 16, 19, pl. 1 (new combination, generic revision, description, comparison with all known material). Gartner et al., 1997:133, fig. 4 (feeding behavior; figure after Bertelsen and Struhsaker, 1977). Love et al., 2005:61 (eastern tropical Pacific). Pietsch, 2005b:215, 222 (reproduction).

### Males and Larvae

Unknown.

### Material

Two metamorphosed females (85 to 365 mm):

Holotype of *Galatheathauma axeli*: ZMUC P92166, 365 mm, *Galathea* station 716, eastern tropical Pacific, 09°23′N, 87°32′W, herring otter trawl, 3570 m, 6 May 1952.

Additional metamorphosed female: SIO, one (85 mm).

### Distinguishing Characters

Length of premaxillae 23.5 to 27.0% (average 25.4%) SL (in specimens 40 mm and larger), anterior premaxillary teeth short: I-1: 3.0, II-1: 1.9, II-2: 4.5, II-3: 4.8, III-1: 1.1, III-2: 1.9, III-3: 2.5, IV-1: 1.0, IV-2: 2.0, V-1: 0.9, V-2: 2.4 mm; esca with a single pair of laterally directed lobes, which in larger specimens are prolonged into tapering, tentacle-like filaments; dorsal-most medial escal appendage wartlike and untapering.

### Description

As given for the genus and family.

### Distribution

The holotype of *Thaumatichthys axeli* was collected in the eastern tropical Pacific at about 9°N, 87°W (Bertelsen and Struhsaker, 1977:28; Fig. 218). The second known specimen (SIO 70-22) is also from the eastern Pacific, but farther north, off the coast of Baja California, at about 31°N, 119°W. Both were taken at abyssal depths, the holotype with a herring trawl at 3570 m, the other with an otter trawl at 3595 to 3695 m.

### *Thaumatichthys* Species

#### Metamorphosed Males and Larvae

*Thaumatichthys pagidostomus*: Bertelsen, 1951:121 (tentatively assigned larvae, description). Brewer, 1973:25 (larval female, eastern tropical Pacific).

*Thaumatichthys binghami*: Bertelsen and Struhsaker, 1977:36 (tentatively assigned larval and metamorphosed males).

*Thaumatichthys* sp.: Bertelsen and Struhsaker, 1977:37 (unidentifiable larvae). Bertelsen, 1990:508 (eastern tropical Atlantic).

#### Material

Four metamorphosed males (12 to 36 mm) and five larvae (3 to 22.5 mm, 4.4 to 34 mm TL):

Metamorphosed males: BMNH, one (12 mm); UW, two (15 to 36 mm); ZMUC, one (31 mm).

Larvae: LACM, one female (22.5 mm, 34 mm TL); ZMUC, one female (6.5 mm, 10 mm TL); ZMUC, three males (3 to 12.5 mm, 4.4 to 18.5 mm TL).

#### Description

For description and locality data, see Bertelsen (1951:121, fig. 80A, B, table 25) and Bertelsen and Struhsaker (1977:27, 37, fig. 15).

#### Distribution

Two of the known larvae were collected in the eastern tropical Atlantic off the Gulf of Guinea, a third is from the Indian Ocean in the southern part of the Mozambique Channel, and the fourth was taken in the eastern tropical Pacific off the coast of Mexico. The fifth larva and the four known metamorphosed males of the genus were collected in the Caribbean Sea, Gulf of Mexico, and off Madeira, which may indicate their conspecificity with *Thaumatichthys binghami* (Bertelsen and Struhsaker, 1977:28, fig. 16).

The three smallest larvae (3.0 to 6.5 mm) were caught in nets fished open in maximum depths of about 20 to 100 m below the surface. The two larger larvae (18.5 to 22.6 mm) were captured in gear fished open at maximum depths of 1500 and 780 m, respectively. Three of the metamorphosed males were collected with pelagic trawls fished open between the surface and 2000 m; the third male (ZMUC P921946) was taken with a bottom trawl at a depth of 1938 m.

## Family Oneirodidae Gill, 1879a
### Dreamers

Figures 10, 14–16, 18, 20, 23, 25, 30, 32, 33, 43, 45, 53, 96–112, 147, 190, 193, 199, 202, 203, 219–229, 240, 245, 248, 250, 253, 254, 263, 264, 291, 295, 296; Tables 1, 8, 9

Type Genus *Oneirodes* Lütken, 1871

Oneirodinae Gill, 1879a:217 (subfamily of family Ceratiidae to contain genus *Oneirodes*).

Oneirodidae Regan, 1925c:562 (family of suborder Ceratioidea to contain genera *Chaenophryne*, *Dolopichthys*, *Lasiognathus*, *Lophodolus*, and *Oneirodes*).

Aceratiidae Parr, 1927:30 (in part; family of suborder Ceratioidea to contain genera *Lipactis*, *Aceratias*, *Rhynchoceratias*, and *Laevoceratias*, all based on free-living males, those of *Rhynchoceratias* subsequently reassigned in part to oneirodid genera *Dolopichthys*, *Chaenophryne*, *Microlophichthys*, *Oneirodes*, and *Pentherichthys* by Bertelsen, 1951).

Eurostrinae Parr, 1930a:20 (in part; subfamily of family Aceratiidae to contain genera *Aceratias* and *Rhynchoceratias*, both based on free-living males, those of *Rhynchoceratias* subsequently reassigned in part to oneirodid genera *Dolopichthys*, *Chaenophryne*, *Microlophichthys*, *Oneirodes*, and *Pentherichthys* by Bertelsen, 1951).

### Distinguishing Characters

Metamorphosed females of the family Oneirodidae are distinguished from those of all other ceratioid families by having a narrow, spatulate, anterodorsally directed process that overlaps the posterolateral surface of the respective sphenotic. The parasphenoid has a pair of anterodorsal extensions that approach or overlap the posterior ventromedial extensions of the respective frontal (absent in *Lophodolus*). The quadrate and articular spines are well developed (small to rudimentary in *Chaenophryne*).

Metamorphosed females are further differentiated in having the following combination of character states: The supraethmoid is present. The frontals are widely separated. The parietals are present. The metapterygoid and mesopterygoid are present. There are 2 hypohyals and 6 (2 + 4) branchiostegal rays. Angular and preopercular spines are absent. The jaws are equal anteriorly. The postmaxillary process of the premaxillae is absent. The anterior-maxillomandibular ligament is well developed. The fourth pharyngobranchial is absent. There is a single ossified basibranchial. Ceratobranchial teeth are absent. Epurals are absent. The hypural plate is entire, without a posterior notch. The pterygiophore of the illicium bears an ossified remnant of the second cephalic spine. The escal bulb has a central lumen and a pore to the outside. The esca is without toothlike denticles. There are 3 pectoral radials. There are 4 to 8 dorsal-fin rays, 4 to 7 (very rarely 3) anal-fin rays, 14 to 30 pectoral-fin rays, and 9 (2 simple + 4 bifurcated + 3 simple) caudal-fin rays (Table 1). The ovaries are paired. One or 2 small pyloric caecae are present or absent.

Although not unique to oneirodids nor characteristic of all members of the family, the following additional features are important in differentiating the family: Each frontal has a prominent, bifurcated ventromedial extension (absent in *Lophodolos*). Sphenotic spines are well developed (absent in *Chaenophryne*). The pterosphenoid is usually present (absent in *Lophodolos*). The hyomandibula usually has a double head (only a single head in *Bertella*). The opercle is bifurcate, the dorsal fork supported by a single rib (except in larval females of *Danaphryne*). An anterior subopercular spine is usually absent (a blunt projection is present in most specimens of *Chaenophryne*, juvenile females of *Lophodolos*, and some larvae and males of *Pentherichthys* and *Dolopichthys*). Quadrate and articular spines are usually well developed (small to rudimentary in *Chaenophryne*). The lower jaw usually has a well-developed symphysial spine (small, blunt to absent in *Chaenophryne*, absent in *Pentherichthys*). The first pharyngobranchial is usually absent (but present in *Spiniphryne* and *Oneirodes*). The second pharyngobranchial is usually well developed and heavily toothed (but reduced and toothless in *Microlophichthys*, *Bertella*, and some species of *Oneirodes*; absent in *Lophodolos* and *Pentherichthys*). Epibranchial teeth are usually absent (present on epibranchial I of some species of *Oneirodes*). An ossified posteroventral process of the coracoid is usually absent (but present in *Spiniphryne* and *Oneirodes*). The pelvic bones are usually rodlike, with or without a slightly expanded distal tip (triradiate or broadly expanded distally in *Chaenophryne*; see Pietsch, 1975b:79, fig. 2). The skin is usually smooth and naked (well-developed dermal spinules are everywhere present in the skin of *Spiniphryne*; minute, widely scattered spinules are present in some species of *Oneirodes*).

Metamorphosed males are distinguished from those of all other ceratioid families in having the following combination of character states: The eyes are directed laterally; they are elliptical in shape, with a short axis, the diameter of the pupil greater than that of the lens. The olfactory organs are large. The anterior nostrils are situated close together, their openings directed anteriorly; the posterior nostrils, usually larger than the eye, are directed laterally. The nasal area is usually pigmented and sometimes slightly inflated. Jaw teeth are absent. The posterior end of the upper denticular is well separated from the anterior end of the pterygiophore of the illicium. Fin-ray counts are as given for metamorphosed females. The skin is smooth and naked (but males of the spiny-skinned oneirodid genus *Spiniphryne* are unknown). The males are free-living and nonparasitic, with two exceptions, both apparently facultative: a single known attached pair of *Bertella idiomorpha* and another of *Leptacanthichthys gracilispinis* (see Reproduction and Early Life History, Chapter Eight).

There is no satisfactory combination of features that serves to adequately diagnose oneirodid larvae. As stated by Bertelsen (1951:71), it is "easier in practice to make the identification [of larvae] from the features characteristic of each of the many genera within the family." To a somewhat lesser extent, the same could also be said for the metamorphosed males and females.

The osteological features cited here and elsewhere in this account of the Oneirodidae are based on examinations of females of nearly all genera and of males of most genera in which they are known (Regan, 1926:16, figs. 9; Regan and Trewavas, 1932:33–37, figs. 37–48; Gregory, 1933:402, 404, 405, figs. 274, 276A, 279; Bertelsen, 1951:71–74, figs. 30, 31, 35, 37, 43, 46, 48–51, 53, 54, 56, 58, 60, 62, 64, 66, 67, 70, 73; Pietsch, 1972a:31, 35, 36, 42–45, fig. 24(4–8), 1974a:4–29, figs. 1–56, 1975b:77–79, figs. 1, 2; Bertelsen and Pietsch, 1975:2–6, figs. 1–3, 1977:181, fig. 7; Pietsch, 1978:2–9, figs. 1–8); specimens of *Dermatias*, *Tyrannophryne*, and *Ctenochirichthys* have so far been unavailable for osteological study.

### Description

Metamorphosed females are highly variable, ranging from short and deep-bodied, more or less globular (e.g., *Chaenophryne* and *Oneirodes*) to elongate and fusiform (e.g., *Dolopichthys* and *Leptacanthichthys*). The head is usually short, approximately 35% SL or less (30% SL in *Dermatias*, nearly 50% SL in some species of *Chaenophryne*). The mouth is large, its opening horizontal to nearly vertical, the cleft extending well past the eye in most genera (terminating anterior to the eye in *Spiniphryne*, *Danaphryne*, *Phyllorhinichthys*, *Puck*, *Ctenochirichthys*, and *Bertella*). The jaw teeth are slender, recurved, and depressible, those in the upper jaw usually somewhat shorter and fewer than those in the lower jaw (the upper-jaw teeth are more numerous than the lower in *Danaphryne*, *Microlophichthys*, *Tyrannophryne*, *Phyllorhinichthys*, *Puck*, *Leptacanthichthys*, and *Ctenochirichthys*). The numbers of premaxillary and dentary teeth are highly variable, the lower jaw with fewer than 20 (e.g., *Phyllorhinichthys* and some species of *Oneirodes*) to nearly 600 (e.g., some species of *Dolopichthys*). Vomerine teeth are usually present (lost with growth in *Bertella* and some species of *Dolopichthys*, absent in *Lophodolos* and *Pentherichthys*). The first epibranchial is free, not bound to the wall of the pharynx by connective tissue. All four epibranchials are closely bound together. The fourth epibranchial and ceratobranchial are bound to the wall of the pharynx, leaving no opening behind the fourth arch. The proximal one-fourth to one-half of the first ceratobranchial is bound to the wall of the pharynx, the distal three-fourths to one-half free. The distal end of the first ceratobranchial is free, not bound by connective tissue to the adjacent third ceratobranchial. The proximal one-quarter to one-half of ceratobranchials II to IV are free, not bound together by connective tissue. Gill filaments are present on the proximal tips of epibranchials II to IV, on the proximal tip of ceratobranchial I, and the full length of ceratobranchials II to IV. A pseudobranch is absent. The length of the illicium is highly variable, extremely short, and nearly fully enveloped by tissue of the esca (e.g., in *Microlophichthys*) to 75% SL (e.g., in some species of *Oneirodes* and *Dolopichthys*). The anterior-most tip of the pterygiophore of the illicium is usually exposed (but hidden in *Tyrannophryne*), emerging at the tip of the snout from between the eyes or further posteriorly. The posterior end of the pterygiophore is usually concealed under the skin (protruding on the dorsal surface of the trunk behind the head in *Oneirodes*). The escal morphology is simple to highly complex. The neuromasts of the acoustico-lateralis system are located at the tips of low, rounded, cutaneous papillae, the pattern of placement as described for other ceratioids (Pietsch, 1969, 1972b, 1974a, 1974b).

For additional description of the males and larvae, see the accounts of the genera below.

The color of females in preservative is dark brown to black over the entire external surface of the body (except for the escal appendages and distal portion of the escal bulb). The oral cavity and viscera, except for the outer surface of the stomach wall, is unpigmented. The external pigmentation of metamorphosed males is as described for females, except that the nasal area is often unpigmented. The subdermal pigmentation of the males and larvae is highly variable (see Bertelsen, 1951:73).

The largest known female of the family is an unidentified, 370-mm specimen of *Oneirodes* (ISH 995/82), collected by the *Walther Herwig* during the 1982 expedition to the Mid-Atlantic Ridge, which differs from all described species of the genus and may prove to be new to science. The right ovary of the 370-mm specimen is approximately 70 mm long (18.9% SL) and 30 mm wide (8.1% SL), and filled with numerous eggs approximately 0.2 to 0.3 mm in diameter. The largest known male of the family measures 16.5 mm.

Diversity

Sixteen genera are currently recognized, differentiated as follows:

Keys to the Genera of the Oneirodidae

FEMALES

The following key (modified from Bertelsen, 1951, and Pietsch, 1974a) will distinguish juvenile and adult female specimens only. Because several genera are known from very few specimens (only three or four juvenile females in *Tyrannophryne*, *Chirophryne*, and *Puck*), the key is tentative and may not include the best diagnostic characters:

1A. Sphenotic spines present; opercle deeply notched posteriorly; pelvic bones rod shaped, with or without slight distal expansion . . . . . . . . . 2
1B. Sphenotic spines absent; opercle not deeply notched posteriorly; pelvic bones triradiate or greatly expanded distally . . . . . . . . . . . . *Chaenophryne* Regan, 1925c, p. 384

NOTE  Five known species; Atlantic, Pacific, and Indian oceans

2A. Pectoral-fin lobe short and broad, shorter than longest pectoral-fin rays . . . . . . . . . . . . . . . . . . . . . . . 3
2B. Pectoral-fin lobe long and narrow, longer than longest pectoral-fin rays . . . . . . . . . . . . . . . . . . . . . . . . 12
3A. Skin covered with close-set dermal spinules . . . . . . . . . . . *Spiniphryne* Bertelsen, 1951, p. 392

NOTE  Two known species; Atlantic, Pacific, and Indian oceans

3B. Skin naked or with minute, widely spaced dermal spinules (visible only with the aid of a microscope in cleared and stained specimens) . . . . . . . . 4
4A. Lower jaw with a symphysial spine; caudal rays without internal pigment . . . . . . . . . . . . . . . . . . . . . . . . . . . 5
4B. Lower jaw without a symphysial spine, ventral margin of dentaries at symphysis concave; caudal rays internally pigmented . . . . *Pentherichthys* Regan and Trewavas, 1932, p. 382

NOTE  A single species; Atlantic, Pacific, and Indian oceans

5A. Illicial apparatus emerging near tip of snout, from between frontal bones . . . . . . . . . . . . . . . . . . . . . . . 6
5B. Illicial apparatus emerging from dorsal surface of head, between or behind sphenotic spines . . . . . . . . . . . . . . . . . . . . . . . . . . . . . *Lophodolos* Lloyd, 1909a, p. 379

NOTE  Two known species; Atlantic, Pacific, and Indian oceans

6A. Depth of caudal peduncle greater than 20% SL . . . . . . . . . . *Dermatias* Smith and Radcliffe, in Radcliffe, 1912, p. 422

NOTE  A single known species; western North Pacific Ocean

6B. Depth of caudal peduncle less than 20% SL . . . . . . . . . . . . . . . . . . . . 7
7A. Dorsal margin of frontal bones strongly convex; subopercle short and broad, ventral end nearly circular . . . . . . . . . . . . . . . . . . . . . . . 8
7B. Dorsal margin of frontal bones nearly straight; subopercle long and narrow, ventral end strongly oval . . . . . . . . . . . . . . . . . . . . . 15
8A. Caudal fin covered with darkly pigmented skin for some distance beyond fin base; anal-fin rays 5, rarely 4 . . . . . . . . . . . . . . . . . . 9
8B. Caudal fin not covered by darkly pigmented skin except at base; anal-fin rays 4, rarely 5 . . . . . . . . . *Oneirodes* Lütken, 1871, p. 394

NOTE  Thirty-five species; nearly cosmopolitan

9A. Lower jaw extremely long, extending posteriorly well beyond base of pectoral fin . . . . . . . . . *Tyrannophryne* Regan and Trewavas, 1932, p. 429

NOTE  A single known species; western and central Pacific Ocean

9B. Lower jaw short, terminating anterior to base of pectoral fin . . . . . 10
10A. Fewer than 20 teeth in lower jaw; esca with 3 stout, nontapering, internally pigmented appendages arising from dorsal surface; smaller specimens (less than about 100 mm) with a pair of unpigmented leaflike appendages on snout . . . . . . . . . . . . . . . . . . . . . . . . . . *Phyllorhinichthys* Pietsch, 1969, p. 427

NOTE  Two known species; Atlantic, Pacific, and western Indian oceans

10B. More than 25 teeth in lower jaw; esca without stout appendages arising from dorsal surface; no leaflike appendages on snout . . . . . . . . . . . 11
11A. Cleft of mouth extending past eye; length of escal bulb more than one-half length of illicial bone; dorsal end of subopercle broad and rounded . . . . . . . . *Microlophichthys* Regan and Trewavas, 1932, p. 425

NOTE  One species (a second species based solely on males); Atlantic, Pacific, and Indian oceans

11B. Cleft of mouth not extending past eye; escal bulb considerably shorter than one-half length of illicial bone; dorsal end of subopercle slender, tapering to a point . . . . . *Danaphryne* Bertelsen, 1951, p. 423

NOTE  A single known species; North Atlantic and western Pacific oceans

12A. Sphenotic, quadrate, and articular spines long, piercing skin; length of pectoral-fin lobe less than 15% SL; pectoral-fin rays 18 to 21 . . . . . . . 13
12B. Sphenotic, quadrate, and articular spines short, in some specimens not piercing skin; length of pectoral-fin lobe greater than 15% SL; pectoral-fin rays 28 to 30 . . . . . *Ctenochirichthys* Regan and Trewavas, 1932, p. 442

NOTE  A single known species; eastern North Atlantic and eastern Pacific oceans

13A. Length of quadrate spine greater than length of articular spine; dorsal profile of frontal bones convex; esca with more than a single appendage, either 5 separate appendages arising from dorsal surface, or 3 dorsal appendages and a lateral filament; anal-fin rays 4 . . . . . . . . . . . . . . . . . . 14
13B. Length of quadrate spine less than length of articular spine; dorsal profile of frontal bones nearly linear; esca with a single appendage arising from dorsal surface; anal-fin rays 5 or 6 . . . . . . . . . . . . . . . . *Leptacanthichthys* Regan and Trewavas, 1932, p. 440

NOTE  A single known species; North Atlantic and Pacific oceans

14A. Length of anterior maxillomandibular ligament greater than one-half length of premaxilla, gape of mouth extending beyond eye; subopercle short and broad, upper end rounded; esca without a lateral filament . . . . . . . . . . . . . . . . . . . . . . . . . *Chirophryne* Regan and Trewavas, 1932, p. 440

NOTE  A single known species; western North Atlantic and Pacific oceans

14B. Length of anterior maxillomandibular ligament less than one-half length of premaxilla, gape of mouth not extending beyond eye; subopercle long and narrow, upper end tapering to a point; esca with a lateral filament . . . . . . . . . . . . . . . . . *Puck* Pietsch, 1978, p. 439

NOTE  A single known species; North Atlantic and Pacific oceans

15A. Hyomandibula with a single head . . . . . . . . . . . . . . . . . . . . . *Bertella* Pietsch, 1973, p. 437

NOTE  A single known species; North Pacific Ocean

15B. Hyomandibula with a double head . . . . . . . . . . . . . *Dolopichthys* Garman, 1899, p. 430

NOTE  Seven recognized species; Atlantic, Pacific, and Indian oceans

MALES

In the following key (slightly modified after Bertelsen, 1951:75), it is uncertain how much diagnostic value can be ascribed to the number of denticular teeth and olfactory lamellae; these presumably increase during juvenile development and therefore may be subject to greater individual variation than indicated here.

1A. Pectoral-fin rays 28 to 30, articulating along dorsal margin of an elongate, pectoral-fin lobe . . . . . . . . . . . . . . . . . . . . . . . . *Ctenochirichthys* Regan and Trewavas, 1932, p. 442

   NOTE  A single known juvenile male, 11.5 mm; eastern North Atlantic

1B. Pectoral-fin rays 14 to 27, articulating along posterior margin of a short and broad pectoral-fin lobe . . . . . 2

2A. Subopercle short and broad, dorsal end rounded . . . . . . . . . . . . . . . 3

2B. Subopercle long and slender, dorsal end tapering more or less to a point . . . . . . . . . . . . . . . . . . . . . . . 4

3A. Skin between nostrils unpigmented; medial surface of subopercle unpigmented; caudal peduncle without subdermal pigment; anal-fin rays 4, rarely 3 or 5; pectoral-fin rays 13 to 19 . . . . . . . . . . . . . . . . *Oneirodes* Lütken, 1871, p. 394

   NOTE  An unknown number of species represented; Atlantic, Pacific, and Indian oceans

3B. Skin between nostrils slightly pigmented; medial surface of subopercle darkly pigmented; caudal peduncle with subdermal pigment; anal-fin rays 5, rarely 4 or 6; pectoral-fin rays 18 to 23 . . . . . . . *Microlophichthys* Regan and Trewavas, 1932, p. 425

   NOTE  Two recognized species; Atlantic, Pacific, and Indian oceans

4A. Posterior nostril contiguous with eye; nasal area unpigmented; opercle deeply notched posteriorly; lower denticular with 4 to 10 teeth . . . . 5

4B. Posterior nostril not contiguous with eye; nasal area pigmented; opercle only slightly concave posteriorly; lower denticular with 17 to 27 teeth . . . . . . . . . . . . . . . . *Chaenophryne* Regan, 1925c, p. 384

   NOTE  An unknown number of species represented; Atlantic, Pacific, and Indian oceans

5A. Lower denticular with 8 to 10 teeth; olfactory lamellae 10 to 11; caudal peduncle with distinct subdermal pigment; caudal-fin rays without internal pigment; pectoral-fin rays 17 to 22 . . . . . . . . . . . . . . *Dolopichthys* Garman, 1899, p. 430

   NOTE  An unknown number of species represented; Atlantic, Pacific, and Indian oceans

5B. Lower denticular with 4 teeth; olfactory lamellae 18; caudal peduncle with faint subdermal pigment; caudal-fin rays internally pigmented; pectoral-fin rays 21 to 27 . . . . . . . . . . . . . . . . . . . . . . . . *Pentherichthys* Regan and Trewavas, 1932, p. 382

   NOTE  A single recognized species; North Atlantic and eastern Pacific

LARVAE

1A. Caudal-fin rays internally pigmented; dorsal-fin rays 6 or 7; anal-fin rays 5 to 7 . . . . . . *Pentherichthys* Regan and Trewavas, 1932, p. 382

   NOTE  A single recognized species; North Atlantic and eastern Pacific oceans

1B. Caudal-fin rays without internal pigment; dorsal-fin rays 5 to 8; anal-fin rays 4 to 6 . . . . . . . . . . . . . . . . . 2

2A. Caudal peduncle pigmented . . . . 3

2B. Caudal peduncle unpigmented . . . . 7

3A. Caudal peduncle with one or two isolated groups of melanophores; tips of caudal-fin rays unpigmented; anal-fin rays 5 or 6, rarely 4 . . . . . . . . . 4

3B. Caudal peduncle without an isolated group of melanophores, body evenly pigmented to base of caudal fin; tips of caudal-fin rays darkly pigmented; anal-fin rays 4 . . . . . . . . . . . . . . . . . . . . . . . *Oneirodes melanocauda* Bertelsen, 1951, p. 416

   NOTE  Five known specimens, 8 to 21 mm TL; western North Atlantic, western Pacific and Indian oceans

4A. Tissue medial to subopercle darkly pigmented; dorsal end of subopercle rounded; branchiostegal pigment faint; body long and slender . . . . 5

4B. Tissue medial to subopercle unpigmented; dorsal end of subopercle slender, tapering to a point; branchiostegal pigment well developed; body short, globose . . . . . . . . . . 6

5A. Caudal peduncle with a single lateral group of melanophores; pectoral-fin rays 18 to 23 . . . . . *Microlophichthys* Regan and Trewavas, 1932, p. 425

   NOTE  Perhaps two species represented, 99 known specimens, 2.4 to 12.5 mm TL; Atlantic, Pacific, and Indian oceans

5B. Caudal peduncle with separate lateral and ventral groups of melanophores; pectoral-fin rays 28 to 30 . . . . . . . . . . . . . . . . *Ctenochirichthys* Regan and Trewavas, 1932, p. 442

   NOTE  A single species, two known specimens, 4.5 to 5.0 mm TL; eastern North Atlantic and eastern Pacific oceans

6A. Caudal peduncle with separate dorsal, lateral, and ventral groups of melanophores . . . . . . . *Dolopichthys* Garman, 1899, p. 430

   NOTE  An unknown number of species represented, 178 specimens, 2.4 to 12 mm TL; Atlantic, Pacific, and Indian oceans

6B. Caudal peduncle with a single lateral group of melanophores . . . . . . . . . . . . . . . . . . *Chaenophryne longiceps* group Bertelsen, 1951, p. 386

   NOTE  A single recognized species, 106 specimens, 3.2 to 19 mm TL; Atlantic, Pacific, and Indian oceans

7A. Pigment of anterior part of body confluent with peritoneal pigment; gill region evenly and heavily pigmented; dorsal end of subopercle short and rounded . . . . . . *Oneirodes* Lütken, 1871, p. 394

   NOTE  An unknown number of species represented, 192 specimens, 3 to 12 mm TL; Atlantic, Pacific, and Indian oceans

7B. Pigment of anterior part of body forming an isolated dorsal group of melanophores; gill region unevenly pigmented or only faintly pigmented; dorsal end of subopercle slender, tapering to a point . . . . . 8

8A. Gill region with a dense, V-shaped group of melanophores . . . . . . . . . . . . . . . . . . . . . . . . . . *Lophodolos* Lloyd, 1909a, p. 379

   NOTE  Probably two species represented, 7.0 to 8.0 mm TL; Atlantic, Pacific, and Indian oceans

8B. Gill region only faintly pigmented . . . . . . . . *Chaenophryne draco* group Bertelsen, 1951, p. 387

   NOTE  An unknown number of species represented, 87 specimens, 3.5 to 13 mm TL; Atlantic, Pacific, and Indian oceans

### Genus *Lophodolos* Lloyd, 1909a (Pugnose Dreamers)

Figures 33, 43, 96–98, 190, 202, 203, 219, 250; Tables 1, 8, 9

Females and Larvae

*Lophodolos* Lloyd, 1909a:167 (type species *Lophodolos indicus* Lloyd, 1909a, by monotypy).

*Lophodolus* Regan, 1925c:563 (emended spelling of *Lophodolos*, followed by numerous subsequent authors, taking the same type species).

*Oneirodes* Murray and Hjort, 1912:104, fig. 90 (in part; erroneous designation; type species *Oneirodes eschrichtii* Lütken, 1871, by monotypy).

*Lophodulus* Bussing, 1965:223 (emended spelling of *Lophodolos*, taking the same type species).

Males

Unknown.

Distinguishing Characters

Metamorphosed females of the genus *Lophodolos* are distinguished from those of all other genera of the family by having the following character states: The dorsal profile of the frontal bones is concave. Ventromedial extensions of the frontals are absent. The posterior end of the frontal makes broad contact with the respective sphenotic and prootic (the pterosphenoid is absent) and extends ventrally to meet the ascending process of the parasphenoid. The sphenotic spines are extraordinarily well developed and curved posteriorly. The pterygiophore of the illicium emerges from between or somewhat behind the sphenotic spines. The symphysial spine of the lower jaw is extremely well developed. The medial ends of the second and third hypobranchials approach each other on the midline (see Pietsch, 1974a:25).

Metamorphosed females of *Lophodolos* are further unique in having the following combination of character states: The ethmoid cartilage and vomer are wide, wider than the distance between the anterolateral tips of the lateral ethmoids and frontals. Vomerine teeth are absent. The nasal foramina are extremely large and nearly circular in shape. The frontals are short, lying posterior to the ethmoid region. The anterior end of the illicial trough is wider and shallower than its posterior end. The symphysial cartilage of the upper jaw is longer than wide. The hyomandibula has a double head. The quadrate spine is extremely well developed, much longer than the articular spine. The posterior margin of the opercle is deeply notched. The subopercle is narrow and elongate, its dorsal end slender and tapering to a blunt point (the posterior margin without indentation), its ventral end narrow and oblong (a small anterior projection is present in some juvenile females). The first and second pharyngobranchials are absent. The caudal-fin rays are without internal pigmentation. The illicium is considerably longer than the length of the esca bulb. The pterygiophore of the illicium is cylindrical throughout its length, the anterior end exposed, the posterior end concealed beneath skin. There are 5 to 8 dorsal-fin rays, the first or anterior-most ray reduced to a small stub. There are 5 to 6 (rarely 4 or 7) anal-fin rays (Table 1). The pectoral-fin lobe is short and broad, shorter than the longest rays of the pectoral fin. There are 17 to 20 (rarely 21) pectoral-fin rays. The coracoid is without a posteroventral process. The pelvic bones are simple, but expanded distally. The skin is naked, without dermal spinules. The darkly pigmented skin of the caudal peduncle extends well past the base of the caudal fin.

The larvae of *Lophodolos* are unique in having the following combination of character states: The body is rather short, its depth about 60% SL. A dark V-shaped patch of pigment is present on the gill cover (or sometimes two crescent-shaped patches of pigment meeting at the base of the lower branchiostegals, an anterior patch along the posterior margin of the opercular region, and a posterior patch along the edge of the gill cover). The dorsal pigment extends posteriorly slightly beyond the anterior margin of the dorsal fin. Peduncular pigment is absent. The ventral part of the peritoneum is unpigmented (Fig. 98).

Description

The body is relatively long, slender, not globular. The snout is short and the mouth large, its cleft extending past the eye. The jaws are equal anteriorly. The angular bone terminates as a well-developed spine. The illicium length is highly variable, 11.1 to 138.0% SL, becoming longer proportionately with growth. The esca bears a pair of unpigmented, bilaterally placed appendages arising from the distal surface.

The teeth are slender, straight, all depressible, and weakly set (easily damaged or lost), arranged in overlapping sets as described for other oneirodids (Pietsch, 1972c:5, fig. 2). The teeth in the lower jaw are larger and more numerous than those in the upper jaw. There are 200 to 280 teeth in the lower jaw. The third pharyngobranchial is well developed, bearing numerous teeth. Epibranchial and ceratobranchial teeth are absent.

The color in preservation is dark brown to black over the entire external surface of the body, except for the bulb and appendages of the esca (escal appendages and the unpigmented distal portion of the escal bulb appear "silvery white" in fresh specimens of *Lophodolos acanthognathus*; Erik Bertelsen, personal communication, 7 January 1974). The oral cavity and guts, except for the outer surface of stomach wall, are unpigmented.

Comments

The original spelling of the generic name *Lophodolos* (Lloyd, 1909a) was reestablished by Pietsch (1974b:7) as the "correct original spelling," as provided by Article 32.2 of the International Code of Zoological Nomenclature.

Diversity

Two species and one nominal form of doubtful validity differentiated as follows:

Key to Females of Species of the Genus *Lophodolos*

1A. Length of illicium less than 25% SL in specimens 30 mm and larger; width of escal bulb 4.4 to 6.7% SL in specimens 25 mm and larger; length of escal appendages 10.2 to 20.9% SL in specimens 25 mm and larger, 8.7 to 22.2% SL (usually greater than 10.0% SL) in specimens less than 25 mm; length of sphenotic spine 4.1 to 9.2% SL (usually greater than 6.0% SL) in specimens 30 mm and larger; length of quadrate spine 2.9 to 6.5% SL (usually greater than 3.5% SL) in specimens 30 mm and larger; dorsal-fin rays 5 to 7 . . . . . . . . . . . . . . . . . . . . . . . . . . . . . . . *Lophodolos acanthognathus* Regan, 1925c, p. 381

NOTE One hundred and forty-nine known specimens, 6 to 73 mm; Atlantic, Pacific, and Indian oceans

1B. Length of illicium greater than 25% SL in specimens 30 mm and larger; width of escal bulb 2.1 to 4.0% SL in specimens 25 mm and larger; length of escal appendages 1.2 to 5.0% SL in specimens 25 mm and larger, 4.2 to 10.5% SL (usually less than 9.0% SL) in specimens less than 25 mm; length of sphenotic spine 1.9 to 6.0% SL (usually less than 5.0% SL) in specimens 30 mm and larger; length of quadrate spine 1.6 to 5.0% SL (usually less than 3.0% SL) in specimens 30 mm and larger; dorsal-fin rays 6 to 8 . . . . . *Lophodolos indicus* Lloyd, 1909a, p. 380

NOTE Thirty-one known specimens, 9.5 to 77 mm; Atlantic, Pacific, and Indian oceans

*LOPHODOLOS INDICUS* LLOYD, 1909A

Figures 96, 97, 219; Table 8

Females

*Lophodolos indicus* Lloyd, 1909a:167 (original description, single specimen). Pietsch, 1974b:8, 13, figs. 1, 2, 4–9 (additional material, description, comparison with all known material; includes *Lophodolos dinema*). Bertelsen and Pietsch, 1977:188 (additional material, eastern Atlantic). Pietsch and Seigel, 1980:395 (additional specimen, Banda Sea). Bertelsen, 1990:502 (synonymy, eastern tropical Atlantic). Pietsch, 1999:2032 (western central Pacific).

*Lophodolus indicus*: Lloyd, 1909b, pl. 45, fig. 7 (holotype figured). Regan, 1925c:563 (comparison with *Lophodolos acanthognathus*). Regan, 1926:30 (brief description after Lloyd, 1909a; comparison with *L. acanthognathus*). Regan and Trewavas, 1932:83 (after Lloyd, 1909a, Regan, 1926). Bertelsen, 1951:108 (description after Lloyd, 1909a, Regan and Trewavas, 1932; comparison with all known material of *Lophodolos*). Grey, 1956a:255–256 (synonymy; distribution).

*Lophodolus dinema* Regan and Trewavas, 1932:83, pl. 4, fig. 3 (original description, single specimen). Bertelsen, 1951:108 (description, comparison with all known material). Grey, 1956a:255 (synonymy, distribution). Nielsen, 1974:96 (listed in type catalog). Parin et al., 1974:123 (western South Atlantic). Pietsch, 1974b:13, 15 (a junior synonym of *Lophodolos indicus*).

Males and Larvae

Unknown.

Material

Thirty-one metamorphosed females (9.5 to 77 mm):

Holotype of *Lophodolos indicus*: ZSI 1024/1, 53 mm, *Investigator* station 307, off Kerala (formerly Travancore), southwest coast of India, 0 to 1624 m.

Holotype of *Lophodolos dinema*: ZMUC P92105, 43 mm, *Dana* station 3716(2), South China Sea, 19°18′N, 120°13′E, 3000 m wire, bottom depth 3225 m, 1400 h, 22 May 1929.

Additional material: ASIZP, one (53 mm); BMNH, one (57 mm); HUMZ, one (75 mm); ISH, five (36 to 75 mm); LACM, four (32.5 to 71 mm); MCZ, five (30 to 64.5 mm); SIO, nine (9.5 to 77 mm); UF, one (23 mm); UW, two (53 to 65 mm).

Distinguishing Characters

Metamorphosed females of *Lophodolos indicus* differ from those of its only congener *L. acanthognathus* in having the following character states: length of illicium greater than 25% SL in specimens 30 mm and larger; width of escal bulb 2.1 to 4.0% SL in specimens 25 mm and larger; length of escal appendages 1.2 to 5.0% SL in specimens 25 mm and larger, 4.2 to 10.5% SL (usually less than 9.0% SL) in specimens less than 25 mm; length of sphenotic spine 1.9 to 6.0% SL (usually less than 5.0% SL) in specimens 30 mm and larger; length of quadrate spine 1.6 to 5.0% SL (usually less than 3.0% SL) in specimens 30 mm and larger; dorsal-fin rays 6 to 8.

Description

Illicium long, 15.2 to 138.0; width of escal bulb small, 2.1 to 5.2; escal appendages short, 1.2 to 10.5 (Figs. 96, 97); sphenotic spines short, 1.9 to 6.0; quadrate spines short, 1.6 to 5.0; dorsal-fin rays 6 to 8; anal-fin rays 5 to 7; pectoral-fin rays 17 to 21 (measurements in percent SL; spine lengths based on specimens greater than 30 mm, fin-ray counts on specimens greater than 20 mm).

Additional description as given for genus and family.

Distribution

The known material of *Lophodolos indicus* is broadly and rather evenly distributed beneath tropical waters of all three major oceans of the world, but appears to be excluded from the western Atlantic where the greatest abundance of its congener, *L. acanthognathus*, is found (Fig. 219). In the eastern Atlantic it extends from about 22°N to the Gulf of Guinea and further south to approximately 18°S, 10°W. The remaining material is rather evenly distributed across the Indian and Pacific oceans between approximately 4°S and 30°N. The holotype was collected off the southwest tip of India.

On the basis of maximum depths reached by fishing gear, metamorphosed specimens of *L. indicus* are vertically distributed between approximately 750 m and an unknown lower limit. All specimens larger than 30 mm (18 individuals) were captured by nets fished below 1000 m; 67% of these were captured by nets fished below 1500 m.

Comments

Large specimens of *Lophodolos indicus* (greater than approximately 30 mm) are easily distinguished from *L. acanthognathus* on the basis of illicial and escal appendage lengths alone (see Key to species above). Smaller specimens are more difficult to identify (see Comments under *L. acanthognathus*).

*Lophodolus dinema* Regan and Trewavas (1932) was described as new solely on the basis of differences in escal morphology. These differences, however, are undoubtedly the result of damage. The esca of the holotype of *L. indicus*, originally described by Lloyd (1909a:167) as being "hard but . . . covered with short, shreddy filaments," has lost the 2 bilaterally placed appendages found in the holotype of *L. dinema* and in all known specimens of *Lophodolos*. Although not examined by me, the poor condition of the esca was confirmed by a sketch made from the holotype of *L. indicus* provided by A. G. K. Menon (personal communication, 23 March 1971) of the Zoological Survey of India. Discrepancies in illicial length (Bertelsen, 1951:107) are also more apparent than real. A plot of illicial length against standard length (see Pietsch, 1974b:4, fig. 1) shows the holotype of *L. dinema* to compare well with the material here considered to be *L. indicus*. In the absence of any significant differences, *L. dinema* is retained within the synonymy of *L. indicus* following the conclusions of Pietsch (1974b).

## *LOPHODOLOS ACANTHOGNATHUS* REGAN, 1925C

Figures 33, 97, 98, 219; Table 8

Females and Larvae

*Oneirodes* n. sp. no. 1: Murray and Hjort, 1912:104, 614, fig. 90 (erroneous designation; specimens referred to *Lophodolos lyra* by Koefoed, 1944, and to *L. acanthognathus* by Nybelin, 1948).

*Lophodolus acanthognathus* Regan, 1925c:563 (original description, two specimens; lectotype ZMUC P92104, designated by Bertelsen, 1951; emended spelling of *Lophodolos*). Regan, 1926:8, 9, 30, pl. 6, fig. 1 (brief description, additional specimen). Schmidt, 1926:262, fig. 2 (after Regan, 1925c, 1926). Norman, 1930:354 (additional specimen, Atlantic). Regan and Trewavas, 1932:9, 11, 12, 83 (description after Regan, 1926; five additional specimens; *Lophodolos lyra* Beebe, 1932b, a synonym of *L. acanthognathus*). Gregory, 1933:405, fig. 277 (osteological comments, specific name misspelled *acanthagnathus*). Beebe, 1937:207 (45 specimens listed from Bermuda). Fowler, 1944:454 (additional material, western North Atlantic). Nybelin, 1948:86, text-fig. 9, table 20 ("*Oneirodes* n. sp. no. 1" of Murray and Hjort, 1912, referred to *L. acanthognathus*; description of an additional specimen; comparison with previous descriptions; geographic, bathymetric distribution). Bertelsen, 1951:107, fig. 64 (synonymy, description, comparison with all known material; *Dana* material listed, comments, in key). Marshall, 1954:114, fig. V, 17a (deep sea biology; after Bertelsen, 1951). Grey, 1955:299 (additional specimen). Grey, 1956a:255 (synonymy, distribution). Maul, 1973:672 (synonymy, eastern North Atlantic). Nielsen, 1974:96 (listed in type catalog).

*Lophodolus lyra* Beebe, 1932b:96, fig. 28 (original description, about 40 specimens). Koefoed, 1944:7, pl. 3, fig. 3 (misidentifications, description; three specimens, *Oneirodes* n. sp. no. 1 of Murray and Hjort, 1912:614). Mead, 1958:133 (type transferred to USNM).

*Lophodolos acanthognathus*: Fowler, 1936:1366, fig. 560 (corrected spelling of *Lophodolos*; brief description after Regan, 1926). Pietsch, 1972a:35, 45 (osteological comments). Pietsch, 1974a:17, 19, 21, 24–28, 82, 86–89, figs. 27, 33, 39E, 45, 51H, 53B, 104 (osteology, relationships). Pietsch, 1974b:8, figs. 3–6, 9 (additional material, description, comparison with all known material). Bertelsen and Pietsch, 1977:188 (additional material, eastern Atlantic). Hubbs et al., 1979:13 (California). Bertelsen, 1986:1392, figs. (eastern North Atlantic). Bertelsen, 1990:502 (synonymy, eastern tropical Atlantic). Nielsen and Bertelsen, 1992:60, fig. 6 (North Atlantic). Andriyashev and Chernova, 1994:102 (listed for Arctic seas). Pietsch, 1999:2032 (western central Pacific). Pietsch, 2002b:1064 (western central Atlantic). Moore et al., 2003:215 (off New England).

*Lophodulus acanthognathus*: Bussing, 1965:223 (emended spelling of *Lophodolos*; two additional specimens, Peru-Chile Trench).

Metamorphosed Males

Unknown.

Material

One hundred and forty-nine metamorphosed females (6 to 73 mm); and five larvae, three males (7 to 8 mm TL), and three females (7 to 8 mm TL):

Lectotype of *Lophodolos acanthognathus*: ZMUC P92104, 12 mm, *Dana* station 1358(5), western North Atlantic, 28°15′N, 56°00′W, 3000 m wire, 1530 h, 2 June 1922.

Paralectotype of *Lophodolos acanthognathus*: BMNH 1925.8.11.15, 17 mm, *Dana* station 1342(1), western North Atlantic, 34°00′N, 70°01′W, 4500 m wire, 1430 h, 15 May 1922.

Holotype of *Lophodolus lyra*: USNM 170949 (originally NYZS 21610), 47 mm, *Gladisfen* net 111, off Bermuda, 32°12′N, 64°36′W, 1463 m, 27 July 1931.

Additional metamorphosed females: BMNH, four (18 to 26 mm); CAS, 32 (6 to 32 mm); FMNH, one (9.5 mm); GNM, one (56 mm); HUMZ, three (58 to 64 mm); ISH, 24 (42 to 73 mm); LACM, four (22.5 to 38 mm); MCZ, 25 (12 to 65 mm); MNHN, two (27 to 52 mm); ROM, 13 (17 to 57 mm); UF, three (20 to 52 mm); USNM, one (10 mm); UW, one (31.5 mm); ZMUB, six (45 to 62 mm); ZMUC, 13 (8.5 to 57 mm); YPM, three (18 to 22 mm).

Larvae: ZMUC, five (7 to 8 mm TL). See Bertelsen (1951:107).

Distinguishing Characters

Metamorphosed females of *Lophodolos acanthognathus* differ from those of its only congener *L. indicus* in having the following character states: length of illicium less than 25% SL in specimens 30 mm and larger; width of escal bulb 4.4 to 6.7% SL in specimens 25 mm and larger; length of escal appendages 10.2 to 20.9% SL in specimens 25 mm and larger, 8.7 to 22.2% SL (usually greater than 10.0% SL) in specimens less than 25 mm; length of sphenotic spine 4.1 to 9.2% SL (usually greater than 6.0% SL) in specimens 30 mm and larger; length of quadrate spine 2.9 to 6.5% SL (usually greater than 3.5% SL) in specimens 30 mm and larger; dorsal-fin rays 5 or 6, rarely 7.

Description

Illicium short, 11.1 to 23.1% SL; width of escal bulb large, 4.2 to 9.0% SL; escal appendages long, 8.7 to 22.2% SL; sphenotic spines long, 4.1 to 9.2% SL; quadrate spines long, 2.9 to 6.5% SL; anal-fin rays 4 to 6; pectoral-fin rays 17 to 21 (spine lengths based on specimens greater than 30 mm, fin-ray counts on specimens greater than 20 mm).

Additional description as given for the genus and family.

Distribution

Most of the material of *Lophodolos acanthognathus* (including all type specimens) has been collected from the western Atlantic Ocean, where it extends from off Greenland and Iceland at about 65°N, to the equator (including the Gulf of Mexico and Caribbean Sea), with a single record in the central South Atlantic at about 40°S, 26°W (Fig. 219). In the Indo-Pacific, it has been collected in the Arabian Sea, Indian Ocean (at approximately 7°N, 59°E), the South China and Celebes seas, and the eastern Pacific, on the equator between 139° and 142°W and from off the coast of Peru. The lectotype was collected from the western North Atlantic at 28°N, 56°W.

On the basis of maximum depths reached by fishing gear, metamorphosed females of *L. acanthognathus* are vertically distributed between 650 m and an unknown lower limit. Nearly all specimens larger than 30 mm were captured by nets fished below 1000 m; about half of these were captured by nets fished below 1500 m.

Comments

Specimens of *Lophodolos acanthognathus* larger than approximately 30 mm can be easily separated from *L. indicus* on the basis of illicial and escal appendage lengths alone (see key to species above). Smaller specimens, especially those smaller than 20 mm, are difficult to identify, and require a combination of meristics and counts, all of which overlap between the two species: illicial and escal appendage lengths, width of the escal bulb, and dorsal-fin ray counts. In some cases, geographic distribution may provide additional data for identification; *L. indicus* apparently does not occur in the western North Atlantic, where approximately 80% of the known material of *L. acanthognathus* has been collected.

The holotype of *L. lyra* Beebe (1932b) compares well with the known material of *L. acanthognathus*; the name is retained within the synonymy of *L. acanthognathus*, following Regan and Trewavas (1932).

On the basis of fin-ray counts, shape of the opercular bones, and a comparison of subdermal pigmentation with juvenile females of *L. acanthognathus*, Bertelsen (1951:106) referred five larvae from the *Dana* collections to the genus *Lophodolos*. Because these larvae were all collected in the North Atlantic where *L. acanthognathus* is most common, they were referred to this species. With no evidence to the contrary, this allocation is followed here.

LOPHODOLOS BIFLAGELLATUS KOEFOED, 1944, NOMEN NUDUM

*Lophodolus biflagellatus* Koefoed, 1944:7 (emended spelling of *Lophodolos*). Pietsch, 1974b:17 *(nomen nudum)*.

Comments

This name was used by Koefoed in a manuscript dated 1918 (not seen by me), and later mentioned in published form (Koefoed, 1944:7) without application to a description or type.

Genus *Pentherichthys* Regan and Trewavas, 1932 (Thickjaw Dreamers)

Figures 33, 99–101, 202, 203, 220, 295; Tables 1, 8

Females

*Dolopichthys* Regan, 1926:28 (in part; five specimens misidentified as *Dolopichthys allector*, reallocated to *Pentherichthys atratus* by Bertelsen, 1951; type species *D. allector* Garman, 1899, by monotypy). Norman, 1930:353, fig. 43 (in part; specimen misidentified as *D. allector*, reallocated to *P. venustus* by Bertelsen, 1951).

*Dolopichthys* (subgenus *Pentherichthys*) Regan and Trewavas, 1932:81, figs. 128, 129 (one of five subgenera of genus *Dolopichthys*; type species *Dolopichthys atratus* Regan and Trewavas, 1932, by subsequent designation of Burton, 1933:62).

*Pentherichthys* Bertelsen, 1951:88, figs. 42, 43 (subgenus *Pentherichthys* given generic status; type species *Dolopichthys atratus* Regan and Trewavas, 1932, by subsequent designation of Burton, 1933:62).

Males

*Rhynchoceratias* Regan, 1926:43 (in part; a single specimen identified as *Rhynchoceratias leuchorhinus* by Regan, 1926; type species *R. brevirostris* Regan, 1925c, by subsequent designation of Fowler, 1936).

Distinguishing Characters

*Pentherichthys* is unique among all known ceratioids in having large melanophores inside the rays of the caudal fin, easily distinguished with the naked eye in larvae and adults of both sexes (see Bertelsen, 1951:103, fig. 61). Metamorphosed females of *Pentherichthys* differ from those of all other oneirodids in having the ethmoid region of the cranium extremely flattened dorsoventrally. The illicial trough is deep and wide, the nasal foramina narrow and elongate (Pietsch, 1974a:16, fig. 25). The symphysial spine of the lower jaw is absent. The dentaries form a thick, broad, posteriorly directed flange immediately lateral to their union on the midline, the ventral margin of the lower jaw at the symphysis concave when viewed anteriorly (Pietsch, 1974a:21, fig. 48). The pterygiophore of the illicium is short, the posterior part broad and dorsoventrally flattened (Pietsch, 1974a: 28, fig. 53C).

Metamorphosed females of *Pentherichthys* are further distinguished from those of all other oneirodid genera in having the following combination of character states: The ethmoid cartilage and vomer are wide, their width about equal to the distance between the anterolateral tips of the lateral ethmoids and frontals. Vomerine teeth are absent. The frontals are short, the anterior end overhanging and extending past the anterior limits of the ethmoid cartilage and vomer, the dorsal margin convex. The ventromedial extensions of the frontals approach each other on the midline, making contact with the parasphenoid. The frontals are separated from the prootics. The pterosphenoid is present. The posterior end of the illicial trough is wider and shallower than its anterior end. The sphenotic spines are well developed. The symphysial cartilage of the upper jaw is considerably wider than long. The hyomandibula has a double head. The quadrate and articular spines are rudimentary. The posterior margin of the opercle is deeply notched. The subopercle is elongate, its dorsal end slender and tapering to a point (the posterior margin is without an indentation), its ventral end oblong (with a small anterior projection in some larvae and males). The first and second pharyngobranchials are absent. The second hypobranchial articulates directly with the second basibranchial. The illicium is considerably longer than the length of the esca bulb. The pterygiophore of the illicium is cylindrical throughout its length, emerging on the snout from between the frontal bones, its anterior end slightly exposed, its posterior end concealed beneath skin. The first ray of the dorsal fin is well developed. There are 6 to 7 dorsal-fin rays and 6 (rarely 5 or 7) anal-fin rays (Table 1). The pectoral-fin lobe is short and broad, shorter than the longest rays of the pectoral fin. There are 21 to 27 pectoral-fin rays. The coracoid lacks a posteroventral process. The pelvic bones are simple, expanded distally. The skin is naked, without dermal spinules. Darkly pigmented skin of the caudal peduncle extends well past the base of the caudal fin.

In addition to having internally pigmented caudal rays, metamorphosed males of *Pentherichthys* differ from those of all other oneirodid genera in having the following combination of character states: The posterior nostril is contiguous with the eye. The nasal area is white. There are 18 olfactory lamellae. There are 7 to 9 upper denticular teeth, fused at their base, and 4 lower denticular teeth (Bertelsen, 1951:102, fig. 62).

In addition to having internally pigmented caudal rays, larvae of *Pentherichthys* are differentiated by having the following combination of character states: The depth of the body anterior to the base of the pectoral fins is about 60% SL. The posterior part of the body is relatively slender. The gill cover has a few scattered melanophores. The dorsal pigmentation consists of a restricted cluster of melanophores, terminating anterior to the origin of the dorsal fin. The posterior part of the caudal peduncle is faintly pigmented. The ventral part of the peritoneum is unpigmented (Bertelsen, 1951:102, fig. 61).

Description

The body of metamorphosed females is relatively long and slender, not globular. The depth of the head is about 45 to 50% SL. The mouth is large, the cleft extending well past the eye. The jaws are equal anteriorly. The illicium length is 12 to 30% SL. The esca bears a small conical or laterally compressed anterior papilla, a median distal crest, and a large tapering posterior appendage. The posterior escal appendage bears in turn several to numerous short filaments along most of its length, and, on each side of its base, a large, simple to complexly branched, anteriorly directed appendage (Fig. 100).

The opercle is strongly bifurcate, the two forks forming an acute angle of about 34 to 47°. The length of the dorsal fork is 50 to 60% length of the ventral fork. The length of the subopercle is about 80 to 82% of the length of the ventral fork of the opercle.

The teeth of the jaws are slender, recurved, and all depressible, arranged in somewhat irregular overlapping transverse sets of up to 7 teeth, increasing in size toward the innermost and most recently developed teeth. The pattern of tooth placement is especially obvious in larger specimens that have higher tooth counts, and, although the pattern is the same in both upper and lower jaws, it is more obvious in the lower jaw. The teeth in the lower jaw are larger and considerably more numerous than those in the upper jaw. The upper and lower jaws bear about 50 and 80 teeth, respectively, in a 19.5-mm specimen, increasing to about 110 and nearly 200 teeth, respectively, in a 119-mm specimen. The third pharyngobranchial is well toothed, bearing about 15 teeth in a 119-mm specimen (Pietsch, 1974a:25, fig. 51G).

For additional description of larvae and males, see Bertelsen (1951:104, 105, figs. 60–62).

Diversity

A single species:

## PENTHERICHTHYS ATRATUS REGAN AND TREWAVAS, 1932

Figures 33, 99–101, 220, 295; Table 8

Females, Males, and Larvae

*Dolopichthys allector*: Regan, 1926:28 (misidentifications; five of nine specimens subsequently described as *Dolopichthys (Pentherichthys) atratus* by Regan and Trewavas, 1932). Norman, 1930:353, fig. 43 (misidentification, single specimen, subsequently described as *D. (P.) venustus* by Regan and Trewavas, 1932).

*Dolopichthys (Pentherichthys) atratus* Regan and Trewavas, 1932:81, fig. 129, pl. 3, fig. 2 (original description, five specimens, lectotype ZMUC P92103, designated by Bertelsen, 1951).

*Dolopichthys (Pentherichthys) venustus* Regan and Trewavas, 1932:81, fig. 130 (original description, single specimen).

*Dolopichthys atratus*: Beebe and Crane, 1947:162 (two additional specimens, eastern tropical Pacific). Nielsen, 1974:91 (listed in type catalog).

*Pentherichthys atratus*: Bertelsen, 1951:105 (subgenus elevated to generic rank, description, comparison of all known material). Briggs, 1960:179 (worldwide distribution). Moore et al., 2003:215 (off New England).

*Pentherichthys venustus*: Bertelsen, 1951:105 (description, comparison of all known material). Pietsch, 1974a:16–21, 24–29, 109, figs. 25, 38, 39H, 47, 48, 51G, 53C (osteology, relationships). Bertelsen and Pietsch, 1977:186, fig. 9 (three additional specimens, eastern South Atlantic). Bertelsen, 1986:1398, fig. (eastern North Atlantic). Bertelsen, 1990:506 (synonymy, eastern tropical Atlantic).

*Ctenochirichthys longimanus*: Bussing, 1965:222 (misidentification, specimen referred to *Pentherichthys atratus* by Pietsch, 1978).

*Pentherichthys* sp.: Maul, 1973:673 (unidentifiable larvae and males, eastern North Atlantic). Pietsch and Seigel, 1980:395 (after Bertelsen, 1951). Bertelsen, 1984:331, fig. 170E (early life history, phylogeny). Pietsch, 1999:2032 (western central Pacific).

Material

Twenty-six metamorphosed females (16.5 to 122 mm); six males, three in metamorphosis (7.5 to 10.5 mm), three metamorphosed (12 to 13.5 mm); and 23 larvae, 10 females and 13 males (3.5 to 16 mm TL):

Lectotype of *Pentherichthys atratus*: ZMUC P92103, 19 mm, *Dana* station 1208(6), Gulf of Panama, 6°48′N, 80°33′W, 2500 m wire, 0810 h, 16 January 1922.

Paralectotypes of *Pentherichthys atratus*: BMNH 1925.8.11.5, 18 mm, *Dana* station 1208(6), Gulf of Panama, 6°48′N, 80°33′W, 2500 m wire, 0810 h, 16 January 1922. BMNH 1925.8.11.7, 21 mm, *Dana* station

1208(15), Gulf of Panama, 6°48'N, 80°33'W, 2600 m wire, 1715 h, 16 January 1922. ZMUC P92160, 16.5 mm, *Dana* station 1206(4), Gulf of Panama, 6°40'N, 80°47'W, 3000 m wire, 1845 h, 14 January 1922. ZMUC P92161, 17.5 mm, *Dana* station 1209(1), Gulf of Panama, 7°15'N, 78°54'W, 3500 m wire, 1845 h, 17 January 1922.

Holotype of *Pentherichthys venustus*: BMNH 1930.1.12.1081, 28 mm, *Discovery* station 297, eastern North Atlantic, 12°08'N, 20°53'W, Young-fish trawl, 0 to 300 m, 28 August 1927.

Additional metamorphosed females: ISH, three (19.5 to 119 mm; 119-mm specimen cleared and stained); LACM, two (19 to 32.5 mm); MCZ, three (24 to 122 mm); MNHN, three (16.5 to 93 mm); NYZS, two (20 to 33 mm); SIO, six (20.5 to 68.5 mm); UW, one (32 mm).

Males SIO, two (7.5 to 9.0 mm); ZMUC, four (10.5 to 13.5 mm).

Larvae MCZ, four (3.5 to 16 mm TL); ZMUC, 19 (5.3 to 15 mm TL; see Bertelsen, 1951:103).

Distinguishing Characters and Description

As given for the genus and family (see Biodiversity, Chapter Three).

Distribution

Metamorphosed females and males of *Pentherichthys atratus* have been collected from numerous, widely scattered localities in the Atlantic and eastern Pacific oceans, but the species is represented in the Indian and western Pacific only by larvae, one from the South China Sea, off the west coast of Luzon, Philippines, and another in the eastern Indian Ocean at about 3°S off the coast of Sumatra (Fig. 220). It has been collected on both sides of the Atlantic between 40°N and 8°S, with one known record in the far eastern South Atlantic off Cape Town, South Africa. In the eastern Pacific it ranges between 22°N and 15°S, from 165°W and eastward into the Gulf of Panama.

Comments

Since Regan and Trewavas (1932) published their monograph on the Ceratioidei of the *Dana* collections, two species of the genus *Pentherichthys* have been recognized, *Pentherichthys venustus* and *P. atratus*. The holotype of the first of these species, a 26-mm juvenile female collected in the North Atlantic by the *Discovery* was tentatively identified by Norman (1930) as *Dolopichthys allector* Garman. While noting that the structure of the esca was different from that of *D. allector*, Norman provided a detailed description and an illustration. Recognizing this specimen as unique, Regan and Trewavas (1932) described it as new and, naming it *D. venustus*, placed it together with another new species, *D. atratus*, in a new subgenus *Pentherichthys*. Bertelsen (1951) later raised *Pentherichthys* to generic status and described the first known males and larvae. He recognized both *P. venustus* and *P. atratus*, but finding no characters to separate the widely distributed larvae into more than a single group, questioned the existence of two species.

Bertelsen and Pietsch (1977) described three additional females (19.5 to 119 mm, the latter specimen cleared and stained and described by Pietsch, 1974a) collected by the *Walther Herwig* in the South Atlantic. Except for the length of the illicium, all counts and measurements of this material compare well with the type material. The illicium of the 19.5- and 92-mm specimens (ISH 2058/71, ISH 131/67) is about 20% SL, nearly the same as that of the 32.5-mm specimen (21% SL) reported by Bussing (1965; misidentified as *Ctenochirichthys longimanus*) and the 33-mm specimen (22% SL) described by Beebe and Crane (1947), but somewhat shorter than that of the 26-mm holotype of *P. venustus* (30% SL). The length of the illicium of the 119-mm, cleared and stained specimen (ISH 130/67) is about 16% SL.

Like most ceratioids, the number of teeth of *Pentherichthys* increases with age. The total number of teeth in the upper jaw is about 50 in the 19.5-mm specimen, about 60 in the 26-mm holotype of *P. venustus*, and about 220 in the 92-mm and 119-mm specimens. Similarly, the number of teeth in the lower jaw increases from about 80 and 100 in the two smaller specimens to nearly 400 in the two larger specimens. The jaw teeth are arranged in a somewhat irregular, oblique, transverse series of up to 7 teeth, increasing in size toward the innermost and least developed.

The esca of the 19.5-mm specimen is well preserved. In the 92-mm specimen most of the skin of the bulb is lost, and in the 119-mm specimen all the escal appendages are shrunken and damaged. Together with the 26-mm holotype of *P. venustus*, they form an ontogenetic series (Bertelsen and Pietsch, 1977). The width of the escal bulb decreases with increasing SL: it is about 11% SL in the 19.5-mm specimen, 8% SL at 26 mm, 4.5% SL at 92 mm, and 2.8% SL at 119 mm. The anterior distal papilla, conical in the two smallest specimens, is compressed and leaf shaped in the 92-mm specimen. In the two smallest specimens the anterior distal papilla is followed by a transparent median crest, pigmented at its base and supported by an unpigmented tapering internal papilla. In the 92-mm specimen the crest is torn and only the internal papilla is left. The length of the large, tapering posterior appendage increases in relative length from nearly 15% SL at 19.5 mm, nearly 20% SL at 26 mm, and 32% SL at 92 mm. Similarly, the filaments on this appendage increase in number with length of fish. Most of the filaments are unbranched; in the two smaller specimens they are marginal on the somewhat flattened appendage, while in the 92-mm specimen they are irregularly spread on the more conical appendage. Some of the proximal filaments are branched, and in all the specimens there is an enlarged anteriorly directed pair of stout branched filaments. They are relatively longest in the holotype of *P. venustus* (about three times the diameter of the bulb), but just slightly longer than the diameter of the bulb in the 19.5- and 92-mm specimens. In all the specimens the posterior appendage as well as its filaments has a very characteristic internal reddish brown color except for a pair of short, stout filaments at the upper margin of the base of the appendage. This pair of unpigmented digitiform filaments is present in all the examined specimens including the holotype of *P. venustus*.

Except for a slightly longer illicium, the holotype of *P. venustus* fits well within the variation described above for the remaining material of the genus. For this reason, and also because of Bertelsen's (1951) doubt of the existence of more than a single species based on a detailed comparison of the widely distributed larvae of the genus, *P. venustus* is hereby recognized as a junior synonym of *P. atratus*.

Genus *Chaenophryne* Regan, 1925c
(Smoothhead Dreamers)

Figures 25, 33, 102–108, 202, 203, 221, 248, 250, 295; Tables 1, 5, 8, 9

Females

*Chaenophryne* Regan, 1925c:564 (type species *Chaenophryne longiceps* Regan, 1925c, by monotypy).

*Himantolophus* Regan, 1926:40 (in part; misidentification; type species *Himantolophus groenlandicus* Reinhardt, 1837, by monotypy).

Males

*Rhynchoceratias* Regan, 1926:44 (in part; type species *Rhynchoceratias brevirostris* Regan, 1925c, by subsequent designation of Fowler, 1936).

*Trematorhynchus* Regan and Trewavas, 1932:91 (in part; type species *Rhynchoceratias leuchorhinus* Regan, 1925c, by subsequent designation of Burton, 1933).

Distinguishing Characters

Metamorphosed females of the genus *Chaenophryne* are distinguished from those of all other genera of the family Oneirodidae by the absence of sphenotic spines. The opercle is only slightly concave posteriorly. The pterygiophore of the illicium is long, 70 to 82% SL (less than 50% SL in other oneirodids; see Pietsch, 1974a:18).

The pelvic bones are triradiate to broadly expanded distally. The bones, especially those closely associated with the external surface of the head, are highly cancellous (a condition not found in any other ceratioid; see Pietsch, 1975b:77).

Metamorphosed females of *Chaenophryne* are further unique in having the following combination of character states: The ethmoid cartilage and vomer are wide, wider than the distance between the anterolateral tips of the lateral ethmoids and frontals. Vomerine teeth are present. The ethmoid region is depressed, the nasal foramina narrow and oval in shape. The frontals are short, their anterior ends overhanging and extending past the anterior limits of the ethmoid cartilage and vomer. The dorsal margin of the frontals is convex. The ventromedial extensions of the frontals approach each other on the midline, making contact with the parasphenoid. The posterior ends of the frontals are separated from the prootics. The pterosphenoid is present. The anterior end of the illicial trough is wider and shallower than its posterior end. The symphysial cartilage of the upper jaw is longer than wide. The lower jaw bears a rudimentary symphysial spine. The hyomandibular has a double head. The quadrate and articular spines are rudimentary. The subopercle is elongate, its dorsal end slender and tapering to a point (its posterior margin without an indentation), its ventral end nearly circular (with a small anterior spine or projection in some specimens). The first pharyngobranchial is absent, but the second pharyngobranchial is well developed. The second hypobranchial articulates directly with the second basibranchial. The caudal-fin rays are without internal pigmentation. The illicium is considerably longer than the length of the esca bulb. The pterygiophore of the illicium is cylindrical throughout its length, emerging on the snout from between the frontal bones, its anterior end usually concealed beneath the skin, its posterior end concealed beneath the skin. The first ray of the dorsal fin is well developed. There are 6 to 8 dorsal-fin rays and 5 or 6 anal-fin rays (Table 1). The pectoral-fin lobe is short and broad, shorter than the longest rays of the pectoral fin. There are 16 to 22 pectoral-fin rays. The coracoid lacks a posteroventral process. The skin is smooth and naked, without dermal spinules. Darkly pigmented skin of the caudal peduncle extends well past the base of the caudal fin.

Metamorphosed males of *Chaenophryne* are distinguished from those of all other genera of the family in having the following combination of character states: The skin between the nostrils and between the posterior nostril and the eye is pigmented. There are 8 to 12 olfactory lamellae. The upper denticular bears 10 to 22, irregularly curved teeth, mutually fused in a semicircular cluster. The lower denticular bears 13 to 31 recurved denticles in two or three irregular series. The skin is black and naked. There are no teeth in the jaws. The shape of the pectoral fin lobe and opercular bones, and the anal-fin ray counts, are as described for females (Bertelsen, 1951:109, figs. 66–69, 72, 73). The males are free-living, apparently never becoming parasitic.

The larvae are unique among oneirodids in having the following combination of character states: The depth of the body is 50 to 65% SL, the length of the head about 45% SL. The dorsal pigment terminates behind the occipital region of the head but extends posteriorly with increasing growth to the posterior margin of the base of the dorsal fin. Pigmentation of the gill cover is weak to moderately strong. Peduncular pigment is present in the *Chaenophryne longiceps* group, but absent in members of the *C. draco* group (Bertelsen, 1951:109, figs. 66–68, 72).

Description

The body of metamorphosed females is short and globular. The mouth is large, its cleft extending past the eye. The illicium length is 20.1 to 47.4% SL, becoming longer proportionately with growth. The anterior end of the pterygiophore of preserved specimens is usually concealed under the skin, within the illicial trough, but apparently capable of considerable forward extension (as in *Oneirodes* and ceratiids; see Bertelsen, 1943, 1951:18). The esca bears a single (*Chaenophryne draco* group) or paired *(C. longiceps)*, conical to elongate, anterior escal appendage or appendages, internally pigmented, with 1 or 3 round translucent windows at its tip. A filamentous medial escal appendage or appendages are present *(C. longiceps)* or absent (*C. draco* group). A posterior escal appendage, consisting of a swollen basal portion and a compressed distal crest, is present, bearing none to numerous filaments anteriorly or distally, with (*C. draco* group) or without *(C. longiceps)* a pair of fringed anterior lobes. One or 2 filamentous, anterolateral escal appendages are present on each side (*C. melanorhabdus*, *C. ramifera*, and *C. quasiramifera*). A paired series of filaments are present, arising laterally from the anterior base of the escal bulb *(C. ramifera* and *C. quasiramifera)*. A subcutaneous internally pigmented appendage emerges from the base of the esca and descends along the anterior margin of the illicial bone (extremely well developed in some specimens of *C. ramifera* and *C. quasiramifera*).

The teeth are slender, recurved, and all depressible, arranged in overlapping sets as described for other oneirodids (Pietsch, 1972c:5, fig. 2, 1974a, 1974b). The teeth in the lower jaw are larger and slightly more numerous than those in the upper jaw. There are 21 to 51 teeth in the upper jaw and 26 to 57 in the lower jaw. Jaw teeth are few and rudimentary in specimens smaller than approximately 18 mm. The vomer bears 4 to 8 teeth, the longest tooth outermost. The second and third pharyngobranchials are well developed and toothed. Epibranchial and ceratobranchial teeth are absent.

For additional description of larvae and males, see Bertelsen (1951:110–116, figs. 66–69, 72, 73).

Diversity

Two species groups differentiated as follows:

Keys to Species Groups of the Genus *Chaenophryne*

FEMALES

1A. Esca with a pair of internally pigmented anterior appendages; medial escal appendage or appendages present; width of escal bulb 5.3 to 11.4% SL in specimens 20 mm and larger; pectoral-fin rays 17 to 22, rarely fewer than 18 .................... ...... *Chaenophryne longiceps* group Bertelsen, 1951, p. 386

NOTE A single known species; Atlantic, Pacific, and Indian oceans

1B. Esca with an unpaired internally pigmented anterior appendage; medial escal appendages absent; width of escal bulb 2.1 to 6.6% SL in specimens 20 mm and larger; pectoral-fin rays 16 to 19, rarely more than 18 ......... *Chaenophryne draco* group Bertelsen, 1951, p. 387

NOTE Four species; Atlantic, Pacific, and Indian oceans

MALES

1A. Upper denticular teeth 17 to 22, lower denticular teeth 23 to 27; pectoral-fin rays 17 to 22 ................. ...... *Chaenophryne longiceps* group Bertelsen, 1951, p. 386
1B. Upper denticular teeth about 10 to 15, lower denticular teeth 15 to 22; pectoral-fin rays 15 to 19 ........ ......... *Chaenophryne draco* group Bertelsen, 1951, p. 387

LARVAE

1A. Caudal peduncle pigmented; pectoral-fin rays 17 to 22 ........... ...... *Chaenophryne longiceps* group Bertelsen, 1951, p. 386
1B. Caudal peduncle unpigmented; pectoral-fin rays 15 to 19 ........... ......... *Chaenophryne draco* group Bertelsen, 1951, p. 387

## CHAENOPHRYNE LONGICEPS GROUP BERTELSEN, 1951

Females, Males, and Larvae

*Chaenophryne longiceps* Group Bertelsen, 1951:72, 110–114, 269, figs. 30, 66, 68–71 (new species group). Pietsch, 1975b:76, table 1 (reallocation of nominal taxa).

### Comments

The *Chaenophryne longiceps* group (Bertelsen, 1951) is retained to include larvae and males readily separated from those of the *Chaenophryne draco* group by larval pigmentation and pectoral-fin ray counts but not divisible into smaller taxonomic units. Within the *C. longiceps* group is also included a single species, *C. longiceps*, based on metamorphosed females that differ from members of the *C. draco* group in the basic structure of the esca and other features listed in the key above. This species is associated with the larvae and males of the *C. longiceps* group by retention in small juvenile specimens of the characteristic larval pigmentation, and by pectoral-fin ray counts (Bertelsen, 1951:110, table 23).

### Diversity

A single species:

## CHAENOPHRYNE LONGICEPS REGAN, 1925C

Figures 103, 104, 106, 107, 221; Tables 5, 8

### Females

*Oneirodes* n. sp. no. 2: Murray and Hjort, 1912:608, 614, fig. 467 (misidentification; specimen referred to *Chaenophryne quadrifilis* by Koefoed, 1944).

*Chaenophryne longiceps* Regan, 1925c:564 (in part, original description, 14 specimens, lectotype ZMUC P92106, designated by Regan and Trewavas, 1932:87). Regan, 1926:31, pl. 6, fig. 2 (in part; description after Regan, 1925c; three additional specimens). Parr, 1927:22, fig. 8 (four additional specimens designated *Chaenophryne longiceps* forma *typica*, var. *quadrifilis* n. var., and var. *quadrifilis*?; description). Regan and Trewavas, 1932:85, 86, figs. 14, 135 (description; lectotype designated as only representative of *C. longiceps*, 13 paralectotypes made types of other species; in key). Bertelsen, 1951:111, 113, figs. 66A, 71 (description; comparison with all known material; in key). Grey, 1956a:256 (synonymy, distribution). Pietsch, 1974a:33 (listed). Nielsen, 1974:89 (listed in type catalog). Pietsch, 1975b:81, 82, figs. 1A, 3–7, 11, 12 (additional material, osteology, description, comparison with all known material). Bertelsen and Pietsch, 1977:186 (additional material, eastern Atlantic). Pietsch and Seigel, 1980:394 (additional male, Banda Sea). Bertelsen, 1984:331, fig. 170D (early life history, phylogeny). Bertelsen, 1986:1386, figs. (eastern North Atlantic). Bertelsen, 1990:498 (synonymy, eastern tropical Atlantic). Nielsen and Bertelsen, 1992:60, fig. 5 (North Atlantic). Stearn and Pietsch, 1995:134, fig. 84 (seven additional females, Greenland). Swinney, 1995a:52, 54 (additional specimen, off Madeira). Stewart and Pietsch, 1998:11, fig. 6 (two additional females, New Zealand). Jónsson and Pálsson, 1999:200, fig. 4 (Iceland). Pietsch, 1999:2032 (western central Pacific). Kukuev and Trunov, 2002:327 (North Atlantic). Pietsch, 2002b:1064 (western central Atlantic). Moore et al., 2003:215 (off New England). Love et al., 2005:60 (eastern North Pacific).

*Himantolophus groenlandicus*: Regan, 1926:40 (in part; misidentification; larval female referred to *Chaenophryne longiceps* group by Bertelsen, 1951).

*Chaenophryne longiceps* var. *quadrifilis* Parr, 1927:22, fig. 8A (original description, two specimens; lectotype YPM 2910 [not BOC 2007 as cited by Parr, 1927], designated by Parr, 1937).

*Chaenophryne crossotus* Beebe, 1932b:83, fig. 21 (original description, single specimen; holotype USNM 170942, originally NYZS 20809). Mead, 1958:132 (type transferred to USNM).

*Chaenophryne bicornis* Regan and Trewavas, 1932:84, 85, fig. 133 (original description, two specimens, lectotype ZMUC P92107, designated by Bertelsen, 1951). Bertelsen, 1951:111, 113, fig. 66C (description, comparison with all known material, lectotype designated, in key). Grey, 1956a:256 (synonymy, distribution). Pietsch, 1974a:33 (listed). Nielsen, 1974:88 (listed in type catalog).

*Chaenophryne crenata* Regan and Trewavas, 1932:84, 86, fig. 134 (original description, single specimen). Bertelsen, 1951:111, 114, fig. 66B (description, comparison with all known material, in key). Grey, 1956a:257 (synonymy, distribution). Pietsch, 1974a:33 (listed). Nielsen, 1974:89 (listed in type catalog).

*Chaenophryne crossata*: Regan and Trewavas, 1932:85, 86 (emendation of specific name; description after Beebe, 1932b; in key).

*Chaenophryne quadrifilis*: Regan and Trewavas, 1932:85, 87, fig. 136 (description based on *Chaenophryne longiceps* var. *quadrifilis* n. var. of Parr, 1927; four specimens). Parr, 1937:63 (listed, lectotype designated). Koefoed, 1944:8, pl. I, figs. 2, 3 (description of additional specimen; *Oneirodes* n. sp. no. 2 of Murray and Hjort, 1912). Bertelsen, 1951:111, 113 (description, comparison with all known material, in key). Grey, 1956a:257 (synonymy, distribution). Maul, 1973:671 (synonymy, eastern North Atlantic). Pietsch, 1974a:33 (listed).

*Chaenophryne haplactis* Regan and Trewavas, 1932:85, 87, fig. 137 (original description, single specimen). Nielsen, 1974:89 (listed in type catalog).

*Chaenophryne longiceps* Group Bertelsen, 1951:71, 110–114, 269, figs. 30, 66, 68–71 (osteological description of larval female, five nominal species grouped, common characters, all available material listed; description of larvae, males, metamorphosing females; in key). Grey, 1956a:256 (synonymy, distribution). Pietsch, 1974a:33 (after Bertelsen, 1951; species listed).

*Chaenophryne crossota*: Bertelsen, 1951:111, 114 (emendation of specific name, description, comparison with all known material, in key). Grey, 1956a:257 (synonymy, distribution). Pietsch, 1974a:33 (listed).

*Chaenophryne* sp.: Kukuev and Trunov, 2002:327, fig. 2b (additional specimen, North Atlantic).

### Males

*Rhynchoceratias leuchorhinus*: Regan, 1926:44 (in part; two males referred to *Chaenophryne longiceps* group by Bertelsen, 1951).

*Trematorhynchus leuchorhinus*: Regan and Trewavas, 1932:91 (in part; two males referred to *Chaenophryne longiceps* group by Bertelsen, 1951).

### Material

Eighty-four metamorphosed females (10 to 245 mm); two females in metamorphosis (12 to 12.5 mm); seven metamorphosing, juvenile, and adult males (11 to 19.5 mm); and 106 larvae (62 females, 42 males, and two undetermined, 3.2 to 19 mm TL):

Lectotype of *Chaenophryne longiceps*: ZMUC P92106, 20 mm, *Dana* station 1203(11), Gulf of Panama, 7°30′N, 79°19′W, 3000 m of wire, 1500 h, 11 January 1922.

Holotype of *Chaenophryne crossotus*: USNM 170942, originally NYZS 20809, 17 mm, Bermuda Oceanographic Expedition, net 1015, 8 mi SE of Nonsuch, 915 m, 15 June 1931.

Lectotype of *Chaenophryne bicornis*: ZMUC P92107, 14 mm, *Dana* station 4005(1), west of Cape Verde Islands, 13°31′N, 18°03′W, 4000 m of wire, 1145 h, 13 March 1930.

Paralectotype of *Chaenophryne bicornis*: BMNH 1932.5.3.24, 14 mm, *Dana* station 4005(2), west of Cape Verde Islands, 13°31′N, 18°03′W, 3500 m of wire, 1145 h, 13 March 1930.

Holotype of *Chaenophryne crenata*: ZMUC P92108, 18 mm, *Dana* station 3714(10), South China Sea, 15°22′N, 115°20′E, 2000 m of wire, 1030 h, 20 May 1929.

Lectotype of *Chaenophryne longiceps* var. *quadrifilis*: YPM 2910 (not BOC 2007), 20.5 mm, *Pawnee*, Third Oceanographic Expedition, station 58, 32°24′N, 64°29′W, 3050 m of wire, 20 April 1927.

Paralectotype of *Chaenophryne longiceps* var. *quadrifilis*: YPM 2911 (not BOC 2027), 20 mm, *Pawnee*, Third Oceanographic

Expedition, station 59, 32°19'N, 64°32'W, 2440 m of wire, 21 April 1927.

Holotype of *Chaenophryne haplactis*: ZMUC P92114, 11 mm, *Dana* station 1152(3), North Atlantic, 30°17'N, 20°44'W, 3000 m of wire, 0730 h, 22 October 1921.

Additional metamorphosed females: BMNH, two (14 to 22.5 mm); BSKU, three (152 to 190 mm); CAS, one (36 mm); CSIRO, three (128 to 148 mm); HUMZ, nine (107 to 245 mm); ISH, six (35 to 103 mm); LACM, two (18 to 40 mm); MCZ, three (17 to 39 mm); MNHN, four (17 to 200 mm); NMNZ, two (90 to 155 mm); NMSZ, one (28.5 mm); NSMT, three (21 to 151 mm); OS, three (25 to 140 mm); ROM, one (170 mm); SAMA, one (161 mm); SIO, six (10 to 122 mm); SIOM, one (23 mm); UF, two (19 to 50 mm); USNM, seven (13.5 to 29.5 mm); UW, five (35 to 90 mm); ZMB, one (25 mm); ZMUC, 10 (70 to 178 mm).

Metamorphosing females, ZMUC, two (12 to 12.5 mm).

Metamorphosing, juvenile, and adult males: BMNH, one (11 mm); LACM, one (12 mm); MCZ, two (14 to 17 mm); USNM, one (19.5 mm); ZMUC, two (15 to 16.5 mm, both with large testes; see Bertelsen, 1951:112).

Larvae: ZMUC, 106 (62 females, 42 males, and two undetermined, 3.2 to 19 mm TL). See Bertelsen (1951:269–270).

Distinguishing Characters

In addition to characters of the esca, which differentiate *Chaenophryne longiceps* from all other species of *Chaenophryne*, metamorphosed females of this species can be further distinguished from *C. draco* and *C. melanorhabdus* by having a smaller ratio between the number of upper- and lower-jaw teeth. The illicium of *C. longiceps* appears to be slightly shorter than that of *C. ramifera* and slightly longer than that of *C. melanorhabdus*. *Chaenophryne longiceps* has fewer dorsal- and anal-fin rays than *C. ramifera*, and a greater number of pectoral-fin rays than all other species of the genus.

Description

Metamorphosed females with esca bearing a pair of elongate internally pigmented (except for distal tip) and bilaterally placed anterior appendages (occasionally sharing a common base as in holotype of *Chaenophryne crenata*; Regan and Trewavas, 1932:86, fig. 134), less than one-tenth to greater than length of escal bulb; 1 to 3 transversely placed, medial escal appendages, bifurcated at midlength to highly filamentous from base, darkly pigmented in some large specimens (102 and 103 mm); medial escal appendages may arise at any point, from distal surface of escal bulb to anterior margin of swollen basal portion of posterior escal appendage; a posterior escal appendage consisting of a swollen basal portion and a membranous distal crest, darkly pigmented in largest known specimens (102 to 170 mm), and bearing 1 to several filaments anteriorly or distally in some specimens; anterolateral appendages and basal series of filaments absent; in most specimens a subcutaneous internally pigmented descending appendage.

Illicial length 22.4 to 40.3% SL; escal bulb width 5.3 to 11.4% SL; total number of teeth in upper jaw 28 to 40, in lower jaw 34 to 57; ratio of number of teeth in upper jaw to number in lower jaw 0.70 to 0.94; vomerine teeth 4 to 8; dorsal-fin rays 6 to 8; anal-fin rays 5 or 6; pectoral-fin rays 17 to 22. Additional description as given for the genus and family.

Distribution

*Chaenophryne longiceps* has a wide geographic distribution, occurring in all three major oceans of the world. It has been collected on both sides of the North Atlantic from the equator to the northern coast of Iceland and off West Greenland as far north as 69°N (Fig. 221). Three specimens have been taken in the Indian Ocean, one from the Arabian Sea (at about 7°N, 65°E), another from the Bay of Bengal (9°N, 90°E), and a third from off Point D'Entrecasteaux, western Australia, at about 36°S. It is also represented by numerous, widely scattered localities throughout the Pacific, extending from the South China Sea, north to Japan and south to New Zealand, with a single record in the Southern Ocean at about 55°S, 159°N. In the eastern Pacific it ranges from the Hawaiian Islands, to the Oregon coast in the north and off Chile (at about 23°S) in the south. The lectotype was collected from the Gulf of Panama.

Based on maximum depths reached by fishing gear, metamorphosed female specimens of *C. longiceps* are vertically distributed between approximately 500 m and an unknown lower limit. About 90% of the known material was captured by gear fished below 850 m.

Comments

Bertelsen (1951) erected the *Chaenophryne longiceps* group to include five nominal species described by Beebe (1932b) and Regan and Trewavas (1932) on the basis of one to four juvenile female specimens less than 25 mm, and on relatively small differences in the morphology of the esca. The greater amount of material available to Pietsch (1975b), providing some understanding of individual and ontogenetic variation, showed that these differences in escal morphology must be regarded as variation exhibited by widely distributed conspecific populations. In the absence of significant differences, these forms were synonymized with *C. longiceps*.

*Chaenophryne haplactis*, known only from the 11-mm holotype, was placed in the *C. draco* group by Bertelsen (1951) on the basis of absence of inner pigment on the caudal peduncle and a low pectoral-fin ray count (15 according to Regan and Trewavas, 1932). Reexamination of the holotype by Bertelsen (personal communication, 16 October 1974) showed the somewhat bleached remains of large stellate melanophores on the peduncle; Pietsch (1975b) counted 18 pectoral rays. Further, the esca of this specimen clearly has the paired anterior appendages characteristic of *C. longiceps* (Regan and Trewavas, 1932:89, fig. 137). For these reasons, *C. haplactis* was removed from the *C. draco* group by Pietsch (1975b) and placed within the synonymy of *C. longiceps*.

## *Chaenophryne draco* Group
Bertelsen, 1951
Females, Males, and Larvae

*Chaenophryne draco* Group Bertelsen, 1951:72, 110, 114–118, 270, figs. 30, 67, 72–75 (new species group). Pietsch, 1975b: 76, 87, table 1 (reallocation of nominal taxa).

Comments

The *Chaenophryne draco* group (Bertelsen, 1951) is retained to include larvae and males readily separated from those of *C. longiceps* by larval pigmentation and pectoral-fin ray counts, but not divisible into smaller taxonomic units. Within the *C. draco* group are also included four species based on metamorphosed females that together differ from *C. longiceps* in the basic construction of the esca and other features listed in the key to species. These three species are associated with the larvae and males of the *C. draco* group by retention in small juvenile specimens of the characteristic larval pigmentation, and by pectoral-fin ray counts (Bertelsen, 1951:110; Table 24).

Diversity

Four species and one nominal form of doubtful validity:

Key to Females of Species of the *Chaenophryne draco* Group

1A. Esca with 1 to 3 anterolateral appendages on each side; ratio of number of teeth in upper and lower jaws 0.76 to 1.30 . . . . . . . . . . . . . . . . . 3
1B. Esca without anterolateral appendages; ratio of number of teeth in upper and lower jaws 1.08 to 1.45 in

specimens 20 mm and larger . . . . . .
. . . . . . . . . . . . . *Chaenophryne draco*
Beebe, 1932b, p. 388

NOTE One hundred and nineteen known specimens, 11 to 123 mm; Atlantic, Pacific, and Indian oceans

2A. Esca with 1 to 12, long, slender, illicial filaments arising laterally from just below base of escal bulb; dorsal-fin rays 8, rarely 7; anal-fin rays 6, rarely 5; pectoral-fin rays 17 to 19, rarely 16 . . . . . . . . . . . . . . . . . . . 3

2B. Esca without illicial filaments arising from below base of escal bulb; dorsal-fin rays 6 or 7, rarely 8; anal-fin rays 5, rarely 6; pectoral-fin rays 16 or 17, rarely 18 . . . . . . . . . . . . . . . . . . .
. . . . . . . *Chaenophryne melanorhabdus*
Regan and Trewavas, 1932, p. 389

NOTE Forty-nine known specimens, 11 to 102 mm; eastern Pacific Ocean

3A. Esca with 1 or 2 long, slender, illicial filaments arising laterally on each side just below base of escal bulb; illicium short, 21.1 to 22.9% SL
. . . . . . . . *Chaenophryne quasiramifera*
Pietsch, 2007, p. 391

NOTE Two known specimens, 45 to 98 mm; Peru-Chile Trench and Nasca Ridge

3B. Esca with a series of 7 to 12, long, slender illicial filaments arising laterally on each side just below base of escal bulb; illicium long, (26.4) 32.8 to 47.4% SL . . . . . . . . . . . . . . . .
. . . . . . . . . . . *Chaenophryne ramifera*
Regan and Trewavas, 1932, p. 390

NOTE Twenty-three known specimens, 13.5 to 87 mm; Atlantic, Pacific, and Indian oceans

## CHAENOPHRYNE DRACO BEEBE, 1932B

Figures 25, 33, 106, 108, 221, 248; Tables 5, 8, 9

### Females

*Chaenophryne longiceps* Regan, 1925c: 564 (in part; original description, 14 specimens; paralectotypes of *C. longiceps* subsequently made types of *C. parviconus*, *C. columnifera*, and *C. melanodactylus* by Regan and Trewavas, 1932). Regan, 1926:31 (in part; after Regan, 1925c).

*Chaenophryne draco* Beebe, 1932b:84, fig. 22 (original description, single specimen, holotype USNM 170943, originally NYZS 22396). Regan and Trewavas, 1932: 85, 89 (description after Beebe, 1932b; in key). Bertelsen, 1951:115, 116 (description, comparison with all known material, in key). Grey, 1956a:258 (synonymy, distribution). Mead, 1958:132 (type transferred to USNM). Pietsch, 1974a:33 (listed). Pietsch, 1975b:81, 87, figs. 1B, 2A, 3–5, 8, 11, 12 (additional material, osteology, description, comparison with all known material). Hubbs et al., 1979:13 (California). Pietsch and Seigel, 1980:395 (two additional females, Banda Sea). Bertelsen and Pietsch, 1983:88, fig. 9 (additional material, Australia). Amaoka, 1984:106, pl. 92C (Japan). Bertelsen, 1986:1385, figs. (eastern North Atlantic). Bertelsen, 1990:498 (synonymy, eastern tropical Atlantic). Swinney, 1995a:52, 54 (seven additional specimens, off Madeira). Stewart and Pietsch, 1998:11, fig. 7 (additional female, New Zealand). Jónsson and Pálsson, 1999:200, figs. 4, 10 (Iceland). Munk, 1999:267, 268, 274, 275, 277–280, fig. 11 (bioluminescence). Pietsch, 1999: 2032 (western central Pacific). Herring, 2000:1275, fig. 1a (misidentification, photograph depicts *C. ramifera*). Anderson and Leslie, 2001:12, fig. 10 (additional female, southern Africa). Herring, 2002:20, fig. 9.6 (misidentification, photograph depicts *C. ramifera*). Kukuev and Trunov, 2002:327 (North Atlantic). Nakabo, 2002: 472, figs. (Japan, in key). Pietsch, 2002b: 1064 (western central Atlantic). Love et al., 2005:60 (eastern North Pacific). Senou et al., 2006:424 (Sagami Sea). Pietsch and Orr, 2007:12, fig. 7B (phylogenetic relationships).

*Chaenophryne parviconus* Regan and Trewavas, 1932:35, 85, 87, figs. 39, 41, 138 (original description, nine specimens, lectotype ZMUC P92110, designated by Bertelsen, 1951; cranial osteology described, figured). Beebe and Crane, 1947:158 (six additional specimens, description; *C. columnifera* and *C. melanorhabdus* synonyms of *C. parviconus*). Bertelsen, 1951:115, 117, fig. 74B, C (in part; description, comparison with all known material, in key; *C. haplactis*, *C. atriconus*, *C. columnifera*, *C. melanodactylus*, *C. macractis*, *C. melanorhabdus*, *C. pterolophus*, and *C. pacis* synonyms of *C. parviconus*). Matsubara, 1955:1360, 1361, fig. 536 (Japan; figure after Imai, 1942). Grey, 1956a:258 (synonymy, distribution). Briggs, 1960:179 (worldwide distribution). Berry and Perkins, 1966:677, fig. 30 (additional specimen). Maul, 1973:670 (synonymy, eastern North Atlantic). Pietsch, 1974a:33, 109 (listed). Nielsen, 1974:90 (listed in type catalog).

*Chaenophryne atriconus* Regan and Trewavas, 1932:85, 87, fig. 139 (original description, single specimen). Nielsen, 1974:88 (listed in type catalog).

*Chaenophryne columnifera* Regan and Trewavas, 1932:85, 88, fig. 140 (original description, three specimens, lectotype ZMUC P92112, designated by Pietsch, 1975b). Nielsen, 1974:89 (listed in type catalog).

*Chaenophryne melanodactylus* Regan and Trewavas, 1932:85, 88, fig. 141 (original description, single specimen). Nielsen, 1974:90 (listed in type catalog).

*Chaenophryne macractis* Regan and Trewavas, 1932:85, 88, fig. 142 (original description, single specimen). Matsubara, 1955:1360, 1361, fig. 536 (in synonymy of *C. parviconus*; figure after Imai, 1942). Nielsen, 1974:90 (listed in type catalog).

### Males and Larvae

Unknown.

### Material

One hundred and nineteen metamorphosed females (11 to 123 mm):

Holotype of *Chaenophryne draco*: USNM 170943 (originally NYZS 22396), 16.5 mm, Bermuda Oceanographic Expedition net 1181, 10 mi SE of Nonsuch, 32°12′N, 64°36′W, 1100 m, 15 August 1931.

Lectotype of *Chaenophryne parviconus*: ZMUC P92110, 12.5 mm, *Dana* station 1209(2), Gulf of Panama, 7°15′N, 78°54′W, 3000 m wire, 1845 h, 17 January 1922.

Paralectotypes of *Chaenophryne parviconus*: BMNH 1925.8.11.19, 15 mm, *Dana* station 1209(3), Gulf of Panama, 7°15′N, 78°54′W, 2500 m wire, 1845 h, 17 January 1922; BMNH 1932.5.3.25, 14.5 mm, *Dana* station 1209(3), Gulf of Panama, 7°15′N, 78°54′W, 2500 m wire, 1845 h, 17 January 1922; BMNH 1932.5.3.26, 13.5 mm, *Dana* station 1208(7), Gulf of Panama, 6°48′N, 80°33′W, 2000 m wire, 0810 h, 16 January 1922.

Holotype of *Chaenophryne atriconus*: ZMUC P92111, 15 mm, *Dana* station 3847(5), off Cocos-Keeling Islands, 12°02′S, 96°43′E, 1500 m wire, 2100 h, 11 October 1929.

Lectotype of *Chaenophryne columnifera*: ZMUC P92112, 14.5 mm, *Dana* station 1208(15), Gulf of Panama, 6°48′N, 80°33′W, 2600 m wire, 1715 h, 16 January 1922.

Paralectotypes of *Chaenophryne columnifera*: BMNH 1925.8.11.17-18, 2 (12 to 13 mm), *Dana* station 1208(15), Gulf of Panama, 6°48′N, 80°33′W, 2600 m wire, 1715 h, 16 January 1922; ZMUC P92113, 13.5 mm, *Dana* station 1209(1), Gulf of Panama, 7°15′N, 78°54′W, 3500 m wire, 1845 h, 17 January 1922.

Holotype of *Chaenophryne melanodactylus*: ZMUC P92116, 15 mm, *Dana* station 1370(13), North Atlantic, 36°36′N, 26°14′W, 3000 m wire, 1150 h, 13 June 1922.

Holotype of *Chaenophryne macractis*: ZMUC P92115, 14 mm, *Dana* station 3561(4), eastern tropical Pacific, 4°20′S, 116°46′W, 2000 m wire, 0900 h, 24 September 1928.

Additional metamorphosed females: AIM, one (97 mm); AMNH, three (25 to 35 mm); AMS three (14 to 63 mm); BMNH, six (12 to 30 mm); CSIRO, five (39 to 121 mm); HUMZ, four (38 to 97 mm); ISH, 16 (18 to 76 mm); LACM, 17 (13 to 42 mm); MCZ, four (12.5 to 18 mm); MNHN, three (16 to 42 mm); MNRJ, one (72 mm); NMNZ, one (123 mm); NMSZ, seven (15 to 21 mm); NSMT, two (52 to 53 mm); NYZS, six (13.5 to 19.5 mm); SAM, one (93 mm); SIO, eight (11 to 51.5 mm); USNM, one (16.5 mm); UW, one (27.5 mm);

ZMUB, one (60 mm); ZMUC, 16 (11 to 117 mm).

### Distinguishing Characters

In addition to the characters of the esca, which separate metamorphosed females of *Chaenophryne draco* from those of all other species of *Chaenophryne*, this species is distinguished by having the highest ratio between the number of upper- and lower-jaw teeth of any species. The illicium of *C. draco* appears to be slightly shorter than that of *C. ramifera* and slightly longer than that of *C. melanorhabdus*. Fewer pectoral-fin rays, and fewer dorsal- and anal-fin rays, help to distinguish *C. draco* from *C. longiceps* and *C. ramifera*, respectively.

### Description

Esca with a single conical to elongate, internally pigmented (except for distal tip) anterior appendage, less than one-seventh to nearly one-third length of escal bulb; medial escal appendages absent; a posterior escal appendage consisting of a swollen basal portion and a somewhat compressed distal crest, with a posterior filament or filaments and a pair of anterior lobes each bearing none to numerous filaments; anterolateral escal appendages and basal series of filaments absent; a subcutaneous, internally pigmented descending appendage present in larger specimens.

Illicium length 24 to 36.4% 8L; escal bulb width 4.3 to 6.7% SL; total number of teeth in upper jaw 35 to 47, in lower jaw 31 to 38; ratio of number of teeth in upper jaw to number in lower jaw 1.08 to 1.45; vomerine teeth 4 to 8; dorsal-fin rays 6 to 8; anal-fin rays 5 or 6 (of 21 specimens counted only one had anal-fin rays 6); pectoral-fin rays 16 to 18.

Additional description as given for the genus and family.

### Distribution

*Chaenophryne draco* is represented by numerous, widely scattered localities in all three major oceans of the world (Fig. 221). It has been collected throughout the North Atlantic, extending from Greenland and Iceland, as far north as about 66°N, to Bermuda (the type locality) and the Cape Verde Islands, with single records from off Rio de Janeiro and the Cape Town, South Africa. It has also been collected on both sides of the Indian Ocean, extending from the equator at approximately 65°E, to the Cocos Islands (12°S, 96°E), and off the southwest corner of western Australia. In the western Pacific it ranges from the South China Sea to Japan in the north and off New South Wales, Tasmania, and New Zealand in the south. In the eastern Pacific it is well represented in the Hawaiian Islands, extending east along the equator to the Gulf of Panama and northern Peru at about 4°S.

Based on maximum depths reached by fishing gear, metamorphosed female specimens of *C. draco* are vertically distributed between approximately 350 m and an unknown lower limit. More than 90% of the known material was captured by gear fished below 700 m. Sufficient material is known from the Gulf of Panama for analysis of vertical data by the procedure outlined by Gibbs (1969). Results indicate a concentration between 700 and 1500 m (see Pietsch, 1975b).

### Comments

Prior to the publication of his monograph on the Ceratioidei, Bertelsen (1951) had not seen the holotype of *Chaenophryne draco* Beebe, 1932b. Beebe's (1932b:85, fig. 22) colorful description and somewhat stylized figure of the esca did not fully agree with descriptions of other species introduced six months later by Regan and Trewavas (1932). These forms were maintained as separate species pending examination of Beebe's type. Later examination by Pietsch (1973) of the holotype of *C. draco* (16.5 mm) confirmed Bertelsen's (1951) prediction that *C. parviconus* (and most of its included synonyms; see Table 5) is a junior synonym of *C. draco*.

## CHAENOPHRYNE MELANORHABDUS REGAN AND TREWAVAS, 1932

Figures 106, 221; Tables 5, 8

### Females

*Chaenophryne melanorhabdus* Regan and Trewavas, 1932:85, 89, fig. 143 (original description, single specimen). Bertelsen, 1951:117, fig. 74E (in synonymy of *Chaenophryne parviconus*). Nielsen, 1974:90 (listed in type catalog). Pietsch, 1975b:81, 90, figs. 1C, 2B, 3–5, 9, 11, 12 (resurrection from synonymy of *C. parviconus*, additional material, osteology, description, comparison with all known material). Peden et al., 1985:9, fig. 37 (eastern North Pacific). Mecklenburg et al., 2002:305, 307 (Alaska, in key). Love et al., 2005:60 (eastern North Pacific).

*Chaenophryne pterolophus* Regan and Trewavas, 1932:85, 89, fig. 144 (original description, single specimen). Nielsen, 1974:90 (listed in type catalog).

*Chaenophryne parviconus*: Grinols, 1966:161, fig. 1 (misidentification, two specimens; description, escae figured, distribution). Pietsch, 1972a:35, 36, 42, 45, fig. 24(6) (misidentification; osteological comments; otolith described, figured). Pietsch, 1974a:33, 109, figs. 26, 37, 39D, 46, 49, 51F, 52B, 56D (misidentification; osteological description, comparison with other oneirodid genera, phylogenetic relationships).

### Males and Larvae

Unknown.

### Material

Forty-nine metamorphosed females (11 to 102 mm):

Holotype of *Chaenophryne melanorhabdus*: ZMUC P92117, 40 mm, *Dana* station 1203(14), Gulf of Panama, 7°30'N, 79°19'W, 2500 m wire, 2030 h, 11 January 1922.

Holotype of *Chaenophryne pterolophus*: ZMUC P92118, 20.5 mm; *Dana* station 1208(16), Gulf of Panama, 6°48'N, 80°33'W, 2100 m wire, 1715 h, 16 January 1922.

Additional material: CAS, one (20 mm); HUMZ, three (56 to 74 mm); LACM, 18 (11 to 97 mm); OS, five (51 to 93 mm); SIO, 12 (13 to 102 mm); USNM, one (63 mm); UW, seven (35 to 77 mm).

### Distinguishing Characters

Metamorphosed females of *Chaenophryne melanorhabdus* are easily separate from those of all other species of *Chaenophryne* by unique features of the esca. In addition, the ratio between the number of upper- and lower-jaw teeth is less than that of *C. draco*, but significantly greater than that of *C. longiceps* and *C. ramifera*. The illicium of *C. melanorhabdus* is shorter than that of its congeners. Finally, fewer pectoral-fin rays, and fewer dorsal-fin and anal-fin rays help to distinguish *C. melanorhabdus* from *C. longiceps* and *C. ramifera*, respectively.

### Description

Esca with a single elongate internally pigmented (except for distal tip) anterior appendage (distal tip bifurcated in some specimens; see Comments below), less than one-fourth to nearly one-third length of escal bulb; medial escal appendages absent; a posterior escal appendage consisting of a swollen basal portion and a somewhat compressed distal crest, with a posterior filament or filaments and a pair of anterior lobes each bearing numerous filaments; a filamentous anterolateral escal appendage on each side; basal series of filaments absent; subcutaneous, internally pigmented, descending escal appendage well developed, often extending full length of illicium.

Illicium length 20.1 to 41.3% SL; escal bulb width 2.1 to 6.3% SL; total number of teeth in upper jaw 21 to 45, in lower jaw 26 to 42; ratio between number of teeth in upper jaw to number of teeth in lower jaw 0.78 to 1.30; vomerine teeth 4 to 7; dorsal-fin rays 6 to 8 (of 19 specimens counted only one had dorsal-fin rays 8); anal-fin rays 5 or 6; pectoral-fin rays 16 to 18.

Additional description as given for the genus and family.

## Distribution

*Chaenophryne melanorhabdus* appears to be restricted to the western continental slope of North and Central America, ranging from approximately 46°N, 125°W in Pacific Subarctic Water, through the mixed transition zone of the California Current, into the eastern Pacific equatorial waters of the Gulf of Panama (type locality), and beyond to the Peru-Chile Trench as far south as about 12°S (Fig. 221).

Based on maximum depths reached by fishing gear, metamorphosed female specimens of *C. melanorhabdus* are vertically distributed between approximately 200 m and an unknown lower limit. About 85% of the known material, including the largest known females was collected by gear fished below 450 m. Sufficient material is known from the transition zone of the California Current for analysis of vertical data by a procedure outlined by Gibbs (1969). Results indicate a concentration between 300 and 1000 m (Pietsch, 1975b).

## Comments

Beebe and Crane (1947:158) synonymized *Chaenophryne melanorhabdus* and *C. columnifera* with *C. parviconus*, considering the material to represent stages of development. Bertelsen (1951:114, 117) agreed with this action and the reasons for it, adding a number of additional forms to the synonymy of *C. parviconus*, including *C. pterolophus*. Pietsch (1975b) reported on eastern Pacific material of *Chaenophryne*, including a complete growth series (12 to 97 mm) that have a pair of filamentous anterolateral escal appendages like those present in *C. melanorhabdus* and *C. pterolophus*. In all other ways this material compared very well with the type material. Consequently, Pietsch (1975b) removed *C. melanorhabdus*, with *C. pterolophus* as a junior synonym, from the synonymy of *C. parviconus* giving it specific status.

Within the known material of *C. melanorhabdus* there are three females (SIO 65-621, 15 mm; SIO 78-37, 39.5 mm; and USNM 213733, 63 mm) from the eastern tropical Pacific that differ from all other specimens in having the distal tip of the anterior escal appendage bifurcated. These specimens most likely represent a new species, but description is deferred pending the discovery of additional material.

## CHAENOPHRYNE RAMIFERA REGAN AND TREWAVAS, 1932

Figures 106, 221; Tables 5, 8

### Females

*Oneirodes* n. sp. no. 3: Murray and Hjort, 1912:609, 614, fig. 468 (misidentification; specimen made holotype of *Chaenophryne pacis* by Koefoed, 1944).

*Chaenophryne ramifera* Regan and Trewavas, 1932:85, 90, fig. 146 (original description, single specimen). Belloc, 1938:305, fig. 29 (after Regan and Trewavas, 1932; comparison with *Chaenophryne intermedia* sp. nov.). Bertelsen, 1951:115, figs. 67D, 75 (description, comparison with all known material, in key; *C. fimbriata* and *C. intermedia* synonyms of *C. ramifera*). Pietsch, 1974a:33 (listed). Nielsen, 1974:90 (listed in type catalog). Parin et al., 1974:123 (western South Atlantic). Pietsch, 1975b:81, 92, figs. 1D, 2C, 3–5, 9, 10–12 (additional material, osteology, description, comparison with all known material). Bertelsen and Pietsch, 1977:186 (additional material, eastern Atlantic). Bertelsen, 1990:499 (synonymy, eastern tropical Atlantic).

*Chaenophryne fimbriata* Regan and Trewavas, 1932:85, 90, fig. 145 (original description, single specimen). Belloc, 1938:305, fig. 27 (after Regan and Trewavas, 1932; comparison with *Chaenophryne intermedia* sp. nov.). Nielsen, 1974:89 (listed in type catalog).

*Chaenophryne intermedia* Belloc, 1938:305, figs. 24, 28 (original description, single specimen).

*Chaenophryne pacis* Koefoed, 1944:9, pl. 2, figs. 6, 7 (original description, single specimen; *Oneirodes* n. sp. no. 3 of Murray and Hjort, 1912). Frøiland, 1979:151 (listed in type catalog).

*Chaenophryne draco*: Herring, 2000:1275, fig. 1a (misidentification). Herring, 2002:200, fig. 9.6 (misidentification after Herring, 2000).

### Males and Larvae

Unknown.

### Material

Twenty-three metamorphosed females (13.5 to 87 mm):

Holotype of *Chaenophryne ramifera*: ZMUC P92119, 17 mm, *Dana* station 3550(6), Gulf of Panama, 7°10′N, 78°15′W, 3000 m wire, 0145 h, 5 September 1928.

Holotype of *Chaenophryne fimbriata*: ZMUC P92120, 16.5 mm, *Dana* station 3917(3), west of Maldives, Indian Ocean, 1°45′N, 71°05′E, 3200 m wire, 1800 h, 5 December 1929.

Holotype of *Chaenophryne intermedia*: MHNLR P316-449, 14 mm, *President Theodore Tissier* station 708, eastern North Atlantic, 14°54′N, 23°15′W, 1000 m wire, 15 May 1936.

Holotype of *Chaenophryne pacis*: ZMUB 4301, 15 mm, *Michael Sars*, North Atlantic Deep-Sea Expedition 1910, station 53, North Atlantic, 34°59′N, 33°01′W, 2600 m wire, 8 to 9 June 1910.

Additional material: ARC, one (32 mm); CAS, one (30 mm); HUMZ, one (87 mm); ISH, five (26 to 55.5 mm); MCZ, five (13.5 to 39 mm); MNHN, two (17 to 28 mm); SIOM, one (16 mm); UF, two (30 to 36.5 mm); ZMUC, one (33.5 mm).

### Distinguishing Characters

Metamorphosed females of *Chaenophryne ramifera* are easily distinguished from those of all other species of *Chaenophryne* by unique features of the esca. The low ratio between the number of upper- and lower-jaw teeth further separates this species from *C. draco* and *C. melanorhabdus*. The illicium of *C. ramifera* is longer than that of its congeners. *Chaenophryne ramifera* has fewer pectoral-fin rays than *C. longiceps*, and a greater number of dorsal-fin and anal-fin rays than all other species.

### Description

Esca with a single elongate internally pigmented anterior appendage, approximately one-fourth to nearly one-third length of escal bulb, trilobed distally with 3 round translucent windows; anterior escal appendage flanked laterally by an elongate unpigmented swelling; medial escal appendages absent; a posterior escal appendage consisting of a swollen basal portion and a somewhat compressed distal crest, with a terminal, anteriorly directed crescent-shaped filament and a pair of anterior lobes each bearing none to several filaments; 2 or 3 filamentous anterolateral escal appendages on each side; a basal series of 7 to 12 filaments on each side; subcutaneous, internally pigmented descending escal appendage extremely well developed in some specimens (see Comments, below), emerging as a free anteriorly and dorsally directed tentacle.

Illicium length 32.8 to 47.4% SL; escal bulb width 4.5 to 6.5% SL; total number of teeth in upper jaw 25 to 51, in lower jaw 33 to 53; ratio between number of teeth in upper jaw to number of teeth in lower jaw 0.76 to 0.98; vomerine teeth 4 to 8; dorsal-fin rays 7 or 8; anal-fin rays 5 or 6; pectoral-fin rays 16 to 19 (of 15 specimens counted only one had pectoral-fin rays 16 and one had pectoral-fin rays 19).

Additional description as given for the genus and family.

### Distribution

*Chaenophryne ramifera* has a wide geographic distribution occurring in all three major oceans of the world (Fig. 221). It extends across the Atlantic from off northern Florida to approximately 35°N, 33°W (the holotype of *C. pacis*), and south to 8°S, with single records from the Gulf of Guinea and off Angola at about 12°S. In the Indo-Pacific, it ranges across the Indian Ocean between 9°N and 2°S, with a single record in the far south at 25°S, 60°E, and continuing eastward along the equator to the Gulf of Panama (type locality) and Peru-Chile Trench at about 10°S.

Based on maximum depth reached by fishing gear, metamorphosed female specimens of *C. ramifera* are distributed vertically between approximately 200 m and an unknown lower limit. Large specimens may be captured at relatively shallow depths: a 35-mm specimen was collected by gear fished above 200 m, a 36.5-mm specimen by gear fished above 550 m. About 85% of the known material was captured by gear fished below 550 m, about 55% by gear fished below 1000 m. Material is not sufficient from any one geographic area for a more analytical treatment of distributional data.

Comments

I have not seen the holotype of *Chaenophryne pacis*, but from Koefoed's (1944:9, figs. 6, 7) description and figures, there can be little doubt that it is a synonym of *C. ramifera*. The anterolateral escal appendages and basal series of filaments have apparently been lost; but the "trilobed stigma" unique to *C. ramifera*, with its 3 translucent windows at the tip of the unpaired anterior escal appendage, is well developed. Thus, *Chaenophryne pacis* is retained within the synonymy of *C. ramifera* following the conclusions of Pietsch (1975b:94).

## CHAENOPHRYNE QUASIRAMIFERA PIETSCH, 2007

Figures 102, 105, 221; Tables 5, 8

### Females

*Chaenophryne* n. sp.: Pietsch and Orr, 2007:3, fig. 2J (phylogenetic relationships).

*Chaenophryne quasiramifera* Pietsch, 2007:164, figs. 1, 2 (original description, two specimens).

### Males and Larvae

Unknown.

### Material

Two metamorphosed females (45 to 98 mm):

Holotype of *Chaenophryne quasiramifera*: SIO 72-180, 98 mm, *Thomas Washington*, cruise Southtow IV/MV 72-II, station MV-72-II-23, Peru-Chile Trench, 20°19.2' S, 71°14.9' W, 3-m Isaacs-Kidd midwater trawl, 0 to 900 m, bottom depth 5856 m, 0023 to 0815 h, 3 May 1972.

Paratype of *Chaenophryne quasiramifera*: SIO 72-186, 45 mm, *Thomas Washington*, cruise Southtow IV/MV-72-II, Station MV-72-II-29, over Nasca Ridge, 16°08.8' S, 75°42.3' W, 3-m Isaacs-Kidd midwater trawl, 0 to 1130 m, bottom depth 5307 m, 2210 to 0655 h, 7 to 8 May 1972.

### Distinguishing Characters

Metamorphosed females of *Chaenophryne quasiramifera* are easily distinguished from those of all other species of the genus in having 1 or 2 elongate illicial filaments arising laterally just below base of escal bulb (comparable only to *C. ramifera*, which has a series of 7 to 12 elongate illicial filaments on each side; it differs further in having the following combination of character states: descending escal appendage extending along anterior margin of illicium, a small proximal portion emerging at articulation of illicium and illicial pterygiophores as a free, anterodorsally directed tentacle (comparable only to some specimens of *C. ramifera*); illicium short, 21.1 to 22.9% SL (comparable only to *C. melanorhabdus*; and a greater number of dorsal- and anal-fin ray counts (D. 8, A. 6; comparable only to *C. ramifera*).

### Description

Esca with a single, elongate, internally pigmented anterior appendage, trilobed distally with 3 tiny, round translucent windows; anterior escal appendage flanked laterally on each side by an elongate unpigmented swelling; medial escal appendages absent; a posterior escal appendage consisting of a swollen basal portion and a somewhat compressed distal crest; 2 or 3 filamentous anterolateral escal appendages on each side (holotype with 3 on each side, mostdistal two sharing a common base, third more proximal in position; paratype with 3 on left, 2 on right, arising from common base); 1 or 2 elongate lateral illicial filaments on each side (holotype with 2 on left, 1 on right; paratype with 2 on each side), each somewhat compressed proximally, gradually tapering and becoming cylindrical distally, proximal portions lightly pigmented; subcutaneous, internally pigmented, descending escal appendage extending proximally along anterior margin of illicium, a small portion (length 5.3 to 2.0% SL) emerging as a free, anterodorsally directed tentacle.

Head length 53.3 to 50.0% SL; least width between frontals 15.1 to 13.3% SL; premaxilla length 42.2 to 8.3% SL; lower-jaw length 58.9 to 58.7; illicium length 21.1 to 22.9% SL; escal-bulb width 5.1 to 3.9% SL; number of teeth in upper jaw 37 to 45, in lower jaw 35 to 36; ratio between number of teeth in upper jaw to number of teeth in lower jaw 1.06 to 1.25; vomerine teeth 5 or 6; dorsal-fin rays 8; anal-fin rays 6; pectoral-fin rays 17 or 18.

Additional description as for the genus and family (see Pietsch, 1975b; Biodiversity, Chapter Three).

### Distribution

*Chaenophryne quasiramifera* is known from only two specimens, both collected in the eastern tropical Pacific Ocean, the holotype from the Peru-Chile Trench, at approximately 20°S, 71°W, in gear fished open between the surface and 900 m, over a bottom depth of 5856 m; the paratype from the Nasca Ridge, 16°S, 75°W, in gear fished open between the surface and 1130 m, over a bottom depth of 5307 m (Fig. 221).

### Comments

*Chaenophryne quasiramifera* can only be confused with *C. ramifera*. The two species share a similar escal morphology, but the relatively simple lateral escal appendages of *C. quasiramifera*, consisting of only 1 or 2 elongate slender filaments on each side, clearly distinguishes this species from *C. ramifera*, which is characterized by having a much more highly complex esca, with 7 to 12 lateral filaments. The type material of *C. quasiramifera* differs further from nearly all known specimens of *C. ramifera* in having a considerably shorter illicium (21.1 to 22.9% SL versus 32.8 to 47.4% SL). The only exception is an 87-mm female of *C. ramifera* (HUMZ 174711), having the typical complex array of lateral escal filaments (10 on each side, some bifurcated near the base), but a surprisingly short illicium (26.4% SL; see Pietsch, 2007, fig. 4). Like the type material of *C. quasiramifera*, this outlier among the known material of *C. ramifera* was also collected from off Peru, but this seems to be no more than coincidental. Three of the five recognized species of *Chaenophryne* have extremely broad geographic ranges, known from multiple localities between approximately 65°N and 25°S in the Atlantic, Pacific, and Indian oceans. All five species are known from the eastern tropical Pacific, including the holotype of *C. ramifera* collected from the Gulf of Panama (see Pietsch, 1975b:94, fig. 11).

*Chaenophryne* is unique among ceratioids in having an internally pigmented, subcutaneous appendage emerging from the base of the esca and descending along the anterior margin of the illicium (first described by Regan and Trewavas, 1932:84, figs. 135, 138–146). In some specimens of *C. ramifera* and both known specimens of *C. quasiramifera*, the descending escal appendage is extremely well developed, extending proximally the full length of the illicium and emerging at the articulation of the illicium and illicial pterygiophore as a free, anterodorsally directed tentacle (Figs. 106D, 248). Although best developed in one of the largest known females of *C. ramifera* (36.5 mm, UF 229702; Fig. 106D), the development of the descending appendage does not appear to be correlated with ontogeny. Of the 23 known specimens of this species, only seven (ranging from 13.5 to 36.5 mm) have an esca with the descending appendage emerging proximally as a free tentacle. In all other known individuals the descending appendage is quite short, only 16.3% of the length of the illicium in a 55.5-mm specimen (ISH 925/1968). Thus, the small, exposed anterodorsally directed tentacle of the two known specimens of *C. quasiramifera*, compared to the enormous development of this structure in the 36.5-mm specimen of *C. ramifera*, most likely bears no taxonomic significance.

### CHAENOPHRYNE GALEATUS KOEFOED, 1944, NOMEN NUDUM

#### Females

*Chaenophryne galeatus* Koefoed, 1944:8 *(nomen nudum)*. Pietsch, 1975b:94 *(nomen nudum)*.

#### Comments

This name was used by Koefoed in a manuscript dated 1918 (not seen by me), and later mentioned in published form (Koefoed, 1944:8) without application to a description or type.

### Genus *Spiniphryne* Bertelsen, 1951 (Spiny Dreamers)

Figures 48, 109–111, 202, 203, 222; Tables 1, 8

#### Females

*Spiniphryne* Bertelsen, 1951:122, fig. 81 (type species *Dolopichthys gladisfenae* Beebe, 1932b, by subsequent designation of Palmer and White, 1953).

*Bertelsenna* Whitley, 1954:30 (unacceptable replacement name for *Spiniphryne* Bertelsen, 1951, therefore taking the same type species *Dolopichthys gladisfenae* Beebe, 1932b).

#### Males and Larvae

Unknown.

#### Distinguishing Characters

Metamorphosed females of *Spiniphryne* are distinguished from those of all other described genera of the family in having the entire body and fins covered with numerous close-set dermal spinules (tiny but obvious without microscopic aid). They are further unique in having the following combination of character states: Vomerine teeth are present. The dorsal margin of the frontals is convex. Sphenotic spines are well developed. The lower jaw bears a well-developed symphysial spine. The hyomandibula has a double head. The quadrate and articular spines are rudimentary. The posterior margin of the opercle is deeply notched. The subopercle is long and narrow, its dorsal end tapering to a point (the posterior margin without indentation), its ventral end nearly circular (without an anterior spine or projection, but irregularly notched unlike all other ceratioids). The second and third pharyngobranchials are well developed. The caudal-fin rays are without internal pigmentation. The illicium is considerably longer than the length of the esca bulb. The pterygiophore of the illicium emerges on the snout from between the frontal bones, its anterior end exposed, its posterior end concealed beneath the skin. The first ray of the dorsal fin is well developed. There are 6 or 7 (of 27 specimens counted, eight had 7) dorsal-fin rays and 5, rarely 4 or 6 (of 27 specimens counted, only one had 4 and one had 6), anal-fin rays (Table 1). The pectoral-fin lobe is short and broad, shorter than the longest rays of the pectoral fin. There are 15 or 16, rarely 17 (of 27 specimens counted, only two had 17), pectoral-fin rays. Darkly pigmented skin of the caudal peduncle extends well past the base of the caudal fin (for internal osteological characters, see Bertelsen and Pietsch, 1975).

#### Description

The body of metamorphosed females is elongate and slender, not globular (Figs. 109, 110). The mouth is moderate in size, the cleft not extending past the eye. The illicium is relatively short, its length 7.8 to 13.8% SL, becoming relatively shorter with increasing standard length. The esca bears 2 distal filaments or bulbous appendages, the filaments simple, the appendages usually covered with small digitiform papillae and clusters of tiny filaments around the base. An anteroposteriorly compressed posterior escal appendage is present, short and simple or large, broad, and divided along its distal margin into 3 or more lobes or several to many slender filaments. A single tiny posterolateral escal filament is present or absent. Filamentous lateral escal appendages are present or absent, often detectable only after considerable manipulation under high-power microscopy.

The teeth are slender and recurved, large and small intermixed in both jaws. The anterior-most two or three teeth of each premaxilla are immobile (evident only in well-preserved specimens), but the remaining teeth of the jaws and vomer are depressible. Teeth in the upper jaw are usually slightly fewer than those in the lower jaw. There are 18 to 47 teeth in the upper jaw, 21 to 52 in the lower jaw (see Pietsch and Baldwin, 2006:404, fig. 5), and 2 to 8 on the vomer. The following measurements are expressed in percent SL: head length 25.0 to 31.8; head width 14.3 to 19.6; head depth 26.6 to 32.7; length of premaxilla 17.6 to 21.9; length of lower jaw 22.6 to 28.6.

The color in life is black, but in preservation red brown to black over the entire external surface of the head, body, and fins, except for the distal surface of the escal bulb. In fresh unfixed specimens examined aboard ship (E. Bertelsen, personal observation, aboard RV *Walther Herwig*, 9 and 16 April 1971; ISH 2131/71, ISH 2734/71), the tips of the distal escal appendages were dark red and the posterior appendage was bright red orange; no part of the esca was silvery (Bertelsen and Pietsch, 1975:9). After 12 hours in formalin the color of the posterior escal appendage remained unchanged, while the distal escal appendages had become brown, probably indicating that the coloration was due to blood.

The subdermal pigmentation of a 12.8-mm adolescent specimen (BMNH 2004.8.24.7) is well preserved (Bertelsen and Pietsch, 1975:10, fig. 6): the anterior part of the body musculature bears relatively large, well-separated, branching melanophores more densely grouped along the back, and extending posteriorly slightly beyond the base of the dorsal fin; the fin bases and caudal peduncle are completely unpigmented; the peritoneum and gill covers are covered with a rather uniform pigmentation, without distinct groupings (pigmentation nearly identical in a 10.8-mm adolescent specimen, BMNH 2004.8.24.6, here referred to *Spiniphryne* sp.).

#### Diversity

Two species differentiated as follows:

Key to Females of the Genus *Spiniphryne*

1A. Esca with a pair of short, slender distal filaments; posterior escal appendage small, simple, without filaments along distal margin; three pairs of long, slender lateral escal filaments on each side; dentary teeth 51 to 52 ..... *Spiniphryne duhameli* Pietsch and Baldwin, 2006, p. 393

   NOTE  Two known specimens 25.5 to 117 mm; eastern Pacific Ocean

1B. Esca with a pair of bulbous distal appendages, each covered with tiny papillae and/or short filaments; posterior escal appendage large, usually extending beyond distal margin of escal bulb, and bearing a series of lobes or filaments along distal margin; no more than 3 slender lateral escal filaments on each side; dentary teeth 21 to 45 .................. ............ *Spiniphryne gladisfenae* Beebe, 1932, p. 393

   NOTE  Twenty-one known specimens, 12.8 to 131 mm; Atlantic, western Pacific, and western Indian oceans

## SPINIPHRYNE GLADISFENAE (BEEBE, 1932)

Figures 48, 109, 111, 222; Table 8

### Females

*Dolopichthys gladisfenae* Beebe, 1932b: 86–88 (original description, single specimen). Mead, 1958:132 (holotype transferred to USNM).

*Centrophryne gladisfenae*: Regan and Trewavas, 1932:84 (new combination, description after Beebe, 1932b).

*Spiniphryne gladisfenae*: Bertelsen, 1951: 122, fig. 81 (new combination, description after Beebe, 1932b; esca, subopercle, pectoral radials, pelvic bone figured). Bertelsen and Pietsch, 1975:1–11, figs. 1–6 (additional material, osteology, description, relationships). Karrer, 1976:375, pl. 2, fig. 5 (off Labrador). Bertelsen and Pietsch, 1977: 172 (*Walther Herwig* material). Pietsch and Seigel, 1980:382 (additional specimen, Banda Sea). Bertelsen, 1990:507 (synonymy, eastern tropical Atlantic). Nielsen and Bertelsen, 1992:60, fig. 8 (North Atlantic). Andriyashev and Chernova, 1994: 102 (listed for Arctic seas and adjacent waters). Pietsch, 1999:2032 (western central Pacific). Pietsch, 2002b:1064 (western central Atlantic). Moore et al., 2003:215 (off New England). Pietsch and Baldwin, 2006:407, figs. 2, 4, 5, 7 (generic revision).

### Males and Larvae

Unknown.

### Material

Twenty-one metamorphosed females (12.8 to 131 mm):

Holotype of *Dolopichthys gladisfenae*: USNM 170944 (originally NYZS 15490), 40 mm, female, Bermuda Oceanographic Expedition, net 639, 9.7 km south of Nonsuch Island, Bermuda, ca. 32°12′N, 64°36′W, 1280 m, 28 May 1930.

Additional material: BMNH, one (12.8 mm); ISH, three (49 to 130 mm); LACM, two (18 to 131 mm); MCZ, nine (14 to 98 mm); NMMBP, one (129 mm); NMNZ, one (67 mm); UW, one (69 mm); ZMB, one (105 mm); ZMUC, one (68 mm).

### Distinguishing Characters

*Spiniphryne gladisfenae* differs from *S. duhameli*, the only other known member of the genus, in details of escal morphology: a pair of bulbous, distal escal appendages, each covered with tiny papillae and/or short filaments; a large posterior escal appendage, usually extending beyond distal margin of escal bulb, and bearing a series of lobes or filaments along distal margin; and no more than 3, slender, lateral escal filaments on each side. *Spiniphryne gladisfenae* differs also in having fewer dentary teeth (21 to 45 versus 51 to 52).

### Description

Esca with 2 bulbous, narrowly based, distal appendages, anterior-most appendage approximately two-thirds to one-half length of posterior-most, each covered with small digitiform papillae and clusters of tiny filaments around the base; bulbous distal prolongation absent; a large (often extending well beyond distal margin of esca), anteroposteriorly compressed, posterior escal appendage (covering opening of escal pore), divided along its distal margin into 3 or more lobes or several to many slender filaments; posterolateral filaments absent; filamentous lateral escal appendages present or absent, when present, unpaired and no more than 3 on a side (e.g., LACM 10970-2, MCZ 161504, MCZ 161524, MCZ 164225, MCZ 164732: 1 tiny lateral filament on one side; MCZ 63337, NMMBP 9031, UW 20824, ZMUC P922146: 1 filament on each side; NMNZ P.39736: 2 filaments on the right side, 3 on the left side), usually detectable only after considerable manipulation under high-power microscopy.

Dorsal-fin rays 6 or 7; anal-fin rays 5, rarely 4 or 6; pectoral-fin rays 15 or 16, rarely 17. Number of teeth in upper jaw 21 to 42, in lower jaw 25 to 45, on vomer 4 to 8. Measurements expressed in percent SL: illicium length 7.8 to 13.3; head length 25.7 to 31.8; head width 14.5 to 19.7; head depth 26.2 to 32.7; length of premaxilla 17.6 to 21.6; length of lower jaw 21.8 to 31.0.

Additional description as given for the genus and family.

### Distribution

Of the 21 known specimens of *Spiniphryne gladisfenae* four have been collected in the eastern Atlantic Ocean between approximately 50°N and 5°S; 12 in the western North Atlantic (including the holotype from off Bermuda) between approximately 64°N and 32°N; a single record in the western Indian Ocean at about 4°S, 41°E; and four in the western Pacific ranging from Taiwan to New Zealand (Fig. 222). All available evidence indicates that this is a deep-dwelling species: all of the specimens, including recently metamorphosed females of only 15 to 16.5 mm, were captured in nets fished at maximum depths greater than 800 m; 95% of the material in nets fished at depths greater than 1000 m; and 52% at depths greater than 1600. An 18-mm female was collected in a closing-net between 650 and 1000 m.

## SPINIPHRYNE DUHAMELI PIETSCH AND BALDWIN, 2006

Figures 109, 222; Table 8

### Females

*Spiniphryne duhameli* Pietsch and Baldwin, 2006:405, figs. 1, 3, 5, 7 (original description, two specimens). Pietsch and Orr, 2007:4, fig. 3B (phylogenetic relationships).

### Males and Larvae

Unknown.

### Material

Two metamorphosed females (25.5 to 117 mm):

Holotype of *Spiniphryne duhameli*: SIO 60-239, female, 117 mm, *Spencer F. Baird*, Tethys Expedition, station 17, central Pacific Ocean, 4°56.5′ to 5°28.0′N, 142°54.5′ to 143°10.0′W, 3-m Isaacs-Kidd midwater trawl, 0 to 2500 m, 0452 to 1304 h, 6 July 1960.

Paratype of *Spiniphryne duhameli*: SIO 70-306, female, 25.5 mm, *Melville*, Antipode Cruise IV, Station 51A-Tr. 1, eastern North Pacific Ocean, 32°00.0′N, 136°12.3′W, 3-m Isaacs-Kidd midwater trawl, 0 to 1400 m, 2355 to 0746 h, 28 to 29 August 1970.

### Distinguishing Characters

*Spiniphryne duhameli* differs from *S. gladisfenae*, the only other known member of the genus, in details of escal morphology: a pair of short, slender, distal escal filaments; a small, simple, posterior escal appendage, without filaments along distal margin; and three pairs of long, slender, lateral escal filaments on each side. *Spiniphryne duhameli* differs also in having a greater number of dentary teeth (51 to 52 versus 21 to 45).

### Description

Esca of holotype with a pair of tiny, slender, distal filaments, anterior-most less than half length of posterior-most, emerging anterior to base of a bulbous, somewhat laterally expanded, distal prolongation; a small (terminating well below distal margin of esca), anteroposteriorly compressed, posterior escal appendage (covering opening of escal pore), with a smooth unbroken distal margin; a tiny posterolateral filament on each side at base of escal bulb; three pairs of lateral escal filaments on each side, with slightly swollen bases, arranged in an oblique row. Esca of paratype damaged, but clearly two distal filamentous appendages, anterior-most about half length of posterior-most, without digitiform papillae and clusters of tiny filaments around the base; all other escal appendages unknown.

Dorsal-fin rays 6, anal-fin rays 5, pectoral-fin rays 15. Number of teeth in upper jaw 42 to 47, in lower jaw 51 to 52, on vomer 4 to 8. Measurements expressed in percent SL: illicium length 9.4 to 12.9; head length 23.1 to 30.6; head width 14.5 to 19.6; head depth 24.8 to 27.8; length of premaxilla 18.8 to 22.4; length of lower jaw 27.4 to 32.2.

Additional description as given for the genus and family.

### Distribution

*Spiniphryne duhameli* is known only from the central and eastern North Pacific Ocean (Fig. 222). Both known specimens were collected with a 3-m Isaacs-Kidd midwater trawl fished open, the holotype between the surface and 2500 m, the paratype between the surface and 1400 m.

### Comments

Like all known ceratioid genera, the esca of *Spiniphryne* has a basic pattern of placement of filaments and appendages that is shared by all contained species: in this case, 2 distal escal appendages; an anteroposteriorly compressed posterior escal appendage; and, in most specimens, extremely slender, unbranched, lateral escal filaments. The esca of *Spiniphryne duhameli*, however, differs strikingly from that of *S. gladisfenae*, its only known congener, in having a pair of slender distal escal filaments (in contrast to bulbous, papilla-covered appendages); a bulbous distal prolongation of the escal bulb; a much smaller posterior escal appendage, without distal lobes or filaments; and lateral escal filaments, with slightly swollen bases, arranged in pairs. In addition, *S. duhameli* has a greater number of jaw teeth, particularly in the lower jaw. This difference is especially great when comparing specimens of comparable standard length.

Unfortunately, the esca of the 25.5-mm paratype of *S. duhameli* is poorly preserved. The true identity of this specimen was in fact not recognized until tooth counts separated it from the material previously identified as *S. gladisfenae*; the escal morphology that remains is limited to the distal surface of the escal bulb, where 2 simple filaments are clearly evident. The geographic distribution of the holotype and paratype, both from the eastern side of the North Pacific (Fig. 222), where *S. gladisfenae* has never been collected (despite ample collecting effort), seems to add further support for their conspecificity. As for the material with missing illicia and escae, tooth counts and geography indicate that all are *S. gladisfenae*, but, with the definitive diagnostic escal characters unavailable, they are here identified only to genus.

Males and larvae of *Spiniphryne* are unknown, but among the known oneirodid larvae, those of *Lophodolos*, *Oneirodes*, and the *Chaenophryne draco* group are similar to the 10.8- and 12.8-mm adolescent females of *Spiniphryne* in lacking pigment on the caudal peduncle. In the shape of the subopercle, these specimens are similar to the larvae of *Oneirodes* but differ distinctly from larvae of *Lophodolos* and the *Chaenophryne draco* group in which the upper end of this bone is slender and tapers to a point (see Bertelsen, 1951). Based on these characters, *Spiniphryne* larvae with 4 anal-fin rays may be confused with those of *Oneirodes*; however, they may be distinguished from the latter by the size and density of the melanophores, which are small and numerous in *Oneirodes* and larger and fewer in *Spiniphryne*. Because metamorphosed males of *Spiniphryne* may be expected to have spiny skin, it should be possible to separate them from other oneirodid males on this character alone. Confirmation of the identification could be made by comparison of the subdermal pigmentation, shape of the subopercle, and in most cases the number of anal-fin rays.

## Genus *Oneirodes* Lütken, 1871
(Common Dreamers)

Figures 10, 18, 20, 23, 30, 45, 53, 112–115, 193, 199, 202, 203, 223–225, 245, 250, 253, 254, 263, 264, 295; Tables 1, 8, 9

### Females

*Oneirodes* Lütken, 1871:56, 72, figs. 1, 2, pl. 2 (type species *Oneirodes eschrichtii* Lütken, 1871, by monotypy). Pietsch, 1974a:33 (generic revision based on all known material).

*Onirodes* Jordan and Gilbert, 1883:848 (erroneous spelling of *Oneirodes*, therefore taking the same type species *Oneirodes eschrichtii* Lütken, 1871; generic description after Gill, 1879a). Alcock, 1899:52, 57 (erroneous spelling; includes *Paroneirodes*; in key).

*Monoceratias* Gilbert, 1915:379, pl. 22, fig. 24 (type species *Monoceratias acanthias* Gilbert, 1915, by original designation and monotypy).

*Dolopichthys* Regan, 1925c:562 (in part; erroneous designations, followed by numerous subsequent authors; type species *Dolopichthys allector* Garman, 1899, by monotypy).

*Oneiroides* Fowler, 1936:1139, 1140, fig. 479 (erroneous spelling of *Oneirodes*, therefore taking the same type species *Oneirodes eschrichtii* Lütken, 1871; description after Günther, 1887).

### Males

*Rhynchoceratias* Regan, 1925c:566 (in part; erroneous designations; 10 specimens, two referred to *Trematorhynchus* by Regan and Trewavas, 1932, subsequently referred to *Oneirodes* by Bertelsen, 1951; type species *Rhynchoceratias brevirostris* Regan, 1925c, by subsequent designation of Fowler, 1936).

*Lipactis* Regan, 1926:43 (in part; six specimens, one from *Dana* station 1152 subsequently referred to *Oneirodes* by Bertelsen, 1951:74; type species *Lipactis tumidus* Regan, 1925c, by monotypy).

*Caranactis* Regan and Trewavas, 1932:58, 59, fig. 86 (type species *Caranactis pumilus* Regan and Trewavas, 1932, by monotypy).

*Trematorhynchus* Regan and Trewavas, 1932:91 (in part; erroneous designations; four males subsequently referred to *Oneirodes* by Bertelsen, 1951; type species *Rhynchoceratias leuchorhinus* Regan, 1925c, by subsequent designation of Burton, 1933).

### Distinguishing Characters

Metamorphosed females of *Oneirodes* are distinguished from those of all other genera of the family by having the posterior end of the pterygiophore of the illicium protruding from the dorsal midline of the trunk behind the head (among ceratioids, shared only with ceratiids). The skin contains minute, widely spaced dermal spinules (visible only microscopically in cleared and stained specimens). Darkly pigmented skin of the caudal peduncle terminates at the base of the caudal fin.

Metamorphosed females of *Oneirodes* are further unique in having the following combination of character states: The ethmoid cartilage and vomer are wide, wider than the distance between the anterolateral tips of the lateral ethmoids and frontals. Vomerine teeth are present. The nasal foramina are large and nearly circular. The frontals are short, lying posterior to the ethmoid region, their dorsal margin convex. The ventromedial extensions of the frontals approach each other on the midline, making contact with the parasphenoid. The frontals are separated from the prootics. The pterosphenoid is present. The posterior end of the illicial trough is wider and shallower than the anterior end. The sphenotic spines are well developed. The symphysial cartilage of the upper jaw is longer than wide. The lower jaw bears a well-developed symphysial spine. The hyomandibula has a double head. The quadrate spine is well developed and distinctly longer than the articular spine. The posterior margin of the opercle is deeply notched. The subopercle is short and broad to long and narrow, its dorsal end rounded to bluntly pointed (the posterior margin convex to indented or deeply notched), its ventral end nearly circular (without an anterior spine or projection). The first pharyngobranchial is present. The second pharyngobranchial is well developed. The second hypobranchial articulates directly with the second basibranchial. The caudal-fin rays are without internal pigmentation. The illicium is considerably longer than the length of the escal bulb. The pterygiophore of the illicium is cylindrical throughout its length, emerging on the snout from between the frontal bones, its anterior end exposed. The first ray of the dorsal fin is well developed. There are 5 to 7 dorsal-fin rays and 4 (very rarely 3 or 5) anal-fin rays (Table 1). The pectoral-fin lobe is short and broad, shorter than the longest rays of the pectoral fin. There are 14 to 18 (very rarely 13 or 19) pectoral-fin rays. The coracoid lacks a posteroventral process. The pelvic bones are simple but somewhat expanded distally.

Metamorphosed males of *Oneirodes* are unique in having the following combination of character states: The septa between the anterior nostrils and between the posterior nostril and the eye are pigmented. The septa between the anterior and posterior nostrils are unpigmented. The medial surface of the subopercle is unpigmented. The caudal peduncle is usually without subdermal pigment (but present in *Oneirodes melanocauda*). There are 6 to 12 olfactory lamellae. There are 6 to 17 upper denticular teeth and 7 to 28 lower denticular teeth. Anal-fin ray counts and the shape of the pectoral-fin lobe and opercular bones are as described for females (Bertelsen, 1951:87, fig. 41). The males are free-living, apparently never becoming parasitic on females.

The larvae of *Oneirodes* are unique in having the following combination of character states: The body is short, its depth about 65% SL. The length of the head is about 55% SL. The anterior part of the body is pigmented. The gill cover is darkly pigmented, the melanophores distributed more or less evenly throughout. The caudal peduncle is unpigmented (Bertelsen, 1951:76, 77, figs. 31, 32, 41).

Description

The body of metamorphosed females is relatively short and globular to moderately fusiform. The mouth is large, the cleft extending past the eye. The length of the illicium is highly variable, 13 to 108% SL. The escal bulb has an internal, arrow-shaped patch of pigment on its dorsal surface. The external morphology of the esca is highly variable but, with perhaps a single exception (*Oneirodes clarkei* Swinney and Pietsch, 1988), the escae of all known species of the genus fall readily into one of three basic escal appendage patterns (Fig. 113):

Escal appendage pattern A: The anterior appendage lacks internal pigment; the terminal papilla bears a single distal spot of pigment; the posterior appendage is cylindrical; the lateral appendage is present or absent; a single pair of anterolateral appendages is present, each represented by a broad membranous flap (Fig. 113A). Pattern A is found in two species: *O. luetkeni* and *O. rosenblatti*.

Escal appendage pattern B: The anterior appendage is internally pigmented; the terminal papilla usually bears a single distal spot of pigment (two distal pigment spots are present in *O. macrosteus*); the posterior appendage is cylindrical or laterally compressed; lateral appendages are present or absent; the anterolateral appendages, if present, are usually represented by a single filamentous pair (four filamentous pairs are present in *O. myrionemus*) (Fig. 113B). With the exception of *O. clarkei* (which is an escal appendage pattern that appears to be intermediate between A and B), pattern B is found in all species of the genus with escae not assigned to pattern A or C.

Escal appendage pattern C: the anterior appendage lacks internal pigment; the terminal papilla usually bears two distal spots of pigment (a single distal pigment spot in *O. schmidti* and *O. mirus*); the posterior appendage is usually anteroposteriorly compressed (cylindrical in *O. schmidti*); lateral appendages are absent; there are usually two pairs of filamentous anterolateral appendages (a single bifurcated pair is present in *O. theodoritissieri*) (Fig. 113C). Pattern C is found only in members of the *O. schmidti* group.

The teeth are slender, recurved, and depressible, those in the lower jaw in overlapping sets as described for other oneirodids (Pietsch, 1972c:5, fig. 2). The pattern of tooth placement is especially obvious in species with high tooth counts, the pattern in the upper jaw apparently the same as that of the lower jaw but not nearly as obvious. Teeth in the lower jaw are larger and more numerous than those in the upper jaw. There are 18 to 65 teeth in the upper jaw, 18 to 160 in the lower jaw. There are 4 to 14 teeth on the vomer, the largest and longest outermost. The first pharyngobranchial is present but reduced, its dorsal end lying free in a connective tissue matrix, with no ossified or ligamentous connection to the medial surface of the suspensorium. The second pharyngobranchial is usually well toothed (toothless in *O. luetkeni*) and slightly more than half as long, and approximately half as wide, as the third pharyngobranchial. The third pharyngobranchial bears numerous teeth.

The following measurements, expressed in percent of standard length, are summarized for females (20 to 213 mm) of all known species: head length 32.1 to 64.3; head width 23.0 to 47.6; head depth 32.2 to 64.3; premaxilla length 22.4 to 39.3; lower-jaw length 34.3 to 57.4; least outside width of frontals 7.5 to 19.0; illicium length 13.0 to 72.3.

For description of males and larvae, see Bertelsen (1951:77, 78).

Diversity

Thirty-five currently recognized species, by far the largest genus of the family as well as the suborder. Not surprisingly, it also has by far the broadest geographic range of any oneirodid and parallels that described for the family as a whole:

Key to Females of Species of the Genus *Oneirodes*

The following key, modified slightly from that of Orr (1991), will differentiate juvenile and adult females only. Males and larvae cannot be identified to species based on females and are thus referred to *Oneirodes* sp. One species, *Oneirodes melanocauda*, based solely on larval material, can be distinguished from all other larvae of the genus by having the tips of the caudal-fin rays darkly pigmented and the caudal peduncle pigmented subdermally (see Bertelsen, 1951:76–79; Pietsch, 1974a:36, 76–77).

1A. Epibranchial of first gill arch toothed ............................ 2
1B. Epibranchial teeth absent ...... 3
2A. Epibranchial of first gill arch with 6 to 17 teeth; a single pair of tooth-bearing pharyngobranchials; escal appendage pattern A (Fig. 113A): anterior appendage without internal pigment, anterolateral appendage represented by a broad membranous flap; ratio of length of dorsal and ventral forks of opercle 0.60 to 0.71 ............... *Oneirodes luetkeni* (Regan, 1925c), p. 397

NOTE Fifty-eight known specimens, 11.5 to 123 mm; eastern tropical Pacific Ocean

2B. Epibranchial of first gill arch with 1 to 5 teeth; two pairs of tooth-bearing pharyngobranchials; escal appendage pattern B (Fig. 113B): anterior appendage internally pigmented, anterolateral appendage absent; ratio of lengths of dorsal and ventral forks of opercle 0.51 to 0.61 ............... *Oneirodes carlsbergi* (Regan and Trewavas, 1932), p. 398

NOTE Thirty known specimens, 18 to 222 mm; tropical Atlantic and Pacific oceans

3A. Escal appendage pattern C (Fig. 113C): anterior appendage without internal pigment, usually two pairs of filamentous anterolateral appendages .......... *Oneirodes schmidti* group, p. 416

NOTE Eight species; Atlantic, Pacific, and Indian oceans

3B. Escal appendage pattern A (Fig. 113A): anterior appendage without internal pigment, a single pair of anterolateral appendages each represented by a broad, membranous flap; or escal appendage pattern B (Fig. 113B): anterior appendage internally pigmented, anterolateral appendages, if present, 1 or 4 filamentous pairs; or escal appendage pattern intermediate between

A and B (*Oneirodes clarkei*) with a cluster of 5 well-developed anterolateral appendages united at base ....... 4

4A. Lower jaw with more than 90 teeth in specimens greater than 45 mm, more than 60 teeth in specimens greater than 25 mm; number of teeth on vomer of specimens greater than 25 mm 8 to 14, usually more than 9 ........................ 5

4B. Lower jaw with fewer than 90 teeth in specimens greater than 45 mm, fewer than 60 teeth in specimens greater than 25 mm; number of teeth on vomer of specimens greater than 25 mm 3 to 9, usually fewer than 8 ........................ 6

5A. Escal appendage pattern A (Fig. 113A): anterior appendage without internal pigment, anterolateral appendage represented by a broad membranous flap; length of illicium less than 35% SL ............ *Oneirodes rosenblatti* Pietsch, 1974a, p. 399

NOTE Twenty-one known specimens, 12.5 to 134 mm; eastern tropical Pacific Ocean

5B. Escal appendage pattern B (Fig. 113B): anterior appendage internally pigmented, anterolateral appendage filamentous; length of illicium 60% SL in 35-mm specimen ............
............ *Oneirodes dicromischus* Pietsch, 1974a, p. 413

NOTE Two known specimens, 21 to 35 mm; western and central Pacific Ocean

6A. Esca with a well-developed lateral appendage ................... 7

6B. Esca with lateral appendage minute or absent ................. 11

7A. Esca with 2 or 3 medial filaments more than twice length of escal bulb ........................ 8

7B. Esca without elongate medial appendages ................. 9

8A. Esca with medial filaments more than six times length of escal bulb (larger specimens) ...... *Oneirodes kreffti* Pietsch, 1974a, p. 405

NOTE Thirty-eight known specimens, 11 to 126 mm; southern latitudes of the Atlantic, Pacific, and Indian oceans

8B. Esca with medial filaments two or three times length of escal bulb ................. *Oneirodes posti* Bertelsen and Grobecker, 1980, p. 405

NOTE Three known specimens, 36.5 to 135 mm; North Atlantic Ocean

9A. Posterior escal appendage equal to or less than one-third length of escal bulb ..................... 10

9B. Posterior escal appendage greater than 1.5 times length of escal bulb (smaller specimens) ............
............ *Oneirodes kreffti* Pietsch, 1974a, p. 405

NOTE Thirty-eight known specimens, 11 to 126 mm; southern latitudes of the Atlantic, Pacific, and Indian oceans

10A. Posterior escal appendage about one-third length of escal bulb; anterior appendage anterodorsally directed, bearing numerous short filaments and 2 unpigmented tapering filaments on anterior margin near distal tip ........ *Oneirodes anisacanthus* (Regan, 1925c), p. 404

NOTE Eleven known specimens, 10.5 to 173 mm; North Atlantic Ocean

10B. Posterior escal appendage minute; anterior appendage narrow, elongate, and anteroventrally directed, bearing a single short distal filament ............ *Oneirodes plagionema* Pietsch and Seigel, 1980, p. 409

NOTE Four known specimens, 25 to 104 mm; Pacific Ocean

11A. Esca with 1 or more well-developed medial appendages ........... 12

11B. Esca with medial appendages minute or absent ................. 23

12A. Esca with anterior and/or posterior appendages laterally compressed ... 13

12B. Esca with anterior and posterior appendages cylindrical ........ 16

13A. Esca with only anterior or posterior appendage compressed; pectoral-fin rays 15 to 19 ............. 14

13B. Esca with both anterior and posterior appendages laterally compressed; pectoral-fin rays 13 to 14 ...........
............... *Oneirodes cristatus* (Regan and Trewavas, 1932), p. 409

NOTE Three known specimens, 20 to 165 mm; Banda and Celebes seas

14A. Esca with only posterior appendage compressed ............. 15

14B. Esca with only anterior appendage compressed ....... *Oneirodes sabex* Pietsch and Seigel, 1980, p. 402

NOTE Twenty-four known specimens, 12 to 189 mm; western Pacific Ocean

15A. Esca with medial appendage represented by tuft of extremely short filaments; anterior appendage simple ............... *Oneirodes pterurus* Pietsch and Seigel, 1980, p. 410

NOTE A single known specimen, 30 mm; Halmahera Sea

15B. Esca with three groups of filamentous medial appendages; anterior appendage filamentous ............
............. *Oneirodes thysanema* Pietsch and Seigel, 1980, p. 414

NOTE Two known specimens, 13 to 26.5 mm; western North Atlantic and Banda Sea

16A. Esca with a single medial appendage, simple throughout its length or bifurcated some distance from its base ....................... 17

16B. Esca with 2 or more highly filamentous medial appendages ....... 20

17A. Esca with medial appendage less than or about equal to length of escal bulb, anterior appendage much longer than length of escal bulb ...... 18

17B. Esca with medial appendage much longer than length of escal bulb; anterior appendage shorter than length of escal bulb ................. 19

18A. Esca with well-developed anterolateral appendages ................
............. *Oneirodes haplonema* Stewart and Pietsch, 1998, p. 414

NOTE Two known specimen, 35 to 116 mm; western South Pacific Ocean

18B. Esca without anterolateral appendages ......... *Oneirodes epithales* Orr, 1991, p. 415

NOTE Two known specimens, 45 to 128 mm; western North Atlantic Ocean

19A. Esca with anterior appendage bearing a fringe of filaments on lateral margin near base and a single unbranched filament on anterior margin ......... *Oneirodes macronema* Regan and Trewavas, 1932, p. 415

NOTE Two known specimens, 16.5 to 27 mm; Caribbean Sea and central North Pacific Ocean

19B. Esca with a trilobed anterior appendage flanked on each side by a cluster of 5 anterolateral appendages united at base ................
............. *Oneirodes clarkei* Swinney and Pietsch, 1988, p. 407

NOTE A single known specimen, 119 mm; eastern North Atlantic Ocean

20A. Esca with posterior appendage highly branched ...............
............ *Oneirodes heteronema* (Regan and Trewavas, 1932), p. 407

NOTE Nine known specimens, 13.5 to 119 mm; eastern tropical Pacific Ocean

20B. Esca with posterior appendage unbranched or bearing only minute filaments .................... 21

21A. Esca with four pairs of filamentous anterolateral appendages; internally pigmented portion of anterior appendage nearly twice length of escal bulb ....... *Oneirodes myrionemus* Pietsch, 1974a, p. 406

NOTE Two known specimens, 43 to 121 mm; eastern North Atlantic Ocean

21B. Esca with or without a single pair of anterolateral appendages; internally pigmented portion of anterior appendage less than length of escal bulb . . . . . . . . . . . . . . . . . . . . . 22

22A. Esca with anterior appendage distally divided into numerous long filaments; posterior appendage usually with 1 or 2 short branches; posterior margin of upper part of subopercle usually indented to deeply notched . . . . . . . . . . . . . . .*Oneirodes bulbosus* Chapman, 1939, p. 403

NOTE  One hundred and fifteen known specimens, 30 to 160 mm; North Pacific Ocean and Bering Sea

22B. Esca with anterior appendage bearing papillae and a few short filaments at distal tip; posterior appendage never branched; posterior margin of upper part of subopercle not indented . . . . . . . *Oneirodes eschrichtii* Lütken, 1871, p. 399

NOTE  One hundred and fifteen known specimens, 10 to 213 mm; Atlantic, Pacific, and Indian oceans

23A. Subopercle long and slender . . . . 24
23B. Subopercle short and broad . . . . 25
24A. Length of illicium greater than 33% SL; esca with an anterolateral appendage and a pigmented anterior appendage . . . . *Oneirodes macrosteus* Pietsch, 1974a, p. 408

NOTE  Twenty-eight specimens, 11.5 to 185 mm; Atlantic Ocean

24B. Length of illicium less than 33% SL; esca without an anterolateral appendage, anterior appendage unpigmented . . . . . *Oneirodes schistonema* Pietsch and Seigel, 1980, p. 412

NOTE  A single known specimen, 74 mm; Banda Sea

25A. Anterior escal appendage bearing 1 to 5 stout papillae along posterior margin, posterior escal appendage branched; ratio of lengths of dorsal and ventral forks of opercle 0.53 to 0.71 . . . . . . . . . . *Oneirodes acanthias* (Gilbert, 1915), p. 410

NOTE  One hundred and fifty-three known specimens, 11.5 to 167 mm; eastern North Pacific Ocean

25B. Anterior escal appendage without stout papillae along posterior margin, posterior escal appendage unbranched; ratio of lengths of dorsal and ventral forks of opercle 0.42 to 0.59 . . . . . . . . . . . . . . . . . . . . . 26

26A. Esca with posterior appendage two to four times length of escal bulb in all known specimens . . . . . . . . . . . . . . . . . . . . . . . . *Oneirodes flagellifer* (Regan and Trewavas, 1932), p. 413

NOTE  Nine known specimens, 10.5 to 22 mm; Indo-west Pacific Ocean

26B. Esca with posterior appendage less than three times length of escal bulb in specimens greater than 70 mm, equal to or less than length of escal bulb in specimens less than 70 mm . . . . . . . . . . . . . . . . . . . . . . . . . . . 27

27A. Anterior escal appendage bearing a compressed papilla and several smaller papillae on distal end, papillae darkly pigmented in specimens greater than 40 mm; posterior margin of upper part of subopercle indented to deeply notched; ratio of lengths of dorsal and ventral forks of opercle 0.42 to 0.54; pectoral-fin rays 15 to 17 . . . . . . *Oneirodes thompsoni* (Schultz, 1934), p. 411

NOTE  One hundred and eight known specimens, 33 to 153 mm; North Pacific Ocean and Bering Sea

27B. Anterior escal appendage usually bearing a compressed papilla and 2 tapering filaments on distal end, papilla and filaments unpigmented; posterior margin of upper part of subopercle not indented; ratio of lengths of dorsal and ventral forks of opercle 0.52 to 0.59; pectoral-fin rays 17 to 19 . . . . . . . . . *Oneirodes notius* Pietsch, 1974a, p. 412

NOTE  Seventeen known specimens, 30 to 150 mm; Southern Ocean

### ONEIRODES LUETKENI (REGAN, 1925C)

Figures 114, 224; Tables 8, 9

#### Females

*Dolopichthys luetkeni* Regan, 1925c:562 (original description, single specimen). Regan, 1926:27, pl. 4, fig. 2 (description, comparison with *Oneirodes eschrichtii*, in key). Parr, 1927:15 (in key; *Dolopichthys heteracanthus* a synonym of *D. luetkeni*). Fowler, 1936:1337 (description after Regan, 1926; in key). Beebe and Crane, 1947:159 (six additional specimens; synonymy, distribution, color; *D. heteracanthus* an immature form of *D. luetkeni*). Nielsen, 1974:94 (listed in type catalog).

*Dolopichthys* (subgenus *Dermatias*) *luetkeni*: Regan and Trewavas, 1932:76, fig. 116 (description; in key).

*Dolopichthys heteracanthus* Regan, 1925c:562 (in part; original description, 10 syntypes, Gulf of Panama). Regan, 1926:28, pl. 5, fig. 1 (in part; misidentifications, description; 21 specimens, 15 of which were subsequently reallocated to *Oneirodes luetkeni* by Bertelsen, 1951; in key). Fowler, 1936:1338, 1339 (description after Regan, 1926; comparison with *Dolopichthys megaceros*, a synonym of *O. eschrichtii*; in key).

*Dolopichthys* (subgenus *Dermatias*) *heteracanthus*: Regan and Trewavas, 1932:77, fig. 117 (misidentifications; description; four additional specimens all referred to *Oneirodes luetkeni* by Bertelsen, 1951; in key).

*Oneirodes luetkeni*: Bertelsen, 1951: 86–87, figs. 31P–S, 40 (new combination; diagnostic characters, available material listed, opercular bones described, figured; in key). Grey, 1956a:248 (synonymy, distribution). Pietsch, 1972a:36, 45 (epibranchial tooth plates). Brewer, 1973:25 (eastern tropical Pacific). Parin et al., 1973:143, figs. 37, 39c (eastern South Pacific). Pietsch, 1974a:38, figs. 19, 28, 50, 61, 106 (additional specimens, description, comparison with all known material). Chirichigno, 1978:58, fig. 27 (additional specimen, description, off Peru). Munk, 1999:267 (bioluminescence).

#### Males and Larvae

Unknown.

#### Material

Fifty-eight metamorphosed females (11.5 to 123 mm):

Holotype of *Oneirodes luetkeni*: ZMUC P9287, 123 mm, *Dana* station 1203(10), 7°30′N, 79°19′W, 3500 m wire, bottom depth 2550 m, 1500 h, 1 November 1922.

Syntypes of *Dolopichthys heteracanthus*: BMNH 1925.8.11.10, 28.5 mm, *Dana* station 1205(2), 6°49′N, 80°25′W, 1000 m wire, 0420 h, 14 December 1922. BMNH 1925.8.11.11, 18 mm, *Dana* station 1208(8), 6°48′N, 80°33′W, 1500 m wire, 0810 h, 16 December 1922.

Additional material: BMNH, one (16.5 mm); HUMZ, 11 (51 to 105 mm); IMARPE, one (39 mm); LACM, seven (15 to 61 mm); NYZS, six (12.5 to 22 mm); SIO, six (14 to 51 mm); SIOM, eight (13 to 95 mm); ZMUC, 15 (11.5 to 60 mm).

#### Distinguishing Characters

A species of *Oneirodes* easily distinguished from all other members of the genus in escal morphology, in lacking teeth on pharyngobranchial II, and in having a relatively short illicium (13.0 to 20.4% SL). In some ways *Oneirodes luetkeni* is similar to *O. carlsbergi*: unlike all other oneirodids, these two species have well-developed teeth on the anterior margin of the epibranchial of the first gill arch. These teeth are present and easily discernible even in the smallest known specimens (11.5 mm). In addition, *O. luetkeni* and *O. carlsbergi*, as well as *O. rosenblatti* and *O. dicromischus* share a relatively high number of jaw teeth.

#### Description

Escal appendage pattern A (Fig. 113A); esca with a stout anterior appendage, distal tip slightly swollen and pigmented in most specimens larger than approximately 60 mm; anterior escal appendage without internal pigment; a truncated terminal escal papilla, with a distal spot of pigment in some specimens, flanked on each side by a medial appendage consisting of 2 or 3 short, unpigmented filaments; an unpigmented, tapering, posterior escal appendage less than length of escal bulb, becoming proportionately longer with growth; usually a pair of lateral escal appendages, each consisting of 3 or 4 short, unpigmented filaments; and a broad, fringed, membranous anterolateral flap.

Subopercle without indentation on posterodorsal margin (Bertelsen, 1951:87, fig. 31P–S); length of ventral fork of opercle 22.8 to 35.7% SL; ratio of lengths of dorsal and ventral forks of opercle 0.60 to 0.71.

Epibranchial I with 6 to 17 teeth on anterior margin (Pietsch, 1974a, fig. 50); teeth absent on pharyngobranchial II; total number of teeth in upper jaw 57 to 134, in lower jaw 51 to 93; number of teeth on vomer 6 to 10; dorsal-fin rays 5 or 6; anal-fin rays 4; pectoral-fin rays 15 to 17.

Measurements in percent of standard length: head length 32.5 to 46.4; head depth 32.5 to 52.6; premaxilla length 30.0 to 39.3; lower-jaw length 42.9 to 57.4; illicium length 13.0 to 20.4.

Additional description as given for the genus and family.

### Distribution

*Oneirodes luetkeni* is known only from the Gulf of Panama and adjacent waters of the eastern Pacific Ocean, extending along the west coast of the Americas to 20°21′N and to 12°20′S (Fig. 224). Vertically this species appears to be concentrated between 700 and 1250 m (see Pietsch, 1974a:39).

## ONEIRODES CARLSBERGI (REGAN AND TREWAVAS, 1932)

Figures 114, 225; Table 8

### Females

*Dolopichthys* (subgenus *Dermatias*) *carlsbergi* Regan and Trewavas, 1932:76, fig. 115 (original description, six specimens; lectotype ZMUC P9285, designated by Bertelsen, 1951).

*Linophryne colletti* Weber, 1913:559 (original description, single specimen, holotype ZMA 101.895; misidentification; teeth present on first epibranchial). Nijssen et al., 1993:225 (listed).

*Oneirodes eschrichtii*: Regan, 1926:26 (in part; misidentifications; two specimens, the holotype of *Oneirodes eschrichtii* and another referred to *O. carlsbergi* by Bertelsen, 1951). Regan and Trewavas, 1932:63 (in part; after Regan, 1926).

*Dolopichthys heteracanthus*: Regan, 1926:28 (in part; misidentifications; 21 specimens, two of which were referred to *Dolopichthys* [subgenus *Dermatias*] *carlsbergi* by Regan and Trewavas, 1932). Norman, 1930:353 (in part; misidentifications; two specimens, the larger referred to *D.* [subgenus *Dermatias*] *carlsbergi*, the smaller to *D.* [subgenus *Dermatias*] *anisacanthus* [= *Oneirodes anisacanthus*], by Regan and Trewavas, 1932).

*Dolopichthys megaceros*: Regan, 1926:29 (in part; misidentifications; nine specimens, one of which was referred to *Dolopichthys* [subgenus *Dermatias*] *carlsbergi* by Regan and Trewavas, 1932).

*Dolopichthys inimicus* Fraser-Brunner, 1935:324, fig. 3 (original description, single specimen).

*Oneirodes inimicus*: Bertelsen, 1951:85 (new combination, diagnostic characters). Grey, 1956a:247 (synonymy, distribution). Maul, 1973:670 (synonymy, eastern North Atlantic; after Bertelsen, 1951).

*Oneirodes carlsbergi*: Bertelsen, 1951:86, figs. 31M–O, 39 (new combination; diagnostic characters, *Dana* material listed, lectotype designated; opercular bones described, figured; in key). Grey, 1956a:247 (synonymy, distribution). Pietsch, 1972a:36, 45 (epibranchial tooth plates). Maul, 1973:670 (synonymy, eastern North Atlantic). Parin et al., 1973:143 (eastern South Pacific). Pietsch, 1974a:39, figs. 62, 107 (additional specimens, description, comparison with all known material). Bertelsen and Pietsch, 1977:172, fig. 1 (*Walther Herwig* material; Atlantic distribution plotted). Pietsch and Seigel, 1980:383 (additional specimen, Banda Sea). Bertelsen, 1986:1394, figs. (includes *Oneirodes inimicus* Fraser-Brunner, 1935; eastern North Atlantic). Bertelsen, 1990:504 (synonymy, eastern tropical Atlantic). Jónsson and Pálsson, 1999:201, fig. 5 (Iceland). Pietsch, 1999:2032 (western central Pacific). Pietsch, 2002b:1064 (western central Atlantic).

*Oneirodes eschrichtii* Group, Bussing, 1965:223 (misidentification; description of additional specimen; comparison with *Dolopichthys brevifilis* [= *Oneirodes eschrichti*]).

*Dolopichthys carlsbergi*: Nielsen, 1974:91 (listed in type catalog).

### Males and Larvae

Unknown.

### Material

Thirty metamorphosed females (18 to 222 mm):

Lectotype of *Oneirodes carlsbergi*: ZMUC P9285, 40 mm, *Dana* station 1206(7), Gulf of Panama, 6°40′N, 80°47′W, 1200 m wire, 1845 h, 15 January 1922.

Paralectotypes of *Oneirodes carlsbergi*: BMNH 1930.1.12.1079, 37 mm, *Discovery*, 13°25′N, 18°22′W, 4.5 m net, 0 to 900 m, 28 October 1925. BMNH 1925.8.11.9, 30.5 mm, *Dana* station 1208(16), 6°48′N, 80°33′W, 2100 m wire, 1715 h, 16 January 1922. BMNH 1932.5.3.15, 21 mm, *Dana* station 3730(1), 16°55′N, 120°02.5′E, 1000 m wire, 2245 h, 15 June 1929. ZMUC P92163, 19.5 mm, *Dana* station 3556(1), 2°52′N, 87°38′W, 2500 m wire, bottom depth 2285 m, 1530 h, 14 September 1928; ZMUC P92162, 19 mm, *Dana* station 1165(8), 12°11′N, 35°49′W, 3000 m wire, 0800 h, 9 November 1921.

Holotype of *Dolopichthys inimicus*: BMNH 1934.8.8.90, 22 mm, Commercial Trawler *Dynevor Castle*, Irish Atlantic Slope, 53°15′N, 12°28′W, 0 to 320 m, July 1934.

Holotype of *Linophryne colletti*: ZMA 101.895, 76 mm, *Siboga* Expedition station 18, Indonesia, Bali Sea, 7°28.2′S, 115°24.6′E, 0 to 1018.

Additional material: BMNH, two (22.5 to 38 mm); CAS, one (70 mm); HUMZ, one (222 mm); ISH, eight (25 to 107 mm); LACM, two (18 to 19 mm); SIO, four (21.5 to 62 mm); SIOM, one (21.5 mm); UF, two (27 to 48 mm); ZMUC, one (159 mm).

### Distinguishing Characters

A species of *Oneirodes* distinguished from all other members of the genus in escal morphology. Although most similar to *Oneirodes luetkeni*, it clearly differs further from the latter and from all other congeners in having the following combination of character states: epibranchial I bearing 1 to 5 teeth; pharyngobranchial II toothed; upper jaw with 29 to 180 teeth; lower jaw with 53 to 160 teeth.

### Description

Escal appendage pattern B (Fig. 113B); esca with a tapering, internally pigmented, anterior appendage, two to more than three times length of escal bulb, becoming proportionately longer with growth, and usually bearing 2 short, unpigmented filaments on anterior margin near distal tip; an unpaired, unpigmented, filamentous medial escal appendage usually consisting of numerous branched filaments, flanked on each side by a similar filamentous medial appendage; a truncated terminal escal papilla, with a distal spot of dark pigment in some specimens; a laterally compressed, crescent-shaped, posterior escal appendage, pigmented on distal margin in some specimens; an unpigmented filamentous lateral escal appendage on each side; anterolateral escal appendages absent.

Subopercle without indention on posterodorsal margin (Bertelsen, 1951:86, fig. 31M–O); length of ventral fork of opercle 23.3 to 30.0% SL; ratio of lengths of dorsal and ventral forks of opercle 0.51 to 0.61.

Epibranchial I with 1 to 5 teeth along anterior margin; teeth present on pharyngobranchial II; total number of teeth in upper jaw 29 to 180, in lower jaw 53 to 160; number of teeth on vomer 4 to 10; dorsal-fin rays 5 to 7; anal-fin rays 4, very rarely 5 (only one specimen had 5 anal-fin rays); pectoral-fin rays 16 to 18.

Measurements in percent of standard length: head length 34.8 to 47.6; head depth 41.9 to 64.3; premaxilla length 26.1 to 38.1; lower-jaw length 40.9 to 54.8; illicium length 15.2 to 35.3.

Additional description as given for the genus and family.

Distribution

Although this species appears to have a relatively narrow circumtropical distribution between approximately 18°N and 8°S, it has not yet been collected in the western Atlantic or Indian Ocean (Fig. 225). In the eastern Atlantic, it ranges only as far west as about 36°W, between 18°N and 5°S. In the western Pacific, there are single records in the Java and Banda seas, and north of Luzon in the Philippine Islands. Several females are known from the relatively nutrient-poor central equatorial Pacific, but many more have been collected in the eastern tropical Pacific between about 8°N and 8°S. One additional specimen, considerably outside the presumed circumtropical range of *Oneirodes carlsbergi*, was taken off the Irish Atlantic Slope (the holotype of *Dolopichthys inimicus* Fraser-Brunner, 1935:324). The lectotype is from the Gulf of Panama.

The available data (maximum depths reached by fishing gear) suggest that *O. carlsbergi* has an extremely wide vertical range compared to that of other species of *Oneirodes*, and that it may be taken at relatively shallow depths. Thirty-six percent of the total known material, including the largest known specimens, was captured by gear fished at maximum depths not exceeding 360 m; 72% was taken by nets fished above 1000 m. Two specimens, 22.5 and 38 mm, were captured by a closing net fished between 690 and 900 m. Data for the Atlantic and Pacific populations were analyzed separately (Pietsch, 1974a), but no significant differences in vertical distribution were found.

*ONEIRODES ROSENBLATTI* PIETSCH, 1974A

Figures 114, 223; Table 8

Females

*Oneirodes rosenblatti* Pietsch, 1974a:41, figs. 63, 64, 108 (original description, 12 specimens).

Males and Larvae

Unknown.

Material

Twenty-one metamorphosed females (12.5 to 134 mm):

Holotype of *Oneirodes rosenblatti*: SIO 69-351, 94 mm, *Piquero*, cruise 8, Gulf of Panama, 3°10′N, 84°10′W, 3-m Isaacs-Kidd midwater trawl, 0950 to 1453 h, 3 July 1969.

Paratypes of *Oneirodes rosenblatti*: LACM 32613-4, 12.5 mm, *Te Vega*, cruise 20, station B-16, 3°46′S, 85°37′W, 6-ft modified Tucker trawl, with opening-closing device, 1000 to 1250 m. SIO 55-246, 2 (67 to 91 mm), *Horizon*, 5°00′N, 78°09′W, 3-m Isaacs-Kidd midwater trawl, 0 to 1436 m, 2050 to 0100 h, 14 to 15 November 1965. SIO 52-384, 84 mm, *Horizon*, Shellback Expedition, 2°09′N, 84°53.5′W, 3-m Isaacs-Kidd midwater trawl, 0 to 1286 m, 29 July 1952. SIO 70-384, 56 mm, *Anton Bruun*, cruise 11, station 183, 8°59′S, 80°37′W, Menzies trawl, 0 to 4501 m, bottom depth 4486 to 4501 m, 4 November 1965. SIO 55-244, 26 mm, *Horizon*, Eastropac Expedition, trawl 12, 3-m Isaacs-Kidd midwater trawl, 0 to 1335 m, 14 November 1955. USNM 201099, 13 mm, *Anton Bruun*, cruise 14, station 570B, 8°33′S, 81°27′W, 3-m Isaacs-Kidd midwater trawl, 0 to 2850 m, 0005 to 0505 h, 14 March 1966. ZIN 49797, 66 mm, *Akademik Kurchatov*, cruise 4, station 282, sample 208, 8°01.5′S, 81°01.2′W, 3-m Isaacs-Kidd midwater trawl, 0 to 1500 m, bottom depth 3870 to 4800 m, 1450 to 1900 h, 28 October 1968. ZIN 49798, 48 mm, *Akademik Kurchatov*, cruise 4, station 295, sample 227, 8°25′S, 81°18′W, 3-m Isaacs-Kidd midwater trawl, 0 to 910 m, 1940 to 0140 h, 1 to 2 November 1968. ZMUC P92188, 32 mm, *Galathea* Expedition station 739, 7°22′N, 79°32′W, herring otter trawl, 0 to 745 m, bottom depth 915 to 975 m, 15 May 1952.

Additional material: HUMZ, eight (68 to 134 mm); SIOM, one (22 mm).

Distinguishing Characters

A species of *Oneirodes* distinguished from all other members of the genus in escal morphology. *Oneirodes rosenblatti* can be confused only with *O. luetkeni*, both of which are characterized by having a relatively high number of teeth in the jaws and on the vomer, and a similar escal morphology. The former, however, is clearly differentiated from *O. luetkeni* by the absence of epibranchial teeth, the presence of teeth on pharyngobranchial II, and a considerably longer illicium.

Description

Escal appendage pattern A (Fig. 113A); esca with a stout, unpigmented anterior appendage, without internal pigment; a pair of medial escal appendages each consisting of 2 or 3 branched filaments, the distal tips of which are pigmented in most specimens larger than approximately 48 mm; a truncated terminal escal papilla, with a distal spot of pigment in some specimens; a stout, unpigmented posterior escal appendage; lateral escal appendages absent; on each side, a large, membranous, oval or crescent-shaped, anterolateral flap, distal half of which is darkly pigmented in most specimens larger than approximately 26 mm.

Subopercular bone relatively long and slender, without indention on posterodorsal margin; length of ventral fork of opercle 20.7 to 26.5% SL; ratio of lengths of dorsal and ventral forks of opercle 0.45 to 0.54.

Epibranchial teeth absent; teeth present on pharyngobranchial II; total number of teeth in upper jaw 50 to 145, in lower jaw 59 to 137; number of teeth on vomer 6 to 14; dorsal-fin rays 5 or 6; anal-fin rays 4; pectoral-fin rays 14 to 16.

Measurements in percent of standard length: head length 32.1 to 40.6; head depth 35.2 to 46.9; premaxilla length 26.8 to 31.2; lower-jaw length 37.2 to 48.3; illicium length 27.7 to 31.2.

Additional description as given for the genus and family.

Distribution

*Oneirodes rosenblatti* is known only from the eastern tropical Pacific Ocean, mostly concentrated in the Gulf of Panama, but extending south into the Peru-Chile Trench as far as about 12°S (Fig. 223). Based on the maximum depths reached by fishing gear, *O. rosenblatti* appears to be a relatively deep dwelling form with a rather wide vertical range. Seventy-five percent of the known material, including the largest known specimens, was captured by gear fished at depths below 1280 m, despite the considerably greater fishing effort made above this depth during most oceanographic work in the eastern tropical Pacific. A single capture (12.5 mm) was made with an opening-closing net between 1000 and 1250 m. Although *O. rosenblatti* occurs between 750 m and perhaps as deep as 3000 m, it seems to be concentrated between 1250 and 2000 m (Pietsch, 1974a).

*ONEIRODES ESCHRICHTII* LÜTKEN, 1871

Figures 10, 18, 20, 112, 114, 223, 245; Tables 8, 9

Females

*Oneirodes eschrichtii* Lütken, 1871:56, figs. 1, 2, pl. 2 (original description, single specimen; in Danish). Lütken, 1872:329, figs. 1, 2, pl. 9 (English translation of Lütken, 1871). Rink, 1877:434 (Greenland). Lütken, 1878a:310, 326, pl. 2, fig. 6 (reference to original description, esca figured, in Danish). Lütken, 1878b:343 (reference to original description, counts, in French). Gill, 1879a:218 (brief description). Gill, 1879b:228 (listed). Jordan, 1885:927 (listed). Günther, 1887:56 (description after Lütken, 1871). Goode and Bean, 1896:492 (description after Gill, 1879a). Jordan and Evermann, 1898:2732 (description after Gill, 1879a). Ehrenbaum, 1901:76 (reference to holotype). Regan, 1926:26, fig. 17 (in part; two specimens, the holotype of *Oneirodes eschrichtii* and another referred to *O. carlsbergi* by Bertelsen, 1951). Regan and Trewavas, 1932:63 (in part; after Regan, 1926). Maul, 1949:34, figs. 13–17 ("sensu lato"; misidentification, specimen referred to *O. anisacanthus* by Pietsch, 1974a; description, osteology, discussion of identity). Einarsson, 1952:91 (biology; after Bertelsen, 1951). Albuquerque, 1954–1956:1064, fig. 436 (after Fowler, 1936; figure after Lütken, 1871). Berry and Perkins, 1966:677 (four additional specimens). Lavenberg and Ebeling, 1967:195, fig. 5 (vertical distribution; name misspelled *esrichtii*). Fitch and Lavenberg, 1968:134 (comparison with *O. acanthias*). Ebeling et al., 1970:31, fig. 4 (ecological groups of deep-sea animals off southern California). Marshall, 1971a:95 (temperature and depth distribution). Maul, 1973: 670 (synonymy, eastern North Atlantic). Nielsen, 1974:96 (listed in type catalog). Pietsch, 1974a:44, figs. 65–70, 109 (review based on all known material). Bertelsen and Pietsch, 1977:174, fig. 1 (*Walther Herwig* material). Hubbs et al., 1979:13 (California). Marshall, 1979:257, 488, fig. 103 (figure after Pietsch, 1974a; distribution). Pietsch and Seigel, 1980:383, fig. 3 (additional specimen, Sulu Sea). Ayling and Cox, 1982:138, fig. (New Zealand). Amaoka, 1984:106, pl. 346-I (Japan). Bertelsen, 1986:1395, figs. (eastern North Atlantic). Pequeño, 1989:43 (Chile). Bertelsen, 1990: 504 (synonymy, eastern tropical Atlantic). Nielsen and Bertelsen, 1992:60, fig. 2, 2b (North Atlantic). Andriyashev and Chernova, 1994:102 (listed for Arctic seas). Stearn and Pietsch, 1995:136, fig. 86 (two additional specimens, Greenland). Swinney, 1995a:52, 55 (three additional specimens, off Madeira). Ellis, 1996:267, fig. (popular account). Munk, 1998:176, figs. 2–4 (escal light-guiding structures). Stewart and Pietsch, 1998:18, fig. 13 (additional specimen, New Zealand). Shedlock et al., 1997:396, fig. 1 (DNA extracted from formalin-fixed specimens). Jónsson and Pálsson, 1999:201, fig. 5 (Iceland). Munk, 1999:267, 268, 272, 274, 275, 277–279, 281, figs. 3A, 5F, 8C (bioluminescence). Pietsch, 1999:2032 (western central Pacific). Sheiko and Fedorov, 2000:24 (Kamchatka). Anderson and Leslie, 2001:16, fig. 12B (additional female, off Namibia). Kukuev and Trunov, 2002:326, fig. 2a (North Atlantic). Mecklenburg et al., 2002:305, 308 (Alaska, in key). Nakabo, 2002:473, figs. (Japan, in key). Pietsch, 2002b:1064 (western central Atlantic). Moore et al., 2003:215 (off New England). Love et al., 2005:61 (eastern North Pacific).

*Onirodes eschrichtii*: Jordan and Gilbert, 1883:848 (emendation of generic name; brief description after Gill, 1878a).

*Oneirodes megaceros* Holt and Byrne, 1908a:93 (original description, single specimen, comparison with *Oneirodes eschrichtii*). Holt and Byrne, 1908b:60 (listed, reference to original description). Murray and Hjort, 1912:94, 614, fig. 81 (specimen thought to resemble *O. megaceros*).

*Dolopichthys megaceros*: Regan, 1926: 29 (new combination, brief description; *Dolopichthys anisacanthus* [= *Oneirodes anisacanthus*], a synonym of *D. megaceros*; additional specimens, comparison with holotype; in key). Parr, 1927:15, 18 (comparison with *D. obtusus* [= *O. eschrichtii*]; in key). Fowler, 1936:1339 (synonymy, description after Regan, 1926, in key). Koefoed, 1944:6, pl. 1, figs. 4a, b, 5, pl. 3, fig. 6 (description of two additional specimens, comparison with holotype). Maul, 1949:40 (a possible synonym of *O. eschrichtii*). Maul, 1961:130 (synonymy, three specimens questionably referred to *D. megaceros*). Wheeler, 1969:585 (reference to original description).

*Dolopichthys* sp.: Regan, 1926:14 (listed). Norman, 1939:115, fig. 41 (in part; two specimens, the smaller, *Discovery* station 186, is *Oneirodes eschrichtii*, the larger, *Discovery* station 193, belongs to the *O. schmidti* group; esca figured).

*Dolopichthys obtusus* Parr, 1927:16, fig. 5 (original description, single specimen; comparison with *Dolopichthys acanthias* [= *Oneirodes acanthias*] and *O. megaceros* [= *O. eschrichtii*]; in key). Maul, 1961:130 (comparison with congeners, a possible synonym of *D. megaceros* [= *O. eschrichtii*]).

*Dolopichthys tentaculatus* Beebe, 1932b: 88, fig. 23 (original description, single specimen; comparison with *Dolopichthys obtusus* [= *Oneirodes eschrichtii*]). Bostelmann, 1934:188 (painting). Beebe, 1937:207 (listed). Mead, 1958:132 (type transferred to USNM). Maul, 1961:130 (comparison with congeners, a possible synonym of *D. megaceros* [= *O. eschrichtii*]).

*Dolopichthys* (subgenus *Dermatias*) *digitatus* Regan and Trewavas, 1932:68, fig. 94 (original description, single specimen).

*Dolopichthys* (subgenus *Dermatias*) *simplex* Regan and Trewavas, 1932:68, fig. 96 (original description, single specimen).

*Dolopichthys* (subgenus *Dermatias*) *pollicifer* Regan and Trewavas, 1932:69, fig. 97 (original description, single specimen).

*Dolopichthys* (subgenus *Dermatias*) *diadematus* Regan and Trewavas, 1932:69, fig. 98 (original description, single specimen).

*Dolopichthys* (subgenus *Dermatias*) *brevifilis* Regan and Trewavas, 1932:69, fig. 99 (original description, single specimen).

*Dolopichthys* (subgenus *Dermatias*) *pennatus* Regan and Trewavas, 1932:69, fig. 100 (original description, single specimen).

*Dolopichthys* (subgenus *Dermatias*) *frondosus* Regan and Trewavas, 1932:70, fig. 101 (original description, single specimen).

*Dolopichthys* (subgenus *Dermatias*) *cirrifer* Regan and Trewavas, 1932:70, fig. 102 (original description, single specimen).

*Dolopichthys* (subgenus *Dermatias*) *tentaculatus*: Regan and Trewavas, 1932:70 (new combination; brief description after Beebe, 1932b, in key).

*Dolopichthys* (subgenus *Dermatias*) *megaceros*: Regan and Trewavas, 1932:71, fig. 103 (new combination; description, reference to holotype, in key).

*Dolopichthys* (subgenus *Dermatias*) *obtusus*: Regan and Trewavas, 1932:71 (new combination; brief description after Parr, 1927, in key).

*Dolopichthys* (subgenus *Dermatias*) *plumatus* Regan and Trewavas, 1932:71, fig. 104 (original description, single specimen).

*Dolopichthys* (subgenus *Dermatias*) *ptilotus* Regan and Trewavas, 1932:73, fig. 107 (original description, single specimen).

*Dolopichthys* (subgenus *Dermatias*) *multifilis* Regan and Trewavas, 1932:73, fig. 108 (original description, single specimen).

*Dolopichthys* (subgenus *Dermatias*) *claviger* Regan and Trewavas, 1932:73, fig. 109 (original description, two specimens, lectotype ZMUC P9270, designated by Bertelsen, 1951).

*Dolopichthys* (subgenus *Dermatias*) *thysanophorus* Regan and Trewavas, 1932:74, fig. 110 (in part; original description, two specimens, lectotype BMNH 1932.5.3.14 [= *Oneirodes flagellifer*], designated by Pietsch, 1974a; paralectotype ZMUC P9281, referred to *O. eschrichtii* by Pietsch, 1974a).

*Dolopichthys hibernicus* Fraser-Brunner, 1935:325, fig. 4 (original description, single specimen). Wheeler, 1969:585 (reference to original description).

*Oneiroides eschrichtii*: Fowler, 1936: 1139, 1140, 1337, fig. 479 (emendation of generic name; description after Günther, 1887; *Oneirodes megaceros* a synonym of *O. eschrichtii*; in key).

*Dolopichthys simplex*: Fowler, 1936: 1365 (listed, in key). Nielsen, 1974:95 (listed in type catalog).

*Dolopichthys diadematus*: Fowler, 1936:1366 (listed, in key). Nielsen, 1974: 92 (listed in type catalog).

*Dolopichthys frondosus*: Fowler, 1936: 1366 (listed, in key). Nielsen, 1974:93 (listed in type catalog).

*Dolopichthys cirrifer*: Fowler, 1936:1366 (listed, in key). Nielsen, 1974:92 (listed in type catalog).

*Dolopichthys ptilotus*: Fowler, 1936: 1366 (listed, in key). Maul, 1961:130 (a possible synonym of *Dolopichthys megaceros* [= *Oneirodes eschrichtii*]). Nielsen, 1974:95 (listed in type catalog).

*Oneirodes eschrichti* Group: Bertelsen, 1951:70, figs. 31E–G, 32–37 (in part; 22 nominal species grouped, common characters, all available material listed; description of larvae, males, females; in key). Grey, 1956a:244 (synonymy, distribution). Maul, 1961:122, figs. 18, 22 (in part; four additional specimens, description, comparison; discussion of possible valid species within *Oneirodes eschrichti* group). Maul, 1962a:17, figs. 7–9 (two additional specimens, description, discussion of proper identity). Bussing, 1965:223 (misidentification, single specimen referred by Pietsch, 1974a, to *O. carlsbergi*; description, comparison with *D. brevifilis* [= *O. eschrichtii*]). Taylor, 1967:2111 (misidentification, single specimen referred by Pietsch, 1974a, to *O. bulbosus*; description; opercular bones described, figured; esca figured).

*Oneirodes pollicifer*: Rass, 1955:334 (misidentification; specimen referred by Pietsch, 1974a, to *Oneirodes eschrichtii*; distribution, Kuril-Kamchatka Trench). Rass, 1967:240 (misidentification; after Rass, 1955).

*Dolopichthys plumatus*: Maul, 1961: 130 (a possible synonym of *Dolopichthys megaceros* [= *Oneirodes eschrichtii*]). Nielsen, 1974:95 (listed in type catalog).

*Dolopichthys digitatus*: Beaufort and Briggs, 1962:249 (description after Regan and Trewavas, 1932, in key). Nielsen, 1974:92 (listed in type catalog).

*Dolopichthys multifilis*: Beaufort and Briggs, 1962:249 (description after Regan and Trewavas, 1932, in key). Nielsen, 1974:94 (listed in type catalog).

*Dolopichthys thysanophorus*: Beaufort and Briggs, 1962:250 (in part; description after Regan and Trewavas, 1932; in key). Nielsen, 1974:95 (listed in type catalog).

*Dolopichthys brevifilis*: Nielsen, 1974:91 (listed in type catalog).

*Dolopichthys claviger*: Nielsen, 1974:92 (listed in type catalog).

*Dolopichthys pennatus*: Nielsen, 1974:95 (listed in type catalog).

*Dolopichthys pollicifer*: Nielsen, 1974: 95 (listed in type catalog).

Males and Larvae

Unknown.

Material

One hundred and fifteen metamorphosed females (10 to 213 mm):

Holotype of *Oneirodes eschrichtii*: ZMUC 64, 153 mm, west coast of Greenland.

Holotype of *Oneirodes megaceros*: NMI SR-497, 61 mm, 51°02′N, 11°36′W, 0 to 1454 m, 10 September 1907.

Holotype of *Dolopichthys obtusus*: YPM 2028, 13 mm, *Pawnee*, Third Oceanographic Expedition, station 59, 32°19′N, 64°32′W, 2440 m wire, 21 April 1927.

Holotype of *Dolopichthys tentaculatus*: USNM 170945 (originally NYZS 23170), 13.5 mm, Bermuda Oceanographic Expedition, station 1271, Bermuda, southeast of Nonsuch Island, 0 to 1097 m, 7 September 1931.

Holotype of *Dolopichthys* (subgenus *Dermatias*) *digitatus*: ZMUC P9272, 21 mm, *Dana* station 3768(1), north of New Guinea, 1°20′S, 138°42′E, 4000 m wire, 1340 h, 24 July 1929.

Holotype of *Dolopichthys* (subgenus *Dermatias*) *simplex*: ZMUC P9279, 26 mm, *Dana* station 4005(2), west of Cape Verde Islands, 13°31′S, 18°03′W, 3500 m wire, 1145 h, 12 March 1930.

Holotype of *Dolopichthys* (subgenus *Dermatias*) *pollicifer*: ZMUC P9277, 11.5 mm, *Dana* station 1208(4), Gulf of Panama, 6°48′N, 80°33′W, 3500 m wire, 0810 h, 16 January 1922.

Holotype of *Dolopichthys* (subgenus *Dermatias*) *diadematus*: ZMUC P9271, 11 mm, *Dana* station 3996(2), South Atlantic, 15°41′S, 5°50′W, 3000 m wire, bottom depth 3620 m, 1800 h, 25 February 1930.

Holotype of *Dolopichthys* (subgenus *Dermatias*) *brevifilis*: ZMUC P9268, 14 mm, *Dana* station 1209(2), Gulf of Panama, 7°15′N, 78°54′W, 3000 m wire, 1845 h, 17 January 1922.

Holotype of *Dolopichthys* (subgenus *Dermatias*) *pennatus*: ZMUC P9275, 58 mm, *Dana* station 1165(8), North Atlantic, 12°11′N, 35°49′W, 3000 m wire, 0800 h, 9 November 1921.

Holotype of *Dolopichthys* (subgenus *Dermatias*) *frondosus*: ZMUC P9273, 23.5 mm, *Dana* station 4006(1), west of Cape Verde, 15°31′N, 18°05′W, 1000 m wire, bottom depth 2425 m, 1830 h, 13 March 1930.

Holotype of *Dolopichthys* (subgenus *Dermatias*) *cirrifer*: ZMUC P9269, 29 mm, *Dana* station 4180(4), North Atlantic, 32°56′N, 23°47′W, 3500 m wire, bottom depth 5225 m, 1115 h, 8 June 1931.

Holotype of *Dolopichthys* (subgenus *Dermatias*) *plumatus*: ZMUC P9276, 10 mm, *Dana* station 1208(6), Gulf of Panama, 6°48′N, 80°33′W, 2500 m wire, 0810 h, 16 January 1922.

Holotype of *Dolopichthys* (subgenus *Dermatias*) *ptilotus*: ZMUC P9278, 10.5 mm, *Dana* station 4009(8), south of Canary Islands, 24°36′N, 17°27′W, 3000 m wire, bottom depth 2425 m, 0700 h, 18 March 1930.

Holotype of *Dolopichthys* (subgenus *Dermatias*) *multifilis*: ZMUC P9274, 17.5 mm, *Dana* station 3686(1), Sulu Sea, 8°34′N, 119°55′E, 1000 m wire, bottom depth 2725 m, 0045 h, 6 April 1929.

Lectotype of *Dolopichthys* (subgenus *Dermatias*) *claviger*: ZMUC P9270, 21.5 mm, *Dana* station 3556(1), between Panama and Galapagos, 21°52′N, 87°38′W, 2500 m wire, bottom depth 2285 m, 1530 h, 14 September 1928.

Paralectotype of *Dolopichthys* (subgenus *Dermatias*) *claviger*: BMNH 1932.5.3.13, 10.5 mm, *Dana* station 3558(2), west of Galapagos, 0°18′S, 99°07′W, 3000 m wire, bottom depth 3360 m, 0830 h, 18 September 1928.

Paralectotype of *Dolopichthys* (subgenus *Dermatias*) *thysanophorus*: ZMUC P9281, 18.5 mm, *Dana* station 3686(7), Sulu Sea, 8°34′N, 119°55′E, 3500 m wire, bottom depth 2725 m, 0745 h, 6 April 1929.

Holotype of *Dolopichthys hibernicus*: BMNH 1934.8.8.91, 23 mm, Commercial Trawler *Dynevor Castle*, 53°15′N, 12°28′W, 0 to 320 m, July 1934.

Additional material: BMNH, two (14.5 to 16 mm); BSKU, one (61 mm); CAS, two (13 to 41.5 mm); HUMZ, four (67 to 172 mm); IFAN, two (30 mm); ISH, 10 (25 to 188 mm); LACM, 11 (10.5 to 71 mm); MCZ, three (11 to 16.5 mm); MMF, two (57 to 95 mm); NMNZ, one (12 mm); ROM, three (101 to 213 mm); NMSZ, three (15.5 to 45.5 mm); NSMT, one (87 mm); SAIAB, one (100 mm); SAM, one (54 mm); SIO, 12 (14 to 121 mm); SIOM, two (21 to 61 mm); USNM, two (14 to 18.5 mm); UW, one (79 mm); ZIN, 10 (standard lengths not recorded); ZMUC, 21 (45 to 104 mm).

Distinguishing Characters

A species of *Oneirodes* distinguished from all other members of the genus in escal morphology. The esca of *Oneirodes eschrichtii* is most similar to that of *O. bulbosus* and *O. anisacanthus*; however, it lacks the numerous tapering distal filaments of the anterior appendage of *O. bulbosus* and the lateral escal appendage of *O. anisacanthus*. Although, the differences in the shape of the subopercular bone between *O. eschrichtii* and *O. bulbosus* (Bertelsen, 1951:83, fig. 31E–G), and the slightly higher jaw-tooth counts of the former, are helpful in distinguishing these two forms, the characters of the esca are the only satisfactory means of separating *O. eschrichtii* from this and most other species of the genus.

Description

Escal appendage pattern B (Fig. 113B); esca with a stout, internally pigmented anterior appendage, not longer than escal bulb, bearing distally a large compressed escal papilla, at the anterior base of which usually arise a pair of filaments shorter than length of anterior appendage; 1 or 2 additional, shorter filaments along anterior margin and several to many along posterior margin of anterior appendage; papilla and distal ends of some filaments of anterior appendage darkly pigmented in most specimens approximately 100 mm and larger (as well as the 61-mm holotype of *Oneirodes megaceros*); a pair of filamentous, medial escal appendages half as long as escal bulb in smaller specimens (about 15 mm) to more than twice the length of escal bulb in specimens of intermediate size (70 mm), less than half as long as escal bulb in larger specimens (150 mm); tips of tapering filaments of medial appendages darkly pigmented in a few large specimens (e.g., 134 mm, 188 mm); terminal escal papilla truncated, with a distal streak of pigment, or conical, with a distal spot of pigment (see Geographic variation, below); an unpigmented, unbranched, tapering posterior escal appendage, 1.5 to three times length of escal bulb in specimens less than approximately 100 mm, becoming proportionately shorter with further growth, less than length of escal bulb in 213-mm specimen; in some specimens, on each side, a dorsolaterally to anterolaterally placed, unpigmented, filamentous appendage shorter than length of medial appendages.

Subopercle without indentation on posterodorsal margin (Bertelsen, 1951:83, fig. 31E–G); length of ventral fork of opercle 20.7 to 31.9% SL; ratio of lengths of dorsal and ventral forks of opercle 0.44 to 0.69.

Epibranchial teeth absent; teeth present on pharyngobranchial II; total number of teeth in upper jaw 24 to 50, in lower jaw 29 to 56; number of teeth on vomer 4 to 8; dorsal-fin rays 5 to 7; anal-fin rays 4; pectoral-fin rays 15 to 19.

Measurements in percent of standard length: head length 33.3 to 46.2; head depth 34.7 to 55.4; premaxilla length 22.4 to 38.9; lower-jaw length 34.3 to 55.7; illicium length 17.8 to 38.6.

Additional description as given for the genus and family.

### Sexual Maturity

The ovaries of several large specimens of *Oneirodes eschrichtii* (121 to 213 mm) are large and tightly packed with eggs. The left ovary of the largest known specimen (213 mm, ROM 27277) is approximately 85 mm long (40% SL). Some of these ripe females were captured considerably outside the larval distribution of ceratioids, described by Bertelsen (1951:224) to be limited to the warmer parts of the oceans between approximately 40°N and 35°S.

### Distribution

*Oneirodes eschrichtii* is the only member of the genus with a nearly cosmopolitan distribution in the Atlantic, Pacific, and Indian oceans (Fig. 223). It is known from both sides of the Atlantic, in the west from off Greenland (as far north as about 66°N) south to Bermuda, and in the east from the Irish Atlantic Slope as far south as 40°S, between 43° and 7°W. It has also been collected in the western Indian Ocean, from the Gulf of Aden and Arabian Sea. In the western Pacific, it has been collected from the Philippines and South China Sea, and north to Japan and the Kuril-Kamchatka Trench at 49°N, 158°E, and south to New South Wales, Tasmania, and New Zealand. In the eastern Pacific it extends from central equatorial waters, at about 150°W, to the west coast of the Americas from 33°N to 34°S, including the Gulf of Panama. Depths of capture are evenly distributed between 750 and 2500 m.

### Geographic Variation

There appears to be a disjunct and morphologically distinct population of *Oneirodes eschrichtii* present in the waters off southern California (Pietsch, 1974a). Morphological differences in the esca are apparent: the medial escal appendages of the southern California population are generally longer and more highly branched than those of specimens collected from other areas; the terminal escal papilla is considerably larger in southern California specimens and bears a distal spot of pigment rather than a distal streak of pigment. Several morphometric characters average higher in the southern California population (see Pietsch, 1974a: 51, figs. 67–70, tables 7, 8).

### Comments

Bertelsen (1951) erected the *Oneirodes eschrichti* group to include 22 nominal species most of which were described by Regan and Trewavas (1932) on the basis of one or two juvenile female specimens less than 25 mm, and on relatively minute differences in the morphology of the esca. With a better understanding of individual and ontogenetic variation since Bertelsen's (1951) monograph, these differences in escal morphology are, in most cases, regarded as variations shown by widely distributed conspecific populations (Pietsch, 1974a). In the absence of significant differences, the maintenance of specific distinction for 18 forms previously included in the *O. eschrichti* group is considered unjustified. These were synonymized with *O. eschrichtii* by Pietsch (1974a).

*Oneirodes thysanophorus* (Regan and Trewavas, 1932) was described on the basis of two specimens (12.5 to 18.5 mm), both collected at the same locality in the Sulu Sea (*Dana* station 3686). The smaller of these specimens (BMNH 1932.5.3.14), the lectotype of *O. thysanophorus*, was referred to *O. flagellifer* by Pietsch (1974a; see Comments under the latter species). The 18.5-mm specimen (ZMUC P9281) has an esca that compares well with that of *O. eschrichtii* and is retained within the material of the latter.

## ONEIRODES SABEX PIETSCH AND SEIGEL, 1980

Figure 225; Table 8

### Females

*Oneirodes eschrichtii*: Pietsch, 1974a:100, 103, fig. 116B (misidentification).

*Oneirodes sabex* Pietsch and Seigel, 1980:387, figs. 9, 10 (original description, 14 specimens). Bertelsen and Pietsch, 1983:85, fig. 6 (two additional specimens, Australia). Pietsch, 1999:2032 (western central Pacific).

*Oneirodes* sp.: Yamakawa, 1984:291, fig. 205 (two specimens, Okinawa Trough).

*Oneirodes appendixus* Ni and Xu, in Ni, 1988:332, fig. 259 (original description, single specimen). Meng et al., 1995:444, fig. 596 (after Ni, 1988).

*Oneirodes sebax*: Stewart and Pietsch, 1998:17, 35, fig. 11 (misspelling of specific name; three additional specimens, New Zealand).

### Males and Larvae

Unknown.

### Material

Twenty-four metamorphosed females (12 to 189 mm):

Holotype of *Oneirodes sabex*: LACM 36116-3, 46 mm, *Alpha Helix* station 84, Banda Sea, 5°04.5′S, 130°12.0′E, 0 to 1500 m, 1400 to 2100 h, 28 April 1975.

Paratypes of *Oneirodes sabex*: AMS I.20315-010, 32.5 mm, *Kapala*, 33°53′S, 152°02′E, 0 to 900 m, bottom depth 1800 m, 1330 to 1990 h, 14 December 1977. AMS I.20314-016, 39 mm, *Kapala*, 33°28′S, 152°33′E, 0 to 900 m, bottom depth 4200 m, 0530 to 1045 h, 14 December 1977. LACM 36087-4, 3 (12 to 26.5 mm), *Alpha Helix* station 135, Halmahera Sea, 0°06.2′S, 128°38.3′E, 820 to 1000 m, 2306 to 0106 h, 18 May 1975. LACM 36068-3, 12 mm, *Alpha Helix* station 25, Banda Sea, 4°16.8′S, 129°34.4′E, 0 to 1800 m, 2010 to 0340 h, 28 March 1975. LACM 36028-5, 13 mm, *Alpha Helix* station 141, Halmahera Sea, 0°05.0′S, 128°52.7′E, 1000 to 1100 m, 0625 to 0855 h, 20 May 1975. LACM 36023-3, 13 mm, *Alpha Helix* station 143, Halmahera Sea, 0°14.5′S, 128°46.7′E, 1250 to 1500 m, 1715 to 1930 h, 20 May 1975. LACM 36089-4, 2 (15 to 17 mm), *Alpha Helix* station 137, Halmahera Sea, 0°08.9′S, 128°40.0′E, 0 to 960 m, 0955 to 1300 h, 19 May 1975. LACM 36051-3, 15 mm, *Alpha Helix* station 38, Banda Sea, 4°40.4′S, 129°39.0′E, 0 to 780 m, 0320 to 0530 h, 12 April 1975. LACM 36088-4, 15.5 mm, *Alpha Helix* station 136, Halmahera Sea, 0°17.4′S, 128°47.5′E, 1000 to 1250 m, 0446 to 0646 h, 19 May 1975. SIO 70-339, 121 mm, *Melville*, cruise Antipode 4, station 69A, trawl 21, Philippine Sea, 19°35′N, 122°57′E, 3-m Isaacs-Kidd midwater trawl 0 to 1450 m, bottom depth 2625 m, 1845 to 0225 h, 15 to 16 September 1970.

Holotype of *Oneirodes appendixus*: ECFR E-1199, 141 mm, 29°50′N, 127°43′E, 423 m, 27 June 1981.

Additional females, AMS, two (12.5 to 50 mm); ASIZP, two (107 to 189 mm); BSKU, two (138 to 149 mm); NMNZ, three (52 to 135 mm).

### Distinguishing Characters

A species of *Oneirodes* distinguished from all other members of the genus in escal morphology: anterior appendage short, laterally compressed, without pigmented internal tube; medial appendages present; posterior appendage short, simple, cylindrical to laterally compressed.

### Description

Escal appendage pattern B (Fig. 113B); esca with anterior appendage strongly compressed, rounded in lateral view, darkly pigmented along distal margin in some specimens; a pair of filamentous medial appendages (so tiny in some specimens that careful observation is required); terminal papilla usually with two distal pigment spots situated on midline, one just behind the other; posterior appendage short, stout, and cylindrical in smaller specimens, laterally compressed in larger specimens; anterolateral appendages absent; escal pore especially large in larger specimens (Pietsch, 1974a, fig. 116B).

Subopercle short and broad, dorsal end tapering to a point, without indentation on posterodorsal margin; length of ventral fork of opercle 24.2 to 27.1% SL; ratio of lengths of dorsal and ventral forks of opercle 0.48 to 0.64. Upper fork of opercle supported by 2 or 3 bony ribs in some larger specimens (e.g., BSKU 29876, 138 mm; BSKU 33971, 149 mm).

Epibranchial teeth absent; teeth present on pharyngobranchial II; number of teeth in upper jaw 26 to 52, in lower jaw 36 to 50; number of teeth on vomer 4 to 10; dorsal-fin rays 5 or 6; anal-fin rays 4; pectoral-fin rays 14 to 17.

Measurements in percent of standard length: head length 32.2 to 39.5; head depth 34.2 to 41.5; premaxilla length 26.0 to 32.1; lower-jaw length 33.2 to 45.3; illicium length 14.4 to 26.5.

Additional description as given for the genus and family.

### Distribution

*Oneirodes sabex* is known only from southeast Asian, Indonesian, eastern Australian, and New Zealand waters: the *Alpha Helix* material, including the holotype and 13 paratypes, was collected in the Halmahera and Banda seas; the 121-mm paratype (SIO 70-339) is from off Luzon, Philippines; the 32.5- and 39-mm paratypes (AMS I. 20315-010, AMS I.20314016) were collected off Sydney, Australia (Fig. 225). The type specimens of *O. appendixus* are from the East China Sea. Several of the *Alpha Helix* specimens were collected with an opening-closing trawl, three (12 to 26.5 mm) between 820 and 1000 m, two (13 to 15.5 mm) between 1000 and 1250 m, and one between 1250 and 1500 m. Of the remaining material, 10 were captured somewhere between the surface and 960 m, and the rest between the surface and 1800 m.

## ONEIRODES BULBOSUS CHAPMAN, 1939

Figure 223; Table 8

### Females

*Oneirodes bulbosus* Chapman, 1939:538, fig. 70 (original description, single specimen; comparison with *Oneirodes eschrichtii*). Clemens and Wilby, 1946:338, fig. 253 (description after Chapman, 1939). Rass, 1955:334, 336 (distribution; Kuril-Kamchatka Trench, Alaska, northern California). Maul, 1961:130 (comparison with other species of *Oneirodes*). Rass, 1967:233, 240, 241 (after Rass, 1955). Ueno, 1971:102 (after Rass, 1955). Pietsch, 1974a:52, figs. 71, 72, 109 (additional material, description, comparison with all known material). Marshall, 1979:490 (distribution). Amaoka, 1983:250, 251, 325, fig. 141 (additional specimen, Okhotsk Sea off Hokkaido). Amaoka, 1984:106, fig. 19, pl. 92-D (Japan). Peden et al., 1985:9, fig. 35 (eastern North Pacific). Amaoka et al., 1995:113, fig. 173 (northern Japan). Sheiko and Fedorov, 2000:24 (Kamchatka). Mecklenburg et al., 2002:305, 310 (Alaska, in key). Nakabo, 2002:473, figs. (Japan, in key). Fedorov et al., 2003:55, fig. 83 (northern Sea of Okhotsk). Love et al., 2005:60 (eastern North Pacific). Orlov and Tokranov, 2005:271, fig. 1 (distribution and size composition).

*Oneirodes eschrichti* Group: Bertelsen, 1951:79 (in part). Taylor, 1967:2111, figs. 3, 4 (misidentification; description of an additional specimen; esca, opercular bones described, figured).

### Males and Larvae

Unknown.

### Material

One hundred and fifteen metamorphosed females (30 to 160 mm):

Holotype of *Oneirodes bulbosus*: USNM 108149, 57 mm, International Fisheries Commission station 1109C, British Columbia, off Graham (or Frederick) Island 53°50′N, 133°54′W, 900 m wire out, 11 March 1934.

Additional material: HUMZ, 26 (35 to 148 mm); NSMT, 20 (21 to 125 mm); SIOM, three (64 to 96 mm); UBC, one (47.5 mm); UW, 37 (30 to 160 mm); ZIN, 27 (47 to 110 mm).

### Distinguishing Characters

A species of *Oneirodes* distinguished from all other members of the genus in escal morphology. *Oneirodes bulbosus* is most similar to *O. eschrichtii* and *O. anisacanthus*. In addition to differences in escal morphology, the shape of the subopercle of most specimens of *O. bulbosus* differs from that of these forms and all other species of the genus, with the exception of *O. thompsoni*. The posterior margin of the upper part of this bone is indented to deeply notched in all specimens of *O. bulbosus* examined except for the holotype. *Oneirodes bulbosus* differs further from *O. anisacanthus* in having slightly fewer teeth in the jaws.

### Description

Escal appendage pattern B (Fig. 113B); esca with a stout, internally pigmented anterior appendage bearing distally numerous tapering filaments, some of which may be twice length of escal bulb and darkly pigmented distally; a pair of highly branched, unpigmented, tapering medial escal filaments, two to nearly five times length of escal bulb in large specimens (81 mm); a short, rounded terminal escal papilla with a distal spot of pigment in specimens 57 mm and larger; and an unpigmented, tapering posterior escal appendage, nearly twice length of escal bulb, and usually bearing 1 or 2 small, unpigmented filaments at one-half to one-third distance from distal tip; lateral and anterolateral escal appendages absent.

Subopercle with posterior margin of dorsal end usually indented to deeply notched; length of ventral fork of opercle 28.6 to 35% SL; ratio of lengths of dorsal and ventral forks of opercle 0.55 to 0.60.

Epibranchial teeth absent; teeth present on pharyngobranchial II; total number of teeth in upper jaw 23 to 39, in lower jaw 24 to 40; number of teeth on vomer 4 to 8; dorsal-fin rays 6 or 7; anal-fin rays 4; pectoral-fin rays 15 to 18.

Measurements in percent of standard length: head length 37.6 to 49.1; head depth 39.1 to 50.8; premaxilla length 29.1 to 38.6; lower-jaw length 44.1 to 56.1; illicial length 23.1 to 38.6.0.

Additional description as given for the genus and family.

### Distribution

*Oneirodes bulbosus* is known throughout the North Pacific and Bering Sea above 50°N, with a single record farther south in the eastern North Pacific at about 43°N, 148°W (Fig. 223). In the west it extends south along the Kuril-Kamchatka Trench as far south as about 38°N off Honshu, Japan. Based on maximum depths reached by gear, it appears to inhabit relatively shallow depths, the bulk of the population concentrated between 600 and 850 m, with perhaps a second peak below approximately 950 m (see Pietsch, 1974a).

## ONEIRODES ANISACANTHUS (REGAN, 1925c)

Figure 224; Table 8

### Females

*Dolopichthys anisacanthus* Regan, 1925c:562 (original description, four specimens, lectotype ZMUC P9267, designated by Bertelsen, 1951). Regan, 1926:29 (synonymized with *Dolopichthys megaceros* [= *Oneirodes eschrichtii*] without comment). Fowler, 1936:1365 (listed, in key). Maul, 1949:39, figs. 13–17 (*O. eschrichtii sensu lato* of Maul, 1949, a possible synonym of *D. anisacanthus*; description, osteology). Bertelsen, 1951:267 (lectotype designated). Maul, 1961:129, figs. 19–21 (*Oneirodes* sp. of *O. eschrichtii* group of Maul, 1961, includes *O. eschrichtii sensu lato* of Maul, 1949; five specimens, the largest referred to *O. anisacanthus* by Pietsch, 1974a, remaining specimens to *O. eschrichtii*). Nielsen, 1974:91 (listed in type catalog). Marshall, 1979:257, fig. 103 (figure after Pietsch, 1974a).

*Dolopichthys megaceros*: Regan, 1926:29 (in part; *Dolopichthys anisacanthus*, a synonym of *D. megaceros*).

*Dolopichthys heteracanthus*: Norman, 1930:353 (in part; misidentifications; two specimens, the larger referred by Regan and Trewavas, 1932, to *Dolopichthys carlsbergi* [= *Oneirodes carlsbergi*], the smaller to *D. anisacanthus* [= *O. anisacanthus*]).

*Dolopichthys* (subgenus *Dermatias*) *anisacanthus*: Regan and Trewavas, 1932:72, text fig. 105, pl. 2, fig. 2 (description, three additional specimens, in key).

*Oneirodes eschrichtii sensu lato*: Maul, 1949:34, figs. 13–17 (description, osteology; a possible synonym of *Dolopichthys anisacanthus*).

*Oneirodes eschrichti* Group: Bertelsen, 1951:79 (in part).

*Oneirodes* sp. of *Oneirodes eschrichti* Group: Maul, 1961:122, figs. 19–21 (in part; *Oneirodes eschrichtii sensu lato* of Maul, 1949; five specimens, the largest referred to *O. anisacanthus* by Pietsch, 1974a, remaining specimens to *O. eschrichtii*; description in part after Maul, 1949).

*Oneirodes anisacanthus*: Pietsch, 1974a:54, figs. 73, 74, 106 (new combination, additional material, description, comparison with all known material). Bertelsen and Pietsch, 1977:174, figs. 1, 2 (*Walther Herwig* material). Bertelsen, 1986:1394, figs. (after Bertelsen and Pietsch, 1977; eastern North Atlantic). Bertelsen, 1990:504 (synonymy, eastern tropical Atlantic). Jónsson and Pálsson, 1999:201, fig. 5 (Iceland). Anderson and Leslie, 2001:15, fig. 12A (additional female, off Durban, South Africa). Pietsch, 2002b:1064 (western central Atlantic).

### Males and Larvae

Unknown.

### Material

Eleven metamorphosed females (10.5 to 173 mm):

Lectotype of *Oneirodes anisacanthus*: ZMUC P9267, 27 mm, *Dana* station 1152(3), North Atlantic, 30°17′N, 20°44′W, 3000 m wire, 1930 h, 22 October 1922.

Additional material: BMNH, three (16 to 47 mm); MMF, one (173 mm); ISH, two (78 to 98 mm); SAM, one (51 mm); UF, two (10.5 to 13 mm); ZMUC, one (39 mm).

### Distinguishing Characters

A species of *Oneirodes* distinguished from all other members of the genus in escal morphology. It is most similar to *Oneirodes eschrichtii* and *O. bulbosus* but differs in having a well-developed lateral escal appendage.

### Description

Escal appendage pattern B (Fig. 113B); esca with a stout anterior appendage as long as or slightly longer than escal bulb, bearing 2 unpigmented, tapering filaments on anterior margin near distal tip, and numerous shorter filaments along anterior, posterior, and lateral margins, with distal tips of some darkly pigmented in specimens 78 mm and larger; proximal half of anterior escal appendage internally pigmented; a series of short, unpigmented filaments arranged along distal midline of escal bulb, usually three distinct groups, the posterior-most group paired; a truncated terminal escal papilla, with a distal streak of pigment in specimens 47 mm and larger; an unpigmented posterior escal appendage, approximately half length of escal bulb, most specimens bearing 2 to 5 short, unpigmented filaments near distal tip; an unpigmented, tapering, unbranched lateral escal appendage less than one-third length of escal bulb in specimens 47 mm and smaller, unbranched and nearly six times length of escal bulb in 78-mm specimen, and branched and nearly seven times length of escal bulb in 173-mm specimen (see Maul, 1949:37, fig. 17); anterolateral escal appendages absent.

Subopercle without indentation on posterodorsal margin; length of ventral fork of opercle 24.9 to 29.6% SL; ratio of lengths of dorsal and ventral forks of opercle 0.44 to 0.60.

Epibranchial teeth absent; teeth present on pharyngobranchial II; total number of teeth in upper jaw 21 to 53, in lower jaw 26 to 54; number of teeth on vomer 5 to 8; dorsal-fin rays 6 or 7; anal-fin rays 4; pectoral-fin rays 15 to 19.

Measurements in percent of standard length: head length 38.2 to 43.6; head depth 38.5 to 42.9; premaxilla length 27.3 to 34.0; lower-jaw length 43.2 to 50.6; illicial length 20.8 to 30.8.

Additional description as given for the genus and family.

### Sexual Maturity

The largest known female of *Oneirodes anisacanthus* (173 mm, MMF 3101), appears to be sexually mature; the right ovary is about 40 mm long (23.1% SL) and contains numerous eggs measuring approximately 0.08 to 0.10 mm in diameter.

### Distribution

*Oneirodes anisacanthus* is wide spread in the Atlantic Ocean, ranging from about 65°N off East Greenland to the Caribbean Sea in the west and Madeira (type locality) and the Gulf of Guinea in the east, with a single record in the far south off Cape Town, South Africa (Fig. 224). A 47-mm female was captured with a closing net between 900 and 1040 m. The remaining vertical distributional data are too few to allow any reasonable conclusions.

### Comments

*Oneirodes anisacanthus* was described originally from four individuals, only one of which actually represents this species: two of the specimens (*Dana* stations 1165 and 1183) were referred by Pietsch (1974a) to *O. eschrichtii*; the specimen from *Dana* station 1256 was described as new by Regan and Trewavas (1932) and remains as the holotype of *O. macronema* (see Pietsch, 1974a:75).

The morphology of the esca of the most recently reported specimen of *O. anisacanthus*, a 98-mm female collected by the *Walther Herwig* in the North Atlantic (see Bertelsen and Pietsch, 1977:174), does not fit the sequence of ontogenetic change described for this species by Pietsch (1974a:55). In previously documented material, the lateral escal appendage is less than one-third the length of the escal bulb in specimens 47 mm and smaller, nearly six times the length of the bulb in a 78-mm specimen, and nearly seven times the length of the bulb in a 173-mm specimen. In contrast, the lateral appendage of the 98-mm specimen is approximately one-third the length of the escal bulb. In all other ways, including counts, measurements, and the shape of the subopercle, the 98-mm specimen compares well with the known material of *O. anisacanthus*.

## ONEIRODES KREFFTI PIETSCH, 1974A

Figure 224; Table 8

### Females

*Oneirodes kreffti* Pietsch, 1974a:57, figs. 75, 76, 107 (original description, three specimens). Bertelsen and Pietsch, 1977:175, figs. 1, 3 (additional specimen, Indian Ocean). Bertelsen and Grobecker, 1980:66 (comparison with *Oneirodes posti*). Bertelsen and Pietsch, 1983:84, fig. 5 (additional specimen, off Sydney, Australia). Pietsch, 1986a:375, fig. 106.1 (after Bertelsen and Pietsch, 1983; southern Africa). Stewart and Pietsch, 1998:17, fig. 10 (four additional specimens, New Zealand). Anderson and Leslie, 2001:17, fig. 12C (*Walther Herwig* material, southern Africa).

*Oneirodes whitleyi* Bertelsen and Pietsch, 1983:85, fig. 7 (original description, three specimens). Roberts, 1991:8 (two additional specimens; Chatham Rise, New Zealand). Stewart and Pietsch, 1998:17, fig. 12 (11 additional specimens, New Zealand). Anderson and Leslie, 2001:17, fig. 12D (three additional females, southern Africa).

### Males and Larvae

Unknown.

### Material

Thirty-eight females, 37 metamorphosed (11 to 126 mm) and one in metamorphosis (8.5 mm):

Holotype of *Oneirodes kreffti*: ISH 1536/71, 50 mm, *Walther Herwig* station 431 III/71, 30°04′S, 5°22′E, CMBT-1600, 0 to 500 m, 2130 to 2252 h, 31 March 1971.

Paratypes of *Oneirodes kreffti*: ISH 1463/71, 53.5 mm, *Walther Herwig* station 427/71, 33°00′S, 7°50′E, CMBT to 1600, 0 to 2000 m, 1925 to 2343 h, 30 March 1971. MCZ 47554, 21 mm, *Anton Bruun*, cruise 3, station 160, trawl 27, APB 7133, 41°07′S, 59°52′E, 3-m Isaacs-Kidd midwater trawl, with Foxton closing device, deep fraction, 150 to 635 m, 1725 to 2105 h, 12 September 1963.

Holotype of *Oneirodes whitleyi*: AMS I.20066-003, 30 mm, east of Brush Island, New South Wales, 35°36′S, 150°55′E, 0 to 650 m, bottom depth 2000 m, 27 October 1977.

Paratypes of *Oneirodes whitleyi*: AMS I.19608-045, 11 mm, off Sydney, New South Wales, 34°05′S, 151°56′E, 0 to 800 m, bottom depth 2410 to 2920 m, 23 March 1971. AMS I.20066-070, 22 mm, east of Brush Island, New South Wales, 35°36′S, 150°55′E, 0 to 650 m, bottom depth 2000 m, 27 October 1977.

Additional material: AMS, three (8.5 to 50 mm); CSIRO, seven (47 to 94 mm); MCZ, one (101 mm); MNHN, one (67 mm); NMNZ, 17 (13 to 111 mm); SAIAB, two (25 to 126 mm); SAM, one (31 mm).

### Distinguishing Characters

A species of *Oneirodes* distinguished from all other members of the genus in escal morphology: a pair of highly branched medial escal appendages, in combination with an elongate cylindrical lateral escal appendage on each side. *Oneirodes kreffti* is most similar to *O. anisacanthus*. These two species are the only members of the genus that possess a well-developed lateral escal appendage. The esca of *O. kreffti*, however, bears a pair of stout, medial appendages that are absent in *O. anisacanthus*. In addition, *O. kreffti* appears to be a more globose form, having a deeper and longer head, and a longer lower jaw than *O. anisacanthus*.

### Description

Escal appendage pattern B (Fig. 113B); esca with a stout, internally pigmented anterior appendage, not longer than escal bulb, typically bearing a conical distal tip, with 2 or more tapering filaments near distal tip and usually several similar filaments along posterior margin; a pair of highly branched, tapering, unpigmented medial escal filaments, less than length of escal bulb in smaller specimens, to more than eight times length of escal bulb in larger specimens; in some specimens medial escal appendages bifurcating from base; a truncated terminal escal papilla with a distal streak of pigment; an unbranched, tapering posterior escal appendage, less than length of escal bulb to more than two times length of escal bulb, sometimes slightly expanded at distal tip; on each side, an unpigmented, tapering lateral escal appendage, unbranched, bifurcated or trifurcated, and considerably longer than medial filaments in larger specimens; anterolateral escal appendage absent; medial and lateral escal appendages laterally compressed in larger specimens.

Subopercle short, dorsal end rounded or tapering to a point, without indentation on posterodorsal margin; length of ventral fork of opercle 28.6 to 31.9% SL; ratio of lengths of dorsal and ventral forks of opercle 0.47 to 0.53.

Epibranchial teeth absent; teeth present on pharyngobranchial II; total of number of teeth in upper jaw 16 to 55, in lower jaw 24 to 50; number of teeth on vomer 4 to 8; dorsal-fin rays 5 or 6; anal-fin rays 4; pectoral-fin rays 16 to 18. Measurements in percent of standard length: head length 38.6 to 64.3; head depth 43.9 to 50.5; premaxilla length 30.7 to 35.5; lower-jaw length 45.5 to 52.3; illicial length 22.3 to 27.3.

Additional description as given for the genus and family.

### Distribution

The known material of *Oneirodes kreffti* has all been taken in southern latitudes of the eastern South Atlantic, Indian, and western Pacific oceans, between 18° and 45°S, and 0° and 180°E (Fig. 224). The northernmost record is off Port Hedland, western Australia, at about 18°S; the southernmost is on the Pacific side of the South Island of New Zealand, at about 45°S. Nearly all the specimens were taken in depths of less than 1000 m, many of them in bottom trawls.

### Comments

*Oneirodes whitleyi* was described by Bertelsen and Pietsch (1983) on the basis of three small females (11 to 30 mm) from off New South Wales. At that time, *O. kreffti* Pietsch (1974a) was known from four specimens, all relatively large (50 to 101 mm) except for a 21-mm paratype, with a somewhat damaged esca. The large increase in material in the past two decades, especially a good number of specimens of intermediate and larger size, clearly shows that the esca undergoes considerable ontogenetic change. The length and degree of branching of the medial and lateral escal appendages increase dramatically between about 35 and 50 mm, filling the previously assumed morphological gap between these two nominal species. Therefore, in the absence of any distinguishing feature, and considering the complete overlap in geographic distribution of the two, *O. whitleyi* is hereby placed in the synonymy of *O. kreffti*.

## ONEIRODES POSTI BERTELSEN AND GROBECKER, 1980

Table 8

### Females

*Oneirodes posti* Bertelsen and Grobecker, 1980:63, fig. 1 (original description, two specimens).

### Males and Larvae

Unknown.

### Material

Three metamorphosed females (36.5 to 135 mm):

Holotype of *Oneirodes posti*: ISH 3184/79, 112 mm, *Anton Dohrn* station 345/79, eastern North Atlantic, 35°24′N, 32°01′W, MT 1600, 0 to 1800 m, 29 April 1979.

Paratype of *Oneirodes posti*: MCZ 51294, 36.5 mm, *Chain* station 105, 52°22′N, 34°51′W, 3-m Isaacs-Kidd midwater trawl, 770 to 850 m, 30 June 1972.

Additional material: MCZ, one (135 mm).

### Distinguishing Characters

A species of *Oneirodes* differing from all previously described species in escal morphology: the heavy black pigmentation of the escal appendages (especially well developed in the 36.5-mm paratype) is unique among species of *Oneirodes*. Most similar to *Oneirodes krefft*, known from the South Atlantic, South Pacific, and Indian Ocean, it shares with the latter species a short anterior escal appendage and a pair of extremely large, branched medial escal appendages. However, the two species differ very distinctly in escal pigmentation and development of the lateral escal appendages.

### Description

Escal appendage pattern B (Fig. 113B); esca with an internally pigmented anterior appendage shorter than half diameter of escal bulb, that of holotype bearing small secondary appendages on anterior and posterior margins, pigmented distally (simple, with only a small posterior spot in paratype); a pair of branched medial filaments, length two to three times diameter of bulb, with stout secondary appendages emerging near posterior margin of base, some proximal secondary appendages bifurcated or trifurcated, distal secondary appendages simple; simple or branched filaments emerging anterolaterally on each side of escal bulb and medially between anterior appendage and medial pair, about 5 in each series, most all heavily pigmented; a truncated, internally pigmented terminal escal papilla with a distal streak of pigment, continuous on each side with a larger proximal, anterolateral pigment spot; posterior escal appendage cylindrical, tapering, unpigmented, length less than one-fourth diameter of escal bulb, with 2 small secondary appendages near distal tip in holotype, simple and about equal to diameter of escal bulb in paratype; except for anterior and posterior escal appendages (and anterior-most anterolateral filaments and distal tips of medial appendages of paratype) all appendages and filaments darkly pigmented throughout length.

Subopercle short and broad, ventral end semicircular, dorsal end tapering in holotype, rounded in paratype, without posterior indentation; length of ventral fork of opercle 25.2 to 29.5% SL; ratio of lengths of dorsal and ventral forks of opercle 0.38 to 0.54.

Epibranchial teeth absent; teeth present on pharyngobranchial II; total number of teeth in upper jaw 30 to 34, in lower jaw 40 to 41; number of teeth on vomer 6 to 8; dorsal-fin rays 5 or 6; anal-fin rays 4; pectoral-fin rays 16 or 17.

Measurements in percent of standard length: head length 36.3 to 42.5; head depth 37.0 to 49.3; premaxilla length 27.4 to 38.5; lower-jaw length 42.6 to 51.3; illicium length 13.7 to 25.6.

### Sexual Maturity

The ovaries of the holotype are about 23 mm long and 12 mm wide, and filled with eggs approximately 0.18 to 0.20 mm in diameter.

### Distribution

All three known specimens were caught between 30° and 40°W in the northern part of the warm-water region of the North Atlantic in pelagic nets fished open in depths of about 800 and 1800 m. The 135-mm specimen (MCZ 162998) was collected with 4000 m of wire out.

### Comments

In agreement with observations of many other ceratioids, the small paratype of *Oneirodes posti* differs from the larger holotype in having a relatively larger head, longer jaws, a shorter, distally more rounded subopercle, and longer sphenotic, quadrate, and articular spines. Despite the great difference in standard length between the 36.5- and 112-mm specimens, they are remarkably similar in tooth counts and in the branching, pigmentation, and relative lengths of the escal appendages. The esca of the 135-mm specimen is damaged and cannot be easily compared in detail with those of the types.

### ONEIRODES MYRIONEMUS PIETSCH, 1974A

Table 8

#### Females

*Oneirodes myrionemus* Pietsch, 1974a:58, figs. 77, 78, 110 (original description, two specimens). Bertelsen and Pietsch, 1977:176, fig. 1 (two additional specimens tentatively identified as *Oneirodes myrionemus*). Bertelsen, 1986:1397, figs. (after Bertelsen and Pietsch, 1977; eastern North Atlantic). Bertelsen, 1990:506 (synonymy, eastern tropical Atlantic). Jónsson and Pálsson, 1999:202, fig. 5 (Iceland).

#### Males and Larvae

Unknown.

#### Material

Two metamorphosed females (43 to 121 mm) and two tentatively identified specimens (76 to 137 mm):

Holotype of *Oneirodes myrionemus*: ISH 3100a/71, 43 mm, *Walther Herwig* station 512/71, 32°47′N, 16°24′W, CMBT-1600, 0 to 1800 m, 1945 to 2348 h, 22 April 1971.

Paratype of *Oneirodes myrionemus*: ISH 3100b/71, 121 mm, *Walther Herwig* station 512/71, data as given for holotype.

Tentatively identified females: ISH, two (76 to 137 mm).

### Distinguishing Characters

In addition to unique differences in escal morphology, *Oneirodes myrionemus* is distinguished from all its congeners by having the following combination of character states: a relatively short head (shorter than that of all other species of the genus, with the exception of *O. rosenblatti*) and a short illicium (shorter than that of all other species of the genus, with the exception of *O. luetkeni*), and a short and broad subopercular bone.

### Description

Escal appendage pattern B (Fig. 113B); esca with a stout, internally pigmented anterior appendage, approximately twice length of escal bulb, bearing distally 2 or 3 unpigmented, tapering filaments (shorter than to more than twice length of anterior appendage), and 1 or 2 branches or papillae (darkly pigmented in 121-mm paratype); numerous, highly filamentous, medial escal appendages in 43-mm holotype; a single unpaired, highly filamentous medial escal appendage in 121-mm paratype; a large, truncated terminal escal papilla with a distal spot of pigment; an unpigmented, tapering posterior escal appendage approximately as long as escal bulb and bearing 2 small filaments in 43-mm holotype; lateral escal appendages absent; 4 highly branched, unpigmented, anterolateral escal appendages on each side.

Subopercle relatively short and broad, without indentation on posterodorsal margin; length of ventral fork of opercle 23.1 to 27.9% SL; ratio of lengths of dorsal and ventral forks of opercle 0.42 to 0.52.

Epibranchial teeth absent; teeth present on pharyngobranchial II; total number of teeth in upper jaw 31 to 33, in lower jaw 38 to 40; number of teeth on vomer 4 to 6; dorsal-fin rays 5 or 6; anal-fin rays 4; pectoral-fin rays 18.

Measurements in percent of standard length: head length 33.1 to 39.5; head depth 40.0 to 51.2; premaxilla length 27.3 to 32.6; lower-jaw length 38.0 to 50.0; illicial length 16.1 to 22.1.

Additional description as given for the genus and family.

### Distribution

Both type specimens of *Oneirodes myrionemus* were captured in the same haul in the eastern North Atlantic at 32°47′N, 16°24′W, between the surface and 1800 m.

### Comments

*Oneirodes myrionemus* was originally described from two specimens (43 to 121 mm) collected by the *Walther Herwig* at the same locality in the eastern North Atlantic (32°N, 16°W). A third specimen (137 mm,

Subopercle without indentation on posterodorsal margin (Pietsch, 1969:367, fig. 3); length of ventral fork of opercle 24.8 to 32.7% SL; ratio of lengths of dorsal and ventral forks of opercle 0.53 to 0.71.

Epibranchial teeth absent; teeth present on pharyngobranchial II; total number of teeth in upper jaw 20 to 55, in lower jaw 24 to 51; number of teeth on vomer 4 to 8; dorsal-fin rays 5 to 7; anal-fin rays 4; pectoral-fin rays 15 to 18.

Measurements in percent of standard length: head length 34.2 to 44.5; head depth 35.2 to 53.6; premaxillary length 27.3 to 33.9; lower-jaw length 38.6 to 51.8; illicium length 18.6 to 28.6.

Additional description as given for the genus and species.

Morphological Variation

The relatively large amount of material of Oneirodes acanthias, including a wide size range of individuals (147 specimens, 11.5 to 167 mm), provides a good understanding of the intraspecific variation and ontogenetic change that occurs within a member of the genus Oneirodes. Head length and depth, illicium and lower-jaw length, and the number of lower-jaw teeth all decrease slightly with growth (see Pietsch, 1974a, figs. 85, 86).

This large series has had its greatest value in elucidating ontogenetic changes that occur in the morphology of the esca, by far the most important taxonomic character complex of the genus. These changes are outlined in the description of the esca of O. acanthias above.

Sexual Maturity

Oneirodes acanthias approaches sexual maturity at a length of approximately 100 mm. The ovaries of a 125-mm specimen (LACM 9664-18) are about 41 mm long (30.5% SL) and contain approximately 100,000 eggs of a single size class, measuring 0.5 to 0.7 mm in diameter.

Distribution

Oneirodes acanthias is found only in the eastern North Pacific Ocean, narrowly restricted to deep water off the coasts of Oregon, California, and Baja California, Mexico (Fig. 224). A statistical analysis of vertical distribution, based on the total geographic range of the species, indicates a concentration between 500 and 1250 m (see Pietsch, 1974a).

ONEIRODES THOMPSONI (SCHULTZ, 1934)

Figures 119, 224; Table 8

Females

Dolopichthys thompsoni Schultz, 1934: 66, figs. 1–4 (original description; single specimen; comparison with Dolopichthys acanthias [= Oneirodes acanthias]). Bertelsen, 1951:85, fig. 38C (D. thompsoni a synonym of O. acanthias). Rass, 1955:334 (a synonym of O. acanthias; distribution).

Oneirodes acanthias: Bertelsen, 1951: 85, fig. 38C (in part; Dolopichthys thompsoni a synonym of Oneirodes acanthias).

Oneirodes thompsoni: Pietsch, 1974a: 68, figs. 87, 88, 106 (new combination, description, comparison with all known material). Marshall, 1979:490 (distribution). Amaoka, 1984:106, fig. 20, pl. 92-E (Japan). Peden et al., 1985:9, fig. 36 (eastern North Pacific). Amaoka et al., 1995:113, fig. 174 (northern Japan). Sheiko and Fedorov, 2000:24 (Kamchatka). Mecklenburg et al., 2002:305, 308 (Alaska, in key). Nakabo, 2002:472, figs. (Japan, in key). Fedorov et al., 2003:55, fig. 84 (northern Sea of Okhotsk). Love et al., 2005:61 (eastern North Pacific). Orlov and Tokranov, 2005:271, fig. 1 (distribution and size composition). Shinohara et al., 2005:420 (off Ryukyu Islands). Pietsch and Orr, 2007:16, fig. 13A (phylogenetic relationships).

Oneirodes sp.: Amaoka, 1983:114, 115, 197, 252, 253, 326, figs. 67, 142 (additional specimens, northeastern Sea of Japan and Okhotsk Sea off Hokkaido).

Males and Larvae

Unknown.

Material

One hundred and eight metamorphosed females (33 to 153 mm):

Holotype of Oneirodes thompsoni: USNM 104495 (originally UW 2890, not UW 2530 as cited by Schultz, 1934), 33 mm, International Fisheries Commission haul 530, Gulf of Alaska, 54°13′N, 159°06′W, 2-m ring net, 0 to 900 m, 0858 h, 3 July 1931.

Additional material: HUMZ, 28 (52 to 153 mm); NSMT, 10 (76 to 145 mm); OS, one (69 mm); SIO, two (37 to 41 mm); SIOM, two (109 to 119 mm); UBC, one (58 mm); UF, two (51 to 142 mm); UW, 46 (44 to 143 mm); ZIN, 15 (70 to 128 mm).

Distinguishing Characters

A species of Oneirodes distinguished from all other members of the genus in escal morphology. Oneirodes thompsoni is most similar to O. acanthias and O. notius. In addition to differences in escal morphology, the shape of the subopercle of O. thompsoni differs from that of these forms and all other species of the genus with the exception of O. bulbosus. The posterior margin of the upper part of this bone is indented to deeply notched in all specimens of O. thompsoni examined. Oneirodes thompsoni further differs from O. notius in having a lower pectoral-fin ray count and a smaller ratio between the lengths of the dorsal and ventral forks of the opercle (0.42 to 0.54 and 0.52 to 0.59 in O. thompsoni and O. notius, respectively).

Description

Escal appendage pattern B (Fig. 113B); esca with a stout, internally pigmented anterior appendage, shorter than escal bulb in smaller specimens (33 mm) to more than three times length of escal bulb in largest known specimen (128 mm), and bearing a large compressed distal papilla, closely associated with 1 to several smaller papillae on posterior margin and usually 2 on anterior margin; papillae of anterior appendage darkly pigmented in most specimens approximately 40 mm and larger; medial escal appendages absent; a truncated or conical terminal escal papilla with a distal streak of pigment; an unpigmented, unbranched, tapering posterior escal appendage, less than length of escal bulb in smaller specimens (69 mm) to more than twice length of escal bulb in largest known specimen (128 mm); lateral and anterolateral escal appendages absent.

Subopercle with posterodorsal margin indented to deeply notched; length of ventral fork of opercle 30.5 to 37.3% SL; ratio of lengths of dorsal and ventral forks of opercle 0.42 to 0.54.

Epibranchial teeth absent; teeth present on pharyngobranchial II, total number of teeth in upper jaw 19 to 42, in lower jaw 18 to 36; number of teeth on vomer 4 to 9; dorsal-fin rays 5 or 6; anal-fin rays 4; pectoral-fin rays 14 to 17.

Measurements in percent of standard length: head length 37.7 to 48.6; head depth 39.1 to 54.1; premaxilla length 28.4 to 34.1; lower-jaw length 41.3 to 54.1; illicium length 23.9 to 39.0.

Additional description as given for the genus and family.

Distribution

Oneirodes thompsoni is everywhere sympatric with its congener O. bulbosus, known only from the North Pacific Ocean and Bering Sea (Fig. 224). The range extends south to approximately 40°N in the east, and to about 37°N off the coast of Honshu, Japan, in the west. Based on maximum depths reached by gear, this species appears to inhabit relatively shallow depths, the bulk of the population concentrated between 600 and 800 m, with a second peak between 950 and 1250 m (Pietsch, 1974a).

Comments

Oneirodes thompsoni was synonymized with O. acanthias by Bertelsen (1951:85, fig. 38) on the basis of similarity in escal morphology and morphometrics. Additional material of both species, however, has shown distinct differences in the esca and in the shape of the subopercular bone (Pietsch, 1974a:68, figs. 87, 88).

### ONEIRODES NOTIUS PIETSCH, 1974A

Figure 223; Table 8

#### Females

*Oneirodes notius* Pietsch, 1974a:70, figs. 30B, 89, 90, 106 (original description, 10 specimens). Bertelsen and Pietsch, 1977: 177 (2 additional specimens). Abe and Iwami, 1979:1, figs. 1–6 (additional specimen). Paulin and Stewart, 1985:28 (New Zealand). Paulin et al., 1989:139 (New Zealand). Pietsch, 1990:212, fig. 1 (after Pietsch, 1974a; Southern Ocean). Trunov, 1999:493 (Atlantic sector of the Subantarctic Ocean). Figueiredo et al., 2002:145, color photograph (misidentification, MZUSP 78220, specimen is *Himantolophus* sp.). Menezes et al., 2003 (misidentification, after Figueiredo et al., 2002).

#### Males and Larvae

Unknown.

#### Material

Seventeen metamorphosed females (30 to 150 mm):

Holotype of *Oneirodes notius*: LACM 11165-9, 132 mm, *Eltanin*, cruise 23, station 1615, 62°13'S, 95°39'W, 3-m Isaacs-Kidd midwater trawl, 0 to 1025 m, bottom depth 4914 m, 0610 to 0919 h, 9 April 1966.

Paratypes of *Oneirodes notius*: ISH 590/71, 150 mm, *Walther Herwig* station 354-II/71, 39°19'S, 48°02'W, CMBT 1600, 0 to 2000 m, 2053 to 0021 h, 6 March 1971. ISH 648/71, 115 mm, *Walther Herwig* station 358-III/71, 39°47'S, 43°30'W, CMBT 1600, 0 to 1015 m, 2035 to 2305 h, 7 March 1971. LACM 10716-6, 106 mm, *Eltanin*, cruise 11, station 949, 65°47'S, 88°48'W, 3-m Isaacs-Kidd midwater trawl, 0 to 1028 m, bottom depth 4526 m, 1845 to 2200 h, 28 January 1964. LACM 10841-4, 60 mm, *Eltanin*, cruise 13, station 1120, 62°05'S, 89°56'W, 3-m Isaacs-Kidd midwater trawl, 0 to 850 m, bottom depth 4721 m, 1825 to 2055 h, 29 May 1964. LACM 11184-6, 54 mm, *Eltanin*, cruise 23, station 1648, 58°14'S, 101°02'W, 3-m Isaacs-Kidd midwater trawl, 0 to 825 m, bottom depth 4685 to 4575 m, 1500 to 1758 h, 19 April 1966. LACM 10875-8, 45 mm, *Eltanin*, cruise 14, station 1204, 55°57'S, 159°23'W, 3-m Isaacs-Kidd midwater trawl, 0 to 1080 m, bottom depth 4145 to 3962 m, 0650 to 1035 h, 10 August 1964. SIO 61-45, 30 mm, *Argo*, Monsoon Expedition, station VII-7, midwater trawl 17, 46°53'S, 179°48'W, 3-m Isaacs-Kidd midwater trawl, 0 to 2000 m, 27 to 28 February 1961.

Additional material: CSIRO, two (92 to 105 mm); ISH, three (39 to 108 mm); ZIN, one (standard length not recorded); ZMUC, one (40 mm).

#### Distinguishing Characters

*Oneirodes notius* is most similar to *Oneirodes acanthias* and *O. thompsoni*. Although the high pectoral-fin ray count of *O. notius* (17 to 19) is helpful in distinguishing this species from its congeners, the characters of the esca are the only satisfactory means of identification. The shape of the frontal bones of *O. notius* differs from those of all other species of *Oneirodes* examined osteologically (see Pietsch, 1974a:17, fig. 30).

#### Description

Escal appendage pattern B (Fig. 113B); esca with a stout, internally pigmented anterior appendage, shorter than escal bulb in smaller specimens (60 mm) to slightly longer than length of bulb in largest specimen (150 mm), usually bearing a compressed papilla and 2 tapering filaments on distal tip; several small tapering filaments along posterior margin of anterior escal appendage of 60-mm specimen; papilla and filaments of anterior escal appendage unpigmented; a pair of minute unpigmented filamentous medial escal appendages present in all but 115-mm specimen; a rounded or truncated terminal escal papilla with a distal streak of pigment; an unpigmented, unbranched tapering posterior escal appendage as long as or shorter than escal bulb; lateral and anterolateral appendages absent.

Subopercle relatively short and broad, without indentation on posterodorsal margin; length of ventral fork of opercle 25.2 to 32.2% SL; ratio of lengths of dorsal and ventral forks of opercle 0.52 to 0.59.

Epibranchial teeth absent; teeth present on pharyngobranchial II; total number of teeth in upper jaw 20 to 37, in lower jaw 22 to 48; number of teeth on vomer 4 to 7; dorsal-fin rays 5 to 7; anal-fin rays 4; pectoral-fin rays 17 to 19.

Measurements in percent of standard length: head length 35.0 to 46.2; head depth 36.3 to 48.8; premaxilla length 25.0 to 35.0; lower-jaw length 38.3 to 50.0; illicium length 19.1 to 28.3.

Additional description as for the genus and family.

#### Distribution

*Oneirodes notius* is restricted to the Southern Ocean, where it extends between 35 and 50°S in the Atlantic and Indian sectors of that ocean, but dips farther south in the Pacific, where most of the material was taken below 45°S and ranging farther south to 75°S, the southernmost locality of any ceratioid (Fig. 223). The holotype was collected at 62°S, 95°W. Based on maximum depths reached by fishing gear, *O. notius* has a relatively wide vertical distribution, from about 700 to 2000 m, with the greatest concentration between 800 and 1100 m.

### ONEIRODES SCHISTONEMA PIETSCH AND SEIGEL, 1980

Table 8

#### Females

*Oneirodes schistonema* Pietsch and Seigel, 1980:389, figs. 11, 12 (original description, single specimen). Pietsch, 1999: 2032 (western central Pacific).

#### Males and Larvae

Unknown.

#### Material

A single metamorphosed female (74 mm):

Holotype of *Oneirodes schistonema*: LACM 36036-3, 74 mm, *Alpha Helix* station 24, Banda Sea, 4°39.1'S, 129°53.7'E, 0 to 2000 m, 0115 to 0745 h, 28 March 1975.

#### Distinguishing Characters

A species of *Oneirodes* unique among all previously described species in escal morphology: anterior appendage branched, unpigmented internally; medial and anterolateral appendages absent; posterior appendage branched.

#### Description

Escal appendage pattern B (Fig. 113B); esca with a stout, unpigmented anterior appendage, less than length of escal bulb, bearing 4 short distal branches; pigmented internal tube of anterior escal appendage absent; medial and anterolateral escal appendages absent; terminal escal papilla without distal pigment spot; posterior escal appendage as large as anterior appendage, bearing 4 short branches near distal end.

Upper end of subopercle long, narrow, and tapering, posterodorsal margin of left subopercle deeply indented; length of ventral fork of opercle 25.3% SL; ratio of lengths of dorsal and ventral forks of opercle 0.48.

Epibranchial teeth absent; teeth present on pharyngobranchial II; total number of teeth in upper jaw 42, in lower jaw 40; number of teeth on vomer 6; dorsal-fin rays 6; anal-fin rays 4; pectoral-fin rays 14.

Measurements in percent of standard length: head length 37.2; head depth 28.4; premaxilla length 26.3; lower-jaw length 41.2; illicium length 28.4.

Additional description as for the genus and family.

#### Distribution

The only known specimen of *Oneirodes schistonema* was collected in the Banda

Sea, with gear fished open between the surface and 2000 m.

### ONEIRODES FLAGELLIFER (REGAN AND TREWAVAS, 1932)

Figures 45, 114, 225; Table 8

#### Females

*Dolopichthys* (subgenus *Dermatias*) *flagellifer* Regan and Trewavas, 1932:74, fig. 111 (original description, single specimen).

*Dolopichthys* (subgenus *Dermatias*) *thysanophorus* Regan and Trewavas, 1932:74, fig. 110 (in part; original description, two specimens: lectotype BMNH 1932.5.3.14, designated by Pietsch, 1974a; paralectotype ZMUC P9281, referred to *Oneirodes eschrichtii* by Pietsch, 1974a).

*Oneirodes flagellifer* Group: Bertelsen, 1951:84, fig. 31J, K (in part; new combination, comparison with *Oneirodes eschrichti* group; opercular bones described, figured; in key). Grey, 1956a:246 (in part; synonymy, distribution).

*Dolopichthys thysanophorus*: Beaufort and Briggs, 1962:250 (in part; description after Regan and Trewavas, 1932).

*Dolopichthys flagellifer*: Nielsen, 1974: 93 (listed in type catalog).

*Oneirodes flagellifer*: Pietsch, 1974a:72, figs. 91, 110 (new combination, description, comparison with all known material). Pietsch and Seigel, 1980:384 (six additional specimens, Sulu Sea). Pietsch, 1999:2032 (western central Pacific).

#### Males and Larvae

Unknown.

#### Material

Nine metamorphosed females (10.5 to 22 mm):

Holotype of *Oneirodes flagellifer*: ZMUC P9280, 22 mm, *Dana* station 3909(3), Indian Ocean off Sri Lanka, 5°21'N, 80°38'E, 3500 m wire, 1900 h, 22 November 1929.

Lectotype of *Dolopichthys* (subgenus *Dermatias*) *thysanophorus*: BMNH 1932.5. 3.14, 12.5 mm, *Dana* station 3686(7), Sulu Sea, 8°34'N, 119°55'E, 3500 m wire, 0745 h, 6 April 1929.

Additional material: LACM, six (10.5 to 15.5 mm); ZMUC, one (22 mm).

#### Distinguishing Characters

A species of *Oneirodes* distinguished from all other members of the genus in escal morphology: anterior escal appendage bearing a distal cluster of filaments; medial escal appendages absent; posterior appendage simple, unusually long and slender; lateral and anterolateral escal appendages absent.

#### Description

Escal appendage pattern B (Fig. 113B); esca with an anterior group of short, unpigmented filaments arising from a common base in 12.5-mm specimen, larger specimens with a stout, internally pigmented anterior escal appendage, slightly more than half length of escal bulb, and bearing numerous unpigmented filaments on distal end, some of which are as long as anterior appendage; medial escal appendages absent; a truncated or rounded terminal escal papilla with a distal streak of pigment in 22-mm specimens; an unpigmented, tapering posterior escal appendage nearly twice as long as escal bulb in 12.5-mm specimen, three to four times length of bulb in two 22-mm specimens; lateral and anterolateral escal appendages absent.

Subopercle without indentation on posterodorsal margin (Bertelsen, 1951:84, fig. 31J, K); length of ventral fork of opercle 28.2 to 28.6% SL; ratio of lengths of dorsal and ventral forks of opercle 0.48 to 0.56.

Epibranchial teeth absent; teeth present on pharyngobranchial II; total number of teeth in upper jaw 17 to 42, in lower jaw 25 to 48; number of teeth on vomer 4 to 7; dorsal-fin rays 5 or 6; anal-fin rays 4; pectoral-fin rays 14 or 15.

Measurements in percent of standard length: head length 40.9 to 46.4; head depth 37.0 to 46.4; premaxilla length 32.3 to 37.0; lower-jaw length 45.8 to 50.0; illicium length 19.3 to 25.0.

Additional description as given for the genus and family.

#### Distribution

*Oneirodes flagellifer* is known from nine specimens all collected in the Indo-Pacific: the holotype from approximately 1750 m in the Indian Ocean off Sri Lanka, the lectotype of *O. thysanophorus* from 1750 m in the Sulu Sea, a third specimen collected in 1951 by the *Galathea* from 3800 m in the South China Sea, and six specimens collected by the *Alpha Helix* in the Sulu Sea (Fig. 225).

#### Comments

*Oneirodes flagellifer* and *O. thysanophorus* were described by Regan and Trewavas (1932) from one and two specimens, respectively. On the basis of similarity in the shape of the opercular bones and escal morphology, Bertelsen (1951:84, fig. 31J, K) placed all three specimens into what he called the *Oneirodes flagellifer* group. The larger of the two syntypes of *O. thysanophorus* (ZMUC P9281) was referred to *O. eschrichtii* by Pietsch (1974a). The 12.5-mm syntype, designated as the lectotype of *O. thysanophorus* by Pietsch (1974a), compares well with the known material of *O. flagellifer*, except for some differences in the escal morphology. The anterior escal appendage of the former consists of a short tuft of filaments arising from a common base, whereas that of the holotype of *O. flagellifer* is stout, internally pigmented, and bears a tuft of filaments distally. That these differences are primarily ontogenetic, as suggested by Bertelsen (1951:84) and supported by Pietsch (1974a:73), is further documented by comparison with six additional specimens (10.5 to 15.5 mm) collected by the *Alpha Helix* in the Sulu Sea (see Pietsch and Seigel, 1980:385). The escal appendages of *O. heteronema* and *O. myrionemus* appear to undergo similar changes with growth, the anterior appendage in the case of the former, the medial appendage in the latter. As no significant differences can be found in the material, *O. thysanophorus* is retained within the synonymy of *O. flagellifer* following the conclusions made by Pietsch (1974a) and Pietsch and Seigel (1980).

### ONEIRODES DICROMISCHUS PIETSCH, 1974A

Table 8

#### Females

*Oneirodes dicromischus* Pietsch, 1974a: 73, figs. 92, 93, 108 (original description, single specimen).

#### Males and Larvae

Unknown.

#### Material

Two metamorphosed females (21 to 35 mm):

Holotype of *Oneirodes dicromischus*: LACM 31463-1, 35 mm, *Caride*, cruise 3, station 59, central Pacific, 00°01'N, 139°06'W, 3-m Isaacs-Kidd midwater trawl, 0 to 840 m, 0816 h, 18 February 1969.

Additional material: MNHN, one (21 mm).

#### Distinguishing Characters

*Oneirodes dicromischus* differs among its congeners in having a relatively long illicium (comparable only to some members of the *O. schmidti* group) high tooth counts (both characters not fully developed in the 21-mm immature specimen), as well as a unique escal morphology. The illicium length and tooth counts of adult specimens alone easily separate this form from all other described species of *Oneirodes*. In addition, *O. dicromischus* appears to be a relatively elongate member of the genus: proportional measurements of head length and depth, and premaxilla and lower-jaw length, lie near the bottom of the range of variation for females of all species of *Oneirodes* combined.

#### Description

Escal appendage pattern B (Fig. 113B); esca with a stout, internally pigmented anterior appendage, the bifurcated distal end of which bears numerous unpigmented tapering filaments; three pairs of filamentous medial escal appendages, most anterior pair the longest, nearly as long as escal bulb; a rounded terminal escal papilla with a distal streak of pigment; an unpigmented tapering posterior escal appendage, slightly expanded and trifurcated at distal tip; lateral escal appendages absent; a stout, unpigmented anterolateral escal appendage bearing a few small filaments.

Subopercle without indentation on posterodorsal margin; length of ventral fork of opercle 21.4 to 23.8% SL; ratio of lengths of dorsal and ventral forks of opercle 0.38 to 0.47.

Epibranchial teeth absent; teeth present on pharyngobranchial II; total number of teeth in upper jaw 20 to 65, in lower jaw 27 to 70; number of teeth on vomer 4 to 8; dorsal-fin rays 6; anal-fin rays 4; pectoral-fin rays 16 or 17.

Measurements in percent of standard length: head length 35.7 to 38.1; head depth 37.1 to 38.6; premaxilla length 25.7 to 31.7; lower-jaw length 38.1 to 38.6; illicium length 35.7 to 60.0.

Additional description as given for the genus and family.

### Distribution

*Oneirodes dicromischus* is known from two specimens, collected in the western and central Pacific on or very near the equator: the holotype north of the Marquesas Islands, captured between the surface and 840 m; and the second specimen in the Line Islands, between the surface and 1030 m.

## ONEIRODES THYSANEMA PIETSCH AND SEIGEL, 1980

Table 8

### Females

*Oneirodes thysanema* Pietsch and Seigel, 1980:389, figs. 13, 14 (original description, two specimens). Pietsch, 1999:2032 (western central Pacific).

### Males and Larvae

Unknown.

### Material

Two metamorphosed females (13 to 26.5 mm):

Holotype of *Oneirodes thysanema*: USNM 207931, 26.5 mm, *Ocean Acre*, cruise 7, station 13N, Bermuda, 32°18'N, 63°30'W, 3-m Isaacs-Kidd midwater trawl, 0 to 1500 m, 1430 to 1730 h, 8 September 1969.

Paratype of *Oneirodes thysanema*: LACM 36073-4, 13 mm, *Alpha Helix* station 94, Banda Sea, 5°01.5'S, 130°04.6'E, 650 to 1000 m, 2245 to 0045 h, 5 May 1975.

### Distinguishing Characters

A species of *Oneirodes* differing from all previously described species in escal morphology: anterior appendage with a series of filaments along posterior margin; medial appendages in three groups; terminal papilla elongate; posterior appendage compressed and branched.

### Description

Escal appendage pattern B (Fig. 113B); esca with a stout, internally pigmented anterior appendage, greater than length of escal bulb, bearing along posterior margin a single branched filament proximally and a series of unbranched filaments distally; medial escal appendages in three groups, a highly filamentous pair lying between a similar but unpaired appendage, and a series of 3 stout papillae situated at base of terminal escal papilla; terminal papilla unusually long, directed posterodorsally; posterior escal appendage as long as anterior appendage, highly compressed, bearing 1 or 2 short, lateral filaments, and a considerably longer branched filamentous anterolateral escal appendage on each side; distal tip of internal tube of anterior escal appendage, and dorsal pigment patch of escal bulb, with a paired circular translucent "eyespot."

Subopercle short and broad, upper end rounded without indentation on posterior margin; length of ventral fork of opercle 27.7 to 30.2% SL; ratio of lengths of dorsal and ventral forks of opercle 0.48 to 0.50.

Epibranchial teeth absent; teeth present on pharyngobranchial II; total number of teeth in upper jaw 32, in lower jaw 30; number of teeth on vomer 4 to 6; dorsal-fin rays 6; anal-fin rays 4; pectoral-fin rays 17.

Measurements in percent of standard length: head length 45.3; head depth 43.4; premaxilla length 30.2; lower-jaw length 47.2; illicium length 26.4.

Additional description as given for genus and family.

### Distribution

The two known specimens of *Oneirodes thysanema* were collected from widely separated localities, the holotype from off Bermuda in the western North Atlantic, with gear fished open between the surface and 1500 m; and the second from the Banda Sea in the Philippine Archipelago, with a closing net between 650 and 1000 m.

## ONEIRODES HAPLONEMA STEWART AND PIETSCH, 1998

Table 8

### Females

*Oneirodes haplonema* Stewart and Pietsch, 1998:15, fig. 9 (original description, single specimen).

### Males and Larvae

Unknown.

### Material

Two metamorphosed females (35 to 116 mm):

Holotype of *Oneirodes haplonema*: NMNZ P-13409, 116 mm, northern Challenger Plateau, 37°31.3'S, 169°31.9'E, 1132 to 1128 m, 23 February 1983.

Additional female: CSIRO, one (35 mm).

### Distinguishing Characters

*Oneirodes haplonema* differs from all other described species of the genus in having the following combination of escal characters: a large, internally pigmented anterior appendage, its width nearly equal to that of illicium, its length including terminal filaments greater than 22% SL; an unpaired, highly filamentous medial escal appendage arising from base of terminal escal papilla; a short, cylindrical posterior escal appendage.

### Description

Escal appendage pattern B (Fig. 113B); esca with a large cylindrical internally pigmented anterior appendage, its width nearly equal to that of illicium, its length, including terminal filaments, 22.4 to 31.4% SL; anterior escal appendage bearing distally a pair of compressed bladelike extensions and 2 elongate tapering filaments, each measuring approximately 9.2 to 14.3% SL; an unpaired highly filamentous medial escal appendage emerging from anterior margin of base of terminal escal papilla; an elongate posteriorly directed terminal papillae bearing a single distal streak of black pigment; a small cylindrical posterior escal appendage, without secondary filaments, its length considerably less than that of medial escal appendage; a highly branched anterolateral escal appendage arising from either side at base of anterior escal appendage, its longest filament about one-third as long as anterior escal appendage; lateral escal appendages absent; anterior escal appendage with distal tip of internal tube bearing a distinct translucent "eyespot."

Subopercle without indentation on posterodorsal margin; length of ventral fork of opercle 27.6 to 31.4% SL; ratio of lengths of dorsal and ventral forks of opercle 0.45 to 0.47.

Epibranchial teeth absent; pharyngobranchial II well toothed; total number of teeth in upper jaw 27 to 34, in lower-jaw teeth 34 to 42; number of teeth on vomer 6; dorsal-fin rays 5; anal-fin rays 4; pectoral-fin rays 16 or 17.

Measurements in percent of standard length: head length 41.4 to 48.6; head depth 44.0 to 51.4; head width 31.0 to 42.9; premaxilla length 34.4 to 36.6; lower-jaw length 45.6 to 48.6; illicium length 16.4 to 27.1.

Additional description as given for the genus and family.

### Distribution

*Oneirodes haplonema* is known only from the Tasman Sea: the holotype from off Auckland, New Zealand, captured with gear fished open between the surface and 1132 m; and a second specimen from off Tasmania, taken between the surface and 960 m.

### Comments

*Oneirodes haplonema* is one of four species of the genus characterized by having a single unpaired medial escal appendage; the other three are *O. macronema* Regan and Trewavas (from the Caribbean Sea, Azores, and the Hawaiian Islands), *O. clarkei* Swinney and Pietsch (from off Madeira), and *O. epithales* Orr (from off Newfoundland). *Oneirodes haplonema* differs from *O. macronema* and *O. clarkei* in having a much longer anterior escal appendage, considerably longer than its illicial length (116 to 137%), but much shorter than the escal bulb in the other species. It differs further from *O. macronema*, and from *O. clarkei* and *O. epithales*, in having a simple unbranched posterior escal appendage.

## ONEIRODES EPITHALES ORR, 1991

Table 8

### Females

*Oneirodes epithales* Orr, 1991:1024, figs. 1, 2 (original description, single specimen; revised key to all known species of *Oneirodes*).

### Males and Larvae

Unknown.

### Material

Two metamorphosed females (45 to 128 mm):

Holotype of *Oneirodes epithales*: ARC 8602571, 128 mm, vessel and cruise N067, western North Atlantic, from off Newfoundland, 41°05′39″N, 56°25′33″W, 0 to 1829 m, 1 September 1986.

Additional female: MCZ, one (45 mm).

### Distinguishing Characters

*Oneirodes epithales* differs from all known congeners in escal morphology: an elongate distally branched anterior escal appendage; a single, short medial escal appendage; lateral escal appendages absent; a tiny anterolateral escal appendage present or absent.

### Description

Escal appendage pattern B (Fig. 113B); esca with an elongate anterior escal appendage, proximal one-half of length internally pigmented, bearing 1 or 2 unpigmented filaments on anterior margin near distal tip (45-mm specimen with about 8 extremely thin filaments more or less equally spaced along posterior margin); 45-mm specimen with 2 tiny, branched medial appendages arising from near base of anterior escal appendage; a short unpaired and unpigmented medial escal appendage emerging near base of terminal papilla, bearing numerous tiny filaments along its length; a rounded terminal escal papilla; a short, unpigmented anteroposteriorly compressed posterior escal appendage, bearing 3 to 6 short, unpigmented lateral filaments along its length; anterolateral escal appendages absent in holotype, a tiny branched anterolateral escal appendage on each side in 45-mm specimen; lateral escal appendages absent; distal tip of internal tube of anterior escal appendage and dorsal pigment patch of escal bulb with a paired circular translucent "eyespot."

Subopercle without indentation on posterodorsal margin; length of ventral fork of opercle 25.0 to 30.0% SL; ratio of lengths of dorsal and ventral forks of opercle 0.46 to 0.55.

Epibranchial teeth absent; pharyngobranchial II well toothed; total number of teeth in upper jaw 36 to 44, in lower jaw 34 to 39; number of teeth on vomer 6; dorsal-fin rays 5; anal-fin rays 4; pectoral-fin rays 15 or 16.

Measurements in percent of standard length: head length 38.0 to 44.4; head depth 44.5 to 49.8; head width 25.8 to 34.7; premaxilla length 30.5 to 35.6; lower-jaw length 43.0 to 51.1; illicium length 15.6.

Additional description as given for the genus and family.

### Distribution

*Oneirodes epithales* is known only from the western North Atlantic, the holotype captured approximately 800 km south of Newfoundland in an open trawl somewhere between the surface and 1829 m; and a second specimen from Georges Bank, depth unknown.

## ONEIRODES MACRONEMA (REGAN AND TREWAVAS, 1932)

Table 8

### Females

*Dolopichthys* (subgenus *Dermatias*) *macronema* Regan and Trewavas, 1932:66, fig. 91 (original description, single specimen).

*Oneirodes schmidti* Group: Bertelsen, 1951:84 (in part).

*Dolopichthys macronema*: Nielsen, 1974:94 (listed in type catalog).

*Oneirodes macronema*: Pietsch, 1974a:75, figs. 96, 97, 110 (new combination, description, comparison with all known material). Grobecker and Pietsch, 1978:547, figs. 1, 2 (second known specimen, description, esca and opercular bones figured). Bertelsen, 1986:1395, figs. (after Grobecker and Pietsch, 1978; eastern North Atlantic). Pietsch, 2002b:1064 (western central Atlantic).

### Males and Larvae

Unknown.

### Material

Two metamorphosed females (16.5 to 27 mm):

Holotype of *Oneirodes macronema*: ZMUC P9282, 27 mm, *Dana* 1256(1), Caribbean Sea, 17°43′N, 64°56′W, 1000 m wire, 1920 h, 4 March 1922.

Additional specimen: LACM, one (16.5 mm). See Grobecker and Pietsch (1978).

### Distinguishing Characters

A species of *Oneirodes* unique among its congeners in escal morphology.

### Description

Escal appendage pattern B (Fig. 113B); esca of 16.5-mm specimen (that of the holotype badly damaged; see Pietsch, 1974a:75, fig. 96) with anterior appendage short and stout, not longer than escal bulb, bearing a fringe of filaments on lateral margin near base and a single unbranched filament on anterior margin; a stout, forked medial escal appendage, more than three times length of escal bulb, bearing 4 or 5 short branches near base; left fork of medial appendage with 2 relatively short branches; an unbranched fingerlike posterior escal appendage shorter than escal bulb; lateral and anterolateral escal appendages absent.

Subopercle short and broad, without indentation on posterodorsal margin; length of ventral fork of opercle 24.2 to 24.7% SL; ratio of lengths of dorsal and ventral forks of opercle 0.37 to 0.45.

Epibranchial teeth absent; teeth present on pharyngobranchial II; total number of teeth in upper jaw 18 to 38, in lower jaw 31 to 44; number of teeth on vomer 6 or 7; dorsal-fin rays 5; anal-fin rays 4; pectoral-fin rays 15 to 17.

Measurements in percent of standard length: head length 40.8 to 45.4; head depth 33.3 to 42.2; premaxilla length 29.6 to 36.3; lower-jaw length 46.3 to 51.5; illicium length 15.1 to 22.2.

Additional description as given for the genus and family.

### Distribution

*Oneirodes macronema* is known from only two specimens, one collected in the Caribbean Sea with 1000 m of wire, and the other from off Oahu, Hawaii, somewhere between the surface and 1000 m.

### Comments

At the time of Pietsch's (1974a) revision of *Oneirodes*, *Oneirodes macronema* was still known only from the poorly preserved holotype and consequently important characters of escal morphology were misinterpreted. In the original illustration of the esca of the holotype (Regan and Trewavas, 1932, fig. 91), the basal part of the medial appendage is shown as unpaired, but Pietsch (1974a:75, fig. 96) described and figured the esca as having "a pair of medial appendages." Later reexamination found the esca to be as originally described, except that the medial appendage is unpaired for a shorter distance than shown by Regan and Trewavas (1932) but about the same distance as in the second known specimen described by Grobecker and Pietsch (1978, fig. 1). In all other ways the second specimen agrees well with the holotype.

Bertelsen (1951:84) included *O. macronema* in the *O. schmidti* group, together with *O. mirus* and *O. schmidti*, because of the long medial escal filaments shared by all three species. The short illicium of the two known specimens of *O. macronema*, however, are considerably outside the range of that of the *O. schmidti* group as restricted by Pietsch (1974a:77). In addition, the shape of the subopercular bone of *O. macronema* does not agree with the relatively elongate subopercle characteristic of all known members of the *O. schmidti* group.

### ONEIRODES MELANOCAUDA BERTELSEN, 1951

Table 8

### Larvae

*Oneirodes melanocauda* Bertelsen, 1951:76, 87, figs. 31L, 41 (original description, four larval specimens). Grey, 1956a:248 (synonymy, distribution). Nielsen, 1974:96 (listed in type catalog). Pietsch, 1974a:76, fig. 108 (additional specimen, Caribbean Sea). Pietsch and Seigel, 1980:385 (after Pietsch, 1974a). Pietsch, 1999:2032 (western central Pacific). Pietsch, 2002b:1064 (western central Atlantic).

### Metamorphosed Females and Males

Unknown.

### Material

Five larval specimens, four females (6.5 to 15.5 mm; and one male, 5 mm:
Holotype of *Oneirodes melanocauda*: ZMUC P9288, female, 15.5 mm, *Dana* station 3688(1), South China Sea, 6°55'N, 114°02'E, 4000 m wire, bottom depth 2900 m, 1700 h, 8 April 1929.
Paratypes of *Oneirodes melanocauda*: ZMUC P92184, female, 9 mm, *Dana* station 3690(3), South China Sea, 8°02'N, 109°36'E, 300 m wire, bottom depth 825 m, 2030 h, 10 April 1929. ZMUC P92186, female, 7 mm, *Dana* station 1250(2), Caribbean Sea, 17°54'N, 67°30'W, 600 m wire, 0330 h, 26 February 1922. ZMUC P92185, male, 5 mm, *Dana* station 3860(20), off west coast of Sumatra, 2°57'S, 99°36'E, 600 m wire, bottom depth 5175 m, 2030 h, 20 October 1929.
Additional material: UF, one female (6.5 mm).

### Distinguishing Characters

*Oneirodes melanocauda* is easily separated from other larvae of the genus by the presence of pigment on the tips of the caudal fin rays, and by subdermal pigment on the caudal peduncle.

### Description

The following description is modified from Bertelsen (1951:87–88): head and body covered with pigment nearly to base of caudal fin; pigment of head darkest on jaws and gill cover; tips of the longest caudal rays pigmented, a few melanophores on caudal fin near base; pigmentation of 5-mm male uniform on body, that of head densest on lower jaw and gill cover, in close agreement with that of 9-mm female but the latter somewhat more densely pigmented, caudal peduncle completely covered, in part with rather large melanophores. Holotype, a 15.5-mm female in metamorphosis, with skin pigmented; eyes small (diameter 4.5% SL); caudal fin with a series of large melanophores near base, distal tips of fin rays darkly pigmented; no clear indication of spines on sphenotic, quadrate, articular, or symphysis of lower jaw; illicium short, undeveloped; esca with a short, posteriorly directed appendage; subopercle relatively larger, ventral fork of opercle relatively shorter, than in other known larvae of *Oneirodes*. Dorsal-fin rays 6; anal-fin rays 4; pectoral-fin rays 16 to 18.

### Distribution

*Oneirodes melanocauda* is represented only by larvae, two specimens from the Caribbean Sea and three from the East Indies.

### Comments

Bertelsen (1951) described *Oneirodes melanocauda* on the basis of four larval specimens possessing several features not found in any other known larvae of *Oneirodes*. These features include the presence of pigment on the tips of the caudal-fin rays, presence of subdermal pigment on the caudal peduncle, apparent absence of sphenotic, quadrate, articular, and symphysial spines, a short illicium (in the 15.5-mm female), a large subopercle, and a short opercular bone (Bertelsen, 1951:87–88, fig. 41). In addition, the esca of the 15.5-mm specimen is poorly developed in contrast to all other known forms of *Oneirodes* in which the characteristic escal morphology of the species is fully formed at a standard length of 10 or 11 mm. There is a good possibility that *O. melanocauda* represents the larvae of an undescribed ceratioid genus.

Careful comparison of the 6.5-mm *Pillsbury* specimen (UF 230270) with the 7-mm *Dana* specimen (ZMUC P92186), both collected in the Caribbean Sea, showed essentially no differences.

### *Oneirodes schmidti* Group Bertelsen, 1951

### Females

*Oneirodes schmidti* Group Bertelsen, 1951:84, fig. 31H, I (in part; three nominal species grouped, in key). Pietsch, 1974a:77 (description; expansion to include *Oneirodes theodoritissieri* Belloc, 1938, and *O. basili* Pietsch, 1974a; exclusion of *O. macronema* Regan and Trewavas, 1932). Grobecker, 1978:568 (expansion to include *O. micronema* Grobecker, 1978). Seigel and Pietsch, 1978:11, 13 (expansion to include *O. alius* Seigel and Pietsch, 1978).

*Dolopichthys* sp.: Norman, 1939:115, fig. 40 (in part; two specimens, the larger referred to *Oneirodes* sp. of *Oneirodes schmidti* group by Pietsch, 1974a). Bertelsen, 1951:80 (in part; specimen of Norman, 1939, listed under *O. eschrichti* group).

*Oneirodes eschrichti* Group Bertelsen, 1951:79–80 (in part; *Dolopichthys* sp. of Norman, 1939, and *Oneirodes theodoritissieri* Belloc, 1938, referred to *O. schmidti* group by Pietsch, 1974a).

### Males and Larvae

Unknown.

### Material

Twenty-six metamorphosed females (10 to 183 mm, see species accounts below).

### Distinguishing Characters

Members of the *Oneirodes schmidti* group are distinguished from all other species of the genus by having escae that fall into escal appendage pattern C (Fig. 113C), a relatively long and narrow subopercle, comparable only to those of *O. macrosteus* and *O. rosenblatti*, and a long illicium (26.5 to 107.7% SL), comparable only to that of *O. macrosteus*.

Description

Subopercle relatively long and narrow, without indentation on posterodorsal margin; length of ventral fork of opercle 23.4 to 28.6% SL; ratio of lengths of dorsal and ventral forks of opercle 0.44 to 0.62.

Epibranchial teeth absent; teeth present on pharyngobranchial II; total number of teeth in upper jaw 33 to 71, in lower 44 to 76; number of teeth on vomer 4 to 8; dorsal-fin rays 5 to 7; anal-fin rays 4; pectoral-fin rays 14 to 16.

Additional description as given for the genus and family.

Distribution

Representatives of the *Oneirodes schmidti* group are known from all three major oceans of the world: four records from the Atlantic, off Newfoundland, Bermuda, the Gulf of Mexico, and French West Africa; four from the western North Pacific, off Japan and Taiwan; 15 from Indonesian waters, ranging from off Sumatra in the eastern Indian Ocean to the Banda and Halmahera seas; and three from the eastern Pacific off southern California and Baja.

Diversity

Eight species, differentiated as follows:

Key to Females of Species of the *Oneirodes schmidti* Group

The following key will identify metamorphosed females of species of the *Oneirodes schmidti* group, with the exclusion of *O. bradburyae*, known only from a single damaged female (23.5 mm) from the Gulf of Mexico (see Comments, below).

1A. Anterior escal appendage absent . . . 
. . . . . . . . . . . . . . . . *Oneirodes pietschi*
Ho and Shao, 2004, p. 420

NOTE  Four known specimen, 41.5 to 117 mm; North Pacific Ocean

1B. Anterior escal appendage present
. . . . . . . . . . . . . . . . . . . . . . . . . 2
2A. Length of all escal appendages less than length of escal bulb, all appendages stout with expanded distal tips, none highly filamentous
. . . . . . . . . . . . *Oneirodes micronema*
Grobecker, 1978, p. 418

NOTE  Two known specimens, 17 to 89 mm; western tropical Pacific Ocean

2B. Length of some escal appendages much greater than or equal to length of escal bulb; if equal in length, highly filamentous, without expanded distal tip . . . . . . . . . . . . . 3
3A. Anterior escal appendage unbranched; anterolateral escal appendage absent . . . . *Oneirodes alius*
Seigel and Pietsch, 1978, p. 417

NOTE  Seven known specimens, 10 to 38 mm; western tropical Pacific Ocean

3B. Anterior escal appendage highly branched; anterolateral escal appendage present . . . . . . . . . . . . . . 4
4A. Ratio of lengths of dorsal and ventral forks of opercle 0.54 to 0.62; distal end of posterior escal appendage anteroposteriorly compressed, the posterior face slightly concave and darkly pigmented . . . . . . . . . . . . . .
. . . . . . . . . . . . . . . . *Oneirodes basili*
Pietsch, 1974a, p. 419

NOTE  Three known specimens, 95 to 159 mm; eastern North Pacific Ocean

4B. Ratio of lengths of dorsal and ventral forks of opercle 0.44 to 0.50; distal end of posterior escal appendage without pigmented, concave posterior surface . . . . . . . . . . . . . . . 5
5A. Anterolateral escal appendage about as long as escal bulb . . . . . . . . . . . .
. . . . . . . . . *Oneirodes theodoritissieri*
Belloc, 1938, p. 420

NOTE  Three known specimens, 58 to 183 mm; eastern tropical Atlantic Ocean

5B. Anterolateral escal appendage more than three times length of escal bulb
. . . . . . . . . . . . . . . . . . . . . . . . . 6
6A. Lower jaw with 49 teeth in 42-mm specimen; anterolateral appendage less than five times length of escal bulb . . . . . . . . . *Oneirodes schmidti*
(Regan and Trewavas, 1932), p. 418

NOTE  Five known specimens, 15.5 to 92 mm; western tropical Pacific Ocean

6B. Lower jaw with 76 teeth in 42-mm specimen; anterolateral appendage greater than standard length . . . . . .
. . . . . . . . . . . . . . . . *Oneirodes mirus*
(Regan and Trewavas, 1932), p. 419

NOTE  A single known specimen, 42 mm; off Sumatra, Indian Ocean

ONEIRODES ALIUS SEIGEL AND PIETSCH, 1978

Figure 223; Table 8

Females

*Oneirodes alius* Seigel and Pietsch, 1978:11, figs. 1, 2 (original description, single specimen). Pietsch and Seigel, 1980:390, fig. 15 (four additional specimens, Halmahera Sea). Orr, 1991:1025 (comparison with *Oneirodes epithales*). Pietsch, 1999:2032 (western central Pacific).

Males and Larvae

Unknown.

Material

Seven metamorphosed females (10 to 38 mm):

Holotype of *Oneirodes alius*: LACM 36026-1, 38 mm, *Alpha Helix* station 122, Halmahera Sea, 0°36.3'S, 129°03.2'E, RMT-8, 575 to 600 m, 2240 to 2340 h, 16 May 1975.

Paratypes of *Oneirodes alius*: LACM 36027-1, 18 mm, *Alpha Helix* station 147, Halmahera Sea, 0°40.0'S, 128°58.5'E, RMT-8, 0 to 1200 m, 0635 to 0935, 21 May 1975. LACM 36028-1, 21 mm, *Alpha Helix* station 141, Halmahera Sea, 0°05.0'S, 128°52.7'E, RMT-8, 1000 to 1100 m, 0625 to 0855, 20 May 1975.

Additional material: LACM, four (10 to 12 mm).

Distinguishing Characters

*Oneirodes alius* is a member of the *O. schmidti* group as diagnosed by Pietsch (1974a:77). In addition to its unique escal morphology, *O. alius* appears to differ from other members of the *O. schmidti* group in having fewer lower-jaw teeth.

Description

Escal appendage pattern C (Fig. 113C); esca with a single elongate tapering anterior appendage, 33.3 to 43.0% SL, without internal pigment; a pair of filamentous medial appendages, half the length of escal bulb in 18-mm paratype, minute in 38-mm holotype; a rounded terminal escal papilla with two bilaterally placed distal pigment spots; a short, unpigmented, anteroposteriorly compressed posterior escal appendage; lateral and anterolateral escal appendages absent.

Subopercle large, elongate, dorsal end tapering to a point, without indentation on posterodorsal margin; length of ventral fork of opercle 30.3 to 32.3% SL; ratio of lengths of dorsal and ventral forks of opercle 0.44 to 52.

Epibranchial teeth absent; teeth present on pharyngobranchial II; total number of teeth in upper jaw 28 to 32, in lower jaw 33 to 37; number of teeth on vomer 6; dorsal-fin rays 5 to 7; anal-fin rays 4; pectoral-fin rays 15 or 16.

Measurements in percent of standard length: head length 50.0; head depth 43.0 to 48.5; premaxilla length 38.0; lower-jaw length 47.5; illicium length 26.5 to 35.5.

Additional description as given for the *Oneirodes schmidti* group, genus, and family.

Distribution

*Oneirodes alius* is known only from the Halmahera Sea (Fig. 223). The holotype, three nontype specimens (11.5 to 12 mm), and the 21-mm paratype were captured with a closing net between 575 and 600 m, 700 and 1000, and 1000 and 1100 m, respectively. The remaining two specimens were captured somewhere between the surface and 1200 m.

Comments

*Oneirodes alius* was originally described from three specimens collected by the *Alpha Helix* in the Halmahera Sea (Seigel and Pietsch, 1978). After the description went to press, four additional specimens were sorted out from *Alpha Helix* stations made in approximately the same localities as the type material. In all respects, these specimens compared well with the type material (Pietsch and Seigel, 1980).

The esca of *O. alius* clearly falls within appendage pattern C as first recognized by Pietsch (1974a). Further, the elongate subopercle of *O. alius* is characteristic of members of the *O. schmidti* group. Within the *O. schmidti* group, *O. alius* appears to have a slightly shorter illicium (26.5 to 35.5% SL compared to 33.3 to 107.7% SL for all other members of the group combined) but differs most strikingly in specific escal characters: it has the well-developed anterior escal appendage of species with escal appendage patterns A and B but lacks the anterolateral escal appendages so well developed in the other members of the *O. schmidti* group. Based on the escal-character complex, *O. alius* is considered the least derived member of the *O. schmidti* group.

ONEIRODES MICRONEMA GROBECKER, 1978

Table 8

Females

*Oneirodes micronema* Grobecker, 1978: 567, figs. 1, 2 (original description, two specimens). Pietsch and Seigel, 1980:393, fig. 18 (reference to original description). Pietsch, 1999:2032 (western central Pacific).

Males and Larvae

Unknown.

Material

Two metamorphosed females (17 to 89 mm):

Holotype of *Oneirodes micronema*: LACM 36039-3, 89 mm, *Alpha Helix*, Southeast Asian Bioluminescence Expedition, station 113, Banda Sea, 5°07.5'S, 130°08.4'E, 650 to 1000 m, 2120 to 2255 h, 13 May 1975.

Paratype of *Oneirodes micronema*: LACM 36043-3, 17 mm, *Alpha Helix*, Southeast Asian Bioluminescence Expedition, station 157, Banda Sea, 4°09.6'S, 130°50.0'E, 0 to 840 m, 24 April 1975.

Distinguishing Characters

A species of *Oneirodes* belonging to the *Oneirodes schmidti* group as diagnosed by Pietsch (1974a). Escal morphology differing from all previously described forms: anterior appendage with lateral filament on each side; a pair of unbranched medial appendages; terminal papilla with pair of pigment spots; posterior appendage slender, with expanded distal tip; a single unbranched anterolateral appendage on each side; lateral appendages absent; all appendages less than length of escal bulb.

Description

Escal appendage pattern C (Fig. 113C); esca with a stout anterior appendage bearing a short unbranched filament on each side; a pair of stout, unbranched medial escal appendages; a rounded terminal escal papilla, with a bilaterally placed pair of distal pigment spots; a slender unbranched posterior escal appendage; a single stout, unbranched anterolateral escal appendage on each side; all appendages unpigmented with slightly expanded distal tips.

Subopercle long and narrow, dorsal end tapering to a point (rounded in juvenile paratype; see Bertelsen, 1951, fig. 31), without indentation on posterodorsal margin; length of ventral fork of opercle 24.8% SL; ratio of length of dorsal and ventral forks of opercle 0.48.

Epibranchial teeth absent; teeth present on pharyngobranchial II; total number of teeth in upper jaw 20 to 56, in lower jaw 32 to 55; number of teeth on vomer 4 to 8; dorsal-fin rays 6; anal-fin rays 4; pectoral-fin rays 14 or 15.

Measurements in percent of standard length: head length 45.3 to 47.1; head depth 34.8 to 35.3; premaxilla length 28.1 to 29.4; lower-jaw length 43.3 to 47.1; illicium length 35.3 to 74.2.

Additional description as given for the *Oneirodes schmidti* group, genus, and family.

Distribution

*Oneirodes micronema* is known from only two specimens, both collected in the Banda Sea. The 89-mm holotype was captured with a closing net between 650 and 1000 m; the 17-mm paratype, somewhere between the surface and 840 m.

Comments

The esca of *Oneirodes micronema* falls clearly within appendage pattern C, along with those of other members of the *O. schmidti* group. The long illicium (74.2% SL in the 89-mm holotype) and elongate subopercle of *O. micronema* further indicate its affinity with members of this group (Pietsch, 1974a). *Oneirodes micronema* differs in escal morphology from others of the *O. schmidti* group: unlike the other species, the escal appendages (except for the anterior appendages) are unbranched, but, more importantly, all are unusually short, shorter than the length of the escal bulb.

ONEIRODES SCHMIDTI (REGAN AND TREWAVAS, 1932)

Table 8

Females

*Dolopichthys* (subgenus *Dermatias*) *schmidti* Regan and Trewavas, 1932:75, fig. 113 (original description; single specimen).
*Oneirodes schmidti* Group: Bertelsen, 1951:84, fig. 311 (in part, new combination). Grey, 1956a:246 (in part; after Bertelsen, 1951; vertical distribution).
*Dolopichthys schmidti*: Nielsen, 1974:95 (listed in type catalog).
*Oneirodes schmidti*: Pietsch, 1974a:78, figs. 98A, 99, 111 (description, comparison with all known material). Pietsch and Seigel, 1980:391, figs. 16, 17 (four additional specimens, Banda Sea). Pietsch, 1999:2032 (western central Pacific).

Males and Larvae

Unknown.

Material

Five metamorphosed females (15.5 to 92 mm):

Holotype of *Dolopichthys* (subgenus *Dermatias*) *schmidti*: ZMUC P9284, 32 mm, *Dana* station 3678(1), Banda Sea, 4°05'S, 128°16'E, 5000 m wire, bottom depth 4700 m, 1840 h, 24 March 1929.

Additional material: LACM, four (15.5 to 92 mm).

Distinguishing Characters

*Oneirodes schmidti* is a member of the *O. schmidti* group as diagnosed above. In addition to differences in escal morphology, *O. schmidti* differs from most other members of the *O. schmidti* group by having a greater number of teeth in the jaws.

Description

Escal appendage pattern C (Fig. 113C); esca with a large complex anterior appendage consisting of a wide compressed base bearing a relatively short unpaired branched filament on posterior margin, 2 extremely long distal filaments (about 16.3 to 17.3% SL) that bifurcate as many as five times, and a stout, bifurcated medial filament, each branch of which becomes highly branched distally; a pair of filamentous, highly branched medial escal appendages less than length of escal bulb; terminal escal papilla with a single distal streak of pigment; a slender, unbranched posterior escal appendage less than length of escal bulb; a relatively short, filamentous, branched anterolateral appendage on each side (the inner pair of stout anterolateral appendages described for the holotype of Oneirodes schmidti by Pietsch, 1974a:78, fig. 99, correspond to the elongate bifurcated filaments that are associated with the anterior appendage of the nontype material).

Subopercle long and narrow, dorsal end tapering, rounded or pointed distally, without indentation on posterodorsal margin; length of ventral fork of opercle 26.6 to 27.7% SL; ratio of length of dorsal and ventral forks of opercle 0.47 to 0.51.

Epibranchial teeth absent; teeth present on pharyngobranchial II; total number of teeth in upper jaw 60 to 67, in lower jaw 56 to 76; number of teeth on vomer 5 or 6; dorsal-fin rays 5 or 6; anal-fin rays 4; pectoral-fin rays 15 to 17.

Measurements in percent of standard length: head length 43.8 to 47.7; head depth 42.4 to 50.0; premaxilla length 35.9 to 40.0; lower-jaw length 50.0 to 52.7; illicium length 37.5 to 107.7.

Additional description as given for the O. schmidti group, genus, and family.

### Distribution

Oneirodes schmidti is known only from the Banda Sea. The holotype was captured in an open trawl fished with 3500 m of wire. The 65- and 92-mm specimens were captured between the surface and 760 m; the 78-mm specimen between the surface and 2000 m. The 15.5-mm specimen was captured in an opening-closing trawl between 650 and 810 m.

### Comments

The 1975 Southeast Asian Bioluminescence Expedition of the *Alpha Helix* provided the first representatives of Oneirodes schmidti since the capture of the holotype by the *Dana* in 1929 (see Pietsch and Seigel, 1980:391). This additional material compares very well with the type specimen in all characters, except for some minor differences in escal morphology that have been added to the diagnosis and description provided above.

## ONEIRODES MIRUS (REGAN AND TREWAVAS, 1932)

Table 8

### Females

*Dolopichthys* (subgenus *Dermatias*) *mirus* Regan and Trewavas, 1932:74, fig. 112 (original description, single specimen).
*Oneirodes schmidti* Group: Bertelsen, 1951:84, fig. 31H (in part). Grey, 1956a:246 (after Bertelsen, 1951; vertical distribution).
*Dolopichthys mirus*: Nielsen, 1974:94 (listed in type catalog).
*Oneirodes mirus*: Pietsch, 1974a:79, figs. 98B, 100, 111 (new combination, description, comparison with all known material).

### Males and Larvae

Unknown.

### Material

A single metamorphosed female (42 mm):
Holotype of *Dolopichthys* (subgenus *Dermatias*) *mirus*: ZMUC P9283, 42 mm, *Dana* station 3828 (10), off west coast of Sumatra, 1°22′N, 96°06.5′E, 3000 m wire, bottom depth 4980 m, 1600 h, 18 September 1929.

### Distinguishing Characters

Oneirodes mirus is a member of the O. schmidti group as diagnosed above. It differs from other members of the O. schmidti group in its escal morphology and is further differentiated from most members of the group by having fewer teeth in the jaws.

### Description

Escal appendage pattern C (Fig. 113C); esca with a stout anterior appendage, shorter than escal bulb, without internal pigment, bearing a single short branch; a pair of unpigmented filamentous medial escal appendages, slightly longer than anterior appendage; a rounded terminal escal papilla with a distal spot of pigment; an anteroposteriorly compressed, unpigmented posterior escal appendage, as long as escal bulb, bearing 2 short branches at midlength; lateral escal appendages absent; two pairs of filamentous anterolateral escal appendages: an inner pair of stout, tapering appendages flanking anterior appendage each longer than SL, lightly pigmented internally except for slightly expanded distal tip, and bearing numerous short filaments; and an outer pair of unpigmented bifurcated appendages, each tapering branch of which is more than three times length of escal bulb and bears 2 short filaments.

Subopercle long and narrow, dorsal end tapering, but rounded distally, without indentation on posterodorsal margin; length of ventral fork of opercle 23.8% SL; ratio of length of dorsal and ventral forks of opercle 0.50.

Epibranchial teeth absent; teeth present on pharyngobranchial II; total number of teeth in upper jaw 33, in lower jaw 49; number of teeth on vomer 6; dorsal-fin rays 6; anal-fin rays 4; pectoral-fin rays 16.

Measurements in percent of standard length: head length 40.4; head depth 47.6; premaxilla length 35.7; lower-jaw length 47.6; illicium length 33.3.

Additional description as given for the O. schmidti group, genus and family.

### Distribution

Oneirodes mirus is known from a single specimen collected in the Indian Ocean, off Sumatra, with 3000 m of wire out.

## ONEIRODES BASILI PIETSCH, 1974A

Table 8

### Females

*Oneirodes* sp. n.: Pietsch, 1972a:42, 43, 45, fig. 24(5) (otolith described, figured).
*Oneirodes basili* Pietsch, 1974a:79, figs. 20, 29, 98C, 101, 111 (original description, three specimens). Hubbs et al., 1979:13 (California). Marshall, 1979:257, fig. 103 (after Pietsch, 1974a). Love et al., 2005:60 (eastern North Pacific).

### Males and Larvae

Unknown.

### Material

Three metamorphosed females (95 to 159 mm):
Holotype of *Oneirodes basili*: LACM 30020-34, 95 mm, *Velero IV* station 11635, 28°08′N, 117°31′W, 3-m Isaacs-Kidd midwater trawl, 0 to 700 m; bottom depth 3520 to 3493 m, 2340 to 0430 h, 20 August 1967.
Paratypes of *Oneirodes basili*: LACM 30028-30, 115 mm, *Velero IV* station 11644, 29°40′N, 118°15′W, 3-m Isaacs-Kidd midwater trawl, 0 to 1400 m, bottom depth 3383 to 3292 m, 1300 to 2117 h, 22 August 1967. LACM 31100-2, 159 mm, *Velero IV* station 13721, 33°06′N, 118°22′W, 3-m Isaacs-Kidd midwater trawl, 0 to 990 m, bottom depth 1152 to 1353 m, 2320 to 0530 h, 18 December 1969.

### Distinguishing Characters

Oneirodes basili is a member of the O. schmidti group as diagnosed above. It differs from other members of the O. schmidti group in its escal morphology. It is further differentiated from O. mirus by having a greater number of upper-jaw teeth, and from O. theodoritissieri by having a greater ratio between the lengths of the dorsal and ventral forks of the opercle.

### Description

Escal appendage pattern C (Fig. 113C); esca with a filamentous, branched anterior appendage, two times length of escal bulb, without internal pigment; spots of dark pigment present on most branches of anterior appendage; medial escal appendage absent; a rounded or conical terminal escal papilla with two bilaterally placed distal spots of pigment; a posterior escal appendage, as long as escal bulb, with an anteroposteriorly compressed distal end, the posterior surface of which is slightly concave and darkly pigmented; lateral escal appendages absent; two pairs of filamentous anterolateral escal appendages: an inner pair of branched tapering appendages flanking anterior appendage, each less than one-half length of escal bulb; and a similar outer pair of appendages equal to or less than length of inner pair.

Subopercle long and narrow, dorsal end tapering to a point, without indentation on posterodorsal margin; length of ventral fork of opercle 25.3 to 28.6% SL; ratio of lengths of dorsal and ventral forks of opercle 0.54 to 0.62.

Epibranchial teeth absent; teeth present on pharyngobranchial II; total number of teeth in upper jaw 49 to 69, in lower jaw 46 to 60; number of teeth on vomer 4 to 8; dorsal-fin rays 5 or 6; anal-fin rays 4; pectoral-fin rays 15 or 16.

Measurements in percent of standard length: head length 38.9 to 43.4; head depth 37.7 to 41.7; premaxilla length 32.2 to 33.3; lower-jaw length 42.1 to 44.3; illicium length 45.9 to 51.6.

Additional description as given for the Oneirodes schmidti group, genus, and family.

### Sexual Maturity

The length of the right ovary of the 159-mm paratype of *Oneirodes basili* measures 33 mm or 20.8% SL; those of the 115-mm paratype are small and undeveloped.

### Distribution

*Oneirodes basili* is known only from off southern California and Baja California, Mexico, collected in midwater trawls fished open at maximum depth of 700 to 1400 m.

## ONEIRODES THEODORITISSIERI BELLOC, 1938

Table 8

### Females

*Oneirodes theodori-tissieri* Belloc, 1938: 303, figs. 23–25 (original description, single specimen).

*Oneirodes eschrichti* Group: Bertelsen, 1951:79 (in part). Grey, 1956a:245 (after Bertelsen, 1951; vertical distribution).

*Oneirodes theodoritissieri*: Aloncle, 1968:691 (listed). Pietsch, 1974a:80, figs. 98D, 102, 111 (comparison with all known material, esca and subopercle figured, additional material, off West Africa). Bertelsen and Pietsch, 1977:178 (*Walther Herwig* specimen, off Cape Verde Islands). Bertelsen, 1990:506 (synonymy, eastern tropical Atlantic).

### Males and Larvae

Unknown.

### Material

Three metamorphosed females (58 to 183 mm):

Holotype of *Oneirodes theodoritissieri*: MHNLR P316-448, 64 mm, *President Theodore Tissier*, cruise 5, station 733, Bissagos, Guinee-Bissau, 11°13′N, 17°26′W, Schmidt net, 1000 m wire, bottom depth 1460 m, 27 May 1936.

Additional material: BMNH, one (58 mm), ISH, one (183 mm).

### Distinguishing Characters

*Oneirodes theodoritissieri* is a member of the *O. schmidti* group as diagnosed above. It differs from other members of the *O. schmidti* group (as well as all other species of the genus) in its unique escal morphology: anterior escal appendage without internal pigment; terminal escal papilla with two distal spots of pigment; posterior escal appendage compressed; lateral appendages absent; a single pair of bifurcated anterolateral appendages. It is further differentiated from *O. schmidti* by having fewer jaw teeth and from *O. basili* by having a smaller ratio between the lengths of the dorsal and ventral forks of the opercle.

### Description

Escal appendage pattern C (Fig. 113C); esca with a filamentous branched anterior appendage, as long as escal bulb, without internal pigment; a pair of filamentous medial appendages as long as escal bulb; a conical terminal papilla with two bilaterally placed distal spots of pigment; an unpigmented anteroposteriorly compressed posterior appendage, as long as escal bulb, the distal half of which bears 6 short branches in 58-mm specimen; lateral appendages absent; a single pair of unpigmented bifurcated anterolateral appendages, each fork highly branched and longer than escal bulb.

Subopercle long and narrow, dorsal end tapering to a point, without indentation on posterodorsal margin; length of ventral fork of opercle 25.9 to 26.6% SL; ratio of lengths of dorsal and ventral forks of opercle 0.44 to 0.50.

Epibranchial teeth absent; teeth present on pharyngobranchial II; total number of teeth in upper jaw 40 to 55, in lower jaw 44 to 58; number of teeth on vomer 5 to 7; dorsal-fin rays 5 or 6; anal-fin rays 4; pectoral-fin rays 15 or 16.

Measurements in percent of standard length: head length 39.6 to 40.6; head depth 38.8 to 40.6; premaxilla length 29.5 to 35.9; lower-jaw length 43.7 to 48.4; illicium length 48.3 to 57.8.

Additional description as given for the *Oneirodes schmidti* group, genus, and family.

### Distribution

*Oneirodes theodoritissieri* is known only from the eastern North Atlantic off Portuguese Guinea and the Cape Verde Islands. The 58-mm specimen was captured with a closing net between 810 and 900 m.

## ONEIRODES PIETSCHI HO AND SHAO, 2004

Table 8

### Females

*Oneirodes pietschi* Ho and Shao, 2004:74, figs. 1–3 (original description, three females).

### Males and Larvae

Unknown.

### Material

Four metamorphosed females (41.5 to 117 mm):

Holotype of *Oneirodes pietschi*: ASIZP 061822, 100 mm, *Ocean Research I* station CD 191, off southwest coast of Taiwan, South China Sea, 21°22.18′N, 118°11.02′E, beam trawl, 1631 to 1635 m, 28 August 2002.

Paratypes of *Oneirodes pietschi*: BSKU 20236, 70 mm, *Sōyō-Maru* station B2, off Miyake-Jima, Izu-Shot Islands, Honshu, Japan, 34°05.0′N, 140°03.04′E, beam trawl, 1250 to 1270 m, 20 June 1972. HUMZ 130128, 117 mm, Hawaiian Islands, North Pacific, 26°59.6′N, 164°29.1′W, midwater trawl, 0 to 400 m, February 1994.

Additional material: ASIZP, one (46 mm).

### Distinguishing Characters

A species of *Oneirodes* belonging to the *Oneirodes schmidti* group, as diagnosed above, differing from all previously described species in lacking an anterior escal appendage. It is further differentiated by having the following details of escal morphology: a pair of simple, elongate, anterolateral escal appendages, without internal pigmentation; a pair of elongate, medial escal appendages, each divided distally into 3 or more short, slender filaments; a simple, elongate, and anteroposteriorly compressed posterior escal appendage, slightly expanded distally; all appendages without pigmentation; terminal escal papilla with or without a pair of small distal pigment spots.

### Description

Escal appendage pattern C (Pietsch, 1974a:34; Fig. 113C): anterior escal appendage absent; anterior margin of escal bulb with a pair of unbranched, cylindrical, anterolateral escal appendages, each without internal pigmentation, about two to three times length of escal bulb; medial escal appendages about half length of anterolateral appendages, those of holotype and 70-mm paratype bearing 3 short, slender filaments, 3 to 5 filaments in 117-mm paratype; posterior escal appendage slightly longer than medial appendages, without internal pigment, anteroposteriorly compressed with distal tip slightly expanded; escal pore relatively large, lying between terminal escal papillae and base of posterior escal appendage in two paratypes, situated closer to base of posterior escal appendage in holotype; distal surface of terminal papilla slightly flattened, with a pair of small pigment spots in two paratypes, more rounded and without pigment spots in holotype

Opercle bifurcate, upper fork broad, bifurcated distally in holotype, blunt in paratypes, lower fork slender and nearly straight; length of lower fork of opercle 23 to 26% SL; ratio of lengths of upper and lower forks of opercle 0.45 to 0.57; subopercle complete on right side of holotype and both sides of 117-mm paratype, broken on left side of holotype and both sides of 70-mm paratype; relatively long and narrow in holotype and 70-mm paratype but short in 117-mm paratype, without indentation on posterodorsal margin, upper part tapering to a point, lower part broad and rounded.

Epibranchial teeth absent; teeth present on pharyngobranchial II; total number of teeth in upper jaw 32 to 58, in lower jaw 34 to 42; number of teeth on vomer 4 to 6; dorsal-fin rays 6; anal-fin rays 4; pectoral-fin rays 14 to 16.

Measurements in percent of standard length: head length 34 to 41; head depth 35 to 43; premaxilla length 27 to 31; lower-jaw length 35 to 42; illicium length 29 to 49.

Additional description as given for the *O. schmidti* group, genus, and family.

Distribution

*Oneirodes pietschi* is represented by four specimens collected from widely separated localities in the North Pacific Ocean: the holotype and the 41.5-mm specimen from the South China Sea, off the southwest coast of Taiwan; and two paratypes from off the Hawaiian Islands and Miyake-Jima, Izu-Shot Islands, Japan. The holotype and 70-mm paratype were collected by bottom trawl at maximum depths of 1270 m and 1635 m, respectively; the 41.5-mm specimen by bottom trawl between 300 and 600; and the 117-mm paratype by midwater trawl somewhere between the surface and 400 m.

## ONEIRODES BRADBURYAE GREY, 1956B

Table 8

Females

*Oneirodes bradburyae* Grey, 1956b:245, fig. 2 (original description, single specimen). Pietsch, 1974a:74, figs. 94, 95, 107 (description, comparison with all known material). McEachran and Fechhelm, 1998:867, 869, fig. (Gulf of Mexico, in key). Pietsch, 2002b:1064 (western central Atlantic).

Males and Larvae

Unknown.

Material

A single known metamorphosed female (23.5 mm):
Holotype of *Oneirodes bradburyae*: USNM 164359, 23.5 mm, *Oregon* station 1028, Gulf of Mexico, 28°28′N, 87°18′W, 0 to 1426 m, 21 April 1954.

Distinguishing Characters

*Oneirodes bradburyae* is tentatively assigned to the *O. schmidti* group as diagnosed above. It appears to differ from all known members of this group in having an extremely long illicium (72% SL). While the illicial lengths of *O. micronema* (35.3 to 74.2% SL) and *O. schmidti* (37.5 to 107.7% SL) range higher than this, they are much shorter at comparable standard lengths (see Comments below).

Description

The esca of the holotype and only known specimen of *Oneirodes bradburyae* is badly damaged. The following is taken from the original description (Grey, 1956b:245): a rather long, fine filament arising from left side of distal end of esca; a shorter club-tipped prolongation on right side; a short, delicate cluster of filaments centrally; all filaments colorless; length of longest filament 15.7% SL.

Subopercle without indentation on posterodorsal margin; length of ventral fork of opercle 24.7% SL; ratio of lengths of dorsal and ventral forks of opercle 0.45.

Epibranchial teeth absent; teeth present on pharyngobranchial II; total number of teeth in upper jaw 42, in lower jaw 54 (jaw-tooth counts are most probably low due to damage); number of teeth on vomer 8; dorsal-fin rays 5; anal-fin rays 4; pectoral-fin rays 14.

Measurements in percent of standard length: head length 43.8; head depth 42.6; premaxilla length 34.0; lower-jaw length 48.9; illicium length 72.3.

Additional description as given for the *O. schmidti* group, genus, and family.

Distribution

*Oneirodes bradburyae* is known from a single specimen collected in the Gulf of Mexico at 28°28′N, 87°18′W, somewhere between the surface and 1426 m.

Comments

The esca of the only known specimen of *Oneirodes bradburyae* is badly damaged: "After the description was written the esca was accidentally dried out and the filaments hardened and shriveled" (Grey, 1956b:245). Nevertheless, the anterior escal appendage is filamentous and clearly lacks an internal pigmented tube; it otherwise appears to be like those of forms included by Pietsch (1974a:77) in the *O. schmidti* group. Furthermore, the subopercle is somewhat elongate, comparable to those of members of the *O. schmidti* group; but, perhaps more significant, the illicium is extremely long (72.3% SL), proportionately longer than that of any other known species of *Oneirodes*.

### *Oneirodes* Species
Females

*Oneirodes* sp.: Pietsch, 1974a:81, 107 (unidentifiable juvenile and adult females). Pietsch and Seigel, 1980:393 (unidentifiable material, Philippine Archipelago). Bertelsen and Pietsch, 1983:87 (unidentifiable material, Australia). Stearn and Pietsch, 1995:138, fig. 88 (six unidentifiable females, Greenland).

Material

Three hundred and eight metamorphosed females (11 to 370 mm):
Distinct forms representing possible new species: AMS I.22810-044, 2 (53 to 133 mm), *Soela*, off Port Hedland, 18°40′S, 116°30′E, Engel trawl, 0 to 736 m, 5 April 1982 (esca unique). ISH 995/82, 370 mm, largest known specimen of the genus, *Walther Herwig* station 401/82, Mid-Atlantic ridge, 49°51.7′N, 16°51.7′W, 0 to 3200 m, 19 June 1982 (esca similar to that of *Oneirodes heteronema*). LACM 36087-5, 13 mm, *Alpha Helix* station 135, Halmahera Sea, 0°06.2′S, 128°38.3′E, 820 to 1000 m, 2306 to 0106 h, 18 May 1975 (illicium 15.4% SL, upper-jaw teeth 14, lower-jaw teeth 21, pectoral-fin rays 15). LACM 36089-5, 14 mm, *Alpha Helix* station 137, Halmahera Sea, 0°08.9′S, 128°40.0′E, 0 to 960 m, 0955 to 1300 h, 19 May 1975 (illicium 14.3% SL, upper-jaw teeth 15, lower-jaw teeth 20, pectoral-fin rays 15; esca like that of LACM 36087-5). LACM 36115-2, 21 mm, *Alpha Helix* station 102, Banda Sea, 4°45.0′S, 129°19.7′E, 0 to 2000 m, 0540 to 1045 h, 7 May 1975 (illicium 23.8% SL, upper-jaw teeth 27, lower-jaw teeth 38, pectoral-fin rays 16). MCZ 57783, 31 mm, *Oceanus*, cruise 55, station MOC 10-158, Florida

Current, 34°29'N, 75°21'W, 0 to 1000 m, 24 February 1979 (esca like that of MCZ 58869). MCZ 58869, 15 mm, *Oceanus*, cruise 118, station MOC 20-013, 38°36'N, 72°3'W, 0 to 1000 m, 25 April 1982 (esca like that of MCZ 57783). SAMA F.6829, 97 mm, *Akebono-Maru 3*, 35°07'S, 133°45'E, 0 to 1012 m, 6 December 1989 (esca similar to that of *O. sabex*, but posterior appendage long and slender). SIO 61-44, 77 mm, Monsoon Expedition MWT 16, 54°22'S, 177°17'W, 0 to 2500 m, 18 February 1961 (illicium length 32.4% SL, upper-jaw teeth about 40, lower-jaw teeth 41, vomerine teeth 7, pectoral-fin rays 16). SIO 61-35, 45 mm, Monsoon Expedition MWT 7, 14°54'S, 70°12'E, 0 to 2000 m, 3 December 1960 (illicium length 80.0% SL, upper-jaw teeth about 120, lower-jaw teeth about 130, vomerine teeth 8, pectoral-fin rays 16, esca badly damaged). SIO 70-334, 13 mm, *Melville*, Antipodes IV, station 67D-Tr 18, 18°59.5'N, 125°29.3'E, 3-m Isaacs-Kidd midwater trawl, 0 to 1350 m, 1314 September 1970 (illicium length 27.7% SL). SIO 75-452, 74 mm, *Melville*, Francis Drake II, station 4, 02°58.7'N, 80°49.9'E, 3-m Isaacs-Kidd midwater trawl, 8 May 1975 (illicium length 24% SL; anterior escal appendage extremely long, 14.8% SL). USNM 207931, 26.5 mm, Ocean Acre 7-13N, 32°18'N, 63°30'W, 3-m Isaacs-Kidd midwater trawl, 0 to 1500 m, 1430 to 1730 h, 8 September 1969 (illicium length 28.3% SL, upper-jaw teeth 30, lower-jaw teeth 34, vomerine teeth 6, pectoral-fin rays 17).

Additional unidentifiable metamorphosed females (the following is only a partial list of the hundreds of unidentifiable female specimens of *Oneirodes* in collections around the world): AMS, four (11 to 105 mm); BMNH, 16 (17 to 82 mm); BSKU, six (94 to 186 mm); CAS, 11 (9 to 15 mm); CSIRO, seven (20 to 166 mm); HUMZ, 19 (38.5 to 182 mm); ISH, 15 (39 to 255 mm); LACM, 94 (9 to 60 mm); MCZ, 39 (10 to 110 mm); MMF, two (55 to 98 mm); MNHN, 11 (standard lengths not recorded); NMNZ, six (11 to 187 mm); NSMT, 21 (27.5 to 158 mm); SIO, 28 (13.5 to 74 mm); SIOM, two (14 to 120 mm); UF, six (10.5 to 57 mm); USNM, seven (11 to 37 mm).

### Comments

These metamorphosed female specimens cannot be readily identified to species. The list includes a few that may represent new species, but description is deferred pending the discovery of additional material.

### Males

*Lipactis tumidus*: Regan, 1926:43 (in part; erroneous designations; six specimens, one from *Dana* station 1152 referred to *Oneirodes eschrichti* group by Bertelsen, 1951:74). Norman, 1930:357 (erroneous designation, specimen subsequently made holotype of *Trematorhynchus exiguus* by Regan and Trewavas, 1932). Fowler, 1936:1349, fig. 567 (in part; after Regan, 1926). Albuquerque, 1954–1956:1062, fig. 434 (after Fowler, 1936).

*Rhynchoceratias leucorhinus*: Regan, 1926:44 (in part; erroneous designations; 17 specimens, two referred to *Trematorhynchus leucorhinus* by Regan and Trewavas, 1932, subsequently referred to *Oneirodes eschrichti* group by Bertelsen, 1951). Albuquerque, 1954–1956:1063, fig. 435 (after Regan, 1926).

*Trematorhynchus leucorhinus*: Regan and Trewavas, 1932:91 (in part; new combination; erroneous designations; four specimens referred to *Oneirodes eschrichti* group by Bertelsen, 1951).

*Trematorhynchus exiguus* Regan and Trewavas, 1932:91, fig. 147 (original description, single specimen; *Lipactis tumidus* of Norman, 1930). Bertelsen, 1951:73 (referred to the Oneirodidae).

*Caranactis pumilus* Regan and Trewavas, 1932:59, fig. 86 (original description, single specimen; referred to *Oneirodes eschrichtii* group by Bertelsen, 1951). Nielsen, 1974:88 (listed in type catalog).

*Oneirodes eschrichti* Group: Bertelsen, 1951:83, figs. 31A, B, 32B, D, E, 35–37 (in part). Grey, 1956a:245 (in part; after Bertelsen, 1951; synonymy, distribution).

*Oneirodes* sp.?: Bertelsen, 1951:88 (ZMUC P92678, 13 mm; male questionably referred to *Oneirodes*).

*Oneirodes* sp.: Pietsch, 1974a:81 (unidentifiable males). Pietsch and Seigel, 1980:394 (unidentifiable males).

### Material

Ninety-seven metamorphosing and metamorphosed males, 8 to 16.5 mm:

Holotype of *Trematorhynchus exiguus*: BMNH 1930.1.12.1102, 10 mm, *Discovery* station 298, 13°01'N, 21°34'W, Young fish trawl, 0 to 1200 m, 29 August 1927 (*Lipactis tumidus* of Norman, 1930).

Holotype of *Caranactis pumilus*: ZMUC P9266, 8.0 mm, *Dana* station 3909(5), Indian Ocean off Sri Lanka, 5°21'N, 80°38'E, 2500 m wire, bottom depth 4120 m, 1900 h, 22 November 1929.

Additional males: LACM, 78 (8.5 to 14 mm); SIO, three (9.5 to 14.5 mm); SIOM, one (9 mm); UF, one (13.5 mm); USNM, two (12.5 to 14.5 mm); ZMUC, 10 (13 to 16.5 mm). Pietsch (1974a:81) and Pietsch and Seigel (1980:394). For a full list of material in the *Dana* collections held by ZMUC, see Bertelsen (1951:88, 267).

### Comments

Of six juvenile and adult males of *Oneirodes* examined by Bertelsen (1951), five appeared to be nearly identical. These were placed in what was then the most common group of species, the *Oneirodes eschrichti* group. The somewhat divergent sixth specimen was designated by Bertelsen (1951) as *Oneirodes* sp.? At present, about 97 juvenile and adult males of *Oneirodes* are known. Despite this large increase in material no characters have been found to identify males to any species based on females. Furthermore, no significant differences within the material could be found in spite of its worldwide distribution. All of this material, including Bertelsen's (1951) *Oneirodes* sp.?, is therefore designated as *Oneirodes* sp. until means of identifying these specimens are found.

### Larvae

Material

One hundred and ninety-two known specimens, 3 to 12 mm TL: AMS, four (7 to 7.5 mm); LACM, 11 (7.5 to 11.5 mm); SWFSC, 12 (lengths not recorded); USNM, two (4.8 to 5 mm); ZMUC, 163 (78 females, 79 males, six undetermined, 3 to 12 mm TL).

For material of the *Dana* collections held by ZMUC, see Bertelsen (1951:267).

Comments

As is the case with nearly all ceratioid larvae, except those belonging to monotypic genera or species groups, these larvae cannot be identified to species.

### Genus *Dermatias* Smith and Radcliffe, in Radcliffe, 1912 (Fattail Dreamers)

Figures 116, 117; Tables 1, 8

Females

*Dermatias* Smith and Radcliffe, in Radcliffe, 1912:206, pl. 17, fig. 3 (type species *Dermatias platynogaster* Smith and Radcliffe, in Radcliffe, 1912, by original designation and monotypy). Bertelsen, 1951:76 (a synonym of *Oneirodes* Lütken). Pietsch, 1974a:33 (concurrence with Bertelsen, 1951).

*Dolopichthys* (subgenus *Dermatias*) Parr, 1927:14 (in key). Regan and Trewavas, 1932:66 (genus *Dolopichthys* broadened to incorporate five subgenera). Bertelsen, 1951:71, 76 (in synonymy of *Oneirodes*).

*Pietschichthys* Kharin, 1989:158, figs. 1, 2 (type species *Pietschichthys horridus* Kharin, 1989, by original designation and monotypy).

Males and Larvae

Unknown.

Distinguishing Characters

Metamorphosed females of *Dermatias* are distinguished from those of all other genera of the family in having an unusually deep caudal peduncle (21.6 to 23.8% SL); a blunt snout and short, highly convex frontals resulting in an extremely short head (29.7 to 30.5% SL); and remarkably few teeth in the jaws (20 to 32 in the upper jaw, 20 to 31 in the lower jaw).

Metamorphosed females of *Dermatias* are further unique in having the following combination of character states: Vomerine teeth are present. The sphenotic spines are well developed (their length 2.9 to 3.6% SL) and directed dorsolaterally. The lower jaw bears a stout symphysial spine. The hyomandibula has a double head. The quadrate spine is well developed (its length 2.5 to 5.0% SL), but the articular spine is less than one-half the length of the quadrate spine. The posterior margin of the opercle is deeply notched. The subopercle is long and narrow, its dorsal end tapering to a point (the posterior margin without indentation), its ventral end oval in shape (without an anterior spine or projection). The caudal-fin rays are without internal pigmentation. The illicium is considerably longer than the length of the esca bulb. The pterygiophore of the illicium emerges on the snout from between the frontal bones, its anterior end exposed, its posterior end concealed beneath the skin. The first ray of dorsal fin is well developed. There are 6 dorsal-fin rays and 4 anal-fin rays (Table 1). The pectoral-fin lobe is short and broad (its length 6.6 to 8.9% SL), shorter than the longest rays of the pectoral fin (16.3 to 19.9% SL). There are 15 or 16 pectoral-fin rays. The skin is smooth and apparently naked, without dermal spinules (but specimens are unavailable for clearing and staining). Darkly pigmented skin of the caudal peduncle extends well past the base of the caudal fin (only three known specimens, 134 to 175 mm; material unavailable for internal osteological examination).

Description

The body of metamorphosed females is moderately short, somewhat fusiform, its depth approximately 45 to 55% SL. The snout is blunt; the head extremely short (29.7 to 30.5% SL). The mouth is small, the opening almost horizontal, and the cleft extending slightly past the eye. The opercular opening is unusually large, 19.4 to 22.5% SL. The illicium length is 15.3 to 17.5% SL. The esca, remarkably similar to those of some species of *Oneirodes* (i.e., those with escal appendage pattern B; Pietsch, 1974a, Fig. 113B), bears a stout, internally pigmented anterior escal appendage approximately as long as the length of the escal bulb; a pair of medial escal appendages, each consisting of a dense cluster of slender filaments, some slightly longer than length of the escal bulb, and each with a darkly pigmented distal tip; a small, rounded terminal papilla, with a single terminal pigment spot; and a simple unpigmented filamentous posterior escal appendage somewhat longer than the length of the escal bulb.

The teeth in the jaws are slender, recurved, and all depressible, those in the lower jaw slightly larger than those of the upper jaw. The longest tooth in upper jaw measures 2.4 to 3.1% SL; the longest in lower jaw is 3.4 to 3.7% SL. There are 20 to 32 teeth in upper jaw and 20 to 31 in the lower jaw. There are 4 to 6 teeth on the vomer. The second and third pharyngobranchials are both well developed and toothed.

The following measurements are expressed in percent SL: head length 29.7 to 30.5; head width 25.5 to 29.1; head depth 33.7 to 36.6; least outside width between frontals 9.1 to 10.9; length of premaxilla 22.4 to 24.0; length of lower jaw 25.2 to 33.6; length of lower fork of opercle 19.7 to 22.5, upper fork 10.9 to 14.9; length of subopercle 8.6 to 11.3.

Diversity

The genus contains a single species:

DERMATIAS PLATYNOGASTER SMITH AND RADCLIFFE, IN RADCLIFFE, 1912

Figures 117, 118; Table 8

Females

*Dermatias platynogaster* Smith and Radcliffe, in Radcliffe, 1912:206–207, pl. 17, fig. 3 (original description, single specimen). Pietsch and Kharin, 2004:123, 124, figs. 1, 3, 4 (resurrection from synonymy of *Oneirodes eschrichtii*). Kharin and Pietsch, 2007:806, figs. 1, 2 (additional specimen, off Queensland, Australia).
*Dolopichthys platynogaster*: Regan, 1926:29–30 (brief description after Smith and Radcliffe, in Radcliffe, 1912; in key). Bertelsen, 1951:79, 81 (one of 22 nominal species placed in the *Oneirodes eschrichti* group). Pietsch, 1974a:44, 53, 102, table 9 (a synonym of *O. eschrichtii*).
*Dolopichthys* (subgenus *Dermatias*) *platynogaster*: Parr, 1927:14 (in key). Regan and Trewavas, 1932:68, fig. 95 (brief description after Smith and Radcliffe, in Radcliffe, 1912; in key).
*Pietschichthys horridus* Kharin, 1989:158–160, figs. 1, 2 (original description, single specimen). Pietsch and Kharin, 2004:123, 125, figs. 2, 4 (a junior synonym of *Dermatias platynogaster*).

Males and Larvae

Unknown.

Material

Three metamorphosed females (134 to 175 mm):

Holotype of *Dermatias platynogaster*: USNM 70269, 134 mm, *Albatross*, Philippine Expedition, station 5463, near Sialat Point Light, off east coast of Luzon, Philippine Islands, western North Pacific, 13°40'57"N, 123°57'45"E, beam trawl, 300 fathoms (549 m), 16 June 1909.

Holotype of *Pietschichthys horridus*: ZIN 47301, 151 mm, Magellan Seamounts, western North Pacific, 19°39'N, 155°21'E, 4.7-m bottom trawl, 1342 m, 13 April 1984.

Additional female: CSIRO, one (175 mm).

Distinguishing Characters and Description

As given for the genus and family.

Distribution

All three known specimens of *Dermatias platynogaster* were collected in the western North Pacific Ocean, with gear fished on the bottom: one off Luzon in the Philippines, taken with a beam trawl in relatively shallow water (549 m); another in the vicinity of the Magellan Seamounts, east of the Mariana Islands, with a bottom trawl fished at a maximum depth of 1342 m; and the third from the Townsville Trough, off Queensland, Australia, with a bottom trawl fished at 1188 m.

Genus *Danaphryne* Bertelsen, 1951 (Dana Dreamers)

Figures 33, 118–120, 202, 203, 226; Tables 1, 8, 9

Larval and Metamorphosed Females

*Danaphryne* Bertelsen, 1951:101, figs. 58, 59 (type species *Dolopichthys* [subgenus *Dermatias*] *nigrifilis* Regan and Trewavas, 1932, by monotypy).

Males

Unknown.

Distinguishing Characters

Metamorphosed females of *Danaphryne* are distinguished from those of all other genera of the family by having an unusually narrow snout, the width of the ethmoid cartilage and vomer considerably less than the distance between the anterolateral tips of the lateral ethmoids and frontals (see Pietsch, 1974a:16, fig. 21).

Metamorphosed females of *Danaphryne* are further unique in having the following combination of character states: Vomerine teeth are present. The nasal foramina are large and nearly circular. The frontals are short, lying posterior to the ethmoid region, their dorsal margin convex. Ventromedial extensions of the frontals approach each other on the midline, making contact with the parasphenoid. The frontals are separated from the prootics. The pterosphenoid is present. The anterior end of the illicial trough is wider and shallower than the posterior end. Sphenotic spines are well developed. The symphysial cartilage of the upper jaw is longer than wide. The lower jaw bears a well-developed symphysial spine. The hyomandibula has a double head. The quadrate spine is well developed, longer than the articular spine. The posterior margin of the opercle is deeply notched. The subopercle is long and narrow, its dorsal end is unusually long and slender, tapering to a point (the posterior margin lacks an indentation), its ventral end nearly circular (without an anterior spine or projection). The first pharyngobranchial is absent. The second pharyngobranchial is well developed. The second hypobranchial articulates directly with the second basibranchial. The caudal-fin rays lack internal pigmentation. The illicium is considerably longer than the length of the esca bulb. The pterygiophore of the illicium is cylindrical throughout length, emerging on the snout from between the frontal bones, its anterior end exposed, its posterior end concealed beneath the skin. The first ray of the dorsal fin is well developed. There are 5 to 7 (rarely 8) dorsal-fin rays and 5 (rarely 4) anal-fin rays (Table 1). The pectoral-fin lobe is short and broad, shorter than the longest rays of the pectoral fin. There are 16 to 19 pectoral-fin rays. The coracoid lacks a posteroventral process. The pelvic bones are simple, expanded distally. The skin is smooth and naked, without dermal spinules. Darkly pigmented skin of the caudal peduncle extends well past base of the caudal fin.

Larval females of *Danaphryne* are distinguished from those of all other genera of the family by having the following combination of character states: The opercle is notched posteriorly, its dorsal fork broad, with 3 supporting ribs. The dorsal end of the subopercle tapers to a point, its ventral end broad and rounded, without a spine or projection on the anterior margin. Subdermal pigment covers the entire head and body except for the snout, pectoral-fin lobe, and caudal peduncle. The pigment is heaviest on the anterior part of the lower jaw and along the ventral and posterior margin of the gill cover (Bertelsen, 1951:123).

### Description

The body of metamorphosed females is relatively short and globular to moderately fusiform. The mouth is moderate, its cleft not extending past the eye. The illicium length is 38.0 to 48.0% SL. The escal bulb is expanded anterodorsally, bearing a distal pair of stout, posteriorly directed tentacle-like filaments (its length 4.5 to 22.5% SL), and a large, laterally compressed posterior escal appendage, with a more or less spherical pigmented swelling on its ventral margin. The escal pore is situated in a cleft between the bases of the tentacle-like filaments, just anterior to the base of the posterior appendage. Internal light-transmitting structures are visible through the more or less transparent distal part of the escal bulb and its appendages. Light reflecting tissue, lying beneath the surface of the anterior tip of the escal bulb, appears as a pair of white spots situated ventrolateral to a small white papilla. The pigmentation, shape, and relative size of the esca and its appendages are variable, but without clear correlation with size or geographic distribution of specimens (Bertelsen and Pietsch, 1977:183).

The teeth are slender, recurved, and all depressible, arranged in an oblique transverse series of up to 5 teeth of increasing size. The teeth in the lower jaw are generally larger but fewer than those in the upper jaw, the number of teeth increasing with size of the specimens. There are 22 teeth in the upper jaw of a 20-mm specimen to approximately 210 in a 105-mm specimen, and 31 teeth in the lower jaw of a 20-mm specimen to approximately 170 in a 105-mm specimen. There are 4 to 14 teeth on the vomer. The second and third pharyngobranchials are well developed and toothed; the epibranchials and ceratobranchials are toothless.

Subdermal larval pigmentation is preserved in the smallest available specimen (20 mm, USNM 216301): the head and body is covered with melanophores except for the caudal peduncle, snout, and pectoral-fin lobes. Concentrations of pigment are present on the margin of the gill cover and anterior part of the lower jaw. In these characters, the 20-mm specimen compares well with the three larvae described by Bertelsen, 1951:123, figs. 82, 83, as "Oneirodid larva, gen. et sp.?".

Additional description as given for the family.

### Diversity

The genus contains a single species:

### DANAPHRYNE NIGRIFILIS (REGAN AND TREWAVAS, 1932)

Figures 33, 118–120, 226; Tables 8, 9

#### Metamorphosed Females and Larvae

*Dolopichthys* (subgenus *Dermatias*) *nigrifilis* Regan and Trewavas, 1932:67, fig. 92 (original description, single specimen).
*Dolopichthys albifilosus* Waterman, 1939b:89, figs. 5, 6 (original description, single specimen). Bertelsen, 1951:102, fig. 59 (a junior synonym of *Dolopichthys nigrifilis*; holotype figured after Waterman, 1939b).
*Danaphryne nigrifilis*: Bertelsen, 1951: 102, figs. 58, 59 (new combination, description; *Dolopichthys albifilosus* a junior synonym of *D. nigrifilis*; holotype and opercular bones figured). Marshall, 1971a:46, fig. 17 (eyes and vision). Pietsch, 1974a: 16–21, 24–28, 32, figs. 21, 31, 39A, 40, 51A, 52 (comparative osteology). Bertelsen and Pietsch, 1977:183, fig. 8 (additional material, eastern Atlantic; description, comparison with all known material, esca figured). Marshall, 1979:132, fig. 49 (after Bertelsen, 1951). Pietsch and Seigel, 1980: 394 (after Bertelsen and Pietsch, 1977). Bertelsen, 1990:500 (synonymy, eastern tropical Atlantic). Stearn and Pietsch, 1995:135, fig. 85 (additional specimen, Greenland). Pietsch, 1999:2032 (western central Pacific). Pietsch, 2002b: 1064 (western central Atlantic). Moore et al., 2003: 215 (off New England).
Oneirodid Larva, Gen. et Sp.?: Bertelsen, 1951:123, figs. 82, 83 (unidentifiable larvae). Bertelsen and Pietsch, 1977:185 (larvae assigned to *Dolopichthys nigrifilis*).
*Dolopichthys nigrifilis*: Nielsen, 1974:94 (listed in type catalog).

#### Males

Unknown.

#### Material

Twenty metamorphosed females (20 to 105 mm) and four larvae (three females, 9 to 17 mm TL; and one male, 9.5 mm):

Holotype of *Dolopichthys nigrifilis*: ZMUC P92102, 24 mm, *Dana* station 3716, South China Sea, 19°18.5′N, 120°13′E, 3000 m wire, 1400 h, 22 May 1929.

Holotype of *Dolopichthys albifilosus*: MCZ 35067, 33 mm, *Atlantis* station 2667, western North Atlantic, 35°40′N, 69°36′W, depth unknown, September 1936.

Additional metamorphosed females: ARC, one (97 mm); BMNH, one (53 mm); BSKU, one (98 mm); HUMZ, one (59 mm); ISH, three (42 to 82 mm; 82-mm specimen cleared and stained); LACM, two (23 to 33 mm); MCZ, two (48 to 105 mm); MNHN, two (22 to 24.5 mm); SIO, one (22 mm); USNM, one (20 mm); UW, two (42.5 to 66 mm); ZMUB, one (76 mm).

Larvae: MCZ, one male (9.5 mm); ZMUC, three females (9 to 17 mm TL).

#### Distinguishing Characters and Description

As given for the genus and family.

#### Description

Additional description as given for the genus and family.

#### Distribution

*Danaphryne nigrifilis* has been collected on both sides of the Atlantic, between approximately 11°N and 41°N, with an additional record from off Greenland at 63°N, 54°W; and from the tropical western Pacific, from the Philippines and South China Sea to the Hawaiian Islands (Fig. 226). The holotype of *D. nigrifilis* is from the South China Sea; that of *D. albifilosus* is from off Bermuda. All of the material, for which depths were recorded, was taken in gear fished open between the surface and 2000 m.

#### Comments

At the time of Bertelsen's (1951) monograph, this species was known from only two specimens, the 24-mm holotype from the South China Sea and the 33-mm holotype of *Dolopichthys albifilosa* described by Waterman (1939b) from the western North Atlantic. Finding no reason to maintain the latter species, Bertelsen (1951) synonymized it with *D. nigrifilis* and placed it in the new genus *Danaphryne*. Additional material reported by Bertelsen and Pietsch (1977) confirmed the earlier assumption that the differences between the two original specimens were well within the individual and ontogenetic variation that might be expected. The subdermal larval pigmentation of the smallest known metamorphosed specimen of *Dolopichthys nigrifilis* (20 mm, USNM 216301) compares well with that of the three larvae (9 to 17 mm TL) described by Bertelsen (1951:123, figs. 82, 83) as "Oneirodid larva, gen. et sp.?" Furthermore, the shape of the opercular bones and fin-ray counts of these larvae are in such good agreement with those of *D. nigrifilis* that they can with little doubt be referred to this species. As confirmed by reexamination (Bertelsen and Pietsch, 1977:185), one of these three larvae has 8 dorsal- and 4 anal-fin rays, one more and one less, respectively, than the numbers observed among the known metamorphosed females of the species.

Genus *Microlophichthys* Regan and Trewavas, 1932 (Shortbait Dreamers)

Figures 33, 121–124, 202, 203, 226; Tables 1, 8

Females, Males, and Larvae

*Dolopichthys* (subgenus *Microlophichthys*) Regan and Trewavas, 1932:77, figs. 118, 119 (one of five subgenera of genus *Dolopichthys*; type species *Dolopichthys microlophus* Regan, 1925c, by subsequent designation of Burton, 1933:62).

*Microlophichthys* Bertelsen, 1951:88, figs. 42, 43 (subgenus *Microlophichthys* given generic status; type species *Dolopichthys microlophus* Regan, 1925c, by subsequent designation of Burton, 1933:62).

Distinguishing Characters

Metamorphosed females of *Microlophichthys* are distinguished from those of all other genera of the family by having the following combination of character states: The ethmoid cartilage and vomer are wide, wider than the distance between the anterolateral tips of the lateral ethmoids and frontals. Vomerine teeth are present. The nasal foramina are large, nearly circular. The frontals are short, lying posterior to the ethmoid region, their dorsal margin convex. Ventromedial extensions of the frontals approach each other on the midline, making contact with the parasphenoid. The frontals are separated from the prootics. The pterosphenoid is present. The anterior end of the illicial trough is wider and shallower than the posterior end. The sphenotic spines are well developed. The symphysial cartilage of the upper jaw is longer than wide. The lower jaw bears a well-developed symphysial spine. The hyomandibula has a double head. The quadrate spine is well developed, longer than the articular spine. The posterior margin of the opercle is deeply notched. The subopercle is short and broad, the dorsal end rounded or tapering to a blunt point (the posterior margin without indentation). The ventral end nearly circular (without an anterior spine or projection). The first pharyngobranchial is absent. The second pharyngobranchial is reduced to a tiny remnant. The second hypobranchial is absent. The caudal-fin rays lack internal pigmentation. The illicium is extremely short, approximately equal to the length of the escal bulb. The pterygiophore of the illicium is cylindrical throughout, emerging on the snout from between the frontal bones, its anterior end just barely exposed, its posterior end concealed beneath the skin. The first ray of the dorsal fin is well developed. There are 5 to 7 dorsal-fin rays and 5 (rarely 4 or 6) anal-fin rays (Table 1). The pectoral-fin lobe is short and broad, shorter than the longest rays of the pectoral fin. There are 18 to 20 (rarely 21 to 23) pectoral-fin rays. The coracoid lacks a posteroventral process. The pelvic bones are simple, with or without a distal expansion. The skin is smooth and naked, without dermal spinules. Darkly pigmented skin of the caudal peduncle extends well past the base of the caudal fin.

The pectoral-fin lobe of males is short and broad, shorter than the longest rays of the pectoral fin. The subopercle is short and broad, its dorsal end rounded. The skin between the nostrils is slightly pigmented. The medial surface of the subopercle is darkly pigmented. The caudal peduncle bears a distinct subdermal group of melanophores (Bertelsen, 1951:75, 89).

The body of larvae is relatively elongate and slender. The dorsal end of the subopercle is rounded, its medial surface with a dark spot of pigment. The branchiostegal pigment is faint. The caudal peduncle is pigmented, the melanophores forming a single isolated group on the lateral surface. The caudal-fin rays lack internal pigmentation (Bertelsen, 1951:75, 76, 89).

Description

The body of metamorphosed females is elongate and slender, not globular. The snout is relatively short, the mouth large, its cleft extending well past the eye. The illicium is extremely short, its length 8.3 to 10.3% SL. The escal bulb has a terminal, truncated escal papilla and a more or less pointed, compressed posterior escal appendage. The terminal escal papilla is darkly pigmented (especially in larger specimens), the pigment continuous with, and spreading out onto, the dorsal surface of the escal bulb, forming a medial patch, more or less expanded on its anterior and posterior ends.

The teeth in the jaws are slender, recurved, and all depressible, arranged in an oblique transverse series of increasing size, those in the lower jaw generally larger but fewer than those in the upper jaw. The number of teeth increase with the size of specimen, the upper jaw with about 160 to 320 teeth, the lower jaw with about 100 to 180 (numerous tiny teeth set extremely close together make counting difficult). There are 4 to 12 teeth on the vomer. The third pharyngobranchial is well developed and toothed. The epibranchials and ceratobranchials are toothless.

The following measurements of metamorphosed females are expressed in percent of standard length: head depth 35.8 to 43.2; head length 31.1 to 39.2; lower-jaw length 34.9 to 46.8; premaxilla length 28.8 to 31.9; least outside width of frontals 7.1 to 11.6.

The posterior nostril of the males is well separated from the eye by a pigmented bridge of skin. There are 13 to 14 olfactory lamellae. There are 8 to 10 upper denticles and 8 lower denticles. There are 6 dorsal-fin rays, 4 or 5 anal-fin rays, and 17 to 19 pectoral-fin rays (Bertelsen, 1951:89).

The larvae have an elongate body, its depth about 45% SL. The length of the head is 50% SL. The gill cover is darkly pigmented, with a concentration of melanophores beneath the subopercle, and beneath and slightly anterior to the ventral fork of the opercle. Dorsal pigmentation is more or less restricted to the head and the anterodorsal part of the body. A broad band of pigment is present on the caudal peduncle, extending anteriorly to or slightly beyond the posterior margins of the bases of the dorsal and anal fins, and posteriorly nearly to base of the caudal fin. The peritoneum is covered with pigment. Larger specimens have a lateral group of melanophores in the skin at the base of the caudal fin. There are 5 to 7 dorsal-fin rays, 4 to 6 (almost always 5) anal-fin rays, and 18 to 23 pectoral-fin rays (Bertelsen, 1951:89).

The body of the smallest known larval specimen (2.4 mm TL) is unpigmented, the notochord straight, with only indistinct signs of the unpaired fins. The gill covers are strongly pigmented (distinguished from larvae of *Oneirodes* only by the more slender shape of the body, and from other oneirodids by the rounded subopercle). The second-smallest larval specimen (2.6 mm TL) shows signs of dorsal pigment on the head and anterior part of the body. A strong but comparatively narrow band of pigment is present on the caudal peduncle and behind the edge of the dorsal fin. The fin rays of the third smallest specimen (4.0 mm TL) are visible, the notochord is reduced; dorsal pigment is still limited to the occipital region, and pigment of the caudal peduncle is somewhat spread. In larger specimens, pigmentation spreads posteroventrally from the dorsal group to the sides of the body, occasionally covering the entire body at lengths of 5.0 to 6.0 mm TL. The area anterodorsal to the anal fin remains without or nearly without pigment until specimens are quite large. The entire body of the largest female larvae (12.5 mm TL) is covered with large stellate melanophores. The dorsal and peduncular pigment appears as slightly stronger concentrations. A lateral group of smaller melanophores is present in the skin covering the posterior part of the caudal peduncle and base of the caudal fin. The latter feature is developed in 6.0-mm TL specimens and larger, but appears only in four of 25 specimens between 6.0 and 10 mm TL and in four of six larger specimens (Bertelsen, 1951:92).

Additional description as given for the family.

### Diversity

The genus contains two species, one based on females, males, and larvae; the other based solely on two adult males:

*MICROLOPHICHTHYS MICROLOPHUS* (REGAN, 1925c)

Figures 33, 121, 122, 124, 126; Table 8

### Females and Larvae

*Dolopichthys microlophus* Regan, 1925c: 563 (original description, single specimen). Regan, 1926:29, fig. 18 (after Regan, 1925c). Nielsen, 1974:94 (listed in type catalog).

*Dolopichthys analogus* Parr, 1927:20, fig. 7 (original description, single specimen).

*Dolopichthys* (subgenus *Microlophichthys*) *microlophus*: Regan and Trewavas, 1932:77, figs. 118, 119 (new combination, description, additional specimen, Caribbean Sea).

*Dolopichthys* (subgenus *Microlophichthys*) *analogus*: Regan and Trewavas, 1932: 78 (new combination; description after Parr, 1927).

*Dolopichthys* (subgenus *Microlophichthys*) *exiguus* Regan and Trewavas, 1932: 78, fig. 121 (original description, single specimen).

*Dolopichthys* (subgenus *Microlophichthys*) *implumis* Regan and Trewavas, 1932:78, fig. 122 (original description, three syntypes). Beebe and Crane, 1947:160, fig. 5 (description, additional material, Costa Rica and Galápagos).

*Microlophichthys microlophus*: Bertelsen, 1951:90, figs. 42–46 (subgenus elevated to generic rank; description of females, males, and larvae based on all known material; *Dolopichthys analogus*, *D. exiguus*, and *D. implumis* junior synonyms of *Microlophichthys microlophus*. Briggs, 1960:179 (worldwide distribution). Parin et al., 1973:144, fig. 37 (eastern South Pacific). Pietsch, 1974a:16, 24, 32, figs. 32, 39B, 41, 51B, 53A, 54, 56A (comparative osteology). Parin, 1975:323 (faunal study). Bertelsen and Pietsch, 1977:185 (four additional females, eastern Atlantic). Pietsch and Seigel, 1980:394 (additional material, a male and two females, Banda and Ceram seas). Bertelsen, 1984: 331, fig. 170F (early life history, phylogeny). Bertelsen, 1986:1392, figs. (eastern North Atlantic). Bertelsen, 1990:503 (synonymy, eastern tropical Atlantic). Pietsch, 1999: 2032 (western central Pacific). Pietsch, 2002b:1064 (western central Atlantic). Moore et al., 2003:215 (off New England). Love et al., 2005:60 (eastern North Pacific).

*Dolopichthys exiguus*: Nielsen, 1974:93 (listed in type catalog).

*Dolopichthys implumis*: Nielsen, 1974: 93 (listed in type catalog).

### Males

*Rhynchoceratias leucorhinus*: Regan, 1926:44 (in part; one of 17 specimens).

### Material

Ninety-four metamorphosed females (11.5 to 112 mm), five males (6.8 to 17 mm), and 110 larvae (2.4 to 12.5 mm TL):

Holotype of *Dolopichthys microlophus*: ZMUC P9292, female, 26 mm, *Dana* station 1159(3), western North Atlantic, 17°55′N, 24°35′W, 3000 m wire, 0450 h, 29 October 1921.

Holotype of *Dolopichthys analogus*: YPM (BOC) 2010, female, 17 mm, *Pawnee*, Third Oceanographic Expedition, station 58, Bermuda, 32°24′N, 64°29′W, 10,000 ft wire, 20 April 1927.

Holotype of *Dolopichthys exiguus*: ZMUC P9289, female, 13 mm, *Dana* station 3904(2), eastern Indian Ocean, 5°18′N, 90°55′E, 3000 m wire, 1745 h, 18 November 1929.

Syntypes of *Dolopichthys implumis*: ZMUC P9290, female, 12 mm, *Dana* station 1208(4), Gulf of Panama, 6°48′N, 80°33′W, 3500 m wire, 16 January 1922; ZMUC P9291, female, 15 mm, *Dana* station 1208(15), Gulf of Panama, 6°48′N, 80°33′W, 2600 m wire, 16 January 1922; BMNH 1932.5.3.18, female, 14 mm, female, *Dana* station 1208(14), Gulf of Panama, 6°48′N, 80°33′W, 3100 m wire, 16 January 1922.

Additional metamorphosed females: BMNH, five (12 to 63 mm); CAS, three (12 to 25 mm); HUMZ, three (80 to 153 mm); ISH, four (61 to 106 mm); LACM, 28 (16 to 97 mm); MCZ, 14 (15 to 112 mm); MNHN, one (65 mm); NSMT, one (34 mm); NYZS, three (11.5 to 14.5 mm); ROM, four (33 to 99 mm, 99-mm specimen cleared and stained); SIO, 10 (15 to 97 mm); TCWC, one (20.5 mm); USNM, three (12.5 to 29 mm); ZMUB, one (107 mm); ZMUC, seven (12 to 48 mm).

Males: LACM, one (17 mm); MCZ, one (15 mm); SIO, one (15 mm); ZMUC, two (6.8 to 12 mm).

Larvae: MCZ, nine (3.5 to 12 mm); SIO, one (8 mm); UF, one male (8.5 mm); ZMUC, 99 (63 females, 32 males, and four undetermined, 2.4 to 12.5 mm TL). See Bertelsen (1951:268).

### Distinguishing Characters

Diagnosis of females and larvae as given for the genus. Males of *Microlophichthys microlophus* differ from those of *M. andracanthus* in lacking spines on the dorsal part of the upper denticular; upper denticular with more than 10 teeth, lower denticular with 8 teeth; and pectoral-fin rays 18 or 19.

### Description

Description of females and larvae as given for the genus. Metamorphosing males with lightly pigmented skin; jaw teeth present; a group of small wartlike dermal papilla containing rudiments of denticles above and below mouth; height of olfactory organ slightly greater than that of eye; length of anterior nostril almost 2.5 times that of posterior nostril and half diameter of eye; posterior nostril situated close to eye; eye with a narrow aphakic space in front of lens; diameter of eye about 12% SL (Bertelsen, 1951:92).

Adult males with shape of subopercle, fin-ray counts, and pigmentation behind subopercle and beneath skin of caudal peduncle conforming well with other known material of the species; jaw teeth absent; premaxilla entirely reabsorbed or nearly so; upper denticular bearing a number (actual count indeterminable due to fusion) of long irregularly curved and fused denticles, directed obliquely forward, with tips directed posteroventrally; lower denticular bearing 8 strong, short denticles placed symmetrically with a single denticle on each side of two longitudinal rows; a narrow bridge of pigmented skin between nostrils; posterior nostril slightly larger than anterior nostril; height of posterior nostril slightly greater than that of eye; skin in narrow space between posterior nostril and margin of eye pigmented; olfactory lamellae badly damaged, count indeterminable; eye with an aphakic space; diameter of eye about 7% SL.

### Sexual Maturity

The two largest known females (106 to 112 mm, ISH 2556/71, MCZ 165969) are nearly mature. The ovary of the 106-mm specimen, about 30 mm long and 20 mm in greatest width, is filled with eggs 0.30 to 0.35 mm in diameter, situated in densely folded lamellae each of which contains a single layer of eggs (Bertelsen and Pietsch, 1977:185). The eggs of the 112-mm specimen are somewhat larger, approximately 0.40 to 0.45, but the specimen has undergone considerable dehydration so they must have been significantly larger in life. The largest of five known metamorphosed males (ZMUC P9290) measures 12 mm (see Bertelsen, 1951:92).

Distribution

*Microlophichthys microlophus* is widely distributed in all three major oceans of the world (Fig. 226). It is present on both sides of the Atlantic, extending from off the Florida coast, north to Bermuda and the central North Atlantic at approximately 54°N, 30°W; and from off West Africa at about 22°N, 23°W, to the Gulf of Guinea and farther south to about 36°S, 5°E. In the Indian Ocean it is known from the Arabian Sea and Bay of Bengal. It is also known from numerous widely scattered records across the Pacific between from Japan, the Philippines, and Banda Sea to the Hawaiian Islands and farther east to the Gulf of Panama and Peru-Chile Trench as far south as about 13°S. It appears to be a rather deep-dwelling species, most records for metamorphosed indicating a range between 800 and 2200 m.

MICROLOPHICHTHYS ANDRACANTHUS BERTELSEN, 1951

Figures 123, 295

Males

*Microlophichthys andracanthus* Bertelsen, 1951:92, fig. 47 (original description, single specimen). Parin et al., 1973:143, fig. 37 (eastern South Pacific, specimen not seen by me). Nielsen, 1974:96 (listed in type catalog). Pietsch, 2005b:220, fig. 11C (reproduction).

Females and Larvae

Unknown.

Material

Two adult males (16 to 17 mm):
Holotype of *Microlophichthys andracanthus*: ZMUC P9293, 17 mm, *Dana* station 1269(1), Caribbean Sea, 17°13′N, 64°58′W, 4500 m wire, 1800 h, 15 March 1922.
Additional material: SIO 70-388, 16 mm, *Te Vega*, cruise 1, station 20, eastern tropical Pacific, 00°44′N, 168°00′W, 21-ft Tucker trawl, 0 to 1050 m, bottom depth 4400 m, 13 August 1963.

Distinguishing Characters

A species of *Microlophichthys* based solely on males, distinguished from those of its only known congener, *Microlophichthys microlophus*, in having a lateral series of spines on each side of the dorsal margin of the upper denticular; upper and lower denticular each with 8 teeth; pectoral-fin rays 17.

Description

Jaws toothless, premaxilla partially or completely reabsorbed; upper denticular with a dorsal bony lamella extending between close-set anterior nostrils and bearing 4 short spines on each side; 8 ventrally directed upper denticular teeth, distal ends turned posteriorly, in a single, horseshoe-shaped horizontal row; lower denticle with 8 short teeth, directed obliquely forward and asymmetrically placed in an irregular double row; eye with an aphakic space, diameter approximately 5% SL; olfactory organ with 13 or 14 lamellae; depth of posterior nostril about 1.5 times that of anterior nostril and twice as great as diameter eye; distance between posterior nostril and anterior margin of eye about two-thirds eye diameter; skin in front of eye and between nostrils pigmented; shape of opercular bones similar to those of males of *Microlophichthys microlophus*; concentrations of pigment present beneath subopercle and beneath skin of caudal peduncle; dorsal-fin rays 6; anal-fin rays 4; pectoral-fin rays 17.

Additional description as given for the genus and family.

Distribution

The two known specimens of *Microlophichthys andracanthus* were collected in the Caribbean Sea (type locality) and in the eastern tropical Pacific between Phoenix Island and the Line Islands.

Comments

The combination of features described above excludes this species from any other known oneirodid genus; however, based on the considerable differences from *Microlophichthys microlophus*, the material could possibly represent an undescribed genus of the family (Bertelsen, 1951:93).

Genus *Phyllorhinichthys* Pietsch, 1969 (Leafysnout Dreamers)

Figures 45, 125–127, 202, 203, 226, 240; Tables 1, 8, 9

Females

*Phyllorhinichthys* Pietsch, 1969:365, figs. 1–4 (type species *Phyllorhinichthys micractis* Pietsch, 1969, by original designation and monotypy).

Males and Larvae

Unknown.

Distinguishing Characters

An oneirodid genus, metamorphosed females of which are distinguished from those of all other described genera of the family in having few teeth in the jaws (upper jaw 10 to 39, lower jaw 8 to 21), a unique escal morphology (1 or 2, stout, internally pigmented anterior appendages; a single internally pigmented distal appendage; and a short, unpigmented, wedge-shaped posterior appendage; a combination of features unknown in any other ceratioid), and a pair of compressed, leaflike appendages on each side of the snout just anteroventral to the eye (length highly variable; see species accounts below).

The metamorphosed females of *Phyllorhinichthys* are further unique in having the following combination of character states: Vomerine teeth are present. The dorsal margin of the frontal bones is strongly convex. Sphenotic spines are well developed. The lower jaw bears a small symphysial spine. The hyomandibula has a double head. The quadrate spine is well developed, much longer than the articular spine. The posterior margin of the opercle is deeply notched. The subopercle is short and broad, its dorsal end rounded in smaller specimens, pointed in larger specimens (the posterior margin without indentation), its ventral end nearly circular (without an anterior spine or projection). The caudal-fin rays lack internal pigmentation. The illicium is short, nearly completely enveloped by the escal bulb (the esca nearly sessile on the snout) to about three times the length of the escal bulb in the largest known specimens. The pterygiophore of the illicium emerges on the snout from between the frontal bones, its anterior end exposed (especially in larger specimens), its posterior end concealed beneath the skin. The first ray of dorsal fin is well developed. There are usually 5 dorsal-fin rays (6 in two of the smallest known specimens, 10.8 to 11.8 mm; see Bertelsen and Pietsch, 1977:181, table 2) and 5 anal-fin rays (Table 1). The pectoral-fin lobe is short and broad, shorter than the longest rays of the pectoral fin. There are 19 to 24 pectoral-fin rays. The skin is smooth and naked, without dermal spinules. Darkly pigmented skin of the caudal peduncle extends well past the base of the caudal fin.

Description

The body of juvenile females is somewhat elongate, but adult females are relatively short, more or less globular. The mouth is small and oblique, the cleft terminating well before the eye. The illicium is short, its length 11.6 to 12.6% SL in the smallest known specimens (8.8 to 18.3 mm), 11.5 to 20.8% SL in larger specimens (52 to 132 mm). The distal half of the escal bulb is internally pigmented, with a heavily pigmented lateral patch on each side. The basal half of the escal bulb lacks internal pigment. The external surface of the escal bulb lacks pigment. An unpigmented wedge of tissue arises from the posteromedial margin of the escal bulb (flanked on each side by a membranous serrate-edged flap of tissue in the 96-mm specimen). The escal pore is situated on the posteromedial margin of the escal bulb, dorsal to the posterior wedge of tissue and just behind the base of the distal escal appendage. A single, internally pigmented, distal escal appendage is present, bearing a silvery distal tip, approximately as long as the escal bulb to nearly 60% SL. One or 2 stout, nontapering, internally pigmented appendages are present, each with a silvery distal tip, arising from the anteromedial or anterolateral margin of the escal bulb. Paired, unpigmented, lateral and medial escal appendages are variously developed depending on the species and standard length (see species accounts below).

A pair of translucent leaflike flaps is placed symmetrically on each side of the snout, flanking the esca, just anteroventral to the eye, each flap with a central nerve fiber extending from the base to the distal tip. Both flaps of each pair are directed anteriorly. In some specimens (LACM 9567-14, ZIN 49815, ARC 8602570) there is a similar, but smaller, anteriorly directed appendage situated on the skin covering the frontal bone, dorsal to the larger paired flaps, a central opaque nerve fiber extending from base to tip. The length of the flaps is highly variable (see species accounts below).

The dorsal end of the subopercle is rounded in the five smallest known specimens (8.8 to 18.3 mm) and in the 52-mm holotype, tapering to a point in larger specimens. The subopercle of larger specimens (118 to 132 mm) is slightly expanded posteriorly (see Bertelsen and Pietsch, 1977:181, fig. 7A, B).

The teeth are slender, all depressible, large and small intermixed in both jaws, those in the lower jaw slightly larger but fewer than those of the upper jaw. The jaw teeth and vomerine teeth of the 8.8-mm specimen are undeveloped. There are 10 to 13 teeth in the upper jaw of smaller specimens (10.8 to 18.3 mm), 24 to 39 in larger specimens; there are 8 to 12 teeth in the lower jaw of smaller specimens (10.8 to 18.3 mm), 12 to 21 in larger specimens. There are 4 to 6 teeth on the vomer. The second and third pharyngobranchials are well developed and toothed. The epibranchials and ceratobranchials are toothless.

The color in preservation is dark brown to black over the entire external surface of the body except for the distal surface of the escal bulb. The holotype has lighter or unpigmented skin on the upper lip, over a considerable area below and including the lower lip, and immediately posterior, anterior, and beneath the snout flaps. The internal pigmentation of the head and anterior part of the body of the four smallest known specimens is diffuse and weak; that of the trunk of the 8.8 to 18.3 mm specimens is continuous and well developed, except for the posterior end of the caudal peduncle. A more or less distinct band of pigment is present on the caudal peduncle (see Bertelsen and Pietsch, 1977:183, fig. 7E).

### Comments

Larvae of *Phyllorhinichthys* are unknown, but the unique pigment pattern of the smallest known specimens (8.8 to 18.3 mm; a band of melanophores on the caudal peduncle and the absence of distinct opercular and dorsal pigment groups), combined with a rounded subopercle and dorsal- and anal-fin ray counts, should distinguish larvae of *Phyllorhinichthys* from those of all other known ceratioids (see Bertelsen and Pietsch, 1977:183).

### Diversity

Two species:

### Key to Females of Species of *Phyllorhinichthys*

1A. Illicium short, 11.5 to 12.6% SL; a pair of simple anterior escal appendages; distal escal appendage short and stout, 6.4 to 15.8% SL . . . . . . . . . . . *Phyllorhinichthys micractis* Pietsch, 1969, p. 428

NOTE  Eleven known specimens, 8.8 to 140 mm; Atlantic, Pacific, and Indian oceans

1B. Illicium long, 16.0 to 20.8% SL; a single anterior escal appendage, bifurcated only at distal tip; distal escal appendage long and slender, 36.4 to 57.6% SL . . . . . . . . . . . . . . . . . . . . . . . . . *Phyllorhinichthys balushkini* Pietsch, 2004, p. 429

NOTE  Seven known specimens, 72 to 168 mm; Atlantic Ocean

### PHYLLORHINICHTHYS MICRACTIS PIETSCH, 1969

Figures 45, 125–127, 226; Tables 8, 9

### Females

*Phyllorhinichthys micractis* Pietsch, 1969:365, figs. 1–4 (original description, single specimen). Pietsch, 1972b:335, figs. 1–6 (additional specimen, eastern tropical Pacific; description, histology of snout flaps and lateral line papillae). Bertelsen and Pietsch, 1977:178, figs. 5–7 (in part; three additional specimens, 118-mm specimen here referred to *Phyllorhinichthys balushkini*; North Atlantic and western Indian Ocean; description; escae, opercular bones, pectoral radials figured). Herring and Morin, 1978:322, fig. 9.12 (in part, species not cited; bioluminescence; figures after Bertelsen and Pietsch, 1977). Bertelsen, 1986:1398, figs. (in part; eastern North Atlantic; after Bertelsen and Pietsch, 1977). Munk, 1992:36, fig. 5 (in part, after Bertelsen and Pietsch, 1977; accessory escal gland). Nielsen and Bertelsen, 1992:60, fig. 7 (misidentification, figure shows *P. balushkini*; North Atlantic). Andriyashev and Chernova, 1994:102 (in part; listed for Arctic seas and adjacent waters). Munk, 1998:181 (in part, after Bertelsen and Pietsch, 1977; escal light-guiding structures). Munk, 1999:267, 278–280, figs. 3F, 10 (in part, after Bertelsen and Pietsch, 1977; bioluminescence). Pietsch, 2002b:1064 (western central Atlantic). Pietsch, 2004:800, figs. 1, 4 (generic revision).

### Males and Larvae

Unknown.

### Material

Eleven metamorphosed females (8.8 to 140 mm):

Holotype of *Phyllorhinichthys micractis*: LACM 9567-14, female, 52 mm, *Velero IV*, cruise 832, station 11187, eastern North Pacific, 22 km off northern end of Guadalupe Island, Mexico, 29°16′30″ to 29°35′30″N, 118°11′30″ to 118°18′30″W, 3-m Isaacs-Kidd midwater trawl fished open between surface and 1050 m, bottom depth 2194 to 3429 m, 1315 to 2103 h, 3 August 1966.

Additional material: ARC, one (120 mm); BMNH, one (10.8 mm); HUMZ, one (108 mm); MCZ, three (8.8 to 18.3 mm); SIO, one (16 mm); ZIN, one (96 mm); ZMUB, one (61 mm); ZMUC, one (140 mm).

### Distinguishing Characters

Metamorphosed females of *Phyllorhinichthys micractis* differ from those of *P. balushkini*, the only other member of the genus, in having a shorter illicium (11.5 to 12.6% SL versus 16.0 to 20.8% SL); 2 anterior escal appendages (versus 1, bifurcated at distal tip); and a considerably shorter, stouter distal escal appendage (6.4 to 15.8% SL versus 36.4 to 57.6% SL).

### Description

Esca varying somewhat among known material in presence and number of appendages and filaments: esca of 8.8 mm specimen with tiny medial appendages, lateral filaments absent; 10.8-mm specimen with medial appendages undeveloped, a single short lateral filament on left side only (see Bertelsen and Pietsch, 1977:178, fig. 6B, C); a simple lateral filament on each side in 11.8-, 18.3-, and 96-mm specimens, but absent in 16-, 52-, and 120-mm specimens; medial appendages present in 16-, 18.3-, 96-, and 120-mm specimens, in each case forming a fanlike series of filaments, all arising from an unpigmented ridge of tissue, oriented anteroposteriorly; 96-mm specimen with 8 filaments on left side, 10 on right (Pietsch, 1972b:337, fig. 3); 120-mm specimen with 4 filaments on each side. All unpigmented tapering filaments with a central opaque nerve fiber extending from base to distal tip; lumen of internal tubes of distal and paired anterior appendages filled with transparent gelatinous tissue, inner walls covered with a silvery reflecting layer.

Snout flaps minute in 8.8 mm specimen; length highly variable in remaining specimens: upper snout flap 8.3% SL in 10.8-mm specimens, 2.1% SL in 11.8-mm specimen, 6.2% SL in 16-mm specimen, 4.9% SL in 18.3-mm specimen, 2.1 to 7.7% SL in 52 to 120 mm specimens, shrunken but present in the 108-mm dried specimen.

Additional description as given for the genus.

Sexual Maturity

The 96-mm specimen (ZIN 49815) appears to be sexually mature. The right ovary, about 26 mm long (27.1% SL), contains approximately 130,000 eggs (count based on extrapolation from wet weight) each about 0.3 mm in diameter.

Distribution

The nine known specimens of *Phyllorhinichthys micractis* were collected from widely separated localities in all three major oceans of the world (Fig. 226): two from the North Atlantic: from off Bermuda, and Newfoundland at about 41°N, 56°W; three from the western Indian Ocean; and four from the North Pacific: from off Japan, the Emperor Seamounts, the Line Islands (just above the equator at about 155°W), and Guadalupe Island, Mexico. All were collected with gear fished open between the surface and 3600 m.

*PHYLLORHINICHTHYS BALUSHKINI* PIETSCH, 2004

Figures 125, 127, 226, 240; Table 8

Females

*Phyllorhinichthys micractis*: Bertelsen and Pietsch, 1977:178, figs. 5 to 7 (misidentification, in part, 118-mm specimen, ISH 536/73). Herring and Morin, 1978:322, fig. 9.12 (bioluminescence, figures after Bertelsen and Pietsch, 1977). Bertelsen, 1986:1398, figs. (misidentification, in part; after Bertelsen and Pietsch, 1977). Nielsen and Bertelsen, 1992:60, fig. 7 (misidentification after Bertelsen and Pietsch, 1977; North Atlantic). Andriyashev and Chernova, 1994:102 (misidentification, in part; listed for Arctic seas and adjacent waters). Munk, 1998:181 (misidentification, in part, after Bertelsen and Pietsch, 1977, and Munk, 1992; escal light-guiding structures). Munk, 1999:267, 278–280, fig. 10 (misidentification, in part, after Bertelsen and Pietsch, 1977, and Munk, 1992, 1998; bioluminescence).

*Phyllorhinichthys balushkini* Pietsch, 2004:800, figs. 1, 4 (original description, five specimens; generic revision).

Males and Larvae

Unknown.

Material

Seven metamorphosed females (72 to 168 mm):

Holotype of *Phyllorhinichthys balushkini*: ISH 536/73, 118 mm, *Walther Herwig*, Overflow 73 Expedition, cruise 51, station 695/73, eastern North Atlantic, 55°43′N, 25°53′W, MT-1600, between the surface and 2600 m, bottom depth 3210 m, 1837 to 2130 hrs, 22 September 1973.

Paratypes of *Phyllorhinichthys balushkini*: BMNH 1998.2.23.51, 120 mm, *Discovery* station 13200-94, eastern North Atlantic, off SW coast of Ireland, Porcupine Abyssal Plain, 48°51′N, 16°26′W, 4847 to 4851 m, 25 July 1997. UW 20826 (formerly VIMS 05726), 86 mm, *Columbus Iselin*, cruise CI-8007, station C043, western North Atlantic, Bahamas, Tongue of the Ocean, 24°55.5′ to 25°03.0′N, 77°44.5′ to 77°43.8′W, 9.15-m otter trawl fished pelagically, with 2000 to 4193 m of wire out, bottom depth 2230 to 2625 m, 0400 to 1000 hrs, 16 September 1980. ZIN 49816, 143 mm, eastern South Atlantic, 32°48′S, 1°48′E, between the surface and 1080 to 1350 m, 19 October 1975. ZMUC P922145, 125 mm, *Walther Herwig*, Mid-Atlantic Ridge Expedition, cruise 52, station 401/82, eastern North Atlantic, 49°51.7′N, 16°51.7′W, PT-1600, between the surface and 3200 m, bottom depth 4800 m, 30 minutes at depth, 16 June 1982.

Additional material: SIO, one (72 mm); ZMUC, one (168 mm).

Distinguishing Characters

Metamorphosed females of *Phyllorhinichthys balushkini* differ from those of *P. micractis*, the only other member of the genus, in having a longer illicium (16.0 to 20.8% SL versus 11.5 to 12.6% SL); a single anterior escal appendage (versus 2), bifurcated only at distal tip; and a considerably longer, more slender, distal escal appendage (36.4 to 57.6% SL versus 6.4 to 15.8% SL). In addition, *P. balushkini* probably has smaller, less well developed snout flaps, but the skin of the snout of three of the five known specimens is torn away, making proper assessment of this character impossible.

Description

Esca of 118-mm holotype best preserved among known material: a single, stout, anterior escal appendage containing a pair of pigmented, internal, light-guiding tubes, diverging from one another and separated only at distal tip; tips of internal tube of anterior escal appendage with circular transparent windowlike opening; internally pigmented distal escal appendage extremely long, 57.6% SL, somewhat damaged, but presumably covered entirely in life with darkly pigmented skin; pigmented internal tube of distal appendage terminating in a pair of small, lenslike, transparent bulbs, partly covered distally by a V-shaped pigmented patch (bright silver in freshly caught specimen; Bertelsen and Pietsch, 1977:179); lumen of internal tubes of distal and anterior appendages filled with completely transparent gelatinous tissue, inner walls covered with a reflecting layer; a paired fan-shaped series of medial escal filaments emerging from dorsal surface of esca between anterior and distal escal appendages, 8 and 9, respectively, all sharing a common base; paired lateral appendages, each consisting of a fan-shaped array of 14 (right side) and 17 (left side) filaments, all sharing a common base.

Snout flaps relatively small; upper and lower flaps about 1.0 to 0.7% SL in 118-mm holotype, 0.9 to 0.7% SL in 120-mm paratype, respectively; skin of snout torn away in three remaining paratypes, presence or absence of snout flaps unknown.

Additional description as given for the genus.

Sexual Maturity

The ovaries of the 125-mm paratype (ZMUC P922145) are about 35-mm long (27% SL) and filled with thousands of tiny eggs, about 0.2 to 0.3 mm in diameter.

Distribution

Six of the seven known specimens of *Phyllorhinichthys balushkini* are from the Atlantic: one from the Tongue of the Ocean off the Bahamas, four from the North Atlantic between approximately 48 to 63°N and 16 to 54°W, and a sixth from the eastern South Atlantic at about 33°S, 2°E (Fig. 226). The seventh specimen was collected in the Pacific sector of the Southern Ocean at 63°S, 179°E. All were collected with gear fished open between the surface and 3200 m.

Genus *Tyrannophryne* Regan and Trewavas, 1932 (Pugnacious Dreamers)

Figure 128; Table 8

Females

*Tyrannophryne* Regan and Trewavas, 1932:83, pl. 4, fig. 1 (type species *Tyrannophryne pugnax* Regan and Trewavas, 1932, by monotypy).

Males and Larvae

Unknown.

### Distinguishing Characters

Metamorphosed females of *Tyrannophryne* (three known specimens, 12 to 50 mm) are distinguished from those of all other genera of the family by having an extremely large oblique mouth, the elements of the upper and lower jaws extending posteriorly far beyond the base of the pectoral fin and opercular opening.

Metamorphosed females of *Tyrannophryne* are further unique in having the following combination of character states: Vomerine teeth are present or absent (apparently lost with growth). Sphenotic spines are well developed. The lower jaw bears a well-developed symphysial spine. The angular is elongate and tapering, forming a long narrow spine. The hyomandibula has a double head. The quadrate spine is small, but longer than the articular spine. The posterior margin of the opercle is deeply notched. The subopercle is short and broad, its dorsal end rounded (the posterior margin without indentation), its ventral end nearly circular (without an anterior spine or projection). The second pharyngobranchial is toothless (absent?). The caudal-fin rays lack internal pigmentation. The illicium is extremely short, almost totally enveloped by tissue of the esca in smaller specimens. The pterygiophore of the illicium emerges on the snout from between the frontal bones, its anterior end just barely exposed, its posterior end concealed beneath the skin. The first ray of the dorsal fin is well developed. There are 5 dorsal-fin rays and 5 anal-fin rays (Table 1). The pectoral-fin lobe is short and broad, shorter than the longest rays of the pectoral fin. There are 18 to 20 pectoral-fin rays. The skin is naked, without dermal spinules. Darkly pigmented skin of the caudal peduncle extends well past the base of the caudal fin (the genus is known from only three metamorphosed females, internal anatomical features are unknown).

### Description

The body of juvenile females is elongate and slender, not globular. The snout is relatively short. The mouth is extremely large, its cleft extending well past the base of the pectoral fin. The lower jaw is extraordinarily long, 67% SL in a 16.5-mm specimen and 68% in a 50-mm specimen. The illicium is extremely short, less than 8.0% SL in the 12-mm holotype, 13.9% SL in a 16.5-mm specimen, and 9.6% SL in a 50-mm specimen. The esca has a medial, distal streak of pigment, forked posteriorly. A terminal escal papilla is absent. A large, laterally compressed posterior escal appendage is present, bearing a small distal cleft and a low anterior crest. Anterior and lateral escal appendages are absent.

The teeth are slender, more or less straight, and all depressible, those in the lower jaw somewhat larger than those in the upper jaw. The total number of teeth in the upper jaw is about 46 to 60 in smaller specimens, about 105 in a 50-mm specimen; those in the lower jaw about 38 to 40 in smaller specimen, about 110 in a 50-mm specimen. There are 2 vomerine teeth in the 12-mm holotype (Regan and Trewavas, 1932:83), but the vomer is toothless in the two larger specimens. The second pharyngobranchial is toothless. The third pharyngobranchial is well developed and heavily toothed. The ceratobranchials and epibranchials are toothless.

### Diversity

A single species:

#### TYRANNOPHRYNE PUGNAX REGAN AND TREWAVAS, 1932 (PUGNACIOUS DREAMER)

Figure 128; Table 8

### Females

*Tyrannophryne pugnax* Regan and Trewavas, 1932:83, pl. 4, fig. 1 (original description, single specimen). Nielsen, 1974:96 (listed in type catalog). Pietsch, 1974a:109 (listed). Pietsch, 1999:2032 (western central Pacific).

### Males and Larvae

Unknown.

### Material

Three metamorphosed females (12 to 50 mm):

Holotype of *Tyrannophryne pugnax*: ZMUC P9294, 12 mm, *Dana* station 3577(7), South Pacific near Tahiti, 18°49′S, 153°10′W, 4000 m wire, 0615 to 0815 h, 19 October 1928.

Additional material: ASIZP, one (50 mm); SIO, one (16.5 mm).

### Distinguishing Characters and Description

As given for the genus and family.

### Distribution

*Tyrannophryne pugnax* is so far known only from the western and central Pacific: the holotype collected in the South Pacific near Tahiti, with an open net somewhere between the surface and about 2000 m; the 16.5-mm specimen from off Bikini Atoll, also taken in an open net, between the surface and about 2100 m; and the 50-mm specimen from off Taiwan by a commercial fishing vessel, with a bottom trawl at a depth of about 400 m.

### Comments

Except for the huge jaws, which are comparable to no other known ceratioid, *Tyrannophryne* greatly resembles *Microlophichthys*, especially in the shape of the opercular bones and the morphology of the esca (Bertelsen, 1951:93).

## Genus *Dolopichthys* Garman, 1899 (Longsnout Dreamers)

Figures 14–16, 25, 33, 129, 132, 135, 190, 202, 203, 227, 228, 295; Tables 1, 8, 9

### Females

*Dolopichthys* Garman, 1899:81, pls. 13–15, figs. 5–7 (type species *Dolopichthys allector* Garman, 1899, by monotypy).

*Oneirodes* Brauer, 1906:316, pl. 15, fig. 6 (in part; erroneous designation subsequently corrected by Regan and Trewavas, 1932; type species *Oneirodes eschrichtii* Lütken, 1871, by monotypy). Gilchrist and von Bonde, 1924:23, pl. 6, fig. 2 (erroneous designation).

*Dolopichthys* (subgenus *Dolopichthys*) Regan and Trewavas, 1932:78 (genus *Dolopichthys* broadened to incorporate five subgenera, amended by Bertelsen, 1951).

### Males

*Rhynchoceratias* Regan, 1926:44 (in part; erroneous designations, two males subsequently referred to *Trematorhynchus* by Regan and Trewavas, 1932; type species *Rhynchoceratias brevirostris* Regan, 1925c, by subsequent designation of Fowler, 1936).

*Trematorhynchus* Regan and Trewavas, 1932:91 (in part; erroneous designations, two males subsequently referred to *Dolopichthys* by Bertelsen, 1951; type species *Trematorhynchus leucorhinus* Regan, 1925c, by subsequent designation of Burton, 1933).

### Distinguishing Characters

Metamorphosed females of *Dolopichthys* are distinguished from those of all other oneirodids by having the following combination of character states: The ethmoid cartilage and vomer are wide, wider than the distance between the anterolateral tips of the lateral ethmoids and frontals. Vomerine teeth are usually present. The ethmoid region is depressed, the nasal foramina narrow and oval in shape. The frontals are long and anterior in position, overhanging and extending past the anterior limits of the ethmoid cartilage and vomer. The dorsal margin of the fronts is nearly linear. Ventromedial extensions of the frontals approach each other on the midline, making contact with the parasphenoid. The frontals are separated from the prootics. The pterosphenoid is present. The anterior end of the illicial trough is wider and shallower than the posterior end.

Sphenotic spines are well developed. The symphysial cartilage of the upper jaw is wider than long. The lower jaw bears a well-developed symphysial spine. The hyomandibula has a double head. The quadrate spine is well developed, considerably longer than the articular spine. The posterior margin of the opercle is deeply notched. The subopercle is long and narrow, its dorsal end slender, tapering to a point (the posterior margin without indentation), its ventral end oval in shape (without an anterior spine or projection). The first pharyngobranchial is absent. The second pharyngobranchial is reduced, but bears as many as 6 teeth. The second hypobranchial is absent. The caudal-fin rays lack internal pigmentation. The illicium is considerably longer than the length of the esca bulb. The pterygiophore of the illicium is cylindrical throughout its length, emerging on the snout from between the frontal bones, its anterior end slightly exposed, its posterior end concealed beneath the skin. The first ray of the dorsal fin is well developed. There are 5 to 7 (rarely 8) dorsal-fin rays and 4 to 6 anal-fin rays (Table 1). The pectoral-fin lobe is short and broad, shorter than the longest rays of the pectoral fin. There are 18 to 21 (rarely 17 or 22) pectoral-fin rays. The coracoid lacks a posteroventral process. The pelvic bones are simple, with or without a distal expansion. The skin is smooth and naked, without dermal spinules. Darkly pigmented skin of the caudal peduncle extends well past the base of the caudal fin.

Metamorphosed males are distinguished from those of all other genera of the family by having the following combination of character states: The opercle is broad, its posterior margin only slightly concave, the dorsal fork as long as and much broader than the ventral fork. The posterior nostril is contiguous to the eye. The nasal area is unpigmented. There are 10 to 11 olfactory lamellae. There are 5 to 8 upper denticular teeth and 8 to 10 lower denticular teeth (Bertelsen, 1951:96).

The larvae are unique in having the following combination of character states: The body is short, its depth 60 to 65% SL. The gill cover is heavily pigmented, especially along its ventral and posterior margins, the melanophores extending to some distance from the distal tips of the branchiostegal rays. The dorsal pigment is restricted to the upper part of the body, reaching posteriorly slightly beyond the base of the dorsal fin in larger specimens. Caudal peduncle pigmentation is separated into dorsal, lateral, and ventral groups. The peritoneum is darkly pigmented (Bertelsen, 1951:96).

Description

The body of metamorphosed females is relatively long and slender, not globular. The depth of the head is 25.5 to 47.7% SL.

The mouth is large, the cleft extending past the eye. The jaws are equal anteriorly. The angular bone terminates as a small spine in some specimens. The length of the illicium is highly variable, 17.8 to 75.3% SL, becoming longer proportionately with growth. The esca bears a large, compressed posterior appendage, consisting of a swollen basal part, darkly pigmented in some specimens, and a distal part consisting of a long tapering anterior filament usually, but not always, connected by a thin membrane to a shorter tapering posterior filament. Dorsal and anterior escal filaments or papillae are present or absent. The distal part of the posterior appendage is darkly pigmented in most specimens, its length greater than approximately 70 mm.

The opercle is strongly bifurcate, the two forks forming an acute angle of 27° to 45°, the length of the dorsal fork 46.2 to 77.1% of the length of the ventral fork.

The teeth are slender, recurved, and all depressible, those in the lower jaw in overlapping sets of up to 8 teeth each, the teeth within each set becoming progressively larger anteriorly, the outer teeth smaller than the inner. The pattern of tooth placement is especially obvious in specimens having high tooth counts. The pattern in the upper jaw apparently the same as the lower but not nearly as obvious. Teeth in the lower jaw are larger and considerably more numerous than those in the upper jaw. The number of teeth in the upper jaw ranges from fewer than 25 teeth to over 400, those in the lower jaw from fewer than 35 to nearly 600. Teeth are very few and rudimentary in specimens smaller than approximately 14 mm. There are 0 to 14 teeth on the vomer, the largest tooth placed outermost. The second pharyngobranchial is less than half the size of the third pharyngobranchial, the former bearing 1 to 4 teeth, the latter with 12 to 19 teeth (pharyngobranchial tooth counts based on three cleared and stained specimens, 56 to 76 mm).

The following measurements, expressed in percent of standard length, are summarized for females (15 to 159 mm) of all species: head length 31.6 to 48.4; head width 15.4 to 39.5; head depth 25.5 to 47.7; lower-jaw length 28.4 to 53.6; premaxilla length 20.4 to 33.3; least outside width of frontal bones 8.7 to 14.8; distance between posterolateral tips of posttemporal bones 15.1 to 25.8; distance from tip of sphenotic spine to posterolateral tip of respective posttemporal bone 10.1 to 20.6; illicium length 25.0 to 75.3.

For additional description of larvae and males, see Bertelsen (1951:98–100).

Diversity

Seven species and two nominal forms of doubtful validity differentiated as follows:

Key to Females of Species of the Genus *Dolopichthys*

1A. Length of illicium greater than 70% SL in specimens more than 70 mm, greater than 50% SL in specimens more than 30 mm, greater than 40% SL in specimens 20 to 30 mm; dorsal midline of escal bulb with 2 darkly pigmented, distally bifurcating papillae . . . . . . . . . . . *Dolopichthys danae* Regan, 1926, p. 434

NOTE  Six known specimens, 20.5 to 115 mm; North Atlantic Ocean

1B. Length of illicium less than 70% SL in specimens more than 70 mm, less than 50% SL in specimens more than 30 mm, less than 40% SL in smaller specimens; dorsal midline of escal bulb without pigmented, distally bifurcating papillae . . . . . . . . . . . . . 2

2A. Number of teeth on vomer 4 or more, usually more than 6; lower jaw with more than 130 teeth in specimens larger than 60 mm, more than 115 in specimens 25 to 60 mm . . . . . . . . 3

2B. Vomer toothless in specimens greater than 35 mm, 2 to 4 vomerine teeth in smaller specimens; lower jaw with fewer than 130 teeth in specimens more than 60 mm, fewer than 115 in smaller specimens . . . . . . . . . . . 4

3A. Lower jaw with more than 300 teeth in specimens greater than 70 mm, more than 150 in specimens 25 to 70 mm, more than 85 in specimens 18 to 25 mm; dorsal midline of escal bulb with a single posteriorly to posterodorsally directed papilla . . . . . . . . . . . . . . . . . *Dolopichthys pullatus* Regan and Trewavas, 1932, p. 432

NOTE  Forty-five known specimens, 10 to 115 mm; Atlantic, Pacific, and Indian oceans

3B. Lower jaw with fewer than 300 teeth in specimens greater than 70 mm, usually fewer than 150 in specimens 25 to 70 mm, fewer than 85 in smaller specimens; dorsal midline of escal bulb with a single dorsally to anterodorsally directed papilla . . . . . . . . . . . . . *Dolopichthys longicornis* Parr, 1927, p. 433

NOTE  Twenty-six known specimens, 14 to 159 mm; Atlantic, Pacific, and Indian oceans

4A. Distal surface of escal bulb with a pair of posteriorly directed, bilaterally placed papillae, each darkly pigmented on dorsal surface and on distal tip . . . . . . . . . . . . . . . . . . 5

4B. Distal surface of escal bulb without posteriorly directed, bilaterally placed papillae . . . . . . . . . . . . . . . 6

5A. Esca with a darkly pigmented, anteriorly directed, medial papilla, flanked on each side by a similar but

posteriorly directed papilla . . . . . . .
. . . . . . . . . . . . . . *Dolopichthys karsteni*
Leipertz and Pietsch, 1987, p. 436

NOTE Eight known specimens, 18 to 99 mm; western North Atlantic Ocean

5B. Esca without an anteriorly directed medial papilla . . . . . . . . . . . . . . . . .
. . . . . . . . . . . . . *Dolopichthys allector*
Garman, 1899, p. 435

NOTE Fifteen known specimens, 16 to 154 mm; Atlantic and eastern Pacific oceans

6A. Dorsal midline of escal bulb with 2 long filaments; lower jaw with fewer than 40 teeth in specimens 22 mm and smaller . . . . . . . . . . . . . . . . . . .
. . . . . . . . . . . . . . *Dolopichthys dinema*
Pietsch, 1972c, p. 436

NOTE Two known specimens, 21 to 22 mm; eastern tropical Atlantic Ocean

6B. Dorsal midline of escal bulb without long filaments; lower jaw with more than 60 teeth in specimens 20 mm and larger . . . . *Dolopichthys jubatus*
Regan and Trewavas, 1932, p. 434

NOTE Eight known females, 15 to 89 mm; Atlantic, Pacific, and Indian oceans

## *DOLOPICHTHYS PULLATUS* REGAN AND TREWAVAS, 1932

Figures 33, 131, 135, 228; Table 8

### Females

*Dolopichthys* (subgenus *Dolopichthys*) *pullatus* Regan and Trewavas, 1932:79, fig. 123, pl. 3, fig. 1 (original description, single specimen).

*Dolopichthys pullatus*: Beebe and Crane, 1947:161, text fig. 6 (description, additional specimen, comparison, comments). Beaufort and Briggs, 1962:252, fig. 59 (description, figure after Regan and Trewavas, 1932). Pietsch, 1972c:7, figs. 1, 3, 4, 11, 12 (additional material, description, comparison with all known material). Brewer, 1973:25 (Gulf of California). Nielsen, 1974:95 (listed in type catalog). Bertelsen and Pietsch, 1977:185 (additional material, eastern Atlantic). Pietsch and Seigel, 1980:394 (additional specimen, Banda Sea). Bertelsen and Pietsch, 1983:87, fig. 8 (additional material, Australia). Paulin and Stewart, 1985:28 (New Zealand). Paulin et al., 1989:139 (New Zealand). Bertelsen, 1990:501 (synonymy, eastern tropical Atlantic). McEachran and Fechhelm, 1998:867, 868 (Gulf of Mexico, in key). Stewart and Pietsch, 1998:13, fig. 8 (additional female, New Zealand). Pietsch, 1999:2032 (western central Pacific). Anderson and Leslie, 2001:14, fig. 11C (additional female, South Africa). Pietsch, 2002b:1064 (western central Atlantic). Love et al., 2005:60 (eastern North Pacific).

*Dolopichthys longicornis*: Bertelsen, 1951:96–98, 100, 101, figs. 53d, 57 (in part; *Dolopichthys pullatus*, a synonym of *D. longicornis*). Bussing, 1965:222, 223 (misidentification, description of an additional specimen here referred to *D. pullatus*; comparisons, comments).

*Dolopichthys allector*: Pietsch, 1972a:35, 36, 42, 43, 45, fig. 24(7) (in part; misidentifications, osteological comments, comparison with other ceratioids, second hypobranchial absent).

### Males and Larvae

Unknown.

### Material

Forty-five metamorphosed females (10 to 115 mm):

Holotype of *Dolopichthys* (subgenus *Dolopichthys*) *pullatus*: ZMUC P 92101, 34 mm, *Dana* station 3680(1), Molucca Sea, 2°22′S, 126°58.5′E, 5000 m wire, 1035 h, 27 March 1929.

Additional material: AMS, four (14 to 73 mm); BMNH, two (28 to 42 mm); CAS, two (10 to 32 mm); ISH, nine (50 to 115 mm); LACM, 10 (15.5 to 93 mm); MCZ, three (18 to 105 mm); NMNZ, one (28 mm); NYZS, one (18 mm); ROM, one (73 mm); SAM, one (63 mm); SIO, eight (14 to 72 mm); UF, two (15 to 21 mm).

### Distinguishing Characters

Metamorphosed females of *Dolopichthys pullatus* differ from those of all other members of the genus in details of escal morphology. Females are further unique in having a considerably greater number of teeth in the lower jaw.

### Description

Escal bulb with a small rounded, anteriorly to anteroventrally directed papilla on anterior midline, continuous internally with a slightly larger posteriorly to posterodorsally directed papilla on dorsal midline; distal tips of anteriorly and posteriorly directed papillae, and internal connection between papillae, darkly pigmented in some specimens (see Comments below); in some specimens, 2 small, unpigmented papillae placed bilaterally on dorsal surface of escal bulb near base of posterior escal appendage; compressed posterior appendage sigmoidal in shape and usually twice as long as escal bulb; swollen basal part of posterior appendage darkly pigmented in specimens 84 mm and larger; connecting membrane between anterior and posterior; filaments of posterior appendage present and pigmented in specimens 84 mm and larger.

Total number of teeth in lower jaw of specimens 14.5 mm and larger, 44 to nearly 600; number of teeth on vomer 4 to 14; dorsal-fin rays 5 to 8; anal-fin rays 4 to 6; pectoral-fin rays 17 to 22.

Measurements in percent of standard length: head length 35.3 to 45.2; head width 19.1 to 30.9; head depth 33.0 to 47.7; premaxilla length 25.9 to 34.2; lower-jaw length 38.7 to 47.5; illicium length 26.4 to 42.4.

Additional description as given for the genus and family.

### Distribution

*Dolopichthys pullatus* is known from both sides of the Atlantic Ocean, including the Gulf of Mexico, between 50°N and 6°S; from the western Indian Ocean off South Africa at about 32°S, extending north and east to the central Arabian Sea; and across the Pacific Ocean from the Molucca Sea, New South Wales, and New Zealand to the Hawaiian Islands, California coast, Gulf of California, and south to the Gulf of Panama, and Chile at approximately 37°S (Fig. 228). Ninety-six percent of the specimens for which data were available were collected by gear fished at maximum depths of 800 m or more; about 66% by gear fished between 1000 and 2000 m. A 58-mm specimen (SIO 70-389) was captured on the bottom at 945 to 960 m.

### Sexual Maturity

The ovaries of the 115-mm specimen (ISH 2732/71) of *Dolopichthys pullatus* are well developed. The left ovary is disk-shaped and approximately 36 mm in diameter (31% SL). The right ovary, in contrast, is oval in shape, approximately 27 mm long (23.5% SL) and 10 mm wide. This difference in shape and size causes most of the volume of the left ovary to be displaced ventrally, thus making more efficient use of the space available for the maturing gonads. Both ovaries are tightly packed with eggs, and as described for the 95-mm holotype of *Dolopichthys mucronatus* (= *D. longicornis*) by Mead et al. (1964), the eggs are of two distinct size groups: ripening eggs, measuring 0.4 to 0.6 mm in diameter, and eggs in early stages of development, approximately 0.2 mm in diameter.

### Comments

Regan and Trewavas (1932) justified their naming of three new species of *Dolopichthys*, *Dolopichthys pullatus*, *D. mucronatus*, and *D. jubatus*, on the basis of differences in the appearance of the esca. However, Bertelsen (1951:26) believed these differences to be "essentially smaller than indicated" and not differing substantially from the esca of *D. longicornis* Parr, 1927. Arranging Regan and Trewavas's types according to size of the specimen, Bertelsen (1951) thought they formed a uniform developmental sequence and accordingly synonymized all three species with *D. longicornis*.

With the advantage of additional material, especially larger sized individuals,

Pietsch (1972c) showed *D. pullatus* to differ from *D. longicornis* in a number of important ways. As in all species of the genus, the escal morphology of *D. pullatus* is diagnostic, differing from that of *D. longicornis* in the placement, shape, and pigmentation of the papillae, and in the size and shape of the posterior escal appendage. The considerably greater number of teeth in the lower jaw of *D. pullatus* easily separates this species from *D. longicornis* and all other members of the group. *Dolopichthys pullatus* is further distinguished from *D. longicornis* by having a slightly longer illicium.

Within the material now considered to be *D. pullatus*, the escae are of two distinct types, differing only in the amount of pigment present on the distal tips of the anteriorly and posteriorly directed papillae, and on the internal connection between the papillae. These parts of the esca are densely pigmented in 10 of the 21 known specimens with well-preserved escae, and lightly pigmented to unpigmented in the remaining 11 individuals. This difference is not ontogenetic, as the size ranges for the two groups (15 to 81 mm for the darkly pigmented type and 14.5 to 105 mm for the lightly pigmented type) are essentially the same. The presence or absence of pigment does not appear to be the result of differences in fixation or preservation or length of time in preservative. Individuals of both escal types occur sympatrically in all three major oceans of the world. The difference in pigmentation may be explained by an expansion and retraction of pigmented tissue layers within the escal bulb. This movement of pigment might well function to control the amount of light that is emitted by the esca. But no anatomical or histological evidence for such a mechanism has been found (see Bioluminescence and Luring, Chapter Six).

### DOLOPICHTHYS LONGICORNIS PARR, 1927

Figure 227; Table 8

#### Females

*Dolopichthys longicornis* Parr, 1927:18, fig. 6 (original description, single specimen). Norman, 1930:354 (misidentifications; two specimens, one subsequently described as *Dolopichthys jubatus* by Regan and Trewavas, 1932; the other, from *Discovery* station 287, was referred to *D. pullatus* by Pietsch, 1972c). Bertelsen, 1951:96, 100, 101, fig. 53e (in part; *D. pullatus*, *D. jubatus*, and a specimen of *D. danae*, synonyms of *D. longicornis*; description, comparison with all known material, in key, comments). Grey, 1956a:253 (in part, synonymy, distribution). Briggs, 1960:179 (worldwide distribution). Mead et al., 1964:588 (egg sizes and numbers). Bussing, 1965:222, 223 (misidentification; description of an additional specimen here referred to *D. pullatus*; comparison, comments). Pietsch, 1972c:12, figs. 1, 3, 5, 11, 12 (additional material, description, comparison with all known material). Maul, 1973:672 (synonymy, eastern North Atlantic). Bertelsen and Pietsch, 1977:185 (additional material, eastern Atlantic). Hubbs et al., 1979:13 (California). Pietsch and Seigel, 1980:394 (additional specimen, Banda Sea). Bertelsen, 1986:1390, figs. (eastern North Atlantic). Pequeño, 1989:43 (Chile). Bertelsen, 1990:501 (synonymy, eastern tropical Atlantic). Nielsen and Bertelsen, 1992:60, fig. 4 (North Atlantic). Andriyashev and Chernova, 1994:102 (listed for Arctic seas). Jónsson and Pálsson, 1999:200, fig. 4 (Iceland). Munk, 1999:267 (bioluminescence). Pietsch, 1999:2032 (western central Pacific). Anderson and Leslie, 2001:13, fig. 11B (*Walther Herwig* material, southern Africa). Pietsch, 2002b:1064 (western central Atlantic). Love et al., 2005:60 (eastern North Pacific).

*Dolopichthys* (subgenus *Dolopichthys*) *mucronatus* Regan and Trewavas, 1932:79, fig. 124 (original description, single specimen).

*Dolopichthys* (subgenus *Dolopichthys*) *longicornis*: Regan and Trewavas, 1932:79 (*Dolopichthys* broadened to incorporate five subgenera, amended by Bertelsen, 1951; brief description after Parr, 1927; in key).

*Dolopichthys mucronatus*: Nielsen, 1974:94 (listed in type catalog).

#### Material

Twenty-six metamorphosed females (14 to 159 mm):

Holotype of *Dolopichthys longicornis*: YPM 2008, 20 mm, *Pawnee*, Third Oceanographic Expedition, station 46, 21°46′N, 72°49′W, 3050 m wire, 4 April 1927.

Holotype of *Dolopichthys* (subgenus *Dolopichthys*) *mucronatus*: ZMUC P92100, 95 mm, *Dana* station 3716(1), 19°18.5′N, 120°13′E, 4000 m wire, 1400 h, 22 May 1929.

Additional material: CAS, two (28 mm); ISH, 10 (60 to 159 mm); LACM, two (16 to 121 mm); MCZ, one (52 mm); SIO, two (14 to 51 mm); SIOM, one (38 mm); UF, one (70 mm); UW, four (21 to 46 mm); ZMUB, one (45 mm).

#### Distinguishing Characters

Metamorphosed females of *Dolopichthys longicornis* can be differentiated from those of all other species of the genus on the basis of jaw and vomerine tooth counts alone. Specimens less than approximately 25 mm, however, may be confused with *D. pullatus*, in which case the diagnostic escal morphology is necessary for positive separation of these two species.

#### Description

Escal bulb with a small rounded anteriorly to anteroventrally directed papilla on anterior midline and a slightly larger dorsally to anterodorsally directed papilla on dorsal midline; papillae simple and unpigmented in specimens less than 98 mm, papillae bifurcated, forming a shallow medial area that is darkly pigmented in specimens 98 mm and larger; in most specimens, 2 small unpigmented papillae placed bilaterally on dorsal surface of escal bulb near base of posterior escal appendage; compressed posterior appendage crescent shaped and approximately as long as escal bulb; swollen basal part of posterior appendage darkly pigmented in specimens 98 mm and larger; connecting membrane between anterior and posterior filaments of posterior appendage present and pigmented in 159-mm specimen.

Total number of teeth in lower jaw of specimens 16 mm and larger, 31 to 286; number of teeth on vomer 4 to 10; dorsal-fin rays 5 to 7; anal-fin rays 4 to 6; pectoral-fin rays 17 to 21.

Measurements in percent of standard length: head length 31.6 to 46.4; head width 16.4 to 39.5; head depth 25.5 to 46.4; premaxilla length 23.3 to 32.1; lower-jaw length 28.4 to 53.6; illicium length 25.0 to 35.7.

Additional description as given for the genus and family.

#### Distribution

*Dolopichthys longicornis* is known from both sides of the Atlantic Ocean as far north as 46°N and 40°S, including the Gulf of Mexico. A single specimen has been collected in the Indian Ocean off Somalia at about 5°N, 52°E (Fig. 227). In the Pacific, it ranges from the South China Sea and Japan to the coast of California between about 41°N and 7°N, including several records in the relatively nutrient poor central water mass. Ninety-three percent of the specimens were captured in gear fished open at maximum depths of 1375 m or more; 73% in gear fished open at maximum depths between 2000 and 2200 m.

#### Comments

*Dolopichthys longicornis*, as restricted by Pietsch (1972c), can be differentiated from the other species of the genus on the basis of jaw and vomerine tooth counts alone. Specimens less than approximately 25 mm, however, may be confused with *D. pullatus*, in which case the diagnostic escal morphology is necessary for positive separation of these two species. Additional differences between *D. longicornis* and *D. pullatus* are discussed under comments on the latter species.

The holotype of *Dolopichthys* (subgenus *Dolopichthys*) *mucronatus* Regan and Trewavas, 1932 (95 mm, ZMUC P 92100), compares well with the material identified as *D. longicornis* in all characters except for a few morphometrics. Head depth, head length, and especially lower-jaw length, expressed in percent of standard length, fall at the extreme lower end of the range of variation of *D. longicornis*. Since, however, the specimen is in poor condition and appears to be stretched out considerably, these relatively low values may be due to an exaggerated standard length. *Dolopichthys mucronatus* is thus retained within the synonymy of *D. longicornis*, following Bertelsen (1951) and Pietsch (1972c).

DOLOPICHTHYS DANAE REGAN, 1926

Figures 131, 228; Table 8

### Females

*Dolopichthys danae* Regan, 1926:29, pl. 4, fig. 1 (original description, single specimen). Regan and Trewavas, 1932:80, fig. 127 (description after Regan, 1926; in key). Albuquerque, 1954–1956:1066, fig. 437 (after Regan, 1926, and Regan and Trewavas, 1932). Bertelsen, 1951:96, 97, 100, 101, figs. 53f, 54d (description after Regan, 1926, and Regan and Trewavas, 1932; comparison with all known material, in key, comments). Pietsch, 1972c:15, figs. 1, 3, 6, 11, 12 (additional material, description, comparison with all known material). Maul, 1973:671 (synonymy, eastern North Atlantic). Nielsen, 1974:92 (listed in type catalog). Bertelsen, 1986:1389, figs. (eastern North Atlantic). Bertelsen, 1990:500 (synonymy, eastern tropical Atlantic). Pietsch, 2002b: 1064 (western central Atlantic).

*Dolopichthys allector*: Maul, 1961:132, fig. 23 (misidentification; description of an additional specimen, comparison with holotype of *Dolopichthys allector*, comments).

### Males and Larvae

Unknown.

### Material

Six metamorphosed females (20.5 to 115 mm):

Holotype of *Dolopichthys danae*: ZMUC P9298, 75 mm, *Dana* station 1171(7), North Atlantic Ocean, 8°19′N, 44°35′W, 6000 m wire, 0800 h, 13 November 1921.

Additional material: BMNH, three (20.5 to 34 mm); MCZ, one (109 mm); MMF, one (115 mm).

### Distinguishing Characters

Metamorphosed females of *Dolopichthys danae* are clearly differentiated from those of all other members of the genus by having a longer illicium and a unique escal morphology. In addition, *D. danae* can be separated from *D. pullatus* and *D. longicornis* by the relative number of teeth in the jaws and on the vomer. Specimens of *D. danae* that have lost the illicial apparatus, however, may be confused with *D. jubatus* or *D. allector*, as all three of these species share, at certain standard lengths, similar tooth counts and morphometrics.

### Description

Escal bulb with 2 distally bifurcating papillae on dorsal midline, each darkly pigmented and well separated from each other; in some specimens, 2 simple, unpigmented, anteriorly directed papillae placed bilaterally on dorsal surface of escal bulb, posterior to bifurcating papillae; swollen basal part of compressed posterior escal appendage darkly pigmented; connecting membrane between anterior and posterior filaments of posterior escal appendage absent; anterior filament of posterior appendage darkly pigmented in 75-mm specimen.

Total number of teeth in lower jaw 53 to nearly 260; number of teeth on vomer 0 to 3; dorsal-fin rays 7; anal-fin rays 5 or 6; pectoral-fin rays 20 or 21.

Measurements in percent of standard length: head length 35.6 to 42.6; head width 16.5 to 26.8; head depth 32.2 to 43.9; premaxilla length 25.2 to 29.3; lower-jaw length 35.6 to 43.8; illicium length 43.9 to 75.3.

Additional description as given for the genus and family.

### Distribution

*Dolopichthys danae* is known from both sides of the North Atlantic Ocean between 40°N and the type locality at 8°N, 44°W (Fig. 228). Two specimens (20.5 to 34 mm) were captured in a closing trawl fished between 600 and 695 m. The 115-mm specimen was taken from the stomach of *Aphanopus carbo* (Maul, 1961). The remaining material was collected with open nets fished at maximum depths of 700 to 2000 m.

### Comments

The 115-mm specimen (MMF 16280A) is questionably referred to *Dolopichthys danae*. An unknown portion of the illicium has been broken off, but the part that remains is 47.8% SL. This illicial length is outside the range of that of all other species except *D. allector*. The high number of teeth in the lower jaw, however, and the presence of teeth on the vomer exclude this specimen from *D. allector*.

Previous authors (Regan and Trewavas, 1932:66, 80; Bertelsen, 1951:97, 98, 100) have indicated that the pigmented posterior appendage of the esca of the holotype of *D. danae* (75 mm) is a character that can be used to separate this species from the other species of the genus. Additional material, however, has shown that all species of the genus, except *D. dinema* (known only from two specimens, 21 and 22 mm), develop dark pigment on the posterior escal appendage with growth, thus eliminating the diagnostic value of this character. Furthermore, it was previously thought (Bertelsen, 1951:97) that *D. danae* could be distinguished from its congeners by the relative diameter of the eye. Examination of most available material revealed the unreliability of this character (Pietsch, 1972c).

DOLOPICHTHYS JUBATUS REGAN AND TREWAVAS, 1932

Figures 25, 129, 227; Tables 8, 9

### Females

*Dolopichthys* (subgenus *Dolopichthys*) *jubatus* Regan and Trewavas, 1932:79, 80, fig. 126 (original description, three specimens, lectotype ZMUC P9299, designated by Pietsch, 1972c).

*Dolopichthys danae*: Regan, 1926:29 (in part; brief description of a specimen questionably referred to *Dolopichthys danae*).

*Dolopichthys megaceros*: Regan, 1926:29 (in part; misidentification, specimen subsequently described as *Dolopichthys jubatus* sp. n. by Regan and Trewavas, 1932).

*Dolopichthys longicornis*: Norman, 1930:354 (in part; misidentification, specimen subsequently described as *Dolopichthys jubatus* sp. n. by Regan and Trewavas, 1932). Bertelsen, 1951:96, 100, 101, fig. 53b, c (*D. jubatus*, a synonym of *D. longicornis*).

*Dolopichthys jubatus*: Pietsch, 1972c:7, 17, figs. 1, 3, 7, 11 (additional material, description, comparison with all known material). Nielsen, 1974:93 (listed in type catalog). Bertelsen and Pietsch, 1977:186 (additional material, eastern Atlantic). Bertelsen, 1986:1389, figs. (eastern North Atlantic). Bertelsen, 1990:501 (synonymy, eastern tropical Atlantic). Anderson and Leslie, 2001:13, fig. 11A (additional female, off Durban, South Africa).

### Males and Larvae

Unknown.

### Material

Eight metamorphosed females (15 to 89 mm):

Lectotype of *Dolopichthys jubatus*: ZMUC P9299, 19.5 mm, *Dana* station 3920(4), Indian Ocean, 1°06′S, 62°25′E, 2500 m wire, 1830 h, 9 December 1929.

Paralectotypes of *Dolopichthys jubatus*: BMNH 1930.1.12.1082, 24 mm, *Discovery*, West Indies, 13°25′N, 18°22′W, 0 to 900 m, 28 October 1925. BMNH 1925.8.11.13, 15.5 mm, *Dana* station 1370(13), off the Azores, 36°36′N, 26°14′W, 3000 m wire, 1150 h, 13 June 1922.

Additional material: BMNH, one (33 mm); ISH, one (89 mm); SAM, one (15 mm); USNM, one (53 mm); ZMUC, one (18.5 mm).

### Distinguishing Characters

Metamorphosed females of *Dolopichthys jubatus* differ from those of all other members of the genus in details of escal morphology. This species is most easily confused with *D. longicornis* but differentiated from the latter by having a slightly longer illicium and considerably fewer teeth in the lower jaw and on the vomer.

Description

Two short distally bifurcating papillae on anterior midline of escal bulb, each devoid of pigment and lying very close together; bilaterally placed papillae on dorsal surface of escal bulb absent; compressed posterior appendage less than length of escal bulb in specimens 33 mm and smaller, more than three times length of escal bulb in 89-mm specimen; swollen basal part of posterior appendage unpigmented in specimens 33 mm and smaller, lightly pigmented in 89-mm specimen; connecting membrane between anterior and posterior filaments of posterior escal appendage absent in specimens 19.5 mm and larger; anterior filament of posterior escal appendage lightly pigmented in 89-mm specimen.

Total number of teeth in lower jaw 58 to approximately 110; number of teeth on vomer 0 to 4; dorsal-fin rays 7 or 8; anal-fin rays 6; pectoral-fin rays 18 to 21.

Measurements in percent of standard length: head length 36.4 to 48.4; head width 24.5 to 33.5; head depth 32.6 to 41.7; premaxilla length 25.1 to 33.0; lower-jaw length 36.0 to 47.9; illicium length 35.5 to 45.2.

Additional description as given for the genus and family.

Distribution

*Dolopichthys jubatus* is known from widely separated localities in the eastern Atlantic (five records ranging from 36°N to about 40°S), the western Indian Ocean (two records, one off South Africa at about 32°S and the other from the central Arabian Sea just below the equator), and the eastern South Pacific Ocean (a single record from off Chile at 39°S, 85°W; Fig. 227). A 33-mm specimen was captured in a closing trawl fished between 690 and 900 m. A 53-mm specimen was taken in an otter trawl fished on the bottom at 1353 m. All the remaining specimens were collected with gear fished open at maximum depths ranging from 800 and 1050 m.

Comments

On the basis of similarity in escal morphology, Bertelsen (1951) synonymized *Dolopichthys jubatus* with *D. longicornis*. The escae of all seven known specimens of *D. jubatus*, however, lack the dorsally directed papillae and the bilaterally placed papillae that are characteristic of the esca of *D. longicornis*. In addition, *D. jubatus* has a longer illicium than *D. longicornis*, and considerably fewer teeth in the lower jaw and on the vomer. For these reasons, *D. jubatus* was resurrected by Pietsch (1972c), and that decision is followed here.

Norman (1930) referred two small ceratioids collected by the *Discovery* to *D. longicornis*. One of these specimens (24 mm, BMNH 1930.1.12.1082) was included in the original description of *D. jubatus* by Regan and Trewavas (1932) and remains as a paralectotype of this species. The other specimen (28 mm, BMNH 1930.1.12.1083) was cleared and stained for osteological study and the skull and pectoral radials erroneously described under the name *D. jubatus* (Regan and Trewavas, 1932:34–37). This latter specimen was referred to *D. pullatus* by Pietsch (1972c).

*DOLOPICHTHYS ALLECTOR* GARMAN, 1899

Figures 14–16, 227; Table 8

Females

*Dolopichthys allector* Garman, 1899:81, pls. 13–15 (original description, single specimen). Gill, 1909:580, figs. 16, 17 (description, figures after Garman, 1899). Regan, 1926:28 (misidentifications; description of nine specimens for which only six can be accounted: five were referred to *Dolopichthys atratus* [= *Pentherichthys atratus*], and one to *D. implumis* [= *Microlophichthys microlophus*] by Regan and Trewavas, 1932). Norman, 1930:353, 354, fig. 43 (misidentification; specimen subsequently described as *D. venustus* sp. n. [= *P. venustus*] by Regan and Trewavas, 1932). Regan and Trewavas, 1932:80 (description after Garman, 1899, and Regan, 1926; in key). Beebe and Crane, 1947:161, text fig. 7 (description of an additional specimen, comparison with holotype). Bertelsen, 1951:100 (description after Garman, 1899; comparison with all known material, in key, comments). Grey, 1956a:251, 252 (synonymy, distribution). Maul, 1961:132, fig. 23 (misidentification; description of a specimen here referred to *D. danae*; comparison with holotype, comments). Pietsch, 1972a:35, 36, 42, 43, 45, fig. 24(7) (misidentifications; osteological comments on specimens here referred to *D. pullatus*). Pietsch, 1972c:7, 19, figs. 1, 3, 8, 11, 12 (additional material, description, comparison with all known material). Bertelsen and Pietsch, 1977:186 (additional material, eastern Atlantic). Bertelsen, 1986:1388, figs. (eastern North Atlantic). Pietsch, 2002b:1064 (western central Atlantic).

Males and Larvae

Unknown.

Material

Fifteen metamorphosed females (16 to 154 mm):

Holotype of *Dolopichthys allector*: MCZ 28735, 61 mm, *Albatross* station 3371, Gulf of Panama, 5°26′20″N, 86°55′00″W, 0 to 1408 m.

Additional material: ARC, two (88 to 120 mm); ISH, six (69 to 154 mm); NYZS, one (18 mm); ROM, one (77 mm); SIO, three (13 to 21 mm); USNM, one (16 mm).

Distinguishing Characters

Metamorphosed females of *Dolopichthys allector* differ from those of all other species of the genus in details of escal morphology. This species is further separated from *D. pullatus* and *D. longicornis* by the relative number of teeth on the vomer. Unlike the other members of the genus, *D. allector* undergoes a complete loss of vomerine teeth at a relatively early growth stage. The relative length of the illicium of *D. allector* further distinguishes this species from *D. pullatus*, *D. longicornis*, and *D. danae*, but not from *D. jubatus*. Larger individuals (greater than 33 mm) of the latter species may lack vomerine teeth and thus be confused with *D. allector*, in which case the diagnostic escal morphology is necessary for positive identification. Finally, specimens of *D. allector* with missing illicia would be difficult to separate from *D. danae* and *D. jubatus* as all three of these species share at certain standard lengths, similar tooth counts and morphometrics.

Description

Two posteriorly directed papillae placed bilaterally on dorsal surface of escal bulb each darkly pigmented on dorsal surface and on distal tip; compressed posterior appendage crescent shaped, unpigmented, and short, equal to or less than length of escal bulb; swollen basal part of posterior appendage unpigmented; connecting membrane between anterior and posterior filaments of posterior escal appendage present; posterior edge of anterior filament of posterior escal appendage darkly pigmented in 101-mm specimen.

Total number of teeth in lower jaw 42 to 128; vomerine teeth absent in specimens 18.5 mm and larger, not more than 2 in smaller specimens; dorsal-fin rays 6 or 7; anal-fin rays 5 or 6; pectoral-fin rays 19 to 21.

Measurements in percent of standard length: head length 33.1 to 42.5; head width 17.6 to 26.0; head depth 26.3 to 37.5; premaxilla length 20.4 to 29.5; lower-jaw length 30.1 to 41.9; illicium length 26.9 to 48.9.

Additional description as given for the genus and family.

Distribution

*Dolopichthys allector* is found only in the Atlantic and eastern Pacific oceans. In the Atlantic, the known records are strangely clustered in the north between about 45° and 53°N, and in the far south between 36 and 40°S (Fig. 227). In the eastern Pacific Ocean, it is known from the type locality in the Gulf of Panama, from off the Galapagos Islands, and from off Chile at about 35°S, 91°W. Eighty-seven percent of the material was captured in gear fished open at maximum depths of 1050 to 2000 m; 73% in gear fished open at maximum depths of 1650 to 2000 m.

### Comments

The 1971 cruise of the RV *Walther Herwig* to the South Atlantic Ocean collected the first unquestionable representatives of *Dolopichthys allector* since the capture of the holotype by the *Albatross* in 1899. This more recent material compares well with the type specimen (despite the poor condition of the latter, which is due partly to desiccation and partly to Garman's, 1899, osteological and myological examination) in all important characters, including escal morphology.

### *DOLOPICHTHYS KARSTENI* LEIPERTZ AND PIETSCH, 1987

Figures 131, 228; Table 8

#### Females

*Dolopichthys karsteni* Leipertz and Pietsch, 1987:406, fig. 1 (original description, three specimens). Pietsch, 2002b: 1064 (western central Atlantic). Moore et al., 2003:215 (off New England).

#### Males and Larvae

Unknown.

#### Material

Eight metamorphosed females (18 to 99 mm):

Holotype of *Dolopichthys karsteni*: MCZ 60991, 99 mm, *Knorr*, cruise 98, station MOC 20-56(0), western North Atlantic, 39°28.0'N, 64°00.6'W, 0 to 1023 m, 30 September 1982.

Paratypes of *Dolopichthys karsteni*: MCZ 61090, 18 mm, *Knorr*, cruise 65, station MOC 10-39(0), western North Atlantic, 39°29.6'N, 69°42.2'W, 500 to 1200 m, 30 April 1977. MCZ 61086, 25 mm, *Oceanus,* cruise 49, station MOC 10-116(2), western North Atlantic, 34°11'N, 74°50'W, 340 to 700 m, 13 August 1978.

Additional material: MCZ, four (22.5 to 29 mm); ZMUB, one (77.5 mm).

#### Distinguishing Characters

Metamorphosed females of *Dolopichthys karsteni* differ from those of all other species of the genus in details of escal morphology.

#### Description

Esca with a single, darkly pigmented, anteriorly directed papilla on dorsal midline of escal bulb, flanked on each side by a similar but posteriorly directed papilla; a slender, compressed, posterior appendage; lightly pigmented along anterior margin, its length approximately equal to length of escal bulb; swollen basal portion of posterior appendage darkly pigmented in holotype.

Total number of teeth in lower jaw 44 to 88; number of teeth on vomer 3 to 6 in five smallest known specimens (18 to 29 mm), vomerine teeth absent in largest known specimen (99 mm); dorsal-fin rays 6 or 7; anal-fin rays 5 or 6; pectoral-fin rays 15 or 16.

Measurements in percent of standard length: head length 40.0 to 46.0; head width 23.0 to 36.0; head depth 35.0 to 44.0; premaxilla length 28.0 to 33.3; lower-jaw length 32.0 to 40.0; illicium length 33.3 to 40.0.

Additional description as given for the genus and family.

#### Distribution

*Dolopichthys karsteni* is known only from the western North Atlantic, extending from off Cape Hatteras north and east to approximately 42°N, 30°W (Fig. 228). All of the material was collected with gear fished open at maximum depth ranging from 700 to 1650 m.

#### Comments

*Dolopichthys karsteni* appears to be most closely related phylogenetically to *D. allector,* with which it shares a pair of posteriorly directed, bilaterally placed escal papillae. It is clearly distinguished from *D. allector,* however, in having an anteriorly directed, medial escal papilla.

### *DOLOPICHTHYS DINEMA* PIETSCH, 1972C

Figures 130, 131, 228; Table 8

#### Females

*Dolopichthys dinema* Pietsch, 1972c:7, 21, figs. 3, 9–11 (original description, two specimens). Bertelsen, 1990:501 (synonymy, eastern tropical Atlantic).

#### Males and Larvae

Unknown.

#### Material

Two metamorphosed females (21 to 22 mm):

Holotype of *Dolopichthys dinema*: BMNH 1970.11.25.1, 21 mm, *Discovery II* station 6662-21, off coast of French Guinea, Africa, 10°50'N, 19°54'W, RMT 8/5, 0 to 700 m, 17 February 1968.

Paratype of *Dolopichthys dinema*: BMNH 1970.11.25.2, 22 mm; *Discovery II* station 6662-21, data as given for the holotype.

#### Distinguishing Characters

Metamorphosed females of *Dolopichthys dinema* differ from those of all other members of the genus in details of escal morphology. In addition, this species has fewer than 40 teeth in the lower jaw, compared to more than 60 teeth in the lower jaw of other species of *Dolopichthys* of similar size. Additional material of larger size will no doubt make these differences more apparent.

#### Description

Two tapering filaments on dorsal midline of escal bulb, each darkly pigmented except for distal one-half to one-fourth of length; posterior filament twice as long as anterior filament, approximately 2.0 mm or 25.0 to 26.7% of illicium length; 2 simple, short, unpigmented tapering appendages placed bilaterally on dorsal surface of escal bulb posterior to long tapering filaments; swollen basal part of compressed posterior escal appendage without pigment; connecting membrane between anterior and posterior filament of posterior escal appendage present.

Total number of teeth in lower jaw 35 to 39; number of teeth on vomer 2; dorsal-fin rays 7; anal-fin rays 6; pectoral-fin rays 19 or 20.

Measurements in percent of standard length: head length 38.1 to 40.9, head width 27.3 to 31.0, head depth 38.6 to 40.5, premaxilla length 27.3 to 28.6, lower-jaw length 36.4 to 40.5, illicium length 35.7 to 36.4.

Additional description as given for the genus and family.

#### Distribution

*Dolopichthys dinema* has been collected only in the eastern North Atlantic Ocean at 10°N, 19°W, and somewhere between the surface and 700 m.

### *DOLOPICHTHYS NIGER* BRAUER, 1902, NOMEN DUBIUM

#### Females

*Dolopichthys niger* Brauer, 1902:292 (original description, two syntypes). Parr, 1927:14 (footnote reference to descriptions of Brauer, 1902, 1906). Regan and Trewavas, 1932:79, fig. 125 (brief description after Brauer, 1902, 1906; in key). Bertelsen, 1951:96, 100, 101, fig. 53a (description after Brauer, 1902, 1906, 1908; comparison with all known material, in key, comments). Grey, 1956a:252 (synonymy, distribution). Marshall, 1966:177, pl. 7 (bioluminescence). Pietsch, 1972c:23 (*nomen dubiun*). Munk, 1999:267 (bioluminescence).

*Oneirodes niger*: Brauer, 1906:316, pl. 15, fig. 6 (new combination). Brauer, 1908:103, 105, 184; pl. 32, figs. 7, 8; pl. 34, figs. 15, 16; pl. 44, figs. 2, 3 (anatomical, histological description of illicial apparatus, esca, and eyes). Regan, 1926:27 (listed as a probable species of *Dolopichthys*). Harvey, 1952:531, 532, fig. 178 (bioluminescence).

Material

A single metamorphosed female (13 mm):

Syntype of *Dolopichthys niger*: ZMB 17707, 13 mm (15.5 mm TL), *Valdivia* station 173, Indian Ocean, 29°06'02"S, 89°39'00"E, 0 to 2500 m, bottom depth 3765 m, 10 January 1899.

Comments

*Dolopichthys niger* was described from two small specimens collected from the Indian Ocean during the German Deep-Sea Expedition of 1898–1899. One of these (13 mm TL from *Valdivia* station 237) was used by Brauer (1908) for his anatomical and histological description of the species. The specimen is not listed in the catalog of the Zoologisches Museum der Humboldt Universität, Berlin (Christine Karrer, personal communication, 9 October 1969), and is most likely lost. A sagittal section through the esca of this specimen, figured by Brauer (1908, pl. 44, fig. 2) looks very much like the esca of *D. pullatus*. Although no further supporting evidence is available, it is likely that this lost syntype of *D. niger* was actually *D. pullatus*.

From figures of the esca of *D. niger* (Brauer, 1906, pl. 15, fig. 6; Regan and Trewavas, 1932:79, fig. 125), Bertelsen (1951) thought the specimens could easily be regarded as younger stages in the same series as that of *D. longicornis* (sensu lato). Brauer (1902, 1906) stated, however, that both specimens of *D. niger* have 8 anal rays. Since no other known specimen of *Dolopichthys* has more than 6 anal rays, Bertelsen (1951) maintained *D. niger* distinct from *D. longicornis*. Examination of the existing syntype of *D. niger* (15.5 mm total length, *Valdivia* station 173) revealed only 6 anal rays (Pietsch, 1972c). The esca has lost much of its original morphology, but appears to come closest to that of *D. longicornis* or *D. jubatus*. The length of the illicium is 3.7 mm (30.0% SL); there are 31 teeth in the lower jaw, and a total of 3 teeth on the vomer. This combination of characters does not exclude the specimen from the known material of any recognized species of *Dolopichthys*. Additional material of all members of the genus, within a size range of 10 to 15 mm, may clarify the situation.

DOLOPICHTHYS CORNUTUS (GILCHRIST AND VON BONDE, 1924), NOMEN DUBIUM

Females

*Oneirodes cornutus* Gilchrist and von Bonde, 1924:23, pl. 6, fig. 2 (original description, single specimen, 12 mm, South African waters). Regan, 1926:27 (listed as a probable species of *Dolopichthys*). Parr, 1927:14 (footnote reference to Regan, 1926). Regan and Trewavas, 1932:63 (footnote reference to original description, comments).

*Dolopichthys cornutus*: Barnard, 1927:1004 (new combination, brief description after Gilchrist and von Bonde, 1924). Smith, 1949:429, fig. 1231 (description, figure after Gilchrist and von Bonde, 1924). Grey, 1956a:252, 253 (synonymy, distribution). Penrith, 1967:187, 188 (holotype lost). Pietsch, 1972c:24 *(nomen dubium)*.

Comments

The holotype of *Dolopichthys cornutus*, and only specimen to which this name has ever been applied, is lost (Penrith, 1967). As far as is known, the facts behind the missing type are as follows: "For a few short years Prof. L. T. Hogben was Professor of Zoology at Cape Town where much of Gilchrist's collection was housed. Hogben was imbued with the spirit of the physiological approach and in a graphic display of his contempt for systematics, threw the types out of a laboratory window. And that was that" (Michael J. Penrith, personal communication, 9 August 1971).

In view of the lack of a type specimen, and an original description that is extremely inadequate and in some instances undoubtedly incorrect, the name *D. cornutus* is here considered a *nomen dubium*.

*Dolopichthys* Species

Unidentifiable Females, Males, and Larvae

*Dolopichthys* sp.: Regan, 1926:5 (unidentifiable females). Regan and Trewavas, 1932:9, 10 (unidentifiable females). Norman, 1939:115, text figs. 40, 41 (misidentifications; two specimens referred to *Oneirodes* Lütken, 1871, by Bertelsen, 1951). Bertelsen, 1951:98, figs. 54a, 54b, 55, 56 (unidentifiable larvae, metamorphosing females, and metamorphosing, juvenile, and adult males; description, comparisons, comments). Grey, 1956a:251 (synonymy, distribution). Pietsch, 1972c:23 (unidentifiable material). Pietsch and Seigel, 1980:394 (unidentifiable male, Banda Sea).

*Rhynchoceratias leuchorhinus*: Regan, 1926:44 (in part; misidentifications, two males subsequently referred to *Trematorhynchus leuchorhinus* by Regan and Trewavas, 1932).

*Trematorhynchus leuchorhinus*: Regan and Trewavas, 1932:91 (misidentifications, two males subsequently referred to *Dolopichthys* sp. by Bertelsen, 1951).

Material

The following females, males, and larvae cannot be identified to species:

Metamorphosed females: ISH, two specimens, 100 to 190 mm (illicia lost); OS 531, 17 mm, *Yaquina*, cruise 6502, station MT-613, 40°18.1'N, 134°06.8'W, 3-m Isaacs-Kidd midwater trawl, 0 to 2400 m, 8 March 1965 (illicium lost; vomerine teeth 3; lower-jaw teeth 63; dorsal-fin rays 6; anal-fin rays 5; pectoral-fin rays 19). SIO 55-229, 16 mm, *Horizon*, 8°31'N, 110°11'W, 3-m Isaacs-Kidd midwater trawl, 0 to 1609 m, 28 October 1955 (illicium lost; vomerine teeth 6; lower-jaw teeth 43; dorsal-fin rays 6; anal-fin rays 5; pectoral-fin rays 19; ZMUC, one specimen, 90 mm (illicium lost).

Females in metamorphosis: ZMUC, two (9.6 to 11 mm).

Males: SIO, three (9.5 to 10.5 mm); ZMUC, five (one in metamorphosis, 8 mm; four juveniles or adults, 10.5 to 13.5 mm).

Larvae: SIO, three (two males, 5.5 to 8 mm; one female, 7 mm); ZMUC, 178 (70 females, 89 males, 19 undetermined, 2.4 to 12 mm TL).

The material of the *Dana* collections held by ZMUC was listed by Bertelsen (1951:268–269).

Comments

Bertelsen's (1951) thorough study failed to reveal any striking differences between males, and no characters were found by which the males could be included in any of the species based on females. Furthermore, the larvae (male and female), easily distinguished from other oneirodid larvae, could not be separated into species. All of this material must therefore remain designated as *Dolopichthys* sp. until means of identifying these specimens is found.

Genus *Bertella* Pietsch, 1973 (Spikehead Dreamers)

Figures 33, 133–136, 202, 203, 226, 291, 296; Tables 1, 8

Females and Parasitic Males

Oneirodidae gen. n. Pietsch, 1972a:35, 42, 43, 45, fig. 24(8) (otolith described and figured).

*Bertella* Pietsch, 1973:193, figs. 1–6 (type species *Bertella idiomorpha* Pietsch, 1973, by original designation and monotypy). Whitley, 1976:46 (not preoccupied by *Bertella* Paetel, 1875, an incorrect subsequent spelling of *Berthella* Blainville, 1825, in Mollusca).

Free-living Males and Larvae

Unknown.

Distinguishing Characters

Metamorphosed females of *Bertella* differ from those of all other described oneirodid genera in having a hyomandibular bone with a single head. They are further unique in having the following combination of character states: The ethmoid cartilage and vomer are wide, wider than the distance between the anterolateral tips of the lateral ethmoids and frontals. Vomerine teeth are present in juvenile specimens, but lost with growth in adults. The ethmoid region is depressed, the nasal foramina narrow and oval in shape. The frontals are long and anterior in position, overhanging and extending past the anterior limits of the ethmoid cartilage and vomer. The dorsal margin of the frontals is nearly linear. Ventromedial extensions of the frontals approach each other on the midline, making contact with the parasphenoid. The frontals are separated from the prootics. The pterosphenoid is present. The anterior end of the illicial trough is wider and shallower than the posterior end. Sphenotic spines are well developed. The symphysial cartilage of the upper jaw is wider than long. The lower jaw bears a well-developed symphysial spine. The quadrate spine is well developed, more than twice, usually three times, as long as the articular spine. The posterior margin of the opercle is deeply notched. The subopercle is long and narrow, its dorsal end slender and tapering to a point (the posterior margin without indentation), its ventral end oval (without an anterior spine or projection). The first pharyngobranchial is absent. The second pharyngobranchial is reduced and toothless. The second hypobranchial is absent. The caudal-fin rays are without internal pigmentation. The illicium is considerably longer than the length of the esca bulb. The pterygiophore of the illicium is cylindrical throughout its length, emerging on the snout from between the frontal bones, its anterior end slightly exposed, its posterior end concealed beneath the skin. The first ray of the dorsal fin is well developed. There are 5 to 6 dorsal-fin rays and 5 (rarely 4) anal-fin rays (Table 1). The pectoral-fin lobe is short and broad, shorter than the longest rays of the pectoral fin. There are 18 to 22 pectoral-fin rays. The coracoid lacks a posteroventral process. The pelvic bones are simple, with or without a distal expansion. The skin is smooth and naked, without dermal spinules. Darkly pigmented skin of the caudal peduncle extends well past the base of the caudal fin.

### Description

The body of metamorphosed females is short and stout, its depth greater than 35% SL. The mouth is moderate in size, its cleft not extending past the eye. The jaws are usually equal anteriorly, the lower jaw extending slightly past the upper in some specimens. The quadrate spine is very well developed and strongly bent posteriorly in specimens greater than 24.5 mm. The distal end of the angular bone is produced into a well-developed spine. The posterior ends of the articular and angular bones are flared out laterally. The posterior-most rays of the dorsal and anal fins are long, extending well past the posterior margin of the hypural plate. The illicial length is 21.4 to 34.5% SL (the illicium of a 37-mm specimen, SIO 73-247, is unusually long, 46.3% SL). The esca bears a single short unpigmented posterior appendage.

The opercle is bifurcated, the two forks forming an acute angle of 37 to 51°; the length of the dorsal fork 46.9 to 76.5% of the length of the ventral fork. The length of the subopercle is 71.9 to 96.3% of the length of the ventral fork of the opercle (the shape of the subopercle does not distinguish this genus from *Dolopichthys*).

The teeth are slender, straight, and depressible, usually in nonoverlapping groups of 2 to 6 (average 3 or 4) in both upper and lower jaws (one specimen, LACM 30561-1, 78 mm, shows no pattern in tooth placement and has considerably more teeth than other individuals). The longest tooth in the lower jaw measures 0.3 mm in a 14-mm specimen and 1.0 mm in an 84-mm specimen. The teeth in the lower jaw are more numerous than those of the upper jaw. There are 40 to 200 teeth in upper jaw and 60 to 250 in the lower jaw. The second pharyngobranchial is reduced and toothless (based on two cleared and stained specimens, 11 to 78 mm). Teeth are present on the third pharyngobranchial. The epibranchials and ceratobranchials are toothless.

The otolith was described and figured by Pietsch (1972a:42, fig. 24(8); *Bertella* referred to as Oneirodidae gen. n.).

The following measurements are expressed in percent of standard length: head depth 35.7 to 42.6; head width 35.1 to 45.3; head length 34.5 to 41.0; lower-jaw length 37.5 to 45.9; premaxilla length 26.3 to 32.4; least outside width of frontals 13.2 to 14.2; distance from tip of sphenotic spine to posterolateral tip of respective posttemporal bone 19.6 to 21.6; distance between posterolateral tips of posttemporal bones 27.4 to 30.6.

### Diversity

A single species:

BERTELLA IDIOMORPHA PIETSCH, 1973

Figures 33, 133–136, 226, 291, 296; Table 8

### Females and Males

Oneirodidae, unidentified: Brewer, 1973:25 (eastern tropical Pacific).
*Bertella idiomorpha* Pietsch, 1973:194, figs. 1–6 (original description, eight specimens). Pietsch, 1974a:16–21, 24–26, 28, 30, 31, 33, 82–89, figs. 24, 36, 39G, 43, 51E, 56B, 103, 104 (comparative osteology, relationships). Hubbs et al., 1979:13 (California). Amaoka, 1984:106, pl. 92F (Japan). Mecklenburg et al., 2002:305, 306 (Alaska, in key). Nakabo, 2002:472, figs. (Japan, in key). Sheiko and Fedorov, 2000:24 (Kamchatka). Love et al., 2005:60 (eastern North Pacific). Pietsch, 2005b:209, 215, 220–222, 232, figs. 8C, 12 (reproduction). Senou et al., 2006:424 (Sagami Sea).

### Larvae

Unknown.

### Material

Thirty-four metamorphosed females (11 to 101 mm) and one attached male (11 mm):

Holotype of *Bertella idiomorpha*: LACM 30601-22, 74 mm, *Velero IV* station 12464, off Guadalupe Island, Mexico, 29°10′00″N, 118°28′15″W, 3-m Isaacs-Kidd midwater trawl, 0 to 940 m, bottom depth 2377 to 3475 m, 1719 to 2240 h, 15 November 1968.

Paratypes of *Bertella idiomorpha*: LACM 30561-1, 78 mm, *Velero IV* station 12006, off San Nicolas Island, southern California, 32°57′30″N, 118°50′40″W, 3-m Isaacs-Kidd midwater trawl, 0 to 900 m, bottom depth 1664 to 1692 m, 2255 to 0350 h, 14 March 1968. LACM 30045-3, 11 mm, *Velero IV* station 11767, Gulf of California, 24°01′29″N, 108°52′11″W, 3-m Isaacs-Kidd midwater trawl, 0 to 1500 m, bottom depth 2560 to 2615 m, 0812 to 1525 h, 16 November 1967. OS 1045, 84 mm, *Yaquina*, Yaloc Expedition, station AK 18-19, haul 854, Gulf of Alaska, 54°37.8′N, 155°33.1′W, 3-m Isaacs-Kidd midwater trawl, 0 to 2900 m, 0600 to 1212 h, 8 July 1966. SIO 60-51, 14 mm, Transpac Expedition, station 59E, eastern North Pacific, 44°06′N, 161°39′W, 13 September 1953. SIOM uncataloged, 57 mm, *Vityaz*, cruise 20, station 3233, sample 33A, western North Pacific, 32°37′N, 150°18′E, 6-m diameter conical ring net fished open with 6000 m wire, bottom depth 5920 to 5880 m, 2225 to 0210 h, 6 to 7 May 1955. ZUMT 52705, 12 mm, 35°02′N, 139°18′E, larvae net, 0 to 1000 m, bottom depth 1200 to 1400 m, 1300 to 1400 h, 27 October 1966. ZIN 49795, 24.5 mm, *Vityaz*, cruise 19, station 3156, sample 55, western North Pacific, 39°57.3′N, 164°52.4′E, 1.6-m diameter conical ring net, 0 to 1000 m, bottom depth 5400 m, 0741 to 0900 h, 28 September 1954.

Additional material: HUMZ, one (61 mm); LACM, one (29.5 mm); MNHN, one (14.5 mm); NSMT, 15 (24 to 72 mm); SIO, four (29 to 101 mm); UW, three (66 to 85 mm; 77-mm female, with 11-mm parasitic male); ZIN, one (98 mm).

### Distinguishing Characters and Description

As given for the genus and family.

### Sexual Maturity

The ovaries of the 84-mm specimen of *Bertella idiomorpha* (OS 1045) are relatively large, approximately 15 mm in length, but appear to be partially spent. About one-third of the volume of each ovary consists of a tight cluster of many small eggs, each approximately 0.3 mm in diameter. This specimen was captured in early July; the condition of the ovaries supports Bertelsen's (1951) findings that the majority of ceratioid species are summer spawners.

Distribution

*Bertella idiomorpha* has so far been collected only in the North Pacific Ocean where it ranges from off Japan and the Bering Sea to the Gulf of California between 24 and 60°N, with a single record on the equator at about 155°W (Fig. 226). Much of the material was taken in relatively shallow depths: 56% of the specimens were collected by gear fished open between 410 and 680 m; and 83% by gear fished open between 410 and 1160 m.

"Long-pectoraled" Oneirodid Genera

Genus *Puck* Pietsch, 1978 (Mischievous Dreamers)

Figures 137–139, 202, 203, 229; Tables 1, 8

Females

*Puck* Pietsch, 1978:10, figs. 1–10, 13A, 18 (type species *Puck pinnata* Pietsch, 1978, by original designation and monotypy).

Males and Larvae

Unknown.

Distinguishing Characters

Metamorphosed females of *Puck* differ from those of all other genera of the family in having the following combination of character states: The ethmoid cartilage and vomer are wide, wider than the distance between the anterolateral tips of the lateral ethmoids and frontals. Vomerine teeth are present. The nasal foramina are large and oval in shape. The frontals are short, lying posterior to the ethmoid region, their dorsal margin strongly convex. Ventromedial extensions of the frontals approach each other on the midline, making contact with the parasphenoid. The frontals are separated from the prootics. The pterosphenoid is present. The anterior end of the illicial trough is wider and shallower than its posterior end. The sphenotic spines are extremely well developed. The symphysial cartilage of the upper jaw is longer than wide. The lower jaw bears a well-developed symphysial spine. The hyomandibula has a double head. The quadrate spine is extremely well developed, nearly six times the length of the articular spine. The posterior margin of the opercle is deeply notched. The subopercle is small, elongate, and narrow throughout its length, its dorsal end tapering to a point (the posterior margin without an indentation), its ventral end rounded (without an anterior spine or projection). The first pharyngobranchial is absent. The second pharyngobranchial is well developed. The second hypobranchial articulates directly with the second basibranchial. The caudal-fin rays lack internal pigmentation. The illicium is considerably longer than the length of the esca bulb. The pterygiophore of the illicium is cylindrical throughout its length, emerging on the snout from between the frontal bones, its anterior end exposed, its posterior end concealed beneath the skin. The first ray of the dorsal fin is well developed. There are 5 dorsal-fin rays and 4 anal-fin rays (Table 1). The pectoral-fin lobe is long and narrow, longer than the longest rays of the pectoral fin. There are 19 to 20 pectoral-fin rays. The coracoid lacks a posteroventral process. The pelvic bones are simple, slightly expanded distally. The skin is naked, without dermal spinules. Darkly pigmented skin of the caudal peduncle extends well past the base of the caudal fin.

Description

The body of metamorphosed females is elongate and stout, not globular. The snout is extremely short, the mouth moderate, opening oblique, the cleft terminating well before the eye. The illicium length is 10.7 to 18.5% SL. The esca bears a stout, rounded, anterodorsally directed and internally pigmented anterior papilla and a similar, and posterodorsally directed, medial escal papilla, without internal pigment. The distal ends of the anterior and medial papillae are darkly pigmented. An unpigmented compressed posterior escal appendage is present, bearing anterodorsally a lump of tissue of uncertain morphology. A tapering unpigmented lateral filament is present on each side, slightly less than the length of the escal bulb.

The opercle is bifurcate, the two forks forming an acute angle of 25°. The length of the ventral fork of the opercle is 22.5 to 23.9% SL, the dorsal fork is 72.7 to 73.6% of the length of the ventral fork. The length of the subopercle is 40.9 to 68.7% of the length of the ventral fork of the opercle.

The teeth are slender, recurved, and all depressible, arranged in an oblique transverse series of increasing size. The number of teeth generally increase with the size of specimens, those of the lower jaw somewhat larger but fewer than those of the upper jaw. There are 35 to 47 teeth in upper jaw and 24 to 25 in the lower jaw. There are 4 to 6 on the vomer. The second and third pharyngobranchials are present and well toothed (Pietsch, 1978:7, fig. 7). The epibranchials and ceratobranchials are toothless.

The following measurements are expressed in percent of standard length: head length 29.6 to 31.5; head depth 35.9 to 44.7; length of premaxilla 24.1 to 27.3; length of lower jaw 34.6 to 38.0; length of pectoral-fin lobe 9.1 to 13.6; length of illicium 10.6 to 18.5; length of sphenotic spine 6.4 to 7.6, quadrate spine 6.2 to 7.6, mandibular spine 1.3 to 1.8.

Diversity

A single species:

PUCK PINNATA PIETSCH, 1978 (MISCHIEVOUS DREAMER)

Figures 137–139, 229; Table 8

Females

*Puck pinnata* Pietsch, 1978:11, figs. 1–10, 13A, 18 (original description, two specimens, comparison with other "long-pectoraled" oneirodid genera). Pietsch and Orr, 2007:18, figs. 17, 18 (phylogenetic relationships).

Males and Larvae

Unknown.

Material

Four metamorphosed females (38 to 81 mm):

Holotype of *Puck pinnata*: LACM 34276-1, female, 46 mm, *Vityaz*, cruise 19, station 3199, sample 123b, western North Pacific, 38°16′N, 152°34′E, 6-m diameter conical ring net fished open with 5350 m of wire, bottom depth 5420 to 5350 m, 0230 to 0545 h, 16 October 1954.

Paratype of *Puck pinnata*: SIO 52-363, female, 66 mm (cleared and stained), *Horizon*, Shellback Expedition, between stations SB 101 and 102, trawl 6, eastern tropical Pacific, 6°53.5′N, 88°32.5′W, 3-m Isaacs-Kidd midwater trawl, fished open between surface and 1464 m, bottom depth about 4000 m, 1 July 1952.

Additional female: ISH, one (38 mm); SIO, one (81 mm).

Distinguishing Characters and Description

As given for the genus and family.

Distribution

*Puck pinnata* is known from only four metamorphosed females: the holotype from the western North Pacific at approximately 38°N, 152°E; the paratype from the eastern tropical Pacific at about 6°N, 88°W; and two previously unreported specimens: one off the Galapagos Islands at 5°S, 87°W, and the other collected during the *Anton Dohrn*, Sargasso Sea Expedition of 1979 from the eastern North Atlantic at about 35°N, 32°W (Fig. 229). It appears to be a relatively deep dwelling species: three of the specimens were taken in nets fished at maximums depth of 1464 to 1800 m; the fourth and largest known specimen (81 mm, SIO 4-35) was collected in an open bottom trawl fished at 4073 m.

## Genus *Chirophryne* Regan and Trewavas, 1932 (Longhand Dreamers)

Figures 32, 137, 140, 141, 202, 203, 229; Tables 1, 8

### Females

*Chirophryne* Regan and Trewavas, 1932:81, figs. 131, 132 (type species *Chirophryne xenolophus* Regan and Trewavas, 1932, by monotypy).

### Males and Larvae

Unknown.

### Distinguishing Characters

Metamorphosed females of *Chirophryne* differ from those of all other oneirodid genera in having the following combination of character states: Vomerine teeth are present. The frontals are short, lying posterior to the ethmoid region, their dorsal margin convex. The sphenotic spines are extremely well developed. The lower jaw bears a small symphysial spine. The hyomandibula has a double head. The quadrate spine is extremely well developed, four to nearly six times longer than the articular spine. The posterior margin of the opercle is deeply notched. The subopercle is short and broad, its dorsal end rounded (the posterior margin without indentation), its ventral end oval in shape (without an anterior spine or projection). The second pharyngobranchial is well developed. The caudal-fin rays are without internal pigmentation. The illicium is considerably longer than the length of the esca bulb. The pterygiophore of the illicium is cylindrical throughout its length, emerging on the snout from between the frontal bones, its anterior end exposed, its posterior end concealed beneath the skin. The first ray of the dorsal fin is well developed. There are 5 to 6 dorsal-fin rays and 4 anal-fin rays (Table 1). The pectoral-fin lobe is long and narrow, longer than the longest rays of the pectoral fin. There are 18 to 19 pectoral-fin rays. The skin is presumably naked, without dermal spinules (but material is unavailable for staining and microscopic examination). Darkly pigmented skin of the caudal peduncle extends well past the base of the caudal fin (known from only two small females, internal anatomical features are unknown).

### Description

The body of metamorphosed females is rather short and stout, somewhat globular. The snout is short, but the mouth is large and oblique, the cleft extending slightly past the eye. The illicium length is 18.2 to 22.7% SL. The escal bulb of the two larger females (22 to 42 mm) bears an unpigmented, tapering anterior appendage connected by a thin membrane to an internally pigmented anterodorsal appendage, darkly pigmented on its distal tip, except for a small circular unpigmented area on each side. A pair of bilaterally placed, unpigmented medial escal appendages (each divided into numerous fine filaments in the 11- and 42-mm specimens) are present, along with an unpigmented, compressed posterior escal appendage bearing distally a rounded lump of tissue that tapers to a point. Lateral escal appendages are absent (see Regan and Trewavas, 1932:82, fig. 132).

The opercle is bifurcated, the two forks forming an acute angle of 35 to 37°. The length of the ventral fork of the opercle is 25.4 to 26.2% SL; the length of the dorsal fork is 52.7 to 55.2% of the length of the ventral fork. The length of the subopercle is 32.5 to 37.3% of the length of the ventral fork of the opercle.

The teeth are as described for the family. The longest tooth in the upper jaw is 0.5 mm; that in the lower jaw 1.0 mm. There are 30 to 39 teeth in upper jaw, 34 to 40 in lower jaw. There are 8 to 10 teeth on vomer. The second and third pharyngobranchials are well developed and toothed. The epibranchials and ceratobranchials are toothless.

The following measurements are expressed in percent of standard length: head length 35.7 to 36.4; head depth 36.4 to 42.9; length of premaxilla 27.3 to 27.7; length of lower jaw 40.5 to 40.9; length of pectoral-fin lobe 13.6 to 16.7; length of illicium 18.2 to 19.0; length of sphenotic spine 6.9 to 10.4, quadrate spine 10 to 11.8, articular spine 2.0 to 2.5.

### Diversity

A single species:

### *CHIROPHRYNE XENOLOPHUS* REGAN AND TREWAVAS, 1932

Figures 32, 137, 140, 141, 229; Table 8

### Females

*Chirophryne xenolophus* Regan and Trewavas, 1932:82, figs. 131, 132 (original description, single specimen). Bertelsen, 1951:75, 94, fig. 50 (redescription of holotype, opercular bones figured, in key). Grey, 1956a:250 (synonymy, distribution). Pietsch, 1974a:31, 32, 89, fig. 58 (second known specimen, esca figured, relationships, in key). Nielsen, 1974:91 (listed in type catalog). Pietsch, 1978:16, figs. 3C, 6D, 13C, 14, 15, 18 (description, comparison with other "long-pectoraled" oneirodid genera). Pietsch and Seigel, 1980:394 (after Pietsch, 1978). Pietsch, 1999:2032 (western central Pacific). Pietsch and Orr, 2007:13, 18, figs. 8B, 17, 18 (phylogenetic relationships).

### Males and Larvae

Unknown.

### Material

Four metamorphosed females (11 to 42 mm):

Holotype of *Chirophryne xenolophus*: ZMUC P9296, 11 mm, *Dana* station 3731(12), South China Sea, 14°37′N, 119°52′E, 2500 m wire; 0200 h, 17 June 1929.

Additional females: MNHN, one (19 mm); SIO, one (22 mm); UW 48243, one (42 mm, cleared and stained).

### Distinguishing Characters and Description

As given for the genus and family.

### Distribution

*Chirophryne xenolophus* is known from four small juvenile females: the holotype collected from off the Philippine Islands in the South China Sea, two from the eastern Pacific Ocean (0°N, 146°W; 32°N, 136°W), and a fourth specimen from the Gulf of Mexico (Fig. 229). The record cited by Pietsch (1978:19, fig. 18) off Japan is erroneous. Like its close relative, *Puck pinnata*, *C. xenolophus* appears to be a relatively deep dwelling species: all the known specimens were collected in nets fished at maximum depths of 1230 to 1400 m.

## Genus *Leptacanthichthys* Regan and Trewavas, 1932 (Lightline Dreamers)

Figures 33, 137, 142–144, 202, 203, 229; Tables 1, 8

### Females and Males

*Dolopichthys* (subgenus *Leptacanthichthys*) Regan and Trewavas, 1932:66, 80, fig. 128 (one of five subgenera of genus *Dolopichthys*; type species *Dolopichthys gracilispinis* Regan, 1925c, by monotypy).

*Leptacanthichthys* Bertelsen, 1951:74, 94, fig. 49 (subgenus *Leptacanthichthys* given generic status; type species *Dolopichthys gracilispinis* Regan, 1925c, by monotypy).

### Larvae

Unknown.

### Distinguishing Characters

Metamorphosed females of *Leptacanthichthys* are distinguished from those of all other known oneirodids by having an extremely well-developed articular spine that is considerably longer than the quadrate spine. Metamorphosed females of *Leptacanthichthys* are further unique in having the following combination of character states: The ethmoid cartilage and vomer are wide, wider than the distance between the anterolateral tips of the lateral ethmoids and frontals. Vomerine teeth are present. The ethmoid region is depressed, the nasal foramina narrow and oval in shape. The frontals are long, anterior in posterior, overhanging and extending past the anterior limits of the ethmoid cartilage and vomer. The dorsal margins of the frontals are nearly linear. Ventromedial extensions of the frontals approach each other on the midline, making contact with the parasphenoid. The frontals are separated from the prootics. The pterosphenoid is present. The anterior end of the illicial trough is wider and shallower than the posterior end. The sphenotic spines are extremely well developed. The symphysial cartilage of the upper jaw is longer than wide. The lower jaw bears a small symphysial spine. The hyomandibula has a double head. The posterior margin of the opercle is deeply notched. The subopercle is short and broad, its dorsal end rounded to bluntly pointed (the posterior margin without indentation), its ventral end rounded (without an anterior spine or projection). The first pharyngobranchial is absent. The second pharyngobranchial is well developed. The second hypobranchial articulates directly with the second basibranchial. The caudal-fin rays lack internal pigmentation. The illicium is considerably longer than the length of the esca bulb. The pterygiophore of the illicium is cylindrical throughout its length, emerging on the snout from between the frontal bones, its anterior end exposed, its posterior end concealed beneath the skin. The first ray of the dorsal fin is well developed. There are 5 or 6 (rarely 4) dorsal-fin rays and 5 anal-fin rays (Table 1). The pectoral-fin lobe is long and narrow, longer than the longest rays of the pectoral fin. There are 18 to 22 pectoral-fin rays. The coracoid lacks a posteroventral process. The pelvic bones are simple, expanded distally. The skin is naked, without dermal spinules. Darkly pigmented skin of the caudal peduncle extends well past base of the caudal fin.

Metamorphosed males of *Leptacanthichthys* are distinguished from those of all other known oneirodids by having the following combination of character states: The pectoral-fin lobe is long and narrow, longer than the longest pectoral-fin rays, articulating along the dorsal margin. The subopercle is short and broad, its dorsal end rounded. There are 6 lower denticular teeth. The skin is naked, without dermal spinules.

Description

The body of metamorphosed females is elongate and slender, not globular. The snout is relatively long, the mouth large, horizontal, the cleft extending past the eye. The illicium length is 19.2 to 24.1% SL. The escal bulb bears a darkly pigmented streak on the dorsal surface and an unpigmented compressed posterior appendage. Anterior and lateral escal appendages are absent.

The opercle is bifurcate, the two forks forming an acute angle of 18° to 27°. The length of the ventral fork of the opercle is 18.1 to 20.9% SL. The length of the dorsal fork of the opercle is 64.4 to 70.5% of the length of the ventral fork. The length of the subopercle is 29.5 to 37.5% of the length of the ventral fork of the opercle.

The teeth are slender, recurved, and all depressible, arranged in an oblique transverse series of increasing size. The number of teeth generally increases with the size of specimens, those in the upper jaw considerably more numerous than those of the lower jaw. There are 52 to 220 teeth in the upper jaw, 44 to 140 in the lower jaw. There are 6 to 16 teeth on the vomer. The second and third pharyngobranchials are present and well toothed. The epibranchials and ceratobranchials are toothless.

The following measurements are expressed in percent of standard length: head length 29.1 to 33.3; head depth 30.8 to 34.9; length of premaxilla 21.8 to 24.6; length of lower jaw 26.7 to 34.1; length of pectoral-fin lobe 7.7 to 13.1; length of illicium 19.2 to 24.1; length of sphenotic spine 3.4 to 4.9, quadrate spine 2.8 to 4.5, articular spine 4.6 to 6.8.

Males are known from a single specimen in late metamorphosis, attached to a sexually mature female (Pietsch, 1976: 785, figs. 2–5): The posterior nostril is well separated from the eye. The gill cover is pigmented, with slightly darker pigmentation along the posterior margin of the subopercle. The dorsal pigment is restricted to the upper part of the body, extending beneath the base of the dorsal fin and just past the anterior base of the anal fin, with a more heavily pigmented dorsal and ventral group of melanophores near the hypural plate. The peritoneum is pigmented.

Diversity

The genus contains a single species:

LEPTACANTHICHTHYS GRACILISPINIS (REGAN, 1925C)

Figures 33, 137, 142–144, 229; Table 8

Males and Females

*Dolopichthys gracilispinis* Regan, 1925c: 563 (original description, two specimens, lectotype ZMUC P9295, designated by Bertelsen, 1951). Regan, 1926:27, 30, pl. 5, fig. 2 (description after Regan, 1925c; in key). Nielsen, 1974:93 (listed in type catalog).

*Dolopichthys (Leptacanthichthys) gracilispinis*: Regan and Trewavas, 1932:66, 80, fig. 128 (new combination; genus *Dolopichthys* broadened to incorporate five subgenera; description after Regan, 1925c, 1926; in key).

*Leptacanthichthys gracilispinis*: Bertelsen, 1951:74, 94, fig. 49 (subgenus *Leptacanthichthys* given generic status, description, in key). Grey, 1956a:250 (synonymy, distribution). Pietsch, 1974a:16–32, 82, 86–89, figs. 22, 34, 39c, 42, 51c, 55, 59, 103, 104 (osteology, relationships, esca figured, in key). Pietsch, 1976:784, figs. 2–5 (attached male described, figured). Pietsch, 1978:20, figs. 3D, 5B, 6E–H, 13D, 16–18 (description, osteology, comparison with other "long-pectoraled" oneirodid genera). Bertelsen, 1986:1391, figs. (additional specimen, eastern North Atlantic). Swinney, 1995a:52, 54 (additional specimen, off Madeira). Pietsch, 1999:2032 (western central Pacific). Pietsch, 2002b:1064 (western central Atlantic). Moore et al., 2003:215 (off New England). Pietsch, 2005b:209, 215, 220, 221, 229, 232 (reproduction). Woodland, 2005:253 (sexual parasitism). Pietsch and Orr, 2007:18, figs. 17, 18 (phylogenetic relationships).

Larvae

Unknown.

Material

Twenty-four metamorphosed females (10.5 to 103 mm), one adult free-living male (13 mm), and one parasitic male in late metamorphosis (7.5 mm):

Lectotype of *Dolopichthys gracilispinis*: ZMUC P9295, 52 mm, *Dana* station 1206(3), Gulf of Panama, 6°40′N, 80°47′W, 3500 m wire, 1845 h, 14 January 1922.

Paralectotype of *Dolopichthys gracilispinis*: BMNH 1925.8.11.14, 43 mm, *Dana* station 1358(5), North Atlantic, 28°15′N, 56°00′W, 3000 m wire, 1530 h, 2 June 1922.

Additional metamorphosed females: ARC, one (42.5 mm); BMNH, one (47 mm); CAS, one (14 mm); ISH, one (72 mm); LACM, three (22 to 56 mm, 56-mm specimen with a 7.5-mm attached male); MCZ, three (12.5 to 55 mm); NMSZ, one (20 mm); OS, one (103 mm); ROM, three (41 to 54 mm); SIO, three (10.5 to 39 mm); ZMUB, four (42 to 61 mm).

Free-living male: BMNH, one (13 mm).

Distinguishing Characters and Description

As given for the genus and family.

Sexual Maturity

The 56-mm female (LACM 33625-2) of *Leptacanthichthys gracilispinis* has a parasitically attached male. Prior to the discovery of this specimen (Pietsch, 1976:784), there was no evidence to conclude that males of oneirodids ever become sexually parasitic. Food items have not been found in the stomach of any oneirodid male (but few specimens have been available for examination), despite the fact that males of at least some genera grow considerably after metamorphosis: for example, metamorphosis stages of *Oneirodes* measure 6.0 to 9.5 mm; whereas, the largest known metamorphosed specimen is 16.5 mm long (Pietsch, 1974a:108). These data indicate that the males probably continue to feed after metamorphosis and are thus not nutritionally dependent on a parasitic association with the female; the especially well developed denticular apparatus of males of this family has been thought to be used for temporary attachment without subsequent tissue fusion. Thus, the discovery of a parasitized female oneirodid is of some considerable interest.

The single known attached male of *Leptacanthichthys* is the first record of sexual parasitism for the genus and family. Its upper denticular bone was embedded in the end of a stalk of tissue, approximately 2.2 mm long, protruding from the belly of the female; the male has since been torn away but the bones of the upper jaw and ethmoid region have remained attached to female tissue. Sagittal sections of the stalk reveal a highly convoluted outer pigmented layer surrounding an inner core of connective tissue. Centrally located within the connective tissue are numerous intercommunicating, thin-walled blood-sinuses much like those described in the area of fusion of parasitic males of other ceratioids (Regan, 1925b:392, 1926:12; Regan and Trewavas, 1932: 15–16, figs. 4, 5; Munk and Bertelsen, 1983:54–73, figs. 4–17).

The male, 7.5 mm long, is in late metamorphosis, with subdermal pigment well developed and skin only lightly pigmented. Juvenile and adult males of all known ceratioids are longer than approximately 10 mm, their subdermal pigment has disappeared, and in all genera but one (*Haplophryne* Regan; see Bertelsen, 1951:167, fig. 112A), the skin is darkly pigmented (Bertelsen, 1951:10).

Examination of the viscera of the male shows a moderately sized liver occupying approximately a third of the volume of the coelomic cavity. The paired testes are small in comparison with those of a ripe male (e.g., see Pietsch and Nafpaktitis, 1971). Serial sections through the entire length of both testes revealed only small undifferentiated spermatocytes. In contrast, the gonads of the female are well developed; the left ovary is filled with approximately 7,500 eggs, measuring approximately 0.4 to 0.6 mm in diameter.

Several facts indicate that the case at hand represents a permanent parasitic association of male and female rather than a temporary attachment: (1) although the male has been torn away from the female, bones of the upper jaw and ethmoid region remain embedded at the tip of a stalk of tissue protruding from the belly of the female, indicating a permanent fusion of tissue between the two sexes; (2) the stalk of female tissue, to which the male was attached, is full of intercommunicating vessels capable of transferring blood and nutrients to the male; (3) as shown by its small size, unpigmented skin, and undeveloped testes, the male is sexually immature; metamorphosed males of families in which parasitic males have been found become attached while immature, whereas temporary attachment should not be expected before the male has reached sexual maturity.

An examination of the ovaries of all known material of *L. gracilispinis* (14 specimens, 10.5 to 72 mm) turned up one additional gravid female (ROM 27284, 54 mm). The ovaries of this specimen are 8.5 mm long (15.7% SL) and contain numerous eggs approximately 0.2 mm in diameter (Pietsch, 1976).

Distribution

*Leptacanthichthys gracilispinis* is well represented in the North Atlantic, with about 16 localities tightly clustered between 28 and 53°N, and 16 and 71°W (Fig. 229). The remaining specimens are from the North Pacific where they range from off Japan and the Hawaiian Islands, to northern California, and south to the equator (at approximately 138°W), and east to the Gulf of Panama. The single known free-living male is from the central North Atlantic at 35°N, 33°W. Ninety-two percent of the known material was captured by gear fished below 1000 m; 62% by gear fished below 1200 m.

Genus *Ctenochirichthys* Regan and Trewavas, 1932 (Combfin Dreamers)

Figures 137, 145–147, 229; Tables 1, 8

Females, Males, and Larvae

*Ctenochirichthys* Regan and Trewavas, 1932:82, pl. 3, fig. 3 (type species *Ctenochirichthys longimanus* Regan and Trewavas, 1932, by monotypy).

Distinguishing Characters

Metamorphosed males and females of the genus *Ctenochirichthys* differ from those of all other oneirodid genera in having an extremely long pectoral-fin lobe (greater than 15% SL in females), bearing a high number of pectoral-fin rays (28 to 30), inserted along the dorsal margin; and an elongate, slender subopercular bone, tapering only slightly, the dorsal end rounded or somewhat squared off.

Metamorphosed females of *Ctenochirichthys* are further unique in having the following combination of character states: Vomerine teeth are present. The dorsal margins of the frontal bones are convex. The sphenotic spines are reduced (in some specimens not piercing the skin). The lower jaw bears a small symphysial spine. The hyomandibula has a double head. The quadrate and articular spines are small (in some specimens not piercing the skin). The posterior margin of the opercle is deeply notched. The subopercle is elongate and slender throughout its length (the posterior margin without indentation), its ventral end elongate and rounded (without an anterior spine or projection). The second pharyngobranchial is well developed. The caudal-fin rays lack internal pigmentation. The illicium is considerably longer than the length of the escal bulb. The pterygiophore of the illicium is cylindrical throughout its length, emerging on the snout from between the frontal bones, its anterior end exposed, its posterior end concealed beneath the skin. The first ray of the dorsal fin is well developed. There are 6 or 7 dorsal-fin rays and 4 or 5 anal-fin rays (Table 1). The pectoral-fin lobe is extremely long and narrow, longer than the longest rays of the pectoral fin. There are 28 to 30 pectoral-fin rays. The skin is presumably naked, without dermal spinules (but material for clearing and staining is unavailable). Darkly pigmented skin of the caudal peduncle extends well past the base of the caudal fin (known from only two metamorphosed females, material is unavailable for internal anatomical examination).

In addition to the long pectoral-fin lobe and high number of pectoral-fin rays, metamorphosed males of the genus *Ctenochirichthys* differ from those of all other oneirodid genera in having 13 olfactory lamellae; 9 upper denticular teeth arranged in a single series; the lower denticular teeth arranged in two rows, the inner row with about 8 teeth, the outer row with 3 teeth; and naked skin, without dermal spinules.

In addition to the high number of pectoral-fin rays, larvae are unique in having the following combination of character states: The body is elongate. The depth of the body and length of the head measures 40 to 45% of SL. A concentration of pigment lies medial to the subopercle. Dorsal pigmentation is restricted to the anterodorsal part of the body. Pigment on the caudal peduncle is separated into dorsal and ventral groups.

Description

The body of metamorphosed females is elongate, not globular. The snout is short, the mouth moderate in size and oblique, the cleft terminating anterior to or beneath the eye. The illicium length is 24.8 to 31.2% SL. The escal bulb bears a short rounded anterior appendage; a darkly pigmented, raised band of tissue extending over the dorsomedial surface and down onto the sides of the bulb, with a circular unpigmented area on each side; and an unpigmented compressed tapering posterior appendage. Lateral escal filaments are absent.

The opercle is bifurcate, the two forks forming an acute angle of 38°. The length of the ventral fork of the opercle is 21.9% SL; the dorsal fork is 52.5% of the length of the ventral fork. The length of the subopercle is 51.2% of the length of the ventral fork of the opercle.

The teeth are slender, recurved, and all depressible, those in the lower jaw slightly larger but fewer than those of the upper jaw. The number of teeth in upper jaw is 17 to 37, in lower jaw 16 to 31. There are 2 to 4 teeth on the vomer. The second and third pharyngobranchials are present and toothed. Epibranchial and ceratobranchial teeth are absent.

### Diversity

A single species:

CTENOCHIRICHTHYS LONGIMANUS REGAN AND TREWAVAS, 1932

Figures 137, 145–147, 229; Table 8

### Females and Larvae

*Dolopichthys heteracanthus*: Regan, 1926:28 (in part).

*Ctenochirichthys longimanus* Regan and Trewavas, 1932:82, pl. 3, fig. 3 (original description, two specimens, lectotype, ZMUC P9297, designated by Bertelsen, 1951). Bertelsen, 1951:75, 94, figs. 51, 52 (redescription of type material, males and larvae described; opercular bones, larval male, juvenile male, and lectotype figured; in key). Grey, 1956a:251 (synonymy; distribution). Bussing, 1965:222 (misidentification, specimen referred to *Pentherichthys atratus* by Pietsch, 1978). Maul, 1973:671 (synonymy, eastern North Atlantic). Pietsch, 1974a:31, 32, 89 (relationships, in key). Nielsen, 1974:91 (listed in type catalog). Pietsch, 1978:14, figs. 3B, 6C, 8B, 11–13B, 18 (comparison with other long-pectoraled genera). Bertelsen, 1986:1387, figs. (eastern North Atlantic). Pequeño, 1989:43 (Chile, record unconfirmed). Pietsch and Orr, 2007:18 (phylogenetic relationships).

### Males

*Trematorhynchus multiradiatus* Beebe and Crane, 1947:166, text fig. 11 (original description, single male specimen). Bertelsen, 1951:95, fig. 52B (redescription, comparison with all known material, figured). Mead, 1958:134 (type transferred to CAS).

### Material

Two females (12.5 to 36.5 mm), a single juvenile male (11.5 mm), and two larval males (4.5 to 5 mm):

Lectotype of *Ctenochirichthys longimanus*: ZMUC P9297, female, 12.5 mm, *Dana* station 3548(2), Gulf of Panama, 7°06′N, 79°55′W, 3000 m wire, 1030 h, 31 September 1928.

Paralectotype of *Ctenochirichthys longimanus*: BMNH 1932.5.3.20, female, 36.5 mm, *Dana* station 1206(5), Gulf of Panama, 6°40′N, 80°47′W, 2500 m wire, 1845 h, 14 January 1922.

Holotype of *Trematorhynchus multiradiatus*: CAS-SU 46491, male, 11.5 mm, eastern Pacific *Zaca* Expedition station 225, net T-1, southwest of Jicaron Island, Panama, 7°08′N, 81°57′W, 1-m diameter conical ring net, 0 to 910 m, 20 March 1938.

Additional material: ZMUC, two larval males (4.5 to 5.0 mm).

### Distinguishing Characters and Description

As given for the genus and family.

### Distribution

*Ctenochirichthys longimanus* is known from five specimens, four of which were collected in the Gulf of Panama. The fifth specimen, a larval male, was taken in the eastern North Atlantic at approximately 34°N, 17°W (Fig. 229). The two metamorphosed females were captured in nets fished at maximum depths of 800 to 1000 m, the single known male, at a maximum depth of 910 m.

### Comments

Based on the unique fin-ray formula and position of the pectoral-fin rays along the dorsal margin of the pectoral-fin lobe, Bertelsen (1951) verified the suggestion of Beebe and Crane (1947:166) that the 11.5-mm specimen described by them under the name *Trematorhynchus multiradiatus* is actually a male of *Ctenochirichthys longimanus*. Using the same characters of the pectoral fin, Bertelsen (1951) was further able to assign two larvae from the *Dana* collections to this species, thus bringing the total number of known specimens to five.

The two known larvae are characterized most strikingly by features of their pigmentation: the unique combination of a dark spot beneath the subopercle and the separation of the pigment on the caudal peduncle into two distinct groupings were found by Bertelsen (1951) to be present as well beneath the skin of the 17-mm lectotype of *C. longimanus*. The distinct shape of the subopercle further confirmed this identity: "As I have found no difference in their pigmentation or other characters which suggest that they represent different species, and as one them as well as the male, mentioned above, was found in the same locality as the two earlier known females, they are all referred to the same and only known species *C. longimanus*" (Bertelsen, 1951:95).

The larvae of *C. longimanus* (two known specimens, 4.5 and 5 mm) are similar to those of *Microlophichthys microlophus* but are easily distinguished by the shape of the subopercle, the number of pectoral-fin rays, and the divided peduncular pigment. Further differentiating the two, the dorsal pigment of *C. longimanus* does not extend forward onto the upper surface of the head, and the sphenotic region, adductor mandibulae, and jaws are unpigmented; the peritoneal pigment does not quite meet on the belly (Bertelsen, 1951:96).

## Family Caulophrynidae Goode and Bean, 1896 (Fanfin Seadevils)

Figures 25, 28, 30, 45, 56, 148–152, 202, 203, 230, 291, 292, 297; Tables 1, 6, 8, 10

Type Genus *Caulophryne* Goode and Bean, 1896

Caulophryninae Goode and Bean, 1896:489 (subfamily of family Ceratiidae to contain genus *Caulophryne*).

Caulophrynidae Regan, 1912:285, 288 (family of division Ceratiiformes to contain genus *Caulophryne*).

### Distinguishing Characters

Metamorphosed females of the family Caulophrynidae are distinguished from those of all other ceratioid families by having extremely long dorsal- and anal-fin rays (greater than 60% SL); only 8 caudal-fin rays; the opercle, subopercle, posttemporal, and ventral portion of the cleithrum greatly reduced; only 2 pectoral radials, the ventral-most radial broadly expanded; the neuromasts of acousticolateralis system located at the tips of extremely long filaments; and the illicium without an expanded escal bulb.

Metamorphosed females are further differentiated by having the following combination of character states: The supraethmoid is present. The frontals lack ventromedial extensions, and meet on the midline anterior to the supraoccipital. The parietals are present, displaced anteriorly. Sphenotic spines are present. The pterosphenoid is absent. The metapterygoid and mesopterygoid are present. The hyomandibula has a double head. There are 2 hypohyals and 6 (2 + 4), rarely 5 (1 + 4), branchiostegal rays. The opercle is bifurcated, but very much reduced. The subopercle is long, narrow, and reduced, with the posterior margin deeply notched, its ventral part without a spine or projection on the anterior margin. Quadrate and articular spines are rudimentary. Angular and preopercular spines are absent. The jaws are equal anteriorly. The lower jaw bears a symphysial spine. A postmaxillary process of premaxilla is absent. The maxillae are reduced. The anterior-maxillomandibular ligament is absent. The first and fourth pharyngobranchials are absent. The second and third pharyngobranchials are well developed and toothed. Ossified hypobranchials and basibranchials are absent. Epibranchial and ceratobranchial teeth are absent. There is a single reduced epural. The posterior margin of the hypural plate is notched in some specimens. The pterygiophore of the illicium bears a small ossified remnant of the second cephalic spine. The escal bulb and associated central lumen and escal pore are absent. The esca lacks toothlike denticles. The coracoid has a well-developed posteroventral process. There are 2 pectoral radials. The pelvic bones are cylindrical, without a distal expansion. There are 6 to 22 (6 in *Robia*, 14 to 22 in *Caulophryne*) dorsal-fin rays, 5 to 19 (5 in *Robia*, 12 to 19 in *Caulophryne*) anal-fin rays, and 14 to 19 pectoral-fin rays (Table 1). Pelvic fins absent (but present in larvae and early metamorphic stages; see below). There are 8 (2 simple + 4 bifurcated + 2 simple) caudal-fin rays (Table 1). The skin is everywhere naked, without a trace of dermal spinules. The ovaries are paired. There are 2 short pyloric caeca.

Metamorphosed males, known from only five specimens, all assigned to *Caulophryne* (two free-living and three parasitically attached to females), have large eyes, with a conspicuous aphakic space anterior to the lens. The olfactory organs are large, the nostrils directed laterally in smaller specimens, the anterior nostrils close set and directed anteriorly in larger specimens. Jaw teeth are absent. The upper denticular has a dorsally directed tapering prolongation, bearing an irregular series of teeth along its ventral margin. The lower denticular is divided into 3 lobes, bearing 5, 8, and 5 teeth, respectively. The pectoral fins are large, their length about 40% SL. The pelvic fins are well developed in young free-living stages, but lost with growth in later stages, including parasitic adults. Fin-ray counts are as given for metamorphosed females. The skin is naked. Males become parasitic, but they are probably facultative (see Reproduction and Early Life History, Chapter Eight).

Larvae are represented by 16 specimens, all assigned to *Caulophryne*. They are characterized by having a short, rounded body, the skin highly inflated. The pectoral fins are unusually large, reaching posteriorly beyond the dorsal and anal fins. The pelvic fins are well developed, with 3 or 4 rays (lost during metamorphosis). Sexual dimorphism is apparently absent, all known specimens with a rudiment of the illicium protruding from the anterodorsal margin of the head. Fin-ray counts are as given for metamorphosed females. Metamorphosis begins at lengths of about 8 to 10 mm (Bertelsen, 1951:35, fig. 11A, B; 1984:326, 328, fig. 167A).

The osteological features of the Caulophrynidae cited here and elsewhere in this account are based only on examination of females of *Caulophryne* (Regan, 1926:16, fig. 7; Regan and Trewavas, 1932:42, figs. 56–58; Bertelsen, 1951:31, fig. 8; Pietsch, 1972a:29, 35, 36, 38, 41–45, fig. 24(3); Pietsch, 1979:4–11, figs. 1–13); males of *Caulophryne* and specimens of the genus *Robia* have so far been unavailable for osteological study.

Description

The body of metamorphosed females is short and globular, its depth approximately 40 to 45% SL. The mouth is large, its opening horizontal to slightly oblique. The lower jaw of most specimens extends posteriorly beyond the base of the pectoral-fin lobe. The oral valve is well developed, lining the inside of both upper and lower jaws. There are two nostrils on each side of the snout situated at the end of a single short tube. The teeth are slender, recurved, and depressible, those in the lower jaw less numerous, but slightly longer than those in the upper jaw. There are 20 to 45 teeth in the upper jaw, 12 to 34 in the lower jaw. There are 1 to 5 vomerine teeth. The first epibranchial and the full length of the first ceratobranchial are free, not bound to the wall of the pharynx by connective tissue. All four epibranchials are closely bound together. The fourth epibranchial and fourth ceratobranchial are bound to the wall of the pharynx, leaving no opening behind the fourth arch. The distal one-third of the first ceratobranchial is bound by connective tissue to the adjacent second ceratobranchial, the proximal two-thirds free. The proximal one-quarter to one-half of ceratobranchials II to IV are free, not bound together by connective tissue. Gill filaments, absent on the epibranchials, are present on the proximal tip of the first ceratobranchial and the full length of ceratobranchials II to IV. A pseudobranch is absent. The length of the illicium is highly variable, 16 to 125.8% SL in *Caulophryne*, 268.3% SL in the only known specimen of *Robia*. The anterior-most tip of the pterygiophore of the illicium is exposed, emerging on the snout, its posterior end concealed beneath the skin. The esca is unpigmented and somewhat translucent, its morphology simple, without elaborate filaments or appendages (*Robia*), or consisting of several branched appendages and/or numerous filaments (*Caulophryne*). In some species the entire length of the illicium is covered with numerous translucent filaments. The neuromasts of the acoustico-lateralis system are located at the tips of elongate cutaneous filaments, the pattern of placement as described for other ceratioids (Pietsch, 1969, 1972b, 1974a, 1974b).

Only two free-living males are known, both assigned to *Caulophryne*. The smaller of the two, a 7.5-mm male in early metamorphosis (ZMUC P92193), has naked skin. The pelvic fins are well developed (each with 4 rays). Jaw teeth and the upper denticular are absent. The lower denticular supports 2 transverse series of teeth. The eyes are large with a conspicuous aphakic space anterior to the lens. The olfactory organs are large, the nostrils directed laterally (Bertelsen, 1951:310). The larger male, an 11-mm specimen in metamorphosis (MCZ 69324), also has naked skin, but the pelvic fins are absent (lost with growth). Jaw teeth are absent. The upper denticular has a dorsally directed tapering prolongation, bearing an irregular series of teeth along its ventral margin. The lower denticular is divided into 3 lobes bearing 5, 8, and 5 teeth, respectively. The olfactory organs are large, the anterior nostrils close set and directed anteriorly. The dorsal- and anal-fin rays counts cannot be determined accurately, but both males have 17 pectoral-fin rays and 8 (2 simple + 4 bifurcated + 2 simple) caudal-fin rays (Table 1).

There are three known parasitic males, 12 to 16 mm, all assigned to *Caulophryne* (BMNH 1930.2.7.1, LACM 36025-1, and MNHN 2001-0140) and all with naked skin and degenerate olfactory organs. There are 19 to 21 dorsal-fin rays, 16 to 17 anal-fin rays, 17 pectoral-fin rays, and 8 (2 simple + 4 bifurcated + 2 simple) caudal-fin rays. Pelvic fins are absent (Regan, 1930b:191; Bertelsen, 1951:31; Pietsch, 1979:20).

Caulophrynid larvae, known from 16 specimens, 3.7 to 11 mm TL, all assigned to *Caulophryne* (ZMUC), have a short body and snout, the latter only about 10% of SL. The skin is highly inflated. The eye is unusually large, its diameter 17% SL in the smallest specimens to 13 to 14% SL in the largest specimens. The nostrils are unusually large, the longest dimension about one-half the diameter of the eye. There are 11 to 20 dorsal-fin rays and 11 to 17 anal-fin rays (the posterior-most rays of the unpaired fins are nearly impossible to distinguish without alizarin staining). The pectoral fins, with 15 to 18 rays, are large, reaching well beyond the dorsal and anal fins, and extending to the base of the caudal fin. The pelvic fins are long and well developed, each with 3 or 4 rays, but become reduced in early metamorphosis, and are lost in juveniles and adults. There are 8 (2 simple + 4 bifurcated + 2 simple) caudal-fin rays (Bertelsen, 1951:31, 35).

The color of preserved females is dark red brown to black over the entire surface of the head, body, fins (except for the esca, and the illicium of some species), and oral cavity. The viscera are unpigmented. The skin of parasitic, and free-living males in metamorphosis, is lightly pigmented.

The largest known female is a 183-mm specimen of *Caulophryne pelagica* collected in the Atlantic Ocean off the Cape Verde Islands (BMNH 2000.1.14.106). The two known free-living males, both in metamorphosis, measure 7.7 and 11 mm (ZMUC P92193, MCZ 69324). The three known parasitic males—one attached to the 142-mm holotype of *C. polynema* (BMNH 1930.2.7.1; see Regan, 1930b), another to a 137-mm specimen of *C. polynema* (MNHN 2001-0140), and the third attached to a 98-mm female identified by Pietsch (1979:18) as *Caulophryne* sp. (LACM 36025-1)—measure 16, 15, and 12 mm, respectively.

Diversity

Two genera differentiated as follows:

Key to Females of Genera of the Family Caulophrynidae

1A. Dorsal-fin rays 14 to 22; anal-fin rays 12 to 19; illicium short, less than 130% SL . . . . . . . . . . . *Caulophryne* Goode and Bean, 1896, p. 445

 NOTE Four recognized species; Atlantic, Pacific, and Indian oceans

1B. Dorsal-fin rays 6; anal-fin rays 5; illicium long, 268% SL in only known specimen . . . . . . . . . . . . . . . *Robia* Pietsch, 1979, p. 450

 NOTE A single known species; Banda Sea

Genus Caulophryne Goode and Bean, 1896 (Fanfin Seadevils)

Figures 25, 28, 30, 45, 56, 148–151, 202, 203, 230, 291, 292, 297; Tables 1, 6, 8, 10

Females, Males, and Larvae

*Caulophryne* Goode and Bean, 1896: 496, pl. 121, fig. 409 (type species *Caulophryne jordani* Goode and Bean, 1896, by monotypy).

*Melanocetus* Brauer, 1902:295 (in part; type species *Melanocetus johnsonii* Günther, 1864, by monotypy). Brauer, 1906:321, pl. 15, fig. 5 (after Brauer, 1902).

*Ceratocaulophryne* Roule and Angel, 1932:500 (type species *Ceratocaulophryne regani* Roule and Angel, 1932, by monotypy). Roule and Angel, 1933:55, pl. 3, figs. 26, 26A (after Roule and Angel, 1932; expanded description, figures).

Distinguishing Characters

Metamorphosed females of the genus *Caulophryne* are distinguished from those of *Robia* (the only other genus of the family) in having a considerably shorter illicium (less than 130% SL) and a greater number of relatively long median-fin rays (dorsal-fin rays 14 to 22, the length of the longest rays greater than 70% SL; anal-fin rays 12 to 19, the length of the longest rays is greater than 60% SL).

Description

The stem of the illicium, bearing slender filaments along its length in some species, is usually darkly pigmented except near the esca (lightly pigmented along the anterior margin in most specimens of *Caulophryne polynema*, unpigmented in *Caulophryne* sp. A). The escal morphology is highly variable (see species accounts). There are 14 to 22 dorsal-fin rays, 12 to 19 anal-fin rays, and 14 to 19 pectoral-fin rays. There are 17 to 46 teeth in the upper jaw and 12 to 34 in the lower jaw. There are 1 to 6 teeth on the vomer. The following measurements are expressed in percent of standard length: head length 27.2 to 41.7; head depth 32.7 to 45.4; head width 21.9 to 30.6; length of premaxilla 28.9 to 44.4; length of lower jaw 31.3 to 50.8; longest tooth in lower jaw 4.7 to 10.1; longest dorsal-fin ray 70 to 248; longest anal-fin ray 60 to 183; length of illicium 15.7 to 125.8.

Additional description as given for the family.

Diversity

Four species differentiated as follows:

Key to Females of Species of the Genus *Caulophryne*

The following key will differentiate juvenile and adult females only. Characters that allow specific identification of free-living males (known from only two specimens, both in metamorphosis) and larvae (16 specimens; see Bertelsen, 1951:35, figs. 8, 11) are unknown.

1A. Stem of illicium naked, without filaments . . . . . . . . . . . . . . . . . . . . . . 2
1B. Stem of illicium with translucent filaments along most of its length . . . . . . . . . . . . . . . . . . . . . 3
2A. Length of illicium 18 to 32% SL in specimens less than 20 mm, greater than 35% SL in specimens greater than 25-mm . . . . . . . . . . . . . . . . . .
 . . . . . . . . . . . . *Caulophryne pelagica* (Brauer, 1902), p. 445

 NOTE Seventeen known specimens, 11 to 183 mm; Atlantic, Pacific, and Indian oceans

2B. Length of illicium 22.5% SL in 169-mm specimen . . . . . . . . . . . . . . . . .
 . . . . . . . . . . . . *Caulophryne bacescui* Mihai-Bardan, 1982, p. 446

 NOTE A single known specimen, 169 mm; Peru-Chile Trench

3A. Stem of illicium with fewer than 15 filaments along length; dorsal-fin rays 16 to 18, rarely 19; anal-fin rays 14 to 17, rarely 18 . . . . . . . . . . . . . . .
 . . . . . . . . . . . . . . *Caulophryne jordani* Goode and Bean, 1896, p. 447

 NOTE Thirty-eight known specimens, 16 to 155 mm; Atlantic, Pacific, and Indian oceans

3B. Stem of illicium with more than 50 filaments along length in specimens greater than 32 mm; dorsal-fin rays 21 or 22, rarely 19 or 20; anal-fin rays 18 or 19, rarely 17 . . . . . . . . . . . . . .
 . . . . . . . . . . . . *Caulophryne polynema* Regan, 1930b, p. 448

 NOTE Fourteen known specimens, 11.5 to 142 mm; Atlantic and eastern Pacific oceans

CAULOPHRYNE PELAGICA (BRAUER, 1902)

Figures 25, 148, 149, 230; Tables 6, 8

Females

*Melanocetus pelagicus* Brauer, 1902:295 (original description, single specimen). Brauer, 1906:321, pl. 15, fig. 5 (after Brauer, 1902). Regan, 1926:22 ("probably a synonym of *C. jordani*"). Harvey, 1952:531, fig. 178 (bioluminescence).

*Caulophryne pelagicus*: Regan, 1912:288 (new combination, listed).

*Caulophryne jordani*: Regan, 1926:22 (in part; specimen later described as holotype of *Caulophryne ramulosa* by Regan and Trewavas, 1932). Bertelsen, 1951:33, 37, fig. 9 (in part, includes all species of *Caulophryne* described to date). Kobayashi et al., 1968:8, figs. 1, 2 (additional specimen, description).

*Caulophryne ramulosa* Regan and Trewavas, 1932:100, 101, pl. 7, fig. 160 (original description, single specimen; holotype previously recorded by Regan, 1926, as *Caulophryne jordani*; in key). Bertelsen, 1951:36, 37 (description, a synonym of *C. jordani pelagica*). Nielsen, 1974:86 (listed in type catalog).

*Caulophryne acinosa* Regan and Trewavas, 1932:100, 101, pl. 8, fig. 161 (original description, single specimen, in key). Bertelsen, 1951:36, 37 (description, a synonym of *Caulophryne jordani pelagica*). Günther and Deckert, 1956:133, fig. 90 (popular account; figure after Regan and Trewavas, 1932). Nielsen, 1974:85 (listed in type catalog).

*Caulophryne pelagica*: Regan and Trewavas, 1932:100, 101, fig. 162 (new orthography, in key; description after Brauer, 1902, 1906; esca figured). Bertelsen, 1951:36, 37, fig. 9 (figure after Brauer, 1906; a subspecies of *Caulophryne jordani*). Pietsch, 1979:14, figs. 16–18, 24 (additional specimens, description, comparison with all known material, in key). Pietsch and Seigel, 1980:380 (after Pietsch, 1979). Pietsch, 1999:2026 (western central Pacific). Nakabo, 2002:467, 1496, figs. (Japan, in key). Love et al., 2005:59 (eastern North

Pacific). Pietsch and Orr, 2007:4, fig. 3C (phylogenetic relationships).

*Caulophryne jordani pelagica*: Bertelsen, 1951:33, 37, fig. 9 (one of three subspecies, description, comparison with all known material, includes *Caulophryne ramulosa* and *C. acinosa*, in key). Grey, 1956a: 233 (synonymy; after Bertelsen, 1951). Kobayashi et al., 1968:9 (additional specimen most similar to *C, jordani pelagica*).

*Caulophryne pietschi* Balushkin and Fedorov, 1985:1035, figs. 1, 2 (original description, single specimen).

Males and Larvae

Unknown.

Material

Seventeen metamorphosed females (11 to 183 mm):

Holotype of *Melanocetus pelagicus*: ZMB 17711, 11 mm, *Valdivia* station 228, Indian Ocean, 2°38′S, 65°59′E, 0 to 2500 m, bottom depth 3460 m.

Holotype of *Caulophryne ramulosa*: ZMUC P9245, 57 mm, *Dana* station 1209(1), Gulf of Panama, 7°15′N, 78°54′W, 3500 m wire, 1845 h, 17 January 1922.

Holotype of *Caulophryne acinosa*: ZMUC P9244, 19 mm, *Dana* station 3920(2), Indian Ocean, 1°06′S, 62°25′E, 3500 m wire, bottom depth 4630 m, 1830 h, 9 December 1929.

Holotype of *Caulophryne pietschi*: ZIN 47196, 150 mm, *Milogradovo* trawl 67, Southern Ocean, south of New Zealand, 48°53′S, 178°39′E, 954 to 960 m, 20 April 1973.

Additional material: BMNH, one (183 mm); HUMZ, three (78 to 93 mm); LACM, two (13 to 67 mm); MCZ, one (28 mm); MNHN, one (20 mm); SIO, three (30 to 97 mm); TFRI, one (80 mm); UW, one (112 mm).

Distinguishing Characters

A species of *Caulophryne*, metamorphosed females of which are distinguished from those of all other described species of the genus in details of escal morphology. Females are further unique in having the following combination of character states: stem of illicium naked, without translucent filaments along length; length of illicium 18 to 125.8% SL; number of teeth in upper jaw 28 to 40; dorsal-fin rays 14 to 17. *Caulophryne pelagica* probably differs further from its congeners in having longer dorsal- and anal-fin rays, but this character is difficult to quantify accurately because the distal ends of the rays are nearly always broken off.

Description

Illicium naked, without translucent filaments, darkly pigmented along entire length except near esca; esca consisting of 5 to 7 elongate filamentous appendages: 2 to 4 appendages more proximal in position, bearing none to several side branches, and 3 or 4 more distal appendages bearing numerous side branches; opaque areas absent.

Dorsal-fin rays 14 to 17; anal-fin rays 12 to 16; pectoral-fin rays 14 to 18; number of teeth in upper jaw 28 to 40, in lower jaw 20 to 34; total number of teeth on vomer 2 to 5.

Measurements in percent SL: length of illicium 18 to 125.8 (18 to 31.6 in specimens less than 20 mm, 35.4 to 43.9 in specimens 28 to 65 mm, 125.8 in 93-mm specimen, 77.0 in 150-mm specimen, 107.6 in 183-mm specimen; longest dorsal-fin ray 105 to 248; longest anal-fin ray 132 to 183; longest tooth in lower jaw 5.3 to 6.8.

Additional description as given for the genus and family.

Distribution

*Caulophryne pelagica*, with 17 known females, is well represented in the Pacific Ocean, ranging from the Halmahera Sea, Taiwan, Japan, and the Bounty Plateau southeast of New Zealand, to the Gulf of Alaska, off Baja California, and into the eastern tropical Pacific off Panama and Ecuador as far south as about 4°S (Fig. 230). There are three records in the western Indian Ocean, one off Somalia and two on the equator west of the Maldives; but only a single known specimen from the Atlantic, collected off the Cape Verde Islands. Several specimens were captured with gear fished at maximum depths of 500 m, and one with gear reaching a depth of 2500 m, but 70% of the known material was taken by gear fished at maximum depths of 900 to 1750 m. Two specimens, one tiny female (13 mm) from the Halmahera Sea, and the single specimen from the Atlantic (183 mm) were collected with closing trawls between 1250 and 1500 m and 1475 and 1500 m, respectively.

Comments

The available evidence indicates that *Caulophryne ramulosa* and *C. acinosa* are junior synonyms of *C. pelagica*. As stated by Bertelsen (1951:36), the illicia of the holotypes of all three species are "well preserved and constructed in a very similar fashion." All three nominal forms agree also in having a slightly longer illicium and fewer median-fin rays than the other species of the genus.

The results of the present study indicate further that *C. pietschi* is also a junior synonym of *C. pelagica*. Balushkin and Fedorov (1985) distinguished *C. pietschi* from previously described species of the genus primarily on the basis of illicial length, but clearly the length of the illicium of the only known specimen (81.3% SL based on personal examination) fits well within the range reported here for *C. pelagica*. Additional features cited by Balushkin and Fedorov (1985), thought to distinguish *C. pietschi* from its congeners, including escal morphology, the presence or absence of filaments on the illicium, the number of teeth in the jaws and on the vomer, and fin-rays counts, are also within the known individual variation described here for *C. pelagica*.

The illicium of the 93-mm specimen (HUMZ 178152) described by Kobayashi et al. (1968) is unusually long (125.8% SL), somewhat outside the range of the remaining known material (18 to 103.8% SL). In all other ways, however, this specimen conforms well to the description of *C. pelagica* as presented here. This specimen may represent an undescribed species, but description must await possible confirmation that might come through examination of additional material.

### *CAULOPHRYNE BACESCUI* MIHAI-BARDAN, 1982

Figure 230; Table 8

Females

*Caulophryne jordani*?: Băcescu, 1966: 34, fig. 4 (misidentification, description). Mayer and Nalbant, 1972:164, fig. 5 (misidentification, description).

*Caulophryne bacescui* Mihai-Bardan, 1982:17, figs. 1–3 (original description, single specimen, after Băcescu, 1966).

Males and Larvae

Unknown.

Material

A single metamorphosed female (169 mm):

Holotype of *Caulophryne bacescui*: NHMB 49922, 169 mm, *Anton Bruun*, cruise 11, Peru-Chile Trench, Peruvian waters, October 1965.

Distinguishing Characters

A species of *Caulophryne*, metamorphosed females of which are distinguished from those of all other described species of the genus in details of escal morphology. Females are further unique in having the following combination of character states: illicium naked, without translucent filaments along length; illicium short, length 22.5% SL; number of teeth in upper jaw 46; dorsal-fin rays 15; anal-fin rays 14.

Description

Illicium naked, without translucent filaments, darkly pigmented along entire length except near esca; esca slightly dilated with a narrowing, distally curved portion, bearing 2 long, translucent and approximately 18 similar filaments of various lengths (Mihai-Bardan, 1982:17, fig. 3C).

Dorsal-fin rays 15; anal-fin rays 14; pectoral-fin rays 17 or 18; number of teeth in upper jaw 46, in lower jaw 32; total number of teeth on vomer 6 (after Mihai-Bardan, 1982).

Measurements in percent of standard length: length of illicium 22.5; dorsal- and anal-fin rays damaged, lengths unknown; longest tooth in lower jaw 4.7 (after Mihai-Bardan, 1982).

Additional description as given for the genus and family.

Distribution

The single known specimen of *Caulophryne bacescui* was collected in the eastern tropical Pacific Ocean, off the coast of Peru, at an unknown depth (Fig. 230).

Comments

*Caulophryne bacescui* differs from its congeners in having a short illicium (22.5% SL), without illicial filaments along its length. While large females of *C. jordani* and *C. polynema*, comparable in size to the only known specimen of *C. bacescui*, have equally short illicia, none lack illicial filaments. The high number of upper-jaw teeth of *C. bacescui* (46) is comparable only to that found in *C. polynema*, while the low numbers of dorsal- and anal-fin rays (15 and 14, respectively) compare best with those of *C. pelagica*.

CAULOPHRYNE JORDANI GOODE AND BEAN, 1896

Figures 28, 45, 150, 230; Tables 6, 8

Females

*Caulophryne jordani* Goode and Bean, 1896:xxxiii, 26*, 496, 541, pl. 121, fig. 409 (original description, single specimen; cited as *Caulophryne setosus* in list of plates, index, and figure caption; see Comments below). Jordan and Evermann, 1898:2735 (description after Goode and Bean, 1896). Jordan and Evermann, 1900, fig. 957 (figure after Goode and Bean, 1896). Gill, 1909:586, fig. 24 (brief description and figure after Goode and Bean, 1896). Regan, 1912:288 (description after Goode and Bean, 1896). Beebe, 1926c:422 (additional specimen, Hudson Gorge). Beebe and Rose, 1926:53, fig. (figured). Regan, 1926:4, 22, figs. 7, 16 (cranial osteology, expanded description; figure after Goode and Bean, 1896, but with numerous filaments added to esca). Beebe, 1929a:19 (after Beebe, 1926c). Regan, 1930b:191, 193 (comparison with *C. polynema*). Regan and Trewavas, 1932:6, 100, 101, fig. 159 (brief description after Goode and Bean, 1896; illicium and esca figured after a sketch made by Barton A. Bean; in key). Roule and Angel, 1933:57 (comparison with *C. regani*). Bertelsen, 1951:33, figs. 8–11 (description, comparison with all known material, includes all species of *Caulophryne* described to date but divided into three subspecies). Marshall, 1966:177, pl. 7 (bioluminescence). Lavenberg and Ebeling, 1967:195, fig. 5 (vertical distribution). Kobayashi et al., 1968:8, figs. 1, 2 (additional specimen, description; referred to *C. pelagica* by Pietsch, 1979). Fitch and Lavenberg, 1968:125, fig. 69 (distinguishing characters, natural history). Briggs, 1960:179 (worldwide distribution). Ueno, 1971:102 (listed). Pietsch, 1972a:29, 35, 36, 38, 41–43, 45, fig. 24(3) (osteology, otolith described, figured). Pietsch, 1976:783 (reproduction). Hubbs et al., 1979:13 (listed for California). Marshall, 1979:415, fig. 145 (lateral-line organs). Pietsch, 1979:15, figs. 1–13, 16, 18, 19, 24 (family revision, comparison with all known material, osteology, relationships, in key). Bertelsen and Pietsch, 1983:80, fig. 1 (additional specimen, New South Wales; description; figures after Beebe and Rose, 1926, and Pietsch, 1979). Fujii, 1983:257, fig. (additional specimen, off Surinam). Amaoka, 1984:105, pl. 92A (misidentification, specimen figured is *C. pelagica*, HUMZ 90932). Pietsch, 1984:321, fig. 164E (early life history, phylogeny). Bertelsen, 1986:1373, 1374, figs. (eastern North Atlantic, in key). Nielsen and Bertelsen, 1992:60, fig. 1 (North Atlantic). Andriyashev and Chernova, 1994:102 (listed for Arctic seas). Meng et al., 1995:438, fig. 586 (China). Stearn and Pietsch, 1995:144, fig. 94 (two additional specimens, Greenland). Amaoka et al., 1995:111, fig. 170 (specimen figured is HUMZ 118865 from off Greenland; see Stearn and Pietsch, 1995:144, fig. 94). Rodríguez-Marín et al., 1996:70, fig. 3 (Flemish Cap bank, east of Newfoundland). Shedlock et al., 1997:396, fig. 1 (DNA extracted from formalin-fixed specimens). Jónsson and Pálsson, 1999:199, fig. 2 (Iceland). Pietsch, 2002b:1057, fig. (western central Atlantic). Moore et al., 2003:214 (off New England). Love et al., 2005:59 (eastern North Pacific). Pietsch, 2005b:215, 216 (reproduction).

*Caulophryne setosus* Goode and Bean, 1896:26*, 541, pl. 121, fig. 409 (name cited in index and list of plates and figures, but clearly an error for *Caulophryne jordani*). Pietsch, 1979:20 *(nomen nudum)*.

*Ceratocaulophryne regani* Roule and Angel, 1932:500 (original description, single specimen). Roule and Angel, 1933:55, pl. 3, figs. 26, 26a (expanded description; figures of holotype, lateral and dorsal views). Roule, 1934:208, 209, fig., color plate 12 (semipopular account). Belloc, 1949:17 (listed). Bertelsen, 1951:36 (a synonym of *Caulophryne jordani polynema*). Kobayashi et al., 1968:8 (after Bertelsen, 1951).

*Caulophryne regani*: Fowler, 1936:1347 (new combination; description after Roule and Angel, 1932, 1933; in key). Albuquerque, 1954–1956:1072 (after Fowler, 1936).

*Caulophryne jordani polynema*: Bertelsen, 1951:33, 37 (one of three subspecies, description, comparison with all known material; includes *Caulophryne regani* Roule and Angel, 1932, a name referred to *C. jordani* by Pietsch, 1979). Grey, 1956a:233 (distribution, synonymy; after Bertelsen, 1951). Maul, 1973:666 (after Bertelsen, 1951).

*Caulophryne jordani jordani*: Bertelsen, 1951:33, 37 (one of three subspecies, description, comparison with all known material, in key). Grey, 1956a:232 (distribution, synonymy; after Bertelsen, 1951).

Males and Larvae

Unknown.

Material

Thirty-eight metamorphosed females (16 to 155 mm):

Holotype of *Caulophryne jordani*: USNM 39265, 26 mm, *Albatross*, off Long Island, New York, 39°27′N, 71°15′W, 0 to 2335 m, 19 September 1887.

Holotype of *Ceratocaulophryne regani*: MOM 91-1529, 38 mm, *Hirondelle II* station 3279, west of the Azores, 38°55′N, 34°07′W, 0 to 3000 m, 23 August 1912.

Additional material: AMS, three (16.5 to 64.5 mm); BMNH, one (36 mm); CSIRO, two (30 to 53 mm); HUMZ, two (123 to 154 mm); ISH, one (130 mm); LACM, three (41 to 54; 41-mm specimen cleared and stained); MCZ, six (16 to 50 mm); MNHN, one (19 mm); MRIR, four (119 to 155 mm); NMNZ, one (61 mm); NSMT, one (100 mm); ORI, one (38 mm); ROM, one (109 mm); SIO, two (66 to 94 mm); USNM, two (34 to 54 mm); UW, two (68 to 84 mm; 68-mm specimen cleared and stained); ZMUC, three (71 to 145 mm).

Distinguishing Characters

A species of *Caulophryne*, metamorphosed females of which are distinguished from those of all other described species of the genus in details of escal morphology. Females are further unique in having the following combination of character states: illicium with 3 to 14 elongate translucent filaments along length; length of illicium 16.8 to 36.8% SL; number of teeth in upper jaw 17 to 39; dorsal-fin rays 16 to 19.

Description

Stem of illicium pigmented except near esca, bearing 3 to 14 elongate (length as great as 87% illicium length), translucent filaments along entire length, all but distal-most filaments arising from posterior margin. Esca with (1) an elongate filamentous anterolateral appendage; (2) 2 distal appendages, each bearing numerous filaments, the anterior-most short and stout, opaque along posterior margin, the posterior-most considerably more slender (separation into 2 distal appendages not as evident in specimens less than 40 mm); and (3) a posterolateral appendage bearing numerous short filaments and with a palmate (usually bilobed) opaque distal tip (illicial and escal filaments absent in 26-mm holotype; esca apparently embellished by Regan, 1926:22, fig. 16).

Dorsal-fin rays 16 to 19; anal-fin rays 14 to 18; pectoral-fin rays 16 to 19; number of teeth in upper jaw 17 to 39, in lower jaw 12 to 26; total number of teeth on vomer 2 to 5.

Measurements in percent of standard length: length of illicium 16.8 to 36.8; longest dorsal-fin ray 78 to 115; longest anal-fin ray 82 to 110; longest tooth in lower jaw 6.1 to 10.1.

Additional description as given for the genus and family.

Distribution

*Caulophryne jordani* is broadly distributed in the North Atlantic ranging from off West Greenland, Iceland, and south to about 5°N; it has so far not been collected in the South Atlantic (Fig. 230). There are two records in the western equatorial Indian Ocean, one off the coast of Somalia and another northwest of the Chagos Archipelago. In the Pacific Ocean it ranges from off New Ireland in the Bismarck Archipelago, off New South Wales, and east to southern California in the north and Peru in the south. It has so far not been collected in the western North Pacific. The species appears to have an unusually broad vertical distribution, with specimens taken as shallow as 275 m and others by gear fished at maximum depths of 3000 m; however, about 70% of the known material was taken by gear fished at maximum depths of 800 to 1625 m. Two specimens taken off West Greenland (123 and 154 mm) were collected in a closing trawl fished at depths ranging from 935 to 1902 m; two specimens from the central North Atlantic (36 and 68 mm) were collected with a closing trawl between 1235 and 1260 m and 1250 and 1510 m, respectively.

Comments

The illicial apparatus of the holotype of *Caulophryne jordani* appears to be somewhat damaged as indicated by the drawing provided by Regan and Trewavas (1932:101, fig. 159) and as noted by Bertelsen (1951:36). Filaments normally found along the length of the illicium and arising from the esca are absent. The posterolateral escal appendage and 2 distal escal appendages, however, remain intact. In these and all other characters, the specimens (with a minor exception, see below) assigned to *C. jordani* by Pietsch (1979) and those reported here for the first time compare very well with the holotype.

Bertelsen (1951:37) believed that *Ceratocaulophryne regani* and *Caulophryne polynema* might be synonyms on the basis of similarity in illicial morphology, "position of the fins, and relative dimensions." This conclusion, however, is not supported here. In illicial and escal morphology, jaw tooth counts, and in dorsal and anal fin-ray counts (16 and 16, not 15 and 14, respectively, as given by Roule and Angel, 1933), the holotype of *C. regani* compares well with the material here recognized as *C. jordani*. *Ceratocaulophryne regani* is therefore recognized as a synonym of *Caulophryne jordani* following the reallocation of Pietsch (1979:16).

The 54-mm specimen of *C. jordani* from the Southern Ocean (LACM 11317-1) differs from other individuals of this species in having a greater number of jaw teeth (see Pietsch, 1979:16, fig. 18).

The name *Caulophryne setosus* appears in a figure caption and is entered in the index and list of plates of Goode and Bean's (1896:26*, 541, pl. 121, fig. 409) original description of *C. jordani*, without application to a description or type. This error was caught by Jordan and Evermann (1898:2735): "plate named *C. setosus*, by slip in proof reading." USNM 39265, the holotype of *C. jordani*, is also listed under the name *setosus* in the "type files" of the Division of Fishes, National Museum of Natural History (Susan L. Jewett, personal communication, 14 September 1976). Although Pietsch (1979:20) set the name aside as a *nomen nudum*, the direct association with Goode and Bean's (1896) figure of the holotype of *C. jordani* is a clear indication of synonymy.

### CAULOPHRYNE POLYNEMA REGAN, 1930B

Figures 230, 291; Tables 6, 8, 10

#### Females and Parasitic Males

*Caulophryne polynema* Regan, 1930b: 191, figs. 1–3 (original description, single female with parasitic male). Parr, 1930b: 131, fig. 4 (parasitic male, figure after Regan, 1930b). Regan and Trewavas, 1932: 100 (description after Regan, 1930b; in key). Fowler, 1936:1347 (description after Regan, 1930b; in key). Bertelsen, 1951:36, fig. 12 (figure after Regan, 1930b; a subspecies of *Caulophryne jordani*). Albuquerque, 1954–1956:1071, figs. 440, 440A (description and figures after Regan, 1930b). Matsubara, 1955:1353, fig. 530B (sexual parasitism; after Regan and Trewavas, 1932). Pietsch, 1976:789 (reproduction). Pietsch, 1979:16, figs. 16, 18, 20, 21, 24 (family revision, comparison with all known material, relationships, in key). Bertelsen, 1986:1373, 1374, fig. (eastern North Atlantic, in key). Bertelsen, 1990:491 (eastern tropical Atlantic, synonymy). Ellis, 1996:268, 269, fig. (popular account). Anderson and Leslie, 2001:3 (after Lloris, 1986). Pietsch, 2002b:1057 (western central Atlantic). Love et al., 2005:59 (eastern North Pacific). Pietsch, 2005b:209, 215, 216, fig. 8 (reproduction).

*Caulophryne jordani*: Bertelsen, 1951: 33, 37, fig. 12 (in part, includes all species of *Caulophryne* described to date). Marshall, 1954:240, figs. 9, 11c (misidentification, semipopular account; figure after Regan, 1930b). Morell, 2005:36–37, color photograph (misidentification, popular account, Monterey Canyon).

*Caulophryne jordani polynema*: Bertelsen, 1951:33, 37, fig. 12 (one of three subspecies, description, comparison with all known material, in key; includes *Caulophryne regani* Roule and Angel, 1932, a name referred to *C. jordani* by Pietsch, 1979). Grey, 1956a:233 (synonymy, distribution; after Bertelsen, 1951). Maul, 1973:666 (after Bertelsen, 1951). Lloris, 1986:249, fig. 130 (additional specimen, off Namibia).

#### Free-living Males and Larvae

Unknown.

#### Material

Fourteen metamorphosed females, 11.5 to 142 mm; and two parasitic males, 15 to 16 mm:

Holotype of *Caulophryne polynema*: BMNH 1930.2.7.1, female, 142 mm, with 16-mm parasitic male, off Funchal Bay, Madeira, "long line, in deep water," 1 February 1929.

Additional material: BMNH, three (11.5 to 110 mm); IIPB, one (100 mm); ISH, one (44 mm); LACM, five (14 to 125 mm); MNHN, one (137-mm female with 15-mm parasitic male); UF, one (79 mm); UW, one (126 mm).

#### Distinguishing Characters

A species of *Caulophryne*, metamorphosed females of which are distinguished from those of all other described species of the genus in details of escal morphology. Females are further unique in having the following combination of character states: illicium with numerous elongate translucent filaments along length; length of illicium 15.7 to 32.1% SL; number of teeth in upper jaw 26 to 45; dorsal-fin rays 19 to 22.

#### Description

Illicium lightly pigmented along anterior margin (fully pigmented in 44-mm specimen; ISH 3000/71), and bearing numerous elongate (as long as 71% illicial length) translucent filaments, most of which arise from posterior margin; more than 50 illicial filaments in specimens larger than 32 mm, as few as 10 in smaller specimens; esca consisting of a posterolateral appendage with a palmate (usually bilobed), opaque, distal tip, and a tapering terminal appendage with an opaque distal tip; escal appendages highly filamentous in specimens larger than 44 mm, number and length of filaments reduced in smaller specimens (most escal filaments lost in holotype, see Regan, 1930b:193, fig. 2).

Dorsal-fin rays 19 to 22; anal-fin rays 17 to 19; pectoral-fin rays 15 to 18; number of teeth in upper jaw 26 to 45, in lower jaw 16 to 32; total number of teeth on vomer 1 to 3.

Measurements in percent of standard length: length of illicium 15.7 to 32.1; longest dorsal-fin ray 70 to 155; longest anal-fin ray 60 to 151; longest tooth in lower jaw 5.6 to 9.5.

Additional description as given for the genus and family.

Distribution

*Caulophryne polynema* is known from widely scattered localities in the Atlantic and eastern Pacific Ocean; it has so far not been taken in the Indian Ocean (Fig. 230). In the Atlantic it ranges from off Florida east to Madeira and south to Cape Verde, Zaire, and Namibia as far south as approximately 28°S. In the eastern Pacific, it has been collected in the Hawaiian Islands and off southern California and Baja. Four specimens were captured with gear fished at maximum depths of 585 to 834 m, and two with gear reaching a depth of 1920 to 2000 m, but all the remaining material was taken by gear fished at maximum depths of 1000 to 1250 m. Two specimens, both from the eastern North Atlantic (11.5 and 100 mm), were collected with a closing trawl between 900 and 1010 m and 1000 and 1250 m, respectively.

Comments

*Caulophryne polynema* was separated by Regan and Trewavas (1932:100) from the other species of the genus because of its higher number of dorsal- and anal-fin rays. Bertelsen (1951:36), however, noted that the high number of fin rays in *Caulophryne* (compared to other ceratioids) has such a large individual variation (dorsal 14 to 22, anal 12 to 19) that "we can scarcely attach much weight to this character." But the greater amount of material now available indicates that despite this variation within the genus, fin-ray counts are of significant taxonomic importance. *Caulophryne polynema* can almost always be separated from its congeners solely on the basis of its greater number of dorsal-fin rays.

The parasitic male attached to the 137-mm female (MNHN 2001-0140) represents the second example of sexual parasitism in *C. polynema* and the third for the family Caulophrynidae. A detailed examination shows no significant differences from the previously described specimens (see Bertelsen, 1951:37, fig. 12; Pietsch, 1979:20, fig. 23, 2005b:216, fig. 8A).

Caulophryne Species

CAULOPHRYNE SP. A

Female and Parasitic Male

*Caulophryne* sp.: Pietsch, 1979:18, 20, figs. 16, 18, 22, 23 (unidentifiable female with parasitic male; description, comparison with all known material; illicium and attached male figured).

*Caulophryne* sp. A: Pietsch and Seigel, 1980:380 (after Pietsch, 1979).

Free-living Males and Larvae

Unknown.

Material

A single metamorphosed female (98 mm), with parasitic male (12 mm): LACM 36025-1, *Alpha Helix*, Southeast Asian Bioluminescence Expedition station 37, Banda Sea, 4°56.3′S, 129°25.5′E, RMT-8, 0 to 2000 m, 1900 to 0155 h, 11 April 1975.

Description

Female with illicium unpigmented, bearing approximately 30 elongate (up to 25% illicium length) translucent filaments along entire length, those on proximal half of illicium arising from posterior margin; esca, a tapering appendage bearing numerous simple filaments and 4 highly branched lateral appendages: three more proximal in position, one more distal; opaque areas absent.

Dorsal-fin rays 17; anal-fin rays 15; pectoral-fin rays 16; number of teeth in upper jaw 22, in lower jaw 14; total number of teeth on vomer 1.

Measurements in percent of standard length: longest tooth in lower jaw 9.2; longest dorsal ray 95+; longest anal ray 95+; length of illicium 27.6.

Male with upper and lower denticulars embedded in an unpigmented conical papilla emerging from belly of female; passageway from outside into pharynx and out through gill openings retained (Pietsch, 1979:20, fig. 23, 2005b:217, fig. 9A); eyes well developed; nostrils degenerate; gills developed; dorsal-, anal-, and pectoral-fin ray counts undetermined; pelvic fins absent; characteristic gelatinous envelope of larvae absent; entire head and body lightly pigmented.

Additional description as given for the genus and family.

Sexual Maturity

The gonads of the female are well developed: the right ovary is 22 mm long (22.4% SL) and contains numerous eggs approximately 0.3 mm in diameter (Pietsch, 1979:18, 20, 2005b:216). The testes of the male are also large, about 2.3 mm long (19.2% SL).

Comments

This female and attached male cannot reasonably be placed within the material of any of the four recognized species of *Caulophryne*. The female has the elongate branched escal appendages characteristic of *Caulophryne pelagica* but also the illicial filaments found only in *C. jordani* and *C. polynema*. The illicium appears to be slightly longer than that of *C. bacescui*, *C. jordani*, and *C. polynema*, but shorter than that of *C. pelagica*. Jaw tooth counts are at the bottom of the range of variation for females of all species combined. Finally, fin-ray counts compare best with *C. jordani*.

CAULOPHRYNE SP. B

Females

*Caulophryne* sp. B: Pietsch and Seigel, 1980:380, fig. 1 (three unidentifiable females).

Males and Larvae

Unknown.

Material

Three juvenile females (10 to 11.5 mm): LACM 36112-1, 10 mm, *Alpha Helix*, Southeast Asian Bioluminescence Expedition station 183, Sulu Sea, 9°24.5′N, 122°12.3′E, RMT-8, 690 to 890 m, 0115 to 0215 h, 6 June 1975. LACM 36111-1, 10.5 mm, *Alpha Helix*, Southeast Asian Bioluminescence Expedition station 184, Sulu Sea, 9°18.0′N, 122°10.0′E, RMT-8, 480 to 550 m, 0355 to 0455 h, 6 June 1975. LACM 36109-2, 11.5 mm, *Alpha Helix*, Southeast Asian Bioluminescence Expedition station 193, Sulu Sea, 9°21.5′N, 122°14.7′E, RMT-8, 800 to 1100 m, 1020 to 1240 h, 7 June 1975.

Comments

These three small females, all from the Sulu Sea, differ significantly from the material of the four recognized species of *Caulophryne* in having an elongate, distally branched lateral appendage on each side of the escal bulb; distal escal filaments, present in all the described species of the genus, are absent. Although these most likely represent a new species, the small size of the specimens and their poor condition do not warrant description at this time (Pietsch and Seigel, 1980:380, fig. 1).

CAULOPHRYNE RADIANS FOWLER, 1936, NOMEN NUDUM

*Caulophryne radians*: Fowler, 1936:1368 (erroneously attributed to Regan).

Comments

This name appears in Fowler's (1936:1368) "Marine Fishes of West Africa," based on the collection of the American Museum Congo Expedition of 1909–1915. Original authorship is attributed to Regan, with *Caulophryne polynema* Regan and Trewavas listed as a synonym. There is no application to a description or type. The name *radians* is otherwise unknown in the Caulophrynidae and just what Fowler might have had in mind cannot now be established.

CAULOPHRYNE RACEMOSA MONOD, 1960, NOMEN NUDUM

*Caulophryne racemosa*: Monod, 1960: 687, fig. 80 (error for *Caulophryne ramulosa* Regan and Trewavas, 1932). Pietsch, 1979:20 *(nomen nudum)*.

Comments

This name appears in a figure that illustrates various stages of specialization of lophiiform pectoral radials. There is no application to a description or type. No doubt Monod (1960) meant to refer to *C. ramulosa*. His figure 80 was taken from Regan and Trewavas's (1932, fig. 58) illustration of the pectoral radials of the holotype of this nominal species.

Genus *Robia* Pietsch, 1979
(Longlure Fanfins)

Figures 152, 230; Table 8

Females

*Robia* Pietsch, 1979:12, figs. 14, 15, 24 (type species *Robia legula* Pietsch, 1979, by original description and monotypy).

Males and Larvae

Unknown.

Distinguishing Characters

Metamorphosed females of the genus *Robia* are distinguished from those of *Caulophryne* (the only other genus of the family) in having a considerably longer illicium (268.3% SL in the only known specimen) and fewer, relatively short median-fin rays (dorsal-fin rays 6, the length of the longest rays less than 65% SL; anal-fin rays 5, the length of the longest rays less than 40% SL).

Description

The illicium, without filaments along its length, is darkly pigmented except near the esca. The esca is translucent, with 2 short lateral appendages and a slightly more opaque distal tip bearing 3 short appendages. There are 6 dorsal-fin rays, 5 anal-fin rays, and 17 pectoral-fin rays (Table 1). There are 33 teeth in upper jaw, 31 in the lower jaw. There are 4 teeth on the vomer. The following measurements are expressed in percent of standard length: head length 39.0; head depth 39.0; head width 23.4; length of premaxilla 37.8; length of lower jaw 47.6; longest tooth in lower jaw 5.1; longest dorsal-fin ray 63.4; longest anal-fin ray 39.0; length of illicium 268.3.

Additional description as given for the family.

Diversity

A single species:

ROBIA LEGULA PIETSCH, 1979
(LONGLURE FANFINS)

Figures 152, 230; Table 8

Females

*Robia legula* Pietsch, 1979:12–14, figs. 14, 15, 24 (original description, single specimen). Pietsch and Seigel, 1980:380 (after Pietsch, 1979). Pietsch, 1999:2026 (western central Pacific).

Males and Larvae

Unknown.

Material

A single known female (41 mm):
Holotype of *Robia legula*: LACM 36024-1, *Alpha Helix*, Southeast Asian Bioluminescence Expedition station 81, Banda Sea, 4°56.5′S, 129°59.5′E, RMT-8 with closing device, 1000 to 1500 m, 0416 to 0616 h, 28 April 1975.

Distinguishing Characters and Description

As given for the genus and family.

Distribution

*Robia legula* is known from a single female specimen collected in the Banda Sea, with an opening-closing trawl fished between 1000 and 1500 m (Fig. 230).

## Family Neoceratiidae Regan, 1926 (Needlebeard Seadevils)

Figures 30, 52, 153, 154, 191, 196, 202, 203, 231, 291, 298, 299; Tables 1, 8, 10

Type Genus *Neoceratias* Pappenheim, 1914

Neoceratiidae Regan, 1926:39 (family of suborder Ceratioidea to contain genus *Neoceratias*).

Distinguishing Characters

Metamorphosed females of the Neoceratiidae are unique among ceratioid families in lacking an illicium and second cephalic spine, as well as the cranial trough, within which the pterygiophore of the illicium lies in all other female ceratioids. The frontals are broad and roughly rectangular in shape. The metapterygoid and mesopterygoid are absent. The preopercle is unusually short and nearly straight. The premaxillae and dentaries are unusually thick and well developed, with rounded edges. A pair of prominent nasal papillae are present on the snout (nostrils and olfactory lamellae are absent). The inner margin of the jaws bears a single inner row of small immobile teeth; the outer margin bears two or three series of long hinged teeth, each with a tiny distal hook. The dorsal and anal fins are mounted on prominent bases (Bertelsen, 1951:156–157, fig. 105).

Metamorphosed females are further differentiated by having the following combination of character states: The supraethmoid is present. The frontals, without ventromedial extensions, meet on the midline. The parietals are present. The sphenotics are conical in shape, without a distal spine. The pterosphenoid is present. The hyomandibula has only a single head. There are 2 hypohyals and 6 (2 + 4), rarely 5 (1 + 4), branchiostegal rays. The opercle is bifurcated and very much reduced. The subopercle, also reduced, is long and narrow, as long as or slightly longer than the lower fork of the opercle, without a posterior notch, its ventral part without a spine or projection on the anterior margin. Quadrate, articular, angular, and preopercular spines are absent. The jaws are subequal, the lower extending anteriorly beyond the upper. The lower jaw lacks a symphysial spine. The postmaxillary process of the premaxilla and the anterior-maxillomandibular ligament are absent. The first, second, and fourth pharyngobranchials are absent; the third is well developed and toothed. There is a single ossified hypobranchial. Ossified basibranchials are absent. Epibranchial and ceratobranchial teeth are absent. Epurals are absent. The posterior margin of the hypural plate is deeply notched. The pterygiophore of the illicium is fully embedded in skin of the head, without ossified remnants of the illicium or second cephalic spine. The esca is absent. The coracoid lacks a posteroventral process. There are 3 pectoral radials. The pelvic bones are small and cylindrical, without a distal expansion. There are 10 to 13 dorsal-fin rays, 10 to 13 anal-fin rays, and 12 to 15 pectoral-fin rays (Table 1). Pelvic fins are absent. The caudal peduncle is unusually long and slender, the caudal fin of females very broad, with the four innermost rays deeply bifurcated. There are 9 (2 unbranched + 4 branched + 3 unbranched) caudal-fin rays, 10 caudal rays in some larvae (see Bertelsen, 1951:159, table 36).

The skin is everywhere naked, dermal spinules absent. The ovaries are paired. There are 2 short pyloric caeca.

Free-living males are unknown, but parasitic males are represented by seven individuals (11.5 to 18 mm). They differ from those of all other ceratioid families in having an unusually slender body. The eyes and olfactory organs are degenerate. The jaws retain a few short teeth on each side. The premaxillae are retained, apparently not reduced. The upper denticular bone is apparently absent, the lower denticular bearing 3 elongate curved projections, each bifurcating distally. Pelvic fins are absent. Fin-ray counts are as given for metamorphosed females. The skin is naked, without a trace of dermal spinules. Sexual parasitism is apparently obligatory (see Reproduction and Early Life History, Chapter Eight).

The larvae (11 known specimens, 3.7 to 9.8 mm) differ from those of all other ceratioid families in having an elongate, slender body, its depth 30 to 40% SL. The length of the head is relatively shorter than those of other ceratioids, 30 to 40% SL. The length of the caudal fin is 20 to 30% SL. The skin is only slightly inflated. The pterygiophore of the illicium is well developed. Sexual dimorphism is apparently absent; all known specimens have an elongate, cylindrical rudiment of the illicium protruding (in a position unique among ceratioid larvae) on tip of the snout, just above the symphysis of the upper jaw. The pectoral fins are relatively small, their length 15 to 20% SL. Pelvic fins are absent. Fin-ray counts are as given for metamorphosed females. Metamorphosis begins at lengths of 8 to 10 mm (Bertelsen, 1951:159, fig. 106A–C; 1984: 326, 328, fig. 167B).

The osteological features cited here and elsewhere in this account of the Neoceratiidae are based on examination of two females, each with a parasitic male: one published account (Bertelsen, 1951: 156, 157, fig. 105; ZMUC P921726, 52 mm), the other (ISH 5546/79, 74 mm) based on notes and rough pencil drawings left behind by Bertelsen and now in the author's possession.

Description

The body of metamorphosed females is elongate and slender, slightly compressed, its depth approximately 18 to 28% SL. The head is short, its length approximately 18 to 25% SL. The mouth is large, its cleft extending posteriorly well past the eye, its opening horizontal to slightly oblique. The oral valve is absent. The eyes are minute, appearing degenerate, their diameter less than 2.5% SL. A pair of large nasal papillae is present on the snout, their length approximately 3.5 to 6.0% SL, unpigmented on their distal ends, nostrils and lamellae absent (function unknown; see Comments, below). The jaws bear an inner row of short, straight, widely spaced immobile teeth, 0 to 6 on each premaxilla, 10 to 21 on each dentary. The outer margins of the jaws bear prominent conical outgrowths, providing articular surfaces for two or three irregular series of long straight hinged teeth, strongly attached by connective tissue and well-developed musculature, and each with a tiny distal hook. The outermost and posterior-most teeth are the largest, the length of some nearly 15% SL. There are 11 to 20 teeth on each premaxilla, 18 to 20 on each dentary (all known material is in very poor condition, making accurate tooth counts difficult). The vomerine teeth are long and slender, one on each side. The third pharyngobranchial is unusually large and well toothed (similar to that of *Gigantactis* in shape and relative size; see Bertelsen et al., 1981:11, figs. 13, 16). The first epibranchial is free, not bound to the wall of the pharynx by connective tissue. All four epibranchials are closely bound together. The fourth epibranchial and ceratobranchial are bound to the wall of the pharynx, leaving no opening behind the fourth arch. The proximal one-half of the first ceratobranchial is bound to the wall of the pharynx. The distal end of the first ceratobranchial is free, not bound by connective tissue to the adjacent second ceratobranchial. The proximal one-quarter of the second ceratobranchial is bound to the third ceratobranchial. The proximal one-half of the third ceratobranchial is bound to the fourth ceratobranchial. Gill filaments are present on the proximal ends of epibranchials II to IV, on the proximal tip of ceratobranchial I, the full length of ceratobranchial II, the distal three-quarters of ceratobranchial III, and the distal one-half of ceratobranchial IV. A pseudobranch is absent. The skin of the head and body bear numerous elongate filaments, undoubtedly bearing neuromasts of the acoustico-lateralis system (not light organs as suggested by Koefoed, 1944:10).

Males are known only as parasites on females. There frontals are relatively smaller and narrower than those of the females. The pterygiophore of the illicium lacks a trace of an illicium or second cephalic spine. Fin-ray counts, the number of branchiostegal rays, and shape of the hyomandibula and opercular bones are comparable to those of the females. The premaxillae are well developed (reduced in most other ceratioid males), each bearing 2 immobile teeth near the symphysis. The upper denticular bone is absent. The dentaries are shorter, but somewhat broader than those of the females, each bearing 3 small immobile teeth. The lower denticular bone is triradiate; each fork is bifurcated distally to form a total of 6 short flattened denticles (Bertelsen, 1951:157–158, fig. 105B).

Bertelsen's (1951:161, figs. 106D, E, 107) description of a parasitic male attached to a 52-mm female (ZMUC P921726) applies well to all known specimens: The male is firmly attached to the female by outgrowths from the lower jaw and snout. The body is long and slender, its depth 20% SL. The length of the head is 26% SL. The length of the caudal rays is about 23% SL. The lower denticular bone is triradiate, firmly anchored in tissue of the female. The olfactory organs are apparently absent, eyes degenerate, covered with weakly pigmented skin and appearing as a deeper-lying irregular mass of pigment. The liver is relatively small, the stomach and intestines undeveloped, but not obviously degenerate. The testes are in development and not remarkably large.

The body of the smallest known larva (3.7 mm) is slender, the tail straight, its skin only slightly inflated. The depth of the body and length of head are 35 to 40% SL. The length of the caudal fin is 25 to 30% SL. The largest known larva (9.8 mm) is somewhat more slender, the body depth and head length about 30% SL. The length of the caudal fin is only about 20% SL. The fin-ray counts, number of branchiostegal rays, and shape of the hyomandibula, opercular, and pelvic bones are comparable to those of metamorphosed specimens (Bertelsen, 1951:156, 160, fig. 106A–C, table 36). The pterygiophore of the illicium is rather well developed, but there is no indication of an ossified illicium. In the two smallest specimens (3.7 to 4.0 mm), a small wartlike structure lies beneath the skin at the tip of the pterygiophore of the illicium; in larger specimens it is cylindrical and elongate, emerging immediately above the symphysis of the upper jaw, projecting dorsally on the snout, and diminishing in size in larger specimens, while the overlying skin becomes increasingly more pigmented. There are no differences detected that might be interpreted as sexual dimorphism (see Bertelsen, 1951:159, 1984:328). The olfactory organs are prominent, similar in relative size in all known specimens. The peritoneum is unpigmented. The body has conspicuous dorsal and ventral streaks of pigment, each several melanophores thick; the pigmentation is strongest between the posterior bases of the dorsal and anal fins (see Bertelsen, 1951:159, fig. 106A–C).

The color of preserved females is dark red brown to black over the entire surface of the head, body, fins, and oral cavity. The viscera are unpigmented. The skin of parasitic males is everywhere more lightly pigmented and semitransparent, allowing the subdermal pigmentation to show through.

The largest known female and male, 108 and 18 mm, respectively (SIO 70-336), are a parasitically attached couple, collected in 1970 by the RV *Melville* in the Philippine Sea off Luzon.

### Diversity

A single genus:

### Genus *Neoceratias* Pappenheim, 1914 (Needlebeard Seadevils)

Figures 30, 52, 153, 154, 196, 202, 203, 231, 291, 298, 299; Tables 1, 8

#### Females, Parasitic Males, and Larvae

*Neoceratias* Pappenheim, 1914:198, fig. 10 (type species *Neoceratias spinifer* Pappenheim, 1914, by monotypy).

#### Distinguishing Characters and Description

As given for the family.

### Diversity

A single known species:

#### *NEOCERATIAS SPINIFER* PAPPENHEIM, 1914 (NEEDLEBEARD SEADEVILS)

Figures 52, 153, 154, 196, 231, 291, 298, 299; Tables 8, 10

#### Females, Parasitic Males, and Larvae

*Neoceratias spinifer* Pappenheim, 1914: 198, fig. 10 (original description, single specimen). Regan, 1926:39, fig. 22 (description after Pappenheim, 1914). Regan, 1927a:4, postcard M 15c (popular account). Regan and Trewavas, 1932:39, 95, fig. 153 (description after Pappenheim, 1914; "may be a male of some unknown Gigantactinid"). Fowler, 1936:1146, fig. 484 (description, figure after Pappenheim, 1914). Gregory and Conrad, 1936:199, 206, figs. 4, 5 (phylogenetic relationships). Koefoed, 1944:9, pl. 2, fig. 5 (second known specimen, North Atlantic). Bertelsen, 1951:158, figs. 105–107 (osteology; description of larvae, a parasitic male, and additional females; comparison of all known material). Grey, 1956a:269 (synonymy, distribution). Wickler, 1961:387, fig. 4 (sexual parasitism). Marshall, 1966:177, pl. 7 (bioluminescence). Marshall, 1971a:70, fig. 30 (brain and gill anatomy). Pietsch, 1976:789, fig. 8 (sexual parasitism). Marshall, 1979:265, 267, 417, 454, figs. 105b, 146, 160 (reduced organ systems, brain and gill arch anatomy, lateral line structures; figures after Bertelsen, 1951). Bertelsen and Pietsch, 1983:93, figs. 14, 15 (additional specimen, with parasitic male; Australia). Bertelsen, 1984:328, fig. 167B (early life history, phylogeny). Jespersen, 1984:37, figs. 1–8 (spermatozoan ultrastructure). Meng et al., 1995:438, fig. 585 (China). Weitzman, 1997:66 (listed among deep-sea families known to occur below 500 m). Stewart and Pietsch, 1998:29 (description after Bertelsen, 1951, and Bertelsen and Pietsch, 1983). Pietsch, 1999:2027 (western central Pacific). Munk, 2000:315, figs. 1–5 (histology of parasitic attachment). Pietsch, 2002b:1058, fig. (western central Atlantic). Moore et al., 2003:214 (off New England). Pietsch, 2005b:210, 216, 225, 232, figs. 8E, 17 (reproduction). Woodland, 2005:253 (sexual parasitism). Pietsch and Orr, 2007:4, 6, 9, figs. 3D, 4L, 6E (phylogenetic relationships).

#### Material

Eighteen metamorphosed females (17 to 108 mm), seven parasitically attached males (8.5 to 18 mm), and 11 larvae (3.7 to 9.8 mm):

Holotype of *Neoceratias spinifer*: ZMB 19383, 25 mm, *Gauss*, Deutschen Südpolar Expedition, tropical Atlantic near St. Helena, 12°11′S, 6°16′W, 0 to 2000 m, 4 September 1903.

Additional females and parasitic males: AMS, one female (77 mm, with 12.5-mm parasitic male); ISH, one female (74 mm, with 12.5-mm parasitic male; female cleared and stained); LACM, two females (23.5 and 42 mm, 42-mm specimen with 8.5-mm parasitic male); MCZ, two females (17 and 21 mm); ORI, one female (31 mm); SIO, seven females (17 to 108 mm; 67.5-mm female with 17.5-mm parasitic male, 108-mm female with 18-mm parasitic male); SIOM, one female (86 mm, with 17.5-mm parasitic male); ZMB, one (43 mm); ZMUC, one female (52 mm, with 15.5-mm parasitic male; cleared and stained).

Larvae: ZMUC, 11 (3.7 to 9.8 mm; 4.0 to 12.5 mm TL). See Bertelsen (1951:158).

#### Distinguishing Characters and Description

As given for the genus and family.

#### Distribution

*Neoceratias spinifer* is known from all three major oceans of the world between about 40°N and 20°S (Fig. 231). In the Atlantic it ranges from off New England (39°N, 69°W) in the west and St. Helena (12°S, 5°W) in the east. There are two known records in the Indian Ocean, a larval specimen from off the northern tip of Madagascar, and an 86-mm female (with a 17.5-mm parasitic male) from the western Bay of Bengal. It is well represented in the Pacific, extending from off Luzon in the Philippine Islands, the Sulu Sea, and Queensland, east to the Hawaiian Islands, with a single record farther east at about 32°N, 136°W. It appears to be a relatively deep living species: none of the material for which data are recorded was collected in gear fished at maximum depths of less than 1000 m.

## Family Gigantactinidae Boulenger, 1904a (Whipnose Seadevils)

Figures 38, 43, 45, 48, 55, 155–166, 197, 202, 203, 232–235, 243, 250, 262, 267–273, 286, 300; Tables 1, 8, 9

### Type Genus *Gigantactis* Brauer, 1902

Gigantactinidae Boulenger, 1904a:188 (family of suborder Pediculati to contain genus *Gigantactis*). Boulenger, 1904b:718, 720 (after Boulenger, 1904a).

Gigactinidae Jordan, 1923:243 (misspelling of family Gigantactinidae).

Aceratiidae Parr, 1927:30 (in part; family of suborder Ceratioidea to contain genera *Lipactis*, *Aceratias*, *Rhynchoceratias*, and *Laevoceratias*, all based on free-living males, that of *Laevoceratias* subsequently reassigned to genus *Gigantactis* by Uwate, 1979, confirmed by Bertelsen et al., 1981).

Cryptorostrinae Parr, 1930a:20 (in part; subfamily of family Aceratiidae to contain *Haplophryne* and *Laevoceratias*, both based on free-living males, that of *Laevoceratias* subsequently reassigned to genus *Gigantactis* by Uwate, 1979, confirmed by Bertelsen et al., 1981).

Laevoceratiidae Regan and Trewavas, 1932:6, 37, 92 (family of order Ceratioidea to contain genera *Laevoceratias* and *Teleotrema*, both based on free-living males, subsequently reassigned to genus *Gigantactis* by Uwate, 1979, and Bertelsen, 1951, respectively).

### Distinguishing Characters

Metamorphosed females of the Gigantactinidae differ from those of all other ceratioid families in lacking a vomer, mesopterygoid, and an ossified scapula; in having the preopercle reduced to a short narrow strut of bone; the interopercle reduced, without ligamentous connection to the angular; the caudal fin emarginate (except in the largest known females of *Gigantactis krefffti* and *G. macronema*), with 9 caudal-fin rays, the ventral-most ray reduced and embedded within skin surrounding the adjacent ray; the pterygiophore of the illicium exceptionally large, the compressed posterior end butting up against the supraoccipital; and 5 pectoral radials (Bertelsen et al., 1981).

Metamorphosed females are further distinguished by having the following combination of character states: The supraethmoid is greatly reduced, usually absent (a tiny rudiment observed in a single specimen; Bertelsen et al., 1981:10, fig. 12). The frontals are narrow, widely separated, and without ventromedial extensions in *Gigantactis*, but absent in *Rhynchactis*. The parietals are greatly reduced in *Gigantactis*, but absent in *Rhynchactis*. Sphenotic spines are absent. The pterosphenoid is absent, the metapterygoid present. The hyomandibula has a single head. There is only 1 hypohyal in *Gigantactis*, but 2 in *Rhynchactis*. There are 6 (2 + 4), very rarely 7 (2 + 5), branchiostegal rays. The opercle is reduced and deeply bifurcated posteriorly to form 2 slender forks, the ventral fork slightly longer than the dorsal fork. The subopercle is long and slender, its ventral part in *Gigantactis* becoming more reduced with age and splitting to form a small projection on the anterior margin and as many as 3 filamentous, posteriorly directed prolongations. Quadrate, articular, angular, and preopercular spines are absent. The jaws are subequal, the upper jaw extending anteriorly beyond the lower jaw. The dentaries are simple, without posterior bifurcation and without a symphysial spine, attached to each other at the symphysis by thick elastic connective tissue. The premaxillae are long and narrow, without ascending or postmaxillary processes, well ossified in *Gigantactis*, but greatly reduced in *Rhynchactis*. The maxillae are reduced to threadlike ossifications or absent. An anterior-maxillomandibular ligament is absent. The first and fourth pharyngobranchials are absent. The second and third pharyngobranchials are extremely large, well ossified, and heavily toothed. The first epibranchial is absent. The anterior half of the third and fourth epibranchials is fused. The first ceratobranchial is reduced, represented only by a short posterior portion and tiny isolated remnants of the anterior portion. The fifth ceratobranchial is absent, except for tiny isolated remnants. Ossified hypobranchials and basibranchials are absent. Epibranchial and ceratobranchial teeth are absent. Remnants of the first ceratobranchial are embedded in tissue of the lateral wall of the pharynx. Epurals are absent. The hypural plate is entire, without a posterior notch. The pterygiophore of the illicium bears a small ossified remnant of the second cephalic spine. The escal bulb has a central lumen and escal pore in *Gigantactis*, but a lumen and pore are absent in *Rhynchactis*. The esca of *Gigantactis* lacks large toothlike denticles. The coracoid is reduced in *Gigantactis* (with an elongate cartilaginous posteroventral process), but unossified in *Rhynchactis* (a posteroventral process absent). Pelvic bones are absent in *Gigantactis*, but short cylindrical remnants are present in *Rhynchactis*. There are 3 to 9 dorsal-fin rays, 3 to 7 anal-fin rays, 14 to 22 pectoral-fin rays, and 9 caudal-fin rays (the ventralmost ray extremely short and covered with skin; Table 1). The caudal-fin rays are all simple in females of *Gigantactis*, but there are 2 simple + 4 bifurcated + 3 simple in males of *Gigantactis* and in both sexes of *Rhynchactis*. Numerous close-set dermal spinules cover the entire head, body, and fins of *Gigantactis*, extending out onto the illicium, and onto the esca in some species (visually obvious in stained specimens of females of all species, without microscopic aid). Minute dermal spinules are present in the skin of the largest known females of *Rhynchactis*, but the skin is smooth and naked in smaller specimens. The ovaries are paired. Pyloric caeca are absent.

Metamorphosed gigantactinid males differ from those of all other ceratioid families in having the following combination of character states: The eyes are minute, their diameter only 3 to 5% SL in most specimens. The olfactory organs are large, their depth 8 to 10% SL in most specimens. The anterior nostrils are close together and directed anteriorly. The premaxillae are degenerate, but the maxillae are well developed. Jaw teeth are absent. The denticular teeth are all or nearly all mutually free. The upper denticular bears 3 to 6 (rarely 2) teeth, without connection to the pterygiophore of the illicium. The lower denticular bone supports 4 to 7 (rarely 3) teeth. The hyomandibula has a single head. There are 6 (rarely 7) branchiostegal rays and 5 pectoral radials. Fin-ray counts are as given for metamorphosed females. Pelvic bones are absent. The skin is naked or densely covered with dermal spinules. The males are free-living, with no evidence that they ever become parasitic, but temporarily attached males have not yet been found (see Reproduction and Early Life History, Chapter Eight).

In addition to the sexual dimorphism common to all metamorphosed ceratioids, gigantactinid males differ further from females of the family in having a symphysial cartilage, a vomer, and a basibranchial ossification; they differ also in having fully developed frontals, parietals, opercular bones, and ceratobranchials (Bertelsen et al., 1981, figs. 11D–F, 14, 15).

Gigantactinid larvae differ from those of all other ceratioid families in having exceptionally large pectoral fins (length 45 to 55% SL), comparable only to those of the Caulophrynidae; they differ further in having the following combination of character states: The body is short, nearly spherical. The skin is highly inflated. Pelvic fins are absent. Fin-ray counts are as given for metamorphosed females. Sexual dimorphism is evident, the females bearing a small, club-shaped illicial rudiment protruding from the head. Metamorphosis begins at 8 to 10 mm, and metamorphosal stages range in size from 9 to 20 mm (Bertelsen et al., 1981, 62; Bertelsen, 1984:326, fig. 168A, B).

The osteological features cited here and elsewhere in this account of the Gigantactinidae are based on examinations of females and males of both *Gigantactis* and *Rhynchactis* (see Regan, 1926:18, fig. 12; Regan and Trewavas, 1932:38–39, figs. 49, 50A; Bertelsen, 1951:145–147, fig. 98; Pietsch, 1972a:29, 34, 35, 41–45, fig. 24(2); Bertelsen et al., 1981:10–19, figs. 5, 11–17).

Description

The body of metamorphosed females is slender and streamlined. The length of the head (measured from the tip of the snout to the base of the pectoral fin) is about 25% SL. The greatest depth of the body is about 25% SL. The length and depth of the caudal peduncle is 20 to 30% SL and 5 to 10% SL, respectively. The caudal fin is deeply cleft in most species, some with one or two dorsal- and ventralmost rays greatly prolonged. The snout of *Gigantactis* is pointed, protruding well beyond the anterior margin of the jaws, but more blunt in *Rhynchactis*. The mouth is large, the opening horizontal, the cleft extending past the eye. An oral valve is absent. The olfactory organs are raised on short cylindrical stalks. The dentition is highly variable: the premaxillary teeth of *Gigantactis* are small, depressible, and relatively few, arranged in one or two series. The dentary teeth are larger, depressible, and more numerous, arranged in two to six, more or less distinct overlapping series, including an external series situated on the outer margin of the jaw. Several anterior-most and lateral-most dentary teeth are greatly enlarged into curved fangs. The jaws of metamorphosed females of *Rhynchactis* are toothless (except for a few premaxillary teeth retained in juveniles). Vomerine teeth are absent in both genera. The upper pharyngeal teeth are extremely large, lying in a forward position in the roof of the mouth. The first epibranchial is absent. Epibranchials II to IV are closely bound together by connective tissue. The fourth epibranchial and ceratobranchial are bound to the wall of the pharynx, leaving no opening behind the fourth arch. A remnant of the first ceratobranchial is free, not bound to the wall of the pharynx. The proximal one-quarter to one-half of ceratobranchials II to IV are closely bound together, displacing gill filaments. Gill filaments are present on epibranchials II to IV and on the distal parts of ceratobranchials II to IV. A discrete pseudobranch is absent, but at least some species have a few gill filaments on the wall of the pharynx adjacent to the second epibranchial. The pterygiophore of the illicium is shorter than the skull and nearly immobile; its short protruding anterior end forms the tip of the snout in *Gigantactis*, but completely concealed under skin of the head in *Rhynchactis*. The illicium is greatly prolonged, its length highly variable among species (ranging from 40 to 100% SL in *Gigantactis longicirra* to 340 to 490% SL in *G. macronema*), reaching its full relative length at a standard length of about 30 mm. *Gigantactis* has an elongate, club- or spindle-shaped escal bulb containing a relatively small, internal, spherical photophore, and bearing appendages in the form of papillae, filaments, and lobes; filaments on the stem of the illicium are present or absent. *Rhynchactis* lacks an escal photophore as well as bioluminescent bacteria (in sharp contrast to all other ceratioid females except those of the Caulophrynidae and Neoceratiidae). The illicium of some specimens of *Rhynchactis* bears elongate slender filaments of a complex structure and unknown function. In both genera, organs of the acoustico-lateralis system are born on short stalks of pigmented skin, more or less distinctly connected in series by narrow unpigmented grooves, the pattern of placement as described for other ceratioids (see Bertelsen et al., 1981:8, 10, fig. 8).

For additional description of males and larvae, see species accounts given below).

The color of females in preservative is uniformly dark brown to black over the entire external surface of the head, body, and fins (except for unpigmented transparent parts of the escal bulb, escal appendages, and stem of the illicium; for ontogenetic and specific variation, see species accounts given below). The oral cavity is more or less pigmented. The skin of metamorphosed males is weakly pigmented in *Rhynchactis*, but varying among species groups of *Gigantactis* from unpigmented to dark brown or black. The subdermal pigment of larvae, males, and juvenile females varies between genera and among species groups of *Gigantactis* from absent (e.g., *G. longicirra*) to dense dorsal and peritoneal groups of melanophores (e.g., *Rhynchactis*).

Gigantactinids are among the largest known ceratioids. While some of the 30 or so females of *Gigantactis* greater than 200 mm (the largest is a 435-mm specimen of *G. elsmani*, ISH 1135/79) have relatively large ovaries, none contain ripe or ripening eggs. Eggs larger than about 0.5 mm in diameter have not been found. At the same time, none of the more than 175 metamorphosed females in collections around the world are parasitized by males, thus it is assumed that males remain free-living. The largest known male, a member of the *Gigantactis* Male Group I, with ripe testes (BMNH 2004.9.12.168), measures 22 mm.

Diversity

Two genera differentiated as follows:

Keys to Genera of the Gigantactinidae

FEMALES

1A. Teeth of lower jaw well developed, in several rows; dorsal-fin rays 5 to 9, rarely 4 or 10; anal-fin rays 4 to 7, rarely 8; escal bulb present . . . . . . .
. . . . . . . . . . . . . . . . . . *Gigantactis* Brauer, 1902, p. 454

NOTE  Eighteen known species; Atlantic, Pacific, and Indian oceans

1B. Lower-jaw teeth absent; dorsal-fin rays 3 or 4, rarely 5; anal-fin rays 3 or 4; escal bulb absent . . . . *Rhynchactis* Regan, 1925c, p. 475

NOTE  Three known species; Atlantic, Pacific, and Indian oceans

MALES

1A. Upper denticular teeth 3, lower denticular teeth 4; dorsal-fin rays 5 to 9, rarely 4 or 10; anal-fin rays 4 to 7, rarely 8; skin spinulose in some species . . . . . . . . . . . . . *Gigantactis* Brauer, 1902, p. 454

1B. Upper denticular teeth 4, lower denticular teeth 6; dorsal-fin rays 3 or 4, rarely 5; anal-fin rays 3 or 4; skin naked . . . . . . . . . . . . . . *Rhynchactis* Regan, 1925c, p. 475

LARVAE

1A. Dorsal group of melanophores weakly developed or absent, when present extending posteriorly to base of dorsal fin but never contiguous with peritoneal pigment; dorsal-fin rays 5 to 9, rarely 4 or 10; anal-fin rays 4 to 7, rarely 8 . . . . *Gigantactis* Brauer, 1902, p. 454

1B. Dorsal group of melanophores strongly developed, terminating in front of base of dorsal fin but contiguous with peritoneal pigment; dorsal-fin rays 3 or 4, rarely 5; anal-fin rays 3 or 4 . . . . . . . . *Rhynchactis* Regan, 1925c, p. 475

Genus *Gigantactis* Brauer, 1902 (Whipnose Seadevils)

Figures 38, 43, 45, 48, 155–161, 197, 202, 203, 232–235, 243, 250, 262, 267–273, 286, 300; Tables 1, 8, 9

Females and Larvae

*Gigantactis* Brauer, 1902:295 (type species *Gigantactis vanhoeffeni* Brauer, 1902, by monotypy).

Males

*Teleotrema* Regan and Trewavas, 1932: 92, fig. 149 (type species *Teleotrema microphthalmus* Regan and Trewavas, 1932, by monotypy).

*Laevoceratias* Parr, 1927:33, fig. 13 (type species *Laevoceratias liparis* Parr, 1927, by monotypy).

Distinguishing Characters

The genus *Gigantactis* is distinguished from *Rhynchactis*, the only other genus of the family, in lacking pelvic bones and by having dorsal-fin rays 5 to 9 (rarely 4 or 10) and anal-fin rays 4 to 7 (rarely 8) (Table 1).

Metamorphosed females of *Gigantactis* differ further from those of *Rhynchactis* in having the following character states: The frontals are present, the parietals present. The premaxillae are well developed, with teeth present throughout their length. The maxillae are represented by threadlike remnants of bone. The dentary bears several rows of strong recurved teeth. There is only a single hypohyal. All rays of the caudal fin are unbranched (in contrast to the males). The skin is spinulose. The snout is produced in front of the mouth (formed primarily by the anterior end of the hypertrophied pterygiophore of the illicium), bearing the illicium on its tip. The esca consists of an expanded luminous bulb.

Metamorphosed males of *Gigantactis* differ further from those of *Rhynchactis* by having the following character states: The diameter of the eye is greater than 3% SL (usually 3.5 to 5% SL). There are 12 (rarely 10 or 11) olfactory lamellae. The depth of the nostrils is rarely greater than 9% SL. There are 3 (rarely 2 or 4) upper denticular teeth and 4 (rarely 3 or 5) lower denticular teeth. All the bases of the denticular teeth are compressed and mutually free. The skin is pigmented or unpigmented, spinulose or naked.

The larvae of *Gigantactis* differ further from those of *Rhynchactis* in having the following character states: A dorsal group of subdermal melanophores is absent or only weakly developed, never contiguous with the peritoneal pigment; when best developed it extends posteriorly behind the base of the dorsal fin. The length of the pectoral fin is approximately 45% SL.

Description

The body of metamorphosed females is elongate and streamlined; the shape is very similar in all species except for some variation in the relative length and depth of the caudal peduncle. The snout is pointed and prolonged. The pterygiophore of the illicium emerges just above and slightly in front of the anterior tip of the upper jaw. The anterior parts of the dorsal and anal fins are raised on more or less distinct triangular keels, supported by the distal ends of well-developed pterygiophores. The caudal peduncle is slender, but broadened posteriorly by a very large hypural plate. The caudal fin is long, differing from all other ceratioids (including males of *Gigantactis* and both sexes of *Rhynchactis*) in having all the rays unbranched. The shape of the caudal fin and the extent of skin coverage on the caudal-fin rays show distinct interspecific variation: the caudal fin is more or less deeply cleft in a number of species; extremely prolonged caudal-fin rays (60 to 100% SL) are present in *Gigantactis longicirra* and members of the *G. gargantua* group (in *G. longicirra* the first and eighth caudal-fin rays are the longest; in the *G. gargantua* group the second and seventh rays are the longest). The proximal one-third to one-half of the caudal fin is covered with undivided skin in some species; in other species, the caudal-fin rays are separated nearly to their bases. Darkly pigmented spiny skin covers each caudal-fin ray, the skin usually compressed, with broad extensions partially connecting the rays. The skin of the head, body, and fins is black, densely covered with small sharp dermal spinules.

With few exceptions, intraspecific variation in illicial length is small, unrelated to standard length in specimens greater than 30 mm. In species with moderately long illicia, the illicial stem is relatively stiff and distinctly compressed proximally, but highly pliable and nearly cylindrical throughout its length in species with extremely long illicia (e.g., members of the *G. macronema* group). The escal bulb is more elongate and slender than in most other ceratioids, more or less club or spindle shaped, and tapering gradually into the stem of the illicium. The greatest diameter of the escal bulb is rarely more than two or three times the minimum diameter of the stem of the illicium. The spherical escal bulb (photophore) is unusually small, its diameter less than one-half that of the escal bulb in most species. The pore of the esca opens on a short protruding tube, with a slightly inflated rim. The skin of the illicial stem and the basal part of the escal bulb is darkly pigmented and spiny in all specimens. The spread of pigment and spinules on the illicium and esca increases with age, in large specimens often completely covering the escal bulb except for a narrow transparent field around the escal pore. The escal appendages are simple or more rarely branched filaments, with variously shaped papillae. Except for the internally transparent distal part of the escal bulb, light-guiding structures are absent. Filaments are present on stem of the illicium in several species.

The premaxillary teeth are small, relatively few, and arranged in one or two longitudinal series, without clear interspecific differences, despite considerable variation in number. More distinct differences are present in the heavy dentition of the lower jaw, varying (in addition to ontogenetic changes) among species in number and relative length of the teeth, and in arrangement in longitudinal series. The series are most distinct in the posterior one-half to two-thirds of the jaw, where two to six series may be traced. The external one or two series consist of enlarged teeth placed on the outside of the jaw (in contrast to all other ceratioids except for *Neoceratias*). All teeth are depressible and strongly curved, especially the large external fangs. The dentaries are connected anteriorly only by elastic ligaments; the right and left elements are mobile relative to one another, their long curved teeth capable of rotating inward from a widely outstretched open position to a situation in which they approach each other within the cavity of the mouth.

Metamorphosed males of *Gigantactis* are similar to those of other ceratioids in the shape of the body and fins, without significant intrageneric differences. The eyes are smaller than those of all other ceratioids (their diameters 3.0 to 5.2% SL, approaching only those of *Centrophryne*, 5.5% SL). The relative size of the eyes differs more or less distinctly between intrageneric groups of males; the metamorphosed specimens of each group with relative eye size appearing to vary independently of the size of specimens, but distinctly larger (6.2 to 7.1% SL) in the known metamorphic stages of males (see Male Group accounts, below).

The skin of the olfactory organs of males is not distinctly inflated. The anterior nostrils are close set, directed anteriorly, and slightly smaller than the posterior nostrils. The posterior nostrils are directed laterally, situated a distance of two or three times the eye diameter in front of the eyes. The length of the series of olfactory lamellae varies between 5.2% and 11% SL. There 10 to 13 olfactory lamellae (one of the specimens has 8 or 9 lamellae, but except for this one exception, none of the characters of the olfactory organs of the metamorphosed males show any significant relationship to size of specimens or to the intrageneric separation into groups). The olfactory organs of metamorphic males are smaller, more lateral in position, and with an incompletely developed number of lamellae.

The upper and lower denticular teeth of metamorphosed males are mutually free, hooked, compressed proximally, and very loosely attached to symphyses of the maxillae and dentaries, respectively. There are 3 (rarely 2 or 4) upper denticular teeth and 4 (rarely 3 or 5) lower denticular teeth. Tooth counts vary without apparent relationship to the separation of specimens into groups.

Metamorphosed males differ significantly in pigmentation and in the occurrence of dermal spines, each of the following combinations occurring: unpigmented and naked, unpigmented and spiny, pigmented and naked, and pigmented and spiny. Each combination shows a more or less distinct relationship to fin-ray counts, eye size, and subdermal pigmentation, and different combinations of these characters form the basis for the separation of *Gigantactis* males into six groups (originally proposed by Bertelsen et al., 1981:59; see accounts given below).

Subdermal pigmentation of the larvae of *Gigantactis* is generally weak to completely absent, but when present subdermal melanophores form (1) an isolated dorsal group, (2) scattered on the dorsal margin of the peritoneum, (3) scattered on the dorsal and ventral margins of the caudal peduncle, and more rarely (4) on the chin and underside of the head. Based on combinations of subdermal pigmentation and the number of dorsal- and anal-fin rays, *Gigantactis* larvae are divided into four groups (following Bertelsen et al., 1981:62; see accounts given below).

Diversity

Eighteen species based on females, five species groups based on males, and four species groups based on larvae:

Key to Females of Species of the Genus *Gigantactis*

The following key will identify metamorphosed females greater than approximately 20 mm. *Gigantactis ovifer* and *G. filibulbosus*, each represented by only a poorly preserved holotype, and recognized in this account as *nomina dubia* (following Bertelsen et al., 1981), are omitted from the key.

1A. Dorsal-fin rays 8 to 10, the first and last distinctly longer than intermediate rays; length of first and eighth caudal-fin rays 60 to 100% SL . . . . .
. . . . . . . . . . . . . *Gigantactis longicirra* Waterman, 1939b, p. 457

NOTE Ten known specimens, 19.5 to 221 mm; Atlantic and eastern tropical Pacific oceans

1B. Dorsal-fin rays 4 to 7, all about equal in length; length of first and eighth caudal-fin rays less than 40% SL . . .
. . . . . . . . . . . . . . . . . . . . . . . . 2

2A. Esca with a darkly pigmented, spinulose distal prolongation (*Gigantactis vanhoeffeni* group) . . . . . . . . . . . . 3

2B. Esca without a darkly pigmented, spinulose distal prolongation . . . . 7

3A. Length of illicium 70 to 120% SL; esca with filaments at base . . . . . . 4

3B. Length of illicium 160 to 200% SL; esca without filaments at base . . . .
. . . . . . . . . . . . . . . *Gigantactis paxtoni* Bertelsen, Pietsch, and Lavenberg, 1981, p. 463

NOTE Eighteen known specimens, 50 to 305 mm; Indo-west Pacific Ocean

4A. Esca without distally flattened papillae . . . . . . . . *Gigantactis gracilicauda* Regan, 1925c, p. 462

NOTE Three known specimens, 21 to 82 mm; North Atlantic Ocean

4B. Esca with distally flattened papillae
. . . . . . . . . . . . . . . . . . . . . . . . . . 5

5A. Illicium with a pair of small papilliform or digitiform appendages on posterior margin below esca; distal prolongation of esca conical and confluent with escal bulb . . . . . . . .
. . . . . . . . . . . *Gigantactis vanhoeffeni* Brauer, 1902, p. 459

NOTE Ninety known specimens, 16.5 to 420 mm; nearly cosmopolitan

5B. Illicium without a pair of appendages below esca; distal prolongation of esca constricted at base . . . . . . . . 6

6A. Esca with distal prolongation nearly cylindrical, more than twice as long as wide, covered with short filaments .............. *Gigantactis meadi* Bertelsen, Pietsch, and Lavenberg, 1981, p. 456

NOTE  Twenty known specimens, 19 to 353 mm; Southern Ocean

6B. Esca with distal prolongation conical, about as long as wide, with filaments restricted to tip ................ .............. *Gigantactis gibbsi* Bertelsen, Pietsch, and Lavenberg, 1981, p. 461

NOTE  Four known specimens, 38 to 114 mm; North Atlantic Ocean

7A. Length of illicium 60 to 120% SL ........................ 8
7B. Length of illicium 130 to 490% SL, rarely less than 200% SL ...... 10
8A. Esca with a posterior pair of appendages at base, fringed in juveniles, divided into branched filaments in older specimens; conical distal prolongation of esca longer than diameter of escal bulb ....... .............. *Gigantactis perlatus* Beebe and Crane, 1947, p. 464

NOTE  Eleven known specimens, 20 to 223 mm; Atlantic, Pacific, and eastern Indian oceans

8B. Esca without posterior pair of appendages; distal part of esca shorter than diameter of escal bulb ..... 9
9A. Base of escal bulb with a posterior medial papilla, but without long filaments; conical distal part of esca with a lateral series of short filaments .............. *Gigantactis kreffti* Bertelsen, Pietsch, and Lavenberg, 1981, p. 458

NOTE  Five known specimens, 44 to 345 mm; eastern South Atlantic and western Pacific oceans

9B. Base of escal bulb without posterior papilla, but with a pair of long, stout filaments; distal part of esca with two pairs of long filaments and several shorter ones .... *Gigantactis elsmani* Bertelsen, Pietsch, and Lavenberg, 1981, p. 465

NOTE  Ten known specimens, 11.5 to 435 mm; Atlantic and Pacific oceans

10A. Esca with distal filaments branched, several filaments emerging from and below base ................ 11
10B. Esca with distal filaments simple, without posterior filaments on or below base ................ 13
11A. Esca with a single anterior filament arising from base ............. .............. *Gigantactis golovani* Bertelsen, Pietsch, and Lavenberg, 1981, p. 466

NOTE  Four known specimens, 25 to 179 mm; Atlantic Ocean

11B. Esca with several anterior filaments arising from base ........... 12

12A. Esca with four or five pairs of distal filaments (large specimens) ....... ............ *Gigantactis gargantua* Bertelsen, Pietsch, and Lavenberg, 1981, p. 466

NOTE  Eleven known specimens, 25 to 408 mm; North Pacific and eastern South Indian oceans

12B. Esca with eight pairs of distal filaments (large specimens) ......... ............ *Gigantactis macronema* Regan, 1925c, p. 468

NOTE  Eleven known specimens, 34 to 354 mm; Atlantic and eastern North Pacific oceans

13A. Esca with a group of anterior filaments arising from base; distal part of escal bulb bearing four or five pairs of stout filaments along posterior margin; second and seventh caudal-fin rays greater than 50% SL (*Gigantactis gargantua* group) ........ 14
13B. Esca without anterior filaments arising from base; filaments of distal part of escal bulb not as above; longest caudal-fin rays less than 40% SL (*Gigantactis macronema* group) .... 16
14A. Esca with less than 10 (only 7 in a 262-mm specimen) filaments arising from base; distal part of escal bulb bearing four pairs of filaments, each gradually tapering and only faintly pigmented at base .............. .............. *Gigantactis herwigi* Bertelsen, Pietsch, and Lavenberg, 1981, p. 468

NOTE  Two known specimens, 105 and 262 mm; Atlantic Ocean

14B. Esca with more than 10 filaments at base; distal part of escal bulb bearing four or five pairs of filaments, each with a pigmented swollen base .... ....................... 15
15A. Distal escal filaments heavily pigmented for more than one-half their length; proximal escal filaments restricted to the anterior margin of the escal bulb ... *Gigantactis watermani* Bertelsen, Pietsch, and Lavenberg, 1981, p. 467

NOTE  Two known specimens, 99 and 305 mm; eastern tropical Atlantic and western South Pacific oceans

15B. Distal escal filaments lightly pigmented for less than one-fifth their length; proximal escal filaments not restricted to the anterior margin of the escal bulb ................ ............ *Gigantactis gargantua* Bertelsen, Pietsch, and Lavenberg, 1981, p. 466

NOTE  Eleven known specimens, 25 to 408 mm; North Pacific and eastern South Indian oceans

16A. Esca with a long distal prolongation; distal escal appendages much longer than escal bulb .............. 17

16B. Esca without distal prolongation, or with only a short rounded distal swelling; distal escal appendages much shorter than escal bulb ........ 18
17A. Length of illicium 340 to 447% SL; distal prolongation of esca truncated, bearing 8 to 20 long distal appendages arranged in nearly symmetrical pairs .................. ............ *Gigantactis macronema* Regan, 1925c, p. 468

NOTE  Eleven known specimens, 34 to 354 mm; Atlantic and eastern North Pacific oceans

17B. Length of illicium 165 to 273% SL; distal prolongation of esca tapering to a point, bearing 10 to 18 long, unpaired distal appendages .............. .............. *Gigantactis savagei* Bertelsen, Pietsch, and Lavenberg, 1981, p. 469

NOTE  Six known specimens, 19 to 150 mm; eastern North Pacific oceans

18A. Esca with a dense cluster of 14 to 16 short lanceolate appendages, each with numerous tiny spherical inclusions, emerging from a distal patch of pigment ........ *Gigantactis ios* Bertelsen, Pietsch, and Lavenberg, 1981, p. 471

NOTE  Four known specimens, 38 to 81 mm; eastern North Atlantic Ocean

18B. Esca with 4 to 10, well-spaced, spatulate, and distally compressed appendages, without spherical inclusions, emerging from a distal patch of pigment ................ 19
19A. Length of illicium 216 to 240% SL; escal bulb with 8 to 10 distal appendages; longest rays of caudal fin 31 to 42% SL ................. ............ *Gigantactis microdontis* Bertelsen, Pietsch, and Lavenberg, 1981, p. 470

NOTE  Twelve known specimens, 19.5 to 310 mm; eastern Pacific Ocean

19B. Length of illicium 330% SL; escal bulb with two pairs of distal appendages; longest rays of caudal fin 50% SL ..... *Gigantactis longicauda* Bertelsen and Pietsch, 2002, p. 471

NOTE  A single known specimen, 114 mm; western North Atlantic Ocean

Key to Males of the Genus *Gigantactis*

1A. Skin spinulose ............... 2
1B. Skin naked .................. 3
2A. Skin darkly pigmented; eyes relatively small, diameter 0.4 to 0.6 mm ................ *Gigantactis* Male Group II, p. 473

NOTE  Twenty-two metamorphosed specimens, 10.5 to 15.5 mm; Atlantic, Pacific, and Indian oceans

2B. Skin unpigmented; eyes relatively large, diameter 0.7 to 0.8 mm ..... ................... *Gigantactis* Male Group III, p. 473

   NOTE  Three metamorphosed specimens, 14 to 16 mm; central Pacific Ocean off Oahu

3A. A distinct V-shaped patch of subdermal pigment on throat; number of dorsal- and anal-fin rays 4 ....... ................... *Gigantactis* Male Group V, p. 473

   NOTE  A single specimen in early metamorphosis, 14.5 mm; central Pacific Ocean off Oahu

3B. Throat without a V-shaped patch of subdermal pigment; number of dorsal- and anal-fin rays 5 to 7 ....... 4

4A. Number of dorsal-fin rays 8 to 10 ......... *Gigantactis longicirra* Waterman, 1939b, p. 457

   NOTE  Four metamorphosed specimens, 12 to 14.5 mm; tropical Atlantic Ocean

4B. Number of dorsal-fin rays 5 to 7 ....................... 5

5A. Number of pectoral-fin rays 18 to 22; eyes relatively large, diameter 0.6 to 0.9 mm; number of olfactory lamellae 11 or 12 ......... *Gigantactis* Male Group I, p. 473

   NOTE  Ten metamorphosed specimens, 15 to 22 mm; Atlantic and Pacific oceans

5B. Number of pectoral-fin rays 15; eyes relatively small, diameter 0.5 mm; number of olfactory lamellae 8 or 9 ................ *Gigantactis* Male Group IV, p. 473

   NOTE  A single specimen in late metamorphosis, 16.5 mm; western tropical Pacific Ocean

### Key to Larvae of the Genus Gigantactis

1A. Dorsal and peritoneal pigment well developed ................... 2

1B. Dorsal and peritoneal pigment weak or absent .................. 3

2A. Dorsal and ventral series of 3 or 4 large subdermal melanophores on caudal peduncle ...... *Gigantactis* Larval Group D, p. 475

2B. No large subdermal melanophores on caudal peduncle .... *Gigantactis* Larval Group A, p. 474

3A. Number of dorsal-fin rays 5 to 7, anal-fin rays 5 to 7 .... *Gigantactis* Larval Group B, p. 474

3B. Number of dorsal-fin rays 8 to 10, anal-fin rays 5 to 8 (Larval Group C) ............. *Gigantactis longicirra* Waterman, 1939b, p. 474

   NOTE  Eight known specimens, 4.7 to 7.5 mm TL; Atlantic and Pacific oceans

## *GIGANTACTIS LONGICIRRA* WATERMAN, 1939B

Figures 38, 156, 158, 197, 200, 233, 267–269, 300; Table 8

### Females, Males, and Larvae

*Gigantactis longicirra* Waterman, 1939b: 82, figs. 1, 2 (original description, single specimen). Waterman, 1948:81, figs. 1–10 (comparative anatomy, comparison with other species of the genus and other ceratioid genera and families). Clarke, 1950: 6, 10, 18, 28 (problems in classifying ceratioid lateral line organs; references to Waterman, 1948). Bertelsen, 1951:150 (comparison with all known material, comments). Grey, 1956a:269 (synonymy, vertical distribution). Bertelsen et al., 1981:26, figs. 1A, 4F, 11A, 11D, 13A, 14A, 15A, 16, 17, 24–26, 66 (additional material, including males and larvae; comparison with all known material). Bertelsen, 1990:514 (eastern tropical Atlantic). Pietsch, 1999:2036 (western central Pacific). Pietsch, 2002b:1068 (western central Atlantic). Moore et al., 2003:216 (off New England). Pietsch and Orr, 2007:15, fig. 12A (gill arches, phylogenetic relationships).

*Gigantactis* Larvae Type C: Bertelsen, 1951:148, 150, 274, fig. 99E, F (11 specimens, description).

*Gigantactis* sp.: Becker et al., 1975:327 (specimen tentatively referred to *Gigantactis longicirra*).

*Gigantactis* Larval Group C: Bertelsen et al., 1981:62, 63, fig. 68 (after Bertelsen, 1951).

### Material

Ten females, nine metamorphosed (34.5 to 222 mm) and one in metamorphosis (19.5 mm); two metamorphosed males (14 and 14.5 mm); and eight larvae (4.7 to 7.5 mm TL).

Holotype of *Gigantactis longicirra*: MCZ 35065, 39 mm, *Atlantis* station 2894, western North Atlantic, 39°06′N, 70°16′W, closing net at 1000 m, bottom depth 2860 m, 20 July 1937.

Additional females: ISH, three (108 to 209 mm, 209-mm specimen cleared and stained); MCZ, one (221 mm); MNRJ, one (222 mm); SIO, two (19.5 and 72.5 mm); SIOM, one (34.5 mm); UW, one (41 mm).

Metamorphosed males: UF, two (14 to 14.5 mm, 14-mm specimen cleared and stained); UW, two (12 to 14 mm).

Larvae: ZMUC, eight (4.7 to 7.5 mm TL).

### Distinguishing Characters

*Gigantactis longicirra* differs from all other species of the genus in having a relatively high number of dorsal-fin rays (8 to 10). Metamorphosed females are further distinguished in having 3 or fewer distal escal filaments, a proximal anterolateral group of short filaments, and a posterior proximal group of long filaments; first and last dorsal-fin rays distinctly longer than intermediate rays; and, in addition, the following combination of character states: length of illicium less than 120% SL (39 to 105% SL); escal bulb without distal prolongation or papillae; dentary teeth relatively long (longest 3.3 to 5.0% SL), in five or six longitudinal series; first and eighth rays of caudal fin prolonged (60 to 100% SL).

Metamorphosed males of *G. longicirra* are further characterized by having the following combination of character states: eyes small, diameter 0.45 and 0.5 mm; olfactory lamellae 11; upper denticular teeth 3; lower denticular teeth 4; skin naked, unpigmented.

Larvae of *G. longicirra* (*Gigantactis* Larval Group C) are further distinguished in lacking dorsal subdermal pigment.

### Description

Stem of illicium without filaments, nearly cylindrical throughout, length highly variable and unrelated to standard length; escal bulb pear shaped in juveniles, elongate and constricted below photophore in adults, without spinules on distal portion; 1 to 3 short, distal filaments. Esca of 39-mm holotype (described and figured by Waterman, 1939b) with escal bulb pear shaped, unpigmented, and naked; a single distal filament and a total of 20 proximal filaments; all filaments unpigmented and unbranched with slight distal swellings. Escal bulb of 118-mm specimen elongate and darkly pigmented, with a distinct constriction below distally placed photophore; skin of proximal part of bulb covered with small spinules; a single unpigmented distal filament bearing a small terminal swelling; all of approximately 25 filaments of proximal anterolateral group unpigmented and arranged in an anterior series (of which the proximal is branched) and a lateral more scattered grouping of 6 or 7 simple filaments; posterior proximal group darkly pigmented and arranged in two symmetrical fan-shaped groups, each consisting of numerous filaments, long and short (some of which are branched), arising from a common base. Esca of 221-mm specimen similar in all major characters to that of 118-mm specimen but differs in having 3 distal filaments, and those of proximal anterolateral group all unbranched and pigmented.

Number of teeth in each premaxilla increasing with standard length from 5 (34.5-mm specimen) to 35 (118-mm specimen) but decreasing in largest specimens due to loss of smallest and oldest teeth; longest premaxillary tooth 1.7 to 2.7% of SL; number of teeth in each dentary increasing with standard length from 9 (34.5 mm) to approximately 60 (209 mm); teeth in posterior part of dentary in five or six longitudinal series in specimens greater than 100 mm; longest dentary teeth 3.3 to 5.2% SL.

Anal-fin rays 6 to 8, pectoral-fin rays 14 to 18; first ray of dorsal fin longest (78% and 80% SL in 108- and 118-mm specimens, respectively; broken in other specimens); caudal-fin rays free nearly to base without remains of connecting membrane, first and eighth rays prolonged 60 to 100% SL, becoming longer with increased SL.

Dermal spinules and lateral line organs on head and body slightly larger than in other members of genus.

For descriptions of males and larvae, see Male Group and Larval Group accounts, below.

### Distribution

*Gigantactis longicirra* has been collected primarily in the Atlantic Ocean, from widely scattered localities that extend from the Gulf of St. Lawrence, south along the New England slope to the Gulf of Mexico, Caribbean Sea, off Venezuela, and east to the Gulf of Guinea, with a single record in the western South Atlantic off Rio de Janeiro (Fig. 233). Two specimens have been taken in the eastern central Pacific at approximately 6°N, 144°W, and 13°N, 127°W. All metamorphosed females were collected by gear fished open at maximum depths of 1000 to 2300 m. The 39-mm holotype was captured with a closing net at 1000 m.

### Comments

Females of *Gigantactis longicirra* are similar to those of a number of other *Gigantactis* species in having a relatively short illicium and long teeth in several rows in the lower jaw. Except for these similarities, the species shows no distinct affinity with any other species of the genus. The esca is unique in several characters: the low number of distal filaments combined with the lack of a distal prolongation of the bulb, and (in adults) the constriction of the bulb below the photophore as well as the posterior proximal pair of darkly pigmented, fan-shaped appendages. The prolongation of the first and last dorsal-fin rays is found in no other species of *Gigantactis*; greatly prolonged caudal rays also occur in *G. gargantua*, *G. watermani*, and *G. herwigi*, but in these species, the second and seventh rays are the longest.

In both metamorphosed males, the testes are relatively short and narrow (0.7 and 0.8 mm in diameter), yet the specimens appear to have passed metamorphosis, as evidenced by their well-developed denticulars, resorbed premaxillae, and close-set olfactory organs (see Male Group accounts, below).

*Gigantactis longicirra* is unique among the recognized species of the genus in having more than 7 dorsal-fin rays, thus the eight larvae with such high dorsal-ray counts referred to "Type C" by Bertelsen (1951) no doubt represent this species. On the other hand, the two larvae with 7 dorsal rays included in this group by Bertelsen (1951) should be removed to the unidentified larvae of Group B (see Larval Group accounts, below). The absence of dorsal subdermal pigment in *G. longicirra* is confirmed in the female metamorphosal stage (SIO 60-241) and in the two metamorphosed males (UF 227411, 227412).

## *GIGANTACTIS KREFFTI* BERTELSEN, PIETSCH, AND LAVENBERG, 1981

Figures 38, 157, 158, 234; Table 8

### Females

*Gigantactis kreffti* Bertelsen et al., 1981:29, figs. 4D, 27–29, 65 (original description, four specimens). Yamakawa, 1982:197, 362, fig. 123 (additional specimen, Kyushu-Palau Ridge). Amaoka, 1984:107, fig. 22, pl. 93-B (Japanese Archipelago). Anderson and Leslie, 2001:23, fig. 15B (reference to paratype, ISH 1262/71; description after Bertelsen et al., 1981; in key). Nakabo, 2002:470, figs. (Japan, in key).

### Males and Larvae

Unknown.

### Material

Five metamorphosed females (44 to 345 mm):

Holotype of *Gigantactis kreffti*: ISH 1099/71, 252 mm, *Walther Herwig* station 406/71, eastern South Atlantic, 39°19′S, 3°15′W, 0 to 2000 m, 19 March 1971.

Paratypes of *Gigantactis kreffti*: ISH 1262/71, 75 mm, *Walther Herwig* station 417/71, eastern South Atlantic, 34°12′S, 16°35′E, 0 to 1550 m, 28 March 1971; ISH 3236/71, 185 mm, *Walther Herwig* station 406/71, eastern South Atlantic, 39°19′S, 3°15′W, 0 to 2000 m, 19 March 1971; ZUMT 52706, 44 mm, western North Pacific, off Japan, 35°06′N, 139°24′E, 0 to 1000 m, bottom depth 1200 to 1400 m.

Additional material: CSIRO, one (345 mm).

### Distinguishing Characters

Metamorphosed females of *Gigantactis kreffti* are distinguished from those of all other species of the genus in having the following combination of character states: length of illicium less than 120% SL (69 to 94% SL); distal escal prolongation with spinules at base and several unpigmented, digitiform filaments on lateral margins; escal papillae absent; a posteromedial papilla at base of escal bulb and a second on illicium below bulb; proximal escal filaments absent; dentary teeth relatively long (longest 2.6 to 5.2% SL), in four or five longitudinal series; rays of caudal fin less than 30% SL.

### Description

Illicium without filaments, nearly cylindrical throughout length, only proximal part slightly compressed; illicium length variable and unrelated to standard length; escal bulb pear shaped with short conical distal prolongation; dermal spinules covering proximal part of bulb, reaching base of distal prolongation on anterior margin. Escal bulb of 252-mm holotype with distal prolongation and oval area surrounding pore of photophore naked and unpigmented; remaining surface of escal bulb darkly pigmented and covered with spinules; pore of photophore protruding from surface of bulb as a short tube; lateral margins of distal prolongation bearing approximately 20 unpigmented digitiform filaments; a stout unpigmented papilla present posteriorly at base of bulb and a similar but slightly smaller papilla on illicium about 65 mm (25% SL) below base of bulb (in the freshly caught unpreserved specimen, the tips of the distal filaments and the papillae were red; beneath the skin below the pore, a bright silvery, circular area was noted). Escae of paratypes similar to that of holotype except for some changes with growth: in smaller specimens (44 and 75 mm) escal bulb more spherical in shape; bulb without pigmentation at 44 mm, pigmentation reaching posterior part of base of bulb at 75 mm; number of distal filaments increasing from 4 or 5 on each side at 44 mm, to 7 or 8 at 75 mm and about 30 at 185 mm; in two smallest specimens, papilla on illicium situated at a distance of about 10% SL below base of bulb (papilla and part of the skin of the illicium are lost in the 185-mm specimen).

Left premaxilla of holotype with 44 teeth, approximately 10 posterior-most teeth curved anteriorly; longest tooth in upper jaw 1.3% SL; number of premaxillary teeth of paratypes increasing from 7 at 44 mm, to 17 at 75 mm and 30 at 185 mm; longest premaxillary tooth 1.3 to 2.5% SL; holotype with about 47 teeth in each dentary, posteriorly arranged in five longitudinal series, one external, one medial, and three internal; number of dentary teeth of paratypes increasing from 15 at 44 mm, to 27 at 75 mm and 42 at 185 mm; four posterior longitudinal series present except in smallest paratype (44 mm) in which second internal series not yet developed.

Dorsal-fin rays 7, anal-fin rays 6, pectoral-fin rays 16 to 18; skin covering caudal fin for some distance beyond fin base; skin of each ray broad, tapering distally, and basally connected by transparent membrane.

### Distribution

*Gigantactis kreffti*, known from five metamorphosed females, has an unusual disjunct distribution with three individuals captured in the South Atlantic off Cape Town, South Africa, one off Japan, and another off Tasmania at about 45°S, 150°E. All material was taken with open gear fished at maximum depths of 1000 to 2000 m.

### Comments

*Gigantactis kreffti* is similar to a number of other species of *Gigantactis* in having a relatively short illicium and long lower-jaw teeth in several posterior series. It is similar to members of the G. *vanhoeffeni* group (including G. *vanhoeffeni*, G. *meadi*, G. *gibbsi*, G. *gracilicauda*, and G. *paxtoni*) in having a relatively spinulose esca with a distal prolongation bearing short filaments, but differs from these species in lacking both spinules on the distal prolongation and filaments on the proximal part of the esca. The relatively short caudal fin with well-developed skin coverage is similar to those of G. *meadi*, G. *gibbsi*, and G. *perlatus*.

## *GIGANTACTIS VANHOEFFENI* BRAUER, 1902

Figures 48, 158, 199, 232, 243, 271, 273; Tables 8, 9

### Females

*Gigantactis vanhoeffeni* Brauer, 1902:296 (original description, two specimens, lectotype ZMHU 17712, designated by Bertelsen et al., 1981; paralectotype lost, used by Brauer, 1908, for anatomical studies). Chun, 1903:567, fig. (brief description after Brauer, 1902). Boulenger, 1904b:720 (brief description). Brauer, 1906:322, pl. 15, figs. 8, 9 (description after Brauer, 1902). Brauer, 1908:103, 184, text fig. 1, pl. 31, figs. 18–24, pl. 32, figs. 1–6, pl. 34, fig. 14, pl. 44, fig. 1 (anatomical, histological description of illicial apparatus, esca, and eyes). Gill, 1909:586, fig. 25 (description, figure after Brauer, 1902, 1906; habits). Regan, 1926:38 (three additional specimens, two subsequently described as new, *Gigantactis sexfilis* and *G. exodon*, by Regan and Trewavas, 1932). Regan and Trewavas, 1932:93, 94 (description after Brauer, 1902, 1906; in key). Fowler, 1936:1345, 1346, fig. 565 (brief description, figure after Brauer, 1902, 1906, and Regan, 1926). Waterman, 1939b:85 (lateral line organs on caudal rays, comparison with *G. longicirra*). Beebe and Crane, 1947:168 (comparison with *G. perlatus*). Bertelsen, 1951:150 (comparison with all known material, comments). Harvey, 1952:533, fig. 179 (bioluminescence; after Brauer, 1908). Grey, 1956a:267 (synonymy, vertical distribution). Marshall, 1966:177, pl. 7 (bioluminescence). Parin and Golovan, 1976:271 (in part, two additional specimens). Hubbs et al., 1979:13 (listed for California). Pietsch and Seigel, 1980:396 (additional material, Philippine Archipelago). Bertelsen et al., 1981:31, figs. 1B, 4A, 5, 6, 8, 12, 19, 21, 30, 31, 64 (family revision, comparison with all known material). Amaoka, 1983:116, 117, 198, figs. 69 (additional specimen, northeastern Sea of Japan). Fujii, 1983:263, fig. (additional specimen, off Surinam). Amaoka, 1984:106, fig. 21, pl. 93-A (Japanese Archipelago). Bertelsen, 1986:1406, figs. (eastern North Atlantic). Ni, 1988:329, fig. 257 (additional specimen, East China Sea). Pequeño, 1989:44 (Chile). Bertelsen, 1990:514 (eastern tropical Atlantic). Nielsen and Bertelsen, 1992:62, fig. 5 (North Atlantic). Andriyashev and Chernova, 1994:103 (listed for Arctic seas). Amaoka et al., 1995:115, fig. 177 (after Amaoka, 1983; northern Japan). Stearn and Pietsch, 1995:133, fig. 83 (three additional specimens, Greenland). Shedlock et al., 1997: 396, 398, figs. 1, 2 (DNA extracted from formalin-fixed specimens). Jónsson and Pálsson, 1999:203, fig. 7 (Iceland). Munk, 1999:267, 268, 274, fig. 5A, 6, 13 (bioluminescence). Pietsch, 1999:2036 (western central Pacific). Anderson and Leslie, 2001:26, fig. 15E (additional specimen, southern Africa; description after Bertelsen et al., 1981; in key). Mecklenburg et al., 2002:313 (Alaska, in key). Nakabo, 2002: 469, 1496, figs. (Japan, in key). Pietsch, 2002b:1068 (western central Atlantic). Moore et al., 2003:217 (off New England). Love et al., 2005:62 (eastern North Pacific). Pietsch and Orr, 2007:16, fig. 13B (phylogenetic relationships).

*Gigantactis exodon* Regan and Trewavas, 1932:93, 94, 95, fig. 151 (original description, single specimen, in key). Fraser-Brunner, 1935:326 (comparison with *Gigantactis filibulbosus*). Waterman, 1939b:84, 85 (comparison with *G. longicirra*). Bertelsen, 1951:150 (comparison with all known material, comments). Grey, 1956a:268 (synonymy, vertical distribution). Pietsch, 1972a:42, 45 (holotype with 5 pectoral radials). Nielsen, 1974:97 (listed in type catalog).

*Gigantactis perlatus*: Parin et al., 1973: 145 (misidentifications, three specimens).

*Gigantactis* sp. 2: Parin et al., 1973:146 (juvenile).

*Gigantactis vahoeffeni*: Bertelsen et al., 1981:23, fig. 23 (erroneous spelling of specific name).

*Gigantactis balushkini* Kharin, 1984: 665, fig. 2 (original description, single specimen).

### Males and Larvae

Unknown.

### Material

Ninety known females, 86 metamorphosed (19 to 420 SL), and three in metamorphosis (16.5 to 21.5 mm):

Lectotype of *Gigantactis vanhoeffeni*: ZMB 17712, 35 mm, *Valdavia* station 239, off Zanzibar, 5°42′S, 43°36′E, open pelagic net, 2500 m, 13 March 1899.

Holotype of *Gigantactis exodon*: ZMUC P92128, 25 mm, *Dana* station 1217(4), Caribbean Sea, west of Jamaica, 18°50′N, 79°07′W, open pelagic net, 2500 m wire, 0630 h, 29 January 1922.

Holotype of *Gigantactis balushkini*: ZIN 46022, 287 mm, *Mys Yunony* trawl 120, western North Pacific, 40°52.8′N, 142°20.8′E, otter trawl, 1210 m, 4 February 1980.

Additional metamorphosed females: ASIZP, five (210 to 420 mm); BSKU, three (250 to 295 mm); SCFR, one (352 mm); CSIRO, one (380 mm); HUMZ, six (147 to 278 mm); ISH, 11 (38 to 270 mm); LACM, 15 (22 to 315 mm); MCZ, one (19 mm); MNHN, one (75 mm); MNRJ, one (308 mm); NMNZ, three (358 to 395 mm); NSMT, two (191 and 261 mm); OS, two (298 and 385 mm); SAM, one (133 mm); SIO, eight (26 to 325 mm); SIOM, 14 (57 to 340 mm); UF, three (33 to 232 mm); UW, four (38 to 295 mm); ZMUB, one (207 mm); ZMUC, four (67.5 to 330 mm).

Females in metamorphosis: LACM, one (17 mm); SIOM, two (16.5 and 21.5 mm).

### Distinguishing Characters

Metamorphosed females of *Gigantactis vanhoeffeni* differ from those of all other species of the genus in having the following combination of character states: length of illicium less than 120% SL (71 to 112% SL in specimens greater than 25 mm); escal bulb with an elongate, spinulose, darkly pigmented distal prolongation; escal bulb and distal prolongation bearing distally flattened papillae; short distal and slender proximal escal filaments present; illicium with a posterior pair of papillae below escal bulb; dentary teeth relatively long (longest 2.6 to 5.0% SL) in three longitudinal series; length of rays of caudal fin less than 45% SL.

Males unknown but probably included in *Gigantactis* Male Group II (see below).

Larvae unknown but probably included in *Gigantactis* Larval Group A (see below).

### Description

Stem of illicium distinctly compressed in specimens greater than 50 mm, depth at base two or three times width, length variable and unrelated to SL; escal bulb gradually tapering into a conical distal prolongation, darkly pigmented proximally, unpigmented and transparent around photophore, and less darkly pigmented distally; bulb covered to distal tip with dermal spinules and bearing several large distally flattened papillae of varying pigmentation, some black, others (especially those in skin covering photophore) nearly transparent; distal filaments short and restricted to tip of distal prolongation, increasing in number from 1 in specimens less than 30 mm to 10 to 30 in specimens greater than 100 mm; 2 or 3 thin filaments on each side of escal bulb immediately below photophore, longest filament reaching to base of distal prolongation; a close-set pair of small appendages arising on posterior margin of illicium some distance below bulb, papilliform and very small in juveniles, larger and somewhat compressed in adults, bifurcate in some of largest specimens; distance from base of paired illicial papillae to tip of distal filaments 10 to 22% SL in specimens less than 100 mm, 7 to 14% SL in larger specimens; 1 to approximately 10 (increasing in number with standard length) short filaments present on posterior margin of distal portion of illicium of specimens greater than 70 mm; in freshly captured unpreserved specimens, paired papillae and tips of distal filaments bright red.

Number of teeth in each premaxilla increasing with standard length from 2 to 8 in smaller specimens to 22 to 53 in larger; longest premaxillary tooth 1.0 to 1.8% SL in specimens greater than 25 mm; number of teeth on each dentary increasing with standard length from 3 to 15 in smaller specimens to 32 to 66 in larger, medial series of small dentary teeth distinctly developed.

Dorsal-fin rays 5 to 7; anal-fin rays 5 or 6, rarely 7; pectoral-fin rays 17 or 18, rarely 19; caudal fin divided between upper and lower lobes nearly to base; skin coverage of caudal fin weakly developed, skin of each ray only slightly compressed; remains of transparent membrane between bases of rays present in some specimens.

Three female metamorphosal stages (16.5 to 21.5 mm, SIOM uncataloged; 17 mm, LACM 36031-1) with illicial length 23.5 to 41% SL; escae with more or less distinct rudiments of papillae and paired posterior appendages; jaw teeth rudimentary, premaxilla with 2 to 4 teeth, dentary with 3 to 5 teeth (representing the medial series and one or two of the first external series); 4 to 8 anterior-most supraorbital lateral-line organs enlarged, the longest about 7% SL; dark subdermal dorsal and peritoneal pigment present.

Distribution

*Gigantactis vanhoeffeni* has a nearly cosmopolitan distribution between approximately 64°N and 43°S (Fig. 232). In the Atlantic it extends from off West Greenland, south to the Gulf of Mexico and Caribbean Sea, and east to the Cape Verde Islands and Gulf of Guinea, with single, isolated records off Rio de Janeiro, the central South Atlantic at about 37°S, 23°W, and off the tip of South Africa. In the Indian Ocean, the known localities are all clustered on the western side, between 18°S and 3°N. In the Pacific it extends from the Sulu Sea, north to Taiwan, Japan, and the Bering Sea, and south to Tasmania and New Zealand. There are numerous records as well that extend across the Pacific to the Hawaiian Islands, and to the Oregon coast in the north, and the Gulf of Panama and off Ecuador in the south. The known material has been taken with gear fished open at maximum depths of 300 to 5300 m, but the species appears to be most commonly found between 700 and 1300.

Comments

*Gigantactis vanhoeffeni* is one of several species of the genus that have a relatively short illicium. It is the type of a group of species (the *G. vanhoeffeni* group, including *G. vanhoeffeni*, *G. meadi*, *G. gibbsi*, *G. gracilicauda*, and *G. paxtoni*) characterized by having a spinulose distal prolongation of the escal bulb bearing short filaments and (except for *G. paxtoni*) having at the base of the bulb a small number of more slender filaments that reach to the base of the distal prolongation. Like *G. meadi*, *G. gibbsi*, and *G. paxtoni*, *G. vanhoeffeni* has distally flattened escal papillae but differs distinctly from these forms in the shape of the distal prolongation of the bulb and in having a posterior pair of papillae on the illicium below the bulb.

Although the characters that distinguish *G. vanhoeffeni* from its nearest relatives are not mentioned in Brauer's (1902) original description (based on two specimens now lost except for badly damaged remains of one), his excellent figure of the esca (Brauer, 1908, pl. 15, fig. 9) shows the characteristic conical shape of its distal prolongation, with the filaments restricted to the tip, as well as the paired appendages on the illicium below the bulb.

Two of the specimens referred to this species have lost the esca, but in both, the characteristic paired papillae are present on the remaining part of the illicium. In one of them (SIOM uncataloged), the loss of the esca seems to have happened before capture. The esca has been broken off just below the photophore, leaving a wound that has healed without sign of regeneration of the lost part. A second break just below the paired appendage has healed in such a way that the remains of the esca are attached to the illicium in a twisted and displaced position.

The esca of the 270-mm specimen (ISH 376/73, representing the most northerly capture (63°N) of a gigantactinid, shows some aberrant features. In addition to the cluster of filaments at the tip of the distal prolongation, it has numerous short, tapering filaments and papillae on the proximal part of the prolongation as well as on the proximal part of the bulb. A pair of filaments on the posterior margin of the illicium below the bulb probably represents the paired appendages characteristic of *G. vanhoeffeni*, but they are neither compressed nor particularly distinct from other filaments present on the illicium in this same region. Although these aberrant features are not present in the three specimens examined of similar or larger size, there is a general tendency among species of *Gigantactis* for the number of distal filaments of the esca as well as those of the illicium to increase with increasing size. For this reason, it is assumed for the present that this variation is due to age, possibly combined with individual differences.

*Gigantactis balushkini* was described by Kharin (1989) as a member of the *G. vanhoeffeni* group, being closest among members of that group to *G. paxtoni* in lacking dermal papillae at the base of the esca, but distinct from the latter in having a considerably shorter illicium (84% SL versus 168 to 198% SL) and in lacking filaments on the head just behind the base of the illicium. It was said to differ further from *G. paxtoni* and all other members of the group in various details of escal morphology. However, despite the poor condition of the holotype, including skin striped from the illicium, a mangled caudal fin, and a desiccated esca, it compares well with the known material of *G. vanhoeffeni*. In fact, the only feature that might distinguish *G. balushkini* from *G. vanhoeffeni* is the absence of secondary escal appendages and filaments, but these have no doubt been lost through desiccation. The type locality of *G. balushkini*, in the western North Pacific at about 40°N, 142°E, compares well with the known distribution of *G. vanhoeffeni*, the only member of the *G. vanhoeffeni* group found in temperate Pacific waters. Based on this body of evidence, *G. balushkini* is placed in synonymy with *G. vanhoeffeni*.

### *GIGANTACTIS MEADI* BERTELSEN, PIETSCH, AND LAVENBERG, 1981

Figures 38, 232, 270, 272; Table 8

Females

*Gigantactis meadi* Bertelsen et al., 1981:33, figs. 4B, 18, 20, 23, 32, 33, 64 (original description, 14 specimens). Pequeño, 1989:44 (Chile). Stewart and Pietsch, 1998:27, fig. 18 (additional specimen, New Zealand). Anderson and Leslie, 2001:24, fig. 16 (additional specimen, southern Africa, description; metamorphosing females tentatively identified by Bertelsen et al., 1981, positively identified; in key).

Males and Larvae

Unknown.

Material

Twenty females, including 15 metamorphosed (35.5 to 353 mm), four in late metamorphosis (19 to 21 mm), and one tentatively assigned to this species (290 mm):

Holotype of *Gigantactis meadi*: MCZ 52572, 306 mm, *Anton Bruun*, cruise 6, station 352A, 33°53′S, 64°55′E, 0 to 350 m, 30 June 1964.

Paratypes of *Gigantactis meadi*: ISH 415/76, 87 mm, *Walther Herwig*, Antarctic Expedition, cruise 20, station 101-I, 47°45′S, 40°05′W, 0 to 2000 m, 5 January 1976; ISH 1004/71, 2 (155 to 201 mm), *Walther Herwig* station 395, 36°49′S, 12°17′W, 0 to 2000 m, 17 March 1971; ISH 965/71, 178 mm, *Walther Herwig* station 384, 39°45′S, 17°40′W, 0 to 2000 m, 13 March 1971; ISH 571/76, 207 mm, *Walther Herwig*, Antarctic Expedition, cruise 20, station 106-II, 39°08′S, 40°00′W, 0 to 1850 m, 8 January 1976; ISH 883/71, 230 mm, *Walther Herwig* station 376, 39°55′S, 26°02′W, 0 to 2000 m, 11 March 1971; ISH 1465/71, 290 mm, *Walther Herwig* station 427, 33°00′S, 7°50′W, 0 to 2000 m, 30 March 1971; LACM 11242-12, 353 mm, *Eltanin* station 1719, 39°58′S, 150°31′W, 0 to 1900 m, bottom depth 5161 m, 16 July 1966; SAM 27811, 21 mm (in late metamorphosis), *Meiring Naudé* station 157, off Durban, 30°06′S, 31°57′E, 0 to 750 m, 18 May 1977; USNM 208032, 35.5 mm, *Eltanin*, cruise 21, station 5, 33°06′S, 83°57′W, 0 to 1050 m, bottom depth 3731 to 3822 m, 28 November 1965.

Additional females: CSIRO, three (255 to 350 mm); NMNZ, one (288 mm); ZIN, one (212 mm).

Additional females in late metamorphosis: SAM, three (19 to 20.5 mm).

Tentatively identified female: SIOM uncataloged, 290 mm, 29°35′S, 14°13′E, 0 to 1300 m (esca lost, length of illicium 93% SL, strong filaments on head at base of illicium).

### Distinguishing Characters

Metamorphosed females of *Gigantactis meadi* differ from those of all other species of the genus (except for *G. paxtoni*) in having filaments on the dorsal surface of the head just behind the base of the illicium. They are further distinguished in having the following combination of character states: length of illicium less than 120% SL (72 to 96% SL); short filaments present along entire posterior margin of illicium; escal bulb with an elongate spinulose and darkly pigmented distal prolongation, slightly constricted at base; escal bulb and distal prolongation bearing distally flattened papillae; short distal and slender proximal escal filaments present; posterior pair of close-set illicial appendages absent; dentary teeth relatively long (longest 2.9 to 3.8% SL) in five or six longitudinal series; rays of caudal fin short (less than 30% SL).

Males unknown but probably included in *Gigantactis* Male Group II (see account, below).

Larvae unknown but probably included in *Gigantactis* Larval Group A (see account, below).

### Description

Proximal part of illicium distinctly compressed, depth at base more than twice width, length variable and unrelated to standard length; escal bulb club shaped, with cylindrical distal prolongation more than twice as long as wide (about three times as long as wide in holotype), spinulose except for area surrounding pore of photophore; small distally flattened papillae present from tip of distal prolongation to some distance below escal bulb; papillae white or only weakly pigmented; 15 to approximately 20 short filaments on distal prolongation, most concentrated at tip, a few (only 2 in holotype) below base distal to photophore; a posterolateral group of filaments at base of escal bulb, some slightly longer than those of distal group; filaments along posterior margin of illicium (34 illicial filaments in holotype, 8 to 42 in paratypes) continuing onto head, forming a cluster (16 cephalic filaments in holotype, 2 to 18 in paratypes).

Number of teeth on each premaxilla 20 in holotype, increasing with standard length from 2 to 8 in smaller specimens to 20 to 27 in larger specimens; longest premaxillary tooth 1.8% SL in holotype, 1.0 to 1.8% in metamorphosed paratypes. Number of dentary teeth 51 in holotype, increasing with standard length from 14 to 24 in smaller specimens to 51 to 81 in larger specimens; teeth on posterior part of lower jaw in five or six longitudinal series: two external series, a distinct medial series of small teeth, and one or two internal series; anterior part of lower jaw of most specimens with a third external series containing largest teeth of jaw, 3.3% SL in holotype, 2.9 to 3.8% SL in paratypes.

Dorsal-fin rays 6 or 7, anal-fin rays 5 or 6, pectoral-fin rays 17 or 18; skin coverage of caudal fin well developed and complete for more than half length of rays; skin of each ray gradually tapering distally and connected by transparent membranes.

Three small females (19 mm, SAM 27807; 20 mm, SAM 27808; 21-mm paratype, SAM 27811) represent an ontogenetic series of late metamorphosal stages: length of illicium 16, 16, and 52% SL, respectively; distal prolongation of esca short, unpigmented, and naked at 19 and 20 mm, elongate, cylindrical, and pigmented, bearing papillae and short filaments at 21 mm; rudimentary dentary teeth in a single series at 19 and 20 mm, arranged in three series (a medial series of 8 teeth, a first external series of 4, and a single tooth representing the second external series) at 21 mm; enlarged lateral-line organs of head absent; dorsal and peritoneal subdermal pigment well developed.

The 290-mm specimen (SIOM uncataloged), tentatively referred to *Gigantactis meadi*, has lost the esca but has apparently retained a complete illicial bone (measuring approximately 93% SL). Cephalic filaments are present at the base of the illicium, and all characters of the teeth and caudal fin agree with the description above.

### Distribution

*Gigantactis meadi* is circumglobal in association with the subtropical convergence of the Southern Ocean, so far collected only between about 30 and 53°S (Fig. 232). It appears to be a relatively deep-living species: all specimens 87 mm and larger were taken by gear fished open at maximum depths of 1850 to 2000 m.

### Comments

*Gigantactis meadi* belongs to the *G. vanhoeffeni* group characterized by having a darkly pigmented spinulose distal prolongation of the escal bulb. In common with three members of this group (*G. vanhoeffeni*, *G. gibbsi*, and *G. paxtoni*), it has distally flattened papillae on the esca. It differs from *G. vanhoeffeni* in having a cylindrical distal prolongation that is somewhat constricted at the base, a cluster of filaments on the head just behind the base of the illicium, and a shorter caudal fin that is covered to a greater extent by skin. It differs from *G. gibbsi* and *G. paxtoni* primarily in the shape and length of the distal escal prolongation. It further differs from *G. paxtoni* in illicial length.

Two females in late metamorphosis (19 mm, SAM 27807; 20 mm SAM 27808), tentatively identified by Bertelsen et al., (1981) as *G. meadi*, are now given full status as representatives of this species, following the recommendation of Anderson and Leslie (2001:25).

## *GIGANTACTIS GIBBSI* BERTELSEN, PIETSCH, AND LAVENBERG, 1981

Figure 234; Table 8

### Females

*Gigantactis gibbsi* Bertelsen et al., 1981:36, figs. 34, 64 (original description, two specimens). Pequeño, 1989:44 (Chile). Bertelsen, 1990:513 (eastern tropical Atlantic). Pietsch, 2002b:1068 (western central Atlantic).

### MALES AND LARVAE
Unknown.

### Material

Four metamorphosed females (38 to 114 mm):

Holotype of *Gigantactis gibbsi*: ZIN 44262, 50 mm, Gulf of Guinea, 2°01′N, 3°56′W, 0 to 465 m.

Paratype of *Gigantactis gibbsi*: USNM 218613, 38 mm, Ocean Acre Expedition, station 12-5C, off Bermuda, 33°00′N, 64°06′W, 0 to 1000 m, 27 August 1971.

Additional females: MCZ, two (41 and 114 mm).

Distinguishing Characters

Metamorphosed females of *Gigantactis gibbsi* differ from those of all other species of the genus in having the following combination of character states: length of illicium less than 120% SL (104 to 118% SL); escal bulb with a short conical, spinulose, darkly pigmented distal prolongation slightly constricted at base; escal bulb bearing distally flattened papillae and short distal and slender proximal filaments; illicium without posterior pair of papillae; dentary teeth in posterior part of jaw in three longitudinal series; length of rays of caudal fin less than 50% SL.

Males unknown but probably included in *Gigantactis* Male Group II (see account, below).

Larvae unknown but probably included in *Gigantactis* Larval Group A (see account, below).

Description

Escal bulb of holotype pear shaped with a short conical distal prolongation, darkly pigmented except at base; skin covering area of photophore and base of distal prolongation unpigmented; dermal spinules present on distal prolongation and bulb except around pore of photophore; distally flattened, unpigmented papillae present on proximal part of bulb; approximately 12 short filaments at tip of distal prolongation; a group of filaments of varying length proximal to bulb, longest about equal to width of bulb. Juvenile paratype with distal prolongation of escal bulb darkly pigmented and spinulose except at base, bearing a short terminal filament; bulb and base of distal prolongation naked and unpigmented; skin torn away from proximal part of bulb and distal part of illicium, with only a few remains of proximal filaments.

Holotype with 13 teeth in each premaxilla (longest tooth 1.4% SL), paratype with 6 (longest 1.3% SL); holotype with 27 teeth in each dentary (longest 2.6% SL), paratype with 23 (longest 2.4% SL); tooth pattern similar to that of *Gigantactis vanhoeffeni* of similar standard length, with three distinct longitudinal series in posterior part of lower jaw; medial series of small teeth well developed.

Dorsal-fin rays 6, anal-fin rays 6, pectoral-fin rays 17 or 18; longest caudal-fin ray (second and seventh) 49% SL in holotype, 26 to 28% SL in paratype; skin coverage of caudal fin relatively well developed.

Distribution

*Gigantactis gibbsi* is known from four specimens all collected in the Atlantic Ocean: the holotype, from equatorial waters of the Gulf of Guinea between the surface and 465 m; the remaining three specimens from off Bermuda and Georges Bank, captured between the surface and 1640 m (Fig. 234).

Comments

*Gigantactis gibbsi* is a typical member of the *G. vanhoeffeni* group having a relatively short illicium, a spinulose distal escal prolongation, and distally flattened papillae on the escal bulb. It differs from *G. vanhoeffeni*, *G. meadi*, and *G. paxtoni* in the shape and length of the distal prolongation and in the length of the longest caudal-fin ray. It further differs from *G. vanhoeffeni* in lacking a pair of papilliform or flattened appendages on the posterior margin of the illicium below the escal bulb, from *G. paxtoni* in illicial length, and from *G. gracilicauda* in having distinct distally flattened escal papillae.

The four specimens on which *G. gibbsi* is based (38 to 114 mm) are juveniles; older specimens of this species could conceivably undergo ontogenetic changes that would indicate conspecificity with some other species of *Gigantactis*. However, a comparison of this material with a complete growth series of *G. vanhoeffeni* (16.5 to 340 mm) indicates that the differences observed in *G. gibbsi* cannot be explained as part of the variation within the better represented *G. vanhoeffeni*. Some of the differences between *G. gibbsi* and the series of specimens representing *G. meadi* could perhaps be explained in this way, but it seems unlikely that the latter species, known only from the Southern Ocean, would occur north of the equator. Finally, it can hardly be assumed that these four juveniles represent *G. gracilicauda* or *G. paxtoni*; besides involving great ontogenetic changes in the shape of the esca, this hypothesis would also imply an ontogenetic loss of either the escal papillae or the proximal escal filaments (Bertelsen et al., 1981).

## *GIGANTACTIS GRACILICAUDA* REGAN, 1925c

Figure 232; Table 8

Females

*Gigantactis gracilicauda* Regan, 1925c: 565 (original description, single specimen). Regan, 1926:18, 19, 38, fig. 12, pl. 10, fig. 2 (cranial osteology; description after Regan, 1925c). Regan and Trewavas, 1932:93, 94 (brief description, in key). Waterman, 1948:90, 93 (osteological comparison with *Gigantactis longicirra*; correction of errors made by Regan, 1926). Bertelsen, 1951:150 (comparison with all known material, comments). Grey, 1956a:267 (synonymy, vertical distribution). Pietsch, 1972a:42, 45 (holotype with 5 pectoral radials). Nielsen, 1974:98 (listed in type catalog). Bertelsen et al., 1981:38, figs. 35, 64 (family revision, comparison with all known material). Bertelsen, 1990:513 (eastern tropical Atlantic). Pietsch, 2002b:1068 (western central Atlantic).

*Gigantactis sexfilis* Regan and Trewavas, 1932:38, 39, 93, 94, figs. 49, 50, 150, pl. 5, fig. 2 (original description, single specimen, osteology of skull and pectoral lobe, in key). Waterman, 1939b:84 (comparison with *Gigantactis longicirra*; holotype with 4 pectoral radials after Regan and Trewavas, 1932). Waterman, 1948:88 (osteology of skull compared to that of *G. longicirra*). Bertelsen, 1951:150 (comparison with all known material, comments). Grey, 1956a:267 (synonymy, vertical distribution). Pietsch, 1972a:42, 45 (holotype with 5 pectoral radials). Nielsen, 1974:98 (listed in type catalog).

*Gigantactis* sp.: Bertelsen, 1951:150, fig. 100 (*Gigantactis vanhoeffeni* of Regan, 1926).

Males and Larvae

Unknown.

Material

Three metamorphosed females (21 to 82 mm):

Holotype of *Gigantactis gracilicauda*: ZMUC P92129, 82 mm, *Dana* station 1183(1), Caribbean Sea, west of station Lucia, 13°47′N, 61°26′W, open pelagic net, 4500 m wire, 1630 h, 24 November 1921.

Holotype of *Gigantactis sexfilis*: ZMUC P92132, 51 mm, *Dana* station 1181(2), tropical Atlantic, 13°07′N, 57°20′W, open pelagic net, 4000 m wire, 1800 h, 21 November 1921.

Additional material: ZMUC, one (21 mm).

Distinguishing Characters

Metamorphosed females of *Gigantactis gracilicauda* differ from those of all other species of the genus in having the following combination of character states: length of illicium less than 120% SL (104 to 107% SL, 86% SL in the 21-mm juvenile); escal bulb with an elongate darkly pigmented spinulose distal prolongation; short distal and slender proximal escal filaments present; distally flattened escal papillae absent; illicium without posterior pair of papillae; dentary teeth in posterior part of jaw in three longitudinal series; rays of caudal fin less than 30% SL.

Males unknown but probably included in *Gigantactis* Male Group II (see account, below).

Larvae unknown but probably included in *Gigantactis* Larval Group A (see account, below).

Description

Stem of illicium slightly compressed laterally in larger specimens; escal bulb gradually tapering to form a darkly pigmented, slender distal prolongation; medial portion of escal bulb containing unpigmented photophore; skin of bulb and distal prolongation covered with spinules but without distinct papillae; distal filaments short, restricted to distal part of tapering prolongation; one or two pairs of slender filaments below escal bulb not reaching base of distal prolongation. Escal bulb of holotype of *Gigantactis gracilicauda* with about 20 short distal filaments, two pairs of slender filaments on posterior margin of illicium just below bulb, and below these, an additional short illicial filament. Escal bulb of holotype of *G. sexfilis* with 4 short distal filaments and a single pair of slender illicial filaments at base of bulb. Esca of juvenile (21 mm) unpigmented except for small spot on distal tip, with a few spinules, but papillae absent; about 8 short distal filaments and two illicial pairs of very thin lateral filaments at base of bulb (some tiny filament-like structure, 0.2 to 0.4 mm long, present on the anterior margin of proximal part of the bulb may be due to abraded skin).

Holotype of *G. gracilicauda* with 10 teeth in each premaxilla (longest 0.7% SL), holotype of *G. sexfilis* with 7 teeth (longest 1.0% SL), juvenile with 4 teeth. Holotype of *G. gracilicauda* with 43 dentary teeth (longest 2.4% SL), holotype of *G. sexfilis* with 23 teeth (longest 2.4% SL), juvenile with approximately 10 (all in development). Tooth pattern similar to that of *G. vanhoeffeni* of similar size, with three distinct longitudinal series in posterior part of lower jaw.

Dorsal-fin rays 5 or 6, anal-fin rays 5 or 6, pectoral-fin rays 18 or 19; longest caudal-fin rays (second and seventh) 26 and 28% SL in holotype of *G. gracilicauda*, both 30% SL in holotype of *G. sexfilis* (both broken in juvenile); skin coverage of caudal fin well developed.

Distribution

*Gigantactis gracilicauda* is known from three individuals all collected in the Atlantic Ocean, from off the coast of Venezuela in the west and western Sahara in the east (Fig. 232). It appears to be a relatively deep-living species: all the specimens were taken by gear fished open at maximum depths 2000 to 2500 m.

Comments

*Gigantactis gracilicauda* is a member of the *G. vanhoeffeni* group, having a relatively short illicium and a darkly pigmented spinulose distal prolongation of the escal bulb, bearing short filaments. It resembles *G. vanhoeffeni* and *G. paxtoni* in the shape of the distal prolongation but differs from these species and other members of the *G. vanhoeffeni* group in lacking distally flattened escal papillae.

The esca of the holotype of *G. gracilicauda* was not described originally by Regan (1925c, 1926) or later by Regan and Trewavas (1932), but its general shape and the presence of some distal filaments is shown in Regan's (1926, pl. 10, fig. 2) illustration of the whole fish. The figure provided by Bertelsen et al. (1981:38, fig. 35A) is to some extent a reconstruction, as the esca is now rather desiccated and shrunken; among the dermal spinules in the wrinkled darkly pigmented skin of the esca are some wartlike protuberances, which could be interpreted as the remains of distally flattened papillae but might just as well be artifacts. The esca of the holotype of *G. sexfilis* (Bertelsen et al., 1981:38, fig. 35B), well described and figured by Regan and Trewavas (1932, fig. 150), is still in good condition; no papillae are present. These distally flattened escal papillae characteristic of several well-represented species of *Gigantactis* (e.g., *G. vanhoeffeni* and *G. meadi*) are developed at metamorphosis and indicate no distinct individual or ontogenetic variation. *Gigantactis gracilicauda* is recognized as a distinct species primarily because these escal papillae are lacking. The differences observed between the holotype of *G. gracilicauda* and the two smaller specimens are within the expected range of variation for this species; for this reason, *G. sexfilis* is regarded as a junior synonym of *G. gracilicauda*.

### GIGANTACTIS PAXTONI BERTELSEN, PIETSCH, AND LAVENBERG, 1981

Figures 38, 233; Table 8

Females

*Gigantactis* sp. 1: Parin et al., 1977:156 (single specimen, off the northwest coast of New Guinea).
*Gigantactis paxtoni* Bertelsen et al., 1981:39, figs. 36–38, 64 (original description, eight specimens). Bertelsen and Pietsch, 1983:92, figs. 12, 13 (additional specimen, Australia). Paulin, 1984:66, fig. 5 (additional specimen, New Zealand). Paulin and Stewart, 1985:27 (New Zealand). Paulin et al., 1989:137, fig. 65.1 (New Zealand). Bertelsen, 1994:141, color plate (popular account). Bertelsen and Pietsch, 1998a:141, color drawing (after Bertelsen, 1994). Stewart and Pietsch, 1998:26, fig. 17 (additional specimens, New Zealand). Pietsch, 1999:2036 (western central Pacific). Stewart and Pietsch, 1998:26, fig. 17 (New Zealand).

Males and Larvae

Unknown.

Material

Eighteen metamorphosed females, 50 to 305 mm:

Holotype of *Gigantactis paxtoni*: AMS I.20314-018, 237 mm, 100 km east of Broken Bay, New South Wales, 33°28′S, 152°33′E, 0 to 900 m over 4200 m, 14 December 1977.

Paratypes of *Gigantactis paxtoni*: AMS I.20070-016, 124 mm, northeast of Cape Howe, New South Wales, 37°24′S, 150°30′E, 0 to 540 m over 3600 m, 1 November 1977. AMS I.20306-007, 142 mm, 65 km east of Broken Bay, New South Wales, 33°31′S, 152°20′E, 0 to 900 m over 1800 to 2900 m, 12 December 1977. AMS I.20314-018, 3 (175 to 228 mm), data as for holotype. SIOM uncataloged, 50 mm, *Vityaz* station 7288, 3°39′N, 131°22′E, 0 to 1500 m (Parin et al., 1977). SIOM uncataloged, 210 mm, *Zwiezda Kryma* station 83, 34°07′S, 44°50′E, 0 to 1260 m, 1976.

Additional females: AMS, one (134 mm); CSIRO, four (200 to 305 mm); NMNZ, five (188 to 295).

Distinguishing Characters

Metamorphosed females of *Gigantactis paxtoni* differ from those of all other species of the genus (except for *G. meadi*) in having filaments on the dorsal surface of the head just behind the base of the illicium. The species is further distinguished by having the following combination of character states: length of illicium 168 to 198% of SL; short filaments present on base of illicium; escal bulb gradually tapering into a conical, spinulose darkly pigmented distal prolongation, length 12 to 28% of SL; escal bulb and distal prolongation bearing low unpigmented papillae; short filaments present on distal prolongation, but absent on base of escal bulb; posterior pair of close-set illicial appendages absent; dentary teeth long (longest tooth 3.4 to 7.1 % SL), in three or four longitudinal series in posterior part of jaw; rays of caudal fin short (27.5 to 35% SL).

Description

Proximal portion of illicium distinctly compressed, depth at base more than twice width; escal bulb gradually tapering into a distal prolongation several times as long as wide (approximately nine times as wide in holotype), spinulose and pigmented except for area surrounding escal pore; small papillae present from tip of distal prolongation to some distance below escal bulb; papillae low (height less than width), unpigmented, and transparent; approximately

10 to 30 short filaments on distal prolongation, none proximal to its base; filaments along posterior margin of proximal part of illicium (except in smallest known specimen) continuing onto head and forming an anterodorsal cluster on snout.

Number of teeth in each premaxilla of holotype 15, increasing with standard length from 7 in 50-mm specimen to 10 to 19 in larger specimens; longest premaxillary tooth 1.3% SL in holotype, 1.0 to 1.8% in paratypes. Number of dentary teeth of holotype 46, increasing from 10 in 50-mm specimen to 28 to 55 in larger specimens; teeth in posterior part of lower jaw in three longitudinal series; anterior part of lower jaw of most specimens with a second external series containing largest teeth of jaw, 5.6% SL in holotype, 3.4 to 7.1% SL in paratypes.

Dorsal-fin rays 6 or 7, anal-fin rays 5 or 6, pectoral-fin rays 18 to 21; skin coverage of caudal fin relatively well developed and complete for about one-third length of rays, skin of each ray gradually tapering and connected by transparent membranes.

### Distribution

Eleven of the 18 known specimens of *Gigantactis paxtoni* were caught off New Zealand and the southeast coast of Australia near the northern boundary of the Subantarctic Water Mass between 33 and 44°S (Fig. 233). Two additional specimens were collected from the western South Indian Ocean and another from the western tropical Pacific off the northwest coast of New Guinea. With the exception of a single individual taken by bottom trawl at 1210 to 1260 m, the material was collected by pelagic gear fished open at maximum depths of 540 to 1500 m (over bottom depths of 1800 to 4200 m).

### Comments

*Gigantactis paxtoni* belongs to the *G. vanhoeffeni* group characterized by having a darkly pigmented spinulose distal prolongation of the escal bulb. It differs from all other members of this group in having a considerably longer illicium, a longer distal prolongation of the escal bulb, and in lacking proximal escal filaments. In common with three members of this group *(G. vanhoeffeni. G. meadi,* and *G. gibbsi), G. paxtoni* has distally flattened escal papillae; these papillae differ from those of the other species, however, in being distinctly lower.

### GIGANTACTIS PERLATUS BEEBE AND CRANE, 1947

Figure 232; Table 8

### Females

*Gigantactis perlatus* Beebe and Crane, 1947:167, text fig. 13, pl. 2, fig. 3 (original description, single specimen). Bertelsen, 1951:150, 151 (comparison of all known material, comments). Grey, 1956a:268 (synonymy, vertical distribution). Mead, 1958:133 (holotype transferred to CAS-SU). Robins and Courtenay, 1958:151 (depth distribution). Parin et al., 1973:145 (misidentifications, three specimens referred to *Gigantactis vanhoeffeni* by Bertelsen et al., 1981). Bertelsen et al., 1981: 41, figs. 1C, 4C, 39, 40, 65 (additional material, family revision, comparison of all known material). Amaoka, 1983:118, 119, 198, fig. 70 (additional specimen, northeastern Sea of Japan). Amaoka, 1984:107, fig. 23, pl. 93-C (Japanese Archipelago). Pequeño, 1989:44 (Chile). Amaoka et al., 1995:115, fig. 178 (after Amaoka, 1983; northern Japan). Pietsch, 1999:2036 (western central Pacific). Anderson and Leslie, 2001:25, fig. 15D (reference to ISH 1466/71; description after Bertelsen et al., 1981; in key). Nakabo, 2002:469, figs. (Japan, in key). Pietsch, 2002b:1068 (western central Atlantic). Moore et al., 2003:216 (off New England).

### Males and Larvae

Unknown.

### Material

Eleven females, nine metamorphosed (23 to 223 mm), and two in metamorphosis (19 to 20 mm):

Holotype of *Gigantactis perlatus*: CAS-SU 46487 (originally NYZS 28621), 32.5 mm, eastern Pacific *Zaca* Expedition station 225 T-1, off Jicaron Island, Panama, 7°08′N, 81°57′W, open pelagic net, 0 to 915 m 20 March 1938.

Additional material: HUMZ, one (187 mm); ISH, one (152 mm); LACM, two (20 to 36 mm); MCZ, two (19 to 23 mm); SIO, two (39 to 222 mm); USNM, two (41 to 223 mm).

### Distinguishing Characters

Metamorphosed females of *Gigantactis perlatus* differ from those of all other species of the genus in having an extremely large distal prolongation of the escal bulb (20% SL in 220-mm specimen), the entire esca densely covered with slightly elongated papillae (not distally flattened papillae as in members of the *G. vanhoeffeni* group). Metamorphosed females are further distinguished in having the following combination of characters: length of illicium less than 120% SL (74 to 111% SL in specimens greater than 30 mm); a pair of posterior escal appendages; dentary teeth large (longest 2.2 to 5.9% SL), arranged in two irregular longitudinal series; rays of caudal fin less than 35% SL.

### Description

Stem of illicium somewhat laterally compressed in largest specimens, length variable and unrelated to standard length; escal bulb with a tapering distal prolongation, increasing in size with standard length from 5 to 6% SL in specimens less than 50 mm to about 20% SL in specimens of approximately 220 mm; entire bulb and distal prolongation without spinules but densely covered with unpigmented, slightly elongated papillae; a few short, paired filaments along posterior margin of distal prolongation; a pair of wing-shaped appendages on posterior margin near base of bulb, with fringed edges in juveniles, divided into numerous branched filaments in older specimens; two pairs of long proximal filaments, larger pair on lateral margin slightly below base of wing-shaped appendages, smaller pair just below and on each side of pore of photophore.

Esca of three smaller specimens (23 to 45 mm) very similar to that of 34-mm holotype (described and figured by Beebe and Crane, 1947, fig. 13), with total length about 10% SL; slightly elongated papillae increasing in number and coverage of bulb with standard length; paired wing-shaped proximal escal appendages simple in 36-mm specimen, fringed on posterior edge in other specimens; 2 or 3 paired filaments along posterior margin of distal prolongation; two proximal pairs of larger filaments. Distal prolongation of three largest specimens (152 to 223 mm) greatly lengthened, total escal length 15% SL at 152 mm, 22 and 27% SL at 222 and 223 mm, respectively; bulb and distal prolongation densely covered with small elongate papillae except for narrow posterior area around tubular opening of pore of photophore; numerous short filaments on distal third of prolongation, in addition to paired filaments on posterior margin; all filaments of distal prolongation except proximal pair unpigmented and simple, proximal pair branched in two largest specimens; paired, wing-shaped appendages divided into numerous short branches in 152-mm specimen, numerous thin filaments in largest specimens. Esca of fresh unpreserved specimens "completely semi-translucent white" (except for darkly pigmented photophore) in holotype (Beebe and Crane, 1947:167); opaque pearly white except for pink and silvery area around pore of photophore in 152-mm specimen; bright red in 223-mm specimen.

Teeth few; number of teeth on each premaxilla 10 in 223-mm specimen, 2 or 3 in all others, longest tooth approximately 1.0% SL; 5 to 10 dentary teeth in specimens 23 to 41 mm (longest about 6% SL), 14 to 18 in specimens 153 to 223 mm (longest 2.2 to 3.0% SL), teeth arranged in two very irregular series throughout length of jaw.

Dorsal-fin rays 5 to 7, anal-fin rays 5 to 7, pectoral-fin rays 16 to 19; longest caudal rays (second and seventh) 22 to

31% SL; skin coverage of caudal fin and thin membranes between rays well developed.

The 20-mm metamorphosal stage (LACM 37518-1) with illicial length 19% SL; esca with relatively short distal prolongation, no papillae, posterior paired appendage and three pairs of filaments rudimentary; edge of premaxilla and dentary in resorption; jaw teeth absent; enlarged cephalic lateral line organs absent; skin and subdermal pigment absent (possibly bleached out).

### Distribution

*Gigantactis perlatus* is known from widely separated localities in the Atlantic, Pacific, and Indian oceans: four specimens from the western North Atlantic; and one each from the South Atlantic off Cape Town, South Africa; the far eastern Indian Ocean near Christmas Island; off Luzon in the South China Sea; off Honshu, Japan; the Hawaiian Islands; Gulf of Panama, and the eastern South Pacific at approximately 36°S, 91°W (Fig. 232). The 36-mm Hawaiian specimen (LACM 36875-2) was collected with a closing net between 670 and 805 m. The remaining material was captured by gear fished open at maximum depths of 800 to 2000 m.

### Comments

Among the species of *Gigantactis* characterized by having relatively short illicia (less than 120% SL), *Gigantactis perlatus* is distinguished by the characters of the esca: the numerous small, slightly elongated papillae (not distally flattened as in members of the *G. vanhoeffeni* group), the extreme development of the distal prolongation, and the characteristic pattern of the filaments and appendages. The presence of paired proximal appendages is shared only with adult specimens of *G. longicirra*; but this species, besides lacking a distal prolongation of the bulb, differs from *G. perlatus* in tooth pattern, median fin-ray counts, and caudal fin-ray lengths. A similar tooth pattern in which few relatively large dentary teeth are arranged in two irregular series is found only in *G. golovani*. The latter species is very distinct from *G. perlatus*, however, in the length of the illicium (greater than 180% SL in *G. golovani*) and in nearly all escal characters.

## *GIGANTACTIS ELSMANI* BERTELSEN, PIETSCH, AND LAVENBERG, 1981

Figure 233; Table 8

### Females

*Gigantactis elsmani* Bertelsen et al., 1981:43, figs. 4E, 41, 42, 65 (original description, two specimens). Amaoka, 1983:118, 119, 199, fig. 71 (additional specimen, northeastern Sea of Japan). Amaoka, 1984:107, fig. 24, pl. 93-D (Japanese Archipelago). Bertelsen, 1990:513 (eastern tropical Atlantic). Fedorov, 1994:414, fig. (additional specimen, description, Okhotsk Sea). Amaoka et al., 1995:116, fig. 179 (after Amaoka, 1983; northern Japan). Sheiko and Fedorov, 2000:24 (Kamchatka). Anderson and Leslie, 2001:23, fig. 15A (reference to tentatively identified female in metamorphosis, MCZ 51269; description after Bertelsen et al., 1981; in key). Balushkin and Fedorov, 2002:8 (after Fedorov, 1994). Nakabo, 2002:470, figs. (Japan, in key).

### Males and Larvae

Unknown.

### Material

Ten females, nine metamorphosed (283 to 435 mm) and one in metamorphosis (11.5 mm) tentatively assigned to this species:

Holotype of *Gigantactis elsmani*: ISH 1360/71, 384 mm, *Walther Herwig* station 459/71, 10°57′S, 11°20′W, 0 to 1900 m, 1818 to 2218 h, 7 April 1971.

Paratype of *Gigantactis elsmani*: LACM 10687-1, 283 mm, *Eltanin* station 904, Southern Ocean, 63°00′S, 114°34′W, 0 to 2932 m, bottom depth 5051 m, 9 January 1964.

Additional material: ASIZP, two (396 to 420 mm); HUMZ, one (350 mm); ISH, two (340 to 435 mm); NSMT, one (322 mm); ZIN, one (403 mm).

Tentatively identified female in metamorphosis: MCZ, one (11.5 mm).

### Distinguishing Characters

Metamorphosed females of *Gigantactis elsmani* differ from those of all other species of the genus in having a single proximal lateral pair and two distal pairs of large escal filaments and in lacking a distinct distal prolongation of the escal bulb. Metamorphosed females are further distinguished in having the following combination of character states: length of illicium less than 130% SL (93 to 126% SL); escal papillae absent; dentary teeth relatively short (longest 3.3% SL), arranged in five or six longitudinal series; longest caudal-fin rays 21 to 32.5% SL.

### Description

Stem of illicium without filaments, proximal part laterally compressed, depth at base about three times width; escal bulb club shaped, without distal prolongation; proximal part of bulb darkly pigmented, covered with spinules extending to or slightly beyond photophore; distal part of bulb above photophore bearing numerous short filaments and one or two pairs of large filaments; a pair of large proximal filaments reaching beyond tip of longest distal filaments. Shape of escal bulb and pattern of escal appendages very similar in four known metamorphosed specimens (238 to 435 mm); proximal pair of long filaments arising in identical position and distal group consisting of 15 to 30 filaments of which four are distinctly enlarged; several elongate filaments broken at various distance from base; relatively longest, most complete filaments found in largest specimen, proximal pair about 30% SL, longest distal pair about 18% SL. Pigmentation of posterior margin of bulb below photophore deeply divided in holotype and largest specimen, forming a rounded indentation in two smaller specimens; spines absent on unpigmented part of bulb in holotype, covering proximal part of bulb in paratype.

Number of teeth on each premaxilla 32 to 58 arranged in two or three overlapping longitudinal series, external and posterior-most teeth turned anteriorly; longest premaxillary tooth 1.0 to 1.4% SL. Number of dentary teeth 35 to 82, longest 2.8 to 3.3% SL; dentary teeth in posterior part of jaw arranged in five longitudinal series, an external, a medial, and three internal.

Dorsal-fin rays 5, anal-fin rays 4 or 5, pectoral-fin rays 16 or 17; longest rays of caudal fin (second and seventh) 21 to 35.5% SL; skin coverage of caudal fin and membranes between caudal rays well developed.

The 11.5-mm metamorphosal stage tentatively referred to this species (MCZ 51269) with length of illicium about 10% SL; esca with naked unpigmented skin, a short conical distal prolongation, and papilliform rudiments of a pair of filaments on base; edge of jaws in resorption, rudiments of teeth absent; on each side a series of 7 or 8 enlarged supraorbital lateral line organs, longest about 10% SL; skin faintly pigmented; no distinct dorsal subdermal pigment (specimen somewhat bleached); differs from holotype and paratype in having dorsal-fin rays 4, pectoral-fin rays 18.

### Distribution

*Gigantactis elsmani* is known from widely scattered localities in the Atlantic and Pacific oceans: two records in the western North Atlantic; one in the South Atlantic at about 12°S, 12°W; two off Taiwan; two from the western North Pacific in the Okhotsk Sea and northeastern Sea of Japan; and one from the far eastern South Pacific at about 63°S, 114°W (Fig. 233). An 11.5-mm specimen in metamorphosis, tentatively referred to this species, is from the South Atlantic off Cape Town, South Africa. All of the known specimens were captured with nonclosing trawls at maximum depths of 1300 to 3000 m.

Comments

The four additional specimens collected since the original description (Bertelsen et al., 1981) are in good agreement with the type material but significantly broaden the range of variation of some of the characters. The comparisons indicate an increase in the relative length of the illicium and caudal fin and in the numbers of jaw teeth with increasing standard length.

The 435-mm specimen (ISH 1135/79) is the largest recorded representative of the genus (the next largest is the 408-mm holotype of *Gigantactis gargantua*, LACM 6903-32) and among ceratioids it is surpassed in standard length only by a few females of *Ceratias*.

Among the species of *Gigantactis* with an illicium less than 120% SL, *G. elsmani* is clearly distinguished by characters of the esca: the lack of a distinct distal prolongation of the escal bulb and the presence of a single proximal and two distal pairs of large filaments. Specimens lacking escae may be identified by using a combination of characters including a low number of fin rays, a relatively high number of posterior longitudinal series of dentary teeth, and well-developed skin coverage on the caudal fin.

*GIGANTACTIS GOLOVANI* BERTELSEN, PIETSCH, AND LAVENBERG, 1981

Figure 234; Table 8

Females

*Gigantactis vanhoeffeni*: Parin and Golovan, 1976:271 (in part, one specimen, misidentification).
*Gigantactis golovani* Bertelsen et al., 1981:44, figs. 4J, 43, 44, 65 (original description, three females). Bertelsen, 1990:513 (eastern tropical Atlantic).

Males and Larvae

Unknown.

Material

Four metamorphosed females, 25 to 179 mm:
Holotype of *Gigantactis golovani*: ISH 2250/71, 179 mm, *Walther Herwig* station 471-III/71, eastern tropical Atlantic, 2°27′S, 19°00′W, MT, 0 to 660 m, 2054 to 2225 h, 10 April 1971.
Paratypes of *Gigantactis golovani*: MCZ 51272, 25 mm, *Atlantis II* station 59, tropical Atlantic, 14°43′N, 25°27′W, 0 to 720 m, 24 November 1970. ZIN 44263, 153 mm, *Zwiezda Kryma* station 279, 10°36′N, 17°38′W, 0 to 1550 m.
Additional female: ISH, one (175 mm).

Distinguishing Characters

Metamorphosed females of *Gigantactis golovani* differ from those of all other species of the genus in having branched distal escal filaments, 3 slender filaments on and below the anterior margin of the escal bulb, and several similar filaments on and below the posterior margin of the escal bulb. They are further distinguished in having the following combination of character states: length of illicium 175 to 200% SL (except in juveniles); escal papillae absent; dentary teeth relatively long (largest 4.0 to 4.5% SL), arranged in two, possibly three, very irregular longitudinal series; rays of caudal fin less than 35% SL.

Description

Stem of illicium nearly cylindrical throughout, length 175 to 199% SL, 72% SL in 25-mm juvenile; escal bulb club shaped, without distal prolongation, naked and unpigmented except at base; a group of 8 to 10 branched distal filaments; a single long filament on anterior margin of base of bulb in holotype and paratype, 2 shorter additional filaments in this position in 175-mm specimen; numerous filaments of different length (some reaching beyond tip of distal escal filaments) on posterior margin of proximal part of bulb and on illicium below bulb. Esca of juvenile paratype with distribution of filaments similar to that of holotype, but number somewhat less, considerably shorter, and unbranched.

Premaxillary teeth 17 to 20 in the three larger specimens (153 to 179 mm), longest 1.5 to 1.8% SL; 5 in smaller paratype (25 mm), longest 2.0% SL; dentary teeth 14 to 18 in larger specimens, 10 in smaller.

Dorsal-fin rays 6 or 7, anal-fin rays 6 or 7, pectoral-fin rays 14 to 17; longest caudal rays (second and seventh) 28 to 32% SL; caudal-fin rays connected by transparent membranes; pigmented skin covering medial caudal-fin rays; rays broad, tapering only at tip.

The 29-mm juvenile (LACM 37517-1), tentatively referred to this species, differing from type material in having only about 4 distal escal filaments, 5 proximal filaments on posterior margin, and none on anterior margin; similar to type material in arrangement of escal filaments, in lacking a distal prolongation of escal bulb, and in dentary tooth pattern, the latter consisting of a few (9) large (six largest 3 to 5% SL) teeth arranged in two (or three) irregular series.

Distribution

*Gigantactis golovani* was only known previously from the eastern tropical Atlantic Ocean (Bertelsen et al., 1981:46, fig. 65). The 175-mm specimen reported here for the first time was collected in the western North Atlantic at approximately 27°N, 52°W (Fig. 234). All the material was captured by gear fished open at maximum depths of 660 to 2000 m.

Comments

The intermediate illicial length of *Gigantactis golovani* (175 to 200% SL, comparable only to *G. paxtoni*, *G. gargantua*, and *G. savagei*, none of which are known from the Atlantic) easily separates this species from the eight species in which the illicium is less than 130% SL, and, at the same time, distinguishes it from all remaining species in which the illicium is considerably longer. In escal characters, *G. golovani* is unique, and in dentary tooth pattern (in which the teeth are relatively long, but few in number and irregularly arranged), *G. golovani* is similar only to *G. perlatus*.

*GIGANTACTIS GARGANTUA* BERTELSEN, PIETSCH, AND LAVENBERG, 1981

Figures 155, 234; Table 8

Females

*Gigantactis* sp. n.: Pietsch, 1972a:42, 43, 45, fig. 24(2) (otolith described, figured).
*Gigantactis gargantua* Bertelsen et al., 1981:46, figs. 1D, 4H, 45, 46, 66 (original description, seven specimens). Amaoka, 1983:120, 121, 199, fig. 72 (additional specimen, northeastern Sea of Japan). Amaoka, 1984:108, pl. 93-E (Japanese Archipelago). Ni, 1988:330, fig. 258 (additional specimen, East China Sea). Amaoka et al., 1995:116, fig. 180 (after Amaoka, 1983; northern Japan). Meng et al., 1995:441, fig. 591 (China). Munk, 1999:282 (illicial movement). Nakabo, 2002:469, figs. (Japan, in key). Love et al., 2005:62 (eastern North Pacific). Pietsch and Orr, 2007:4, fig. 3E (phylogenetic relationships).

Males and Larvae

Unknown.

Material

Eleven females, 10 metamorphosed (49 to 408 mm) and one in metamorphosis (25 mm):
Holotype of *Gigantactis gargantua*: LACM 6903-32, 408 mm, *Velero IV*, eastern Pacific, San Clemente Basin, 32°16′N, 117°43′W, 0 to 1250 m, bottom depth 1775 m, 21 February 1966.
Paratypes of *Gigantactis gargantua*: LACM 30415-27, 25 mm, *Velero IV*, eastern Pacific, near Guadalupe Island, 28°44′N, 118°10′W, 0 to 1850 m; LACM 32749-3, 49 mm, central Pacific, leeward Oahu, 21°20 to 30′N, 158°20 to 30′W, 0 to 1000 m, 1 March 1971; LACM 30997-2, 105 mm, *Velero IV*, 31°32′N, 118°29′W, 0 to 1300 m, 14 August 1969; LACM 30996-16, 106 mm, *Velero IV*, eastern Pacific, off southern California, 31°54′N, 118°39′W, 0 to 500 m, 14 August 1969; LACM 9748-28, 166 mm, *Velero IV*, eastern Pacific, San Clemente Basin, 32°13′N, 117°47′W, 0 to 835 m, bottom depth 1756 m, 15 April

1966; SIOM uncataloged, 325 mm, 31°30.8'S, 95°27.2'E, 0 to 1400 m.

Additional females: ECFR, one (184 mm); HUMZ, one (340 mm); NMV, one (385 mm); OS, one (325 mm).

Distinguishing Characters

Metamorphosed females of *Gigantactis gargantua* differ from those of all other species of the genus in having the following combination of character states: length of illicium 134 to 354% SL; four or five pairs of large, distal escal filaments; 30 to 50 proximal filaments centered on anterior margin of escal bulb; escal papillae absent; dentary teeth relatively short (longest 2.3% SL), arranged posteriorly in four longitudinal series; second and seventh caudal-fin rays extremely long (30 to 47% and 54 to 76% SL, respectively).

Males unknown but probably included in *Gigantactis* Male Group I (see account, below).

Description

Illicium usually without filaments (some filaments on distal part of illicium of 325-mm specimen, SIOM uncataloged), distinctly compressed proximally, depth near base more than twice width in holotype; length of illicium 134 to 216% SL in five metamorphosed specimens from eastern Pacific, 354% SL in specimen from Indian Ocean (325 mm, SIOM uncataloged). Escal bulb club shaped, with a short distal prolongation; posterior surface of distal prolongation darkly pigmented, with a distal and two or three lateral pairs of swellings forming bases for four or five pairs (two on distal pair of swellings) of large unpigmented filaments, more or less branched in large specimens; pore of photophore on a pigmented papilla; proximal part of bulb spinulose, unpigmented, with numerous filaments of different length (longest reaching beyond tip of distal prolongation), longest and most dense on anterior margin. Esca of 408-mm holotype differing from those of paratypes: proximal right swelling of distal prolongation small, deformed, with only a short filament (possibly in regeneration); division into six pairs of distal filaments less distinct, nearly all branched, some bifurcated near close-set bases; proximal filaments more numerous, group centered on anterior margins surrounding base of bulb. Most escal filaments of type material slightly swollen at tip.

Premaxillary teeth increasing in number with standard length from 7 in smallest paratype to 59 in holotype (longest tooth 0.7 to 0.9% SL). Dentary teeth increasing in number with standard length from 16 in smallest paratype (longest 1.6% SL) to 75 in holotype (longest 2.3% SL), arranged posteriorly in four longitudinal series, some teeth of second external series present anteriorly in the jaw of large specimen.

Dorsal-fin rays 5 to 7, anal-fin rays 6, pectoral-fin rays 19 to 22; skin coverage of proximal part of caudal fin well developed.

The 25-mm metamorphosal stage (LACM 30415-27) with length of illicium 50% SL; esca with short distal prolongation and indistinct rudiments of distal filaments; teeth in early development; no enlarged lateral line organs on head; dorsal subdermal pigment faint.

Distribution

*Gigantactis gargantua* is known from 11 specimens: seven collected in the eastern North Pacific Ocean from the Hawaiian Islands, off northern Oregon at about 46°N, and off southern California (Fig. 234). The remaining specimens were taken in the northeastern Sea of Japan; the East China Sea; off Kangaroo Island, South Australia; and the eastern South Indian Ocean at about 35°S, 93°E. All of the specimens were taken by open gear fished at maximum depths of 500 to 1300 m.

Comments

*Gigantactis gargantua*, *G. herwigi*, and *G. watermani* form a group of closely related forms here referred to as the *G. gargantua* group. These three species are distinguished from all other members of the genus in having four to five pairs of large distal filaments arising from swollen bases, a group of proximal filaments centered on the anterior margin of the bulb, and especially long second and seventh caudal-fin rays. *Gigantactis gargantua* differs from *G. watermani* and *G. herwigi* in details of the escal morphology, particularly in having a greater number of proximal filaments (about 20 to 50 compared with 12 and 7 in *G. watermani* and *G. herwigi*, respectively).

The material forming the *G. gargantua* group is quite similar in nearly all characters, but three species are recognized within the group for the following reasons: the six metamorphosed specimens from the Pacific and Indian oceans, described by Bertelsen et al. (1981) as *G. gargantua*, agree completely in the escal characters that separate them from the two Atlantic specimens; in light of the small intraspecific and ontogenetic variation in these characters observed in other species of *Gigantactis*, it seems unlikely that the differences between the two Atlantic specimens could be explained in this way. The variation in illicial length is unusually great in *G. gargantua* compared to that in most other species of *Gigantactis*, even within the representatives of the eastern Pacific population. For this reason, the extreme length of the illicium observed in the specimen from the Indian Ocean (SIOM uncataloged) does not warrant specific distinction. For the same reason, the large difference in illicial length between the holotypes of *G. watermani* and *G. herwigi* may not be a diagnostic means of distinguishing these two species.

### GIGANTACTIS WATERMANI BERTELSEN, PIETSCH, AND LAVENBERG, 1981

Figures 45, 157, 235; Table 8

Females

*Gigantactis watermani* Bertelsen et al., 1981:49, figs. 47 to 49, 66 (original description, single specimen). Bertelsen, 1990:515 (eastern tropical Atlantic).

Males and Larvae

Unknown.

Material

Two metamorphosed females (99 to 305 mm):

Holotype of *Gigantactis watermani*: ISH 2330/71, 99 mm, *Walther Herwig* station 478/71, eastern tropical Atlantic, 1°04'N, 18°22'W, 0 to 2100 m, 1842 to 2245 h, 12 April 1971.

Additional female: MNHN, one (305 mm).

Distinguishing Characters

Metamorphosed females of *Gigantactis watermani* differ from those of all other species of the genus in having the following combination of character states: length of illicium 203 to 231% SL; five pairs of large distal escal filaments; 12 to 26 proximal filaments centered on anterior margin of escal bulb; escal papillae absent; dentary teeth short (longest 2.5 to 2.6% SL), arranged posteriorly in four longitudinal series; second caudal-fin ray extremely long (53 to 70% SL), seventh caudal-fin ray broken in holotype (remaining portion measures 30% SL), 59% SL in 305-mm specimen.

Description

Stem of illicium without filaments, proximal part not distinctly compressed; escal bulb club shaped with a darkly pigmented distal prolongation approximately five times as long as diameter of bulb, with proximal one-half greatly swollen; distal prolongation bearing five pairs of stout tapering filaments, all but distal-most pair densely covered with black pigment except for narrow tapering tip; opening of pore of photophore present in darkly pigmented skin at base of distal prolongation, not raised on a papilla; 12 narrow unpigmented filaments on anterior margin of base of bulb, longest reaching base of distal-most pair of distal filaments; base of bulb proximal to filaments pigmented and spinulose.

Premaxillary teeth 16 to 84 (longest 1.1 to 1.5% SL); dentary teeth 43 to 132, arranged in four longitudinal series in posterior part of jaw, some teeth of second external series present anteriorly.

Dorsal-fin rays 5 or 6, anal-fin rays 5, pectoral-fin rays 18 or 19 in holotype, 22 on both sides in 305-mm specimen.

Distribution

*Gigantactis watermani* is only known from two females, the 99-mm holotype collected in the eastern tropical Atlantic at approximately 1°N, 18°W, and a 305-mm specimen from the western tropical Pacific, off New Caledonia (Fig. 235). The holotype was captured with gear fished open between the surface and 2100 m. The second specimen was taken somewhere between the surface and 1383 m.

Comments

*Gigantactis watermani* is a member of the *G. gargantua* group, distinguished from other members of the genus in having four to five pairs of large distal filaments arising from swollen bases, a group of proximal filaments centered on the anterior margin of the escal bulb, and especially long second and seventh caudal-fin rays. It is distinguished from the other two members of this group (*G. gargantua* and *G. herwigi*) in having a more elongate distal escal prolongation and larger and more heavily pigmented proximal portions of the distal escal filaments. It is further distinguished from *G. gargantua* in having the proximal filaments restricted to the anterior margin of the escal bulb.

GIGANTACTIS HERWIGI BERTELSEN, PIETSCH, AND LAVENBERG, 1981

Figure 235; Table 8

Females

*Gigantactis herwigi* Bertelsen et al., 1981:49, figs. 4G, 50, 51, 66 (original description, single specimen).

Males and Larvae

Unknown.

Material

Two metamorphosed females (105 to 262 mm):

Holotype of *Gigantactis herwigi*: ISH 972/68, 262 mm, *Walther Herwig* station 17/68, tropical Atlantic, 4°43′S, 26°39′W, MT, 0 to 2000 m, 1155 to 1215 h, 4 February 1968.

Tentatively assigned specimen: UW, one (105 mm).

Distinguishing Characters

Metamorphosed females of *Gigantactis herwigi* differ from those of all other species of the genus in having the following combination of character states: length of illicium 373% SL; four pairs of large distal escal filaments; 7 proximal filaments centered on anterior margin of escal bulb; escal papillae absent; dentary teeth short (longest 1.6% SL), arranged posteriorly in four longitudinal series; second and seventh caudal-fin rays prolonged (70 and 95% SL, respectively).

Description

Stem of illicium without filaments, proximal part slightly compressed laterally; escal bulb club shaped, without distinct prolongation, bearing four pairs of large unbranched distal filaments, slightly swollen and pigmented posteriorly at base; opening of pore of photophore on pigmented papilla; proximal part of bulb spinulose with 7 narrow unpigmented filaments on anterior margin of base of bulb, longest reaching beyond base of distal filaments.

Premaxillary teeth 34 (longest 0.7% SL); dentary with approximately 65 teeth, arranged posteriorly in four series, some teeth of second external series present anteriorly.

Distribution

*Gigantactis herwigi* is known from the holotype collected on the equator at about 26°W, somewhere between the surface and 2000 m, and a second specimen from the Gulf of Mexico at some unknown depth (Fig. 235).

Comments

*Gigantactis herwigi* is a member of the *G. gargantua* group. It differs from other members of this group in lacking a distinct prolongation of the escal bulb. It differs further from *G. watermani* in having distal filaments that are less swollen and pigmented only near the base (swollen and darkly pigmented for more than half their length in *G. watermani*), and in having unpigmented proximal filaments (pigmented in *G. watermani*). It differs further from *G. gargantua* (known only from the Pacific and Indian oceans) in having all distal filaments of the esca unbranched and proximal escal filaments (7 compared to about 30 in *G. gargantua*) restricted to the anterior margin of the bulb (surrounding the bulb in *G. gargantua*).

GIGANTACTIS MACRONEMA REGAN, 1925C

Figures 38, 157, 159, 233; Table 8

Females

*Gigantactis macronema* Regan, 1925c: 565 (original description, single specimen). Regan, 1926:38, pl. 11 (description after Regan, 1925c). Schmidt, 1926:265, fig. 5 (after Regan, 1925c, 1926). Regan, 1927a:4, postcard M 13 (popular account). Regan and Trewavas, 1932:93, 94 (description after Regan, 1925c, 1926; in key). Waterman, 1939b:84 (comparison with *Gigantactis longicirra*, largest known gigantactinid). Waterman, 1948:130 (comparison with *G. longicirra*). Bertelsen, 1951:150, fig. 101 (comparison with all known material, comments). Marshall, 1954:269, 270, fig. XI, 2 (deep sea biology; figure after Regan, 1926). Grey, 1956a:267 (synonymy, vertical distribution). Günther and Deckert, 1956:132, fig. 89 (popular account; figure after Regan, 1926). Robins and Courtenay, 1958:151 (comparison with additional specimen designated *Gigantactis* sp.). Courtenay, 1959: 221, fig. (after Robins and Courtenay, 1958). Fitch and Lavenberg, 1968:135, fig. 74 (distinguishing characters, natural history, age and growth). Pietsch, 1972a:29, 34, 35, 41, 42, 45 (comments on osteology, holotype with 5 pectoral radials). Nielsen, 1974:98 (listed in type catalog). Hubbs et al., 1979:13 (listed for California). Bertelsen et al., 1981:50, figs. 1E, 4I, 22, 52, 53, 64 (additional material, revision of family). Fujii, 1983:264, fig. (additional specimen, off Surinam). Munk, 1999:281 (bioluminescence). Anderson and Leslie, 2001:24, fig. 15C (reference to ISH 1596/71; description after Bertelsen et al., 1981; in key). Pietsch, 2002b:1068 (western central Atlantic). Love et al., 2005:62 (eastern North Pacific).

Males and Larvae

Unknown.

Material

Eleven metamorphosed females, 34 to 354 mm:

Holotype of *Gigantactis macronema*: ZMUC P92130, 98 mm, *Dana* station 1365(9), central North Atlantic, 31°47′N, 41°41′W, 5000 m wire, 1030 h, 8 June 1922.

Additional material: ISH, two (135 to 232 mm); LACM, four (34 to 354 mm); MCZ, one (141 mm, cleared and stained); NSMT, 1 (116 mm); SIO, two (35 to 61 mm).

Distinguishing Characters

Metamorphosed females of *Gigantactis macronema* differ from those of all other species of the genus in having the following combination of character states: length of illicium 340 to 447% SL; escal bulb with a lightly pigmented truncated distal prolongation; 6 to 20 long distal filaments; escal papillae absent; proximal escal filaments absent; dentary teeth short (longest 1.3 to 3.4% SL, average 2.3% SL), arranged posteriorly in two longitudinal series; length of caudal-fin rays less than

45% SL; skin coverage of proximal part of caudal fin weakly developed; caudal-fin rays free nearly to base.

Males unknown but probably included in *Gigantactis* Male Group I (see account, below).

Larvae unknown but probably included in *Gigantactis* Larval Group B (see account, below).

Description

Stem of illicium very slender, nearly cylindrical throughout length, without filaments except in largest known specimen (354 mm, from off California) in which distal one-sixth of stem is covered with filaments; variation in illicial length unrelated to standard length; escal bulb elongate and club shaped, increasing gradually in width from illicium toward unpigmented area surrounding photophore; skin of bulb spinulose except area distal to photophore; pore of photophore raised on papilla, pigmented at tip except in 135-mm specimen from western North Atlantic; distal prolongation truncated, faintly pigmented on posterior margin, length more than twice width at base except in largest specimen where it is somewhat shorter; pigmentation of prolongation faint and restricted to posterior side except in 135-mm specimen in which it is covered with dark pigment; number of distal filaments tending to increase with length of specimens: about four pairs in a 35-mm specimen, eight pairs in three of four specimens of 93 to 232 mm in which they are retained, and approximately 20 more irregularly placed filaments in 354-mm specimen (but only three pairs in 135-mm specimen); esca of 98-mm holotype poorly preserved, with remains of at least 8 filaments (see Comments, below).

Number of teeth on each premaxilla increasing with standard length from 3 to 5 in smaller specimens to 23 in largest, those in posterior part of jaw turned forward; longest premaxillary tooth 0.6 to 1.6% SL. Number of dentary teeth varying between 18 and 31 in six largest specimens, those in posterior part of jaw arranged in two longitudinal series.

Dorsal-fin rays 5 or 6, anal-fin rays 5 or 6, pectoral-fin rays 17 to 20; skin coverage of proximal part of caudal fin less developed than in most other species, rays cylindrical, free nearly to base, without connecting membranes; longest caudal-fin rays (second and seventh) 26 to 42% SL, only slightly longer than intermediate rays.

Distribution

*Gigantactis macronema* is known from the central North and eastern South Atlantic, from the eastern Pacific just north and west of the Hawaiian Islands, and from off the coasts of southern California and Baja California (Fig. 233). The material was captured by open gear fished at maximum depths of 650 to 2500 m.

Comments

*Gigantactis macronema* is one of a number of species of *Gigantactis* that have an illicium that is more than twice the standard length and that lack filaments on the proximal part of the escal bulb. These species, here referred to as the *G. macronema* group, are further characterized by having relatively small dentary teeth (those in the posterior part of the jaw in 2, rarely 3, longitudinal series) and a caudal fin with weakly developed skin coverage and without greatly prolonged rays (the longest, usually 30 to 35% SL). *Gigantactis macronema* differs from the other three members of the *G. macronema* group in its combination of escal characters and extreme illicial length. Specimens with missing escae can be referred to the species group only.

The esca of the holotype of *G. macronema* was not described in the original description (Regan, 1925c). A well-developed distal prolongation of the escal bulb and some slender distal filaments, however, are shown in an illustration provided by Regan (1926, pl. 11). The esca is now desiccated and shrunken so that the position and the number of filaments of the distal prolongation are difficult to ascertain. Because of the poor condition of the holotype, the allocation of the available material to this species is open to some doubt. Additional material is needed to show whether the differences from the specimens previously referred to *G. macronema* observed in the 135-mm specimen from the western North Atlantic (ISH 5550/79) recorded here for the first time represent intraspecific variation or might be of specific order.

*GIGANTACTIS SAVAGEI* BERTELSEN, PIETSCH, AND LAVENBERG, 1981

Figures 159, 233; Table 8

Females

*Gigantactis savagei* Bertelsen et al., 1981:53, figs. 54, 55, 67 (original description, three specimens). Love et al., 2005: 62 (eastern North Pacific).

Males and Larvae

Unknown.

Material

Six females: two metamorphosed (56 to 150 mm), one in metamorphosis (19 mm), and three (33 to 44 mm) tentatively assigned to this species: Holotype of *Gigantactis savagei*: LACM 9706-41, 150 mm, *Velero IV* station 11169, Baja California, Cortez Bank, 31°40′N, 120°23′W, 0 to 650 m 31 July 1966.

Paratypes of *Gigantactis savagei*: LACM 37080-1, 56 mm, *Teuthis* station 33, central Pacific, leeward Oahu, 21°20 to 30′N, 158°20 to 30′W, 0 to 1250 m, 22 April 1971; LACM 37520-1, 19 mm, *Teuthis* station 141, central Pacific, leeward Oahu, 21°20 to 30′N, 158°20 to 30′W, 0 to 985 m.

Tentatively identified females: LACM, two (33 to 44 mm); SIO, one (38 mm).

Distinguishing Characters

Metamorphosed females of *Gigantactis savagei* differ from those of all other species of the genus in having the following combination of character states: length of illicium 165 to 268% SL; escal bulb with a darkly pigmented tapering distal prolongation; 10 to 18 narrow distal filaments; escal papillae absent; proximal escal filaments absent; dentary teeth short (longest 1.7 to 2.1 % SL), arranged posteriorly in two or three longitudinal series; length of caudal-fin rays less than 35% SL; skin coverage of proximal part of caudal fin weakly developed, caudal-fin rays free nearly to base.

Larvae unknown but probably included in *Gigantactis* Larval Group B (see account, below).

Description

Illicium without filaments, length variable and unrelated to standard length, nearly cylindrical throughout length; escal bulb club shaped, proximal part pigmented and spinulose, without filaments; tapering distal prolongation darkly pigmented (except area under swollen bases of filaments), distal filaments not arranged in a distinct pattern, largest and most numerous at tip, each with a series of small swellings throughout length; 18 distal filaments in holotype (eight longest present at tip), 10 in metamorphosed paratype (five longest at tip); tissue between oval swellings very thin and flexible.

Number of premaxillary teeth 7 to 14 (longest 0.8 to 1.4% SL); number of dentary teeth 17 or 18.

Dorsal-fin rays 5 or 6, anal-fin rays 5 or 6, pectoral-fin rays 18 to 20; skin coverage of proximal part of caudal fin less developed than in most other species, caudal-fin rays free nearly to base, skin nearly cylindrical, without connecting membranes; longest caudal-fin rays (second and seventh) 30 to 35% SL, only slightly larger than intermediate rays.

The 19-mm metamorphosal stage (LACM 37520-1) with relatively shorter illicium; esca with conical distal prolongation and 4 distal filaments; some remains of enlarged supraorbital lateral line organs; skin and subdermal pigment absent (possibly bleached away).

## Distribution

*Gigantactis savagei* is known only from the North Pacific Ocean extending from the Hawaiian Islands to the coast of southern California and Baja California, with a single tentatively identified specimen from the western North Pacific at about 42°N, 167°E (Fig. 233). It appears to be a relatively shallow-living species: the holotype was captured in open gear fished above 700 m; the remaining specimens were collected by open gear fished at maximum depths of approximately 1000 to 1250 m.

## Comments

*Gigantactis savagei* is a member of the *G. macronema* group. It differs from the other four members of this group (*G. macronema, G. microdontis, G. ios*, and *G. longicauda*), as well as from all other members of the genus, in escal pigmentation, in the size and shape of the distal prolongation of the escal bulb, and in the structure and position of the distal escal filaments. *Gigantactis savagei* further differs from *G. macronema* in having a significantly shorter illicium.

The three juvenile specimens (33 to 44 mm), tentatively referred to this species, compare well with members of the *G. macronema* group in illicial length, in lacking proximal escal filaments, and in having relatively short dentary teeth in two longitudinal series. In other escal characters, however, they are not in complete agreement with the material here placed within the *G. macronema* group. The largest (44 mm) has 9 short distal escal filaments, each with a series of swellings as is characteristic of *G. savagei*, but the bulb is distally unpigmented and without a distal prolongation. In each of the two smaller specimens, the bulb has a somewhat pigmented short conical distal prolongation with short rudiments of distal filaments (approximately 7 in the 38-mm specimen, indistinct rudiments in the 33-mm specimen), and 2 additional filaments at the base. From a comparison with specimens of *G. macronema* and *G. microdontis* of a similar standard length, it seems less likely that they represent one of these species than developmental stages of *G. savagei*.

## GIGANTACTIS MICRODONTIS BERTELSEN, PIETSCH, AND LAVENBERG, 1981

Figures 38, 234; Table 8

### Females

*Gigantactis* sp. 2: Parin et al., 1973:146 (19.5-mm metamorphosal stage).
*Gigantactis microdontis* Bertelsen et al., 1981:54, figs. 56–58, 67 (original description, seven specimens). Love et al., 2005:62 (eastern North Pacific).

### Males and Larvae

Unknown.

### Material

Twelve females, 11 metamorphosed (25.5 to 310 mm) and one in late metamorphosis (19.5 mm):

Holotype of *Gigantactis microdontis*: MCZ 52574, 66 mm, *Anton Bruun*, cruise 18B, station 739B, off Peru, 15°12′S, 75°44′W, 0 to 700 m, bottom depth 1060 m, 25 August 1966.

Paratypes of *Gigantactis microdontis*: LACM 32776-2, 25.5 mm, central North Pacific, leeward Oahu, 21°20 to 30′N, 158°20 to 30′W, 0 to 925 m, 6 July 1970; LACM 32791-3, 38.5 mm, central North Pacific, leeward Oahu, 21°20 to 30′N 158°20 to 30′W, 0 to 1175 m, 22 September 1970; LACM 30284-29, 44 mm, *Velero IV*, eastern North Pacific, Guadalupe Island, 28°48′N, 118°10′W, 0 to 650 m, 24 August 1968 (esca poorly preserved); LACM 9693-34, 118 mm, *Velero IV* station 11628, eastern North Pacific, San Clemente Basin, 31°45′N, 118°45′W, 0 to 600 m, 18 August 1967 (esca previously drawn, but specimen now lost); LACM 32204-2, 127 mm, *Velero IV* station 13083, eastern North Pacific, Guadalupe Island, 28°20′N, 118°18′W, 0 to 650 m, 18 June 1969 (esca poorly preserved); SIOM uncataloged, 19.5 mm, 12°30′S, 87°45′W, 0 to 100 m.

Additional females: FMNH, one (39.5 mm); OS, one (310 mm); SIO, two (93 to 299 mm); UW, one (125 mm).

### Distinguishing Characters

Metamorphosed females of *Gigantactis microdontis* differ from those of all other species of the genus in having exceptionally short dentary teeth (longest 1.1 to 1.6% SL). They further differ in having the following combination of character states: length of illicium 216 to 240% SL (in specimens 38.5 mm and larger); escal bulb with a short distal prolongation pigmented on distal surface; 8 to 10 short distal filaments; escal papillae absent; proximal escal filaments absent; dentary teeth arranged posteriorly in two longitudinal series; length of caudal-fin rays less than 45% SL; skin coverage of proximal part of caudal fin weakly developed, caudal-fin rays free nearly to base.

Larvae unknown but probably included in *Gigantactis* Larval Group B (see account, below).

### Description

Illicium without filaments, nearly cylindrical throughout length; escal bulb club-shaped, pigmented only near base and on distal part of short distal prolongation; distal patch of pigment oblong in shape, tapering posteriorly toward pore of photophore; 10 distal filaments in holotype, 9 in paratype, 8 in 310-mm specimen (OS 13915), arising from surface of bulb just inside edge of distal pigment patch in two nearly parallel, lateral series that meet anteriorly; filaments short, approximately diameter of bulb in holotype, less in paratype, spatulate and compressed distally; most distal filaments lost in 44-mm and 127-mm specimens (but in illicial length, in the shape of the escal bulb, and remains of the distal prolongation, they are in good agreement with the above description), those remaining differing in being compressed nearly from base; esca of 118-mm specimen (Bertelsen et al., 1981:54, fig. 56c; sketched before being lost) without a distinct distal prolongation, but bearing 10 filaments of similar shape; late metamorphosal stage and juvenile (19.5 and 25 mm, respectively) with illicium short (33% and 43% SL, respectively), esca unpigmented with about 8 short rudimentary filaments close set on distal surface of bulb.

Number of premaxillary teeth 8 to 16 (longest 0.8 to 1.0% SL); number of dentary teeth 16 to 51.

Dorsal-fin rays 4 to 6, anal-fin rays 4 to 6, pectoral-fin rays 17 to 19; skin coverage of proximal part of caudal fin less developed than in most other species, rays free nearly to base, skin nearly cylindrical without connecting membranes; longest caudal-fin rays (second and seventh) 30 to 43% SL, only slightly longer than intermediate rays.

The 19.5-mm SIOM specimen is in late metamorphosis with nine to 10 anteriormost supraorbital lateral-line organs enlarged, largest 10% SL; skin very faintly pigmented, no distinct subdermal pigmentation (possibly bleached away).

### Distribution

*Gigantactis microdontis* is known only from the eastern Pacific Ocean, extending from the Hawaiian Islands, to off the coasts of Oregon and California in the north, and off the coast of Peru as far south as about 15°S (Fig. 234). It appears to be a rather shallow-living species; all specimens were collected by gear fished open above 1200 m; southern California and Peruvian specimens by open gear fished above 700 m. The Oregon specimen was collected with a bottom trawl at about 915 m.

### Comments

*Gigantactis microdontis* is a member of the *G. macronema* group, separated from all other species of the genus by the characters listed for this group (see species account for *G. macronema*); it differs from the other members of this group (*G. macronema, G. savagei, G. ios*, and *G. longicauda*) in pigmentation, shape and size of the distal prolongation of the escal bulb, and structure and position of the distal escal filaments. It differs further from *G. macronema* in illicial length, from *G. savagei* in the length of the longest dentary tooth, and from *G. longicauda* in illicial length and caudal-fin ray length.

GIGANTACTIS IOS BERTELSEN, PIETSCH, AND LAVENBERG, 1981

Figure 232; Table 8

### Females

*Gigantactis ios* Bertelsen et al., 1981:56, figs. 59, 57 (original description, single specimen). Bertelsen, 1990:514 (eastern tropical Atlantic). Swinney, 1995a:52, 56 (two additional specimens, off Madeira). Swinney, 1995b:39, 40, fig. 1 (after Swinney, 1995a; description). Jónsson and Pálsson, 1999:203, figs. 7, 11 (Iceland).

### Males and Larvae

Unknown.

### Material

Four metamorphosed female (38 to 81 mm):

Holotype of *Gigantactis ios*: BMNH 1977.9.13.1, 57 mm, *Discovery* station 7856-48, eastern North Atlantic, southwest of Madeira, 29°49′N, 23°00′W, RMT, 1005 to 1250 m, 4 April 1972.

Additional females: NMSZ, two (38 to 51 mm); SIO, one (81 mm).

### Distinguishing Characters

Metamorphosed females of *Gigantactis ios* differ from those of all other species of the genus in having the following combination of character states: length of illicium 247 to 256% SL; escal bulb without distal prolongation; a dense group of 14 to 16 short distal filaments; escal papillae absent; proximal escal filaments absent; dentary teeth short (longest 1.1 to 1.4% SL), arranged posteriorly in two longitudinal series; length of caudal-fin rays less than 37% SL; skin coverage of proximal part of caudal fin weakly developed, caudal-fin rays free nearly to base.

Larvae unknown but probably included in *Gigantactis* Larval Group B (see account, below).

### Description

Illicium without filaments; escal bulb club-shaped; pigment confined to base of bulb except for a small distal spot; distal escal filaments lanceolate with tiny internal transparent bulbs.

Premaxillary teeth 10 to 16 (longest 0.8 to 1.0% SL); dentary teeth 16 to 32.

Dorsal-fin rays 5, anal-fin rays 5, pectoral-fin rays 17 or 18; skin coverage of proximal part of caudal fin weakly developed, rays free nearly to base, skin nearly cylindrical without connecting membranes; longest caudal-fin ray (first or seventh) 31 to 36% SL, only slightly longer than other rays.

### Distribution

*Gigantactis ios* is known from four metamorphosed females all collected in the eastern North Atlantic Ocean off Madeira and south to about 20°N, 20°S (Fig. 232). The holotype was taken with a closing trawl between 1005 and 1250 m. The other three specimens were captured in open trawls fished at maximum depths of 1200 to 1360 m (Swinney, 1995b).

### Comments

*Gigantactis ios* is a member of the *G. macronema* group, separated from all other species of the genus by the characters listed for this group (see species account for *G. macronema*); it differs from the other members of this group (*G. macronema*, *G. savagei*, *G. microdontis*, and *G. longicauda*) in escal pigmentation, shape and size of the distal prolongation of the escal bulb, and number, structure and position of the distal escal filaments.

Swinney's (1995a) description of two additional females from the eastern North Atlantic tentatively identified by him as *G. ios* casts some doubt on the validity of this species and of the similar species *G. microdontis*. As pointed out by Swinney (1995b:42), the escae of his specimens, well preserved only in the larger of the two (51 mm, NMSZ 1994.050.1), appears to be intermediate between those of *G. ios* and *G. microdontis* as described by Bertelsen et al. (1981). The distal escal appendages, 14 in all, are short and lanceolate and contain tiny spherical inclusions just as described for the holotype of *G. ios* (the 8 to 10 distal escal appendages of all seven known specimens of *G. microdontis* lack these little inclusions), but in contrast to those of *G. ios* and very much like those of *G. microdontis*, the distal appendages are neatly arranged in a U-shaped pattern emerging from the periphery of a circular or teardrop-shaped patch of pigment. All of the known material of *G. microdontis* is from the eastern Pacific while *G. ios* is known only from the eastern North Atlantic, but despite this disjunct distribution, there is a good chance that these two species are really one and the same. Additional material, however, is required to resolve this question (Swinney, 1995b:43).

GIGANTACTIS LONGICAUDA BERTELSEN AND PIETSCH, 2004

Table 8

### Females

*Gigantactis longicauda* Bertelsen and Pietsch, 2004:958, figs. 1–3 (original description, single specimen).

### Males and Larvae

Unknown.

### Material

A single metamorphosed female (114 mm):

Holotype of *Gigantactis longicauda*: ISH 5539/79, 114 mm, *Anton Dohrn*, Sargasso Sea Expedition, cruise 210/92, station 98/79, western North Atlantic, 23°46′N, 58°59′W, MT-1600, 500 to 1200 (–0) m, bottom depth 5600 m, 28 March 1979.

### Distinguishing Characters

Metamorphosed females of *Gigantactis longicauda* differ from those of all other species of the genus in having the following combination of character states: length of illicium 330% SL; escal bulb nearly spherical, with a short, truncated distal prolongation, pigmented distally, and bearing two pairs of short, spatulate appendages; stem of illicium without filaments; dentary teeth short (longest tooth 1.7% SL), arranged posteriorly in two longitudinal series; longest caudal rays 50% SL.

### Description

Stem of illicium without filaments, its proximal part not distinctly compressed. Length of distal escal prolongation less than half diameter of escal bulb; two pairs of distal appendages well separated from each other, emerging from border of a nearly circular distal patch of pigment.

Number of teeth on left premaxilla 10, on left dentary 22.

Dorsal-fin rays 5; anal-fin rays 5; pectoral-fin rays 16; caudal-fin rays cylindrical and free nearly to base; first, second, seventh, and eighth caudal rays nearly equal in length (about 50% SL), third and sixth rays 44% SL, fourth and fifth rays about 35% SL.

### Distribution

*Gigantactis longicauda* is known only from the holotype collected in the western North Atlantic in an open midwater trawl, with a primary fishing depth of 500 to 1200 m.

### Comments

According to the dentition, length of the illicium, and shape of the caudal fin, *Gigantactis longicauda* is a typical member of the *G. macronema* group as diagnosed by Bertelsen et al. (1981, table 1). Among the four species already assigned to this group (*G. macronema*, *G. savagei*, *G. ios*, and *G. microdontis*), *G. longicauda* is most similar in escal morphology to *G. microdontis*, known from seven specimens, all collected in the eastern Pacific Ocean.

It differs most strikingly from *G. microdontis* in the length of the illicium (330% SL versus 216 to 240% SL), the number of escal appendages (4 versus 8 to 10), and the length of the longest caudal-fin rays (50% versus 31 to 42% SL).

As a cautionary note, the holotype of *G. longicauda* as well as the known specimens of *G. savagei*, *G. microdontis*, and *G. ios*, are small juveniles (none greater than 150 mm) that may not have yet developed the full array of illicial and escal appendages. Among species of the *G. macronema* group, only *G. macronema* is represented by large adult individuals, yet illicial filaments are absent in all except for the largest known specimen, a 354-mm female, in which the distal one-sixth of the illicium is covered with numerous elongate filaments. Thus, the absence of illicial and escal filaments and appendages in the other species of the group might be size related and therefore not diagnostic.

### GIGANTACTIS OVIFER REGAN AND TREWAVAS, 1932, NOMEN DUBIUM

#### Females

*Gigantactis ovifer* Regan and Trewavas, 1932:93, 95, fig. 152 (original description, single specimen). Fraser-Brunner, 1935:326 (comparison with *Gigantactis filibulbosus*). Waterman, 1939b:85 (comparison with *G. longicirra*). Bertelsen, 1951:150 (comparison with all known material, comments). Grey, 1956a:268 (synonymy, vertical distribution). Pietsch, 1972a:42, 45 (holotype with 5 pectoral radials). Nielsen, 1974:98 (listed in type catalog). Bertelsen et al., 1981:57 *(nomen dubium)*.

#### Males and Larvae

Unknown.

#### Material

A single metamorphosed female (30 mm):
Holotype of *Gigantactis ovifer*: ZMUC P92131, 30 mm, *Dana* station 3731(12), South China Sea, 14°37′N, 119°52′E, 2500 m wire, 0200 h, 17 June 1929.

#### Comments

The esca of the holotype of *Gigantactis ovifer* is damaged. The "two very short terminal appendages" described by Regan and Trewavas (1932:95, fig. 152) represent fragments of torn tissue. The relatively long dentary teeth of this specimen (longest 3.4% SL), and the presence of a few teeth that appear to represent a third longitudinal series, only excludes it from the *G. macronema* group. Because it remains doubtful that its illicial length (90% SL) represents the relative length as an adult, and because in other morphometric and meristic characters it is similar to several species of *Gigantactis* recognized here, *G. ovifer* is regarded as a *nomen dubium*.

### GIGANTACTIS FILIBULBOSUS FRASER-BRUNNER, 1935, NOMEN DUBIUM

#### Females

*Gigantactis filibulbosus* Fraser-Brunner, 1935:326, fig. 5 (original description, single specimen). Waterman, 1939b:85 (comparison with *Gigantactis longicirra*). Robins and Courtenay, 1958:151 (depth distribution). Wheeler, 1969:585 (reference to original description). Bertelsen et al., 1981:57 *(nomen dubium)*.
*Gigantactis filibolosus*: Bertelsen, 1951:150 (misspelling of specific name, comparison with all known material, comments).
*Gigantactis filibulosus*: Bertelsen, 1951:151, 152, table 31 (misspelling of specific name, comparison with all known material, comments). Maul, 1973:675 (listed).

#### Males and Larvae

Unknown.

#### Material

A single metamorphosed female (25 mm):
Holotype of *Gigantactis filibulbosus*: BMNH 1934.8.8.92, 25 mm, *Dynevor Castle*, Irish Atlantic slope, 53°15′N, 12°28′W, 0 to 320 m, July 1934.

#### Comments

The holotype and only specimen ever referred to this species is a juvenile with an undeveloped illicium (length 84% SL) that gives no indication of what its length might have been when fully developed. The escal bulb is shrunken, with no remains of "a slender filament, expanded at tip, on each side of its distal end," as described and figured by Fraser-Brunner (1935, fig. 5). Among the species recognized here, an esca somewhat similar to this description is found only in *Gigantactis microdontis*; however, this species has about 10 escal filaments. Even if we assume that the pair of filaments present in *G. filibulbosus* represents remains of a larger number, the length of the dentary teeth of the holotype (which according to the description and figure reach a length of at least 3% SL) fall well outside the range recorded for *G. microdontis*. Further, since the juvenile holotype has lost the distal portion of the caudal fin and is similar to several *Gigantactis* species in proportions and meristic characters, *G. filibulbosus* is regarded a *nomen dubium*.

### Gigantactis Males

Characters used to separate *Gigantactis* larvae into groups (median fin-ray counts and the concentration of subdermal, dorsal, and peritoneal pigment; Bertelsen, 1951:148) are of limited use in separating the males (see Bertelsen et al., 1981). The two specimens here referred to *Gigantactis longicirra* on the basis of their high dorsal-fin counts agree with larval "Type C" of Bertelsen (1951:150, fig. 99E, F) in having weakly developed pigmentation. In the remaining material, the density of subdermal melanophores is relatively great but varies without significant correlation to other characters. This variation is due to ontogenetic changes, but it is also probably caused by different preservation and storage times allowing for an unequal loss of pigment.

The size of the olfactory organs, used to distinguish other ceratioid males (Bertelsen, 1951), also proved to be of little importance in distinguishing *Gigantactis* males. The thin fragile skin covering the olfactory lamellae is torn or deformed in most of the specimens so that the only reliable measurement of size is the length of the series of olfactory lamellae (this length varied from 0.8 to 1.8 mm, without any significant relationship to other characters); but even in the best preserved males, the size of the nostrils of left and right sides differs significantly. There is the impression, however, that the anterior nostrils are slightly smaller and more narrow than the posterior nostrils and have a depth of approximately one-half the length of the series of olfactory lamellae.

The final results of a thorough investigation (Bertelsen et al., 1981) provided five characters that can be used to separate *Gigantactis* males into groups: (1) eye diameter, (2) dermal pigmentation, (3) subdermal pigmentation (used to separate a single specimen), (4) presence or absence of skin spines, and (5) fin-ray counts (the only character found in males that is also useful in distinguishing species based on females). In evaluating their significance, changes in these characters during metamorphosis were considered in view of the fact that (1) the skin of larvae is unpigmented and naked, (2) the eyes of larvae are normally developed and relatively large, and (3) the olfactory organs of larvae are small (with few olfactory lamellae) and occupy a lateral position. To judge the relative stage of development, the testes of each specimen were examined and the maximum diameter of the largest of the pair was measured (since each testis is more or less pear shaped, tapering into a stalk of variable length, no useful length measurement could be obtained). On this basis, the material of *Gigantactis* was separated into six groups: *G. longicirra* (two specimens) and five additional groups that are for convenience called Group I through Group V. Group I includes the material referred to as the "naked type" by Bertelsen (1951:152, fig. 102C). Group II includes *G. microphthalmus* (Regan and Trewavas, 1932:92, fig.

149; Bertelsen, 1951:152). The remaining three groups include material described by Bertelsen et al. (1981).

### GIGANTACTIS MALE GROUP I

Figure 160

*Laevoceratias liparis* Parr, 1927:33, fig. 13 (original description, single specimen). Regan and Trewavas, 1932:93 (after Parr, 1927). Bertelsen, 1951:70, fig. 29 (tentatively referred to Diceratiidae, reallocated to Gigantactinidae by Uwate, 1979; figure after Parr, 1927).

*Gigantactis* Male Group I: Bertelsen et al., 1981:59, figs. 61A, 68 (description, comparison with all known material).

#### Material

Ten metamorphosed males (15 to 22 mm):

Holotype of *Laevoceratias liparis*: YPM 2013, 17 mm, *Pawnee*, Third Oceanographic Expedition, station 33, 24°11′N, 75°37′W, 2440 m wire, 22 March 1927.

Additional material: BMNH, one (22 mm); GNM, one (19.5 mm); LACM, three (16.5 to 19 mm, 17.5 mm specimen stained); SIO, two (17 to 21.5 mm); USNM, one (15 mm); ZMUC, one (19 mm).

#### Distinguishing Characters

Group I males differ from other *Gigantactis* males in having the following combination of character states: eyes relatively large, diameter 0.6 to 0.9 mm (average 0.79 mm); olfactory lamellae 11 or 12; upper denticular teeth 3 (rarely 2); lower denticular teeth 4 (rarely 3); skin naked, pigmented; dorsal-fin rays 5 or 6; anal-fin rays 5 to 7; pectoral-fin rays 18 to 22 (Bertelsen et al., 1981, table 19).

#### Comments

The material of Group I differs from other known specimens in length distribution: eight of the 10 known specimens are larger than the males of all other groups, ranging from 17 to 22 mm. In two of the smallest known specimens of Group I (15 mm, USNM 218615; 17 mm, SIO 68-490), the anterior nostrils are somewhat separated, but otherwise their metamorphosis seems to be complete. The greatest diameter of the testes of the largest known specimen (22 mm, BMNH 2004.9.12.168) is 2.7 mm; in the remaining material, this measurement varies disproportionately with standard length from 0.8 to 2.2 mm.

### GIGANTACTIS MALE GROUP II

Figures 160, 286

*Teleotrema microphthalmus* Regan and Trewavas, 1932:93, fig. 149 (original description, single specimen). Nielsen, 1974:98 (listed in type catalog).

*Gigantactis microphthalmus*: Bertelsen, 1951:146, 152, 153, figs. 102C, 103E (new combination; comparison with all known material).

*Gigantactis* Male Group II: Pietsch and Seigel, 1980:396 (Philippine Archipelago). Bertelsen et al., 1981:60, figs. 61B, 68 (description, comparison with all known material). Pietsch, 2005b, fig. 224, 15C (reproduction). Pietsch and Orr, 2007:6, fig. 4M (phylogenetic relationships).

#### Material

Twenty-two metamorphosed males (10.5 to 15.5 mm):

Holotype of *Teleotrema microphthalmus*: ZMUC P92127, 16 mm, *Dana* station 4003(2), eastern tropical Atlantic, west of Sierra Leone, 8°26′N, 15°11′W, 5000 m wire, 1130 h, 9 March 1930.

Additional material: BMNH, three (13 to 15.5 mm); LACM, eight (12 to 14 mm, 13.5-mm specimen cleared and stained); MCZ, two (10.5 to 12 mm); SIO, five (12.5 to 15.5 mm); SIOM, one (13 mm); ZMUC, two (11.5 to 14.5 mm, 14.5-mm specimen cleared and stained).

#### Distinguishing Characters

Group II males differ from other *Gigantactis* males in having the following combination of characters: eyes relatively small, diameter 0.45 to 0.60 (average 0.54 mm); olfactory lamellae 11 or 12 (two specimens with 10 and 13 lamellae, respectively); upper denticular teeth 3 (rarely 4); lower denticular teeth 4 (rarely 3 or 5); skin densely covered with spines, darkly pigmented; dorsal-fin rays 5 or 6; anal-fin rays 5 to 7; pectoral-fin rays 16 to 18 (Bertelsen et al., 1981, table 19).

#### Comments

In the material of Group II, the diameter of the testes ranges from 0.4 to 1.9 mm, with some tendency for larger males to have larger testes (diameter less than 1.0 mm in all specimens less than 14 mm, greater than 1.5 mm in all material more than 14 mm). All known specimens are postmetamorphosal stages.

### GIGANTACTIS MALE GROUP III

*Gigantactis* Male Group III: Bertelsen et al., 1981:60, fig. 68 (description, comparison with all known material).

#### Material

Three metamorphosed males: LACM, two (14 to 15 mm); SIO, one (16 mm).

#### Distinguishing Characters

Group III males differ from other *Gigantactis* males in having the following combination of character states: eyes relatively large, diameter 0.7 and 0.8 mm; olfactory lamellae 10 or 11; denticular teeth present in only the 15-mm male, 3 upper and 4 lower; skin spinulose, unpigmented; dorsal-fin rays 5 or 6; anal-fin rays 5 or 6; pectoral-fin rays 16 to 18 (Bertelsen et al., 1981, table 19).

#### Comments

The diameter of the testes of the Group III males ranges from 1.3 to 1.5 mm. Both specimens are postmetamorphosal stages.

### GIGANTACTIS MALE GROUP IV

*Gigantactis* Male Group IV: Pietsch and Seigel, 1980:396 (Philippine Archipelago). Bertelsen et al., 1981:61, fig. 68 (description, comparison with all known material).

#### Material

A single male in metamorphosis: LACM, one (16.5 mm).

#### Distinguishing Characters

The Group IV male differs from other *Gigantactis* males in having relatively low pectoral-fin ray counts (15, 15). This specimen differs further in having the following combination of character states: eyes small, diameter 0.5 mm; olfactory organs well separated with 8 to 9 lamellae; upper denticular teeth 3; lower denticular teeth 4; skin naked, darkly pigmented; dorsal-fin rays 6; anal-fin rays 5 (Bertelsen et al., 1981, table 19).

#### Comments

The diameter of the testes of the Group IV male is 0.4 mm. The development of the olfactory organs and testes indicate that the specimen is a late metamorphosal stage.

### GIGANTACTIS MALE GROUP V

*Gigantactis* Male Group V: Bertelsen et al., 1981:61, figs. 62, 68 (description, comparison with all known material).

#### Material

A single male in metamorphosis: LACM, one (14.5 mm).

#### Distinguishing Characters

The Group V male differs from other *Gigantactis* males in having a distinct V-shaped patch of subdermal pigment on the throat and in having relatively low dorsal- and anal-fin ray counts (only 4 in both fins). This specimen differs further in having the following combination of character states: eyes large, diameter 1.0 mm; olfactory organs in development, not contiguous, with approximately 8 lamellae; denticulars in development with 2 upper teeth and 4 lower teeth; skin naked, unpigmented; pectoral-fin rays 21 (Bertelsen et al., 1981, table 19).

Comments

Despite the relatively large testes of the Group V male (approximately 2 mm long and 1.2 mm wide), the development of the premaxillae (unresorbed), eyes, olfactory organs, and denticulars indicates that the specimen is an early metamorphosal stage.

DISCUSSION

Except for *Gigantactis longicirra*, it is not possible to satisfactorily refer *Gigantactis* males to species based on females. However, there seems little doubt that the best represented Group II contains the males of the most common species, *G. vanhoeffeni*, having the same fin-ray counts. The males of other members of the *G. vanhoeffeni* group are probably included in Group II as well. The high number of pectoral-fin rays characteristic of Group I indicates that these males correspond to members of the *G. macronema* and *G. gargantua* groups. The fin-ray counts of Group III males are shared with a number of species based on females, thus providing no indication of their identity. The exceptionally low number of pectoral rays in Group IV males (15) corresponds only to *G. golovani*, but since both known females of this species are from the Atlantic and the single Group IV male is from the western Pacific, this evidence seems too slight to conclude that these forms represent the same species. Similarly, the low number of dorsal- and anal-fin rays (4) in the Group V male corresponds only to *G. elsmani*, *G. savagei*, and *G. microdontis*. Females of *G. savagei* and *G. microdontis* and the Group V male are known only from the eastern Pacific, yet none of the known females have more than 19 pectoral-fin rays, while the Group V male has 21. The presence of a well-developed supraethmoid, and the fact that the subdermal pigment of the Group V male differs significantly from that of all known gigantactinid larvae, males, and juvenile females, possibly indicates that it represents an undescribed genus.

The most conspicuous difference between the groups of *Gigantactis* males is the presence or absence of skin spines. Since this difference is correlated with standard length (spines present in all the smaller specimens, but absent in all the larger specimens), it might be only an ontogenetic difference. This conclusion is clearly contradicted by the difference between Group I and Group II males in pectoral-fin ray counts. Furthermore, a decrease in actual eye diameter from the metamorphosal stages (see below) to the small spiny specimens, followed by an increase in the larger naked specimens, is very unlikely. Finally, a comparison of the size of gonads shows that some of the naked specimens have smaller and probably less advanced testes than most of the spiny specimens.

*Gigantactis* Larvae

Figure 161

Material

In addition to the 233 larvae (2.5 to 14 mm TL) listed by Bertelsen (1951:148, 274–275), Bertelsen et al. (1981) examined another 66 specimens (4 to 15 mm TL): MCZ, 18; SWFSC, 48.

Comments

Bertelsen (1951) divided *Gigantactis* larvae into three types (here referred to as "groups"): Group A, with dorsal and peritoneal pigment well developed, dorsal-fin rays 5 to 7, anal-fin rays 5 to 7; Group B, with dorsal and peritoneal pigment weak or absent, dorsal-fin rays 4 to 6, and anal-fin rays 4 to 6; and Group C, with pigment as in Group B, but dorsal-fin rays 7 to 10, anal-fin rays 5 to 8. The groups, however, could not be sharply distinguished. Several specimens were found to be intermediate in pigmentation, and the chosen limit between Groups B and C in fin-ray counts was doubtful. In analyzing additional larvae made available since 1951 (Bertelsen et al., 1981), it was impossible to judge whether differences in the strength of pigmentation in larvae obtained from different sources might be due to variation in the degree of bleaching caused by differences in preservation and storing time. Furthermore, much of the recently collected material is small (2.5 to 5 mm TL), representing young stages in which the separation between Groups A and B is especially uncertain. No larvae with more than 7 dorsal-fin rays were found. For these reasons, the attempt to separate *Gigantactis* larvae into distinct groupings failed. Nevertheless, combined with the much greater information on the genus now available, reexamination of the larvae by Bertelsen et al. (1981) revealed some new facts:

GIGANTACTIS LARVAL GROUP A

*Gigantactis* Larvae Type A: Bertelsen, 1951:148, fig. 99A–D, I–J (new grouping, description). Maul, 1973:672 (after Bertelsen, 1951; eastern North Atlantic).

*Gigantactis* Larval Group A: Bertelsen et al., 1981, 62, 63, fig. 68 (after Bertelsen, 1951; comparison with other larvae of *Gigantactis*).

Comments

Among the identified female metamorphosal stages (representing seven of the 16 recognized species of *Gigantactis*), only those of *Gigantactis vanhoeffeni* and *G. meadi* have a well-developed dorsal group of subdermal pigment. Furthermore, distinct remains of this pigment were found under the darkly pigmented skin of the juvenile specimens of *G. gibbsi* (38 mm, USNM 218613) and *G. gracilicauda* (21 mm, ZMUC P921535), the two remaining species of the *G. vanhoeffeni* group. This indicates that Group A contains the larvae of species belonging to the *G. vanhoeffeni* group and further infers that this type of pigmentation is unique to the group. If this is correct, five of the eight unidentified female metamorphosal stages listed by Bertelsen et al. (1981:63; LACM 32768-3, SIO 61-48, SIO 73-158, SAM 27810, and ZMUC P921655) also belong to the *G. vanhoeffeni* group. Because approximately one-half of the identified *Gigantactis* belong to this species group, this possibility is not too unlikely.

GIGANTACTIS LARVAL GROUP B

*Gigantactis* Larvae Type B: Bertelsen, 1951:148, 150, fig. 99G, H (new grouping, description).

*Gigantactis* Larval Group B: Bertelsen et al., 1981, 62, 63, fig. 68 (after Bertelsen, 1951; comparison with other larvae of *Gigantactis*).

Comments

In the identified female metamorphosal stages representing *Gigantactis perlatus*, *G. elsmani*, *G. gargantua*, *G. savagei*, and *G. microdontis*, the dorsal subdermal pigment is either absent or (as in *G. gargantua*) very faint. Because none of the specimens have darkly pigmented skin, the possibility that this lack of subdermal pigment is due to bleaching cannot be excluded. However, it seems probable that the larvae of these species are included among those of Group B (see also *Gigantactis* Male Group I).

GIGANTACTIS LARVAL GROUP C

*Gigantactis* Larvae Type C: Bertelsen, 1951:148, 150, fig. 99E, F (new grouping, description).

*Gigantactis* Larval Group C: Bertelsen et al., 1981, 62, 63, fig. 68 (after Bertelsen, 1951; comparison with other larvae of *Gigantactis*).

Comments

Because *longicirra* is unique among the recognized species of the genus in having more than 7 dorsal-fin rays, the eight larvae with such high dorsal-ray counts referred to Group C by Bertelsen (1951) no doubt represent this species. On the other hand, the two larvae with 7 dorsal rays included in this group by Bertelsen (1951) are removed to the unidentified larvae of Group B. The absence of dorsal subdermal pigment in *G. longicirra* is confirmed in the identified female metamorphosal stage and in the two metamorphosed males.

### GIGANTACTIS LARVAL GROUP D

*Gigantactis* Larval Group D: Bertelsen et al., 1981, 62, 63, fig. 68 (after Bertelsen, 1951; comparison with other larvae of *Gigantactis*).

### Comments

Two small females (both 9 mm; MCZ 54041 and ZMUC P921605) differ slightly in subdermal pigmentation from other specimens examined. According to the presence of numerous tiny melanophores in the skin and a developing external illicium, they represent early metamorphosal stages. Besides having a well-developed dorsal group of subdermal melanophores, they both have a short dorsal and ventral series of 3 to 4 large melanophores on the caudal peduncle (see also Bertelsen, 1951, fig. 59c). All *Gigantactis* larvae examined lack pigment on the caudal peduncle, whereas metamorphosal stages and juveniles of Group A may have diffuse pigment extending posteriorly to the caudal peduncle. In no other specimens examined is pigment concentrated in large well-separated melanophores. For this reason, it seems probable that the Group D specimens represent one of the species for which no metamorphosal stages have yet been identified.

### Genus *Rhynchactis* Regan, 1925c (Toothless Seadevils)

Figures 55, 162–166, 235, 286; Tables 1, 8

### Females, Males, and Larvae

*Rhynchactis* Regan, 1925c:565 (type species *Rhynchactis leptonema* Regan, 1925c, by monotypy).

### Distinguishing Characters

The genus *Rhynchactis* is distinguished from *Gigantactis* by having pelvic bones and by having 3 or 4 (rarely 5) dorsal-fin rays and 3 or 4 anal-fin rays. In addition, metamorphosed females differ in having the following character states: The frontals and parietals are absent. Each premaxilla is represented by an anterior remnant bearing 0 to 2 teeth (0 to 4 premaxillary teeth in total). The maxillae are absent (but present in larvae). The dentaries are toothless or bear only minute teeth. There are 2 hypohyals. There are 9 (2 simple + 4 branched + 3 simple) caudal-fin rays. The skin is covered with minute spinules in larger specimens, but juveniles are naked. The snout is truncated, bearing the illicium slightly behind its tip. A bulbous, terminal, escal light organ is absent.

Metamorphosed males are distinguished from those of *Gigantactis* by having the following character states: The diameter of the eye is 2.4% SL in the largest known specimens (17 to 18.5 mm). There are 13 to 15 olfactory lamellae. The depth of the nostrils is 10 to 12% SL. There are 4 to 6 upper denticular teeth and 6 or 7 lower denticular teeth. At least some of the denticular teeth are paired, with broad conical bases. There are 17 to 19 pectoral-fin rays. The skin is naked and weakly pigmented (subdermal pigment as in the larvae, see below).

The larvae of *Rhynchactis* differ from those of *Gigantactis* in having the following character states: The dorsal group of subdermal melanophores is extremely dense and contiguous with the peritoneal pigment, but never extending back to the base of the dorsal fin. The length of the pectoral fin is approximately 50 to 55% SL.

### Description

The illicium of metamorphosed females is longer than the body and of approximately constant proportion to standard length during growth. Lower-jaw teeth are absent. The dentary is greatly reduced in all metamorphosed specimens. Usually 1 or 2, small curved teeth are present on the anterior part of each reduced premaxilla of smaller specimens (27 to 60 mm), but premaxillary teeth are absent in larger specimens.

A dense pavement of white papillae-like glands cover the inner surface of the upper and lower jaw, each gland with a more or less distinct central groove and outlined by pigmented skin (Bertelsen et al., 1981, fig. 9). Histologically each gland consist of a simple short tube with a more or less pigmented wall and covered internally with large glandular cells that nearly fill the lumen of the tube (Bertelsen et al., 1981, fig. 10).

There are 17 or 18 pectoral-fin rays (but as many as 20 have been recorded in larvae; see Bertelsen et al., 1981:67, table 21). The caudal fin is distinctly divided into upper and lower portions containing 4 and 5 caudal-fin rays, respectively. The longest caudal-fin rays (the second and seventh) measure 34 to 45% SL in six specimens in which they appear complete. The skin of larger females (110 to 130 mm) is densely covered with minute spinules, the largest spinules with a length and basal diameter of nearly 0.1 mm. Spinules are apparently absent in all the smaller specimens.

For description of males and larvae, see the generic diagnosis, above.

### Diversity

Three species:

### Key to Females of the Species of *Rhynchactis*

1A. Length of illicium 143 to 158% SL, without secondary filaments . . . . . .
. . . . . . . . . . . *Rhynchactis leptonema*
Regan, 1925c, p. 475

NOTE Five known specimens, 42 to 118 mm; western Atlantic, and western and central Pacific

1B. Length of illicium 109 to 210% SL, with 11 to 20 secondary filaments . .
. . . . . . . . . . . . . . . . . . . . . . . 2

2A. Illicial filaments long, the five longest 6 to 15% SL, some with bifid tip; length of illicium 109 to 144% SL, length proximal to emergence of secondary filaments 43 to 72% of total illicial length . . . . . . . . . . . . . . .
. . . . . . . . . . . *Rhynchactis macrothrix*
Bertelsen and Pietsch, 1998b, p. 476

NOTE Ten known specimens, 27 to 152 mm; Atlantic, western Pacific, and western Indian oceans

2B. Illicial filaments short, the five longest 2 to 4% SL, all simple; length of illicium 210% SL, length proximal to emergence of secondary filaments 86% of total illicial length . . . . . . .
. . . . . . . . . . . *Rhynchactis microthrix*
Bertelsen and Pietsch, 1998b, p. 476

NOTE A single known specimen, 113 mm; western Indian Ocean

### RHYNCHACTIS LEPTONEMA REGAN, 1925C

Figure 235; Table 8

### Females

*Rhynchactis leptonema* Regan, 1925c: 565 (original description, single specimen). Regan, 1926:38, pl. 10, fig. 1 (description after Regan, 1925c). Regan and Trewavas, 1932:95 (after Regan, 1925c, 1926). Waterman, 1939b:84 (comparison with *Gigantactis longicirra*). Bertelsen, 1951:153, fig. 104 (diagnostic characters, description of a juvenile male and 23 larvae, *Dana* material listed). Grey, 1956a:269 (synonymy, vertical distribution). Pietsch, 1972a:42, 45 (holotype with 5 pectoral radials). Nielsen, 1974:98 (listed in type catalog). Becker et al., 1975:327 (additional specimen). Parin et al., 1977:107 (additional larval stage). Bertelsen et al., 1981:66, figs. 2, 3, 7, 9, 10, 11B, C, E, 13B, 14B, 15B, 63, 69 (osteology, generic revision, comparison with all known material). Bertelsen, 1984:329, fig. 168B (metamorphic male, early life history, phylogeny). Bertelsen, 1990:514 (eastern tropical Atlantic). Bertelsen and Pietsch, 1998b:587, figs. 1, 7. Pietsch, 1999:2036 (western central Pacific). Pietsch, 2002b: 1068 (western central Atlantic).

### Males and Larvae

Unknown.

### Material

Five metamorphosed females (42 to 118 mm):

Holotype of *Rhynchactis leptonema*: ZMUC P92133, 42 mm, *Dana* station 1171(11), tropical western Atlantic, 8°19′N, 44°35′W, 3000 m of wire, 0800 h, 13 November 1921.

Additional material: ASIZP, three (88 to 141 mm); HUMZ, one (118 mm).

### Distinguishing Characters

Length of illicium 143 to 158% SL (complete in the 118-mm specimen, apparently complete in the 42-mm holotype); secondary illicial filaments absent.

### Description

Longest caudal-fin rays 36 to 45% SL; dorsal-fin rays 3 or 4; anal-fin rays 4; pectoral-fin rays 18; each premaxilla with a single small tooth in 42-mm holotype, premaxillae toothless in 118-mm specimen; skin spines apparently absent in holotype, distinct in the 118-mm specimen.

Additional description as given for the genus and family.

### Distribution

The known specimens of *Rhynchactis leptonema* are from widely separated localities, the holotype from the western tropical Atlantic at approximately 8°N, 44°W, another from off Hawaii at about 20°N, 168°W, and the remaining specimens from off Taiwan (Fig. 235). All were collected with nonclosing gear, the holotype fished from a maximum depth of about 1500 m, the second specimen from a maximum depth of 400 m, and the Taiwanese material with a shrimp trawl fished just over the bottom at 300 to 600 m.

### Comments

The observation by Bertelsen et al. (1981) that a juvenile female *Rhynchactis* had secondary illicial filaments in contrast to all other known females was the first indication that the genus might contain more than one species. Since that time, the discovery of additional material (described by Bertelsen and Pietsch, 1998b) has shown considerably more variation in illicial morphology such that, among the four known specimens with secondary illicial filaments, three are remarkably similar in a number of characters in which they differ most from the fourth. Overlapping in length distribution as well, there seems no reason to doubt that these two groups represent distinct species (Bertelsen and Pietsch, 1998b).

Although it might be argued that secondary illicial filaments develop with growth and that the 42-mm holotype *Rhynchactis leptonema* might thus be conspecific with *R. macrothrix* or *R. microthrix* (see below), the recent discovery of a large, extremely well-preserved specimen of *R. leptonema* (HUMZ 129796) lacking any traces of secondary illicial filaments makes it evident that all three species must be recognized.

### RHYNCHACTIS MACROTHRIX BERTELSEN AND PIETSCH, 1998B

Figures 163, 164, 235; Table 8

#### Females

*Rhynchactis macrothrix* Bertelsen and Pietsch, 1998b:587, figs. 2–4, 7 (original description, three specimens). Pietsch, 2002b:1068 (western central Atlantic). Pietsch, 2005b:216, 225 (reproduction).

#### Males and Larvae

Unknown.

#### Material

Ten metamorphosed females (27 to 152 mm):

Holotype of *Rhynchactis macrothrix*: ISH 605/74, 110 mm, *Anton Dohrn*, Gate Expedition, station 39/74, central Atlantic, 7°55′N, 32°41′W, 0 to 2000 m, bottom depth 4840 m, 21 July 1974.

Paratypes of *Rhynchactis macrothrix*: BMNH 2004.7.9.17, 27 mm, *Discovery* station 8281-3, 31°47′N, 63°39′W, 910 to 1000 m, 13 March 1973; SIOM uncataloged, 130 mm, *Vityaz* station 17-2603, 14°57′S, 43°37′E, 0 to 1300 m.

Additional material: ASIZP, six (70 to 152 mm); UW, one (34 mm).

#### Distinguishing Characters

Length of illicium 109 to 144% SL, distal tip bearing 3 or 4 small unpigmented filaments, some with a tiny distal swelling; distal 28 to 57% of illicial length bearing a series of 11 to 20 secondary filaments, the five longest filaments measuring 6 to 15% SL; each secondary filament with 1 or 2 unpigmented terminal filaments, some with a tiny distal swelling. In two largest specimens (110 and 130 mm), secondary filaments darkly pigmented to base of terminal filament(s), divided on each side by a narrow but well-defined longitudinal transparent band.

#### Description

Longest caudal-fin rays 34 to 37% SL; dorsal-fin rays 4; anal-fin rays 3; pectoral-fin rays 17 or 18; in 27-mm specimen, 2 teeth on each rudimentary premaxilla, premaxillary teeth absent in two larger specimens; skin spines distinct in two larger specimens (clearly evident on the few remains of skin of 130-mm specimen), not visible in 27-mm specimen.

Sections of one of the secondary illicial filaments of the 110-mm holotype were prepared for light microscopy. They did not show much more than the sections of the 27-mm specimen described by Bertelsen et al. (1981, fig. 3), except for (1) the presence of a densely pigmented wall on both sides of the transparent band; (2) a greater density of the strongly stained nuclei of the central core, which contains distinct blood vessels and nerves; and (3) a similar concentration of strongly stained nuclei in the tiny swellings of the tips of the terminal filaments of the secondary illicial filaments.

Additional description as given for the genus and family.

#### Distribution

*Rhynchactis macrothrix* is known from widely scattered localities in the Atlantic, western Indian, and western Pacific oceans (Fig. 235). In the Atlantic, it is represented by three specimens: the holotype from central equatorial waters at about 8°N, 32°W; the 27-mm paratype from off Bermuda; and a third individual from the Gulf of Mexico. The 130-mm paratype is from off the west coast of Madagascar in the western Indian Ocean. The six remaining specimens are all from off Taiwan, collected by shrimp trawls fished just over the bottom at 300 to 600 m. The 27-mm paratype was captured with a closing trawl between 910 and 1000 m. The remaining specimens were taken in trawls fished open at maximum depth of approximately 1300 to 2000 m.

### RHYNCHACTIS MICROTHRIX BERTELSEN AND PIETSCH, 1998B

Figure 235; Table 8

#### Females

*Rhynchactis microthrix* Bertelsen and Pietsch, 1998b:588, figs. 5–7 (original description, a single specimen).

#### Males and Larvae

Unknown.

#### Material

A single metamorphosed female (113 mm):

Holotype of *Rhynchactis microthrix*: MCZ 57516, 113 mm, *Anton Bruun*, cruise 6, station 337B, western Indian Ocean, 00°14′S, 65°03′E, 0 to 2250 m, 28 May 1964.

#### Distinguishing Characters

Length of illicium 210% SL, its tip with 4 tiny filaments; an irregular series of 19 filaments on the distal 14% of length of illicium, the five longest 2 to 4% SL, all gradually tapering from base to tip.

#### Description

Longest caudal-fin rays 40% SL; dorsal-fin rays 4; anal-fin rays 4; pectoral-fin rays

18; premaxillary teeth absent; skin spines well developed.

Additional description as given for the genus and family.

Distribution

The holotype and only known specimen of *Rhynchactis microthrix* was captured in the western Indian Ocean, on the equator at about 65°E, between the surface and approximately 2250 m (Fig. 235).

*Rhynchactis* Species

The following lists of material examined include unidentifiable males and larvae, and metamorphosed females that have more or less incomplete illicia and lack secondary illicial filaments:

Metamorphosed Females

Ten specimens, 17 to 126 mm (three previously recorded as *Rhynchactis leptonema* by Bertelsen et al., 1981): FMNH, one (40.5 mm); SIOM, one (32 mm); ISH, two (60 to 126 mm, 60-mm specimen cleared and stained); MCZ, one (61 mm); SIO, two (20 to 30 mm); UW, two (17 to 17.5 mm); VIMS, one (35 mm).

Metamorphosed Males

Seven specimens, 13 to 20 mm: SIOM, one (13 mm); LACM, one (17 mm, cleared and stained); SIO, two (15 to 16.5 mm); UW, two (16 to 20 mm); ZMUC, one (18.5 mm).

Larvae

Twenty-seven specimens, 2.5 to 20 mm: SIOM, one (20 mm); LACM, one (18 mm); ORI, one (10.5 mm); SIO, one (5 mm); ZMUC, 23 (2.5 to 7 mm, 7-mm specimen cleared and stained).

## Family Linophrynidae Regan, 1925c (Leftvent Seadevils)

Figures 1, 4, 13, 20, 25, 31, 35, 40, 45, 47, 50, 51, 53, 54, 179–185, 190, 194, 202, 203, 206, 237–241, 247, 249, 251, 257–259, 297, 301, 306–309; Tables 1, 7–10

Type Genus *Linophryne* Collett, 1886

Aceratiidae Brauer, 1906:323 (family of order Pediculati to contain genus *Aceratias* Brauer, 1902, and three species, all based on free-living males, subsequently reassigned to Linophrynidae by Regan and Trewavas, 1932). Parr, 1927:30 (in part; family of suborder Ceratioidea to contain genera *Lipactis*, *Aceratias*, *Rhynchoceratias*, and *Laevoceratias*, all based on free-living males, those of *Aceratias* subsequently reassigned to Linophrynidae by Regan and Trewavas, 1932). Koefoed, 1944:13 (family of suborder Ceratioidea to contain genera *Aceratias* and *Hyaloceratias*, based on larvae and free-living males, subsequently reassigned to family Linophrynidae by Bertelsen, 1951).

Aoerratiidae Gill, 1909:568 (misspelling of Aceratiidae; family of order Pediculati to contain genus *Aceratias*, following Brauer, 1902, 1906).

Linophrynidae Regan, 1925c:562 (family of suborder Ceratioidea to contain genera *Edriolychus*, *Haplophryne*, *Borophryne*, and *Linophryne*).

Photocorynidae Regan, 1925c:562 (family of suborder Ceratioidea to contain genus *Photocorynus*).

Eurostrinae Parr, 1930a:20 (in part; subfamily of family Aceratiidae to contain genera *Aceratias* and *Rhynchoceratias*, both based on free-living males, those of *Aceratias* subsequently reassigned to Linophrynidae by Regan and Trewavas, 1932).

Cryptorostrinae Parr, 1930a:20 (in part; subfamily of family Aceratiidae to contain genera *Haplophryne* and *Laevoceratias*, both based on free-living males, those of *Haplophryne* reassigned to family Linophrynidae by Regan and Trewavas, 1932, and subsequently to genera *Edriolychnus* [= *Haplophryne*] and *Linophryne* by Bertelsen, 1951).

Distinguishing Characters

The family Linophrynidae differs from all other ceratioid families in having 3 (rarely 4) dorsal-fin rays, 3 (rarely 2 or 4) anal-fin rays, 5 (rarely 4) branchiostegal rays, and a sinistral anus.

Metamorphosed females are further differentiated by having the following combination of character states: The supraethmoid is present. The frontals lack ventromedial extensions and are separated from each other in most genera but meet on the midline in *Photocorynus*. The parietals are present. The sphenotics are overlapped by the anterolateral margin of the pterotic. Sphenotic spines are well developed. The pterosphenoid and mesopterygoid are absent. The metapterygoid is present. The hyomandibula has a single head. There are 2 hypohyals and 5 (1 + 4), rarely 4 (0 + 4), branchiostegal rays. The opercle is bifurcated, its posterior margin moderately concave. The subopercle is long and extremely slender, without a spine or projection on the anterior margin. Quadrate and articular spines are absent. The angular is produced to form a spine in some taxa. The preopercle usually bears one or more spines (preopercle spines are absent in *Acentrophryne*). The jaws are more or less equal anteriorly, the lower extending slightly beyond the upper in same taxa. The lower jaw usually has a symphysial spine (absent in the *Linophryne* subgenus *Stephanophryne*). The postmaxillary process of the premaxilla is absent. The anterior-maxillomandibular ligament is weak to absent. The first pharyngobranchial is usually absent (present but rudimentary in *Photocorynus*). The second and third pharyngobranchials are well developed and toothed. The fourth pharyngobranchial is absent. The first epibranchial is well developed in *Photocorynus* and *Haplophryne*, but somewhat reduced in *Borophryne* and *Linophryne*. The fifth ceratobranchial is usually absent (a small rudiment is present in *Photocorynus*). Ossified hypobranchials are usually absent (a single ossified hypobranchial is present in *Photocorynus*). An ossified basibranchial is present or absent. Epibranchial and ceratobranchial teeth are absent. Epurals are usually absent (a single small element is apparently present in some specimens of *Photocorynus*). The posterior margin of the hypural plate is usually deeply notched (entire in *Photocorynus*). The pterygiophore of the illicium lacks a remnant of the second cephalic spine. The escal bulb has a central lumen and a pore. The esca lacks large toothlike denticles. The posteroventral process of the coracoid is absent. There are 3 pectoral radials. Pelvic bones are absent. There are 13 to 19 pectoral-fin rays and 9 (2 simple + 4 bifurcated + 3 simple ) caudal-fin rays, the ninth very short in *Photocorynus*, about one-half the length of the eighth caudal-fin ray in all other genera (Table 1). The skin is everywhere naked, dermal spinules are absent. The ovaries are paired. Pyloric caecae are absent.

Metamorphosed males of the family Linophrynidae are further differentiated by having the following combination of character states: The eyes are prominent, tubular, and directed anteriorly. The olfactory organs are relatively large, the anterior nostrils well separated, but directed more or less anteriorly. The denticular bones are short and stout, the upper denticular bearing 3 to 7 teeth, the lower denticular with 2 to 13 teeth. There is no connection between the pterygiophore of the illicium and the upper denticular bone. Fin-ray counts are as given for metamorphosed females. The skin is naked, without dermal spinules. The eyes and olfactory organs of parasitic stages are somewhat degenerated. Sexual parasitism is apparently obligatory (see Reproduction and Early Life History, Chapter Eight).

The larvae of the Linophrynidae are further differentiated by having the following combination of character states: The body is slender, more or less elongate, the skin highly inflated. Subdermal pigmentation of the body is either absent or distributed along the lateral margins, never on the dorsal part of the body. The pectoral fins are small, not reaching beyond the dorsal and anal fins. Pelvic fins are absent. Fin-ray counts are as given for metamorphosed females. Sexual dimorphism is well developed, the females bearing a distinct illicial rudiment (females of *Linophryne* also have a rudiment of the hyoid barbel, present as an opaque, wartlike thickening of the skin in specimens greater than about 10 mm). Metamorphosis is delayed in at least some taxa, the larvae attaining lengths of 17.5 to 22 mm. Specimens in metamorphosis measure 15 to 32 mm (Bertelsen, 1951:135, 140, figs. 90, 93; 1984:329, fig. 168C–E).

The osteological features cited here and elsewhere in this account of the Linophrynidae are based on examination of females, males, and larvae of *Photocorynus*, *Haplophryne*, *Borophryne*, and *Linophryne* (Regan, 1926:16, figs. 6, 8; Beebe, 1929b:23–34, figs. 2–5; Parr, 1930a: 14–20; Regan and Trewavas, 1932:43–48, figs. 59–70; Gregory, 1933:402–403, 407–408, fig. 280; Bertelsen, 1951:161–162, fig. 108; Pietsch, 1972a:35, 36, 41, 43–45); specimens of *Acentrophryne* have been unavailable for internal osteological study.

### Description

The body of metamorphosed females is short, more or less oval in shape to globular. The length of the head, measured from the tip of the snout to the base of the pectoral fin, is 50 to 60% SL. The snout is relatively short, usually less than 20% SL. The mouth is large, the opening horizontal or somewhat oblique, the cleft extending past the eye. The oral valve is well developed. The nostrils are set on rounded papillae. The dentition is highly variable among genera: the teeth of *Acentrophryne*, *Borophryne*, and *Linophryne* are long and few, arranged in several oblique longitudinal series; those of *Haplophryne* and *Photocorynus* are considerably shorter and more numerous, arranged in several series (for details, see the generic accounts, below). The vomer is well toothed in *Acentrophryne*, *Borophryne*, and *Linophryne*, but toothless in *Photocorynus* and *Haplophryne*. The first epibranchial is free, not bound to the wall of the pharynx by connective tissue. All four epibranchials are closely bound together by connective tissue. The fourth epibranchial and ceratobranchial are bound to the wall of the pharynx, leaving no opening behind the fourth arch. The proximal one-half of the first ceratobranchial is bound to the wall of the pharynx, but its distal one-half is free, not bound by connective tissue to the adjacent second ceratobranchial. The proximal one-quarter of the second ceratobranchial is bound to the third ceratobranchial and the proximal one-half of third ceratobranchial is bound to the fourth ceratobranchial. Gill filaments are present on the full length of second ceratobranchial, the distal three-quarters of the third ceratobranchial, and the distal one-half of the fourth ceratobranchial. A pseudobranch is absent. The illicium is relatively short, the stem shorter than the diameter of the escal bulb in *Haplophryne*, *Photocorynus*, *Borophryne*, and several species of *Linophryne*, but reaching 35 to 40% SL in other species of *Linophryne* and about 70% SL in *Acentrophryne*. The pterygiophore of the illicium is short, its anterior end hidden beneath the skin of the head or slightly protruding on the snout. A bioluminescent hyoid barbel is present in *Linophryne*, but absent in all other genera. The neuromasts of the acoustico-lateralis system are located at the tips of low cutaneous papillae, the pattern of placement as described for other ceratioids (Pietsch, 1969, 1972b, 1974a, 1974b).

Free-living and parasitic males are known for all genera, except *Acentrophryne*. The eyes of metamorphosed free-living males are relatively large (their diameter 6 to 9% SL), more or less tubular, and directed anteriorly, without an aphakic space. The olfactory organs are inflated and moderately to strongly enlarged. The anterior nostrils are directed anteriorly. There are 3 to 13 olfactory lamellae, but the number is variable among genera. Jaw teeth are present in *Photocorynus* and *Haplophryne*, but absent in *Borophryne* and *Linophryne*. The upper denticular bone is simple or bears a short dorsal prolongation, not reaching the tip of the illicial pterygiophore. The lower denticular bone is simple or (in *Haplophryne*) divided into a bilateral pair. Each denticular bears a single series of 3 to 7 upper and 2 to 13 lower denticular teeth. Sphenotic spines are prominent in the males of *Borophryne* and in *Linophryne* subgenus *Linophryne*, but absent in all other taxa. A hyoid barbel is absent in the males of all genera. The skin is smooth and naked, without dermal spinules. The denticular teeth, eyes, and olfactory organs of parasitic males appear somewhat degenerated; the belly is greatly inflated in mature specimens.

The larvae are more elongate than those of most other ceratioids, the length of the head generally about 45% SL. The skin of the head and body is moderately inflated. The pectoral fins are relatively short. A dorsal group of subdermal melanophores is absent. A small rudiment of a hyoid barbel is present in the largest known female larvae of *Linophryne*. Pointed sphenotic spines are present in *Borophryne* and the *Linophryne* subgenus *Linophryne*, in contrast to the larvae of all other ceratioids.

The color of metamorphosed females is usually dark brown to black over the entire surface of the body, except for the escal appendages, the distal parts of the escal bulb, the barbel of *Linophryne*, and fin rays. The skin is everywhere unpigmented in *Haplophryne* (in contrast to all other metamorphosed ceratioid females). The free-living males of *Linophryne* are dark brown to black, but unpigmented in all other linophrynid genera. The parasitic males of *Borophryne* and *Linophryne* are dark brown to black, but those of *Photocorynus* and *Haplophryne* are unpigmented.

The largest known female of the family is a 275-mm specimen of *Linophryne lucifer* (ZMUC P922443, with a 29-mm parasitic male); several additional specimens of *Linophryne* are larger than 180 mm. The largest recorded females of the other four genera of the family range from 50 to 159 mm: 69 mm in *Photocorynus* (SIO 53-356), 159 mm in *Haplophryne* (NMNZ P.21248), 105 mm in *Acentrophryne* (HUMZ 175257), and 101 mm in *Borophryne* (LACM 30053-10). The largest known free-living male of *Photocorynus spiniceps* is 8.6 mm (ZMUC P921727), 16- to 21-mm maximum known lengths in the other genera. The largest known parasitic males are 7.3 mm in *Photocorynus* (ZMUC P92134), 15 mm in *Haplophryne* (AMS I.21365-8), 22 mm in *Borophryne* (LACM 30053-10), and 30 mm in *Linophryne* (IMB).

### Diversity

The family consists of five genera, three of which contain only a single recognized species, while the fourth, *Acentrophryne*, contains two species, and the fifth, *Linophryne*, includes 22 species divided among three subgenera:

### Keys to Genera of the Linophrynidae

FEMALES

1A. Epiotic and posttemporal spines present; preopercle with 5 or 6 spines ..................... *Photocorynus* Regan, 1925b, p. 479

   NOTE  A single species; Atlantic, Pacific, and Indian oceans

1B. Epiotic and posttemporal spines absent; preopercle with 0 or 1 spine .......................... 2

2A. Preopercular spines with 2 to 5 radiating cusps; skin unpigmented; teeth small and numerous ............ ..................... *Haplophryne* Regan, 1912, p. 481

   NOTE  A single species; Atlantic, Pacific, and Indian oceans

2B. Preopercular spine simple or absent; skin darkly pigmented; teeth large and few ................... 3

3A. Preopercular spine absent ........ ................... *Acentrophryne* Regan, 1926, p. 484

   NOTE  Two species; eastern Pacific Ocean

3B. Preopercular spine present ...... 4

4A. Hyoid barbel absent .... *Borophryne* Regan, 1925c, p. 486

   NOTE  A single species; eastern Pacific Ocean

4B. Hyoid barbel present .... *Linophryne* Collett, 1886, p. 487

   NOTE  Twenty-two species; Atlantic, Pacific, and Indian oceans

MALES

The following key will differentiate free-living metamorphosed males of *Photocorynus*, *Haplophryne*, *Borophryne*, and *Linophryne*. Males of *Acentrophryne* are unknown.

1A. Premaxillae well developed; several teeth in jaws; denticular teeth weak, not meeting in front of mouth; skin unpigmented ................ 2
1B. Premaxillae more or less resorbed; teeth in jaws few or absent; denticular teeth well developed, meeting in front of mouth; pigmentation of skin present or absent ............. 3
2A. Teeth in jaws shorter than denticular teeth; lower denticular bone undivided; subdermal pigment absent .................... *Photocorynus* Regan, 1925b, p. 479

NOTE  A single species; Atlantic, Pacific, and Indian oceans

2B. Longest teeth in jaws longer than denticular teeth; a pair of lower denticular bones; two lateral series of subdermal melanophores on body .................... *Haplophryne* Regan, 1912, p. 481

NOTE  A single species; Atlantic, Pacific, and Indian oceans

3A. Pigmentation of skin faint or absent; subdermal pigment absent; sphenotic spines present .... *Borophryne* Regan, 1925c, p. 486

NOTE  A single species; eastern Pacific Ocean

3B. Skin darkly pigmented, brown or black; subdermal pigment usually well developed, rarely weak or absent; sphenotic spines present or absent ................ *Linophryne* Collett, 1886, p. 487

NOTE  Twenty-two species; Atlantic, Pacific, and Indian oceans

LARVAE

The larvae of *Borophryne*, *Linophryne* subgenus *Linophryne*, and *Linophryne* subgenus *Stephanophryne* can be differentiated with confidence, but those of *Haplophryne* and *Linophryne* subgenus *Rhizophryne* together form an inseparable group referred to as "Hyaloceratias" (see Bertelsen, 1951:189). Larvae of *Acentrophryne* are unknown; those of *Photocorynus* are unknown as well, but their morphology can be predicted by features displayed by recently metamorphosed males.

1A. Sphenotic spines long and pointed ........................... 2
1B. Sphenotic spines very short or absent ........................... 3
2A. Body short, depth more than 60% SL; subdermal pigment absent .... .................... *Borophryne* Regan, 1925c, p. 486

NOTE  A single species; eastern Pacific Ocean

2B. Body more elongate, depth less than 60% SL; a distinct group of subdermal melanophores on caudal peduncle, rarely faint or absent ....... .... *Linophryne* subgenus *Linophryne* Bertelsen, 1982, p. 498

NOTE  Fourteen species; Atlantic, Pacific, and Indian oceans

3A. Subdermal pigment restricted to peritoneum ........... *Photocorynus* Regan, 1925b, p. 479

NOTE  A single species; Atlantic, Pacific, and Indian oceans

3B. Two lateral series of melanophores on body .................... 4
4A. Groups of melanophores on posterior angle of lower jaw and posterior tip of peritoneum .............. *Linophryne* subgenus *Stephanophryne* Bertelsen, 1982, p. 489

NOTE  A single species; Pacific and Indian oceans

4B. Posterior angle of lower jaw and posterior tip of peritoneum without melanophores ................ *Hyaloceratias* (an inseparable mix of larvae belonging to genus *Haplophryne* and *Linophryne* subgenus *Rhizophryne*), pp. 481, 491

NOTE  Atlantic, Pacific, and Indian oceans

## Genus *Photocorynus* Regan, 1925b (Spinyhead Seadevils)

Figures 25, 30, 167–169, 202, 203, 236, 297, 301–303; Tables 1, 8, 10

### Females and Males

*Photocorynus* Regan, 1925b:394 (type species *Photocorynus spiniceps* Regan, 1925b, by monotypy). Regan, 1926:21 (*Photocorynus* assigned to family Photocorynidae). Bertelsen, 1951:161 (*Photocorynus* reassigned to family Linophrynidae).

### Larvae

Unknown.

### Distinguishing Characters

Metamorphosed females of the genus *Photocorynus* are distinguished from those of all other ceratioids by having a spine on the epiotic, a pair of spines on the posttemporal, and 5 or 6 spines on the preopercle. They differ further from those of all other genera of the family in having the following combination of character states: The frontals meet on the midline, each with a well-developed anterodorsal spine. The maxillae are moderately strong, the jaws with numerous short teeth arranged in several series. Vomerine teeth are absent. The first pharyngobranchial is present. The ceratohyal lacks an anterodorsal process. The posterior margin of the hypural plate is entire. The ninth caudal-fin ray is extremely short, the length of the illicium less than 10% SL. The esca is nearly sessile on the snout, without appendages. A hyoid barbel is absent. The second and third pectoral radials are subequal. The skin is darkly pigmented.

Free-living juvenile males of *Photocorynus* are unique in lacking cranial and preopercular spines. The preopercle is not strongly curved. The epiotic region of the skull is highly elevated. There are 8 or 9 teeth on each side of the upper and lower jaws, all shorter than the denticular teeth. The denticular bones are small and placed slightly behind the tip of the jaws. The denticular teeth are slender, 3 on the upper denticular and 3 or 4 on lower denticular. The eyes are slightly tubular, their diameter 8 to 9% SL. The olfactory organs are moderately enlarged and inflated, nearly as large as the eyes. There are 3 olfactory lamellae. The skin is unpigmented.

Adult males become parasitic. Their skin is very faintly pigmented. The olfactory organs and to a lesser extent the eyes are degenerated. Jaw teeth are lost except in the lower jaw of one known specimen.

The larvae are unknown, but extrapolating from the features displayed by recently metamorphosed males, they most probably lack sphenotic spines and subdermal pigmentation (except on the peritoneum) and have relatively small pectoral fins.

### Description

The body of metamorphosed females is relatively short and oval to globular, the length and depth of the head about 50% SL. The sphenotics, epiotics, posttemporals, preopercles, anterodorsal margins of the frontals, heads of the palatines, symphysis of the lower jaw, all with well-developed spines. The posteroventral margin of the articulars is notched to form a pair of short spines. The positions, numbers, and relative length of the spines of the head is somewhat variable: the left posttemporal bears only a single spine in one known specimen; in all specimens examined, the 5 or 6 preopercular spines include a dorsal and ventral pair, while the position of 1 or 2 median spines varies in relation to that of the pairs.

The nostrils are set on low papilla. The lateral-line organs are stalked and unpigmented. The escal bulb is unpigmented and globular or slightly oval in shape, nearly sessile on the blunt tip of the pterygiophore of the illicium. The diameter of the escal bulb is 3.5 to 4.5% SL. Escal filaments and appendages are absent. The inner wall of the spherical photophore is black except for a relatively narrow distal transparent field.

The teeth are slender, recurved, depressible, and relatively short. The longest teeth are 2 to 3% SL and placed in the anterior part of the lower jaw. There are about 30 teeth on each premaxilla in a 24-mm specimen and about 80 in a 45-mm specimen; there are about 45 teeth on each dentary in a 24-mm specimen and about 110 in a 45-mm specimen. Teeth in the lower jaw are placed in a very regular pattern of oblique diagonal series: five to 10 overlapping longitudinal series, each with teeth increasing in length posteriorly; more numerous transverse series each with teeth increasing in length medially. The pattern of placement of the premaxillary teeth is similar but less regular. There are 3 or 4 dorsal- and anal-fin rays, and 15 to 17 pectoral-fin rays (Table 1).

The color of metamorphosed females in preservation is light brown, the skin semitransparent in smaller specimens (16 to 26 mm), but more densely pigmented in larger specimens (45 to 69 mm). Subdermal pigmentation is absent except on the peritoneum.

The body of free-living and parasitic males is elongate, the posterior part of the head strongly arched. Sphenotic spines are absent. The dorsal body contour is concave. There are 8 or 9 short recurved teeth on each side of the upper and lower jaws, arranged in one or two series, the largest teeth about one-half the length of the denticular teeth. The slightly tubular eyes are directed anteriorly, their diameter about 8% SL. The anterior nostrils are directed anteriorly but are well separated and oval in shape, their vertical diameter about two-thirds that of the posterior nostrils. The posterior nostrils are directed anterolaterally and are well separated from the eyes. The olfactory organs and to some extent the eyes are degenerated in parasitic males. There are 3 or 4 dorsal- and anal-fin rays, and 15 to 17 pectoral-fin rays. The skin is unpigmented in free-living males, but faintly pigmented in parasitic males. Subdermal pigment is restricted to the peritoneum.

Females attain a maximum known length of 69 mm (SIO 53-356); the largest parasitic male measures 7.3 mm (ZMUC P92134) and the largest free-living male 8.6 mm (ZMUC P921727).

### Diversity

A single species:

#### PHOTOCORYNUS SPINICEPS REGAN, 1925B

Figures 25, 30, 167–169, 236, 297, 301–303; Tables 8, 10

### Females and Males

*Photocorynus spiniceps* Regan, 1925b: 393, figs. 6, 7 (original description, a single female, with parasitic male). Beebe, 1926c:375, fig. 59 (figured). Regan, 1926: 12, 16, 21, figs. 2, 6, pl. 1, fig. 1 (description, and figures in part, after Regan, 1925b; skull "less specialized than that of any other ceratioid, and more nearly approaches that of *Lophius* in essentials"; allocated to new family Photocorynidae). Schmidt, 1926:268, fig. 8 (after Regan, 1925b, 1926). Regan, 1927b:4, postcards M-18, M-20a (popular account). Regan and Trewavas, 1932:23, 24, 43, 102, figs. 16B, 59, 60 (lateral line, osteology, additional specimen; description after Regan, 1925b, 1926). Fowler, 1944:528 (Gulf of Panama). Bertelsen, 1951:161, 165, figs. 108A, 109, 110 (description, additional material; osteology of a male, first record of free-living males, genus reallocated to family Linophrynidae; figure of Regan, 1926, corrected). Grey, 1956a:270 (distribution). Idyll, 1964:198, fig. 10-4 (popular account). Marshall, 1966, pl. 7 (figured). Parin et al., 1973:146, fig. 37 (additional female, eastern South Pacific). Nielsen, 1974:100 (listed in type catalog). Pietsch, 1976:788 (sexual parasitism). Paulin et al., 1989:138, fig. 66.2 (New Zealand). Bertelsen, 1990:519 (eastern tropical Atlantic). Swinney, 1995a:52, 57, fig. 2 (additional specimen, off Madeira). Pietsch, 2002b:1070 (western central Atlantic). Pietsch, 2005b:216, 226–229, 232, figs. 9E, 19D, 24 (reproduction; world's smallest sexually mature vertebrate). Guinness World Records, 2007:41, fig. (smallest vertebrate). Pietsch and Orr, 2007:4, 14, figs. 3F, 11 (phylogenetic relationships).

### Larvae

Unknown.

### Material

Thirty-one metamorphosed females (16 to 69 mm), three free-living metamorphosed males (6.5 to 8.6 mm), and four parasitic males (6.2 to 7.3 mm):

Holotype of *Photocorynus spiniceps*: ZMUC P92134, female, 46 mm, with 7.3-mm parasitic male, *Dana* station 1209(3), Gulf of Panama, 7°15'N, 78°54'W, 2500 m of wire, 17 January 1929.

Additional females and parasitic males: BMNH, six (20.5 to 36 mm); CAS, one (45 mm); ISH, one (50.5 mm, with 7.0-mm parasitic male); LACM, one (29.5 mm); MCZ, one (20 mm); NMSZ, one (20.5 mm); NSMT, one (50 mm); SIO, six (16.5 to 69 mm, 46-mm female with 6.2-mm parasitic male, 49-mm female with 6.5-mm parasitic male); SIOM, six (20.5 to 29 mm); TCWC, two (29 to 35 mm); UF, two (16 to 22 mm); ZMUC, two (24 to 25 mm).

Free-living males: BMNH, one (8.5 mm); ZMUC, two (6.5 to 8.6 mm).

### Distinguishing Characters and Description

As given for the genus and family.

### Reproduction and Sexual Maturity

The 31 metamorphosed females of *Photocorynus spiniceps* include four with parasitic males. One of these, a 7.3-mm specimen attached on the face just above the right eye of the 46-mm holotype of *P. spiniceps* (ZMUC P92135), was described and figured by Regan (1925b; see also Bertelsen, 1951:166, fig. 110). The remaining three couples were described by Pietsch (2005b:228): ISH 1913/71, a 50.5-mm female, with a 7.0-mm male attached to the belly; SIO 70-326, a 46-mm female, with a 6.2-mm male attached just behind the head above the pectoral fin, slightly to the right of the dorsal midline; and SIO 70-346, a 49-mm female, with a 6.5-mm male in almost the same position, but slightly to the left of the dorsal midline. The ovaries of the ISH specimen are large and contain ripening eggs 0.4 to 0.5 mm in diameter, while the three remaining parasitized females appear to be immature with egg diameters of less than 0.1 mm. The belly of the 7.3-mm male attached to the holotype "is greatly inflated by the testes" (Bertelsen, 1951:166, fig. 110C). The remaining three parasitic males are also mature or at least close to maturity, their testes distinctly enlarged. Those of the two SIO specimens (6.2 to 6.5 mm) are especially well developed, their greatest length about 1.4 mm and 1.3 mm (22.6 and 20.0% SL, respectively), each containing large amounts of developing spermatozoa, but without evidence of flagellated sperm. Together with the 7.5-mm parasitic male of *Leptacanthichthys gracilispinis* described above, the parasitic males of *P. spiniceps* are the smallest known within the suborder Ceratioidei and, if regarded as adults, which histological evidence seems to indicate in the case of *Photocorynus*, they are among the world's smallest known sexually mature vertebrates as defined in terms of length, volume, and weight (see Winterbottom and Emery, 1981; see also Roberts, 1986; Weitzman and Vari, 1988; Kottelat and Vidthayanon, 1993; Watson and Walker, 2004; Kottelat et al., 2006).

### Distribution

Until recently, *Photocorynus spiniceps* was known only from the Gulf of Panama and off Peru, but the available material now includes records from Hawaii, the western North Pacific, the Indian Ocean, and both sides of the Atlantic between latitudes of 32°N and 13°S. Included is a new record from off Japan (SIO 53-356) that represents both the largest known individual (69 mm) as well as the northernmost record (35°N) for the species (Fig. 236). Four specimens were captured in closing nets at depths of 990 to 1420 m; the remaining material was caught in open nets, none fished at maximum fishing depths less than 1000 m.

## Genus *Haplophryne* Regan, 1912 (Ghostly Seadevils)

Figures 6, 31, 170–172, 202, 203, 236, 250, 284, 297, 301, 304; Tables 1, 8–10

### Females

*Edriolychnus* Regan, 1925b:398, figs. 8, 9 (type species *Edriolychnus schmidti* Regan, 1925b, by monotypy).

*Haplophryne* Munk and Bertelsen, 1983:52 (*Haplophryne* Regan, 1912, a senior synonym of *Edriolychnus* Regan, 1925b, taking *Aceratias mollis* Brauer, 1902, as type species).

### Males

*Aceratias* Brauer, 1902:296 (in part; type species *Aceratias macrorhinus* by subsequent designation of Jordan, 1920:497).

*Haplophryne* Regan, 1912:289 (type species *Aceratias mollis* Brauer, 1902, by monotypy). Munk and Bertelsen, 1983:52 (*Haplophryne* Regan, 1912, a senior synonym of *Edriolychnus* Regan, 1925b, therefore taking *Aceratias mollis* Brauer, 1902, as type species).

*Halpophryne* Waterman, 1939b:89 (misspelling of *Haplophryne* Regan, 1912, therefore taking the same type species *Aceratias mollis* Brauer, 1902).

### Larvae

*Hyaloceratias* Koefoed, 1944:16 (in part, type species *Hyaloceratias parri* Koefoed, 1944, by monotypy). Bertelsen, 1951:180, 189, fig. 123 (an artificial, inseparable assemblage of larvae of *Haplophryne* and *Linophryne*).

*Haplophryne* Munk and Bertelsen, 1983:52 (*Haplophryne* Regan, 1912, a senior synonym of *Edriolychnus* Regan, 1925b, taking *Aceratias mollis* Brauer, 1902, as type species).

### Distinguishing Characters

Metamorphosed females of *Haplophryne* are distinguished from those of all other ceratioids by having unpigmented skin and a single compressed preopercular spine consisting of 2 to 5 radiating cusps. They differ further from those of all other genera of the family in having the following combination of character states: The frontals are widely separated, each with an anterodorsal spine. Epiotic and posttemporal spines are absent. The maxillae is reduced and extremely slender. The teeth are relatively short and numerous, placed in several overlapping oblique longitudinal series. Vomerine teeth are absent. The first pharyngobranchial is absent. The ceratohyal lacks an anterodorsal process. The posterior margin of the hypural plate is notched. The ninth caudal-fin ray is about one-half the length of the eighth ray. The illicium is extremely short, the escal bulb sessile on the snout. There is only a single escal appendage. A hyoid barbel is absent. The second pectoral radial is broader than the third.

The free-living males of *Haplophryne* are distinguished from those of all other genera of the family by having the following combination of character states: The sphenotic spines are weak, nearly absent. The preopercle is slender and angled at midlength. The epiotic region of the skull is only moderately elevated. The premaxillae are well developed. There are 20 to 24 teeth on each side of the upper and lower jaws, arranged in four series. The longest jaw teeth are longer than the denticular teeth. There is a transverse series of 3 to 6 upper denticular teeth, more or less completely fused at the base to form a small upper denticular bone. A pair of lower denticular bones are widely spaced near tip of lower jaw, each with 1 to 3 denticular teeth. All the denticular teeth are short and not meeting in front of the closed mouth. The olfactory organs are moderately enlarged and inflated, nearly as large as the eyes. The posterior nostril is well separated from the eye. There are 5 to 7 olfactory lamellae. The skin is everywhere unpigmented.

The larvae of *Haplophryne* are distinguished from those of all other genera of the family by having the following character states: Sphenotic spines are absent. There are two series of melanophores along the sides of the body, coming together at the base of the caudal fin to form a single more or less dense cluster. The pectoral fins are relatively small. Except in later metamorphic stages, they cannot be distinguished from those of *Linophryne* subgenus *Rhizophryne*, and are thus included in an artificial assemblage of larvae here termed "*Hyaloceratias*," as first described and defined by Bertelsen (1951:180, 189).

### Description

The body of metamorphosed females is short, its depth at the base of the pectoral fin 55 to 70% SL. The length of the head is 50 to 60% SL. The sphenotics, frontals, and preopercles all bear strong prominent spines. The preopercular spine is compressed, its distal one-half normally divided into 2 to 5 short broad cusps (in one specimen the spine is undivided on the left side), the upper- and lowermost spine more or less strongly curved. The lower jaw bears a symphysial spine. The ventral edge of the articulars is straight.

The nostrils of females are set on elongate papilla. The lateral-line organs are stalked and unpigmented. The pterygiophore of the illicium is short, completely embedded in the skin of the head. The escal bulb is sessile and spherical or distally flattened, with a diameter of about 10% SL, and completely unpigmented except for the inner wall of the photophores. There is a single short posterior escal appendage, divided distally into 2 to 6 short branches, simple in some juveniles.

The teeth of females are slender, recurved, and depressible, the longest about 5% SL. In the larger specimens, there are somewhere between 50 and 80 teeth on each side of the upper and lower jaws, arranged in several diagonal series, 6 to at least 10 oblique longitudinal series, and a greater number of oblique transverse series, increasing with the size of specimens.

The skin of females is totally unpigmented, the peritoneum black. Secondary subdermal pigmentation spreads posteriorly with increasing size of specimens from an anterior concentration on the dorsal surface of the trunk. It also spreads anteriorly from the base of the caudal fin, gradually obscuring the pattern of larval melanophores, and completely covering the body musculature of adults, except for the myosepta. The skin of free-living and parasitic males is also unpigmented.

The body of free-living and parasitic males is elongate. The roof of the skull is not strongly arched. Sphenotic spines are absent or represented only by a small blunt knob. The dorsal contour of the body is straight to slightly convex. The teeth are recurved and depressible. There are 20 to 24 on each side of the upper and lower jaws, arranged in four very distinct oblique longitudinal series, each with the length of teeth increasing posteriorly. The longest teeth are 3 to 4% SL and two to four times the length of the denticular teeth. Some jaw teeth are lost in larger parasitic males.

The eyes of males are directed anteriorly; they are slightly tubular, with diameters of 6 to 7% SL. The olfactory organs are situated on the sides of the blunt snout, well separated from the eyes and inflated, their greatest diameter about 6% SL. The anterior nostrils are directed anteriorly, about one-half the size of the posterior nostrils. The olfactory organs and eyes are degenerated in the largest parasitic males (see Munk and Bertelsen, 1983:66).

The skin of free-living and parasitic males is unpigmented. Subdermal pigment is present on the peritoneum and in two series of melanophores along the sides of the body: a dorsal series frequently consisting of a single row of melanophores, but sometimes 2 or 3 melanophores in width; and a ventral series of melanophores 2 to several melanophores in width. The dorsal and ventral series fuse to form a single group of melanophores at the base of the caudal fin.

The body shape and patterns of jaw-teeth and subdermal pigmentation of the larvae are as described for free-living males. The numbers of teeth and melanophores increase with the size of specimens. Except for metamorphic stages, the larvae are inseparable from those of the *Linophryne* subgenus *Rhizophryne* (see Comments, p. 483).

Fin-ray counts are shared by females and males: the dorsal and anal fins each contain 3 rays, the pectoral-fin 15 or 16 rays (the specimen reported by Roule and Angel, 1930:122, to have 4 dorsal, 4 anal rays, and 13 or 14 pectoral rays has not reexamined). The caudal-fin contains 9 rays, the ninth ray about one-half the length of the eighth.

Females reach a maximum known length of 159 mm (NMNZ P.21248); free-living males, 16 mm (MCZ 32308); and parasitic males, 15 mm (AMS I.21365-8).

Diversity

A single species:

### HAPLOPHRYNE MOLLIS (BRAUER, 1902)

Figures 6, 31, 170–172, 236, 250, 284, 297, 301, 304; Tables 8–10

Females and Parasitic Males

*Edriolychnus schmidti* Regan, 1925b:398, figs. 8, 9 (original description, single female with parasitic male). Jordan, 1926:367, fig. (sexual parasitism, after Regan, 1925b). Regan, 1926:14, 25, fig. 5, pl. 3, fig. 2 (two additional females, description). Schmidt, 1926:269, fig. 9 (after Regan, 1925b, 1926). Regan, 1927b:4, postcards M-19, M-20b (popular account). Roule and Angel, 1930:122, pl. 6, figs. 156–158 (additional female). Regan and Trewavas, 1932:15, 104, figs. 3–6, pl. 9, fig. 2 (additional female with three parasitic males, description). Fowler, 1936:1342, fig. 562 (West Africa). Fowler, 1944:528 (Gulf of Panama). Fowler, 1949:160 (Oceania). Bertelsen, 1951:168, figs. 108B, 111A, B, 112–115, 124A (description, additional specimens, comparison of all known material; *Edriolychnus* sp., *E. macracanthus*, *E. radians*, and *E. roulei* junior synonyms of *E. schmidti*). Albuquerque, 1954–1956:1075, figs. 443, 443A (Portugal). Matsubara, 1955:1353, 1354, figs. 529, 530C, 531 (sexual parasitism; after Regan and Trewavas, 1932). Grey, 1956a:270 (distribution). Günther and Deckert, 1956:179, fig. 122 (popular account). Briggs, 1960:179 (worldwide distribution). Maul, 1961:155, figs. 27A, C, 32 (three additional females, two in metamorphosis and one metamorphosed). Wickler, 1961:387, fig. 5 (sexual parasitism). Maul, 1962b:45 (additional female). Mead et al., 1964:588 (reproduction). Blache et al., 1970:452, fig. 1147 (eastern Atlantic). Maul, 1973:677 (eastern North Atlantic). Parin et al., 1973:146, fig. 37 (two additional females, eastern South Pacific). Nielsen, 1974:99 (listed in type catalog). Olsson, 1974:225, figs. 1–6 (endocrine organs of a parasitic male). Pietsch, 1976:782, 788 (three additional females with parasitic males). Pietsch and Seigel, 1980:397 (expected in Philippine Archipelago). Herring, 1983:202, fig. 3d, table 9 (bioluminescence emission spectra). Paulin and Stewart, 1985:27 (New Zealand). Pequeño, 1989:44 (Chile).

*Edriolychnus* sp.: Regan and Trewavas, 1932:104, fig. 163 (female in metamorphosis). Beebe, 1937:207 (two additional females).

*Edriolychnus macracanthus* Regan and Trewavas, 1932:104 (original description, three females including two referred to *Edriolychnus schmidti* by Regan, 1926; no type designation). Fowler, 1944:528 (Gulf of Panama). Nielsen, 1974:99 (listed in type catalog).

*Edriolychnus radians* Regan and Trewavas, 1932:17, 104, fig. 64A, pl. 9, fig. 1 (original description, single female with parasitic male). Fowler, 1936:1368 (after Regan and Trewavas, 1932; West Africa). Nielsen, 1974:99 (listed in type catalog).

*Edriolychnus roulei* Regan and Trewavas, 1932:105, fig. 64B (original description, based on single female described as *Edriolychnus schmidti* by Roule and Angel, 1930). Fowler, 1936:1368 (after Regan and Trewavas, 1932; West Africa).

*Haplophryne mollis*: Munk and Bertelsen, 1983:49, 52, figs. 1–17 (a senior synonym of *Edriolychnus schmidti*, histology of parasitic attachment). Bertelsen and Pietsch, 1983:94, figs. 17, 18 (seven additional females, five with parasitic males). Bertelsen, 1986:1408, figs. (eastern North Atlantic). Bertelsen, 1990:516 (eastern tropical Atlantic). Nielsen and Bertelsen, 1992:62, fig. 6 (North Atlantic). Herring and Munk, 1994:747, figs. 1–11 (histology of escae, observations on control of bioluminescence). Swinney, 1995a:52, 56, fig. 1 (five additional specimens, off Madeira). McEachran and Fechhelm, 1998:858, 859, 860 (Gulf of Mexico, in key). Stewart and Pietsch, 1998:30, 37, fig. 19 (six additional females and 11 parasitic males, New Zealand). Munk, 1999:267, 268, 271, 272, 274, 277, 281, figs. 3B, 4D (bioluminescence). Pietsch, 1999:2037 (western central Pacific). Pietsch, 2002b:1070 (western central Atlantic). Moore et al., 2003:217 (listed). Pietsch, 2005b:210, 213, 214, 216, 226, 227, 232, figs. 6, 9C, 19B, 20, 21 (reproduction). Pietsch and Orr, 2007:4, 9, fig. 3G, 6H (phylogenetic relationships).

Free-living Males

*Aceratias mollis* Brauer, 1902:297 (original description, single male). Brauer, 1906:324, pl. 16, fig. 10 (description of holotype after Brauer, 1902, allocation to new family Aceratiidae). Murray and Hjort, 1912:615 (after Brauer, 1902, 1906).

*Aceratias macrorhinus indicus*: Murray and Hjort, 1912:87, 90, 96, 609, 615, 618, 625, 627, 744, figs. 68, 470, 536 (in part, misidentifications). Whitley, 1934, unpaged (listed). Koefoed, 1944:13, 15, pl. 2, fig. 3 (in part, four misidentified males referred to as "juvenes").

*Haplophryne mollis*: Regan, 1912:289 (new combination). Regan and Trewavas, 1932:20, 111, fig. 171 (two males tentatively referred to *Edriolychnus*). Albuquerque, 1954–1956:1077, fig. 444 (after Fowler, 1936; Portugal). Munk and Bertelsen, 1983:49, figs. 1–17 (histology of parasitic attachment, *Edriolychnus schmidti* a junior synonym of *Haplophryne mollis*).

*Haplophryne simus* Borodin, 1930b:285 (original description, single male in metamorphosis). Borodin, 1931:83, pl. 4, fig. 1 (after Borodin, 1930b; "nearest to *Aceratias mollis*").

*Haplophryne* sp. (? *Edriolychnus* male): Regan and Trewavas, 1932:111, fig. 171 (additional free-living male).

*Edriolychnus schmidti* (? = *Haplophryne mollis*): Bertelsen, 1951:168, figs. 108B, 111A, B, 112A (additional free-living males, osteology).

Larvae

*Haplophryne mollis*: Regan, 1916:148, pl. 10, fig. 2 (single female larva, probably *Haplophryne mollis*). Regan, 1926:25, pl. 3, fig. 3 (in part; two specimens reallocated by Bertelsen, 1951: a metamorphic female assigned to *Linophryne* subgenus *Rhizophryne* and a larva assigned to *Hyaloceratias*). Idyll, 1964:255 (popular account). Bertelsen, 1984:328, fig. 167D (metamorphic male, early life history, phylogeny).

*Haplophryne* sp.: Regan and Trewavas, 1932:20, 111, figs. 11A, 171 (male larva, possibly *Haplophryne mollis*; viscera described, figured). Beebe, 1937:207 (three additional specimens).

*Haplophryne triregium* Whitley and Phillipps, 1939:236 (original description, single female specimen, based on *Haplophryne mollis* of Regan, 1916:148; not the same as *Aceratias mollis* of Brauer, 1902). Whitley, 1956:413 (listed; New Zealand). Whitley, 1968:89 (listed).

*Hyaloceratias parri* Koefoed, 1944:16, pl. 1, figs. 7, 8 (in part; original description, 13 larvae, no type designation). Frøiland, 1979:151 (listed in type catalog).

*Hyaloceratias* sp.: Bertelsen, 1951:180, 189, figs. 123, 124B (16 larvae of Regan, 1916, Regan and Trewavas, 1932, and Koefoed, 1944, plus 220 new records; an assemblage of larvae containing *Edriolychnus schmidti*, now recognized as *Haplophryne mollis* and *Linophryne* subgenus *Rhizophryne*, and possibly young larvae of *Linophryne corymbifera*, a name now recognized as a junior synonym of *L. indica*).

Material

Eighty-eight metamorphosed and metamorphic females (13 to 159 mm), 43 parasitic males (8.9 to 15 mm), 20 free-living metamorphosed and metamorphic males (10 to 16 mm), and about 250 "*Hyaloceratias*" larvae (2.6 to 17.5 mm):

Holotype of *Aceratias mollis*: ZMB 17713, male, 13 mm, *Valdivia* station 175, Indian Ocean, 26°3′S, 93°43′E, open pelagic net, 0 to 2200 m, 12 January 1899.

Holotype of *Edriolychnus schmidti*: ZMUC P92135, female, 48 mm, with 10-mm parasitic male, *Dana* station 1183(1), Caribbean Sea, west of St. Lucia, 13°47′N, 61°26′W, open pelagic net, 4500 m of wire, 1630 h, 24 November 1921.

Holotype of *Haplophryne simus*: MCZ 32308, male in metamorphosis, 16 mm, *Atlantis* station 319, Sargasso Sea, 34°50′N, 64°20′W, 0 to 1500 m, 29 August 1929.

Syntypes of *Edriolychnus macracanthus*: BMNH 1925.8.11.4, female, 40 mm, *Dana* station 1203(10), Gulf of Panama, 7°30′N, 79°19′W, open pelagic net, 3500 m of wire, 1500 h, 11 January 1922. ZMUC P92136, female, 26 mm, *Dana* station 1156(5), North Atlantic, 25°11′N, 20°57′W, open pelagic net, 6000 m of wire, 1810 h, 25 October 1921. ZMUC P92137, female, 33 mm, *Dana* station 3613(8), western Pacific, off New Caledonia, 22°43′S, 166°05.8′E, open pelagic net, 3000 m of wire, 0530 h, 28 November 1928.

Holotype of *Edriolychnus radians*: ZMUC P92138, female, 34 mm, with 11-mm parasitic male, *Dana* station 3980(9), South Atlantic, 23°26′S, 3°56′E, open pelagic net, 3000 m of wire, 0910 h, 17 February 1930.

Holotype of *Edriolychnus roulei*: MOM, female, 35 mm TL, Prince Albert 1st of Monaco, expedition of 1911, station 3131, North Atlantic, between Madeira and Azores, 0 to 3500 m, 16 August 1911.

Holotype of *Haplophryne triregium*: BMNH 1916.3.20.184, larval female, 10 mm TL, British Antarctic (Terra-Nova) Expedition of 1910, station 127, off Three Kings Islands, New Zealand, on the surface, 25 August 1911.

Syntypes of *Hyaloceratias parri*: 11 larvae, 4.5 to 17.5 mm (6 to 23 mm TL), *Michael Sars*, North Atlantic Deep-Sea Expedition: ZMUB 4279, one, station 53, 34°59′N, 33°59′W, 100 m of wire, 8 to 9 June 1910. ZMUB 4280, four, station 64, 34°44′N, 47°52′W, 300 m of wire, 24 June 1910. ZMUB 4281, one, station 64, 34°44′N, 47°52′W, 200 m of wire, 24 June 1910. ZMUB 4282, one, station 45, 28°42′N, 20°00′W, 300 m of wire, 28 to 29 May 1910. ZMUB 4283, two, station 45, 28°42′N, 20°00′W, 300 m of wire, 28 to 29 May 1910. ZMUB 4284, one, station 51, 31°20′N, 35°07′W, 200 m of wire, 5 to 6 June 1910. ZMUB 4307, one, station 49B, 29°08′N, 25°16′W, 370 m of wire, 1 June 1910.

Additional females and parasitic males: AIM, one (63 mm); AMS, seven (29 to 62 mm, 39-mm specimen with 12.5-mm parasitic male; 46-mm specimen with 11-mm parasitic male; 48-mm specimen with two parasitic males, 10.5 to 12 mm; 52.5-mm specimen with two parasitic males, 11.5 to 12 mm; 62-mm specimen with two parasitic males, 12 to 15 mm); BMNH, one (40 mm, with two parasitic males, 10.5 to 11 mm); CSIRO, none (24 to 60 mm; 54-mm specimen with 9-mm male; 60-mm specimen with 14-mm male); FMNH, one (24 mm); ISH, eight (23 to 41 mm); LACM, two (33 to 70 mm, 70-mm specimen with two parasitic males, 11 to 12.5 mm); MCZ, 14 (13 to 47, 47-mm, with parasitic male, about 13 mm); MMF, four (21 to 31 mm); MNHN, three (33.5 to 48 mm; 45- and 48-mm females each with a 12-mm parasitic male); NMNZ, nine (43.5 to 159, 45-mm specimen with 13.6-mm parasitic male; 51-mm specimen with 11.6-mm parasitic male; 54-mm specimen with six males, 8.9 to 10.5 mm; 58-mm specimen with two parasitic males, 11.6 to 14.2; 60-mm specimen with 11-mm parasitic male; 70-mm specimen with two parasitic males, 11.4 to 11.5 mm; 159-mm specimen with two males, 9.8 to 10.8 mm); NMSZ, five (25 to 30 mm); NMV, four (37 to 57 mm, 52-mm specimen with two parasitic males, 10 to 10.3 mm); NSMT, one (38.5 mm); SIO, one (40 mm); SIOM, two (27 to 28 mm); NOC, two (35-mm female with two parasitic males, 10.5 to 11.5 mm; 38-mm female with two parasitic males, 9.5 to 10 mm); UF, four (28 to 48.5 mm); ZMUC, four (37 to 53 mm, 50-mm female with three parasitic males, 11 to 12 mm).

Additional free-living males: SIO, four (10 to 13 mm); ZMUB, four (12 to 14 mm); ZMUC, 10 (11.5 to 14 mm).

For a list of "*Hyaloceratias*" larvae (2.6 to 10 mm) in the *Dana* Collections, see Bertelsen, 1951:276.

### Distinguishing Characters and Description

As given for the genus and family.

### Reproduction and Sexual Maturity

The 88 known females of *Haplophryne mollis* include 25 specimens that bear one to six parasitic males; three others bear a scar on various parts of the body that represents a lost male (Bertelsen and Pietsch, 1983:97, fig. 18B; Munk and Bertelsen, 1983:63, fig. 13; see Table 10). Such scars are not known in any other ceratioid (white circular scars, one each on four specimens and four species of *Himantolophus*, are most probably the result of prior attachment of parasitic copepods; see Bertelsen and Krefft, 1988:25–26). While the scars and most of the parasitic males are found on the belly of the females, several males are attached in various places on the head; especially remarkable is the position of a 12-mm male (AMS I.21365-008) attached to the distal surface of the esca of the female (Bertelsen and Pietsch, 1983:97, fig. 18C). One of the two parasitic males attached to the latter specimen (AMS I.21365-008) is somewhat larger (15 mm) and distinctly more bulky than the largest known free-living male, indicating growth based on true parasitism (Bertelsen and Pietsch, 1983:97, fig. 17B). Several of the parasitic males have large, apparently ripe testes, but no females with eggs larger than 0.15 mm in diameter have been reported.

### Distribution

*Haplophryne mollis* is known from all three major oceans of the world, the majority of the localities scattered over the North Atlantic between about 55°N and 40°S, including records in the Gulf of Mexico and the Caribbean Sea. In the Pacific Ocean, most specimens are from off eastern Australia, New Caledonia, and New Zealand, but numerous specimens have also been taken in the central North Pacific in and around the Hawaiian Islands, with one additional record from the Gulf of Panama (Fig. 236). The species is represented in the Indian Ocean by the holotype of *Aceratias mollis* collected at approximately 26°S, 94°E, and another free-living male taken off Cape Leeuwin, western Australia.

All specimens were caught in non-closing pelagic trawls. Of the 88 known females, at least 20, including six with parasitic males, were collected by trawls fished at maximum depths of 550 to 900 m. The remaining material, for which depth data are known, all came from nets fished at maximum fishing depths of 1000 to 2000 m. With the exception of one specimen captured in less than 200 m, the 20 known free-living males came from hauls with maximum depths of more than 1500 m.

### Comments

Females of the genus were divided into four species by Regan and Trewavas (1932) on the basis of differences in the branching of the preopercular spine. These differences, however, were shown by Bertelsen (1951) to represent intraspecific variation and the considerably larger number of specimens now available (additional material reported by Bertelsen and Pietsch, 1983, and Stewart and Pietsch, 1998, plus the new records presented here) seems to confirm that the genus is monotypic.

The pattern of subdermal larval melanophores distributed in series along the sides of the body is shared by *Haplophryne* and two of the three subgenera of *Linophryne*, *Rhizophryne*, and *Stephanophryne* (Bertelsen, 1951, 1982). Except in the youngest stages, larvae of the subgenus *Stephanophryne* can be distinguished by the presence of additional groups of melanophores, but no characters have been found to separate larvae of *Haplophryne* and subgenus *Rhizophryne*. As first proposed by Bertelsen (1951:180, 189), this assemblage of larvae is termed "*Hyaloceratias*," using the name introduced by Koefoed (1944). Males in metamorphosis can be distinguished by the fact that at the stage when the denticular teeth of *Linophryne* appear the pigmentation of their skin is in development, in difference to the permanently unpigmented skin of *Haplophryne*. Females of *Linophryne* in metamorphic and late larval stages can be distinguished from *Haplophryne* by having a papilliform rudiment of the hyoid barbel, unique to the females of the first-mentioned genus.

Genus *Acentrophryne* Regan, 1926
(Fangtooth Seadevils)

Figures 173–175, 236; Tables 1, 8

Females

*Acentrophryne* Regan, 1926:23, pl. 1, fig. 2 (type species *Acentrophryne longidens* Regan, 1926, by monotypy).

Males and Larvae

Unknown.

Distinguishing Characters

Metamorphosed females of *Acentrophryne* differ from those of all other genera of the family in lacking a spine on the preopercle. They differ further in having the following combination of character states: Frontal, epiotic, and posttemporal spines are absent. Jaw teeth are few in number (6 to 26 on each premaxilla, 9 to 16 on each dentary), extremely long (longest about 20% SL), and arranged in two or three overlapping, oblique longitudinal series. There are 2 to 6 vomerine teeth. The ninth caudal-fin ray is about one-half the length of the eighth. The length of the illicium is 35.7 to 70.5% SL. The esca bears a single, unpigmented distal appendage. A hyoid barbel is absent. The skin is uniformly black except for distal parts of the esca. Material is unavailable for internal anatomical examination.

Description

The body of metamorphosed females is short and globular, its depth 50 to 60% SL. The length of the head is 46 to 67% SL, the length of the lower jaw, 32 to 57% SL. The caudal peduncle is unusually short, the dorsal and anal fins terminating nearly at the base of the caudal-fin rays. The sphenotic spines are prominent. The dorsolateral margin of the frontals is smooth, without spines or projections. The lower jaw is without a symphysial spine. The posterior tip of the angular forms a sharp spine. The anterior end of the pterygiophore of the illicium protrudes slightly from the snout. The relative length of the illicium increases with growth: it measures 35.7% SL in a 42-mm specimen, 67.9% SL in a 56-mm specimen, and 70.5% SL in a 105-mm specimen. The escal bulb is relatively large (its width 6.2 to 8.2% SL) and bears a single, unpigmented distal appendage, 2.3 to 35.7% SL. Additional escal appendages and filaments are absent. The skin of the head and body is uniformly black to dark red brown, except for the distal part of the escal bulb and escal appendage.

The teeth are relatively few: there are only 6 or 7 on each premaxilla and 9 on each dentary in a 42-mm specimen; 15 on each premaxilla and 10 to 12 on each dentary in a 56-mm specimen; and 25 or 26 teeth on each premaxilla and 16 on each dentary in a 105-mm specimen. The longest premaxillary tooth measures 5.0 to 14.9% SL, the longest dentary tooth, 8.8 to 19.6% SL. The dorsal and anal fins each contain 3 rays, the pectoral fin, 16 to 19 rays. Pelvic fins are absent. There are 9 caudal-fin rays, the ninth ray (the ventralmost) about one-half the length of the eighth caudal-fin ray (Table 1).

Females reach a maximum known length of 105 mm (HUMZ 175257); males and larvae are unknown.

Diversity

Two species:

Key to Metamorphosed Females of Species of *Acentrophryne*

1A. Distal appendage of esca short (2.3 to 2.5% SL), illicium short (35.7 to 54.9% SL), head wide (35.7 to 37.3% SL), pectoral-fin rays 18 or 19 . . . . . . . . . . . *Acentrophryne longidens* Regan, 1926, p. 484

NOTE Two known specimens, 42 to 50 mm; Gulf of Panama and Pacific coast of Costa Rica

1B. Distal appendage of esca long (27.6 to 35.7% SL), illicium long (63.6 to 70.5% SL), head narrow (25.0 to 28.1% SL), pectoral-fin rays 16 . . . . . . . . . . . . *Acentrophryne dolichonema* Pietsch and Shimazaki, 2005, p. 485

NOTE Three known specimens, 55 to 105 mm; off Peru

ACENTROPHRYNE LONGIDENS REGAN, 1926

Figures 174, 236; Table 8

Females

*Acentrophryne longidens* Regan, 1926:23, pl. 1, fig. 2 (original description, single specimen). Regan and Trewavas, 1932:106 (after Regan, 1926). Fowler, 1944:528 (Gulf of Panama). Beebe and Crane, 1947:170, fig. 15 (additional specimen, questionably referred to *Acentrophryne longidens*). Bertelsen, 1951:192 (after Regan, 1926, and Beebe and Crane, 1947). Grey, 1956a:278 (distribution). Nielsen, 1974:98 (holotype listed in type catalog). Pietsch and Lavenberg, 1980:906, figs. 1, 2 (first known fossil ceratioid, Late Miocene of California). Pietsch, 2005b: 216, 226 (reproduction). Pietsch and Shimazaki, 2005:247, fig. 1 (generic revision).

*Borophryne apogon*: Nigrelli, 1947:183 (misidentification based on AMNH 51440, 42 mm; histology of stomach tumor).

Males and Larvae

Unknown.

Material

Two metamorphosed females (42 to 50 mm):

Holotype of *Acentrophryne longidens*: ZMUC P921981, 50 mm, *Dana* station 1203(14), eastern tropical Pacific, Gulf of Panama, 7°30′N, 79°19′W, open conical ring-trawl, 2500 m wire, bottom depth 2550 m, 2030 h, 11 January 1922.

Additional specimen: AMNH 51440 (formerly NYZS 28411), 42 mm, eastern Pacific *Zaca* Expedition station 210 T-10, south of Cape Blanco, Costa Rica, 9°11′N, 85°08′W, 1-m-diameter tow-net, 500 fathoms (910 m), 27 February 1938.

Distinguishing Characters

Metamorphosed females of *Acentrophryne longidens* differ from those of *A. dolichonema* in having a much shorter distal escal appendage (2.3 to 2.5% SL versus 27.6 to 35.7% SL), a shorter illicium (35.7 to 54.9% SL versus 63.6 to 70.5% SL), a wider head (35.7 to 37.3% SL versus 25.0 to 28.1% SL), and a greater number of pectoral-fin rays (18 or 19 versus 16).

Description

Metamorphosed females with body short, globular, head length 46.0 to 57.0% SL, head depth 47.6 to 58.8% SL, head width 35.7 to 37.3% SL; length of premaxilla 38.1 to 52.9% SL; length of lower jaw 32.1 to 55.9% SL; length of illicium 35.7 to 54.9% SL; width of escal bulb 6.2 to 8.2% SL; a short distal escal appendage, with smooth rounded distal tip, showing no signs of damage, length 2.3 to 2.5% SL. Jaw teeth relatively few, 6 to 14 in each premaxilla, 9 to 11 in each dentary; longest premaxillary tooth 5.0 to 14.9% SL, longest dentary tooth 8.8 to 19.6% SL; dorsal- and anal-fin rays 3; pectoral-fin rays 18 or 19; caudal-fin rays 9, ninth ray (ventral-most) 18.6 to 22.6% SL.

Additional description as given for the genus and family.

Distribution

The two known specimens of *Acentrophryne longidens* were collected in the

Gulf of Panama at about 7°'N, 79°'W, and from off the coast of Costa Rica at 9°'N, 85°'W, with open nets fished at maximum depths of about 1250 to 1280 m (Fig. 236).

### ACENTROPHRYNE DOLICHONEMA PIETSCH AND SHIMAZAKI, 2005

Figures 173, 236; Table 8

#### Females

*Acentrophryne dolichonema* Pietsch and Shimazaki, 2005:248, fig. 2 (original description, three specimens; generic revision).

#### Males and Larvae

Unknown.

#### Material

Three metamorphosed females (55 to 105 mm):

Holotype of *Acentrophryne dolichonema*: HUMZ 175257, 105 mm, *Shinkai-Maru*, eastern tropical Pacific, off Peru, 8°10.6 to 11.9'S, 80°32.0 to 32.4'W, Bacalao Trawl net-586 MSK (nonclosing otter trawl), bottom depth 1061 to 1105 m, 27 April 2000.

Paratypes of *Acentrophryne dolichonema*: HUMZ 167353, 56 mm, *Shinkai-Maru*, eastern tropical Pacific, off Peru, 9°28.4 to 27.0'S, 79°33.5 to 34.8'W, Bacalao Trawl net-586 MSK (nonclosing otter trawl), bottom depth 201 to 223 m, 6 August 1999; HUMZ 189134, 55 mm, *Humboldt*, eastern tropical Pacific, off Peru, 8°44.00 to 42.98'S, 80°06.66 to 07.16'W, Nordsea Balloon Trawl (nonclosing otter trawl), bottom depth 966 m, 9 October 2003.

#### Distinguishing Characters

Metamorphosed females of *Acentrophryne dolichonema* differ from those of *A. longidens* in having a much longer distal escal appendage (27.6 to 35.7% SL versus 2.3 to 2.5% SL), a longer illicium (63.6 to 70.5% SL versus 35.7 to 54.9% SL), a narrower head (25.0 to 28.1% SL versus 35.7 to 37.3% SL), and fewer pectoral-fin rays (16 versus 18 or 19).

#### Description

Metamorphosed females with body short, globular, head length 56.4 to 66.7% SL, head depth 53.6 to 54.5% SL, head width 25.0 to 28.1% SL; length of premaxilla 45.7 to 51.8% SL; length of lower jaw 50.9 to 57.1% SL; length of illicium 63.6 to 70.5% SL; width of escal bulb 6.7 to 7.1% SL; an elongate distal escal appendage, constricted at base, tapering gradually to form a smooth rounded distal tip, length 27.6 to 35.7% SL. Jaw teeth relatively few, 15 to 26 in each premaxilla, 10 to 16 in each dentary; longest premaxillary tooth 9.5 to 11.0% SL, longest dentary tooth 15.7 to 18.1% SL; dorsal- and anal-fin rays 3; pectoral-fin rays 16; caudal-fin rays 9, ninth ray (ventral-most) 10.5 to 17.3% SL.

Additional description as given for the genus and family.

#### Distribution

The three known specimens of *Acentrophryne dolichonema* were captured in close proximity to each other, off the coast of Peru, between approximately 8° and 9°S, and 79° and 80°W. All were captured by bottom trawls at depths ranging from 201 to 1105 m (Fig. 236).

#### Comments

*Acentrophryne dolichonema* is distinguished from *A. longidens* by having a much longer distal escal appendage, but it might be argued that the two known specimens of *A. longidens* are simply *A. dolichonema* that have lost the distal escal appendage at some time prior to capture. However, it seems highly unlikely that this could have happened without leaving some evidence of damage, even given ample time for healing and possible tissue regeneration. The short, papilla-like, distal escal appendage of both specimens of *A. longidens* is nearly identical, tapering quickly to form a smooth, rounded distal termination. The two species are further separated by a difference in illicial length. Because the two specimens of *A. longidens* are small (42 to 50 mm) and all known individuals of *A. dolichonema* are somewhat larger (55 to 105 mm), it might be argued further that *A. longidens* represents juvenile specimens of the latter. However, if this were true it would necessitate an extremely rapid ontogenetic increase in illicial length, the evidence for which is lacking in all other lophiiform fishes for which adequate material has been studied. These two characters, coupled with distinct differences in the width of the head and in pectoral-fin ray counts, strongly indicate the existence of two species. It should be pointed out also that the two species probably occupy different habitats: the known material of *A. longidens* was taken in pelagic nets fished far above the bottom, whereas all three specimens of *A. dolichonema* were captured in bottom trawls.

### †*Acentrophryne* Species

Figure 175

#### Females

*Acentrophryne longidens*: Pietsch and Lavenberg, 1980:906, figs. 1, 2 (first known fossil ceratioid, Late Miocene of California).

#### Males and Larvae

Unknown.

#### Material

A single known fossil female from the Late Miocene of California: LACM, Department of Vertebrate Paleontology, 117685, accession no. A-6839-78-167, 28 mm, finely laminated, light to medium brown diatomite, Clarendonian, Late Miocene (7.2 to 8.6 million years before present), Puente Formation, Locality No. 6908, Chalk Hill, Hacienda Heights, Los Angeles County, California, August 1977.

#### Comments

Although lophioid and antennarioid fishes are well known from Pliocene and Eocene deposits—*Lophius budegassa* Arambourg, 1927, Lower Pliocene of Algeria; *L. brachysomus* Agassiz, 1835, 1844, Eocene of Monte Bolca, Italy; *Eosladenia caucasica* Bannikov, 2004, Middle Eocene of the Northern Caucasus; *Histionotophorus bassani* (Zigno, 1887), Monte Bolca (for a review of Monte Bolca material, see Eastman, 1904; see also Bannikov, 2004:420; Carnevale and Pietsch, 2006:454; Carnevale et al., 2008:996)—the fossil of *Acentrophryne* from the Late Miocene Puente Formation of southern California, described by Pietsch and Lavenberg (1980), is the first known fossilized ceratioid. Easily recognized as a female ceratioid by the presence of an illicium and the absence of pelvic fins, the fossil clearly belongs to the family Linophrynidae in having 3 dorsal- and three anal-fin rays, 3 pectoral radials, and a single-headed hyomandibular bone; and to the genus *Acentrophryne* in the absence of a preopercular spine and in having an elongate illicium (Pietsch and Lavenberg, 1980). Because *Acentrophryne* was considered monotypic at the time, the fossil was initially described as *Acentrophryne longidens*, but with a second species now recognized (*A. dolichonema* Pietsch and Shimazaki, 2005), the fossil, lacking the information required for specific identification, is here referred to as *Acentrophryne* sp.

Between January and November 1993, about a dozen additional ceratioid fossils were collected from this same formation during earth-moving activities associated with construction of the Metro Rail Red Line, Wilshire Boulevard/Vermont Avenue Subway Station, in Los Angeles. Unfortunately, this material, donated by the Los Angeles Metropolitan Transportation Authority, is rather poorly preserved and currently remains unidentified in the collections of the Department of Vertebrate Paleontology (LACM). The Puente Formation and other southern California diatomite deposits are well known for their meso- and bathypelagic fishes (see Jordan, 1925, and Crane, 1966).

## Genus *Borophryne* Regan, 1925c
(Greedy Seadevils)

Figures 176–178, 202, 203, 236, 297, 301, 305;
Tables 1, 8, 10

### Females, Males, and Larvae

*Borophryne* Regan, 1925c:564 (type species *Borophryne apogon* Regan, 1925c, by monotypy).

### Distinguishing Characters

Metamorphosed females of *Borophryne* are very similar to those of *Linophryne* but differ in lacking a hyoid barbel. They differ from those of all other genera of the family in having the following combination of character states: The frontals are widely separated, their dorsal margin with a conspicuous, rounded, laterally compressed protuberance. Epiotic and posttemporal spines are absent. There is a single preopercular spine (a feature shared only with *Linophryne*). The maxillae are reduced and extremely slender. The teeth in the jaws are unusually well developed and few in number (fewer than 30 on each premaxilla, fewer than 20 on each dentary), relatively long (the longest teeth measuring 25% SL), and arranged in three overlapping oblique longitudinal series. There are 0 to 4 vomerine teeth. The first pharyngobranchial is absent. The ceratohyal lacks an anterodorsal process. The posterior margin of the hypural plate is notched. The ninth caudal-fin ray is about one-half the length of the eighth ray. The length of the illicium is about 20% SL. The esca bears distal filaments and a bifurcated terminal appendage. The second pectoral radial is broader than the third. A hyoid barbel is absent. The skin is uniformly black except on the distal parts of the esca and, except in some largest specimens, the distal parts of the fins.

Free-living metamorphosed males of *Borophryne* differ from those of other genera of the family in having the following combination of character states: The sphenotic spines are well developed. The premaxillae are degenerate, more or less completely resorbed. The upper and lower jaws are toothless. The denticles are short and strong, connected basally and meeting in front of the mouth when the jaws are closed. There are 3 to 7 teeth on the upper denticular. The lower denticular bears an anteriorly directed median tooth and a row of 2 to 6 more or less posteriorly directed teeth on each side (but the denticles are often placed asymmetrically, their size irregular among known material). The olfactory organs are unpigmented and strongly inflated, the length of the posterior nostril almost twice the diameter of the eye. There are 7 to 9 olfactory lamellae. The skin is light brown, but weakly pigmented, and more or less transparent, closely surrounding the body.

Larvae of *Borophryne* are distinguished from those of other linophrynids in having the following combination of character states: The body is relatively short, its greatest depth about 45 to 50% SL. The skin is highly inflated, unpigmented, and more or less transparent. The eye diameter is approximately 14% SL in smaller specimens, decreasing to about 8% in larger specimens.

### Description

The body of metamorphosed females is short and globular, its depth about 40 to 55% SL. The length of head is about 50 to 60% SL, the lower jaw 40 to 50% SL. The caudal peduncle is unusually short. The frontals bear a conspicuous, rounded, laterally compressed dorsal protuberance. The sphenotic spines are well developed in young specimens, but become proportionately smaller with growth and covered with skin in the largest known specimens. The preopercle is angled posteriorly, its posterior margin with a strong pointed spine. A symphysial spine is present on the lower jaw. The angular and posteroventral part of the articular form a sharp spine.

The pterygiophore of the illicium protrudes slightly from the snout, its exposed length rarely more than the diameter of the escal bulb. The length of the illicium is about 20% SL. The escal bulb is large, its diameter about 10% SL. The esca bears a large bifurcated terminal appendage and numerous fine filaments on each side. The skin of the head and body is uniformly black or dark red brown. The fins are unpigmented in most smaller specimens, but pigmentation at the base of the fins gradually spreads with growth to cover the distal tips of the rays.

The dentition of females, including placement and the number and length of teeth, is very similar to that found in species of *Linophryne* (Bertelsen, 1980a: 34; see Fig. 35): there are three, longitudinal, overlapping series of teeth on each premaxilla and dentary; the number of teeth in each series tends to increase with increasing length of specimens; the length of teeth in each series increases posteriorly. There are 7 to 28 teeth on each premaxilla and 5 to 17 on each dentary. The dorsal and anal fins each have 3 rays. There are 15 to 18 pectoral-fin rays and 9 caudal-fin rays, the ninth ray about one-half the length of the eighth (Table 1).

For description of free-living males and larvae, see the Distinguishing Characters above.

The largest known specimen is a 101-mm female (LACM 30053-10), with two parasitically attached males (16 to 22 mm). The largest of 62 known free-living males is 17.5 mm (LACM 30059-5). Of the 13 known parasitic males, two have reached lengths of 20 and 22 mm (SIO 68-112, LACM 30053-10), indicating considerable growth after attachment (see Table 10). Both of the latter males have enlarged testes and show distinct degeneration of olfactory organs and eyes.

### Diversity

A single species:

#### BOROPHRYNE APOGON REGAN, 1925c

Figures 176–178, 236, 297, 301, 305; Tables 8, 10

### Females, Males, and Larvae

*Borophryne apogon* Regan, 1925c:564 (original description, four specimens, lectotype ZMUC P92147 by subsequent designation of Bertelsen, 1951). Regan, 1926:23, pl. 2, fig. 1 (description after Regan, 1925c). Regan and Trewavas, 1932:18, 20, 106, figs. 7, 8, 10A (two additional specimens, each with a parasitic male). Fowler, 1944:528 (Gulf of Panama). Beebe and Crane, 1947:151, 171, pl. 2, fig. 4 (five additional specimens, two with a parasitic male: NYZS 28707 apparently lost; NYZS 28708, now AMNH 211332). Nigrelli, 1947:183 (misidentification, specimen is *Acentrophryne longidens*; histology of stomach tumor). Bertelsen, 1951:193, figs. 125, 126 (additional material, assignment of larvae and free-living males, comparison of all known material, lectotype designated). Harvey, 1952:530 (bioluminescence). Matsubara, 1955: 1353, fig. 530D (sexual parasitism, figure after Regan and Trewavas, 1932). Grey, 1956a:279 (distribution). Günther and Deckert, 1956:134, fig. 91 (popular account). Bruun, 1958:104 (restricted distribution, Gulf of Panama). Idyll, 1964:196, fig. 10-3 (popular account). Blache et al., 1970:453, fig. 1149 (erroneously listed for the eastern Atlantic). Pietsch, 1972a:29, 35, 45 (osteological notes). Brewer, 1973: 26 (36 additional specimens, eastern tropical Pacific). Parin et al., 1973:1467, fig. 37 (four additional females, eastern South Pacific). Nielsen, 1974:98 (listed in type catalog). Matthews et al., 1977:12, 14, tables 1, 2 (misidentification; specimen from tuna stomach, western North Atlantic). Bertelsen, 1984:328, fig. 167G (early life history, phylogeny). Ackerman and Murawski, 1997:88, color photograph (female with two parasitic males, LACM 30053-10). Beltrán-León and Herrera, 2000:284, fig. 92 (larvae, Pacific Colombia). Love et al., 2005:59 (eastern North Pacific). Pietsch, 2005b:210, 216, 226, 227, figs. 9B, 19A (reproduction). Pietsch and Orr, 2007:6, 9, figs. 4O, 6I (phylogenetic relationships).

### Material

Thirty-nine metamorphosed females (15 to 101 mm), 62 free-living males (11.0 to 17.5 mm), 13 parasitic males (9.5 to 22 mm), and 14 larvae (4.0 to 11.0 mm):

12A. Hyoid barbel with a proximal pair of lateral branches and 4 primary distal branches, each dividing into several filaments; esca with 2 blunt distal prolongations in front of each other and a pair of fringed filaments ........ ............ *Linophryne* (subgenus *Linophryne*) *bipennata* Bertelsen, 1982, p. 503

NOTE  A single known specimen, 37.5 mm; eastern South Indian Ocean

12B. Hyoid barbel without proximal lateral branches, 4 primary distal branches, only one of them undivided; esca without distal prolongation .... 13

13A. Esca with several branched filaments: 1 distal filament and a series of lateral filaments ......... *Linophryne* (subgenus *Linophryne*) *macrodon* Regan, 1925c, p. 504

NOTE  Ten known specimens, 19 to 91 mm; western North Atlantic and Gulf of Panama

13B. Esca with a single ribbonlike filament on each side of distal part of escal bulb, distal filament absent ....... ............ *Linophryne* (subgenus *Linophryne*) *maderensis* Maul, 1961, p. 504

NOTE  Four known specimens, 29 to 105 mm; eastern North Atlantic Ocean

## LINOPHRYNE ARCTURI (BEEBE, 1926B)

Figures 47, 183, 237, 241, 257; Tables 7–9

### Females

*Diabolidium arcturi* Beebe, 1926b:80, fig. (original description, single female; "the little sea devil of the *Arcturus*"; apparently the earliest observation of bioluminescence in a ceratioid). Beebe, 1926c: 360–362, fig. 1 (popular account, after Beebe, 1926b). Mead, 1958:132 (holotype transferred to CAS).

*Linophryne arcturi*: Regan and Trewavas, 1932:107 (new combination; description after Beebe, 1926b). Beebe and Crane, 1947:173, fig. 16 (description of holotype, bioluminescence). Bertelsen, 1951:174, 183 (after Beebe and Crane, 1947; in key). Harvey, 1952:530 (bioluminescence, after Beebe and Crane, 1947). Grey, 1956a:275 (distribution). Bertelsen, 1982:65, 97, figs. 6, 7 (first records from the Atlantic, comparison of all known material, in key). Bertelsen, 1990:518 (eastern tropical Atlantic). Munk, 1999:267 (bioluminescence).

*Diabolicum arcturi*: Munk, 1999:267 (misspelling of generic name).

### Males and Larvae

Unknown.

### Material

Three metamorphosed females (28.6 to 51 mm):

Holotype of *Diabolidium arcturi*: CAS-SU 46505 (formerly NYZS 6333), 28.5 mm, *Arcturus* station 74 T-70, south of Cocos Island, Costa Rica, 4°50′N, 87°00′W, open pelagic net, 915 m (500 fathoms), 2 June 1925.

Additional material: BMNH, one (34 mm); ISH, one (51 mm).

### Distinguishing Characters

Metamorphosed females of *Linophryne arcturi* differ from those of all other species of the genus in having the following escal and hyoid barbel characters: escal bulb with a compressed, flange-shaped anterior appendage; a stout, cylindrical distal escal appendage with terminal filament; base of distal escal appendage with a pair of structures on posterior margin, each consisting of a hemispherical protuberance with a digitiform anterior prolongation; a compressed posterior escal appendage. Barbel 118 to 357% SL, with several short, unpigmented distal branches bearing sessile and stalked photophores; stem of barbel pigmented with or without short simple lateral branches, each with a distal photophore.

### Description

Frontal spines absent; spines of sphenotic, preopercle, angular, and symphysis of lower jaw sharply pointed; teeth on each premaxillary 9 to 14, arranged in four series; teeth on each dentary 7 to 9, in three series, longest tooth 13 to 18% SL; a single pair of vomerine teeth in two smaller specimens, two pairs in largest known specimen; dorsal-fin rays 3; anal-fin rays 3; pectoral-fin rays 16 or 17.

Length of illicium 26 to 39% SL; escal appendages similar in basic pattern, but differing in shape in three known specimens: anterior escal appendage (1) a simple compressed flange, with smooth edge in 28.6-mm holotype, (2) compressed with deeply notched serrated edge in 51-mm specimen, and (3) distally prolonged into a pair of fringed flaps in 34-mm specimen; distal escal appendage (1) pigmented except on tip, simple cylindrical with a simple distal filament in 28.6-mm holotype, (2) with scattered melanophores, a short, pointed posterolateral branch, and a distal filament with tiny lateral branches in 51-mm specimen, and (3) pigmented at base with several fringed flaps on anterior margin and distal tip; largest specimen differing further from two smaller specimens in having a filament on tip of pair of structures at posterior base of distal appendage, and a short filament on each side of dorsal part of escal bulb.

Hyoid barbels of three known specimens also highly variable: total length of barbel 118 to 160% SL in two smaller females, 360% SL in 51-mm specimen; barbel of holotype bearing a number of short lateral branches irregularly spaced along length of stem, somewhat longer and more close set distally, but without forming a distinct tassel; barbel of two smaller specimens with a distal tassel very similar to those of *Linophryne coronata*; tassel of 34-mm specimen consisting of 3 proximal branches, a medial pair of branches, and 3 primary distal branches, each branch subdivided and bearing stalked and sessile photophores; tassel of 51-mm specimen consisting of a tapering proximal pair of primary branches, with a number of small photophores on slender stalks, and a distal cluster of 30 to 40 stalked photophores of varying size on a short stout base; stem of barbel of 34-mm specimen simple, without lateral branches, that of 51-mm specimen with short lateral branches throughout its length, each with a photophore on distal tip.

Subdermal pigmentation of holotype somewhat bleached; remaining specimens with slight but distinct concentration of subdermal pigment on caudal peduncle.

### Distribution

The holotype of *Linophryne arcturi* was caught in the eastern tropical Pacific off Cocos Island, Costa Rica, the other two specimens in the eastern Atlantic: the ISH specimen from off Ascension Island and the BMNH specimen from south of Madeira (Fig. 237). All were caught in open nets with maximum fishing depths of 650 to 1900 m.

### Comments

The three specimens here referred to *Linophryne arcturi* vary significantly in details of both the esca and hyoid barbel. Additional material is needed to determine whether they represent different closely related species or belong to one species with remarkably wide intraspecific variation.

## LINOPHRYNE CORONATA PARR, 1927

Figures 20, 47, 180, 183, 238; Tables 7–10

### Females and Parasitic Males

*Linophryne coronata* Parr, 1927:13, fig. 4 (original description, single specimen). Regan and Trewavas, 1932:107, 110 (after Parr, 1927). Bertelsen, 1951:174, 176 (*Linophryne longibarbata* Borodin, 1930a, and *L. coronata diphlegma* Parr, 1934, synonyms of *L. coronata*; in key). Grey, 1956a: 272 (distribution). Ponomarenko, 1959:83, fig. (additional female with remnant of parasitic male). Fitch and Lavenberg, 1968: 131, fig. 72 (distinguishing characters, natural history). Bertelsen, 1973:68 (comparison with *L. lucifer* and *L. sexfilis*). Bertelsen,

1976:10, figs. 2, 4 (description of female with parasitic male). Bertelsen, 1982:56, 97, figs. 3 to 6 (four additional females, comparison of all known material, in key). Bertelsen, 1986:1410, 1411, figs. (eastern North Atlantic, in key). Jónsson et al., 1986a:186 (additional female with parasitic male, southwest Iceland). Bertelsen, 1990: 518 (eastern tropical Atlantic). Nielsen and Bertelsen, 1992:62, fig. 8 (North Atlantic). Andriyashev and Chernova, 1995:103 (southern Greenland). Stearn and Pietsch, 1995:142, fig. 92 (three additional females, Greenland). Munk, 1998:182, figs. 7, 8 (escal light-guiding structures). Jónsson and Pálsson, 1999:204, fig. 8 (Iceland). Munk, 1999:267, 268, 271, 274, 278, 279, fig. 5C (bioluminescence). Pietsch, 2002b:1070 (western central Atlantic). Love et al., 2005:59 (eastern North Pacific). Pietsch, 2005b:211, 228 (reproduction). Bañón et al., 2006:385, figs. 1, 2 (new record, off Galician Bank, Portugal).

*Linophryne longibarbata* Borodin, 1930a: 87 (original description, single specimen). Borodin, 1931:83 (additional specimen, off Bermuda). Regan and Trewavas, 1932:107 (after Borodin, 1930a, 1931).

*Linophryne coronata coronata*: Parr, 1934:50, figs. 15, 16 (new subspecies based on holotype of *Linophryne coronata* Parr, 1927).

*Linophryne coronata longibarbata*: Parr, 1934:52, figs. 17, 18 (new subspecies based on holotype of *Linophryne longibarbata* Borodin, 1930a).

*Linophryne coronata diphlegma* Parr, 1934:54, figs. 19, 20 (original description, single specimen). Bertelsen, 1951:176 (differences among subspecies due to individual and ontogenetic variation).

*Linophryne coronata diplegma*: Bertelsen, 1976:11 (misspelling of subspecific name).

*Linophryne* aff. *coronata* A: Bertelsen, 1982:62, figs. 5A, C, 6 (description, single female from eastern tropical Atlantic, BMNH 2004.11.6.86, tentatively referred to *Linophryne coronata*).

*Linophryne* aff. *coronata* B: Bertelsen, 1982:62, fig. 5B, D, 6 (description, single female from eastern North Pacific, SIO 60-282, tentatively referred to *Linophryne coronata*).

*Linophryne* aff. *coronata*: Swinney, 1995a:52, 57 (additional specimen tentatively referred to *Linophryne coronata*, off Madeira). Swinney, 1995b:45, fig. 3 (after Swinney, 1995a; description, esca and hyoid barbel figured).

Free-living Males and Larvae

Unknown.

Material

Twenty-two metamorphosed females (29.5 to 225 mm), including three specimens (29.5 to 51.5 mm) tentatively referred to this species; and four parasitic males (18 to 26 mm, one represented only by partial remains):

Holotype of *Linophryne coronata*: YPM 2005, 33 mm, *Pawnee* station 39, Bahamas, 22°43′N, 74°23′W, open pelagic net, 8000 ft of wire, 29 March 1927.

Holotype of *Linophryne longibarbata*: MCZ 32307, 75 mm, *Atlantis*, cruise 1929, station 325, north Sargasso Sea, 37°00′N, 67°12′W, 1500 m, 29 April 1929.

Holotype of *Linophryne coronata diphlegma*: YPM 3207, 33 mm, *Atlantis* station 1478, Bahamas, 25°39′N, 77°18′W, triangular pelagic trawl, 1050 to 1100 m, 20 to 21 February 1933.

Additional material: HUMZ, three (150 to 169 mm); IIPB, one (172-mm female, with 16-mm parasitic male); ISH, one (36 mm); LACM, one (34.5 mm); MCZ, one (35 mm); MRIR, two (219-mm female, with 26-mm parasitic male; 290 mm TL female, with parasitic male); PINRO, one (225-mm female, with partial remains of a parasitic male); ROM, two (69 to 105 mm); SIO, one (65 mm); ZMUC, three (85 to 210 mm).

Tentatively assigned material: BMNH, one (36 mm); NMSZ, one (29.5 mm); SIO, one (51.5 mm).

Distinguishing Characters

Metamorphosed females of *Linophryne coronata* differ from those of other species of the genus in details of escal and barbel morphology: escal bulb with a distal prolongation, conical at base and distally more or less deeply bifurcated; a second, shorter pair of blunt, conical posterolateral appendages; posterior appendage short, conical or compressed; additional escal appendages absent. Barbel with long undivided stem and a distal tassel of branches; total length 90 to 276% SL (321 to 358% in tentatively referred specimens); stem of barbel without filaments or lateral branches (except in tentatively referred specimens); distal tassel more or less distinctly divided into 3 or 4 basal and 3 terminal branches; most branches compressed and pointed, tapering into a filament with a photophore on tip and with a single series of sessile or stalked photophores along one edge; length of tassel 8.5 to 28% SL (35 to 61% in tentatively referred specimens).

Juvenile females with some concentration of subdermal pigment on caudal peduncle.

Parasitic males with pointed sphenotic spines.

Description

Frontal spines absent; angular spine and that at symphysis of lower jaw relatively short; sphenotic and preopercular spines shortest in largest specimens (as in other species of the genus); two to three pairs of vomerine teeth in specimens greater than 35 mm; number of jaw teeth increasing with standard length from 11 to 14 to 17 to 21 on each side of upper jaw, from 5 to 10 to 10 to 15 in lower jaw; longest tooth (posterior-most in anteriormost series of dentary) 12 to 21% SL; dorsal-fin rays 3; anal-fin rays 2 or 3; pectoral-fin rays 13 or 14. A concentration of subdermal pigmentation on caudal peduncle discernible in smallest specimens.

Length of illicium 19 to 35% SL, shortest in juvenile holotype (19%) and in largest known specimen (24%); external pigmentation of esca increasing with age, completely covering escal bulb and appendages in largest known specimen (except for a narrow field behind the distal escal appendages; see Bertelsen, 1976, fig. 2B); morphology of escal and barbel appendages varying without distinct relation to size of specimens; length of distal prolongation of escal bulb varying from about 0.5 to about 1.5 times diameter of escal bulb, its distal bifurcation ranging from an indistinct low cleft to separation into a pair of short conical branches; position of smaller pair of appendages (in relation to base of escal prolongation) varying between lateral and posterior; some specimens with both pairs of distal appendages have nipplelike tips; posterior appendage varying from a small, blunt or pointed papilla to a compressed, somewhat larger flap. Branching of distal tassel of barbel ranging from blade shaped with a single series of more or less sessile photophores along compressed edge, to truncated, with a distal concentration of photophores, and to brush shaped, consisting of a cluster of stalked photophores.

Barbels of two specimens differing to such a degree from those of other specimens as well as from each other that they may represent separate species (for which reason they have been referred to as *Linophryne* aff. *coronata* A and B) (see Bertelsen, 1982:2, fig. 5); both differ from other specimens in having a longer barbel (321 and 358% SL versus 90 to 276%), having a number of lateral branches on stem of barbel (versus none), and having some branches of terminal tassel prolonged and branched (versus shorter and unbranched); *L.* aff. *coronata* A from eastern tropical Atlantic with lateral branches branched and restricted to distal half of stem; *L.* aff. *coronata* B from eastern North Pacific with lateral branches of barbel simple and distributed throughout length of stem of barbel.

Two largest females (219 and 225 mm) with parasitic males, one represented by only a small remnant (see Bertelsen, 1976:11); the other well preserved, 26 mm, with pointed sphenotic spine, pattern of larval subdermal melanophores obliterated by secondary pigmentation.

Distribution

Most of the known material of *Linophryne coronata* was collected from the western North Atlantic, where records extend from off Greenland and Iceland, as far north as 67°N, south to Newfoundland, Bermuda, and the Bahamas (Fig. 238). In the eastern North Atlantic it is known from off Portugal at about 42°N, and off West Africa at about 11°N. Three specimens have been taken in the eastern Pacific: one from off southern California, another in central water at about 27°N, 138°W, and a third from the equator at about 110°W. The two largest females (219 to 225 mm) were caught in commercial bottom trawls at 90 and 370 to 480 m, respectively; another was taken in a pelagic closing net between 1000 and 1250 m; and the remaining specimens in open pelagic nets with maximum fishing depths of 1000 to 3000 m.

### LINOPHRYNE LUCIFER COLLETT, 1886

Figures 13, 183, 237, 308, 309; Tables 7, 8, 10

#### Females and Parasitic Males

*Linophryne lucifer* Collett, 1886:138, pl. 15 (original description, single female). Günther, 1887:57 (description after Collett, 1886). Lütken, 1894:79 (after Collett, 1886). Goode and Bean, 1896:496, pl. 121, fig. 408 (details of capture; description after Collett, 1886). Jordan and Evermann, 1896:510 (after Collett, 1886). Gill, 1909: 585, fig. 23 (semipopular account, figure after Collett, 1886). Regan, 1926:24 (after Collett, 1886). Fowler, 1936:1141, 1341, fig. 480 (West Africa). Bertelsen, 1951:174, 176 (after Collett, 1886; in key). Grey, 1956a: 272 (distribution). Brandes and Kotthaus, 1960:72 (additional specimen, UMB 4792, Iceland-Faroe Ridge). Krefft, 1963:83 (additional specimen, ISH 54/61, Anton Dohrn Bank). Wellershaus, 1963:163, figs. 1, 2 (additional specimen, Denmark Strait). Idyll, 1964:274 (popular account). Bertelsen and Krefft, 1965:294, figs. 1–4 (three additional females, North Atlantic; description, figures of escae and barbels). Jónsson, 1967b:316, figs. 1, 2 (female with parasitic male, East Greenland). Pethon, 1969:13 (listed in type catalog). Ejsymont, 1970:59, figs. 1–5 (additional female, Flemish Cap). Jónsson, 1970:278 (additional female, southwest Iceland). Bertelsen, 1973:68 (comparison with *Linophryne sexfilis* and *L. coronata*). Bertelsen, 1976:8, figs. 1, 4 (description of female with parasitic male recorded by Jónsson, 1967b). Bertelsen, 1982:51, 97, figs. 1, 6 (three additional females, one with parasitic male; description, comparison with all known material; first record from Indian Ocean; in key). Jónsson et al., 1984: 205 (additional female, Iceland). Jónsson et al., 1986a:186 (additional female, West Iceland). Nielsen and Bertelsen, 1992:62, fig. 7 (North Atlantic). Stearn and Pietsch, 1995:143, fig. 93 (additional female, Greenland). Ellis, 1996:268, fig. (popular account). Jónsson and Pálsson, 1999: 204, fig. 8 (Iceland). Pietsch, 2005b:211, 228, fig. 23A (reproduction).

*Linophryne lucifera*: Regan and Trewavas, 1932:107, 109 (emendation of specific name, description after Collett, 1886; in key). Jónsson et al., 1979:229 (additional female, Iceland). Albuquerque, 1954–1956:1073, 1074, fig. 441 (Portugal, in key). Maul, 1973:676 (eastern North Atlantic). Bertelsen, 1986:1409, 1411, figs. (eastern North Atlantic, in key). Jónsson et al., 1986c:628 (two additional specimens, Iceland). Andriyashev and Chernova, 1994:103 (southern Greenland).

*Linophryne digitopogon* Balushkin and Trunov, 1988:62, fig. 1 (original description, single female). Heemstra, 1995:377, fig. 109a.2 (after Balushkin and Trunov, 1988). Anderson and Leslie, 2001:28, fig. 18 (after Balushkin and Trunov, 1988, and Heemstra, 1995).

#### Free-living Males and Larvae

Unknown.

#### Material

Twenty-nine females, 28 metamorphosed (30 to 275 mm) and one in late metamorphosis (22 mm); six with parasitic males (24 to 29 mm):

Holotype of *Linophryne lucifer*: ZMUO J.5560, 31.5 mm, off Madeira, ca. 36°N, 20°W, found floating on the surface, May 1877.

Holotype of *Linophryne digitopogon*: ZIN 48199, 30 mm, *Gizhiga*, eastern South Atlantic, off Childs Bank, 30°38′S, 14°50′E, 0 to 920 m, 17 February 1974.

Additional material: AUS, one (193 mm); BMNH, one (135 mm); BSKU, one (130 mm); IMB, two (165 to 214 mm); ISH, one (175 mm); MRIR, three (180-mm TL female, with 40-mm TL parasitic male; 190-mm TL female, with 30-mm TL parasitic male; 230-mm female, with 24-mm parasitic male); ROM, two (22 to 71 mm); SIOM, one (190-mm female, with 24.5-mm parasitic male); SMNS, one (197.5 mm); UMB, two (146 to 164.5 mm); ZMUC, 12 (85 to 275 mm; 174-mm female, with 23.5-mm parasitic male; 275-mm female, with 29-mm parasitic male).

#### Distinguishing Characters

Metamorphosed females of *Linophryne lucifer* differ from those of all other species of the genus in details of escal and hyoid barbel morphology: escal bulb with a conical distal prolongation bearing short, stout filaments; posterior escal appendage small, conical; additional escal appendages absent. Barbel with a single stem, divided distally into a pair of compressed pointed branches, each bearing a series of photophores on one edge, and with a pair of short, simple lateral branches on distal half of stem, each with a few photophores at distal tip.

Parasitic males of *Linophryne lucifer* with pointed sphenotic spines; juvenile females and parasitic males with a concentration of subdermal melanophores on caudal peduncle.

#### Description

Frontal spines absent; sphenotic and preopercular spines, and symphysial spine of lower jaw well developed in small specimens, shorter and less pointed in adults; each premaxilla with 10 to 20 teeth arranged in four series, each dentary with 8 to 13 teeth in three series; longest tooth in lower jaw 10 to 25.5% SL; total number of teeth on vomer 2 or 3; dorsal-fin rays 3; anal-fin rays 2 or 3 (rarely 2); pectoral-fin rays 14 to 16.

Length of illicium 10 to 16.5% SL, except holotype 28.5%; escal bulb nearly spherical, tapering gradually to form a short, conical distal prolongation, bearing 1 to 8 filaments, varying in numbers unrelated to standard length; escal filaments short and stout, shorter than 25% diameter of escal bulb in most specimens, reaching about 50% of escal bulb in some specimens; a small blunt or conical posterior appendage below escal pore; additional escal appendages absent.

Length of barbel 68 to 75% SL in specimens 49 to 214 mm, shorter in 22-mm metamorphic specimen and in largest female (230 mm, barbel 41% SL); distal pair of branches 6 to 17% SL, longest in 49-mm holotype, shortest in largest known specimen; branches slender and pointed, blade shaped, and of equal length in most specimens, asymmetrical in a few large specimens, and with an additional short intermediate branch in one specimen; photophores normally in a single series of 20 to 35 along one edge of each branch; a pair of short, simple lateral branches on distal half of barbel, placed at somewhat varying distance above base of terminal branches; one specimen with two close-set pairs; each lateral branch bearing a few distal photophores, usually 1 born at distal tip of a distal filament.

Body of 22-mm metamorphic female and parasitic males finely covered with subdermal pigment, with a distinct but weak concentration of melanophores on caudal peduncle; parasitic males with well-developed pointed sphenotic spines, but without characters distinguishing them from other males of the subgenus.

#### Distribution

Except for a single specimen from the eastern South Indian Ocean, collected in a bottom trawl at 1250 m (which may represent an undescribed species, see Comments, below), all known material was taken in the North Atlantic, from off

Iceland and East and West Greenland (Fig. 237). A few additional specimens were taken farther south, extending from off Newfoundland to Madeira. Nearly all the specimens, including all the largest known females, were caught in high latitudes by commercial bottom trawls fished at depths of about 300 to 600 m. Two smaller specimens (22 to 71 mm) were taken with open pelagic trawls fished at a maximum depth of 1000 m.

Comments

The specimen from the Indian Ocean, which is the only record of *Linophryne lucifer* outside the Atlantic, differs from the description given above in some details of the esca and barbel and might represent a separate closely related species (see Bertelsen, 1982:53, fig. 1).

*LINOPHRYNE SEXFILIS* BERTELSEN, 1973

Figure 240; Tables 7–9

Females

*Linophryne sexfilis* Bertelsen, 1973:65, figs. 1, 2 (original description, single specimen). Hansen and Herring, 1977:104, 107, fig. 1e (bioluminescence). Herring and Morin, 1978:322, fig. 9.12 (bioluminescence, after Hansen and Herring, 1977). Bertelsen, 1982:92 (after Bertelsen, 1973; species referred to *Linophryne* subgenus *Linophryne*). Bertelsen, 1986:1410, 1413, figs. (eastern North Atlantic, in key). Munk, 1999:267 (bioluminescence).

Males and Larvae

Unknown.

Material

A single metamorphosed female (38.5 mm):

Holotype of *Linophryne sexfilis*: BMNH, 1976.11.4.1, 38.5 mm, *Discovery* station 7856-50, North Atlantic, 30°04′N, 23°00′W, RMT 8, 1250 to 1000 m, 5 April 1972.

Distinguishing Characters

Metamorphosed females of *Linophryne sexfilis* differ from those of all other species of the genus in the following details of escal and hyoid barbel morphology: escal bulb with a cylindrical distal prolongation, divided distally into 6 simple tapering filaments; posterior appendage tiny, papilliform; additional escal appendages absent. Barbel simple, undivided, bearing a distal tassel of appendages consisting of a simple, cylindrical, distally unpigmented and slightly expanded prolongation of primary stem of barbel, surrounded at base by 3 shorter, unpigmented, compressed pointed branches, each bearing a series of sessile photophores along outer margin.

Description

Frontal spines absent; lower jaw with a short symphysial spine; sphenotic, preopercular, and angular spines well developed, sharp, piercing skin; vomer with two pairs of teeth; each premaxilla with 16 teeth in four series (1 + 3 + 5 + 7), each dentary with 11 teeth in three series (2 + 4 + 5); posterior-most tooth of most series in development; longest dentary tooth (second tooth of first series) 18.6% SL; dorsal-fin rays 3; anal-fin rays 3; pectoral-fin rays 16 or 17.

Length of illicium 22% SL; stem of illicium and slightly oval escal bulb darkly pigmented; length of distal escal appendage 33.5% SL; distal escal prolongation and filaments unpigmented; cylindrical base of escal prolongation bearing 3 stout tapering filaments surrounding a more narrow continuation of base, again dividing into 3 somewhat shorter filaments; all filaments ending at about same distance from base of escal prolongation.

Total length of barbel 72% SL, undivided darkly pigmented stem 45% SL; distal third of unpigmented, rod-shaped central branch slightly expanded, containing a central opaque thread ending in 2 subdermal photophores; 3 blade-shaped branches at base about half as long, each bearing 8 to 10 sessile photophores along outer compressed edge; caudal peduncle covered with secondary subdermal pigment, obliterating eventual pattern of larval pigmentation.

Distribution

The single known specimen of *Linophryne sexfilis* was caught in the eastern North Atlantic, northwest of the Canary Islands, in an open midwater trawl fished at a maximum depth of 1250 m.

*LINOPHRYNE ARGYRESCA* REGAN AND TREWAVAS, 1932

Figures 297, 306; Table 8

Females and Parasitic Males

*Linophryne argyresca* Regan and Trewavas, 1932:19, 108, figs. 9, 167, 168, pl. 10, fig. 3 (original description, single female with attached male). Parr, 1934: Bertelsen, 1951:175, 185, fig. 121E (additional description, comparison with all known material, in key). Matsubara, 1955:1354, fig. 530E (sexual parasitism; figure after Regan and Trewavas, 1932). Grey, 1956a:277 (distribution). Günther and Deckert, 1956:135, fig. 93 (popular account). Idyll, 1964:279, fig. 13-11 (sexual parasitism, popular account). Nielsen, 1974:99 (listed in type catalog). Bertelsen, 1982:54, 98, figs. 2, 6 (detailed description of holotype, comparison of all known material, in key). Pietsch, 2005b:211, fig. 9D (reproduction).

Free-living Males and Larvae

Unknown.

Material

A single metamorphosed female (61 mm), with parasitic male (12 mm):

Holotype of *Linophryne argyresca*: ZMUC P92142, 61 mm, with 12-mm parasitic male, *Dana* station 3904(1), Indian Ocean, 5°18′N, 90°55′W, 3500 m wire, 1745 h, 18 November 1929.

Distinguishing Characters

Metamorphosed females of *Linophryne argyresca* differ from those of all other species of the genus in the following details of escal and hyoid barbel morphology: escal bulb with a large, oblong, internally silvery distal prolongation (as long as diameter of bulb), with a terminal cluster of about 7 branched filaments inside a circle of 7 shorter filaments; on each side at base of distal prolongation a small, distally fringed lobe; posterior escal appendage compressed, with fringed edge. Primary stem of barbel unbranched, darkly pigmented, length 18% SL, with an unpigmented, slightly tapering, filamentous distal prolongation, flanked on each side by a similar filamentous appendage; length of each lateral branch about 12% SL; distal end of all three terminal filaments bearing a few internal photophores.

In addition, the single metamorphosed female of *Linophryne argyresca* differs from all other females of *Linophryne*, except for *L. quinqueramosa* and some specimens of *L. algibarbata* (see Bertelsen, 1980b:239), in having a pair of small, pointed frontal spines piercing the skin above and slightly in front of the eyes.

Parasitic male with pointed sphenotic spines and a distinct, but weak concentration of subdermal pigment on caudal peduncle.

Description

Frontal spines well developed, sharp, piercing skin; lower jaw with a short symphysial spine; sphenotic, preopercular, and angular spines well developed, sharp, piercing skin; vomer with two pairs of teeth (plus 1 tiny tooth lying just in front of pair on right side); each premaxilla with 20 teeth in four series (2 + 3 + 5 + 10), left dentary with 11 teeth, right dentary with 12 teeth, each in three series (3 + 3-4 + 5); longest premaxillary tooth (second tooth of first series) 6.7% SL; longest dentary tooth (third tooth of first series) 12.2% SL; dorsal-fin rays 3; anal-fin rays 3; pectoral-fin rays 16.

Length of illicium about 22% SL; length of escal bulb about 14% SL; length of distal prolongation of esca about 7% SL.

Parasitic male without apparent degeneration of eyes or olfactory organs, and without swelling of abdomen; attached to belly of female below and slightly in front of anus (see Regan and Trewavas, 1932, fig. 9, pl. 10, fig. 3); a relatively weak concentration of subdermal pigment on caudal peduncle (see Bertelsen, 1951:184, fig. 121E; similar concentration of pigment on peduncle of female, but hardly distinguishable, being mixed with light, evenly distributed secondary pigmentation).

Sexual Maturity

The ovaries of the holotype are small and immature, about 6 mm in length and width, and contain small oocytes, each about 0.05 mm in diameter.

Distribution

The only known specimen of Linophryne argyresca was collected in the eastern tropical Indian Ocean somewhere between the surface and about 1750 m.

Comments

Among the known species of the genus, Linophryne argyresca is most similar to L. sexfilis in the development of the distal prolongation of the escal bulb, in the pattern of distal escal filaments, as well as in the branching of the hyoid barbel (2 or 3 lateral branches emerging from the base of a simple distal prolongation of the stem). It differs most distinctly from L. sexfilis and all other species of the genus in the shape of the posterior escal appendage, the presence of fringed lateral escal lobes, and in the number and morphology of the distal branches of the barbel.

LINOPHRYNE ESCARAMOSA
BERTELSEN, 1982

Figure 183; Table 8

Females

Linophryne escaramosa Bertelsen, 1982: 69, 95, figs. 8, 13 (original description, single specimen, in key).

Males and Larvae

Unknown.

Material

A single known metamorphosed female (36.5 mm):

Holotype of Linophryne escaramosa: LACM 42296-1, 36.5 mm, Hawaiian Islands off Oahu, 21°10′ to 20′N, 158°10′ to 20′W, 3-m Isaacs-Kidd midwater trawl, oblique tow, 0 to 1350 m, 1134 to 1418 h, 27 May 1974.

Distinguishing Characters

Metamorphosed females of Linophryne escaramosa differ from those of all other members of the genus in the following details of escal morphology (barbel missing, morphology unknown): esca with an oblong distal prolongation of bulb, as long as diameter of escal photophore, with internal light-guiding structure bifurcated at tip and bearing branched distal filaments; a series of 6 or 7 elongate branched filaments on each side of escal bulb, the proximal-most emerging from base of bulb; a small conical posterior appendage behind escal pore. A dense concentration of subdermal pigment on caudal peduncle.

Description

Barbel and surrounding skin lost in the otherwise well-preserved holotype: body proportions, dentition, and development of spines well within variation observed in other species of genus; 12 teeth on each premaxilla, arranged in four series (1 + 2 + 4 + 5), longest premaxillary tooth (the first in the first series) 13% SL; 10 teeth on each dentary, arranged in three series (2 + 4 + 4), longest dentary tooth (the second in the first series) 18% SL, last dentary tooth of second and third series in development; 3 vomerine teeth (plus a small tooth in development on left side); dorsal-fin rays 3; anal-fin rays 3; pectoral-fin rays 16; caudal peduncle covered with a dense layer of subdermal pigment, extending anteriorly to middle of bases of dorsal and anal-fins.

Esca unpigmented except for photophore; base of distal prolongation of bulb cylindrical; distal filaments, one lying just anterior to the other, as long as diameter of bulb, each with 4 or 5 short simple lateral branches; length of lateral filaments three or four times diameter of bulb, tending to form a double series on each side, each bearing 6 to about 18 short simple secondary filaments.

Distribution

The only known specimen of Linophryne escaramosa was collected off Oahu, Hawaiian Islands, somewhere between the surface and 1350 m.

Comments

Although the missing barbel of the holotype of Linophryne escaramosa precludes a full diagnostic description, a combination of escal characters separates this species distinctly from all other previously described species of the genus. The series of branched filaments on each side of the escal bulb are similar to those of L. polypogon, L. racemifera, and L. macrodon, and all four species have a distal escal filament; but, in contrast to L. escaramosa, the distal escal filament of these three otherwise similar species is not raised on a distal prolongation of the bulb. In the several species that possess such a prolongation of the bulb, it is developed in the smallest known metamorphosed specimens. Because the largest known specimens of L. racemifera and L. macrodon are considerably larger than the holotype of L. escaramosa, it is most unlikely that the difference in this character from L. escaramosa could be due to a difference in age. Furthermore, in contrast to L. escaramosa and most other species of the genus, L. polypogon, L. racemifera, and L. macrodon lack a posterior appendage placed directly behind the escal pore.

LINOPHRYNE BIPENNATA
BERTELSEN, 1982

Table 8

Females

Linophryne bipennata Bertelsen, 1982:71, 98, figs. 9, 13 (original description, single specimen, in key). Pietsch, 1999:2037 (western central Pacific).

Males and Larvae

Unknown.

Material

A single metamorphosed female (37.5 mm):

Holotype of Linophryne bipennata: SIO 61-31, 37.5 mm, Argo, Monsoon Expedition station II-7, trawl 3, Java Trench, 12°05.9′S, 115°26.2′E, 3-m Isaacs-Kidd midwater trawl, 0 to 2000 m, bottom depth 4970 m, 28 to 29 October 1960.

Distinguishing Characters

Metamorphosed females of Linophryne bipennata differ from those of all other species of Linophryne in details of escal and hyoid barbel morphology: escal bulb with a low distal prolongation, length about one-half diameter of bulb, distally divided to form 2 short, distally rounded appendages (posterior appendage larger than anterior appendage); on each side of base of distal prolongation a single compressed filament, length about three times diameter of bulb, each bearing 2 to 4 short tapering lateral filaments; a low papilliform posterior appendage behind escal pore. Barbel about 25% SL in total length, with 2 lateral branches emerging from posterior proximal margin, and distally divided at about half length to form 3 primary branches; all branches repeatedly divided, each terminal branch with a series of internal photophores and 1 or more distal filaments. A dense concentration of subdermal pigment on caudal peduncle.

Description

Body proportions, and development of spines and dentition within the variation observed in other species of the genus: 12 teeth on each premaxilla, arranged in three series (1 + 2 + 3 + 6); longest premaxillary tooth (the first in the first series) 13% SL; 11 teeth on each dentary, arranged in three series (2 + 3 + 6); longest dentary tooth (the second of the first series) 19% SL; last premaxillary tooth of fourth series and last dentary tooth of third series in development; 3 vomerine teeth; dorsal-fin rays 3; anal-fin rays 3; pectoral-fin rays 16 or 17; caudal peduncle covered with dense layer of subdermal pigment, extending to anterior base of dorsal and anal rays.

Esca unpigmented except for internal photophore; compressed filaments with broad bases tapering into narrow shafts proximal to compressed distal part; 2 short, tapering lateral branches on each side of left filament, 2 similar branches on one side of right filament; posterior appendage a low hemispherical papilla. Barbel unpigmented except at base; left proximal lateral branch trifurcated, right proximal lateral branch simple; anterior branch of trifurcated distal part of barbel with a small proximal lateral branch and a bifurcated distal part; median branch bifurcated, each secondary branch with 2 or 3 small tertiary branches; posterior branch simple with 2 short lateral branches; all terminal branches with several internal photophores and filaments, some with proximal internal photophores.

Distribution

The only known specimen of *Linophryne bipennata* was taken in the Java Trench somewhere between the surface and 2000 m.

Comments

Among the known species of the genus, *Linophryne bipennata* is most similar to *L. maderensis* in having a single pair of compressed escal filaments; it is further similar to *L. maderensis*, as well as to *L. macrodon*, in the pattern of distal branching of the barbel. It differs significantly from these species, however, in having a distal prolongation of the escal bulb and a pair of lateral branches emerging from the proximal part of the stem of the barbel. Known only from the eastern Indian Ocean, *L. bipennata* differs further from *L. maderensis* and *L. macrodon* in geographic distribution, the former known only from the eastern North Atlantic, the later known only from the Gulf of Panama and the western North Atlantic.

## *LINOPHRYNE MADERENSIS* MAUL, 1961

Figure 238; Tables 7, 8, 10

### Females and Parasitic Males

*Linophryne maderensis* Maul, 1961:151, figs. 30, 31A, B (original description, three females). Maul, 1973:676 (eastern North Atlantic). Bertelsen, 1982:73, 98, figs. 10, 13, 18 (detailed description of type material, comparison of all known material, in key). Bertelsen, 1986:1409, 1412, figs. (eastern North Atlantic, in key). Pietsch, 2005b:211, 216 (reproduction).

### Free-living Males and Larvae

Unknown.

### Material

Four metamorphosed females (29 to 105 mm) and one parasitic male (15 mm):
Holotype of *Linophryne maderensis*: MMF 9094, 34 mm, off Madeira, from stomach of *Aphanopus carbo*, 10 August 1956.
Paratypes of *Linophryne maderensis*: MMF 9911, 38 mm, off Madeira, from stomach of *Aphanopus carbo*, 16 October 1956; MMF 15119A, 29 mm, off Madeira, from stomach of *A. carbo*, 15 September 1958.
Additional material: ZMB, one (105-mm female, with 15-mm parasitic male).

### Distinguishing Characters

Metamorphosed females of *Linophryne maderensis* differ from those of all other species of the genus in having the following escal and hyoid barbel characters: a single unbranched filamentous lateral appendage emerging from each side of escal bulb; additional escal appendages and distal prolongation of escal bulb absent. Total length of barbel (retained only in holotype and 105-mm specimen) about 60% SL, with proximal half simple, gradually dividing distally into a proximal branch, a medial branch, and a pair of distal branches; each branch tapering to form a slender filament or subdividing into 2 to 4 distal filaments, each filament with 1 to 5 internal photophores.

A distinct concentration of subdermal pigment on caudal peduncle.

### Description

Dorsal profile of frontals angular, but without spine; symphysial spine of lower jaw short; sphenotic, preopercular, and angular spines slender and sharp (as in other *Linophryne* of similar size); teeth in each premaxillary 11 to 13, arranged in four series; teeth in each dentary 8 to 11, arranged in three series; longest tooth in lower jaw 19% SL; a single pair of vomerine teeth in 29-mm paratype, vomer of other specimens with 3 irregularly spaced teeth; dorsal-fin rays 3; anal-fin rays 3; pectoral-fin rays 16 or 17.

Length of illicium 30.5 to 36% SL; escal bulb smooth, ovoid, with escal pore close to apex; bases of lateral escal filaments (retained in all specimens) arising in an acute angle from surface of bulb slightly above its greatest diameter; right lateral escal filament of holotype slender and untapering, length about 1.5 times diameter of escal bulb (about 16% SL), its distal tip split into 3 or 4 tiny threads, possibly representing remains of distal branching.

Length of hyoid barbel of holotype 60% SL, undivided part 26% SL, darkly pigmented except for distal part of branches; proximal branch simple, with a single terminal photophore; medial branch dividing to form a cluster of 4 distal filaments each with a series of 3 to 5 photophores; one of two distal pairs of branches simple, except for a tiny lateral branch at base of unpigmented part, bearing a series of 4 photophores; other distal pair of branches divided into 3 filaments each with 3 or 4 photophores; photophores unstalked, appearing as pearls on a string.

Subdermal pigmentation with numerous minute melanophores on caudal peduncle (see Maul, 1961:152, fig. 30).

### Distribution

Three of the four known specimens of *Linophryne maderensis* were collected from the stomachs of *Aphanopus carbo* (family Trichiuridae) caught in a long-line fishery off Madeira (Parin, 1986; Fig. 238). The fourth specimen was taken just north of Madeira at approximately 37°N, 19°W, between 0 and 880 m (Peter Bartsch, personal communication, 13 August 2007).

## *LINOPHRYNE MACRODON* REGAN, 1925C

Figures 20, 40, 239, 249, 258; Tables 7, 8, 10

### Females and Parasitic Males

*Linophryne macrodon* Regan, 1925c:564 (original description, single female). Regan, 1926:24, pl. 2, fig. 2 (after Regan, 1925c). Parr, 1927:7 (in key). Regan and Trewavas, 1932:107 (after Regan, 1925c, 1926). Fowler, 1944:528 (Gulf of Panama). Bertelsen, 1951:175, 185, figs. 120B, 121C (description, new figures of esca and subdermal pigmentation; material assigned to *Linophryne polypogon* group, in key). Grey, 1956a:276 (distribution). Idyll, 1964:198, fig. 10-4 (popular account). Parin et al., 1973:146, fig. 37 (additional female, eastern South Pacific). Nielsen, 1974:100 (listed in type catalog). Bertelsen, 1982:76, 98, figs. 11–13 (additional specimens, comparison of all known material, in key). Pietsch, 2002b:1070 (western central Atlantic). Pietsch, 2005b:211, 228 (reproduction). Pietsch and Orr, 2007:4, fig. 3H (phylogenetic relationships).

*Linophryne brevibarbis* Parr, 1927:7, fig. 1 (original description, single female). Regan and Trewavas, 1932:108 (after Parr, 1927). Parr, 1934:43 (in key). Bertelsen, 1951:175, 185 (after Parr, 1927; assigned to *Linophryne polypogon* group). Grey, 1956a:276 (distribution). Günther and Deckert, 1956:180, fig. 124C (evolution of dorsal-fin spines, semipopular account).

*Linophryne polypogon* Group: Pietsch, 1976:788 (female with parasitic male, western Atlantic).

Free-living Males and Larvae

Unknown.

Material

Ten metamorphosed females (19 to 91 mm), including two tentatively assigned specimens (28 to 28.5 mm); and one parasitic male (21.5 mm):

Holotype of *Linophryne macrodon*: ZMUC P92144, 37 mm, *Dana* station 1208(4), Gulf of Panama, 6°48'N, 80°33'W, open pelagic net, 3500 m of wire, 0810 m, 16 January 1922.

Holotype of *Linophryne brevibarbis*: YPM 2001, 25 mm, *Pawnee* station 58, off Bermuda, 32°24'N, 64°29'W, open pelagic net, 10,000 ft wire, 20 April 1927.

Additional material: BMNH, one (26 mm); UF, one (91-mm female with 21.5-mm parasitic male); USNM, one (24 mm); ZMUB, one (19 mm); ZMUC, two (40 to 41 mm).

Two tentatively assigned specimens: MCZ, one (28.5 mm); UW, one (28 mm).

Distinguishing Characters

Metamorphosed females of *Linophryne macrodon* differ from those of all other species of the genus in having the following escal and hyoid barbel characters: escal bulb without distal prolongation; a single distal filament and three pairs of lateral filaments, all branched except in juveniles; anterior and posterior escal appendages absent. Length of barbel 24 to 80% SL, with stem unbranched and distal half to two-thirds gradually dividing into 3 primary branches, each terminating into 1 to 4 tapering, branched or simple filaments, each with a series of few to 25 to 30 internal photophores. A dense concentration of subdermal melanophores on caudal peduncle.

Males with pointed sphenotic spines.

Description

Dorsal profile of frontals angular, but without spine; sphenotic, preopercular, and angular spines similar to those of females of similar size of other species of subgenus; teeth in each premaxillary 7 to 17, arranged in four series; teeth in each dentary 8 to 11, arranged in three series, longest tooth in fully metamorphosed specimens 14 to 19% SL; one or two pairs of vomerine teeth; dorsal-fin rays 3; anal-fin rays 3; pectoral-fin rays 15 to 17.

Metamorphosed females with length of illicium 26 to 37% SL, shorter in metamorphic specimens; distal escal filament simple in holotype, with a rudimentary pair of filaments at base in three metamorphic females, two pairs of filaments in three largest specimens; six of seven specimens with lateral escal filaments very similar, most branching to form as many as 6 secondary filaments, length of each (median and longest pair of filaments one to 1.5 times escal diameter) slightly increasing with increasing standard length; sixth specimen (41-mm) with filaments somewhat longer (longest about three times diameter of escal bulb) and more highly branched, each forming 8 to about 25 secondary filaments.

Length of barbel increasing from 24 to 32% SL in metamorphic specimens to 65 to 80% SL in metamorphosed specimens; divided portion of stem of barbel 10 to 14% SL in metamorphic specimens, 18 to 27% SL in metamorphosed specimens; branching very similar in all specimens; proximal posterior branch simple with a distal series of 2 to 6 internal photophores; metamorphic specimens with anterior branch simple except for tiny rudiments of lateral branches; metamorphosed specimens with anterior branch divided distally into 3 or 4 secondary branches, some further divided to form a cluster of 7 to 10 slender tapering filaments, each with a series of few to about 20 photophores; distal pair of primary branches of unequal length, bearing a few rudimentary lateral branches in metamorphic specimens, of nearly equal length in metamorphosed specimens, each divided distally to form a terminal cluster of 2 to 5 slender filaments, longest with a series of 25 to 30 photophores.

A dense concentration of subdermal pigment on caudal peduncle.

Single known parasitic male, 21.5 mm, with pointed sphenotic spines, attached to largest known female, 91 mm, with immature eggs each about 0.1 mm in diameter.

Distribution

The holotype of *Linophryne macrodon* was caught in the Gulf of Panama, but all the additional specimens were taken in the western North Atlantic between about 24° and 40°N, west of 63°W (Fig. 239). The 41-mm specimen was taken in a bottom trawl at about 300 m, but all the remaining specimens were caught in various nonclosing nets, with maximum fishing depths exceeding 1000 m.

## LINOPHRYNE POLYPOGON REGAN, 1925C

Figures 181, 182, 237; Table 8

Females

*Linophryne polypogon* Regan, 1925c:565 (original description, single female). Regan, 1926:24, pl. 2, fig. 3 (after Regan, 1925c). Regan and Trewavas, 1932:108 (after Regan, 1926). Parr, 1934:43 (in key). Fowler, 1936:1142 (West Africa). Bertelsen, 1951:174, 183, figs. 120A, 121A (additional description of holotype, comparison with all known material, in key). Albuquerque, 1954–1956:1073, 1074, fig. 442 (Portugal, in key). Grey, 1956a:276 (distribution). Maul, 1973:676 (eastern North Atlantic). Nielsen, 1974:100 (listed in type catalog). Bertelsen, 1980b:243, figs. 5, 6 (redescription of holotype, comparison with all known material). Bertelsen, 1986:1409, 1412, figs. (eastern North Atlantic, in key). Ni, 1988:334, fig. 260 (additional specimen, East China Sea). Meng et al., 1995:443, fig. 595 (after Ni, 1988). Bertelsen, 1994:140, color plate (popular account). Bertelsen and Pietsch, 1998a:140, color photograph (after Bertelsen, 1994). Pietsch and Orr, 2007:4, fig. 3I (phylogenetic relationships).

Males and Larvae

Unknown.

Material

Three metamorphosed females (32 to 97 mm):

Holotype of *Linophryne polypogon*: ZMUC P92145, 32 mm, *Dana* station 1142(6), 33°25'N, 16°58'W, open pelagic net, 5000 m of wire, 15 October 1921.

Additional material: BMNH, one (33 mm); ECFR, one (97 mm).

Distinguishing Characters

Metamorphosed females of *Linophryne polypogon* differ from those of all other species of the genus in details of escal and barbel morphology: escal bulb without distal prolongation; 2 distal filaments and a series of 5 to 8 branched filaments on each side; anterior and posterior escal appendages absent. Barbel divided at base into numerous (14 to 19) branches, each bearing numerous tiny filaments and a single distal photophore.

Caudal peduncle with a concentration of subdermal melanophores.

Description

Based primarily on 32-mm holotype (data for 97-mm specimen, courtesy of Ni Yong, personal communication, 10 October 1985): frontals with relatively high rounded projections; sphenotic, preopercular, and angular spines well developed; symphysial spine of lower jaw short; teeth in each premaxillary 9 to 13, arranged in four series; teeth in each dentary 8 or 9, arranged in three series, longest tooth 20% SL; dorsal-fin rays 3; anal-fin rays 3; pectoral-fin rays 16 or 17.

Length of illicium 31% SL; 2 distal escal filaments, emerging side by side, anterior filament longest, nearly equal to diameter of escal bulb, with two pairs of short lateral secondary branches, posterior filament simple; each lateral escal filament with a somewhat flattened stem and bearing 0 to 6 short lateral branches, single or arranged in pairs; longest lateral filament 1 to 1.5 times diameter of escal bulb.

Barbel of holotype with 19 branches, each tapering gradually to form long slender filaments; 16 branches of about equal length (40 to 50% SL); 2 shorter (12% SL) branches forming an anterior pair, a single branch (about 25% SL) placed behind anterior pair perhaps broken. Barbel of 97-mm specimen with 14 branches, longest branch 86% SL, shortest branch 8% SL. Stalked photophores distributed throughout length of branches of barbel except for a short distance from base.

Posterior half of caudal peduncle with a well-defined group of subdermal melanophores.

### Distribution

The three known specimens of *Linophryne polypogon* are from widely separated localities (Fig. 237): two were collected in the North Atlantic, the holotype from off Madeira in an open pelagic net, with 5000 m of wire out (corresponding to a depth of about 2500 m), and the BMNH specimen from off the Azores in a closing net between 1200 and 1295 m. The third specimen is from the East China Sea, taken with open gear fished at a maximum depth of 1055 m.

### *LINOPHRYNE RACEMIFERA* REGAN AND TREWAVAS, 1932

Figures 25, 45, 47, 183, 194, 239; Table 8

### Females

*Linophryne racemifera* Regan and Trewavas, 1932:108, figs. 165, 166, pl. 10, fig. 2 (original description, single female). Bertelsen, 1951:175, 184, fig. 121B (additional description of holotype, in key). Grey, 1956a:277 (distribution). Nielsen, 1974:100 (listed in type catalog). Bertelsen, 1982:80, figs. 13, 14 (two additional specimens, comparison with all known material). Bertelsen, 1990:518 (eastern tropical Atlantic). Swinney, 1995a:52, 57 (two additional specimens, off Madeira). Swinney, 1995b:46, figs. 4, 5 (after Swinney, 1995a; description, esca and hyoid barbel figured). Pietsch, 2002b:1070 (western central Atlantic). Love et al., 2005:59 (eastern North Pacific). Pietsch and Orr, 2007:13, fig. 9 (phylogenetic relationships).

### Males and Larvae

Unknown.

### Material

Five metamorphosed females (25.5 to 81 mm):

Holotype of *Linophryne racemifera*: ZMUC P92146, 52 mm, *Dana* station 4009(6), western North Atlantic, south of Canaries, 24°36.5′N, 17°27′W, open pelagic net, 4000 m of wire, 0700 h, 18 March 1930.

Additional material: LACM, one (58 mm); NMSZ, two (25.5 to 27 mm); USNM, one (81 mm).

### Distinguishing Characters

Metamorphosed females of *Linophryne racemifera* differ from those of all other species of the genus in details of escal and hyoid barbel morphology: escal bulb without distal prolongation; a single distal filament and a series of about 10 branched filaments on each side; in some specimens a posterior escal filament at or near base of bulb; anterior escal appendage absent. Total length of barbel 25 to 33% SL, distal part with a series of 5 photophores, proximal photophores set on short stalks.

A dense concentration of subdermal melanophores on caudal peduncle.

### Description

Dorsal profile of frontals slightly angular, without spine; sphenotic, preopercular, and angular spines similar to those of females of other species of subgenus; teeth in each premaxillary 16 to 24, arranged in four series; teeth in each dentary 10 to 16, arranged in three series, longest tooth 12 to 17% SL; dorsal-fin rays 3; anal-fin rays 3; pectoral-fin rays 15 to 17.

Length of illicium 25 to 26% SL; distal filament of esca with 8 to 10 lateral branches in holotype and largest known specimen, simple in 58-mm specimen; lateral branches well separated at base in 58-mm specimen, tending to fuse at base to form pairs in other specimens; length of each lateral filament three or four times diameter of escal bulb, gradually dividing into 5 to 10 slender filaments; a bifid posteromedial filament present in two nontype specimens, absent in holotype. Distal tip of barbel of holotype with 5 photophores, each set on a short stalk, but only proximal-most photophore protruding through skin as a tiny free appendage; barbel of 58-mm specimen with 2 tiny branches, each without distinct photophores, emerging proximal to 3 photophores contained within distal tip of barbel. A dense concentration of subdermal pigment on caudal peduncle.

### Distribution

The five known specimens of *Linophryne racemifera* are from widely separated localities: three in the eastern North Atlantic from off Madeira and the Canary Islands, a fourth in the equatorial western Atlantic off French Guiana, and the fifth in the eastern Pacific off southern California (Fig. 239). Two of the specimens (58 to 81 mm) came from nets fished at maximum depths of 210 and 500 m; two others (25.5 to 27 mm) at maximum fishing depths of 1360 to 1500 (Swinney, 1995a). The holotype was collected in an open pelagic net with 4000 m of wire.

### *LINOPHRYNE TREWAVASAE* BERTELSEN, 1978

Figure 307; Tables 7, 8, 10

### Females and Parasitic Males

Anglerfish: Robison, 1976:40, color photograph (popular account; specimen subsequently made holotype of *Linophryne trewavasae* by Bertelsen, 1978).

*Linophryne* sp. B: Hansen and Herring, 1977:107 (bioluminescence).

*Linophryne trewavasae* Bertelsen, 1978:26, figs. 1, 2 (original description, single female with parasitic male). Pietsch and Seigel, 1980:397 (Philippine Archipelago; after Bertelsen, 1978). Pietsch, 1999:2037 (western central Pacific). Pietsch, 2005b:211 (reproduction).

### Free-living Males and Larvae

Unknown.

### Material

A single metamorphosed female (73.5 mm), with parasitic male (10.7 mm):

Holotype of *Linophryne trewavasae*: LACM 36116-5, female, 73.5 mm, with parasitic male, 10.7 mm, Southeast Asian Bioluminescence Expedition, *Alpha Helix* station 84, 5°04.5′S, 130°12.0′E, RMT-8 oblique haul, 0 to 1500 m, 28 April 1975.

### Distinguishing Characters

Metamorphosed females of *Linophryne trewavasae* differ from those of all other species of the genus in details of escal and hyoid barbel morphology: escal bulb egg shaped, without distal appendages, bearing three pairs of stout filaments on posterolateral surface of proximal half of bulb, median pair distally compressed and fringed. Barbel short (about 20% SL), undivided, with distal series of white tubercles; frontal spines small and blunt.

Parasitic male with sphenotic spines pointed, olfactory lamellae 8, caudal peduncle with distinct concentration of subdermal pigment.

### Description

Metamorphosed females with frontal, sphenotic, preopercular, articular, angular, and symphysial spine of lower jaw present but all relatively short, blunt, and concealed by skin, except for those of sphenotic and preopercle. Jaw teeth arranged in overlapping oblique series, length of teeth in each series increasing posteriorly (as described in *Linophryne sexfilis*; see Bertelsen, 1973); teeth in each premaxilla 14, arranged in four series; 11 teeth in each dentary, arranged in three series; two pairs of vomerine teeth, those of anterior pair about 1.0 mm long, longest tooth of posterior pair 4.0 mm. Dorsal-fin rays 3; anal-fin rays 3; pectoral-fin rays 17, 17.

Length of illicium 29.8% SL; length of escal bulb 9.8% SL; diameter of escal bulb 5.0% SL; length of longest escal filament 10.0% SL; escal darkly pigmented except for transparent distal one-third, through which a black photophore appears; pore of photophore opening slightly behind tip of bulb surrounded by circular disc of reflecting silvery tissue; lateral surface of proximal one-half of bulb bearing 3 stout filaments on each side, forming symmetrical pairs of which lowermost is placed at base of bulb; all escal filaments pigmented and approximately equal in length (about 10% SL), those of upper and lower pair cylindrical and tapering, those of median pair distally compressed and fringed. Barbel relatively short, length 19.3% SL, unbranched and pigmented throughout length; posterior surface of distal one-third of barbel with 7 white tubercles raised on short papillae, and a more irregular group of about 6 tubercles more or less embedded in transparent tip. Color of female uniformly brownish black except for white tips of esca, barbel, and caudal fin rays (see Robison, 1976).

Parasitic male attached to the belly of female, about 10 mm below and slightly in front of sinistral anus, by separate outgrowths from snout and tip of lower jaw; sphenotic spines sharply pointed, piercing skin, upper denticular triangular, bearing 3 hooked denticles and a horseshoe-shaped lower denticular, with indications of 6 denticles embedded in tissue connecting female and male (determined by X ray); eyes without evidence of degeneration, diameter 0.95 mm (9% SL), diameter of lens 0.8 mm (7.5 % SL); olfactory lamellae 8 on each side; greatest diameter of posterior nostrils about 10% SL about twice that of anterior nostrils; caudal peduncle with distinct concentration of relatively large subdermal melanophores; fin-ray counts as for female; testes about 2 mm long and 1.5 mm wide.

### Sexual Maturity

The ovaries of the holotype are about 12 mm long, each containing numerous eggs with diameters of less than 0.05 mm.

### Distribution

*Linophryne trewavasae* is known only from the holotype collected in the Banda Sea, with a net fished open between the surface and 1500 m.

### Comments

*Linophryne trewavasae* shows no close resemblance to any of the other recognized species of the genus. The combination of an escal bulb with lateral filaments, a proximally unbranched barbel and a concentration of subdermal pigment on the caudal peduncle is shared with five species: *L. macrodon* Regan, 1925c; *L. brevibarbis* Parr, 1927; *L. bicornis* Parr, 1927; *L. racemifera* Regan and Trewavas, 1932; and *L. maderensis* Maul, 1961. *Linophryne bicornis* and *L. maderensis* resemble *L. trewavasae* furthermore in having few (one to three) pairs of lateral escal filaments and in lacking a distal escal filament, but both species differ distinctly from *L. trewavasae* in barbel characters; the unbranched barbel of *L. bicornis* bears a distal, bilaterally symmetrical tassel of stalked tubercles and long unbranched filaments; in *L. maderensis* (as well as *L. macrodon* and *L. brevibarbis*) the distal one-third to one-half of the barbel is repeatedly branched. On the other hand, *L. racemifera* resembles *L. trewavasae* in having an unbranched barbel with a short series of tubercles on the tip, but differs in escal characters, having numerous branched lateral filaments and a well-developed distal filament.

No characters were found to distinguish the male of *L. trewavasae* from other males of the genus, except for those of *L. arborifera* and *L. corymbifera*, which have distinctive subdermal pigment patterns and lack sphenotic spines (see Bertelsen, 1951:178, 182, figs. 116, 117; 1978:30, fig. 3). The low number of olfactory lamellae observed in *L. trewavasae* (8) has been found previously only in a single free-living male, *L. masculina* (Parr, 1934), and in the parasitic males of *L. lucifer* and *L. coronata*; as in the latter forms, however, the low count might be due to degeneration after attachment, as suggested by Bertelsen (1976:10, 13).

### LINOPHRYNE BICORNIS PARR, 1927

Figures 183, 237; Tables 7, 8, 10

#### Females and Parasitic Males

*Linophryne bicornis* Parr, 1927:10, fig. 2 (original description, single female). Regan and Trewavas, 1932:107, 111 (after Parr, 1927). Bertelsen, 1951:184 (comparison with *Linophryne brevibarbis*). Grey, 1955:300, fig. 56 (additional specimen, off Bermuda). Grey, 1956a:275 (distribution). Behrmann, 1977:93, figs. 1–4 (female with parasitic male). Bertelsen, 1982:82, fig. 15 (revision, additional female with parasitic male). Pietsch, 2002b:1070 (western central Atlantic). Moore et al., 2003:217 (listed). Pietsch, 2005b:211, 228 (reproduction).

#### Free-living Males and Larvae

Unknown.

#### Material

Five metamorphosed females (27 to 185 mm) and three parasitic males (18 to 30 mm):

Holotype of *Linophryne bicornis*: YPM 2030, 27 mm, *Pawnee* station 59, Bermuda, 32°19′N, 64°32′W, open pelagic net, about 2500 m of wire, 21 April 1927.

Additional material: FMNH, one (28.5 mm); IMB, one (185-mm female with 30-mm parasitic male); MCZ, one (101-mm female with 19-mm parasitic male); SIOM, one (180-mm female with 18-mm parasitic male).

#### Distinguishing Characters

Metamorphosed females of *Linophryne bicornis* differ from those of all other species of the genus in details of escal and barbel morphology: escal bulb without distal prolongation or filaments, anterior and posterior appendages absent; a stout, tapering filament on each side of upper part of escal bulb, each one to 1.5 times diameter of bulb; 0 to 2 minute conical papillae below each lateral filament. Primary stem of barbel undivided, length 20 to 32% SL; distal end divided into a pair of short branches, each with a lateral series of 3 or 4 long filaments, decreasing in length distally and gradually continuing as a cluster of stalked photophores on tip and medial side of each branch; each filament with a distal series of internal photophores; length of proximal and longest filaments 35 to 65% SL; a small branch with internal photophores in cleft between the pair of primary branches.

Caudal peduncle with a dense concentration of subdermal pigment in 27-mm holotype and in parasitic males.

Parasitic males with pointed sphenotic spines.

#### Description

Dorsal profile of frontals nearly straight; sphenotic, preopercular, and angular spines sharp and well developed in juveniles, blunt and shorter in adults; juveniles with 8 or 9 premaxillary teeth, 6 or 7 dentary teeth, and a pair of vomerine teeth; adults with 17 or 18 premaxillary teeth, 11 to 16 dentary teeth, and 2 pairs of vomerine teeth; dorsal-fin rays 3; anal-fin rays 3; pectoral-fin rays 15 or 16.

Length of illicial 24.5 to 35% SL; tip of escal bulb smoothly rounded in adults, with a low transverse keel in 28.5-mm specimen (possibly caused by shrinkage); stout pair of tapering filaments on upper part of bulb, simple except for 2 tiny lateral branches on right filament of 185-mm specimen; a single pair of conical papillae on proximal part of bulb of 27-mm holotype, two pairs in adults, none in 28.5-mm specimen.

Barbel very similar in all three specimens in which it is retained; distal pair of branches truncated and short, length measured to base of filaments 3 to 5% SL; 28.5-mm specimen with medial branch about same length, with slender stem bearing two pairs of pear-shaped photophores (Grey, 1955:300, fig. 56), medial branch of 180-mm specimen much shorter, with about 10 internal photophores; medial branch of 185-mm specimen undeveloped or lost; all barbel filaments and stalked photophores unpigmented; dense pigmentation of stem of barbel extending onto 2 primary branches of adults, terminating proximal to bifurcation in juvenile specimen.

Caudal peduncle (examined in the holotype and the 18-mm parasitic male) with strong concentration of subdermal pigment.

Sphenotic spines of parasitic male small, but sharply pointed.

### Distribution

Three of the five known females of *Linophryne bicornis* were caught in the western North Atlantic, one from off New England at about 39°N, 72°W, collected at maximum fishing depths of 992 to 1098 m; and two off Bermuda, at maximum fishing depths of 1100 to 1200 m (Fig. 237). The 185-mm female, with a parasitic male, was caught off Newfoundland in a commercial bottom trawl between 620 and 660 m; the 180-mm parasitized female, in the eastern South Indian Ocean, with a bottom trawl between 1220 and 1272 m; and the third attached pair, off New England, in a pelagic trawl between 992 and 1098 m.

### LINOPHRYNE ALGIBARBATA WATERMAN, 1939B

Figures 238, 308, 309; Tables 7, 8, 10

#### Females and Parasitic Males

*Linophryne algibarbata* Waterman, 1939b:85, figs. 3, 4, (original description, single female). Koefoed, 1944:13, pl. 2, fig. 1 (additional female, description). Bertelsen, 1951:174, 175 (reexamination of female recorded by Koefoed, 1944; a male referred to *Aceratias macrorhinus* by Koefoed, 1944, reassigned to *L. algibarbata*; in key). Grey, 1956a:271 (distribution). Behrmann, 1974:364, figs. 1, 2 (female with parasitic male). Bertelsen, 1976:13, figs. 3, 4 (description of parasitized female recorded by Behrmann, 1974). Bertelsen, 1980b:237, figs. 2, 3, 6 (additional material, comparison of all known material). Stearn and Pietsch, 1995:141, fig. 91 (additional female, Greenland). Jónsson and Pálsson, 1999:203, fig. 8 (Iceland). Pietsch, 2002b:1070 (western central Atlantic). Moore et al., 2003:217 (listed). Pietsch, 2005b:211, 228, fig. 23B (reproduction).

#### Free-living Males

*Aceratias macrorhinus*: Whitley, 1934, unpaged (listed). Koefoed, 1944:15 (in part; male referred to *Linophryne algibarbata* by Bertelsen, 1951). Beaufort and Briggs, 1962:261, fig. 67 (after Regan and Trewavas, 1932; Indo-Australian Archipelago).

#### Larvae

Unknown.

#### Material

Nine metamorphosed females (28 to 182 mm) and two parasitic males (23 to 29 mm):

Holotype of *Linophryne algibarbata*: MCZ 35066, 28 mm, *Atlantis* station 2894, 39°06′N, 70°16′W, closing net in 400 m, 20 July 1937.

Additional material: HUMZ, one (135 mm); IINH, one (155-mm female with 23-mm parasitic male); IMB, one (182-mm female with 29-mm parasitic male); MCZ, one (26.5 mm); ROM, two (28.5 to 30 mm); USNM, one (28 mm); ZMUB, one (30 mm).

#### Distinguishing Characters

Metamorphosed females of *Linophryne algibarbata* differ from those of all other species of the genus in details of escal and barbel morphology: escal bulb globular, that of small specimens surrounded distally and laterally by a blunt ridge; escal filaments or other appendages absent. Barbel divided at base into 4 primary branches of nearly equal length, 40 to 86% SL, each with a varying number of secondary and tertiary branches, more or less prolonged into filaments, each with a distal photophore.

Caudal peduncle with a concentration of subdermal melanophores.

Males with pointed sphenotic spines.

#### Description

Juvenile females (28 to 30 mm) with dorsal profile of frontals angular, usually raised into a short sharp spine; sphenotic, preopercular, and angular spines well developed, symphysial spine of lower jaw short; adult females (135 to 182 mm) with all spines short and blunt, frontal spines absent; juveniles with 4 to 8 teeth on each premaxilla, arranged in four series; 5 to 8 teeth on each dentary, arranged in three series; longest tooth 10 to 23% SL; adults with 14 premaxillary teeth, 7 or 8 dentary teeth; juveniles with a pair of vomerine teeth, adults with two pairs; length of illicium 39 to 60% SL (50 to 60% in specimens 30 to 182 mm); dorsal-fin rays 3; anal-fin rays 2 or 3; pectoral-fin rays 15 or 16.

Esca and distal part of illicial stem of juveniles unpigmented, pigment extending onto base of escal bulb in adults; shape of esca very similar in three juveniles in which it is retained; transverse blunt ridge of escal bulb slightly raised into lateral lobes, most distinct in 30-mm specimen reported by Koefoed (1944); escal bulb of adults smoothly rounded without transverse ridge. Length of barbel 40 to 86% SL, longest in holotype, but without distinct relation to size of specimens; 4 primary branches of barbel nearly equal in length in all specimens, tending to form two pairs, situated side by side; each primary branch bearing throughout their length 15 to 25 secondary branches some of which divide into tertiary branches; filamentous tip of each branch with a swelling containing a photophore of various size, the longest at ultimate tip of each of 4 primary branches.

Posterior two-thirds of caudal peduncle covered with subdermal pigment; well preserved in three juvenile females, somewhat obscured by secondary pigment in parasitic males.

Parasitic males with sharply pointed sphenotic spines.

#### Distribution

The nine known females of *Linophryne algibarbata* were all collected in the North Atlantic Ocean, several from off New England at about 39 to 40°N, 67 to 70°W; and the rest from off Iceland as far north as about 67°N (Fig. 238). The 28-mm holotype was caught in a closing net at 400 m, the remaining juveniles in open nets with maximum fishing depths of 1000 to 2200 m. Two of the three adults (155 to 182 mm), both from high latitudes, were caught in commercial bottom trawls in less than 500 m; the third (135 mm) in gear fished open between the surface and 1075 m.

### FREE-LIVING MALES OF THE SUBGENUS LINOPHRYNE

#### Males

*Aceratias macrorhinus* Brauer, 1902:296 (original description, single specimen). Brauer, 1906:324, pl. 16, figs. 4, 5 (after Brauer, 1902). Brauer, 1908:110, 184, text figs. 2, 8 (anatomy of illicial pterygiophore and eyes). Regan, 1912:289 (after Brauer, 1906). Regan, 1926:45, fig. 27 (additional specimens). Regan, 1927a:4, postcard

M-15b (popular account). Regan and Trewavas, 1932:20, 113, fig. 10B (additional specimens). Whitley, 1934, unpaged (listed). Koefoed, 1944:15, pl. 3, fig. 5 (additional specimens). Bertelsen, 1951:186 (placed in *Linophryne macrorhinus* group).

*Haplophryne hudsonius* Beebe, 1929a:19 (original description, single specimen, reference to Beebe, 1929b). Beebe, 1929b:23, figs. 2–5 (detailed description, osteology). Borodin, 1931:84 (comparison with *Haplophryne simus* Borodin, 1930b). Gregory, 1933:402, 403, 407–410, fig. 280 (osteology). Mead, 1958:133 (type transferred to USNM).

*Anomalophryne hudsonia*: Regan and Trewavas, 1932:112 (new combination; new generic name for *Haplophryne hudsonius* Beebe, 1929a, 1929b).

*Aceratias edentula* Beebe, 1932b:102, fig. 31 (original description, single specimen, jaw mechanism). Bertelsen, 1951:187 (placed in *Linophryne macrorhinus* group). Mead, 1958:131 (type transferred to USNM).

*Borophryne masculina* Parr, 1934:56, fig. 21 (original description, single specimen). Bertelsen, 1951:186 (reassigned to *Linophryne*, perhaps a male of *L. coronata*). Bertelsen, 1976:11 (after Bertelsen, 1951).

*Linophryne masculina*: Bertelsen, 1951: 175, 176, 186 (new combination; *Borophryne masculina* Parr, 1934, referred to *Linophryne*, suspected to be male of *Linophryne coronata*; in key). Grey, 1956a: 277 (distribution).

*Linophryne algibarbata*: Bertelsen, 1951:175 (tentative reallocation of specimen referred to *Aceratias macrorhinus* by Koefoed, 1944).

*Linophryne macrorhinus* Group: Bertelsen, 1951:186, figs. 121D, 122 (additional material, comparison of all known material). Grey, 1956a:277 (distribution). Maul, 1973:677 (eastern North Atlantic). Bertelsen, 1982:85 (males grouped and allocated to *Linophryne* subgenus *Linophryne*).

*Halophryne hudsonius*: Bertelsen, 1951: 186 (misspelling of generic name; specimen referred to *Linophryne macrorhinus* group).

Material

About 30 known specimens (13 to 21 mm):

Holotype of *Aceratias macrorhinus*: ZMB 17714, 21 mm, *Valdivia* station 73, off Angola, West Africa, 9°31′S, 9°46′E, 2000 m, 7 October 1898.

Holotype of *Haplophryne hudsonius*: USNM 170954 (originally NYZS 7696), 15 mm, Eleventh Expedition of the Department of Tropical Research, New York Zoological Society, SS *Wheeler* station 114, T-6, Hudson Gorge, 39°15′N, 72°00′W, 1100 m (600 fathoms), 7 July 1928.

Holotype of *Aceratias edentula*: USNM 170951 (originally NYZS 20751), 19.5 mm, Bermuda Oceanographic Expedition, New York Zoological Society, off Nonsuch Island, Bermuda, 32°12′N, 64°36′W, 1830 m (1000 fathoms), 2 June 1931.

Holotype of *Borophryne masculina*: YPM 3208, 19.5 mm, *Atlantis* station 1478, Bahamas, 25°39′N, 77°18′W, 0 to 1100 m, 20 to 21 February 1933.

Additional material: See Bertelsen (1951:186; 1982:85).

Distinguishing Characters

Metamorphosed free-living males of *Linophryne* subgenus *Linophryne* differ from those of all other subgenera of the genus as well as other genera of the family in having the following combination of character states: head with pointed sphenotic spines; premaxillaries degenerate; jaw teeth absent (some larval teeth retained in smallest metamorphosed specimens); denticular bones well developed with 3 (rarely 4) upper and 5 to 9 lower denticular teeth; eyes well developed, slightly tubular and directed anteriorly; olfactory organs inflated, larger than eyes, with 8 to 13 olfactory lamellae; skin dark brown or black, subdermal pigment consisting of small melanophores restricted to, or concentrated on, caudal peduncle, not arranged in lateral series along body.

Description

As given in the diagnosis above and in the general description of the genus and family.

Distribution

Recorded from all major oceans of the world between about 35°N and 35°S and, like other metamorphosed ceratioid males, rarely collected in hauls with a maximum fishing depth of less than 1000 m.

Comments

A detailed comparison of the parasitic males of known species of the genus failed to reveal any differences that might allow specific identification of the free-living males (Bertelsen, 1982:86).

LARVAE AND EARLY FEMALE METAMORPHIC STAGES OF THE SUBGENUS *LINOPHRYNE*

*Haplophryne mollis*: Norman, 1930: 352 (misallocation, single metamorphic female).

*Cryptolychnus paucidens* Regan and Trewavas, 1932:105, pl. 8, fig. 2 (original description based on a female recorded by Norman, 1930).

*Allector paucidens*: Fowler, 1936:1368 (new combination; see Myers, in Fowler, 1936:1368).

*Linophryne macrorhinus* Group: Bertelsen, 1951:186, fig. 122 (in part; including metamorphosed males now referred to subgenus *Linophryne*).

*Haplophryne macrorhinus* Group: Maul, 1961:142, fig. 27B (additional metamorphic female).

*Linophryne* sp.: Maul, 1961:142, figs. 28A, B (two additional metamorphic females).

Material

About 75 larvae of both sexes (3 to 11 mm), and four females in metamorphosis (22 to 32 mm):

Holotype of *Cryptolychnus paucidens*: BMNH 1930.1.12.1078, female, 32 mm, Gulf of Guinea, 2°49′30″S, 9°25′30″W, Young-fish trawl, 0 to 1000 m, 19 August 1927.

Additional material: MCZ, 22 (lengths not recorded); MMF, three metamorphic females (22 to 26 mm); SIO (number and lengths not recorded); ZMUC, 40 (3 to 11 mm; see Bertelsen, 1951:187, 275–276).

Distinguishing Characters

Larvae of *Linophryne* subgenus *Linophryne* differ from those of other subgenera of the genus as well as other genera of the family in having the following combination of character states: sphenotics with pointed spines; subdermal pigment restricted to caudal peduncle, consisting of a varying number of tiny melanophores. Metamorphic females differ further from those of other linophrynid genera in having a papilliform rudiment of a hyoid barbel.

Description

As given in the diagnosis above and in the general description of the genus and family.

Distribution

Larvae of the subgenus *Linophryne* have been collected in all three major oceans of the world. Like other larvae of the family, they tend to be most numerous at depths of 100 to 200 m.

# REALLOCATION OF NOMINAL SPECIES OF THE CERATIOIDEI BASED ON FEMALES

| Nominal Species | Currently Recognized Taxon |
|---|---|
| *Acentrophryne dolichonema* Pietsch and Shimazaki, 2005 | *Acentrophryne dolichonema* Pietsch and Shimazaki, 2005 |
| *Acentrophryne longidens* Regan, 1926 | *Acentrophryne longidens* Regan, 1926 |
| *Aegoeonichthys appelii* Clarke, 1878 | *Himantolophus appelii* (Clarke, 1878) |
| *Allector cheloniae* Heller and Snodgrass, 1903 | *Linophryne* subgenus *Rhizophryne* sp. |
| *Bertella idiomorpha* Pietsch, 1973 | *Bertella idiomorpha* Pietsch, 1973 |
| *Borophryne apogon* Regan, 1925c | *Borophryne apogon* Regan, 1925c |
| *Bufoceratias shaoi* Pietsch et al., 2004 | *Bufoceratias shaoi* Pietsch et al., 2004 |
| *Caulophryne acinosa* Regan and Trewavas, 1932 | *Caulophryne pelagica* (Brauer, 1902) |
| *Caulophryne bacescui* Mihai-Bardan, 1982 | *Caulophryne bacescui* Mihai-Bardan, 1982 |
| *Caulophryne pietschi* Balushkin and Fedorov, 1985 | *Caulophryne pelagica* (Brauer, 1902) |
| *Caulophryne polynema* Regan, 1930b | *Caulophryne polynema* Regan, 1930b |
| *Caulophryne racemosa* Monod, 1960 | *Nomen nudum* |
| *Caulophryne radians* Fowler, 1936 | *Nomen nudum* |
| *Caulophryne ramulosa* Regan and Trewavas, 1932 | *Caulophryne pelagica* (Brauer, 1902) |
| *Caulophryne setosus* Goode and Bean, 1896 | *Nomen nudum* |
| *Centrophryne spinulosa* Regan and Trewavas, 1932 | *Centrophryne spinulosa* Regan and Trewavas, 1932 |
| *Ceratias bispinosus* Günther, 1887 | *Diceratias bispinosus* (Günther, 1887) |
| *Ceratias carunculatus* Günther, 1887 | *Cryptopsaras couesii* Gill, 1883b |
| *Ceratias holboelli* Krøyer, 1845 | *Ceratias holboelli* Krøyer, 1845 |
| *Ceratias mitsukurii* Tanaka, 1908 | *Cryptopsaras couesii* Gill, 1883b |
| *Ceratias murrayi* Bertelsen, 1990 | *Nomen nudum* |
| *Ceratias uranoscopus* Murray, in Thomson, 1877 | *Ceratias uranoscopus* Murray, in Thomson, 1877 |
| *Ceratocaulophryne regani* Roule and Angel, 1932 | *Caulophryne jordani* Goode and Bean, 1896 |
| *Chaenophryne atriconus* Regan and Trewavas, 1932 | *Chaenophryne draco* Beebe, 1932b |
| *Chaenophryne bicornis* Regan and Trewavas, 1932 | *Chaenophryne longiceps* Regan, 1925c |
| *Chaenophryne columnifera* Regan and Trewavas, 1932 | *Chaenophryne draco* Beebe, 1932b |

*(Continued)*

| Nominal Species | Currently Recognized Taxon |
|---|---|
| *Chaenophryne crenata* Regan and Trewavas, 1932 | *Chaenophryne longiceps* Regan, 1925c |
| *Chaenophryne crossotus* Beebe, 1932b | *Chaenophryne longiceps* Regan, 1925c |
| *Chaenophryne draco* Beebe, 1932b | *Chaenophryne draco* Beebe, 1932b |
| *Chaenophryne fimbriata* Regan and Trewavas, 1932 | *Chaenophryne ramifera* Regan and Trewavas, 1932 |
| *Chaenophryne galeatus* Koefoed, 1944 | *Nomen nudum* |
| *Chaenophryne haplactis* Regan and Trewavas, 1932 | *Chaenophryne longiceps* Regan, 1925c |
| *Chaenophryne intermedia* Belloc, 1938 | *Chaenophryne ramifera* Regan and Trewavas, 1932 |
| *Chaenophryne longiceps* Regan, 1925c | *Chaenophryne longiceps* Regan, 1925c |
| *Chaenophryne longiceps* var. *quadrifilis* Parr, 1927 | *Chaenophryne longiceps* Regan, 1925c |
| *Chaenophryne macractis* Regan and Trewavas, 1932 | *Chaenophryne draco* Beebe, 1932b |
| *Chaenophryne melanodactylus* Regan and Trewavas, 1932 | *Chaenophryne draco* Beebe, 1932b |
| *Chaenophryne melanorhabdus* Regan and Trewavas, 1932 | *Chaenophryne melanorhabdus* Regan and Trewavas, 1932 |
| *Chaenophryne pacis* Koefoed, 1944 | *Chaenophryne ramifera* Regan and Trewavas, 1932 |
| *Chaenophryne parviconus* Regan and Trewavas, 1932 | *Chaenophryne draco* Beebe, 1932b |
| *Chaenophryne pterolophus* Regan and Trewavas, 1932 | *Chaenophryne melanorhabdus* Regan and Trewavas, 1932 |
| *Chaenophryne quasiramifera* Pietsch, 2007 | *Chaenophryne quasiramifera* Pietsch, 2007 |
| *Chaenophryne ramifera* Regan and Trewavas, 1932 | *Chaenophryne ramifera* Regan and Trewavas, 1932 |
| *Chirophryne xenolophus* Regan and Trewavas, 1932 | *Chirophryne xenolophus* Regan and Trewavas, 1932 |
| *Corynolophus globosus* Tanaka, 1918b | *Himantolophus sagamius* (Tanaka, 1918a) |
| *Corynolophus sagamius* Tanaka, 1918a | *Himantolophus sagamius* (Tanaka, 1918a) |
| *Corynophorus compressus* Osório, 1912 | *Himantolophus compressus* (Osório, 1912) |
| *Cryptolychnus paucidens* Regan and Trewavas, 1932 | *Linophryne* subgenus *Linophryne* sp. |
| *Cryptolynchnus micractis* Regan and Trewavas, 1932 | *Linophryne* subgenus *Rhizophryne* sp. |
| *Cryptopsaras couesii* Gill, 1883b | *Cryptopsaras couesii* Gill, 1883b |
| *Cryptosparas atlantidis* Barbour, 1941b | *Cryptopsaras couesii* Gill, 1883b |
| *Cryptosparas normani* Regan and Trewavas, 1932 | *Cryptopsaras couesii* Gill, 1883b |
| *Cryptosparas pennifer* Regan and Trewavas, 1932 | *Cryptopsaras couesii* Gill, 1883b |
| *Cryptosparas valdiviae* Regan and Trewavas, 1932 | *Cryptopsaras couesii* Gill, 1883b |
| *Ctenochirichthys longimanus* Regan and Trewavas, 1932 | *Ctenochirichthys longimanus* Regan and Trewavas, 1932 |
| *Dermatias platynogaster* Smith and Radcliffe, in Radcliffe, 1912 | *Dermatias platynogaster* Smith and Radcliffe, in Radcliffe, 1912 |
| *Diabolidium arcturi* Beebe, 1926b | *Linophryne arcturi* (Beebe, 1926b) |
| *Diceratias glomerulosus* Regan, 1926 | *Diceratias bispinosus* (Günther, 1887) |
| *Diceratias pileatus* Uwate, 1979 | *Diceratias pileatus* Uwate, 1979 |
| *Diceratias trilobus* Balushkin and Fedorov, 1986 | *Diceratias trilobus* Balushkin and Fedorov, 1986 |
| *Dolopichthys albifilosus* Waterman, 1939b | *Danaphryne nigrifilis* (Regan and Trewavas, 1932) |
| *Dolopichthys allector* Garman, 1899 | *Dolopichthys allector* Garman, 1899 |
| *Dolopichthys analogus* Parr, 1927 | *Microlophichthys microlophus* (Regan, 1925c) |
| *Dolopichthys anisacanthus* Regan, 1925c | *Oneirodes anisacanthus* (Regan, 1925c) |
| *Dolopichthys atratus* Regan and Trewavas, 1932 | *Pentherichthys atratus* (Regan and Trewavas, 1932) |
| *Dolopichthys brevifilis* Regan and Trewavas, 1932 | *Oneirodes eschrichtii* Lütken, 1871 |

| Nominal Species | Currently Recognized Taxon |
|---|---|
| *Dolopichthys carlsbergi* Regan and Trewavas, 1932 | *Oneirodes carlsbergi* (Regan and Trewavas, 1932) |
| *Dolopichthys cirrifer* Regan and Trewavas, 1932 | *Oneirodes eschrichtii* Lütken, 1871 |
| *Dolopichthys claviger* Regan and Trewavas, 1932 | *Oneirodes eschrichtii* Lütken, 1871 |
| *Dolopichthys cristatus* Regan and Trewavas, 1932 | *Oneirodes cristatus* (Regan and Trewavas, 1932) |
| *Dolopichthys danae* Regan, 1926 | *Dolopichthys danae* Regan, 1926 |
| *Dolopichthys diadematus* Regan and Trewavas, 1932 | *Oneirodes eschrichtii* Lütken, 1871 |
| *Dolopichthys digitatus* Regan and Trewavas, 1932 | *Oneirodes eschrichtii* Lütken, 1871 |
| *Dolopichthys dinema* Pietsch, 1972c | *Dolopichthys dinema* Pietsch, 1972c |
| *Dolopichthys exiguus* Regan and Trewavas, 1932 | *Microlophichthys microlophus* (Regan, 1925c) |
| *Dolopichthys flagellifer* Regan and Trewavas, 1932 | *Oneirodes flagellifer* (Regan and Trewavas, 1932) |
| *Dolopichthys frondosus* Regan and Trewavas, 1932 | *Oneirodes eschrichtii* Lütken, 1871 |
| *Dolopichthys gladisfenae* Beebe, 1932b | *Spiniphryne gladisfenae* (Beebe, 1932b) |
| *Dolopichthys gracilispinis* Regan, 1925c | *Leptacanthichthys gracilispinis* (Regan, 1925c) |
| *Dolopichthys heteracanthus* Regan, 1925c | *Oneirodes luetkeni* (Regan, 1925c) |
| *Dolopichthys heteronema* Regan and Trewavas, 1932 | *Oneirodes heteronema* (Regan and Trewavas, 1932) |
| *Dolopichthys hibernicus* Fraser-Brunner, 1935 | *Oneirodes eschrichtii* Lütken, 1871 |
| *Dolopichthys implumis* Regan and Trewavas, 1932 | *Microlophichthys microlophus* (Regan, 1925c) |
| *Dolopichthys inimicus* Fraser-Brunner, 1935 | *Oneirodes carlsbergi* (Regan and Trewavas, 1932) |
| *Dolopichthys jubatus* Regan and Trewavas, 1932 | *Dolopichthys jubatus* Regan and Trewavas, 1932 |
| *Dolopichthys karsteni* Leipertz and Pietsch, 1987 | *Dolopichthys karsteni* Leipertz and Pietsch, 1987 |
| *Dolopichthys longicornis* Parr, 1927 | *Dolopichthys longicornis* Parr, 1927 |
| *Dolopichthys luetkeni* Regan, 1925c | *Oneirodes luetkeni* (Regan, 1925c) |
| *Dolopichthys macronema* Regan and Trewavas, 1932 | *Oneirodes macronema* (Regan and Trewavas, 1932) |
| *Dolopichthys microlophus* Regan, 1925c | *Microlophichthys microlophus* (Regan, 1925c) |
| *Dolopichthys mirus* Regan and Trewavas, 1932 | *Oneirodes mirus* (Regan and Trewavas, 1932) |
| *Dolopichthys mucronatus* Regan and Trewavas, 1932 | *Dolopichthys longicornis* Parr, 1927 |
| *Dolopichthys multifilis* Regan and Trewavas, 1932 | *Oneirodes eschrichtii* Lütken, 1871 |
| *Dolopichthys niger* Brauer, 1902 | *Nomen dubium* |
| *Dolopichthys nigrifilis* Regan and Trewavas, 1932 | *Danaphryne nigrifilis* (Regan and Trewavas, 1932) |
| *Dolopichthys obtusus* Parr, 1927 | *Oneirodes eschrichtii* Lütken, 1871 |
| *Dolopichthys pennatus* Regan and Trewavas, 1932 | *Oneirodes eschrichtii* Lütken, 1871 |
| *Dolopichthys plumatus* Regan and Trewavas, 1932 | *Oneirodes eschrichtii* Lütken, 1871 |
| *Dolopichthys pollicifer* Regan and Trewavas, 1932 | *Oneirodes eschrichtii* Lütken, 1871 |
| *Dolopichthys ptilotus* Regan and Trewavas, 1932 | *Oneirodes eschrichtii* Lütken, 1871 |
| *Dolopichthys pullatus* Regan and Trewavas, 1932 | *Dolopichthys pullatus* Regan and Trewavas, 1932 |
| *Dolopichthys schmidti* Regan and Trewavas, 1932 | *Oneirodes schmidti* (Regan and Trewavas, 1932) |
| *Dolopichthys simplex* Regan and Trewavas, 1932 | *Oneirodes eschrichtii* Lütken, 1871 |
| *Dolopichthys tentaculatus* Beebe, 1932b | *Oneirodes eschrichtii* Lütken, 1871 |
| *Dolopichthys thompsoni* Schultz, 1934 | *Oneirodes thompsoni* (Schultz, 1934) |
| *Dolopichthys thysanophorus* Regan and Trewavas, 1932 | *Oneirodes flagellifer* (Regan and Trewavas, 1932) |

*(Continued)*

| Nominal Species | Currently Recognized Taxon |
|---|---|
| *Dolopichthys venustus* Regan and Trewavas, 1932 | *Pentherichthys atratus* (Regan and Trewavas, 1932) |
| *Edriolychnus macracanthus* Regan and Trewavas, 1932 | *Haplophryne mollis* (Brauer, 1902) |
| *Edriolychnus radians* Regan and Trewavas, 1932 | *Haplophryne mollis* (Brauer, 1902) |
| *Edriolychnus roulei* Regan and Trewavas, 1932 | *Haplophryne mollis* (Brauer, 1902) |
| *Edriolychnus schmidti* Regan, 1925b | *Haplophryne mollis* (Brauer, 1902) |
| *Galatheathauma axeli* Bruun, 1953 | *Thaumatichthys axeli* (Bruun, 1953) |
| *Gigantactis balushkini* Kharin, 1984 | *Gigantactis vanhoeffeni* Brauer, 1902 |
| *Gigantactis elsmani* Bertelsen et al., 1981 | *Gigantactis elsmani* Bertelsen et al., 1981 |
| *Gigantactis exodon* Regan and Trewavas, 1932 | *Gigantactis vanhoeffeni* Brauer, 1902 |
| *Gigantactis filibulbosus* Fraser-Brunner, 1935 | *Gigantactis* sp., nomen dubium |
| *Gigantactis gargantua* Bertelsen et al., 1981 | *Gigantactis gargantua* Bertelsen et al., 1981 |
| *Gigantactis gibbsi* Bertelsen et al., 1981 | *Gigantactis gibbsi* Bertelsen et al., 1981 |
| *Gigantactis golovani* Bertelsen et al., 1981 | *Gigantactis golovani* Bertelsen et al., 1981 |
| *Gigantactis gracilicauda* Regan, 1925c | *Gigantactis gracilicauda* Regan, 1925c |
| *Gigantactis herwigi* Bertelsen et al., 1981 | *Gigantactis herwigi* Bertelsen et al., 1981 |
| *Gigantactis ios* Bertelsen et al., 1981 | *Gigantactis ios* Bertelsen et al., 1981 |
| *Gigantactis kreffti* Bertelsen et al., 1981 | *Gigantactis kreffti* Bertelsen et al., 1981 |
| *Gigantactis longicauda* Bertelsen and Pietsch, 2002 | *Gigantactis longicauda* Bertelsen and Pietsch, 2002 |
| *Gigantactis longicirra* Waterman, 1939b | *Gigantactis longicirra* Waterman, 1939b |
| *Gigantactis macronema* Regan, 1925c | *Gigantactis macronema* Regan, 1925c |
| *Gigantactis meadi* Bertelsen et al., 1981 | *Gigantactis meadi* Bertelsen et al., 1981 |
| *Gigantactis microdontis* Bertelsen et al., 1981 | *Gigantactis microdontis* Bertelsen et al., 1981 |
| *Gigantactis ovifer* Regan and Trewavas, 1932 | *Gigantactis* sp., nomen dubium |
| *Gigantactis paxtoni* Bertelsen et al., 1981 | *Gigantactis paxtoni* Bertelsen et al., 1981 |
| *Gigantactis perlatus* Beebe and Crane, 1947 | *Gigantactis perlatus* Beebe and Crane, 1947 |
| *Gigantactis savagei* Bertelsen et al., 1981 | *Gigantactis savagei* Bertelsen et al., 1981 |
| *Gigantactis sexfilis* Regan and Trewavas, 1932 | *Gigantactis gracilicauda* Regan, 1925c |
| *Gigantactis vanhoeffeni* Brauer, 1902 | *Gigantactis vanhoeffeni* Brauer, 1902 |
| *Gigantactis watermani* Bertelsen et al., 1981 | *Gigantactis watermani* Bertelsen et al., 1981 |
| *Haplophryne trireguim* Whitley and Phillipps, 1939 | *Haplophryne mollis* (Brauer, 1902) |
| *Himantolophus albinares* Maul, 1961 | *Himantolophus albinares* Maul, 1961 |
| *Himantolophus azurlucens* Beebe and Crane, 1947 | *Himantolophus azurlucens* Beebe and Crane, 1947 |
| *Himantolophus borealis* Kharin, 1984 | *Himantolophus borealis* Kharin, 1984 |
| *Himantolophus cornifer* Bertelsen and Krefft, 1988 | *Himantolophus cornifer* Bertelsen and Krefft, 1988 |
| *Himantolophus crinitus* Bertelsen and Krefft, 1988 | *Himantolophus crinitus* Bertelsen and Krefft, 1988 |
| *Himantolophus danae* Regan and Trewavas, 1932 | *Himantolophus danae* Regan and Trewavas, 1932 |
| *Himantolophus groenlandicus* Reinhardt, 1837 | *Himantolophus groenlandicus* Reinhardt, 1837 |
| *Himantolophus kainarae* Barbour, 1942b | *Himantolophus sagamius* (Tanaka, 1918a) |
| *Himantolophus macroceras* Bertelsen and Krefft, 1988 | *Himantolophus macroceras* Bertelsen and Krefft, 1988 |
| *Himantolophus macroceratoides* Bertelsen and Krefft, 1988 | *Himantolophus macroceratoides* Bertelsen and Krefft, 1988 |
| *Himantolophus mauli* Bertelsen and Krefft, 1988 | *Himantolophus mauli* Bertelsen and Krefft, 1988 |

| Nominal Species | Currently Recognized Taxon |
|---|---|
| *Himantolophus melanolophus* Bertelsen and Krefft, 1988 | *Himantolophus melanolophus* Bertelsen and Krefft, 1988 |
| *Himantolophus melanophus* Jónsson and Pálsson, 1999 | *Himantolophus melanophus* Bertelsen and Krefft, 1988 |
| *Himantolophus multifurcatus* Bertelsen and Krefft, 1988 | *Himantolophus multifurcatus* Bertelsen and Krefft, 1988 |
| *Himantolophus nigricornis* Bertelsen and Krefft, 1988 | *Himantolophus nigricornis* Bertelsen and Krefft, 1988 |
| *Himantolophus paucifilosus* Bertelsen and Krefft, 1988 | *Himantolophus paucifilosus* Bertelsen and Krefft, 1988 |
| *Himantolophus pseudalbinares* Bertelsen and Krefft, 1988 | *Himantolophus pseudalbinares* Bertelsen and Krefft, 1988 |
| *Himantolophus ranoides* Barbour, 1942b | *Himantolophus groenlandicus* Reinhardt, 1837 |
| *Himantolophus reinhardti* Lütken, 1878a | *Himantolophus groenlandicus* Reinhardt, 1837 |
| *Hyaloceratias parri* Koefoed, 1944 | *Haplophryne mollis* (Brauer, 1902) |
| *Lasiognathus amphirhamphus* Pietsch, 2005a | *Lasiognathus amphirhamphus* Pietsch, 2005a |
| *Lasiognathus ancistrophorus* Maul, 1962b | *Lasiognathus saccostoma* Regan, 1925c |
| *Lasiognathus beebei* Regan and Trewavas, 1932 | *Lasiognathus beebei* Regan and Trewavas, 1932 |
| *Lasiognathus intermedius* Bertelsen and Pietsch, 1996 | *Lasiognathus intermedius* Bertelsen and Pietsch, 1996 |
| *Lasiognathus saccostoma* Regan, 1925c | *Lasiognathus saccostoma* Regan, 1925c |
| *Lasiognathus waltoni* Nolan and Rosenblatt, 1975 | *Lasiognathus waltoni* Nolan and Rosenblatt, 1975 |
| *Linophryne algibarbata* Waterman, 1939b | *Linophryne algibarbata* Waterman, 1939b |
| *Linophryne andersoni* Gon, 1992 | *Linophryne andersoni* Gon, 1992 |
| *Linophryne arborifer* Regan, 1925c | *Linophryne arborifera* Regan, 1925c |
| *Linophryne arborifera* Beebe, 1932a | *Linophryne arborifera* Regan, 1925c |
| *Linophryne argyresca* Regan and Trewavas, 1932 | *Linophryne argyresca* Regan and Trewavas, 1932 |
| *Linophryne bicornis* Parr, 1927 | *Linophryne bicornis* Parr, 1927 |
| *Linophryne bipennata* Bertelsen, 1982 | *Linophryne bipennata* Bertelsen, 1982 |
| *Linophryne brevibarbata* Beebe, 1932b | *Linophryne brevibarbata* Beebe, 1932b |
| *Linophryne brevibarbis* Parr, 1927 | *Linophryne macrodon* Regan, 1925c |
| *Linophryne colletti* Weber, 1913 | *Oneirodes carlsbergi* (Regan and Trewavas, 1932) |
| *Linophryne coronata diphlegma* Parr, 1934 | *Linophryne coronata* Parr, 1927 |
| *Linophryne coronata diplegma*: Bertelsen, 1976 | *Linophryne coronata* Parr, 1927 |
| *Linophryne coronata* Parr, 1927 | *Linophryne coronata* Parr, 1927 |
| *Linophryne corymbifera* Regan and Trewavas, 1932 | *Linophryne indica* (Brauer, 1902) |
| *Linophryne densiramus* Imai, 1941 | *Linophryne densiramus* Imai, 1941 |
| *Linophryne digitopogon* Balushkin and Trunov, 1988 | *Linophryne lucifer* Collett, 1886 |
| *Linophryne escaramosa* Bertelsen, 1982 | *Linophryne escaramosa* Bertelsen, 1982 |
| *Linophryne eupogon* Regan and Trewavas, 1932 | *Linophryne arborifera* Regan, 1925c |
| *Linophryne longibarbata* Borodin, 1930a | *Linophryne coronata* Parr, 1927 |
| *Linophryne lucifer* Collett, 1886 | *Linophryne lucifer* Collett, 1886 |
| *Linophryne lucifera* Regan and Trewavas, 1932 | *Linophryne lucifer* Collett, 1886 |
| *Linophryne macrodon* Regan, 1925c | *Linophryne macrodon* Regan, 1925c |
| *Linophryne maderensis* Maul, 1961 | *Linophryne maderensis* Maul, 1961 |
| *Linophryne parini* Bertelsen, 1980b | *Linophryne parini* Bertelsen, 1980b |
| *Linophryne pennibarbata* Bertelsen, 1980a | *Linophryne pennibarbata* Bertelsen, 1980a |

*(Continued)*

| Nominal Species | Currently Recognized Taxon |
| --- | --- |
| *Linophryne polypogon* Regan, 1925c | *Linophryne polypogon* Regan, 1925c |
| *Linophryne quinqueramosus* Beebe and Crane, 1947 | *Linophryne quinqueramosus* Beebe and Crane, 1947 |
| *Linophryne racemifera* Regan and Trewavas, 1932 | *Linophryne racemifera* Regan and Trewavas, 1932 |
| *Linophryne sexfilis* Bertelsen, 1973 | *Linophryne sexfilis* Bertelsen, 1973 |
| *Linophryne trewavasae* Bertelsen, 1978 | *Linophryne trewavasae* Bertelsen, 1978 |
| *Lophodolos indicus* Lloyd, 1909a | *Lophodolos indicus* Lloyd, 1909a |
| *Lophodolus acanthognathus* Regan, 1925c | *Lophodolus acanthognathus* Regan, 1925c |
| *Lophodolus biflagellatus* Koefoed, 1944 | *Nomen nudum* |
| *Lophodolus dinema* Regan and Trewavas, 1932 | *Lophodolos indicus* Lloyd, 1909a |
| *Lophodolus lyra* Beebe, 1932b | *Lophodolus acanthognathus* Regan, 1925c |
| *Mancalias bifilis* Regan and Trewavas, 1932 | *Ceratias tentaculatus* (Norman, 1930) |
| *Mancalias kroyeri* Koefoed, 1944 | *Ceratias uranoscopus* Murray, in Thomson, 1877 |
| *Mancalias sessilis* Imai, 1941 | *Ceratias uranoscopus* Murray, in Thomson, 1877 |
| *Mancalias tentaculatus* Norman, 1930 | *Ceratias tentaculatus* (Norman, 1930) |
| *Mancalias uranoscopus triflos* Roule and Angel, 1933 | *Ceratias uranoscopus* Murray, in Thomson, 1877 |
| *Mancalias uranoscopus typhlos* Bertelsen, 1951 | *Ceratias uranoscopus* Murray, in Thomson, 1877 |
| *Mancalias xenistius* Regan and Trewavas, 1932 | *Ceratias uranoscopus* Murray, in Thomson, 1877 |
| *Mancalius uranoscopus typhlops* Beebe and Crane, 1947 | *Ceratias uranoscopus* Murray, in Thomson, 1877 |
| *Melanocetus bispinossus* Günther, 1880 | *Nomen nudum* |
| *Melanocetus cirrifer* Regan and Trewavas, 1932 | *Melanocetus johnsonii* Günther, 1864 |
| *Melanocetus eustalus* Pietsch and Van Duzer, 1980 | *Melanocetus eustalus* Pietsch and Van Duzer, 1980 |
| *Melanocetus ferox* Regan, 1926 | *Melanocetus johnsonii* Günther, 1864 |
| *Melanocetus johnsonii* Günther, 1864 | *Melanocetus johnsonii* Günther, 1864 |
| *Melanocetus johnstoni* Filhol, 1885 | *Melanocetus johnsonii* Günther, 1864 |
| *Melanocetus krechi* Brauer, 1902 | *Melanocetus johnsonii* Günther, 1864 |
| *Melanocetus megalodontis* Beebe and Crane, 1947 | *Melanocetus johnsonii* Günther, 1864 |
| *Melanocetus murrayi* Günther, 1887 | *Melanocetus murrayi* Günther, 1887 |
| *Melanocetus niger* Regan, 1925 | *Melanocetus niger* Regan, 1925 |
| *Melanocetus pelagicus* Brauer, 1902 | *Caulophryne pelagica* (Brauer, 1902) |
| *Melanocetus polyactis* Regan, 1925c | *Melanocetus polyactis* Regan, 1925c |
| *Melanocetus rossi* Balushkin and Fedorov, 1981 | *Melanocetus rossi* Balushkin and Fedorov, 1981 |
| *Melanocetus rotundatus* Gilchrist, 1903 | *Melanocetus johnsonii* Günther, 1864 |
| *Melanocetus tumidus* Parr, 1927 | *Melanocetus murrayi* Günther, 1887 |
| *Melanocetus vorax* Brauer, 1902 | *Melanocetus murrayi* Günther, 1887 |
| *Miopsaras myops* Gilbert, 1905 | *Ceratias* sp., *nomen dubium* |
| *Monoceratias acanthias* Gilbert, 1915 | *Oneirodes acanthias* (Gilbert, 1915) |
| *Neoceratias spinifer* Pappenheim, 1914 | *Neoceratias spinifer* Pappenheim, 1914 |
| *Oneirodes alius* Seigel and Pietsch, 1978 | *Oneirodes alius* Seigel and Pietsch, 1978 |
| *Oneirodes appendixus* Ni and Xu, in Ni, 1988 | *Oneirodes sabex* Pietsch and Seigel, 1980 |
| *Oneirodes basili* Pietsch, 1974a | *Oneirodes basili* Pietsch, 1974a |
| *Oneirodes bradburyae* Grey, 1956b | *Oneirodes bradburyae* Grey, 1956b |

| Nominal Species | Currently Recognized Taxon |
|---|---|
| *Oneirodes bulbosus* Chapman, 1939 | *Oneirodes bulbosus* Chapman, 1939 |
| *Oneirodes clarkei* Swinney and Pietsch, 1988 | *Oneirodes clarkei* Swinney and Pietsch, 1988 |
| *Oneirodes cornutus* Gilchrist and von Bonde, 1924 | *Dolopichthys* sp., *nomen dubium* |
| *Oneirodes dicromischus* Pietsch, 1974a | *Oneirodes dicromischus* Pietsch, 1974a |
| *Oneirodes epithales* Orr, 1991 | *Oneirodes epithales* Orr, 1991 |
| *Oneirodes eschrichtii* Lütken, 1871 | *Oneirodes eschrichtii* Lütken, 1871 |
| *Oneirodes haplonema* Stewart and Pietsch, 1998 | *Oneirodes haplonema* Stewart and Pietsch, 1998 |
| *Oneirodes kreffti* Pietsch, 1974a | *Oneirodes kreffti* Pietsch, 1974a |
| *Oneirodes macrosteus* Pietsch, 1974a | *Oneirodes macrosteus* Pietsch, 1974a |
| *Oneirodes megaceros* Holt and Byrne, 1908a | *Oneirodes eschrichtii* Lütken, 1871 |
| *Oneirodes melanocauda* Bertelsen, 1951 | *Oneirodes melanocauda* Bertelsen, 1951 |
| *Oneirodes micronema* Grobecker, 1978 | *Oneirodes micronema* Grobecker, 1978 |
| *Oneirodes myrionemus* Pietsch, 1974a | *Oneirodes myrionemus* Pietsch, 1974a |
| *Oneirodes notius* Pietsch, 1974a | *Oneirodes notius* Pietsch, 1974a |
| *Oneirodes plagionema* Pietsch and Seigel, 1980 | *Oneirodes plagionema* Pietsch and Seigel, 1980 |
| *Oneirodes pietschi* Ho and Shao, 2004 | *Oneirodes pietschi* Ho and Shao, 2004 |
| *Oneirodes posti* Bertelsen and Grobecker, 1980 | *Oneirodes posti* Bertelsen and Grobecker, 1980 |
| *Oneirodes pterurus* Pietsch and Seigel, 1980 | *Oneirodes pterurus* Pietsch and Seigel, 1980 |
| *Oneirodes rosenblatti* Pietsch, 1974a | *Oneirodes rosenblatti* Pietsch, 1974a |
| *Oneirodes sabex* Pietsch and Seigel, 1980 | *Oneirodes sabex* Pietsch and Seigel, 1980 |
| *Oneirodes schistonema* Pietsch and Seigel, 1980 | *Oneirodes schistonema* Pietsch and Seigel, 1980 |
| *Oneirodes sebax* Stewart and Pietsch, 1998 | *Oneirodes sabex* Pietsch and Seigel, 1980 |
| *Oneirodes theodoritissieri* Belloc, 1938 | *Oneirodes theodoritissieri* Belloc, 1938 |
| *Oneirodes thysanema* Pietsch and Seigel, 1980 | *Oneirodes thysanema* Pietsch and Seigel, 1980 |
| *Oneirodes whitleyi* Bertelsen and Pietsch, 1983 | *Oneirodes kreffti* Pietsch, 1974a |
| *Paroneirodes glomerosus* Alcock, 1890 | *Diceratias bispinosus* Günther, 1887 |
| *Parrichthys merrimani* Barbour, 1942b | *Ceratias* sp., *nomen dubium* |
| *Photocorynus spiniceps* Regan, 1925b | *Photocorynus spiniceps* Regan, 1925b |
| *Phrynichthys thele* Uwate, 1979 | *Bufoceratias thele* (Uwate, 1979) |
| *Phrynichthys wedli* Pietschmann, 1926 | *Bufoceratias wedli* (Pietschmann, 1926) |
| *Phyllorhinichthys balushkini* Pietsch, 2004 | *Phyllorhinichthys balushkini* Pietsch, 2004 |
| *Phyllorhinichthys micractis* Pietsch, 1969 | *Phyllorhinichthys micractis* Pietsch, 1969 |
| *Pietschichthys horridus* Kharin, 1989 | *Dermatias platynogaster* Smith and Radcliffe, in Radcliffe, 1912 |
| *Puck pinnata* Pietsch, 1978 | *Puck pinnata* Pietsch, 1978 |
| *Reganichthys giganteus* Bigelow and Barbour, 1944a | *Ceratias holboelli* Krøyer, 1845 |
| *Reganula gigantea* Bigelow and Barbour, 1944b | *Ceratias holboelli* Krøyer, 1845 |
| *Rhynchactis leptonema* Regan, 1925c | *Rhynchactis leptonema* Regan, 1925c |
| *Rhynchactis macrothrix* Bertelsen and Pietsch, 1998b | *Rhynchactis macrothrix* Bertelsen and Pietsch, 1998b |
| *Rhynchactis microthrix* Bertelsen and Pietsch, 1998b | *Rhynchactis microthrix* Bertelsen and Pietsch, 1998b |

*(Continued)*

| Nominal Species | Currently Recognized Taxon |
|---|---|
| *Robia legula* Pietsch, 1979 | *Robia legula* Pietsch, 1979 |
| *Spiniphryne duhameli* Pietsch and Baldwin, 2006 | *Spiniphryne duhameli* Pietsch and Baldwin, 2006 |
| *Spiniphryne gladisfenae* Beebe, 1932b | *Spiniphryne gladisfenae* Beebe, 1932b |
| *Thaumatichthys binghami* Parr, 1927 | *Thaumatichthys binghami* Parr, 1927 |
| *Thaumatichthys pagidostomus* Smith and Radcliffe, 1912 | *Thaumatichthys pagidostomus* Smith and Radcliffe, 1912 |
| *Typhloceratias firthi* Barbour, 1942b | *Ceratias* sp., *nomen dubium* |
| *Typlopsaras shufeldti* Gill, 1883b | *Ceratias uranoscopus* Murray, in Thomson, 1877 |
| *Tyrannophryne pugnax* Regan and Trewavas, 1932 | *Tyrannophryne pugnax* Regan and Trewavas, 1932 |

# REALLOCATION OF NOMINAL SPECIES OF THE CERATIOIDEI BASED ON FREE-LIVING MALES

| Nominal Species | Currently Recognized Taxon |
|---|---|
| Originally allocated to family Aceratiidae Brauer, 1906 | |
| *Aceratias macrorhinus indicus* Brauer, 1902 | *Linophryne indica* (Brauer, 1902) |
| *Aceratias macrorhinus* Brauer, 1902 | *Linophryne* subgenus *Linophryne* sp. |
| *Aceratias mollis* Brauer, 1902 | *Haplophryne mollis* (Brauer, 1902) |
| *Lipactis tumidus* Regan, 1925c (original material represents two currently recognized taxa) | *Himantolophus brevirostris* group? Bertelsen and Krefft, 1988 |
| | *Oneirodes* sp. |
| *Rhynchoceratias brevirostris* Regan, 1925c | *Himantolophus brevirostris* group |
| *Rhynchoceratias leucorhinus* Regan, 1925c (original material represents seven currently recognized taxa) | *Centrophryne spinulosa* Regan and Trewavas, 1932 |
| | *Chaenophryne draco* group Bertelsen, 1951 |
| | *Chaenophryne longiceps* group Bertelsen, 1951 |
| | *Oneirodes* sp. |
| | *Dolopichthys* sp. |
| | *Microlophichthys microlophus* (Regan, 1925c) |
| | *Pentherichthys atratus* (Regan and Trewavas, 1932) |
| *Rhynchoceratias oncorhynchus* Regan, 1925c | *Himantolophus brevirostris* group Bertelsen and Krefft, 1988 |
| *Rhynchoceratias rostratus* Regan, 1925c | *Himantolophus rostratus* group Bertelsen and Krefft, 1988 |
| *Rhynchoceratias acanthirostris* Parr, 1927 | *Melanocetus murrayi* Günther, 1887 |
| *Rhynchoceratias latirhinus* Parr, 1927 | *Melanocetus murrayi* Günther, 1887 |
| *Laevoceratias liparis* Parr, 1927 | *Gigantactis* Male Group I Bertelsen et al., 1981 |
| *Haplophryne hudsonius* Beebe, 1929b | *Linophryne* subgenus *Linophryne* sp. |
| *Haplophryne simus* Borodin, 1930b | *Linophryne* subgenus *Rhizophryne* sp.? |
| *Rhynchoceratias longipinnis* Parr, 1930a | *Melanocetus murrayi* Günther, 1887 |
| *Aceratias edentula* Beebe, 1932b | *Linophryne* subgenus *Linophryne* sp. |
| Originally allocated to family Melanocetidae Gill, 1879b | |
| *Centrocetus spinulosus* Regan and Trewavas, 1932 | *Melanocetus johnsonii* Günther, 1864 |
| *Xenoceratias longirostris* Regan and Trewavas, 1932 | *Melanocetus* sp. |
| *Xenoceratias micracanthus* Regan and Trewavas, 1932 | *Melanocetus johnsonii* Günther, 1864 |
| *Xenoceratias heterorhynchus* Regan and Trewavas, 1932 | *Melanocetus johnsonii* Günther, 1864 |
| *Xenoceratias laevis* Regan and Trewavas, 1932 | *Melanocetus johnsonii* Günther, 1864 |
| *Xenoceratias brevirostris* Regan and Trewavas, 1932 | *Melanocetus johnsonii* Günther, 1864 |
| *Xenoceratias braueri* Koefoed, 1944 | *Melanocetus johnsonii* Günther, 1864 |
| *Xenoceratias regani* Koefoed, 1944 | *Melanocetus murrayi* Günther, 1887 |
| *Xenoceratias nudus* Beebe and Crane, 1947 | *Melanocetus* sp. |

*(Continued)*

| Nominal Species | Currently Recognized Taxon |
|---|---|
| Originally allocated to family Himantolophidae Gill, 1861 | |
| *Rhynchoceratias altirostris* Regan and Trewavas, 1932 | *Himantolophus brevirostris* group Bertelsen and Krefft, 1988 |
| Originally allocated to family Diceratiidae Regan and Trewavas, 1932 | |
| *Caranactis pumilus* Regan and Trewavas, 1932 | *Oneirodes* sp. |
| Originally allocated to family Oneirodidae Gill, 1879a | |
| *Trematorhynchus exiguus* Regan and Trewavas, 1932 | *Oneirodes* sp. |
| *Trematorhynchus obliquidens* Regan and Trewavas, 1932 | *Chaenophryne draco* group Bertelsen, 1951 |
| *Trematorhynchus phyllodon* Parr, 1934 | Oneirodidae ? |
| *Trematorhynchus adipatus* Beebe and Crane, 1947 | *Chaenophryne* sp. |
| *Trematorhynchus moderatus* Beebe and Crane, 1947 | *Chaenophryne* sp. |
| *Trematorhynchus multilamellatus* Beebe and Crane, 1947 | Unidentifiable |
| *Trematorhynchus multiradiatus* Beebe and Crane, 1947 | *Ctenochirichthys longimanus* Regan and Trewavas, 1932 |
| *Trematorhynchus paucilamellatus* Beebe and Crane, 1947 | *Oneirodes* sp. |
| *Microlophichthys andracanthus* Bertelsen, 1951 | *Microlophichthys andracanthus* Bertelsen, 1951 |
| Originally allocated to family Laevoceratiidae Regan and Trewavas, 1932 | |
| *Teleotrema microphthalmus* Regan and Trewavas, 1932 | *Gigantactis* Male Group II Bertelsen et al., 1981 |
| Originally allocated to family Linophrynidae Regan, 1925c | |
| *Nannoceratias denticulatus* Regan and Trewavas, 1932 | *Linophryne* subgenus *Rhizophryne* sp. |
| *Borophryne masculina* Parr, 1934 | *Linophryne* subgenus *Linophryne* sp. |

# SYMBOLIC CODES FOR INSTITUTIONAL COLLECTIONS

The systematic revision presented in this volume is based on metamorphosed stages of some 6,310 females and 785 males, comprising as far as possible all available material worldwide, borrowed from or examined at the following 90 institutions:

| | |
|---|---|
| AIM: | Auckland Institute and Museum, Auckland, New Zealand |
| AMNH: | American Museum of Natural History, New York |
| AMS: | Australian Museum, Sydney, Australia |
| ARC: | Atlantic Reference Centre, Huntsman Marine Laboratory, St. Andrews, New Brunswick, Canada |
| ARIK: | Atlantic Research Institute of Fisheries and Oceanography, Kaliningrad, Russia |
| ASIZP: | Academia Sinica, Institute of Zoology, Taipei, Taiwan, Republic of China |
| AUS: | Agricultural University, Szczecin, Poland |
| BMNH: | The Natural History Museum, London (formerly the British Museum of Natural History), England |
| BPBM: | Bernice P. Bishop Museum, Honolulu, Hawaii |
| BSKU: | Kochi University, Department of Biology, Faculty of Science, Kochi City, Japan |
| CAS: | California Academy of Sciences, San Francisco |
| CAS-SU: | Stanford University, material now housed at the California Academy of Sciences, San Francisco |
| CMA: | Cabrillo Marine Aquarium, San Pedro, California |
| CMC: | Canterbury Museum, Christchurch, New Zealand |
| CSIRO: | Commonwealth Scientific and Industrial Research Organization, Hobart, Tasmania |
| ECFR: | East China Sea Fisheries Research Institute, Chinese Academy of Fisheries Science, Shanghai, China |
| FMNH: | Field Museum of Natural History, Chicago |
| GNM: | Göteborgs Musei Zoologiska Avdelning, Sweden |
| HSU: | Humboldt State University, Arcata, California |
| HUMZ: | Hokkaido University, Laboratory of Marine Biodiversity, Graduate School of Fisheries Sciences, Hakodate, Japan |
| IEOS: | Instituto Español de Oceanografia, Santande, Spain |
| IFAN: | Institut Français d'Afrique Noire, Dakar, Senegal |
| IFM: | Leibniz-Institut für Meereswissenschaften an der Universität Kiel, Kiel, Germany |
| IINH: | Icelandic Institute of Natural History, Reykjavik, Iceland |
| IIPB: | Instituto de Investigaciones Pesquéras de Barcelona, Barcelona, Spain |
| IMARPE: | Instituto del Mar del Perú, Callao, Peru |
| IMB: | Institut für Meeresforschung, Bremerhaven, Germany (now part of the Alfred Wegner Institut, Bremerhaven) |
| IMBV: | Institute of Marine Biology, Vladivostok, Russia |
| INM: | Ibaraki Nature Museum, Ibaraki, Japan |
| ISH: | Institut für Seefischerei, Hamburg, Germany, collections now housed at the Zoological Museum, University of Hamburg |
| LACM: | Natural History Museum of Los Angeles County, Los Angeles |
| MACN: | Museo Argentino de Ciencias Naturales, Buenos Aires, Argentina |
| MBL: | Museu Nacional de História Natural (Museu Bocage), Universidade de Lisboa, Lisbon, Portugal |
| MCZ: | Museum of Comparative Zoology, Harvard University, Cambridge, Massachusetts |
| MHNLR: | Musée d'Histoire Naturelle de La Rochelle, La Rochelle, France |
| MMA: | Museo Marítimo de Asturias, Luanco, Asturias, Spain |
| MMF: | Museu Municipal do Funchal, Madeira, Portugal |
| MNHN: | Muséum National d'Histoire Naturelle, Paris, France |
| MNHNC: | Museo Nacional de Historia Natural, Santiago, Chile |
| MNRJ: | Museu Nacional, Rio de Janeiro, Brazil |
| MOM: | Musée Oceanographique de Monaco, Monaco |
| MRIR: | Marine Research Institute, Reykjavik, Iceland |
| MTF: | Museum of Thorshavn, Faeroe Islands, Denmark |

| | | | |
|---|---|---|---|
| MZUSP: | Museu de Zoologia da Universidade de São Paulo, São Paulo, Brazil | SIOM: | Shirshov Institute of Oceanography, Moscow, Russia (formerly IOAN or IOM, Institute of Oceanography, Russian Academy of Sciences, Moscow) |
| NHMB: | "Grigore Antipa" Natural History Museum, Bucharest, Romania | SMNS: | Staatliches Museum für Naturkunde, Stuttgart, Germany |
| NHRM: | Swedish Museum of Natural History, Stockholm, Sweden | SWFSC: | Southwest Fisheries Science Center, National Marine Fisheries Service, La Jolla, California |
| NMI: | National Museum of Ireland, Dublin, Ireland | TCWC: | Texas Cooperative Wildlife Collection, Texas A & M University, College Station, Texas |
| NMMBP: | National Museum of Marine Biology and Aquarium, Checheng, Pingtung, Taiwan, Republic of China | TFRI: | Taiwan Fisheries Research Institute, Taipei, Taiwan, Republic of China |
| NMNS: | National Museum of Natural Science and Technology, Taichung, Taiwan, Republic of China | UBC: | University of British Columbia, Vancouver, British Columbia, Canada |
| NMNZ: | National Museum of New Zealand, Wellington, New Zealand | UF: | Florida State Museum, University of Florida, Gainesville, Florida (including material formerly housed at the University of Miami Marine Laboratory) |
| NMSZ: | National Museums of Scotland, Edinburgh, Scotland | | |
| NMV: | National Museum of Victoria, Melbourne, Australia | | |
| NMW: | Naturhistorisches Museum, Vienna, Austria | | |
| NOC: | National Oceanography Centre, Southampton, England (formerly SOC, Southampton Oceanography Centre, Southampton, England) | UMB: | Übersee-Museum, Bremen, Germany |
| | | USNM: | National Museum of Natural History, Washington, D.C. |
| NSMN: | Niigata Science Museum, Niigata, Japan | UW: | University of Washington, Seattle, Washington |
| NSMT: | National Science Museum, Tokyo, Japan | VIMS: | Virginia Institute of Marine Science, Gloucester Point, Virginia |
| NYZS: | New York Zoological Society, New York (collections now housed at California Academy of Sciences, San Francisco) | WAM: | Western Australian Museum, Perth, Australia |
| | | YPM: | Bingham Oceanographic Collections, Peabody Museum of Natural History, Yale University, New Haven, Connecticut (formerly BOC) |
| ORI: | Ocean Research Institute, University of Tokyo, Tokyo, Japan | | |
| OS: | Oregon State University, Department of Oceanography, Corvallis, Oregon | ZIN: | Zoological Institute, Russian Academy of Sciences, St. Petersburg, Russia |
| PINRO: | Polar Research Institute of Marine Fisheries, Murmansk, Russia | ZMA: | Zoological Museum, University of Amsterdam, Amsterdam, Netherlands |
| ROM: | Royal Ontario Museum, Toronto, Canada | ZMB: | Zoologisches Museum der Humboldt-Universität, Berlin, Germany |
| SAIAB: | South African Institute for Aquatic Biodiversity, Grahamstown, South Africa (formerly RUSI: Rhodes University, J. L. B. Smith Institute of Ichthyology) | ZMH: | Universität Hamburg, Zoologisches Institut und Museum, Hamburg, Germany |
| SAM: | South African Museum, Cape Town, South Africa | ZMUB: | Zoological Museum, University of Bergen, Bergen, Norway |
| SAMA: | South Australian Museum, Adelaide, Australia | ZMUC: | Zoological Museum, University of Copenhagen, Copenhagen, Denmark |
| SCFR: | South China Sea Fisheries Research Institute, Chinese Academy of Fisheries Science, Kwangzhou, China | ZMUO: | Zoological Museum, University of Oslo, Oslo, Norway |
| SCMT: | Science College Museum of Tokyo, Tokyo, Japan | ZSI: | Zoological Survey of India, Calcutta, India |
| SFU: | Shanghai Fisheries University, Shanghai, China | ZUMT: | Department of Zoology, University Museum, University of Tokyo, Tokyo, Japan |
| SIO: | Scripps Institution of Oceanography, University of California, La Jolla, California | | |

# GLOSSARY

ABYSSOPELAGIC   That part of the oceanic zone below 4000 m.

ACANTHOPTERGII   A superorder of derived actinopterygian fishes containing a huge assemblage of spiny-rayed teleosts, some 13 orders and 270 families, the interrelations of which are essentially unknown.

ACOUSTICO-LATERALIS SYSTEM   A complicated network of sensory canals (containing mechanoreceptors, i.e., neuromasts) arranged on the head and body of fishes (and larval and permanently aquatic amphibians) that serves to transduce and relay waterborne vibrations to nerve cells, allowing for detection of disturbances in the surrounding environment; also called *lateral-line system*.

ALLOPATRIC   In biogeography, distributions that do not overlap.

ANTENNARIOIDEI   A suborder of anglerfishes containing the frogfishes and handfishes, four families, 15 genera, and about 54 species of laterally compressed, shallow to moderately deep-water, benthic forms.

APHAKIC SPACE   A space between the pupil and lens resulting in an elongation of the eye and a consequent increase in the sensitivity of the eye to light as well as an increase in binocular overlap, thus aiding in depth perception; an adaptation that has evolved convergently in numerous relatively unrelated groups of deep-sea fishes.

APOMORPHY   In cladistic analysis, a derived character state.

AUTAPOMORPHY   In cladistic analysis, a derived character state present in only a single taxon.

BATHYPELAGIC   That part of the oceanic zone that extends from 1000 to 4000 m, below the extent of solar radiation, often called the aphotic zone.

BATRACHOIDIFORMES   One of some 44 orders of bony or ray-finned fishes (class Actinopterygii) that contains all the toadfishes of the world (a single family, 22 genera, and about 78 species), thought by some to be the sister group of the Lophiiformes.

BENTHIC   All those phenomena and things that are strictly associated with the bottom, regardless of depth.

BENTHOPELAGIC   All those phenomena and things that are closely associated with the bottom, regardless of depth.

BRANCHIOSTEGAL RAY   One of a series of elongate, flattened or cylindrical, riblike elements that supports the membranous outer margin of the gill cover of teleost fishes; interconnected by muscles, the branchiostegals open and close relative to one another, thus increasing the efficiency of the pumping mechanism responsible for moving water across the gills; they are also responsible for feeding mechanisms that rely on the production of suction.

CARUNCLE   A spherical, esca-like light organ associated with the anterior-most two or three soft dorsal-fin rays found only in members of the ceratioid family Ceratiidae.

CAUDAL PEDUNCLE   The narrowest part of the body of a fish, nearly always located just anterior to the base of the caudal fin, responsible for transmitting the power of the lateral undulation of the locomotory trunk muscles to the tail.

CERATIOIDEI   A suborder of anglerfishes containing the seadevils, comprising 11 families, 35 genera, and 160 currently recognized species of globose to elongate, mesopelagic, bathypelagic, and abyssal-benthic forms.

CHAUNACOIDEI   A suborder of anglerfishes containing the gapers, coffinfishes, and sea toads, a single family, at least two genera, and as many as 14 species of globose, deep-water benthic forms, many of which remain to be described.

CLADE   A branch of a cladogram containing all the descendents of a common ancestor (see *monophyletic*).

CLADISTICS   A method of phylogenetic analysis first described in detail by Willi Hennig (1950, 1966), in which monophyletic groups are formed by shared derived characters, or synamorphies.

CLADOGRAM   A branching diagram (tree) showing the sequence of evolutionary divergence of organisms through time, based on cladistic analysis; a product of systematic research.

CLASSIFICATION   The ordering of plants and animals into groups based on their similarity and relationship.

CLASSIFICATIONS   Concise lists of organisms, grouped or ranked according to the pattern of branching seen in a branching diagram (cladogram); a product of systematic research.

CONGENERIC   Belonging to the same genus.

CONSPECIFIC   Belonging to the same species.

DENTICULAR BONES   Tooth-bearing elements, homologous to dermal spinules of the anterior tip of the snout and lower

**DERIVED** Modified relative to the primitive condition.

**ENDEMIC** In biogeography, a taxon found only in a particular geographic locality.

**EPAXIAL** That portion of the body that develops dorsal to the axis of the body formed by the vertebral column.

**EPIPELAGIC** That part of the ocean that extends from the surface to 200 m, a depth that corresponds on average to the margin of the continental slope, which in turn is approximately equivalent to the lower limit of photosynthesis, often called the ephotic zone.

**ESCA** The Latin word for "bait"; a fleshy structure borne on the distal tip of the illicium of anglerfishes; bioluminescent in females of nearly all deep-sea anglerfishes.

**ETYMOLOGY** From the Greek word *etymos*, the true, original, literal root-meaning of a word, often derived by analyzing its individual component parts.

**EXTRINSIC MUSCLES** Muscles that originate separately or away from, but insert on, the bony element in question.

**HEMIBRANCH** A gill arch with only one row of gill lamellas.

**HERMAPHRODITISM** With reference to ceratioid anglerfishes, the result of tissue fusion between a conspecific male and female, forming a single organism having both male and female reproductive organs.

**HOLOBRANCH** A gill arch with two rows of gill lamellae, one row on each side.

**HOLOTYPE** A single specimen designated as such by an author in an original description and set aside as the best example of the species in question, thus serving as a reference standard for that taxon.

**HOMOLOGOUS** Characters in different taxa that are structurally similar due to common evolutionary origin.

**HOMOPLASY** Nonhomologous similarity due to convergent, parallel, or reversed evolution.

**HYOID APPARATUS** In teleost fishes, a series of bony elements (including the urohyal, dorsal and ventral hypohyals, epihyal, ceratohyal, and interhyal) that forms in part the floor of the mouth, supports the gills, and provides articular surface for the branchiostegal rays.

**HYOID BARBEL** A cutaneous structure emanating from the chin of the ceratioid genera *Centrophryne* and *Linophryne*: nonluminescent, relatively tiny, and lost with growth in both females and males of the single known species of the former; bioluminescent and elaborate in adult females of all 22 species of the latter.

**HYPAXIAL** That portion of the body that develops ventral to the axis of the body formed by the vertebral column.

**HYPURAL** One of a series of bony elements of the tail skeleton of a fish, which in anglerfishes are fused together to form a solid, roughly triangular bony plate (hypural plate) that supports the rays of the caudal fin.

**ILLICIUM** The luring apparatus of anglerfishes: the modified first dorsal-fin spine situated on the tip of the snout or on top of the head of female ceratioids that bears the bioluminescent bait or esca.

**INNERVATION** The distribution of nerves to an organ of the body.

**INTRINSIC MUSCLE** A muscle that both originates and inserts on the bony element in question.

**LECTOTYPE** A single specimen selected from a series of syntypes (or a specimen known to have been used by the original describer if no type was identified) by a subsequent author to serve in place of a holotype.

**LOPHIIFORMES** One of some 44 orders of bony or ray-finned fishes (class Actinopterygii) that contains all the anglerfishes of the world.

**LOPHIOIDEI** A suborder of anglerfishes containing the goosefishes and monkfishes, a single family, four genera, and 25 living species of shallow- to deep-water, dorsoventrally flattened forms.

**MELANOPHORE** A pigment cell containing melanin, a brown, black, or rust-colored pigment.

**MESOPELAGIC** That part of the ocean that extends from 200 to 1000 m, the greater depth corresponding to the limit of penetration of solar radiation, often called the twilight zone or disphotic zone.

**METAMORPHOSIS** A stage in early development during which an animal goes through a relatively abrupt and radical change in structure and function.

**MONOPHYLETIC** A natural group of organisms that includes all the descendents of a common ancestor (i.e., a clade).

**NEUROMAST** A mechanoreceptor of the acoustico-lateralis system.

**NOMEN DUBIUM** A name representing a nominal species based on one or more unidentifiable specimens (e.g., damaged specimens that have lost all diagnostic characters), or a questionable taxon for which a type specimen or specimens have been lost or never retained; a name not certainly applicable to any known species.

**NOMEN NUDUM** A name appearing in the literature without application to a type, description, or illustration; a name without indication.

**OGCOCEPHALOIDEI** A suborder of anglerfishes containing the batfishes, a single family of 10 general and about 68 species of dorsoventrally flattened, deep-water benthic forms.

**ONTOGENETIC** Pertaining to developmental change through the life cycle of an organism.

**OPERCULAR APPARATUS** In teleost fishes, a series of bony elements (including the preopercle, opercle, subopercle, and interopercle) that forms a major part of the respiratory pump and suction-feeding apparatus, and also serves to cover and protect the gills.

**OUTGROUP** A closely related taxon outside the study group used in cladistic analysis to provide information about the direction of character-state change.

**PAIRED FINS** Fins that insert off the midline, the pectorals and pelvics.

**PARACANTHOPTERYGII** A superorder of actinopterygian fishes containing the trout-perches, cods, cusk-eels, toadfishes, and anglerfishes, thought by most to be an artificial, unnatural assemblage.

**PARALECTOTYPE** One of two or more syntypes remaining after a lectotype has been selected from a type series by a subsequent describer.

**PARATYPE** A specimen used in the description of a new species in addition to the specimen designated as the holotype.

**PARSIMONY** In cladistic analysis, the principle that determines the choice of hypotheses requiring the fewest ad hoc assumptions about character convergence, parallelism, and reversal.

**PECTORAL GIRDLE** In teleost fishes, bony elements (including the scapula, coracoid, cleithrum, supracleithrum, and one or more postcleithra) that support the pectoral fin.

PECTORAL RADIALS   In teleost fishes, those bony elements that form the pectoral-fin lobe and provide articular support for the rays of the pectoral fin, consisting in ceratioids of two to five separate ossifications.

PEDICULATI   Now obsolete ordinal name first used by British ichthyologist Albert Günther (1861) to contain all the anglerfishes, a junior synonym of Lophiiformes.

PELAGIC   All those phenomena and things that lie within the water column, in contrast to the bottom.

PERCOMORPHA   The most-derived teleostean clade, containing all the perches and perchlike fishes, a huge assemblage of some nine orders and 245 families, the interrelationships of which are essentially unknown.

PERITONIUM   The smooth, transparent membrane that lines the gut cavity and is folded inward over the abdominal and pelvic viscera.

PHARYNGEAL TEETH   In teleost fishes, teeth borne on upper and lower parts of the gill arches that permit the separation of grabbing (outer jaw) and chewing (pharyngeal jaw) functions of the mouth; the upper pharyngeal teeth are extremely well developed in ceratioids.

PHARYNX   The throat or more generally the mouth.

PHOTOPHORE   A light-producing structure.

PHYLOGENY   The evolutionary relationships among taxa based on their descent from a common ancestor.

PLESIOMORPHY   In cladistic analysis, a primitive character state.

POLYTOMY   In cladistic analysis, the branching of three or more clades from a single node; an unresolved branching point.

PRIMITIVE   Unmodified relative to the derived condition.

PSEUDOBRANCH   A small structure situated on the inner wall of the opercular cavity of many fishes, composed of filaments like those of the gills, and apparently involved in providing oxygenated blood to the eye.

PTERYGIOPHORE   With respect to ceratioid anglerfishes, the bony element that supports and allows for movement of the illicium of females; and which is closely associated with the upper denticular bone of males, serving in part to close the denticular jaws when biting and holding fast to a conspecific female.

SEXUAL DIMORPHISM   The existence of structures that differ morphologically between conspecific males and females.

SEXUAL PARASITISM   In some ceratioid anglerfishes, a symbiotic relationship between conspecific genders in which the male benefits to the disadvantage of the female in terms of nutrition.

SINISTRAL   Of, related to, or inclined to the left, as describing the position of the anus in members of the ceratioid family Linophrynidae.

SISTER GROUP   In cladistic analysis, a taxon most closely related to another by common ancestry.

STANDARD LENGTH   The length of a fish, excluding the rays of the tail or caudal fin, measured from the anterior-most tip of the upper jaw to the posterior-most margin of the bony structures that support the caudal-fin rays (except where noted, all fish lengths in this volume are standard length, abbreviated SL; total length, abbreviated TL, is occasionally used for small larvae that have not yet developed the caudal fin).

SUSPENSORIUM   A functional unit of the teleost head skeleton (consisting of the hyomandibula, preopercle, symplectic, pterygoids, palatine, and quadrate) that functions primarily to support or "suspend" the upper and lower jaws.

SYMBIOSIS   In reference to anglerfishes, the intimate and apparently obligatory association of bioluminescent bacteria that live within the esca of female ceratioids.

SYMPATRIC   In biogeography, two or more taxa that occupy the same geographic locality.

SYMPHYSIAL SPINE   A ventrally directed protuberance, long and sharp in some ceratioids, formed by the distal ends of the dentaries (tooth-bearing elements of the lower jaw) where they meet on the ventral midline.

SYMPLESIOMORPHY   In cladistic analysis, a shared primitive character.

SYNAPOMORPHY   In cladistic analysis, a shared derived character.

SYNTYPE   One of a series of two or more specimens used in the description of a new species in which no single specimen is given holotype status.

SYSTEMATICS   The study of biological diversity, or, more specifically, the ordering of the diversity of nature through construction of a classification that can serve as a general reference system.

TAXON   Any of the formal categories used in classifying organisms; the plural is *taxa*.

TAXONOMY   That branch of the biological sciences that deals with the discovery, recognition, definition, and naming of groups of organisms.

TELEOST   A member of the division Teleostei containing all the so-called bony or ray-finned fishes, with about 26,840 living species, more than all other vertebrate species combined.

TETRAODONTIFORMES   One of some 44 orders of bony or ray-finned fishes (class Actinopterygii) that contains all the spikefishes, triggerfishes, boxfishes, puffers, and molas of the world (nine families, approximately 101 genera and 357 species), thought by some to be closely related to the Lophiiformes.

TOTAL LENGTH   The length of a fish, measured from the anterior-most tip of the upper jaw to the tip of the longest caudal-fin ray (except where noted, all fish lengths in this volume are standard length, abbreviated SL; total length, abbreviated TL, is occasionally used for small larvae that do not have a fully developed caudal skeleton).

TROPHIC   Related to nutrition or mode of feeding.

UNPAIRED FINS   Fins that insert on the midline, the dorsal, anal, and caudal.

VOMERINE TEETH   Teeth borne on the vomer, an unpaired bony element of the cranium that forms the anterior-most part of the roof of the mouth.

# REFERENCES

Abe, T. 1967. Records from northern Japan of two females of *Ceratias holboelli* each parasitized by a male. *Proc. Japan Acad.*, 43:797–800.

Abe, T., and M. Funabashi. 1992. A record of an adult female of the deep sea ceratioid anglerfish, *Cryptopsaras couesi* Gill, with four parasitic adult males from off Ibaraki Prefecture, Japan. *UO*, 41:cover, 1–3.

Abe, T., and T. Iwami. 1979. A record of the ceratioid anglerfish *Oneirodes notius* Pietsch caught along with the Antarctic krill. *Bull. Biogeogr. Soc. Japan*, 34(1):1–7.

Abe, T., and C. Nakamura. 1954. A record of an adult female with a supposedly parasitic male of *Cryptopsaras couesi* from the Pacific coast of northern Japan. *Japan. J. Ichthyol.*, 3(2):95–96.

Ackerman, J., and D. A. Murawski. 1997. Parasites looking for a free lunch. *Nat. Geogr.*, 192(4):74–91.

Agassiz, J. L. R. 1835. *Revue critique des poissons fossiles figurés dans L'Ittiolitologia Veronese (extraits de la 4me livr. des Recherches sur les Poissons Fossiles)*. Imprimerie de Petitpierre et Prince, Neuchâtel, Switzerland. [Reprinted in German in *Neues Jahrbuch*, 1835.]

Agassiz, J. L. R. 1844. *Recherches sur les Poissons Fossiles*, Vol 5. Published by the author, printed in Neuchâtel, Switzerland. [Dating after Brown, 1890.]

Agassiz, L. 1846. *Nomenclatoris Zoologici. Index universalis, continens nomina systematica classium, ordinum, familiarum et generum animalius omnium, tam viventium quam fossilius, secundum ordinem alphabeticum unicum disposita. . . .* Jent and Gassmann, Soloduri.

Albuquerque, R. M. 1954–1956. Peixes de Portugal e ilhas adjacentes. Chaves para a sua determinaçao, com uma nota prefacial por J. A. Serra. *Portugaliae Acta Biologica*, Lisboa, Ser. B, Vol. 5, xvi + 1164 pp.

Alcock, A. W. 1890. Natural history notes from H.M. Indian Marine Survey steamer "Investigator," Commander R. F. Hoskyn, R. N., commanding. No. 16. On the bathybial fishes collected in the Bay of Bengal during the season 1889–90. *Ann. Mag. Nat. Hist.*, Ser. 6, 6(26):197–222.

Alcock, A. W. 1896. Natural history notes from H.M. Indian Marine Survey steamer "Investigator," Commander C. F. Oldham, R. N., commanding. Series 2. No. 23. A supplementary list of the marine fishes of India, with descriptions of 2 new genera and 8 new species. *J. Asiatic Soc. Bengal*, 65, Pt. 2, Natural Science, 3:302–338.

Alcock, A. W. 1899. *A descriptive catalogue of the Indian deep-sea fishes in the Indian Museum, being a revised account of the deep-sea fishes collected by the Royal Indian Marine Survey ship "Investigator."* Printed by Order of the Trustees, Calcutta.

Alcock, A. W. 1900. *Illustrations of the Zoology of the Royal Indian Marine Surveying Steamer "Investigator,"* pt. 7, Fishes. Printed by Order of the Trustees, Calcutta, pls. 27–35.

Alcock, A. W. 1902. *A Naturalist in Indian Seas. Or, Four Years with the Royal Indian Marine Survey Ship "Investigator."* John Murray, London.

Alexander, R. M. 1970. Mechanics of the feeding action of various teleost fishes. *J. Zool.*, London, 162:145–156.

Aloncle, H. 1968. Catalogues des types de poissons téléostéens en collection au Museum de La Rochelle. *Bull. Mus. Nat. Hist. Nat.*, Ser. 2, 40(4):683–691.

Amanieu, M., and C. Cazaux. 1962. Animaux rares observes dans la region d'Arcachon en 1961–1962. *P. V. Soc. Linn. Bordeaux*, 99:74–86.

Amaoka, K. 1983. [Ceratioid families]. pp. 114–121, 197–199, 250–253, 325–326, in: K. Amaoka, K. Nakaya, H. Araya, and T. Yasui (editors), *Fishes from the North-Eastern Sea of Japan and the Okhotsk Sea off Hokkaido. The Intensive Research of Unexploited Fishery Resources on Continental Slopes*. Japan Fisheries Resource Conservation Association, Tokyo.

Amaoka, K. 1984. Suborder Ceratioidei. pp. 105–108, plates 92–93, in: H. Masuda, K. Amaoka, C. Araga, T. Uyeno, and T. Yoshino (editors), *The Fishes of the Japanese Archipelago*, 2 vols. Tokai University Press, Tokyo.

Amaoka, K. 1990. [Family Ceratiidae]. p. 202, in: K. Amaoka, K. Matsuura, T. Inada, M. Takeda, H. Hatanaka, and K. Okada (editors), *Fishes Collected by the R/V Shinkai Maru around New Zealand*. Japan Marine Fishery Resources Research Center, Tokyo.

Amaoka, K., K. Nakaya, and M. Yabe. 1995. *The Fishes of Northern Japan*. Kita-Nihon Kaiyo Center Co., Sapporo, Hokkaido, Japan.

Anderson, M.E., and R.W. Leslie. 2001. Review of the deep-sea anglerfishes (Lophiiformes: Ceratioidei) of southern Africa. *Ichthyol. Bull., J.L.B. Smith Inst. Ichthyol.*, 70, 32 pp.

Andriyashev, A.P., and N.V. Chernova. 1994. Annotated list of fishlike vertebrates and fish of the Arctic seas and adjacent waters. *Vopr. Ikhtiol.*, 34(4):435–456. [In Russian, English translation in *J. Ichthyol.*, 35(1):81–123, 1995.]

Annandale, N., and J.T. Jenkins. 1910. Plectognathi and Pediculati. Part 3, pp. 7–21, in: Report on the fishes taken by the Bengal Fisheries Steamer "Golden Crown." *Mem. Indian Mus.*, Calcutta, 3(1).

Arambourg, C. 1927. Les poissons fossiles d'Oran. *Mater. Carte Geol. Algerie*, Ser. 1, Paleontol., 6:1–298.

Arrington, D.A., K.O. Winemiller, W.F. Loftus, and S. Akin. 2002. How often do fishes "run on empty"? *Ecology*, 83(8):2145–2151.

Arronte, J.C., and T.W. Pietsch. 2007. First record of *Himantolophus mauli* (Lophiiformes: Himantolophidae) on the slope off Asturias, central Cantabrian Sea, eastern North Atlantic Ocean. *Cybium*, 31(1):85–86.

Ayling, T., and G.J. Cox. 1982. *Collins Guide to the Sea Fishes of New Zealand*. Collins, Auckland.

Băcescu, M.C. 1966. Colaborarea țării noastre la explorarea vieții animale din zonele hadale ale Pacificului sud-estic. *Natura, Seria Geografie-Geologie*, 6:27–36.

Backus, R.H., G.W. Mead, R.L. Haedrich, and A.W. Ebeling. 1965. The mesopelagic fishes collected during Cruise 17 of the R/V CHAIN with a method for analyzing faunal transects. *Bull. Mus. Comp. Zool.*, Harvard, 145:139–157.

Backus, R.H., J.E. Craddock, R.L. Haedrich, and D.L. Shores. 1970. The distribution of mesopelagic fishes in the Equatorial and western North Atlantic Ocean. *J. Mar. Res.*, 28(2): 179–201.

Baird, R.C. 1971. The systematics, distribution, and zoogeography of the marine hatchetfishes (family Sternoptychidae). *Bull. Mus. Comp, Zool.*, 142(1):1–128.

Ballintijn, C.M., and G.M. Hughes. 1965. The muscular basis of the respiratory pumps in the trout. *J. Exp. Biol.*, 43: 349–362.

Balushkin, A.V., and V.V. Fedorov. 1981. On finding the deepwater anglerfishes (*Melanocetus rossi* sp. n. and *Oneirodes notius*) in the Ross Sea (Antarctica). *Biol. Morya*, 2(2):79–82. [In Russian, with English abstract.]

Balushkin, A.V., and V.V. Fedorov. 1985. *Caulophryne pietschi* sp. nov., a new species of moss anglerfish (Caulophryidae) from the notal regions of the southwestern Pacific Ocean. *Vopr. Ikhtiol.*, 25(6):1035–1037. [In Russian, English translation in *J. Ichthyol.*, 26(1):151–154, 1986.]

Balushkin, A.V., and V.V. Fedorov. 1986. A new species of diceratiid deepsea anglerfish, *Diceratias trilobus* sp. n. (Fam. Diceratiidae, Ceratioidei), from the coast of Japan. *Vopr. Ikhtiol.*, 26(5):855–856. [In Russian, English translation in *J. Ichthyol.*, 27(1):136–138, 1987.]

Balushkin, A.V., and V.V. Fedorov. 2002. New contributions to the fish fauna of the Southern Ocean. *Contributions from the Zoological Institute*, 4. Russian Academy of Sciences, St. Petersburg. [In Russian, with English abstract.]

Balushkin, A.V., and L.A. Trunov. 1988. A new species of deep-sea anglerfish, *Linophryne digitopogon* sp. n. (Linophrynidae; Ceratioidei; Pisces), from the south east Atlantic. *Biol. Morya*, 1988(6):62–65. [In Russian, with English abstract.]

Bannikov, A.F. 2004. The first discovery of an anglerfish (Teleostei, Lophiidae) in the Eocene of the Northern Caucasus. *Paleontol. Zhur.*, 38(4):67–72. [In Russian, English translation in *Paleontol. J.*, 38(4):420–425, 2004.]

Bañón, R., T.W. Pietsch, and C.-G. Piñeiro. 2006. New Record of *Linophryne coronata* (Lophiiformes, Linophrynidae) from the north-eastern Atlantic Ocean. *Cybium*, 30(4):385–386.

Barbour, T. 1941a. *Ceratias mitsukurii* in M.C.Z. *Copeia*, 1941(3):175.

Barbour, T. 1941b. Notes on pediculate fishes. *Proc. New Engl. Zool. Club*, 19:7–14.

Barbour, T. 1942a. The northwestern Atlantic species of frog fishes. *Proc. New Engl. Zool. Club*, 19:21–40.

Barbour, T. 1942b. More concerning ceratioid fishes. *Proc. New Engl. Zool. Club*, 21:77–86.

Barnard, K.H. 1927. A monograph of the marine fishes of South Africa. Part II. (Teleostei–Discocephali to end. Appendix.) *Ann. So. Afr. Mus.*, 21(2):419–1065.

Barnett, M.A. 1983. Species structure and temporal stability of mesopelagic fish assemblages in the Central Gyres of the North and South Pacific Ocean. *Mar. Biol.*, 74:245–256.

Barnett, M.A. 1984. Mesopelagic fish zoogeography in the central tropical and subtropical Pacific Ocean: Species composition and structure at representative locations in three ecosystems. *Mar. Biol.*, 82:199–208.

Bassot, J.-M. 1966. On the comparative morphology of some luminous organs. pp. 557–610, in: F.H. Johnson and Y. Haneda (editors), *Bioluminescence in Progress: Proceedings of the Luminescence Conference Sponsored by the Japan Society for the Promotion of Science and by the National Science Foundation, under the United States–Japan Cooperative Science Program, 12–16 September 1965, Hakone National Park, Kanagawa-ken, Japan*. Princeton University Press, Princeton, New Jersey.

Beaufort, L.F. de, and J.C. Briggs. 1962. *The Fishes of the Indo-Australian Archipelago XI. Scleroparei, Hypostomides, Pediculati, Plectognathi, Opisthomi, Discocephali, Xenopterygii*. E.J. Brill, Leiden.

Becker, V.E., Y.N. Shcherbachev, and V.M. Tchuvasov. 1975. Deep-sea pelagic fishes of the Caribbean Sea, Gulf of Mexico, and Puerto-Rican Trench. *Trans. Shirshov Inst. Oceanol.*, 100:289–336.

Beebe, W. 1926a. Two dramatic secrets recently given up by the sea. *Bull. N.Y. Zool. Soc.*, 29(2):77–79.

Beebe, W. 1926b. A new ceratioid fish. Preliminary description of a new genus and species. *Bull. N.Y. Zool. Soc.*, 29(2):80.

Beebe, W. 1926c. *The Arcturus Adventure, an Account of the New York Zoological Society's First Oceanographic Expedition*. G.P. Putnam's Sons, New York and London, published under the auspices of the New York Zoological Society.

Beebe, W. 1929a. Deep sea fish of the Hudson Gorge. Taken at station 113 of the Arcturus and station 114 of the Eleventh Expedition of the Department of Tropical Research of the New York Zoological Society. *Zoologica*, New York, 12(1):1–19.

Beebe, W. 1929b. *Haplophryne hudsonius*: A new species; description and osteology. *Zoologica*, New York, 12(2):21–36.

Beebe, W. 1930. The Bermuda Oceanographic Expedition. Twelfth expedition of the Department of Tropical Research of the New York Zoological Society. *Bull. N.Y. Zool. Soc.*, 33(2):35–66.

Beebe, W. 1932a. The depths of the sea. Strange life forms a mile below the surface. *Nat. Geogr. Mag.*, 61(1):65–88.

Beebe, W. 1932b. Nineteen new species and four post-larval deep-sea fish. *Zoologica*, New York, 13(4):47–107.

Beebe, W. 1933. On the *Antares* to the West Indies. Narrative of the fifteenth and seventeenth expeditions of the Department of Tropical Research. *Bull. N.Y. Zool. Soc.*, 36(4):97–115.

Beebe, W. 1934a. *Half Mile Down*. Harcourt, Brace and Co., New York.

Beebe, W. 1934b. Three new deep-sea fish seen from the bathysphere. *Bull. N.Y. Zool. Soc.*, 37(6):190–193.

Beebe, W. 1934c. A half mile down. Strange creatures, beautiful and grotesque as figments of fancy, reveal themselves at windows of the bathysphere. *Nat. Geogr. Mag.*, 66(6):661–704.

Beebe, W. 1937. Preliminary list of Bermuda deep-sea fish. Based on the collections from fifteen hundred metre-net hauls, made in an eight-mile circle south of Nonsuch Island, Bermuda. *Zoologica*, New York, 22(3):197–208.

Beebe, W. 1938. *Ceratias*—siren of the deep. Through perpetual darkness the tiny male seeks out his mate and attaches himself to her spiny skin for life. *Bull. N.Y. Zool. Soc.*, 41(2):50–53.

Beebe, W., and J. Crane. 1947. Eastern Pacific expeditions of the New York Zoological Society. XXXVII. Deep-sea ceratioid fishes. *Zoologica*, New York, 31(11):151–182.

Beebe, W., and R. Rose. 1926. "The Arcturus Adventure." Excerpts of several chapters from the story of the Arcturus Oceanographic Expedition. *Bull. N.Y. Zool. Soc.*, 29(2):42–59.

Behrmann, G. 1974. *Linophryne algibarbata*, ein seltener Tiefseeangler mit angewachsenem Zwergmännchen. *Natur u. Museum*, 104(12):364–366.

Behrmann, G. 1977. Ein neuer Fund des Tiefseeanglers *Linophryne bicornis* (Pisces: Ceratioidea) aus dem Atlantik. *Veröff. Inst. Meeresforsch. Bremerh.*, 16:93–98.

Belloc, G. 1938. Résultats des croisières scientifiques du navire "President Théodore-Tissier." Liste des poissons pélagiques et bathypélagiques capturés au cours de la cinquième croisière avec diagnoses préliminaires de deux espèces nouvelles. *Rev. Trav. Pêches marit.*, Paris, 11(3):281–313.

Belloc, G. 1949. Catalogue des types de poissons du Musée Océanographique de Monaco. *Bull. Inst. Océanogr.*, Monaco, 958, 23 pp.

Beltrán-León, B.S., and R.R. Herrera. 2000. *Estadios tempranos de peces del Pacífico Colombiano*, Vol 1. Instituto Nacional de Pesca y Agricultura, Buenaventura, Republica de Colombia.

Berg, L.S. 1940. Classification of fishes, both recent and fossil. *Trav. Zool. Inst. Akad. Nauk SSSR*, 5:87–517 [Reprint, J.W. Edwards, Ann Arbor, Michigan, 1947.]

Berry, F.H., and H.C. Perkins. 1966. Survey of pelagic fishes of the California current area. *Fish. Bull.*, U.S. Fish Wildl. Serv., 65(3):625–682.

Bertelsen, E. 1943. Notes on the deep-sea angler-fish *Ceratias holbölli* Kr. based on specimens in the Zoological Museum of Copenhagen. *Vidensk. Medd. fra Dansk Naturh. Foren.*, 107:185–206.

Bertelsen, E. 1951. The ceratioid fishes. Ontogeny, taxonomy, distribution and biology. *Dana Rept.*, 39, 276 pp.

Bertelsen, E. 1973. A new species of deep-sea angler fish, *Linophryne sexfilis* (Pisces, Ceratioidei). *Steenstrupia*, Copenhagen, 3(7):65–69.

Bertelsen, E. 1976. Records of parasitic males in three species of *Linophryne* (Pisces, Ceratioidei). *Steenstrupia*, Copenhagen, 4(2):7–18.

Bertelsen, E. 1978. Notes on linophrynids. IV. A new species of deepsea anglerfish of the genus *Linophryne* and the first record of a parasitic male in *Linophryne corymbifera* (Pisces, Ceratioidei). *Steenstrupia*, Copenhagen, 5(3):25–32.

Bertelsen, E. 1980a. Notes on Linophrynidae. V. A revision of the deepsea anglerfishes of the *Linophryne arborifer*-group (Pisces, Ceratioidei). *Steenstrupia*, Copenhagen, 6(6):29–70.

Bertelsen, E. 1980b. Notes on Linophrynidae. VI. A new species of deepsea anglerfish of the genus *Linophryne*, with notes on other *Linophryne* species with multi-stemmed barbels (Pisces, Ceratioidei). *Steenstrupia*, Copenhagen, 6(15):233–249.

Bertelsen, E. 1981. Notes on Linophrynidae. VII. New records of the deepsea anglerfish *Linophryne indica* (Brauer, 1902), a senior synonym for *Linophryne corymbifera* Regan and Trewavas, 1932 (Pisces, Ceratioidei). *Steenstrupia*, Copenhagen, 7(1):1–14.

Bertelsen, E. 1982. Notes on Linophrynidae. VIII. A review of the genus *Linophryne*, with new records and descriptions of two new species. *Steenstrupia*, Copenhagen, 8(3):49–104.

Bertelsen, E. 1983. First records of metamorphosed males of the families Diceratiidae and Centrophrynidae (Pisces, Ceratioidei). *Steenstrupia*, Copenhagen, 8(16):309–315.

Bertelsen, E. 1984. Ceratioidei: Development and relationships. pp. 325–334, in: Moser, H.G., W.J. Richards, D.M. Cohen, M.P. Fahay, A.W. Kendall, Jr., and S.L. Richardson (editors), *Ontogeny and Systematics of Fishes*, Spec. Publ. No. 1. Amer. Soc. Ichthyol. Herpetol., Lawrence, Kansas.

Bertelsen, E. 1986. [Ceratioidei, Caulophrynidae, Melanocetidae, Himantolophidae, Diceratiidae, Oneirodidae, Thaumatichthyidae, Centrophrynidae, Ceratiidae, Gigantactinidae, and Linophrynidae]. pp. 1371–1414, in: P.J.P. Whitehead, M.-L. Bauchot, J.-C. Hureau, J. Nielsen, and E. Tortonese (editors), *Fishes of the North-eastern Atlantic and Mediterranean*, Vol. 3. United Nations Educational Scientific and Cultural Organization, Paris.

Bertelsen, E. 1990. [Caulophrynidae, Melanocetidae, Himantolophidae, Diceratiidae, Oneirodidae, Thaumatichthyidae, Centrophrynidae, Ceratiidae, Gigantactinidae, and Linophrynidae]. pp. 491–519, in: J.C. Quéro, J.-C. Hureau, C. Karrer, A. Post, and L. Saldanha (editors), *Check-list of the Fishes of the Eastern Tropical Atlantic*, Vol. 1. Junta Nacional de Investigação Científica e Tecnológica, Lisbon.

Bertelsen, E. 1994. Anglerfishes. pp. 137–141, in: J.R. Paxton and W.N. Eschmeyer (editors), *Encyclopedia of Fishes, A Comprehensive Illustrated Guide by International Experts*. University of New South Wales Press, Sydney, New South Wales, Australia.

Bertelsen, E., and D.B. Grobecker. 1980. A new species of the ceratioid anglerfish genus *Oneirodes* (Pisces: Lophiiformes) from the eastern North Atlantic. *Arch. FischWiss.*, 31(2):63–66.

Bertelsen, E., and J. Grøntved. 1949. The light organs of a bathypelagic fish. *Vidensk. Medd. fra Dansk Naturh. Foren.*, 111:163–167.

Bertelsen, E., and G. Krefft. 1965. On a rare ceratioid fish, *Linophryne lucifer* Collett, 1886. *Vidensk. Medd. fra Dansk naturh. Foren.*, 128:293–301.

Bertelsen, E., and G. Krefft. 1988. The ceratioid family Himantolophidae (Pisces, Lophiiformes). *Steenstrupia*, Copenhagen, 14(2):9–89.

Bertelsen, E., J.G. Nielsen, and D.G. Smith. 1989. Suborder Saccopharyngoidei: Families Saccopharynidae,

Eurypharygidae, and Monognathidae. pp. 636–655, in: E.G. Böhlke (editor), *Fishes of the Western North Atlantic*, Memoir 1, Pt. 9, Vol. 1. Sears Foundation for Marine Research, Yale University, New Haven, Connecticut.

Bertelsen, E., and T.W. Pietsch. 1975. Results of the research cruises of FRV "Walther Herwig" to South America. XXXVIII. Osteology and relationships of the ceratioid anglerfish genus *Spiniphryne* (family Oneirodidae). *Arch. FischWiss.*, 26(1):1–11.

Bertelsen, E., and T.W. Pietsch. 1977. Results of the research cruises of the FRV "Walther Herwig" to South America. XLVII. Ceratioid anglerfishes of the family Oneirodidae collected by the FRV "Walther Herwig." *Arch. FischWiss.*, 27(3):171–189.

Bertelsen, E., and T.W. Pietsch. 1983. The ceratioid anglerfishes of Australia. *Rec. Aust. Mus.*, 35(2):77–99.

Bertelsen, E., and T.W. Pietsch. 1984. Results of the research cruises of FRV "Walther Herwig" to South America. LXIII. A resurrection of the ceratioid anglerfish *Ceratias tentaculatus* (Norman, 1930), with notes on the occurrence of the species of *Ceratias* in the Atlantic Ocean (Pisces: Lophiiformes). *Arch. FischWiss.*, 35(1/2):43–51.

Bertelsen, E., and T.W. Pietsch. 1996. Revision of the deep-sea anglerfish genus *Lasiognathus* (Lophiiformes: Thaumatichthyidae), with the description of a new species. *Copeia*, 1996(2):401–409.

Bertelsen, E., and T.W. Pietsch. 1998a. Anglerfishes. pp. 137–141, in: J.R. Paxton and W.N. Eschmeyer (editors), *Encyclopedia of Fishes, A Comprehensive Illustrated Guide by International Experts*, 2nd edition. Weldon Owen Pty Limited, McMahons Point, New South Wales, Australia.

Bertelsen, E., and T.W. Pietsch. 1998b. Revision of the deepsea anglerfish genus *Rhynchactis* Regan (Lophiiformes: Gigantactinidae), with descriptions of two new species. *Copeia*, 1998(3):583–590.

Bertelsen, E., and T.W. Pietsch. 2002. A new species of deep-sea anglerfish of the genus *Gigantactis* (Lophiiformes: Gigantactinidae) from the Western North Atlantic Ocean. *Copeia*, 2002(4):958–961.

Bertelsen, E., T.W. Pietsch, and R.J. Lavenberg. 1981. Ceratioid anglerfishes of the family Gigantactinidae: Morphology, systematics, and distribution. *Nat. Hist. Mus. Los Angeles Co., Contrib. Sci.*, 332, vi + 74 pp.

Bertelsen, E., and J.-C. Quéro. 1981. Capture au large du maroc de *Centrophryne spinulosa* Regan et Trewavas, 1932 (Pisces, Lophiiformes, Centrophrynidae), espèce nouvelles pour l'Atlantique nord-est. *Cybium*, 13(2):192.

Bertelsen, E., and P.J. Struhsaker. 1977. The ceratioid fishes of the genus *Thaumatichthys*: Osteology, relationships, distribution, and biology. *Galathea Rept.*, 14:7–40.

Bigelow, H.B., and T. Barbour. 1944a. A new giant ceratioid fish. *Proc. New Engl. Zool. Club*, 23:9–15.

Bigelow, H.B., and T. Barbour. 1944b. *Reganula gigantea* to replace *Reganichthys giganteus*. *Copeia*, 1944(2):123.

Blache, J., J. Cadenat, and A. Stauch. 1970. Clés de détermination des poissons de mer signalés dans l'Atlantique Oriental (entre le 20ᵉ parallèle N. et le 15ᵉ parallèle S.). *Faune Tropicale*, Office de la Recherche Scientifique et Technique Outre Mer, 18, 479 pp.

Blacker, R.W. 1967. English observations on rare fishes in 1965. *Cons. Intern. Expl. Mer, Ann. Biol., Copenh.*, 22(1965):186–187.

Blacker, R.W. 1968. English observations on rare fishes in 1966. *Cons. Intern. Expl. Mer, Ann. Biol., Copenh.*, 23(1966):212–213.

Blacker, R.W. 1972. English observations on rare fishes in 1970. *Cons. Intern. Expl. Mer, Ann. Biol., Copenh.*, 27(1970):193–195.

Blacker, R.W. 1973. English observations on rare fishes in 1971. *Cons. Intern. Expl. Mer, Ann. Biol., Copenh.*, 28(1971):221–222.

Blacker, R.W. 1977. English observations on rare fish in 1975. *Cons. Intern. Expl. Mer, Ann. Biol., Copenh.*, 32(1975):184–185.

Blacker, R.W. 1980. English observations on rare fish in 1978. *Cons. Intern. Expl. Mer, Ann. Biol., Copenh.*, 35(1978):249–250.

Blainville, H.M.D. de. 1825. *Manuel de Malacologie et de Conchyliologie*. F.G. Levrault, Paris, vol. 1.

Bleeker, P. 1865. *Atlas ichthyologique des Indes Orientales Néêrlandaises, publié sous les auspices du Gouvernement Colónial Néêrlandais*. Vol. 5, Baudroies, Ostracions, Gymnodontes et Balistes. Frédéric Muller, Amsterdam.

Böhlke, J.E. 1966. Order Lyomeri. pp. 603–628, in: G.W. Mead, H.B. Bigelow, C.M. Breder, D.M. Cohen, D. Merriman, Y.H. Olsen, W.C. Schroeder, L.P. Schultz, and J. Tee-Van (editors), *Fishes of the Western North Atlantic*, Memoir 1, Pt. 5. Sears Foundation for Marine Research, Yale University, New Haven, Connecticut.

Bolin, R.L., and G.S. Myers. 1950. Station records of the Crocker-Stanford Deep-sea Expedition, coast of California, September 1938. *Stanford Ichthyol. Bull.*, 3:203–208.

Borodin, N. 1930a. Some more new deep-sea fishes. *Proc. N. Engl. Zool. Club*, 11:87–92.

Borodin, N. 1930b. A new deep sea fish. *Occ. Pap. Boston Soc. Nat. Hist.*, 5:285–286.

Borodin, N. 1931. Atlantic deep-sea fishes. *Bull. Mus. Comp. Zool., Harvard*, 72(3):55–89.

Bostelmann, E. 1934. Silver-whip anglerfish *Dolopichthys tentaculatus* Beebe. From six hundred fathoms. *Bull. N.Y. Zool. Soc.*, 37(6):188.

Boulenger, G.A. 1904a. A synopsis of the suborders and families of teleostean fishes. *Ann. Mag. Nat. Hist.*, Ser. 7, 13(75):161–190.

Boulenger, G.A. 1904b. Systematic account of Teleostei. *The Cambridge Natural History*, MacMillan and Co., London, 7:421–760.

Bowen, B.K. 1963. The angler-fish *Ceratias holboelli* from Western Australian waters. *J. Roy. Soc. W. Aust.*, 46(3):91–92.

Bradbury, M.G. 1967. The genera of batfishes (family Ogcocephalidae). *Copeia*, 1967(2):399–422.

Bradbury, M.G. 1980. A revision of the fish genus *Ogcocephalus*, with descriptions of new species from the western North Atlantic Ocean (Ogcocephalidae; Lophiiformes). *Proc. Calif. Acad. Sci.*, 42(7):229–285.

Bradbury, M.G. 1988. Rare fishes of the deep-sea genus *Halieutopsis*: A review with descriptions of four new species (Lophiiformes: Ogcocephalidae). *Fieldiana Zool.* (n.s.), 44:1–22.

Bradbury, M.G. 1999. A review of the fish genus *Dibranchus*, with descriptions of new species and a new genus, *Solocisquama* (Lophiiformes: Ogcocephalidae). *Proc. Calif. Acad. Sci.*, 51(5):259–310.

Brandes, C.-H., and A. Kotthaus. 1960. Rare fishes: B. Records of the Institut für Meeresforschung and the Abteilung Fischereibiologie der Biologischen Anstalt Helgoland, Bremerhaven. *Cons. Intern. Expl. Mer, Ann. Biol., Copenh.*, 15(1958):72.

Brandes, C.-H., A. Kotthaus, and G. Krefft. 1953. Rare fishes. *Cons. Intern. Expl. Mer, Ann. Biol., Copenh.*, 9(1952):47–48.

Brandes, C.-H., A. Kotthaus, and G. Krefft. 1956. Rare fishes from distant northern seas. *Cons. Intern. Expl. Mer, Ann. Biol., Copenh.*, 11(1954):29–30.

Brandes, C.-H., A. Kotthaus, and G. Krefft. 1957. Rare fishes from distant northern seas: Germany. *Cons. Intern. Expl. Mer, Ann. Biol., Copenh.*, 12(1955):54–55.

Brauer, A. 1902. Diagnosen von neuen Tiefseefischen, welche von der Valdivia-Expedition gesammelt sind. *Zool. Anz.*, 25, 668(4):277–298.

Brauer, A. 1904. Über die Leuchtorgane der Knochenfische. *Verhand. Deut. Zool. Gesells.*, 1904(1):16–35.

Brauer, A. 1906. Die Tiefsee-fische. I. Systematischer Teil. *Wiss. Ergebn. Deut. Tiefsee-Exped. "Valdivia,"* 15(1):1–432.

Brauer, A. 1908. Die Tiefsee-fische. II. Anatomischer Teil. *Wiss. Ergebn. Deut. Tiefsee-Exped. "Valdivia,"* 15(2):1–266.

Bremer, K. 1988. The limits of amino acid sequence data in angiosperm phylogenetic reconstruction. *Evolution*, 42:795–803.

Brewer, G.D. 1973. Midwater fishes from the Gulf of California and the adjacent eastern tropical Pacific. *Nat. Hist. Mus. L.A. Co., Contri. Sci.*, 242, 47 pp.

Briggs, J.C. 1960. Fishes of worldwide (circumtropical) distribution. *Copeia*, 1960(3):171–180.

Brinton, E. 1962. The distribution of pacific euphausiids. *Bull. Scripps Inst. of Oceanog., Univ. Calif.*, 8(2):51–270.

Brown, W.H. 1890. Dates of publication of "Recherches sur les poissons fossiles," by L. Agassiz. pp. xxv–xxix, in: A.S. Woodward and C.D. Sherborn, *A Catalogue of British Fossil Vertebrata*. Dulau and Co., London.

Brusca, R.C., and G.J. Brusca. 1990. *Invertebrates*. Sinauer Associates, Inc., Sunderland, Massachusetts.

Bruun, A.F. 1953. Dybhavets dyreliv. pp. 153–198, In: *Galatheas jordomsejling 1950–1952, den Danske dybhavsekspeditions virke og resultater skildret of deltagerne.* Under redaktion af A.F. Bruun, S. Greve, H. Mielche, and R. Spärck, med forord af H.K.H. Prins Axel. J.H. Schultz Forlag, Copenhagen. [In Danish.]

Bruun, A.F. 1956. Animal life of the deep sea bottom. pp. 149–195, in: A.F. Bruun, S. Greve, H. Mielche, and R. Spärck (editors), translated from the Danish by R. Spink, *The Galathea Deep Sea Expedition 1950–1952, Described by Members of the Expedition*. George Allen and Unwin, Ltd., London.

Bruun, A.F. 1958. On the restricted distribution of two deep-sea fishes, *Borophryne apogon* and *Stomias colubrinus*. *J. Mar. Res.*, 17:103–112.

Büchner, S. 1973. Rudolph von Willemoes-Suhm (1847–1875), ein Andenken an einen jungen holsteinischen Forschungsreisenden. *Steinburger Jahrbuch*, Selbstverlag, Kreisverein des Schleswig-Holsteinischen Heimatbundes, Itzehoe, 1973:23–32.

Burne, R.H., and J.R. Norman. 1943. Charles Tate Regan 1878–1943. *Obituary Notices of Fellows of The Royal Society*, printed and published for the Royal Society by Harrison and Sons, London, 4:411–426.

Burton, M. 1933. *Zoological Record*, for 1932. 69(13):1–62.

Bussing, W.A. 1965. Studies of the midwater fishes of the Peru-Chile Trench. Amer. Geophys. Union, *Biol. Antarctic Seas II, Antarctic Res. Ser.*, 5:185–227.

Carnevale, G., and T.W. Pietsch. 2006. Filling the gap: A fossil frogfish, genus *Antennarius* (Teleostei: Lophiiformes: Antennariidae), from the Miocene of Algeria. *J. Zoology*, London, 163:1–10.

Carnevale, G., T.W. Pietsch, G.T. Takeuchi, and R.W. Huddleston. 2008. Fossil ceratioid anglerfishes (Teleostei: Lophiiformes) from the Miocene of the Los Angeles Basin, California. *J. Paleon.*, 82(5):996–1008.

Cartes, J.E., F. Maynou, F. Sarda, J.B. Company, D. Lloris, and S. Tudela. 2004. The Mediterranean deep-sea ecosystems: An overview of their diversity, structure, functioning and anthropogenic impacts. pp. 9–38, in: S. Tudela and F. Simard (editors), *The Mediterranean Deep-Sea Ecosystems: An Overview of Their Diversity, Structure, Functioning and Anthropogenic Impacts, with a Proposal for Conservation*. IUCN, Malaga, and WWF, Rome.

Caruso, J.H. 1975. Sexual dimorphism of the olfactory organs of lophiids. *Copeia*, 1975(2):380–381.

Caruso, J.H. 1981. The systematics and distribution of the lophiid anglerfishes. I. A revision of the genus *Lophiodes*, with the description of two new species. *Copeia*, 1981(3):522–549.

Caruso, J.H. 1983. The systematics and distribution of the lophiid anglerfishes. II. Revisions of the genera *Lophiomus* and *Lophius*. *Copeia*, 1983(1):11–30.

Caruso, J.H. 1985. The systematics and distribution of the lophiid anglerfishes. III. Intergeneric relationships. *Copeia*, 1985(4):870–875.

Caruso, J.H. 1986. [Lophiidae]. pp. 1362–1363, in: P.J.P. Whitehead, M.-L. Bauchot, J.-C. Hureau, J. Nielsen, and E. Tortonese (editors), *Fishes of the North-eastern Atlantic and Mediterranean*, Vol. 3. United Nations Educational Scientific and Cultural Organization, Paris.

Caruso, J.H. 1989a. Systematics and distribution of the Atlantic chaunacid anglerfishes (Pisces: Lophiiformes). *Copeia*, 1989(1):153–165.

Caruso, J.H. 1989b. A review of the Indo-Pacific members of the deep-water chaunacid anglerfish genus *Bathychaunax*, with the description of a new species from the Eastern Indian Ocean (Pisces: Lophiiformes). *Bull. Mar. Sci.*, 45(3):574–579.

Caruso, J.H., and H.R. Bullis, Jr. 1976. A review of the lophiid angler fish genus *Sladenia*, with a description of a new species from the Caribbean Sea. *Bull. Mar. Sci.*, 26(1):59–64.

Caruso, J.H., and R.D. Suttkus. 1979. A new species of lophiid angler fish from the western North Atlantic. *Bull. Mar. Sci.*, 29(4):491–496.

Chapman, W.M. 1939. Eleven new species and three new genera of oceanic fishes collected by the International Fisheries Commission from the northeastern Pacific. *Proc. U.S. Nat. Mus.*, 86(3062):501–542.

Chen, J.T.F., M.-C. Liu, and S.-C. Lee. 1967. A review of the pediculate fishes of Taiwan. *Tunghai Univ. Biol. Bull.*, 33(Ichthyol. Ser. 7):1–23.

Childress, J.J. 1971. Respiratory rate and depth of occurrence of midwater animals. *Limnol. Oceanogr.*, 16:104–106.

Chirichigno, N. 1978. Nuevas adiciones a la ictiofauna marina del Perú. *Inf. Inst. Mar Perú*, 46:1–109.

Chun, C. 1903. *Aus den Tiefen des Weltmeeres von Carl Chun. Schilderungen von der Deutschen Tiefsee-Expedition*. G. Fischer, Jena.

Clarke, F.E. 1878. On two new fishes. *Trans. N.Z. Inst.*, 10(30):243–246.

Clarke, M.R.. and P.J. Herring. 1971. Animal miscellany: A survey of oceanic families. pp. 164–204, in: P.J. Herring and M.R. Clarke (editors), *Deep Oceans*. Praeger Publishers, New York and Washington, D.C.

Clarke, R. 1950. The bathypelagic angler fish *Ceratias holbölli* Kröyer. *Disc. Rept.*, 26:1–32.

Clarke, R. 1956. Sperm whales of the Azores. *Disc. Rept.*, 28:237–298.

Clemens, W.A., and G.V. Wilby. 1946. Fishes of the Pacific coast of Canada. *Bull. Fish. Res. Bd. Canada*, 38:1–368.

Collett, R. 1886. On a new pediculate fish from the sea off Madeira. *Proc. Zool. Soc.*, London, 1886:138–143.

Collette, B.B. 1966. A review of the venomous toadfishes, subfamily Thalassophryninae. *Copeia*, 1966(4):846–864.

Collette, B.B. 1968. *Daector schmitti*, a new species of venomous toadfish from the Pacific coast of Central America. *Proc. Biol. Soc. Washington*, 81:155–160.

Collette, B.B. 1995. *Potamobatrachus trispinosus*, a new freshwater toadfish (Batrachoididae) from Rio Tocantins, Brazil. *Ichthyol. Expl. Freshwaters*, 6(4):333–336.

Combs, C.L. 1973. *Structure and probable feeding function of the batfish esca*. Unpublished M.S. thesis, Florida State University, Tallahassee, Florida.

Cope, E.D. 1872. Observations on the systematic relations of the fishes. *Proc. Amer. Assoc. Adv. Sci.*, August 1871, pp. 317–343.

Costa, M.J. 1980. *Phrynichthys wedli* Pietschmann, 1926 (Pisces, Diceratiidae), poisson nouveau pour les côtes du Portugal. *Cybium, Ser. 3*, 1980(8):89–90.

Courtenay, W.R., Jr. 1959. The rare angler. *Sea Frontiers*, 5(4):221–224.

Cowles, D.L., and J.J. Childress. 1995. Aerobic metabolism of the anglerfish *Melanocetus johnsoni*, a deep-pelagic marine sit-and-wait predator. *Deep-Sea Res.*, 42(9):1631–1638.

Craddock, J.E., and G.W. Mead. 1970. Midwater fishes from the Eastern South Pacific Ocean. *Sci. Res. S.E. Pac. Exped., Anton Bruun Rept.*, 3:3–46.

Crane, J.M. 1966. Late tertiary radiation of viperfish (Chauliodontidae) based on a comparison of recent and Miocene species. *Nat. Hist. Mus. Los Angeles Co., Contrib. Sci.*, 115:1–37.

Crane, J.M. 1968. Bioluminescence in the batfish *Dibranchus atlanticus*. *Copeia*, 1968(2):401–411.

Cuvier, G.L. C.F.D. 1829. *Le régne animal distribué d'après son organisation, pour servir de base à l'histoire naturelle des animaux et d'introduction à l'anatomie comparée*, nouvelle édition, revue et augmentée, Vol. 2. Déterville, Paris.

Dahlgren, U. 1928. The bacterial light organ of *Ceratias*. *Science*, 68:65–66.

DeWitt, H.H., P.C. Heemstra, and O. Gon. 1990. Notothenliidae. pp. 279–331, in: O. Gon and P.C. Heemstra (editors), *Fishes of the Southern Ocean*. J.L.B. Smith Institution of Ichthyology, Grahamstown, South Africa.

Donnelly J., and J.V. Gartner, Jr. 1990. Notes on two rare species of anglerfish *Himantolophus* (Ceratioidei: Himantolophidae) collected in the eastern Gulf of Mexico. *Northeast Gulf Sci.*, 11(1):77–78.

Drioli, M., and I.E. Vignes. 1992. Presencia de *Ceratias tentaculatus* (Norman, 1930) (Pisces, Ceratioidei) en Aguas Argentina, a los 39°30'L.S. y 54°57.4'L.W. *Rev. Mus. Argentino Cienc. Nat. "Bernardino Rivadavia," Inst. Nac. Invest. Cienc. Nat., Zoologia*, 16(2):19–26.

Du Buit, M.-H., J. Gueguen, D. Latrouite, and J.-C. Quéro. 1980. Observations française sur les poissons rares en 1978. *Cons. Intern. Expl. Mer, Ann. Biol., Copenh.*, 35(1978): 250–252.

Du Buit, M.-H., C. Ozouf-Costaz, and J.-C. Quéro. 1989. Observations a concarneau de *Cryptopsaras couesi* et *Ceratias* sp. (Pisces, Lophiiformes, Ceratiidae), espèces nouvelles pour la faune ichtyologique Française. Leur distribution en atlantique nord-est. *Cybium*, 13(2):192.

Eastman, C.R. 1904. Descriptions of Bolca fishes. *Bull. Mus. Comp. Zool.*, Harvard, 46(1):1–36.

Eaton, T.H., Jr., C.A. Edwards, M.A. McIntosh, and J.P. Rowland. 1954. Structure and relationships of the anglerfish, *Lophius americanus*. *J. Elisha Mitchell Sci. Soc.*, 70(2):205–218.

Ebeling, A.W. 1962. Melamphaidae. I. Systematics and zoogeography of the species in the bathypelagic fish genus *Melamphaes* Günther. *Dana Rept.*, 58, 164 pp.

Ebeling, A.W., R.M. Ibara, R.J. Lavenberg, and F.J. Rohlf. 1970. Ecological groups of deep-sea animals off Southern California. *Nat. Hist. Mus. Los Angeles Co., Sci. Bull.*, 6:1–43.

Ehrenbaum, E. 1901. Die Fische. Faune Arctica, cine Zusammenstellung der arktischen Tierformen, mit besonderer Berucksichtigung des Spitzbergen-Gebietes auf Grund der Ergebnisse der Deutschen Expedition in das Nordliche Eismeer in Jahre 1898. *Faune Arctica*, 2(1):65–168.

Ehrenbaum, E. 1936. Naturgeschichte und wirtschaftliche Bedeutung der Seefische Nordeuropas. Vol. 2, in: H. Lübbert and E. Ehrenbaum (editors), *Handbuch der Seefischerei Nordeuropas*. Schweizerbart'sche Verlagsbuchhandlung, Stuttgart.

Einarsson, H. 1952. Saedjöflarnir. *Sérprentun úr Náttúrufraeðingnum*, 22:90–96.

Ejsymont, E. 1970. Record of the ceratioid fish, *Linophryne lucifer*, in the NW Atlantic. *Acta Ichthyol. Piscatoria*, 1:59–65.

Ellis, R. 1996. *Deep Atlantic: Life, Death, and Exploration in the Abyss*. Knopf, New York.

Ellis, R. 2005. *Singing Whales and Flying Squid: The Discovery of Marine Life*. Lyons Press, Guilford, Connecticut, xvii + 269 pp.

Endo, H., and G. Shinohara. 1999. A new batfish, *Coelophrys bradburyae* (Lophiiformes: Ogcocephalidae) from Japan, with comments on the evolutionary relationships of the genus. *Ichthyol. Res.*, 46(4):359–365.

Eschmeyer, W.N. 1990. *Catalog of the Genera of Recent Fishes*. California Academy of Sciences, San Francisco vi + 697 pp.

Everly, A.W. 2002. Stages of development of the goosefish, *Lophius americanus*, and comments on the phylogenetic significance of the development of the luring apparatus in Lophiiformes. *Environ. Biol. Fishes*, 64:393–417.

Fast, T.N. 1957. The occurrence of the deep-sea anglerfish, *Cryptopsaras couesi*, in Monterey Bay, California. *Copeia*, 1957(3):237–240.

Fedorov, V.V. 1994. *Gigantactis elsmani*, first report of a species of the family Gigantactinidae (Lophiiformes) from the Sea of Okhotsk. *Vopr. Ikhtiol.*, 34(3):414–415. [In Russian, English translation in *J. Ichthyol.*, 34(8):132–134, 1994.]

Fedorov, V.V., I.A. Chereshnev, M.V. Nazarkin, A.V. Shestakov, and V.V. Volobuev 2003. *Catalog of Marine and Freshwater Fishes of the Northern Part of the Sea of Okhotsk*. Dalnauka, Vladivostok.

Field, J.G. 1966. Contributions to the functional morphology of fishes. Part II. The feeding mechanism of the angler-fish, *Lophius piscatorius* Linnaeus. *Zool. Afric.*, 2:45–67.

Figueiredo, J.L. de., A.P. dos Santos, N. Yamaguti, R.Á. Bernardes, and C.L. Del Bianco Rossi-Wongtschowski. 2002.

*Peixes da Zona Econômica Exclusiva da Região Sudeste-Sul do Brasil. Levantamento com rede de meia água.* Editora de Universidade de São Paulo, São Paulo, Brazil.

Filhol, H. 1885. *La vie au fond des mers, les explorations sous-marines et les voyages du* Travailleur *et du* Talisman. Masson, Paris.

Fitch, J.E. 1973. The second record of the giant seadevil, *Ceratias holboelli*, from California, with notes on its life history. *Bull. So. Calif. Acad. Sci.*, 72:164.

Fitch, J.E., and R.J. Lavenberg. 1968. *Deep-Water Fishes of California*. University of California Press, Berkeley and Los Angeles.

Fonseca, N.C. 1968. Nuevos registros para la ictiofauna marina del Peru. *Bol. Inst. Mar Peru*, Callao, 1(8):377–504.

Fowler, H.W. 1936. The marine fishes of West Africa based on the collections of the American Museum Congo Expedition, 1909–1915. *Bull. Amer. Mus. Nat. Hist.*, 70(2):607–1493.

Fowler, H.W. 1944. The fishes. Results of the fifth George Vanderbilt Expedition, 1941. *Monogr. Acad. Nat. Sci., Philad.*, 6:57–529.

Fowler, H.W. 1949. Fishes of Oceania, Suppl. 3. *Mem. Bernice P. Bishop Mus.*, 12(2):37–186.

Fraser-Brunner, A. 1935. New or rare fishes from the Irish Atlantic Slope. *Proc. Roy. Irish Acad.*, 42B(9):319–326.

Frøiland, Ø. 1979. Fish types in the Zoological Museum, University of Bergen. *Sarsia*, 64:143–154.

Fujii, E. 1983. [Caulophrynidae, Melanocetidae, Diceratiidae, Himantolophidae, Gigantactinidae, and Ceratiidae]. pp. 257–265, in: T. Uyeno, K. Matsuura, E. Fujii (editors), *Fishes Trawled off Suriname and French Guiana*. Japan Marine Fishery Resources Research Center, Tokyo.

Gaimard, J.P. 1852. *Voyages de la Commission scientifique du Nord, en Scandinavie, en Laponie, au Spitzberg et au Feröe, pendant les années 1838, 1839 et 1840 sur la corvette la Recherche, commandée par M. Fabvre. . . . Pub. par ordre du roi sous la direction de M. Paul Gaimard. . . . Atlas de Zoologie*, Sect. 7, Poissons, Pl. 9. A. Bertrand, Paris.

Garman, S. 1899. Report on an exploration off the west coasts of Mexico, Central and South America, and off the Galapagos Islands, in charge of Alexander Agassiz, by the U.S. Fish Commission steamer "Albatross," during 1891, Lieut. Commander Z.L. Tanner, U.S.N., commanding. XXVI. The fishes. *Mem. Mus. Comp. Zool.*, Harvard, 24, 431 pp. + 97 pls.

Gartner, Jr., J.V., R.E. Crabtree, and K.J. Sulak. 1997. Feeding at depth. pp. 115–193, in: D.J. Randall and A.P. Farrell (editors), *Deep-Sea Fishes*. Academic Press, San Diego.

Gibbs, R.H., Jr. 1969. Taxonomy, sexual dimorphism, vertical distribution, and evolutionary zoogeography of the bathypelagic fish genus *Stomias* (Stomiatidae). *Smithson. Contrib. Zool.*, 31:1–25.

Gilbert, C.H. 1905. The deep-sea fishes. Pt. 2, pp. 577–713, in: D.S. Jordan and B.W. Evermann (editors), The aquatic resources of the Hawaiian Islands. *Bull. U.S. Fish Comm.*, 23(2).

Gilbert, C.H. 1915. Fishes collected by the United States Fisheries steamer "Albatross" in southern California in 1904. *Proc. U.S. Nat. Mus.*, 48:305–380.

Gilchrist, J.D.F. 1903. Descriptions of new South African fishes. *Mar. Invest. So. Afr.*, 2:203–211.

Gilchrist, J.D.F., and W.W. Thompson. 1917. A catalogue of the sea fishes recorded from Natal, Part 2. *Ann. Durban Mus.*, 1(4):291–431.

Gilchrist, J.D.F., and C. Von Bonde. 1924. Deep-sea fishes procured by the S.S. "Pickle," Part II. *Fish. Mar. Biol. Survey Rept.*, 3(7):1–24.

Gill, T.N. 1861. Catalogue of the fishes of the eastern coast of North America, from Greenland to Georgia. *Proc. Acad. Nat. Sci. Philad.*, 13(Suppl.):1–63.

Gill, T.N. 1863. Descriptions of some new species of Pediculati, and on the classification of the group. *Proc. Acad. Nat. Sci. Philad.*, 15:88–92.

Gill, T.N. 1872. Arrangement of the Families of Fishes, or Classes Pisces, Marsipobranchii, and Leptocardii. *Smithson. Misc. Coll.*, 247, xlvi + 49 pp.

Gill, T.N. 1873. Catalogue of the fishes of the east coast of North America. *Rept. U.S. Fish Comm.*, 1(19):779–822.

Gill, T.N. 1879a. Synopsis of the pediculate fishes of the eastern coast of extratropical North America. *Proc. U.S. Nat. Mus.*, 1:215–221.

Gill, T.N. 1879b. Note on the Ceratiidae. *Proc. U.S. Nat. Mus.*, 1:227–231.

Gill, T.N. 1883a. Supplementary note on the pediculati. *Proc. U.S. Nat. Mus.*, 5:551–556.

Gill, T.N. 1883b. Deep-sea fishing fishes. *Forest and Stream*, Nov. 8, 1883, p. 284.

Gill, T.N. 1909. Angler fishes: Their kinds and ways. *Smithson. Inst., Annu. Rept.* 1908(1909):565–615.

Girard, A.A. 1893. Etude sur un poisson des grandes profondeurs du genre *Himantolophus* dragué sur les côtes du Portugal. *Bol. Soc. Geogr. Lisboa*, Ser. 11, 9:603–610.

Golvan, Y.-J. 1962. Catalogue systématique des noms de genre de poissons actuels de la X$^e$ édition du "Systema naturae" de Charles Linné jusqu'à la fin de l'année 1959. *Ann. Parasito. Hum. et Comp.*, 37(6):1–227.

Gon, O. 1992. A new deep-sea anglerfish of the genus *Linophryne* (Teleostei: Ceratioidei) from the central equatorial Pacific Ocean. *Micronesica*, 25(2):137–143.

Goode, G.B. 1881. Fishes from the deep water on the south coast of New England obtained by the United States Fish Commission in the summer of 1880. *Proc. U.S. Nat. Mus.*, 3(177):467–486.

Goode, G.B., and T.H. Bean. 1896. Oceanic ichthyology, a treatise on the deep-sea and pelagic fishes of the world, based chiefly upon the collections made by the steamers Blake, Albatross, and Fish Hawk in the northwestern Atlantic. *U.S. Nat. Mus., Spec. Bull.*, 2:1–553.

Gordon, M. 1955. *Histrio*: The fish on the Sargasso merry-go-round. *Aquarium*, 24:386–393.

Gosline, W.A. 1960. Contributions toward a classification of modern isospondylopus fishes. *Bull. Brit. Mus. (Nat. Hist.), Zoology*, 6:327–365.

Gosline, W.A. 1971. *Functional Morphology and Classification of Teleost Fishes*. University of Hawaii Press, Honolulu.

Greenfield, D.W. 1998. *Halophryne hutchinsi*: A new toadfish (Batrachoididae) from the Philipiine Islands and Pulau Waigeo, Indonesia. *Copeia*, 1998(3):696–701.

Greenwood, P.H. 1994. Ethelwynn Trewavas, 5 Nov. 1900–16 Aug. 1993. *Copeia*, 1994(2):565–569.

Greenwood, P.H., D.E. Rosen, S.H. Weitzman, and G.S. Myers. 1966. Phyletic studies of teleostean fishes, with a provisional classification of living forms. *Bull. Amer. Mus. Nat. Hist.*, 131:339–456.

Gregory, W.K. 1928. Studies on the body-form of fishes. *Zoologica*, New York, 8(6):325–421.

Gregory, W. K. 1933. Fish skulls: A study of the evolution of natural mechanisms. *Trans. Amer. Phil. Soc.*, 23(2): 75–481.

Gregory, W. K. 1951. *Evolution Emerging: A Survey of Changing Patterns from Primeval Life to Man*, Vol. 1, text, Vol. 2, atlas. The Macmillan Company, New York.

Gregory, W. K., and G. M. Conrad. 1936. The evolution of the pediculate fishes. *Amer. Nat.*, 70(728):193–208.

Grey, M. 1955. Notes on a collection of Bermuda deep-sea fishes. *Fieldiana, Zoology*, 37:265–302.

Grey, M. 1956a. The distribution of fishes found below a depth of 2000 meters. *Fieldiana, Zoology*, 36(2):73–337.

Grey, M. 1956b. New records of deep-sea fishes, including a new species, *Oneirodes bradburyae*, from the Gulf of Mexico. *Copeia*, 1956(4):242–246.

Grey, M. 1959. Descriptions of newly discovered western Atlantic specimens of *Diceratias bispinosus* Günther and *Paroneirodes wedli* (Pietschman). *Copeia*, 1959(3):225–228.

Grinols, R. B. 1966. Addition of adult anglerfish, *Chaenophryne parviconus* Regan and Trewavas (Pisces: Oneirodidae), to the eastern subarctic Pacific Ocean. *Calif. Fish Game*, 52(3): 161–165.

Grobecker, D. B. 1978. A new species of anglerfish, genus *Oneirodes* (Oneirodidae), from the Banda Sea. *Copeia*, 1978(4):567–568.

Grobecker, D. B., and T. W. Pietsch. 1978. Second specimen of the rare deep-sea anglerfish *Oneirodes macronema*. *Copeia*, 1978(3):547–548.

Grobecker, D. B., and T. W. Pietsch. 1979. High-speed cinematographic evidence for ultrafast feeding in antennariid anglerfishes. *Science*, 205:1161–1162.

Gudger, E. W. 1945a. The frogfish, *Antennarius scaber*, uses its lure in fishing. *Copeia*, 1945(2):111–113.

Gudger, E. W. 1945b. The angler-fishes, *Lophius piscatorius* et *americanus*, use the lure in fishing. *Amer. Nat.*, 79(785): 542–548.

Guinness World Records 2007. Guinness World Records, Ltd., a HIT Entertainment, Ltd. Company, London.

Guinness World Records 2009. Guinness World Records, Ltd., a HIT Entertainment, Ltd. Company, London.

Günther, A. C. L. G. 1861. *Catalogue of the Acanthopterygian Fishes in the Collection of the British Museum*, Vol. 3. Trustees of the British Museum, London.

Günther, A. C. L. G. 1864. On a new genus of pediculate fish from the sea of Madeira. *Proc. Zool. Soc. London*, 1864(6):301–303.

Günther, A. C. L. G. 1880. *An Introduction to the Study of Fishes*. Black, Edinburgh.

Günther, A. C. L. G. 1887. Report on the deep-sea fishes collected by H.M.S. *Challenger* during the years 1873–1876. *Rept. Sci. Res. Voy. Challenger, Zoology*, 22, 335 pp.

Günther, K., and K. Deckert. 1956. *Creatures of the Deep Sea*. Charles Scribner's Sons, New York.

Haneda, Y. 1968. Observations on the luminescence of the deep sea luminous angler fish, *Himantolophus groenlandicus*. *Sci. Rept. Yokosuka City Mus.*, 14:1–6.

Hansen, K. 1970. On the luminous organs in the barbels of some stomiatoid fishes. *Vidensk. Meddr. Dansk naturh. Foren.*, 133:69–84.

Hansen, K., and P. J. Herring. 1977. Dual bioluminescent systems in the anglerfish genus *Linophryne* (Pisces: Ceratioidea). *J. Zool.*, London, 182:103–124.

Hardy, A. C., and E. R. Günther. 1935. The plankton of the South Georgia whaling grounds and adjacent waters, 1926–1927. *Disc. Rept.*, 11:1–456.

Harvey, E. N. 1922. The production of light by the fishes, *Photoblepharon* and *Anomalops*. *Carnegie Inst. Wash. Dept. Mar. Biol. Publ.*, 312:43–60.

Harvey, E. N. 1931. Stimulation by adrenalin of the luminescence of deep-sea fish. *Zoologica*, New York, 12(6):67–69.

Harvey, E. N. 1940. *Living Light*. Princeton University Press, Princeton, New Jersey.

Harvey, E. N. 1952. *Bioluminescence*. Academic Press, New York.

Haygood, M. G. 1993. Light organ symbioses in fishes. *Crit. Rev. Microbiol.*, 19(4):191–216.

Haygood, M. G., and D. L. Distal. 1993. Bioluminescent symbionts of flashlight fishes and deep-sea anglerfishes form unique lineages related to the genus *Vibrio*. *Nature*, 363:154–156.

Haygood, M. G., D. L. Distal, and P. J. Herring. 1992. Polymerase chain reaction and 16S rRNA gene sequences from the luminous bacterial symbionts of two deep-sea anglerfishes. *J. Mar. Biol. Assoc. U.K.*, 72:149–159.

Heemstra, P. C. 1995. Linophrynidae. p. 377, in: M. M. Smith and P. C. Heemstra (editors), *Smiths' Sea Fishes*, 2nd edition. Southern Book Publishers, Johannesburg, South Africa.

Heller, E., and R. E. Snodgrass. 1903. Papers from the Hopkins Stanford Galapagos Expedition, 1898–1899. XV. New fishes. *Proc. Wash. Acad. Sci.*, 5:189–229.

Hennig, W. 1950. *Grundzüge einer Theorie der Phylogenetischen Systematik*. Deutscher Zentralverlag, Berlin.

Hennig, W. 1966. *Phylogenetic Systematics*. University of Illinois Press, Urbana.

Herring, P. J. 1983. The spectral characteristics of luminous marine organisms. *Proc. Roy. Soc. London*, B, 220(1219): 183–217.

Herring, P. J. 1993. Light genes will out. *Nature*, 363:110–111.

Herring, P. J. 2000. Species abundance, sexual encounter and bioluminescent signaling in the deep sea. *Phil. Trans. Roy. Soc. London*, B, 355:1273–1276.

Herring, P. J. 2002. *The Biology of the Deep Ocean*. Oxford University Press, Oxford, England.

Herring, P. J., and J. G. Morin. 1978. Bioluminescence in fishes. pp. 273–329, in: P. J. Herring (editor), *Bioluminescence in Action*. Academic Press, London, New York, San Francisco.

Herring, P. J., and O. Munk. 1994. The escal light gland of the deep-sea anglerfish *Haplophryne mollis* (Pisces: Ceratioidei), with observations on luminescence control. *J. Mar. Biol. Assn. U.K.*, 74:747–763.

Herring, P. J., E. A. Widder, and O. Munk. 1997. Flashing anglerfish; an unexpected signalling system. *Abstracts of the Eighth Deep Sea Biology Symposium*, Monterey, California, 1997:52.

Hjortaa, H. 1961. "Galatheas" wunderfisch—*Galatheathauma axeli*. *Der Präparator, Zeitschrift für Museumstechnik*, 7(1): 169–171.

Ho, H.-C., and K.-T. Shao. 2004. New species of deep-sea ceratioid anglerfish, *Oneirodes pietschi* (Lophiiformes: Oneirodidae), from the North Pacific Ocean. *Copeia*, 2004(1):74–77.

Hobart, W. L. (editor) 1999. The U.S. Fish Commission Steamer *Albatross*: A History. *Mar. Fish. Rev.*, 61(4):viii + 1–91.

Holcroft, N.I. 2004. A molecular test of alternative hypotheses of tetraodontiform (Acanthomorpha: Tetraodontiformes) sister group relationships using data from the RAG1 gene. *Mol. Phylog. Evol.*, 32:749–760.

Holcroft, N.I. 2005. A molecular analysis of the interrelationships of tetraodontiform fishes (Acanthomorpha: Tetraodontiformes). *Mol. Phylog. Evol.*, 34:525–544.

Holt, E.W.L., and L.W. Byrne. 1908a. New deep-sea fishes from the southwest coast of Ireland. *Ann. Mag. Nat. Hist.*, 1(8): 86–95.

Holt, E.W.L., and L.W. Byrne. 1908b. Second report on the fishes of the Irish Atlantic Slope. IV. Recent additions to the British and Irish list. *Rept. Fish. Ireland*, 1906, Pt. 2, Sci. Invest., 5:51–63.

Holt, E.W.L., and L.W. Byrne. 1909. Biological observations. I. Fishes. *Mem. Challenger Soc.*, 1:195.

Hopkins, T.L., and R.C. Baird. 1975. Net feeding in mesopelagic fishes. *U.S. Fish. Bull.*, 73:908–914.

Hopkins, T.L., and R.C. Baird. 1977. Aspects of the feeding ecology of oceanic midwater fishes. pp. 325–360, in: N.R. Anderson and B.J. Zahuranec (editors), *Oceanic Sound Scattering Prediction*. Plenum Press, New York.

Hove, J.R., L.M. O'Bryan, M.S. Gordon, P.W. Webb, and D. Weihs. 2001. Boxfishes (Teleostei: Ostraciidae) as a model system for fishes swimming with many fins: Kinematics. *J. Exp. Biol.*, 204:1459–1471.

Hubbs, C.L., W.I. Follett, and L.J. Demspter. 1979. List of the fishes of California. *Occ. Pap. Calif. Acad. Sci.*, 133, 51 pp.

Hulet, W.H., and G. Musil. 1968. Intracellular bacteria in the light organ of the deep-sea angler fish, *Melanocetus murrayi*. *Copeia*, 1968(3):506–512.

Idyll, C.P. 1964. *Abyss: The Deep-Sea and the Creatures That Live in It*. Thomas Y. Crowell Co., New York.

Iglésias, S.P. 2005. *Himantolophus pseudalbinares* Bertelsen and Krefft, 1988, a junior synonym of *H. albinares* Maul, 1961 (Himantolophidae), with the first record from the Pacific Ocean. *Cybium*, 29(2):191–194.

Imai, S. 1941. Seven new deep-sea fishes obtained in Sagami Sea and Suruga Bay. *Japan. J. Zool.*, 9(2):233–250.

Imai, S. 1942. On some deep-sea angler-fishes obtained in Sagami Bay and Suruga Bay. *J. Dept. Agri., Kyūsyū Imp. Univ.*, 7(2):37–48.

Jespersen, A. 1984. Spermatozoans from a parasitic dwarf male of *Neoceratias spinifer* Pappenheim, 1914. *Vidensk. Medd. fra Dansk Naturh. Foren.*, 145:37–42.

Johnson, R.K., and D.M. Cohen. 1974. Revision of the chiasmodontid fish genera *Dysalotus* and *Kali*, with descriptions of two new species. *ArchFischwiss.*, 25(1/2):13–46.

Jones, D.H. 1972. A rare deep-sea angler from Weather Station "Alfa." *Mar. Obser.*, 42:24–26.

Jónsson, G. 1967a. "Sjaldgaefir fiskar," sem Fiskideild og Hafrannsóknastofnunarinnar hafa borizt 1955–1966. *Ægir*, 60:31–34, 55–58.

Jónsson, G. 1967b. Sjaldsédur fiskur veidist vid Graenland. *Ægir*, 60:316–317.

Jónsson, G. 1968a. Rare fishes recorded by the Marine Research Institute in Reykjavik during 1966. *Cons. Intern. Expl. Mer, Ann. Biol., Copenh.*, 23(1966):211.

Jónsson, G. 1970. Rare fishes recorded by the Marine Research Institute in Reykjavik during 1969. *Cons. Intern. Expl. Mer, Ann. Biol., Copenh.*, 26(1969):277–278.

Jónsson, G. 1973. Rare fishes: Rare fishes recorded by the Marine Research Institute in Reykjavik during 1971. *Cons. Intern. Expl. Mer, Ann. Biol., Copenh.*, 28(1971):219–220.

Jónsson, G. 1975. Rare fishes: Icelandic observations on rare fish in 1973. *Cons. Intern. Expl. Mer, Ann. Biol., Copenh.*, 30(1973):207.

Jónsson, G., J. Magnússon, and J.V. Magnússon. 1977. Rare fishes: Icelandic observations on rare fish in 1975. *Cons. Intern. Expl. Mer, Ann. Biol., Copenh.*, 32(1975):180–182.

Jónsson, G., J. Magnússon, and J.V. Magnússon. 1978. Rare fishes: Icelandic observations on rare fish in 1976. *Cons. Intern. Expl. Mer, Ann. Biol., Copenh.*, 33(1976):180–183.

Jónsson, G., J. Magnússon, and J.V. Magnússon. 1979. Rare fishes: Icelandic observations on rare fish in 1977. *Cons. Intern. Expl. Mer, Ann. Biol., Copenh.*, 34(1977):228–230.

Jónsson, G., J. Magnússon, and J.V. Magnússon. 1980. Rare fishes: Icelandic observations on rare fish in 1978. *Cons. Intern. Expl. Mer, Ann. Biol., Copenh.*, 35(1978):247–248.

Jónsson, G., J. Magnússon, and J.V. Magnússon. 1981. Rare fishes: Icelandic observations on rare fish in 1979. *Cons. Intern. Expl. Mer, Ann. Biol., Copenh.*, 36(1979):192–193.

Jónsson, G., J. Magnússon, and J.V. Magnússon. 1983. Rare fishes: Icelandic observations on rare fish in 1980. *Cons. Intern. Expl. Mer, Ann. Biol., Copenh.*, 37(1980):234–235.

Jónsson, G., J. Magnússon, and J.V. Magnússon. 1984. Rare fishes: Icelandic observations on rare fish in 1981. *Cons. Intern. Expl. Mer, Ann. Biol., Copenh.*, 38(1981):204–205.

Jónsson, G., J. Magnússon, and J.V. Magnússon. 1985. Rare fishes: Icelandic observations on rare fish in 1982. *Cons. Intern. Expl. Mer, Ann. Biol., Copenh.*, 39(1982):184–185.

Jónsson, G., J. Magnússon, and J.V. Magnússon. 1986a. Rare fishes: Icelandic observations on rare fish in 1983. *Cons. Intern. Expl. Mer, Ann. Biol., Copenh.*, 40(1983):185–187.

Jónsson, G., J. Magnússon, and J.V. Magnússon. 1986b. Rare fishes: Icelandic observations on rare fish in 1984. *Cons. Intern. Expl. Mer, Ann. Biol., Copenh.*, 41(1984):158–160.

Jónsson, G., J. Magnússon, and V. Vilhelmsdóttir. 1986c. Sjaldsédar fisktegundir á Íslandsmidum árid 1985. *Ægir*, 79:626–629.

Jónsson, G., J. Magnússon, and V. Vilhelmsdóttir. 1987. Sjaldsédir fiskar árid 1986. *Ægir*, 80:358–360.

Jónsson, G., and J. Pálsson. 1999. Fishes of the suborder Ceratioidei (Pisces: Lophiiformes) in Icelandic and adjacent waters. *Rit Fiskideildar*, 16:197–207.

Jordan, D.S. 1885. A catalogue of the fishes known to inhabit the waters of North America, north of the Tropic of Cancer, with notes on the species discovered in 1883 and 1884. pp. 789–973, in: *Report of the Commissioner of Fish and Fisheries for the Year Ending June 30, 1885*, Vol. 13, Washington, D.C. [separate in 1885; full report published in 1887.]

Jordan, D.S. 1905. *A Guide to the Study of Fishes*, Vol. 2. Holt, New York.

Jordan, D.S. 1920. The genera of fishes. Part IV. From 1881 to 1920, thirty-nine years, with the accepted type of each, a contribution to the stability of scientific nomenclature. *Stanf. Univ. Publ., Univ. Ser.*, 43:411–576 + i–xviii.

Jordan, D.S. 1923. A classification of fishes. Including families and genera as far as known. *Stanf. Univ. Publ., Univ. Ser., Biol. Sci.*, 3(2):77–243 + 10 pp.

Jordan, D.S. 1925. The fossil fishes of the Miocene of southern California. *Stanf. Univ. Publ., Univ. Ser., Biol. Sci.*, 4(1):1–51.

Jordan, D.S. 1926. The male sea devil and his ways. *Sci. Monthly*, 22:367.

Jordan, D.S., and B.W. Evermann. 1896. A check list of the fishes and fish-like vertebrates of North and Middle America. pp. 207–584, in: *Report of the Commissioner of Fish and Fisheries for the Year Ending June 30, 1895*, Vol. 21, Append. 5, Washington, D.C.

Jordan, D.S., and B.W. Evermann. 1898. The fishes of North and Middle America, a descriptive catalogue of the species of fish-like vertebrates found in the waters of North America, north of the isthmus of Panama, Part 3. *Bull. U.S. Nat. Mus.*, 47(3):xxiv + 2183–3136.

Jordan, D.S., and B.W. Evermann. 1900. The fishes of North and Middle America, a descriptive catalogue of the species of fish-like vertebrates found in the waters of North America, north of the isthmus of Panama, Part 4. *Bull. U.S. Nat. Mus.*, 47(3):cii + 3137–3314 + 392 plates.

Jordan, D.S., and C.H. Gilbert. 1883. Synopsis of the fishes of North America. Contributions to North American ichthyology, based primarily on the collections of the United States National Museum. *Bull. U.S. Nat. Mus.*, 16, lvi + 1018 pp.

Jordan, D.S., and E.K. Jordan. 1922. A list of the fishes of Hawaii, with notes and descriptions of new species. *Mem. Carnegie Mus.*, 10(1):1–92.

Jordan, D.S., and M. Sindo. 1902. A review of the pediculate fishes or anglers of Japan. *Proc. U.S. Nat. Mus.*, 24:361–381.

Jordan, D.S., S. Tanaka, and J.O. Snyder. 1913. A catalogue of the fishes of Japan. *J. Coll. Sci., Tokyo Imp. Univ.*, 33(1):1–497.

Karrer, C. 1973a. Über das Vorkommen von Fischarten im Nordwestatlantik (Neufundland-Baffinland). *Fischerei-Forsch. Wiss. Schriftenr.*, Rostock, 11:73–90.

Karrer, C. 1973b. Über Fische aus dem Südostatlantik. *Mitt. Zool. Mus. Berlin*, 49(1):191–257.

Karrer, C. 1975. Über Fische aus dem Südostatlantik, Teil 2. *Mitt. Zool. Mus. Berlin*, 51(1):63–82.

Karrer, C. 1976. Über Fischarten aus der Davisstrasse und Labradorsee. *Mitt. Zool. Mus. Berlin*, 52(2):371–376.

Kharin, V.E. 1984. Two new species of deepwater anglerfishes (Ceratioidei: Himantolophidae, Gigantactinidae) from the North Pacific. *Vopr. Ikhtiol.*, 24(4):663–667. [In Russian, English translation in *J. Ichthyol.*, 24(3):112–117, 1984.]

Kharin, V.E. 1989. A new genus and species of deep water angler fish of the family Oneirodidae from the northwestern Pacific. *Vopr. Ikhtiol*, 29(1):158–160. [In Russian, English translation in *J. Ichthyol.*, 29(4):156–158, 1989.]

Kharin, V.E. 2006a. *Himantolophus sagamius* (Himantolophidae), a new fish species for the fauna of Russia. *Vopr. Ikhtiol.*, 46(2):281–282. [In Russian, English translation in *J. Ichthyol.*, 46(3):274–275, 2005.]

Kharin, V.E. 2006b. On the species composition and distribution of ceratiid anglers (Ceratiidae) in Russian and adjacent waters. *Vopr. Ikhtiol.*, 46(3):420–423. [In Russian, English translation in *J. Ichthyol.*, 46(5):409–412, 2006.]

Kharin, V.E. 2007. Finding a rare species of leftvent seadevils *Linophryne indica* (Linophrynidae) near the Russian Economic Zone. *Vopr. Ikhtiol.*, 47(2):266–268. [In Russian, English translation in *J. Ichthyol.*, 47(2):198–200, 2006.]

Kharin, V.E., and P.G. Milovankin. 2007. A new occurrence of the deepsea anglerfish *Cryptopsaras couesii* (Osteichthys: Lophiiphormes: Ceratiidae) in Russian waters. *Vopr. Ikhtiol.*, 47(1):119–120. [In Russian, English translation in *J. Ichthyol.*, 47(1):112–113, 2007.]

Kharin, V.E., and T.W. Pietsch. 2007. New finding of a rare deep-sea ceratioid anglerfish, *Dermatias platynogaster* Smith and Radcliffe (Lophiiformes: Oneirodidae). *J. Ichthyol.*, 47(9):806–808.

Kitchell, J.F. 1983. Energetics. pp. 312–338, in: P.W. Webb and D. Weihs (editors), *Fish Biomechanics*. Praeger Publishers, New York.

Klepadlo, C., P.A. Hastings, and R.H. Rosenblatt. 2003. Pacific footballfish, *Himantolophus sagamius* (Tanaka) (Teleostei: Himantolophidae), found in the surf zone at Del Mar, San Diego County, California, with notes on its morphology. *Bull. So. Calif. Acad. Sci.*, 102(3):99–106.

Knudsen, M., P. Jespersen, and A.V. Tåning. 1934. Introduction to the reports from the Carlsberg Foundation's Oceanographical Expedition Round the World 1928–30. *Dana Rept.*, 1, 130 pp. + 7 pls.

Kobayashi, K., T. Ueno, H. Omi, and K. Abe. 1968. Records of some rare deep-sea anglerfishes obtained from the waters of Hokkaido. *Bull. Fac. Fish., Hokkaido Univ.*, 19(1):7–18.

Koefoed, E. 1944. Pediculati from the "Michael Sars" North Atlantic Deep-sea Expedition 1910. *Rept. Sci. Res. "Michael Sars" Exped.* 4, 2(1):1–18.

Korsmeyer, K.E., J.F. Steffensen, and J. Herskin. 2002. Energetics of median and paired fin swimming, body and caudal swimming, and gait transition in parrotfish (*Scarus schlegeli*) and triggerfish (*Rhinecanthus aculeatus*). *J. Exp. Biol.*, 253:1253–1263.

Kottelat, M., R. Britz, T.H. Hui, and K.-E. Witte. 2006. *Paedocypris*, a new genus of Southeast Asian cyprinid fish with a remarkable sexual dimorphism, comprises the world's smallest vertebrate. *Proc. Roy. Soc. B*, 273:895–899.

Kottelat, M., and C. Vidthayanon. 1993. *Boraras micros*, a new genus and species of minute freshwater fish from Thailand (Teleostei: Cyprinidae). *Ichthyol. Expl. Freshwaters*, 4(2):161–176.

Kotthaus, A. 1979. Fische des Indischen Ozeans. Ergebnisse der ichthyologischen Untersuchungen während der Expedition des Forschungsschiffes "Meteor" in den Indischen Ozean, Oktober 1964 bis Mai 1965. A. Systematischer Teil, XXI, Diverse Ordnungen. *"Meteor" Forsch.-Ergebn., Reihe D*, 28:6–54.

Kramp, P.L. 1953. Pelagisk dyreliv. pp. 77–96, In: *Galatheas jordomsejling 1950–1952, den Danske dybhavsekspeditions virke og resultater skildret af deltagerne*. Under redaktion af A.F. Bruun, S. Greve, H. Mielche, and R. Spärck, med forord af H.K.H. Prins Axel. J.H. Schultz Forlag, Copenhagen. [In Danish.]

Kramp, P.L. 1956. Pelagic fauna. pp. 65–86, in: A.F. Bruun, S. Greve, H. Mielche, and R. Spärck (editors), *The Galathea Deep Sea Expedition 1950–1952, Described by Members of the Expedition*, translated from the Danish by R. Spink. George Allen and Unwin, Ltd., London.

Krefft, G. 1954. Ein Tiefseeangler aus isländischen Gewässern. *Die Fischwirtschaft*, 6(4):78–79.

Krefft, G. 1960. Rare fish. A. Records of the Institut für Seefischerei, Hamburg. *Cons. Intern. Expl. Mer, Ann. Biol., Copenh.*, 15(1958):70–72.

Krefft, G. 1961. Rare fish: Germany. *Cons. Intern. Expl. Mer, Ann. Biol., Copenh.*, 16(1959):90–91.

Krefft, G. 1963. Rare fish: Germany. *Cons. Intern. Expl. Mer, Ann. Biol., Copenh.*, 18(1961):82–83.

Krefft, G. 1967. Rare fish: German observations on rare fish in 1965. *Cons. Intern. Expl. Mer, Ann. Biol., Copenh.*, 22(1965):183–186.

Krøyer, H.N. 1845. Ichthyologiske bidrag (Fortsaettelse). 10. *Ceratias holboelli* Kr. *Naturh. Tidsskr.*, 1844–1845, 2(1):639–649.

Kubota, T. 1973. Four links of food chains from the lancetfish, *Alepisaurus ferox*, to zooplankton in Suruga Bay, Japan. *J. Fac. Mar. Sci. Techn., Tokai Univ.*, 7:231–243.

Kubota, T., and T. Uyeno. 1970. Food habits of lancetfish *Alepisaurus ferox* (Order Myctophiformes) in Suruga Bay, Japan. *Japan J. Ichthyol.*, 17(1):22–28.

Kujawa, S. 1964. Szczególnie rzadki okaz ryby głębinowej w zbiorach Muzeum Morskiego Instytutu Rybackiego w Gdyni. [A rare abyssal fish in the Museum of the Sea Fisheries Institute, Gdynia.] *Przegląd Zoologiczny*, 8(3):301–303.

Kukuev, E.I., and I.A. Trunov. 2002. Composition of the meso- and bathypelagic fish fauna of the Irminger Current zone and adjacent areas. *Vopr. Ikhtiol.*, 42(3):322–329. [In Russian, English translation in *J. Ichthyol.*, 42(5):377–384, 2002.]

Kuronuma, K. 1941. Notes on rare fishes taken off the Pacific coast of Japan. *Bull. Biogeogr. Soc. Japan*, 11(8):37–67.

Lancraft, T.M., and B.H. Robison. 1980. Evidence of postcapture ingestion by midwater fishes in trawl nets. *U.S. Fish. Bull.*, 77(3):713–715.

Landesman, J. 1990. A voracious fisher of the deep. *Tidelines*, Cabrillo Marine Museum, San Pedro, California, 10(1):1–8.

Last, P.R., D.C. Gledhill, and B.H. Holmes. 2007. A new handfish, *Brachionichthys australis* sp. nov. (Lophiiformes: Brachionichthyidae), with a redescription of the critically endangered spotted handfish, *B. hirsutus* (Lacepède). *Zootaxa*, 1666:53–68.

Last, P.R., E.O.G. Scott, and F.H. Talbot. 1983. *Fishes of Tasmania*. Tasmanian Fisheries Development Authority, Old Wharf, Hobart, Tasmania.

Lavenberg, R.J., and A.W. Ebeling. 1967. Distribution of midwater fishes among deep-water basins of the southern California shelf. pp. 185–201, in: *Proc. Symp. Biol. Calif. Islands*. Santa Barbara Botanical Garden., Santa Barbara, California.

Lea, R.N. 1988. Family Himantolophidae added to the ichthyofauna of the temperate Eastern North Pacific. *Calif. Fish Game*, 74(3):180–182.

Le Danois, Y. 1964. Étude anatomique et systématique des Antennaires, de l'Ordre des Pédiculates. *Mém. Mus. Nat. Hist. Nat., Paris, n.s., Sér. A, Zool.*, 31(1):1–162.

Le Danois, Y. 1978. La topographie du système latéro-muqueux d' *Himantolophus groenlandicus* Reinhardt (Pisces, Pediculati, Ceratioidei). *Vie Milieu (AB)*, 28/29:307–321.

Lee, S.-C. 1988. Fishes of lophiiformes (Pediculati) of Taiwan. *Bull. Inst. Zool., Acad. Sinica*, 27(1):13–26.

Lee, S.-C. 1993. Lophiiformes. pp. 180–185, in: S.-C. Shen (editor), *Fishes of Taiwan*. Department of Zoology, National Taiwan University, Taipei.

Leim, A.H., and W.B. Scott. 1966. Fishes of the Atlantic Coast of Canada. *Bull. Fish. Res. Bd. Canada*, 155, 485 pp.

Leipertz, S.L., and T.W. Pietsch. 1987. A new species of ceratioid anglerfish of the genus *Dolopichthys* (Pisces: Lophiiformes) from the western North Atlantic Ocean. *Copeia*, 1987(2):406–409.

Leisman, G., D.H. Cohn, and K.H. Nealson. 1980. Bacterial origin of luminescence in marine animals. *Science*, New York, 208:1271–1273.

Liem, K.F. 1970. Comparative functional anatomy of the Nandidae (Pisces: Teleostei). *Fieldiana, Zool.*, 56:1–166.

Lloris, D. 1981. Contribución al conocimiento de la ictiofauna del SO africano: *Phrynichthys wedli* Pietschmann, 1926 (Lophiiformes, Diceratiidae). *Res. Exp. Cient.*, 9:53–58.

Lloris, D. 1986. Ictiofauna Demersal y aspectos biogeográficos de la costa sudoccidental de África (SWA/Namibia). *Monogr. Zool. Mar., Instituto de Ciencias del Mar Barcelona*, 1: 9–432.

Lloyd, R.E. 1909a. A description of the deep-sea fish caught by the R.I.M.S. Ship "Investigator" since the year 1900, with supposed evidence of mutation in *Malthopsis*. *Mem. Indian Mus.*, Calcutta, 2(3):139–180.

Lloyd, R.E. 1909b. Illustrations of the zoology of the Royal Indian Marine Survey Ship Investigator under the command of Commander WE. G. Beauchamp, R.I.M. Fishes. Part X. *Mem. Indian Mus.*, Calcutta, 2(3), pls. 44–50.

Longhurst, A. 2007. *Ecological Geography of the Sea*, 2nd edition. Academic Press, Boston.

Love, M.S., C.W. Mecklenburg, T.A. Mecklenburg, and L.K. Thorsteinson. 2005. *Resource Inventory of Marine and Estuarine Fishes of the West Coast and Alaska: A Checklist of North Pacific and Arctic Ocean Species from Baja California to the Alaska-Yukon Border*, OCS Study MMS 2005-030 and USGS/NBII 2005-001. U.S. Department of the Interior, U.S. Geological Survey, Biological Resources Division, Seattle, Washington.

Lozano y Rey, L. 1960. Ictyologia iberica. IV. Peces fisoclistas (ordenes Equeneiformes y Gobiformes), Pediculados y Asimetricos. *Mem. R. Acad. Cienc. Exact. Fis. Nat. Madr., Ser., Cienc. Nat.*, 14:617.

Luck, D.G., and T.W. Pietsch. 2008. In-situ observations of a deep-sea ceratioid anglerfish of the genus *Oneirodes* (Lophiiformes: Oneirodidae). *Copeia*, 2008(2):446–451.

Lütken, C.F. 1871. *Oneirodes eschrichtii* Ltk. en ny grønlandsk Tudsefisk. *Oversigt over det Kongl. Danske Vidensk. Selsk. Forhandl.*, 1871:56–74. [In Danish, French summary in the same volume, pp. 9–18.]

Lütken, C.F. 1872. On *Oneirodes eschrichtii*, Lütken, a new lophioid fish from Greenland. *Ann. Mag. Nat. Hist., Ser. 4*, 9(35):329–344.

Lütken, C.F. 1878a. Til Kundskab om to arktiske slaegter af Dybhavs-Tudsefiske: *Himantolophus* og *Ceratias*. *Danske Vidensk. Selsk. Skr., 5, R. Nat. Math. Afd.*, 11(5):307–338.

Lütken, C.F. 1878b. Contributions pour servir à l'histoire de deux genres de poissons de la famille des Baudroies, *Himantolophus* et *Ceratias*, habitant les grandes profondeurs des mers arctiques. *Danske Vidensk. Selsk. Skr., 5, R. Nat. Math. Afd.*, 11(5):339–348.

Lütken, C.F. 1887. Fortsatte bidrag til Kundskab om de arktiske Dybhavs-Tudsefiske, saerligt Slaegten *Himantolophus*. *Danske Vidensk. Selsk. Skr., 6, R. Nat. Math. Afd.*, 4(5):325–334.

Lütken, C.F. 1894. En bemaerkning I anledning af fangsten af en *Himantolophus Reinhardti* I havet vest for Portugal. *Vidensk. Meddr. Dansk naturh. Foren.*, 1894:78–81.

Macdonald, A.G. 1975. *Physiological Aspects of Deep Sea Biology*. Cambridge University Press, Cambridge, England.

Machida, Y., and T. Yamakawa. 1990. Occurrence of the deep-sea diceratiid anglerfish *Phrynichthys wedli* in the East China Sea. *Proc. Japan. Soc. Syst. Zool.*, 42:60–65.

Mackintosh, N.A. 1937. The seasonal circulation of the Antarctic macro-plankton. *Disc. Rept.*, 16:365–412.

Marshall, N.B. 1954. *Aspects of Deep-Sea Biology*. Hutchinsons, London.

Marshall, N.B. 1966. *The Life of Fishes*. World Publishing Co., Cleveland and New York.

Marshall, N.B. 1967a. The olfactory organs of bathypelagic fishes. pp. 57–70, in: N.B. Marshall (editor), *Aspects of Marine Zoology, Symp. Zool. Soc. London*, 19. Published for the Zoological Society of London by Academic Press, London.

Marshall, N.B. 1967b. The organization of deep-sea fishes. pp. 473–479, in: F.M. Bayer, C.P. Idyll, J.I. Jones, F.F. Koczy, A.A. Myrberg, C.R. Robins, F.G.W. Smith, G.L. Vos, E.J.F. Wood, and A.C. Jensen (editors), *Proceedings of the International Conference on Tropical Oceanography, 17–24 November 1965, Miami Beach, Florida, Studies in Tropical Oceanography, Miami*, Vol. 5, Institute of Marine Sciences, University of Miami, Miami, Florida.

Marshall, N.B. 1971a. *Explorations in the Life of Fishes*. Harvard University Press, Cambridge, Massachusetts.

Marshall, N.B. 1971b. Animal ecology. pp. 205–224, in: P.J. Herring and M.R. Clarke (editors), *Deep Oceans*. Praeger Publishers, New York and Washington, D.C.

Marshall, N.B. 1974. Midwater fishes of the dark ocean. *Spectrum, Brit. Sci. News*, 118:2–5.

Marshall, N.B. 1979. *Developments in Deep-Sea Biology*. Blandford Press, Ltd., Poole, United Kingdom.

Marshall, N.J. 1996. The lateral line systems of three deep-sea fish. *J. Fish Biol.*, 49(Suppl. A):239–258.

Martin, W.E. 1975. *Hydrichthys pietschi*, new species (Coelenterata) parasitic on the fish *Ceratias holboelli*. *Bull. So. Calif. Acad. Sci.*, 74(1):1–5.

Marzuola, C. 2002. Upside way down: Video turns fish story on its head. *Science News*, 162:262.

Matsubara, K. 1955. *Fish Morphology and Hierarchy*, Vol. 2. Ishizaki-Shoten, Tokyo.

Matthews, F.D., D.M. Damkaer, L.W. Knapp, and B.B. Collette. 1977. Food of western North Atlantic tunas *(Thunnus)* and lancetfishes *(Alepisaurus)*. NOAA Techn. Rept. NMFS SSRF-706, 19 pp.

Maul, G.E. 1949. Alguns peixes notáveis. *Bol. Mus. Mun. Funchal*, 4(11):22–42.

Maul, G.E. 1961. The ceratioid fishes in the collection of the Museu Municipal do Funchal (Melanocetidae, Himantolophidae, Oneirodidae, Linophrynidae). *Bol. Mus. Mun. Funchal*, 14(50):87–159.

Maul, G.E. 1962a. On a small collection of ceratioid fishes from off Dakar and two recently acquired specimens from stomachs of *Aphanopus carbo* taken in Madeira (Melanocetidae, Himantolophidae, Diceratiidae, Oneirodidae, Ceratiidae). *Bol. Mus. Mun. Funchal*, 16(54):5–27.

Maul, G.E. 1962b. Report on the fishes taken in Madeiran and Canarian waters during the summer-autumn cruises of the "Discovery II" 1959 and 1961. I. The ceratioid fishes (Melanocetidae, Himantolophidae, Oneirodidae, Gigantactinidae, Linophrynidae). *Bol. Mus. Mun. Funchal*, 16(56):33–46.

Maul, G.E. 1973. [Caulophrynidae, Melanocetidae, Himantolophidae, Diceratiidae, Oneirodidae, Ceratiidae, Gigantactinidae, and Linophrynidae]. pp. 666–677, in: J.-C. Hureau and T. Monod (editors), *Check-list of the Fishes of the North-eastern Atlantic and Mediterranean*, Vol. 1. United Nations Educational Scientific and Cultural Organization, Paris.

Maurin, C., F. Lozano Cabo, and M. Bonnet. 1970. Inventaire faunistique des principales espèces ichthyologiques fréquentant les côtes nord-ouest africaines. *Rapp. P.-v. Réun. Cons. Perm. Intern. Explor. Mer.*, 159:15–21.

Mayer, R.F., and T.T. Nalbant. 1972. Additional species of fishes in the fauna of Peru Trench. Results of the 11th Cruise of the R/V "Anton Bruun," 1965. *Rev. Roum. Biol.-Zoologie*, Bucharest, 17(3):159–165.

MBARI. 1997. *MBARI's first decade: A retrospective*. Monterey Bay Aquarium Research Institute, Moss Landing, California.

McEachran, J.D., and J.D. Fechhelm. 1998. *Fishes of the Gulf of Mexico*. Vol. 1, Myxiniformes to Gasterosteiformes. University of Texas Press, Austin.

McFall-Ngai, M.J., and E.G. Ruby. 1991. Symbiotic recognition and subsequent morphogenesis as early events in an animal-bacterial mutualism. *Science*, New York, 254:1491–1494.

McGinnis, R.F. 1982. Biogeography of lanternfishes (family Myctophidae) south of 30°S. Biology of the Antarctic Seas XII, *Antarctic Research Series*, Vol. 35. American Geophysical Union, Washington, D.C.

Mead, G.W. 1958. A catalog of the type specimens of fishes formerly in the collections of the Department of Tropical Research, New York Zoological Society. *Zoologica*, New York, 43(11):131–134.

Mead, G.W., E. Bertelsen, and D.M. Cohen. 1964. Reproduction among deep-sea fishes. *Deep-Sea Res.*, 11:569–596.

Mecklenburg, C.W., T.A. Mecklenburg, and L.K. Thorsteinson. 2002. *Fishes of Alaska*. American Fisheries Society, Bethesda, Maryland.

Meléndez, R., and I. Kong. 1997. Himantolophid fishes from Chile (Pisces: Lophiiformes). *Revista de Biol. Mar. Oceanogr.*, 32(1):11–15.

Menezes, N.A., P.A. Buckup, J.L. de. Figueiredo, and R.L. de Moura. 2003. *Catálogo das Espécies de Peixes Marinhos do Brasil*. Museu de Zoologia da Universidade de São Paulo, São Paulo, Brazil.

Meng, Q., J. Su, and X. Miao. 1995. *Systematics of Fishes*. Contemporary Scientific and Technological Works: Agriculture. China Agricultural Press, Beijing.

Menon, A.G.K, and K.V. Rama Rao. 1975. A catalogue of type-specimens of fishes described in the biological collections of the R.I.M.S. "Investigator" during 1884–1926. *Matsya, Bull. Indian Soc. Ichthyol.*, 1:31–48.

Menon, A.G.K, and G.M. Yazdani. 1963. Catalogue of type-specimens in the Zoological Survey of India. Part 2. Fishes. *Rec. Zool. Surv. India*, 61(1–2):91–190.

Merrett, N.R., J. Badcock, and P.J. Herring. 1973. The status of *Benthalbella infans* (Pisces: Myctophoidei), its development, bioluminescence, general biology, and distribution in the western North Atlantic. *J. Zool.*, London, 170:1–48.

Mihai-Bardan, A. 1982. *Caulophryne bacescui*, a new species of anglerfishes from the Peruvian waters, eastern South Pacific (Pisces, Caulophrynidae). *Rev. Roum. Biol., Biol. Anim.*, Bucharest, 27(1):17–21.

Mills, E.L. 1989. *Biological Oceanography: An Early History, 1870–1960*. Cornell University Press, Ithaca and London.

Minchin, D. 1988. A record of the deep-sea anglerfish, *Cryptopsaras couesi* Gill, from the north-eastern Atlantic. *J. Fish Biol.*, 32:313.

Miya, M., T.P. Satoh, and M. Nishida. 2005. The phylogenetic position of toadfishes (order Batrachoidiformes) in the higher ray-finned fish as inferred from partitioned Bayesian analysis of 102 whole mitochondrial genome sequences. *Biol. J. Linn. Soc.*, 85:289–306.

Miya, M., H. Takeshima, H. Endo, N.B. Ishiguro, J.G. Inoue, T. Mukai, T.P. Satoh, M. Yamaguchi, A. Kawaguchi, K. Mabuchi, S.M. Shirai, and M. Nishida. 2003. Major patterns of higher teleostean phylogenies: A new perspective based on 100 complete mitochondrial DNA sequences. *Mol. Phylog. Evol.*, 26:121–138.

Monod, T. 1960. A propos du pseudobrachium des *Antennarius* (Pisces, Lophiformes). *Bull. Inst. Français Afrique Noire*, Ser. A, 22:620–698.

Montgomery, J., and N. Pankhurst. 1997. Sensory physiology. pp. 325–349, in: D.J. Randall and A.P. Farrell (editors), *Deep-Sea Fishes*. Academic Press, San Diego.

Moore, H.B. 2002. *Marine Ecology*. John Wiley and Sons, New York.

Moore, J. 2002. Upside-down swimming behavior in a whipnose anglerfish (Teleostei: Ceratioidei: Gigantactinidae). *Copeia*, 2002:1144–1146.

Moore, J., K.E. Hartel, J.E. Craddock, and J. Galbraith. 2003. An annotated list of deepwater fishes from off the New England region, with new area records. *Northeastern Nat.*, 10(2):159–248.

Morell, V. 2005. Way down deep: There's a place in California where the sun never shines: Monterey Canyon. *Nat. Geo.*, June, 205(6):36–55.

Morozov, E.G., K. Trulsen, M.G. Velarde, and V.I. Vlasenko. 2002. Internal tides in the Strait of Gibraltar. *J. Phys. Oceanogr.*, 32(11):3193–3206.

Munk, O. 1964. The eyes of some ceratioid fishes. *Dana Rept.*, 62, 17 pp.

Munk, O. 1966. Ocular anatomy of some deep-sea teleosts. *Dana Rept.*, 70, 62 pp.

Munk, O. 1988. Glandular tissue of escal light organ in the deep-sea anglerfish *Oneirodes eschrichti* (Pisces, Ceratioidei). A light and electron microscopic study. *Vidensk. Meddr. Dansk Naturh. Foren.*, 147:93–120.

Munk, O. 1992. Accessory escal gland (AEG) in some deep-sea anglerfishes. *Acta Zool.*, Stockholm, 73(1):33–37.

Munk, O. 1998. Light guides of the escal light organs in some deep-sea anglerfishes (Pisces: Ceratioidei). *Acta Zool.*, Stockholm, 79(3):175–186.

Munk, O. 1999. The escal photophore of ceratioids (Pisces; Ceratioidei)—a review of structure and function. *Acta Zool.*, Stockholm, 80(4):265–284.

Munk, O. 2000. Histology of the fusion area between the parasitic male and the female in the deep-sea anglerfish *Neoceratias spinifer* Pappenheim, 1914 (Teleostei, Ceratioidei). *Acta Zool.*, Stockholm, 81(4):315–324.

Munk, O., and E. Bertelsen. 1980. On the esca light organ and its associated light-guiding structures in the deep-sea anglerfish *Chaenophryne draco* (Pisces, Ceratioidei). *Vidensk. Meddr. Dansk naturh. Foren.*, 142:103–129.

Munk, O., and E. Bertelsen. 1983. Histology of the attachment between the parasitic male and the female in the deep-sea anglerfish *Haplophryne mollis* (Brauer, 1902) (Pisces, Ceratioidei). *Vidensk. Meddr. Dansk naturh. Foren.*, 144: 49–74.

Munk, O., K. Hansen, and P.J. Herring. 1998. On the development and structure of the escal light organ of some melanocetid deep sea anglerfishes (Pisces Ceratioidei). *J. Mar. Biol. Assn. U.K.*, 78:1321–1335.

Munk, O., and P.J. Herring. 1996. An early stage in development of escae and caruncles in the deep-sea anglerfish *Cryptopsaras couesi* (Pisces: Ceratioidei). *J. Mar. Biol. Assn. U.K.*, 76:517–527.

Munro, I.S.R. 1967. *The Fishes of New Guinea*. Dept. of Agric., Stock and Fisheries, Port Moresby, New Guinea.

Murray, J., and J. Hjort. 1912. *The Depths of the Ocean. A General Account of the Modern Science of Oceanography Based Largely on the Scientific Researches of the Norwegian Steamer* Michael Sars *in the North Atlantic*. Macmillian and Co., London.

Nakabo, T. 2002. Ceratioidei. pp. 467–476, 1496–1497, in: T. Nakabo (editor), *Fishes of Japan, with Pictorial Keys to the Species*, English edition, Vols. 1 and 2. Tokai University Press, Tokyo.

Ni, Y. 1988. Lophiiformes. pp. 316–336, in: Deep Water Fishes of the East China Sea Editorial Subcommittee, *The Deep Water Fishes of the East China Sea*. Xue Lin Publishing House, Shanghai, China. [In Chinese, with English abstracts.]

Ni, Y., H. Wu, and S. Li. 1989. Two anglerfishes Diceratiidae new to Chinese fauna. *So. China Sea Fish. Res.*, 1:87–94. [In Chinese, with English abstract.]

Nielsen, J.G. 1974. *Fish Types in the Zoological Museum of Copenhagen*. Zoological Museum, University of Copenhagen, Copenhagen, Denmark.

Nielsen, J.G. 1994. Erik Bertelsen, 8 Aug. 1912–18 March 1993. *Copeia*, 1994(2):564–565.

Nielsen, J.G., and E. Bertelsen. 1985. The gulper-eel family Saccopharyngidae (Pisces, Anguilliformes). *Steenstrupia*, Copenhagen, 11(6):157–206.

Nielsen, J.G., and E. Bertelsen. 1992. *Fiskar i Nordur-Atlantshaft*. MM Publ., Reykjavik.

Nigrelli, R.F. 1947. Spontaneous neoplasms in fishes. II. Fibro-carcinoma-like growth in the stomach of *Borophryne apogon* Regan, a deep-sea ceratioid fish. *Zoologica*, New York, 31(12):183–184.

Nijssen, H., L. van Tuijl, and I.J.H. Isbrücker. 1993. Revised catalogue of the type specimens of recent fishes in the Institute of Taxonomic Zoology (Zoölogisch Museum), University of Amsterdam, The Netherlands. *Bull. Zool. Mus., Univ. Amsterdam*, 13(18):211–260.

Nobre, A. 1935. *Fauna marinha de Portugal*. I. Vertebrados. (Mamíferos, Reptis e Peixes). Companhia Editora do Minho, Barcelona, Portugal.

Nolan, R.S, and R.H. Rosenblatt. 1975. A review of the deep-sea angler fish genus *Lasiognathus* (Pisces: Thaumatichthyidae). *Copeia*, 1975(1):60–66.

Norman, J.R. 1930. Oceanic fishes and flatfishes collected in 1925–1927. *Disc. Rept.*, 2:261–370.

Norman, J.R. 1939. Fishes. *Sci. Rept. John Murray Exped.*, London, 7(1):1–115.

Norman, J.R., and P.H. Greenwood. 1975. *A History of Fishes*, 3rd edition. John Wiley and Sons, New York.

Nybelin, O. 1948. Fishes collected by the "Skagerak" Expedition in the eastern Atlantic 1946. *Kungl. Vetensk. Vitterh.-Samh. Handl., F. 6, Ser. B*, 5(16):1–93.

Ochiai, A., and F. Mitani. 1956. A revision of the pediculate fishes of the genus *Malthopsis* found in the waters of Japan (family Ogcocephalidae). *Pac. Sci.*, 10:271–285.

O'Day, W.T. 1974. Bacterial luminescence in the deep-sea anglerfish *Oneirodes acanthias* (Gilbert, 1915). *Nat. Hist. Mus. L.A. Co., Contri. Sci.*, 255, 12 pp.

Ogilby, J.D. 1907. Some new pediculate fishes. *Proc. Roy. Soc. Queensland*, 20:17–25.

Okada, Y.K. 1961. Notes on the parasitic male of *Cryptopsaras couesi*. *Proc. Japan Acad.*, 37(3):155–157.

Olsson, R. 1974. Endocrine organs of a parasitic male deep-sea angler-fish, *Edriolychnus schmidti*. *Acta Zool.*, Stockholm, 55:225–232.

Orlov, A.M., and A.M. Tokranov. 2005. Distribution and size composition of anglers of the genus *Oneirodes* (Oneirodidae) off the Northern Kurils and Southeastern Kamchatka. *Vopr. Ikhtiol.*, 45(2):285–288. [In Russian, English translation in *J. Ichthyol.*, 45(3):271–274, 2005.]

Orr, J.W. 1991. A new species of the ceratioid anglerfish genus *Oneirodes* (Oneirodidae) from the western North Atlantic, with a revised key to the genus. *Copeia*, 1991(4): 1024–1031.

Oshima, M. 1934. Occurrence of *Cryptopsaras couesii* in the Japanese water. *Rec. Oceanogr. Work Japan*, 6(1):108–109.

Osório, B. 1909. Contribuicao para o conhecimento da fauna bathypelágica visinha das costas de Portugal. *Mem. Mus. Bocage*, 1:1–35.

Osório, B. 1912. Nova Contribuicao para o conhecimento da fauna bathypelágica visinha das costas de Portugal. *Mem. Mus. Bocage*, 4:89–93.

Paetel, F. 1875. *Die bisher veröffentlichten Familien und Gattungsnamen der Mollusken*. Verlag von Gebrüder Paetel, Berlin, vol. 4, 229 pp.

Palmer, G., and E.I. White. 1953. Pisces. *Zool. Rec.* 88:1–85.

Pappenheim, P. 1914. Die Fische der Deutschen Südpolar-Expedition 1901–1903. II. Die Tiefseefische. *Deut. Südpolar-Exped.*, 15(7):161–200.

Parin, N.V. 1970. *Ichthyofauna of the Epipelagic Zone*. Israel Program for Scientific Translations, Jerusalem, iii + 206 pp.

Parin, N.V. 1975. Change of pelagic ichthyocoenoses along the section of the equator in the Pacific Ocean between 97 and 155°W. *Trans. Shirshov Inst. Oceanogr.*, 102:313–334. [In Russian, with English abstract.]

Parin, N.V. 1986. Trichiuridae. pp. 976–980, in: P.J.P. Whitehead, M.-L. Bauchot, J.-C. Hureau, J. Nielsen, and E. Tortonese (editors), *Fishes of the North-eastern Atlantic and Mediterranean*, Vol. 2. United Nations Educational Scientific and Cultural Organization, Paris.

Parin, N.V., A.P. Andriashev, O.D. Borodulina, and V.M. Tchuvassov. 1974. Midwater fishes of the southwestern Atlantic Ocean. *Akad. Nauk CCCP*, Moscow, 98:76–140. [In Russian, with English abstract.]

Parin, N.V., V.E. Becker, O.D. Borodulina, E.S. Karmovskaya, B.I. Fedoryako, J.N. Shcherbachev, G.N. Pokhilskaya, and V.M. Tchuvassov. 1977. Midwater fishes in the western tropical Pacific Ocean and the seas of the Indo-Australian Archipelago. *Trans. Shirshov Inst. Oceanogr.*, 107:68–188. [In Russian, with English abstract.]

Parin, N.V., V.E. Becker, O.D. Borodulina, and V.M. Tchuvassov. 1973. Deep-sea pelagic fishes of the southeastern Pacific Ocean. *Trans. Shirshov Inst. Oceanogr.*, 94:71–172. [In Russian, with English abstract.]

Parin, N.V., and G.A. Golovan. 1976. Pelagic deep-sea fishes of the families characteristic of the open ocean collected over the continental slope off West Africa. *Trans. Shirshov Inst. Oceanogr.*, 104:251–276. [In Russian, with English abstract.]

Parr, A.E. 1927. Scientific results of the Third Oceanographic Expedition of the "Pawnee" 1927. Ceratioidea. *Bull. Bingh. Oceanogr. Coll., Yale Univ.*, 3(1):1–34.

Parr, A.E. 1930a. On the osteology and classification of the pediculate fishes of the genera *Aceratias*, *Rhynchoceratias*, *Haplophryne*, *Laevoceratias*, *Allector*, and *Lipactis*, with taxonomic and osteological description of *Rhynchoceratias longipinna*, new species, and a special discussion of the rostral structures of the Aceratiidae. *Occ. Pap. Bingh. Oceanog. Coll., Yale Univ.*, 3:1–23.

Parr, A.E. 1930b. On the probable identity, life-history and anatomy of the free-living and attached males of the ceratioid fishes. *Copeia*, 1930(4):129–135.

Parr, A.E. 1932. On a deep-sea devilfish from New England waters and the peculiar life and looks of its kind. *Bull. Bost. Soc. Nat. Hist.*, 63:3–16.

Parr, A.E. 1934. Report on experimental use of a triangular trawl for bathypelagic collecting, with an account of the fishes obtained and a revision of the family Cetominidae. *Bull. Bingh. Oceanogr. Coll., Yale Univ.*, 4(6):1–34.

Parr, A.E. 1937. Concluding report on fishes, with species index for articles 1–7 (fishes of the Third Oceanographic Expedition of the "Pawnee"). *Bull. Bingh. Oceanogr. Coll., Yale Univ.*, 3(7):1–79.

Patterson, C., and D.E. Rosen. 1989. The paracanthopterygii revisited: Order and disorder. pp. 5–36, in: D.M. Cohen (editor), Papers on the systematics of gadiform fishes, *Nat. Hist. Mus. Los Angeles Co., Sci. Ser.*, 32.

Paulin, C.D. 1984. Six families of fishes new to the New Zealand fauna. *N.Z.J. Zool.*, 11:63–70.

Paulin, C.D., and A.L. Stewart. 1985. A list of New Zealand teleost fishes held in the National Museum of New Zealand. *Nat. Mus. N.Z., Misc. Ser.*, 12, 63 pp.

Paulin, C.D., A.L. Stewart, C.D. Roberts, and P.J. MacMillan. 1989. New Zealand fish, a complete guide. *Nat. Mus. N.Z., Misc. Ser.*, 19, 279 pp.

Paxton, J.R. 1998. Squirrelfishes and their allies. pp. 160–164, in: J.R. Paxton and W.N. Eschmeyer (editors), *Encyclopedia of Fishes, A Comprehensive Illustrated Guide by International Experts*, 2nd edition. Weldon Owen Pty Limited, McMahons Point, New South Wales, Australia.

Paxton, J.R., and R.J. Lavenberg. 1973. Feeding mortality in a deep sea angler fish (*Diceratias bispinosus*) due to a macrourid fish (*Ventrifossa* sp.). *Aust. Zool.*, 18(1):47–51.

Peden, A.E., W. Ostermann, and L.J. Pozar. 1985. Fishes obtained at Canadian Weathership Ocean Station Papa (50°N, 145°W), with notes on the Trans-Pacific Cruise of the CSS Endeavor. *Brit. Columbia Prov. Mus., Hertitage Rec.*, 18, vi + 50 pp.

Penrith, M.J. 1967. Ceratioid angler-fishes from South Africa. *J. Nat. Hist.*, 1:185–188.

Pequeño, G. 1989. Peces de Chile: Lista sistematica revisada y comentada. *Rev. Biol. Mar.*, Valparaiso, 24(2):1–132.

Pethon, P. 1969. List of type specimens of fishes, amphibians and reptiles in the Zoological Museum, University of Oslo. *Rhizocrinus, Occ. Pap. Zool. Mus., Univ. Oslo*, 1(1):1–17.

Pierantoni, U. 1917. Gli organi simbiotici e la luminescenza batterica dei Cefalopode. *Pub. Staz. Zool.*, Napoli, 20:105.

Pietsch, T.W. 1969. A remarkable new genus and species of deep-sea anglerfish (family Oneirodidae) from off Guadalupe Island, Mexico. *Copeia*, 1969(2):365–369.

Pietsch, T.W. 1972a. A review of the monotypic deep-sea anglerfish family Centrophrynidae: Taxonomy, distribution, and osteology. *Copeia*, 1972(1):17–47.

Pietsch, T.W. 1972b. A second specimen of the deep-sea anglerfish, *Phyllorhinichthys micractis* (family Oneirodidae), with a histological description of the snout flaps. *Copeia*, 1972(2):335–340.

Pietsch, T.W. 1972c. Ergebnisse der Forschungsreisen des FFS "Walther Herwig" nach Südamerika. XIX. Systematics and distribution of ceratioid fishes of the genus *Dolopichthys* (family Oneirodidae), with the description of a new species. *Arch. FischWiss.*, 23(1):1–28.

Pietsch, T.W. 1973. A new genus and species of deep-sea anglerfish (Pisces: Oneirodidae) from the northern Pacific Ocean. *Copeia*, 1973(2):193–199.

Pietsch, T.W. 1974a. Osteology and relationships of ceratioid anglerfishes of the family Oneirodidae, with a review of the genus *Oneirodes* Lütken. *Nat. Hist. Mus. L.A. Co., Sci. Bull.*, 18, 113 pp.

Pietsch, T.W. 1974b. Systematics and distribution of ceratioid anglerfishes of the genus *Lophodolos* (family Oneirodidae). *Breviora*, 425:1–19.

Pietsch, T.W. 1975a. Precocious sexual parasitism in the deep-sea ceratioid anglerfish *Cryptopsaras couesi* Gill. *Nature*, 256:38–40.

Pietsch, T.W. 1975b. Systematics and distribution of ceratioid anglerfishes of the genus *Chaenophryne* (family Oneirodidae). *Bull. Mus. Comp. Zool., Harvard*, 147(2):75–100.

Pietsch, T.W. 1976. Dimorphism, parasitism and sex: Reproductive strategies among deepsea ceratioid anglerfishes. *Copeia*, 1976(4):781–793.

Pietsch, T.W. 1978. A new genus and species of deep-sea anglerfish from the eastern North Pacific Ocean, with a review of the allied genera *Leptacanthichthys*, *Chirophryne*, and *Ctenochirichthys* (family Oneirodidae). *Nat. Hist. Mus. Los Angeles Co., Contrib. Sci.*, 297:1–25.

Pietsch, T.W. 1979. Ceratioid anglerfishes of the family Caulophrynidae with the description of a new genus and species from the Banda Sea. *Nat. Hist. Mus. Los Angeles Co., Contrib. Sci.*, 310:1–25.

Pietsch, T.W. 1981. The osteology and relationships of the anglerfish genus *Tetrabrachium*, with comments on lophiiform classification. *U.S. Fish. Bull.*, 79(3):387–419.

Pietsch, T.W. 1984. Lophiiformes: Development and relationships. pp. 320–325, in: H.G. Moser, W.J. Richards, D.M. Cohen, M.P. Fahay, A.W. Kendall, Jr., and S.L. Richardson (editors), *Ontogeny and Systematics of Fishes*, Spec. Publ. No. 1. Amer. Soc. Ichthyol. Herpetol., Lawrence, Kansas.

Pietsch, T.W. 1986a. [Order Lophiiformes: Introduction and key to families; families Ceratiidae, Oneirodidae, Melanocetidae, Himantolophidae, and Diceratiidae]. pp. 362–363, 373–377, in: M.M. Smith and P.C. Heemstra (editors), *Smiths' Sea Fishes*. Macmillan South Africa, Johannesburg.

Pietsch, T.W. 1986b. Systematics and distribution of bathypelagic anglerfishes of the family Ceratiidae (Order: Lophiiformes). *Copeia*, 1986(2):479–493.

Pietsch, T.W. 1990. [Ceratiidae, Oneirodidae, and Melanocetidae]. pp. 210–214, in: O. Gon and P.C. Heemstra (editors), *Fishes of the Southern Ocean*. J.L.B. Smith Institution of Ichthyology, Grahamstown, South Africa.

Pietsch, T.W. 1999. [Caulophrynidae, Neoceratiidae, Melanocetidae, Himantolophidae, Diceratiidae, Oneirodidae, Thaumatichthyidae, Centrophrynidae, Ceratiidae, Gigantactinidae, and Linophrynidae]. pp. 2026–2037, in: *FAO Species Identification Sheets for Fishery Purposes. Western Central Pacific (Fishing Area 71 and the southwestern part of Area 77)*, Vol. 3, Batoid fishes, Chimaeras, and Bony fishes, Pt. 1 (Elopidae to Linophrynidae). Food and Agriculture Organization of the United Nations, Rome.

Pietsch, T.W. 2002a. Seadevils or deep-sea anglerfishes, family Ceratiidae. pp. 274–276, in: B.B. Collette and G. Klein-MacPhee (editors), *Bigelow and Schroeder's Fishes of the Gulf of Maine*, 3rd edition. Smithsonian Institution Press, Washington and London.

Pietsch, T.W. 2002b. [Caulophrynidae, Melanocetidae, Himantolophidae, Diceratiidae, Oneirodidae, Thaumatichthyidae, Centrophrynidae, Ceratiidae, Gigantactinidae, Neocertiidae, and Linophrynidae]. pp. 1057–1070, in: *FAO Species Identification Guide for Fishery Purposes. The Living Marine Resources of the Western Central Atlantic (Fishing Area 31)*, Vol. 2, Bony Fishes, Pt. 1 (Acipenseridae to Grammatidae). Food and Agriculture Organization of the United Nations, Rome.

Pietsch, T.W. 2004. Revision of the deep-sea anglerfish genus *Phyllorhinichthys* Pietsch (Lophiiformes: Ceratioidei: Oneirodidae), with the description of a new species from the Atlantic Ocean. *Copeia*, 2004(4):797–803.

Pietsch, T.W. 2005a. A new species of the ceratioid anglerfish genus *Lasiognathus* Regan (Lophiiformes: Thaumatichthyidae) from the eastern North Atlantic off Madeira. *Copeia*, 2005(1):77–81.

Pietsch, T.W. 2005b. Dimorphism, parasitism, and sex revisited: Modes of reproduction among deep-sea ceratioid anglerfishes (Teleostei: Lophiiformes). *Ichthyol. Res.*, 52(3):207–236.

Pietsch, T.W. 2007. A new species of the ceratioid anglerfish genus *Chaenophryne* Regan (Lophiiformes: Oneirodidae) from the eastern tropical Pacific Ocean. *Copeia*, 2007(1):163–168.

Pietsch, T.W., and Z.H. Baldwin. 2006. A revision of the deep-sea anglerfish genus *Spiniphryne* Bertelsen (Lophiiformes: Ceratioidei: Oneirodidae), with description of a new species from the eastern North Pacific Ocean. *Copeia*, 2006(3):403–410.

Pietsch, T.W., A.V. Balushkin, and V.V. Fedorov. 2006. New records of the rare deep-sea anglerfish *Diceratias trilobus* Balushkin and Fedorov (Lophiiformes: Ceratioidei: Diceratiidae) from the western Pacific and eastern Indian oceans. *J. Ichthyol.*, 46(Suppl. 1):S97–S100.

Pietsch, T.W., and D.B. Grobecker. 1978. The compleat angler: Aggressive mimicry in an antennariid anglerfish. *Science*, 201(4353):369–370.

Pietsch, T.W., and D.B. Grobecker. 1980. Parental care as an alternative reproductive strategy in antennariid anglerfishes. *Copeia*, 1980(3):551–553.

Pietsch, T.W., and D.B. Grobecker. 1987. *Frogfishes of the World: Systematics, Zoogeography, and Behavioral Ecology*. Stanford Univ. Press, Stanford, California.

Pietsch, T.W., and D.B. Grobecker. 1990. Frogfishes: Masters of aggressive mimicry, these voracious carnivores can gulp prey faster than any other vertebrate predator. *Sci. Amer.*, 262(6):96–103.

Pietsch, T.W., H.-C. Ho, and H.-M. Chen. 2004. Revision of the deep-sea anglerfish genus *Bufoceratias* Whitley (Lophiiformes: Ceratioidei: Diceratiidae), with description of a new species from the Indo-West Pacific Ocean. *Copeia*, 2004(1):98–107.

Pietsch, T.W., and V.E. Kharin. 2004. *Pietschichthys horridus* Kharin, 1989, a junior synonym of *Dermatias platynogaster* Smith and Radcliffe, in Radcliffe, 1912 (Lophiiformes: Oneirodidae), with a revised key to oneirodid genera. *Copeia*, 2004(1):122–127.

Pietsch, T.W., and R.J. Lavenberg. 1980. A fossil ceratioid anglerfish from the Late Miocene of California. *Copeia*, 1980(4):906–908.

Pietsch, T.W., and B.G. Nafpaktitis. 1971. A male *Melanocetus johnsoni* attached to a female *Centrophryne spinulosa* (Pisces: Ceratioidei). *Copeia*, 1971(2):322–324.

Pietsch, T.W., and J.W. Orr. 2007. Phylogenetic relationships of deep-sea anglerfishes of the suborder Ceratioidei (Teleostei: Lophiiformes) based on morphology. *Copeia*, 2007(1):1–34.

Pietsch, T.W., and J.E. Randall. 1987. First Indo-Pacific occurrence of the deepsea ceratioid anglerfish, *Diceratias pileatus* (Lophiiformes: Diceratiidae). *Japan. J. Ichthyol.*, 33(4):419–421.

Pietsch, T.W., and J.A. Seigel. 1980. Ceratioid anglerfishes of the Philippine Archipelago, with descriptions of five new species. *U.S. Fish. Bull.*, 78(2):379–399.

Pietsch, T.W., and M. Shimazaki. 2005. Revision of the deep-sea anglerfish genus *Acentrophryne* Regan (Lophiiformes: Ceratioidei: Linophrynidae), with the description of a new species from off Peru. *Copeia*, 2005(2):146–151.

Pietsch, T.W., and J.P. Van Duzer. 1980. Systematics and distribution of ceratioid anglerfishes of the family Melanocetidae, with description of a new species from the eastern North Pacific Ocean. *U.S. Fish. Bull.*, 78(1):59–87.

Pietschmann, V. 1926. Ein neuer Tiefseefisch aus der Ordnung der Pediculati. *Anz. Akad. Wiss. Wien*, 63(11):88–89.

Pietschmann, V. 1930. *Phrynichthys wedli* Pietschm., nov. gen. et spec., ein Tiefsee Pediculate. *Ann. Nat. Mus. Wien*, 44:419–422.

Ponomarenko, V.P. 1959. Rare deep-sea fish of the North Atlantic. *Priroda*, Moscow, 2:83–85.

Potthoff, T. 1984. Clearing and staining techniques. pp. 35–37, in: H.G. Moser, W.J. Richards, D.M. Cohen, M.P. Fahay, A.W. Kendall, Jr., and S.L. Richardson (editors), *Ontogeny and Systematics of Fishes*, Spec. Publ. No. 1. Amer. Soc. Ichthyol. Herpetol., Lawrence, Kansas.

Quéro, J.-C., and J.-J. Vayne. 1988. Les petits et grands pêcheurs abyssaux (Pisces, Ceratiidae) pêchés dans les eaux Européennes. *Mésogée*, 48:173–181.

Quigley, D.T.G, and K. Flannery. 1997. First record of the Atlantic football fish *Himantolophus groenlandicus* (Reinhardt, 1837) (Pisces: Lophiiformes, Ceratioidea, Himantolophidae) from Irish waters. *Irish Nat. J.*, 25(11/12):442–444.

Radcliffe, L. 1912. Scientific results of the Philippine cruise of the fisheries steamer "Albatross," 1907–1910. No. 16. New pediculate fishes from the Philippine Islands and contiguous water. *Proc. U.S. Nat. Mus.*, 42:199–214.

Rasquin, P. 1958. Ovarian morphology and early embryology of the pediculate fishes *Antennarius* and *Histrio*. *Bull. Amer. Mus. Nat. Hist.*, 114(4):331–371.

Rass, T.S. 1955. [The deepwater fishes of the Kuril-Kamchatka Trench.] *Proc. Trudy Inst. Oceanol. Akad. Nauk, USSR*, 12:328–339. [In Russian.]

Rass, T.S. 1967. Some regularities in distribution of deep-sea fishes. Part 2, Chapter 5, pp. 228–246, in: *The Pacific Ocean*, Vol. 7, Biology of the Pacific Ocean, Pt. 3, Fishes of the Open Waters. Nauka, Moscow. [In Russian.]

Rauther, M. 1941. Einige Beobachtungen über den tentakelapparat von *Ceratias holbölli* (Teleostei, Pediculati). *Zool. Anz. Bd.*, 136:131–140.

Regan, C.T. 1912. The classification of the teleostean fishes of the order Pediculati. *Ann. Mag. Nat. Hist.*, Ser. 8, 9(28):277–289.

Regan, C.T. 1913. A deep-sea angler-fish *Melanocetus johnsoni*. *Proc. Zool. Soc. London*, 1913:1096–1097.

Regan, C.T. 1916. Larval and post-larval fishes. *Brit. Antarctic ("Terra Nova") Exped., 1910, Nat. Hist. Rept., Zool.*, 1(4):125–156.

Regan, C.T. 1925a. A rare angler fish (*Ceratias holbolli*) from Iceland. *Naturalist*, 1925:41–42.

Regan, C.T. 1925b. Dwarfed males parasitic on the females in oceanic angler-fishes (Pediculati, Ceratioidea). *Proc. Roy. Soc., B*, 97:386–400.

Regan, C.T. 1925c. New ceratioid fishes from the N. Atlantic, the Caribbean Sea, and the Gulf of Panama, collected by the "Dana." *Ann. Mag. Nat. Hist.*, Ser. 8, 8(62):561–567.

Regan, C.T. 1926. The pediculate fishes of the suborder Ceratioidea. *Dana Oceanogr. Rept.* 2, 45 pp.

Regan, C.T. 1927a. *Oceanic Angler-fishes*. Series 1. Trustees of the British Museum of Natural History, London, 4 pp. of text, five postcards.

Regan, C.T. 1927b. *Oceanic Angler-fishes*. Series 2. Trustees of the British Museum of Natural History, London, 4 pp. of text, five postcards.

Regan, C.T. 1930a. Angler-fishes. *Proc. Roy. Inst. Great Brit.*, 26(2):211–217.

Regan, C.T. 1930b. A ceratioid fish (*Caulophryne polynema*, sp. n.), female with male, from off Madeira. *J. Linn. Soc.*, London, 37:191–195.

Regan, C.T., and E. Trewavas. 1932. Deep-sea anglerfish (Ceratioidea). *Dana Rept.*, 2, 113 pp.

Reinhardt, J.C.H. 1837. Ichthyologiske bidrag til den grønlandske fauna. *Kgl. Danske Vidensk. Selsk. Naturvid. Math. Afhandl.*, 4(7):83–196.

Reinhardt, J.C.H. 1857. Fortegnelse over Grønlands Pattedyr, Fugle og Fiske. pp. 3–27, in: H.J. Rink, *Grønland: Geografisk og Statistisk Beskrevet*, Bd. 2, Det søndre Inspectorat med Afbildninger, Kaart og Naturhistoriske Tillaeg. Andr. Fred. Høst, København.

Rice, D.W. 1972. First record of the giant seadevil *Ceratias holboelli* from California, based upon a specimen from the stomach of a sperm whale. *Bull. So. Calif. Acad. Sci.*, 71(3):158.

Rink, H.J. 1877. *Danish Greenland, its People and its Products, with Illustrations by the Eskimo and a Map*. Henry S. King and Co., London.

Roberts, C.D. 1991. Fishes of the Chatham Islands, New Zealand; a trawl survey and summary of the ichthyofauna. *N.Z.J. Mar. Freshwater Res.*, 25:1–19.

Roberts, T.R. 1986. *Danionella translucida*, a new genus and species of cyprinid fish from Burma, one of the smallest living vertebrates. *Environ. Biol. Fish.*, 16:231–241.

Robins, C.R., and W.R. Courtenay, Jr. 1958. A deep sea ceratioid angler-fish of the genus *Gigantactis* from Florida. *Bull. Mar. Sci. Gulf Carib.*, 8(2):146–151.

Robison, B.H. 1976. Deep-sea fishes. *Nat. Hist.*, New York, 85(7):38–45.

Rodríguez-Marín, E., E. De Cárdenas, and F. Saborido. 1996. New record of two species of suborder Ceratioidei on Flemish Cap, northwest Atlantic. *NAFO Sci. Counc. Stud.*, 27:69–71.

Rondelet, G. 1554. *Libri de piscibus marinis, in quibus verae piscium effigies expressae sunt*. Matthiam Bonhomme, Lyon.

Rosen, D.E. 1985. An essay on euteleostean classification. *Amer. Mus. Novitates*, 2827:1–57.

Rosen, D.E., and C. Patterson. 1969. The structure and relationships of the Paracanthopterygian fishes. *Bull. Amer. Mus. Nat. Hist.*, 141:357–474.

Roule, L. 1919. Poissons provenant des campagnes du yacht Princesse-Alice (1891–1913) et du yacht Hirondelle II (1914). *Résult. Camp. Sci. Prince Albert I*, 52:1–190.

Roule, L. 1934. *Les Poissons et le Monde Vivant des Eaux. Études Ichthyologiques et Philosophiques.* Vol. 7, L'Abime des Grands Fonds Marins. Librairie Delagrave, Paris.

Roule, L., and F. Angel. 1930. Larves et alevins de poissons provenant des croisières du Prince Albert I<sup>er</sup> de Monaco. *Résult. Camp. Sci. Prince Albert I,* 79:1–148.

Roule, L., and F. Angel. 1932. Notice préliminaire sur un nouveau genre de poisson abyssal provenant des collections du Musée Océanographique de Monaco. *Bull. Mus. Oceanogr. Monaco, Ser. 2,* 4(5):500.

Roule, L., and F. Angel. 1933. Poissons provenant des campagnes du Prince Albert I<sup>er</sup> de Monaco. *Résult. Camp. Sci. Prince Albert I,* 86:1–115.

Ruby, R. G., and M. J. McFall-Ngai. 1992. A squid that glows at night: Development of an animal-bacterial mutualism. *J. Bacteriol.,* 174:4865–4870.

Saemundsson, B. 1908. Oversigt over Islands fiske. *Skr. Kommn. Havunders., Kbhvn.,* 5:1–140.

Saemundsson, B. 1922. Zoologiske meddelelser fra Island. XIV. 11 Fiske, ny for Island, og supplerende om andre, tidligere kendte. *Vidensk. Medd. Dansk Naturh. Foren.,* 74:159–201.

Saemundsson, B. 1927. Zoologiske meddelelser fra Island. XV. 6 Fiske, nye for Island, og Tilføjelser om andre, tidligere kendte. *Vidensk. Medd. Dansk Naturh. Foren.,* 84:151–187.

Saemundsson, B. 1939. Zoologiske meddelelser fra Island. XVII. 6 Fiske, nye for Island, og Tilføjelser om andre, tidligere kendte. *Vidensk. Medd. Dansk Naturh. Foren.,* 102:183–212.

Saemundsson, B. 1949. Marine pisces. *The Zoology of Iceland,* Copenhagen and Reykjavik, 4(72):1–150.

Saruwatari, T., T. W. Pietsch, A. M. Shedlock, I. Oohara, K. Itaya, T. Chiyotani, M. Abe, K. Ishihara, A. Sakai, A. Sugawara, T. Inagaki, K. Ishida, T. Komatsu, T. Sakai, and T. Kobayashi. 2001. Sibling analysis between female and parasitic males of a ceratioid anglerfish (*Cryptopsaras couesi,* Ceratiidae, Teleostei). *DNA Polymorphism,* 9:82–85. [In Japanese.]

Sassa, C., K. Kawaguchi, T. Kinoshita, and C. Watanabe. 2002. Assemblages of vertical migratory mesopelagic fish in the transitional region of the western North Pacific. *Fish. Oceanogr.,* 11(4):193–204.

Schmidt, J. 1926. "Dana" expeditionens tudsefisk. *Naturens Verden,* 6:261–271.

Schnakenbeck, W. 1936. Fang eines eigenartigen Fisches. *Der Fischmarkt,* Hamburg, 4(7):189–190.

Schroeder, E. H. 1963. North Atlantic temperatures at a depth of 200 meters. *Serial Atlas Mar. Environ.,* Folio 2. American Geographical Society, New York.

Schultz, L. P. 1934. A new ceratiid fish from the Gulf of Alaska. *Copeia,* 1934(2):66–68.

Schultz, L. P. 1957. The frogfishes of the family Antennariidae. *Proc. U.S. Nat. Mus.,* 107:47–105.

Schwab, I. R. 2004. A notch in time. *J. Ophthalmol.,* 88:1234.

Seigel, J. A., and T. W. Pietsch. 1978. A new species of the ceratioid anglerfish genus *Oneirodes* (Pisces: Lophiiformes) from the Indo-west Pacific. *Copeia,* 1978(1):11–13.

Senou, H., K. Matsuura, and G. Shinohara. 2006. Checklist of fishes in the Sagami Sea, with zoogeographical comments on shallow water fishes occurring along the coastlines under the influence of the Kuroshio Current. *Mem. Nat. Sci. Mus.,* Tokyo, 41:389–542.

Shedlock, A. M., M. G. Haygood, T. W. Pietsch, and P. Bentzen. 1997. Enhanced DNA extraction and PCR amplification of mitochondrial genes from formalin-fixed museum Specimens. *BioTechniques,* 22(3):394–400.

Shedlock, A. M., T. W. Pietsch, M. G. Haygood, P. Bentzen, and M. Hasegawa. 2004. Molecular systematics and life history evolution of anglerfishes (Teleostei: Lophiiformes): Evidence from mitochondrial DNA. *Steenstrupia,* Copenhagen, 28(2):129–144.

Sheiko, B. A., and V. V. Fedorov. 2000. [Cephalaspidomorpha, Chondrichthyes, Holocephali, and Osteichthyes]. pp. 7–69, in: *Catalog of Vertebrates of Kamchatka and Adjacent Waters.* Kamchatskiy Petchatniy Dvor, Petropavlovsk-Kamchatsky. [In Russian.]

Shimazaki, M., and K. Nakaya. 2004. Functional anatomy of the luring apparatus of the deep-sea ceratioid anglerfish *Cryptopsaras couesii* (Lophiiformes: Ceratioidae). *Ichthyol. Res.,* 51:33–37.

Shinohara, G., T. Sato, Y. Aonuma, H. Horikawa, K. Matsuura, T. Nakabo, and K. Sato. 2005. Annotated checklist of deep-sea fishes from the waters around the Ryukyu Islands, Japan. pp. 385–452, in: K. Hasegawa, G. Shinohara, and M. Takeda (editors), *Deep-Sea Fauna and Pollutants in the Nansei Islands,* National Science Museum Monographs, No. 29. National Science Museum, Tokyo.

Shoemaker, H. H. 1958. A female ceratioid angler, *Cryptopsaras couesi* Gill, from the Gulf of Mexico, bearing three parasitic males. *Copeia,* 1958(2):143–145.

Simmons, M. P., and M. Miya. 2004. Efficiently resolving the basal clades of a phylogenetic tree using Bayesian and parsimony approaches: A case study using mitogenomic data from 100 higher teleost fishes. *Mol. Phylog. Evol.,* 31:351–362.

Smale, M. J., G. Watson, and T. Hecht. 1995. Otolith atlas of southern African marine fishes. *J.L.B. Smith Inst. Ichthyol., Ichthyol. Mongr.,* 1, 253 pp., 149 pls.

Smith, H. M., and L. Radcliffe. 1912. Scientific results of the Philippine cruise of the fisheries steamer "Albatross," 1907–1910, No. 20. Description of a new family of pediculate fishes from Celebes. *Proc. U.S. Nat. Mus.,* 42:579–581.

Smith, J. L. B. 1949. *The Sea Fishes of Southern Africa.* Central News Agency, Cape Town, South Africa.

Stearn, D., and T. W. Pietsch. 1995. [Ceratiidae, Gigantactinidae, Oneirodidae, Melanocetidae, Linophrynidae, and Caulophrynidae]. pp. 131–144, in: O. Okamura, K. Amaoka, M. Takeda, K. Yano, K. Okada, and S. Chikuni (editors), *Fishes Collected by the R/V Shinkai Maru around Greenland.* Japan Marine Fishery Resources Research Center, Tokyo.

Stehmann, M. F. W. 1997. Gerhard Krefft (1912–1993) and post-World War II collection building in ichthyology at the Institut für Seefischerei, Hamburg: A melding of applied and basic research. pp. 121–132, in: T. W. Pietsch and W. D. Anderson, Jr. (editors), *Collection Building in Ichthyology and Herpetology,* Special Publ. No. 3. Amer. Soc. Ichthyol. Herpetol., Lawrence, Kansas.

Stehmann, M. F. W., and P. A. Hulley 1994. Gerhard Krefft, 30 March 1912—20 March 1993. *Copeia,* 1994(2): 558–564.

Stewart, A. L., and T. W. Pietsch. 1998. The ceratioid anglerfishes (Lophiiformes: Ceratioidei) of New Zealand. *J. Roy. Soc. N.Z.,* 28(1):1–37.

Struhsaker, P. 1962. The ceratioid fish *Melanocetus johnsoni* off the southeastern coast of the United States and a morphological observation. *Copeia,* 1962(4):841–842.

Swainson, W. 1839. *The Natural History and Classification of Fishes, Amphibians, and Reptiles, or Monocardian Animals*, Vol. 2. Longman, Orme, Brown, Green, and Longman's, London.

Swinney, G.N. 1995a. The first record of *Photocorynus spiniceps* Regan (Teleostei, Lophiiformes, Ceratoioidei, Linophrynidae) from the north-eastern Atlantic, with notes on a collection of female anglerfish from off Madeira. *Bol. Mus. Mun. Funchal (Hist. Nat.)*, Madeira, 47(261):51–62.

Swinney, G.N. 1995b. Ceratioid anglerfish of the families Gigantactinidae and Linophrynidae (Lophiiformes, Ceratioidea) collected off Madeira, including two species new to the North-eastern Atlantic. *J. Fish Biol.*, 47(1):39–49.

Swinney, G.N., and T.W. Pietsch. 1988. A new species of the ceratioid anglerfish genus *Oneirodes* (Pisces: Lophiiformes) from the eastern North Atlantic off Madeira. *Copeia*, 1988(4):1054–1056.

Tanaka, S. 1908. Notes on some rare fishes of Japan, with descriptions of two new genera and six new species. *J. Coll. Sci., Imp. Univ., Tokyo*, 23(13):1–24.

Tanaka, S. 1911. *Paraceratias mitsukurii* (Tanaka) (Ceratiidae). pp. 30–32, in: S. Tanaka, *Figures and Descriptions of the Fishes of Japan, including Riukiu Islands, Bonin Islands, Formosa, Kurile Islands, Korea, and Southern Sakhalin*, Vol. 2. Tokyo Printing Co., Tokyo.

Tanaka, S. 1918a. *Corynolophus sagamius*, n. sp. (Ceratiidae). pp. 491–494, in: S. Tanaka, *Figures and Descriptions of the Fishes of Japan, including Riukiu Islands, Bonin Islands, Formosa, Kurile Islands, Korea, and Southern Sakhalin*, Vol. 27. Tokyo Printing Co., Tokyo.

Tanaka, S. 1918b. *Corynolophus globosus*, n. sp. (Ceratiidae). pp. 529–532, in: S. Tanaka, *Figures and Descriptions of the Fishes of Japan, including Riukiu Islands, Bonin Islands, Formosa, Kurile Islands, Korea, and Southern Sakhalin*, Vol. 29. Tokyo Printing Co., Tokyo.

Tarakanov, E.S., and A.V. Balushkin. 1987. Systematics of the anglerfishes of the genus *Ceratias*. *Biologiya Morya*, 5:32–39. [In Russian with English abstract, English translation in *Soviet J. Mar. Biol.*, 13(5):265–271, 1988.]

Taylor, F.H.C. 1967. Unusual fishes taken by midwater trawl off the Queen Charlotte Islands, British Columbia. *J. Fish. Res. Bd. Canada*, 24(10):2101–2115.

Thompson, W.W. 1918. Catalogue of fishes of the Cape Province. *Mar. Biol. Rept., So. Afri.*, 4:75–177.

Thomson, C.W. 1877. *The Voyage of the "Challenger." The Atlantic: A Preliminary Account of the General Results of the Exploring Voyage of H.M.S. "Challenger" During the Year 1873 and the Early Part of the Year 1876*, Vol 2. Macmillan, London.

Thurman, H.V. 1975. *Introductory Oceanography*. C.E. Merrill, Columbus, Ohio.

Trunov, I.A. 1974. Notes on the fish fauna of the southeastern Atlantic. A rare species of anglerfish [*Paroneirodes wedli* (Pietschmann, 1926)]. *Vopr. Ikhtiol.*, 14(1):168–170. [In Russian, English translation in *J. Ichthyol.*, 14(1):145–147, 1974.]

Trunov, I.A. 1981. The ichthyofauna of the submarine bank Valdivia (south-eastern Atlantic). *Bull. Soc. Natur. Moscou, Biol.*, 86:51–64.

Trunov, I.A. 1999. New data on species of fish from subantarctic and Antarctic waters of the Atlantic Ocean. *Vopr. Ikhtiol.*, 39(4):460–468. [In Russian, English translation in *J. Ichthyol.*, 39(7):488–497, 1999.]

Trunov, I.A. 2001. The first finding of the blue-lighted anglerfish *Himantolophus azurlucens* (Himantolophidae) in the Atlantic Ocean. *Vopr. Ikhtiol.*, 41(4):560–561. [In Russian, English translation in *J. Ichthyol.*, 41(7):543–544, 2001.]

Tsukahara., H., S. Matsui, T. Honda, Y. Nonogami, and T. Ozawa. 1974. *Data on Fish Collected with Larva Net. Preliminary Report of the Hakuho Maru Cruise KH-73-2*. Ocean Research Institute, Tokyo.

Ueno, T. 1966. [Fishes of Hokkaido. 22. Anglerfishes, frogfishes, and deep-sea anglerfishes (Ceratioidei)]. *Hokusuishi Geppo*, 23(11):532–541. [In Japanese.]

Ueno, T. 1971. List of the marine fishes from the waters of Hokkaido and its adjacent regions. *Sci. Rept. Hokkaido Fish. Exp. Sta.*, 13, 46(3):61–102. [In Japanese.]

Ueno, T., and K. Abe. 1967. [Deep-sea anglerfishes taken from off Kushiro, Hokkaido, Japan.] *Hokusuishi Geppo*, 24(8):274–279. [In Japanese.]

Utrecht, W.L. van. 1957. Een "Diepzeevis" uit de maag van een potvis. *Visserij-Nieuws*, 1957(5):72–73.

Uwate, K.R. 1979. Revision of the anglerfish family Diceratiidae, with descriptions of two new species. *Copeia*, 1979(1):129–144.

Vaillant, L. 1888. *Expéditions Scientifiques du "Travailleur" et du "Talisman" pendant les années 1880, 1881, 1882, 1883. Poissons*. G. Masson, Paris.

Valenciennes, A. 1837. Acanthoptérygiens a pectorales pédiculées. pp. 335–507, in: G. Cuvier and A. Valenciennes, *Histoire Naturelle des Poissons*, Vol. 12. Levrault, Paris and Strasbourg.

Wagner, H.-J. 2001. Sensory brain areas in mesopelagic fishes. *Brain Behav. Evol.*, 57:117–133.

Waite, E.R. 1912. Notes on New Zealand fishes, No. 2. *Trans. N.Z. Inst.*, Wellington, 44(10):194–202.

Waterman, T.H. 1939a. Studies of deep-sea angler-fishes (Ceratioidea). I. An historical survey of our present state of knowledge. *Bull. Mus. Comp. Zool., Harvard*, 85(3):65–81.

Waterman, T.H. 1939b. Studies of deep-sea angler-fishes (Ceratioidea). II. Three new species. *Bull. Mus. Comp. Zool., Harvard*, 85(3):82–94.

Waterman, T.H. 1948. Studies on deep-sea angler-fishes (Ceratioidea). III. The comparative anatomy of *Gigantactis longicirra* Waterman. *J. Morphol.*, 82(2):81–150.

Watson, W., and H.J. Walker. 2004. The world's smallest vertebrate, *Schindleria brevipinguis*, a new paedomorphic species in the family Schindleriidae (Perciformes: Gobioidei). *Rec. Aust. Mus.*, 56:139–142.

Weber, M. 1913. Die Fische der Siboga-Expedition. *Siboga-Exped. Monogr.*, 57, xii + 710 pp.

Weitzman, S.H. 1997. Systematics of deep-sea fishes. pp. 43–77, in: D.J. Randall and A. P. Farrell (editors), *Deep-Sea Fishes*. Academic Press, San Diego.

Weitzman, S.H., and R. Vari. 1988. Miniaturization in South American freshwater fishes: An overview and discussion. *Proc. Biol. Soc. Wash.*, 101(2):444–465.

Wellershaus, S. 1963. Über zwei Tiefseefische in der marinbiologischen Sammlung des Instituts für Meeresforschung in Bremerhaven: *Linophryne lucifer* Collett und *Trigonolampa miriceps* Regan and Trewavas. *Veröffentl. Inst. Meeresf. Bremerh.*, 8:163–166.

Westneat, M.W., M.E. Hale, M.J. McHenry, and J.H. Long, Jr. 1998. Mechanics of the fast-start: Muscle function and the role of intramuscular pressure in the escape behavior of *Amia calva* and *Polypterus palmas*. *J. Exp. Biol.*, 201:3041–3055.

Wheeler, A.C. 1969. *The Fishes of the British Isles and North-west Europe*. Macmillan, London, Melbourne, and Toronto.

Whitley, G.P. 1931. New names for Australian fishes. *Aust. Zool.*, 6(4):310–334.

Whitley, G.P. 1934. Supplement to the check-list of the fishes of New South Wales. In: A.R. McCulloch, *The Fishes and Fish-like Animals of New South Wales*, 3rd edition. Royal Zoological Society of New South Wales, Sydney. [following title page, 12 pages, unpaged.]

Whitley, G.P. 1949. The handfish. *Aust. Mus. Mag.*, 9(12):398–403.

Whitley, G.P. 1954. New locality records for some Australian fishes. *Proc. Roy. Zool. Soc. N.S.W.*, 1952–53:23–30.

Whitley, G.P. 1956. Name-list of New Zealand fishes. pp. 397–414, in: D.H. Graham, *A Treasury of New Zealand Fishes*, 2nd edition. Reed, Wellington.

Whitley, G.P. 1958. Descriptions and records of fishes. *Proc. Roy. Zool. Soc. N.S.W.*, 1956–57:28–51.

Whitley, G.P. 1968. A check-list of the fishes recorded from the New Zealand region. *Aust. Zool.*, 15(1):1–100.

Whitley, G.P. 1970. Ichthyological quiddities. *Aust. Zool.*, 15(3):242–247.

Whitley, G.P. 1976. More fish genera scrutinized. *Aust. Zool.*, 19(1):45–50.

Whitley, G.P., and W.J. Phillipps. 1939. Descriptive notes on some New Zealand fishes. *Trans. Proc. Roy. Soc, N.Z.*, 69(2):228–236.

Whitmee, S.J. 1875. On the habits of the fishes of the genus *Antennarius*. *Proc. Zool. Soc.*, London, 35(7):543–546.

Wickler, W. 1961. Über die Paarbildung der Tiefsee-angler. *Natur und Volk, Ber. Senckenb. Naturf. Gesells.*, 91(11):381–390.

Willemoes-Suhm, R. von. 1876. Von der Challenger-Expedition. Briefe von R.v. Willemoes-Suhm an C. Th. E.v. Siebold. VI. H.M.S. Challenger, auf der Fahrt von Japan nach den Sandwich-Inseln, im Juli 1875. *Zeit. Wiss. Zoöl.*, 26(4):lxxvii–xci.

Williamson, H.C. 1911. Occurrence of *Himantolophus reinhardti*. *Rept. Fishery Bd. Scotland*, 1909, 3:51.

Wilson, D.P. 1937. The habits of the angler-fish *Lophius piscatorius* L., in the Plymouth Aquarium. *J. Mar. Biol. Assoc. U.K.*, 21(2):477–496.

Winterbottom, R. 1974. A descriptive synonymy of the striated muscles of the Teleostei. *Proc. Acad. Nat. Sci., Phil.*, 125(12):225–317.

Winterbottom, R., and A.R. Emery. 1981. A new genus and two new species of gobiid fishes (Perciformes) from the Chagos Archipelago, central Indian Ocean. *Environ. Biol. Fishes*, 6(2):139–149.

Wolff, T. 1960. Strejflys over dybhavets dyreliv. Alle tiders kvantitativt og kvalitativt rigeste trawltræk på dybt vand: Galathea Station 716, nord for Panamabugten, 3570 m dybde. *Naturens Verden*, Juni 1960, pp. 161–191.

Wolff, T. 1961. Animal life from a single abyssal trawling. *Galathea Rept.*, 5:129–162.

Wolff, T. 1967. *Danish Expeditions on the Seven Seas*. Rhodos International Science and Art Publishers, Copenhagen, Denmark.

Woodland, D.J. 2005. Parasitic marine fishes. pp. 250–258, 497–500, in: K. Rohde (editor), *Marine Parasitology*. CSIRO Publishing, Collingwood, Australia.

Yamakawa, T. 1982. [Gigantactinidae and Ceratiidae]. pp. 197–199, 362–363, in: O. Okamura, K. Amaoka, and F. Mitani (editors), *Fishes of the Kyushu-Palau Ridge and Tosa Bay. The Intensive Research of Unexploited Fishery Resources on Continental Slopes*. Japan Fisheries Resource Conservation Association, Tokyo.

Yamakawa, T. 1984. [Melanocetidae, Diceratiidae, and Oneirodidae]. pp. 288–291, 384–385, in: O. Okamura and T. Kitajima (editors), *Fishes of the Okinawa Trough and the Adjacent Waters. The Intensive Research of Unexploited Fishery Resources on Continental Slopes*. Japan Fisheries Resource Conservation Association, Tokyo.

Yamanoue, Y., M. Miya, K. Matsuura, N. Yagishita, K. Mabuchi, H. Sakai, M. Katoh, M. Nishida. 2007. Phylogenetic position of tetraodontiform fishes within the higher teleosts: Bayesian inference based on 44 whole mitochondrial genome sequences. *Mol. Phylog. Evol.*, 45:89–101.

Yasaki, Y. 1928. On the nature of the luminescence of the knight-fish, *Monocentris japonicus* (Houttuyn). *J. Exp. Zool.*, 50(3):495–505.

Yasaki, Y., and Y. Haneda. 1936. Uber einen neuen Typus von Leuchtorgan im Fische *(Acropoma japonimcum)*. *Proc. Imp. Acad. Japan*, 12:55–57.

Young, R.E., and C.F.E. Roper. 1977. Intensity regulation of bioluminescence during countershading in living midwater animals. *U.S. Fish. Bull.*, 75(2):239–252.

Zigno, A. De. 1887. Nuove aggiunte alia ittiofauna dell' epoca Eocena. *Mem. R. Istit. Veneto*, 23:81.

# ILLUSTRATION CREDITS

Sources for the illustrations used in this volume are usually given in the legends for the figures themselves; where more detailed credits are required, they are given below, cited by figure number. Unless indicated otherwise, all drawings reproduced from *Steenstrupia*, *Dana Reports*, and *Galathea Reports* were made by E. Bertelsen. Illustrations for which no attribution is cited, either in the legends or here, are by the author.

Title page illustration, figures 3–5, 24–27, 30, 31, 34–38, 41, 42, 44, 45–47, 50, 51, 54–56, 58–60, 63, 66, 67, 69–74, 77, 82–84, 87, 88, 90, 93–95, 98–99, 101, 107, 108, 115, 120–125, 128, 132, 145, 147, 149, 153, 154, 157–166, 168, 169, 171, 172, 174, 177–179, 182–185, 194–198, 205–207, 247, 257, 274–278, 280–286, 288, 292, 294, 295, 297–301, 306–309: images that appear in various issues of *Steenstrupia*, *Dana Reports*, and *Galathea Reports*, all publications of the Zoological Museum, University of Copenhagen. Used with permission.

Figure 1: Marshall, 1974, *Spectrum, Brit. Sci. News*, 118, unnumbered figure, p. 2.

Figures 6, 42, 245, 248, 304: images that appear in issues of *Vidensk. Meddr. Dansk naturh. Foren.*; courtesy of C.A. Reitzels Boghandel & Forlag A/S, Copenhagen. Used with permission.

Figure 7: Photo by Geert Brovad of a Krøyer painting, from the original in the archives of the Zoological Museum, University of Copenhagen. Used with permission.

Figures 9–11: Lütken, 1878, *Danske Vidensk. Selsk. Skr.*, 5, R. Nat. Math. Afd., 11(5), pls. 1, 2, bound in between pp. 338 and 339; Lütken, 1887, *Danske Vidensk. Selsk. Skr.*, 6, R. Nat. Math. Afd., 4(5), unnumbered pl., foldout following p. 334: courtesy of Eileen C. Mathias and the Ewell Sale Stewart Library, The Academy of Natural Sciences of Philadelphia. Used with permission.

Figure 17: Image Resources, The Natural History Museum, London. Used with permission.

Figure 18: Brusca and Brusca, 1990, *Invertebrates*, fig. 18A, p. 406; courtesy of Richard C. Brusca, Linda VandenDolder, and Sinauer Associates, Inc. Used with permission.

Figure 20: Parr, 1927, *Bull. Bingh. Oceanic Coll.*, Yale University, 3(1):9, 12, 17, figs. 1, 4, 5. Used with permission.

Figures 21, 22: Photographs courtesy of Peter Rask Møller, Jørgen Nielsen, and the Zoological Museum, University of Copenhagen. Used with permission.

Figures 23, 28, 33, 38, 45, 112–114, 137, 138, 140–142, 144, 146, 151, 152, 157–160, 162–165, 193, 197, 200, 253, 270–273, 297, 300: images that appear in various issues of the *Science Bulletin* and *Contributions in Science*, scientific publications of the Natural History Museum of Los Angeles County, Los Angeles; courtesy of Managing Editor K. Victoria Brown. Used with permission.

Figures 32, 45, 60, 62, 65, 76, 88, 102, 105, 109, 114, 117, 125, 126, 130, 131, 134–136, 144, 173, 174, 181, 186, 188, 191, 199, 202, 203, 244, 262–264, 290, 293: images that appear in publications of the American Society of Ichthyologists and Herpetologists; courtesy of Secretary Maureen A. Donnelly and the American Society of Ichthyologists and Herpetologists. Used with permission.

Figure 40: Image taken from the *Bull. Bost. Soc. Nat. Hist.*, 63:8, fig. 3; courtesy of Kimberely Altomere, Michael Morrison, and the Museum of Science, Boston. Used with permission.

Figure 43: Images that appear in *FAO Species Identification Sheets for Fishery Purposes*, Food and Agriculture Organization of the United Nations, Rome; copyright FAO; courtesy of Michel Lamboeuf. Used with permission.

Figure 44: Norman, 1930, *Disc. Rept.*, 2:365, fig. 46; courtesy of Paul Cooper and the General Library, The Natural History Museum, London. Used with permission.

Figures 49, 241, 260, 310: Images published by William Beebe in various issues of *Zoologica* and the *Bulletin of the New York Zoological Society*, copyright now held by the Wildlife Conservation Society, New York; courtesy of Deborah A. Behler and the Wildlife Conservation Society. Used with permission.

Figure 52: Marshall, 1971a, *Explorations in the Life of Fishes*, p. 70, fig. 30, Harvard University Press; courtesy of Scarlett R. Huffman and Harvard University Press. Used with permission.

Figure 53: Marshall, 1967a, *Symp. Zool. Soc. London*, 19:59, fig. 1; courtesy of Jennifer Jones and Elsevier Limited. Used with permission.

Figure 70: Maul, 1961, *Bol. Mus. Mun. Funchal*, 14(50):97, fig. 2; courtesy of Manuel Biscoito and the Museu Municipal do Funchal, Madeira. Used with permission.

Figures 78, 279–281, 284–286, 291–298, 301, 302, 304, 309: Pietsch, 2005b, *Ichthy. Res.*, 52(3):212–228, figs. 2–24; courtesy of Seishi Kimura, Yoshiaki Kai, and the Ichthyological Society of Japan. Used with permission.

Figures 89, 176: Bruun et al. (editors), 1956, *The Galathea Deep Sea Expedition 1950–1952, Described by Members of the Expedition*, unnumbered figures on pp. 86, 175; courtesy of George Allen and Unwin, Ltd. Used with permission.

Figures 91, 92, 109, 170, 180, 249, 258: images of deep-sea anglerfishes in the collection of the Department of Ichthyology, Museum of Comparative Zoology; courtesy of Karsten Hartel, Linda Ford, and the Museum of Comparative Zoology, Harvard University. Used with permission.

Figures 96, 97, 103, 106, 118, 267–269: images taken from various issues of *Breviora* and the *Bulletin of the Museum of Comparative Zoology*; courtesy of Karsten Hartel, Linda Ford, and the Museum of Comparative Zoology, Harvard University. Used with permission.

Figures 100, 110, 111, 119, 125, 127: images published in various issues of *Arch. FischWiss.*; courtesy of Ralf Thiel, Walter W. Kühnhold, and the *Journal of Applied Ichthyology*. Used with permission.

Figure 116: from the picture collection, Fish Division, U.S. National Museum of Natural History; courtesy of Lisa Palmer and the Smithsonian Institution, Washington, D.C. Used with permission.

Figures 129, 139, 143, 150, 287: previously unpublished drawings by artists employed by the Zoological Museum, University of Copenhagen; courtesy of Peter Rask Møller, Jørgen Nielsen, and the Zoological Museum, University of Copenhagen. Used with permission.

Figures 133, 242, 305: National Geographic Image Collection; courtesy of Ashley Morton, Susan Riggs, Mimi Dornack, and the National Geographic Society. Used with permission.

Figure 156: Courtesy of H. J. Walker, R. H. Rosenblatt, and the Scripps Institution of Oceanography. Used with permission.

Figure 175: Courtesy of Gary T. Takeuchi and the Department of Vertebrate Paleontology, Natural History Museum of Los Angeles County, Los Angeles, California.

Figure 186: Collette, 1995, *Ichthyol. Explor. Freshwaters*, 6(4):335, fig. 2; courtesy of Bruce B. Collette, Friedrich Pfeil, and IEF. Used with permission.

Figure 187: Gregory, 1933, *Trans. Am. Phil. Soc.*, 23(2), plate 2, foldout opposite p. 411; courtesy of Shannon L. Ryder and the Krieger Publishing Company. Used with permission.

Figure 189: Gregory and Conrad, 1936, *Am. Nat.*, 70(728):206, fig. 5; courtesy of Judy Choi and the University of Chicago Press. Used with permission.

Figure 190: Gregory, 1951, *Evolution Emerging*, p. 312, fig. 9.155; courtesy of The Macmillan Company. Used with permission.

Figures 201, 246, 298: images appearing in issues of *Acta Zoologica*, Stockholm; courtesy of Laura Wilson and Wiley Blackwell. Used with permission.

Figure 240: Herring and Moran, 1978, *Bioluminescence in Action*, p. 322, fig. 9.12; courtesy of Peter Herring, Laura Gould, and Elsevier Limited. Used with permission.

Figure 250: Herring, 1983, *Proc. Roy. Soc. London*, B, 220(1219):204, fig. 3; courtesy of Peter J. Herring, Jennifer Kren, and the Royal Society of London. Used with permission.

Figures 251, 259: Hansen and Herring, 1977, *J. Zool., London*, 182:106, 111, figs. 2, 3; courtesy of Peter Herring, Laura Wilson, and Wiley Blackwell. Used with permission.

Figure 255: Haygood, 1993, *Crit. Rev. Microbiol.*, 19(4):198, fig. 6; courtesy of Dan Short, Taylor and Francis, Inc., and the Copyright Clearance Center. Used with permission.

Figures 263, 264: Courtesy of Bruce H. Robison and the Monterey Bay Aquarium Research Institute, Moss Landing, California. Used with permission.

Figure 266: Paxton and Lavenberg, 1973, *Aust. Zool.*, 18(1):48, fig. 1; courtesy of Dan Lunney and the *Australian Zoologist*. Used with permission.

Figure 289: Pietsch, 1975a, *Nature*, 256:38, fig. 1; courtesy of Jessica Rutt and Macmillan Publishers, Ltd. Used with permission.

# INDEX

A number in boldface type indicates the first page of a major discussion or critical description. Those followed by "t" or "k" denote tables or keys to the identification of a taxon, respectively. Ceratioid families, genera, and species that are recognized here as valid taxa are indicated in boldface to distinguish them from taxa that are regarded as synonyms. Personal names are indexed as they appear in the text, but names of those that appear only as authors of taxa, collectors of specimens, or authors in the References are not indexed.

abductor, pectoral musculature, 254; fig. 268
*abyssicola, Bathyteuthis*, 250
abyssal-benthic, 213, 272, 314, 315
abyssopelagic zone, fig. 1
*acanthias, Dolopichthys*, 410
*acanthias, Dolopichthys (Dermatias)*, 410
*acanthias, Dolopichthys (Monoceratias)*, 410
*acanthias, Monoceratias*, 109, 410
***acanthias*, Oneirodes**, 10, 108, 197, 199, 218, 219, 235t, 256, 275, 397, **410**, 411; figs. 23, 193, 224, 253, 254
*acanthirostris, Rhynchoceratias*, 76t, 367
*acanthirostris, Xenoceratias*, 367
***acanthognathus*, Lophodolos**, 96, 97, 214, 215, 380k, **381**; figs. 33A, 97A, 98B, 190, 219
*acanthognathus, Lophodolus*, 381
*acanthognathus, Lophodulus*, 382
*Acanthoptérygiens à Pectorales Pédiculées*, 173
Acanthopterygii, 175
accessory escal glands, 234, **237**
***Acentrophryne***, 28t, 150, 151k, 155, **157**, 164, 181, 225, 300, 478k, **484**; fig. 175
*Aceratias*, 149, 481, 487
Aceratiidae, 11, 17, 149, 154, 277, 278, 315, 319, 331, 359, 376, 477
*acinosa, Caulophryne*, 128, 129, 130t, 445; fig. 149
acoustico-lateralis system, **29**, 128, 130, 137, 262, 268; figs. 29, 30
adductor mandibulae, 265, 443
*adipatus, Trematorhynchus*, 100t
adrenalin, 231, 233, 244, 246
Aegaeonichthyinae, 331
*Aegaeonichthys*, 331, 333
*Aegeonichthys*, 331, 333
*Aegoeonichthys*, 331, 333
*Aegonichthys*, 331, 333
Aeschynichthyidae, 352
*Aeschynichthys*, 354
Agassiz, A, 10
*Agyropelecus*, fig. 1
*Albatross*, U.S. Fish Commission Steamer, 10, 89, 128, 436
*albifilosus, Dolopichthys*, 112, 424; fig. 118
**albinares group, *Himantolophus***, 64t, 70, 210, 333k, **341**
***albinares*, Himantolophus**, 63, 64t, 70, 207, 210, 234, 235t, 248, 250, 342k, **342**; figs. 34A, B, 213

*Alepisaurus ferox*, 275, 498
algal, stomach contents, 274
***algibarbata*, Linophryne**, 15, 163t, 171, 227, 228, 245, 287t, 302, 498k, **508**, 509; figs. 238, 308B, 309B
***alius*, Oneirodes**, 111, 218, 220, 417k, **417**; fig. 223
*Allector*, 487
***allector*, Dolopichthys**, 10, 13, 97, 118, 120, 220, 249, 250, 382, 432k, 434, **435**; figs. 14–16, 227
Alpha Helix, Southeast Asian Bioluminescence Expedition, 130, 136, 403, 413, 418, 419
*altirostris, Rhynchoceratias*, 64t, 350
*Amacrodon*, 81, 82, 373
*amazonica, Thalassophryne*, fig. 186C
ambush predation, 119
*americanus, Lophius*, 175; fig. 40B
amphipods, 249, 250, 261, 262
***amphirhamphus*, Lasiognathus**, 83, 85, 370k, **372**; figs. 85, 86, 217
*Amphitretus*, fig. 1
anal opening, sinistral, **30**, 150, 189; fig. 31
*analogus, Dolopichthys*, 426
*analogus, Dolopichthys (Microlophichthys)*, 426
*ancistrophorus, Lasiognathus*, 82, 83, 371
***andersoni*, Linophryne**, 148, 162, 163t, 166, 227, , 492k, **492**
***andracanthus*, Microlophichthys**, 114, 115, 220, **427**; figs. 123, 295D
Andresen, P., fig. 13
anguilliform locomotion, 256
*anisacanthus, Dolopichthys*, 108, 404
*anisacanthus, Dolopichthys (Dermatias)*, 398, 404
***anisacanthus*, Oneirodes**, 106, 108, 218–220, 396k, 398, **404**; fig. 224
*Anomalophryne*, 487
Anomalopidae, fig. 255
Antarctic Polar Front, 212, 222
Antarctic Silverfish, 364
*antarcticum, Pleuragramma*, 364
Antennariidae, 177, 178k, 179, 249, 259, 260; fig. 188
Antennarioidea, 315
Antennarioidei, 3, 33, 157, 173, 177, 178k, 179, 181; figs. 2B–E, 188, 202, 203
*Antennarius*, 8, 13, 263; fig. 189
*Antennarius commerson*, figs. 2B, 2C

*Antennarius maculatus*, 249; fig. 265
*Antennarius striatus*, fig. 2D
*Antigonia capros*, 175
Aoerratiidae, 477
aphakic space, 19, 42, 289; fig. 51
*Aphanopus carbo*, 162, 275
aphotic zone, fig. 1
***apogon*, Borophryne**, 148, 159, 162, 199, 254, 287t, 301, 305, 484, **486**; figs. 176–178, 236, 297D, 301C, 305
**appelii group, *Himantolophus***, 64t, 69, 333k, 334k, **339**, 350
*appelii, Aegaeonichthys*, 339
*appelii, Aegoeonichthys*, 339; fig. 12
*appelii, Aegonichthys*, 339
***appelii*, Himantolophus**, 10, 13, 63, 64t, 65, 69, 207, 209, 210, 275, **339**, **350**; figs. 12, 68, 72D, 212
*appendixus, Oneirodes*, 402
*arborifer, Linophryne*, 496; fig. 190
arborifer group, *Linophryne*, 497
arborifera group, *Linophryne*, 497; figs. 5, 50C, 51C, 54
*arborifera, Haplophryne*, 497
***arborifera*, Linophryne**, 162, 163t, 166, 226, 228, 229, 231, 236t, 239, 245, 246, 247, 287t, 300, 491k, 493, 494, **496**; figs. 35, 179, 182C, 184, 206, 239, 247, 259
*arborifera, Linophryne arborifera*, 496
Arctic Polar Front, 212, 222
*arcturi, Diabolicum*, 499
*arcturi, Diabolidium*, 163t, 229, 236t, 499
***arcturi*, Linophryne**, 163t, 170, 228, 229, 236t, 247, 498k, **499**; figs. 47C, E, 183D, E, 237, 241, 257
***argyresca*, Linophryne**, 162, 163t, 170, 227, 287t, 498k, **502**; figs. 297E, 306
*atlantidis, Cryptosparas*, 58t, 330
*Atlantis*, research vessel, 112, 233
*Atolla*, fig. 1
*atratus, Dolopichthys*, 97, 383
*atratus, Dolopichthys (Pentherichthys)*, 383
***atratus*, Pentherichthys**, 97, 98, **383**; figs. 33B, 99–101, 220, 295A
*atriconus, Chaenophryne*, 99, 100t, 388
*axeli, Galatheathauma*, 82, 83, 270, 373, 376; figs. 89, 92
***axeli*, Thaumatichthys**, 82, 87, 270, 294, 374k, **376**; figs. 36C, E, 37, 89, 92, 218, 275

549

*axeli, Thaumatichthys (Galatheathauma)*, 376
***azurlucens, Himantolophus***, 63, 64t, 70, 210, 235t, 247, 250, 254, 346k, **348**; figs. 49, 260

***bacescui, Caulophryne***, 129, 130, 130t, 134, 445k, **446**; fig. 230
bacteria, bioluminescent, 4, 7, 37, 40, 138, 179, 234, 237–239, **241**, 244, 246, 251, 269; figs. 240, 253–255
bacterial luciferase, 234
*balushkini, Gigantactis*, 138, 459
***balushkini, Phyllorhinichthys***, 116, 236t, 428k, **429**; figs. 125B, 127A, 226
barbel, hyoid, of *Centrophryne*, 40, 55, 56, 164, 245, 319, **321**; fig. 60; of *Linophryne*, 7, 40, 42, 162, 164, 165, 229, **245**, 247, 251, 314; figs. 47, 184, 240, 257–259; of stomiiforms, 246, 247
Barbugede Tudsefiske, 8
bare-bellied toadfishes, 8
***basili, Oneirodes***, 106, 111, 417k, **419**
bass, largemouth, fig. 40A
*bassani, Histionotophorus*, 157, 485
batfishes, 6; fig. 2G
*Bathyceratias trilychnus*, 63, 229, 235t, 328; fig. 242
bathypelagic, 3, 8, 17, 51, 197, 200, 210, 228, 250, 253, 256, 260, 261, 314, 315; fig. 1
*Bathypterois*, fig. 1
bathysphere, of Beebe, 229, 256, 331; fig. 242
*Bathyteuthis abyssicola*, 250
Batrachoidea, 175
Batrachoidei, fig. 187
Batrachoididae, 173; fig. 186
Batrachoidiformes, 173, 175, 195; fig. 186
*Batrachus*, 173
Bean, T. H., 10
**Bearded Seadevils**, 152k, **162**, **487**
Beebe, W., 11, 229, 245, 247, 248, 254, 256, 277, 331; figs. 241, 242
***beebei, Lasiognathus***, 82, 84, 85, 212, 370k, **370**; figs. 88A, B, 217
Belloc, G., 11
*Benthalbella infans*, 251
benthic ambulatory function, 51, 181, 315
benthic life style, 13, 15, 17, 51, 85, 90, 173, 181, 211–213, 228, 261, 270, 272
*Benthodytes typica*, 272
benthopelagic, 228; fig. 1
***Bertella***, 28t, 90, 94k, **121**, 214, **221**, 295, 296, 305, 307, 378k, **437**; figs. 133–136, 202, 203
Bertelsen, E., ix, 11, 18–22, 122, 148, 173, 179, 180, 197, 200, 233, 278–281, 302; figs. 21, 22, 37
*Bertelsenna*, 103, 392
*Berthella*, 437
Berycidae, 29
*bicornis, Chaenophryne*, 99, 100t, 386
***bicornis, Linophryne***, 163t, 170, 227, 228, 287t, 302, 498k, **507**; figs. 183F, 237
*bifilis, Mancalias*, 58t, 327
*biflagellatus, Lophodolos*, 97, **382**
*biflagellatus, Lophodolus*, 382
Bingham, H. P., fig. 20
Bingham Oceanographic Foundation, 17; fig. 20
*binghami, Amacrodon*, 82, 375
***binghami, Thaumatichthys***, 81, 82, 87, 212, 213, 271, 294, 374k, **375**, 376; figs. 36A, B, 37, 40D, E, 45D, 90, 91, 93, 218, 276, 277
binocular vision, 19, 42, 303; fig. 51

bioluminescence, ix, 4, 9, 21, 58, 61, 162, **229**, 235t, 262, 268, 277, 303; figs. 240, 241, 250, 251, 255, 261
***bipennata, Linophryne***, 163t, 170, 227, 499k, **503**
*bispinossus, Melanocetus*, 76t, 366
*bispinosus, Ceratias*, 354
*bispinosus, Ceratias (Diceratias)*, 71, 75, 354
***bispinosus, Diceratias***, 10, 72, 75, 211, 354k, **354**; fig. 215
Black Scabbard Fish, 162, 274
**Black Seadevils**, 53k, **75**, 79, **359**
Blennioidei, 175
blood vessels, intra-escal, **234**; fig. 248
Blue Planet video series, 294
Blue-lighted Anglerfish, 254
body shape, of females, **30**; of larvae, **48**, 190, 254; figs. 55, 56
*borealis, Himantolophus*, 64t, 70, 210, 342k, **345**; fig. 214
Borodin, N. A., 11
***Borophryne***, 17, 28t, 150, 152k, 155, **158**, 164, 165, 225, 298, 300–303, 478k, 479k, **486**; figs. 202, 203
Bothidae, 150
boxfishes, 175
Brachionichthyidae, 177, 178k; fig. 188
*brachysomus, Lophius*, 157, 485
***bradburyae, Oneirodes***, 111, 417k, **421**
brain, 45; figs. 52, 53
branchiostegal rays, **25**; fig. 16B
Brauer, A. B., 11, 17, 148, 233, 234; figs. 243, 256
*braueri, Xenoceratias*, 76t, 362
***brevibarbata, Linophryne***, title page, 163t, 166, 228, 239, 287t, 246, 491k, **493**, 497; figs. 4, 238, 251
*brevibarbis, Linophryne*, 163t; fig. 20C
*brevifilis, Dolopichthys*, 398, 401
*brevifilis, Dolopichthys (Dermatias)*, 400
***brevirostris group, Himantolophus***, 64t, 334k, **350**; figs. 40F, 41, 73A–C
*brevirostris, Rhynchoceratias*, 64t, 321, 350, 382, 384, 394, 430; fig. 73B
*brevirostris, Xenoceratias*, 76t, 362; fig. 82B
British Broadcasting Corporation, 294
Bruun, A., 82, 159, 199, 270, 272
*budegassa, Lophius*, 15, 157, 485
*Bufichthys*, 72
***Bufoceratias***, 17, 28t, 72, 74k, **75**, 84, 199, 261, 291, 315, 354k, **356**; figs. 43B, 76, 202, 203
***bulbosus, Oneirodes***, 106, 107, 218, 219, 397k, **403**; fig. 223
burst swimming, 256

Caproidae, 175
*capros, Antigonia*, 175
*Caranactis*, 394
*Caratias*, 323
*carbo, Aphanopus*, 162, 275
*Carcharodon*, fig. 1
Carlsberg Foundation, 11
*carlsbergi, Dolopichthys*, 398
*carlsbergi, Dolopichthys (Dermatias)*, 107, 398
***carlsbergi, Oneirodes***, 106, 107, 197, 200, 218–220, 395k, **398**; figs. 114E, 225
carnivorous plants, 81
caruncles, 40, 48, 58, 188, 229, 231, **244**, 303, 314, 317, 322; figs. 44A, B, 240, 250, 256
*carunculatus, Ceratias*, 58t, 330
*carunculatus, Cryptosparas*, 330
*catodon, Physeter*, 275
*caucasica, Eosladenia*, 157, 485
caudal skeleton, **25**, 187; figs. 11, 15, 28, 200

***Caulophryne***, 19, 28t, 129, 130, 130t, **133**, 133k, 260, 285t, 297, 307, 445k, **445**; figs. 30C, 148–151, 202, 203, 292D, E, 297A
**Caulophrynidae**, 19, 21, 28t, 53k, 128, 133, 133k, 141, 194, **221**, 234, 254, 261, 262, 285t, 297, 305, 314, 316k, 317k, **443**; figs. 148–152, 189, 191, 202, 203
Caulophryninae, 443
*Centrocetus*, 360
***Centrophryne***, 28t, **57**, 84, 103, 143, 303, **320**; figs. 57–60, 202, 203
**Centrophrynidae**, 19, 21, 28t, 53k, **55**, 103, 192, **201**, 250, 280, **281**, 306, 317k, **319**; figs. 57–60, 191, 202, 203
cephalopods, 234, 261
Ceratiadae, 315
*Cératiades*, 9, 315
Ceratiaeformes, 322
Ceratianae, 322
***Ceratias***, 9, 23, 28t, 55, **62**, 62k, 84, 229, 234, 237, 244, 251, 267, 280, 281, 282t, 283t, 290, 291, 301–304, 307, 323k, **323**; figs. 61–63, 44A, 48B, 50A, 51A, 63, 202, 203, 286, 288
**Ceratiidae**, 9, 11, 21, 28t, 52k, **57**, 58t, 192, **201**, 235t, 242, 263, **281**, 282t, 303, 314, 316k, 317k, **322**; figs. 61–67, 189, 191, 202, 203
Ceratiiformes, 315, 331, 359
Ceratiina, 315
Ceratiinae, 322
Ceratina, 315
Ceratioidae, 322
Ceratioidea, 315, 331, 359, 368, 376
*Ceratocaulophryne*, 445
Cetomimoidei, 3, 173
***Chaenophryne draco* group**, 99, 100t, 102, 379k, 385k, **387**; figs. 108, 295B
***Chaenophryne longiceps* group**, 99, 100t, 102, 379k, 385k, **386**; fig. 107
***Chaenophryne***, 21, 28t, 90, 91, 92k, **98**, 100t, 214, **216**, 220, 237, 247, 248, 260, 378k, 379k, **384**; figs. 102–108, 202, 203, 250
chaetognaths, 250, 260, 263
*Challenger*, HMS, 10, 71, 229
*Chauliodus*, fig. 1
Chapman, W. M., 11
Chaunacidae, 173, 177; fig. 188
Chaunacoidei, 3, 33, 177, 178k, 181; figs. 2F, 39A, 188B, 202, 203
*Chaunax umbrinus*, fig. 2F
*cheloniae, Allector*, 497
chemical attractant, 3, 7, 237, 270; communicant, 250
Chiasmodontidae, 99
Chironecteoidei, 322, 331
***Chirophryne***, 21, 28t, 90, 95k, 111, 123, **125**, 378k, **440**; figs. 137B, 140, 141, 202, 203
cinematographic analysis, of feeding, 263; of locomotion, 256, 259; figs. 262–264
*cirrhifer, Stephanolepis*, 175
*cirrifer, Dolopichthys*, 400
*cirrifer, Dolopichthys (Dermatias)*, 400
*cirrifer, Melanocetus*, 76, 76t, 362
cladistic analysis, 180, 181, 306; figs. 188, 202, 203
Clarke, F. E., 10; fig. 12
***clarkei, Oneirodes***, 108, 396k, **407**
*claviger, Dolopichthys*, 401
*claviger, Dolopichthys (Dermatias)*, 400
coelenterates, 261
coffinfishes, fig. 2F
Collett, R., 10, 148, 245

550 INDEX

*colletti*, *Linophryne*, 398
*columnifera*, *Chaenophryne*, 99, 100t, 388
**Combfin Dreamers**, 94k, **128**, **442**
*commerson*, *Antennarius*, fig. 2B, C
**Common Dreamers**, 92k, **105**, **394**
*compressus*, *Corinophorus*, 341
*compressus*, *Corynophorus*, 64t, 70, 349
***compressus*, *Himantolophus***, 63, 64t, 70, 210, 346k, **349**
Conrad, G. M., 3
copepods, 250, 260, 262
***cornifer* group, *Himantolophus***, 64t, 70, 210, 333k, **346**
***cornifer*, *Himantolophus***, 64t, 70, 207, 210, 346k, **346**; figs. 45B, 69B, 72E, 213
*cornutus*, *Dolopichthys*, 121, **437**
*cornutus*, *Oneirodes*, 437
*coronata*, *Linophryne coronata*, 500
***coronata*, *Linophryne***, 163t, 170, 226–228, 236t, 287t, 288t, 302, 498k, **499**; figs. 20B, 47B, D, 180, 183B, 238
*Coryceus*, 260
*corymbifera*, *Linophryne*, 149, 163t, 236t, 489
*Corynolophus*, 333
*Corynophorus*, 333
*couesi*, *Ceratias*, 330
*couesii*, *Ceratias* (*Cryptopsaras*), 330
***couesii*, *Cryptopsaras***, 10, 57, 58t, 63, 197, 199, 207, 210, 229, 233, 235t, 241, 248, 251, 254, 275, 280, 283t–285t, 289, 290, 298, 304, **329**; figs. 24, 44B, 48C, 50B, 51B, D–G, 64–67, 210, 240, 250, 255, 256, 261, 282, 283, 286, 289, 290, 291A
*couesii*, *Cryptopsarus*, 330
*couesii*, *Cryptosaras*, 330
*couesii*, *Cryptosparas*, 330
counter-shading, bioluminescent, 229, **248**; fig. 261
Crane, J., 11
*crenata*, *Chaenophryne*, 99, 100t, 386
***crinitus*, *Himantolophus***, 64t, 68, 207, 209, 334k, **338**; figs. 72B, 212
*cristatus*, *Dolopichthys*, 409
*cristatus*, *Dolopichthys* (*Dermatias*), 108, 409
***cristatus*, *Oneirodes***, 108, 396k, **409**
*crocodilus*, *Lampanyctus*, 261
*crossota*, *Chaenophryne*, 99, 100t, 386
*crossotus*, *Chaenophryne*, 386
crustaceans, 250, 261–263, 275
*Cryptolychnus*, 487
***Cryptopsaras***, 28t, 62, 62k, **63**, 84, 130, 229, 231, 237, 244, 260, 281, 290, 291, 302, 303, 307, 323k, **328**; figs. 64–67, 202, 203, 240
*Cryptopsarus*, 328
Cryptorostrinae, 452, 477
*Cryptosaras*, 328
*Cryptosarus*, 328
*Cryptosparas*, 328
C-start escape response, 255, 256
***Ctenochirichthys***, 21, 28t, 90, 94k, 111, 123, **128**, 181, 378k, 379k, **442**; figs. 137D, 145–147
Cuvier, G., 173
*Cyclothone*, fig. 1

*Daector schmitti*, fig. 186D
Dahlgren, U., 234
**Dana Dreamers**, 93k, **112**, **423**
*Dana*, collections, 11, 18, 21, 63, 138, 152, 162, 197, 260, 300; expeditions, 11, 21, 197, 207, 253, 254, 277; fig. 207; Royal Danish Research Ship, 11, 89, 113, 233
***danae*, *Dolopichthys***, 118, 119, 431k, **434**; figs. 131E, 190, 228

***danae*, *Himantolophus***, 63, 64t, 68, 208, 334k, **337**; fig. 71C
***Danaphryne***, 28t, 90, 93k, **112**, 378k, **423**; figs. 118–120, 202, 203
deep-sea toadfishes, 8
Deep-sea Swallower, 275
*Delaware*, research vessel, 246
***densiramus*, *Linophryne***, 163t, 166, 226, 228, 288t, 491k, **494**; figs. 47A, 238
denticular apparatus, of males, 3, 40, **45**, 181, 190, 263, 277–279, 289, 291, 294, 300, 304, 305, 314; figs. 5, 6, 41B, C, 54, 283, 293, 294, 299
denticular teeth, of males, 23, **45**, 263, 289, 294, 305, 314, 315; figs. 3, 5, 41B, C, 54, 283, 293, 294
*denticulatus*, *Nannoceratias*, 163t, 497
depressor dorsalis muscle, of the illicium, 40, 267; fig. 46
***Dermatias***, 28t, 89, 90, 92k, **111**, 118, 181, 291, 378k, **422**; figs. 116–117
descending escal appendage, of *Chaenophryne*, 237; figs. 106D, 248
*Diabolicum*, 487
*Diabolidium*, 487
*diadematus*, *Dolopichthys*, 400
*diadematus*, *Dolopichthys* (*Dermatias*), 400
*Diceratia*, 352
***Diceratias***, 17, 28t, 71, 72, 74k, **75**, 84, 199, 260, 353k, **354**; figs. 75, 202, 203
**Diceratiidae**, 21, 28t, 53k, **71**, 74, 138, 192, 197, **211**, 242, 244, **291**, 305, 316k, 317k, **328**; figs. 44C, 75–77, 189, 191, 202, 203, 292B
*dicromischus*, *Oneirodes*, 110, 396k, **413**
*digitatus*, *Dolopichthys*, 401
*digitatus*, *Dolopichthys* (*Dermatias*), 400
*digitopogon*, *Linophryne*, 163t, 501
dilatator operculi muscle, 265
***dinema*, *Dolopichthys***, 120, 432k, **436**; figs. 130, 131D, 228
*dinema*, *Lophodolus*, 96, 380
*Dionaea*, 81
*diphlegma*, *Linophryne coronata*, 163t, 500
*diplegma*, *Linophryne coronata*, 500
*Discovery*, research vessel, 97, 239, 246, 254, 294
disphotic zone, fig. 1
distribution, benthic, 13, 15, 17, 51, 85, 90, 173, 181, 211–213, 228, 314; of juveniles and adults, 199; of larvae, 198; relative to physical, chemical, and biological parameters, 200; seasonal, 197; vertical, 198
*doliatus*, *Siganus*, 175
*dolichonema*, *Acentrophryne*, 148, 157, 158, 484k, **485**; figs. 173, 236
***Dolopichthys***, 21, 28t, 82, 89, 90, 94k, 103, 106, 111, 114, **118**, 121, 123, 126, 214, **220**, 233, 250, 260, 262, 378k, 379k, 394, **430**; figs. 129–132, 190, 202, 203, 295E
**Doublespine Seadevils**, 53k, **71**, 74k, 75, **352**
**Doublewart Seadevils**, 62, 62k, **323**
***draco* group, *Chaenophryne***, 99, 100t, 102, 379k, 385k, **387**; figs. 108, 295B
***draco*, *Chaenophryne***, 99, 100t, 102, 216, 234, 235t, 236, 237, 388k, **388**, 390; figs. 25C, 33C, 106B, 221, 248
**Dreamers**, 54k, **87**, **376**
***duhameli*, *Spiniphryne***, 103, 392k, **393**; figs. 109B, 222
dwarfism, of males, ix, 3, 11, 18, 23, 51, 152, 173, 189, 315, 277; fig. 17
Dybhavs-Tudsefiske, fig. 10
*Dysalotus*, 99

*edentula*, *Aceratias*, 163t, 509
*Edriolychnus*, 17, 149, 154, 481
egg raft, 177, 198, 300; fig. 206
elasmobranchs, 55
***elsmani*, *Gigantactis***, 145, 224, 456k, **465**; fig. 233
emission spectra, 238, 251; fig. 250
Endo, H., 294
endocrine glands, 280
endocrinology, 4, 8
Eocene, of Monte Bolca, 157, 485
*Eosladenia caucasica*, 157, 485
epaxial musculature, 369
epipelagic, 198–200, 210, 213, 291; fig. 1
***epithales*, *Oneirodes***, 110, 396k, **415**
epurals, 27, 187; fig. 28
erector dorsalis muscle, of the illicium, 40, 267; fig. 46
esca, definition of, 6, 15, 37, 314; diagnostic value of, 6, 9, 23, 37; external structure, 9, **37**, 188, 249; figs. 10C, E, F, 14E–G, 45, 59, 66, 72, 240, 245A, 247, 248A, 249; internal structure, 7, 188, 229, **234**; figs. 201, 240, 243, 244, 245B, 246, 248B–F; presence in larvae, 19, 237; species-specificity, 37, 250
escal, accessory glands, 234, **237**; appendages, 6, 9, 23, 24, **37**, 249, 251, 262, 263, 314; figs. 113, 240, 245, 247, 248; bacteria, 4, 7, 37, 40, 138, 179, 234, 237–239, **241**, 244, 246, 251, 269; figs. 240, 253–255; denticles, of thaumatichthyids, 81–87, 188, 314; figs. 86, 88, 93; descending appendage, of *Chaenophryne*, 237; figs. 106D, 248; innervation, **234**, 263; light-guiding structures, 7, 39, 164, **237**, 239, 241, 244, 249, 262; figs. 201, 240, 245B, 246, 247; light-reflecting structures, 7, 39, 234, **237**, 241, 249, 424, 429, 507; musculature, **234**, 239, 241, 251; pore, 37, 40, 188, 234, 237, 241, 244, 251, 314; figs. 201, 240, 245, 246
escape response, 259; C-start, 255, 256
***escaramosa*, *Linophryne***, 163t, 170, 227, 498k, **503**; fig. 183G
Eschricht, D. F., 8, 9, 89
*eschrichtii* group, *Oneirodes*, 106, 111, 398, 400, 403, 404, 416, 420, 422
***eschrichtii*, *Oneirodes***, 9, 13, 89, 105, 106, 107, 111, 218–220, 234, 235t, 237, 249, 256, 275, 397k, 398, **399**, 402; figs. 10F, 18A, 18B, 20A, 112, 114C, 223, 245
*eschrichtii*, *Oneiroides*, 400
*eschrichtii*, *Onirodes*, 400
euphotic zone, fig. 1
*eupogon*, *Linophryne*, 163t, 496
*eupogon*, *Linophryne arborifera*, 496
*Euprymna scolopes*, 241
Eurostrinae, 319, 331, 359, 376, 477
Eurypharyngidae, 275
*eustales*, *Melanocetus*, 366
***eustalus*, *Melanocetus***, 76t, 77, 80, 212, 235t, 360k, **366**; figs. 79, 80B, 216, 252
*exiguus*, *Dolopichthys*, 426
*exiguus*, *Dolopichthys* (*Microlophichthys*), 426
*exiguus*, *Trematorhynchus*, 422
exocrine glands, 237
*exodon*, *Gigantactis*, 138, 459
expatriates, geographic, 208, 210, 211, 227, 228
extrinsic muscles, of the illicium, 40, 249, 251, 263, 267, 269; figs. 15A, 42, 46, 273

eyes, bowl-shaped, 19, 42, 61, 189, 192; of females, 19, 42, 262, 274; fig. 51F, G; of larvae, 19, 42; fig. 51D–F; of males, 3, 19, **42**, 189, 192, 250, 302; figs. 5, 50, 51A–E, 286, 294; telescopic, tubular, 19, 42, 150, 165, 189, 192, 300; fig. 51

facultative sexual parasitism, 48, 91, 130 **195**, 280, **305**, 306t; figs. 202, 203
**Fanfin Seadevils**, 53k, 133k, **133**, **443**
**Fangtooth Seadevils**, 151k, **157**, **484**
**Fattail Dreamers**, 92k, **111**, **422**
feeding mechanism, gape-and-suck, **263**; of Gigantactinidae, **267**; gripping, 263; snagging, 31, 269; snapping, 263, 273; suction, 263; of Thaumatichthyidae, 270; trapping, 31, 273, 274
*ferox*, *Alepisaurus*, 275, 498
*ferox*, *Melanocetus*, 76, 76t, 236t, 362, 364, 366
fiber optics, 234; fig. 240
Filhol, H., 13; fig. 19
*filibolosus*, *Gigantactis*, 472
*filibulbosus*, *Gigantactis*, 138, 146, **472**
*filibulosus*, *Gigantactis*, 472
*fimbriata*, *Chaenophryne*, 99, 100t, 390
fins, fin rays, **27**, 42, 187, 314, 316; figs. 11, 15, 16C, 28
*firthi*, *Typhloceratias*, 58t, 63, 328
fishing-frog, 13, 15
*flagellifer* group, *Oneirodes*, 106, 413
*flagellifer*, *Dolopichthys*, 413
*flagellifer*, *Dolopichthys* (*Dermatias*), 110, 413
***flagellifer***, ***Oneirodes***, 106, 110, 218, 220, 397k, **413**; figs. 45E, 114F, 225
flatfishes, 150
food, diet, 250, 251; of females and larvae, **260**; of males, **263**
**Footballfishes**, 54k, **63**, 68, **331**
fosforescens, 229
fossils, **157**, **485**; fig. 175
*fraenatus*, *Sufflamen*, 175
Fraser-Brunner, A., 11
frogfishes, 51, 253; fig. 2B–E
*frondosus*, *Dolopichthys*, 400
*frondosus*, *Dolopichthys* (*Dermatias*), 400
frontal bones, **30**

*Galathea* Deep Sea Expedition, 17, 82, 272
*Galatheathauma*, 82, 368, 373
Galatheathaumatidae, 368
*galeatus*, *Chaenophryne*, 100t, 102, **392**
gape-and-suck feeding, **263**
gapers, fig. 2F
***gargantua***, ***Gigantactis***, 145, 224, 225, 251, 268, 269, 456k, **466**; figs. 155, 234
*gargantua* group, *Gigantactis*, 456k
Garman, S. W., 10, 13, 15, 37, 89, 249; figs. 14–16
gastropods, 261
**Ghostly Seadevils**, 151k, **154**, **481**
*gibbsi*, *Gigantactis*, 144, 224, 225, 456k, **461**; fig. 234
Gigactinidae, 452
**Gigantactinidae**, 11, 21, 22, 28t, 52k, 136, **138**, 141, 141k, 194, **222**, 236t, 250, 254, 263, 278, **298**, 303, 305, 314, 316k, 317k, **452**; figs. 155–166, 189, 191, 202, 203
**Gigantactis**, 17, 22, 23, 28t, 84, 105, 136, **138**, **141**, 141k, 142, 233, 234, 237, 249–251, 254–256, 259–263, 265, 266, **267**, 298, 303, 316k, 454k, **454**; figs. 43D, 155–161, 202, 203, 250, 262, 287
*gigantea*, *Reganula*, 58t, 325
*giganteus*, *Reganichthys*, 58t, 201, 325
Gilbert, C. H., 10
gills, function in males, 298, 301

gill arches, **33**, 45, 55; figs. 16A, 39, 41D, 197–199, 269
gill filaments, **33**, 187; fig. 199
Gill, T. N., 9, 10, 51, 63, 173
*gladisfenae*, *Centrophryne*, 55, 103, 393
*gladisfenae*, *Dolopichthys*, 55, 102, 103, 392, 393
***gladisfenae***, ***Spiniphryne***, 103, 392k, **393**; figs. 48G, 109A, 110, 111, 222
*Glaucosoma*, 323
Glaucosomatidae, 323
*globosus*, *Corynolophus*, 63, 64t, 336
*globosus*, *Himantolophus*, 336
*glomerosus*, *Diceratias*, 357
*glomerosus*, *Oneirodes*, 354
*glomerosus*, *Onirodes*, 354
*glomerosus*, *Paroneirodes*, 71, 72, 354, 357
*glomerulosus*, *Diceratias*, 354
*glomerulosus*, *Paroneirodes*, 354, 357
Gnathophausidae, 261
gobies, 249
***golovani***, ***Gigantactis***, 145, 223–225, 456k, **466**; fig. 234
Gondogeneia, 249
Goode, G. B., 10
goosefishes, figs. 2A, 40B
***gracilicauda***, ***Gigantactis***, 138, 144, 223–225, 455k, **462**; fig. 232
*gracilispinis*, *Dolopichthys*, 123, 128, 441
*gracilispinis*, *Dolopichthys* (*Leptacanthichthys*), 441
***gracilispinis***, ***Leptacanthichthys***, 91, 127, 285t, 295, 296, 301, **441**; figs. 33F, 137C, 142–144, 229
**Greedy Seadevils**, 152k, **158**, **486**
Gregory, W. K., 3; figs. 187
*grimaldii*, *Opisthoproctus*, 241
***groenlandicus* group**, ***Himantolophus***, 64t, 68, 208, 209, 334k, **334**
***groenlandicus***, ***Himantolophus***, 8, 13, 63, 64t, 65, 68, 207–210, 234, 235t, 250, 251, 259, 291, 334k, **334**, 384, 386; figs. 9, 10E, 11, 34C, 71D, 72A, 198, 211
Guinness World Records, 3
Gulf of Panama, 11, 199
gulper eel, 275
Günther, A. C. L. G., 8, 9; fig. 8
gustatory structures, 263

hagfishes, 55
Halibatrachi, 9, 314, 315
*Halieutaea retifera*, fig. 2G
*Halophryne hutchinsi*, fig. 186B
Halpophryne, 481
Haneda, Y., 251
handfishes, fig. 2E
*haplactis*, *Chaenophryne*, 99, 100t, 386
***haplonema***, ***Oneirodes***, 110, 396k, **414**
**Haplophryne**, 28t, 149, 150, 151k, **154**, 158, 166, 225, 238, 290, 296–298, 300–303, 307, 314, 478k, 479k, **481**; figs. 202, 203
Haygood, M., 241
head skeleton, **24**; of females, figs. 11, 14D, 15, 23–25, 192, 193–196, 276, 277; of larvae, fig. 27; of males, figs. 26, 41, 283, 293, 294, 299
head spines, **30**; fig. 32B
hemibranchs, 33, 315
hermaphroditism, ix, 3, 307
***herwigi***, ***Gigantactis***, 146, 223, 456k, **468**; fig. 235
*heteracanthus*, *Dolopichthys*, 397, 398, 404, 407, 443; fig. 190
*heteracanthus*, *Dolopichthys* (*Dermatias*), 397
*heteronema*, *Dolopichthys*, 407

*heteronema*, *Dolopichthys* (*Dermatias*), 108, 407
***heteronema***, ***Oneirodes***, 106, 108, 218, 219, 396k, **407**; figs. 114G, 225
*heterorhynchus*, *Xenoceratias*, 76t, 362; fig. 82D
*hibernicus*, *Dolopichthys*, 400
**Himantolophidae**, 9, 17, 21, 28t, 54k, 56, **63**, 192, **207**, 235t, 263, **291**, 305, 314, 316k, 317k; **331**; figs. 68–74, 189, 191, 202, 203
Himantolophiformes, 331
Himantolophinae, 331
***Himantolophus***, 9, 23, 28t, 64t, **68**, 105, 198, 210, 229, 237, 250, 260, 291, 303, **333**; figs. 68–74, 202, 203, 292A
*Histionotophorus bassani*, 157, 485
*Histioteuthis*, fig. 1
*Histrio histrio*, 260
*histrio*, *Pterophryne*, 260
Hogben, L. T., 437
***holboelli***, ***Ceratias holboelli***, 3, 8, 11, 57, 58t, 61, 62, 152, 201, 206, 207, 235t, 251, 261, 275, 277, 279, 280, 282t, 324k, **324**; figs. 3, 42, 45A, 61, 62A, 209, 250, 280, 281, 287, 310
*holboellii*, *Caratias*, 325
Holbøll, C. P., 8
*holbolli*, *Ceratias*, 325
holobranchs, 33, 315
holothurian, 261, 271, 272, 274
*Horizon*, research vessel, 118
hormonal communication, 51
*horridus*, *Pietschichthys*, 111, 423
*hudsonia*, *Anomalophryne*, 509
*hudsonius*, *Halophryne*, 509
*hudsonius*, *Haplophryne*, 163t, 509
*hutchinsi*, *Halophryne*, fig. 186B
*Hyaloceratias*, 157, 166, 260, 479k, 481, 488, 489
*Hydrichthys pietschi*, 324
hydroid, 324
hyoid barbel, of *Centrophryne*, 40, 55, 56, 164, 245, 319, **321**; fig. 60; of *Linophryne*, 7, 40, 42, 162, 164, 165, 229, **245**, 247, 251, 314; figs. 47, 184, 240, 257–259; of stomiiforms, 246, 247
hypaxial musculature, 266

***idiomorpha***, ***Bertella***, 91, 123, 221, 285t, 295, 297, **438**; figs. 33H, 133, 134, 135A, 136, 226, 291C, 296
Idyll, C. P., 253
illicium, definition of, 6, 15, 36, 314
illicial apparatus, 23, **33**, 233, 249; figs. 14D, 42, 46, 273; evolution of, fig. 40; innervation of, 249; musculature of, 249, 262, 263, **267**; figs. 42, 46, 273, 283; presence in larvae, 19, 190; presence in males, figs. 56, 283; pterygiophore of, **33**, 46, 190, 244, 249–251, 262, 263, 269, 314; figs. 41–43, 283, 293
Imai, S., 11
immunology, 4, 8
*implumis*, *Dolopichthys*, 426
*implumis*, *Dolopichthys* (*Microlophichthys*), 426
*in situ* observations, **249**, 253, **255**; figs. 262–264
inclinator dorsalis muscle, of the illicium, 40, 251, 267; figs. 42, 46, 273, 283
***indica***, ***Linophryne***, 149, 163t, 166, 226, 228, 236t, 246, 288t, **489**; figs. 239, 307B
***indica***, ***Linophryne*** (***Stephanophryne***), 488k, **489**
*indicus*, *Aceratias macrorhinus*, 149, 163t, 482, 489, 497
*indicus*, *Aceratias*, 149, 489, 497

*indicus*, **Lophodolos**, 96, 214, 215, 380k, **380**; figs. 96, 97B, 219
*indicus*, *Lophodulus*, 380
*infans*, *Benthalbella*, 251
inflation, body of larvae, 48, 190, 254; figs. 55, 56
*inimicus*, *Dolopichthys*, 219, 398
*inimicus*, *Oneirodes*, 106, 398
innervation, escal, **234**, 263; illicial, 249
intermandibularis muscle, 265, 266
*intermedia*, *Chaenophryne*, 99, 100t, 390
***intermedius***, ***Lasiognathus***, 85, 212, 370k, **371**; figs. 88C, D, 217
intrinsic muscles, of the illicium, 40, 249, 267, 274; figs. 46, 273
*Investigator*, research vessel, 96
***ios***, ***Gigantactis***, 146, 224, 225, 456k, **471**; fig. 232
isotherms, 219; fig. 207

*Jason*, remote underwater vehicle, 253, 255; fig. 262
jaws, of females, **31**; figs. 25, 34, 35, 270, 271; of males, figs. 6, 26, 41C, 54, 282
jet-propulsion, **259**; fig. 265
Johnson, J. Y., 8; fig. 8
*johnsonii*, *Ceratias*, 362
*johnsoni*, *Melanocetes*, 362
***johnsonii***, ***Melanocetus***, 9, 13, 75, 76, 76t, 77, 80, 199, 212, 235t, 241, 254, 255, 259, 261, 280, 281, 285t, 294, 361k, **361**; figs. 8, 19, 25A, 26, 32A, 45C, 78, 80A, 81A, 83, 192, 216, 255
*Johnson Sea-Link II*, submersible, fig. 29,
*johnstoni*, *Melanocetus*, 362
*jordani*, *Caulophryne jordani*, 128, 447
***jordani***, ***Caulophryne***, 10, 128, 129, 130t, 135, 221, 297, 445k, **447**; figs. 28, 150
***jubatus***, ***Dolopichthys***, 118, 119, 221, 432k, **434**; figs. 25D, 129, 227
*jubatus*, *Dolopichthys* (*Dolopichthys*), 119, 434
juvenile stage, definition of, 314

*kainarae*, *Himantolophus*, 64t, 336
*Kali*, 99
***karsteni***, ***Dolopichthys***, 120, 432k, **436**; figs. 131C, 228
Koefoed, E., 11, 245
*krechi*, *Melanocetus*, 76, 76t, 362, 367
Krefft, G., 148; fig. 22
***kreffti***, ***Gigantactis***, 144, 224, 456k, **458**; figs. 38B, 157C, 158C, 234
***kreffti***, ***Oneirodes***, 108, 218, 219, 396k, **405**; fig. 224
Krøyer, H. N., 8; fig. 7
Krøyer, P. S., fig. 7
*kroyeri*, *Mancalias*, 58t, 326

*laevis*, *Xenoceratias*, 76t, 362; fig. 82C
*Laevoceratias*, 454
Laevoceratiidae, 18, 19, 21, 315, 452; fig. 189
*Lampanyctus crocodilus*, 261
lampreys, 55
lancetfish, 275
lanternfishes, 275
larvae, body inflation of, 48, 190, 254, 314; figs. 55, 56; distribution of, 200; fig. 207; hump-backed, 48, 61, 192, 317; figs. 63, 67; locomotion, **254**; pectoral fins, 49, 190, 254; figs. 55, 56; pelvic fins, 49, 190; fig. 56A, B; pigmentation, **49**, 314
larval stage, definition of, 314
***Lasiognathus***, 22, 28t, 58, 82, 83, **84**, 84k, 86, 90, 212, 251, **270**, 294, 314, 369k, **369**; figs. 85–88, 195, 202, 203

lateral-line system, **29**, 128, 130, 137, 262, 268; figs. 29, 30
*latirhinus*, *Rhynchoceratias*, 76t, 367
*latirhinus*, *Xenoceratias*, 367
Lavenberg, R. J., 22
*lavenbergi*, *Saccopharynx*, 275; fig. 278
**Leafysnout Dreamers**, 93k, **115**, **427**
**Leftvent Seadevils**, 54k, **148**, **477**
***legula***, ***Robia***, 130, 136, **450**; figs. 152, 230
***Leptacanthichthys***, 21, 28t, 90, 94k, 111, 118, 123, **126**, 295, 296, 305, 307, 378k, **440**; figs. 137C, 142–144, 202, 203
*leptonema*, *Rhynchactis*, 138, 139, 147, 224, 475k, **475**; fig. 235
*leuchorhinus*, *Rhynchoceratias*, 55, 100t, 382, 384, 394, 437
*leuchorhinus*, *Trematorhynchus*, 55, 437
*leucorhinus*, *Rhynchoceratias*, 321, 364, 422, 426
*leucorhinus*, *Trematorhynchus*, 422, 430
levator arcus palatini muscle, 265
levator maxillae superioris muscle, 273
levator operculi muscle, 265
levatores externi muscles, 266
levatores interni muscles, 266
lie-in-wait predation, 256, 259
light-guiding structures, 7, 39, 164, **237**, 241, 244, 249, 262; figs. 201, 240, 245B, 246, 247
light-reflecting structures, 7, 39, 234, **237**, 241, 249, 424, 429, 507
**Lightline Dreamers**, 94k, **126**, **440**
*Limacina*, 260
*Linocetus*, 360
***Linophryne***, 7, 23, 28t, 105, 148–150, 152k, 155, 157, 158, 160, **162**, 163t, 225, 226, 228, 234, 237, 246, 247, 251, 263, 288t, 300–303, 307, 478k, 479k, **487**; figs. 1, 53A, 185, 202, 203, 240, 301D
***Linophrynidae***, 7, 19, 21, 28t, 54k, **148**, 151k, 194, 197, 198, **225**, 236t, 260, 263, 286t, 296, **298**, 303, 314, 317k, **477**; figs. 167–185, 189–191, 202, 203
*Liocetus*, 359, 360
*Lipactis*, 63, 333, 394
*liparis*, *Laevoceratias*, 21, 138, 359, 473
**Longhand Dreamers**, 95k, **125**, **440**
*longibarbata*, *Linophryne coronata*, 500
*longibarbata*, *Linophryne*, 163t, 500
*longicauda*, *Gigantactis*, 138, 146, 223, 456k, **471**
*longiceps*, *Chaenophryne*, 99, 100t, 102, 216, 217, 237, 384, **386**; figs. 103, 106A, 221
**longiceps group**, ***Chaenophryne***, 99, 100t, 102, 379k, 385k, **386**; fig. 107
***longicirra***, ***Gigantactis***, 15, 138, 140, 144, 224, 455k, 457k, **457**; figs. 38A, 156, 158A, 197, 200, 233, 267–269, 300
***longicornis***, ***Dolopichthys***, 118, 119, 220, 236t, 431k, 432, **433**, 434; fig. 227
*longicornis*, *Dolopichthys* (*Dolopichthys*), 433
***longidens***, ***Acentrophryne***, 148, 157, 158, 484k, **484**; figs. 174, 236
***longimanus***, ***Ctenochirichthys***, 123, 128, 383, **343**; figs. 137D, 145–147, 229
*longipinnis*, *Rhynchoceratias*, 17, 76t, 367; fig. 293
*longipinnis*, *Xenoceratias*, 367
*longirostris*, *Melanocetus*, 77, 81, 361k, **368**
*longirostris*, *Xenoceratias*, 76t, 81, 368
Long-lure fanfins, 135k, **135**, **450**
Longnose Lancetfish, 275
long-pectoraled oneirodid genera, 21, 111, **123**, 188, **439**; figs. 137–147

**Longsnout Dreamers**, 94k, **118**, **430**
Lophichthyidae, 177, 178k; fig. 188
Lophiidae, 45, 177, 249; figs. 188, 189
**Lophiiformes**, 3, 23, 51, 175, 181, 195, 198, 249, 254, 259, 261, 263, 281, 300, 315, 368; figs. 2, 186, 188
*Lophiodes reticulatus*, fig. 2A
Lophioidae, 322, 331
Lophioidea, 9, 315
*Lophioidea apoda*, 9, 314
Lophioidei, 3, 33, 157, 173, 177, 178k, 179, 181, 229; figs. 2A, 188, 202, 203
Lophioideorum (*Halibatrachorum*) *apodum*, 314
*Lophioides apodes*, 314
*Lophius*, 8, 13, 179, 197
*Lophius americanus*, 175; fig. 40B
*Lophius brachysomus*, 157, 485
*Lophius budegassa*, 157, 485
*Lophius piscatorius*, 13, 15, 263
***Lophodolos***, 21, 28t, 89, 91, 91k, **96**, 98, 111, **214**, 378k, 379k, **379**; figs. 43A, 96–98, 202, 203, 250
*Lophodolos*, 379
*Lophodulus*, 379
***lucifer***, ***Linophryne***, 10, 148, 150, 162, 163t, 166, 170, 226–228, 245, 261, 302, 498k, **501**; figs. 13, 183A, 237, 308A, 309A
*lucifera*, *Linophryne*, 163t, 501
Lütken, C. F., 9, 229, 249; figs. 9, 10
*luetkeni*, *Dolopichthys*, 107, 397; fig. 190
*luetkeni*, *Dolopichthys* (*Dermatias*), 397
***luetkeni***, ***Oneirodes***, 107, 218, 219; 236t, 395k, **397**; figs. 113A, 114A, 224
luring, apparatus, 3, 177, 179, 277; behavior, ix, 13, 17, 249, 250, 262, 271
*lyra*, *Lophodolos*, 381

*macracanthus*, *Edriolychnus*, 482; fig. 171A
*macracanthus*, *Xenoceratias*, 76t, 362
*macractis*, *Chaenophryne*, 99, 100t, 388
***macroceras***, ***Himantolophus***, 64t, 70, 210, 346k, **347**; figs. 70B, 214
***macroceratoides***, ***Himantolophus***, 64t, 70, 210, 346k, **348**; figs. 70A, 72F
***macrodon***, ***Linophryne***, 163t, 170, 228, 246, 248, 288t, 302, 499k, **504**; figs. 20C, 40C, 239, 249, 258
*macronema*, *Dolopichthys*, 415
*macronema*, *Dolopichthys* (*Dermatias*), 110, 415
***macronema***, ***Gigantactis***, 138, 146, 224, 250, 456k, **468**; figs. 38E, 157A, 159A, B, 233
**macronema** group, *Gigantactis*, 456k
***macronema***, ***Oneirodes***, 110, 396k, **415**
macrophagy, 261, 263
*macrorhinus*, *Aceratias*, 17, 149, 163t, 166, 481, 508
**macrorhinus** group, *Haplophryne*, 509
**macrorhinus** group, *Linophryne*, 509
macrosmatic, 45
***macrosteus***, ***Oneirodes***, 108, 218, 219, 397k, **408**; figs. 114D, 225
***macrothrix***, ***Rhynchactis***, 139, 147, 224, 298, 475k, **476**; figs. 163, 164, 235
Macrouridae, 29
*maculatus*, *Antennarius*, 249; fig. 265
***maderensis***, ***Linophryne***, 148, 162, 163t, 170, 228, 499k, **504**; fig. 238
*Mancalias*, 322, 323
*Mancalius*, 323
*Manta*, fig. 1
Marshall, N. B., 197, 259
*masculina*, *Borophryne*, 163t, 509
*masculina*, *Linophryne*, 509

INDEX 553

mate location, **302**
***mauli*, *Himantolophus*,** 64t, 70, 207, 210, 342k, **343**; figs. 69C, 70C, 212
***meadi*, *Gigantactis*,** 144, 223–225, 456k, **460**; figs. 38C, 232, 270, 272
Mediterranean Sea, 15, 37, 197, 200, 212, 316
*megaceros*, *Dolopichthys*, 398, 400, 404, 434
*megaceros*, *Dolopichthys* (*Dermatias*), 400
*megaceros*, *Oneirodes*, 400
*megalodontis*, *Melanocetus*, 76t, 254, 362
***melanocauda*, *Oneirodes*,** 110, 379k, 395k, **416**
Melanocetes, 360
**Melanocetidae,** 17, 21, 28t, 53k, **75**, 192, **212**, 235t, 263, 278, 280, 285t, 288t, **293**, 305, 316k, 317k, **359**; figs. 78–84, 189–191, 202, 203
Melanocetinae, 359
***Melanocetus*,** 9, 28t, 76t, **79**, 199, 234, 237, 238, 247, 250, 260, 293, 305, **360**, 445; figs. 1, 30A, 78–84, 202, 203, 292C
Melanocoetidae, 359
Melanocoetus, 360
*melanodactylus*, *Chaenophryne*, 99, 100t, 388
***melanolophus*, *Himantolophus*,** 64t, 70, 207, 210, 340k, **341**; figs. 72C, 214
*melanophus*, *Himantolophus*, 64t, 341
melanophores, subdermal, 49
***melanorhabdus*, *Chaenophryne*,** 99, 100t, 102, 216, 217, 237, 388k, **389**; figs. 104, 106C, 221
Melanostomiinae, barbels of, 246
*Melville*, research vessel,
*merrimani*, *Parrichthys*, 58t, 63, 328
mesopelagic, 8, 197, 200, 210, 228, 250, 253, 256, 260, 261, 291, 314, 315; fig. 1
metamorphosis stage, definition of, 314
*mexicanus*, *Triphoturus*, 275
*Michael Sars*, North Atlantic Deep-sea Expedition, 11
*micracanthus*, *Xenoceratias*, 76t, 362; fig. 82E
*micractis*, *Allector*, 497
*micractis*, *Cryptolychnus*, 497; fig. 185B
***micractis*, *Phyllorhinichthys*,** 115, 116, 236t, 250, 428k, **428**; figs. 45F, 125A, 126, 127B, C, 226
***microdontis*, *Gigantactis*,** 146, 224, 225, 456k, **470**; figs. 38F, 234
***Microlophichthys*,** 28t, 90, 93k, **114**, 118, 214, **220**, 260, 378k, 379k, **425**; figs. 121–124, 202, 203
*microlophus*, *Dolopichthys*, 426
*microlophus*, *Dolopichthys* (*Microlophichthys*), 426
***microlophus*, *Microlophichthys*,** 114, 115, 220, 250, **426**; figs. 33E, 121, 122, 124, 226
***micronema*, *Oneirodes*,** 111, 417k, **418**
*microphthalmus*, *Gigantactis*, 473
*microphthalmus*, *Teleotrema*, 138, 473
*Micropterus salmoides*, fig. 40A
microsmatic, 45
***microthrix*, *Rhynchactis*,** 139, 148, 223, 475k, **476**; fig. 235
migration, ontogenetic vertical, 198; fig. 205
mimicry, 13, 249, 262; fig. 18
Miocene, of California, 157, 485
Miopsaras, 323
*mirus*, *Dolopichthys*, 419
*mirus*, *Dolopichthys* (*Dermatias*), 111, 419
***mirus*, *Oneirodes*,** 106, 111, 417k, **419**
**Mischievous Dreamers,** 95k, **124**, **439**
*mitsukurii*, *Ceratias*, 58t, 330
*mitsukurii*, *Ceratias* (*Paraceratias*), 330
*mitsukurii*, *Cryptosarus*, 330
*mitsukurii*, *Cryptosparas*, 330

*mitsukurii*, *Paraceratias*, 330
Miya, M., 175
*moderatus*, *Trematorhynchus*, 100t
*Mola*, fig. 1
molecular, evidence for relationships, 175, **195**, 306, 307
*mollis*, *Aceratias*, 149, 154, 481, 482
*mollis*, *Aceratias macrorhinus*,
***mollis*, *Haplophryne*,** 17, 149, 154, 157, 236t, 250, 277, 280, 286t, 287t, 298, 301, 305, **482**, 497, 509; figs. 6, 31, 170–172, 236, 250, 297C, 301B, 304
mollusks, 437
monkfishes, fig. 2A
Monoceratias, 394
monophyly, 51, 179, 181, 190, 191, 192, 306
Monte Bolca, 157, 485
Monterey Bay Aquarium Research Institute, 136, 256
Moore, J., 255
*mucronatus*, *Dolopichthys*, 118, 433
*mucronatus*, *Dolopichthys* (*Dolopichthys*), 433
*multifilis*, *Dolopichthys*, 401
*multifilis*, *Dolopichthys* (*Dermatias*), 400
***multifurcatus*, *Himantolophus*,** 64t, 70, 210, 342k, **344**; fig. 69A
multiple attachment, of males, 4, 281, 290, 307
*multiradiatus*, *Trematorhynchus*, 443
Munk, O., 234, 280
Murray, J., 10
***murrayi*, *Melanocetus*,** 10, 76, 76t, 77, 80, 199, 212, 229, 235t, 285t, 294, 360k, 361k, **366**, 367; figs. 39B, 81D, 84, 216, 244, 291B, 293
*murrayi*, *Melanocetus* (*Liocetus*), 366
*muscipula*, *Dionaea*, 81
musculature, of feeding mechanism, **263**; figs. 267, 270-272, 277; illicial, **40**, 249, **267**; figs. 42, 46, 267, 273; intra-escal, **234**, 239, 241, 251; of jaws, figs. 270, 271, 277; of pectoral fin, fig. 268; of pharyngeals, 266; figs. 269, 270, 272
*mutsukurii*, *Cryptosarus*,
Myctophidae, 199, 275
myctophiform, 17
*myops*, *Miopsaras*, 58t, 62, 328
Myopsaras, 323
Myosparas, 323
***myrionemus*, *Oneirodes*,** 108, 396k, **406**

Nafpaktitis, B. G., 21, 253
Nannoceratias, 487
nasal papilla, 42, 136, 189; fig. 52
**Needlebeard Seadevils,** 52k, **136**, **450**
*Neoceratias*, 19, 28t, **137**, 142, 263, 298, 303, 305, 307, **452**; figs. 30D, 153, 154, 202, 203
**Neoceratiidae,** 21, 28t, 52k, **136**, 141, 194, 197, **222**, 234, 261, 285t, **298**, 314, 315, 316k, 317k, **450**; figs. 153, 154, 189, 191, 202, 203
nereid polychaete, 13; fig. 18C
*Nesogobius*, 249
net contamination, 260
net feeding, 260
New York Zoological Society, 157, 254
*Nezumia*, fig. 1
*niger*, *Dolopichthys*, 118, 121, 236t, **436**
*niger*, *Melanocetus*, 76, 76t, 77, 80, 212, 361k, 362, 364, **365**; figs. 81C, 216
*niger*, *Oneirodes*, 236t, 436
***nigricornis* group, *Himantolophus*,** 64t, 69, 210, 333k, **340**

***nigricornis*, *Himantolophus*,** 64t, 69, 210, 340k, **341**; fig. 214
***nigrifilis*, *Danaphryne*,** 15, 113, 233, 236t, **424**; figs. 33D, 118–120, 226
*nigrifilis*, *Dolopichthys*, 112, 113, 233, 236t, 424
*nigrifilis*, *Dolopichthys* (*Dermatias*), 424
*nomen dubium*, definition of, 313
*nomen nudum*, definition of, 313
nonparasitic temporary sexual attachment, 48, **195**, **305**, 306t
Norman, J. R., 11, 97
*normani*, *Cryptosparas*, 58t, 330
**Northern Giant Seadevil,** 3, 62, **324**; fig. 3
nostrils, of females, 42; of males, 3, 42, 189, 303; figs. 5, 286, 294
***notius*, *Oneirodes*,** 110, 199, 218, 219, 355, 397k, **412**; fig. 223
*nudus*, *Melanocetus*, 77, 81, 361k, **368**
*nudus*, *Xenoceratias*, 76t, 81, 368
nutritional parasites, 279

obligatory sexual parasitism, 46, 137, **195**, **303**, 305, 306t; figs. 202, 203
obliqui dorsales muscle, 266
*obliquidens*, *Trematorhynchus*, 100t
*obtusus*, *Dolopichthys*, 400; fig. 20A
*obtusus*, *Dolopichthys* (*Dermatias*), 400
ocular degeneration, 42, 166, 298, 486, 488, 490, 503, 507
Ogcocephalidae, 45, 177, 179, 249; figs. 188, 189
Ogcocephaloidei, 3, 33, 173, 177, 178, 179k, 181; figs. 2G, 188B, 202, 203
*ogcorrhynchus*, *Rhynchoceratias*, 350
olfactory organs, of females, 19, 42; lamellae, 42, 43; of larvae, 19; lobes of the brain, fig. 53; of males, 19, 42, 189, 250, 302, 303; figs. 5, 53
omnivorous, 274
*onchorhynchus*, *Rhynchoceratias*, 350
*oncorhynchus*, *Rhynchoceratias*, 64t, 350, 497; fig. 73C
***Oneirodes*,** 9, 21, 23, 28t, 58, 71, 72, 90, 91, 92k, 103, **105**, 111, 118, 199, 200, 214, **217**, 223, 233, 247, 249–251, 256, 259, 260, 262, 267, 275, 295, 305, 378k, 379k, **394**, 430; figs. 30B, 53B, 112–115, 202, 203, 250, 263, 264, 295C
**Oneirodidae,** 17, 21, 28t, 54k, 55, 82, 91k, **87**, 103, 192, 194, **214**, 235t, 237, 278, 285t, **294**, 306, 314, 317k, 368, **376**; figs. 96–147, 189–191, 202, 203
Oneirodinae, 376
Oneiroides, 394
Onirodes, 354, 394
ontogenetic vertical migration, 198; fig. 205
ontogeny, 19; fig. 3; of denticular apparatus of males, 45, 279; fig. 54; of escal photophore, **237**
opercular bones, **25**; figs. 16B, 27
Ophiiformes, 175
*Opisthoproctus grimaldii*, 241
oral glands, 31, 138; fig. 165
Ostraciidae, 259
ostracods, 261
ovaries, anatomy of, 55, 177, 281, 319
***ovifer*, *Gigantactis*,** 138, 146, **472**

*pacis*, *Chaenophryne*, 99, 100t, 390
***pagidostomus*, *Thaumatichthys*,** 10, 81, 82, 87, 212, 213, 272, 374k, **375**, 376; figs. 36D, 37, 48F, 218
pair formation, 250
Pappenheim, P., 136, 137
Paracanthopterygii, 175

*Paraceratias*, 328
*Paraneirodes*, 354
parasitism, sexual, ix, 11, 17, 21, **46**, 48, 51, 61, 124, 137, 173, 179, **195**, **277**, 282–288t, 315; figs. 3, 31, 202, 203, 280, 281, 284, 287, 289–291, 298; discovery of, 17, 152, 277; figs. 17, 280, 287; facultative, 48, 91, 195, **305**, 306t; figs. 202, 203; obligatory, 46, 137, 195, **303**, 305, 306t; figs. 202, 203
*parini*, **Linophryne**, 163t, 166, 227, 492k, **492**; fig. 183C
**Paroneirodes**, 71, 72, 354, 356
Parr, A. E., 11, 17, 277–279; fig. 20
*parri*, *Hyaloceratias*, 481, 482
*Parrichthys*, 323
*parviconus*, *Chaenophryne*, 99, 100t, 236t, 388, 389
*paucidens*, *Allector*, 509
*paucidens*, *Cryptolychnus*, 509
***paucifilosus***, ***Himantolophus***, 64t, 68, 207, 209, 235t, 334k, **338**; fig. 213
Pawnee, expeditions of, 17; fig. 20
***paxtoni***, ***Gigantactis***, 145, 224, 225, 455k, **463**; figs. 38D, 233
pectoral fins, of larvae, 190
pectoral radials, **27**
pectoral-fin oscillation, synchronous, 256, 259
Pediculati, 8, 13, 137, 173, 175, 179, 315, 322, 359, 368; figs. 12, 187
pelagic zone, fig. 1
*pelagica*, *Caulophryne jordani*, 128, 446
***pelagica***, ***Caulophryne***, 128, 129, 130, 130t, 133, 134, 221, 445k, **445**; figs. 25B, 45G, 148, 230
*pelagicus*, *Caulophryne*, 445
*pelagicus*, *Melanocetus*, 130t, 134, 445
pelvic bones, **27**, 188; fig. 41E
pelvic fins, of larvae, 49, 188, 190; fig. 56A, B; of males, 188; figs. 56C, 292D
Penaeidae, 261
*pennatus*, *Dolopichthys*, 401
*pennatus*, *Dolopichthys* (*Dermatias*), 400
***pennibarbata***, ***Linophryne***, 163t, 166, 228, 491k, **495**; figs. 47G, H, 182A, 237
*pennifer*, *Cryptopsaras*, 330
*pennifer*, *Cryptopsaras couesii*, 330
*pennifer*, *Cryptosparas*, 58t, 330
***Pentherichthys***, 28t, 90, 91k, **97**, **118**, 378k, 379k, **382**; figs. 99–101, 202, 203
Perciformes, 195
Percomorpha, 175
***perlatus***, ***Gigantactis***, 138, 145, 224, 225, 456k, 459, **464**; fig. 232
pharyngeals, movement of, 263, 269; musculature of, 266, 272; teeth of, 31, 269; figs. 33, 39A, 272
pheromones, 7, 237, 289
phosphorescence, 3, 229, 245
*Photobacterium phosphoreum*, 241; fig. 255
Photocorynidae, 19, 179, 477; fig. 189
***Photocorynus***, 28t, 150, 151k, **152**, 155, 158, 225, 298, 300–302, 478k, 479k, **479**; figs. 30E, 202, 203
photogenic cells, granules, 7, 247; figs. 240, 259
photosynthesis, fig. 1
photophore, 229; fig. 1; definition, 37, 234; ectodermally derived, 7, 247, 251; escal, 40, 233, **234**, 237, 242, 244, 261, 262; of hyoid barbel, 42, 164, 165, **245**, 247, 251; mesodermally derived, 7, 247, 251
*Phrynichthys*, 72, 356
*Phyllophryne scortea*, 249
***Phyllorhinichthys***, 28t, 90, 93k, **115**, 237, 378k, **427**; figs. 125–127, 202, 203, 240

*Physalia*, fig. 1
*Physeter catodon*, 275
*pictus*, *Chaunax*, fig. 39A
*pietschi*, *Caulophryne*, 130, 130t, 446
*pietschi*, *Hydrichthys*, 324
***pietschi***, ***Oneirodes***, 111, 417k, **420**
Pietschichthys, 111, 422
Pietschmann, V., 11
pigmentation, of larvae, 19, 23, 63, **49**; of metamorphosed specimens, 23, 37, 316
***pileatus***, ***Diceratias***, 74, 75, 211, 261, 293, 354k, **355**; figs. 48D, 215
***pinnata***, ***Puck***, 124, 125, **439**; figs. 137A, 138, 139, 229
piscatorius, *Lophius*, 13, 15, 263
placenta-like attachment, 279
***plagionema***, ***Oneirodes***, 108, 396k, **409**
***platynogaster***, ***Dermatias***, 10, 111, 112, **423**; figs. 116, 117
*platynogaster*, *Dolopichthys*, 423
*platynogaster*, *Dolopichthys* (*Dermatias*), 423
*Pleuragramma antarcticum*, 364
Pleuronectidae, 150
Pleuronectiformes, 150
Pliocene, of Algeria, 157, 485
*plumatus*, *Dolopichthys*, 401
*plumatus*, *Dolopichthys* (*Dermatias*), 400
*politus*, *Sympterichthys*, fig. 2E
*pollicifer*, *Dolopichthys*, 401
*pollicifer*, *Dolopichthys* (*Dermatias*), 400
*pollicifer*, *Oneirodes*, 401
***polyactis***, ***Melanocetus***, 76, 76t, 77, 80, 212, 361k, **364**; figs. 81B, 190, 216
polychaetes, 261; nereid, 13; fig. 18C
***polynema***, ***Caulophryne***, 128, 129, 130t, 133, 135, 221, 285t, 297, 445k, **448**; figs. 230, 291D
*polynema*, *Caulophryne jordani*, 129, 447, 448
*polypogon* group, *Linophryne*, 505
***polypogon***, ***Linophryne***, 163t, 170, 228, 246, 498k, **505**; figs. 181, 182B, 237
pontogeneiid amphipod, 249
population sizes, 253, 256, 303
***posti***, ***Oneirodes***, 108, 396k, **405**
postmaxillary process, 31
*Potamobatrachus trispinosus*, fig. 186A
**Prickly Seadevils**, 53k, **55**, 57, **319**
protractor hyoidei muscle, 266
protractor muscles, 251
protrusible upper jaw, 31, 263
***pseudalbinares***, ***Himantolophus***, 64t, 70, 210, 342k, **344**; fig. 213
pseudobranch, 33, 315
*pterolophus*, *Chaenophryne*, 99, 100t, 389
*Pterophryne histrio*, 260
pteropods, 260
***pterurus***, ***Oneirodes***, 108, 396k, **410**
pterygiophore, illicial, **33**, 46, 187, 249–251, 262, 263, 269, 314; figs. 41C, 42, 46, 273, 293
*ptilotus*, *Dolopichthys*, 400
*ptilotus*, *Dolopichthys* (*Dermatias*), 400
***Puck***, 21, 28t, 90, 95k, 111, 123, **124**, 378k, **439**; figs. 137A, 138, 139, 202, 203
Puente Formation, Late Miocene, 157, 485; fig. 175
pufferfishes, 175
**Pugnacious Dreamers**, 93k, **116**, **429**
***pugnax***, ***Tyrannophryne***, 116, 118, **430**; fig. 128
**Pugnose Dreamers**, 91k, **96**, **379**
***pullatus***, ***Dolopichthys***, 118, 119, 220, 221, 250, 431k, **432**; figs. 33G, 131A, B, 135B, 228
*pullatus*, *Dolopichthys* (*Dolopichthys*), 119, 432
*pumilus*, *Caranactis*, 72, 359, 394, 422

*quadrifilis*, *Chaenophryne*, 99, 100t, 386
*quadrifilis*, *Chaenophryne longiceps forma typica*, 386
*quadrifilis*, *Chaenophryne longiceps* var., 386
***quasiramifera***, ***Chaenophryne***, 100t, 102, 216, 217, 237, 388k, **391**; figs. 102, 105, 221
***quinqueramosa***, ***Linophryne***, 166, 227, 492k, **493**

***racemifera***, ***Linophryne***, 163t, 170, 228, 498k, **506**; figs. 25F, 45I, 47F, 183H, 194, 239
*racemosa*, *Caulophryne*, 130t, 135, 450, **450**
Radcliffe, L., 10, 89
*radians*, *Caulophryne*, 130t, 135, **449**
*radians*, *Edriolychnus*, 482; fig. 171B
***ramifera***, ***Chaenophryne***, 99, 100t, 102, 216, 217, 237, 388k, **390**; figs. 106D, 221
*ramulosa*, *Caulophryne*, 128, 130t, 445; fig. 149
*ranoides*, *Himantolophus*, 63, 64t, 335
rattail, fig. 1
*Regalecus*, fig. 1
Regan, C. T., 11, 17, 23, 152, 175, 277, 279, 280; fig. 17
*regani*, *Caulophryne*, 128, 129, 130t, 447
*regani*, *Ceratocaulophryne*, 447
*regani*, *Xenoceratias*, 76t, 367
*Reganichthys giganteus*, 201, 323
*Reganula*, 323
Reinhardt, J. C. H., 8, 9
*reinhardti*, *Corynolophus*, 335, 336
*reinhardti*, *Himantolophus*, 9, 63, 64t, 335; figs. 9, 10A–C, 71D
remotely operated vehicles, *in situ* observations from, **249**, 253, **255**; figs. 262–264
*reticulatus*, *Lophiodes*,
*retifera*, *Halieutaea*, fig. 2G
retractor dorsalis muscle, 266, 267
retractor muscles, 251
***Rhizophryne***, 157, 162, 163t, 164–166, 479k, 488k, 489k, **491**; fig. 185A
***Rhynchactis***, 22, 28t, 138, 139, 141k, 142, **146**, 254, 261, 263, 265, 266, **269**, 298, 303, 314, 316k, 317k, 454k, **475**; figs. 55, 162–166, 202, 203, 286
*Rhynchoceratias*, 63, 320, 333, 360, 382, 384, 394, 430
***Robia***, 28t, 133, 133k, **135**, 181, 262, 297, 317k, 445k, **450**; fig. 152
Robison, B. H., 136
Roper, C. F. E., 248
***rosenblatti***, ***Oneirodes***, 107, 218, 219, 396k, **399**; figs. 113A, 114B, 223
*rossi*, *Melanocetus*, 76t, 77, 80, 360k, **364**
*rostratus*, *Himantolophus*, 349
*rostratus*, *Rhynchoceratias*, 63, 64t, 349, 364; fig. 73D
***rostratus* group**, ***Himantolophus***, 64t, 334k, **349**; fig. 73E, F
*rotundatus*, *Melanocetus*, 76t, 362
*rotundatus*, *Melanocoetus*, 362
*roulei*, *Edriolychnus*, 482

***sabex***, ***Oneirodes***, 107, 218, 220, 396k, **402**; fig. 225
Saccopharynidae, 275
*Saccopharynx lavenbergi*, 275; fig. 278
***saccostoma***, ***Lasiognathus***, 82–84, 212, 370k, **371**; figs. 87, 190, 217, 274
Saemundsson, B., 277, 281; figs. 279, 280, 287
*sagamius*, *Corynolophus*, 63, 64t, 68, 336

INDEX 555

*sagamius*, *Himantolophus*, 64t, 68, 207, 208, 235t, 251, 275, 334k, **336**; fig. 211
*Sagitta*, 260
*salmoides*, *Micropterus*, fig. 40A
*Sargassum*, 272
Sargassum fish, 260
***savagei*, *Gigantactis***, 146, 224, 225, 456k, **469**; figs. 159C, 233
Scaridae, 259
scavengers, 261
***schistonema*, *Oneirodes***, 110, 397k, **412**
Schmidt, J., 11
***schmidti* group, *Oneirodes***, 106, 395k, 415, **416**; fig. 113C
*schmidti*, *Dolopichthys* (*Dermatias*), 111, 418
***schmidti*, *Edriolychnus***, 154, 277, 280, 418, 481, 482
***schmidti*, *Oneirodes***, 106, 111, 417k, **418**
*schmitti*, *Daector*, fig. 186D
*schufeldti*, *Ceratias* (*Typlopsaras*), 326
Schultz, L. P., 11
*scolopes*, *Euprymna*, 241
Scopelarchidae, 251
Scopeloidei, 148, 261
Scorpaenidae, 72
*scortea*, *Phyllophryne*, 247
sculling behavior, 259
sea-bats, 179
sea-mice, 179
sea toads, fig. 2F
sea urchins, 261
*sebax*, *Oneirodes*, 402
second cephalic ray, 34, 40
second dorsal-fin spine, **34**, 40, 48, 61, 188, 229, **242**, 315; figs. 44, 46, 283
Seefischerei, Institute für, fig. 22
Seitenorganen, 229
Sepiola, 234
*sessilis*, *Ceratias holboelli*, 326
*sessilis*, *Mancalias*, 58t, 326
*setosus*, *Caulophryne*, 130t, 135, 447
sex change, 281
sex ratio, 281
*sexfilis*, *Gigantactis*, 138, 462
***sexfilis*, *Linophryne***, 163t, 170, 227, 236t, 246, 498k, **502**; fig. 240
sexual dimorphism, 3, 11, 17, 21, 23, 45, 48, 51, 78, 133, 179, 181, 190, 194, 314, 315; fig. 3
sexual maturity, 304
sexual parasitism, ix, 11, 17, 21, **46**, 48, 51, 61, 124, 137, 173, 179, **195**, **277**, 282–288t; figs. 3, 31, 202, 203, 280, 281, 284, 287, 289–291, 298; discovery of, 17, 152, 277; figs. 17, 280, 287; facultative, 48, 91, 195, **305**, 306t; figs. 202, 203; obligatory, 46, 137, 195, **303**, 305, 306t; figs. 202, 203
***shaoi*, *Bufoceratias***, 72, 75, 261, 357k, **358**; figs. 76B, 215
Shoemaker, H. H., 280
**Shortbait Dreamers**, 93k, **114**, **425**
*shufeldti*, *Ceratias*, 326
*shufeldti*, *Mancalias*, 326
*shufeldti*, *Typhlopsarus*, 326
*shufeldti*, *Typlopsaras*, 58t, 326
Siebold, C. T. E. von, 229
*Siganus doliatus*, 175
*simplex*, *Dolopichthys*, 400
*simplex*, *Dolopichthys* (*Dermatias*), 400
*simus*, *Haplophryne*, 163t, 482
sinistral anus, **30**, 150, 189; fig. 31
sister groups, 175, 177, 190, 192, 193, 195
Smith, H. M., 10, 89
**Smoothhead Dreamers**, 92k, **98**, **384**

snagging, snapping, feeding mechanisms, 31, 263, 269, 273
**Snaggletooth Seadevils**, 84k, **369**
**Southern Giant Seadevil**, 62, **327**
sperm whale, 275, 328
**Spikehead Dreamers**, 94k, **121**, **437**
spines and spinules, dermal, of females, 42, 84, 188; figs. 10B, 48, 49; of males, 46, 190; fig. 54
***spiniceps*, *Photocorynus***, 148, 151, 250, 277, 286t, 298, 300, 301, 304, **480**; figs. 25E, 107–109, 236, 297B, 301A, 302, 303
***spinifer*, *Neoceratias***, 136, 137, 152, 154, 222, 285t, 298, **452**; figs. 52, 153, 154, 196, 231, 291E, 298, 299
***Spiniphryne***, 21, 28t, 55, 74, 84, 90, 91, 92k, **102**, 317k, 378k, **392**; figs. 109–111, 202, 203
*spinulosa*, *Centrocetus*, 363
***spinulosa*, *Centrophryne***, 55, 57, 103, 201, 280, 281, 294, **321**; figs. 48A, 57–60, 208, 285
*spinulosus*, *Centrocetus*, 76t, 362; fig. 82A
**Spiny Dreamers**, 92k, **102**, **392**
**Spinyhead Seadevils**, 151k, **152**, **479**
Spitzdrüse, 234
squid, 241, 250, 262
standard length, definition of, 313; shrinkage in preservation, 313
**Stargazing Seadevil**, 62, **326**
*Stephanolepis cirrhifer*, 175
***Stephanophryne***, 150, 162, 163t, 164–166, 479k, 488k, **489**
sternohyoideus muscle, 266
*Sternoptyx*, 229
stomiiforms, barbels of, 246, 247
*striatus*, *Antennarius*, fig. 2D
Struhsaker, P. J., 21
submersibles, *in situ* observations from, **249**, 253, **255**; figs. 262–264
subtropical convergence, 209, 210, 461
suction feeding, **263**
*Sufflamen fraenatus*, 175
supracarinales anterior muscle, of the illicium, 40, 251, 267; figs. 42, 46, 273, 283
suspensorium, 24; figs. 24, 25
symbiotic bacteria, 4, 7, 37, 40, 138, 179, 234, 237–239, **241**, 244, 246, 251, 269; figs. 240, 253–255
*Sympterichthys politus*, fig. 2E
*Synanceia*, 72, 356
*Synaphobranchus*, fig. 1

Tansei-Maru, research vessel, 294
*Tarachtichthys*, fig. 1
teeth, of females, **31**, 185; figs. 34–38; of gill arches, 31, 33, 45, 56, 186; figs. 33, 39A, 198; of males, 23, **45**, 83, 189, 190, 315; figs. 3, 5, 6, 41B, C, 54, 283, 293, 294; of pharyngeals, 31, 45, 185, 186; figs. 33, 39A
*Teleotrema*, 454
temporary nonparasitic sexual attachment, 48, **195**, **305**, 306t
***tentaculatus*, *Ceratias***, 22, 57, 58t, 62, 201, 206, 207, 235t, 275, 282t, 324k, **327**; figs. 62C, 209
*tentaculatus*, *Ceratias holboelli*, 327
*tentaculatus*, *Ceratias holbolli*, 327
*tentaculatus*, *Dolopichthys*, 400
*tentaculatus*, *Dolopichthys* (*Dermatias*), 400
*tentaculatus*, *Mancalias*, 58t, 62, 327
Tetrabrachiidae, 177, 178k; fig. 188
Tetraodontiformes, 175, 195
*Tetrapturus*, fig. 1
*Thalassophryne amazonica*, fig. 186C

**Thaumatichthyidae**, 28t, 52k, **81**, 83, 90, 192, 197, **212**, 263, 270, **294**, 306, 316k, 317k, **368**; figs. 85–95, 190, 191, 202, 203
***Thaumatichthys***, 17, 21, 28t, 82, 83, 84, 84k, **85**, 90, 173, 197, 212, 213, 261, 263, **270**, 294, 314, 315, 369k, **373**; figs. 27, 43C, 89–95, 202, 203, 294
*thele*, *Bufoceratias*, 75, 211, 261, 293, 357k, **358**; fig. 215
*thele*, *Phrynichthys*, 72, 75, 358
***theodoritissieri*, *Oneirodes***, 106, 111, 417k, **420**
**Thickjaw Dreamers**, 91k, **97**, **382**
*thompsoni*, *Dolopichthys*, 411
***thompsoni*, *Oneirodes***, 109, 218, 219, 397k, **411**; figs. 199A, 224
Thomson, C. W., 10, 13
Three-starred Seadevil, 63, 229; fig. 242
*Thunnus*, 275
***thysanema*, *Oneirodes***, 110, 396k, **414**
*thysanophorus*, *Dolopichthys*, 413
*thysanophorus*, *Dolopichthys* (*Dermatias*), 400, 413
*thysanophorus*, *Oneirodes*, 106, 402, 413
*Tiburon*, remote underwater vehicle, 253, 256; figs. 263, 264
toadfishes, 173, 175; figs. 186, 187
**Toady Seadevils**, 74k, **75**, **356**
**Toothless Seadevils**, 141k, **146**, **475**
transversi dorsales muscle, 266
trap-mouthed wonder fish, 81, 272
trapping, feeding mechanism, 31, 273, 274
*Trematorhynchus*, 384, 394, 430
Trewavas, E., 11
***trewavasae*, *Linophryne***, 163t, 171, 227, 246, 288t, 498k, **506**; fig. 307A
*triflos*, *Mancalias* (*Ceratias*) *uranoscopus*, 326
*triflos*, *Mancalias uranoscopus*, 58t, 326
triggerfishes, 175
***trilobus*, *Diceratias***, 72, 75, 211, 261, 354k, **356**; figs. 2H, 75, 215, 266
*trilychnus*, *Bathyceratias*, 63, 229, 235t, 331; fig. 242
*Triphoturus mexicanus*, 275
Triplestar Seadevil, 63, 229; fig. 242
**Triplewart Seadevils**, 62k, **63**, **328**
*triregium*, *Haplophryne*, 482
*trispinosus*, *Potamobatrachus*, fig. 186A
true anglers, 179
*Tudsefiske*, 8
*tumidus*, *Lipactis*, 64t, 350, 394, 422; fig. 73A
*tumidus*, *Melanocetus*, 76t, 367
tuna, 275; fig. 1
twilight zone, fig. 1
*Typhloceratias*, 323
*typhlops*, *Mancalius uranoscopus*, 326
*Typhlopsarus*, 323
*typhlos*, *Mancalias uranoscopus*, 326
*Typhlosparas*, 323
*typica*, *Benthodytes*, 272
*Typlopsaras*, 323
***Tyrannophryne***, 28t, 90, 93k, **116**, 181, 263, 378k, **429**; fig. 128

*umbrinus*, *Chaunax*, fig. 2F
***uranoscopus*, *Ceratias***, 10, 13, 57, 58t, 62, 206, 207, 235t, 282t, 324k, **326**; figs. 62B, 209
*uranoscopus*, *Ceratias* (*Mancalias*), 326
*uranoscopus*, *Mancalias*, 326
*uranoscopus*, *Mancalius*, 326
*vahoeffeni*, *Gigantactis*, 459

*Valdivia*, collections, 233, expedition, 11, 138, German research vessel, 11, 233

*valdiviae*, *Cryptosparas*, 58t, 330
*Vampyroteuthis*, fig. 1
**vanhoeffeni**, ***Gigantactis***, 138, 144, 224, 225, 236t, 455k, **459**, 466; figs. 48H, 158B, 199B, 232, 243, 271, 273
*vanhoeffeni* group, *Gigantactis*, 455k
vascularization, of male-female attachment, 3, 154, 279, 280; of hyoid barbel, 247; intra-escal, **234**
*Velero IV*, research vessel, 275
*Ventrifossa johnboborum*, 261; fig. 266
Venus Flytrap, 81
*venustus*, *Dolopichthys*, 97,
*venustus*, *Dolopichthys* (*Pentherichthys*), 383
*venustus*, *Pentherichthys*, 97, 383
vertebrae, 187; figs. 11, 15B, 28, 200
vertical migration, ontogenetic, 198; fig. 205
*Vibrio*, 241; fig. 255
video analysis, of feeding, 263; of locomotion, 256, 259; figs. 262–264

viperfish, 260
viscera, 296; figs. 16D, E, 288
vision, acumen, 302, 303; binocular, 19, 42, 303; fig. 51; stereoscopic, 268
*Vityaz*, Russian Research Vessel, fig. 21
*vorax*, *Melanocetus*, 76t, 366

*Walther Herwig*, research cruises of, 21, 113, 259, 313
Walton, I., 13
**waltoni**, ***Lasiognathus***, 83, 85, 212, 370k, **370**; fig. 217
**Warty Seadevils**, 52k, **57**, **322**
Waterman, T. H., 11, 112, 233, 254, 278, 279; fig. 267
**watermani**, ***Gigantactis***, 145, 223, 456k, **467**; figs. 45H, 157B, 235
**wedli**, ***Bufoceratias***, 75, 211, 261, 357k, **357**; figs. 29, 44D, 48E, 76A, 215
*wedli*, *Diceratias*, 357

*wedli*, *Paraneirodes*, 357
*wedli*, *Paroneirodes*, 357
*wedli*, *Phrynichthys*, 72, 75, 357, 358
whalefishes, 3, 173
**Whipnose Seadevils**, 52k, 141k, **141**, **452**
*whitleyi*, *Oneirodes*, 405
Willemoes-Suhm, R. von, 9, 229
**Wolftrap Seadevils**, 52k, **81**, **368**
**Wonderfishes**, 81, 84k, **85**, **373**

*xenistius*, *Ceratias holboelli*, 326
*xenistius*, *Mancalias*, 58t, 326
*Xenoceratias*, 360
**xenolophus**, ***Chirophryne***, 123, 126, **440**; figs. 32B, 137B, 140, 141, 229

Young, R. E., 248

*Zaca*, Eastern Pacific Expedition, 157
Zeiformes, 175

| | |
|---|---|
| Composition: | Aptara, Inc. |
| Text: | 8/10.75 Stone Serif |
| Display: | Stone Serif, Akzidenz Grotesk |
| Printer and Binder: | Imago, Inc. |